de Gruyter Expositions in Mathematics 21

Editors

O. H. Kegel, Albert-Ludwigs-Universität, Freiburg
V. P. Maslov, Academy of Sciences, Moscow
W. D. Neumann, The University of Melbourne, Parkville
R. O. Wells, Jr., Rice University, Houston

de Gruyter Expositions in Mathematics

1 The Analytical and Topological Theory of Semigroups, *K. H. Hofmann, J. D. Lawson, J. S. Pym* (Eds.)

2 Combinatorial Homotopy and 4-Dimensional Complexes, *H. J. Baues*

3 The Stefan Problem, *A. M. Meirmanov*

4 Finite Soluble Groups, *K. Doerk, T. O. Hawkes*

5 The Riemann Zeta-Function, *A. A. Karatsuba, S. M. Voronin*

6 Contact Geometry and Linear Differential Equations, *V. E. Nazaikinskii, V. E. Shatalov, B. Yu. Sternin*

7 Infinite Dimensional Lie Superalgebras, *Yu. A. Bahturin, A. A. Mikhalev, V. M. Petrogradsky, M. V. Zaicev*

8 Nilpotent Groups and their Automorphisms, *E. I. Khukhro*

9 Invariant Distances and Metrics in Complex Analysis, *M. Jarnicki, P. Pflug*

10 The Link Invariants of the Chern-Simons Field Theory, *E. Guadagnini*

11 Global Affine Differential Geometry of Hypersurfaces, *A.-M. Li, U. Simon, G. Zhao*

12 Moduli Spaces of Abelian Surfaces: Compactification, Degenerations, and Theta Functions, *K. Hulek, C. Kahn, S. H. Weintraub*

13 Elliptic Problems in Domains with Piecewise Smooth Boundaries, *S. A. Nazarov, B. A. Plamenevsky*

14 Subgroup Lattices of Groups, *R. Schmidt*

15 Orthogonal Decompositions and Integral Lattices, *A. I. Kostrikin, P. H. Tiep*

16 The Adjunction Theory of Complex Projective Varieties, *M. C. Beltrametti, A. J. Sommese*

17 The Restricted 3-Body Problem: Plane Periodic Orbits, *A. D. Bruno*

18 Unitary Representation Theory of Exponential Lie Groups, *H. Leptin, J. Ludwig*

19 Blow-up in Quasilinear Parabolic Equations, *A. A. Samarskii, V. A. Galaktionov, S. P. Kurdyumov, A. P. Mikhailov*

20 Semigroups in Algebra, Geometry and Analysis, *K. H. Hofmann, J. D. Lawson, E. B. Vinberg* (Eds.)

Compact Projective Planes

With an Introduction to Octonion Geometry

by

Helmut Salzmann
Dieter Betten
Theo Grundhöfer
Hermann Hähl
Rainer Löwen
Markus Stroppel

Walter de Gruyter · Berlin · New York 1995

Authors

Helmut Salzmann
Mathematisches Institut
der Universität Tübingen
Auf der Morgenstelle 10
D-72076 Tübingen

Dieter Betten, Hermann Hähl
Mathematisches Seminar
der Universität Kiel
Ludewig-Meyn-Str. 4
D-24118 Kiel

Theo Grundhöfer
Mathematisches Institut
Universität Würzburg
Am Hubland
D-97074 Würzburg

Rainer Löwen
Institut für Analysis
TU Braunschweig
Pockelsstraße 14
D-38106 Braunschweig

Markus Stroppel
Fachbereich Mathematik
TH Darmstadt
Schloßgartenstraße 7
D-64289 Darmstadt

1991 Mathematics Subject Classification: 51-02; 51H10, 51H25, 51A40, 51N30
Keywords: Projective plane, topological geometry, motion groups, collineation groups

∞ Printed on acid-free paper which falls within the guidelines of the ANSI to ensure permanence and durability.

Library of Congress Cataloging-in-Publication Data

Compact projective planes : with an introduction to octonion geometry / by H. Salzmann ... [et al.].
      p. cm. — (De Gruyter expositions in mathematics, ISSN 0938-6572 ; 21)
  Includes bibliographical references and index.
  ISBN 3-11-011480-1 (alk. paper)
  1. Projective planes. I. Salzmann, H. II. Series.
QA554.C66 1995
516'.5—dc20
                                    95-14979
                                       CIP

Die Deutsche Bibliothek — *Cataloging-in-Publication Data*

Compact projective planes : with an introduction to octonion geometry / by Helmut Salzmann ... – Berlin ; New York : de Gruyter, 1995
  (De Gruyter expositions in mathematics ; 21)
  ISBN 3-11-011480-1
NE: Salzmann, Helmut; GT

© Copyright 1995 by Walter de Gruyter & Co., D-10785 Berlin.
All rights reserved, including those of translation into foreign languages. No part of this book may be reproduced or transmitted in any form or by any means, electronic or mechanical, including photocopy, recording, or any information storage or retrieval system, without permission in writing from the publisher.
Typeset using the authors' TeX files: Lewis & Leins, Berlin.
Printing: Ratzlow-Druck, Berlin. Binding: Lüderitz & Bauer GmbH, Berlin.
Cover design: Thomas Bonnie, Hamburg.

Preface

The old and venerable subject of geometry has been changed radically by the famous book of Hilbert [1899, 30] on the foundations of geometry. It was Hilbert's aim to give a simple axiomatic characterization of the real (Euclidean) geometries. He expressed the necessary continuity assumptions in terms of properties of an order. Indeed, the real projective plane is the only desarguesian ordered projective plane where every monotone sequence of points has a limit, see the elegant exposition in Coxeter [61].

However, the stipulation of an order excludes the geometries over the complex numbers (or over the quaternions or octonions) from the discussion. In order to include these geometries, the order properties have been replaced by topological assumptions (like local compactness and connectedness), see Kolmogoroff [32], Köthe [39], Skornjakov [54], Salzmann [55, 57], Freudenthal [57a,b]. This is the historical origin of topological geometry in the sense of this book.

Topological geometry studies incidence geometries endowed with topologies which are compatible with the geometric structure. The prototype of a topological geometry is a topological projective plane, that is, a projective plane such that the two geometric operations of joining distinct points and intersecting distinct lines are continuous (with respect to given topologies on the point set and on the line set). Only few results can be proved about topological planes in general. In order to obtain deeper results, and in order to stay closer to the classical geometries, we concentrate on compact, connected projective planes. Planes of this type exist in abundance; topologically they are very close to one of the four classical planes treated in Chapter 1, but they can deviate considerably from these classical planes in their incidence-geometric structure. Most theorems in this book have a 'homogeneity hypothesis' requiring that the plane in question admits a collineation group which is large in some sense. Of course, there is a multitude of possibilities for the meaning of 'large'. It is a major theme here to consider compact projective planes with collineation groups of large topological dimension. This approach connects group theory and geometry, in the spirit of F. Klein's Erlangen program. We shall indeed use various methods to describe a geometry in group-theoretic terms, see the remarks after (32.20). Usually, the groups appearing in our context turn out to be Lie groups.

*

In this book we consider mainly projective (or affine) planes. This restriction is made for conciseness. Let us comment briefly on some other types of incidence geometries (compare Buekenhout [95] for a panorama of incidence geometry). Topological projective spaces have been considered by Misfeld [68], Kühne–Löwen [92] and others, see also Groh [86a,b]. It is a general phenomenon that spatial geometries are automatically much more homogeneous than plane geometries; we just mention the validity of Desargues' theorem (and its consequences) in each projective space, and the more recent classification of all spherical buildings of rank at least 3 by Tits. This phenomenon effectuates a fundamental dichotomy between plane geometry and space geometry. Stable planes are a natural generalization of topological projective planes, leading to a rich theory, compare (31.26). Typical examples are obtained as open subgeometries of topological projective planes. The reader is referred to Grundhöfer–Löwen [95] and Steinke [95] for surveys on locally compact space geometries (including stable planes) and circle geometries, respectively.

Projective planes are the same thing as generalized triangles, and the generalized polygons are precisely the buildings of rank 2. A theory of topological generalized polygons and of topological buildings is presently developing, see Burns–Spatzier [87], Knarr [90] and Kramer [94] for fundamental results in this direction, compare also Grundhöfer–Löwen [95] Section 6. These geometries are of particular interest in differential geometry, see Thorbergsson [91, 92]. Another connection with differential geometry is provided by the study of symmetric planes, see Löwen [79a,b, 81b], Seidel [90a, 91], H. Löwe [94, 95], Grundhöfer–Löwen [95] 5.27ff.

*

Now we give a rough description of the contents of this book (see also the introduction of each chapter).

In Chapter 1, we consider in detail the classical projective planes over the real numbers, over the complex numbers, over Hamilton's quaternions, and over Cayley's octonions; these classical division algebras are denoted by $\mathbb{R}$, $\mathbb{C}$, $\mathbb{H}$, $\mathbb{O}$. The four classical planes are the prime examples (and also the most homogeneous examples, as it turns out) of compact, connected projective planes. We describe the full collineation groups of the classical planes, as well as various interesting subgroups, like motion groups with respect to polarities. In the case of the octonion plane, this comprises a complete and elementary description of some exceptional Lie groups (and of their actions on the octonion plane), including proofs of their simplicity; e.g. the full collineation group has type E_6, and the elliptic motion group is the compact group of type F_4.

Chapter 2 is a brief summary of notions and results concerning projective and affine planes, coordinates and collineations. It is meant as a reference for known facts which entirely belong to incidence geometry.

In Chapter 3 we study planes on the point set $\mathbb{R}^2$, with lines which are homeomorphic to the real line $\mathbb{R}$, so that each line is a curve in $\mathbb{R}^2$. This chapter is the most intuitive part of the book. If the parallel axiom is satisfied, that is, if we have an affine plane, then we can form the usual projective completion, which leads to a (topologically) 2-dimensional compact projective plane. These planes have been studied by Salzmann in 1957–1967 with remarkable success. He proved that the full collineation group Σ of such a plane is a Lie group of dimension at most 8, that the real projective plane $\mathcal{P}_2\mathbb{R}$ is characterized by the condition $\dim \Sigma > 4$, and that the Moulton planes are the only planes of this type with $\dim \Sigma = 4$. Furthermore, he explicitly classified all 2-dimensional compact projective planes with $\dim \Sigma = 3$. All these classification results are proved in Chapter 3.

In Chapter 4 we begin a systematic study of topological projective planes in general. Most results require compactness, and some results (like contractibility properties) are based on connectedness assumptions. Note that, for every prime p, the plane over the p-adic numbers $\mathbb{Q}_p$ provides an example of a compact, totally disconnected plane. We show that the four classical planes studied in Chapter 1 are precisely the compact, connected Moufang planes; Moufang planes are defined by a very strong homogeneity condition, which implies transitivity on triangles (and even on quadrangles). Furthermore, we prove that the group of all continuous collineations of a compact projective plane is always a locally compact group (with respect to the compact-open topology).

Chapter 5 deals with the algebraic topology of compact, connected projective planes of finite topological dimension. As Löwen has shown, the point spaces of these planes have the very same homology invariants as their classical counterparts considered in Chapter 1; moreover, the lines are homotopy equivalent to spheres. We obtain that the topological dimension of a line in such a plane is one of the numbers $1, 2, 4, 8$; the (point sets of the) corresponding planes have topological dimensions $2, 4, 8, 16$. The topological resemblance to classical planes has strong geometric consequences, which are discussed in Section 55 and in Chapters 6–8. In fact, these results determine a subdivision of the whole theory into four cases. In order to understand Chapters 6–8, it suffices to be acquainted with the main *results* of Chapter 5; the *methods* of proof in that chapter are not used in other chapters.

In Chapter 6 we consider compact, connected projective planes which are homogeneous in some sense. As indicated above, the idea of homogeneity plays a central rôle in this book. We prove that a compact, connected projective plane which admits an automorphism group transitive on points is isomorphic to one of the four classical planes treated in Chapter 1. This is a remarkable result; it says that for compact, connected projective planes, the Moufang condition is a consequence of transitivity on points. Furthermore, we consider groups of axial collineations and transitivity conditions for these groups, and we study planes which admit a classical motion group. Often, these homogeneity conditions are strong enough to allow an explicit classification of the planes in question. In Section 65 we employ the topological dimension of the automorphism group as a measure of homogeneity (this idea is fully developed in Chapters 3, 7, 8), and Section 66 is a short report on groups of projectivities in our context.

In Chapters 7 and 8 we determine all compact projective planes of dimension 4, 8 or 16 which admit an automorphism group of sufficiently large topological dimension. This approach leads first to the classical planes over $\mathbb{C}$, $\mathbb{H}$, $\mathbb{O}$, and then the most homogeneous non-classical planes appear in a systematic fashion. In contrast to Chapter 3, deeper methods are required, and proper translation planes arise.

In Chapter 7 we study compact projective planes of topological dimension 4; these planes are the topological relatives of the complex projective plane $\mathcal{P}_2\mathbb{C}$. We prove that the automorphism group Σ of such a plane $\mathcal{P}$ is a (real) Lie group of dimension at most 16, and that the complex projective plane is characterized by the condition $\dim \Sigma > 8$. This result is one of the highlights of the theory of 4-dimensional planes. If $\dim \Sigma \geq 7$, then $\mathcal{P}$ is a translation plane (up to duality) or a shift plane. All translation planes $\mathcal{P}$ with $\dim \Sigma \geq 7$ and all shift planes $\mathcal{P}$ with $\dim \Sigma \geq 6$ have been classified explicitly; Chapter 7 contains a classification of the translation planes with $\dim \Sigma \geq 8$ and of the shift planes with $\dim \Sigma \geq 7$. Finally, we show in Section 75 that only the complex plane admits a complex analytic structure.

The theory of 4-dimensional compact planes is distinguished from the theory of higher-dimensional compact planes, regarding both the phenomena and the methods. For instance, the class of shift planes appears only in low dimensions. In higher dimensions, special tools connected with the recognition and handling of low-dimensional manifolds are not available.

Chapter 8 deals with compact projective planes of topological dimension 8 or 16, that is, with the relatives of the quaternion plane $\mathcal{P}_2\mathbb{H}$ or of the octonion plane $\mathcal{P}_2\mathbb{O}$. In some parts of this chapter, the results are only surveyed, with references to the literature. Again, classification results on planes admitting automorphism groups of large dimension constitute the main theme. It turns out that such planes are often translation planes up to duality. They carry a vector space structure, whence special tools become available. Accordingly, the theory of translation planes and the classification of the most homogeneous ones form a theory on their own. Fundamental results of this theory are developed in Section 81; in Section 82, the classification of all 8-dimensional compact translation planes $\mathcal{P}$ satisfying $\dim \operatorname{Aut} \mathcal{P} \geq 17$ and of all 16-dimensional compact translation planes $\mathcal{P}$ satisfying $\dim \operatorname{Aut} \mathcal{P} \geq 38$ is presented.

In the following sections, classification results of this kind are extended to 8- and 16-dimensional compact planes in general. For reasons of space, the results often are not proved in their strongest form. Salzmann [81a, 90] proved, on the basis of Hähl [78], that the quaternion plane $\mathcal{P}_2\mathbb{H}$ is the only compact projective plane of dimension 8 such that $\dim \operatorname{Aut} \mathcal{P} > 18$. In Section 84, this result is proved under the stronger assumption $\dim \operatorname{Aut} \mathcal{P} \geq 23$. Similarly, the octonion plane is known to be the only compact projective plane $\mathcal{P}$ of dimension 16 such that $\dim \operatorname{Aut} \mathcal{P} > 40$, see Salzmann [87], Hähl [88]. In Section 85, we characterize the octonion plane by the stronger condition that $\dim \operatorname{Aut} \mathcal{P} \geq 57$. The proofs of these characterization results make use of the corresponding characterization results for translation planes.

In Section 86, we construct and characterize the compact Hughes planes of dimensions 8 and 16. They form two one-parameter families of planes with rather singular properties. Section 87 contains basic results indicating a viable route towards an extension of the classification results presented here. This should help the reader to go beyond the limitations of the exposition here; moreover, it may serve as a guide to future research. Among other things, these results explain the special rôle played in the classification by translation planes on the one hand and by Hughes planes on the other hand.

The final Chapter 9 is an appendix. Here we collect a number of results from topology, and we give a systematic outline of Lie theory, as required in this book. In this chapter we usually do not give proofs, but rather refer to the literature (with an attempt to give references also for folklore results). The topics covered in Chapter 9 include the topological characterization of Lie groups (Hilbert's fifth problem), and the structure and the classification of (simple) Lie groups; in fact, we require only results for groups of dimension at most 52. Furthermore we report on real linear representations of almost simple Lie groups, and we list all irreducible representations of these groups on real vector spaces of dimension at most 16. Finally, we deal with various classification results on (not necessarily compact) transformation groups.

*

This book gives a systematic account of many results which are scattered in the literature. Some results are presented in improved form, or with simplified proofs, others only in weakened versions. A few of the more recent results are only mentioned, because their proofs appear to be too complex to be included here. However, we hope to provide a convenient introduction to compact, connected projective planes, as well as a sound foundation for future research in this area.

Many colleagues and friends have offered helpful advice, or have read parts of the typescript and contributed improvements. In particular, the authors would like to thank Richard Bödi, Sven Boekholt, Michael Dowling, Karl Heinrich Hofmann, Norbert Knarr, Linus Kramer, Helmut Mäurer, Kai Niemann, Joachim Otte, Burkard Polster, Barbara Priwitzer, Eberhard Schröder, Jan Stevens, Peter Sperner, and Bernhild Stroppel. We are also indebted to the Oberwolfach Institute, which gave us the opportunity to hold several meetings devoted to the work on this book. Finally, we would like to thank de Gruyter Verlag and, in particular, Dr. Manfred Karbe, for support and professional advice.

Tübingen, September 1994   The authors

Contents

Preface v

Chapter 1
The classical planes 1

11 The classical division algebras 4
12 The classical affine planes 25
13 The projective planes over $\mathbb{R}$, $\mathbb{C}$, and $\mathbb{H}$ 42
14 The planes over $\mathbb{R}$, $\mathbb{C}$, and $\mathbb{H}$ as topological planes 56
15 Geometry of a projective line 65
16 The projective octonion plane $\mathcal{P}_2\mathbb{O}$ 77
17 The collineation group of $\mathcal{P}_2\mathbb{O}$ 88
18 Motion groups of $\mathcal{P}_2\mathbb{O}$ 103

Chapter 2
Background on planes, coordinates and collineations 131

21 Projective and affine planes 131
22 Coordinates, ternary fields 135
23 Collineations 137
24 Lenz–Barlotti types 142
25 Translation planes and quasifields 144

Chapter 3
Geometries on surfaces 148

31 $\mathbb{R}^2$-planes 149
32 Two-dimensional compact projective planes 163
33 Towards classification 178
34 The Moulton planes 185
35 Skew hyperbolic planes 196
36 Skew parabola planes 202
37 Planes over Cartesian fields 205
38 Flexibility, rigidity and related topics 212

Chapter 4
Compact projective planes 216

41 The topology of locally compact planes 217
42 Compact, connected planes 225
43 Ternary fields for compact planes 230
44 Automorphism groups 235

Chapter 5
Algebraic topology of compact, connected planes 244

51 General properties 245
52 Assuming that lines are manifolds 257
53 Conditions implying that lines are manifolds 266
54 Lines are homology manifolds 276
55 Geometric consequences 283

Chapter 6
Homogeneity 309

61 Axial collineations 311
62 Planes admitting a classical motion group 323
63 Transitive groups 339
64 Transitive axial groups 349
65 Groups of large dimension 368
66 Remarks on von Staudt's point of view 370

Chapter 7
Four-dimensional planes 372

71 Automorphism groups 373
72 Characterizing $\mathcal{P}_2\mathbb{C}$ 383
73 Four-dimensional translation planes 393
74 Four-dimensional shift planes 420
75 Analytic planes 445

Chapter 8
Eight- and sixteen-dimensional planes 449

81 Translation planes 452
82 Classification of translation planes 477
83 Stiffness 519
84 Characterizing $\mathcal{P}_2\mathbb{H}$ 535
85 Characterizing $\mathcal{P}_2\mathbb{O}$ 547

86 Hughes planes 555
87 Principles of classification 581

Chapter 9
Appendix: Tools from topology and Lie theory 590

91 Permutation groups 590
92 Topological dimension and remarks on general topology 592
93 Locally compact groups and Lie groups 597
94 Lie groups and their structure 602
95 Linear representations 616
96 Transformation groups 630

Bibliography 643
Notation 679
Index 683

Chapter 1

The classical planes

In this introductory chapter, the classical examples of topological projective planes will be presented and studied. These are the planes over the following coordinate domains: the field $\mathbb{R}$ of real numbers, the field $\mathbb{C}$ of complex numbers, the skew field $\mathbb{H}$ of quaternions, and the alternative field $\mathbb{O}$ of octonions. In a later chapter, see (42.7), it will turn out that these classical planes are the only locally compact, connected topological planes which either satisfy Desargues' law (valid in the planes over the fields $\mathbb{R}$, $\mathbb{C}$, and $\mathbb{H}$), or at least possess the Moufang property (which holds in the plane over $\mathbb{O}$, as well).

A careful elementary study of the latter plane, the plane over the octonions $\mathbb{O}$, is a particular objective of this chapter. This comprises the investigation of the group Σ of its collineations, and of certain subgroups of Σ, including the elliptic motion group Φ. The group Σ is known to be the real exceptional simple Lie group $E_6(-26)$, and Φ is the compact exceptional simple Lie group of type F_4. One of the aims of our presentation is to give a detailed study of these groups which is mainly based on incidence geometry, and which makes only marginal references to the theory of simple Lie groups (just what we need for the identification of these groups among the simple Lie groups, for instance). In particular, the simplicity of these groups is proved without recourse to Lie group theory. It seems to us that this approach offers a pleasant road to an intimate understanding of $E_6(-26)$ and its distinguished subgroups.

It should be said here that the material of this chapter is classical and well known. Distinctive features of our presentation are, we believe, the particular blend of arguments and techniques in dealing with the octonion plane, and our systematic use of methods from incidence geometry.

The subject is rooted in incidence geometry on the one hand, and it has aspects which are important for topology and Lie group theory on the other hand. There may be readers whose interest is primarily on one side and who are less familiar with the other side. The presentation in this chapter is therefore intended to be accessible with few prerequisites from either side. In particular, Sections 12 and 13 about the affine planes over $\mathbb{R}$, $\mathbb{C}$, $\mathbb{H}$, and $\mathbb{O}$, and about the projective planes over $\mathbb{R}$, $\mathbb{C}$, and $\mathbb{H}$, may serve at the same time as a concrete introduction to some basic notions about affine and projective planes and their collineations, before one turns to Chapter 2, which is a short and rather abstract summary on projective planes in general. We trust that the more experienced reader will find it easy to skip the extra explanations implied by this approach.

In Section 11, the Cayley–Dickson process is applied to the field $\mathbb{R}$ in order to construct the algebras $\mathbb{C}$, $\mathbb{H}$, and $\mathbb{O}$. We present their characteristic properties and study their automorphism groups. Section 12 is concerned with the affine planes over these algebras. For the planes over $\mathbb{R}$, $\mathbb{C}$, and $\mathbb{H}$, a complete description of their (affine) collineation groups is given. For the plane over $\mathbb{O}$, this is less easy due to non-associativity, and a full treatment is therefore postponed until Sections 15–18. However, first results will be given in the octonion case, as well. They concern special collineations, which are in close connection with the Moufang identities in $\mathbb{O}$, and the description of all collineations fixing the coordinate axes; the latter topic is closely related to the triality principle. It should be pointed out that in our presentation the mentioned algebraic laws, viz. the Moufang identities and the triality principle, are obtained as a by-product of the geometrical reasoning, whereas they are usually proved by algebraic means and then employed for geometrical conclusions, among other things. Section 13 is devoted to a description of the projective planes over $\mathbb{R}$, $\mathbb{C}$, and $\mathbb{H}$ by homogeneous coordinates (a tool not available for the non-associative algebra $\mathbb{O}$) and to a study of their collineation groups. The fundamental theorem of projective geometry describing all collineations is proved, and the elliptic and hyperbolic motion groups are presented. In Section 14, these planes are studied as topological planes; the topological structure is introduced using homogeneous coordinates.

A step of major conceptual and technical importance is the study of the geometry of a projective line in Section 15. A projective line is represented as a quadric of index 1 in a certain real vector space. This interpretation is valid over $\mathbb{O}$ as well as over $\mathbb{R}$, $\mathbb{C}$, and $\mathbb{H}$. It provides a description in algebraic terms of the Möbius geometry ('conformal geometry') on a projective line, which was introduced and used by Tits [53] in his fundamental paper on the octonion plane. The Möbius geometry helps to understand the stabilizer of a line in the collineation group, because the action of that stabilizer on the given line respects this additional geometric structure. This fact is an important tool for the study of the collineation group in the octonion plane.

Sections 16 through 18 are devoted to a close study of the projective plane over $\mathbb{O}$. In Section 16, 'Veronese coordinates' are introduced as a substitute for homogeneous coordinates, which are not available over $\mathbb{O}$. (Veronese coordinates could equally well be used for the planes over $\mathbb{R}$, $\mathbb{C}$, and $\mathbb{H}$, so that a unified treatment would result.) The topological properties of the octonion plane can now be derived in virtually the same way as for the other classical planes.

In Section 17, the group Σ of all collineations of the octonion plane is determined. We show that Σ is generated by elations, and we deduce that it is a simple group. Incidentally, we obtain that all collineations are induced by appropriate linear transformations acting on Veronese coordinates; this result is an octonion analogue of the fundamental theorem of projective geometry over (not necessarily commutative) fields. Then we study the stabilizer of a triangle. In affine language, such a stabilizer has already been described in Section 12 as the stabilizer of the coordinate axes in the affine collineation group. We now show that its maximal compact, connected subgroup, which is a normal subgroup, is the universal covering group $\mathrm{Spin}_8\mathbb{R}$ of $\mathrm{SO}_8\mathbb{R}$, and we exhibit the triality automorphism of $\mathrm{Spin}_8\mathbb{R}$. In the stabilizer of a degenerate quadrangle, we accordingly find the universal covering group $\mathrm{Spin}_7\mathbb{R}$ of $\mathrm{SO}_7\mathbb{R}$. One thus obtains a concrete geometric understanding of these groups and of the homogeneous space $\mathrm{Spin}_7\mathbb{R}/\mathrm{G}_2 \approx \mathbb{S}_7$.

In Section 18, we study the groups of collineations which commute with the standard elliptic polarity or with the standard hyperbolic polarity of the octonion plane, the so-called elliptic and hyperbolic motion groups. We prove their simplicity by a geometric argument; using their dimensions, they are then easily identified as real simple Lie groups of exceptional type F_4. The elliptic motion group turns out to be a maximal compact subgroup of the full collineation group; this fact finally allows us to recognize the latter among the simple Lie groups as a real form of type E_6. A crucial step in the study of the motion groups is the analysis of the stabilizer of a point. For an arbitrary point in the elliptic case and for an interior point in the hyperbolic case, the stabilizer is isomorphic to the universal covering group $\mathrm{Spin}_9\mathbb{R}$ of $\mathrm{SO}_9\mathbb{R}$; the covering map $\mathrm{Spin}_9\mathbb{R} \to \mathrm{SO}_9\mathbb{R}$ has a very simple geometric description. As by-products of this analysis one obtains that the homogeneous space $\mathrm{F}_4/\mathrm{Spin}_9\mathbb{R}$ is homeomorphic to the point space of the projective octonion plane, and that $\mathrm{Spin}_9\mathbb{R}/\mathrm{Spin}_7\mathbb{R} \approx \mathbb{S}_{15}$. The section closes with a classification of all polarities of the octonion plane, up to equivalence; besides the standard elliptic polarity and the standard hyperbolic polarity, there appears just one further possibility.

We close this summary by a list of references to places in this chapter where descriptions of various classical or exceptional groups or further information about them may be found.

$SO_3\mathbb{R}$	(11.22 through 25), (11.29)	$SU_2\mathbb{C}$	(11.26)
		$U_2\mathbb{C}$	(13.14)
$O_3\mathbb{R}$	(13.13)	$SU_3\mathbb{C}$	(11.34 and 35)
$SO_4\mathbb{R}, O_4\mathbb{R}$	(11.22 and 23)	$PU_3\mathbb{C}$	(13.13 and 15)
$SO_8\mathbb{R}, O_8\mathbb{R}$	(11.22 and 23), (12.18)	$PU_3(\mathbb{C}, 1)$	(13.13 and 17)
$SO_9\mathbb{R}$	(18.8)	$U_2\mathbb{H}$	(13.14), (18.9)
		$PU_3\mathbb{H}$	(13.13 and 15)
$\mathrm{Spin}_3\mathbb{R}$	(11.26)	$PU_3(\mathbb{H}, 1)$	(13.13 and 17)
$\mathrm{Spin}_5\mathbb{R}$	(18.9)		
$\mathrm{Spin}_7\mathbb{R}$	(17.14 and 15)	$PGL_2\mathbb{F}$	(12.12), (15.6)
$\mathrm{Spin}_8\mathbb{R}$	(17.13 and 16)	$PGL_3\mathbb{F}$	(13.4)
$\mathrm{Spin}_9\mathbb{R}$	(18.8, 13 and 16)		($\mathbb{F} \in \{\mathbb{R}, \mathbb{C}, \mathbb{H}\}$)
$O_3(\mathbb{R}, 1)$	(13.13 and 17), (15.6)	$G_2 = \mathrm{Aut}\,\mathbb{O}$	(11.30 through 33), (17.15)
$PSO_4(\mathbb{R}, 1)$	(15.6)	$F_4 = F_4(-52)$	(18.10, 15 and 16)
$PSO_6(\mathbb{R}, 1)$	(15.6)	$F_4(-20)$	(18.23 and 26)
$O'_9(\mathbb{R}, 1)$	(18.22)	$E_6(-26)$	(18.19)
$PSO_{10}(\mathbb{R}, 1)$	(15.6)		

11 The classical division algebras

In this section, we apply the Cayley–Dickson process to the field $\mathbb{R}$ of real numbers in order to construct the field $\mathbb{C}$ of complex numbers, the skew field $\mathbb{H}$ of quaternions and the (non-associative) alternative field $\mathbb{O}$ of octonions, and we derive the characteristic properties of these algebras. They will be called the classical division algebras; notice that, in our terminology, a division algebra is not necessarily associative. The multiplication of these algebras provides useful descriptions of certain orthogonal groups. This will be explained and applied for a study of the automorphism groups $\mathrm{Aut}\,\mathbb{H}$ and $\mathrm{Aut}\,\mathbb{O}$. Little will be said about $\mathrm{Aut}\,\mathbb{C}$, as this is more a matter of field theory.

11.1 The Cayley–Dickson process serves to construct a sequence $\mathbb{F}_m$ of $\mathbb{R}$-algebras, each furnished with an involutory antiautomorphism $a \mapsto \bar{a}$, called *conjugation*. The construction proceeds inductively in the following way: One starts with

$$\mathbb{F}_0 = \mathbb{R}, \quad \bar{a} = a \quad \text{for } a \in \mathbb{R};$$

then, assuming that $\mathbb{F}_{m-1}$ ($m \geq 1$) with its conjugation has been constructed, one puts

$$\mathbb{F}_m := \mathbb{F}_{m-1} \times \mathbb{F}_{m-1},$$

with addition, multiplication, and conjugation defined by

$$(a, b) + (c, d) := (a + c, b + d)$$
$$(a, b)(c, d) := (ac - \bar{d}b, da + b\bar{c})$$
$$\overline{(a, b)} := (\bar{a}, -b)$$

(for $a, b, c, d \in \mathbb{F}_{m-1}$). Obviously, the dimension of $\mathbb{F}_m$ over $\mathbb{R}$ is 2^m.

For $m = 1$, this is the familiar definition of the field $\mathbb{C}$ of complex numbers, $\mathbb{F}_2 =: \mathbb{H}$ is the algebra of Hamilton's *quaternions*, and $\mathbb{F}_3 =: \mathbb{O}$ is the algebra of *octonions* (or Cayley numbers). The further steps in the ladder will not be of interest to us, because they lead to algebras having zero divisors (11.17).

Via the map $\mathbb{F}_{m-1} \to \mathbb{F}_m : a \mapsto (a, 0)$ we may identify $\mathbb{F}_{m-1}$ with a subalgebra of $\mathbb{F}_m$. In this way, $\mathbb{R} = \mathbb{F}_0$ is a central subfield of all the algebras $\mathbb{F}_m$, and we have inclusions

$$\mathbb{R} \subset \mathbb{C} \subset \mathbb{H} \subset \mathbb{O}.$$

It is easily verified by induction that conjugation is indeed an antiautomorphism of $\mathbb{F}_m$, and that its fixed elements are precisely the real numbers:

11.2. $\{ x \in \mathbb{F}_m \mid \bar{x} = x \} = \mathbb{R}$. □

11.3 The norm form. For $x = (a, b) \in \mathbb{F}_m$, $a, b \in \mathbb{F}_{m-1}$, one computes that

$$x\bar{x} = (a, b)(\bar{a}, -b) = (a\bar{a} + \bar{b}b, 0)$$
$$\bar{x}x = (\bar{a}, -b)(a, b) = (\bar{a}a + \bar{b}b, 0).$$

By induction on m, one obtains that $x\bar{x} = \bar{x}x$, and that for $x \neq 0$ this is a positive element of the subfield $\mathbb{R}$. This positive real number will also be written as

$$\|x\|^2 := x\bar{x} = \bar{x}x.$$

The map $x \mapsto \|x\|^2$ is a positive definite quadratic form (the so-called *norm form*) on the $\mathbb{R}$-vector space $\mathbb{F}_m$. The associated bilinear form is

$$\langle x \mid y \rangle := \|x + y\|^2 - \|x\|^2 - \|y\|^2 = \bar{x}y + \bar{y}x,$$

as is easily computed. It is a positive definite inner product; note that

$$\langle x \mid x \rangle = 2\|x\|^2.$$

Since conjugation is involutory, one has

$$\|\bar{x}\|^2 = \|x\|^2$$
$$\langle \bar{x} \mid \bar{y} \rangle = \langle x \mid y \rangle.$$

Our initial calculation shows that the norms in $\mathbb{F}_m$ and in $\mathbb{F}_{m-1}$ are connected as follows:
$$\|(a,b)\|^2 = \|a\|^2 + \|b\|^2$$
for $a, b \in \mathbb{F}_{m-1}$. In particular, the subspaces $\mathbb{F}_{m-1} \times \{0\}$ and $\{0\} \times \mathbb{F}_{m-1}$ of $\mathbb{F}_m = \mathbb{F}_{m-1} \times \mathbb{F}_{m-1}$ are orthogonal with respect to the inner product.

11.4 Inverses. For a nonzero element $x \in \mathbb{F}_m$, one easily verifies that
$$x^{-1} := \|x\|^{-2}\bar{x}$$
is a two-sided multiplicative inverse, and that
$$\|x^{-1}\|^2 = (\|x\|^2)^{-1} .$$

11.5 $\mathbb{F}_m$ as a quadratic algebra. By (11.2), one immediately obtains that for every $x \in \mathbb{F}_m$
$$x + \bar{x} \in \mathbb{R}, \quad \text{so that} \quad \bar{x} \in \mathbb{R} + \mathbb{R}x .$$
Consequently,
$$x^2 = (x + \bar{x})x - \|x\|^2 \in \mathbb{R} + \mathbb{R}x .$$
In particular, every element of $\mathbb{F}_m$ satisfies a quadratic equation with real coefficients.

11.6 Pure elements. The subspace
$$\operatorname{Pu} \mathbb{F}_m := \{\, x \in \mathbb{F}_m \mid \bar{x} = -x \,\} \leq \mathbb{F}_m$$
of *pure* elements is the orthogonal complement of $\mathbb{R} = \mathbb{R} \cdot 1$ with respect to the inner product. One infers directly from the quadratic equation in (11.5) that this subspace can also be described as
$$\operatorname{Pu} \mathbb{F}_m = \{\, x \in \mathbb{F}_m \mid x^2 \in \mathbb{R}, \ x^2 \leq 0 \,\} .$$
More precisely,

(1) $\qquad x^2 = -\|x\|^2 \quad \text{if, and only if, } x \in \operatorname{Pu} \mathbb{F}_m .$

For $u, v \in \operatorname{Pu} \mathbb{F}_m$, we have

(2) $\qquad \langle u \mid v \rangle = 0 \iff uv = -vu \iff uv \in \operatorname{Pu} \mathbb{F}_m ;$

this is clear from the definitions, since $\overline{uv} = \bar{v}\bar{u} = (-v)(-u) = vu$.

We now deal with the associativity properties of our algebras. $\mathbb{F}_1 = \mathbb{C}$ is associative and commutative, and $\mathbb{F}_2 = \mathbb{H}$ is associative; $\mathbb{F}_3 = \mathbb{O}$ is not associative, but weak forms of associativity survive, which we shall now consider. Our discussion will also cover the facts mentioned about $\mathbb{C}$ and $\mathbb{H}$. For $m \geq 4$, little associativity is left in $\mathbb{F}_m$, see (11.17); the following fact is nevertheless quite general.

11.7 Mono-associativity. *For $x \in \mathbb{F}_m \setminus \mathbb{R}$, the span $\mathbb{R} + \mathbb{R}x$ of 1 and x is an associative and commutative subalgebra of $\mathbb{F}_m$, and this subalgebra is isomorphic to $\mathbb{C}$.*

Proof. The span $A = \mathbb{R} + \mathbb{R}x$ intersects the hyperplane $\operatorname{Pu}\mathbb{F}_m$ in a 1-dimensional subspace $\mathbb{R}u$ with $u \in \operatorname{Pu}\mathbb{F}_m$ and $\|u\|^2 = 1$; then $\{1, u\}$ is a basis of A. By (11.6), we have $u^2 = -\|u\|^2 = -1 \in \mathbb{R}$. One now easily verifies that A is a subalgebra, which is associative and commutative. Clearly, it is isomorphic to $\mathbb{C}$. □

The key to stronger associativity properties of $\mathbb{F}_m$ for $m \leq 3$ is

11.8 Alternativity. *Assume that $\mathbb{F}_{m-1}$ is associative. Then $\mathbb{F}_m$ has the following property:*

(1) $$\bar{x}(xy) = (\bar{x}x)y = \|x\|^2 y$$

for all $x, y \in \mathbb{F}_m$, and hence, upon conjugation,

(2) $$(yx)\bar{x} = y(x\bar{x}) = \|x\|^2 y.$$

Equivalently, $\mathbb{F}_m$ is alternative; i.e., the following identities hold:

$$x(xy) = x^2 y \quad \text{and} \quad (yx)x = yx^2.$$

Proof. Let $x = (a, b)$, $y = (c, d)$ for $a, b, c, d \in \mathbb{F}_{m-1}$. By direct computation and the associativity of $\mathbb{F}_{m-1}$ one obtains that $\bar{x}(xy) = (\bar{a}ac - \bar{a}\bar{d}b + \bar{a}\bar{d}b + c\bar{b}b, d a\bar{a} + b\bar{c}\bar{a} - b\bar{c}\bar{a} + b\bar{b}d)$. Since $a\bar{a} = \bar{a}a = \|a\|^2$ and $b\bar{b} = \bar{b}b = \|b\|^2$ are scalars, and since $\|a\|^2 + \|b\|^2 = \|x\|^2$, we conclude that $\bar{x}(xy) = (\|a\|^2 c + \|b\|^2 c, \|a\|^2 d + \|b\|^2 d) = \|x\|^2 (c, d) = \|x\|^2 y$. This proves (1), and (2) follows upon conjugation. The alternative laws are identical to (1) and (2) if x is a pure element, i.e., if $x = -\bar{x}$. The general case is dealt with by decomposing x into a scalar in $\mathbb{R}$ and a pure element. □

From (11.8), we now derive the associativity of $\mathbb{H}$ together with the following weaker associativity property of $\mathbb{O}$, which includes alternativity:

11.9 Biassociativity. *For all $x \in \mathbb{O} \setminus \mathbb{R}$ and $y \in \mathbb{O} \setminus (\mathbb{R} + \mathbb{R}x)$, the span $\mathbb{R} + \mathbb{R}x + \mathbb{R}y + \mathbb{R}xy$ is an associative subalgebra of $\mathbb{O}$ isomorphic to $\mathbb{H}$.*

Remark. Taken together with (11.7), this implies that $\mathbb{O}$ is *biassociative* in the sense that any two elements $x, y \in \mathbb{O}$ belong to an associative subalgebra (containing 1). By (11.4 and 5), then, this subalgebra also contains $\bar{x}$, $\bar{y}$, $x^{-1} = \|x\|^{-2}\bar{x}$, and $y^{-1} = \|y\|^{-2}\bar{y}$. As a consequence, brackets are of no importance in multiple products whose factors are among x, y, $\bar{x}$, $\bar{y}$, x^{-1}, and y^{-1}, or are scalars in $\mathbb{R}$.

Proof of (11.9). Intersecting $\mathbb{R} + \mathbb{R}x$ and $\mathbb{R} + \mathbb{R}x + \mathbb{R}y$ with the hyperplane $\mathrm{Pu}\,\mathbb{O}$, we obtain elements $u, v \in \mathrm{Pu}\,\mathbb{O}$ for which $\|u\|^2 = 1 = \|v\|^2$ and $\langle u \mid v \rangle = 0$, and such that $x = r_0 + r_1 u$, $y = s_0 + s_1 u + s_2 v$, for suitable scalars $r_\nu, s_\nu \in \mathbb{R}$ ($\nu = 0, 1, 2$) with $r_1 \neq 0, s_2 \neq 0$. By (11.6), we have $u^2 = -\|u\|^2 = -1$, and therefore $\mathbb{R} + \mathbb{R}x + \mathbb{R}y + \mathbb{R}xy = \mathbb{R} + \mathbb{R}u + \mathbb{R}v + \mathbb{R}uv$. It thus suffices to prove the following statement, which is formulated so as to give information about $\mathbb{F}_2 = \mathbb{H}$, as well.

11.10 Proposition. *Assume that $2 \leq m \leq 3$, and let $u, v \in \mathrm{Pu}\,\mathbb{F}_m$ be such that*

(1) $\qquad\qquad \|u\|^2 = 1 = \|v\|^2 \quad and \quad \langle u \mid v \rangle = 0 \,.$

Then the product $w := uv$ satisfies

(2) $\qquad\qquad w \in \mathrm{Pu}\,\mathbb{F}_m \,, \quad \langle u \mid w \rangle = 0 = \langle v \mid w \rangle, \quad and \quad \|w\|^2 = 1 \,.$

Moreover, we have the following multiplication table:

	u	v	w
u	-1	w	$-v$
v	$-w$	-1	u
w	v	$-u$	-1

.

The span $\mathbb{R} + \mathbb{R}u + \mathbb{R}v + \mathbb{R}w$ is an associative subalgebra of $\mathbb{F}_m$ isomorphic to $\mathbb{H}$.

Terminology. A triple u, v, w with these properties is called a *Hamilton triple*.

Proof. At a first stage, we shall assume in addition that $\mathbb{F}_m$, $m = 2, 3$, is already known to be alternative. This assumption will later prove to be innocuous. From (1) and (11.6(2)), we infer $u^2 = -1 = v^2$, and $w \in \mathrm{Pu}\,\mathbb{F}_m$. Assuming alternativity one obtains

$$uw = u(uv) = u^2 v = -v \quad \text{and} \quad wv = (uv)v = uv^2 = -u \,.$$

By (11.6(2)) again, it now follows that u, v, and w are mutually orthogonal with respect to $\langle \mid \rangle$ and anticommuting. Using alternativity once more we obtain that $w^2 u = w(wu) = wv = -u$. Since w^2 is a scalar by (11.6(1)), this means that $w^2 = -1$ and $\|w\|^2 = 1$. Thus the multiplication table is as asserted, and it follows that

$$A := \mathbb{R} + \mathbb{R}u + \mathbb{R}v + \mathbb{R}w$$

is a subalgebra. Next we remark that A is *flexible*, which means that

(3) $$a(ba) = (ab)a$$

for all $a, b \in A$. This follows immediately from alternativity by expanding the two sides of the equation $(a+b)((a+b)a) = (a+b)^2 a$.

For the proof of associativity of A, it now suffices to show that triple products composed of u, v and w are associative. If two of the factors of such a triple product coincide, then this is a consequence of alternativity and flexibility. As to triple products with different factors, we have, for instance,

$$(uv)w = w^2 = -1 = u^2 = u(vw) \quad \text{and} \quad (vu)w = -w^2 = 1 = -v^2 = v(uw).$$

All the other triple products with different factors are obtained from these by cyclic permutation of u, v, and w, under which the corresponding equalities remain valid because of the symmetry of our multiplication table. Thus A is associative.

We must now dispose of our supplementary hypothesis of alternativity. In $\mathbb{F}_2$, this is easy. Since $\mathbb{F}_1 = \mathbb{C}$ is associative, $\mathbb{F}_2$ is alternative by (11.8), and so the above arguments are valid in $\mathbb{F}_2$. In this case, $1, u, v, w$ span the 4-dimensional algebra $\mathbb{F}_2 = \mathbb{H}$, since they are linearly independent by (1) and (2); thus $\mathbb{F}_2 = A$. In particular, from the above, $\mathbb{F}_2 = \mathbb{H}$ is now known to be associative.

Using (11.8) again, we infer that also $\mathbb{F}_3 = \mathbb{O}$ is alternative. Thus our previous arguments apply here, as well, with the result that the subalgebra A spanned by 1 and by elements u, v, w of the specified kind is associative. Any two such algebras are isomorphic, since multiplication is entirely determined by the given multiplication table. As $\mathbb{H}$ is also spanned by such a basis, every such subalgebra of $\mathbb{F}_3 = \mathbb{O}$ is isomorphic to $\mathbb{H}$. □

Since, by (11.4), there are multiplicative inverses, associativity of one of our algebras implies that it satisfies all the axioms of a (not necessarily commutative) field. From the preceding discussion, in particular from (11.10), we thus infer the following.

11.11. $\mathbb{C} = \mathbb{F}_1$ and $\mathbb{H} = \mathbb{F}_2$ *are associative*; $\mathbb{C}$ *is a commutative field, and* $\mathbb{H}$ *is a skew field.*

(The non-commutativity of $\mathbb{F}_2$ is obvious from the multiplication table in (11.10).) □

It was already noted in the proof of (11.10) that, by (11.8), we now know $\mathbb{F}_3 = \mathbb{O}$ to be alternative. This has the following consequence.

11.12. *For $m \leq 3$, the algebra $\mathbb{F}_m$ has no zero divisors, and therefore is a division algebra in the sense that, for $a \neq 0$, the $\mathbb{R}$-linear maps $x \mapsto ax$ and $x \mapsto xa$ are bijective.*

Indeed, if $xy = 0$, then $\|x\|^2 y = \bar{x}(xy) = 0$, so that $x = 0$ or $y = 0$. □

We summarize the properties of $\mathbb{O}$ which we have now obtained (11.8, 9 and 12).

11.13. $\mathbb{O}$ *is an* **alternative field***, i.e., an alternative division algebra, and it is biassociative.* □

11.14 Multiplicativity of the norm. *If $m \leq 3$, i.e., if $\mathbb{F}_m$ is $\mathbb{R}$, $\mathbb{C}$, $\mathbb{H}$, or $\mathbb{O}$, then*

(1) $$\|xy\|^2 = \|x\|^2 \|y\|^2$$

for $x, y \in \mathbb{F}_m$. Moreover, for each $a \in \mathbb{F}_m$,

(2) $$\langle x \mid \bar{a}y \rangle = \langle ax \mid y \rangle \quad \text{and} \quad \langle x \mid y\bar{a} \rangle = \langle xa \mid y \rangle.$$

In particular, for $p \in \operatorname{Pu} \mathbb{F}_m$, we have

(3) $$\langle px \mid x \rangle = 0.$$

Remark. For $m \geq 4$, these equalities are not true, see (11.17).

Proof. By (11.9) and (11.11), all the algebras in question are at least biassociative. Therefore, $\|xy\|^2 = \overline{(xy)}(xy) = \bar{y}(\bar{x}x)y = \|x\|^2 \bar{y}y = \|x\|^2 \|y\|^2$, proving (1). For (2), we may assume $a \neq 0$. Put $z = a^{-1}y$, so that $az = a(a^{-1}y) = y$. Using (1), we obtain $\langle ax \mid y \rangle = \langle ax \mid az \rangle = \|ax + az\|^2 - \|ax\|^2 - \|az\|^2 = \|a\|^2(\|x+z\|^2 - \|x\|^2 - \|z\|^2) = \|a\|^2 \langle x \mid z \rangle = \langle x \mid \|a\|^2 z \rangle = \langle x \mid (\bar{a}a)z \rangle = \langle x \mid \bar{a}(az) \rangle = \langle x \mid \bar{a}y \rangle$. The first equality of (2) is thus proved. The second is obtained by applying conjugation, which preserves the inner product (11.3). Identity (3) follows from (2), since $\bar{p} = -p$. □

The following result is a motivation ex post of the Cayley–Dickson process (11.1).

11.15 Proposition: Constructing $\mathbb{O}$ from quaternion subfields. *Let H be a subalgebra of $\mathbb{O}$ isomorphic to $\mathbb{H}$, and let $z \in \operatorname{Pu} \mathbb{O}$ be of unit length $\|z\|^2 = 1$ and orthogonal to H. Then the 4-dimensional $\mathbb{R}$-linear subspace Hz is orthogonal to H, so that the $\mathbb{R}$-vector space $\mathbb{O}$ decomposes into the direct sum*

$$\mathbb{O} = H \oplus Hz.$$

For $a, b, c, d \in H$, we have

$$(a + bz)(c + dz) = (ac - \bar{d}b) + (da + b\bar{c})z.$$

Proof. Clearly, H is invariant under conjugation (11.5). By (11.14(2)), we have $\langle a \mid bz \rangle = \langle \bar{b}a \mid z \rangle = 0$, so that H and Hz are orthogonal. Using alternativity (11.8) we obtain, for $x, y, w \in \mathbb{O}$,

$$\|x+y\|^2 \cdot w = (x+y)((\bar{x}+\bar{y})w) = (\|x\|^2 + \|y\|^2)w + x(\bar{y}w) + y(\bar{x}w).$$

Now $\|x+y\|^2 = \|x\|^2 + \|y\|^2$ whenever $\langle x \mid y \rangle = 0$, so that

(∗)   if $\langle x \mid y \rangle = 0$, then $x(\bar{y}w) = -y(\bar{x}w)$.

In particular, for $w = 1$,

(∗∗)   if $y \in \operatorname{Pu}\mathbb{O}$ and $\langle x \mid y \rangle = 0$, then $xy = y\bar{x}$.

Using these pieces of information together with alternativity, we may now evaluate the terms obtained from expanding the left hand side of the asserted product formula as follows.

$$(bz)c = \bar{c}(bz) = \bar{c}(z\bar{b}) = -\bar{z}(c\bar{b}) = -(b\bar{c})\bar{z} = (b\bar{c})z,$$
$$a(dz) = a(z\bar{d}) = -\bar{z}(\bar{a}\bar{d}) = z(\overline{da}) = (da)z,$$
$$(bz)(dz) = -\bar{d}((\bar{z}\bar{b})z) = -\bar{d}((b\bar{z})z) = -\bar{d}(b\|z\|^2) = -\bar{d}b. \quad \square$$

11.16 Cayley triples are triples $u, v, z \in \operatorname{Pu}\mathbb{O}$, $\|u\|^2 = \|v\|^2 = \|z\|^2 = 1$ such that u and v are mutually orthogonal, and such that z is orthogonal to u, v and uv.

This notion is symmetric in the sense that a permutation of a Cayley triple is another Cayley triple. For example, by (11.15 and 10), we obtain the following identities.

(1) $u(vz) = (vu)z = -(uv)z$,

(2) $(vz)u = -(vu)z$.

Thus, in particular, $u(vz) = -(vz)u$. From (11.6(2)) one now infers that the pure element vz is orthogonal to u. In passing, we note that (1) shows $\mathbb{O}$ to be non-associative.

By (11.10 and 9), for a given Cayley triple u, v, z, the pure octonions u, v and $w := uv$ form a Hamilton triple, and together with 1 they span a subalgebra H isomorphic to $\mathbb{H}$. Furthermore, according to (11.15), the elements 1, u, v, w, z, uz, vz, wz constitute a basis of $\mathbb{O}$ as a vector space over $\mathbb{R}$. From (11.10 and 15), it is easy to compute a multiplication table for this basis; even without explicit computation, the following assertion is obvious.

For all Cayley triples u, v, z one obtains the same multiplication table with respect to the basis 1, u, v, w, z, uz, vz, $(uv)z$.

In particular, this shows that, for any two Cayley triples, the $\mathbb{R}$-linear transformation of $\mathbb{O}$ mapping the basis obtained in this way from the first Cayley triple onto the basis corresponding to the second Cayley triple is an automorphism of $\mathbb{O}$. We thus have proved the following.

For any two Cayley triples of $\mathbb{O}$, there is a unique automorphism of $\mathbb{O}$ mapping the first Cayley triple onto the second.

This observation may be used to reduce the effort in computing the multiplication table mentioned above.

The *standard Cayley triple* and the associated basis of $\mathbb{O}$ are obtained as follows. Let $i \in \mathbb{C}$ be an 'imaginary unit', that is, $i \in \operatorname{Pu}\mathbb{C}$ with $i^2 = -1$. In $\mathbb{H} = \mathbb{C} \times \mathbb{C}$, one usually considers the basis

$$1 = (1, 0), \quad i = (i, 0), \quad j := (0, 1), \quad k := ij = (i, 0)(0, 1) = (0, i).$$

The triple i, j, k is clearly a Hamilton triple as in (11.10). In $\mathbb{O} = \mathbb{H} \times \mathbb{H}$, by identifying $\mathbb{H}$ with the subalgebra $\mathbb{H} \times \{0\}$, we find these elements again, namely

$$i \stackrel{\wedge}{=} (i, 0), \quad j \stackrel{\wedge}{=} (j, 0), \quad k \stackrel{\wedge}{=} (k, 0).$$

They are elements of $\operatorname{Pu}\mathbb{O}$. Putting

$$l = (0, 1) \in \operatorname{Pu}\mathbb{O},$$

we note that i, j, l is a Cayley triple, and we compute the following products.

$$\begin{aligned} il &= (i, 0)(0, 1) = (0, i), \\ jl &= (j, 0)(0, 1) = (0, j), \\ kl &= (k, 0)(0, 1) = (0, k). \end{aligned}$$

A discussion and presentation of the complete multiplication table for the basis $1, i, j, k, l, il, jl, kl$ of $\mathbb{O}$ may be found for instance in Porteous [81] Chap. 14, p. 277 ff. Following an idea of Freudenthal [85] 1.5.13, p. 19, one may represent this multiplication table graphically as shown in Figure 11a, using the projective plane with 7 points.

Figure 11a is to be interpreted as follows: If basis elements a, b, c are on a line of this projective plane, then $ab = \pm c$, the sign depending on whether or not the cyclic order of (a, b, c) matches with the orientation indicated in the diagram. For example, $ij = k$ but $ji = -k$. (When comparing this to Freudenthal, loc. cit., one should note that Freudenthal uses a different basis, so that the orientations implicit in his diagram do not completely agree with ours. Also, there is a misprint; $e_4 e_7$ should be e_3.)

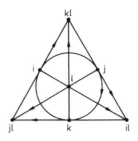

Figure 11a

11.17 Warning. If $m \geq 4$, then $\mathbb{F}_m$ has zero divisors. Consequently, $\mathbb{F}_m$ is not alternative, and the norm in $\mathbb{F}_m$ is not multiplicative.

It suffices to show this for $\mathbb{F}_4 \subseteq \mathbb{F}_m$. Let u, v, z be a Cayley triple of $\mathbb{O}$ as discussed in (11.16). In $\mathbb{F}_4 = \mathbb{O} \times \mathbb{O}$, we then compute $(z, -u)(vz, uv) = \big(z(vz) - (uv)u, (uv)z + u(vz)\big) = \big(-z(zv) + (vu)u, 0\big) = \big((-z^2 + u^2)v, 0\big) = (0, 0)$; here, we have applied (11.10) and the alternativity of $\mathbb{O}$. Thus, there are zero divisors; hence, clearly, the norm cannot be multiplicative. If $\mathbb{F}_m$ were alternative, there could be no zero divisors by the argument of (11.12).

11.18 Notes. Biassociativity as a consequence of alternativity can be derived much more generally for alternative rings. This result is commonly ascribed to Artin, see Zorn [31].

There are further weak associativity properties of $\mathbb{O}$, which involve quadruple products of three elements, the so-called Moufang identities. Usually they are derived from alternativity by algebraic means. We shall obtain them as a by-product from geometric considerations, see (12.14 and 15) and the references given there.

We have seen that among the algebras $\mathbb{F}_m$ only the first few have satisfactory associativity properties, are division algebras (without zero divisors), and have multiplicative norms. These facts are special cases of much more general results: A theorem of Frobenius [1878] says that $\mathbb{R}$, $\mathbb{C}$, and $\mathbb{H}$ are the only fields which contain $\mathbb{R}$ as central subfield and have finite dimension over $\mathbb{R}$. By a theorem of Zorn [33], the octonions form the only non-associative alternative field of finite dimension over $\mathbb{R}$.

For modern proofs of these results see Herstein [64] Chap. 7 Sect. 3, p. 326 ff, Palais [75], Ebbinghaus et al. [90, 92] Chap. 8 §2 and Chap. 9 §3. Much more generally, there is a complete structure theory for non-associative alternative fields, culminating in the theorem of Bruck–Kleinfeld–Skornyakov, which is presented, for instance, in the following books: Kleinfeld [63], Pickert [75] Chap. 6, Theorem 13, p. 175 and Theorem 15, p. 177, Schafer [66] Theorem 3.17, p. 56, Zhevlakov et al. [82] 7.3 Corollary 2, p. 152; for further references, in particular to the original papers, one may consult Grundhöfer–Salzmann [90] XI.7.8, p. 322.

The algebras $\mathbb{R}$, $\mathbb{C}$, $\mathbb{H}$, and $\mathbb{O}$ can also be characterized as being the only finite-dimensional composition algebras over $\mathbb{R}$, i.e., finite-dimensional algebras admitting a positive definite quadratic form which is multiplicative; this is a theorem of Hurwitz [1898], see also Freudenthal [85] 1.5.14 p. 19, Harvey–Lawson [82] Appendix IVA, p. 140 ff and Theorem $A.12$, p. 143, Ebbinghaus et al. [90, 92] Chap. 10 §1, Curtis [90] VD, p. 156 ff. For generalizations see Zhevlakov et al. [82] Chap. 2, p. 22 ff.

These facts are of an algebraic nature. Adams, Atiyah, Bott, Hirzebruch, Kervaire, and Milnor proved the following (much broader) result by using methods of algebraic topology: Finite-dimensional (not necessarily associative) division algebras over $\mathbb{R}$ can only exist in dimensions 1, 2, 4, and 8, that is, in the dimensions of the classical examples $\mathbb{R}$, $\mathbb{C}$, $\mathbb{H}$, and $\mathbb{O}$. See e.g. Bott–Milnor [58], Milnor [58], Kervaire [58]. An even more general, purely topological result determining the possible dimensions of spheres which admit a multiplication with (homotopy) unit will play an important rôle in Chap. 5, cf. (52.5 and 8). A survey by Hirzebruch of these topics may be found in Ebbinghaus et al. [90, 92] Chap. 11. For a recent proof using mainly analytic means see Gilkey [87].

There are variations of the Cayley–Dickson process. When these are carried out over other scalar fields than $\mathbb{R}$, higher stages of the process may remain free of zero divisors, thus producing (non-alternative) division algebras of dimension at least 16, in contrast with the above and (11.17). In this way, one obtains, for instance, a 16-dimensional division algebra over $\mathbb{Q}$. See Schafer [45] and R.B. Brown [67].

The deviation in $\mathbb{F}_m$ from commutavity and associativity is measured by the following results.

11.19 The center of $\mathbb{F}_m$. *For $m > 1$, the only elements of $\mathbb{F}_m$ commuting with all elements are the scalars in $\mathbb{R} \subset \mathbb{F}_m$; i.e., the center of $\mathbb{F}_m$ is $\mathbb{R}$.*

Proof. Let $c, d \in \mathbb{F}_{m-1}$ be such that $(c, d) \in \mathbb{F}_m = \mathbb{F}_{m-1} \times \mathbb{F}_{m-1}$ commutes with each element of $\mathbb{F}_m$. Then, in particular, $(-\bar{d}, \bar{c}) = (0, 1)(c, d) = (c, d)(0, 1) = (-d, c)$, so that $c = \bar{c} \in \mathbb{R}$, $d = \bar{d} \in \mathbb{R}$ (11.2). Moreover, for arbitrary $z \in \mathbb{F}_{m-1}$, we have $(zc, dz) = (z, 0)(c, d) = (c, d)(z, 0) = (cz, d\bar{z})$, so that $d(z - \bar{z}) = 0$. Since d is a real scalar, it must be zero, whence $(c, d) = (c, 0) \in \mathbb{R}$. □

11.20 The kernel of $\mathbb{F}_m$ is defined as

$$\operatorname{Ker} \mathbb{F}_m = \{\, a \in \mathbb{F}_m \mid x(ya) = (xy)a \text{ for all } x, y \in \mathbb{F}_m \,\} \, .$$

The fields $\mathbb{F}_m$, $m \leq 2$ coincide with their kernels. We shall show that

$$\operatorname{Ker} \mathbb{F}_m = \mathbb{R} \quad \text{for } m \geq 3 \, ,$$

so that, in particular,

$$\operatorname{Ker} \mathbb{O} = \mathbb{R} \, .$$

Remarks. 1) We are particularly interested in the case of $\mathbb{F}_3 = \mathbb{O}$. The geometric significance of the result will be discussed in (12.13), see also (23.11) and (25.4).

2) The same result can be obtained more generally for arbitrary 8-dimensional real division algebras by means of algebraic topology, see Buchanan–Hähl [77].

Proof of (11.20). Let $m \geq 3$. It is clear that $\operatorname{Ker} \mathbb{F}_m$ contains the scalar field $\mathbb{R}$. Conversely, let $a \in \operatorname{Ker} \mathbb{F}_m$, and write $a = (c, d) \in \mathbb{F}_{m-1} \times \mathbb{F}_{m-1} = \mathbb{F}_m$. For $u, v \in \mathbb{F}_{m-1}$ and for

$$x = (u, 0), \ y = (v, 1) \in \mathbb{F}_{m-1} \times \mathbb{F}_{m-1} = \mathbb{F}_m,$$

by definition of the multiplication in $\mathbb{F}_m$ via the Cayley–Dickson process, one has

(1)
$$\begin{aligned}(xy)a &= (uv, u)(c, d) = \bigl((uv)c - \bar{d}u, d(uv) + u\bar{c}\bigr) \\ x(ya) &= (u, 0)(vc - \bar{d}, dv + \bar{c}) = \bigl(u(vc) - u\bar{d}, (dv)u + \bar{c}u\bigr).\end{aligned}$$

Because of $a \in \operatorname{Ker} \mathbb{F}_m$, these two products coincide. With $v = 0$ and u arbitrary, we obtain that $\bar{c}$ and $\bar{d}$ belong to the center of $\mathbb{F}_{m-1}$, which is $\mathbb{R}$ by (11.19), so that $c, d \in \mathbb{R}$. Now, with arbitrary u and v, a comparison of the second components in (1) yields

(2)
$$d(uv) = (dv)u = d(vu).$$

Since $\mathbb{H} = \mathbb{F}_2 \subseteq \mathbb{F}_{m-1}$ is not commutative, we may choose u and v in such a way that $uv \neq vu$; it then follows from (2) that the scalar d must be 0, so that $a = (c, d) = (c, 0) \in \mathbb{R}$. □

Orthogonal groups

In the sequel, let $\mathbb{F}$ be one of the algebras $\mathbb{F}_m$ for $0 \leq m \leq 3$, i.e., one of the algebras $\mathbb{R}, \mathbb{C}, \mathbb{H}$, or $\mathbb{O}$. Using the multiplication in $\mathbb{F}$, one may give convenient descriptions of certain groups of orthogonal transformations acting on the $\mathbb{R}$-vector space $\mathbb{F} = \mathbb{R}^n$, where

$$n = \dim_\mathbb{R} \mathbb{F} = 2^m \in \{1, 2, 4, 8\}.$$

11.21. The group of $\mathbb{R}$-linear transformations of $\mathbb{F}$ which preserve the norm form $\|\ \|^2$, or, equivalently, which are orthogonal with respect to the inner product $\langle\ |\ \rangle$ on $\mathbb{F}$ (11.3), is denoted by

$$\mathrm{O}_n\mathbb{R} = \bigl\{ C : \mathbb{F} \to \mathbb{F} \bigm| C \text{ is } \mathbb{R}\text{-linear}, \forall x \in \mathbb{F} : \|Cx\|^2 = \|x\|^2 \bigr\}.$$

The normal subgroup

$$\mathrm{SO}_n\mathbb{R} = \{ C \in \mathrm{O}_n\mathbb{R} \mid \det C = 1 \}$$

has index 2. Together with the scalar homotheties, these groups generate the group of 'similitudes'

$$\mathrm{GO}_n\mathbb{R} := \{\, \mathbb{F} \to \mathbb{F} : x \mapsto r \cdot Cx \mid r \in \mathbb{R} \setminus \{0\},\, C \in \mathrm{O}_n\mathbb{R} \,\},$$

and the group of 'direct similitudes'

$$\mathrm{GO}_n^+\mathbb{R} := \{\, \mathbb{F} \to \mathbb{F} : x \mapsto r \cdot Cx \mid r \in \mathbb{R} \setminus \{0\},\, C \in \mathrm{SO}_n\mathbb{R} \,\}.$$

Note that $\mathrm{GO}_1\mathbb{R} = \mathrm{GO}_1^+\mathbb{R} = \{\, x \mapsto rx \mid r \in \mathbb{R} \setminus \{0\} \,\}$.

11.22 Lemma.

(a) *For each $a \in \mathbb{F} \setminus \{0\}$, the transformations $x \mapsto ax$ and $x \mapsto xa$ of $\mathbb{F} = \mathbb{R}^n$ belong to $\mathrm{GO}_n^+\mathbb{R}$. If $\|a\|^2 = 1$, they belong to $\mathrm{O}_n\mathbb{R}$, and even to $\mathrm{SO}_n\mathbb{R}$ for $\mathbb{F} \in \{\mathbb{C}, \mathbb{H}, \mathbb{O}\}$.*

(b) *The group $\mathrm{SO}_n\mathbb{R}$ is generated by the transformations*

$$\mathbb{F} \to \mathbb{F} : x \mapsto axa, \quad \text{where } a \in \mathbb{F},\, \|a\|^2 = 1.$$

(c) *The group $\mathrm{GO}_n^+\mathbb{R}$ is generated by the transformations*

$$\mathbb{F} \to \mathbb{F} : x \mapsto raxa, \quad \text{where } a \in \mathbb{F},\, \|a\|^2 = 1 \text{ and } r \in \mathbb{R} \setminus \{0\}.$$

Note that in the non-commutative case the transformations in (b) and in (c) do not constitute a group by themselves. For more precise information in the associative case, see (11.23).

Proof. Since the norm is multiplicative (11.14), the $\mathbb{R}$-linear transformations in (a) change the square of the norm by a factor $\|a\|^2$; hence, upon multiplying by $\|a\|^{-1}$, one obtains an element of $\mathrm{O}_n\mathbb{R}$. If $\mathbb{F} \in \{\mathbb{C}, \mathbb{H}, \mathbb{O}\}$, every element belongs to a subfield isomorphic to $\mathbb{C}$, see (11.7), and therefore is a square; by alternativity (11.8), it follows that the given transformations are squares of transformations of the same kind, and thus have positive determinant over $\mathbb{R}$. For $\mathbb{F} = \mathbb{R}$ use the fact that $\mathrm{GO}_1\mathbb{R} = \mathrm{GO}_1^+\mathbb{R}$.

Assertion (b) is trivial for $\mathbb{F} = \mathbb{R}$, since $\mathrm{SO}_1\mathbb{R} = \{\mathrm{id}\}$. In general, the transformations considered in (b) belong to $\mathrm{SO}_n\mathbb{R}$ by (a). In order to show that they generate $\mathrm{SO}_n\mathbb{R}$, we use reflections in hyperplanes. (These reflections do *not* belong to $\mathrm{SO}_n\mathbb{R}$!) For any $a \in \mathbb{F}$ with $\|a\|^2 = 1$, the reflection in the hyperplane $a^\perp$ orthogonal to a is the mapping

$$\varrho_a : x \mapsto x - 2a \cdot \frac{\langle x \mid a \rangle}{\langle a \mid a \rangle} = x - 2a \cdot \frac{(\bar{x}a + \bar{a}x)}{2\|a\|^2} = x - a(\bar{x}a + \bar{a}x) = -a\bar{x}a\,;$$

note that $a(\bar{a}x) = \|a\|^2 x = x$ by alternativity (11.8). As is well known, every element of $\mathrm{SO}_n\mathbb{R}$ is the product of an even number of such hyperplane reflections, that is, the product of transformations having the form $\varrho_a\varrho_b$ for $a, b \in \mathbb{F}$ with $\|a\|^2 = 1 = \|b\|^2$, see e.g. Porteous [81] p. 159–160. Now $x^{\varrho_a\varrho_b} = -b(\overline{-a\bar{x}a})b = b(\bar{a}x\bar{a})b$, so that $\varrho_a\varrho_b$ can also be written as the composition of the transformations $x \mapsto \bar{a}x\bar{a}$ and $x \mapsto bxb$. □

In contrast to the preceding result, the following does not hold in the case $\mathbb{F} = \mathbb{O}$ for lack of associativity.

11.23 Corollary. *For $\mathbb{F} \in \{\mathbb{C}, \mathbb{H}\}$ and $n = \dim_\mathbb{R} \mathbb{F}$, the group $\mathrm{GO}_n^+\mathbb{R}$ consists precisely of the transformations*

$$\mathbb{F} \to \mathbb{F} : x \mapsto b^{-1}xa,$$

where $a, b \in \mathbb{F} \setminus \{0\}$, and the group $\mathrm{SO}_n\mathbb{R}$ consists of these transformations for $\|a\|^2 = 1 = \|b\|^2$.

Proof. By (11.22a), these transformations belong to $\mathrm{SO}_n\mathbb{R}$ and to $\mathrm{GO}_n^+\mathbb{R}$, respectively, and they form subgroups of $\mathrm{SO}_n\mathbb{R}$ and $\mathrm{GO}_n^+\mathbb{R}$. (Here, the associativity of $\mathbb{F}$ is required.) On the other hand, these subgroups contain the transformations of (11.22b) (put $b = a^{-1}$), and of (11.22c) (put $b = (ra)^{-1}$), which generate $\mathrm{SO}_n\mathbb{R}$ and $\mathrm{GO}_n^+\mathbb{R}$, respectively. □

11.24. Using $\mathbb{F} = \mathbb{H}$, we may also obtain a description of $\mathrm{SO}_3\mathbb{R}$ from (11.23). The subspace $\mathrm{Pu}\,\mathbb{H} \cong \mathbb{R}^3$ of the $\mathbb{R}$-vector space $\mathbb{H}$ is the orthogonal complement of $\mathbb{R} = \mathbb{R} \cdot 1$ in $\mathbb{H}$, see (11.6). The group of $\mathbb{R}$-linear transformations of $\mathbb{H}$ which are orthogonal with respect to the inner product and which fix 1 may therefore be identified with the group of orthogonal transformations of $\mathrm{Pu}\,\mathbb{H} \cong \mathbb{R}^3$. Abusing notation somewhat, we denote both groups by $\mathrm{O}_3\mathbb{R}$. The normal subgroup $\mathrm{SO}_3\mathbb{R}$ of elements of determinant 1 has index 2. Now the transformation $\mathbb{H} \to \mathbb{H} : x \mapsto b^{-1}xa$ in (11.23) fixes 1 if, and only if $b = a$, whence the following result.

Corollary. *$\mathrm{SO}_3\mathbb{R}$ consists precisely of the transformations*

$$\mathrm{int}(a) : \mathbb{H} \to \mathbb{H} : x \mapsto a^{-1}xa,$$

where $a \in \mathbb{H}$ satisfies $\|a\|^2 = 1$. □

The notation $\mathrm{int}(a)$ reflects the fact that this transformation is an inner automorphism of the skew field $\mathbb{H}$.

The preceding result shows that $\mathrm{SO}_3\mathbb{R}$ is an epimorphic image of the unit sphere

$$\mathbb{S}_3 = \{\, a \in \mathbb{H} \mid \|a\|^2 = 1 \,\}$$

of $\mathbb{H} = \mathbb{R}^4$, which, by multiplicativity of the norm (11.14), is a subgroup of the multiplicative group $\mathbb{H}^\times$.

11.25 Corollary. *The kernel of the epimorphism* $\mathrm{int} : \mathbb{S}_3 \to \mathrm{SO}_3\mathbb{R} : a \mapsto \mathrm{int}(a)$ *is* $\{1, -1\}$.

Proof. An element $a \in \mathbb{S}_3$ belongs to the kernel if, and only if $a^{-1}xa = x$ for all $x \in \mathbb{H}$, or, equivalently, if a belongs to the center of $\mathbb{H}$, which is $\mathbb{R}$, cf. (11.19); now $\mathbb{R} \cap \mathbb{S}_3 = \{1, -1\}$. □

Remark. In other words, $\mathbb{S}_3$ is a two-fold covering group of $\mathrm{SO}_3\mathbb{R}$. Since the sphere $\mathbb{S}_3$ is simply connected, this shows that $\mathbb{S}_3$ is (isomorphic to) the *universal covering group* of $\mathrm{SO}_3\mathbb{R}$, cf. (94.2), which is denoted systematically by $\mathrm{Spin}_3\mathbb{R}$. This group shall now be interpreted as a $\mathbb{C}$-linear group.

11.26 An isomorphism $\mathrm{Spin}_3\mathbb{R} \cong \mathrm{SU}_2\mathbb{C}$. We consider $\mathbb{H} = \mathbb{C} \times \mathbb{C}$ as a right vector space over the subfield $\mathbb{C} \,\hat{=}\, \mathbb{C} \times \{0\}$. Thus, scalar multiplication of $(a, b) \in \mathbb{C} \times \mathbb{C} = \mathbb{H}$ by the scalar $c \in \mathbb{C}$ is given by

$$(a, b)c = (a, b)(c, 0) = (ac, b\bar{c}).$$

The elements $1 \,\hat{=}\, (1, 0)$ and $j = (0, 1)$ of $\mathbb{H} = \mathbb{C} \times \mathbb{C}$ form a $\mathbb{C}$-basis. Under the standard Hermitian form with respect to this basis, the inner product of $x = x_1 + jx_2$ and $y = y_1 + jy_2$, for $x_1, x_2, y_1, y_2 \in \mathbb{C}$, is, by definition, $x_1\overline{y_1} + x_2\overline{y_2}$. Now note that, for $c \in \mathbb{C}$, we have $cj = j\bar{c}$ as $ij = -ji$, see also (11.15(∗)). Therefore, $\|x\|^2$ may be expressed in the following way.

$$(*) \quad \begin{aligned} \|x\|^2 &= (x_1 + jx_2)(\overline{x_1} - jx_2) = x_1\overline{x_1} + jx_2\overline{x_1} - j\overline{x_1}x_2 - j^2 \overline{x_2}x_2 \\ &= x_1\overline{x_1} + \overline{x_2}x_2 \, ; \end{aligned}$$

this is the inner product of x by itself under the standard Hermitian form. For $a \in \mathbb{H}$, the transformation

$$\lambda_a : \mathbb{H} \to \mathbb{H} : x \mapsto ax$$

is $\mathbb{C}$-linear by associativity. Moreover, if $a \in \mathbb{S}_3$, the norm is preserved, see (11.14), so that λ_a is unitary with respect to the standard Hermitian form.

By $\mathrm{U}_2\mathbb{C}$ we mean the group of all unitary transformations of the 2-dimensional $\mathbb{C}$-vector space $\mathbb{H}$, and $\mathrm{SU}_2\mathbb{C}$ is the normal subgroup consisting of the unitary transformations having (complex) determinant 1.

Proposition. *The group* $\{\lambda_a \mid a \in \mathbb{S}_3\}$, *which clearly is isomorphic to* $\mathbb{S}_3 = \mathrm{Spin}_3\mathbb{R}$, *coincides with* $\mathrm{SU}_2\mathbb{C}$.

Proof. For $a_1, a_2 \in \mathbb{C}$ and $a = a_1 + ja_2 = a_1 + \overline{a_2}j$, we have $aj = a_1j + \overline{a_2}j^2 = -\overline{a_2} + j\overline{a_1}$, so that the matrix of λ_a is

$$\begin{pmatrix} a_1 & -\overline{a_2} \\ a_2 & \overline{a_1} \end{pmatrix}.$$

It has determinant 1 and is unitary if, and only if, $a_1\overline{a_1} + a_2\overline{a_2} = 1$. By (*), this is equivalent to $a \in \mathbb{S}_3$. Hence, $\{\lambda_a \mid a \in \mathbb{S}_3\}$ is a subgroup of $\mathrm{SU}_2\mathbb{C}$, and the two groups coincide since both are sharply transitive on $\mathbb{S}_3$. □

Automorphisms

11.27. We now study the group $\mathrm{Aut}\,\mathbb{F}_m$ of automorphisms of the ring $\mathbb{F}_m$. That is, we consider automorphisms with respect to addition and multiplication in $\mathbb{F}_m$; linearity over $\mathbb{R}$ is not presupposed. As is well known, the only automorphism of $\mathbb{F}_0 = \mathbb{R}$ is the identity. The only *continuous* automorphisms of the field $\mathbb{F}_1 = \mathbb{C}$ are the identity and conjugation, because every continuous automorphism fixes the elements of $\mathbb{R}$ (the closure of the prime field $\mathbb{Q}$) and maps i onto i or $-i$. However, $\mathbb{C}$ has a multitude of other automorphisms, see the references in (44.11). Concerning information about their topological (mis)behaviour, cf. (55.22a) and the references given there.

For $m \geq 2$, the picture becomes simpler again:

11.28 Proposition. *Let $m \geq 2$. Then an automorphism $\alpha \in \mathrm{Aut}\,\mathbb{F}_m$ fixes every element of the center $\mathbb{R}$ of $\mathbb{F}_m$, and so is $\mathbb{R}$-linear. Moreover, α leaves $\mathrm{Pu}\,\mathbb{F}_m$ invariant, commutes with conjugation, and is orthogonal with respect to the inner product of $\mathbb{F}_m$.*

Proof. The center $\mathbb{R}$ of $\mathbb{F}_m$ (11.19) is invariant under α, and $\alpha|_\mathbb{R} = \mathrm{id}$ since $\mathbb{R}$ has no other automorphism. This is equivalent to $\mathbb{R}$-linearity. The invariance of $\mathrm{Pu}\,\mathbb{F}_m = \{x \in \mathbb{F}_m \mid x^2 \leq 0\}$, see (11.6), is now immediate. Therefore conjugation, which on $\mathrm{Pu}\,\mathbb{F}_m$ induces $-\mathrm{id}$, commutes with α. It follows that $\|x^\alpha\|^2 = x^\alpha \overline{x^\alpha} = x^\alpha \bar{x}^\alpha = (x\bar{x})^\alpha = (\|x\|^2)^\alpha = \|x\|^2$, whence α is orthogonal. □

We now determine $\mathrm{Aut}\,\mathbb{H}$.

11.29. Because of associativity of $\mathbb{H}$, conjugation by an element $a \in \mathbb{H} \setminus \{0\}$ is an automorphism of the skew field $\mathbb{H}$, the *inner automorphism* $\mathrm{int}(a) : \mathbb{H} \to \mathbb{H} : x \mapsto a^{-1}xa$. Let $\mathrm{Int}\,\mathbb{H}$ be the group of all these inner automorphisms. Since $\|a\|$ belongs to the center $\mathbb{R}$ of $\mathbb{H}$, we have $\mathrm{int}\left(\frac{1}{\|a\|}a\right) = \mathrm{int}(a)$. Now $\frac{1}{\|a\|}a$ has norm 1, so that

$$\mathrm{Int}\,\mathbb{H} = \{\mathrm{int}(a) \mid a \in \mathbb{H}, \|a\|^2 = 1\}.$$

According to (11.28 and 24), we already know that $\operatorname{Aut}\mathbb{H}$ is contained in the group $O_3\mathbb{R}$ of orthogonal transformations of $\mathbb{H} = \mathbb{R}^4$ fixing 1, and that $\operatorname{Int}\mathbb{H} = SO_3\mathbb{R}$.

Proposition. $\operatorname{Aut}\mathbb{H} = \operatorname{Int}\mathbb{H} = SO_3\mathbb{R}$.

Proof. One merely has to show that $\operatorname{Aut}\mathbb{H} \subseteq O_3\mathbb{R}$ cannot be bigger than $SO_3\mathbb{R} = \operatorname{Int}\mathbb{H}$. Now, $SO_3\mathbb{R}$ has index 2 in $O_3\mathbb{R}$, whence, if $\operatorname{Aut}\mathbb{H}$ were bigger than $SO_3\mathbb{R}$, we would have $\operatorname{Aut}\mathbb{H} = O_3\mathbb{R}$. In particular, conjugation (which on $\operatorname{Pu}\mathbb{H}$ induces $-\operatorname{id} \in O_3\mathbb{R}$) would have to be an element of $\operatorname{Aut}\mathbb{H}$, but conjugation is an *anti-automorphism* and not an automorphism, as $\mathbb{H}$ is not commutative. □

Finally, we study $\operatorname{Aut}\mathbb{O}$. We begin by stating a few transitivity properties which follow from the sharp transitivity of $\operatorname{Aut}\mathbb{O}$ on the set of Cayley triples, compare (11.16).

11.30 Lemma.

(a) $\operatorname{Aut}\mathbb{O}$ *acts transitively on* $\{(u, v) \mid u, v \in \operatorname{Pu}\mathbb{O}, u \perp v, \|u\|^2 = 1 = \|v\|^2\}$. *In other words,* $\operatorname{Aut}\mathbb{O}$ *is transitive on the 6-sphere* $\{u \in \operatorname{Pu}\mathbb{O} \mid \|u\|^2 = 1\}$, *and the stabilizer* $(\operatorname{Aut}\mathbb{O})_i$ *is transitive on the unit sphere of the orthogonal space of i in* $\operatorname{Pu}\mathbb{O}$, *that is, on the 5-sphere* $\{u \in \operatorname{Pu}\mathbb{O} \mid \|u\|^2 = 1, u \perp i\}$.
(b) *The stabilizer* $(\operatorname{Aut}\mathbb{O})_{i,j}$ *is sharply transitive on the unit sphere of* $\{0\} \times \mathbb{H} \subseteq \mathbb{H} \times \mathbb{H} = \mathbb{O}$, *that is, on the 3-sphere* $\{(0, b) \in \mathbb{H} \times \mathbb{H} \mid \|b\|^2 = 1\}$.

Proof. Concerning (a), note that, for any two pure orthogonal elements u, v of unit length, there is a Cayley triple having u, v as the first two elements. The 3-sphere in (b) consists precisely of the elements $z \in \operatorname{Pu}\mathbb{O}$ with $\|z\|^2 = 1$ and such that i, j, z is a Cayley triple; this is because i, j, ij span $\operatorname{Pu}\mathbb{H} \times \{0\}$, and $\{0\} \times \mathbb{H}$ is the orthogonal space of $\operatorname{Pu}\mathbb{H} \times \{0\}$ in $\operatorname{Pu}\mathbb{O}$, cf. (11.3 and 6). □

11.31 Lemma.

(a) *The automorphism group* $\Lambda = \operatorname{Aut}\mathbb{O}$ *acts transitively on the set* $\mathcal{H}$ *of subalgebras H of $\mathbb{O}$ with $H \cong \mathbb{H}$, so that the stabilizers Λ_H of such subalgebras are conjugate. These stabilizers cover* Λ.
(b) *The stabilizer of* $\mathbb{H} = \mathbb{H} \times \{0\} \subseteq \mathbb{H} \times \mathbb{H} = \mathbb{O}$ *is*

$$\Lambda_\mathbb{H} = \{(x, y) \mapsto (a^{-1}xa, b^{-1}ya) \mid a, b \in \mathbb{H}, \|a\|^2 = 1 = \|b\|^2\},$$

where $(x, y) \in \mathbb{O} = \mathbb{H} \times \mathbb{H}$. *It is isomorphic to* $SO_4\mathbb{R}$.
(c) *The stabilizer of i and j is*

$$\Lambda_{i,j} = \{(x, y) \mapsto (x, b^{-1}y) \mid b \in \mathbb{H}, \|b\|^2 = 1\}.$$

(d) *All involutions of Λ are conjugate, and Λ is generated by them.*
(e) $\operatorname{Aut}\mathbb{O} \subseteq SO_8\mathbb{R}$.

Note. More precisely, it can be proved that every element in Aut $\mathbb{O}$ is a product of at most two involutions, see Wonenburger [69].

Proof. (a) A subalgebra $H \cong \mathbb{H}$ intersects Pu $\mathbb{O}$ in a 3-dimensional subspace, and therefore contains pure elements u, v of unit length which are orthogonal. By (11.10), H is the span of 1, u, v, and uv. It follows from (11.30a) that $\Lambda =$ Aut $\mathbb{O}$ is transitive on the set $\mathcal{H}$ of such subalgebras. In particular, the stabilizers Λ_H for $H \in \mathcal{H}$ are conjugate, cf. (91.1a).

We must show in addition that every element $\lambda \in \Lambda$ leaves some subalgebra $H \in \mathcal{H}$ invariant. Under the orthogonal action of λ, the 7-dimensional invariant subspace Pu $\mathbb{O}$ (11.28) decomposes into 1- and 2-dimensional invariant subspaces. In particular, there is an invariant 2-dimensional subspace spanned by orthogonal pure vectors u, v of norm 1. The subalgebra H spanned by 1, u, v, uv is isomorphic to $\mathbb{H}$, see (11.10), and clearly is invariant under λ.

(b and c) By definition of the multiplication in $\mathbb{O} = \mathbb{H} \times \mathbb{H}$, one readily verifies that the right-hand side in assertion (b) is a subgroup M of Aut $\mathbb{O}$. The stabilizer $M_{i,j}$ fixes every element of the subalgebra $\mathbb{H} \times \{0\}$, which is spanned by 1, i, j, and $k = ij$. Thus, $M_{i,j}$ consists of the transformations described in assertion (b) with $a = \pm 1$, cf. (11.25), and is the group on the right-hand side of assertion (c).

In particular, $M_{i,j}$ acts transitively on the unit sphere of $\{0\} \times \mathbb{H}$. Now, by (11.30), the larger group $\Lambda_{i,j} \supseteq M_{i,j}$ is sharply transitive on this unit sphere. So, we have

(1) $$\Lambda_{i,j} = M_{i,j}.$$

Obviously, M leaves $\mathbb{H} \cong \mathbb{H} \times \{0\}$ invariant, and induces the full automorphism group Int $\mathbb{H} = SO_3\mathbb{R}$ (11.29). In particular, M is transitive on the set of pairs of pure orthogonal elements of length 1 contained in $\mathbb{H} \cong \mathbb{H} \times \{0\} \subseteq \mathbb{H} \times \mathbb{H} = \mathbb{O}$, and this set is invariant under $\Lambda_\mathbb{H}$ by (11.28). From (1), we therefore infer that $\Lambda_\mathbb{H} = M$, using the general principle (91.3).

Restriction to $\{0\} \times \mathbb{H}$ provides an epimorphism of $\Lambda_\mathbb{H}$ onto $SO_4\mathbb{R}$ according to the description of $SO_4\mathbb{R}$ given in (11.23) (with $\mathbb{F} = \mathbb{H}$). We even obtain an isomorphism; indeed, if $b^{-1}ya = y$ for all $y \in \mathbb{H}$, then $b = a$.

(d) Let $\iota \in \Lambda$ be an involution. Since ι is orthogonal (11.28), the $\mathbb{R}$-vector space $\mathbb{O}$ is the orthogonal sum of the eigenspaces F_+ and F_- of ι corresponding to the eigenvalues 1 and -1, and ι is uniquely determined by its fixed space F_+, which is a subalgebra of $\mathbb{O}$.

Now, multiplication by an element $a \in F_- \setminus \{0\}$ is a vector space isomorphism between F_+ and F_-, so that F_+ has dimension 4 over $\mathbb{R}$. By (11.9 and 10), the subalgebra F_+ is isomorphic to $\mathbb{H}$. The fact that Λ is transitive on the set

$\mathcal{H}$ of such subalgebras according to (a) now implies that all involutions are conjugate.

It is well known that $\Lambda_\mathbb{H} \cong SO_4\mathbb{R}$ is generated by its involutions, see e.g. Dieudonné [71] Chap. II §6 no. 1), p. 51. The same then holds for the conjugates Λ_H, $H \in \mathcal{H}$, and for their union Λ, see (a).

(e) From (11.28), we know that $\mathrm{Aut}\,\mathbb{O} \subseteq O_8\mathbb{R}$, and (a) and (b) imply that every automorphism has determinant 1. □

11.32 Theorem. $\mathrm{Aut}\,\mathbb{O}$ *is a simple group.*

Proof. Let $\mathsf{N} \neq \{\mathrm{id}\}$ be a normal subgroup of $\Lambda = \mathrm{Aut}\,\mathbb{O}$. We must show that $\mathsf{N} = \mathrm{Aut}\,\mathbb{O}$. By (11.31d), it suffices to prove that N contains an involution. From (11.31a) we infer that N intersects some and therefore each of the conjugate subgroups Λ_H, $H \in \mathcal{H}$ non-trivially. Now every non-trivial normal subgroup of $SO_4\mathbb{R} \cong \Lambda_\mathbb{H}$ contains an involution, as is well known, and may be seen as follows.

According to (11.23), write

$$SO_4\mathbb{R} = \left\{\, \mathbb{H} \to \mathbb{H} : y \mapsto b^{-1}ya \,\middle|\, a, b \in \mathbb{H},\ \|a\|^2 = 1 = \|b\|^2 \,\right\}.$$

The two subgroups

$$\mathsf{A} = \left\{\, \mathbb{H} \to \mathbb{H} : y \mapsto ya \,\middle|\, a \in \mathbb{H},\ \|a\|^2 = 1 \,\right\}$$

and

$$\mathsf{B} = \left\{\, \mathbb{H} \to \mathbb{H} : y \mapsto b^{-1}y \,\middle|\, b \in \mathbb{H},\ \|b\|^2 = 1 \,\right\}$$

are normal subgroups, and $\mathsf{AB} = SO_4\mathbb{R}$. The centralizer $\mathrm{Cs}\,\mathsf{B}$ of B is A (and vice versa). If N is a normal subgroup of $SO_4\mathbb{R}$ such that $\mathsf{N} \cap \mathsf{B} = \{\mathrm{id}\}$, then $\mathsf{N} \subseteq \mathrm{Cs}\,\mathsf{B} \subseteq \mathsf{A}$. As A and B are isomorphic to $\mathbb{S}_3 = \{\, a \in \mathbb{H} \mid \|a\|^2 = 1 \,\}$, it now suffices to ascertain that every nontrivial normal subgroup H of $\mathbb{S}_3$ contains an involution. For this, we consider the epimorphism $\mathrm{int} : \mathbb{S}_3 \to SO_3\mathbb{R}$ with kernel $\{1, -1\}$ of (11.25). If H contains the involution -1, the proof is finished. If not, then H is mapped isomorphically onto a non-trivial normal subgroup of $SO_3\mathbb{R}$ by int. As $SO_3\mathbb{R}$ is simple (see e.g. Artin [57] p. 178), we then have $\mathrm{int}(H) = SO_3\mathbb{R}$, so that H must contain involutions since $SO_3\mathbb{R}$ does. (In fact, the latter case does not occur, but that does not affect the argument.) □

The following result connects our discussion with the theory of simple Lie groups; it will not be used in this chapter in an essential way.

11.33 Theorem. $\mathrm{Aut}\,\mathbb{O}$ *is a compact, connected simple Lie group of dimension* 14. *By the classification of simple Lie groups, it is therefore isomorphic to the exceptional compact Lie group* $G_2 = G_2(-14)$.

Proof. By (11.28), $\operatorname{Aut}\mathbb{O}$ is a subgroup of $O_8\mathbb{R}$. It is closed in the usual topology of $O_8\mathbb{R} \subseteq \operatorname{GL}_8\mathbb{R}$, as multiplication in $\mathbb{O}$ is continuous. Hence $\operatorname{Aut}\mathbb{O}$ is compact since $O_8\mathbb{R}$ is (see for instance Porteous [81] Prop. 17.8, p. 337). As a closed linear group, $\operatorname{Aut}\mathbb{O}$ is a Lie group, cf. (94.3).

Simplicity has been proved in (11.32). It follows that $\operatorname{Aut}\mathbb{O}$ is connected, because the connected component of the identity of any topological group is a normal subgroup, and because $\operatorname{Aut}\mathbb{O}$ contains non-trivial connected subsets, see (11.31b and c).

The dimension of $\operatorname{Aut}\mathbb{O}$ may be computed by applying the dimension formula for stabilizers (96.10) to the transitive actions described in (11.30), as follows. By (11.31c or 30b), we have $\dim(\operatorname{Aut}\mathbb{O})_{i,j} = \dim\mathbb{S}_3 = 3$. As $(\operatorname{Aut}\mathbb{O})_{i,j}$ is a stabilizer of the transitive action of $(\operatorname{Aut}\mathbb{O})_i$ on the 5-sphere of pure unit octonions orthogonal to i, it follows that $\dim(\operatorname{Aut}\mathbb{O})_i = \dim(\operatorname{Aut}\mathbb{O})_{i,j} + 5 = 8$. By transitivity of $\operatorname{Aut}\mathbb{O}$ on the 6-dimensional unit sphere of $\operatorname{Pu}\mathbb{O}$, we conclude that $\dim\operatorname{Aut}\mathbb{O} = \dim(\operatorname{Aut}\mathbb{O})_i + 6 = 14$.

Now, according to the classification of almost simple Lie groups, cf. (94.32 and 33), there is just one compact almost simple Lie group of dimension 14, viz. the exceptional Lie group $G_2 = G_2(-14)$. □

Remarks. The classification of almost simple Lie groups also yields that $\operatorname{Aut}\mathbb{O}$ is simply connected; we shall prove this independently in (17.15c).

In the above proof, the only information needed about the stabilizer $(\operatorname{Aut}\mathbb{O})_i$ was its dimension. In fact, it is not difficult to determine this stabilizer completely, as we shall see now.

11.34 Automorphisms of $\mathbb{O}$ which are $\mathbb{C}$-linear. Biassociativity (11.13) of $\mathbb{O}$ implies that $\mathbb{O}$ is a left vector space over the subfield $\mathbb{C}$ spanned by 1 and i, scalar multiplication being multiplication within $\mathbb{O}$. Explicitly, scalar multiplication of $(x, y) \in \mathbb{H} \times \mathbb{H} = \mathbb{O}$ by the scalar $c \in \mathbb{C} \subseteq \mathbb{H} = \mathbb{H} \times \{0\}$ is given by

(1) $$c(x, y) := (c, 0)(x, y) = (cx, yc).$$

An automorphism in $(\operatorname{Aut}\mathbb{O})_i$ has the following properties: it fixes 1, is $\mathbb{C}$-linear, and is orthogonal with respect to the standard inner product of $\mathbb{O} = \mathbb{R}^8$; for the latter property, see (11.28). We shall show that $(\operatorname{Aut}\mathbb{O})_i$ consists precisely of the transformations which have these properties and in addition have *complex determinant* 1.

For convenience of notation, we first identify $\mathbb{O}$ and $\mathbb{C}^4$ in a suitable way. It is easy to see that the vectors 1, j, $l = (0, 1)$, and $jl = (0, j)$ belonging to the standard basis of $\mathbb{O} = \mathbb{H} \times \mathbb{H}$ form a $\mathbb{C}$-basis of $\mathbb{O}$. We note that these vectors are orthonormal with respect to the standard inner product of $\mathbb{O} = \mathbb{R}^8$. If we identify $\mathbb{O}$ and $\mathbb{C}^4$ via this basis, then these vectors are also orthonormal with respect to the standard Hermitian form on $\mathbb{C}^4 = \mathbb{O}$. We note that with these identifications we obtain that $O_8\mathbb{R} \cap GL_4\mathbb{C} = U_4\mathbb{C}$, where $U_4\mathbb{C}$ is the group of unitary transformations of $\mathbb{O} \cong \mathbb{C}^4$ with respect to the standard Hermitian form.

The properties of $(\mathrm{Aut}\,\mathbb{O})_i$ stated above now can be expressed by saying that $(\mathrm{Aut}\,\mathbb{O})_i$ is contained in the stabilizer of $1 \in \mathbb{O}$ in $U_4\mathbb{C}$. With a slight abuse of notation, this stabilizer will be denoted by $U_3\mathbb{C}$, as it induces the identity on $\mathbb{C} = \mathbb{R} + \mathbb{R}i \cong \mathbb{C} \times \{0\}$, and acts in the usual way on the orthogonal space $\mathbb{C}^\perp \cong \{0\} \times \mathbb{C}^3$. The normal subgroup consisting of the elements of $U_3\mathbb{C}$ with *complex* determinant 1 will be denoted by $SU_3\mathbb{C}$.

Proposition. $(\mathrm{Aut}\,\mathbb{O})_i = SU_3\mathbb{C}$.

Proof. The group $SU_3\mathbb{C}$ is transitive on the unit sphere of $\mathbb{C}^\perp$, and so is $\Lambda_i := (\mathrm{Aut}\,\mathbb{O})_i$ by (11.30a). It therefore suffices to show that $\Lambda_i \subseteq SU_3\mathbb{C}$ and that

(2) $$\Lambda_{i,j} = (SU_3\mathbb{C})_j,$$

see (91.3). By (11.31c and 26), the stabilizer $\Lambda_{i,j}$ acts trivially on $\mathbb{H} \times \{0\} \subseteq \mathbb{H} \times \mathbb{H} = \mathbb{O}$ and induces the group $SU_2\mathbb{C}$ on $\{0\} \times \mathbb{H} \cong \{0\} \times \mathbb{C}^2$, so that (2) is clear. It remains to prove that $\Lambda_i \subseteq SU_3\mathbb{C}$. Every $\mathbb{C}$-linear map has an eigenvector, hence Λ_i is the union of all stabilizers $\Lambda_{i,\mathbb{C}x}$ with $0 \neq x \in \mathbb{C}^\perp$. The action of Λ_i on the set of such subspaces $\mathbb{C}x$ is transitive by (11.30a); according to the principle of conjugate stabilizers (91.1a), this implies that Λ_i acts transitively on the set of the corresponding stabilizers $\Lambda_{i,\mathbb{C}x}$ by conjugation. Thus, because of $\Lambda_i \subseteq U_3\mathbb{C}$, the proof is finished if we show that $\Lambda_{i,\mathbb{C}j} \subseteq SU_3\mathbb{C}$.

Now the subalgebra of $\mathbb{O}$ generated by i and $\mathbb{C}j$ is $\mathbb{H} \cong \mathbb{H} \times \{0\}$, so that $\Lambda_{i,\mathbb{C}j} = \Lambda_{\mathbb{H},i}$. The explicit description (11.31b) of $\Lambda_\mathbb{H}$ shows that $\Lambda_{i,\mathbb{C}j}$ is the product of $\Lambda_{i,j} = \{(x,y) \mapsto (x, b^{-1}y) \mid b \in \mathbb{H}, \|b\|^2 = 1\}$ by a subgroup of $\Lambda_{i,l}$, namely $\{(x,y) \mapsto (a^{-1}xa, a^{-1}ya) \mid a \in \mathbb{C}, \|a\|^2 = 1\}$. From (11.30a), we know that the stabilizer $\Lambda_{i,l}$ is conjugate in $\Lambda_i \subseteq U_3\mathbb{C}$ to $\Lambda_{i,j} = (SU_3\mathbb{C})_j$, see (2). We conclude that $\Lambda_{i,l} \subseteq SU_3\mathbb{C}$ and that $\Lambda_{i,\mathbb{C}j} \subseteq \Lambda_{i,j}\Lambda_{i,l} \subseteq SU_3\mathbb{C}$. □

The following converse of the preceding result will only be used in later chapters.

11.35 Proposition. *Every closed, connected subgroup of* $\mathrm{Aut}\,\mathbb{O}$ *which is locally isomorphic to* $SU_3\mathbb{C}$ *is conjugate to* $(\mathrm{Aut}\,\mathbb{O})_i$.

Proof. Let Δ be a subgroup of this kind; it is a Lie group by (94.3). As $SU_3\mathbb{C}$ is connected and simply connected (see e.g. Porteous [81] Prop. 17.22, p. 340 and Husemoller [66] 12.3, p. 93), there is a surjective covering homomorphism $SU_3\mathbb{C} \to \Delta$, which may be viewed as a representation of $SU_3\mathbb{C}$ on $\mathbb{O} = \mathbb{R}^8$. (For the notion of covering homomorphism, see (94.2).) According to (95.3), $\mathbb{R}^8$

is the direct sum of the subspace F of fixed vectors and of irreducible subspaces. Now, $1 \in F$, so that $\dim F \geq 1$, and, up to equivalence, the only irreducible representation of $SU_3\mathbb{C}$ of dimension ≤ 7 is the usual representation of $SU_3\mathbb{C}$ on $\mathbb{C}^3 = \mathbb{R}^6$, see (95.10). Thus F is 2-dimensional, $F = \mathbb{R} + \mathbb{R}u$ for a suitable $u \in \mathrm{Pu}\,\mathbb{O}$ with $\|u\|^2 = 1$. Up to conjugation with an automorphism of $\mathbb{O}$ mapping i to u (11.30a), we may assume that $F = \mathbb{R} + \mathbb{R}i = \mathbb{C}$; then $\Delta \subseteq (\mathrm{Aut}\,\mathbb{O})_i$. By assumption, $\dim \Delta = 8 = \dim(\mathrm{Aut}\,\mathbb{O})_i$, cf. (11.34) or the proof of (11.33). We conclude that $\Delta = (\mathrm{Aut}\,\mathbb{O})_i$ by connectedness, cf. (93.12). □

12 The classical affine planes

This section is concerned with the classical planes over $\mathbb{R}$, $\mathbb{C}$, $\mathbb{H}$, and $\mathbb{O}$ from an affine point of view. We introduce the affine planes over these division algebras; the corresponding projective planes can then be obtained by projective completion, i.e., by adjunction of a line at infinity. Our main objects of study are the affine collineations, which may also be viewed as collineations of the projective completion fixing the line at infinity. A special rôle is played by collineations having an axis and a center; these notions are introduced and illustrated by a concrete description of all such collineations for particular axes and centers. The group of collineations of the affine plane over a division algebra $\mathbb{F}$ can be easily determined if $\mathbb{F}$ is a field like $\mathbb{R}$, $\mathbb{C}$, or $\mathbb{H}$ in our context here. Then, the affine collineations are just the semilinear affine transformations of the vector space $\mathbb{F} \times \mathbb{F}$; this is the so-called fundamental theorem of affine geometry. On the line at infinity, the affine collineations induce the fractional semilinear transformations.

The plane over $\mathbb{O}$ has no comparable vector space structure, as $\mathbb{O}$ is not associative. Consequently, for the octonion plane, the description of the collineation group is much more involved. A full discussion will therefore take place separately in Sections 16–18. Here, we restrict ourselves to first results. We construct special affine collineations of particular interest; they will be obtained as products of certain easily accessible central collineations (shears). The geometric fact that they are collineations readily translates into well-known algebraic properties of $\mathbb{O}$, the Moufang identities. These collineations fix the coordinate axes, and the group of all collineations doing so will be determined completely, starting from these special collineations. The description obtained for this group is in terms of orthogonal transformations of the coordinate axes and of the line at infinity. As a direct algebraic interpretation of these geometric results, we obtain the so-called triality principle for the group $SO_8\mathbb{R}$.

12.0 General assumption. Throughout this section, $\mathbb{F}$ shall denote one of the Cayley–Dickson division algebras $\mathbb{R}$, $\mathbb{C}$, $\mathbb{H}$, or $\mathbb{O}$.

12.1 The affine plane $\mathcal{A}_2\mathbb{F}$ over $\mathbb{F}$ is constructed as follows. The set of points is $\mathbb{F} \times \mathbb{F}$, and the following subsets of $\mathbb{F} \times \mathbb{F}$ are called lines:

$$[s, t] = \{ (x, sx + t) \mid x \in \mathbb{F} \} \quad \text{for } s, t \in \mathbb{F},$$
$$[c] \ = \{c\} \times \mathbb{F} \quad \text{for } c \in \mathbb{F}.$$

We say that the line $[s, t]$ has *slope* s, and the line $[c]$ has *slope* ∞.

Because $\mathbb{F}$ is a division algebra, it is easily verified that this structure has the following properties:

1) For two points $(x_1, y_1) \neq (x_2, y_2)$ there is a unique line joining them (i.e., containing both of them), namely the line $[s, y_1 - sx_1]$, where s is uniquely determined by the equation

(1) $$s(x_2 - x_1) = y_2 - y_1$$

if $x_1 \neq x_2$, and the line $[x_1]$ if $x_1 = x_2$.

2) Two lines of different slopes have a unique point of intersection. In fact, we have $[c] \cap [s, t] = \{(c, sc + t)\}$, and for $s_1 \neq s_2$ we have $[s_1, t_1] \cap [s_2, t_2] = \{(x, s_1 x + t_1)\}$, where x is uniquely determined by the equation

(2) $$(s_1 - s_2)x = t_2 - t_1.$$

3) Two different lines are disjoint if they have the same slope, and by property 2) this condition is also necessary. Two lines will be called *parallel* if they have the same slope, i.e., if they are disjoint or equal.

4) The so-called *parallel axiom* holds: For each line L and each point (x, y), there is a unique line which passes through (x, y) and is parallel to L. Depending on whether the slope of L is ∞ or $s \in \mathbb{F}$, this parallel is given by $[x]$ or by $[s, y - sx]$, respectively.

Properties 1) – 4) say that $\mathcal{A}_2\mathbb{F}$ satisfies the axioms of an *affine plane*, see (21.8).

Since $\mathbb{F}$ is biassociative (11.11 and 13), the solutions of equations (1) and (2) may be written down explicitly as

(1') $$s = (y_2 - y_1)(x_2 - x_1)^{-1}$$
(2') $$x = (s_1 - s_2)^{-1}(t_2 - t_1).$$

12.2 The projective completion $\overline{\mathcal{A}_2\mathbb{F}}$ of the affine plane $\mathcal{A}_2\mathbb{F}$ is obtained by adjoining 'points at infinity', one for each parallel class. All the lines of the parallel class are thought to intersect in the respective point at infinity. Furthermore, one adds a 'line at infinity' passing through just these points at infinity.

For $s \in \mathbb{F} \cup \{\infty\}$, the point at infinity on the lines of slope s (which form a parallel class) will be denoted by (s), and the line at infinity is

$$[\infty] := \{ (s) \mid s \in \mathbb{F} \cup \{\infty\} \} .$$

By this construction, we obtain a geometry in which any two lines $L_1 \neq L_2$ always have a unique point of intersection $L_1 \wedge L_2$. (If L_1, L_2 are lines of the affine plane $\mathcal{A}_2 \mathbb{F}$, their intersection point in the projective completion $\overline{\mathcal{A}_2 \mathbb{F}}$ is a point at infinity if, and only if, in $\mathcal{A}_2 \mathbb{F}$ the lines L_1, L_2 are parallel.) The property that for any two points $p_1 \neq p_2$ there is a unique line $p_1 p_2$ joining them is preserved in this extension process. Thus, the projective completion $\overline{\mathcal{A}_2 \mathbb{F}}$ is a *projective* plane in the sense of definition (21.1).

Of course, this construction applies quite generally to any affine plane, see Hughes–Piper [73] Theorem 3.10, p. 83.

For certain points and lines of the projective completion $\overline{\mathcal{A}_2 \mathbb{F}}$, we shall systematically use special names as indicated in the diagram of Figure 12a.

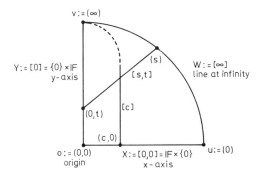

Figure 12a

12.3 Collineations. A *collineation* of an affine or projective plane is a bijection of the set of points onto itself mapping lines onto lines.

As for the latter condition, it suffices to postulate that lines are mapped *into* lines, see (23.2).

A collineation of an affine plane maps every parallel class of lines to a parallel class of lines; one therefore obtains a unique extension to a collineation of the projective completion by permuting the points at infinity accordingly. Conversely, if a collineation of the projective completion leaves the line at infinity invariant, then it arises in this way from an affine collineation. Henceforth, we shall not distinguish between an affine collineation and its projective extension. We remark that, besides these collineations, the projective completion may have collineations which move the line at infinity, cf. (13.5) and (17.6).

It is straightforward that for a collineation φ, for points $p_1 \neq p_2$ and for lines $L_1 \neq L_2$ one has

$$(p_1 p_2)^\varphi = p_1^\varphi p_2^\varphi \quad \text{and} \quad (L_1 \wedge L_2)^\varphi = L_1^\varphi \wedge L_2^\varphi .$$

The collineations of a given plane form a group under composition. For a subgroup Δ of this group, for a point p and a line L, the stabilizers Δ_p and Δ_L are the subgroups consisting of all collineations in Δ which fix p or leave L invariant, respectively.

12.4 Axial collineations. We now present a notion which is basic for the study of collineation groups of projective planes, together with some standard results, cf. also (23.7 ff).

(a) Consider a point p and a line L. We say that a collineation φ has *axis* L if φ fixes every point on L; dually, φ has *center* p if φ leaves every line through p invariant. A collineation of a *projective* plane has an axis if, and only if, it has a center (see Hughes–Piper [73] Theorem 4.9, p. 94; in the present chapter, we shall not use this fact). The center may be on the axis or not.

(b) For a group Δ of collineations, $\Delta_{[p,L]}$ denotes the subset of all collineations in Δ with center p and axis L; it is obviously a subgroup. For $\gamma \in \Delta$ one easily verifies the following

Conjugation formula: $\qquad \gamma^{-1} \Delta_{[p,L]} \gamma = \Delta_{[p^\gamma, L^\gamma]} ,$

cf. also (91.1a).

(c) *Uniqueness properties.* A collineation $\delta \in \Delta_{[p,L]}$ having a fixed point q outside $L \cup \{p\}$ must be the identity. Indeed, by joining q to the points of L, one sees that q is a center of δ, too; therefore, every point not on pq is a fixed point, being the intersection of two different lines through the centers p and q, and the same argument applied to such a fixed point instead of q finally shows that $\delta = \mathrm{id}$.

Dually, the only collineation in $\Delta_{[p,L]}$ leaving a line $M \neq L$ with $p \notin M$ invariant is the identity, because then M is an axis as well, its points being the points of intersection of the invariant line M and the invariant lines through the center p.

In particular, if a non-identical collineation has a center and an axis, these are uniquely determined.

We now give concrete descriptions of collineations of $\overline{\mathcal{A}_2}\mathbb{F}$ having a center and an axis. We first deal with such collineations for which center and axis are incident, so-called *elations*. The other case, in which center and axis are not incident, will be taken up later (12.13).

12 The classical affine planes

12.5 Proposition: Elations of $\overline{\mathcal{A}_2\mathbb{F}}$.

(a) *The collineations of $\overline{\mathcal{A}_2\mathbb{F}}$ having the line at infinity $W = [\infty]$ as axis and a center on W are precisely the translations*

$$\tau_{a,b} : (x, y) \mapsto (x+a, y+b); \quad \tau_{a,b}|_W = \mathrm{id}$$

with $a, b \in \mathbb{F}$. The translation group

$$\mathsf{T} = \{\, \tau_{a,b} \mid a, b \in \mathbb{F} \,\}$$

is commutative and sharply transitive on the affine point set $\mathbb{F} \times \mathbb{F}$.

(b) *The collineations of $\overline{\mathcal{A}_2\mathbb{F}}$ with axis $Y = [0]$ and center $v = (\infty)$ are precisely the shears*

$$\sigma_a : (x, y) \mapsto (x, y+ax), \quad (s) \mapsto (s+a), \quad (\infty) \mapsto (\infty)$$

for $a \in \mathbb{F}$; they form a commutative group which is sharply transitive on $W \setminus \{v\}$.

(c) *The collineations of $\overline{\mathcal{A}_2\mathbb{F}}$ with axis $X = [0, 0]$ and center $u = (0)$ are precisely the collineations*

$$\sigma'_a : (x, y) \mapsto (x+ay, y), \quad (s) \mapsto ((s^{-1}+a)^{-1}) \quad \text{for } s \neq 0, -a^{-1},$$
$$(0) \mapsto (0), \quad (-a^{-1}) \mapsto (\infty), \quad (\infty) \mapsto (a^{-1})$$

for $a \in \mathbb{F} \setminus \{0\}$ (together with $\sigma'_0 = \mathrm{id}$).

Note that (c) is obtained from (b) by conjugation with the following collineation, see the conjugation formula in (12.4b).

12.6. *The reflection*

$$(x, y) \mapsto (y, x), \quad (s) \mapsto (s^{-1}) \quad \text{for } s \neq 0,$$
$$(0) \mapsto (\infty), \quad (\infty) \mapsto (0)$$

is a collineation of $\overline{\mathcal{A}_2\mathbb{F}}$ with axis $[1, 0]$ and center (-1).

Proof of (12.5 and 6). 1) We first consider the restrictions of the maps in (12.5a and b) and in (12.6) to the affine plane with point set $\mathbb{F} \times \mathbb{F}$. One may easily verify by direct computation that these restrictions transform affine lines into lines and therefore are collineations of the affine plane $\mathcal{A}_2\mathbb{F}$. For (12.5a and b), this only requires the distributive laws; for (12.6), one also uses biassociativity (11.9). The necessary computations show that lines are mapped as follows:

(1) $\qquad [s, t]^{\tau_{a,b}} = [s, t+b-sa]; \quad [c]^{\tau_{a,b}} = [c+a]$

(2) $\qquad [s, t]^{\sigma_a} = [s+a, t]; \quad [c]^{\sigma_a} = [c],$

and that the reflection of (12.6) interchanges

(3)
$$[s, t] \text{ with } [s^{-1}, -s^{-1}t] \text{ for } s \neq 0,$$
$$[0, t] \text{ with } [t].$$

The effect of these collineations on the slopes of lines shows that their projective extensions act on the line at infinity W in the specified way.

From (3), one infers that the collineation defined in (12.6) leaves all the lines of slope -1 invariant, so that it has center (-1). It is clear that this collineation has axis $[1, 0] = \{ (x, x) \mid x \in \mathbb{F} \}$.

Thus, (12.6) is proved. Concerning (12.5), it now suffices to prove (a) and (b), because (c) then follows from (b), as was noted above.

2) The collineation $\tau_{a,b}$ has axis W. If $a \neq 0$ and if $s \in \mathbb{F}$ is the solution of $sa = b$, then by (1) all the lines $[s, t]$, $t \in \mathbb{F}$, are invariant under $\tau_{a,b}$. Now these are precisely the affine lines through the point (s), so that $\tau_{a,b}$ has center (s). For $a = 0$, the center is (∞), since then all the lines $[c]$ are invariant according to equation (1).

Commutativity of T and transitivity on $\mathbb{F} \times \mathbb{F}$ are obvious. We use a standard transitivity argument in order to show that T contains every collineation with axis W and center p on W. For fixed $p \in W$, the group $\mathsf{A}_{[p,W]}$ of all these collineations contains $\mathsf{T}_{[p,W]}$. Now, for any line $M \neq W$ through p, the latter group is transitive on $M \setminus \{p\}$; indeed, for two different points $q_1, q_2 \in M \setminus \{p\}$, the translation $\tau \in \mathsf{T}$ mapping q_1 to q_2 leaves $M = pq_1 = pq_2$ invariant, so that, by the uniqueness properties (12.4c), the center of τ belongs to M and therefore equals $p = M \wedge W$. Again by the uniqueness properties (12.4c), the group $\mathsf{A}_{[p,W]}$ is now seen to be sharply transitive on $M \setminus \{p\}$, so that $\mathsf{A}_{[p,W]} = \mathsf{T}_{[p,W]}$; see also (23.9). This completes the proof of (12.5a).

3) It is obvious that σ_a has axis $Y = \{0\} \times \mathbb{F}$, and from (2) one sees that all the lines $[c]$ of slope ∞, in other words the lines through $v = (\infty)$, are invariant, so that σ_a has center v. Transitivity of the group $\{ \sigma_a \mid a \in \mathbb{F} \}$ on $W \setminus \{v\}$ is obvious. By the same argument as above, it follows that this group coincides with the group $\mathsf{A}_{[v,Y]}$ of all collineations with axis Y and center v, since the latter group is seen to be sharply transitive on $W \setminus \{v\}$ by virtue of the uniqueness properties (12.4c). Thus, (12.5b) is verified. □

Next, we prove some general statements about collineations having non-collinear fixed points.

12.7 Lemma: Stabilizer of a triangle. *The collineations of $\overline{\mathcal{A}_2}\mathbb{F}$ fixing the points $o = (0, 0)$, $u = (0)$, and $v = (\infty)$ are precisely the transformations of the form*

$$(x, y) \mapsto (x^\alpha, y^\beta), \quad (s) \mapsto (s^\gamma), \quad (\infty) \mapsto (\infty)$$

with automorphisms α, β, γ of the additive group of $\mathbb{F}$ satisfying

(∗) $$(sx)^\beta = s^\gamma x^\alpha$$

for all $s, x \in \mathbb{F}$. In particular, these collineations are completely determined by their actions on the coordinate axes $\mathbb{F} \times \{0\}$ and $\{0\} \times \mathbb{F}$.

Proof. A collineation of this kind leaves the coordinate axes $X = ou$ and $Y = ov$ invariant and maps parallels of X and Y to parallels of X and Y, respectively; this means that in the image of (x, y) the first coordinate is independent of y and the second one is independent of x, so that the collineation is of the stated form with bijections α, β, γ of $\mathbb{F}$ fixing 0. We now find the algebraic conditions for such a transformation to be a collineation. Consider the point $(x, sx + t)$ on the line $[s, t]$ joining $(0, t)$ and (s). The corresponding condition for the image points, requiring that the point $(x^\alpha, (sx + t)^\beta)$ lies on the line joining $(0, t^\beta)$ and (s^γ), that is, on the line $[s^\gamma, t^\beta]$, is expressed algebraically by

(1) $$(sx + t)^\beta = s^\gamma x^\alpha + t^\beta.$$

The special case $t = 0$, which is just (∗), allows us to write (1) as $(sx + t)^\beta = (sx)^\beta + t^\beta$, thus exhibiting the additivity of β, which in turn, once more via (∗), implies the additivity of α and γ.

Conversely, if (∗) holds and if β is additive, then (1) is valid, and we have a collineation. □

12.8 Corollary: Stabilizer of a quadrangle. *The collineations of $\overline{\mathcal{A}_2}\mathbb{F}$ fixing the points $o = (0, 0)$, $u = (0)$, $v = (\infty)$, and $e = (1, 1)$ are precisely the transformations*

$$(x, y) \mapsto (x^\alpha, y^\alpha), \quad (s) \mapsto (s^\alpha), \quad (\infty) \mapsto (\infty) \quad \text{with } \alpha \in \operatorname{Aut} \mathbb{F}.$$

Proof. These collineations are the transformations given in (12.7) which, in addition, satisfy $1^\alpha = 1 = 1^\beta$. From (12.7(∗)) it then follows that also $1^\gamma = 1$, that $\alpha = \beta = \gamma$, and that, consequently, α is an automorphism of $\mathbb{F}$. □

12.9 Remark. *The collineations in (12.8) coincide with the affine collineations fixing the points o, $e_1 = (1, 0)$, and $e_2 = (0, 1)$.*

Indeed, if we consider them as collineations of the projective completion $\overline{\mathcal{A}_2}\mathbb{F}$, the line at infinity W remains invariant; now $u = (oe_1) \wedge W$, $v = (oe_2) \wedge W$, $e = (e_1 v) \wedge (e_2 u)$, and conversely $e_1 = (ve) \wedge (ou)$, $e_2 = (ue) \wedge (ov)$. □

In the field case $\mathbb{F} \in \{\mathbb{R}, \mathbb{C}, \mathbb{H}\}$, the affine collineation group can now be determined completely.

12.10 Fundamental theorem of affine geometry. *Let* $\mathbb{F} \in \{\mathbb{R}, \mathbb{C}, \mathbb{H}\}$. *The collineations of the affine plane* $\mathcal{A}_2\mathbb{F}$ *are precisely the mappings*

$$(x, y) \mapsto \varphi(x^\alpha, y^\alpha) + (a, b)$$

for $a, b \in \mathbb{F}$, $\alpha \in \mathrm{Aut}\,\mathbb{F}$, *and* $\varphi \in \mathrm{GL}_2\mathbb{F}$ (*the group of linear transformations of* $\mathbb{F} \times \mathbb{F}$ *considered as a* right $\mathbb{F}$-*vector space*). *They form the group* $\mathsf{A} = \mathrm{A}\Gamma\mathrm{L}_2\mathbb{F}$, *which is the semidirect product* $\mathsf{A} = \mathsf{T} \rtimes \mathsf{A}_o$ *of the translation group* T *described in* (12.5) *by the stabilizer of the origin*

$$\mathsf{A}_o = \Gamma\mathrm{L}_2\mathbb{F} = \mathrm{Aut}\,\mathbb{F} \ltimes \mathrm{GL}_2\mathbb{F}\,.$$

Remark. This theorem and the proof given below are in fact valid over any (not necessarily commutative) field $\mathbb{F}$, cf. (23.6) and the references given there.

Proof of (12.10). Let A denote the group of all affine collineations. As T is the group of all collineations with axis W and center on W, see (12.5), the conjugation formula (12.4b) implies that T is a normal subgroup of A. By transitivity of T on the affine point set, the Frattini argument (91.2a) shows that $\mathsf{A} = \mathsf{A}_o \cdot \mathsf{T} = \mathsf{T} \cdot \mathsf{A}_o$.

The group $\mathrm{GL}_2\mathbb{F}$ consists of collineations because in the field case the lines are just the one-dimensional affine subspaces of the *right* $\mathbb{F}$-vector space $\mathbb{F} \times \mathbb{F}$, as is immediate from their definition (12.1), so that linear transformations map lines to lines. Now $\mathrm{GL}_2\mathbb{F}$ is sharply transitive on the set of bases of $\mathbb{F} \times \mathbb{F}$, whence $\mathsf{A}_o = \mathsf{A}_{o,(1,0),(0,1)} \cdot \mathrm{GL}_2\mathbb{F}$ by the Frattini argument. Finally, the stabilizer $\mathsf{A}_{o,(1,0),(0,1)}$ is described by $\mathrm{Aut}\,\mathbb{F}$ according to (12.9 and 8). □

We now determine how these affine collineations act on the line at infinity W. Since $\mathsf{A} = \mathsf{T} \cdot \mathsf{A}_o$, and since the translation group T acts trivially on W, only the collineations fixing o need to be considered.

12.11. *Let* $\mathbb{F} \in \{\mathbb{R}, \mathbb{C}, \mathbb{H}\}$. *The stabilizer* A_o *consists of all the mappings*

$$\gamma_{\alpha,\varphi} : (x, y) \mapsto \varphi(x^\alpha, y^\alpha)$$

for $\alpha \in \mathrm{Aut}\,\mathbb{F}$ *and* $\varphi \in \mathrm{GL}_2\mathbb{F}$. *If* φ *is described by a matrix as*

$$\varphi : \mathbb{F}^2 \to \mathbb{F}^2 : \begin{pmatrix} x \\ y \end{pmatrix} \mapsto \begin{pmatrix} a & b \\ c & d \end{pmatrix} \begin{pmatrix} x \\ y \end{pmatrix},$$

then the projective extension of the collineation $\gamma_{\alpha,\varphi}$ *acts on the line at infinity as the fractional semilinear transformation*

$$\gamma_{\alpha,\varphi}|_W : W \to W : (s) \mapsto \left((c + d \cdot s^\alpha)(a + b \cdot s^\alpha)^{-1}\right)\,.$$

This should be understood also for $s = 0$ and $s = \infty$, with the usual conventions about the rôle of 0 and ∞ in such expressions, e.g. with $0^{-1} = \infty$, $\infty^{-1} = 0$.

In order to obtain the asserted description of $\gamma_{\alpha,\varphi}|_W$, we decompose the given collineation as $\gamma_{\alpha,\varphi} = \gamma_{\alpha,\mathrm{id}} \gamma_{\mathrm{id},\varphi}$. The projective extension of $\gamma_{\alpha,\mathrm{id}}$ is the collineation described in (12.8) mapping (s) to (s^α). The projective extension of $\gamma_{\mathrm{id},\varphi}$ maps (s) to (s'), where s' is the slope of the image of the line $[s, 0] = \{(x, sx) \mid x \in \mathbb{F}\} = (1, s)\mathbb{F}$. The image line is $((a+bs), (c+ds))\mathbb{F} = (1, s')\mathbb{F}$, so that $s' = (c+ds)(a+bs)^{-1}$. The description of $\gamma_{\alpha,\varphi}|_W$ now follows by composition. □

We thus have obtained the following result.

Fundamental theorem for the projective line. *Let $\mathbb{F} \in \{\mathbb{R}, \mathbb{C}, \mathbb{H}\}$. The group $\mathsf{A}|_W$ of transformations induced on the line at infinity W by the group A of collineations of $\mathcal{A}_2\mathbb{F}$ (via projective extension) is the product*

$$\mathsf{A}|_W = \mathrm{Aut}\,\mathbb{F} \cdot \mathrm{PGL}_2\mathbb{F}$$

of the group

$$\mathrm{PGL}_2\mathbb{F} = \left\{(s) \mapsto \left((c+ds)(a+bs)^{-1}\right) \,\Big|\, \begin{pmatrix} a & b \\ c & d \end{pmatrix} \in \mathrm{GL}_2\mathbb{F}\right\}$$

of fractional linear transformations by the group

$$\{(s) \mapsto (s^\alpha), (\infty) \mapsto (\infty) \mid \alpha \in \mathrm{Aut}\,\mathbb{F}\} \cong \mathrm{Aut}\,\mathbb{F}.$$

Note. In (15.6), we shall present a version of the preceding theorem which is valid for $\mathbb{F} = \mathbb{O}$, as well.

Addenda.
(a) *The last-mentioned group is the stabilizer of (0), (1), and (∞) in $\mathsf{A}|_W$.*
(b) *The stabilizer of (0) and (∞) in $\mathrm{PGL}_2\mathbb{F}$ consists of the transformations $(s) \mapsto (dsa^{-1})$ for $a, d \in \mathbb{F}^\times$. These are precisely the maps $(s) \mapsto (Bs)$ where B belongs to the group $\mathrm{GO}_n^+\mathbb{R}$ defined in (11.21).*
(c) *The fractional linear group $\mathrm{PGL}_2\mathbb{F}$ is triply transitive on W. It is sharply triply transitive if, and only if $\mathbb{F}$ is commutative.*
(d) *$\mathrm{Aut}\,\mathbb{F}$ is trivial for $\mathbb{F} = \mathbb{R}$. In the case $\mathbb{F} = \mathbb{H}$, the transformation $(s) \mapsto (s^\alpha)$ for $\alpha \in \mathrm{Aut}\,\mathbb{F}$ belongs to $\mathrm{PGL}_2\mathbb{H}$. Thus,*

$$\mathsf{A}|_W = \mathrm{PGL}_2\mathbb{F} \quad \text{for } \mathbb{F} = \mathbb{R}, \mathbb{H}.$$

Proof of Addenda. We begin with assertion (b). The first part of the assertion is immediate from the explicit description of $\mathrm{PGL}_2\mathbb{F}$. The second part has been proved in (11.23).

We now turn to (c). The stabilizer of (∞) in $\mathrm{PGL}_2\mathbb{F}$ contains the transformations $(s) \mapsto (s+c)$ for $c \in \mathbb{F}$ and hence is transitive on $W \setminus (\infty)$. Under $(s) \mapsto (s^{-1})$, the points (0) and (∞) are interchanged. Hence, the group $\mathrm{PGL}_2\mathbb{F}$ is doubly transitive on W, and (b) shows that it is even triply transitive. Also from (b), we infer that the stabilizer Λ of the triple (0), (1), (∞) in $\mathrm{PGL}_2\mathbb{F}$ is induced by the inner automorphisms of $\mathbb{F}$, so that this stabilizer is trivial precisely if $\mathbb{F}$ is commutative. Thus, (c) is proved. At the same time, we have obtained that this stabilizer is contained in the subgroup of $\mathsf{A}|_W$ corresponding to $\mathrm{Aut}\,\mathbb{F}$, which proves (a). Moreover, the latter subgroup reduces to $\Lambda \subseteq \mathrm{PGL}_2\mathbb{F}$ whenever all automorphisms of $\mathbb{F}$ are inner, so that then $\mathsf{A}|_W = \mathrm{PGL}_2\mathbb{F}$. This is the case for $\mathbb{F} = \mathbb{H}$ by (11.29), whence (d). □

From now on, we include the case $\mathbb{F} = \mathbb{O}$ in our discussion again. Our next topic is a concrete description of collineations with *non*-incident center and axis. Such collineations are called *homologies*.

12.13 Proposition: Homologies of $\overline{\mathcal{A}_2}\mathbb{F}$.

(a) *The collineations of $\overline{\mathcal{A}_2}\mathbb{F}$ having the line at infinity $W = [\infty]$ as axis and fixing the origin $o = (0, 0)$ are precisely the mappings*

$$\mu_a : (x, y) \mapsto (xa, ya); \quad \mu_a|_W = \mathrm{id}$$

$$\text{for } 0 \neq a \in \begin{cases} \mathbb{F} & \text{if } \mathbb{F} \in \{\mathbb{R}, \mathbb{C}, \mathbb{H}\} \\ \mathrm{Ker}\,\mathbb{O} = \mathbb{R} & \text{if } \mathbb{F} = \mathbb{O}. \end{cases}$$

They have center o. These collineations are called homotheties.
  In the field case $\mathbb{F} \in \{\mathbb{R}, \mathbb{C}, \mathbb{H}\}$, the group of these homotheties is transitive on $X \setminus \{o, X \wedge W\}$. For $\mathbb{F} = \mathbb{O}$, this is not so.

(b) *The collineations of $\overline{\mathcal{A}_2}\mathbb{F}$ with axis $X = [0, 0]$ and center $v = (\infty)$ are precisely the mappings*

$$(x, y) \mapsto (x, ay), \quad (s) \mapsto (as)$$

for $0 \neq a \in \mathbb{F}$ if $\mathbb{F} \in \{\mathbb{R}, \mathbb{C}, \mathbb{H}\}$, and for $0 \neq a \in \mathbb{R}$ if $\mathbb{F} = \mathbb{O}$.

(c) *The collineations of $\overline{\mathcal{A}_2}\mathbb{F}$ with axis $Y = [0]$ and center $u = (0)$ are precisely the mappings*

$$(x, y) \mapsto (ax, y), \quad (s) \mapsto (sa^{-1})$$

for $0 \neq a \in \mathbb{F}$ if $\mathbb{F} \in \{\mathbb{R}, \mathbb{C}, \mathbb{H}\}$, and for $0 \neq a \in \mathbb{R}$ if $\mathbb{F} = \mathbb{O}$.

Remark. The difference between $\mathbb{O}$ and the fields $\mathbb{R}, \mathbb{C}, \mathbb{H}$ exhibited here reflects the non-associativity of $\mathbb{O}$, as will become clear in the proof. Synthetically, this difference is expressed by the fact that Desargues' theorem for non-incident centers and axes, which holds in planes over (not necessarily commutative) fields, is not valid in the octonion plane. The fundamental interrelations between transitivity properties of groups of homologies, the validity of Desargues' theorem, and associativity properties of the coordinate domain are due to Baer [42] Theorem 5.1, p. 146 and Theorem 6.2, p. 151, cf. also (23.22) and the books mentioned there.

Proof of (12.13). (a) The collineations in question are precisely the transformations of the form stated in (12.7) with $\gamma = \mathrm{id}$, i.e., the collineations

$$\delta : (x, y) \mapsto (x^\alpha, y^\beta); \quad \delta|_W = \mathrm{id}$$

with bijections α, β of $\mathbb{F}$ satisfying

(∗) $$(sx)^\beta = s \cdot x^\alpha$$

for all $s, x \in \mathbb{F}$. With $x = 1$ and $a := 1^\alpha$ this gives $s^\beta = sa$; and putting $s = 1$ we obtain $\beta = \alpha$. Thus, δ is of the form

(1) $$\delta : (x, y) \mapsto (xa, ya),$$

and (∗) requires that

(2) $$(sx)a = s(xa)$$

for all $s, x \in \mathbb{F}$.

Now, if $\mathbb{F} \in \{\mathbb{R}, \mathbb{C}, \mathbb{H}\}$, then every element a satisfies (2) because of associativity; if $\mathbb{F} = \mathbb{O}$, then (2) holds precisely for the elements a of the kernel $\mathrm{Ker}\, \mathbb{O} = \mathbb{R}$, see (11.20). Thus, the first part of assertion (a) is proved.

For $\mathbb{F} \in \{\mathbb{R}, \mathbb{C}, \mathbb{H}\}$, the easy direction of the preceding argument would suffice, namely the verification that the homotheties μ_a are indeed collineations for all $a \in \mathbb{F} \setminus \{0\}$. The proof can then be completed by the following standard transitivity argument. Obviously, in these cases, the set $\{ \mu_a \mid a \in \mathbb{F} \setminus \{0\} \}$ of homotheties is transitive on $X \setminus \{o, X \wedge W\} = \mathbb{F} \times \{0\} \setminus \{(0, 0)\}$; one now observes that, by the uniqueness properties (12.4c), the group $\mathrm{A}_{[o,W]}$ of all homologies with center o and axis W is sharply transitive there and thus consists precisely of the homotheties as stated. See also (23.9).

A further proof for $\mathbb{F} \in \{\mathbb{R}, \mathbb{C}, \mathbb{H}\}$ can be obtained from the fundamental theorem of affine geometry (12.10). The collineations fixing the origin are given in (12.11) together with their actions on W; the collineation $\gamma_{\alpha,\varphi}$ described there has axis W if, and only if $c + d \cdot s^\alpha = s(a + b \cdot s^\alpha)$ for all $s \in \mathbb{F}$. By putting in turn $s = 0, s = 1$, and $s = -1$ one sees that this is equivalent to $c = 0 = b$, $d = a$ and $s^\alpha = a^{-1}sa$ for all $s \in \mathbb{F}$. In this situation, $\gamma_{\alpha,\varphi}$ maps (x, y) to $\varphi(x^\alpha, y^\alpha) = (a \cdot x^\alpha, a \cdot y^\alpha) = (a \cdot (a^{-1}xa), a \cdot (a^{-1}xa)) = (xa, ya)$; this proves our proposition anew. Note that, in the non-commutative case $\mathbb{F} = \mathbb{H}$ at hand, $\gamma_{\alpha,\varphi}$ is not a linear transformation (of the affine point set $\mathbb{H}^2$ considered as a *right* $\mathbb{H}$-vector space)!

(b) may be proved analogously. The collineations with axis X and center v are the mappings

$$(x, y) \mapsto (x, y^\beta), \quad (s) \mapsto (s^\gamma)$$

with bijections β, γ of $\mathbb{F}$ satisfying $(sx)^\beta = s^\gamma x$ for all $s, x \in \mathbb{F}$. With $s = 1$ and $c := 1^\gamma$, we obtain $x^\beta = cx$. Putting $x = 1$, we infer $\gamma = \beta$. Thus, the collineations in question are the transformations $(x, y) \mapsto (x, cy)$, where $c \neq 0$ satisfies

(3) $\qquad\qquad\qquad c(sx) = (cs)x \quad \text{for all } s, x \in \mathbb{F}.$

For $\mathbb{F} \in \{\mathbb{R}, \mathbb{C}, \mathbb{H}\}$, the latter condition is trivially fulfilled because of associativity. For $\mathbb{F} = \mathbb{O}$, by applying conjugation to (3), we obtain the equivalent condition $(\bar{x}\bar{s})\bar{c} = \bar{x}(\bar{s}\bar{c})$ for all $s, x \in \mathbb{F}$, which says that $\bar{c}$, and hence c, belongs to the kernel $\mathbb{R}$ of $\mathbb{O}$.

In the field cases $\mathbb{F} \in \{\mathbb{R}, \mathbb{C}, \mathbb{H}\}$, analogous variations of the proof as in (a) are possible, of course.

(c) is obtained from (b) by conjugation; one applies the conjugation formula (12.4b) to the reflection $(x, y) \mapsto (y, x)$, $(s) \mapsto (s^{-1})$ of (12.6), which interchanges the axis X with Y and the center u with v. □

Collineations of $\mathcal{A}_2\mathbb{O}$, Moufang identities, and triality

Finally, we proceed to a more detailed study of the stabilizer ∇ of the triangle $o = (0, 0)$, $u = (0)$, and $v = (\infty)$ in the collineation group of the projective octonion plane $\overline{\mathcal{A}_2}\mathbb{O}$. The group ∇ can also be viewed as the stabilizer of the coordinate axes $X = ou$ and $Y = ov$ in the group of collineations of the affine plane $\mathcal{A}_2\mathbb{O}$. This stabilizer is an easily accessible part of the collineation group. A study of the whole collineation group of the projective plane $\overline{\mathcal{A}_2}\mathbb{O}$ will be the subject of separate sections (17 and 18).

First, we exhibit special collineations which generate ∇ and which are closely related to the Moufang identities (12.15). By an idea of Salzmann [59c], these collineations can be constructed very easily as compositions of elations which are known from (12.5). When translating the geometric fact that these transformations are collineations into algebraic language, one immediately obtains the Moufang identities. They express other facets of the weak associativity properties of $\mathbb{O}$ besides alternativity (11.13). Actually, they are inherent in alternativity, although this is not entirely obvious. Here, this fact may be obtained by inspection of our argument. For purely algebraic proofs, see Pickert [75] 6.1 no. 2, p. 160, Schafer [66] III.1, p. 28, Zhevlakov et al. [82] Lemma 2.7, p. 35.

With the special collineations mentioned above at our disposal, it is then easy to determine ∇. The structure of ∇ is closely related to the triality principle (12.18).

As a special trait of our presentation, triality does not appear as an algebraic fact having geometric applications, but is intertwined with geometric phenomena right from the start.

Our arguments would be valid more generally for any alternative field instead of $\mathbb{O}$. In particular, they work over (not necessarily commutative) fields, as well; but then, the collineations in question may also be obtained directly from the fundamental theorem of affine geometry (12.10), see (12.19), and the Moufang identities are trivially satisfied because of associativity.

12.14 Proposition. *The following mappings are collineations of $\overline{\mathcal{A}_2\mathbb{O}}$, for every $a \in \mathbb{O} \setminus \{0\}$:*

$$\gamma_a : (x, y) \mapsto (a^{-1}x, ay), \quad (s) \mapsto (asa)$$
$$\gamma'_a : (x, y) \mapsto (axa, ya), \quad (s) \mapsto (sa^{-1})$$
$$\gamma''_a : (x, y) \mapsto (xa, aya), \quad (s) \mapsto (as).$$

Proof. We use the elations $\sigma_a : (x, y) \mapsto (x, y + ax)$, $\sigma'_a : (x, y) \mapsto (x + ay, y)$ from (12.5b and c) and check that γ_a is the following product of such elations:

$$(*) \qquad \gamma_a = \sigma'_1 \, \sigma_{a-1} \, \sigma'_{-a^{-1}} \, \sigma_{a-a^2} \, .$$

The biassociativity of $\mathbb{O}$ (11.9) will be essential for the necessary calculations. Under the given product of elations, an affine point is mapped as follows:

$$(x, y) \mapsto (x + y, y)$$
$$\mapsto (x + y, y + (a-1)(x+y)) = (x + y, a(x+y) - x)$$
$$\mapsto \left(x + y - a^{-1}(a(x+y)) + a^{-1}x, a(x+y) - x\right) = \left(a^{-1}x, a(x+y) - x\right)$$
$$\mapsto \left(a^{-1}x, a(x+y) - x + (a - a^2)(a^{-1}x)\right) = \left(a^{-1}x, a(x+y) - x + x - ax\right)$$
$$= (a^{-1}x, ay).$$

The line $[s, 0]$ joining $(0,0)$ and $(1, s)$ is mapped onto the line joining the image points $(0, 0)$ and (a^{-1}, as); this line is $[asa, 0]$, since $(asa)a^{-1} = as$ by biassociativity, so that indeed $(a^{-1}, as) \in [asa, 0]$. The projective extension thus maps (s) to (asa), and $(*)$ is proved.

According to (12.7), the fact that γ_a is a collineation is expressed algebraically by the identity $a(sx) = (asa)(a^{-1}x)$. Substituting $z = a^{-1}x$, $az = x$ yields the first *Moufang identity*

$$(1) \qquad a(s(az)) = (asa)z\,.$$

Applying conjugation on both sides of (1) and renaming, one obtains the second Moufang identity

$$(2) \qquad ((ba)x)a = b(axa)\,.$$

With $s = ba$, $sa^{-1} = b$ this is transformed into the identity $(sx)a = (sa^{-1})(axa)$, from which in turn, by applying (12.7) backwards, it follows that γ'_a is a collineation. An immediate verification with the help of biassociativity shows that $\gamma''_a = \gamma_a \gamma'_a$, so that this is a collineation as well. By (12.7) once again, the latter fact is equivalent to the third Moufang identity

(3) $$a(sx)a = (as)(xa) \, .$$

We collect these identities as a corollary to the proof.

12.15 The Moufang identities. *For all $a, b, c \in \mathbb{O}$, the following hold:*

$$a(b(ac)) = (aba)c$$
$$((ab)c)b = a(bcb)$$
$$(ab)(ca) = a(bc)a \, .$$ □

12.16 Lemma. *The group of collineations generated by the collineations γ_a and γ'_a of (12.14) for $a \in \mathbb{O} \setminus \{0\}$ fixes the vertices of the triangle o, u, v and is transitive on the set $(\mathbb{O} \setminus \{0\}) \times (\mathbb{O} \setminus \{0\})$ of affine points not incident with the sides of this triangle.*

Proof. For $a, b \in \mathbb{O} \setminus \{0\}$, we show that the point $(1, 1)$ may be mapped to the point (a, b) by a product of collineations of the specified kind. Indeed, since ba belongs to a subfield of $\mathbb{O}$ isomorphic to $\mathbb{C}$, see (11.7), there is $c \in \mathbb{O}$ with $ba = c^3$. Now, using alternativity (11.8), one easily verifies that $\gamma_{c^{-1}} \gamma'_c \gamma_b$ maps $(1, 1)$ to (a, b). □

In the following proposition, we use the notation for orthogonal groups introduced in (11.21).

12.17 Proposition: The stabilizer of the coordinate axes.

(a) *The group ∇ of collineations of $\overline{\mathcal{A}_2 \mathbb{O}}$ leaving the points o, u, and v fixed consists of the transformations*

$$(A, B \,|\, C) : (x, y) \mapsto (Ax, By), \quad (s) \mapsto (Cs) \, ,$$

where A, B, C belong to $\mathrm{GO}_8^+ \mathbb{R}$ and satisfy the **triality condition**

(∗) $$B(s \cdot x) = Cs \cdot Ax \quad \text{for all } s, x \in \mathbb{O} \, .$$

(b) *The group ∇ is the direct product of the subgroup consisting of the transformations*

$$\mu_{r,t} : (x, y) \mapsto (rx, ty), \quad (s) \mapsto (tr^{-1} \cdot s) \quad \text{for } 0 < r, t \in \mathbb{R}$$

by the subgroup

$$S\nabla = \{ (A, B \mid C) \in \nabla \mid A, B, C \in SO_8\mathbb{R} \} \,.$$

(c) *The projection homomorphisms* $\mathrm{pr}_\nu : S\nabla \to SO_8\mathbb{R}$ *for* $\nu = 1, 2, 3$ *defined by* $\mathrm{pr}_\nu(A_1, A_2 \mid A_3) = A_\nu$ *are surjective. Their kernels are of order 2 and are generated by the reflections*

$$\begin{aligned}
\iota_v &= \mu_{1,-1} : (x, y) \mapsto (x, -y), & (s) &\mapsto (-s) & &\text{with center } v, \text{ axis } X = [0, 0]\\
\iota_u &= \mu_{-1,1} : (x, y) \mapsto (-x, y), & (s) &\mapsto (-s) & &\text{with center } u, \text{ axis } Y = [0]\\
\iota_o &= \mu_{-1,-1} : (x, y) \mapsto (-x, -y), & (s) &\mapsto (s) & &\text{with center } o, \text{ axis } W = [\infty] \,.
\end{aligned}$$

These reflections are the non-trivial elements of the center $\mathbb{Z}_2 \times \mathbb{Z}_2$ *of* $S\nabla$.

Proof. 1) According to (12.7), the transformations of the form $(A, B \mid C)$ for $A, B, C \in GO_8^+\mathbb{R}$ are collineations if, and only if, the triality condition (∗) is satisfied; they obviously fix o, u, and v. Let Ψ be the subgroup of ∇ consisting of these transformations. By (11.22), the collineations given in (12.14) belong to Ψ. From (12.16), we know that Ψ is transitive on $(\mathbb{O} \setminus \{0\}) \times (\mathbb{O} \setminus \{0\})$. The Frattini argument (91.2a) yields that $\nabla = \nabla_e \cdot \Psi$, where $e = (1, 1)$. According to (12.8), the stabilizer ∇_e of o, u, v, e consists of the collineations $(x, y) \mapsto (x^\alpha, y^\alpha)$, $(s) \mapsto (s^\alpha)$ for $\alpha \in \mathrm{Aut}\,\mathbb{O}$. Now $\mathrm{Aut}\,\mathbb{O}$ is contained in $SO_8\mathbb{R}$ by (11.31e). Thus $\nabla_e \subseteq \Psi$ and $\nabla = \Psi$, as asserted in (a).

2) We now prove (b). For an element $(A, B \mid C)$ of ∇, let $A = rA'$, $B = tB'$, $C = qC'$ with $0 < r, t, q \in \mathbb{R}$ and $A', B', C' \in SO_8\mathbb{R}$. Then, in the equation

$$tB'(s \cdot x) = qC's \cdot rA'x$$

obtained from (∗), we consider the norms of both sides. As A', B', C' are orthogonal maps and as the norm is multiplicative (11.14), we see that $t^2 \|s\|^2 \|x\|^2 = q^2 \|s\|^2 r^2 \|x\|^2$, whence $q = tr^{-1}$. This shows that $(A, B \mid C) = \mu_{r,t} \cdot (A', B' \mid C')$.

3) Surjectivity of pr_ν follows from the fact that the collineations γ_a, γ'_a, and γ''_a constructed in (12.14) for $a \in \mathbb{O} \setminus \{0\}$ belong to $S\nabla$ if $\|a\|^2 = 1$. Indeed, the corresponding transformations $x \mapsto ax$, $x \mapsto xa$, and $x \mapsto axa$ belong to $SO_8\mathbb{R}$ and generate this group by (11.22).

The kernel of pr_3 consists of homologies with axis W and center o, i.e., of real homotheties according to (12.13a). Since $S\nabla$ induces an orthogonal group on the coordinate axes, it follows that $\ker \mathrm{pr}_3 = \{(x, y) \mapsto (\pm x, \pm y)\}$. Analogously, $\ker \mathrm{pr}_1 = S\nabla_{[v, X]}$ and $\ker \mathrm{pr}_2 = S\nabla_{[u, Y]}$ can be obtained from the description of homologies in (12.13b and c).

It is obvious that the reflections ι_v, ι_u, and ι_o belong to the center of $S\nabla$. On the other hand, since the center of $SO_8\mathbb{R}$ and the kernel of $\mathrm{pr}_1 : S\nabla \to SO_8\mathbb{R}$ both have order 2, the center of $S\nabla$ cannot have more than four elements, so that it consists of id, ι_v, ι_u, and ι_o. □

Reformulating certain aspects of the preceding geometric result (12.17) in algebraic terms, one obtains the following well-known statement.

12.18 Triality principle. *Among the triples of transformations $A, B, C \in SO_8\mathbb{R}$ satisfying*

(∗) $$B(s \cdot x) = Cs \cdot Ax \quad \text{for all } s, x \in \mathbb{O},$$

each of A, B, or C may take on every value in $SO_8\mathbb{R}$, and, in such a triple, each of the elements A, B, C determines the other two uniquely up to sign.

This is an immediate consequence of the information in (12.17c) about the surjective homomorphisms pr_ν and their kernels. By (12.7), the triality condition (∗) just expresses the fact that $(A, B \,|\, C)$ is a collineation.

For other approaches to the triality principle, see e.g. van der Blij–Springer [60] and Harvey [90] p. 275 ff.

12.19 Remarks. 1) The study of the stabilizer ∇ and of certain of its subgroups will be continued in Section 17, see (17.11 through 16). For instance, the homomorphisms pr_ν will be used to show that $S\nabla$ is the universal (two-fold) covering group $\mathrm{Spin}_8\mathbb{R}$ of $SO_8\mathbb{R}$. The discussion of the latter topic in (17.13) does not make essential use of Sections 13–16, so the reader may continue right there if he wishes.

2) Over $\mathbb{R}$ and $\mathbb{H}$ instead of $\mathbb{O}$, the group ∇ may be described in complete analogy with (12.17). Over $\mathbb{C}$, the collineations analogous with those given in (12.17) only constitute the $\mathbb{C}$-linear part of ∇.

Over $\mathbb{F} \in \{\mathbb{R}, \mathbb{H}\}$, even the proof of (12.17) carries over verbatim. On the other hand, the result may be obtained directly from the fundamental theorem (12.10) of affine geometry, according to which the affine collineations fixing the coordinate axes $X = \mathbb{F} \times \{0\}$ and $Y = \{0\} \times \mathbb{F}$ are the maps

$$(x, y) \mapsto (ax^\alpha, dy^\alpha),$$

for $a, d \in \mathbb{F}^\times$ and $\alpha \in \mathrm{Aut}\,\mathbb{F}$. The only automorphism of $\mathbb{R}$ is the identity, and the automorphisms of $\mathbb{H}$ are precisely the inner automorphisms (11.25). Thus, the collineations in question, with their actions on the line at infinity according to (12.11), are the mappings of the form

$$(x, y) \mapsto (axc, dyc), \quad (s) \mapsto (dsa^{-1}).$$

(This collineation is composed of the collineation $(x, y) \mapsto (c^{-1}xc, c^{-1}yc)$,

$(s) \mapsto (c^{-1}sc)$ obtained from the inner automorphism $x \mapsto c^{-1}xc$, and of the collineation $(x, y) \mapsto (acx, dcy)$, whose action on the line at infinity is $(s) \mapsto (dcs(ac)^{-1}) = (dcsc^{-1}a^{-1})$.) Now, the transformations $\mathbb{H} \to \mathbb{H} : x \mapsto axc$ constitute the group $\mathrm{GO}_4^+ \mathbb{R}$, see (11.23), so that we arrive at a description of ∇ for the plane over $\mathbb{H}$ which is analogous to (12.17).

Over $\mathbb{C}$, the collineations corresponding to the multitude of field automorphisms, see the references in (44.11), are not covered by this description; except conjugation, they are all discontinuous.

12.20 Note. For $\mathbb{F} \in \{\mathbb{R}, \mathbb{C}, \mathbb{H}\}$, the group $\mathsf{A}|_W$ of transformations induced on the line at infinity by the group of all affine collineations was discussed in (12.12) and was shown to be the product $\mathrm{Aut}\,\mathbb{F} \cdot \mathrm{PGL}_2\mathbb{F}$. For the case $\mathbb{F} = \mathbb{O}$, we shall not enter into a detailed discussion of $\mathsf{A}|_W$ at this point; the usual definitions of $\mathrm{GL}_2\mathbb{F}$ and $\mathrm{PGL}_2\mathbb{F}$ for a field $\mathbb{F}$ do not make sense if $\mathbb{F}$ is replaced by $\mathbb{O}$. Instead, we shall give a different geometric description of the group $\mathsf{A}|_W$ in Section 15, which is valid uniformly for $\mathbb{F} \in \{\mathbb{R}, \mathbb{C}, \mathbb{H}\}$ and for $\mathbb{F} = \mathbb{O}$, see (15.6). At this stage, we shall merely add a few remarks related to the results obtained so far; these remarks will not be used in the sequel.

From what we know, it is not difficult to deduce that $\mathsf{A}|_W$ is generated by the transformations $\sigma_a|_W$ and $\sigma'_a|_W$ for $a \in \mathbb{O}$, where σ_a and σ'_a are the shears described in (12.5b and c). Indeed, the subgroup Ω generated by these transformations acts 2-transitively on W, so that it suffices to show that Ω contains the stabilizer of (0) and (∞). This stabilizer is the group $\nabla|_W$, where, as before, ∇ is the stabilizer of the points o, (0), and (∞) in the group of all collineations; recall that $\mathsf{A}|_W = \mathsf{A}_o|_W$ because of transitivity of the group of translations with axis W. Now ∇ was discussed in (12.17); arguing as in step 3) of the proof there, one sees that $\nabla|_W$ is generated by the restrictions $\gamma_a|_W$, $a \in \mathbb{O}\setminus\{0\}$, of the collineations described in (12.14). These, in turn, have been constructed explicitly as products of shears.

If $\mathbb{O}$ is replaced with an arbitrary Cayley division algebra K, then Ω need not coincide with $\mathsf{A}|_W$ any more, but still is an important normal subgroup. As shown by Timmesfeld [94], the group Ω encodes the entire geometry of the plane over K. One of the main themes of his paper are characterizations of the subgroup of A generated by the shears σ_a and σ'_a for $a \in K$. Timmesfeld calls that group $\mathrm{SL}_2 K$, thus giving a meaning to this otherwise undefined symbol. In the case of a field K, this is in accordance with the usual meaning and with the description of $\mathsf{A}|_W$ using the group $\mathrm{PGL}_2 K$ as in (12.12).

The following fact will be useful in later sections when we continue the study of the collineation group of the octonion plane.

12.21 Lemma. *The group A of all collineations of the affine octonion plane $\mathcal{A}_2\mathbb{O}$ is the semidirect product of the translation group T described in (12.5) by the* stabilizer A_o *of the origin* $o = (0, 0)$, *and* A_o *consists of* $\mathbb{R}$-*linear transformations of* $\mathbb{O} \times \mathbb{O}$.

Proof. Since T is a normal subgroup of A and is transitive on the affine point set $\mathbb{O} \times \mathbb{O}$, we have $\mathsf{A} = \mathsf{A}_o \cdot \mathsf{T} = \mathsf{T} \cdot \mathsf{A}_o$ exactly as in (12.10), by the Frattini argument (91.2a). If one is willing to use basic information about translation planes, see (25.5), the linearity assertion is obtained readily: the collineations in A_o are known to be semilinear over the kernel of a coordinatizing quasifield, which in our case is $\mathbb{O}$. Now the kernel of $\mathbb{O}$ is $\mathbb{R}$ (11.20) and has no automorphism except the identity, so that semilinearity implies linearity.

Without using this information, a direct argument for the linearity of A_o may be given as follows. The subgroup Λ of A_o generated by the shears $(x, y) \mapsto (x, y + ax)$, $(s) \mapsto (s + a)$, $(\infty) \mapsto (\infty)$ of (12.5b) and by the reflection $(x, y) \mapsto (y, x)$, $(s) \mapsto (s^{-1})$ for $s \neq 0$, $(0) \leftrightarrow (\infty)$, obviously consists of $\mathbb{R}$-linear collineations. Moreover, Λ acts 2-transitively on the line at infinity, so that $\mathsf{A}_o = \Lambda \cdot \mathsf{A}_{o,u,v}$ by the Frattini argument. Now, finally, $\mathsf{A}_{o,u,v} = \nabla$ also consists of $\mathbb{R}$-linear transformations by (12.17a). □

13 The projective planes over $\mathbb{R}$, $\mathbb{C}$, and $\mathbb{H}$

The projective planes $\overline{\mathscr{A}_2 \mathbb{F}}$ over $\mathbb{F} \in \{\mathbb{R}, \mathbb{C}, \mathbb{H}, \mathbb{O}\}$ are highly homogeneous structures. For instance, the group of all collineations is transitive both on the set of points and on the set of lines, including the points at infinity and the line at infinity, see (13.5) and (17.2). This fact is obscured, however, by the representation of $\overline{\mathscr{A}_2 \mathbb{F}}$ as the projective completion of $\mathscr{A}_2 \mathbb{F}$; that description, given in (12.2), assigns a special rôle to the elements at infinity. The classical method to remedy this defect is to introduce homogeneous coordinates. For the planes over the fields $\mathbb{F} \in \{\mathbb{R}, \mathbb{C}, \mathbb{H}\}$, this is easy and will be our first objective here. Due to the non-associativity of $\mathbb{O}$, an equally homogeneous description of the octonion projective plane is of needs more complicated, and will be put off to Section 16. The present section continues with a proof of the fundamental theorem of projective geometry, which describes the group of all collineations of the projective plane over $\mathbb{F} \in \{\mathbb{R}, \mathbb{C}, \mathbb{H}\}$ in terms of homogeneous coordinates. Finally, we examine some distinguished subgroups of those collineation groups, namely the elliptic motion groups and the hyperbolic motion groups. They can be defined as the groups of all (linear) collineations which commute with certain polarities. Our interest focusses on the transitivity properties of these groups.

13.0 General assumption. In this section, $\mathbb{F}$ shall denote one of the fields $\mathbb{R}$, $\mathbb{C}$, or $\mathbb{H}$. Many of our arguments would work for an arbitrary field.

13 The projective planes over $\mathbb{R}$, $\mathbb{C}$, and $\mathbb{H}$

13.1 The projective plane $\mathcal{P}_2\mathbb{F}$ over $\mathbb{F}$ is defined as follows. Consider $\mathbb{F}^3$ as a right vector space over $\mathbb{F}$. The 1-dimensional subspaces of $\mathbb{F}^3$ will be called 'points'; the 'lines' will be the 2-dimensional subspaces. Incidence between points and lines is given by inclusion. The set of points will be denoted by $P_2\mathbb{F}$, and the set of lines by $\mathcal{L}_2\mathbb{F}$.

In this definition, the rôles of points and of lines are completely symmetric. If one wishes to think of lines as subsets of the set of points, one has to identify a line with the *point row* consisting of the points incident with it.

Elementary linear algebra shows that this geometry $\mathcal{P}_2\mathbb{F}$ is indeed a projective plane in the sense of definition (21.1). We shall see soon (13.3) that in fact $\mathcal{P}_2\mathbb{F}$ is isomorphic to the projective completion $\overline{\mathcal{A}_2\mathbb{F}}$ constructed in (12.2).

13.2 Homogeneous coordinates in $\mathcal{P}_2\mathbb{F}$. A point $p \in P_2\mathbb{F}$, i.e., a 1-dimensional subspace of $\mathbb{F}^3$, is spanned by some nonzero vector $(x_1, x_2, x_3) \in \mathbb{F}^3$,

$$p = (x_1, x_2, x_3)\mathbb{F}.$$

The point p determines the coordinates $x_1, x_2, x_3 \in \mathbb{F}$ uniquely up to a common nonzero factor from the right; these coordinates are called *homogeneous coordinates* of p.

In order to introduce homogeneous coordinates for lines, we recall that the 2-dimensional subspaces of $\mathbb{F}^3$ are precisely the kernels of nonzero linear forms of $\mathbb{F}^3$. Assume that the 2-dimensional subspace U is the kernel of the linear form

$$(a_1 \ a_2 \ a_3) : \mathbb{F}^3 \longrightarrow \mathbb{F} : \begin{pmatrix} x_1 \\ x_2 \\ x_3 \end{pmatrix} \longmapsto (a_1 \ a_2 \ a_3) \begin{pmatrix} x_1 \\ x_2 \\ x_3 \end{pmatrix} = \sum_{\nu=1}^{3} a_\nu x_\nu,$$

$$U = \mathrm{Ker}\,(a_1 \ a_2 \ a_3),$$

where $(a_1, a_2, a_3) \in \mathbb{F}_3 \setminus \{0\}$. Then U determines the coefficients $a_\nu \in \mathbb{F}$ of such a linear form uniquely up to a common factor from the left. These coefficients are called homogeneous coordinates of the line U. It is obvious that incidence of points and lines can be expressed in homogeneous coordinates as follows.

The point $p = (x_1, x_2, x_3)\mathbb{F}$ *is incident with the line* $\mathrm{Ker}\,(a_1 \ a_2 \ a_3)$ *if, and only if,*

$$(*) \qquad (a_1 \ a_2 \ a_3) \begin{pmatrix} x_1 \\ x_2 \\ x_3 \end{pmatrix} = \sum_{\nu=1}^{3} a_\nu x_\nu = 0.$$

13.3 An isomorphism $\overline{\mathcal{A}_2\mathbb{F}} \cong \mathcal{P}_2\mathbb{F}$. In $\mathcal{P}_2\mathbb{F}$, we single out the line

$$W := \mathbb{F} \times \mathbb{F} \times \{0\} = \mathrm{Ker}\,(0 \ 0 \ 1) \in \mathcal{L}_2\mathbb{F}.$$

The affine plane $\mathcal{A}_2\mathbb{F}$ as described in (12.1) can be located within $\mathbb{F}^3$ on the 2-dimensional affine subspace

$$A := \mathbb{F} \times \mathbb{F} \times \{1\}$$

parallel to W by means of the bijection

$$\mathbb{F} \times \mathbb{F} \to A : (x, y) \mapsto (x, y, 1).$$

This bijection maps the lines of $\mathcal{A}_2\mathbb{F}$ as defined in (12.1) onto the 1-dimensional affine subspaces of A.

Now we interpret this geometry on A in the projective plane $\mathcal{P}_2\mathbb{F}$, see Figure 13a. The vectors in A span the 1-dimensional linear subspaces of $\mathbb{F}^3$ not contained in W; these are the points of $\mathcal{P}_2\mathbb{F}$ not incident with the line W. A line L in A (a 1-dimensional affine subspace) spans a 2-dimensional linear subspace $U \neq W$ of $\mathbb{F}^3$, which is a line of $\mathcal{P}_2\mathbb{F}$; the points on this line are the 1-dimensional linear subspaces containing some vector on L, and the 1-dimensional subspace $U \cap W$. Denoting by $P_2\mathbb{F} \setminus W$ the set of points of $\mathcal{P}_2\mathbb{F}$ not incident with W, we have a bijection

$$\psi : \mathbb{F} \times \mathbb{F} \to P_2\mathbb{F} \setminus W : (x, y) \mapsto (x, y, 1)\mathbb{F},$$

and ψ is a collineation of the affine plane $\mathcal{A}_2\mathbb{F}$ onto an affine plane with point set $P_2\mathbb{F} \setminus W$ whose lines are the elements of $\mathcal{L}_2\mathbb{F} \setminus \{W\}$ (disregarding their points on W); this affine plane will be denoted by $\mathcal{P}_2\mathbb{F}^W$.

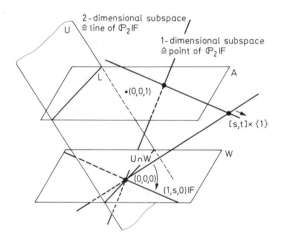

Figure 13a

We shall write down the inverse map ψ^{-1} explicitly, and incidentally prove the bijectivity of ψ once again. To do this, we use the fact that one is allowed to multiply homogeneous coordinates of a point by a common scalar from the right.

Thus, for a point $(x_1, x_2, x_3)\mathbb{F}$ not belonging to W, i.e., with $x_3 \neq 0$, we have $(x_1, x_2, x_3)\mathbb{F} = (x_1 x_3^{-1}, x_2 x_3^{-1}, 1)\mathbb{F}$. Hence,

(1) $$P_2\mathbb{F} \setminus W = \{(x, y, 1)\mathbb{F} \mid x, y \in \mathbb{F}\},$$

and the inverse of ψ is the map

$$\psi^{-1} : P_2\mathbb{F} \setminus W \to \mathbb{F} \times \mathbb{F} : (x_1, x_2, x_3)\mathbb{F} \mapsto (x_1 x_3^{-1}, x_2 x_3^{-1}).$$

By an analogous normalization we obtain that the points on $W = \mathbb{F} \times \mathbb{F} \times \{0\}$ can be written as

(2) $$(1, s, 0)\mathbb{F} \quad \text{for } s \in \mathbb{F} \quad \text{and} \quad (0, 1, 0)\mathbb{F}.$$

In the affine plane $\mathcal{P}_2\mathbb{F}^W$, two different lines $L_1, L_2 \in \mathcal{L}_2\mathbb{F} \setminus \{W\}$ are parallel if, and only if, their point of intersection in the projective plane $\mathcal{P}_2\mathbb{F}$ lies on W. Thus, $\mathcal{P}_2\mathbb{F}$ is (isomorphic to) the projective completion of the affine plane $\mathcal{P}_2\mathbb{F}^W$, with W having the rôle of the line at infinity. To put it more precisely, the collineation $\psi : \mathcal{A}_2\mathbb{F} \to \mathcal{P}_2\mathbb{F}^W$ extends uniquely to a collineation $\overline{\psi}$ of the projective completion $\overline{\mathcal{A}_2\mathbb{F}}$ onto the projective plane $\mathcal{P}_2\mathbb{F}$. We now give a complete coordinate description of that collineation.

A correspondence between affine and homogeneous coordinates.
The collineation $\overline{\psi}$ of the projective completion $\overline{\mathcal{A}_2\mathbb{F}}$ onto $\mathcal{P}_2\mathbb{F}$ is given by the following maps between points and lines, respectively, of $\overline{\mathcal{A}_2\mathbb{F}}$ and $\mathcal{P}_2\mathbb{F}$.

$$(x, y) \mapsto (x, y, 1)\mathbb{F} \qquad [s, t] \to \text{Ker}(-s \ \ 1 \ \ -t)$$
$$(s) \mapsto (1, s, 0)\mathbb{F} \qquad [c] \to \text{Ker}(1 \ \ 0 \ \ -c)$$
$$(\infty) \mapsto (0, 1, 0)\mathbb{F} \qquad [\infty] \to \text{Ker}(0 \ \ 0 \ \ 1) = W.$$

Indeed, from (1) and (2) it is immediate that the left column describes a bijection between the point sets of $\overline{\mathcal{A}_2\mathbb{F}}$ and $\mathcal{P}_2\mathbb{F}$. Furthermore, one has to check that this bijection maps the point row of every line onto the point row of the image line specified in the right column. For example, in $\mathcal{A}_2\mathbb{F}$ the point (x, y) is incident with the line $[s, t]$ if, and only if $y = sx + t$, i.e., $-sx + y - t = 0$; this is equivalent to the condition that $\overline{\psi}(x, y) = (x, y, 1)\mathbb{F}$ is incident with the line $\text{Ker}(-s \ \ 1 \ \ -t)$ of $\mathcal{P}_2\mathbb{F}$, see (13.2(*)). That line also contains the point $(1, s, 0)\mathbb{F}$ of W corresponding to the point (s) at infinity of $[s, t]$. The other verifications are just as easy. □

The isomorphism $\overline{\mathcal{A}_2\mathbb{F}} \cong \mathcal{P}_2\mathbb{F}$ described above may serve to illustrate the term 'point at infinity'. The 1-dimensional linear subspace $(1, s, 0)\mathbb{F} = \overline{\psi}((s))$ is parallel to the line $[s, t] \times \{1\} = \{(x, sx + t, 1) \mid x \in \mathbb{F}\}$ in $A \subseteq \mathbb{F}^3$ for every $t \in \mathbb{F}$. Every projective point corresponding to a point of this affine line can be rewritten as $(x, sx + t, 1)\mathbb{F} = (1, s + tx^{-1}, x^{-1})\mathbb{F}$ for $x \neq 0$. If x 'tends to infinity', this projective point 'converges' to the point $(1, s, 0)\mathbb{F}$ corresponding to the point at infinity (s), see Figure 13a. This will be made more precise in Section 14.

Collineations

13.4 Linear collineations. A linear transformation $A \in \mathrm{GL}_3\mathbb{F}$ of the right vector space $\mathbb{F}^3$ maps 1-dimensional subspaces of $\mathbb{F}^3$ to 1-dimensional subspaces and therefore induces a bijection of the point set of $\mathscr{P}_2\mathbb{F}$, given by

$$[A] : \mathrm{P}_2\mathbb{F} \to \mathrm{P}_2\mathbb{F} : x\mathbb{F} \mapsto (Ax)\mathbb{F},$$

where $x \in \mathbb{F}^3$. Furthermore, $[A]$ is a collineation of $\mathscr{P}_2\mathbb{F}$, because A maps 2-dimensional subspaces of $\mathbb{F}^3$ to 2-dimensional subspaces. The collineations which are obtained in this way are called *linear collineations*; they form a subgroup

$$\mathrm{PGL}_3\mathbb{F} := \{ [A] \mid A \in \mathrm{GL}_3\mathbb{F} \}$$

of the group $\mathrm{Aut}\,\mathscr{P}_2\mathbb{F}$ of all collineations of $\mathscr{P}_2\mathbb{F}$.

As to transitivity properties of $\mathrm{PGL}_3\mathbb{F}$, we consider *non-degenerate quadrangles* of $\mathscr{P}_2\mathbb{F}$, i.e., quadruples of points no three of which are collinear. One such quadrangle is the *standard quadrangle* $e_1\mathbb{F}$, $e_2\mathbb{F}$, $e_3\mathbb{F}$, $(e_1 + e_2 + e_3)\mathbb{F}$ where e_1, e_2, e_3 is the standard basis of $\mathbb{F}^3$. If p_1, p_2, p_3, p_4 is an arbitrary non-degenerate quadrangle, then the 1-dimensional subspaces p_1, p_2, p_3 generate $\mathbb{F}^3$ because they are not collinear; thus, there is a basis b_1, b_2, b_3 such that $p_\nu = b_\nu\mathbb{F}$ ($\nu = 1, 2, 3$). A vector $x \in \mathbb{F}^3$ generating the 1-dimensional subspace p_4 may be represented as $x = \sum_{\nu=1}^{3} b_\nu \lambda_\nu$ with $\lambda_\nu \in \mathbb{F}$; and the scalars λ_ν are all nonzero because p_4 is not collinear with any two of p_1, p_2, p_3. Let $A \in \mathrm{GL}_3\mathbb{F}$ be the linear transformation mapping the standard basis e_1, e_2, e_3 onto $b_1\lambda_1, b_2\lambda_2, b_3\lambda_3$; then the collineation $[A]$ maps the standard quadrangle onto p_1, p_2, p_3, p_4. Thus we have obtained the following result.

13.5 Proposition: Homogeneity. $\mathrm{PGL}_3\mathbb{F}$ *is transitive on the set of non-degenerate quadrangles.* □

13.6 Fundamental theorem of projective geometry. *The collineations of $\mathscr{P}_2\mathbb{F}$ are precisely the transformations induced by semilinear bijections of $\mathbb{F}^3$. Explicitly, these are the transformations $[A, \alpha]$ defined by*

$$[A, \alpha] : \mathrm{P}_2\mathbb{F} \to \mathrm{P}_2\mathbb{F} : \begin{pmatrix} x_1 \\ x_2 \\ x_3 \end{pmatrix} \cdot \mathbb{F} \mapsto A \begin{pmatrix} x_1^\alpha \\ x_2^\alpha \\ x_3^\alpha \end{pmatrix} \cdot \mathbb{F},$$

where A is a regular 3×3-matrix over $\mathbb{F}$, and $\alpha \in \mathrm{Aut}\,\mathbb{F}$.

Proof. By (13.5) and with the help of the Frattini argument (91.2a), the proof reduces to showing that the stabilizer of the standard quadrangle

$$e_1\mathbb{F} = (1,0,0)\mathbb{F}, \quad e_2\mathbb{F} = (0,1,0)\mathbb{F}, \quad e_3\mathbb{F} = (0,0,1)\mathbb{F}, \quad (e_1+e_2+e_3)\mathbb{F} = (1,1,1)\mathbb{F}$$

in $\mathrm{Aut}\,\mathcal{P}_2\mathbb{F}$ consists precisely of the transformations $[\mathrm{id}, \alpha]$, $\alpha \in \mathrm{Aut}\,\mathbb{F}$.

By means of the collineation $\overline{\psi} : \overline{\mathcal{A}_2\mathbb{F}} \cong \mathcal{P}_2\mathbb{F}$ described in (13.3), a collineation of $\mathcal{P}_2\mathbb{F}$ fixing the standard quadrangle can be written as $\overline{\psi} \circ \delta \circ \overline{\psi}^{-1}$, where δ is a collineation of $\overline{\mathcal{A}_2\mathbb{F}}$ that fixes the corresponding quadrangle formed by the points $u = (0)$, $v = (\infty)$, $o = (0,0)$, and $e = (1,1)$. By (12.8), the collineations of $\overline{\mathcal{A}_2\mathbb{F}}$ fixing those points are precisely the transformations $\delta : (x,y) \mapsto (x^\alpha, y^\alpha)$, $(s) \mapsto (s^\alpha)$, $(\infty) \mapsto (\infty)$ for all $\alpha \in \mathrm{Aut}\,\mathbb{F}$. For these, it is easy to show that $\overline{\psi} \circ \delta \circ \overline{\psi}^{-1} = [\mathrm{id}, \alpha]$ by checking that $\overline{\psi} \circ \delta = [\mathrm{id}, \alpha] \circ \overline{\psi}$. □

13.7 Remarks. Not only the collineations coming from automorphisms of $\mathbb{F}$, but in fact all collineations of $\mathcal{A}_2\mathbb{F}$ as determined by the fundamental theorem of affine geometry (12.10) are easily translated into homogeneous coordinates via the collineation $\overline{\psi} : \overline{\mathcal{A}_2\mathbb{F}} \to \mathcal{P}_2\mathbb{F}$ of (13.3). In this way, the collineation

$$\begin{pmatrix} x \\ y \end{pmatrix} \mapsto \begin{pmatrix} c_{11} & c_{12} \\ c_{21} & c_{22} \end{pmatrix} \begin{pmatrix} x^\alpha \\ y^\alpha \end{pmatrix} + \begin{pmatrix} a \\ b \end{pmatrix}$$

of $\mathcal{A}_2\mathbb{F}$ yields the collineation $[C, \alpha]$ of $\mathcal{P}_2\mathbb{F}$ with

$$C = \begin{pmatrix} c_{11} & c_{12} & a \\ c_{21} & c_{22} & b \\ 0 & 0 & 1 \end{pmatrix} \in \mathrm{GL}_3\mathbb{F}.$$

For a diagonal matrix

$$D = \begin{pmatrix} d & & \\ & d & \\ & & d \end{pmatrix}, \quad 0 \neq d \in \mathbb{F},$$

the collineation $[D] = [D, \mathrm{id}]$ is the same as $[\mathrm{id}, \mathrm{int}\,d]$, where $\mathrm{int}\,d$ is the inner automorphism $\mathrm{int}\,d : \mathbb{F} \to \mathbb{F} : x \mapsto dxd^{-1}$; indeed, $[D, \mathrm{id}]$ maps $(x_1, x_2, x_3)\mathbb{F}$ to $(dx_1, dx_2, dx_3)\mathbb{F} = (dx_1 d^{-1}, dx_2 d^{-1}, dx_3 d^{-1})\mathbb{F}$.

Thus, if all automorphisms of $\mathbb{F}$ are inner automorphisms, then every collineation is induced by a linear transformation of $\mathbb{F}^3$. Here, this applies to $\mathbb{R}$ and $\mathbb{H}$, see (11.26 and 28).

13.8 Corollary. *For $\mathbb{F} \in \{\mathbb{R}, \mathbb{H}\}$, the full collineation group of $\mathcal{P}_2\mathbb{F}$ is equal to $\mathrm{PGL}_3\mathbb{F}$.* □

13.9 Warning. For $\mathbb{F} = \mathbb{C}$, conjugation is an automorphism, and therefore the map

$$(x_1, x_2, x_3)\mathbb{C} \mapsto (\overline{x_1}, \overline{x_2}, \overline{x_3})\mathbb{C}$$

is an involutory collineation of $\mathcal{P}_2\mathbb{C}$, whose fixed points are precisely the points of $\mathcal{P}_2\mathbb{R}$ viewed as a subplane of $\mathcal{P}_2\mathbb{C}$. In the case $\mathbb{F} = \mathbb{H}$, the analogous definition would not even give a well-defined map. For example, with the standard Hamilton triple i, j, k of $\mathbb{H}$ (11.16), one has $(j, k, 0)\mathbb{H} = (j, ij, 0)\mathbb{H} = (1, i, 0)\mathbb{H}$ and $(\bar{j}, \bar{k}, \bar{0})\mathbb{H} = (-j, -k, 0)\mathbb{H} = (j, k, 0)\mathbb{H} = (1, i, 0)\mathbb{H} \neq (1, -i, 0)\mathbb{H} = (\bar{1}, \bar{i}, \bar{0})\mathbb{H}$. On the other hand, conjugation does give rise to a polarity of $\mathcal{P}_2\mathbb{F}$ in all cases, see (13.12).

Polarities and their motion groups

13.10 The dual plane $\mathcal{P}_2^*\mathbb{F}$. The symmetry of the rôles of points and lines in the axioms of a projective plane $\mathcal{P}$ (21.1) allows us to interchange the notions of points and lines; in this way, we obtain another projective plane, called the *dual* plane $\mathcal{P}^*$. This yields the following *duality principle*: If a statement is true in all projective planes, then its *dual* statement, obtained by interchanging the words 'point' and 'line', is equally true.

Now we redescribe the construction of the projective plane $\mathcal{P}_2\mathbb{F}$ in a way which is particularly suited to determine its dual plane; the latter will be denoted by $\mathcal{P}_2^*\mathbb{F}$. The points of $\mathcal{P}_2\mathbb{F}$ are the 1-dimensional subspaces of $\mathbb{F}^3$ as a *right* vector space, and via homogeneous coordinates (13.2) the lines correspond to the 1-dimensional subspaces of $\mathbb{F}^3$ as a *left* vector space; recall that the homogeneous coordinates of a line are determined up to a common factor from the left. Incidence is described by equation (13.2(∗)). Thus, we obtain the dual plane $\mathcal{P}_2^*\mathbb{F}$ instead of $\mathcal{P}_2\mathbb{F}$ when we reverse the order of multiplication in $\mathbb{F}$, passing to the opposite field $\mathbb{F}^{\text{op}}$:

$$\mathcal{P}_2^*\mathbb{F} \cong \mathcal{P}_2(\mathbb{F}^{\text{op}}).$$

Note that in our context, for $\mathbb{F} \in \{\mathbb{R}, \mathbb{C}, \mathbb{H}\}$, we have that $\mathbb{F}^{\text{op}} \cong \mathbb{F}$ and therefore $\mathcal{P}_2\mathbb{F} \cong \mathcal{P}_2^*\mathbb{F}$; in general, this is true if, and only if, $\mathbb{F}$ admits an antiautomorphism.

13.11 Definitions. A *duality* of a projective plane $\mathcal{P}$ with point set P and line set $\mathcal{L}$ is a collineation of $\mathcal{P}$ onto its dual plane $\mathcal{P}^*$. A duality can be described as a bijection π of the disjoint union $P \dot\cup \mathcal{L}$ onto itself exchanging P and $\mathcal{L}$ and having the following property:

p is incident with L if, and only if, L^π is incident with p^π

for all $p \in P$ and $L \in \mathcal{L}$. This condition is equivalent to

$$(pq)^\pi = p^\pi \wedge q^\pi \quad \text{and} \quad (L \wedge M)^\pi = L^\pi M^\pi$$

for distinct points p, q and distinct lines L, M. A *polarity* is a duality which as a bijection of $P \dot\cup \mathcal{L}$ is involutory, i.e., satisfies

$$(p^\pi)^\pi = p \quad \text{and} \quad (L^\pi)^\pi = L$$

for all $p \in P$ and $L \in \mathcal{L}$. The line p^π is called the *polar* of p, and the point L^π is the *pole* of L. A collineation γ of $\mathcal{P}$ is said to *commute* with the polarity π if for every point p one has $p^{\gamma\pi} = p^{\pi\gamma}$. This is equivalent to $L^{\gamma\pi} = L^{\pi\gamma}$ for every line L. The collineations commuting with a given polarity form a subgroup of the group of all collineations.

13.12 The standard elliptic polarity and the standard hyperbolic polarity of $\mathcal{P}_2\mathbb{F}$, $\mathbb{F} \in \{\mathbb{R}, \mathbb{C}, \mathbb{H}\}$ are constructed using the following two Hermitian forms on $\mathbb{F}^3$:

$$(x \mid y)_\pm := \overline{x_1}y_1 + \overline{x_2}y_2 \pm \overline{x_3}y_3$$

for $x = (x_1, x_2, x_3)$, $y = (y_1, y_2, y_3) \in \mathbb{F}^3$.

By mapping 1- and 2-dimensional subspaces of $\mathbb{F}^3$ to their orthogonal spaces with respect to either $(\ \mid\ )_+$ or $(\ \mid\ )_-$, one clearly obtains polarities π^+ and π^- of $\mathcal{P}_2\mathbb{F}$, respectively, which will be called the *standard elliptic polarity* and the *standard hyperbolic polarity*. In homogeneous coordinates, they have the following description:

$$\pi^\pm : P_2\mathbb{F} \to \mathcal{L}_2\mathbb{F} : (x_1, x_2, x_3)\mathbb{F} \mapsto \operatorname{Ker}(\overline{x_1}\ \overline{x_2}\ \pm\overline{x_3})$$
$$\mathcal{L}_2\mathbb{F} \to P_2\mathbb{F} : \operatorname{Ker}(x_1\ x_2\ x_3) \mapsto (\overline{x_1}, \overline{x_2}, \pm\overline{x_3})\mathbb{F}.$$

Note that these maps are well-defined, since conjugation is an antiautomorphism of $\mathbb{F}$. For later use, we remark that obviously

(1) $$\pi^- = \pi^+ \iota_o,$$

where ι_o is the linear collineation

$$\iota_o = \left[\begin{pmatrix} 1 & & \\ & 1 & \\ & & -1 \end{pmatrix}\right] = \left[\begin{pmatrix} -1 & & \\ & -1 & \\ & & 1 \end{pmatrix}\right].$$

In affine coordinates, ι_o is given by $(x, y) \mapsto (-x, -y)$, cf. (13.7). In other words, ι_o is the unique involutory homology having the point $o = (0, 0)$ as center and the line at infinity as axis (12.13).

We now want to determine the (linear) collineations which commute with these polarities. Among them are the collineations induced by linear transformations U of $\mathbb{F}^3$ which are *unitary* with respect to the Hermitian form $(\ \mid\ )_\pm$, i.e., satisfy

(*) $$(Ux \mid Uy)_\pm = (x \mid y)_\pm \quad \text{for all } x, y \in \mathbb{F}^3.$$

The unitary transformations form a subgroup of $\operatorname{GL}_3\mathbb{F}$, which is denoted by $\operatorname{U}_3\mathbb{F}$ in the case of $(\ \mid\ )_+$ and by $\operatorname{U}_3(\mathbb{F}, 1)$ for $(\ \mid\ )_-$. If $\mathbb{F} = \mathbb{R}$, it is more customary to call these transformations orthogonal and to write $O_3\mathbb{R}$ and $O_3(\mathbb{R}, 1)$ instead of $U_3\mathbb{R}$ and $U_3(\mathbb{R}, 1)$, respectively.

13.13 Proposition: Motion groups of $\mathcal{P}_2\mathbb{F}$. *The elliptic motion group and the hyperbolic motion group of $\mathcal{P}_2\mathbb{F}$, defined as the groups of all linear collineations which commute with the standard elliptic or hyperbolic polarity, respectively, coincide with the groups*

$$PU_3\mathbb{F} = \{\,[A] \mid A \in U_3\mathbb{F}\,\} \quad \text{and} \quad PU_3(\mathbb{F}, 1) = \{\,[A] \mid A \in U_3(\mathbb{F}, 1)\,\}$$

of collineations induced by the respective unitary transformations.

Proof. We give a direct proof adapted to our special situation; for a more conceptual approach in a general setting see Baer [52] IV.5 Prop. 1, p. 144 ff.

To simplify notation, write (|) for either (|)$_+$ or (|)$_-$, and let π stand for π^+ or π^-, and PU_3 for $PU_3\mathbb{F}$ or $PU_3(\mathbb{F}, 1)$, accordingly. By definition, $(x\mathbb{F})^\pi = x^\perp$, the orthogonal space of $x \in \mathbb{F}^3$ with respect to (|). Let Φ be the motion group belonging to π. Obviously, $PU_3 \subseteq \Phi$. We have to show the converse inclusion.

It is well known that every element $A \in GL_3\mathbb{F}$ has a (uniquely determined) adjoint $A^* \in GL_3\mathbb{F}$ satisfying

$$(A^*x \mid y) = (x \mid Ay)$$

for all $x, y \in \mathbb{F}^3$, see e.g. Porteous [81] Prop. 11.26, p. 207 ff. Using the adjoint, we find for $y \in \mathbb{F}^3$ that

$$(Ay)^\perp = \{\,x \mid x \in \mathbb{F}^3,\ (x \mid Ay) = 0\,\} = \{\,x \mid x \in \mathbb{F}^3,\ (A^*x \mid y) = 0\,\}$$
(1)
$$= \{\,A^{*-1}z \,\Big|\, z \in \mathbb{F}^3,\ (z \mid y) = 0\,\}$$
$$= A^{*-1}(y^\perp)\,.$$

Now let A be such that $[A] \in \Phi$, that is, $(Ay)^\perp = A(y^\perp)$ for all $y \in \mathbb{F}^3$. By (1), this translates into

(2) $$A^*A(y^\perp) = y^\perp$$

for all $y \in \mathbb{F}^3$. Since every line of $\mathcal{P}_2\mathbb{F}$ is of the form $y^\perp = (y\mathbb{F})^\pi$ for a suitable $y \in \mathbb{F}^3$, condition (2) says that the collineation of $\mathcal{P}_2\mathbb{F}$ induced by A^*A fixes every line and, hence, every point. In other words, A^*A leaves every 1-dimensional subspace of $\mathbb{F}^3$ invariant, and thus is of the form

(3) $$A^*A = c \cdot \mathrm{id}$$

for a suitable $c \in \mathbb{F}^\times$. Moreover, in the case $\mathbb{F} = \mathbb{H}$, it follows at this point already that c must belong to the subfield $\mathbb{R}$, because this subfield is the center of $\mathbb{H}$.

However, for other reasons, the fact that $c \in \mathbb{R}$ will presently be obtained in general, the case $\mathbb{F} = \mathbb{C}$ included.

From the definition of (|), it is immediate that the 2-dimensional subspace $\mathbb{F}^2 \times \{0\}$ is positive definite with respect to (|) not only in the elliptic case, but also in the hyperbolic case, in the sense that, for $0 \neq x \in \mathbb{F}^2 \times \{0\}$, the value $(x \mid x) \in \mathbb{R}$ is positive. If we choose $x \neq 0$ from the non-trivial intersection of the 2-dimensional subspaces $\mathbb{F}^2 \times \{0\}$ and $A^{-1}(\mathbb{F}^2 \times \{0\})$, then $0 < (x \mid x)$ and $0 < (Ax \mid Ax) = (A^*Ax \mid x) = (cx \mid x) = \bar{c}(x \mid x)$. It follows that $c \in \mathbb{R}$ and $\bar{c} = c > 0$.

Now, it is important that the real scalars c and $b = 1/\sqrt{c}$ belong to the center of $\mathbb{F}$ and are fixed under conjugation. The first property implies that the linear transformation $B = bA$ induces the same collineation of $\mathcal{P}_2\mathbb{F}$ as A. Using (3), we obtain, moreover, that $(Bx \mid By) = (bAx \mid bAy) = b^2(A^*Ax \mid y) = c^{-1}(cx \mid y) = c^{-1}c(x \mid y) = (x \mid y)$ for all $x, y \in \mathbb{F}^3$, which means that B is unitary. Thus, $[A] = [B] \in \mathrm{PU}_3$. □

On the subspace $\mathbb{F}^2 = \mathbb{F}^2 \times \{0\} \leq \mathbb{F}^3$, both Hermitian forms $(\mid)_+$ and $(\mid)_-$ of (13.12) induce the same Hermitian form $(a, b) \mapsto \overline{a_1}b_1 + \overline{a_2}b_2$ for $a = (a_1, a_2)$, $b = (b_1, b_2) \in \mathbb{F}^2$. The group $\mathrm{U}_2\mathbb{F}$ of unitary transformations of $\mathbb{F}^2$ with respect to this Hermitian form appears in the following description of stabilizers of the elliptic and hyperbolic motion groups.

13.14 Proposition. *The stabilizer Φ_o of the point $o = (0, 0, 1)\mathbb{F}$ in the elliptic motion group $\mathrm{PU}_3\mathbb{F}$ and the stabilizer of o in the hyperbolic motion group $\mathrm{PU}_3(\mathbb{F}, 1)$ coincide.*

The stabilizer Φ_o leaves the line $W = \mathrm{Ker}\,(0\ 0\ 1)$ invariant and is transitive on the set of its points. On the point set of the affine plane $\mathcal{P}_2\mathbb{F}^W \cong \mathcal{A}_2\mathbb{F}$, identified with $\mathbb{F}^2$ according to (13.3), Φ_o acts as the group

$$\left\{ \binom{x}{y} \mapsto U\binom{xc}{yc} \,\middle|\, U \in \mathrm{U}_2\mathbb{F},\, c \in \mathbb{F},\, \|c\|^2 = 1 \right\}.$$

Addendum. In the commutative cases $\mathbb{F} \in \{\mathbb{R}, \mathbb{C}\}$, the homotheties $\mu_c : \binom{x}{y} \mapsto \binom{xc}{yc}$ for $c \in \mathbb{F}$, $\|c\|^2 = 1$ belong to $\mathrm{U}_2\mathbb{F}$. For $\mathbb{F} = \mathbb{H}$, the homothety μ_c is not a linear transformation of the *right* $\mathbb{H}$-vector space $\mathbb{H}^2$ unless $c \in \mathbb{R}$. The group of all homotheties μ_c with $\|c\|^2 = 1$ is isomorphic to $\mathrm{Spin}_3\mathbb{R}$ in this case (11.25). Thus

$$\Phi_o \cong \begin{cases} \mathrm{O}_2\mathbb{R} \\ \mathrm{U}_2\mathbb{C} \\ \mathrm{U}_2\mathbb{H} \cdot \mathrm{Spin}_3\mathbb{R} \end{cases} \text{ for } \mathbb{F} = \begin{cases} \mathbb{R} \\ \mathbb{C} \\ \mathbb{H} \end{cases}.$$

Proof of (13.14). The orthogonal space of $(0, 0, 1)\mathbb{F}$ with respect to both Hermitian forms $(\mid)_\pm$ is the 2-dimensional subspace $\mathbb{F}^2 \times \{0\} = \mathrm{Ker}\,(0\ 0\ 1)$ representing the line W. A transformation in $U_3\mathbb{F}$ or in $U_3(\mathbb{F}, 1)$ leaving $(0, 0, 1)\mathbb{F}$ invariant therefore leaves $W = \mathbb{F}^2 \times \{0\}$ invariant, as well. Since the Hermitian forms $(\ |\ )_+$ and $(\ |\ )_-$ coincide on $\mathbb{F}^2 \times \{0\}$, the transformations in question in both the elliptic and the hyperbolic case are precisely those which are represented by the matrices

(1) $$\begin{pmatrix} U & \\ & a \end{pmatrix} \text{ with } U \in U_2\mathbb{F} \text{ and } a \in \mathbb{F},\ \bar{a}a = 1.$$

The group of these transformations is transitive on the set of 1-dimensional subspaces of $\mathbb{F}^2 \times \{0\}$, that is, on the set of points of the line W. A point not on W with affine coordinates $(x, y) \in \mathbb{F}^2$ has homogeneous coordinates $(x, y, 1)\mathbb{F}$; the image point under the collineation induced by transformation (1) has homogeneous coordinates

$$\begin{pmatrix} U\begin{pmatrix} x \\ y \end{pmatrix} \\ a \end{pmatrix} \mathbb{F} = \begin{pmatrix} U\begin{pmatrix} x \\ y \end{pmatrix}a^{-1} \\ 1 \end{pmatrix} \mathbb{F} = \begin{pmatrix} U\begin{pmatrix} xa^{-1} \\ ya^{-1} \end{pmatrix} \\ 1 \end{pmatrix} \mathbb{F}$$

and affine coordinates $U\begin{pmatrix} xc \\ yc \end{pmatrix}$, where $c = a^{-1}$. Thus our proposition is proved.

If $\mathbb{F}$ is commutative, then the transformation $\begin{pmatrix} x \\ y \end{pmatrix} \mapsto \begin{pmatrix} xc \\ yc \end{pmatrix} = \begin{pmatrix} c & \\ & c \end{pmatrix}\begin{pmatrix} x \\ y \end{pmatrix}$ is linear, and for $\bar{c}c = 1$ it obviously belongs to $U_2\mathbb{F}$. □

In the sequel, we determine various orbits of the motion groups.

13.15. *The standard elliptic motion group* $PU_3\mathbb{F}$ *of* $\mathcal{P}_2\mathbb{F}$, $\mathbb{F} \in \{\mathbb{R}, \mathbb{C}, \mathbb{H}\}$, *is flag transitive, i.e., it acts transitively on the set of all incident point–line pairs.*

Proof. For a given flag $p \subseteq L$, there is a basis (b_1, b_2, b_3) of $\mathbb{F}^3$ which is orthonormal with respect to the positive definite Hermitian form $(\ |\ )_+$ and satisfies $p = b_1\mathbb{F}$, $L = b_1\mathbb{F} + b_2\mathbb{F}$. The assertion now follows because $U_3\mathbb{F}$ is (sharply) transitive on the set of orthonormal bases. □

The Hermitian form $(\ |\ )_-$ differs from $(\ |\ )_+$ in that there are 'isotropic' vectors $x \neq 0$ satisfying $(x\ |\ x)_- = 0$. For the point $p = x\mathbb{F}$ of $\mathcal{P}_2\mathbb{F}$, this is equivalent to saying that p is incident with its polar under the standard hyperbolic polarity π^-. This situation is covered by the following general notion.

13.16 Definition. Let π be a polarity of a projective plane. A point p is said to be *absolute* if p is incident with its polar p^π; dually, an *absolute line* is a line L which is incident with its pole L^π.

Obviously, the set of absolute points and the set of absolute lines is invariant under every collineation commuting with the polarity π.

13.17 Proposition: Orbits of the hyperbolic motion group.

(a) *The set of absolute points of the standard hyperbolic polarity π^- of $\mathcal{P}_2\mathbb{F}$, $\mathbb{F} \in \{\mathbb{R}, \mathbb{C}, \mathbb{H}\}$, is*

$$Q = \{ (x, y, 1)\mathbb{F} \mid x, y \in \mathbb{F}, \|x\|^2 + \|y\|^2 = 1 \} .$$

The set of exterior points, i.e., of non-absolute points which are incident with some absolute line, is

$$E = \{ (x, y, 1)\mathbb{F} \mid x, y \in \mathbb{F}, \|x\|^2 + \|y\|^2 > 1 \}$$
$$\cup \{ (x, y, 0)\mathbb{F} \mid 0 \neq (x, y) \in \mathbb{F}^2 \} .$$

The set of interior points, i.e., of points which are not incident with any absolute line, is

$$I = \{ (x, y, 1)\mathbb{F} \mid x, y \in \mathbb{F}, \|x\|^2 + \|y\|^2 < 1 \} .$$

(b) *The group $\mathrm{PU}_3(\mathbb{F}, 1)$ is transitive on Q, E and I. It is even flag transitive on the 'interior hyperbolic plane' (the geometry consisting of the interior points and of the lines through interior points).*

(c) *The lines through interior points are precisely the polars of exterior points.*

Remarks. 1) It is clear that the hyperbolic motion group cannot act flag transitively on the 'exterior hyperbolic plane' whose lines are all the lines through exterior points, because through every exterior point there are different kinds of lines (absolute lines, non-absolute lines without interior points, and lines with interior points), which cannot be transformed into each other by hyperbolic motions.

2) Via the collineation $\overline{\mathcal{A}_2\mathbb{F}} \cong \mathcal{P}_2\mathbb{F}$ of (13.3), the set Q of absolute points corresponds to the unit sphere in the affine point set $\mathbb{F} \times \mathbb{F} \cong \mathbb{R}^{2n}$ ($n = 1, 2, 4$). The point set I of the interior hyperbolic plane becomes the 'interior' ($=$ bounded) complementary component of the unit sphere, and E corresponds to the exterior component, together with the points at infinity.

Proof of (13.17). By (13.14), the orbits of the stabilizer Φ_o of $o = (0, 0, 1)\mathbb{F}$ in the hyperbolic motion group $\mathrm{PU}_3(\mathbb{F}, 1)$ are the subsets S_r corresponding to the spheres of positive radius $r \in \mathbb{R}$ in $\mathbb{F} \times \mathbb{F}$,

$$S_r = \{ (x, y, 1)\mathbb{F} \mid x, y \in \mathbb{F}, \|x\|^2 + \|y\|^2 = r \} ,$$

and the point row $\{ (x, y, 0)\mathbb{F} \mid (0, 0) \neq (x, y) \in \mathbb{F}^2 \}$ of the line W at infinity.

The set Q of absolute points, the set E of exterior points, and the set I of interior points are invariant under the hyperbolic motion group $\mathrm{PU}_3(\mathbb{F}, 1)$. The explicit descriptions of Q, E and I given in (a) and the transitivity assertions stated in (b) now may be inferred from the following facts, which shall be established subsequently.

1) The point $(1, 0, 1)\mathbb{F} \in S_1$ is absolute.
2) The point $(0, 0, 1)\mathbb{F}$ is an interior point, and its orbit under $\mathrm{PU}_3(\mathbb{F}, 1)$ contains points from every sphere S_r of radius $r < 1$.
3) The point $(0, 1, 0)\mathbb{F}$ on W is an exterior point, and its orbit under $\mathrm{PU}_3(\mathbb{F}, 1)$ contains points from every sphere S_r of radius $r > 1$.

Flag transitivity on the interior hyperbolic plane is then obtained from known transitivity properties of Φ_o. Indeed, Φ_o is transitive on the set of lines through $o = (0, 0, 1)\mathbb{F} \in I$, as Φ_o is transitive on the points of W.

In the following proofs of assertions 1)–3), we use the explicit description of π^- in homogeneous coordinates given in (13.12).

Proof of 1). $(1, 0, 1)\mathbb{F} \subseteq \mathrm{Ker}\,(1 \ \ 0 \ \ -1) = (1, 0, 1)\mathbb{F}^{\pi^-}$.

Proof of 2). The lines through $(0, 0, 1)\mathbb{F}$ are of the form $\mathrm{Ker}\,(a_1 \ \ a_2 \ \ 0)$; none of them contains its pole $\mathrm{Ker}\,(a_1 \ \ a_2 \ \ 0)^{\pi^-} = (\overline{a_1}, \overline{a_2}, 0)\mathbb{F}$, because $a_1\overline{a_1} + a_2\overline{a_2} > 0$. Hence, these lines are not absolute, and $(0, 0, 1)\mathbb{F}$ is an interior point. The matrix

$$(*) \qquad \begin{pmatrix} 1 & & \\ & \sqrt{1+t^2} & t \\ & t & \sqrt{1+t^2} \end{pmatrix} = \begin{pmatrix} 1 & & \\ & \cosh\tau & \sinh\tau \\ & \sinh\tau & \cosh\tau \end{pmatrix}$$

with $t \in \mathbb{R}$, $\tau = \ln(t + \sqrt{1+t^2})$ belongs to $\mathrm{U}_3(\mathbb{F}, 1)$ and maps $(0, 0, 1)\mathbb{F}$ to $(0, t, \sqrt{1+t^2})\mathbb{F}$. For $t > 0$, this point can be rewritten as $(0, r^{-1}, 1)\mathbb{F}$, where

$$r = \sqrt{1+t^{-2}},$$

and $t > 0$ can be chosen such that r^{-1} is any preassigned number between 0 and 1.

Proof of 3). The point $(0, 1, 0)\mathbb{F}$ is not contained in its polar $(0, 1, 0)\mathbb{F}^{\pi^-} = \mathrm{Ker}\,(0 \ \ 1 \ \ 0)$, i.e., it is not absolute. It is contained in the line $\mathrm{Ker}\,(1 \ \ 0 \ \ 1)$, which is absolute: $\mathrm{Ker}\,(1 \ \ 0 \ \ 1)^{\pi^-} = (1, 0, -1)\mathbb{F} \subseteq \mathrm{Ker}\,(1 \ \ 0 \ \ 1)$. The matrix $(*)$ maps $(0, 1, 0)\mathbb{F}$ to $(0, \sqrt{1+t^2}, t)\mathbb{F}$. For $t > 0$, this is the point $(0, r, 1)\mathbb{F}$, where r is defined as above; for a suitable $t > 0$, the radius r takes on any given value greater than 1.

Thus, (a) and (b) are proved. As to assertion (c), note that the interior point $o = (0, 0, 1)\mathbb{F}$ is contained in the line $\mathrm{Ker}\,(0 \ \ 1 \ \ 0)$, which is the polar of the exterior point $(0, 1, 0)\mathbb{F}$. Assertion (c) now follows, because, according to (b), the hyperbolic motion group is transitive both on E and on the set of lines through interior points, and because hyperbolic motions respect the polar relation. $\square$

13.18 Notes. Here we record further polarities which, together with the standard elliptic polarity and the standard hyperbolic polarity, represent all equivalence classes of polarities of $\mathcal{P}_2\mathbb{F}$ except the discontinuous polarities of $\mathcal{P}_2\mathbb{C}$.

According to a theorem of Birkhoff and von Neumann, cf. Baer [52] IV.1 Prop. 2, p. 103 and IV.3 Theorem 1, p. 111 or Taylor [92] Theorem 7.1, p. 53, every polarity of $\mathcal{P}_2\mathbb{F}$ may be derived from some non-degenerate Hermitian form $(x, y) \mapsto f(x, y)$ on $\mathbb{F}^3$ accompanied by an involutory antiautomorphism α of $\mathbb{F}$. (Notice that the other possibility appearing in the general form of that theorem, which is formulated for projective spaces of arbitrary dimension, is excluded for projective planes. It concerns the case of a non-degenerate alternating bilinear form, instead of a Hermitian form; in particular, the underlying field then is necessarily commutative. However, every alternating form on a vector space of odd dimension is degenerate.)

For $\mathbb{F} = \mathbb{R}$, there is no (anti-)automorphism except $\alpha = \mathrm{id}$. For $\mathbb{F} = \mathbb{C}$, we are not interested in the multitude of possibilities arising from non-continuous (anti-)automorphisms, so we only consider the cases $\alpha = \mathrm{id}$ and $\alpha = \mathrm{conjugation}$ (11.26).

For $\mathbb{F} \in \{\mathbb{R}, \mathbb{C}\}$ it is then easy to see (essentially by Sylvester's theorem) that, up to equivalence, there is only one polarity other than the standard elliptic and hyperbolic polarities of $\mathcal{P}_2\mathbb{R}$ and $\mathcal{P}_2\mathbb{C}$, namely the polarity ϱ of $\mathcal{P}_2\mathbb{C}$ obtained from the standard bilinear form

$$f(x, y) = \sum_{\nu=1}^{3} x_\nu y_\nu$$

of $\mathbb{C}^3$, see e.g. Lewis [82] Sect. 3, p. 256 and Sect. 4, p. 261. The absolute points of ϱ form a conic C, with a connected complement $P_2\mathbb{C} \setminus C$. This property distinguishes ϱ from π^+, which has no absolute points, and from π^-, whose set of non-absolute points is disconnected (13.17).

For $\mathbb{F} = \mathbb{H}$, the classification of polarities up to equivalence may be obtained from Dieudonné [71] Chap. I §§6,8 together with Dieudonné [52] §19, p. 383; the result can also be found in Lewis [82]. For $\alpha = \mathrm{conjugation}$ there are just the standard elliptic polarity and the standard hyperbolic polarity (Lewis [82] Sect. 5, p. 263; cf. also Baer [52] Chap. IV Appendix I, Application 2, p. 130, Dieudonné [71] p. 16 (end of §8)).

A polarity of $\mathcal{P}_2\mathbb{H}$ accompanied by any other involutory antiautomorphism may also be described using a non-degenerate *skew*-Hermitian form g accompanied by conjugation, and up to equivalence there is just one possibility (Lewis [82] Sect. 6, p. 264), given by

$$g(x, y) = \sum_{\nu=1}^{3} \overline{x_\nu} \cdot i \cdot y_\nu .$$

For the corresponding polarity, one can again show that the set A of absolute points is non-empty and that the set of non-absolute points is connected. In contrast to the complex case, however, A is not what one would like to call a conic, because non-absolute lines carry more than two absolute points.

The two further polarities presented here, one of $\mathcal{P}_2\mathbb{C}$ and the other of $\mathcal{P}_2\mathbb{H}$, are analogous to the *standard planar polarity* of the octonion projective plane discussed in (18.28) ff. The method of classification presented there for $\mathbb{F} = \mathbb{O}$ also works for $\mathbb{F} \in \{\mathbb{R}, \mathbb{C}, \mathbb{H}\}$, see (18.31). This offers an alternative approach to the classification of polarities as indicated here.

14 The planes over $\mathbb{R}$, $\mathbb{C}$, and $\mathbb{H}$ as topological planes

In this section, we study the topological properties of the projective planes over $\mathbb{F} \in \{\mathbb{R}, \mathbb{C}, \mathbb{H}\}$. Primarily, we shall show that they are examples illustrating the subject matter of this book, topological projective planes. This means that the geometric operations in $\mathcal{P}_2\mathbb{F}$ are continuous with respect to a topology introduced in a rather standard way. The proof of this result will work for an arbitrary topological field in place of $\mathbb{F}$.

More generally, we shall consider the point sets $P_d\mathbb{F}$ of the projective spaces of arbitrary finite dimension d. We introduce their natural topologies, and we show that they are manifolds. The section closes with a description of the Hopf maps associated with the fields $\mathbb{F}$.

14.1 The topology on a projective space over $\mathbb{F}$. For $d \in \mathbb{N}$, we consider $\mathbb{F}^{d+1}$ as a *right* vector space over $\mathbb{F}$. The 'points' of *projective d-space over $\mathbb{F}$* are the 1-dimensional subspaces of $\mathbb{F}^{d+1}$, hence the point set of this space is

$$P_d\mathbb{F} := \{ x\mathbb{F} \mid 0 \neq x \in \mathbb{F}^{d+1} \}.$$

Of course, $\mathbb{F}^{d+1}$ is also a left vector space over $\mathbb{F}$, the set of whose 1-dimensional subspaces will be denoted by

$$P_d^*\mathbb{F} := \{ \mathbb{F}x \mid 0 \neq x \in \mathbb{F}^{d+1} \}.$$

If $\mathbb{F}$ is not commutative, then $P_d\mathbb{F}$ and $P_d^*\mathbb{F}$ differ. The latter space is identified in a natural way with the set

$$\mathcal{H}_d\mathbb{F} := \{ H \leq \mathbb{F}^{d+1} \mid \dim H = d \}$$

of hyperplanes of $\mathbb{F}^{d+1}$, since the hyperplanes are precisely the kernels of nonzero linear forms, and since a linear form is determined by its kernel up to a scalar factor from the left. Thus, there is a bijection

(1) $$P_d^*\mathbb{F} \to \mathcal{H}_d\mathbb{F} : \mathbb{F}x \mapsto \mathrm{Ker}\,(\begin{matrix} x_1 & x_2 & \cdots & x_{d+1} \end{matrix})\,,$$

where $x = (x_1, x_2, \ldots, x_{d+1}) \in \mathbb{F}^{d+1} \setminus \{0\}$, and where $(\begin{matrix} x_1 & x_2 & \cdots & x_{d+1} \end{matrix})$ is to be interpreted as the matrix representation of a linear form with respect to the standard basis.

For $d = 2$, this agrees with the construction of the projective plane $\mathcal{P}_2\mathbb{F}$ as in (13.1 and 2), with point set $P_2\mathbb{F}$ and line set $\mathcal{L}_2\mathbb{F} = \mathcal{H}_2\mathbb{F}$.

In order to introduce topologies on $P_d\mathbb{F}$ and $\mathcal{H}_d\mathbb{F} \cong P_d^*\mathbb{F}$, we use the natural topology on $\mathbb{F} = \mathbb{R}^n$ ($n = 1, 2, 4$), defined by the norm $\|a\|^2 = \bar{a}a$, and we consider the product topology on $\mathbb{F}^{d+1}$. The spaces $P_d\mathbb{F}$ and $P_d^*\mathbb{F}$ will be endowed with the *quotient topologies*, cf. (92.19), determined by the canonical maps

$$\vartheta : \mathbb{F}^{d+1} \setminus \{0\} \to P_d\mathbb{F} : x \mapsto x\mathbb{F}$$

and

$$\vartheta^* : \mathbb{F}^{d+1} \setminus \{0\} \to P_d^*\mathbb{F} : x \mapsto \mathbb{F}x\,.$$

A topology on $\mathcal{H}_d\mathbb{F}$ is obtained by transfer via the bijection (1). This topology may also be viewed as the quotient topology determined by the map

$$\mathrm{Ker} : \mathbb{F}^{d+1} \setminus \{0\} \to \mathcal{H}_d\mathbb{F} : (x_1, x_2, \ldots, x_{d+1}) \mapsto \mathrm{Ker}\,(\begin{matrix} x_1 & x_2 & \cdots & x_{d+1} \end{matrix})\,.$$

With respect to these topologies, the maps ϑ, ϑ^*, and Ker are open maps. In order to see this for ϑ, one has to prove that, for an open subset $U \subseteq \mathbb{F}^{d+1} \setminus \{0\}$, the image $\vartheta(U)$ is open in $P_d\mathbb{F}$. By definition of the quotient topology, this is equivalent to showing that the inverse image $\vartheta^{-1}(\vartheta(U))$ is open in $\mathbb{F}^{d+1} \setminus \{0\}$. Now, $\vartheta^{-1}(\vartheta(U)) = \bigcup\{\vartheta^{-1}(\vartheta(x)) \mid x \in U\} = \bigcup\{x\mathbb{F}^\times \mid x \in U\} = U\mathbb{F}^\times = \bigcup\{Ua \mid a \in \mathbb{F}^\times\}$ is the union of the open sets Ua. For ϑ^* and Ker, the assertion is obtained analogously. (The same argument proves a similar general assertion about the quotient topology of orbit spaces, see (96.2). The projective space $P_d\mathbb{F}$ is the orbit space of $\mathbb{F}^{d+1} \setminus \{0\}$ under the action of the multiplicative group $\mathbb{F}^\times$ by scalar multiplication from the right, and similarly for $P_d^*\mathbb{F}$.)

By definition, the spaces $P_d\mathbb{F}$ and $P_d^*\mathbb{F} \cong \mathcal{H}_d\mathbb{F}$ are interchanged if we switch the right and left vector space structures of $\mathbb{F}^{d+1}$. Equivalently, we may replace the field $\mathbb{F}$ with the opposite field $\mathbb{F}^{\mathrm{op}}$, in which the order of multiplication is reversed. Therefore, the duality principle of (13.10) extends as follows: Any true statement about $P_d\mathbb{F}$ depending only on properties of $\mathbb{F}$ which are shared by $\mathbb{F}^{\mathrm{op}}$ (e.g., topological properties) is equally valid for $P_d^*\mathbb{F}$.

Note. A description of these topologies has been given by Pontryagin in 1938, see Pontryagin [86] Sect. 27 Example 48, p. 183. There, the topology on $\mathcal{H}_d\mathbb{F}$ is introduced by a different method, which, more generally, is used to define topologies on the Grassmann manifolds consisting of all k-dimensional subspaces of $\mathbb{F}^{d+1}$, for arbitrary $k \in \{1, ..., d\}$. See also (64.3) and Kühne–Löwen [92], where it is shown in 2.8 that the resulting topology on $\mathcal{H}_d\mathbb{F}$ is the same as the topology defined here.

14.2 Linear transformations. A linear transformation $A \in \mathrm{GL}_{d+1}\mathbb{F}$ of the right vector space $\mathbb{F}^{d+1}$ maps 1-dimensional subspaces to 1-dimensional subspaces, and therefore induces a bijection

$$[A] : \mathrm{P}_d\mathbb{F} \to \mathrm{P}_d\mathbb{F} : x\mathbb{F} \mapsto (Ax)\mathbb{F},$$

where $x \in \mathbb{F}^{d+1} \setminus \{0\}$; this generalizes (13.4). This bijection is a homeomorphism with respect to the quotient topology, since it is obtained from the homeomorphism A of $\mathbb{F}^{d+1}$ by passing to quotients, cf. (92.20).

In accordance with (13.4), the group of all these transformations will be denoted by

$$\mathrm{PGL}_{d+1}\mathbb{F} := \{ [A] \mid A \in \mathrm{GL}_{d+1}\mathbb{F} \}.$$

The linear transformation A maps the hyperplane $\mathrm{Ker}\,(\,x_1 \;\cdots\; x_{d+1}\,)$ onto the hyperplane $\mathrm{Ker}\,(\,(x_1 \;\cdots\; x_{d+1})A^{-1}\,)$, for $(x_1, ..., x_{d+1}) \in \mathbb{F}^{d+1} \setminus \{0\}$. The resulting bijection

$$\mathcal{H}_d\mathbb{F} \to \mathcal{H}_d\mathbb{F} : \mathrm{Ker}\,(\,x_1 \;\cdots\; x_{d+1}\,) \mapsto \mathrm{Ker}\,(\,(x_1 \;\cdots\; x_{d+1})A^{-1}\,)$$

also is a homeomorphism, since it is obtained from the homeomorphism $(x_1, ..., x_{d+1}) \mapsto (x_1, ..., x_{d+1})A^{-1}$ of $\mathbb{F}^{d+1}$ by passing to quotients.

The group $\mathrm{GL}_{d+1}\mathbb{F}$ carries a natural topology as a subset of the $\mathbb{F}$-vector space of all $(d+1) \times (d+1)$-matrices. The corresponding projective group $\mathrm{PGL}_{d+1}\mathbb{F}$ is endowed with the quotient topology defined by the natural map

$$\eta : \mathrm{GL}_{d+1}\mathbb{F} \to \mathrm{PGL}_{d+1}\mathbb{F} : A \mapsto [A].$$

We assert that these two groups act continuously on $\mathrm{P}_d\mathbb{F}$, which means that the maps

(1) $$\mathrm{GL}_{d+1}\mathbb{F} \times \mathrm{P}_d\mathbb{F} \to \mathrm{P}_d\mathbb{F} : (A, x\mathbb{F}) \mapsto (Ax)\mathbb{F}$$

and

(2) $$\mathrm{PGL}_{d+1}\mathbb{F} \times \mathrm{P}_d\mathbb{F} \to \mathrm{P}_d\mathbb{F} : ([A], x\mathbb{F}) \mapsto (Ax)\mathbb{F}$$

are continuous, cf. (96.1). This will result from the obvious continuity of the map

(3) $$\mathrm{GL}_{d+1}\mathbb{F} \times (\mathbb{F}^{d+1} \setminus \{0\}) \to \mathbb{F}^{d+1} \setminus \{0\} : (A, x) \mapsto Ax,$$

14 The planes over $\mathbb{R}$, $\mathbb{C}$, and $\mathbb{H}$ as topological planes

with the help of the following commutative diagram by the universal property of identifying maps (quotient maps), cf. (92.19).

$$
\begin{array}{ccc}
\mathrm{GL}_{d+1}\mathbb{F} \times (\mathbb{F}^{d+1} \setminus \{0\}) & \longrightarrow & \mathbb{F}^{d+1} \setminus \{0\} \\
\mathrm{id} \times \vartheta \downarrow & & \downarrow \vartheta \\
\mathrm{GL}_{d+1}\mathbb{F} \times \mathrm{P}_d\mathbb{F} & \longrightarrow & \mathrm{P}_d\mathbb{F} \\
\eta \times \mathrm{id} \downarrow & & \downarrow \mathrm{id} \\
\mathrm{PGL}_{d+1}\mathbb{F} \times \mathrm{P}_d\mathbb{F} & \longrightarrow & \mathrm{P}_d\mathbb{F}
\end{array}
$$

The horizontal maps are the maps (3), (1), (2) above. The vertical maps are open, since ϑ and η are; in particular, they are identifying (92.20b). Thus, by the universal property, the continuity of the first row in turn implies the continuity of the second and third.

14.3 Lemma. *For each hyperplane H of $\mathbb{F}^{d+1}$, the set*

$$\mathrm{P}_d\mathbb{F} \setminus H := \{ x\mathbb{F} \mid x \notin H \}$$

of points not contained in H is open in $\mathrm{P}_d\mathbb{F}$ and homeomorphic to $\mathbb{F}^d$.

Proof. Since $\mathrm{PGL}_{d+1}\mathbb{F}$ is transitive on the set of hyperplanes and consists of homeomorphisms of $\mathrm{P}_d\mathbb{F}$, see (14.2), it suffices to prove the lemma for a particular hyperplane, e.g., for

$$W := \mathbb{F}^d \times \{0\} \leq \mathbb{F}^{d+1} .$$

The complement $\mathrm{P}_d\mathbb{F} \setminus W$ is open in $\mathrm{P}_d\mathbb{F}$ in the quotient topology, since its inverse image under the canonical map ϑ of (14.1) is the open subset $\mathbb{F}^{d+1} \setminus W$ of $\mathbb{F}^{d+1} \setminus \{0\}$. Therefore, the topology induced on $\mathrm{P}_d\mathbb{F} \setminus W$ can also be described as the quotient topology defined by the restriction $\vartheta|_{\mathbb{F}^{d+1}\setminus W}$. The map

$$\psi : \mathbb{F}^d \to \mathrm{P}_d\mathbb{F} \setminus W : (x_1, \ldots, x_d) \mapsto (x_1, \ldots, x_d, 1)\mathbb{F}$$

is a continuous bijection, the inverse map being

$$\psi^{-1} : \mathrm{P}_d\mathbb{F} \setminus W \to \mathbb{F}^d : (y_1, \ldots, y_d, y_{d+1})\mathbb{F} \mapsto (y_1 y_{d+1}^{-1}, \ldots, y_d y_{d+1}^{-1}) ;$$

note that $(y_1, \ldots, y_d, y_{d+1})\mathbb{F} \in \mathrm{P}_d\mathbb{F} \setminus W$ if, and only if $y_{d+1} \neq 0$. The inverse map ψ^{-1} is continuous by the universal property (92.19) of the canonical map $\vartheta|_{\mathbb{F}^{d+1}\setminus W}$, because the composition $\psi^{-1} \circ \vartheta : \mathbb{F}^{d+1} \setminus W \to \mathbb{F}^d : (y_1, \ldots, y_d, y_{d+1}) \mapsto (y_1 y_{d+1}^{-1}, \ldots, y_d y_{d+1}^{-1})$ is clearly continuous. Thus, ψ is a homeomorphism. $\square$

Note. In (13.3), we considered ψ in the special case $d = 2$ of a projective plane. In higher dimensions d, a projective space has an even richer geometric structure giving rise to various geometric operations. If the underlying field is a topological field, these geometric operations have good continuity properties. Here, this subject will be studied only in the special case of projective planes. Regarding the general case, see Misfeld [68], Groh [86a,b], Zanella [89], Kühne–Löwen [92], and, as far as the special needs of this book are concerned, (64.3). The basic result in this respect for the case of a projective plane will now be proved.

14.4 Theorem. $\mathcal{P}_2\mathbb{F}$ *is a topological projective plane, i.e., joining distinct points by a line and intersecting distinct lines are continuous operations.*

Formally, this means that, with the quotient topologies on the point set $P_2\mathbb{F}$ and on the line set $\mathcal{L}_2\mathbb{F} = \mathcal{H}_2\mathbb{F} \cong P_2^*\mathbb{F}$ as defined in (14.1), the maps

$$\vee : P_2\mathbb{F} \times P_2\mathbb{F} \setminus \{(p, p) \mid p \in P_2\mathbb{F}\} \to \mathcal{L}_2\mathbb{F} : (p, q) \mapsto pq,$$
$$\wedge : \mathcal{L}_2\mathbb{F} \times \mathcal{L}_2\mathbb{F} \setminus \{(L, L) \mid L \in \mathcal{L}_2\mathbb{F}\} \to P_2\mathbb{F} : (L, M) \mapsto L \wedge M$$

are continuous.

Note. These maps, moreover, have good differentiability properties, see the note in (14.9) and (75.4), where also an alternative for the proof of the present theorem is mentioned. Here, we shall give two proofs, one right now and the other in (14.6).

First proof of Theorem (14.4). The group $\mathrm{PGL}_3\mathbb{F}$ acts transitively on the set of pairs of distinct points (13.5), and it consists of homeomorphisms of $P_2\mathbb{F}$, which induce homeomorphisms on $\mathcal{L}_2\mathbb{F} \cong \mathcal{H}_2\mathbb{F}$ (14.2). Hence, it suffices to verify the continuity of the join map $\vee$ in the neighbourhood of two particular points, e.g., the points $o = (0, 0, 1)\mathbb{F}$ and $p_1 = (1, 0, 1)\mathbb{F}$. These belong to the open subset $P_2\mathbb{F} \setminus W$ of $P_2\mathbb{F}$; see (14.3) with $d = 2$. Therefore, we may use the homeomorphism ψ obtained in the proof of (14.3), and argue in the affine plane $\mathbb{F}^2 = \mathbb{F} \times \mathbb{F}$. Let U_0 and U_1 be disjoint neighbourhoods of 0 and 1 in $\mathbb{F}$; then $V_0 = \psi(U_0 \times U_0)$ and $V_1 = \psi(U_1 \times U_0)$ are (disjoint) neighbourhoods of $o = \psi(0, 0)$ and $p_1 = \psi(1, 0)$ in $P_2\mathbb{F}$. It suffices to establish the continuity of the join map $\vee$ on $V_0 \times V_1$.

In the affine plane $\mathcal{A}_2\mathbb{F}$ with point set $\mathbb{F} \times \mathbb{F}$, the line joining $(x_0, y_0) \in U_0 \times U_0$ and $(x_1, y_1) \in U_1 \times U_0$ is the line $[s, t]$, where

$$s = (y_1 - y_0)(x_1 - x_0)^{-1} \quad \text{and} \quad t = y_0 - sx_0;$$

note that $x_1 \neq x_0$ because of $U_0 \cap U_1 = \emptyset$. Using the isomorphism $\overline{\mathcal{A}_2\mathbb{F}} \cong \mathcal{P}_2\mathbb{F}$ afforded by ψ in (13.3), one obtains that the line joining $\psi(x_0, y_0)$ and $\psi(x_1, y_1)$ in $\mathcal{P}_2\mathbb{F}$ is $\mathrm{Ker}\,(-s \ \ 1 \ \ -t)$ with s, t as given above; this can also be verified directly. That line depends continuously on $\psi(x_0, y_0)$ and $\psi(x_1, y_1)$, since ψ is a homeomorphism, and since the canonical map $\mathrm{Ker} : \mathbb{F}^3 \setminus \{0\} \to \mathcal{H}_2\mathbb{F} \cong \mathcal{L}_2\mathbb{F}$ is continuous. Thus, continuity of the join operation is proved. Continuity of intersection simply follows by duality. Indeed, intersection is the join operation in the dual plane, which may be considered as the plane over the opposite field $\mathbb{F}^{op}$, see (13.10) and the end of (14.1). □

We note that the arguments of the preceding proof are valid over an arbitrary topological (not necessarily commutative) field. Our second proof of the same result in (14.6) is valid if there are compactness properties available. These shall now be established for the fields in which we are mainly interested.

14.5 Proposition. *$P_d\mathbb{F}$ and $P_d^*\mathbb{F}$ are compact Hausdorff spaces for $\mathbb{F} \in \{\mathbb{R}, \mathbb{C}, \mathbb{H}\}$.*

Remark. More generally, this is true for any non-discrete locally compact topological field $\mathbb{F}$.

Proof. By (14.3), $P_d\mathbb{F}$ is a Hausdorff space, every pair of points being contained in the complement of a suitable hyperplane. We choose a norm ν on $\mathbb{F}^{d+1}$; see comments below. The unit sphere $S = \{ x \in \mathbb{F}^{d+1} \mid \nu(x) = 1 \}$ is a compact subset of $\mathbb{F}^{d+1} \setminus \{0\}$. Every 1-dimensional subspace $x\mathbb{F}$, $0 \neq x \in \mathbb{F}^{d+1} \setminus \{0\}$, meets S; indeed, there is an element $a \in \mathbb{F} \setminus \{0\}$ such that $\nu(x) = \|a\|$ and hence $xa^{-1} \in S \cap x\mathbb{F}$. Thus, $P_d\mathbb{F}$ is compact, being the image of the compact set S under the canonical map ϑ, which is continuous. For $P_d^*\mathbb{F}$, the assertion follows by duality, see (14.1).

Specifically, for $\mathbb{F} \in \{\mathbb{R}, \mathbb{C}, \mathbb{H}\}$ one might use the Euclidean norm $\nu(x) = \sqrt{\sum_{i=1}^{d+1} \|x_i\|^2}$. Another possibility is the maximum norm $\nu(x) = \max_i \|x_i\|$, which has the advantage of making the proof work for an arbitrary (non-discrete) locally compact field $\mathbb{F}$, with $\| \ \|$ being an absolute value which describes the topology of $\mathbb{F}$, see Samuel [88] 2.8 Theorem 45, p. 87 ff. Compare also Lenz [65] Kap. XI Satz 2.5, p. 341. □

14.6. *Second proof of Theorem* (14.4). As $P_2\mathbb{F}$ and $\mathscr{L}_2\mathbb{F} \cong P_2^*\mathbb{F}$ are compact Hausdorff spaces (14.5), we may apply the criterion (43.1) for the continuity of the geometric operations. Accordingly, it suffices to prove that the flag space $\mathbf{F} \subseteq P_2\mathbb{F} \times \mathscr{L}_2\mathbb{F}$, consisting of all incident pairs (p, L), is a closed subset. Now, the canonical maps $\vartheta : \mathbb{F}^3 \setminus \{0\} \to P_2\mathbb{F}$ and $\mathrm{Ker} : \mathbb{F}^3 \setminus \{0\} \to \mathscr{L}_2\mathbb{F}$ are open (14.1), and therefore the map $\vartheta \times \mathrm{Ker} : (\mathbb{F}^3 \setminus \{0\})^2 \to P_2\mathbb{F} \times \mathscr{L}_2\mathbb{F}$ is open, as well. Thus, it suffices to verify that $(\vartheta \times \mathrm{Ker})^{-1}(\mathbf{F})$ is closed in $(\mathbb{F}^3 \setminus \{0\})^2$. But this is obvious: $(\vartheta \times \mathrm{Ker})^{-1}(\mathbf{F}) = \{ (x, y) \in (\mathbb{F}^3 \setminus \{0\})^2 \mid x\mathbb{F} \subseteq \mathrm{Ker}(y_1 \ y_2 \ y_3) \} = \{ (x, y) \in (\mathbb{F}^3 \setminus \{0\})^2 \mid \sum y_i x_i = 0 \}$. □

14.7 Corollary. *In the projective plane $\mathcal{P}_2\mathbb{F}$, the point row of each line is a closed subset of $P_2\mathbb{F}$ homeomorphic to the one-point compactification $\mathbb{F} \cup \{\infty\}$ of $\mathbb{F}$.*

Specifically, for the line $W = \mathbb{F}^2 \times \{0\} = \mathrm{Ker}\,(0\ 0\ 1)$, *a homeomorphism of* $\mathbb{F} \cup \{\infty\}$ *onto the point row of* W *is given by*

$$s \mapsto (1, s, 0)\mathbb{F} \,\widehat{=}\, (s), \quad \infty \mapsto (0, 1, 0)\mathbb{F} \,\widehat{=}\, (\infty) \,.$$

Proof. We use the same symbol for a line and its point row. Every line L is homeomorphic to W. A homeomorphism is given by central projection from a point c not incident with L or W, that is, by the map $L \to W : p \mapsto pc \wedge W$. Indeed, this map is a bijection with inverse map $W \to L : w \mapsto wc \wedge L$, and both these maps are continuous by (14.4). Of course, this argument is valid in every topological projective plane, see (41.2).

The restriction of the quotient map $\vartheta : \mathbb{F}^3 \setminus \{0\} \to \mathrm{P}_2\mathbb{F}$ to the closed subset $(\mathbb{F}^2 \times \{0\}) \setminus \{0\}$ is the quotient map defining the topology of $\mathrm{P}_1\mathbb{F}$, so that we may identify W with $\mathrm{P}_1\mathbb{F}$. In particular, W is compact by (14.5), and W is the one-point compactification of $W \setminus \{(0, 1, 0)\mathbb{F}\}$. Now the case $d = 1$ of the proof of (14.3) shows that the map $\alpha : \mathbb{F} \cup \{\infty\} \to W$ specified in the assertion restricts to a homeomorphism $\alpha' : \mathbb{F} \to W \setminus \{(0, 1, 0)\mathbb{F}\}$. The map α is the canonical extension of α' to the one-point compactifications, hence α is a homeomorphism as well. □

We now consider properties which are specific for the projective spaces over $\mathbb{R}$, $\mathbb{C}$, and $\mathbb{H}$.

14.8 Proposition. *For* $\mathbb{F} \in \{\mathbb{R}, \mathbb{C}, \mathbb{H}\}$, *the projective space* $\mathrm{P}_d\mathbb{F}$ *is a compact, connected topological manifold of dimension* $n \cdot d$, *where* $n = \dim_\mathbb{R} \mathbb{F} \in \{1, 2, 4\}$. *The dual space* $\mathrm{P}_d^*\mathbb{F} \cong \mathcal{H}_d\mathbb{F}$ *is homeomorphic to* $\mathrm{P}_d\mathbb{F}$ *via the homeomorphism*

$$\pi : \mathrm{P}_d\mathbb{F} \to \mathcal{H}_d\mathbb{F} : (x_1, \ldots, x_{d+1})\mathbb{F} \mapsto \mathrm{Ker}\,(\overline{x_1}\ \overline{x_2}\ \cdots\ \overline{x_{d+1}}) \,.$$

Proof. By (14.5), $\mathrm{P}_d\mathbb{F}$ is a compact Hausdorff space. Every point belongs to an open subset of the form $\mathrm{P}_d\mathbb{F} \setminus H$, where H is a suitable hyperplane of $\mathbb{F}^{d+1}$. By (14.3), these subsets are homeomorphic to $\mathbb{F}^d = \mathbb{R}^{n \cdot d}$. Thus, $\mathrm{P}_d\mathbb{F}$ is a manifold of dimension $n \cdot d$. Note that the topology of $\mathrm{P}_d\mathbb{F}$ has a countable basis (a property which is often required in the definition of a topological manifold), since in fact $\mathrm{P}_d\mathbb{F}$ may be covered by finitely many of these subsets $\mathrm{P}_d\mathbb{F} \setminus H$, with $H \in \mathcal{H}_d\mathbb{F}$. As these subsets are connected and meet each other, $\mathrm{P}_d\mathbb{F}$ is connected, as well.

The map π is well-defined, since conjugation is an antiautomorphism, and π is bijective. (If $d = 2$, then π is the standard elliptic polarity considered in (13.12).) The map $\mathbb{F}^{d+1} \to \mathbb{F}^{d+1} : (x_1, \ldots, x_{d+1}) \mapsto (\overline{x_1}, \ldots, \overline{x_{d+1}})$ clearly is a homeomorphism, and π is obtained from it by passing to quotients with respect to the identifying maps $\vartheta : \mathbb{F}^{d+1} \setminus \{0\} \to \mathrm{P}_d\mathbb{F}$ and $\mathrm{Ker} : \mathbb{F}^{d+1} \setminus \{0\} \to \mathcal{H}_d\mathbb{F}$ of (14.1). By the universal property (92.20), it follows that π is a homeomorphism with respect to the quotient topologies. □

14.9 Corollary. *The point set* $P_2\mathbb{F}$ *and the line set* $\mathcal{L}_2\mathbb{F}$ *of the projective plane* $\mathcal{P}_2\mathbb{F}$ *over* $\mathbb{F} \in \{\mathbb{R}, \mathbb{C}, \mathbb{H}\}$ *are compact, connected topological manifolds of dimension* $2n$, $n \in \{1, 2, 4\}$. *The point row of each line of* $\mathcal{P}_2\mathbb{F}$ *is a closed subset of* $P_2\mathbb{F}$ *homeomorphic to the n-sphere* $\mathbb{S}_n$.

Proof. The first part is a special case of (14.8); the second part follows from (14.7), since the one-point compactification of $\mathbb{F} \cong \mathbb{R}^n$ is homeomorphic to $\mathbb{S}_n$. □

Note. As is outlined in (75.4), these manifolds moreover carry natural differentiable structures, in which the geometric operations are differentiable, so that the structure of a differentiable projective plane results.

14.10 Visualization of $\mathcal{P}_2\mathbb{R}$. For $\mathbb{F} = \mathbb{R}$, a pictorial representation of the projective plane may be obtained in the following way.

As in (14.5), every point p of $\mathcal{P}_2\mathbb{R}$, i.e., every 1-dimensional subspace of $\mathbb{R}^3$, intersects the unit sphere $\mathbb{S}_2$ of $\mathbb{R}^3$, and the intersection is a pair of antipodal points. If p is not contained in the line represented by the 2-dimensional subspace $W = \mathbb{R}^2 \times \{0\}$, then p intersects the upper hemisphere D^+ of $\mathbb{S}_2$ in exactly one point. If p is incident with W, the intersection with D^+ is a pair of antipodal points lying on the equator $E = \mathbb{S}_2 \cap W$. Thus, the point set $P_2\mathbb{R}$ can be visualized as the upper hemisphere D^+ with antipodal points on the equator E identified, a so-called 'cross cap'.

In this picture, the equator E covers the line W twice. A line given by a 2-dimensional subspace $L \neq W$ of $\mathbb{R}^3$ is represented as the great half circle $L \cap D^+$ with end points identified. In particular, a point row homeomorphic to the circle $\mathbb{S}_1$ results, in accordance with (14.9).

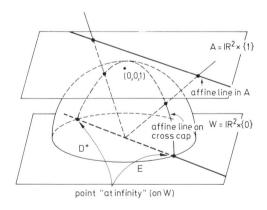

Figure 14a

Let $\mathcal{P}_2\mathbb{R}^W$ be the affine plane obtained by removing the line W (and considering it as the line at infinity). In (13.3), this affine plane was represented on the

2-dimensional affine subspace $A = \mathbb{R}^2 \times \{1\}$ of $\mathbb{R}^3$. This representation corresponds with the description on D^+ via central projection from the origin.

14.11 The Hopf map. In the affine plane over $\mathbb{F} \in \{\mathbb{R}, \mathbb{C}, \mathbb{H}\}$ with point set $\mathbb{F}^2 \cong \mathbb{R}^{2n}$, $n = \dim_\mathbb{R} \mathbb{F} \in \{1, 2, 4\}$, consider the unit sphere

$$\mathbb{S}_{2n-1} = \left\{ (x, y) \in \mathbb{F}^2 \mid \|x\|^2 + \|y\|^2 = 1 \right\}.$$

Via the isomorphism between the projective completion $\overline{\mathcal{A}_2}\mathbb{F}$ and $\mathcal{P}_2\mathbb{F}$ given in (13.3), the line at infinity W can be considered as a subset of $\mathrm{P}_2\mathbb{F}$, which is homeomorphic to $\mathbb{S}_n$ according to (14.9).

By central projection of $\mathbb{S}_{2n-1}$ from the origin $(0, 0)$ onto W, we obtain the *Hopf map*

$$\mathbb{S}_{2n-1} \to W \approx \mathbb{S}_n : (x, y) \mapsto \big((0,0) \vee (x, y)\big) \wedge W = \begin{cases} (yx^{-1}) & \text{for } x \neq 0, \\ (\infty) & \text{for } x = 0. \end{cases}$$

The image point, in affine notation as in (12.2), represents the slope of the line joining $(0, 0)$ and (x, y). The continuity of this map is a consequence of the continuity of the geometric operations in the topological projective plane $\mathcal{P}_2\mathbb{F}$, see (14.4).

A fiber of the Hopf map, i.e., the inverse image of a point of W, is the intersection of $\mathbb{S}_{2n-1}$ with a line through $(0, 0)$. These lines are just the 1-dimensional $\mathbb{F}$-linear subspaces of $\mathbb{F}^2$, and therefore are n-dimensional $\mathbb{R}$-linear subspaces. Their intersections with $\mathbb{S}_{2n-1}$ are spheres of dimension $n - 1$. The fiber containing the point $(x, y) \in \mathbb{S}_{2n-1}$ may be explicitly described as

(∗) $$\left\{ (x\lambda, y\lambda) \mid \lambda \in \mathbb{F}, \ \|\lambda\|^2 = 1 \right\}.$$

This is because, for $\lambda \in \mathbb{F}$, we have $(x\lambda, y\lambda) \in \mathbb{S}_{2n-1}$ if, and only if $\|\lambda\|^2 = 1$.

The Hopf map over the octonions $\mathbb{O}$ can be defined quite analogously, see (16.15). The reader should be warned, however, that over $\mathbb{O}$ the description (∗) of a fiber is not valid.

Note. The domain $\mathbb{S}_{2n-1}$ of the Hopf map may be replaced by the larger set of *all* points $\neq (0, 0)$ of the *projective* plane. The resulting map, viz. the central projection from $(0, 0)$ onto the line W, is studied in (51.23) for arbitrary topological projective planes. By an easy geometric construction, the central projection is shown to be a locally trivial fiber bundle. The fibers are the projective lines through $(0, 0)$ with the point $(0, 0)$ removed. In the classical situation here, the fibers thus are homeomorphic to $\mathbb{R}^n$, see (14.9), and it is not difficult to see that the fiber bundle in question is even a vector bundle, of which the Hopf map is the associated sphere bundle.

15 Geometry of a projective line

We shall now endow a line of the projective plane $\mathcal{P}_2\mathbb{F}$, $\mathbb{F} \in \{\mathbb{R}, \mathbb{C}, \mathbb{H}, \mathbb{O}\}$, with the additional metric structure of a 'conformal space', that is, of a projective quadric of index 1. Specifically, this will be carried out for the line W at infinity of $\overline{\mathcal{A}_2\mathbb{F}} \cong \mathcal{P}_2\mathbb{F}$, and the quadric will be defined by a quadratic form q of index 1 on the $\mathbb{R}$-vector space $\mathbb{F} \times \mathbb{R}^2$. This will serve to interpret the group of transformations induced on the line W by the collineations of $\mathcal{A}_2\mathbb{F}$ as a conformal group acting on the quadric; in the case $\mathbb{F} = \mathbb{C}$, this interpretation is confined to $\mathbb{C}$-linear collineations. Put more precisely, we show that the group in question is induced on W considered as a quadric by the special orthogonal group $\mathrm{SO}(\mathbb{F} \times \mathbb{R}^2, q)$.

In the cases where $\mathbb{F}$ is a field, the same group was described in (12.12) in a different way, namely as the group of all fractional linear transformations of $\mathbb{F} \cup \{\infty\}$. While the latter description is restricted to the field cases, the new description as a conformal action covers those cases and the octonion case in a uniform manner.

In connection with the homogeneous presentation of the projective octonion plane via Veronese coordinates in Section 16, the metric structure on W will suggest itself very naturally (16.2), and will provide us with valuable insight into the structure of the collineation group of the octonion plane. In particular, stabilizers of lines in the elliptic and hyperbolic motion groups correspond to certain familiar subgroups of $\mathrm{SO}(\mathbb{F} \times \mathbb{R}^2, q)$, which we shall study at the end of this section, as a preparation for Section 18. There, this correspondence will be a crucial step in our study of those motion groups in the octonion plane.

15.1 The metric extension of $\mathbb{F}$ of index 1. The following constructions and notational conventions will be used throughout this section.

(a) For $\mathbb{F} \in \{\mathbb{R}, \mathbb{C}, \mathbb{H}, \mathbb{O}\}$, we consider the $\mathbb{R}$-vector space

$$\widehat{\mathbb{F}} := \mathbb{F} \times \mathbb{R} \times \mathbb{R}$$

of dimension $n + 2$, where

$$n = \dim_{\mathbb{R}} \mathbb{F} \in \{1, 2, 4, 8\}.$$

We define a quadratic form q of index 1 on $\widehat{\mathbb{F}}$,

$$q : \widehat{\mathbb{F}} \to \mathbb{R} : (x, r_1, r_2) \mapsto \bar{x}x - r_1 r_2 = \|x\|^2 - r_1 r_2,$$

where $x \in \mathbb{F}$, $r_\nu \in \mathbb{R}$. The quadric Q associated with q is the following subset of the real projective space $\mathrm{P}\widehat{\mathbb{F}} = \{\mathbb{R}z \mid z \in \widehat{\mathbb{F}} \setminus \{0\}\}$ of 1-dimensional $\mathbb{R}$-linear subspaces of $\widehat{\mathbb{F}}$:

$$Q = \left\{ \mathbb{R}z \mid z \in \widehat{\mathbb{F}} \setminus \{0\}, \ q(z) = 0 \right\} \subseteq \mathrm{P}\widehat{\mathbb{F}}.$$

(b) We endow the real projective space $P\widehat{\mathbb{F}} = P_{n+1}\mathbb{R}$ with its standard quotient topology (14.1), induced by the canonical map $\vartheta : \widehat{\mathbb{F}} \setminus \{0\} \to P\widehat{\mathbb{F}} : z \mapsto \mathbb{R}z$. The inverse image $\vartheta^{-1}(Q) = \{ z \mid q(z) = 0 \}$ of the quadric Q is obviously closed in $\widehat{\mathbb{F}} \setminus \{0\}$. Hence, Q is closed in $P\widehat{\mathbb{F}}$ and therefore compact (14.5). We show that Q is homeomorphic to the one-point compactification $\mathbb{F} \cup \{\infty\}$ of $\mathbb{F} = \mathbb{R}^n$ via the map

(1) $$\mathbb{F} \cup \{\infty\} \to Q : s \mapsto \mathbb{R}(s, \|s\|^2, 1), \ \infty \mapsto \mathbb{R}(0, 1, 0).$$

It is immediate that the image is contained in Q, and it suffices to verify that the restriction

(1') $$\mathbb{F} \to Q \setminus \{\mathbb{R}(0, 1, 0)\} : s \mapsto \mathbb{R}(s, \|s\|^2, 1)$$

is a homeomorphism. The continuity of (1') is clear; we shall prove that this map has a continuous inverse. Now $\mathbb{R}(0, 1, 0)$ is the only point of Q contained in the hyperplane $H = \mathbb{F} \times \mathbb{R} \times \{0\}$; indeed, $q(x, r, 0) = \bar{x}x$ is zero only if $x = 0$, and $\mathbb{R}(0, r, 0) = \mathbb{R}(0, 1, 0)$ for $r \neq 0$. In the proof of (14.3) it has been shown that the map

(2) $$\mathbb{F} \times \mathbb{R} \to P\widehat{\mathbb{F}} \setminus H : (x, r) \mapsto \mathbb{R}(x, r, 1)$$

is a homeomorphism. (When specializing (14.3) to the present situation, note that $\mathbb{F} \times \mathbb{R}$ here is to be interpreted as $\mathbb{R}^d$, where $d = n+1$, and that $\mathbb{F}$ in (14.3) is taken to be $\mathbb{R}$.) The point $\mathbb{R}(x, r, 1)$ is in Q if, and only if $q(x, r, 1) = \bar{x}x - r = 0$, that is, $r = \|x\|^2$. Thus, it is clear from the homeomorphism (2) that the map $\mathbb{R}(x, r, 1) \mapsto x$ induces an inverse of (1') and is continuous.

(c) We now consider the line at infinity W in the projective completion $\overline{\mathcal{A}_2 \mathbb{F}}$ of the affine plane over $\mathbb{F}$,

$$W = \{ (s) \mid s \in \mathbb{F} \} \cup \{(\infty)\}.$$

In what follows, W shall be identified with the quadric Q as suggested by the bijection (1); thus we identify

(s) and $\mathbb{R}(s, \|s\|^2, 1)$,

(∞) and $\mathbb{R}(0, 1, 0)$.

Via this bijection, W inherits a topology from Q. Now in the cases $\mathbb{F} \in \{\mathbb{R}, \mathbb{C}, \mathbb{H}\}$, we have already put a topology on W in a different way in Section 14. There, the projective plane $\mathcal{P}_2 \mathbb{F}$ has been endowed with the structure of a topological plane, so that W carries a topology as a line in this projective plane. We shall now convince ourselves that this topology coincides with the topology obtained from Q. Indeed, with respect to both topologies, W is homeomorphic to $\mathbb{F} \cup \{\infty\}$; for W as a line of $\mathcal{P}_2 \mathbb{F}$, such a homeomorphism is given in (14.7). This homeomorphism and the homeomorphism (1) above show that, under the identification of W and Q, the given topologies match.

15.2 Conformal transformations. Consider the group

$$O(\widehat{\mathbb{F}}, q) = \left\{ A \in \mathrm{GL}(\widehat{\mathbb{F}}) \;\middle|\; \forall z \in \widehat{\mathbb{F}} : q(Az) = q(z) \right\}.$$

It consists of the $\mathbb{R}$-linear transformations of $\widehat{\mathbb{F}}$ which are orthogonal with respect to the quadratic form q. The normal subgroup

$$\mathrm{SO}(\widehat{\mathbb{F}}, q) = \left\{ A \in O(\widehat{\mathbb{F}}, q) \;\middle|\; \det A = 1 \right\}$$

has index 2, since orthogonal transformations have determinant ± 1, as is well known, and since the map $(x, r_1, r_2) \mapsto (x, r_2, r_1)$ belongs to $O(\widehat{\mathbb{F}}, q)$ and has determinant -1.

The groups of transformations induced by $O(\widehat{\mathbb{F}}, q)$ and $\mathrm{SO}(\widehat{\mathbb{F}}, q)$ on the space $P\widehat{\mathbb{F}}$ of 1-dimensional subspaces will be denoted by $PO(\widehat{\mathbb{F}}, q)$ and $PSO(\widehat{\mathbb{F}}, q)$, respectively. These groups leave the quadric $W = Q$ invariant and induce transformation groups on W, which will be denoted, respectively, by

$$PO_{n+2}(1)^W := PO(\widehat{\mathbb{F}}, q)|_W$$

and

$$PSO_{n+2}(1)^W := PSO(\widehat{\mathbb{F}}, q)|_W.$$

The notation was chosen so as to indicate the dimension ($= n+2$) of the vector space $\widehat{\mathbb{F}}$ and the index ($= 1$) of the quadratic form q. The elements of $PO_{n+2}(1)^W$ are sometimes called *conformal* transformations, and the elements of $PSO_{n+2}(1)^W$ *direct conformal* transformations. In the case $\mathbb{F} = \mathbb{R}$, $n = 1$ we have

$$PSO_3(1)^W = PO_3(1)^W,$$

because on $\widehat{\mathbb{R}} = \mathbb{R}^3$ the reflection $-\mathrm{id}$ has determinant -1, so that $O(\widehat{\mathbb{R}}, q)$ is generated by $-\mathrm{id}$ together with $\mathrm{SO}(\widehat{\mathbb{R}}, q)$. In the other cases $\mathbb{F} \in \{\mathbb{C}, \mathbb{H}, \mathbb{O}\}$, where the dimension of $\widehat{\mathbb{F}}$ is even, the normal subgroup $PSO_{n+2}(1)^W$ has index 2 in $PO_{n+2}(1)^W$, as we shall see in (15.5).

Next we record some well known facts about quadrics and corresponding orthogonal groups, together with ad hoc proofs.

15.3 Lemma. *Let A be an $\mathbb{R}$-linear automorphism of $\widehat{\mathbb{F}} = \mathbb{F} \times \mathbb{R}^2$ such that the transformation $[A]$ induced on the projective space $P\widehat{\mathbb{F}}$ maps the quadric $W = Q$, defined by the quadratic form q, onto itself. Then A is a scalar multiple of a transformation in $O(\widehat{\mathbb{F}}, q)$, and $[A] \in PO(\widehat{\mathbb{F}}, q)$.*

This is a consequence of the fact that Q determines the quadratic form up to a scalar factor. More precisely, by assumption, the quadratic form $x \mapsto q(Ax)$ vanishes on the same vectors as q, so that there is a constant $r \in \mathbb{R}$ such that

(1) $$q(Az) = r \cdot q(z) \quad \text{for all } z \in \widehat{\mathbb{F}},$$

see e.g. Samuel [67] and [88] 3.1 Theorem 46, p. 91. In our context here, $U = \{(x, r_1, -r_1) \mid x \in \mathbb{F},\ r_1 \in \mathbb{R}\}$ is a subspace of $\widehat{\mathbb{F}}$ of codimension 1 on which q is positive definite, and if $z \neq 0$ is chosen in $U \cap A^{-1}(U) \neq \{0\}$, then it follows from (1) that $r > 0$. Thus $C := \frac{1}{\sqrt{r}} A \in \mathrm{O}(\widehat{\mathbb{F}}, q)$, and $[A] = [C] \in \mathrm{PO}(\widehat{\mathbb{F}}, q)$. □

The following lemma, whose proof is independent of the preceding one, is concerned with the special case $[A]|_W = \mathrm{id}$.

15.4 Lemma. *The 1-dimensional subspaces belonging to the quadric $W = Q$ generate $\widehat{\mathbb{F}} = \mathbb{F} \times \mathbb{R}^2$, and the $\mathbb{R}$-linear transformations of $\widehat{\mathbb{F}}$ which leave every element of W invariant are precisely the scalar homotheties.*

Proof. Consider an orthonormal basis $e_1, \ldots, e_n$ of the $\mathbb{R}$-vector space $\mathbb{F}$ with respect to the inner product on $\mathbb{F}$. The vectors $(e_\nu, 1, 1)$, $(0, 1, 0)$ and $(0, 0, 1)$ form a basis of $\widehat{\mathbb{F}}$, and they generate 1-dimensional subspaces belonging to W. If a linear transformation φ of $\widehat{\mathbb{F}}$ leaves all 1-dimensional subspaces belonging to W invariant, then these vectors are eigenvectors of φ. All the corresponding eigenvalues coincide, since for the same reason the vector $(\sum_{\nu=1}^n e_\nu, n+1, \frac{n}{n+1}) = \sum_{\nu=1}^n (e_\nu, 1, 1) + (0, 1, 0) - \frac{n^2}{n+1}(0, 0, 1)$ is an eigenvector of φ, as well. □

15.5 Remark. It follows from (15.4) that $-\mathrm{id}$ generates the kernel of the epimorphism $\mathrm{O}(\widehat{\mathbb{F}}, q) \to \mathrm{PO}_{n+2}(1)^W$ given by restriction to W. For $\mathbb{F} \in \{\mathbb{C}, \mathbb{H}, \mathbb{O}\}$, where $\dim_\mathbb{R} \widehat{\mathbb{F}}$ is even, this kernel is contained in $\mathrm{SO}(\widehat{\mathbb{F}}, q)$, hence $\mathrm{PSO}_{n+2}(1)^W$ has index 2 in $\mathrm{PO}_{n+2}(1)^W$, as asserted in (15.2).

Our main concern in this section is the following conformal interpretation of the group of transformations induced on W by the collineations of $\overline{\mathcal{A}_2\mathbb{F}} \cong \mathcal{P}_2\mathbb{F}$ leaving W invariant, or, equivalently, by the collineations of the affine plane $\mathcal{A}_2\mathbb{F}$. For $\mathbb{F} = \mathbb{C}$, only linear collineations will be considered (see below). We regard W as the line at infinity of the affine plane $\mathcal{A}_2\mathbb{F}$. Let

$$A_\mathbb{F} = \begin{cases} \text{group of } \textit{all} \text{ affine collineations of } \mathcal{A}_2\mathbb{F} & \text{for } \mathbb{F} \in \{\mathbb{R}, \mathbb{H}, \mathbb{O}\} \\ \text{group of } \mathbb{C}\text{-affine collineations of } \mathcal{A}_2\mathbb{C} & \text{for } \mathbb{F} = \mathbb{C}. \end{cases}$$

For $\mathbb{F} = \mathbb{C}$, this group is the semidirect product of the group of translations by $\mathrm{GL}_2\mathbb{C}$, see (12.10); in particular, we exclude the collineations $(x, y) \mapsto (x^\alpha, y^\alpha)$ belonging to a non-identical automorphism α of the field $\mathbb{C}$.

15.6 Fundamental theorem for the projective line. *The group of transformations of the line W induced by $A_\mathbb{F}$ coincides with the group of direct conformal transformations introduced in* (15.2),

$$A_\mathbb{F}|_W = \mathrm{PSO}_{n+2}(1)^W .$$

In the cases $\mathbb{F} \in \{\mathbb{R}, \mathbb{C}, \mathbb{H}\}$, the group $A_\mathbb{F}|_W$ can also be described as the group $\mathrm{PGL}_2 \mathbb{F}$ of fractional linear transformations introduced in (12.12); for $\mathbb{F} = \mathbb{C}$, this is due to our particular choice of $A_\mathbb{F}$. Thus, the preceding theorem has the following consequence.

Corollary. *For $\mathbb{F} = \mathbb{R}, \mathbb{C}, \mathbb{H}$, and corresponding values $n = 1, 2, 4$, we have the following equality of transformation groups of the projective line W:*

$$\mathrm{PGL}_2 \mathbb{F} = \mathrm{PSO}_{n+2}(1)^W .$$

Remarks. 1) We offer the following a priori reasons why, in the case $\mathbb{F} = \mathbb{C}$, the collineations involving a non-identical automorphism of $\mathbb{C}$ are excluded from the group $A_\mathbb{F}$, which is considered in the theorem stated above. Automorphisms other than the identity and complex conjugation are highly discontinuous, see (11.27), (44.11) and (55.22a), and complex conjugation has negative determinant over $\mathbb{R}$. In contrast with this, the automorphisms of $\mathbb{H}$ and $\mathbb{O}$ are all $\mathbb{R}$-linear and have determinant 1 over $\mathbb{R}$, see (11.29 and 31e).

2) In Tits [53] 4.11, p. 321, where the result of (15.6) for $\mathbb{F} = \mathbb{O}$, $n = 8$ may be found, there is a little imprecision, the group in question being given as $\mathrm{PO}_{10}(1)^W$ instead of $\mathrm{PSO}_{10}(1)^W$.

Because of the transitivity properties of the two groups in theorem (15.6), the assertion can be obtained by a comparison of stabilizers. The proof will be accomplished in (15.9), after some preparations. First, we describe the stabilizer of two points of W in $\mathrm{PSO}_{n+2}(1)^W$. We use the inner product $\langle \ | \ \rangle$ on $\mathbb{F} = \mathbb{R}^n$ and the associated norm $\| \ \|^2$ (11.3), together with the corresponding groups of orthogonal transformations and similitudes, as studied in (11.21 through 23); we keep the notation introduced there.

15.7 Proposition. *The stabilizer of the 1-dimensional subspaces $(0), (\infty) \in W$ of $\widehat{\mathbb{F}}$ in $O(\widehat{\mathbb{F}}, q)$ consists of the transformations*

$$(x, r_1, r_2) \mapsto (Cx, tr_1, t^{-1}r_2)$$

with $C \in O_n \mathbb{R}$ and $t \in \mathbb{R} \setminus \{0\}$. On W, they induce the transformations

$$(s) \mapsto (t \cdot Cs), \quad (\infty) \mapsto (\infty) .$$

Consequently, the stabilizer of (0) *and* (∞) *in* $\mathrm{PO}_{n+2}(1)^W = \mathrm{PO}(\widehat{\mathbb{F}}, q)|_W$ *is*

$$\mathrm{GO}_n^W := \{ W \to W : (s) \mapsto (Bs),\ (\infty) \mapsto (\infty) \mid B \in \mathrm{GO}_n\mathbb{R} \},$$

and the stabilizer of these points in $\mathrm{PSO}_{n+2}(1)^W$ *is the group*

$$\mathrm{GO}_n^{+W} := \{ W \to W : (s) \mapsto (Bs),\ (\infty) \mapsto (\infty) \mid B \in \mathrm{GO}_n^+\mathbb{R} \}.$$

Proof. If $\delta \in \mathrm{O}(\widehat{\mathbb{F}}, q)$ leaves the 1-dimensional subspaces $\mathbb{R}(0, 0, 1) \cong (0)$ and $\mathbb{R}(0, 1, 0) \cong (\infty)$ invariant, then $\mathbb{F} \times \{(0, 0)\}$ is also invariant, since this is the orthogonal space with respect to q of the span of the two 1-dimensional subspaces. Therefore, δ has the form

$$\delta : (x, r_1, r_2) \mapsto (Cx, t_1 r_1, t_2 r_2),$$

with $t_1, t_2 \in \mathbb{R} \setminus \{0\}$ and $C \in \mathrm{O}_n\mathbb{R}$, since on $\mathbb{F} \times \{(0, 0)\}$ the quadratic form is given by the norm form $\| \ \|^2$ on $\mathbb{F}$. The fact that δ leaves q invariant is equivalent to $t_2 = t_1^{-1}$. In particular, δ belongs to $\mathrm{SO}(\mathbb{F}, q)$ if, and only if $C \in \mathrm{SO}_n\mathbb{R}$. Finally, $(s)^\delta = \mathbb{R}\left(s, \|s\|^2, 1\right)^\delta = \mathbb{R}\left(Cs, t_1\|s\|^2, t_1^{-1}\right) = \mathbb{R}\left(t_1 Cs, t_1^2\|s\|^2, 1\right) = \mathbb{R}\left(t_1 Cs, \|t_1 Cs\|^2, 1\right) = (t_1 Cs)$. □

15.8 Lemma. $\mathrm{PSO}_{n+2}(1)^W$ *contains the transformations*

(1) $W \to W : (s) \mapsto (s + a),\ (\infty) \mapsto (\infty)$ *for* $a \in \mathbb{F}$, *and*

(2) $W \to W : (s) \mapsto (s^{-1}),\ (0) \mapsto (\infty),\ (\infty) \mapsto (0)$.

Proof. Transformation (1) is induced on the quadric $W \cong Q \subseteq \mathrm{P}\widehat{\mathbb{F}}$ by the following linear map T of $\widehat{\mathbb{F}} = \mathbb{F} \times \mathbb{R}^2$:

$$T(x, r_1, r_2) = (x + r_2 a, \langle x \mid a \rangle + r_1 + r_2 \|a\|^2, r_2)$$
$$= (x + r_2 a, \|x + a\|^2 - \|x\|^2 - \|a\|^2 + r_1 + r_2 \|a\|^2, r_2).$$

Indeed, T maps $(s) \cong \mathbb{R}(s, \|s\|^2, 1)$ onto $\mathbb{R}(s + a, \|s + a\|^2, 1) \cong (s + a)$. Also, T preserves the quadratic form q:

$$q(T(x, r_1, r_2)) = \|x + r_2 a\|^2 - \langle x \mid r_2 a \rangle - \|r_2 a\|^2 - r_1 r_2$$
$$= \|x + r_2 a\|^2 - (\|x + r_2 a\|^2 - \|x\|^2 - \|r_2 a\|^2) - \|r_2 a\|^2 - r_1 r_2$$
$$= \|x\|^2 - r_1 r_2 = q(x, r_1, r_2).$$

Thus, transformation (1) belongs to $\mathrm{PO}_{n+2}(1)^W$. Moreover, it is the square of the map $(s) \mapsto (s + \frac{1}{2}a)$, which for the same reason belongs to $\mathrm{PO}_{n+2}(1)^W$, as well. Since the subgroup $\mathrm{PSO}_{n+2}(1)^W$ has index 2 in $\mathrm{PO}_{n+2}(1)^W$, it therefore contains transformation (1).

Transformation (2) is induced by the linear map $R : (x, r_1, r_2) \mapsto (\bar{x}, r_2, r_1)$; indeed, R maps $(s) \cong \mathbb{R}(s, \|s\|^2, 1)$ onto $\mathbb{R}(\bar{s}, 1, \|s\|^2) = \mathbb{R}(\|s\|^{-2}\bar{s}, \|s\|^{-2}, 1) = \mathbb{R}(s^{-1}, \|s^{-1}\|^2, 1) \cong (s^{-1})$. Clearly, R belongs to $O(\widehat{\mathbb{F}}, q)$, and, for $\mathbb{F} \in \{\mathbb{C}, \mathbb{H}, \mathbb{O}\}$, the determinant of R is 1. This proves our assertion about transformation (2) in these cases. For $\mathbb{F} = \mathbb{R}$, $n = 1$, use the fact that $\mathrm{PSO}_3(1)^W = \mathrm{PO}_3(1)^W$, see (15.2). □

15.9 *Proof of Theorem* (15.6): $A_\mathbb{F}|_W = \mathrm{PSO}_{n+2}(1)^W$.
The groups $\mathrm{PSO}_{n+2}(1)^W$ and $A_\mathbb{F}|_W$ both contain the transformations (1) and (2) of (15.8), which are induced on W by the following elements of the affine collineation group $A_\mathbb{F}$, see (12.5b and 6): the shears $(x, y) \mapsto (x, y + ax)$, and the reflection $(x, y) \mapsto (y, x)$. Hence, both groups contain the group generated by these transformations. This group is 2-transitive on W: indeed, the group of the transformations (1) fixes (∞) and is transitive on $W \setminus \{(\infty)\}$, and transformation (2) maps (∞) to (0). Therefore, it suffices to show that the stabilizer GO_n^{+W} of the two points $u = (0)$ and $v = (\infty)$ in $\mathrm{PSO}_{n+2}(1)^W$, see (15.7), coincides with the stabilizer of these two points in $A_\mathbb{F}|_W$, cf. (91.3). Since the subgroup $T \leq A_\mathbb{F}$ of translations (12.5a) fixes every point of W and is transitive on the affine point set, the stabilizer of u and v in $A_\mathbb{F}|_W$ is induced by the stabilizer $\nabla = (A_\mathbb{F})_{o,u,v}$, where $o = (0, 0)$. For $\mathbb{F} = \mathbb{O}$, the latter has been determined in (12.17), and it has been shown that $\nabla|_W = \mathrm{GO}_8^{+W}$, indeed. For $\mathbb{F} \in \{\mathbb{R}, \mathbb{C}, \mathbb{H}\}$, we have $A_\mathbb{F}|_W = \mathrm{PGL}_2\mathbb{F}$ according to (12.12), and the stabilizer of the points (0) and (∞) in this group is GO_n^{+W}, see Addendum (b) of (12.12). Thus, by (15.7), that stabilizer coincides indeed with the stabilizer of the same points in $\mathrm{PSO}_{n+2}(1)^W$. □

In the remainder of this section, we study certain subgroups of $\mathrm{PSO}_{n+2}(1)^W$ which are isomorphic to $\mathrm{SO}_{n+1}\mathbb{R}$ or $\mathrm{SO}_{n+1}(\mathbb{R}, 1)$. This material is standard; it is presented here within our notational frame for the convenience of the reader, but it will be needed in Section 18 only, for the study of motion groups of the octonion plane. Indeed, for $\mathbb{F} \in \{\mathbb{R}, \mathbb{C}, \mathbb{H}, \mathbb{O}\}$, the groups in question are related to the elliptic motion group or to the hyperbolic motion group of the plane over $\mathbb{F}$, respectively. In this book, however, that relationship will be explained and used only in the octonion case, except for a few indications in (18.9) concerning some of the other cases. For the fields $\mathbb{R}$, $\mathbb{C}$, and $\mathbb{H}$, we have indeed seen easier ways to deal with the respective motion groups in Section 13.

15.10 The groups $\mathrm{O}_{n+1}{}^W$ **and** $\mathrm{O}_{n+1}(1)^W$.
1) These groups are induced on W by the stabilizers in $\mathrm{PO}(\widehat{\mathbb{F}}, q)$ of two 1-dimensional subspaces of $\widehat{\mathbb{F}}$ which *do not belong* to W. Those subspaces are spanned by the vectors

$$h^\pm = (0, 1, \pm 1) \in \mathbb{F} \times \mathbb{R} \times \mathbb{R} = \widehat{\mathbb{F}}.$$

Their orthogonal spaces are

$$T^\pm = \{ (x, r, \mp r) \mid x \in \mathbb{F}, r \in \mathbb{R} \} \ .$$

As $q(h^\pm) \neq 0$, we have an orthogonal sum decomposition

$$\widehat{\mathbb{F}} = T^\pm \oplus \mathbb{R} h^\pm \ .$$

Every element of $O(\widehat{\mathbb{F}}, q)$ leaving $\mathbb{R} h^\pm$ invariant induces an orthogonal transformation on $T^\pm$. Conversely, every map $B \in O(T^\pm, q)$ gives rise to a transformation

$$\widehat{B} : \widehat{\mathbb{F}} \to \widehat{\mathbb{F}} : z + rh^\pm \mapsto Bz + rh^\pm \ ,$$

($z \in T^\pm$, $r \in \mathbb{R}$), which belongs to $O(\widehat{\mathbb{F}}, q)$ and fixes $h^\pm$. Regarding the induced transformation $[\widehat{B}]$ of the projective space $P\widehat{\mathbb{F}}$, we claim that the maps

(1)
$$\begin{array}{ccccc} O(T^\pm, q) & \to & O(\widehat{\mathbb{F}}, q)_{h^\pm} & \to & PO(\widehat{\mathbb{F}}, q)_{\mathbb{R} h^\pm} \\ B & \mapsto & \widehat{B} & \mapsto & [\widehat{B}] \end{array}$$

are isomorphisms. For the first map, this is obvious, and so is the injectivity of the second map. In order to prove its surjectivity, assume that $A \in O(\widehat{\mathbb{F}}, q)$ leaves $\mathbb{R} h^\pm$ invariant. Since $q(h^\pm) \neq 0$, the orthogonal transformation A maps $h^\pm$ either to itself or to $-h^\pm$. In the first case, A belongs to $O(\widehat{\mathbb{F}}, q)_{h^\pm}$. In the second case, the same is true for $-A$ instead of A; but $[-A] = [A]$. Thus, surjectivity of the second map is proved. If $\det A = 1$ and if n is even, then $\det(-A) = \det A = 1$. Hence, the restrictions of the isomorphisms (1)

(2)
$$SO(T^\pm, q) \to SO(\widehat{\mathbb{F}}, q)_{h^\pm} \to PSO(\widehat{\mathbb{F}}, q)_{\mathbb{R} h^\pm}$$

are isomorphisms as well for n even.

We remark that the quadratic form q is positive definite on T^+ and that the restriction of q to T^- has index 1. Consequently, we have the following isomorphisms.

(3)
$$\begin{aligned} O(T^+, q) &\cong O_{n+1} \mathbb{R} \ , \\ O(T^-, q) &\cong O_{n+1}(\mathbb{R}, 1) \ . \end{aligned}$$

2) We now consider the actions of these groups on the quadric W. Let

$$\begin{aligned} O_{n+1}{}^W &:= PO(\widehat{\mathbb{F}}, q)_{\mathbb{R} h^+}|_W \\ O_{n+1}(1)^W &:= PO(\widehat{\mathbb{F}}, q)_{\mathbb{R} h^-}|_W \ . \end{aligned}$$

These are subgroups of $PO_{n+2}(1)^W = PO(\widehat{\mathbb{F}}, q)|_W$. The notation adopted here reflects the isomorphisms (1) and (3); also note that $PO(\widehat{\mathbb{F}}, q)$ acts effectively on W

by (15.4). According to the isomorphisms (1), we have

$$O_{n+1}{}^W = \{ [A]|_W \mid A \in O(\widehat{\mathbb{F}}, q)_{h^+} \} = \{ [\widehat{B}]|_W \mid B \in O(T^+, q) \},$$
$$O_{n+1}(1)^W = \{ [A]|_W \mid A \in O(\widehat{\mathbb{F}}, q)_{h^-} \} = \{ [\widehat{B}]|_W \mid B \in O(T^-, q) \}.$$

We obtain subgroups $SO_{n+1}{}^W$ and $SO_{n+1}(1)^W$ by insisting that the elements A and B which appear in the above definition be of determinant one. With these subgroups, we have

(4)
$$PSO(\widehat{\mathbb{F}}, q)_{\mathbb{R}h^+}|_W = \begin{cases} SO_{n+1}{}^W & \text{for } n \text{ even} \\ O_2{}^W & \text{for } n = 1 \end{cases}$$
$$PSO(\widehat{\mathbb{F}}, q)_{\mathbb{R}h^-}|_W = \begin{cases} SO_{n+1}(1)^W & \text{for } n \text{ even} \\ O_2(1)^W & \text{for } n = 1. \end{cases}$$

For n even, this is obvious from the isomorphisms (2). For n odd, one has $PSO(\widehat{\mathbb{F}}, q) = PO(\widehat{\mathbb{F}}, q)$.

3) We note that the groups $SO_{n+1}{}^W$, $SO_{n+1}(1)^W$, and the group GO_n^{+W} introduced in (15.7) have the following subgroup in common:

$$SO_n{}^W := \{ W \to W : (s) \mapsto (Cs), (\infty) \mapsto (\infty) \mid C \in SO_n\mathbb{R} \}.$$

Indeed, the transformations in this subgroup are induced by the linear transformations $(x, r_1, r_2) \mapsto (Cx, r_1, r_2)$ of $\widehat{\mathbb{F}} = \mathbb{F} \times \mathbb{R} \times \mathbb{R}$ for $C \in SO_n\mathbb{R}$, which clearly belong to $SO(\widehat{\mathbb{F}}, q)$, and which fix h^+ and h^-.

We now provide information about the actions of these groups on W.

15.11. *The group $SO_{n+1}{}^W$ acts transitively on W.*

In fact, we show that this group acts on W exactly in the same way as $SO(T^+, q)$ acts on the unit sphere

$$ST^+ = \{ z \in T^+ \mid q(z) = 1 \}$$
$$= \{ (x, r, -r) \mid x \in \mathbb{F}, r \in \mathbb{R}, \|x\|^2 + r^2 = 1 \} \approx \mathbb{S}_n.$$

First, the map $SO(T^+, q) \to SO_{n+1}{}^W : B \mapsto [\widehat{B}]|_W$ is an isomorphism; this follows from the isomorphism (1) in (15.10) and from the definition of $SO_{n+1}{}^W$ in step 2) there. Secondly, we shall prove that the map

(1) $$T^+ \to P\widehat{\mathbb{F}} : z \mapsto \mathbb{R}(z + h^+)$$

induces a bijection $ST^+ \to W$. It is then clear that this bijection is equivariant with respect to the isomorphism $B \mapsto [\widehat{B}]|_W$, compare the definition of $\widehat{B}$ in step 1) of

(15.10), so that (15.11) will be proved. In order to establish the asserted bijection $ST^+ \to W$, recall that W has been identified with the quadric of q. Since q is positive definite on T^+, no element of W is contained in T^+. Therefore, every 1-dimensional subspace in W intersects the affine hyperplane $T^+ + h^+$ in a unique vector $z + h^+$. The given subspace is then generated by $z + h^+$. Now $z \in T^+$ is orthogonal to h^+, so that $q(z + h^+) = q(z) + q(h^+) = q(z) - 1$ is zero if, and only if $z \in ST^+$. □

15.12. *The group* $O_{n+1}(1)^W$ *leaves the* $(n-1)$*-sphere*

$$W_1 = \{\,(s) \mid s \in \mathbb{F}, \|s\|^2 = 1\,\} \subseteq W$$

invariant, and $SO_{n+1}(1)^W$ *acts transitively on the complement* $W \setminus W_1$, *which consists of the two connected components*

and
$$W_- = \{\,(s) \mid s \in \mathbb{F}, \|s\|^2 < 1\,\}$$
$$W_+ = \{\,(s) \mid s \in \mathbb{F}, \|s\|^2 > 1\,\} \cup \{(\infty)\}\,.$$

Every element of $SO_{n+1}(1)^W$ *either fixes or interchanges the components. The subgroup of index 2 formed by the elements of the first kind is precisely the commutator subgroup* $O'_{n+1}(1)^W$ *of* $O_{n+1}(1)^W$.

For $n \geq 2$, *the action of* $SO_{n+1}(1)^W$ *on* W_1 *is triply transitive. If* $n > 2$, *then this holds for* $O'_{n+1}(1)^W$, *as well.*

Remark. The case $n = 1$ is somewhat particular, W_1 consisting of two points only; the stabilizer of these two points in $O_2(1)^W$ is precisely $SO_2(1)^W$.

Proof of (15.12). We use the bilinear form β_q on $\widehat{\mathbb{F}} = \mathbb{F} \times \mathbb{R}^2$ associated with the quadratic form q. For

$$z = (x, r_1, r_2),\ z' = (x', r'_1, r'_2),$$

where $x, x' \in \mathbb{F}$, $r_\nu, r'_\nu \in \mathbb{R}$, we have

$$\beta_q(z, z') = q(z + z') - q(z) - q(z')$$
$$= \|x + x'\|^2 - (r_1 + r'_1)(r_2 + r'_2) - \|x\|^2 + r_1 r_2 - \|x'\|^2 + r'_1 r'_2$$
$$= \langle x \mid x' \rangle - r_1 r'_2 - r'_1 r_2\,;$$

here, $\langle \ \mid\ \rangle$ denotes the inner product in $\mathbb{F}$, see (11.3). In particular, for $(s) \in W$, $(s) \triangleq \mathbb{R}(s, \|s\|^2, 1)$, we have $\beta_q(h^-, (s, \|s\|^2, 1)) = \|s\|^2 - 1$. Thus, we obtain:

$$W_1 = \{\,\mathbb{R}z \mid q(z) = 0,\ \beta_q(h^-, z) = 0\,\}\,.$$

Clearly, $O_{n+1}(1)^W = PO(\widehat{\mathbb{F}}, q)_{\mathbb{R}h^-}|_W$ leaves W_1 invariant. We shall now show that the group $O_{n+1}(1)^W$ is transitive on $W \setminus W_1$ and that it is triply transitive on W_1 for $n \geq 2$, and afterwards turn to $SO_{n+1}(1)^W$ and $O'_{n+1}(1)^W$.

Let $z, z' \in \widehat{\mathbb{F}} \setminus \{0\}$ be such that $\mathbb{R}z, \mathbb{R}z' \in W \setminus W_1$. Then we have

(1) $$q(z) = 0 = q(z'),$$

and $\beta_q(h^-, z) \neq 0 \neq \beta_q(h^-, z')$. Replacing z' by a suitable scalar multiple, if necessary, we may achieve that

(2) $$\beta_q(h^-, z) = \beta_q(h^-, z').$$

According to (1) and (2), we obtain a linear q-isometry of $\mathbb{R}h^- + \mathbb{R}z$ onto itself which fixes h^- and sends z to z'. This extends to an isometry of $\widehat{\mathbb{F}}$ by Witt's theorem, see e.g. Taylor [92] Theorem 7.4, p. 57. Therefore, the group $O(\widehat{\mathbb{F}}, q)_{h^-}$, inducing $O_{n+1}(1)^W$ on W, is transitive on $W \setminus W_1$.

We now prove that $O_{n+1}(1)^W$ is triply transitive on W_1 for $n \geq 2$. Consider a triple $z_1, z_2, z_3 \in \widehat{\mathbb{F}} \setminus \{0\}$ such that the corresponding points $\mathbb{R}z_\nu$ are different and belong to W. Then $q(z_\nu) = 0$ for $\nu = 1, 2, 3$. Hence $\beta_q(z_\mu, z_\nu) \neq 0$ for $\mu \neq \nu$, or else β_q would be identically 0 on the 2-dimensional subspace spanned by z_μ and z_ν, contradicting the fact that the index of β_q is 1. It follows that z_1, z_2, z_3 are linearly independent. Moreover, replacing these vectors by suitable scalar multiples, if necessary, we may arrange that

$$\beta_q(z_1, z_2) = 1 = \beta_q(z_2, z_3) \quad \text{and} \quad \beta_q(z_1, z_3) = \varepsilon \in \{1, -1\}.$$

The Gram matrix of the restriction of β_q to the 3-dimensional subspace S spanned by z_1, z_2, z_3 with respect to this basis has determinant 2ε; in particular, $\beta_q|_S$ is regular. We shall decide about the sign of ε using the fact that the sign of the determinant of the Gram matrix of $\beta_q|_S$ does not depend on the specific basis. The quadratic form q is positive definite on the 2-dimensional intersection $S \cap T^+$ with the hyperplane T^+. Hence, for a vector $w \in S \setminus \{0\}$ belonging to the orthogonal space of $S \cap T^+$ we have $q(w) < 0$, or else β_q would be positive definite on S, contradicting the fact that $q(z_\nu) = 0$. The Gram matrix of $\beta_q|_S$ with respect to the basis of S consisting of w and an orthogonal basis of $S \cap T^+$ thus has negative determinant, so that $\varepsilon = -1$.

In addition, if the points $\mathbb{R}z_\nu$ belong to W_1, we have $\beta_q(z_\nu, h^-) = 0$ for $\nu = 1, 2, 3$. Then, the vectors z_1, z_2, z_3, h^- are linearly independent, and their inner products under β_q are 0, 1, or -1, as described. If another such triple z_1', z_2', z_3' is given, then by Witt's theorem there is an element of $O(\widehat{\mathbb{F}}, q)$ fixing h^- and sending z_ν to z_ν' for $\nu = 1, 2, 3$. Thus, $O_{n+1}(1)^W$ is triply transitive on W_1.

The assertion about the connected components of $W \setminus W_1$ refers to the topology on the quadric $W \cong Q \subseteq \mathbb{P}\widehat{\mathbb{F}}$ as described in (15.1). Under the homeomorphism established there between $W \cong Q$ and the one-point compactification $\mathbb{F} \cup \{\infty\}$, see (15.1(2)), the subset W_1 corresponds to the unit sphere $\{ s \in \mathbb{F} \mid \|s\|^2 = 1 \} = \mathbb{S}_{n-1}$ in $\mathbb{F} = \mathbb{R}^n$, and W_- and W_+ correspond to the interior and the exterior of $\mathbb{S}_{n-1}$ in $\mathbb{F} \cup \{\infty\} \approx \mathbb{S}_n$; these are indeed the connected components of the complement of $\mathbb{S}_{n-1}$. The elements of $\mathrm{O}(\widehat{\mathbb{F}}, q)$ induce homeomorphisms of the real projective space $\mathrm{P}\widehat{\mathbb{F}}$, see (14.2), which restrict to homeomorphisms of W. Thus, $\mathrm{O}_{n+1}(1)^W$ consists of homeomorphisms, hence permutes the connected components W_+ and W_- of $W \setminus W_1$.

For a transfer of transitivity properties to the subgroups $\mathrm{SO}_{n+1}(1)^W$ and $\mathrm{O}'_{n+1}(1)^W$, we shall use certain special elements of $\mathrm{O}(\widehat{\mathbb{F}}, q)_{h^-}$, which we study now. The transformation

$$P : \widehat{\mathbb{F}} \to \widehat{\mathbb{F}} : (x, r_1, r_2) \mapsto (-\bar{x}, r_1, r_2)$$

belongs to $\mathrm{O}(\widehat{\mathbb{F}}, q) \setminus \mathrm{SO}(\widehat{\mathbb{F}}, q)$ and fixes h^-. It induces an element $[P]|_W$ of $\mathrm{O}_{n+1}(1)^W \setminus \mathrm{SO}_{n+1}(1)^W$ fixing $\mathbb{R}(0, 0, 1) \cong (0) \in W \setminus W_1$.

For $2 \leq \nu \leq n$, we consider moreover an orthogonal $\mathbb{R}$-linear involution J of $\mathbb{F}$ such that the eigenspace N of J for the eigenvalue -1 has dimension ν. Then the transformation

$$R_\nu : \widehat{\mathbb{F}} \to \widehat{\mathbb{F}} : (x, r_1, r_2) \mapsto (Jx, -r_2, -r_1)$$

belongs to $\mathrm{O}(\widehat{\mathbb{F}}, q)$, has determinant $(-1)^{\nu+1}$ and fixes h^-. For $s \in N$, this transformation maps the point $\mathbb{R}(s, \|s\|^2, 1) \cong (s)$ to $\mathbb{R}(s, 1, \|s\|^2) = \mathbb{R}\left(\frac{s}{\|s\|^2}, \frac{1}{\|s\|^2}, 1\right) = \mathbb{R}\left(\bar{s}^{-1}, \|\bar{s}^{-1}\|^2, 1\right) \cong (\bar{s}^{-1})$. In particular, the transformation $[R_\nu]|_W$ induced by R_ν interchanges the components of $W \setminus W_1$ and fixes the points of the $(\nu-1)$-sphere $\{(s) \mid s \in N, \|s\|^2 = 1\} \subseteq W_1$ corresponding to the unit sphere of $N \leq \mathbb{F}$.

Using composition with the transformation $[P]|_W \in \mathrm{O}_{n+1}(1)^W \setminus \mathrm{SO}_{n+1}(1)^W$ fixing $(0) \in W \setminus W_1$, we now see that transitivity of $\mathrm{O}_{n+1}(1)^W$ on $W \setminus W_1$ carries over to $\mathrm{SO}_{n+1}(1)^W$. The subgroup of $\mathrm{SO}_{n+1}(1)^W$ formed by the elements that leave the components of $W \setminus W_1$ invariant is a normal subgroup of index 2 in $\mathrm{SO}_{n+1}(1)^W$, which contains the commutator subgroup $\mathrm{O}'_{n+1}(1)^W$ of $\mathrm{O}_{n+1}(1)^W$. That subgroup coincides with the commutator subgroup, because $\mathrm{O}'_{n+1}(\mathbb{R}, 1)$ is known to have index 2 in $\mathrm{SO}_{n+1}(\mathbb{R}, 1)$, see Dieudonné [71] Chap. II §8 no. 2), p. 56.

If $n \geq 2$, we may use composition with the transformation $[R_2]|_W$, which belongs to $\mathrm{O}_{n+1}(1)^W \setminus \mathrm{SO}_{n+1}(1)^W$ and fixes many points of W_1, in order to obtain that $\mathrm{SO}_{n+1}(1)^W$ is triply transitive on W_1 because $\mathrm{O}_{n+1}(1)^W$ is. Finally, if $n > 2$, then this property carries over to $\mathrm{O}'_{n+1}(1)^W$, as can be seen by composition with the transformation $[R_3]|_W \in \mathrm{SO}_{n+1}(1)^W$, which then is available, and which interchanges the components of $W \setminus W_1$ and fixes many points of W_1. □

15.13 Notes. The method of expressing the geometric structure of the projective line over $\mathbb{F}$ by considering it as a projective quadric of $\mathbb{F} \times \mathbb{R}^2$ is adopted from Märer [73] p. 43, and was used by Grundhöfer [83] §3 for the study of projectivities of Moufang planes. Märer's paper is concerned with Möbius geometry (inversive geometry). Now, the groups studied in the present section are closely related to that topic, as we shall indicate. The quadric $W \cong Q$ can be considered as an n-sphere in $(n+1)$-space, cf. (15.11), and the Möbius geometry on this sphere is given by the system of spheres of smaller dimensions obtained as the intersections of W with affine subspaces. The group $\mathrm{PO}_{n+2}(1)^W$ introduced in (15.2) is precisely the group of all automorphisms of that geometry, and the group $\mathrm{O}_{n+1}(1)^W$ of (15.10) is the stabilizer of a sphere of dimension $n-1$, namely of W_1. From complex analysis, the term 'Möbius transformations' for the elements of the group $\mathrm{PGL}_2\mathbb{C} = \mathrm{PSO}_4(1)^W$, see (15.6), is familiar. There are also close connections to hyperbolic geometry. For instance, the Klein model of hyperbolic $(n+1)$-space can be realized on the bounded component of $(\widehat{\mathrm{P}\mathbb{F}} \setminus T^+) \setminus Q$, the quadric $Q \cong W$ being the space of ends of this hyperbolic space, and $\mathrm{PO}(\widehat{\mathbb{F}}, q)$ is the group of automorphisms of this geometry. There are also models of hyperbolic geometry within Möbius geometry. Specifically, in the situation of (15.12), hyperbolic n-space can be projected onto each of the two hemispheres W_+ and W_- of W; in this way, the groups considered there are closely related to the automorphism group of hyperbolic n-space. For systematic information about these topics, see e.g. Berger [77, 87] Chapters 18 and 19, Iversen [92], Schröder [91] part I, §4 and part II, §10, and Wilker [81].

16 The projective octonion plane $\mathcal{P}_2\mathbb{O}$

In Section 13, we described the classical projective planes over $\mathbb{R}$, $\mathbb{C}$, and $\mathbb{H}$ in terms of homogeneous coordinates. We now present a different method of introducing coordinates in these planes. These Veronese coordinates, as we call them, might well replace homogeneous coordinates in the planes over $\mathbb{R}$, $\mathbb{C}$, and $\mathbb{H}$, and they have the advantage of working equally well over the octonions $\mathbb{O}$. For the sake of clarity, the construction will be carried out over $\mathbb{O}$ only. However, Veronese coordinates for the planes over $\mathbb{R}$, $\mathbb{C}$, and $\mathbb{H}$ may be obtained simply by restriction to these subfields of $\mathbb{O}$.

In Section 14, the planes over $\mathbb{R}$, $\mathbb{C}$, and $\mathbb{H}$ proved to be topological projective planes in a natural way. At the end of the present section, we shall obtain an equally natural topological structure for the projective octonion plane via Veronese coordinates.

16.1 Veronese coordinates. (a) Consider the real vector space

$$V := \mathbb{O}^3 \times \mathbb{R}^3$$

of dimension 27. The elements of V will be written in the form

$$(x_\nu; \xi_\nu)_\nu = (x_1, x_2, x_3; \xi_1, \xi_2, \xi_3) \quad \text{with} \quad x_\nu \in \mathbb{O}, \ \xi_\nu \in \mathbb{R}, \ \nu = 1, 2, 3 \ .$$

We are particularly interested in the subset

$$H \subseteq V$$

of all Veronese vectors, which we define now. A vector $(x_\nu; \xi_\nu)_\nu$ is called a *Veronese vector* if the following relation among its coordinates is satisfied for every cyclic permutation (λ, μ, ν) of $(1, 2, 3)$:

(∗) $\qquad\qquad \xi_\lambda \overline{x_\lambda} = x_\mu x_\nu \quad \text{and} \quad \|x_\lambda\|^2 = \xi_\mu \xi_\nu \,.$

The choice of this subset will be motivated in (16.8). It is clear that, for every Veronese vector $w \in H \setminus \{0\}$, the 1-dimensional subspace $\mathbb{R}w$ is entirely contained in H.

We now define a geometry having precisely these 1-dimensional subspaces as points. It will turn out (16.3) that this geometry is a projective plane isomorphic to the projective completion $\overline{\mathcal{A}_2 \mathbb{O}}$ of the affine plane over $\mathbb{O}$ as defined in (12.1 and 2). Consequently, this geometry will be called the projective plane $\mathcal{P}_2 \mathbb{O}$ over $\mathbb{O}$. Comparing this with the construction of the projective plane over $\mathbb{F} \in \{\mathbb{R}, \mathbb{C}, \mathbb{H}\}$ in (13.1), we note two differences. In the first place, we consider 1-dimensional subspaces over $\mathbb{R}$, not over any larger domain, and secondly, not all 1-dimensional subspaces are honoured by being called points, but only those spanned by Veronese vectors. Thus, the set of points will be the following proper subset of the projective space $PV = P_{26}\mathbb{R}$ of all 1-dimensional $\mathbb{R}$-linear subspaces of $V = \mathbb{R}^{27}$:

$$P_2 \mathbb{O} := \{ \mathbb{R}w \mid w \in H \setminus \{0\} \} \subset PV \,.$$

(b) Lines will be introduced using the following positive definite symmetric bilinear form β on V:

$$\beta\big((x_\nu; \xi_\nu)_\nu \mid (y_\nu; \eta_\nu)_\nu\big) = \sum_{\nu=1}^{3} \big(\langle x_\nu \mid y_\nu \rangle + \xi_\nu \eta_\nu\big) \,,$$

where $\langle x_\nu \mid y_\nu \rangle = \overline{x_\nu} y_\nu + \overline{y_\nu} x_\nu$ is the inner product on $\mathbb{O}$ as in (11.3). The quadratic form corresponding to β is given by

$$\tfrac{1}{2}\beta\big((x_\nu; \xi_\nu)_\nu \mid (x_\nu; \xi_\nu)_\nu\big) = \tfrac{1}{2} \cdot \sum_{\nu=1}^{3} (2\overline{x_\nu} x_\nu + \xi_\nu^2) = \sum_{\nu=1}^{3} (\|x_\nu\|^2 + \tfrac{1}{2}\xi_\nu^2) \,;$$

it is important that the real parts have only half the weight of the octonion parts.

(c) The lines of our geometry will be particular hyperplanes of the $\mathbb{R}$-vector space $V = \mathbb{O}^3 \times \mathbb{R}^3$, namely those which are perpendicular to some Veronese vector $w \in H \setminus \{0\}$ (or, equivalently, to the point $p = \mathbb{R}w \in P_2\mathbb{O}$) with respect to the bilinear form β. In other words, we consider orthogonal spaces

$$w^\perp = p^\perp = \{ z \in V \mid \beta(z \mid w) = 0 \} \,,$$

and we define the set of all lines by

$$\mathcal{L}_2\mathbb{O} := \{\, p^\perp \mid p \in P_2\mathbb{O} \,\} \;.$$

This is a proper subset of the space $\mathcal{H}V = \mathcal{H}_{26}\mathbb{R}$ of all hyperplanes of $V = \mathbb{R}^{27}$.

A point $p \in P_2\mathbb{O}$ is said to be incident with a line $q^\perp$, for $q \in P_2\mathbb{O}$, if, and only if, $p \subseteq q^\perp$.

We have now defined a geometry $\mathcal{P}_2\mathbb{O} = (P_2\mathbb{O}, \mathcal{L}_2\mathbb{O}, \subseteq)$, and next we want to compare it with the projective completion $\overline{\mathcal{A}_2\mathbb{O}}$. First, we shall establish that the lines in these two geometries have metric structures that match. For $\overline{\mathcal{A}_2\mathbb{O}}$, the metric structure of a line was introduced in Section 15; in Veronese coordinates, a corresponding structure can be obtained even more naturally, as the following observation shows.

16.2 Geometric structure of a line. *Consider the line* $W = (0, 0, 0; 0, 0, 1)^\perp$ *of* $P_2\mathbb{O}$. *The set of points incident with* W *is*

$$\{\, \mathbb{R}(0, 0, s; \|s\|^2, 1, 0) \mid s \in \mathbb{O} \,\} \cup \{\mathbb{R}(0, 0, 0; 1, 0, 0)\} \;.$$

This is the quadric of the quadratic form q *of index* 1 *defined on the* 10-*dimensional* $\mathbb{R}$-*linear subspace*

$$V_3 = \{0\} \times \{0\} \times \mathbb{O} \times \mathbb{R}^2 \times \{0\} \quad \leq \quad \mathbb{O}^3 \times \mathbb{R}^3 = V$$

by

$$q(0, 0, x; r_1, r_2, 0) = \bar{x}x - r_1 r_2 \;.$$

Remark. Ignoring the zero coordinates, we see that the point set of W actually is the quadric with which the line at infinity of $\overline{\mathcal{A}_2\mathbb{O}}$ was identified in (15.1c). Moreover, we shall see in (16.3) that this identification extends to an isomorphism between the projective planes $\overline{\mathcal{A}_2\mathbb{O}}$ and $\mathcal{P}_2\mathbb{O}$.

Proof of (16.2). By definition, a point $\mathbb{R}(x_\nu; \xi_\nu)_\nu \in P_2\mathbb{O}$ is incident with W if, and only if $\xi_3 = 0$. One distinguishes the cases $\xi_2 = 0$ and $\xi_2 \neq 0$; in the latter case, upon multiplication by a scalar if necessary, one may assume that $\xi_2 = 1$. In both cases, it now follows directly from the conditions (16.1(∗)) defining $P_2\mathbb{O}$ that the point in question has the asserted form. □

16.3 A correspondence between affine and Veronese coordinates. *An isomorphism of the projective completion* $\overline{\mathcal{A}_2\mathbb{O}}$ *onto* $\mathcal{P}_2\mathbb{O}$ *is given by the following maps* between points and lines, respectively, of $\overline{\mathcal{A}_2\mathbb{O}}$ and $\mathcal{P}_2\mathbb{O}$:

$$(x, y) \mapsto \mathbb{R}(x, \bar{y}, y\bar{x}; \|y\|^2, \|x\|^2, 1) \qquad [s, t] \mapsto (\bar{s}t, -\bar{t}, -s; 1, \|s\|^2, \|t\|^2)^\perp$$
$$(s) \mapsto \mathbb{R}(0, 0, s; \|s\|^2, 1, 0) \qquad [c] \mapsto (-c, 0, 0; 0, 1, \|c\|^2)^\perp$$
$$(\infty) \mapsto \mathbb{R}(0, 0, 0; 1, 0, 0) \qquad [\infty] \mapsto (0, 0, 0; 0, 0, 1)^\perp ,$$

for $x, y, s, t, c \in \mathbb{O}$. In particular, $\mathcal{P}_2\mathbb{O}$ is a projective plane.

Proof. By testing the condition (16.1(∗)), which defines Veronese vectors, it is easy to verify that the 1-dimensional subspaces and hyperplanes of $V = \mathbb{O}^3 \times \mathbb{R}^3$ appearing in the proposition are points and lines of $\mathcal{P}_2\mathbb{O}$.

Next, one has to show that every point of $P_2\mathbb{O}$ is among the image points. Let $\mathbb{R}(x_\nu; \xi_\nu)_\nu \in P_2\mathbb{O}$. The case $\xi_3 = 0$ has been dealt with in (16.2). If $\xi_3 \neq 0$, we may assume that $\xi_3 = 1$. Then condition (16.1(∗)) says that $\overline{x_3} = x_1 x_2$, $\xi_1 = \|x_2\|^2$, and $\xi_2 = \|x_1\|^2$, which proves our claim.

Now it is easy to see that the given point map is a bijection between the point sets of $\overline{\mathcal{A}_2\mathbb{O}}$ and of $\mathcal{P}_2\mathbb{O}$. In the same way, one may verify that every line of $\mathcal{P}_2\mathbb{O}$ appears exactly once in the list of our proposition.

It remains to be shown that the points of a line in $\overline{\mathcal{A}_2\mathbb{O}}$ are mapped onto the points of the specified image line of $\mathcal{P}_2\mathbb{O}$. For $[\infty] = \{ (s) \mid s \in \mathbb{O} \} \cup \{(\infty)\}$, this is clear by (16.2). As for $[c]$, the line $(-c, 0, 0; 0, 1, \|c\|^2)^\perp$ contains the image point $\mathbb{R}(x, \bar{y}, y\bar{x}; \|y\|^2, \|x\|^2, 1)$ of (x, y) if, and only if $(x, y) \in [c]$, since $0 = \langle -c \mid x \rangle + \|x\|^2 + \|c\|^2 = \|x - c\|^2$ is equivalent to $x = c$. Furthermore, this line contains the image of (∞), but it does not contain the image of (s) for $s \in \mathbb{O}$. The line $(\bar{s}t, -\bar{t}, -s; 1, \|s\|^2, \|t\|^2)^\perp$ contains the image $\mathbb{R}(0, 0, a; \|a\|^2, 1, 0)$ of (a) if, and only if $0 = \langle -s \mid a \rangle + \|a\|^2 + \|s\|^2 = \|a - s\|^2$, i.e., if $a = s$. It contains the image $\mathbb{R}(x, \bar{y}, y\bar{x}; \|y\|^2, \|x\|^2, 1)$ of (x, y) if, and only if $(x, y) \in [s, t]$, since $0 = \langle x \mid \bar{s}t \rangle - \langle \bar{y} \mid \bar{t} \rangle - \langle y\bar{x} \mid s \rangle + \|y\|^2 + \|x\|^2 \|s\|^2 + \|t\|^2 = \langle sx \mid t \rangle - \langle y \mid t \rangle - \langle y \mid sx \rangle + \|y\|^2 + \|sx\|^2 + \|t\|^2 = \|sx + t - y\|^2$ is equivalent to $sx + t - y = 0$. Here, we have used the multiplicative properties (11.14) of the norm $\| \ \|^2$ and of the inner product $\langle \ \mid \ \rangle$. □

16.4 Convention. In the sequel, the projective planes $\overline{\mathcal{A}_2\mathbb{O}}$ and $\mathcal{P}_2\mathbb{O}$ will be identified via the isomorphism given in (16.3). We shall freely use both affine and Veronese coordinates for points and lines, depending on which notation is more practical for the respective purposes.

16.5 Remark. Restricting the coordinates in (16.3) to a subfield $\mathbb{F}$ of $\mathbb{O}$ which is $\mathbb{R}, \mathbb{C}$, or $\mathbb{H}$, one obtains a collineation of the projective completion $\overline{\mathcal{A}_2\mathbb{F}}$ of the affine plane over $\mathbb{F}$ onto a subplane of $\mathcal{P}_2\mathbb{O}$. This yields a description of $\mathcal{P}_2\mathbb{F} \cong \overline{\mathcal{A}_2\mathbb{F}}$ in Veronese coordinates.

Up to now, we only know collineations of $\mathcal{P}_2\mathbb{O} \cong \overline{\mathcal{A}_2\mathbb{O}}$ which leave the line at infinity $[\infty]$ invariant (Section 12). It is fortunate that there is a collineation moving $[\infty]$, the triality collineation, which is very easy to describe in Veronese coordinates. This will be the key to transitivity properties of the collineation group of $\mathcal{P}_2\mathbb{O}$, see Section 17.

16.6 The triality collineation. The conditions (16.1($*$)) determining the set $H \subseteq V = \mathbb{O}^3 \times \mathbb{R}^3$ of Veronese vectors are obviously invariant under simultaneous cyclic permutation of both the octonion coordinates and the real coordinates. Thus, the transformation realizing such a permutation,

$$\tilde{\tau} : V \to V : (x_1, x_2, x_3; \xi_1, \xi_2, \xi_3) \mapsto (x_2, x_3, x_1; \xi_2, \xi_3, \xi_1),$$

leaves H invariant and therefore induces a bijection

$$\tau : P_2\mathbb{O} \to P_2\mathbb{O} : \mathbb{R} \cdot w \mapsto \mathbb{R} \cdot \tilde{\tau}(w).$$

Since $\tilde{\tau}$ is obviously orthogonal with respect to the inner product β on V defined in (16.1), it also maps lines of $\mathcal{P}_2\mathbb{O}$ onto lines. Thus, τ is a collineation of $\mathcal{P}_2\mathbb{O}$, the *triality collineation*. By definition, we have $\tau^3 = \mathrm{id}$ and therefore $\tau^{-1} = \tau^2$.

We translate the action of τ into affine coordinates, observing the identifications of (16.3 and 4).

$$(x, y)^\tau = (y^{-1}, xy^{-1}) \quad \text{for } y \neq 0, \quad (x, 0)^\tau = (x),$$
$$(s)^\tau = (0, s^{-1}) \quad \text{for } s \neq 0, \quad (0)^\tau = (\infty), \quad (\infty)^\tau = (0, 0),$$

cf. also Figure 16a. In particular, the points $(0, 0)$, (0), (∞) are permuted cyclically by τ, and the same is true for the sides $W = [\infty]$, $Y = [0]$, and $X = [0, 0]$ of this triangle.

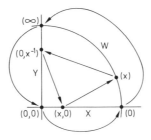

Figure 16a

Proof. $(x, y)^\tau = \left(\mathbb{R}(x, \bar{y}, y\bar{x}; \|y\|^2, \|x\|^2, 1)\right)^\tau = \mathbb{R}(\bar{y}, y\bar{x}, x; \|x\|^2, 1, \|y\|^2)$. If $y \neq 0$, we may multiply the Veronese coordinates of the image point by $\|y\|^{-2}$, which belongs to the center $\mathbb{R}$ of $\mathbb{O}$. Now $\|y\|^{-2} = (y\bar{y})^{-1} = y^{-1}\bar{y}^{-1}$ and also $\|y\|^{-2} = (\bar{y}y)^{-1} = y^{-1}\bar{y}^{-1} = \overline{y^{-1}y^{-1}} = \|y^{-1}\|^2$. Hence, $\|y\|^{-2}\bar{y} = y^{-1}$, $\|y\|^{-2}(y\bar{x}) = (\|y\|^{-2}y)\bar{x} = y^{-1}\bar{x} = \overline{xy^{-1}}$, and, by alternativity, $\|y\|^{-2}x = x\|y\|^{-2} = x(y^{-1}\bar{y}^{-1}) = (xy^{-1})\bar{y}^{-1}$. Using the multiplicativity of the norm (11.14), we thus obtain $(x, y)^\tau = \mathbb{R}\bigl(y^{-1}, \overline{xy^{-1}}, \overline{(xy^{-1})y^{-1}}; \|xy^{-1}\|^2, \|y^{-1}\|^2, 1\bigr) = (y^{-1}, xy^{-1})$. Next, $(x, 0)^\tau = \bigl(\mathbb{R}(x, 0, 0; 0, \|x\|^2, 1)\bigr)^\tau = \mathbb{R}(0, 0, x; \|x\|^2, 1, 0) = (x)$, and $(0)^\tau = (\mathbb{R}(0, 0, 0; 0, 1, 0))^\tau = \mathbb{R}(0, 0, 0; 1, 0, 0) = (\infty)$. The images of the other points can now be obtained using the relation $\tau^2 = \tau^{-1}$. □

16.7 The standard elliptic polarity of $\mathcal{P}_2\mathbb{O}$. (a) This is the polarity defined by the maps

$$\pi : \mathcal{P}_2\mathbb{O} \to \mathcal{L}_2\mathbb{O} : p \mapsto p^\perp, \quad \mathcal{L}_2\mathbb{O} \to \mathcal{P}_2\mathbb{O} : p^\perp \mapsto p = (p^\perp)^\perp \, .$$

From the symmetry of the orthogonality relation, it is clear that this is a polarity indeed.

(b) In affine coordinates as given by (16.3 and 4), this polarity maps the points as follows:

$$(x, y)^\pi = \begin{cases} [-\overline{xy^{-1}}, -\overline{y^{-1}}] & \text{for } y \neq 0 \\ [-\overline{x^{-1}}] & \text{for } y = 0,\, x \neq 0 \\ [\infty] = W & \text{for } (x, y) = (0, 0) \end{cases}$$

$$(s)^\pi = \begin{cases} [-\overline{s^{-1}}, 0] & \text{for } s \neq 0 \\ [0] & \text{for } s = 0 \end{cases}$$

$$(\infty)^\pi = [0, 0] \, .$$

Indeed, the orthogonal space of the point with affine coordinates (x, y) is

$$(x, \bar{y}, y\bar{x}; \|y\|^2, \|x\|^2, 1)^\perp \, .$$

For $y \neq 0$, the same computations as in the proof of (16.6) show that this hyperplane may also be written as $\bigl(\overline{(xy^{-1})y^{-1}}, y^{-1}, \overline{xy^{-1}}; 1, \|xy^{-1}\|^2, \|y^{-1}\|^2\bigr)^\perp$; in affine notation, this is the line $[-\overline{xy^{-1}}, -\overline{y^{-1}}]$. For $x \neq 0$, we have $(x, 0)^\pi = (x, 0, 0; 0, \|x\|^2, 1)^\perp = (\|x\|^{-2}x, 0, 0; 0, 1, \|x\|^{-2})^\perp = (\bar{x}^{-1}, 0, 0; 0, 1, \|\bar{x}^{-1}\|^2)^\perp = [-\bar{x}^{-1}] = [-\overline{x^{-1}}]$, and for $s \neq 0$ we similarly obtain $(s)^\pi = (0, 0, s; \|s\|^2, 1, 0)^\perp = (0, 0, \bar{s}^{-1}; 1, \|\bar{s}^{-1}\|^2, 0)^\perp = [-\overline{s^{-1}}, 0]$. Finally, $(0, 0)^\pi = (0, 0, 0; 0, 0, 1)^\perp = [\infty]$, $(0)^\pi = (0, 0, 0; 0, 1, 0)^\perp = [0]$, and $(\infty)^\pi = (0, 0, 0; 1, 0, 0)^\perp = [0, 0]$.

(c) Consider the projective plane $\mathcal{P}_2\mathbb{F}$ over $\mathbb{F} \in \{\mathbb{R}, \mathbb{C}, \mathbb{H}\}$ as a subplane of $\mathcal{P}_2\mathbb{O}$ according to (16.5). Then, by restriction, the polarity π of $\mathcal{P}_2\mathbb{O}$ induces the standard elliptic polarity π^+ of $\mathcal{P}_2\mathbb{F}$, which has been defined in (13.12) in terms of homogeneous coordinates.

In order to see this, we express π^+ in affine coordinates, using the correspondence (13.3) between affine and homogeneous coordinates. According to (13.12), the point $(x, y, 1)\mathbb{F}$ corresponding to the affine point (x, y) is mapped to the line $\mathrm{Ker}\,(\bar{x}\ \bar{y}\ 1)$. If $y \neq 0$, we may multiply the coordinates of this line by $\bar{y}^{-1}$ from the left, obtaining that $\mathrm{Ker}\,(\bar{x}\ \bar{y}\ 1) = \mathrm{Ker}\,(-\bar{y}^{-1}\bar{x}\ 1\ \bar{y}^{-1})$ represents the line $[\overline{-xy^{-1}}, \overline{-y^{-1}}]$ in affine coordinates. For $y = 0$, $x \neq 0$, we similarly have $\mathrm{Ker}\,(\bar{x}\ 0\ 1) = \mathrm{Ker}\,(1\ 0\ \bar{x}^{-1})$; in affine coordinates, this is the line $[\overline{-x^{-1}}]$. Finally, π^+ maps the point $(1, s, 0)\mathbb{F} \,\hat{=}\, (s)$ for $s \neq 0$ to $\mathrm{Ker}\,(1\ \bar{s}\ 0) = \mathrm{Ker}\,(\bar{s}^{-1}\ 1\ 0) \,\hat{=}\, [\overline{-s^{-1}}, 0]$, the point $(0, 0, 1)\mathbb{F} \,\hat{=}\, (0, 0)$ to $\mathrm{Ker}\,(0\ 0\ 1) \,\hat{=}\, [\infty]$, and the point $(0, 1, 0)\mathbb{F} \,\hat{=}\, (\infty)$ to $\mathrm{Ker}\,(0\ 1\ 0) \,\hat{=}\, [0, 0]$. This agrees with the description of π in affine coordinates as given in (b).

16.8 Notes. In an indirect way, a correspondence between Veronese and homogeneous coordinates of $\mathcal{P}_2\mathbb{F}$, $\mathbb{F} \in \{\mathbb{R}, \mathbb{C}, \mathbb{H}\}$ was indicated in (16.5). The embedding into $\mathcal{P}_2\mathbb{O}$ was used, and affine coordinates appeared as an intermediate step. We shall now sketch a more direct description of that correspondence. This will in fact show how Veronese coordinates arise from a very natural construction. (See also Faulkner–Ferrar [84] §5, p. 251 ff, Brada–Pecaut-Tison [87].)

Homogeneous coordinates describe the points of $\mathcal{P}_2\mathbb{F}$ as the 1-dimensional subspaces of the right $\mathbb{F}$-vector space $\mathbb{F}^3$. For each point p, consider the orthogonal projection pr_p of $\mathbb{F}^3$ onto the 1-dimensional subspace p with respect to the standard positive definite Hermitian form $(\ |\ )_+$ on $\mathbb{F}^3$, see (13.12). This defines a 1-1-correspondence between the point set and the set of all idempotent self-adjoint endomorphisms of $\mathbb{F}^3$ having rank 1, or equivalently, trace 1; note that, for idempotent endomorphisms, trace and rank are equal. With respect to the standard basis of $\mathbb{F}^3$, the projection pr_p hence is represented by a Hermitian matrix of the form

$$A = \begin{pmatrix} \xi_1 & x_3 & \bar{x}_2 \\ \bar{x}_3 & \xi_2 & x_1 \\ x_2 & \bar{x}_1 & \xi_3 \end{pmatrix}$$

for suitable $x_\nu \in \mathbb{F}$, $\xi_\nu \in \mathbb{R}$ such that $\xi_1 + \xi_2 + \xi_3 = \mathrm{tr}A = 1$.

By a direct, if somewhat tedious computation, one obtains that A is idempotent if, and only if, the conditions (∗) of (16.1) hold, which say that $(x_\nu; \xi_\nu)_\nu$ is a Veronese vector. Thus, if we identify $\mathbb{F}^3 \times \mathbb{R}^3$ with the space $\mathbb{H}_3\mathbb{F}$ of Hermitian 3×3-matrices as suggested by the way in which the matrix A is written, then the set of all Veronese vectors as introduced in (16.1) corresponds to the set of all real multiples of idempotent self-adjoint matrices of trace 1.

We contend that for each point p the matrix A representing the projection pr_p contains the Veronese coordinates of this point p. Let $p = \mathbb{F}w \le \mathbb{F}^3$ for a suitable $w \in \mathbb{F}^3 \setminus \{0\}$. Then, the orthogonal projection pr_p of $\mathbb{F}^3$ onto p is the map $z \mapsto w \cdot (w\,|\,w)_+^{-1} \cdot (w\,|\,z)_+$. If we write w as $w = (a_1, a_2, a_3)$ with suitable $a_1, a_2, a_3 \in \mathbb{F}$ such that $a_1\bar{a}_1 + a_2\bar{a}_2 + a_3\bar{a}_3 = 1$, then the entries of the matrix of pr_p are obtained as $a_\mu \bar{a}_\nu$, for $\mu, \nu \in \{1, 2, 3\}$. The Veronese vector corresponding

84   1 The classical planes

to this matrix indeed belongs to the point p. For instance, if $w = (x, y, 1)$, that is, if p is the point with affine coordinates (x, y), then the matrix in question is

$$\frac{1}{\|x\|^2 + \|y\|^2 + 1} \begin{pmatrix} \|x\|^2 & x\bar{y} & x \\ y\bar{x} & \|y\|^2 & y \\ \bar{x} & \bar{y} & 1 \end{pmatrix} ;$$

its entries correspond to the Veronese coordinates associated with the affine point (x, y) according to (16.3).

Via our identification $\mathbb{F}^3 \times \mathbb{R}^3 \cong H_3\mathbb{F}$, the inner product β on $\mathbb{F}^3 \times \mathbb{R}^3$, as defined in (16.1), corresponds to the inner product $(A, B) \mapsto \frac{1}{2}(\operatorname{tr} AB + \overline{\operatorname{tr} AB})$ on the space of Hermitian 3×3-matrices. Now, two 1-dimensional subspaces of $\mathbb{F}^3$ are orthogonal if, and only if, the product AB of the orthogonal projections A, B onto these subspaces is 0, and then A and B are orthogonal in $H_3\mathbb{F}$. From this, one may deduce directly that the descriptions of the standard elliptic polarity on $\mathcal{P}_2\mathbb{F}$ in Veronese coordinates on the one hand and in homogeneous coordinates on the other hand agree. Indeed, in both descriptions, the polarity is given by orthogonality with respect to the appropriate inner product. This approach is more conceptual than the one used in (16.7c).

Although there is no 3-dimensional vector space over $\mathbb{O}$, Hermitian 3×3-matrices over $\mathbb{O}$ are meaningful. They form the so-called exceptional Jordan algebra over $\mathbb{R}$ (of dimension 27). This is the starting point for an investigation of the octonion plane (including generalizations) based on the algebraic theory of Jordan algebras, see Jordan [49], Freudenthal [85], Jacobson [59, 60, 61], Springer [60, 62], Springer–Veldkamp [63], Veldkamp [68], Faulkner–Ferrar [84]. The history and a synopsis of the contributions of Freudenthal, Springer, Veldkamp and others may be found in Veldkamp [91]. See also Freudenthal [65].

Another type of homogeneous coordinates taken from $\mathbb{O}^3$ was proposed by Porteous [81] p. 285 and developed further by Aslaksen [91]. In Faulkner [89], a quite different, geometric construction of $\mathcal{P}_2\mathbb{O}$ starting from quaternion projective 3-space $\mathcal{P}_3\mathbb{H}$ is given.

Our suggestion to use the term 'Veronese coordinates' is motivated by the close relationship to the Veronese embeddings of classical projective planes into Euclidean spaces. See Cecil–Ryan [85] Chap. 1 Sect. 9, p. 87 ff for a comprehensive study of the rôle of these embeddings in differential geometry.

The projective octonion plane $\mathcal{P}_2\mathbb{O}$ as a topological plane

16.9 Topologies on $P_2\mathbb{O}$ and $\mathcal{L}_2\mathbb{O}$. By construction, $P_2\mathbb{O}$ is contained in the space $P_{26}\mathbb{R}$ of 1-dimensional subspaces of $V = \mathbb{O}^3 \times \mathbb{R}^3 = \mathbb{R}^{27}$, and $\mathcal{L}_2\mathbb{O}$ is contained in the space $\mathcal{H}_{26}\mathbb{R}$ of hyperplanes of V. According to (14.1), the spaces $P_{26}\mathbb{R}$ and $\mathcal{H}_{26}\mathbb{R} \approx P_{26}^*\mathbb{R}$ have natural topologies; $P_2\mathbb{O}$ and $\mathcal{L}_2\mathbb{O}$ will be endowed with the induced topologies. By (14.5), the spaces $P_{26}\mathbb{R}$ and $\mathcal{H}_{26}\mathbb{R}$ are compact Hausdorff spaces. We show that this carries over to the point space and the line space of the octonion plane.

$P_2\mathbb{O}$ and $\mathcal{L}_2\mathbb{O}$ are compact Hausdorff spaces.

Proof. The topology of the projective space $P_{26}\mathbb{R}$ is the quotient topology determined by the map

$$\vartheta : V \setminus \{0\} = \mathbb{R}^{27} \setminus \{0\} \to P_{26}\mathbb{R} : w \mapsto \mathbb{R}w .$$

We know that $\vartheta^{-1}(P_2\mathbb{O})$ is the set $H \setminus \{0\}$ of nonzero Veronese vectors, see (16.1). From the conditions (16.1(∗)) describing the elements of H it is obvious that H is closed in $\mathbb{R}^{27}$, so that $P_2\mathbb{O}$ is closed in $P_{26}\mathbb{R}$, and therefore is compact. By the following lemma, compactness carries over to $\mathcal{L}_2\mathbb{O}$. □

16.10 Lemma. *The standard elliptic polarity π provides a homeomorphism between $P_2\mathbb{O}$ and $\mathcal{L}_2\mathbb{O}$.*

Proof. We first show that the map $\perp : \mathbb{R}^{27} \setminus \{0\} \to \mathcal{H}_{26}\mathbb{R} : w \mapsto w^\perp$ is continuous and open, where $w^\perp$ is the orthogonal space of w with respect to the inner product β defined in (16.1). Consider the map $\tilde{\pi} : \mathbb{R}^{27} \setminus \{0\} \to \mathbb{R}^{27} \setminus \{0\}$ which to $w \in \mathbb{R}^{27} \setminus \{0\}$ associates the coefficient vector of the linear form $z \mapsto \beta(w \mid z)$ with respect to the standard basis. Obviously, $\tilde{\pi}$ is an $\mathbb{R}$-linear isomorphism and hence a homeomorphism. Moreover, $\perp$ is obtained by composition of $\tilde{\pi}$ with the continuous and open map $\mathrm{Ker} : \mathbb{R}^{27} \setminus \{0\} \to \mathcal{H}_{26}\mathbb{R}$ considered in (14.1). In particular, $\perp$ is an identifying map, cf. (92.20b). The standard elliptic polarity is the restriction of the bijective map $P_{26}\mathbb{R} \to \mathcal{H}_{26}\mathbb{R} : p \mapsto p^\perp$. This bijection is a homeomorphism by the general principle (92.20a) for quotient topologies, being the bottom map of the following commutative diagram

$$\begin{array}{ccc} \mathbb{R}^{27} \setminus \{0\} & \xrightarrow{\mathrm{id}} & \mathbb{R}^{27} \setminus \{0\} \\ \vartheta \downarrow & & \downarrow \perp \\ P_{26}\mathbb{R} & \longrightarrow & \mathcal{H}_{26}\mathbb{R} \end{array},$$

in which ϑ and $\perp$ are identifying maps. □

The following results are analogous to (14.4 and 3).

16.11 Theorem. *$P_2\mathbb{O}$ is a topological projective plane.*

Note. One even obtains a differentiable projective plane, see the note after (16.14) and (75.4), where also an alternative for the proof of the present theorem is mentioned.

Proof of (16.11). We argue as in the second proof of theorem (14.4), see (14.6). By compactness and by the general criterion (43.1), it suffices to prove that the flag space $\mathbf{F} := \{ (p, L) \in P_2\mathbb{O} \times \mathcal{L}_2\mathbb{O} \mid p \text{ incident with } L \}$ is closed in $P_2\mathbb{O} \times \mathcal{L}_2\mathbb{O} \subseteq P_{26}\mathbb{R} \times \mathcal{H}_{26}\mathbb{R}$. The latter product has the quotient topology determined by the map

$$\vartheta \times \bot : (\mathbb{R}^{27} \setminus \{0\}) \times (\mathbb{R}^{27} \setminus \{0\}) \to P_{26}\mathbb{R} \times \mathcal{H}_{26}\mathbb{R} : (w, z) \mapsto (\mathbb{R}w, z^\bot) \,;$$

indeed, by (14.1) and by the first part of the proof of (16.10), both ϑ and $\bot$ are continuous and open maps, so that $\vartheta \times \bot$ is continuous and open, too. Therefore, it suffices to verify that the inverse image $(\vartheta \times \bot)^{-1}(\mathbf{F})$ is closed. Now, by the construction of the octonion plane from the set H of Veronese vectors (16.1), it is clear that $(\vartheta \times \bot)^{-1}(\mathbf{F}) = \{ (w, z) \in H \times H \mid \beta(w \mid z) = 0, w \neq 0 \neq z \}$. This set is closed in $(\mathbb{R}^{27} \setminus \{0\})^2$, since $H \setminus \{0\}$ is closed in $\mathbb{R}^{27} \setminus \{0\}$. □

16.12 Lemma. *The set $P_2\mathbb{O} \setminus W$ of points not incident with the line $W = (0, 0, 0; 0, 0, 1)^\bot = \mathbb{O}^3 \times \mathbb{R}^2 \times \{0\}$ is open in $P_2\mathbb{O}$ and homeomorphic to $\mathbb{O} \times \mathbb{O}$. In fact, the map*

$$\psi : \mathbb{O} \times \mathbb{O} \to P_2\mathbb{O} \setminus W : (x, y) \mapsto \mathbb{R}(x, \bar{y}, y\bar{x}; \|y\|^2, \|x\|^2, 1)$$

induced by the isomorphism $\overline{\mathcal{A}_2\mathbb{O}} \cong \mathcal{P}_2\mathbb{O}$ of (16.3) is a homeomorphism.

Proof. A line L of $\mathcal{P}_2\mathbb{O}$ is given by a certain hyperplane of $V = \mathbb{R}^{27}$. According to (14.3), $P_{26}\mathbb{R} \setminus L$ is open in $P_{26}\mathbb{R}$, so that $P_2\mathbb{O} \setminus L$ is open in $P_2\mathbb{O}$. (This is equivalent to saying that the point row of L is closed in $P_2\mathbb{O}$; incidentally, this is a general fact in topological projective planes, see (41.2b).)

The map ψ obviously is continuous. For the continuity of the inverse, we consider the continuous map

(1) $$V \setminus W \to \mathbb{O} \times \mathbb{O} : (x_\nu; \xi_\nu)_\nu \mapsto (x_1 \xi_3^{-1}, \overline{x_2 \xi_3^{-1}}) \,;$$

note that $(x_\nu; \xi_\nu)_\nu \notin W$ if, and only if $\xi_3 \neq 0$. The open subset $P_{26}\mathbb{R} \setminus W$ of $P_{26}\mathbb{R}$ has the quotient topology determined by the canonical map $\vartheta|_{V \setminus W} : V \setminus W \to P_{26}\mathbb{R} \setminus W$; and the map (1) factors to yield the well-defined map $P_{26}\mathbb{R} \setminus W \to \mathbb{O} \times \mathbb{O} : \mathbb{R}(x_\nu; \xi_\nu)_\nu \mapsto (x_1 \xi_3^{-1}, \overline{x_2 \xi_3^{-1}})$. By the universal property of the quotient topology (92.19), this map is continuous. Its restriction to $P_2\mathbb{O} \setminus W$ is ψ^{-1}. □

16.13 Proposition. *Every line of $\mathcal{P}_2\mathbb{O}$, when considered as a point row in $P_2\mathbb{O}$, is homeomorphic to the one-point compactification $\mathbb{O} \cup \{\infty\}$ of $\mathbb{O} \cong \mathbb{R}^8$, i.e., to the*

16 The projective octonion plane $\mathcal{P}_2\mathbb{O}$

8-sphere $\mathbb{S}_8$. *Specifically, the map*

$$s \mapsto \mathbb{R}(0, 0, s; \|s\|^2, 1, 0) \mathrel{\widehat{=}} (s) \quad \text{for } s \in \mathbb{O}, \qquad \infty \mapsto \mathbb{R}(0, 0, 0; 1, 0, 0) \mathrel{\widehat{=}} (\infty)$$

is a homeomorphism of $\mathbb{O} \cup \{\infty\}$ *onto the point row of the line* $W \mathrel{\widehat{=}} [\infty]$.

Proof. The point rows of every two lines are homeomorphic to each other by central projection in the topological projective plane $\mathcal{P}_2\mathbb{O}$, compare the proof of (14.7) or the general result (41.2). Thus, it suffices to prove the statement about W.

In (16.2), W was considered as the quadric of a quadratic form on the subspace $V_3 = \{0\} \times \{0\} \times \mathbb{O} \times \mathbb{R}^2 \times \{0\}$ of $V = \mathbb{O}^3 \times \mathbb{R}^3$. The projective spaces PV_3 and PV corresponding to these $\mathbb{R}$-linear spaces carry the standard topologies, defined as the quotient topologies with respect to the canonical maps $V_3 \setminus \{0\} \to PV_3$ and $V \setminus \{0\} \to PV$. Since $V_3 \setminus \{0\}$ is closed in $V \setminus \{0\}$, it is immediate that the standard topology on PV_3 coincides with the topology induced from PV. In particular, the topology on W as a point row in $P_2\mathbb{O} \subseteq PV$ coincides with the topology induced by the projective subspace PV_3, in which W is embedded as a quadric of a particular kind. For such a quadric, the homeomorphism of our assertion has been established in (15.1b). □

16.14 Corollary. $P_2\mathbb{O}$ *and* $\mathcal{L}_2\mathbb{O}$ *are compact, connected, 16-dimensional manifolds.*

Proof. Compactness has been proved in (16.9). As a general fact in topological projective planes, the set of points and the set of lines are manifolds whenever the point rows of lines have this property, see (52.1). In the present context, the assertion can be verified directly as follows, using the triality collineation τ described in (16.6).

That collineation is induced by the linear transformation $\tilde{\tau}$ of $V = \mathbb{R}^{27}$; it therefore is a homeomorphism of $P_2\mathbb{O} \subseteq PV = P_{26}\mathbb{R}$. (Note that, by (14.2), $\tilde{\tau}$ induces a homeomorphism $[\tilde{\tau}]$ of PV, which in turn induces τ by restriction to $P_2\mathbb{O}$.) According to (16.12), the subsets $P_2\mathbb{O} \setminus W$, $P_2\mathbb{O} \setminus W^\tau$, and $P_2\mathbb{O} \setminus W^{(\tau^2)}$ are homeomorphic to $\mathbb{R}^{16}$. They cover $P_2\mathbb{O}$, since the lines W, $W^\tau = [0] = Y$, and $W^{(\tau^2)} = [0, 0] = X$, see (16.6), do not have a point in common. Thus, $P_2\mathbb{O}$ is a manifold of dimension 16. It is connected, because the connected subsets $P_2\mathbb{O} \setminus W$, $P_2\mathbb{O} \setminus X$, and $P_2\mathbb{O} \setminus Y$ have non-trivial intersection. The line space $\mathcal{L}_2\mathbb{O}$ is homeomorphic to $P_2\mathbb{O}$, see (16.10). □

Note. It can be shown along the lines of the preceding proof that the manifolds $P_2\mathbb{O}$ and $\mathcal{L}_2\mathbb{O}$ moreover carry natural differentiable structures, in which the geometric operations are differentiable, so that the structure of a differentiable projective plane results. This was worked out by Otte [93]; for some indications, cf. (75.4).

16.15 The octonion Hopf map. Within $\mathcal{P}_2\mathbb{O}$, one can now describe the corresponding Hopf map analogously to the description in (14.11) of the Hopf maps over $\mathbb{R}$, $\mathbb{C}$, and $\mathbb{H}$ in the respective projective planes. We consider the unit sphere in the point set $\mathbb{O} \times \mathbb{O} = \mathbb{R}^{16}$ of the affine octonion plane $\mathcal{A}_2\mathbb{O}$,

$$\mathbb{S}_{15} = \{ (x, y) \in \mathbb{O} \times \mathbb{O} \mid \|x\|^2 + \|y\|^2 = 1 \},$$

and we identify $\overline{\mathcal{A}_2\mathbb{O}}$ and $\mathcal{P}_2\mathbb{O}$ according to (16.3 and 4). By central projection of $\mathbb{S}_{15}$ from the origin $(0, 0)$ onto the line $[\infty] \cong W$ at infinity in $\overline{\mathcal{A}_2\mathbb{O}}$, we obtain the Hopf map

$$\mathbb{S}_{15} \to W \approx \mathbb{S}_8 : (x, y) \mapsto \big((0, 0) \vee (x, y)\big) \wedge W = \begin{cases} (yx^{-1}) & \text{for } x \neq 0 \\ (\infty) & \text{for } x = 0. \end{cases}$$

The continuity of this map is a consequence of the continuity of the geometric operations in the topological projective plane $\mathcal{P}_2\mathbb{O}$.

A fiber of the Hopf map is the intersection of $\mathbb{S}_{15}$ with a line through the origin of the affine plane $\mathbb{O} \times \mathbb{O}$. Every such line is an 8-dimensional $\mathbb{R}$-linear subspace of $\mathbb{O} \times \mathbb{O} = \mathbb{R}^{16}$, and its intersection with $\mathbb{S}_{15}$ is the unit sphere of this subspace. Thus, the fibers of the Hopf map are 7-spheres.

As in (14.11), we note that the Hopf map is a locally trivial fiber bundle, and, in fact, the sphere bundle of a vector bundle which may be constructed within the topological projective plane $\mathcal{P}_2\mathbb{O}$, see (51.23).

Warning. In contrast to the situation over $\mathbb{R}$, $\mathbb{C}$, or $\mathbb{H}$, as described in (14.11), the fiber of a point $(x, y) \in \mathbb{S}_{15}$ under the octonion Hopf map does *not* consist of the points $(x\lambda, y\lambda)$, where $\lambda \in \mathbb{O}$, $\|\lambda\|^2 = 1$. Indeed, these points may belong to different fibers, because in general, for lack of associativity, xy^{-1} is different from $(x\lambda)(\lambda^{-1}y^{-1}) = (x\lambda)(y\lambda)^{-1}$.

17 The collineation group of $\mathcal{P}_2\mathbb{O}$

In this section, it will be shown that the group Σ of all collineations of $\mathcal{P}_2\mathbb{O}$ is generated by elations, and that every collineation is induced by a linear transformation of the vector space $V = \mathbb{O}^3 \times \mathbb{R}^3$; recall that $P_2\mathbb{O}$ is embedded via Veronese coordinates in the projective space PV of this vector space. The first result implies that Σ is a simple group, and the second result (which might be considered as an analogue of the fundamental theorem of projective geometry) shows that Σ is a Lie group, whose dimension will be computed to be 78. The identification of Σ among the simple Lie groups of dimension 78 will be obtained in the next section (18.19), when more is known about subgroups of Σ.

17 The collineation group of $\mathcal{P}_2\mathbb{O}$

By means of elations one may easily show that Σ is transitive on the set of non-degenerate triangles, and our previous study in (12.14 ff) of the stabilizer ∇ of a triangle is helpful for the discussion of the whole group Σ. In the second part of the present section, we shall continue the study of ∇ using Veronese coordinates. Important subgroups of ∇ will be identified as the universal covering groups $\mathrm{Spin}_8\mathbb{R}$ and $\mathrm{Spin}_7\mathbb{R}$ of $\mathrm{SO}_8\mathbb{R}$ and $\mathrm{SO}_7\mathbb{R}$; and these groups contain the stabilizer of a non-degenerate quadrangle, which is isomorphic to $\mathrm{Aut}\,\mathbb{O} \cong G_2$. In this way, we obtain concrete geometric descriptions of these groups and of some related homogeneous spaces.

The material presented in this and the next section is contained in the fundamental papers by Tits [53, 54]. There are differences of approach and presentation, though. For example, Veronese coordinates are an important technical tool here; the geometrical properties which they embody, see (16.2), substitute the arguments from conformal geometry used by Tits. Furthermore, we refrained from using the fact that Σ is a Lie group in an essential way, whereas Tits does so freely. Concerning the pioneer work of Freudenthal and further developments of this subject, we refer to the notes in (16.8) and (17.17).

To a great extent, the arguments of this section can be adapted for arbitrary non-desarguesian Moufang planes. This was worked out by Kreh [92] on the basis of a preliminary version of this chapter; we owe him some improvements.

17.1 Conventions and notation. In this section, we make no distinction between the projective completion $\overline{\mathcal{A}_2\mathbb{O}}$ of the affine octonion plane and the projective plane $\mathcal{P}_2\mathbb{O}$ as described in (16.1). We constantly use the identification $\overline{\mathcal{A}_2\mathbb{O}} \cong \mathcal{P}_2\mathbb{O}$ made in (16.3 and 4) and freely employ both affine and Veronese coordinates.

By Σ, we shall always denote the group of all collineations of $\mathcal{P}_2\mathbb{O}$. For temporary use, we introduce the subgroup E of Σ generated by the elations (the axial collineations with centers on the axes), cf. (12.4 and 5). This subgroup is frequently considered in projective geometry, often under the name of *little projective group*. For the octonion plane, it will soon turn out (17.7) that E coincides with Σ.

Generally, E is a normal subgroup of Σ, as is clear from the formula (12.4b) on conjugate axial collineations. By definition, E contains the group of all elations with axis $W = [\infty]$, i.e. the group $\mathsf{T} = \{\,\tau_{a,b} \mid a, b \in \mathbb{O}\,\}$ of all translations described in (12.5a).

In the following proposition, the triality collineation τ of (16.6), which moves the line $W = [\infty]$, will be employed to derive transitivity properties of E.

17.2 Proposition. E *is generated by* T *and its conjugates* $\mathsf{T}^\tau = \tau^{-1}\mathsf{T}\tau$ *and* $\mathsf{T}^{(\tau^2)}$, *and* E *acts transitively on the set of non-degenerate triangles of* $\mathcal{P}_2\mathbb{O}$.

Proof. For the moment, the subgroup of E generated by T, T^τ, and $\mathsf{T}^{(\tau^2)}$ will be denoted by Ξ. We first study transitivity properties of Ξ and then show that $\Xi = \mathsf{E}$.

The translation group T fixes every point of W and is transitive on $\mathcal{P}_2\mathbb{O} \setminus W$, cf. (12.5a). Analogous transitivity properties follow by conjugation with τ and τ^2. As $W^\tau = [0] = Y$ and $W^{(\tau^2)} = [0, 0] = X$, we obtain that $\mathsf{T}^\tau \leq \Xi_{(\infty)}$ is transitive on $\mathcal{P}_2\mathbb{O} \setminus Y$ and that $\mathsf{T}^{(\tau^2)}$ is transitive on $\mathcal{P}_2\mathbb{O} \setminus X$. Consequently, Ξ is transitive on $\mathcal{P}_2\mathbb{O}$, and $\Xi_{(\infty)}$ is transitive on $\mathcal{P}_2\mathbb{O} \setminus \{(\infty)\}$, so that Ξ is doubly transitive on $\mathcal{P}_2\mathbb{O}$. Thus, for transitivity of Ξ on the set of non-degenerate triangles, it is sufficient that the stabilizer of two points, (0) and (∞), say, is transitive on the set of all points which are not collinear with the two given points. But this is clear from transitivity of $\mathsf{T} \leq \Xi_{[W]}$ on $\mathcal{P}_2\mathbb{O} \setminus W$.

In particular, Ξ is transitive on the set of lines. Now, for $\varphi \in \Xi$, the subgroup T^φ is the group of all elations with axis W^φ, by the conjugation formula (12.4b). Hence, Ξ contains all elations with arbitrary axes, so that $\Xi = \mathsf{E}$ by definition. □

The last argument also shows that, for every line L, the group of elations with axis L is transitive on $\mathcal{P}_2\mathbb{O} \setminus L$. In current terminology, see (24.8), this is expressed as follows.

17.3 Corollary. *$\mathcal{P}_2\mathbb{O}$ is a Moufang plane.* □

17.4 Theorem. E *is a simple group.*

Note. The proof given below is valid for arbitrary Moufang planes. The structure of the argument goes back to Iwasawa's proof of the simplicity of the special projective groups, cf. Iwasawa [51b] and Huppert [67] Chap. II 6.12 and 6.13, p. 182–183. The transfer to Moufang planes appears to be due to Dembowski, see Lenz [65] Chap. IX Satz 2.3, p. 254, H. Lüneburg [76] Lemma 8, p. 387.

Proof of (17.4). Let $\mathsf{N} \neq \{\mathrm{id}\}$ be a normal subgroup of E; we have to show that $\mathsf{N} = \mathsf{E}$. It follows immediately from (17.2) that E is doubly transitive on the set $\mathcal{L}_2\mathbb{O}$ of lines. Therefore, the normal subgroup N is transitive on $\mathcal{L}_2\mathbb{O}$; this is a standard argument, see (91.4). In particular, every elation is conjugate by an element of N to an element of the group T of elations with axis W, so that the subgroup $\mathsf{NTN} = \mathsf{TN}$ contains all elations and hence coincides with E by definition. Consequently, $\mathsf{E}/\mathsf{N} = \mathsf{TN}/\mathsf{N} \cong \mathsf{T}/(\mathsf{T} \cap \mathsf{N})$ is commutative, as T is commutative (12.5a). In other words, N contains all commutators of E.

Now we show that each elation is a commutator in E (which then proves that $\mathsf{N} = \mathsf{E}$). As an immediate consequence of (17.2), we know that E is flag transitive (i.e., transitive on the set of incident point–line pairs). Hence, every elation is conjugate in E to an elation with axis W and center (∞), that is, to a translation of the form $\tau_{0,b} : (x, y) \mapsto (x, y + b)$, see (12.5a). Now one easily verifies that $\tau_{0,b}$ is the commutator $\sigma_1^{-1}\tau_{b,0}\sigma_1\tau_{b,0}^{-1}$ obtained from the translation $\tau_{b,0} : (x, y) \mapsto (x+b, y)$ and from the shear $\sigma_1 : (x, y) \mapsto (x, y+x)$, which is an elation with axis Y, see (12.5b). □

17.5 Lemma. *For $a \in \mathbb{O} \setminus \{0\}$, the collineations*

$$\gamma_a : (x, y) \mapsto (a^{-1}x, ay), \quad (s) \mapsto (asa)$$
$$\gamma_a' : (x, y) \mapsto (axa, ya), \quad (s) \mapsto (sa^{-1})$$

of (12.14) *belong to* E.

Proof. In (12.14), the collineation γ_a is constructed as a product of elations, so that $\gamma_a \in$ E. We now verify that $\gamma_a' = \tau\gamma_a\tau^{-1}$, where τ is the triality collineation; this will prove that $\gamma_a' \in$ E, since E is a normal subgroup of Σ.

The collineations $\tau\gamma_a\tau^{-1}$ and γ_a' fix the points $(0, 0)$, (0), and (∞). Such collineations are determined by their actions on the affine coordinate axes, see (12.7). Hence, it suffices to establish that $\tau\gamma_a\tau^{-1}$ and γ_a' coincide there. From the explicit description of τ in affine coordinates (16.6), we obtain, for $x \in \mathbb{O}$ and $y \in \mathbb{O} \setminus \{0\}$, that indeed $(x, 0)^{\tau\gamma_a\tau^{-1}} = (x)^{\gamma_a\tau^{-1}} = (axa)^{\tau^{-1}} = (axa, 0)$, and $(0, y)^{\tau\gamma_a\tau^{-1}} = (y^{-1}, 0)^{\gamma_a\tau^{-1}} = (a^{-1}y^{-1}, 0)^{\tau^{-1}} = (0, ya)$. (A direct proof that γ_a' and $\tau\gamma_a\tau^{-1}$ coincide everywhere would need more algebraic effort and the use of the Moufang identities (12.15).) □

17.6 Theorem. E *is transitive on the set of non-degenerate quadrangles.*

Proof. Since, by (17.2), it is already known that E is transitive on the set of non-degenerate triangles, it suffices to prove that the stabilizer in E of a specific triangle, e.g., of the triangle $(0, 0)$, (0), (∞), is transitive on the set of points not incident with the sides of the triangle. This transitivity property was already established in (12.16), using the collineations of (12.14) and (17.5), which belong to E. □

17.7 Theorem. $\Sigma =$ E; *in words: Every collineation of $\mathcal{P}_2\mathbb{O}$ is a product of elations.*

Proof. The transitivity of E on the set of non-degenerate quadrangles implies that Σ is the product of E by the stabilizer $\Sigma_{o,e,u,v}$ of the quadrangle consisting of the points $o = (0, 0)$, $u = (0)$, $v = (\infty)$, and $e = (1, 1)$, by the Frattini argument (91.2a). Hence, it suffices to show that $\Sigma_{o,u,v,e} \subseteq$ E.

According to (12.8) and (11.32), the stabilizer $\Sigma_{o,u,v,e} \cong \operatorname{Aut} \mathbb{O}$ is simple. Therefore, our assertion follows if we establish that the normal subgroup $E \cap \Sigma_{o,u,v,e}$ of $\Sigma_{o,u,v,e}$ is non-trivial. Now, for $a \in \mathbb{O} \setminus \{0\}$, the collineation $\gamma_a \gamma_{a^2}' \in$ E (17.5) can be easily computed using biassociativity; (x, y) is mapped to (axa^2, aya^2), and

(s) to (asa^{-1}). If $a^3 = 1 \neq a$, this collineation fixes o, u, v, and e, but it is not the identity, since a does not belong to the center $\mathbb{R}$ of $\mathbb{O}$. □

Note. For arbitrary Moufang planes, the preceding theorems are *not* generally valid. It is true, however, that the group Σ of *all* collineations is transitive on the set of non-degenerate quadrangles, cf. Pickert [75] Chap. 7 Theorem 14, p. 195 and Ciftci–Kaya–Ferrar [88]. The arguments of our proof of (17.6), together with (12.16), are similar to those of the latter paper. Pickert [59], Salzmann [59c], and Springer [62] Prop. 20, p. 466 have determined precisely those Moufang planes for which (17.6) is true. In all other Moufang planes, Σ and E differ.

For the second part of the proof of the following theorem, we are indebted to Kreh [92]. Recall that the point set $\mathrm{P}_2\mathbb{O}$ is the set of 1-dimensional $\mathbb{R}$-linear subspaces of $V = \mathbb{O}^3 \times \mathbb{R}^3$ which are generated by Veronese vectors, see (16.1).

17.8 Theorem.

(a) *Every collineation of $\mathcal{P}_2\mathbb{O}$ is induced by a linear transformation of $\mathbb{O}^3 \times \mathbb{R}^3$. More precisely, the collineations are exactly the maps having the form*

$$[\tilde{\varphi}] : \mathrm{P}_2\mathbb{O} \to \mathrm{P}_2\mathbb{O} : \mathbb{R} \cdot w \mapsto \mathbb{R} \cdot \tilde{\varphi}(w),$$

where $\tilde{\varphi}$ is a linear transformation of $V = \mathbb{O}^3 \times \mathbb{R}^3$ mapping the set $H \subseteq V$ of Veronese vectors onto itself.

(b) *In (a), one may assume in addition that $\det \tilde{\varphi} = 1$. With the subgroup*

$$\widetilde{\Sigma} = \{ \tilde{\varphi} \in \mathrm{SL}(V) \mid \tilde{\varphi}(H) = H \},$$

the map $\widetilde{\Sigma} \to \Sigma : \tilde{\varphi} \mapsto [\tilde{\varphi}]$ is an isomorphism.

Proof. 1) First, we show that every collineation is induced by a linear transformation as described in (a). It is enough to consider the translations and the triality collineation τ, because $\Sigma = \mathsf{E}$ is generated by these collineations (17.2). The triality collineation, according to its very definition (16.6), is induced by the linear transformation $\tilde{\tau}$. We shall now verify that the translation $\tau_{a,b} : (x, y) \mapsto (x+a, y+b)$ is induced by the linear transformation $\tilde{\tau}_{a,b}$ of $\mathbb{O}^3 \times \mathbb{R}^3$ which maps $(x_\nu; \xi_\nu)_\nu$ to

$$(x_1 + \xi_3 a, x_2 + \xi_3 \bar{b}, x_3 + b\overline{x_1} + \overline{x_2}\,\bar{a} + \xi_3 b\bar{a};$$
$$\xi_1 + \langle \overline{x_2} \mid b \rangle + \xi_3 \|b\|^2, \xi_2 + \langle x_1 \mid a \rangle + \xi_3 \|a\|^2, \xi_3).$$

Indeed, $\tilde{\tau}_{a,b}$ maps the 1-dimensional subspace $\mathbb{R}(x, \bar{y}, y\bar{x}; \|y\|^2, \|x\|^2, 1) \in \mathrm{P}_2\mathbb{O}$, which by (16.3) corresponds with the affine point (x, y), to

$$\mathbb{R}(x+a, \bar{y}+\bar{b}, y\bar{x}+b\bar{x}+y\bar{a}+b\bar{a}; \|y\|^2+\langle y \mid b \rangle+\|b\|^2, \|x\|^2+\langle x \mid a \rangle+\|a\|^2, 1)$$
$$= \mathbb{R}(x+a, \overline{y+b}, (y+b)\overline{(x+a)}; \|y+b\|^2, \|x+a\|^2, 1),$$

17 The collineation group of $\mathcal{P}_2\mathbb{O}$

which in affine coordinates is the point $(x + a, y + b)$.

The subspaces $\mathbb{R}(0, 0, s; \|s\|^2, 1, 0)$ and $\mathbb{R}(0, 0, 0; 1, 0, 0)$ corresponding to the points at infinity (s) and (∞) are clearly fixed under $\tilde{\tau}_{a,b}$.

That $\tilde{\tau}$ maps H onto itself is immediate from its definition (16.6). For $\tilde{\tau}_{a,b}$ the same is true just because the induced collineation $\tau_{a,b}$ is a bijection of $P_2\mathbb{O}$.

2) Conversely, we have to show that a linear transformation $\tilde{\varphi}$ of $V = \mathbb{O}^3 \times \mathbb{R}^3$ satisfying $\tilde{\varphi}(H) = H$ induces a collineation. For a given line L of $\mathcal{P}_2\mathbb{O}$, we must find a line L' such that $[\tilde{\varphi}]$ maps the points of L into L'. We choose two distinct points p, q on L. By the transitivity properties (17.2) and by step 1), there are linear transformations $\tilde{\varrho}, \tilde{\sigma}$ of V satisfying $\tilde{\varrho}(H) = H = \tilde{\sigma}(H)$ and inducing collineations ϱ and σ such that

$$\tilde{\varrho}(p) = (0) = \mathbb{R}(0, 0, 0; 0, 1, 0), \quad \tilde{\varrho}(q) = (\infty) = \mathbb{R}(0, 0, 0; 1, 0, 0),$$
$$\tilde{\sigma}(\tilde{\varphi}(p)) = (0), \quad \tilde{\sigma}(\tilde{\varphi}(q)) = (\infty).$$

Then, the transformation

$$\tilde{\psi} = \tilde{\sigma} \circ \tilde{\varphi} \circ \tilde{\varrho}^{-1}$$

satisfies $\tilde{\psi}(H) = H$ as well, and we have $\tilde{\psi}((0)) = (0)$ and $\tilde{\psi}((\infty)) = (\infty)$. Moreover, the collineation ϱ maps the points of the line $L = p \vee q$ into the line W at infinity joining (0) and (∞), and the collineation σ^{-1} maps the points of the line W into the line $L' := \tilde{\varphi}(p) \vee \tilde{\varphi}(q)$. Hence, its remains to show that $\tilde{\psi}$ maps points of W into W, for then it is clear that $\tilde{\varphi} = \tilde{\sigma}^{-1} \circ \tilde{\psi} \circ \tilde{\varrho}$ maps the points of L into L'.

As the 1-dimensional subspaces (0) and (∞) are invariant under $\tilde{\psi}$, we have

$$\tilde{\psi}(0, 0, 0; 1, 0, 0) = (0, 0, 0; \alpha_1, 0, 0) \quad \text{and} \quad \tilde{\psi}(0, 0, 0; 0, 1, 0) = (0, 0, 0; 0, \alpha_2, 0)$$

for some $\alpha_1, \alpha_2 \in \mathbb{R} \setminus \{0\}$. The points of $W \setminus \{(\infty)\}$ have the form $(s) \triangleq \mathbb{R}(0, 0, s; \|s\|^2, 1, 0)$. For fixed $s \in \mathbb{O}$, let

$$\tilde{\psi}(0, 0, s; \|s\|^2, 1, 0) = (x_1, x_2, x_3; \xi_1, \xi_2, \xi_3) \in H.$$

For arbitrary $\gamma \in \mathbb{R} \setminus \{0\}$, the vector $z = (0, 0, s; \gamma\|s\|^2, \gamma^{-1}, 0)$ also belongs to H, since it represents the point $\mathbb{R}(0, 0, \gamma s; \gamma^2\|s\|^2, 1, 0) \triangleq (\gamma s)$. This vector can be written as a linear combination

$$z = (0, 0, s; \|s\|^2, 1, 0) + (\gamma - 1)\|s\|^2(0, 0, 0; 1, 0, 0) + (\gamma^{-1} - 1)(0, 0, 0; 0, 1, 0).$$

Its image under $\tilde{\psi}$ is

$$(x_1, x_2, x_3; \xi_1, \xi_2, \xi_3) + (\gamma - 1)\|s\|^2(0, 0, 0; \alpha_1, 0, 0) + (\gamma^{-1} - 1)(0, 0, 0; 0, \alpha_2, 0)$$
$$= (x_1, x_2, x_3; \xi_1 + (\gamma - 1)\|s\|^2\alpha_1, \xi_2 + (\gamma^{-1} - 1)\alpha_2, \xi_3).$$

By assumption, this vector belongs to H, as well, so that $\left(\xi_2 + (\gamma^{-1} - 1)\alpha_2\right)\xi_3 = \|x_1\|^2$. As this holds for every $\gamma \neq 0$, we obtain that $\xi_3 = 0$; this means that the point spanned by $\tilde{\psi}(0,0,s;\|s\|^2,1,0)$ is contained in the line $(0,0,0;0,0,1)^\perp = W$. Thus, (a) is proved.

3) For (b), first note that if a collineation is induced by a linear transformation $\tilde{\varphi}$, then the same collineation is induced by $(\sqrt[27]{\det \tilde{\varphi}})^{-1} \cdot \tilde{\varphi}$, which has determinant 1. Thus, $\widetilde{\Sigma} \to \Sigma$ is surjective. The proof is finished if we show that a linear transformation $\tilde{\varphi}$ of determinant 1 which induces the identity on $P_2\mathbb{O}$ is the identity. This is immediate from the following statement.

17.9 Lemma. *The linear transformations of $V = \mathbb{O}^3 \times \mathbb{R}^3$ which induce the identity on $P_2\mathbb{O}$ are just the scalar homotheties $r \cdot \mathrm{id}$ for $r \in \mathbb{R} \setminus \{0\}$.*

Proof. By (16.2), the point row of the line W is a quadric belonging to the quadratic form q of index 1 on the subspace $V_3 = \{0\} \times \{0\} \times \mathbb{O} \times \mathbb{R}^2 \times \{0\}$ of $V = \mathbb{O}^3 \times \mathbb{R}^3$. By (15.4), W generates V_3; hence, a linear transformation $\tilde{\varphi}$ of V inducing the identity on $P_2\mathbb{O}$ leaves V_3 invariant and induces a scalar homothety on V_3. The same is true for the subspaces $V_1 = \mathbb{O} \times \{0\} \times \{0\} \times \{0\} \times \mathbb{R}^2$ and $V_2 = \{0\} \times \mathbb{O} \times \{0\} \times \mathbb{R} \times \{0\} \times \mathbb{R}$, as one may see by applying the same argument to $\tilde{\tau}^{-1} \circ \tilde{\varphi} \circ \tilde{\tau}$ and $\tilde{\tau}^{-2} \circ \tilde{\varphi} \circ \tilde{\tau}^2$ and using $V_2 = \tilde{\tau}(V_3)$, $V_1 = \tilde{\tau}^2(V_3)$; here again, $\tilde{\tau}$ is the linear transformation inducing the triality collineation τ, cf. (16.6). Thus, $\tilde{\varphi}|_{V_\nu} = r_\nu \cdot \mathrm{id}$ for every $\nu \in \{1,2,3\}$ with suitable $r_\nu \in \mathbb{R} \setminus \{0\}$. Finally, $r_1 = r_2 = r_3$, because the subspaces V_ν have pairwise non-trivial intersection, and the assertion follows since these subspaces generate V. □

We now translate the information which we have obtained about Σ into the language of Lie groups. For our further investigations of subgroups of Σ the result will *not* be used systematically; we prefer to keep the discussion on a purely incidence geometric level wherever this is adequate.

17.10 Theorem. *The group Σ of all collineations of $\mathcal{P}_2\mathbb{O}$ is a simple (center free) Lie group of dimension 78.*

Note. Later, we shall show that, in fact, Σ is the exceptional real simple Lie group $E_6(-26)$, see (18.19).

Proof of (17.10). By (17.8), Σ is naturally isomorphic to a group $\widetilde{\Sigma}$ of linear transformations of $V = \mathbb{O}^3 \times \mathbb{R}^3$. We endow Σ with the topology obtained from this isomorphism. The action of Σ on $P_2\mathbb{O}$ is continuous, see (14.2).

By its description in (17.8), the group $\widetilde{\Sigma}$ is (topologically) closed in $\mathrm{GL}(V)$, because the set H of Veronese vectors is closed in $\mathbb{O}^3 \times \mathbb{R}^3$, as is immediate from its definition (16.1). Hence, $\Sigma \cong \widetilde{\Sigma}$ is a Lie group, cf. (94.3). The simplicity of $\Sigma = \mathrm{E}$, see (17.7), has been proved in (17.4).

17 The collineation group of $\mathcal{P}_2\mathbb{O}$

In order to compute the dimension of Σ, we use the fact (17.6) that Σ is transitive on the set of non-degenerate quadrangles. For the quadrangle o, u, v, e, the following orbits under stabilizers in Σ result: $v^{\Sigma_u} = P_2\mathbb{O} \setminus \{u\}$, $o^{\Sigma_{u,v}} = P_2\mathbb{O} \setminus uv$, $e^{\Sigma_{o,u,v}} = P_2\mathbb{O} \setminus (uv \cup ou \cup ov)$. As the point rows uv, ou, ov are closed in $P_2\mathbb{O}$ by (16.10), these orbits are open in the 16-dimensional manifold $P_2\mathbb{O}$, see (16.12). Applying the dimension formula (96.10) four times, we obtain

$$\dim \Sigma = 4 \cdot 16 + \dim \Sigma_{o,u,v,e} .$$

Now, in affine coordinates, the elements of $\Sigma_{o,u,v,e}$ are the maps $(x, y) \mapsto (x^\alpha, y^\alpha)$, $(s) \mapsto (s^\alpha)$ for $\alpha \in \mathrm{Aut}\,\mathbb{O}$, see (12.8). It is straightforward that these collineations are induced by the linear transformations $(x_\nu; \xi_\nu)_\nu \mapsto (x_\nu^\alpha; \xi_\nu)_\nu$ of $\mathbb{O}^3 \times \mathbb{R}^3$; in verifying this via the correspondence (16.3) between affine and Veronese coordinates, one uses the fact that $\alpha \in \mathrm{Aut}\,\mathbb{O}$ commutes with conjugation and is norm-preserving (11.28). Hence, as a topological group, $\Sigma_{o,u,v,e}$ is isomorphic to $\mathrm{Aut}\,\mathbb{O}$; in particular, it is a Lie group of dimension 14, see (11.33). Thus, finally, $\dim \Sigma = 64 + 14 = 78$. □

The stabilizer ∇ of a non-degenerate triangle in Σ has already been studied in (12.17 ff). We now take up this thread again in order to complete our previous results by a description of ∇ in Veronese coordinates, and we determine the isomorphism types of various subgroups of ∇.

17.11 The stabilizer of a triangle in Veronese coordinates.

1) We consider the non-degenerate triangle consisting of the points

$$(0, 0) \cong \mathbb{R}(0, 0, 0; 0, 0, 1) = o$$
$$(0) \quad \cong \mathbb{R}(0, 0, 0; 0, 1, 0) = u$$
$$(\infty) \cong \mathbb{R}(0, 0, 0; 1, 0, 0) = v$$

and its stabilizer

$$\nabla := \Sigma_{o,u,v} .$$

In (12.17), the group ∇ has been described in affine coordinates as the direct product of the subgroup consisting of the transformations

$$\mu_{r,t} : (x, y) \mapsto (rx, ty), \quad (s) \mapsto (tr^{-1} \cdot s) ,$$

for $0 < r, t \in \mathbb{R}$, by the subgroup $S\nabla$ whose elements are the collineations

$$(A, B \,|\, C) : (x, y) \mapsto (Ax, By), \quad (s) \mapsto (Cs)$$

with $A, B, C \in \mathrm{SO}_8\mathbb{R}$ such that

(*) $\qquad\qquad B(s \cdot x) = Cs \cdot Ax \quad$ for all $s, x \in \mathbb{O}$.

According to (17.8), each of these collineations is induced by a unique linear transformation of $\mathbb{O}^3 \times \mathbb{R}^3$ of determinant 1. Our aim is to make this explicit. We freely use the identification made in (16.3 and 4) between affine and Veronese coordinates.

2) It is straightforward to verify that the collineation $\mu_{r,t}$ is induced by the following transformation $\tilde{\mu}_{r,t}$ of $\mathbb{O}^3 \times \mathbb{R}^3$:

$$\tilde{\mu}_{r,t} : (x_\nu; \xi_\nu)_\nu \mapsto (rt)^{-2/3} \cdot (rx_1, tx_2, rtx_3; t^2\xi_1, r^2\xi_2, \xi_3) ;$$

the factor $(rt)^{-2/3}$ is chosen so as to obtain a transformation of determinant 1.

We now prove a statement concerning linear transformations of $\mathbb{O}^3 \times \mathbb{R}^3$ which induce collineations of the type $(A, B \,|\, C) \in S\nabla$ as described above.

3) *For $R_\nu \in SO_8\mathbb{R}$, $\nu \in \{1, 2, 3\}$, the linear transformation (R_1, R_2, R_3) of $\mathbb{O}^3 \times \mathbb{R}^3$ defined by*

$$(R_1, R_2, R_3) : (x_\nu; \xi_\nu)_\nu \mapsto (R_\nu x_\nu; \xi_\nu)_\nu$$

induces a collineation if, and only if, the condition

($\tilde{*}$) $\qquad\qquad R_3(y \cdot \bar{x}) = \overline{R_2 \bar{y}} \cdot R_1 x \quad \text{for all } x, y \in \mathbb{O}$

is satisfied. For the induced collineation, denoted by $[R_1, R_2, R_3]$, we find that

$$[R_1, R_2, R_3] = (A, B \,|\, C) ,$$

where

(1) $\qquad\qquad A = R_1, \quad By = \overline{R_2 \bar{y}} \text{ for } y \in \mathbb{O}, \quad C = R_3 .$

Indeed, (R_1, R_2, R_3) maps the point $\mathbb{R}(x, \bar{y}, y\bar{x}; \|y\|^2, \|x\|^2, 1) \cong (x, y)$ to

$$\mathbb{R}(R_1 x, R_2 \bar{y}, R_3(y\bar{x}); \|y\|^2, \|x\|^2, 1) = \mathbb{R}(R_1 x, R_2 \bar{y}, R_3(y\bar{x}); \|R_2\bar{y}\|^2, \|R_1 x\|^2, 1) .$$

If (R_1, R_2, R_3) induces a collineation, this 1-dimensional subspace consists of Veronese vectors (16.1), whence ($\tilde{*}$). Conversely, if ($\tilde{*}$) is satisfied, this is the point with affine coordinates $(R_1 x, \overline{R_2 \bar{y}})$. Furthermore, the point $\mathbb{R}(0, 0, s; \|s\|^2, 1, 0) \cong (s)$ is mapped to $\mathbb{R}(0, 0, R_3 s; \|s\|^2, 1, 0) = \mathbb{R}(0, 0, R_3 s; \|R_3 s\|^2, 1, 0) \cong (R_3 s)$. Thus, (R_1, R_2, R_3) induces the collineation $(A, B \,|\, C)$ as specified by (1).

4) We now want to see that, in fact, all collineations of $S\nabla$ may be obtained in this way. This amounts to showing that conditions ($\tilde{*}$) and ($*$) are equivalent for elements of $SO_8\mathbb{R}$ as specified by (1). We use conjugation by the linear transformation $\tilde{\tau}$ of $\mathbb{O}^3 \times \mathbb{R}^3$ which induces the triality collineation τ and is given by cyclic permutation of Veronese coordinates (16.6). Obviously,

(2) $\qquad\qquad \tilde{\tau} (R_1, R_2, R_3) \tilde{\tau}^{-1} = (R_2, R_3, R_1) .$

For $A, B, C \in SO_8\mathbb{R}$ and corresponding transformations R_1, R_2, R_3 according to (1), condition (∗) directly translates into $\overline{R_2(y \cdot \bar{x})} = R_3 x \cdot R_1 \bar{y}$, or, equivalently, $R_2(y \cdot \bar{x}) = \overline{R_1 \bar{y}} \cdot \overline{R_3 x}$ for all $x, y \in \mathbb{O}$. By step 3), this condition says that (R_3, R_1, R_2) induces a collineation. From (2), we infer that (R_1, R_2, R_3) induces a collineation, as well, which, according to step 3), is just $(A, B \,|\, C)$.

We may now summarize these results as follows.

Proposition. *Let $\widetilde{\nabla}$ denote the group of all linear transformations of $V = \mathbb{O}^3 \times \mathbb{R}^3$ having determinant 1 and inducing a collineation that belongs to ∇. This group $\widetilde{\nabla}$ is the direct product of the subgroup*

$$\{\, \tilde{\mu}_{r,t} \mid 0 < r, t \in \mathbb{R} \,\}$$

by the subgroup

$$S\widetilde{\nabla} = \{\, (R_1, R_2, R_3) \mid R_\nu \in SO_8\mathbb{R} \text{ satisfy } (\tilde{*}) \,\} \,;$$

the latter subgroup induces the group of collineations

$$S\nabla = \{\, (A, B \,|\, C) \mid A, B, C \in SO_8\mathbb{R} \text{ satisfy } (*) \,\} \,.$$ □

Remark. Obviously, $S\widetilde{\nabla}$ is the intersection of $\widetilde{\nabla}$ with $O(V, \beta)$, the orthogonal group of the inner product β defined in (16.1b).

In the considerations leading to the preceding proposition, the following automorphism of $S\nabla$ has already made its appearance.

17.12 Proposition: The triality automorphism of $S\nabla$. *Conjugation with the triality collineation τ induces an automorphism of $S\nabla$, which is given by the map*

$$S\nabla \to S\nabla : [R_1, R_2, R_3] \mapsto [R_2, R_3, R_1] \,.$$

This automorphism permutes the non-trivial elements of the center $\mathbb{Z}_2 \times \mathbb{Z}_2$ of $S\nabla$ cyclically.

Proof. As τ permutes the vertices of the standard triangle ouv, see (16.6), conjugation by τ is an automorphism of the stabilizer of this triangle. Correspondingly, conjugation by the linear transformation $\tilde{\tau}$ of $V = \mathbb{O}^3 \times \mathbb{R}^3$ inducing τ is an automorphism of $\widetilde{\nabla}$. Now $\tilde{\tau}$ belongs to $O(V, \beta)$, cf. (16.6); therefore, conjugation by $\tilde{\tau}$ leaves $S\widetilde{\nabla} = \widetilde{\nabla} \cap O(V, \beta)$ invariant. On the level of collineations, we thus obtain the triality automorphism of $S\nabla$; its explicit description is immediate from equation (2) in step 4) of the preceding proof (17.11). We now consider the actions of $S\nabla$ on the sides of the triangle ouv. The non-trivial elements of the kernels of these actions are the three non-trivial elements of the center of $S\nabla$, see (12.17c). Since τ permutes the sides of ouv cyclically, it is clear that conjugation by τ permutes the corresponding kernels accordingly. □

We have seen in (12.17c) that the action of SV on the sides of the triangle ouv is described by epimorphisms

$$\mathrm{pr}_\nu : S\nabla \to SO_8\mathbb{R}$$

for $\nu \in \{1, 2, 3\}$ with kernels $\mathbb{Z}_2$. These epimorphisms have an important rôle in the following result.

17.13 Proposition. *$S\nabla$ is compact and connected, and therefore is the (simply connected) universal covering group $\mathrm{Spin}_8\mathbb{R}$ of $SO_8\mathbb{R}$ (via any of the epimorphisms pr_ν). Every proper normal subgroup of $S\nabla$ is contained in the center $\mathbb{Z}_2 \times \mathbb{Z}_2$.*

Remarks. 1) The epimorphisms pr_ν are the three so-called half-spin representations of $\mathrm{Spin}_8\mathbb{R} \cong S\nabla$, cf. (17.16) and (95.10).

2) The Dynkin diagram of the Lie algebra D_4 of $\mathrm{Spin}_8\mathbb{R}$ and $SO_8\mathbb{R}$ is shown in Figure 17a.

Figure 17a

We mention without proof that the automorphism of this Dynkin diagram corresponding to the triality automorphism (17.12) of $S\nabla \cong \mathrm{Spin}_8\mathbb{R}$ is a 1/3 - turn.

Proof of (17.13). $S\nabla$ is isomorphic to the group $\widetilde{S\nabla}$ of linear transformations of $\mathbb{O}^3 \times \mathbb{R}^3$ described in (17.11). From this description, it is clear that $\widetilde{S\nabla}$ is topologically closed in $(SO_8\mathbb{R})^3$. It follows that $\widetilde{S\nabla}$ and $S\nabla$ are compact.

Connectedness will follow from the assertion about normal subgroups. We shall deduce this assertion from the corresponding known properties of $SO_8\mathbb{R}$ using the epimorphism pr_3, whose kernel is generated by the reflection $\iota_o : (x, y) \mapsto (-x, -y)$ with axis W, cf. (12.17c). The only proper normal subgroup of $SO_8\mathbb{R}$ is the center $\{\mathrm{id}, -\mathrm{id}\}$, see e.g. Artin [57] Theorem 5.3, p. 178; its preimage $\mathrm{pr}_3^{-1}\{\mathrm{id}, -\mathrm{id}\}$ has four elements and consists of the center Z of $S\nabla$. Now let N be a normal subgroup of $S\nabla$. Its image under the epimorphism pr_3 is a normal subgroup of $SO_8\mathbb{R}$. Hence, we either have $\mathrm{pr}_3(N) \subseteq \{\mathrm{id}, -\mathrm{id}\}$, in which case $N \subseteq \mathrm{pr}_3^{-1}\{\mathrm{id}, -\mathrm{id}\} = Z$, or $\mathrm{pr}_3(N) = SO_8\mathbb{R}$. In the latter case, $S\nabla = \mathrm{pr}_3^{-1}(\mathrm{pr}_3(N)) = N \cdot \ker \mathrm{pr}_3 = N \cup N \cdot \iota_o$, and N has index at most 2 in $S\nabla$. We shall prove in a moment that ι_o is a square in $S\nabla$. It then follows that $\iota_o \in N$, and $N = S\nabla$.

In order to show that ι_o is a square, we recall the following collineations from (12.14):

$$\gamma_a : (x, y) \mapsto (a^{-1}x, ay), \quad (s) \mapsto (asa) \, ;$$

if $a \in \mathbb{S}_7 = \{\, a \in \mathbb{O} \mid \|a\|^2 = 1 \,\}$, they belong to $\mathrm{S}\nabla$ according to (12.17b), because the transformations $x \mapsto ax$ and $x \mapsto axa$ of $\mathbb{O} = \mathbb{R}^8$ then belong to $\mathrm{SO}_8\mathbb{R}$, see (11.22). If, in particular, $a \in \mathbb{S}_7 \cap \mathrm{Pu}\,\mathbb{O}$, i.e., if $a^2 = -\bar{a}a = -\|a\|^2 = -1$, then $(\gamma_a)^2 = \iota_o$, as is immediate from biassociativity (11.9).

This completes the proof of the assertion about normal subgroups. As to connectedness, we consider the connected component $\mathrm{S}\nabla^1$ of $\mathrm{S}\nabla$ containing id, and we have to show that $\mathrm{S}\nabla^1 = \mathrm{S}\nabla$. Now $\mathrm{S}\nabla^1$ is a normal subgroup of $\mathrm{S}\nabla$; therefore, by the above, all we need to know is that $\mathrm{S}\nabla^1$ has more than four elements. But, since the sphere $\mathbb{S}_7$ is connected, the collineations γ_a for $a \in \mathbb{S}_7$ form a connected subset of $\mathrm{S}\nabla$ containing $\mathrm{id} = \gamma_1$; hence, all these collineations belong to $\mathrm{S}\nabla^1$.

Thus, $\mathrm{S}\nabla$ is connected. By compactness, $\mathrm{pr}_3 : \mathrm{S}\nabla \to \mathrm{SO}_8\mathbb{R}$ is a closed map, hence the natural topology of $\mathrm{SO}_8\mathbb{R}$ coincides with the quotient topology determined by pr_3, and this implies that the homomorphism pr_3 is also an open map, cf. (92.20b) and (93.16). We now may summarize these properties by saying that $\mathrm{S}\nabla$ is a connected, two-sheeted covering group of $\mathrm{SO}_8\mathbb{R}$, see (94.2). As the fundamental group of $\mathrm{SO}_8\mathbb{R}$ is $\mathbb{Z}_2$, cf. Steenrod [51] 22.9, p. 118, there is only one such covering group, and it is simply connected; in other words, it is the universal covering group, which in the case of $\mathrm{SO}_8\mathbb{R}$ is usually denoted by $\mathrm{Spin}_8\mathbb{R}$. □

We now investigate the stabilizer of a degenerate quadrangle. We obtain such a group from the stabilizer ∇ of the standard triangle ouv by keeping the additional point (1) on the line at infinity $W = uv$ fixed. An explicit description of the stabilizer $\nabla_{(1)}$ can be extracted from the description of ∇ given in (12.17). The most interesting part of ∇ is the subgroup $\mathrm{S}\nabla \cong \mathrm{Spin}_8\mathbb{R}$; in the following proposition, we therefore study the corresponding stabilizer $\mathrm{S}\nabla_{(1)}$ more closely.

Its action on the line at infinity W will be described using the following notation. The stabilizer of $1 \in \mathbb{O} = \mathbb{R}^8$ in $\mathrm{SO}_8\mathbb{R}$ leaves the orthogonal complement $\mathrm{Pu}\,\mathbb{O} = \mathbb{R}^7$ of 1, see (11.6), invariant and induces the group $\mathrm{SO}_7\mathbb{R}$ on it; somewhat negligently, this stabilizer itself will be denoted by $\mathrm{SO}_7\mathbb{R}$.

17.14 Proposition: The stabilizer of a degenerate quadrangle. *The group $\mathrm{S}\nabla_{(1)}$ consists of the collineations which in affine coordinates are of the form*

$$(A, A \mid C) : (x, y) \mapsto (Ax, Ay), \quad (s) \mapsto (Cs) \, ,$$

with $A \in \mathrm{SO}_8\mathbb{R}$ and $C \in \mathrm{SO}_7\mathbb{R}$ satisfying

(∗) $$A(s \cdot x) = Cs \cdot Ax$$

for all $s, x \in \mathbb{O}$. The group $S\nabla_{(1)}$ is connected, and the restriction map

$$\mathrm{pr}_3 : S\nabla_{(1)} \to SO_7\mathbb{R} : (A, A \,|\, C) \mapsto C$$

is a two-sheeted covering homomorphism whose kernel is generated by the reflection $\iota_o : (x, y) \mapsto (-x, -y)$. Hence, $S\nabla_{(1)}$ is isomorphic to the (simply connected) universal covering group $\mathrm{Spin}_7\mathbb{R}$ of $SO_7\mathbb{R}$. The kernel $\ker \mathrm{pr}_3 = \{\mathrm{id}, \iota_o\}$ is the center of $S\nabla_{(1)}$, and is the only non-trivial, proper normal subgroup.

The restriction map $\mathrm{pr}_1 : S\nabla_{(1)} \to SO_8\mathbb{R} : (A, A\,|\,C) \mapsto A$ maps $S\nabla_{(1)}$ isomorphically onto a subgroup of $SO_8\mathbb{R}$.

The non-trivial orbits of $S\nabla_{(1)}$ on the line $Y = ov = [0]$ are the 7-spheres $S(\varrho) = \{(0, y) \mid \|y\|^2 = \varrho^2\}$ for $0 < \varrho \in \mathbb{R}$.

Proof. The given description of $S\nabla_{(1)}$ is obtained from the description (12.17) of $S\nabla$ in the following way. An element $(A, B\,|\,C) \in S\nabla$, with $A, B, C \in SO_8\mathbb{R}$ satisfying the triality condition (12.17(∗)), fixes (1) if, and only if $C(1) = 1$. The triality condition, $B(s \cdot x) = Cs \cdot Ax$ for all $s, x \in \mathbb{O}$, when evaluated for $s = 1$, then gives that $B = A$.

The surjectivity of the restriction homomorphism $\mathrm{pr}_3 : S\nabla \to SO_8\mathbb{R}$ stated in (12.17) implies that it induces an epimorphism of $S\nabla_{(1)}$ onto $SO_7\mathbb{R}$.

Using this epimorphism, the further proof may now proceed exactly as in (17.13). One uses the fact that $SO_7\mathbb{R}$ is simple, see e.g. Artin [57] Theorem 5.3, p. 178. In order to make the proof of (17.13) work analogously, we only have to provide the following two facts: firstly, the reflection ι_o is a square not only in $S\nabla$, as was proved in (17.13), but even in $S\nabla_{(1)}$, and secondly, the connected component $(S\nabla_{(1)})^1$ of $S\nabla_{(1)}$ has (infinitely) many elements. This needs some more care than was necessary in (17.13) because the point (1) has to be kept fixed. In the proof of (17.13), we first remarked that $S\nabla$ contains the collineations $\gamma_a : (x, y) \mapsto (a^{-1}x, ay)$, $(s) \mapsto (asa)$ for $a \in \mathbb{S}_7 = \{a \in \mathbb{O} \mid \|a\|^2 = 1\}$. By composition with the reflection $\iota_u : (x, y) \mapsto (-x, y)$, $(s) \mapsto (-s)$ with center u and axis $Y = ov$, we obtain the following elements of $S\nabla$:

$$\gamma_a \iota_u : (x, y) \mapsto (-a^{-1}x, ay), \quad (s) \mapsto (-asa).$$

We now make the restriction $a \in \mathbb{S}_7 \cap \mathrm{Pu}\,\mathbb{O}$, i.e., $a^2 = -\bar{a}a = -\|a\|^2 = -1$. Then, these collineations leave the point (1) fixed. Moreover, since ι_u belongs to the center of $S\nabla$, the square of any of these collineations equals the square of γ_a; for $a \in \mathbb{S}_7 \cap \mathrm{Pu}\,\mathbb{O}$, this square is ι_o, as we remarked in the proof of (17.13).

The claim that the connected component $(S\nabla_{(1)})^1$ of $S\nabla_{(1)}$ containing id has infinitely many elements can also be obtained from the collineations $\gamma_a \iota_u$ for $a \in \mathbb{S}_7 \cap \mathrm{Pu}\,\mathbb{O}$, although the identity is not among them. The unit sphere $\mathbb{S}_7 \cap \mathrm{Pu}\,\mathbb{O}$ of $\mathrm{Pu}\,\mathbb{O} = \mathbb{R}^7$ is connected; hence, all these collineations belong to one and the same connected component of $S\nabla_{(1)}$, which is a coset of the connected component of id.

With these ingredients, one obtains analogously to the proof of (17.13) that $\ker \mathrm{pr}_3$ is the only non-trivial, proper normal subgroup of $S\nabla_{(1)}$, and that $S\nabla_{(1)}$ can be considered as the universal covering group $\mathrm{Spin}_7 \mathbb{R}$ of $\mathrm{SO}_7 \mathbb{R}$. Since ι_o is central, the center of $S\nabla_{(1)}$ equals $\ker \mathrm{pr}_3$. It remains to verify the last two statements of the proposition.

Injectivity of pr_1 is immediate from condition (*), which uniquely determines C for any given A.

The sphere $S(\varrho)$ is invariant under $S\nabla_{(1)}$, since that group acts on Y by orthogonal transformations. Transitivity on $S(\varrho)$ is related to transitivity of Σ on the set of non-degenerate quadrangles (17.6) in the following way. For $y \in \mathbb{O}$ with $\|y\|^2 = \varrho^2$, there is $\sigma \in \nabla = \Sigma_{o,u,v}$ mapping (y, y) to (ϱ, ϱ). Since the latter points both belong to the line through o of slope 1, this line is invariant under σ, and hence its point at infinity, the point (1), is fixed. It is immediate that $(0, y)^\sigma = (0, \varrho)$. Furthermore, $\|y\|^2 = \varrho^2$ implies that $\sigma \in S\nabla$ because by (12.17) we have $\nabla = \{\mu_{r,t} \mid 0 < r, t \in \mathbb{R}\} \cdot S\nabla$, and $\mu_{r,t} \neq \mathrm{id}$ changes the norms of coordinates. Thus, $\sigma \in S\nabla_{(1)}$. $\square$

According to (12.8), the stabilizer in Σ of the non-degenerate quadrangle with vertices o, u, v, and $e := (1, 1)$ consists of the collineations

$$(x, y) \mapsto (x^\alpha, y^\alpha), \quad (s) \mapsto (s^\alpha), \quad (\infty) \mapsto (\infty) \quad \text{for } \alpha \in \mathrm{Aut}\,\mathbb{O}.$$

We now relate the discussion of $\mathrm{Aut}\,\mathbb{O} \cong G_2$ in (11.30 through 33) with the present context.

17.15 The stabilizer of a non-degenerate quadrangle.

(a) $\Sigma_{o,u,v,e} \cong \mathrm{Aut}\,\mathbb{O} \cong G_2$ *is contained in* $S\nabla$; *therefore, the following stabilizers coincide:*

$$\Sigma_{o,u,v,e} = S\nabla_e = S\nabla_{(1),(0,1)}\,.$$

The latter description shows that we deal with a stabilizer of the transitive action of $S\nabla_{(1)} \cong \mathrm{Spin}_7 \mathbb{R}$ on the 7-sphere $S(1)$, see (17.14). This has the following consequences.

(b) $\mathrm{Spin}_7 \mathbb{R}/G_2 \approx \mathbb{S}_7$.
(c) *In particular,* $\mathrm{Aut}\,\mathbb{O}$ *is simply connected.*

Proof. For abbreviation, we put $\Lambda = \Sigma_{o,u,v,e}$. From (11.31e), we know that $\mathrm{Aut}\,\mathbb{O} \subseteq \mathrm{SO}_8 \mathbb{R}$, so that $\Lambda \subseteq S\nabla$ according to (12.17), and $\Lambda = S\nabla_e$, the converse inclusion $S\nabla_e \subseteq \Lambda$ being trivial. The identity $S\nabla_e = S\nabla_{(1),(0,1)}$ is a simple geometric fact; indeed, $e = (o \vee (1)) \wedge ((0, 1) \vee u)$, $(1) = oe \wedge uv$, and $(0, 1) = ov \wedge eu$. Assertion (b) follows from the transitive action of $S\nabla_{(1)} \cong \mathrm{Spin}_7 \mathbb{R}$ on the 7-sphere by a standard argument, cf. (96.9a). The exact homotopy sequence associated with that action, see (96.12), shows that Λ is simply connected, since $\mathrm{Spin}_7\mathbb{R}$ has this property and since the second homotopy group $\pi_2(\mathbb{S}_7)$ is trivial. □

For later use, see (62.9), we finally prove here that the stabilizers of $S\nabla$ at points on the sides of the triangle ouv can be given a purely (Lie) group theoretic characterization. We use results of the present section and facts about representations of compact Lie groups.

17.16 Proposition. *The group $\mathrm{Spin}_8\mathbb{R}$ contains precisely three conjugacy classes of closed, connected proper subgroups Ω of dimension at least 21. They are permuted cyclically by the triality automorphism (17.12). Under the isomorphism $S\nabla \cong \mathrm{Spin}_8\mathbb{R}$ of (17.13), one class corresponds to the set of stabilizers $S\nabla_w$ for $w \in W^* := W \setminus \{u, v\}$. In particular, all these subgroups Ω are isomorphic to $\mathrm{Spin}_7\mathbb{R}$.*

Proof. 1) We identify $\mathrm{Spin}_8\mathbb{R}$ with $S\nabla$. By (17.6), ∇ is transitive on the sides of the triangle ouv minus the vertices. We have $\nabla \cong S\nabla \times \mathbb{R}^2$ by (12.17). Therefore, all stabilizers $S\nabla_w$ for $w \in W^*$ are conjugate. If the stabilizers on another side of the triangle were in the same conjugacy class within $S\nabla$, then $S\nabla_w$ would fix a non-degenerate quadrangle and would be a subgroup of the 14-dimensional group G_2, cf. (17.6 and 15) and (11.31), contrary to $\dim S\nabla_w = 21$, see (17.14).

2) However, under the triality collineation τ and its inverse, the stabilizers on the other sides of the triangle (minus the vertices) are conjugate to the stabilizers $S\nabla_w$, $w \in W^*$, because τ permutes the sides of the triangle cyclically (16.6). In particular, all the information obtained in (17.14) for the groups $S\nabla_w$ and their actions carries over to these other classes of stabilizers.

3) By the description of $S\nabla$ in (12.17), the group Ω acts on the sides of the triangle ouv as a closed, connected subgroup of $\mathrm{SO}_8\mathbb{R}$ in its natural action on $\mathbb{R}^8 \cup \{\infty\}$, and the resulting epimorphism $\Omega \to \mathrm{SO}_8\mathbb{R}$ is either an isomorphism onto this subgroup or a two-fold covering homomorphism. Now, the closed, connected subgroups of $\mathrm{SO}_8\mathbb{R}$ having large dimension can be determined without difficulty, see (95.12); if such a subgroup has dimension at least 21, then it is isomorphic to one of the groups $\mathrm{SO}_8\mathbb{R}$, $\mathrm{Spin}_7\mathbb{R}$, and $\mathrm{SO}_7\mathbb{R}$. Moreover, in the first two cases, the action on $\mathbb{R}^8$ is irreducible, whereas in the last case it is equivalent to the action obtained from the standard action of $\mathrm{SO}_7\mathbb{R}$ on $\mathbb{R}^7 \times \mathbb{R}$.

If the group induced by Ω were $\mathrm{SO}_8\mathbb{R}$, then $\dim \Omega = \dim \nabla$, see (17.13), and consequently, by connectedness, $\Omega = \nabla$, cf. (93.12), contrary to our assumptions.

If Ω induces a group isomorphic to $\mathrm{Spin}_7\mathbb{R}$ on one of the sides of the triangle, then the corresponding epimorphism $\Omega \to \mathrm{Spin}_7\mathbb{R}$ is an isomorphism, since $\mathrm{Spin}_7\mathbb{R}$ is simply connected and consequently does not admit a non-trivial covering. In other words, the action of Ω on the respective side of the triangle is effective. We now show that this cannot happen for all three sides at the same time. The central involution $\iota \in \Omega \cong \mathrm{Spin}_7\mathbb{R}$ has a fixed point other than a vertex on some side of the triangle by simple geometric reasons, see (23.17). Now the fixed points of ι on this side correspond to a linear subspace of $\mathbb{R}^8$, which is Ω-invariant since ι belongs to the center. Thus, either the action of Ω on this side is not irreducible, or ι acts trivially on this side, and the action of Ω is not effective.

Hence, on one of the sides of the triangle, Ω induces the group $\mathrm{SO}_7\mathbb{R}$ in its natural action on $\mathbb{R}^7 \times \mathbb{R}$. In particular, Ω fixes a point p other than a vertex on this side, $\Omega \subseteq \mathrm{S}\nabla_p$. Now recall from step 2) that $\mathrm{S}\nabla_p \cong \mathrm{Spin}_7\mathbb{R}$. Again by connectedness and dimension, we conclude that $\Omega = \mathrm{S}\nabla_p$. □

17.17 Notes. As basic references for the subject of this section, we have already cited Freudenthal [85] (a reprint of a set of lecture notes from the early fifties) and Tits [53, 54]. It was already said in (16.8) that the theory of planes over general Cayley algebras and of their collineation groups evolved in close connection with the algebraic theory of (exceptional) Jordan algebras. Among the contributions in this spirit, we mention again Jacobson [59, 61] and Springer [60, 62], where far-reaching generalizations of the results of the present section may be found. Faulkner [70] develops a uniform theory covering the cases of both odd and even characteristic. Further references may be found in (16.8).

Timmesfeld [94] shows how the plane over a Cayley algebra K and its little projective group (of type E_6) can be comprehended entirely within a geometric study of the group which he calls $\mathrm{SL}_2 K$ (the group generated by the shears of (12.5b and c)). In our approach, we have also studied it; on the line at infinity, this group induces the group $\mathsf{A}|_W = \mathrm{PSO}_{10}(\mathbb{R}, 1)$ considered in (15.6), see (12.20).

18 Motion groups of $\mathcal{P}_2\mathbb{O}$

This section is devoted to the study of polarities of $\mathcal{P}_2\mathbb{O}$ and of the corresponding motion groups. We mainly occupy ourselves with the standard elliptic polarity, which is already known from Section 16, and with the standard hyperbolic polarity. The elliptic motion group $\Phi = \Phi^+$ and the hyperbolic motion group Φ^- are simple groups. This well-known fact is proved here by elementary geometric considerations, without recourse to the structure theory of Lie groups. Then these motion groups are located in the list of simple Lie groups as the exceptional compact Lie group $\mathrm{F}_4 = \mathrm{F}_4(-52)$ in the elliptic case, and as the group $\mathrm{F}_4(-20)$ in the hyperbolic case. The elliptic motion group turns out to be a maximal compact subgroup of the group Σ of all collineations of $\mathcal{P}_2\mathbb{O}$; this allows us to recognize Σ among the simple Lie groups of dimension 78 as the exceptional real simple Lie group $\mathrm{E}_6(-26)$. The section closes with a classification of all polarities of $\mathcal{P}_2\mathbb{O}$, showing that up to equivalence there is just one further polarity besides the standard elliptic and the standard hyperbolic polarity. This classification relies on knowledge about the collineation group; it could be carried out in this manner over $\mathbb{R}$, $\mathbb{C}$, and $\mathbb{H}$, as well, as an alternative to the approach of (13.18) via Hermitian forms in the field case.

Like the standard elliptic polarity (16.7), the standard hyperbolic polarity may be derived from a non-degenerate symmetric bilinear form on $V = \mathbb{O}^3 \times \mathbb{R}^3$. The first steps of the investigation consist in showing that the corresponding motion groups are induced by, and are isomorphic to certain subgroups $\widetilde{\Phi}^+$ and $\widetilde{\Phi}^-$ of the orthogonal groups belonging to the respective bilinear forms, and that $\widetilde{\Phi}^+$ and $\widetilde{\Phi}^-$ each leave a certain 26-dimensional subspace of the 27-dimensional $\mathbb{R}$-vector space $\mathbb{O}^3 \times \mathbb{R}^3$ invariant. (In fact, the linear actions of $\widetilde{\Phi}^+$ and $\widetilde{\Phi}^-$ on the respective subspaces yield the first fundamental representations of $F_4(-52)$ and $F_4(-20)$, see (18.17).)

In studying the stabilizer $\Phi_L^\pm$ of a line L in the respective motion group, we have two notions of orthogonality at our disposal, the one corresponding to the given polarity, and the one which is inherent in the geometry of a line (16.2). The interplay between these two notions of orthogonality will be used in order to determine $\Phi_L^\pm$. As a result, the stabilizer $\Phi_p = \Phi_L$ of a polar point–line pair (p, L) in the elliptic motion group Φ satisfies $\Phi_L|_L \cong SO_9\mathbb{R}$ and is isomorphic to the (two-fold) universal covering group $Spin_9\mathbb{R}$ of $SO_9\mathbb{R}$, the obvious restriction map $\Phi_L \to \Phi_L|_L$ being a covering homomorphism, see (18.8). This provides a very simple geometric description of $Spin_9\mathbb{R}$. From the transitivity properties of $\Phi = F_4$ which we shall establish, it follows that $P_2\mathbb{O}$ can be viewed as the homogeneous space $F_4/Spin_9\mathbb{R}$. In the hyperbolic motion group, the stabilizers of interior points are the same as in the elliptic motion group; this can be used to obtain the homeomorphism $Spin_9\mathbb{R}/Spin_7\mathbb{R} \approx \mathbb{S}_{15}$.

For the planes over $\mathbb{R}$, $\mathbb{C}$, or $\mathbb{H}$ in place of $\mathbb{O}$, it would be possible to give a completely analogous treatment of the motion groups, essentially by restriction of the arguments of the present section to the corresponding subplanes of $\mathcal{P}_2\mathbb{O}$. We have refrained from carrying this out systematically; indeed, in Section 13 we have seen easier ways to deal with motion groups of the planes over the fields $\mathbb{R}$, $\mathbb{C}$, and $\mathbb{H}$. At a few places, however, it is particularly interesting to reconsider those motion groups under the aspect of the present section. In this way, one may recognize, for instance, that the point stabilizers of the elliptic complex or quaternion motion group contain normal subgroups isomorphic to $Spin_3\mathbb{R}$ or to $Spin_5\mathbb{R}$, respectively. On the other hand, we know from Section 13 that those motion groups are unitary groups. By combining these two points of view, we shall obtain a very easy and natural explanation of the isomorphisms $SU_2\mathbb{C} \cong Spin_3\mathbb{R}$ and $U_2\mathbb{H} \cong Spin_5\mathbb{R}$, see (18.9).

18.0 Basic notions. In this section, we always use the construction of the projective octonion plane $\mathcal{P}_2\mathbb{O}$ by means of Veronese vectors and of the inner product β on $\mathbb{O}^3 \times \mathbb{R}^3$ as explained in (16.1). We freely employ both Veronese or affine coordinates, on the basis of the correspondence (16.3) between these two systems.

The *standard elliptic polarity* π of $\mathcal{P}_2\mathbb{O}$ as defined in (16.7) maps a point $p \in P_2\mathbb{O}$ to the line represented by the orthogonal space $p^\perp$ of p in $V = \mathbb{O}^3 \times \mathbb{R}^3$ with respect to the inner product β. In order to define a further polarity, we use the reflection

$$\iota_o = \mu_{-1} = \mu_{-1,-1} : (x, y) \mapsto (-x, -y),\ (s) \mapsto (s),$$

which is induced by the following linear transformation of $\mathbb{O}^3 \times \mathbb{R}^3$:

$$\tilde{\iota}_o = \tilde{\mu}_{-1,-1} : (x_1, x_2, x_3; \xi_1, \xi_2, \xi_3) \mapsto (-x_1, -x_2, x_3; \xi_1, \xi_2, \xi_3),$$

see (12.13a and 17c) and part 2) of (17.11). As $\tilde{\iota}_o$ is obviously orthogonal with respect to β, the collineation ι_o commutes with the standard elliptic polarity π. Therefore,

$$\pi^- := \pi\iota_o = \iota_o\pi$$

is a polarity, as well (13.11); indeed, $(\pi^-)^2 = \pi\iota_o\iota_o\pi = \pi^2 = \mathrm{id}$. It is called the *standard hyperbolic polarity*.

We remark that this agrees with the notion of the standard hyperbolic polarity of $\mathcal{P}_2\mathbb{F}$ for $\mathbb{F} \in \{\mathbb{R}, \mathbb{C}, \mathbb{H}\}$. More precisely, if $\mathcal{P}_2\mathbb{F}$ is considered as a subplane of $\mathcal{P}_2\mathbb{O}$ according to (16.5), then the restriction of π^- to $\mathcal{P}_2\mathbb{F}$ is the standard hyperbolic polarity of $\mathcal{P}_2\mathbb{F}$ as defined in (13.12); this is immediate from the corresponding statement for the standard elliptic polarity in (16.7c) together with equation (1) in (13.12).

The standard hyperbolic polarity may be described by the non-degenerate symmetric bilinear form β^- of $\mathbb{O}^3 \times \mathbb{R}^3$ defined as follows:

$$\beta^-(w\,|\,z) = \beta(w\,|\,\tilde{\iota}_o z).$$

Indeed $\tilde{\iota}_o$, being orthogonal with respect to β and involutory, is also self-adjoint, so that β^- is indeed symmetric. Furthermore, a point q is incident with the line $p^{\pi^-} = p^{\iota_o\pi}$, if and only if, $0 = \beta\left(p^{\iota_o}\,|\,q\right) = \beta\left(\tilde{\iota}_o(p)\,|\,q\right) = \beta\left(p\,|\,\tilde{\iota}_o(q)\right) = \beta^-(p\,|\,q)$, so that p^{π^-} is the orthogonal space of p with respect to β^-.

Below we note how $\beta^+ := \beta$ and β^- are expressed in coordinates. For β^+, this is just the definition (16.1b); the corresponding expression for β^- then follows from the definition of β^- above and from the description of $\tilde{\iota}_o$.

$$\beta^\pm\left((x_\nu; \xi_\nu)_\nu\,\big|\,(y_\nu; \eta_\nu)_\nu\right) = \pm\langle x_1\,|\,y_1\rangle \pm \langle x_2\,|\,y_2\rangle + \langle x_3\,|\,y_3\rangle + \sum_{\nu=1}^{3} \xi_\nu\eta_\nu.$$

In the sequel, it will be useful to know that on the set H of Veronese vectors the quadratic forms belonging to $\beta^+ = \beta$ and to β^- take on a particularly simple form. Indeed, from the definition of H in (16.1(*)) we obtain for

$$w = (x_\nu; \xi_\nu)_\nu \in H$$

that

(*) $$\begin{aligned}\beta^\pm(w \mid w) &= \pm 2\|x_1\|^2 \pm 2\|x_2\|^2 + 2\|x_3\|^2 + \xi_1^2 + \xi_2^2 + \xi_3^2 \\ &= \pm 2\xi_2\xi_3 \pm 2\xi_1\xi_3 + 2\xi_1\xi_2 + \xi_1^2 + \xi_2^2 + \xi_3^2 \\ &= (\pm \xi_1 \pm \xi_2 + \xi_3)^2 \, .\end{aligned}$$

We now consider absolute points of these polarities. The point $p = \mathbb{R}w$, $w \in H \setminus \{0\}$, is absolute with respect to the standard elliptic or hyperbolic polarity if, and only if p is contained in its own orthogonal space with respect to $\beta^\pm$, i.e., if w is self-orthogonal. In the elliptic case, this does not occur as $\beta^+ = \beta$ is positive definite. For the standard hyperbolic polarity, it follows from (*) that $p = \mathbb{R}w$ is π^--absolute if, and only if $\xi_3 = \xi_1 + \xi_2$. Passing to affine notation, we conclude that the points $(s) = \mathbb{R}(0, 0, s; \|s\|^2, 1, 0)$ and $(\infty) = \mathbb{R}(0, 0, 0; 1, 0, 0)$ at infinity are not π^--absolute, and that $(x, y) = \mathbb{R}(x, \bar{y}, y\bar{x}; \|y\|^2, \|x\|^2, 1)$ is π^--absolute if, and only if $\|x\|^2 + \|y\|^2 = 1$. Thus we have proved:

18.1 Proposition. *The standard elliptic polarity π has no absolute points. The set of absolute points of the standard hyperbolic polarity π^- is the 15-dimensional sphere*

$$Q = \{ (x, y) \in \mathbb{O} \times \mathbb{O} \mid \|x\|^2 + \|y\|^2 = 1 \}$$

(the unit sphere in the affine point set $\mathbb{O} \times \mathbb{O}$). □

18.2 Motion groups. The groups of all collineations commuting with the polarities of (18.0) will be denoted by $\Phi = \Phi^+$ (for the standard elliptic polarity π) and by Φ^- (for the standard hyperbolic polarity π^-); they will be called the *elliptic motion group* Φ and the *hyperbolic motion group* Φ^- of $\mathcal{P}_2\mathbb{O}$.

By (17.8), every collineation is induced by a unique linear transformation of $V = \mathbb{O}^3 \times \mathbb{R}^3$ having determinant 1, so let

$$\widetilde{\Phi} = \widetilde{\Phi}^+ := \{ \tilde{\varphi} \in \mathrm{SL}(V) \mid \tilde{\varphi}|_{P_2\mathbb{O}} \in \Phi \}$$
$$\widetilde{\Phi}^- := \{ \tilde{\varphi} \in \mathrm{SL}(V) \mid \tilde{\varphi}|_{P_2\mathbb{O}} \in \Phi^- \} \, .$$

The natural maps $\widetilde{\Phi} \to \Phi$ and $\widetilde{\Phi}^- \to \Phi^-$ are isomorphisms. The subgroups $\widetilde{\Phi}$ and $\widetilde{\Phi}^-$ of $\mathrm{GL}(V)$ have obvious topologies; on Φ and Φ^- we shall consider the topologies obtained by transfer via the natural maps.

Our polarities are defined by the symmetric bilinear forms $\beta = \beta^+$ and β^-. It is quite natural that, by the following result, the groups $\widetilde{\Phi}$ and $\widetilde{\Phi}^-$ turn out to be subgroups of the corresponding proper orthogonal groups $SO(V, \beta) \cong SO_{27}\mathbb{R}$ and $SO(V, \beta^-)$, respectively, namely

$$\widetilde{\Phi} = \widetilde{\Phi}^+ = \{\, \widetilde{\varphi} \in SO(V, \beta) \mid \widetilde{\varphi}(H) = H \,\}$$
$$\widetilde{\Phi}^- = \{\, \widetilde{\varphi} \in SO(V, \beta^-) \mid \widetilde{\varphi}(H) = H \,\} \,.$$

More generally, we prove the following.

Proposition. *Let ϱ be a polarity of $\mathcal{P}_2\mathbb{O}$ which is described by a non-degenerate symmetric bilinear form η of $V = \mathbb{O}^3 \times \mathbb{R}^3$ in the sense that the polar of a point $p \in P_2\mathbb{O}$ is the orthogonal space of p with respect to η. Let $\widetilde{\Psi} \le SL(V)$ consist of those transformations which induce collineations of $\mathcal{P}_2\mathbb{O}$ commuting with ϱ. Then*

$$\widetilde{\Psi} = \{\, \widetilde{\psi} \in SO(V, \eta) \mid \widetilde{\psi}(H) = H \,\} \,.$$

In particular, $\widetilde{\Psi}$ is a closed subgroup of $SO(V, \eta)$.

Proof. If $\widetilde{\psi} \in SO(V, \eta)$ and if $\widetilde{\psi}(H) = H$, then $\widetilde{\psi}$ induces a collineation (17.8), which by orthogonality clearly commutes with the polarity ϱ.

Conversely, let $\widetilde{\psi} \in \widetilde{\Psi}$ induce a collineation ψ on $P_2\mathbb{O}$ which commutes with ϱ. Let $\widetilde{\psi}^\circ$ denote the adjoint of $\widetilde{\psi}$ relative to η, that is, the unique linear transformation satisfying $\eta(w \mid \widetilde{\psi}^\circ z) = \eta(\widetilde{\psi}w \mid z)$ for all $w, z \in V$. For a point $p \in P_2\mathbb{O}$, by definition of the polarity ϱ, a vector $w \in V$ belongs to the line p^ϱ (considered as a hyperplane of V) if, and only if $\eta(w \mid p) = 0$. By definition of the adjoint, this is equivalent to $\eta(\widetilde{\psi}w \mid \widetilde{\psi}^{\circ\,-1}(p)) = 0$, so that $\widetilde{\psi}(p^\varrho)$ is the orthogonal space of $\widetilde{\psi}^{\circ\,-1}(p)$ with respect to η. On the other hand, because of $\widetilde{\psi}(p^\varrho) = (p^\varrho)^\psi = (p^\psi)^\varrho = (\widetilde{\psi}(p))^\varrho$, this is also the orthogonal space of $\widetilde{\psi}(p)$. We infer that $\widetilde{\psi}^{\circ\,-1}(p) = \widetilde{\psi}(p)$ for all $p \in P_2\mathbb{O}$, in other words, $\widetilde{\psi}\widetilde{\psi}^\circ$ induces the identity on $P_2\mathbb{O}$. By (17.9), we therefore have $\widetilde{\psi}\widetilde{\psi}^\circ = r \cdot \mathrm{id}$ for a suitable $r \in \mathbb{R} \setminus \{0\}$. Now $\det \widetilde{\psi}^\circ = \det \widetilde{\psi} = 1$, hence $r^{27} = 1 = r$, and $\widetilde{\psi} = \widetilde{\psi}^{\circ\,-1}$, so that for all $w, z \in V$ we have $\eta(\widetilde{\psi}w \mid \widetilde{\psi}z) = \eta(w \mid \widetilde{\psi}^\circ\widetilde{\psi}z) = \eta(w \mid z)$, in other words, $\widetilde{\psi}$ is orthogonal with respect to η. Thus the asserted description of $\widetilde{\Psi}$ is proved.

As the set H of Veronese vectors is closed in V by its very definition (16.1(∗)), it is then clear that $\widetilde{\Psi}$ is closed in $SO(V, \eta)$. □

18.3 Lemma. *In $V = \mathbb{O}^3 \times \mathbb{R}^3$, we consider the vectors*

$$e^+ = (0, 0, 0; 1, 1, 1), \quad e^- = (0, 0, 0; -1, -1, 1) \,,$$

and their orthogonal spaces

$$U^+ = \{ (x_\nu; \xi_\nu)_\nu \mid \xi_1 + \xi_2 + \xi_3 = 0 \}, \quad U^- = \{ (x_\nu; \xi_\nu)_\nu \mid -\xi_1 - \xi_2 + \xi_3 = 0 \}.$$

The subspace U^+ is invariant under $\widetilde{\Phi} = \widetilde{\Phi}^+$, and U^- is invariant under $\widetilde{\Phi}^-$.

Remarks. U^+ is the orthogonal space of e^+ with respect to the inner product β, but also with respect to the symmetric bilinear form β^- defining the standard hyperbolic polarity (18.0), and likewise for U^-. The hyperplanes U^+ and U^- are *not* lines as e^+ and e^- are not Veronese vectors. The linear form

$$\mathbb{O}^3 \times \mathbb{R}^3 \to \mathbb{R} : (x_\nu; \xi_\nu)_\nu \mapsto \xi_1 + \xi_2 + \xi_3,$$

of which U^+ is the kernel, is sometimes called the *trace* form. When $\mathbb{O}^3 \times \mathbb{R}^3$ is viewed as the space of Hermitian matrices with coefficients in $\mathbb{O}$ (16.8), this is indeed the trace in the ordinary sense.

Proof of (18.3). Write $\beta^+ := \beta$. For $w = (x_\nu; \xi_\nu)_\nu \in \mathbb{O}^3 \times \mathbb{R}^3$ we have $\beta^\pm(w \mid e^\pm) = \pm \xi_1 \pm \xi_2 + \xi_3$. If $w \in H$, then it follows from (18.0(*)) that $e^\pm$ and $-e^\pm$ belong to the set $\{ e \in V \mid \forall w \in H : \beta^\pm(w \mid e)^2 = \beta^\pm(w \mid w) \}$. Now this set obviously is invariant under $\widetilde{\Phi}^\pm \subseteq \mathrm{SO}(V, \beta^\pm)$. We show that it consists of $e^\pm$ and $-e^\pm$ alone; then the proof is finished, since $U^\pm$ is the orthogonal space of $e^\pm$. So let $e = (e_\nu; \varepsilon_\nu)_\nu \in \mathbb{O}^3 \times \mathbb{R}^3$ satisfy the condition

(1) $$\beta^\pm(w \mid e)^2 = \beta^\pm(w \mid w) \quad \text{for all } w \in H.$$

We evaluate this condition for Veronese vectors w corresponding to the points with affine coordinates $(x, 0)$, $(0, y)$, and (s) for $x, y, s \in \mathbb{O}$, see (16.3). For $w = (x, 0, 0; 0, \|x\|^2, 1)$, condition (1) says

(2) $$\left(\pm \langle x \mid e_1 \rangle + \|x\|^2 \varepsilon_2 + \varepsilon_3 \right)^2 = \pm 2\|x\|^2 + \|x\|^4 + 1 = (\|x\|^2 \pm 1)^2.$$

This condition is particularly simple if $\langle x \mid e_1 \rangle = 0$; as $\|x\|^2$ is arbitrary, we infer that

$$\varepsilon_2 = \pm \varepsilon_3 \in \{1, -1\}.$$

On the other hand, if in (2) we put $x = re_1$ with $r \in \mathbb{R}$, we obtain the polynomial equation $\left(\pm 2r\|e_1\|^2 + r^2\|e_1\|^2 \varepsilon_2 + \varepsilon_3 \right)^2 = (r^2\|e_1\|^2 \pm 1)^2$ for r, whose linear terms show that $e_1 = 0$. If, in the same way, we evaluate condition (1) for $w = (0, \bar{y}, 0; \|y\|^2, 0, 1)$, we obtain $\varepsilon_1 = \pm \varepsilon_3$ and $e_2 = 0$. Finally, using $w = (0, 0, s; \|s\|^2, 1, 0)$ we conclude that $e_3 = 0$. Thus, indeed, $e \in \{e^\pm, -e^\pm\}$. □

The following is an easy consequence of the definitions:

18.4 Proposition. *The stabilizers of o in the elliptic motion group and in the hyperbolic motion group coincide:*

$$\Phi_o^- = \Phi_o = \Phi_o^+ .$$

Proof. Since the polar $o^\pi = (0,0)^\pi = W$ (16.7) is invariant under ι_o, it is also the polar of o under the standard hyperbolic polarity $\pi^- = \pi\iota_o$; hence, both stabilizers leave W invariant and are contained in $\Sigma_{o,W}$. Now every element of $\Sigma_{o,W}$ commutes with ι_o, as ι_o is the only involutory homology with center o and axis W (12.13). Thus, for elements of $\Sigma_{o,W}$, the conditions of commuting with π or with $\pi^- = \pi\iota_o$ are equivalent. □

The elliptic motion group

We now specialize our discussion to the elliptic motion group Φ, for the sake of definiteness. The hyperbolic case, which can be treated in close analogy, will be presented afterwards, see (18.20 ff). The facts stated next are immediate consequences of our description of elliptic motions by orthogonal transformations (18.2).

18.5 Fact. *The triality collineation τ belongs to Φ.*

Indeed, the transformation $\tilde{\tau} \in \mathrm{GL}(V)$, which by definition (16.6) induces τ, is orthogonal with respect to β. □

18.6 Fact. *The stabilizer of the vertices $o = (0,0)$, $u = (0)$, and $v = (\infty)$ of the standard triangle in the elliptic motion group is the group*

$$\Phi_{o,u,v} = S\nabla \cong \mathrm{Spin}_8\mathbb{R}$$

described in (12.17) and in (17.11 and 13).

Proof. Let $\widetilde{\nabla}$ be the subgroup of $\mathrm{SL}(V)$ inducing the stabilizer $\nabla = \Sigma_{o,u,v}$. As was noted at the end of (17.11), one has $\widetilde{\nabla} \cap \mathrm{SO}(V, \beta) = S\widetilde{\nabla}$. By (18.2), it follows that $\Phi_{o,u,v} = \Phi \cap \nabla$ is the group $S\nabla$ induced by $S\widetilde{\nabla}$. □

According to (12.17), the stabilizer ∇ is the product of $S\nabla = \Phi_{o,u,v}$ by the subgroup consisting of the collineations

$$\mu_{r,t} : (x, y) \mapsto (rx, ty), \ (s) \mapsto (tr^{-1} \cdot s)$$

for $0 < r, t \in \mathbb{R}$. We state how the latter collineations behave with regard to the standard elliptic polarity.

18.7 Lemma. $\pi \mu_{r,t} = \mu_{r,t}^{-1} \pi$.

Proof. The linear transformation $\tilde{\mu} = \tilde{\mu}_{r,t}$ of $\mathbb{O}^3 \times \mathbb{R}^3$ inducing the collineation $\mu = \mu_{r,t}$, see (17.11 step 2), is obviously self-adjoint with respect to the inner product β. Therefore, $\tilde{\mu}$ maps the orthogonal space $p^\perp$ of a point $p \in P_2\mathbb{O}$ onto $\tilde{\mu}^{-1}(p)^\perp$; indeed, for $w \in p^\perp$ we have $\beta\left(\tilde{\mu}w \mid \tilde{\mu}^{-1}(p)\right) = \beta(w \mid p) = 0$. Thus, $p^{\pi\mu} = (p^\perp)^\mu = \tilde{\mu}^{-1}(p)^\perp = p^{\mu^{-1}\pi}$. □

We now determine the structure of the stabilizer of the line W in the elliptic motion group Φ. As W is the polar o^π of the point $o = (0,0)$, see (16.7b), we have of course

$$\Phi_W = \Phi_o.$$

The main task is to identify the subgroup corresponding to Φ_W in the group $\Sigma_W|W$ of transformations of W induced by *all* collineations fixing W. Now, according to the fundamental theorem (15.6) for the projective line, $\Sigma_W|W$ is the group $PSO_{n+2}(1)^W = PSO_{10}(1)^W$ of direct conformal transformations of W described in (15.2). We now show that Φ_W induces the subgroup $SO_{n+1}{}^W = SO_9{}^W$ introduced in (15.10).

18.8 Theorem. *The group* $\Phi_o|W$ *of transformations of the point row W which are induced by elements of* $\Phi_W = \Phi_o$ *is*

$$\Phi_o|W = SO_9{}^W.$$

In particular, Φ_o is transitive on W. The kernel of the restriction epimorphism

$$\mathrm{pr} : \Phi_o \to SO_9{}^W : \varphi \mapsto \varphi|W$$

has two elements, the non-trivial element being the reflection with axis W and center o,

$$\iota_o = \mu_{-1} : (x, y) \mapsto (-x, -y), \ (s) \mapsto (s).$$

The kernel $\ker \mathrm{pr} = \{\mathrm{id}, \iota_o\}$ *is the center and the only non-trivial, proper normal subgroup of* Φ_o.

The group Φ_o is a compact and connected Lie group; it is isomorphic to the universal covering group $\mathrm{Spin}_9\mathbb{R}$ *of* $SO_9\mathbb{R}$.

Proof. 1) First, we prove that $\Phi_o|W \subseteq SO_9{}^W$. We recall from (16.2) that W is a quadric in the subspace

$$V_3 = \{0\} \times \{0\} \times \mathbb{O} \times \mathbb{R}^2 \times \{0\}$$

of $V = \mathbb{O}^3 \times \mathbb{R}^3$. In order to make notation match with Section 15, we identify V_3 and
$$\mathbb{O} \times \mathbb{R}^2 =: \hat{\mathbb{O}}$$
in the obvious way. Then, as in (15.1), W is the projective quadric of the quadratic form q on $\hat{\mathbb{O}}$ given by
$$q(x, r_1, r_2) = \bar{x}x - r_1 r_2 .$$
According to (15.4), V_3 is generated by W. Therefore, for $\varphi \in \Phi_W$, a linear transformation $\tilde{\varphi} \in \tilde{\Phi}$ of $\mathbb{O}^3 \times \mathbb{R}^3$ inducing φ leaves V_3 invariant. By (18.3), the 9-dimensional subspace
$$T_3^+ := U^+ \cap V_3 = \{ (0, 0, x_3; r, -r, 0) \mid x_3 \in \mathbb{O}, r \in \mathbb{R} \}$$
is also invariant under $\tilde{\varphi}$. It is the orthogonal space in V_3 of the vector $h_3^+ := (0, 0, 0; 1, 1, 0) \in V_3$ with respect to the inner product β, hence $\tilde{\varphi}(h_3^+)$ equals h_3^+ or $-h_3^+$, as $\tilde{\varphi} \in \tilde{\Phi}$ is orthogonal. Under our identification $V_3 = \hat{\mathbb{O}}$, the subspace T_3^+ is mapped to
$$T^+ = \{ (x, r, -r) \mid x \in \mathbb{O}, r \in \mathbb{R} \} \leq \hat{\mathbb{O}} ,$$
the vector h_3 is identified with
$$h^+ = (0, 1, 1) \in \hat{\mathbb{O}} ,$$
and the restriction $\tilde{\varphi}|_{V_3}$ corresponds to a linear transformation ϱ of $\hat{\mathbb{O}}$ satisfying $\varrho(h^+) \in \{h^+, -h^+\}$.

On the other hand, according to the fundamental theorem for the projective line (15.6), the restriction $\varphi|_W$ of the collineation φ is also induced by an orthogonal transformation $\sigma \in \mathrm{SO}(\hat{\mathbb{O}}, q)$ with respect to q. Since both ϱ and σ induce the same transformation of the quadric W, namely $\varphi|_W$, we know from (15.4) that σ is a scalar multiple of ϱ; in particular, σ leaves the 1-dimensional subspace $\mathbb{R}h^+$ invariant. Thus, $\varphi|_W = [\sigma]|_W$ belongs to the group $\mathrm{PSO}(\hat{\mathbb{O}}, q)_{\mathbb{R}h^+}|_W = \mathrm{SO}_9{}^W$ described in (15.10(4)).

2) We now prove the converse inclusion $\Phi_o|_W \supseteq \mathrm{SO}_9{}^W$. Let $\bar{\sigma} \in \mathrm{SO}_9{}^W \subseteq \mathrm{PSO}_{10}(1)^W$. By the fundamental theorem for the projective line (15.6), there is a collineation $\varphi \in \Sigma_W$ such that $\varphi|_W = \bar{\sigma}$. Our aim is to show that φ can be modified so as to become an element of Φ_o, without changing its action on W. Composing φ with a translation (with axis W) if necessary, we may assume that $o^\varphi = o$, so that $\varphi \in \Sigma_{o,W}$.

According to the definition of $\mathrm{SO}_9{}^W$ in (15.10 step 2), $\bar{\sigma}$ is induced by an element $\sigma \in \mathrm{SO}(\hat{\mathbb{O}}, q)$ satisfying $\sigma(h^+) = h^+$. The orthogonal space of h^+ with respect to the bilinear form β_q on $\widehat{\mathbb{O}}$ associated with the quadratic form q is invariant under the orthogonal transformation σ. Now it is fortunate that this orthogonal space is the same subspace T^+ which is also the orthogonal space of h^+ with respect to the symmetric bilinear form obtained by restricting the inner product β to $V_3 \cong \widehat{\mathbb{O}}$. Furthermore, on T^+ the restrictions of β_q and of β coincide. (Due to the special position of T^+ in $\widehat{\mathbb{O}}$, this is true, even though q is not positive definite on $\widehat{\mathbb{O}}$, whereas β is.) Thus, σ is orthogonal not only with respect to q, but also with respect to β.

Now, for a point p on W, the point $p^\pi \wedge W$ is the unique point p' on W satisfying $p' \subseteq p^\perp$. Since σ is β-orthogonal, it follows that $(p')^{\bar\sigma} \subseteq (p^{\bar\sigma})^\perp$. Hence, we obtain

(1) $\qquad p^{\pi\varphi} \wedge W = (p^\pi \wedge W)^\varphi = (p^\pi \wedge W)^{\bar\sigma} = p^{\bar\sigma\pi} \wedge W = p^{\varphi\pi} \wedge W$.

As p is on W, the line p^π is incident with the point $W^\pi = o$, and so are the lines $p^{\pi\varphi}$ and $p^{\varphi\pi}$, because $\varphi \in \Sigma_{o,W}$. Therefore, we have $p^{\pi\varphi} = o \vee (p^{\pi\varphi} \wedge W)$ and $p^{\varphi\pi} = o \vee (p^{\varphi\pi} \wedge W)$, and (1) gives $p^{\pi\varphi} = p^{\varphi\pi}$. Hence, the collineation $\pi\varphi\pi\varphi^{-1} : P_2\mathbb{O} \xrightarrow{\pi} \mathcal{L}_2\mathbb{O} \xrightarrow{\varphi} \mathcal{L}_2\mathbb{O} \xrightarrow{\pi} P_2\mathbb{O} \xrightarrow{\varphi^{-1}} P_2\mathbb{O}$ has axis W. Also, it fixes o; indeed, $o^{\pi\varphi\pi\varphi^{-1}} = W^{\varphi\pi\varphi^{-1}} = W^{\pi\varphi^{-1}} = o^{\varphi^{-1}} = o$. The collineations having axis W and fixing o have been determined in (12.13); they are precisely the homotheties $\mu_t : (x, y) \mapsto (tx, ty)$, $(s) \mapsto (s)$ for $t \in \mathbb{R} \setminus \{0\}$. Thus $\pi\varphi\pi\varphi^{-1} = \mu_t$, $\varphi\pi\varphi^{-1} = \pi\mu_t$ for a suitable t. Now let $r := \sqrt{|t|}$ and $\varphi' := \mu_r\varphi$. The collineations φ and φ' induce the same transformation on W, and by (18.7) we have

$$\varphi'\pi\varphi'^{-1} = \mu_r\varphi\pi\varphi^{-1}\mu_r^{-1} = \mu_r\pi\mu_t\mu_r^{-1} = \pi\mu_r^{-1}\mu_t\mu_r^{-1} = \pi\mu_\varepsilon, \quad \text{with } \varepsilon \in \{1, -1\}$$

depending on whether $t > 0$ or $t < 0$. If $\varepsilon = 1$, $\mu_\varepsilon = \mathrm{id}$, then φ' commutes with π, i.e., $\varphi' \in \Phi_o$, as desired.

We now show that $\varepsilon = -1$ cannot occur. In that case, $\varphi'\pi\varphi'^{-1} = \pi\mu_{-1}$ would be the standard hyperbolic polarity π^- of (18.0), which has absolute points (18.1), so that the conjugate $\pi = \varphi'^{-1}\pi^-\varphi'$ would have absolute points, as well. Indeed, if p is an absolute point of π^-, then $p^{\varphi'}$ is an absolute point of $\varphi'^{-1}\pi^-\varphi'$, as is easily verified. But π does not have absolute points (18.1).

Thus, $\Phi_o|_W = \mathrm{SO}_9{}^W$ is proved. Transitivity of Φ_o on W follows by (15.11).

3) The stabilizer Φ_o is compact, since it is closed in $\Phi \cong \widetilde{\Phi}$, and since $\widetilde{\Phi}$ is closed in the compact group $\mathrm{SO}(V, \beta)$, see (18.2). Being (isomorphic to) a closed subgroup of a linear group, Φ_o is a Lie group, cf. (94.3). The proof may now proceed exactly as in (17.13) for the group $\mathrm{S}\nabla$, which is a subgroup of Φ_o, see (18.6). One uses the epimorphism $\mathrm{pr} : \Phi_o \to \Phi_o|_W = \mathrm{SO}_9{}^W$ and the fact that $\mathrm{SO}_9\mathbb{R}$ is simple, see e.g. Artin [57] Theorem 5.3, p. 178. Analogously as in the proof of (17.13) we obtain that $\ker \mathrm{pr} = \{\mathrm{id}, \iota_o\}$ is the center and the only non-trivial, proper normal subgroup of Φ_o, that, in particular, Φ_o is connected, and that Φ_o can be considered as the universal covering group of $\mathrm{SO}_9{}^W$ via the covering map pr. Thus, Φ_o is isomorphic to the universal covering group $\mathrm{Spin}_9\mathbb{R}$ of $\mathrm{SO}_9\mathbb{R}$. $\square$

18.9 Restriction to $\mathcal{P}_2\mathbb{H}$ and to $\mathcal{P}_2\mathbb{C}$. As a digression, we shall sketch here how restriction of the preceding discussion to the subplanes $\mathcal{P}_2\mathbb{H}$ and $\mathcal{P}_2\mathbb{C}$ and their elliptic motion groups, combined with the description of these motion groups by unitary groups in Section 13, yields isomorphisms

$$U_2\mathbb{H} \cong \mathrm{Spin}_5\mathbb{R} \quad \text{and} \quad SU_2\mathbb{C} \cong \mathrm{Spin}_3\mathbb{R}.$$

It is well known that there are such isomorphisms, see e.g. Porteous [81] Cor. 13.60, p. 266 in connection with Cor. 11.57 p. 220 (note that Porteous writes Sp(n) for $U_n\mathbb{H}$). We feel that the way suggested here for obtaining such isomorphisms, as an analogue to the isomorphism $\Phi_o \cong \mathrm{Spin}_9\mathbb{R}$ established in (18.8) for the octonion plane, is very natural.

Let $\mathbb{F} \in \{\mathbb{C}, \mathbb{H}\}$ and $n = \dim_\mathbb{R} \mathbb{F} \in \{2, 4\}$, and let Φ denote the elliptic motion group of $\mathcal{P}_2\mathbb{F}$. In (16.5 and 7) it has been shown that we may freely switch between affine, homogeneous or Veronese coordinates for the discussion of this situation. The stabilizer Φ_o has been determined in (13.14) in affine coordinates. It is a product $\Delta \cdot \Lambda$, where Δ is the normal subgroup

$$\Delta = \begin{cases} SU_2\mathbb{C} & \text{for } \mathbb{F} = \mathbb{C} \\ U_2\mathbb{H} & \text{for } \mathbb{F} = \mathbb{H}, \end{cases}$$

which in affine coordinates acts in the usual way on the *right* $\mathbb{F}$-vector space $\mathbb{F}^2$, and where

$$\Lambda = \left\{ \begin{pmatrix} x \\ y \end{pmatrix} \mapsto \begin{pmatrix} xc \\ yc \end{pmatrix} \;\middle|\; c \in \mathbb{F},\, \|c\|^2 = 1 \right\}.$$

Since Λ acts trivially on W (12.13a), we have $\Phi_o|_W = \Delta|_W$. The kernel of the restriction map

$$\mathrm{pr} : \Delta \to \Delta|_W = \Phi_o|_W$$

is

$$\ker \mathrm{pr} = \{\mathrm{id}, \mu_{-1}\}.$$

Indeed, its elements are homologies having axis W and fixing o, hence (12.13) they are of the form

$$\mu_a : \begin{pmatrix} x \\ y \end{pmatrix} \mapsto \begin{pmatrix} xa \\ ya \end{pmatrix}$$

for $0 \neq a \in \mathbb{F}$. Now, in order to be an element of Δ, in the first place μ_a must be $\mathbb{F}$-linear with respect to scalar multiplication from the *right*; this is the case if, and only if a belongs to the center of $\mathbb{F}$. In the second place, then, μ_a must be unitary; this is equivalent to $\|a\|^2 = 1$. In the case $\mathbb{F} = \mathbb{C}$, for μ_a to belong to Δ, there is the third condition $1 = \det \mu_a = a^2$, so that finally $a = \pm 1$. For $\mathbb{F} = \mathbb{H}$, we have obtained that a is an element of the center $\mathbb{R}$ of $\mathbb{H}$ (11.19) having norm 1, so that here, too, the only possibilities are $a = \pm 1$.

In Veronese coordinates over $\mathbb{F}$, on the other hand, one may carry out exactly the same analysis as was done for $\mathbb{O}$ in (18.8), with the result that $\Phi_o|_W \cong \mathrm{SO}_{n+1}\mathbb{R}$. From the two-fold covering homomorphism $\mathrm{pr} : \Delta \to \Phi_o|_W$ discussed above, we therefore learn that Δ is isomorphic to the universal covering group $\mathrm{Spin}_{n+1}\mathbb{R}$ of $\mathrm{SO}_{n+1}\mathbb{R}$. For this last step, one needs the fact that Δ is connected. This is well known, see Porteous [81] Prop. 17.22, p. 340. It can also be proved exactly as was done for Φ_o in (18.8); one notes that the non-trivial element μ_{-1} of ker pr is the square of the matrix $\begin{pmatrix} i & \\ & -i \end{pmatrix} \in \Delta$ and concludes as in the proof of (17.13).

Let us point out finally that the proof of $\Phi_o|_W \cong \mathrm{SO}_{n+1}\mathbb{R}$ can also be carried out without passing to Veronese coordinates for the entire projective plane; it suffices to consider W as the quadric of the metric extension $\widehat{\mathbb{F}} = \mathbb{F} \times \mathbb{R}^2$ endowed with the quadratic form q of index 1 as in Section 15. The analysis in step 2) of the proof of (18.8) shows that a transformation $\bar{\sigma} \in \mathrm{PSO}_{n+2}(1)^W$ of the point row W is induced by a collineation φ belonging to Φ_o if, and only if, for every point p on W, one has

(1) $\qquad (p^\pi \wedge W)^{\bar{\sigma}} = p^{\bar{\sigma}\pi} \wedge W ,$

in other words, if $\bar{\sigma}$ commutes with the bijection

$$W \to W : p \mapsto p^\pi \wedge W .$$

It is now easy to calculate this bijection explicitly in affine coordinates and then to show that, if W is considered as a quadric of $\widehat{\mathbb{F}}$, this bijection is induced by the following orthogonal involution of $\widehat{\mathbb{F}}$:

$$\varepsilon : \widehat{\mathbb{F}} \to \widehat{\mathbb{F}} : (x, r_1, r_2) \mapsto (-x, r_2, r_1) .$$

With an orthogonal transformation $\sigma \in \mathrm{SO}(\widehat{\mathbb{F}}, q)$ inducing $\bar{\sigma}$ on W, condition (1) now can be expressed by demanding that $\sigma \circ \varepsilon$ and $\varepsilon \circ \sigma$ induce the same transformation on W; according to (15.4), then, they only differ by a scalar, and our condition is equivalent to σ leaving the 1-eigenspace of the orthogonal involution ε invariant. This eigenspace is spanned by the vector $h^+ = (0, 1, 1) \in \widehat{\mathbb{F}} = \mathbb{F} \times \mathbb{R}^2$. Thus, $\Phi_o|_W = \mathrm{PSO}(\widehat{\mathbb{F}}, q)_{\mathbb{R}h^+}|_W = \mathrm{SO}_{n+1}{}^W$, see (15.10).

We now return to the octonion plane.

18.10 Corollary. *The elliptic motion group Φ of $\mathcal{P}_2\mathbb{O}$ is flag transitive.*

Proof. By (18.8), the stabilizer Φ_W is transitive on W, so that $\Phi_o = \Phi_W$ is transitive on the set $\mathcal{L}_o$ of lines through o. Hence, it suffices to show that Φ is transitive on the set of points. By conjugation with the triality collineation τ, which belongs to Φ, see (18.5), we infer that $\Phi_W{}^\tau$ is transitive on the line $W^\tau = Y \in \mathcal{L}_o$ (16.6). Now, any point can be moved to a point of Y by an element of Φ_o, and the image point can then be mapped to o by an element of $\Phi_W{}^\tau$. $\square$

By transitivity of Φ on the set of lines, the next statement, which will be useful for proving simplicity of Φ, follows immediately from theorem (18.8), where the special case of the line W was already dealt with.

18.11 Corollary. *For any line L, there is exactly one non-trivial collineation in Φ with axis L; it is involutory and has center $p = L^\pi$ (the pole of L). This collineation will be called the reflection ι_p at p.*

The subgroup $\{\mathrm{id}, \iota_p\}$ generated by ι_p is the center and the only non-trivial, proper normal subgroup of $\Phi_p = \Phi_L$. □

18.12 Proposition. *On the affine point set $\mathbb{O} \times \mathbb{O} = \mathbb{R}^{16}$, the stabilizer $\Phi_o = \Phi_W \cong \mathrm{Spin}_9\mathbb{R}$ acts as an orthogonal group with respect to the norm $\|(x,y)\|^2 = \|x\|^2 + \|y\|^2$. It acts transitively on every sphere with center o (a sphere of dimension 15).*

Note that the sphere of radius 1 is the set Q of absolute points of the standard hyperbolic polarity π^-, see (18.1).

Proof. By (12.21), Φ_o consists of $\mathbb{R}$-linear transformations of $\mathbb{O} \times \mathbb{O}$. Therefore, it suffices to show that Φ_o leaves Q invariant and acts transitively on it. By (18.4), the elements of Φ_o commute not only with the standard elliptic polarity π, but also with the standard hyperbolic polarity $\pi^- = \pi \iota_o$; hence, the set Q of absolute points of π^- is invariant under Φ_o. Furthermore, Φ_o acts transitively on the set of lines through o, see (18.8), so that by suitable elements of Φ_o any two points on Q can be transformed into points on $X = \mathbb{O} \times \{0\}$ having norm 1. Now for two such points there is an element of $S\nabla \subseteq \Phi_o$ (18.6) mapping one of the points to the other, because $S\nabla$ induces the group $\mathrm{SO}_8\mathbb{R}$ on $\mathbb{O} \times \{0\}$ by (12.17c). Thus, Φ_o is transitive on Q. □

18.13 Addendum. *The stabilizer of the point $(1,0) \in Q$ in $\Phi_o \cong \mathrm{Spin}_9\mathbb{R}$ coincides with the stabilizer of that point in the group $S\nabla$ analyzed in (12.17) and (17.11 ff):*

$$\Phi_{o,(1,0)} = S\nabla_{(1,0)}.$$

This stabilizer is isomorphic to $\mathrm{Spin}_7\mathbb{R}$. For the coset space corresponding to the transitive action of $\Phi_o \cong \mathrm{Spin}_9\mathbb{R}$ on $Q \approx \mathbb{S}_{15}$ we therefore obtain

$$\mathrm{Spin}_9\mathbb{R}/\mathrm{Spin}_7\mathbb{R} \approx \mathbb{S}_{15}.$$

Proof. $\Phi_{o,(1,0)}$ fixes the lines $o \vee (1,0) = X = [0,0]$ and $o^\pi = W$ and the points $X \wedge W = u = (0)$ and $X^\pi = (\infty) = v$, see (16.7). Therefore $\Phi_{o,(1,0)} = \Phi_{o,u,v,(1,0)} = S\nabla_{(1,0)}$ by (18.6). From the principle of conjugate stabilizers (91.1a), we infer that $S\nabla_{(1,0)} = \tau S\nabla_{(1)} \tau^{-1}$, where τ is the triality collineation (16.6), which normalizes $S\nabla$, see (17.12), and maps $(1,0)$ to (1). In particular, $S\nabla_{(1,0)}$ is isomorphic to $\mathrm{Spin}_7\mathbb{R}$, since $S\nabla_{(1)}$ is, cf. (17.14). The homeomorphism $\mathrm{Spin}_9\mathbb{R}/\mathrm{Spin}_7\mathbb{R}$ is obtained by a standard argument for transitive actions, cf. (96.9a). □

The proof of the following well-known result is one of the principal objectives of this section.

18.14 Theorem. *The elliptic motion group Φ is simple.*

Proof. 1) A non-trivial normal subgroup of Φ has no fixed point, since the set of its fixed points is invariant under Φ, cf. (91.1d), and since Φ is transitive on $P_2\mathbb{O}$ by (18.10).

2) Let N be a normal subgroup of Φ. We have to show that either $\mathsf{N} = \Phi$ or $\mathsf{N} = \{\mathrm{id}\}$. The connected component N^1 of the identity in N is a normal subgroup of Φ, too. If $\mathsf{N}^1 = \{\mathrm{id}\}$, then N is totally disconnected and therefore (93.18) is centralized by the group Φ_o, which is connected (18.8). Now o is the only fixed point of Φ_o, see (18.10), and hence a fixed point of N, cf. (91.1c). By step 1), we infer that in this case $\mathsf{N} = \{\mathrm{id}\}$. Thus, by passing to the connected component if necessary, we may henceforth assume that N is connected and that $\mathsf{N} \neq \{\mathrm{id}\}$.

3) Then, the orbit o^N is connected. We want to show that it is open in $P_2\mathbb{O}$. By homogeneity, it suffices to prove that o^N contains a neighbourhood of o. We shall obtain such a neighbourhood using the fact that the orbit o^N of the normal subgroup N is invariant under the stabilizer Φ_o, cf. (91.1d), and that the orbits of Φ_o in the affine point set $\mathbb{O} \times \mathbb{O}$ are the spheres with center o, see (18.14). We consider the compact ball $B_1 = \{(x, y) \in \mathbb{O} \times \mathbb{O} \mid \|x\|^2 + \|y\|^2 \leq 1\}$. The continuous mapping $\varrho : (x, y) \mapsto \|x\|^2 + \|y\|^2$ of B_1 onto the unit interval can be continuously extended to $P_2\mathbb{O}$ by putting $\varrho(p) = 1$ for $p \notin B_1$. Since $o^\mathsf{N} \neq \{o\}$ by step 1), we see that the connected subset $\varrho(o^\mathsf{N})$ contains a neighbourhood J of 0 in the unit interval, and $\varrho^{-1}(J)$ is a neighbourhood of o. We may assume that $1 \notin J$; then $\varrho^{-1}(J) \subseteq o^\mathsf{N}$, since $J \subseteq \varrho(o^\mathsf{N})$ and since for $0 < r < 1$ the preimage $\varrho^{-1}(r)$, which is a sphere, is an orbit of Φ_o.

4) Φ permutes the orbits of the normal subgroup N, cf. (91.1d), and acts transitively on $P_2\mathbb{O}$ by (18.10); it therefore follows from step 3) that *every* N-orbit is open. But $P_2\mathbb{O}$ is connected (16.14), so that there is just one N-orbit; in other words, N is transitive on $P_2\mathbb{O}$.

5) The stabilizer N_p of a point p is a normal subgroup of Φ_p; therefore, by (18.11), we either have $\mathsf{N}_p = \Phi_p$ or $\mathsf{N}_p \leq \{\mathrm{id}, \iota_p\}$.

6) We consider these cases separately for $p = o$.

Case 1: If $\mathsf{N}_o = \Phi_o$, then transitivity of N immediately implies that $\Phi = \mathsf{N}$, cf. (91.3).

Case 2: If $\mathsf{N}_o \leq \{\mathrm{id}, \iota_o\}$, then we also consider the reflection ι_v at $v = (\infty)$, which has axis $v^\pi = X = [0, 0]$, cf. (18.11) and (16.7). In particular, ι_v fixes o and $u = (0)$; by our assumption about N_o, it does not belong to N. Hence, according to step 5), we have $\mathsf{N}_v = \{\mathrm{id}\}$, and N is sharply transitive on $\mathrm{P}_2\mathbb{O}$. In particular, there is a unique collineation $\eta \in \mathsf{N}$ mapping v to u. Now $\iota_v \eta \iota_v^{-1}$ does so, as well, so that $\iota_v \eta \iota_v^{-1} = \eta$, $\iota_v = \eta^{-1} \iota_v \eta$. But this contradicts the uniqueness of the center v of ι_v, since $\eta^{-1} \iota_v \eta$ has center $v^\eta = u$, cf. (12.4). □

18.15 Theorem. *The elliptic motion group Φ is a compact, connected simple Lie group of dimension 52, hence it is isomorphic to the exceptional Lie group $\mathrm{F}_4 = \mathrm{F}_4(-52)$. In particular, Φ is simply connected.*

Proof. By (18.2), Φ is isomorphic to the closed subgroup $\widetilde{\Phi}$ of the compact group $\mathrm{SO}(V, \beta) \leq \mathrm{GL}(V)$, hence Φ is compact, and a Lie group (94.3). It is a simple group by (18.14). In particular, it is connected, since the connected component of the identity, which is a normal subgroup, is non-trivial; the latter fact can be obtained in exactly the same way as in the proof of (17.13) for the group $\mathrm{S}\nabla$, which is a subgroup of Φ, cf. (18.6). By (18.10), Φ acts transitively on the point set $\mathrm{P}_2\mathbb{O}$, which therefore is homeomorphic to the homogeneous space Φ/Φ_o, cf. (96.9a). Now recall that $\mathrm{P}_2\mathbb{O}$ is a 16-dimensional manifold; furthermore, we know from (18.8) that $\Phi_o \cong \mathrm{Spin}_9\mathbb{R}$ is a Lie group of dimension 36. Thus, $16 = \dim \Phi/\Phi_o = \dim \Phi - \dim \Phi_o = \dim \Phi - 36$, cf. (94.3), whence $\dim \Phi = 52$. We now use the classification of simple Lie groups, see (94.32 and 33), according to which the exceptional Lie group $\mathrm{F}_4 = \mathrm{F}_4(-52)$ is the only compact simple Lie group of dimension 52.

The classification yields, in particular, that Φ is simply connected. Without the classification, the latter fact can be seen from the transitive action of Φ on $\mathrm{P}_2\mathbb{O} \approx \Phi/\Phi_o$. One uses the exact homotopy sequence (96.12) and the fact that $\Phi_o \cong \mathrm{Spin}_9\mathbb{R}$ and $\mathrm{P}_2\mathbb{O}$ are simply connected (51.28). □

We record explicitly the description of $\mathrm{P}_2\mathbb{O}$ as a homogeneous space which we have just obtained.

18.16 Corollary. $\mathrm{F}_4/\mathrm{Spin}_9\mathbb{R}$ *is homeomorphic to* $\mathrm{P}_2\mathbb{O}$. □

18.17 Remark. According to (18.3), the linear group $\widetilde{\Phi}$ inducing the elliptic motion group Φ leaves the orthogonal space U^+ of $e^+ = (0, 0, 0; 1, 1, 1)$ in $\mathbb{O}^3 \times \mathbb{R}^3$ invariant. One thus obtains a linear representation of $\mathrm{F}_4 \cong \Phi \cong \widetilde{\Phi}$ of dimension 26. This is the first fundamental representation of F_4.

18.18 Proposition. *The elliptic motion group Φ is a maximal compact subgroup of the group Σ of all collineations of $\mathrm{P}_2\mathbb{O}$.*

Proof. Let K be a compact subgroup of Σ containing Φ; we want to show that $K = \Phi$. By transitivity of Φ on the set of lines (18.10) it suffices to show that $K_W = \Phi_W$, cf. (91.3).

We consider the transformation groups $K_W|_W$ and $\Phi_W|_W = SO_9{}^W$ (18.8) induced by these stabilizers on the point row W. From the fundamental theorem for the octonion projective line (15.6) we know that $\Sigma_W|_W = PSO_{10}(1)^W$; thus, $\Phi_W|_W = SO_9{}^W$ is a maximal compact subgroup of $\Sigma_W|_W$, whence $K_W|_W = \Phi_W|_W$. It follows that $K_W = \Phi_W$ as asserted, once we know that the kernel $K_{[W]}$ of the action of K_W on W coincides with $\Phi_{[W]}$. This in turn is an immediate consequence of the fact proved next that $\Phi_{[W]}$ is a maximal compact subgroup of $\Sigma_{[W]}$.

The Frattini argument (91.1a) tells us that $\Sigma_{[W]}$ is the semidirect product of the translation group $T \cong \mathbb{R}^{16}$ by the stabilizer of o in $\Sigma_{[W]}$. According to (12.13), this stabilizer is the group $\Sigma_{[o,W]} = \{ \mu_r \mid r \in \mathbb{R} \setminus \{0\} \}$ of real homotheties. From (18.8), we know that $\Phi_{[W]} = \{id, \mu_{-1}\}$, and clearly this is a maximal compact subgroup of $\Sigma_{[o,W]}$. Now, to finish the proof, we use the fact that $T \cong \mathbb{R}^{16}$ does not have non-trivial compact subgroups. Under the continuous homomorphism $\Sigma_{[W]} = \Sigma_{[o,W]} \cdot T \to \Sigma_{[o,W]}$ obtained by factoring out T, the image of a compact subgroup Ψ of $\Sigma_{[W]}$ containing $\Phi_{[W]}$ equals $\Phi_{[W]}$, so that $\Phi_{[W]} \leq \Psi \leq \Phi_{[W]} \cdot T$. As the compact subgroup $T \cap \Psi$ of T is trivial, we conclude that $\Psi = \Phi_{[W]}$. □

18.19 Theorem. *The group Σ of all collineations of $\mathcal{P}_2\mathbb{O}$ is isomorphic to the real exceptional simple Lie group* $E_6(-26)$.

Proof. From (17.10) and (18.18 and 15), we know that Σ is a simple Lie group of dimension 78 having a maximal compact subgroup isomorphic to $F_4 = F_4(-52)$. According to the classification of simple Lie groups, the only such group is $E_6(-26)$. (For references concerning this classification, see (94.32).) □

Note. Springer [62] p. 247 gives a different proof of the preceding result (in a much more general context). He shows that Σ is distinguished among all simple Lie groups of the same dimension by having an outer automorphism of order 2. Such an automorphism can be obtained via conjugation by the standard elliptic polarity π. Indeed, if this were an inner automorphism, which could also be obtained via conjugation by a collineation σ, then the stabilizer Σ_W of the line W would fix the point $W^{\pi\sigma}$, but Σ_W has no fixed points, see e.g. (17.2).

The hyperbolic motion group

We now discuss the hyperbolic motion group Φ^- defined in (18.0 and 2). The methods will be analogous to those used above for the elliptic motion group; this will allow for a less detailed presentation.

As the reflection ι_o fixes the points $o = (0,0)$, $u = (0)$, and $v = (\infty)$, their polars under the standard elliptic polarity π and the standard hyperbolic polarity $\pi^- = \iota_o \pi$ coincide. In (16.7), these polars have been computed; we recall the result.

18.20.
$$o^{\pi^-} = o^\pi = W = [\infty] = uv$$
$$u^{\pi^-} = u^\pi = Y = [0] = ov$$
$$v^{\pi^-} = v^\pi = X = [0,0] = ou.$$

In other words, o, u, and v constitute a *polar triangle* with respect to both π and π^-.

The absolute points of the standard hyperbolic polarity π^- have already been determined in (18.1); we now supplement this by describing the interior and the exterior points. The result is analogous to (13.17).

18.21 Proposition. *The set of absolute points of the standard hyperbolic polarity π^-, written in affine coordinates, is*

$$Q = \left\{ (x, y) \in \mathbb{O} \times \mathbb{O} \mid \|x\|^2 + \|y\|^2 = 1 \right\}.$$

The set of exterior points, i.e., of non-absolute points which are incident with some absolute line, is

$$E = \left\{ (x, y) \in \mathbb{O} \times \mathbb{O} \mid \|x\|^2 + \|y\|^2 > 1 \right\} \cup W.$$

The set of interior points, i.e., of points which are not incident with any absolute line, is

$$I = \left\{ (x, y) \in \mathbb{O} \times \mathbb{O} \mid \|x\|^2 + \|y\|^2 < 1 \right\}.$$

The sets I and E are open, connected subsets of $P_2\mathbb{O}$.

Proof. By (18.12), the action of the stabilizer $\Phi_o^- = \Phi_o$ (18.4) on the affine point set $\mathbb{O} \times \mathbb{O}$ preserves the norm form $\|(x,y)\|^2 = \|x\|^2 + \|y\|^2$ and is transitive on the spheres with center o. In particular, Φ_o^- is transitive on Q; consequently, Φ_o^- is also transitive on the set of absolute lines, which is just the image of Q under π^-. Therefore, in order to establish the asserted description of the exterior points, it suffices to consider the points of a single absolute line, e.g., of the polar of the absolute point $(1, 0)$. By (16.7), this polar is the line $(1, 0)^{\pi^-} = (1, 0)^{\iota_o \pi} = (-1, 0)^\pi = [1]$. Under Φ_o^-, an affine point (x, y) may be transformed into a point $(1, y')$ of $[1]$ if, and only if $\|x\|^2 + \|y\|^2 = 1 + \|y'\|^2$, and it is not absolute if, and only if $y' \neq 0$, or equivalently, if $\|x\|^2 + \|y\|^2 > 1$. The points of the line W at infinity are all exterior, since $v = (\infty) \in [1]$ is exterior and since $\Phi_o^- = \Phi_o$ is transitive on W, see (18.8). Thus E is as stated, and the description of $I = P_2\mathbb{O} \setminus (Q \cup E)$ follows.

120   1 The classical planes

The affine point set $\mathbb{O} \times \mathbb{O} = P_2\mathbb{O} \setminus W$ is open in $P_2\mathbb{O}$ (16.12), and the open unit disk I of $\mathbb{O} \times \mathbb{O} = \mathbb{R}^{16}$ clearly is connected. The set E is the complement in $P_2\mathbb{O}$ of the closed unit disk $\{ (x, y) \in \mathbb{O} \times \mathbb{O} \mid \|x\|^2 + \|y\|^2 \leq 1 \}$, which is compact, hence E is open in $P_2\mathbb{O}$. Finally, we prove that E is connected. A projective line, and a projective line minus a point are connected point sets, the latter being homeomophic to $\mathbb{R}^8$, see (16.13). Consequently, for an absolute line L, the set $(L \setminus \{L^\pi\}) \cup W$, composed of two connected point sets intersecting in $L \wedge W$, is connected, and so is the union E of all these sets. □

18.22 The stabilizer of an exterior point. The stabilizer $\Phi_o^- = \Phi_o$ (18.4) of the interior point o is already known from our investigation of the elliptic motion group. By analogous methods, we now study the stabilizer of an exterior point, e.g., of $v = (\infty)$. The stabilizer Φ_v^- leaves the polar $v^\pi = X$ invariant; in fact

$$\Phi_v^- = \Phi_X^- .$$

In order to describe the action of this stabilizer on the point row X, we relate X to the line W which was used in Section 15 as the standard projective line. Specifically, we identify the point rows X and W via the bijection provided by the triality collineation τ (16.6). In (15.10 and 12), with $n = 8$, subgroups $O_9(1)^W \supseteq SO_9(1)^W \supseteq O_9'(1)^W$ of $PSO_{10}(1)^W$ have been described. The transformation groups of the point row X obtained from these groups via conjugation by the bijection $\tau : X \to W$ will be denoted by

$$O_9(1)^X \supseteq SO_9(1)^X \supseteq O_9'(1)^X .$$

These groups can also be described more directly. We shall only state and sketch this here, because it will not be used in the sequel. Analogously to (16.2), where the point row W is interpreted as a quadric in the subspace $V_3 = \{0\} \times \{0\} \times \mathbb{O} \times \mathbb{R}^2 \times \{0\}$ of $V = \mathbb{O}^3 \times \mathbb{R}^3$, the point row X is the quadric of the quadratic form q_1 defined on the subspace $V_1 = \mathbb{O} \times \{0\} \times \{0\} \times \{0\} \times \mathbb{R}^2$ by $q_1(x_1, 0, 0; 0, \xi_2, \xi_3) = \overline{x_1}x_1 - \xi_2\xi_3$. Analogously to (15.10(4)), we have $O_9(1)^X = PO(V_1, q_1)_{Rh_1^-}|_X$, where $h_1^- = (0, 0, 0; 0, 1, -1) \in V_1$, and similarly for the subgroup $SO_9(1)^X$. The group $O_9'(1)^X$ is the commutator subgroup of $O_9(1)^X$, analogously to (15.12). This group now describes the action of Φ_v^- on X.

Theorem. *The transformation group $\Phi_v^-|_X$ of the point row X which is induced by $\Phi_v^- = \Phi_X^-$ is*

$$\Phi_v^-|_X = O_9'(1)^X .$$

The kernel of the epimorphism

$$\mathrm{pr} : \Phi_v^- \to O_9'(1)^X : \varphi \mapsto \varphi|_X$$

has two elements, the non-trivial element being the reflection with axis X and center v

$$\iota_v = \mu_{1,-1} : (x,y) \mapsto (x,-y), \ (s) \mapsto (-s),$$

see (12.17c and 18.11). The kernel $\ker \operatorname{pr} = \{\operatorname{id}, \iota_v\}$ is the center of Φ_v^- and the only non-trivial, proper normal subgroup. In particular, Φ_v^- is connected.

The stabilizer Φ_v^- leaves the sets

$$X \cap I = \{(x,0) \mid \|x\|^2 < 1\} \quad \text{and} \quad X \cap E = \{(x,0) \mid \|x\|^2 > 1\} \cup \{(0)\}$$

of interior and exterior points of X invariant and acts transitively on each of them; moreover, this stabilizer is triply transitive on the set

$$X \cap Q = \{(x,0) \mid \|x\|^2 = 1\}$$

of absolute points on X.

Proof. We shall proceed in close analogy to the proof in the elliptic case (18.8). Here, we are dealing with the stabilizer of the line X, whereas there we were concerned with the line W. Therefore, in order to make the analogy more direct, we shall prove the statement into which our theorem is transformed by conjugation with the triality collineation τ (16.6). The group Φ^- will be replaced by the conjugate

$$\Phi^= = \tau^{-1} \Phi^- \tau,$$

which is easily seen to be the motion group of the conjugate polarity

$$\pi^= = \tau^{-1} \pi^- \tau,$$

i.e., the group of all collineations commuting with $\pi^=$. The point v and the line X will be replaced by $v^\tau = o$ and by $X^\tau = W$, that is, we study the stabilizer $\Phi_o^=$ and its action on $W = o^{\pi^=}$. We shall show that $\Phi_o^=|_W = O_9'(1)^W$, and that the kernel of this action is generated by $\tau^{-1} \iota_v \tau = \iota_{v^\tau} = \iota_o$.

According to (18.0), the polarity π^- is obtained from the non-degenerate symmetric bilinear form β^- of $\mathbb{O}^3 \times \mathbb{R}^3$ in the sense that the polar of a point $p \in P_2\mathbb{O}$ is the orthogonal space of p with respect to β^-. Hence, the polarity $\pi^=$ is obtained from the following non-degenerate symmetric bilinear form $\beta^=$:

$$\beta^=(w \mid z) = \beta^- \left(\tilde{\tau}^{-1} w \mid \tilde{\tau}^{-1} z \right),$$

where $\tilde{\tau}$ is the linear transformation inducing the triality collineation τ, see (16.6). In coordinates, $\beta^=$ is described as follows:

$$\beta^= \left((x_\nu; \xi_\nu)_\nu \mid (y_\nu; \eta_\nu)_\nu \right) = - \langle x_1 \mid y_1 \rangle + \langle x_2 \mid y_2 \rangle - \langle x_3 \mid y_3 \rangle + \sum_{\nu=1}^{3} \xi_\nu \eta_\nu.$$

From the general result in (18.2) we know that the motion group $\Phi^=$ is induced by the following group of $\beta^=$-orthogonal transformations of $\mathbb{O}^3 \times \mathbb{R}^3$:

$$\widetilde{\Phi}^= = \{\, \widetilde{\varphi} \in \mathrm{O}(V, \beta^=) \mid \widetilde{\varphi}(H) = H \,\}\,.$$

Of course, $\widetilde{\Phi}^=$ is conjugate under $\widetilde{\tau}$ to the group $\widetilde{\Phi}^- \subseteq \mathrm{SO}(V, \beta^-)$ inducing Φ^-. Now, $\widetilde{\Phi}^-$ leaves the subspace U^- described in (18.3) invariant; hence, the image of U^- under $\widetilde{\tau}$,

$$U^= = \{\, (x_\nu; \xi_\nu)_\nu \mid -\xi_1 + \xi_2 - \xi_3 = 0 \,\}\,,$$

is invariant under $\widetilde{\Phi}^=$.

One now proceeds in direct analogy with the proof of (18.8). One considers the intersection of $U^=$ with the subspace $V_3 = \{0\} \times \{0\} \times \mathbb{O} \times \mathbb{R}^2 \times \{0\}$, and one identifies V_3 with $\widehat{\mathbb{O}} = \mathbb{O} \times \mathbb{R}^2$ in the obvious way; then, this intersection becomes

$$T^- = \{\, (x, r, r) \mid x \in \mathbb{O},\, r \in \mathbb{R} \,\}\,.$$

It is the orthogonal space of

$$h^- = (0, 1, -1) \in \widehat{\mathbb{O}}$$

for the bilinear form β_q associated with the standard quadratic form q on $\widehat{\mathbb{O}}$ determining the geometry of W, see (15.1), but also for the symmetric bilinear form obtained by restriction of $\beta^=$ to $V_3 \cong \widehat{\mathbb{O}}$. Furthermore, on T^- the symmetric bilinear forms β_q and $-\beta^=$ coincide. (The minus sign slightly disturbs the analogy with (18.8), but that does not affect the argument.)

One may now conclude exactly as in (18.8) that $\Phi_o^=|_W$ is contained in the stabilizer of $\mathbb{R}h^-$ in $\mathrm{PSO}(V, q)$, which is $\mathrm{SO}_9(1)^W$ according to (15.10(4)). We even want to prove that $\Phi_o^=|_W \subseteq \mathrm{O}_9'(1)^W$; therefore we distinguish interior and exterior points on W with respect to the polarity $\pi^= = \tau^{-1}\pi^-\tau$. The set of interior points is the τ-image of the set $X \cap I$ of interior points on X for the polarity π^-. By (18.21) and (16.6), this image set is

$$(X \cap I)^\tau = \{\, (s) \mid \|s\|^2 < 1 \,\} = W_-\,;$$

W_- is the notation of (15.12). Analogously, the set of exterior points on W for $\pi^=$ is

$$(X \cap E)^\tau = \{\, (s) \mid \|s\|^2 > 1 \,\} \cup \{(\infty)\} = W_+\,.$$

Now $\Phi_o^=$ preserves these sets, so that $\Phi_o^=|_W$ is contained in the subgroup $\mathrm{O}_9'(1)^W$ of $\mathrm{SO}_9(1)^W$ according to (15.12).

For the proof of the converse inclusion, one also needs to know the behaviour of the collineations $\mu_{r,t}$ of (12.17) under conjugation by the polarity in question. The triality collineation τ commutes with π (18.5), and by definition we have $\pi^- = \pi \iota_o$, so that

(1) $$\pi^= = \tau^{-1}\pi\iota_o\tau = \pi\tau^{-1}\iota_o\tau = \pi\iota_{o^\tau} = \pi\iota_u \, .$$

Furthermore, the reflection $\iota_u = \mu_{-1,1}$ commutes with $\mu_{r,t}$. From the fact (18.7) that $\pi\mu_{r,t} = \mu_{r,t}^{-1}\pi$ we now derive the analogous statement for $\pi^=$, namely $\pi^=\mu_{r,t} = \pi\iota_u\mu_{r,t} = \pi\mu_{r,t}\iota_u = \mu_{r,t}^{-1}\pi\iota_u = \mu_{r,t}^{-1}\pi^=$. Using this exactly as in step 2) of (18.8), one obtains that every element $\bar{\sigma} \in O_9'(1)^W$ is induced by a collineation φ' fixing o and W and satisfying

(2) $$\varphi'\pi^=\varphi'^{-1} = \pi^=\mu_\varepsilon \quad \text{with } \varepsilon \in \{1, -1\} \, .$$

In order to finish the proof of the identity $\Phi_o^=|_W = O_9'(1)^W$, one has to show that the case $\varepsilon = -1$ cannot occur. According to (15.12), since $\varphi'|_W = \bar{\sigma} \in O_9'(1)^W$, the set W_- is invariant under φ', so that $(0)^{\varphi'}$ is an interior point of the polarity $\pi^=$. In particular, the image $L^{\varphi'}$ of a line L passing through (0) is not absolute, i.e., it is not incident with $L^{\varphi'\pi^=}$. Consequently, L is not incident with $L^{\varphi'\pi^=\varphi'^{-1}} = L^{\pi^=\mu_\varepsilon}$, see (2). Now, for $L = [0, 1]$, our computations of certain π-images in (16.7) together with equation (1) give $[0, 1]^{\pi^=} = [0, 1]^{\pi\iota_u} = (0, -1)^{\iota_u} = (0, -1)$, whence $[0, 1]^{\pi^=\mu_\varepsilon} = (0, -\varepsilon)$. For $\varepsilon = -1$, the latter point would be incident with the line $[0, 1]$, contradicting the observation made above.

From (18.4 and 6), we know that $\Phi_{o,u,v}^- = S\nabla$, and according to (17.12) this group is invariant under conjugation by τ, so that it coincides with the stabilizer $\Phi_{o,u,v}^=$. One may now conclude exactly as in (17.13) for the subgroup $S\nabla$ that the kernel of the restriction homomorphism $\mathrm{pr} : \Phi_o^= \to \Phi_o^=|_W = O_9'(1)^W$ is $\{\mathrm{id}, \iota_o\}$, that it is the center and the only non-trivial, proper normal subgroup of $\Phi_o^=$, and that $\Phi_o^=$ is connected. One employs the simplicity of $O_9'(1)^W \cong O_9'(\mathbb{R}, 1)$, a proof of which may be found in Dieudonné [71] Chap. II §9C), p. 59 together with §8 no. 2b), p. 56.

As for the transitivity assertions of our theorem, we finally remark that $W_- = (X \cap I)^\tau$ and $W_+ = (X \cap E)^\tau$ are orbits of $O_9'(1)^W = \Phi_o^=|_W$, and that this group is triply transitive on $W \cap Q = \{ (s) \mid \|s\|^2 = 1 \} = W_1$, see (15.12). Under transformation by τ, we obtain the corresponding transitivity properties of Φ_v^-, as asserted. □

18.23 Corollary. *The hyperbolic motion group Φ^- is flag transitive on the 'interior hyperbolic plane' consisting of the set I of interior points and of all the lines passing through interior points. It is transitive on the set E of exterior points, and doubly transitive on the set $Q \approx \mathbb{S}_{15}$ of absolute points.*

Remarks. 1) Of course, Φ^- cannot act flag transitively on the 'exterior hyperbolic plane' whose lines are all the lines through exterior points (exactly for the same reason as in (13.17), see the corresponding remark there).

2) Tits [53] 5.7 and 6.1, p. 324 ff derives this result from the classification of polarities, whereas we are rather proceeding the other way round, see the remark following the proof of (18.29).

Proof of (18.23). From (18.12), we know that $\Phi_o^- = \Phi_o$. By (18.8), this subgroup acts transitively on W and, consequently, on the set $\mathcal{L}_o$ of lines passing through o. In particular, every point orbit of Φ^- contains a point of X. Transitivity of Φ^- on I and E now follows from transitivity of $\Phi_v^-|_X = O_9'(1)^X$ on $X \cap I$ and $X \cap E$, see (18.22). Flag transitivity on the interior hyperbolic plane then is immediate from transitivity of Φ_o^- on $\mathcal{L}_o$.

The fact that Φ^- is doubly transitive on Q will follow from transitivity on the set of lines passing through interior points and from the fact that Φ_v^- is doubly transitive (in fact even triply transitive) on $X \cap Q$, once we have shown that a line L joining two points of Q contains an interior point. Since $W \cap Q = \emptyset$, we certainly have $L \neq W$, and by transitivity of Φ^- on E, we may assume that $L \wedge W = (0)$. We have to show that a line $[0, t]$ through (0) either contains an interior point, or has at most one point in common with Q. By (18.21), the point (x, t) on $[0, t]$ is an interior point if, and only if $\|x\|^2 + \|t\|^2 < 1$. Hence, if none of these points is an interior point, we have that $\|t\|^2 \geq 1$. For $\|t\|^2 > 1$, there are no absolute points, either, and if $\|t\|^2 = 1$, then $(0, t)$ is the only absolute point on the line $[0, t]$. □

18.24 Corollary. *The lines through interior points are precisely the polars of exterior points. The stabilizer Φ_L^- of such a line L and the stabilizer Φ_q^- of an exterior point q are connected two-fold covering groups of $O_9'(\mathbb{R}, 1)$.*

Proof. This has been proved for the special cases $L = X = v^\pi$ and $q = v$ in (18.22); the assertion in general follows by the transitivity properties of Φ^- derived in (18.23). □

18.25 Theorem. *The hyperbolic motion group Φ^- is simple.*

Proof. The argument is similar and in many steps identical to the proof of the corresponding result for the elliptic case, (18.14). The complications of our proof, when compared to Tits [53] 6.6, p. 327, are due to the fact that we do not use the structure theory of Lie groups, because our aim is to give an elementary proof.

1) A normal subgroup which fixes an interior point consists of the identity alone. Indeed, the set of its fixed points is invariant under Φ^-, cf. (91.1d), and Φ^- is transitive on the set I of interior points (18.23), so that I consists of fixed points. Now, I is open; hence, a collineation which fixes every point of I is the identity, because every point can be represented as the intersection of two lines carrying many points from I.

2) Let N be a normal subgroup of Φ^-; we have to show that $N = \{\mathrm{id}\}$ or $N = \Phi^-$. Exactly as in the elliptic case (18.14), we may assume that N is connected and $N \neq \{\mathrm{id}\}$.

3) The orbit o^N is open in $P_2\mathbb{O}$; this follows as in the elliptic case, since $\Phi_o^- = \Phi_o$ by (18.4).

4) By transitivity of Φ^- on the open connected set I (18.21 and 23), it follows from step 3) that N, too, is transitive on I.

5) Since Φ^- is also transitive on the set E of exterior points (18.23), the principle of conjugate stabilizers (91.1a) yields that the stabilizer N_q of an exterior point q is conjugate to N_v, $v = (\infty)$, and analogously the stabilizer N_p of an interior point p is conjugate to N_o. Concerning these normal subgroups of Φ_v^- and of $\Phi_o^- = \Phi_o$, we obtain the following alternatives from (18.22 and 8): either $N_v = \Phi_v^-$ or $N_v \leq \{\mathrm{id}, \iota_v\}$, and either $N_o = \Phi_o^-$ or $N_o \leq \{\mathrm{id}, \iota_o\}$.

6) We now consider the possibilities for N_o separately.

Case 1: If $N_o = \Phi_o^-$, then transitivity of N on I implies that $N = \Phi^-$, cf. (91.3).

Case 2: If $N_o \leq \{\mathrm{id}, \iota_o\}$, then $N_v = \{\mathrm{id}\}$, since ι_v fixes o and hence does not belong to N by assumption. In the same way, we now see that the reflection ι_o, which fixes v, does not belong to N, so that $N_o = \{\mathrm{id}\}$. Thus, the situation is as follows: N is sharply transitive on I, and all the stabilizers in N of exterior points are trivial. It shall be shown in a moment that this implies sharp transitivity of N on E, as well. In particular, there is exactly one element $\eta \in N$ such that $v^\eta = u$. From this, one may then derive a contradiction exactly as in the elliptic case.

It remains to be shown that, in the situation at hand, N would have to be transitive on E. We first derive further properties of N from the sharply transitive action on I, using the continuous bijective map

$$\alpha : N \to I : \eta \mapsto o^\eta.$$

At this point we meet extra difficulties due to our policy not to use Lie group theory in an essential way. In particular, we do not assume N to be an analytic subgroup, hence we do not yet know much about the topology of N. Therefore, we construct a large compact subset in N; on that set, α then induces a homeomorphism. We start from the fact that $\Phi_o^- = \Phi_o$ is compact and connected (18.8). For fixed $\eta \in N \setminus \{\mathrm{id}\}$, the subset

$$M = \left\{ \varphi_2^{-1} \left(\varphi_1^{-1} \eta \varphi_1 \eta^{-1} \right) \varphi_2 \mid \varphi_1, \varphi_2 \in \Phi_o \right\}$$

of Φ^- therefore is compact and connected as well, being a continuous image of $\Phi_o \times \Phi_o$, and M is contained in the normal subgroup N. Since M is invariant under conjugation by elements of $\Phi_o = \Phi_o^-$, the image $\alpha(M) = o^M$ is the union of the Φ_o-orbits of its elements, and these orbits are 15-spheres centered at o, cf. (18.12). Obviously, $o \in \alpha(M)$, but $\alpha(M)$ contains more points, e.g., the subset $\{o^{\varphi^{-1}\eta\varphi\eta^{-1}} \mid \varphi \in \Phi_o\} = o^{\eta\Phi_o\eta^{-1}}$, which is a bijective image of the non-trivial Φ_o-orbit of the point $o^\eta \ne o$. By a connectedness argument as in the elliptic case, see step 3) of the proof of (18.14), it follows that $\alpha(M) = o^M$ contains a neighbourhood of o. In particular, the homeomorphic subset $M \subseteq N$ contains an open subset B which is homeomorphic to an open ball of dimension 16.

Now we return to E, which is a connected open subset of $P_2\mathbb{O}$ (18.21), and hence is a 16-dimensional manifold (16.14). The free action of N on E gives rise to continuous and bijective maps of N onto each of its orbits in E. The image of the 16-dimensional ball $B \subseteq N$ under each of these maps is open in E by domain invariance (51.18 and 19), so that, as a consequence of homogeneity, each orbit of N in E is open. Since E is connected, there is only one orbit, i.e., N is transitive on E. □

18.26 Corollary. *The hyperbolic motion group Φ^- is a connected, simply connected, simple Lie group of dimension 52; it is isomorphic to the real exceptional Lie group $F_4(-20)$.*

Proof. By (18.2), Φ^- is isomorphic to the closed subgroup $\widetilde{\Phi}^-$ of $GL(V)$, so that Φ^- is a Lie group (94.3). It is simple by (18.25). In particular, Φ^- is connected, since the connected component of the identity is non-trivial. By (18.23), Φ^- acts transitively on the set I of interior points, which is the open unit disk in $\mathbb{O} \times \mathbb{O} = \mathbb{R}^{16}$ (18.21). The stabilizer $\Phi_o^- = \Phi_o \cong \mathrm{Spin}_9\mathbb{R}$ (18.4 and 8) is simply connected and has dimension 36. The exact homotopy sequence (96.12) associated with this transitive action now shows that Φ^- is simply connected, as well. Furthermore, I is homeomorphic to the manifold Φ^-/Φ_o of dimension $\dim \Phi^- - \dim \Phi_o$, cf. (96.9a) and (94.3a), whence $\dim \Phi^- = \dim I + \dim \Phi_o = 16 + 36 = 52$.

Now one applies the classification of simple Lie groups, cf. (94.32 ff). The only simple Lie groups of dimension 52 are $F_4(-52)$, $F_4(-20)$, and $F_4(4)$. The group $F_4(-52)$ is compact, whereas Φ^- is not. A maximal compact subgroup of $F_4(4)$ has dimension 24, but the compact subgroup $\Phi_o^- = \Phi_o$ of Φ^- has dimension 36. Thus the only remaining possibility is $F_4(-20)$. □

18.27 Motion groups over $\mathbb{R}$, $\mathbb{C}$, and $\mathbb{H}$. It is well known that the elliptic motion groups $PU_3\mathbb{F}$ of the planes $\mathcal{P}_2\mathbb{F}$ for $\mathbb{F} \in \{\mathbb{R}, \mathbb{C}, \mathbb{H}\}$ and the hyperbolic motion groups $PU_3(\mathbb{F}, 1)$ for $\mathbb{F} \in \{\mathbb{C}, \mathbb{H}\}$, see (13.13), are simple, as well; in the real hyperbolic motion group $PO_3(\mathbb{R}, 1)$, the commutator subgroup $PO_3'(\mathbb{R}, 1)$ is a simple subgroup of index 2. Indeed, these groups are among the almost simple Lie groups, cf. (94.32 and 33), and they are easily seen to be center free, hence they are simple, cf. (94.21).

In the hyperbolic case, there are purely algebraic arguments not using Lie theory which establish the corresponding results quite generally over arbitrary fields, with the exception of a few small finite fields. In this way, the real hyperbolic group is covered for instance by Dieudonné [71] Chap. II §9A) no. 3), p. 58, and the complex and quaternion hyperbolic group by Dieudonné [71] Chap. II §5, p. 49 ff together with Chap. II §4A,B), p. 45 ff. For a direct proof in the real elliptic case, see Artin [57] Theorem 5.3, p. 178.

On the other hand, one may obtain proofs valid over $\mathbb{R}$, $\mathbb{C}$, and $\mathbb{H}$ by restricting the arguments of the present section, in particular (18.14 and 25), to $\mathcal{P}_2\mathbb{F}$ as a subplane of $\mathcal{P}_2\mathbb{O}$, much in the same way as in (18.9), where the point stabilizers of the elliptic motion groups were considered.

Classification of polarities

At the end of this section, we want to prove that, up to equivalence, $\mathcal{P}_2\mathbb{O}$ has only one further polarity besides the standard elliptic and hyperbolic polarities (18.0). A representative of this third class of polarities shall now be described.

18.28 Construction: The standard planar polarity. We consider the following involutory automorphism of the octonion algebra $\mathbb{O} = \mathbb{H} \times \mathbb{H}$, see (11.31):

$$\varepsilon : \mathbb{H} \times \mathbb{H} \to \mathbb{H} \times \mathbb{H} : (a, b) \mapsto (a, -b) \ .$$

According to (12.8), that automorphism gives rise to an involutory collineation δ of $\mathcal{P}_2\mathbb{O}$,

$$\delta : (x, y) \mapsto (x^\varepsilon, y^\varepsilon), \ (s) \mapsto (s^\varepsilon) \ .$$

This collineation belongs to $S\nabla = \Phi_{o,u,v}$, see (12.17b) and (18.6); in particular, it commutes with the standard elliptic polarity π, so that $\pi\delta$ is a polarity again. We call it the *standard planar polarity*. (The name reflects the fact that δ is a planar involution in the sense of (23.14)).

18.29 Theorem: Classification of polarities. *Every polarity of the octonion plane is equivalent to the standard elliptic polarity, to the standard hyperbolic polarity, or to the standard planar polarity.*

This theorem is due to Tits [53, 54]. He gave an explicit proof only for the so-called Hermitian (elliptic or hyperbolic) polarities, and a sketch of proof for the general result. Our proof is different from his. An important step relies on the following lemma.

18.30 Lemma. *Every involution in Φ is conjugate either to the reflection ι_o or to the planar involution δ of* (18.28).

Remark. In particular, it does not matter which planar involution is used to define a planar polarity.

Proof. Let $\vartheta \in \Phi$ be an involution. According to a general result on involutory collineations of projective planes, see (23.17), either ϑ is an axial collineation, or it is planar, i.e., it fixes a non-degenerate quadrangle.

First, we deal with the case that ϑ is an axial collineation. Up to conjugation, we may assume that the axis of ϑ is W, by transitivity of Φ on the set of lines (18.10). Then ϑ is the unique non-identical elliptic motion ι_o with this axis, see (18.8).

Now let ϑ be a planar involution. Again, flag transitivity of Φ allows us to assume that W is a fixed line of ϑ, and that $v = (\infty)$ is a fixed point. Then ϑ also fixes the points $o = W^\pi$ and $v^\pi \wedge W = [0,0] \wedge W = (0) = u$. By planarity, ϑ has a further fixed point p which is not collinear with any two of the three points o, u, and v. Using transitivity of the full collineation group Σ on the set of non-degenerate quadrangles (17.6), we find a collineation $\gamma \in \nabla = \Sigma_{o,u,v}$ such that $p^\gamma = e = (1,1)$; then $\gamma^{-1}\vartheta\gamma \in \Sigma_{o,u,v,e}$. By our analysis of ∇ in (12.17), we have a decomposition $\gamma = \psi\mu$ where $\psi \in S\nabla$ and $\mu = \mu_{r,t}$ for suitable $r, t \in \mathbb{R} \setminus \{0\}$. The collineations in $\Sigma_{o,u,v,e}$ are described by automorphisms of $\mathbb{O}$ (12.8), which are $\mathbb{R}$-linear (11.28); in particular, $\gamma^{-1}\vartheta\gamma$ fixes $(r,t) = e^\mu$, whence $\psi^{-1}\vartheta\psi = \mu\gamma^{-1}\vartheta\gamma\mu^{-1}$ also belongs to $\Sigma_{o,u,v,e}$. Now any two involutions in Aut $\mathbb{O}$ are conjugate (11.31), so that $\psi^{-1}\vartheta\psi$ is conjugate within $\Sigma_{o,u,v,e}$ to the involution δ of (18.28). Because of the facts $\Sigma_{o,u,v,e} \subseteq S\nabla$ (17.15), $\psi \in S\nabla$, and $S\nabla \subseteq \Phi$ (18.6), our claim is proved. □

Proof of (18.29). Let ϱ be any polarity. First we collect a few facts which hold for polarities in general. An absolute point of ϱ is incident with just one absolute line, namely its polar, cf. Hughes–Piper [73] XII.2 Lemma 12.1, p. 239. So if there is a line Z all of whose points are absolute, then the absolute lines are just the polars of the points of Z, and the absolute points are precisely the points of Z. (In fact, for polarities of $\mathcal{P}_2\mathbb{O}$ such a line will not exist.) In particular, there is always a point a which is not absolute; in case a line Z of the mentioned kind exists, we assume in addition that $a \neq Z^\varrho$. There is a non-absolute point b on the line a^ϱ. Let $c = a^\varrho \wedge b^\varrho$; then abc is easily seen to be a *polar triangle* with respect to ϱ, i.e., a triple of non-collinear points such that the polar of any of the points is the line joining the other two.

For a collineation γ, the equivalent polarity $\gamma^{-1}\varrho\gamma$ has $a^\gamma b^\gamma c^\gamma$ as a polar triangle. Therefore, up to equivalence, transitivity of the group of collineations on the set of triangles (17.2) allows us to assume that the triangle ouv is a polar triangle of ϱ. This triangle is a polar triangle of the standard elliptic polarity, as well (18.20), so that the composition $\varrho\pi$ is a collineation γ fixing o, u, v, and we have

$$\varrho = \gamma\pi.$$

By (12.17), $\gamma \in \nabla = \Sigma_{o,u,v}$ may be written as $\gamma = \psi \mu_{r,t}$ where $\psi \in S\nabla$ and $r, t \in \mathbb{R}$, $r > 0$, $t > 0$, whence $\gamma = \psi \mu^2 = \mu \psi \mu$ with $\mu = \mu_{\sqrt{r},\sqrt{t}}$; note that μ commutes with $\psi \in S\nabla$. From (18.7), we know that $\pi \mu = \mu^{-1} \pi$. Now we consider the following polarity ϱ', which is equivalent to ϱ:

$$\varrho' := \mu^{-1} \varrho \mu = \mu^{-1} \gamma \pi \mu = \mu^{-1}(\mu \psi \mu) \mu^{-1} \pi = \psi \pi = \pi \psi.$$

(For the last identity, recall from (18.6) that $\psi \in S\nabla$ belongs to the elliptic motion group Φ.) Now, using the properties of a polarity, we infer that $\mathrm{id} = \varrho' \varrho' = (\psi \pi)(\pi \psi) = \psi \pi^2 \psi = \psi^2$.

If $\psi = \mathrm{id}$, then $\varrho' = \pi$ is the standard elliptic polarity. If ψ is an involution, then, by (18.30), there is an elliptic motion $\varphi \in \Phi$ such that $\varphi^{-1} \psi \varphi$ is the reflection ι_o or the planar involution δ, and the equivalent polarity $\varphi^{-1} \varrho' \varphi = \varphi^{-1} \pi \psi \varphi = \pi \varphi^{-1} \psi \varphi$ is the standard hyperbolic polarity $\pi^- = \pi \iota_o$ or the standard planar polarity $\pi \delta$. □

In comparison with the approach by Tits [53, 54], we have mainly used our knowledge about collineation groups and motion groups for the classification of polarities, whereas Tits proceeds in another order and uses parts of the classification of polarities to derive properties of the motion groups. For the classification of polarities, he studies the involution which is induced by a polarity on a line (cf. the last part of (18.9) where this involution was considered for the standard elliptic polarity). An important tool is the fact that this involution belongs to the group of conformal transformations (15.2) and therefore can easily be identified in specific cases. For the proof of that fact, Tits relates the involution in question to projectivities (21.6) and shows that the group of projectivities of a line onto itself is the group of direct conformal transformations. The latter result corresponds to the fundamental theorem for the projective line (15.6).

18.31 Note: Polarities of $\mathcal{P}_2 \mathbb{F}$, $\mathbb{F} \in \{\mathbb{R}, \mathbb{C}, \mathbb{H}\}$. By restriction to the subplane $\mathcal{P}_2 \mathbb{F}$ of $\mathcal{P}_2 \mathbb{O}$, the preceding arguments give an analogous classification of the polarities of $\mathcal{P}_2 \mathbb{F}$. From another point of view, this classification has already been discussed in (13.18); there, the problem was reduced to the classification of Hermitian forms, whereas the present argument relies entirely on the analysis of the collineation group. For $\mathbb{F} = \mathbb{R}$, there is no planar polarity. In the case $\mathbb{F} = \mathbb{C}$, the method described here produces the continuous polarities only.

18.32 Note: Absolute points and motions of planar polarities. Let ϱ denote the standard planar polarity of $\mathcal{P}_2 \mathbb{F}$, $\mathbb{F} \in \{\mathbb{C}, \mathbb{H}, \mathbb{O}\}$, and put $n = \dim_\mathbb{R} \mathbb{F} \in \{2, 4, 8\}$. It can be shown that the set Q_ϱ of absolute points of ϱ is a sphere of dimension $3n/2 - 1$, and that the group Ψ of those (continuous) collineations of $\mathcal{P}_2 \mathbb{F}$ that commute with ϱ is doubly transitive on Q_ϱ and transitive on $\mathrm{P}_2 \mathbb{F} \setminus Q_\varrho$. The connected component of Ψ is $\mathrm{PU}_4(\mathbb{F}', 1)^1$, where $\mathbb{F}' \in \{\mathbb{R}, \mathbb{C}, \mathbb{H}\}$ is the fixed field of the involutory automorphism ε of $\mathbb{F}$ which is used to define the standard planar polarity. Explicitly, these are the groups $\mathrm{PSO}_4(\mathbb{R}, 1)$ for $\mathbb{F} = \mathbb{C}$, $\mathrm{PSU}_4(\mathbb{C}, 1)$ for $\mathbb{F} = \mathbb{H}$, and $\mathrm{PU}_4(\mathbb{H}, 1)$ for $\mathbb{F} = \mathbb{O}$. This can be seen from the classification of doubly transitive actions of Lie groups by Tits [55], cf. (96.17). Of course, such an argument is not very geometric. A direct geometric proof could be obtained using the hints by Tits [54] p. 33 and the construction of $\mathcal{P}_2\mathbb{F}$ from the hyperbolic polarity of the projective space $\mathcal{P}_3\mathbb{F}'$ by Faulkner [89]. For $\mathbb{F} = \mathbb{C}$, the result is familiar; the standard complex planar polarity is the polarity corresponding to the standard bilinear form on $\mathbb{C}^3$, see (13.18), and $\mathrm{PO}_3(\mathbb{C}) \cong \mathrm{PSO}_4(\mathbb{R}, 1)$.

18.33 Notes. References to the literature have already been given in (16.8) and (17.17). We mention again the following papers dealing in particular with the subject of this section, polarities and motion groups: Freudenthal [85], Tits [53, 54], Jacobson [60], Springer [62], Springer–Veldkamp [63], Veldkamp [68], Faulkner [70]. As pointed out in earlier notes, most of these papers place these topics in a more general context, and use more sophisticated methods. The elliptic motion group of $\mathcal{P}_2\mathbb{O}$ is also discussed in Harvey [90] p. 289 ff.

The techniques developed by Tits [53, 54] for the classification of polarities of $\mathcal{P}_2\mathbb{O}$ were used by Rutar [82] for more general Cayley algebras. The elliptic motion group Φ coincides with the isometry group of $\mathcal{P}_2\mathbb{O}$ as a Riemannian manifold. Concerning the octonion plane as an object of differential geometry, see Besse [78] Chap. 3, G p. 86 ff, Brada–Pécaut-Tison [87] and the references given in the latter paper. A detailed study of the orbits of Φ_o in $\mathcal{P}_2\mathbb{F}$ for $\mathbb{F} \in \{\mathbb{R}, \mathbb{C}, \mathbb{H}, \mathbb{O}\}$ and of their properties as homogeneous spaces from the point of view of Riemannian geometry was made by Tetzlaff [89].

Chapter 2

Background on planes, coordinates and collineations

This chapter is meant as a reference for incidence geometry; it is a compilation of fundamental geometric facts and theorems, which are used in this book. Furthermore we fix the notation for geometric (and some algebraic) objects. We do not deal with topology here. A reader with a good background in geometry might skip this chapter entirely, or just use it to absorb the notation. For proofs we usually refer to the literature; Hughes–Piper [73] or Pickert [75] can serve as a general reference for Chapter 2, compare also Dembowski [68].

In Section 21 we review the definition and some basic properties of projective and affine planes, and Section 22 explains the coordinatization of arbitrary affine planes, leading to the concept of ternary fields. Collineations and collineation groups are the topic of Section 23, which is important for many classification results in later chapters. In Section 24, we give a rough description of the Lenz–Barlotti hierarchy of projective planes, and in Section 25 we take a closer look at translation planes and their ternary fields.

21 Projective and affine planes

An *incidence structure* is a triple $\mathcal{P} = (P, \mathcal{L}, \mathbf{F})$ of sets with $\mathbf{F} \subseteq P \times \mathcal{L}$. The elements of P are the *points*, the elements of $\mathcal{L}$ are the *lines*, and the elements of $\mathbf{F}$, the incident point–line pairs, are called *flags* (or *incidences*). A set of points is *collinear* if there exists a line incident with all these points.

21.1 Definition. A *projective plane* is an incidence structure $\mathcal{P} = (P, \mathcal{L}, \mathbf{F})$ which satisfies the following axioms:

(i) Any two distinct points p, q are incident with a unique line $p \vee q$, the line joining p and q (also denoted by pq).
(ii) Any two distinct lines L, M are incident with a unique point $L \wedge M$, their intersection point.
(iii) There exists a non-degenerate quadrangle in $\mathcal{P}$, i.e. a set of four points no three of which are collinear.

Often (but not always) the lines are considered as subsets of P, i.e. each line L is identified with its *point row* $\{\, p \in P \mid (p, L) \in \mathbf{F}\,\}$; then one writes $p \in L$ instead of $(p, L) \in \mathbf{F}$ (and one also writes $\mathcal{P} = (P, \mathcal{L})$). For every point p, we denote by $\mathcal{L}_p = \{\, L \in \mathcal{L} \mid (p, L) \in \mathbf{F}\,\}$ the *pencil* of p.

Interchanging the rôles of points and lines in a projective plane $\mathcal{P} = (P, \mathcal{L}, \mathbf{F})$, we obtain the *dual plane* $\mathcal{P}^* = (\mathcal{L}, P, \mathbf{F}^*)$, with $(L, p) \in \mathbf{F}^*$ if, and only if, $(p, L) \in \mathbf{F}$; it is easy to see that $\mathcal{P}^*$ is again a projective plane.

21.2 Classical examples. Let K be a (not necessarily commutative) field, and consider K^3 as a right vector space over K. Denote by $P = P_2 K = P(K^3)$ the set of all one-dimensional subspaces of K^3, and by $\mathcal{L}$ the set of all two-dimensional subspaces. Then $\mathcal{P}_2 K := (P, \mathcal{L}, \subseteq)$, with inclusion $\subseteq$ as incidence, is a projective plane, the *projective plane over K*. If we identify $\mathcal{L}$ with the set of one-dimensional subspaces of the dual space of K^3 (i.e. K^3 as a left K-space), then we can use homogeneous coordinates $(x_1, x_2, x_3)K$ for points and $K(a_1, a_2, a_3)$ for lines such that incidence holds if, and only if, $\sum_{i=1}^{3} a_i x_i = 0$, compare (13.1 and 2). For the projective plane over the octonions $\mathbb{O}$ see (12.2), (16.3).

21.3 Desarguesian planes. The *Desargues configuration* is the incidence structure $\mathcal{D} = \left(\binom{\Omega}{2}, \binom{\Omega}{3}, \subseteq\right)$ where $\Omega = \{1, 2, 3, 4, 5\}$ and $\binom{\Omega}{i} = \{\, X \mid X \subseteq \Omega, |X| = i\,\}$. It may be regarded as a configuration of two triangles which are perspective (with *center* $\{1, 2\}$) and axial (with *axis* $\{3, 4, 5\}$).

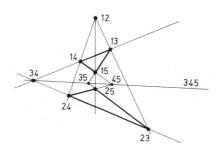

Figure 21a

A projective plane $\mathcal{P}$ is said to be *desarguesian*, if the Desargues configuration 'always closes' in $\mathcal{P}$; this means that the following 'configurational' proposition holds: Label ten distinct points and ten distinct lines of $\mathcal{P}$ by the elements of $\binom{\Omega}{2}$ and $\binom{\Omega}{3}$, respectively, and assume that 29 of the 30 incidences of $\mathcal{D}$ hold in $\mathcal{P}$. Then also the remaining incidence of $\mathcal{D}$ holds in $\mathcal{P}$ (it does not matter which incidence is singled out as the conclusion, because $\mathcal{D}$ has a flag-transitive collineation group). This means that the configuration $\mathcal{D}$ occurs in $\mathcal{P}$ in all conceivable ways (possibly with additional incidences, like the incidence of center and axis). A projective plane $\mathcal{P} = (P, \mathcal{L})$ is (p, L)-*desarguesian*, if each Desargues configuration with center $p \in P$ and axis $L \in \mathcal{L}$ closes in $\mathcal{P}$, see Pickert [75] 3.2 p. 73f, Hughes–Piper [73] p. 108 for details.

Complicated synthetic arguments based on the Desargues configuration can often be replaced by simple algebraic computations, due to the following characterization, which was discovered by Hilbert [1899] [30] §§24–27, see also Wiener [1891] p. 47; for modern proofs compare Hughes–Piper [73] Chap. VI, Pickert [75] 4.4 p. 110.

21.4 Theorem. *The desarguesian projective planes are precisely the planes $\mathcal{P}_2 K$ where K is a (not necessarily commutative) field.*

21.5 Pappian planes. A projective plane $\mathcal{P}$ is *pappian*, if the Pappos configuration

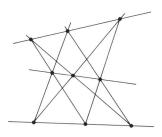

Figure 21b

always closes in $\mathcal{P}$. By a theorem of Hessenberg, *every pappian plane is desarguesian* (cp. Seidenberg [76], Herzer [72], Pickert [75] 5.2 p. 144f.). *The pappian projective planes are precisely the projective planes $\mathcal{P}_2 K$ where K is a commutative field.*

21.6 Projectivities. (a) Let (L, p) be a non-incident line–point pair of a projective plane $\mathcal{P}$. Denote by $\pi(L, p)$ the bijection of L onto $\mathcal{L}_p$ given by $x \mapsto x \vee p$, and by $\pi(p, L)$ its inverse; these bijections are called *perspectivities*. The products

$$\pi(L_1, p_1, L_2, p_2, \ldots, p_{n-1}, L_n) := \pi(L_1, p_1)\pi(p_1, L_2)\pi(L_2, p_2) \cdots \pi(p_{n-1}, L_n)$$

are the *projectivities* of L_1 onto L_n. Dually, one obtains bijections between pencils, and these bijections are also called projectivities.

The projectivities of a line L onto itself form a permutation group $\Pi = \Pi(\mathcal{P})$, the isomorphism type of which does not depend on the choice of L in $\mathcal{P}$. It is easy to see that Π is always triply transitive on L: the projectivities of the form $\pi(L, a, M, x, L)$ with a, b, M as indicated in Figure 21c fix b and act (sharply) doubly transitively on $L \setminus \{b\}$.

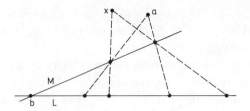

Figure 21c

(b) If $\mathcal{P} = \mathcal{P}_2 K$ is the desarguesian plane over the skew field K, then every projectivity is induced by a linear mapping, hence $\Pi(\mathcal{P})$ is isomorphic to the group $PGL_2 K$ in its natural action on the projective line $K \cup \{\infty\}$, compare (12.12). This group is sharply triply transitive precisely if K is commutative. For non-desarguesian planes $\mathcal{P}$, the groups $\Pi(\mathcal{P})$ tend to be quite large, see Section 66.

21.7 Subplanes. The projective plane $\mathcal{P}' = (P', \mathcal{L}', \mathbf{F}')$ is a *subplane* of the projective plane $\mathcal{P} = (P, \mathcal{L}, \mathbf{F})$ if, and only if, $P' \subseteq P$, $\mathcal{L}' \subseteq \mathcal{L}$ and $\mathbf{F}' = \mathbf{F} \cap (P' \times \mathcal{L}')$. A proper subplane $\mathcal{P}'$ of $\mathcal{P}$ is a *Baer subplane* if every point of $\mathcal{P}$ is on a line of $\mathcal{P}'$ and every line of $\mathcal{P}$ carries a point of $\mathcal{P}'$. It is easy to see that each Baer subplane is a maximal subplane (cp. Hughes–Piper [73] Thm. 4.6 p. 93).

Examples: If K is a field, then $\mathcal{P}_2 K'$ is a subplane of $\mathcal{P}_2 K$ for every subfield K' of K. In the chain $\mathcal{P}_2 \mathbb{R}$, $\mathcal{P}_2 \mathbb{C}$, $\mathcal{P}_2 \mathbb{H}$, $\mathcal{P}_2 \mathbb{O}$, compare (11.1) and (16.3), each plane is a Baer subplane of its successor. See also (23.18) on Baer involutions.

21.8 Definition. An *affine plane* is an incidence structure $\mathcal{A} = (A, \mathcal{L}, \mathbf{F})$ satisfying the following axioms:

(i) Any two distinct points are on a unique line.
(ii) For every pair $(p, L) \in P \times \mathcal{L}$ there exists a unique line $M \in \mathcal{L}_p$ such that either $M = L$ or M and L have no point in common; we call M the *parallel* to L through p.
(iii) There exists a triangle in $\mathcal{A}$, i.e. a set of three points which are not collinear.

21.9 Affine parts and projective completion. If $\mathcal{P} = (P, \mathcal{L}, \in)$ is a projective plane and if $W \in \mathcal{L}$ is an arbitrary line, then *the affine part*

$$\mathcal{P}^W := (P \setminus W, \mathcal{L} \setminus \{W\}, \in)$$

is an affine plane (here lines are considered as point sets). Conversely, every affine plane $\mathcal{A} = (A, \mathcal{L}, \mathbf{F})$ has a *projective completion* $\overline{\mathcal{A}}$, which is the projective plane obtained from $\mathcal{A}$ by adding a new line W (called the 'line at infinity') and a set of new points on W (called the 'points at infinity') which correspond to the parallel classes of $\mathcal{A}$; compare (12.2) or Hughes–Piper [73] 3.10 p. 83, Pickert [75] p. 11. Hence every affine plane has the form $\mathcal{P}^W$.

22 Coordinates, ternary fields

The introduction of coordinates connects planes with algebraic structures. A ternary field is the algebraic equivalent of an affine plane with a distinguished triangle.

22.1. Let $\mathcal{A} = (A, \mathcal{L}, \mathbf{F})$ be an affine plane, and fix a quadrangle o, u, v, e such that u, v are points at infinity. Let K be the set of affine points on the line oe, and define $\tau : K^3 \to K$ by

$$\tau(s, x, t) = (xv \wedge L)u \wedge oe,$$

where L is the line through $tu \wedge ov$ parallel to $o(su \wedge ev)$.

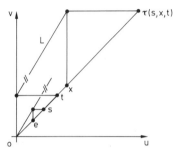

Figure 22a

We coordinatize $\mathcal{A}$ by identifying each affine point p with the pair

$$(pv \wedge oe, pu \wedge oe) \in K^2;$$

then the line L may be described as $L = \{ (x, \tau(s, x, t)) \mid x \in K \}$. Furthermore, (K, τ) is a ternary field in the sense of the following definition (with $0 = o$ and $1 = e$).

22.2 Definition. A *ternary field* (K, τ), or *ternary* for short, is a set K with a ternary operation $\tau : K^3 \to K$ and two distinguished elements $0, 1 \in K$ with $0 \neq 1$ such that the following conditions are satisfied:

(i) $\tau(0, x, t) = \tau(s, 0, t) = t$ and $\tau(s, 1, 0) = \tau(1, s, 0) = s$ for all $s, t, x \in K$.
(ii) For $s, x, y \in K$ there exists a unique $t = t(s, x, y) \in K$ with $\tau(s, x, t) = y$.
(iii) For $s_1, s_2, t_1, t_2 \in K$ with $s_1 \neq s_2$ there exists a unique $x = x(s_1, t_1, s_2, t_2)$ with $\tau(s_1, x, t_1) = \tau(s_2, x, t_2)$.
(iv) For $x_1, x_2, y_1, y_2 \in K$ with $x_1 \neq x_2$ there exist uniquely determined elements $s = s(x_1, y_1, x_2, y_2)$ and $t \in K$ with $y_i = \tau(s, x_i, t)$ for $i = 1, 2$.

Every field K yields a ternary field by the definition $\tau(s, x, t) = sx + t$.

22.3. The procedure of coordinatization can be reversed: every ternary field $K = (K, \tau)$ yields an affine plane $\mathcal{A}_K = (K^2, \mathcal{L}, \in)$ with

$$\mathcal{L} = \{\, [s, t] \mid s, t \in K \,\} \cup \{\, [c] \mid c \in K \,\},$$

where

$$[s, t] = \{\, (x, \tau(s, x, t)) \mid x \in K \,\} \text{ and } [c] = \{c\} \times K,$$

compare (12.1). The line $[s, t]$ is said to have *slope* s, and the line $[c]$ has *slope infinity*. The axioms (iii, iv) say that lines with different slopes intersect in a unique point and that any two points are joined by a unique line. These axioms are referred to as *planarity conditions*.

If K is constructed from $\mathcal{A}$ as in (22.1), then $\mathcal{A}$ is isomorphic to $\mathcal{A}_K$ via $p \mapsto (pv \wedge oe, pu \wedge oe)$. This is expressed by saying that $\mathcal{A}$ is coordinatized by K. We denote by (s) the common point at infinity of all lines $[s, t]$ with $t \in K$, and by (∞) the point at infinity of the lines $[c]$, compare (21.9).

22.4. Every ternary field (K, τ) gives rise to two binary operations which are defined by

$$x + t = \tau(1, x, t) \quad \text{and} \quad s \cdot x = \tau(s, x, 0).$$

The structure $(K, +, \cdot)$ is a 'double loop', compare Grundhöfer–Salzmann [90]. The division operations $\backslash$ and $/$ are defined by the equations

$$x \cdot (x \backslash y) = y = (y/x) \cdot x$$

for $x, y \in K$, $x \neq 0$.

In general, the ternary operation τ is not determined by $(K, +, \cdot)$, cp. Grundhöfer–Salzmann [90], Section 4, for examples. A ternary field is called *linear*, if $\tau(s, x, t) = s \cdot x + t$ for all $s, x, t \in K$, that is, if the ternary operation τ can be reconstructed from the double loop $(K, +, \cdot)$. Sections 24 and 25 contain more information about special types of ternary fields.

Ternary fields, especially linear ones, are useful for constructing planes. The underlying double loop is used for investigating topological properties of a line in a topological plane, see (42.8), (51.4 and 28), Grundhöfer–Salzmann [90].

Looking at projective or affine planes via ternary fields has the following disadvantage: the ternary field coordinatizing a given plane depends in an essential way on the choice of a quadrangle in the plane; examples are provided by planes over semifields which are not alternative, compare (64.16).

23 Collineations

23.1 Definition. A *collineation* (or an *isomorphism*) between incidence structures $(P, \mathcal{L}, \mathbf{F})$ and $(P', \mathcal{L}', \mathbf{F}')$ is a pair (α, β) of bijections $\alpha : P \to P'$ and $\beta : \mathcal{L} \to \mathcal{L}'$ such that $(p, L) \in \mathbf{F}$ if, and only if, $(p^\alpha, L^\beta) \in \mathbf{F}'$. The existence of an isomorphism is expressed by saying that the two incidence structures are *isomorphic*.

For affine or projective planes, lines may be identified with point sets. Hence β is determined uniquely by α, and we describe L^β as $L^\alpha = \{x^\alpha \mid x \in L\}$. Every collineation α of an affine or projective plane satisfies $(pq)^\alpha = p^\alpha q^\alpha$ and $(L \wedge M)^\alpha = L^\alpha \wedge M^\alpha$ for distinct points p, q and distinct lines L, M.

A *duality* of a projective plane $\mathcal{P}$ is a collineation of $\mathcal{P}$ onto the dual plane $\mathcal{P}^*$ of $\mathcal{P}$. A duality δ with $\delta^2 = 1\!\!1$ is called a *polarity*, compare (13.11).

23.2. A bijection α between the point sets of two affine (or projective) planes is a collineation if, and only if, α maps lines *into* lines (i.e. α preserves the relation of collinearity). Indeed, if the image L^α of a line L is properly contained in the line M, then M also contains the image L_1^α of some other line L_1. Every point is the intersection of two lines that meet L and L_1, hence α maps every point into M, a contradiction.

23.3. The *collineation group* or *automorphism group* $\mathrm{Aut}\,\mathcal{P}$ of an incidence structure $\mathcal{P} = (P, \mathcal{L}, \mathbf{F})$ is the group of all collineations of $\mathcal{P}$ onto itself. Let Δ be a subgroup of $\mathrm{Aut}\,\mathcal{P}$. The following notation is used for points p and lines L:

$$\Delta_p = \{\delta \in \Delta \mid p^\delta = p\} \text{ and } \Delta_L = \{\delta \in \Delta \mid L^\delta = L\} \text{ are the stabilizers,}$$

$$p^\Delta = \{p^\delta \mid \delta \in \Delta\} \text{ and } L^\Delta = \{L^\delta \mid \delta \in \Delta\} \text{ are the orbits, and}$$

$$\Delta_{[L]} = \bigcap_{p \in L} \Delta_p, \quad \Delta_{[p]} = \bigcap_{L \in \mathcal{L}_p} \Delta_L$$

are the pointwise stabilizer of L and the linewise stabilizer of p, respectively.

23.4. Every collineation between two affine planes $\mathcal{A}, \mathcal{A}'$ has a unique extension to the projective completions $\overline{\mathcal{A}}, \overline{\mathcal{A}'}$ (since parallel classes of lines are mapped onto parallel classes of lines); in particular, $\mathrm{Aut}(\mathcal{P}^W) \cong (\mathrm{Aut}\,\mathcal{P})_W$ for every line W of a projective plane $\mathcal{P}$.

23.5. Let $\mathcal{P}, \mathcal{P}'$ be projective planes which are coordinatized by the ternary fields $(K, \tau), (K', \tau')$ with respect to quadrangles o, u, v, e and o', u', v', e'. The collineations of $\mathcal{P}$ onto $\mathcal{P}'$ mapping (o, u, v, e) onto (o', u', v', e') correspond bijectively to the isomorphisms of (K, τ) onto (K', τ'). In particular, the stabilizer

(Aut $\mathcal{P}$)$_{o,u,v,e}$ of the ordered quadrangle (o, u, v, e) is isomorphic to the automorphism group of (K, τ), i.e. the group of all bijections $\alpha : K \to K$ which satisfy $\tau(s^\alpha, x^\alpha, t^\alpha) = \tau(s, x, t)^\alpha$ for every $s, x, t \in K$. Indeed, the map

$$(\text{Aut } \mathcal{P})_{o,u,v,e} \to \text{Aut}(K, \tau) : \alpha \mapsto \alpha|_{oe}$$

is an isomorphism; its inverse is given by the map $\alpha \mapsto (\alpha, \alpha)$, where (α, α) maps an affine point (x, y) to (x^α, y^α). See Pickert [75] 1.24 p. 37, compare also (12.9).

23.6 Examples. If F is a (not necessarily commutative) field, then Aut $\mathcal{P}_2 F = P\Gamma L_3 F$, compare (13.6), Hughes–Piper [73] p. 30, Pickert [75] 4.7 p. 112, H. Lüneburg [66], Taylor [92] 3.1. The subgroup PGL$_3 F$ acts transitively on the set of non-degenerate quadrangles. The automorphism group of the affine plane over F is the group $A\Gamma L_2 F = F^2 \rtimes \Gamma L_2 F$, compare (12.10), Taylor [92] 2.3. For Aut $\mathcal{P}_2 \mathbb{O}$ see Section 17.

23.7 Axial collineations. Let α be a collineation of a projective plane $\mathcal{P} = (P, \mathcal{L})$. We say that α has *axis* $L \in \mathcal{L}$ if α fixes every point on L; dually, α has *center* $p \in P$ if α fixes every line through p. In fact, α has an axis if, and only if, it has a center (see e.g. Hughes–Piper [73] Theorem 4.9 p. 94 or Pickert [75] 3.4 p. 65). The collineations with center and axis are called *axial collineations* (some authors use the terms *central collineations* or *perspectivities*).

An axial collineation with center p and axis L is called a *homology* if $p \notin L$, and an *elation* if $p \in L$. Together, the homologies and elations of a projective plane $\mathcal{P}$ generate the *projective group* of $\mathcal{P}$, and the elations generate the *little projective group* of $\mathcal{P}$.

From an affine point of view, an elation is called a *translation* if its axis L is the line at infinity; if the axis L is an affine line and if the center is at infinity, then the elation is called a *shear*. A homology with center p and axis L is a *homothety* if L is the line at infinity, and it is a *strain* if L is an affine line and p is at infinity.

23.8. An axial collineation α is already determined by its center p, its axis L, and any point-pair (q, q^α) with $q \notin L \cup \{p\}$, see Hughes–Piper [73] Theorem 4.7 p. 93, Pickert [75] p. 65/66, or (12.4c). If α is an elation, then $p = L \wedge qq^\alpha$, hence α is determined by L and (q, q^α).

Let $\alpha \neq \mathbb{1}$ be an axial collineation. We infer from the above that the center p and the axis L of α are uniquely determined by α. Furthermore, p and the points on L are the only fixed points of α (and dually for the fixed lines of α). This means that the group of all axial collineations with center p and axis L acts freely (and effectively) on each line through p which is distinct from L. As a consequence, we obtain:

23.9. Let Θ be a set of collineations with center p and axis L, and let $M \neq L$ be a line through p. If Θ is transitive on $M \setminus \{p, M \wedge L\}$, then Θ is the group of all collineations with center p and axis L.

23.10. Let Δ be a group of collineations of a projective plane $\mathcal{P} = (P, \mathcal{L})$. We use the following notation:

$$\Delta_{[p,L]} = \Delta_{[p]} \cap \Delta_{[L]} = \{\delta \in \Delta \mid \delta \text{ has center } p \text{ and axis } L\}$$

for $p \in P$, $L \in \mathcal{L}$, and

$$\Delta_{[L,L]} = \bigcup_{p \in L} \Delta_{[p,L]}$$

is the group of all elations in Δ with axis L. This union is indeed a group, see e.g. Hughes–Piper [73] Theorem 4.13 p. 96, Pickert [75] p. 65. The group $\Delta_{[p,p]}$ is defined dually. By (23.8), the group $\Delta_{[L,L]}$ acts freely on the set $P \setminus L$ of affine points of $\mathcal{P}^L$.

23.11 Lemma. *Let K be the ternary field which coordinatizes $\mathcal{P} = (P, \mathcal{L})$ with respect to the quadrangle o, u, v, e, and let $\Gamma = \operatorname{Aut} \mathcal{P}$. Then the groups $\Gamma_{[o,uv]}$, $\Gamma_{[u,uv]}$ and $\Gamma_{[v,uv]}$, acting on the affine points $P \setminus uv$, are described by*

$$\Gamma_{[o,uv]} = \left\{ (x, y) \mapsto (x \cdot c, y \cdot c) \;\middle|\; \begin{array}{l} 0 \neq c \in K \text{ such that for all } s, x, t \in K \\ \tau(s, x, t) \cdot c = \tau(s, x \cdot c, t \cdot c) \end{array} \right\},$$

$$\Gamma_{[u,uv]} = \left\{ (x, y) \mapsto (x + a, y) \;\middle|\; \begin{array}{l} a \in K \text{ such that for all } s, x, t \in K \\ \tau(s, x + a, t) = \tau(s, x, \tau(s, a, t)) \end{array} \right\},$$

$$\Gamma_{[v,uv]} = \left\{ (x, y) \mapsto (x, y + b) \;\middle|\; \begin{array}{l} b \in K \text{ such that for all } s, x, t \in K \\ \tau(s, x, t) + b = \tau(s, x, t + b) \end{array} \right\}.$$

Proof. It is easy to verify that the maps on the right hand side are collineations of the asserted type. For the converse, we remark that every element of $\Gamma_{[o,uv]}$ is of the form $(x, y) \mapsto (x^\alpha, y^\beta)$ for suitable bijections α, β of K onto itself. The line $[s, t]$ is mapped onto $[s, t^\beta]$, hence $\tau(s, x, t)^\beta = \tau(s, x^\alpha, t^\beta)$ for every $s, x, t \in K$. Specializing $t = 0$ gives $(s \cdot x)^\beta = s \cdot x^\alpha$, as $0^\beta = 0$, hence $\alpha = \beta$ and $s^\beta = s \cdot c$ with $c = 1^\alpha$. For elements of $\Gamma_{[v,uv]}$ we have $\alpha = 1$ and $\tau(s, x, t)^\beta = \tau(s, x, t^\beta)$, which leads to $x^\beta = x + 0^\beta$ upon specializing $s = 1, t = 0$. Finally, the elements of $\Gamma_{[u,uv]}$ have the shape $(x, y) \mapsto (x^\alpha, y)$, and the preimage of a line $[s, t]$ is of the form $[s, t']$, hence $\tau(s, x, t') = \tau(s, x^\alpha, t)$ for $s, x, t \in K$. Now $x = 0$ gives $t' = \tau(s, a, t)$ with $a = 0^\alpha$, and $s = 1, t = 0$ gives $x^\alpha = x + \tau(1, a, 0) = x + a$. □

23.12. Let $\Delta \leq \Gamma = \operatorname{Aut} \mathcal{P}$. If $\gamma \in \Gamma$, then the action of γ by conjugation satisfies

$$(\Delta_{[p,L]})^\gamma = (\Delta^\gamma)_{[p^\gamma, L^\gamma]}$$

for $p \in P, L \in \mathcal{L}$; in other words, the γ-conjugate of a collineation with center p and axis L has center p^γ and axis L^γ (this is an immediate consequence of (91.1a)). In particular, if $\Delta_{[p,L]} \neq \mathbb{1}$, then the normalizer $\mathrm{Ns}_\Gamma (\Delta_{[p,L]})$ is contained in $\Gamma_{p,L}$.

23.13. If p, q are distinct points on a line L such that $\Delta_{[p,L]} \neq \mathbb{1} \neq \Delta_{[q,L]}$, then the group $\Delta_{[L,L]}$ is abelian, and all its elements have the same order (see Hughes–Piper [73] Theorem 4.14 p. 97 or Pickert [75] p. 199).

23.14. A *planar collineation* of a projective plane $(P, \mathcal{L})$ is a collineation α such that the set F of fixed points contains a non-degenerate quadrangle. Then F is (the point set of) a subplane. If this subplane is a Baer subplane, then α is called a *Baer collineation*. Such a Baer collineation acts freely on $P \setminus F$, since Baer subplanes are maximal subplanes (21.7).

23.15. A collineation group Δ of a projective plane $\mathcal{P}$ is said to be *straight on* $\mathcal{P}$ if every point orbit of Δ is contained in some line.

If a cyclic group Δ is straight on $\mathcal{P}$, then it is also straight on the dual plane $\mathcal{P}^*$, i.e. every line orbit is contained in some pencil, see Baer [46] Lemma 1. The group $\Gamma_{[c]}$ of all axial collineations with center c in a desarguesian plane shows that this is not true for straight groups in general.

The proof of Baer [46] Lemma 1 yields

23.16 Lemma. *Let Δ be a group of collineations of a projective plane $\mathcal{P} = (P, \mathcal{L})$. Assume that Δ is straight on $\mathcal{P}$ and on the dual plane $\mathcal{P}^*$. Then either Δ acts trivially on some Baer subplane of $\mathcal{P}$, or $\Delta = \Delta_{[p,L]}$ for some pair $(p, L) \in P \times \mathcal{L}$.*

23.17 Corollary. *If α is a collineation of a projective plane $(P, \mathcal{L})$ with $\alpha^2 = \mathbb{1} \neq \alpha$, then α is either an axial collineation or a Baer collineation (a 'Baer involution').*

For a direct proof of (23.17) see Hughes–Piper [73] Theorems 4.3 and 4.4 p. 91/92 or Pickert [75] p. 72.

23.18. If β is an involutory automorphism (23.5) of a ternary field K, then β yields a Baer involution α which acts as $(x, y)^\alpha = (x^\beta, y^\beta)$ on the affine points of the plane over K.

For example, the automorphisms $\beta : (a, b) \mapsto (a, -b)$ of $\mathbb{C} = \mathbb{R}^2$, $\mathbb{H} = \mathbb{C}^2$, $\mathbb{O} = \mathbb{H}^2$ yield Baer involutions α of the planes over $\mathbb{C}, \mathbb{H}, \mathbb{O}$, compare (11.31), (18.28).

23.19. Every involutory axial collineation α is already determined by its axis and any point-pair (p, p^α) with $p^\alpha \neq p$. Indeed, the product $\alpha\alpha'$ of two collineations α, α' of this type fixes p, p^α and every point on the axis, hence $\alpha\alpha' = \mathbb{1}$, see the second paragraph of (23.8), and $\alpha = \alpha'$.

23.20. A *reflection* is a homology of order 2. If α, β are reflections with common axis L and distinct centers p, q, then $\alpha\beta$ is an elation with center $pq \wedge L$ (and axis L), compare Hughes–Piper [73] Lemma 4.21 p. 101, Pickert [75] p. 212.

23.21. Let Δ be a collineation group of a projective plane $\mathcal{P}$, let p be a point and L a line. We say that Δ is (p, L)-*transitive*, and that $\Delta_{[p,L]}$ is *linearly transitive*, if $\Delta_{[p,L]}$ acts transitively on $M \setminus \{p, M \wedge L\}$ for every line $M \in \mathcal{L}_p \setminus \{L\}$. The plane $\mathcal{P}$ is said to be (p, L)-*homogeneous*, if $\mathcal{P}$ admits a collineation group which is (p, L)-transitive.

23.22 Theorem (Baer). *A projective plane $\mathcal{P}$ is (p, L)-homogeneous if, and only if, $\mathcal{P}$ is (p, L)-desarguesian.*

Proofs can be found in Baer [42] 5.1 and 6.2, Hall [59] 20.2.4 p. 353, Pickert [75] 3.20 p. 76 and Hughes–Piper [73] Theorem 4.29 p. 108.

23.23. If a collineation group Δ is (p, pq)-transitive and (q, pq)-transitive with distinct points p, q, then Δ is (x, pq)-transitive for every point x on pq, see Hughes–Piper [73] Theorem 4.19 p. 100, Pickert [75] 3.11 p. 67. This condition is equivalent to the transitivity of $\Delta_{[pq,pq]}$ on the points not on pq, and this is the defining condition for translation planes, compare (24.6), (25.1).

The following lemma, which gives a group-theoretic description of flag-homogeneous incidence structures, is well known (compare Dembowski [68] 1.2.17).

23.24 Lemma. *Let Δ be a flag-transitive collineation group of an incidence structure $\mathcal{P}$, and let (p, L) be a flag. Then $\mathcal{P}$ is isomorphic to the coset geometry $(\Delta/\mathsf{A}, \Delta/\mathsf{B}, \mathbf{F})$, where $\mathsf{A} = \Delta_p$, $\mathsf{B} = \Delta_L$ and $(\mathsf{A}\delta, \mathsf{B}\delta') \in \mathbf{F} \iff \mathsf{A}\delta \cap \mathsf{B}\delta' \neq \emptyset$.*

Proof. Evaluation at p (or at L) gives a bijection of Δ/A onto the point set of $\mathcal{P}$ (or of Δ/B onto the line set of $\mathcal{P}$). A pair $(p^\delta, L^{\delta'})$ is a flag of $\mathcal{P}$ precisely if $(p^\delta, L^{\delta'}) = (p^\gamma, L^\gamma)$ for some $\gamma \in \Delta$, and this condition on γ is equivalent to $\gamma \in \mathsf{A}\delta \cap \mathsf{B}\delta'$. □

24 Lenz–Barlotti types

24.1. Let $\mathcal{P} = (P, \mathcal{L}, \mathbf{F})$ be a projective plane, and let Δ be a subgroup of the group Γ of all collineations of $\mathcal{P}$. The *Lenz–Barlotti type* of Δ is determined by the set

$$C = \{ (p, L) \in P \times \mathcal{L} \mid \Delta \text{ is } (p, L)\text{-transitive} \}.$$

The enumeration of all a priori possible configurations C arising in this fashion yields the Lenz–Barlotti hierarchy of projective planes. Here we concentrate mainly on the Lenz types, i.e. we consider only the flags in C and the corresponding groups of elations, and we sometimes ignore the refinement (due to Barlotti) which can be obtained by considering homologies. For more details see Pickert [75] Anhang 6 p. 334f, Dembowski [68] 3.1.20 p. 123f, Hughes–Piper [73] p. 153f, Hall [59] 20.4, 20.5.

24.2. First we give the definitions of the *Lenz types*. The dual of type IV has been omitted; all other Lenz types are self-dual. By o, u, v we denote a suitable triangle.

I $\quad C \cap \mathbf{F} = \emptyset$
II $\quad C \cap \mathbf{F} = \{(v, uv)\}$
III $\quad C \cap \mathbf{F} = \{ (x, xu) \mid x \in ov \}$
IV $\quad C \cap \mathbf{F} = \{ (x, uv) \mid x \in uv \}$
V $\quad C \cap \mathbf{F} = \{ (x, uv) \mid x \in uv \} \cup \{ (v, X) \mid X \in \mathcal{L}_v \}$
VII $\quad C \cap \mathbf{F} = \mathbf{F}$

For the missing type VI see (24.9). The proof that the preceding list is complete uses (23.12 and 23) as essential tools.

24.3. We say that $\mathcal{P}$ is of Lenz type (at least) N if $\mathcal{P}$ admits a collineation group of type N. Extend the triangle o, u, v to a non-degenerate quadrangle o, u, v, e. The following list contains the names for the special ternary fields which coordinatize the planes $\mathcal{P}$ of given Lenz type with respect to o, u, v, e:

I   ternary field
II   Cartesian field
III Cartesian field with special properties,
     compare Spencer [60]
IV   quasifield
V   semifield
VII alternative field

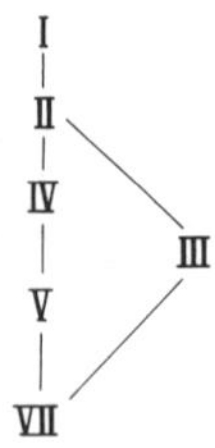

Figure 24a

The special algebraic properties of these ternary fields (see below) can be obtained from (23.11), compare also Hughes–Piper [73] Theorem 6.2 p. 129 and Theorem 6.5 p. 132.

24.4 Type II. The planes of type II are coordinatized by Cartesian fields. These Cartesian fields $(K, +, \cdot)$ may be characterized as follows: $(K, +)$ is a group with neutral element 0, the binary operation $\cdot : K^2 \to K$ satisfies $x \cdot 1 = x = 1 \cdot x$ for a distinguished element $1 \neq 0$, and, for $a \neq b$, each map $x \mapsto -a \cdot x + b \cdot x$ or $x \mapsto x \cdot a - x \cdot b$ is bijective, compare (22.2iii,iv). The corresponding ternary field (K, τ) with $\tau(s, x, t) = s \cdot x + t$ is also called a Cartesian field. The group $(K, +)$ is isomorphic to the group of all translations with center v and axis uv, see (23.11).

24.5 Type III. This type splits into two Lenz–Barlotti types, III.1 and III.2, which are defined by $C = \{(x, xu) \mid x \in ov\}$ and $C = \{(x, xu) \mid x \in ov\} \cup \{(u, ov)\}$ respectively. Examples of type III.2 are provided by the Moulton planes, see (31.25b), Section 34, and (64.18). Compare also Spencer [60], Kalhoff [89, 90].

24.6 Type IV. The planes of type (at least) IV are called *translation planes* (with *translation line* uv). They are coordinatized by quasifields $Q = (Q, +, \cdot)$, i.e. by Cartesian fields which satisfy the distributive law $x \cdot (y + z) = x \cdot y + x \cdot z$. See Section 25 for a discussion of translation planes and quasifields.

We mention two of the Lenz–Barlotti types which refine Lenz type IV. A generic translation plane has Lenz–Barlotti type IVa.1, which is defined by $C = C \cap \mathbf{F} = \{(x, uv) \mid x \in uv\}$. Type IVa.2 is defined by

$$C = \{(x, uv) \mid x \in uv\} \cup \{(u, X) \mid X \in \mathcal{L}_v\} \cup \{(v, X) \mid X \in \mathcal{L}_u\} \;;$$

the planes of this type are called *nearfield planes*, see (25.7).

24.7 Type V. The planes of Lenz type (at least) V are those translation planes (with translation line uv) which are also dual translation planes (with translation point v). They are precisely the planes which can be coordinatized by *semifields*, i.e., by quasifields satisfying both distributive laws. In other terminology, semifields are also referred to as (non-associative) division algebras or division rings.

In every plane over a semifield, the maps $(x, y) \mapsto (x, a \cdot x + y)$ form the (linearly transitive) group of all shears (23.7) with center $v = (\infty)$ and axis $ov = [0]$. Every translation plane with at least one linearly transitive group of shears has Lenz type at least V (one can apply the dual of (23.23), since uv is the axis of further linearly transitive groups of elations). Compare also (25.4 and 8).

24.8 Type VII. A plane of Lenz type (at least) VII is called a *Moufang plane*. This type splits into two Lenz–Barlotti types, VII.1 and VII.2. Using the notation from (24.1), type VII.2 is defined by $C = P \times \mathcal{L}$, i.e. the plane $\mathcal{P}$ is desarguesian and therefore coordinatized by a (not necessarily commutative) field K, see (23.22) and (21.4).

Type VII.1 is defined by $C = \mathbf{F}$; then $\mathcal{P}$ is a translation plane with respect to any line. These planes are coordinatized by *alternative fields*, i.e. by semifields with the property that any two elements are contained in a subfield (or, equivalently, by semifields which satisfy the alternative laws $x \cdot (x \cdot y) = x^2 \cdot y$ and $(x \cdot y) \cdot y = x \cdot y^2$). An alternative field which is not a field is always a Cayley–Dickson algebra of dimension 8 over its center, see Pickert [75] 6.13 p. 175 and p. 178, Zhevlakov et al. [82] 7.3. The octonions $\mathbb{O}$ are a typical example.

The collineation group of a Moufang plane $\mathcal{P}$ acts transitively on the set of ordered quadrangles in $\mathcal{P}$, see Pickert [75] 7.14 p. 195, Ciftci–Kaya–Ferrar [88], or (23.6) for the special case of desarguesian planes. Thus all ternary fields (alternative fields) coordinatizing $\mathcal{P}$ are isomorphic (23.5).

24.9. Type VI, which requires the existence of several translation lines (up to duality), has been omitted. This is due to the following theorem, which implies that there exists no projective plane of Lenz type exactly VI.

Theorem (Skornjakov–San Soucie). *Every projective plane which has two distinct translation lines is a Moufang plane.*

For a proof see Hughes–Piper [73] p. 140–151.

25 Translation planes and quasifields

25.1 Definition. An (affine) *translation plane* is an affine plane $\mathcal{A}$ such that the group of all translations (23.7) of $\mathcal{A}$ acts transitively (and hence sharply transitively) on the points of $\mathcal{A}$. This means that the line W at infinity is a translation line of the projective completion $\overline{\mathcal{A}}$, compare (24.6) and (23.23), or equivalently, that for every $c \in W$, the group of all translations with center c and axis W is linearly transitive.

25.2 Definition. A *quasifield* is a set Q with two binary operations $+$ and $\cdot$ and two distinguished elements $0, 1 \in Q$ with $0 \neq 1$ such that the following conditions are satisfied:

(i) $(Q, +)$ is a group with neutral element 0.
(ii) $(Q \setminus \{0\}, \cdot)$ is a loop, i.e. $x \cdot 1 = x = 1 \cdot x$ for $x \in Q$, and all maps $x \mapsto a \cdot x$ and $x \mapsto x \cdot a$ with $0 \neq a \in Q$ are bijections of $Q \setminus \{0\}$ onto itself.
(iii) The distributive law $x \cdot (y + z) = x \cdot y + x \cdot z$ holds.
(iv) $0 \cdot x = 0$ for all $x \in Q$.
(v) The planarity condition holds, i.e. for $a \neq b$ the (additive) map $x \mapsto a \cdot x - b \cdot x : Q \to Q$ is surjective (hence bijective).

This definition agrees with the definition given in (24.6). In older terminology, quasifields are also called Veblen–Wedderburn systems. The translation planes are precisely the affine planes which can be coordinatized by quasifields. If Q is a quasifield, then $\mathsf{T} = \{(x, y) \mapsto (x + a, y + b) \mid a, b \in Q\} \cong (Q, +) \oplus (Q, +)$ is the translation group of the plane coordinatized by Q, compare (23.11), Hughes–Piper [73] Lemma 7.10 p. 165. Therefore, the additive group $(Q, +)$ of a quasifield is always abelian, see (23.13).

Examples of quasifields can be found in Sections 64, 73, 82 and in Grundhöfer–Salzmann [90].

25.3 Vector space structure. The *kernel* of a quasifield $(Q, +, \cdot)$ is defined by

$$K = \left\{ k \in Q \;\middle|\; \begin{array}{l} (x + y) \cdot k = x \cdot k + y \cdot k \\ (x \cdot y) \cdot k = x \cdot (y \cdot k) \end{array} \text{ for all } x, y \in Q \right\}.$$

The kernel K is a (not necessarily commutative) field, and Q is a right vector space over K (the multiplication by scalars is obtained by restricting the quasifield multiplication). Furthermore, all maps $x \mapsto a \cdot x$ with $a \in Q$ are K-linear. If Q has finite dimension over K, then the planarity condition (v) is a consequence of the other conditions in (25.2), see Hughes–Piper [73] Thm. 7.3 p. 160, H. Lüneburg [80] 5.3.

Due to this vector space structure, translation planes are more easily tractable than arbitrary planes.

25.4 Axial collineations. Let Q be a quasifield. Denote by $o = (0, 0)$ the origin of the affine plane over Q and by $u = (0)$, $v = (\infty)$ the points at infinity of the x-axis and of the y-axis, respectively (22.3). Let Γ be the collineation group of the affine (or projective) plane coordinatized by Q.

The geometric interpretation of the kernel K of Q is given by the isomorphism $K^\times \to \Gamma_{[o,uv]}$ which maps $k \in K^\times$ to the homology $(x, y) \mapsto (x \cdot k, y \cdot k)$, see (23.11), Hughes–Piper [73] Lemma 7.9 p. 164, H. Lüneburg [80] 5.4.

Furthermore, the *nuclei*

$$N_m(Q) = \{ c \in Q \mid (x \cdot c) \cdot y = x \cdot (c \cdot y) \text{ for all } x, y \in Q \} \text{ and}$$

$$N_l(Q) = \{ c \in Q \mid (c \cdot x) \cdot y = c \cdot (x \cdot y) \text{ for all } x, y \in Q \}$$

describe the collineation groups

$$\Gamma_{[u,ov]} = \{ (x, y) \mapsto (c \cdot x, y) \mid 0 \neq c \in N_m(Q) \} \text{ and}$$

$$\Gamma_{[v,ou]} = \{ (x, y) \mapsto (x, c \cdot y) \mid 0 \neq c \in N_l(Q) \} \;;$$

for examples see (11.20), (64.20b) and (82.17). Generalizing the situation of (24.7), we introduce the *distributor*

$$D(Q) = \{\, d \in Q \mid (d+x) \cdot y = d \cdot y + x \cdot y \text{ for all } x, y \in Q \,\} \,,$$

which describes the group

$$\cdot \, \Gamma_{[v,ov]} = \{\, (x, y) \mapsto (x, d \cdot x + y) \mid d \in D(Q) \,\}.$$

For proofs of these facts see H. Lüneburg [80], Sections 3 and 5 (with slightly different notation), André [54b], Dembowski [68] 3.1.30 p. 134.

25.5 Collineation group. Let Q be a quasifield with kernel K, and let Γ be the collineation group of the affine plane over Q. Then $\Gamma \leq A\Gamma L(Q \oplus Q, K)$, in fact $\Gamma = \mathsf{T} \rtimes \Gamma_o$, where $\mathsf{T} = \{\, (x, y) \mapsto (x+a, y+b) \mid a, b \in Q \,\}$ is the translation group and $\Gamma_o \leq \Gamma L(Q \oplus Q, K)$ consists of semi-linear transformations of the vector space $Q \oplus Q$ over K, see H. Lüneburg [80] 1.10. These facts generalize (12.10) and (17.8).

25.6 Spreads. A *spread* of an abelian group A is a set S of proper subgroups of A such that $A = \bigcup S$ and $A = X \oplus Y$ for all $X, Y \in S$ with $X \neq Y$.

Let $\mathcal{A}$ be the translation plane coordinatized by a quasifield Q. We identify the set of affine points with the translation group $\mathsf{T} \cong Q \oplus Q$. Then the lines of $\mathcal{A}$ through $(0, 0)$ form a spread of the abelian group $Q \oplus Q$. These are the lines $[s, 0] = \{\, (x, s \cdot x) \mid x \in Q \,\}$ and $[0] = \{0\} \times Q$. In fact, these lines are subspaces of $Q \oplus Q$ considered as a vector space over the kernel of Q. In terms of the translation group T, the spread is described as $S = \{\, \mathsf{T}_{[p]} \mid p \in W \,\}$, where W is the line at infinity of $\mathcal{A}$.

Conversely, every spread S of an abelian group A yields a translation plane $\mathcal{A} = (A, \mathcal{L}, \in)$ with $\mathcal{L} = \{\, a + X \mid a \in A, X \in S \,\}$. The ring of all endomorphisms φ of A which satisfy $X^\varphi \subseteq X$ for every $X \in S$ coincides with $\Gamma_{[o,uv]} \cup \{0\}$, compare (25.4), and hence is isomorphic to the kernel of any quasifield coordinatizing $\mathcal{A}$. See André [54b], Bruck–Bose [64, 66], H. Lüneburg [80] Section 1 for more information.

25.7 Nearfields. A (planar) *nearfield* is a quasifield with associative multiplication. The nearfields coordinatize precisely the planes of Lenz–Barlotti type at least IVa.2 as defined in (24.6), see Pickert [75] 3.46 p. 103 or use (25.4). For examples and more information on nearfields see Wähling [87] and (64.19 and 20).

25.8 Semifields. A *semifield* is a quasifield satisfying both distributive laws, compare (24.7). As a consequence of the second distributive law, each nucleus (25.4) of a semifield is closed under addition (and not only under multiplication, as in quasifields in general). Hence each nucleus of a semifield is a (not necessarily commutative) field. The intersection of the nuclei, the kernel and the center of a semifield Q is a commutative field F, and Q is an F-algebra. Semifields are also referred to as (non-associative) division algebras or division rings. For examples see (64.16), (82.1, 16 and 21) and Grundhöfer–Salzmann [90].

Chapter 3

Geometries on surfaces

In this chapter we study 'real' planes: we start with the point set $\mathbb{R}^2$, and we consider suitable systems of curves, called lines, in $\mathbb{R}^2$.

The axioms for the line system imposed in Section 31 require that any two points are joined by a unique line, but not the dual condition, nor any kind of parallel axiom. This means that the so-called $\mathbb{R}^2$-planes of Section 31 are not in general affine planes (they are never projective planes). We deal with these geometries because they provide very concrete and intuitive examples of topological geometries.

The following result was proved by Salzmann [69a] [67b] 2.7, Löwen [95]: Let S be a surface (i.e., a topological manifold of dimension 2) which is endowed with a 'line' system $\mathcal{L}$ of closed subsets which are homeomorphic to $\mathbb{R}$ or to the circle $\mathbb{S}_1$, such that any two points of S are joined by a unique line. Then suitable natural continuity conditions (compare (31.22)) imply that $(S, \mathcal{L})$ is either an $\mathbb{R}^2$-plane in the sense of Section 31, or $(S, \mathcal{L})$ is a 2-dimensional compact projective plane (see Section 32), or S is homeomorphic to the Möbius strip. In the last case, some lines are homeomorphic to $\mathbb{R}$ and others are homeomorphic to $\mathbb{S}_1$. We shall not explicitly consider this last case, we just mention that examples of this type can be obtained by deleting a point or a suitable closed disk from the point set of a 2-dimensional compact projective plane; for a concrete example, consider the exterior of the quadric Q in (35.1), see also Betten [68]. Related characterizations of $\mathbb{R}^2$-planes and of 2-dimensional compact projective planes have been obtained by Busemann [55] Chap. I, in the context of metric spaces.

The topology of these geometries is intimately connected with orders (on each line). In fact, the order properties in Hilbert's axiomatic characterization of the real geometries can be replaced by topological properties (of the corresponding order topology), see (32.5 and 8), (42.10 and 11), Freudenthal [57a,b]. The present chapter contains several characterizations of the real projective plane (and of other projective planes) as a topological geometry with special properties. Giving preference to topological arguments, we shall not deal explicitly with order geometry; for this topic, references can be found in (31.4) and (32.24). The connection between topology and order is a special feature of (locally compact) planes of topological dimension 2. Planes of higher topological dimension, like the planes over $\mathbb{C}$, $\mathbb{H}$ or $\mathbb{O}$, cannot be ordered. These planes will be studied in Chapters 4 through 8.

In Section 32 we consider those $\mathbb{R}^2$-planes which are affine planes. The projective completion of such a plane is a compact projective plane of topological dimension 2; the point space of the completion is a surface homeomorphic to the point space of the real projective plane. We derive a number of fundamental properties of these planes and of their collineation groups. The (full) collineation groups turn out to be Lie groups (32.21b). The real projective plane $\mathcal{P}_2\mathbb{R}$ is characterized as the only compact projective plane of dimension 2 admitting a collineation group which is transitive on the points (32.23). See (63.8) for the generalization to compact projective planes of higher dimensions.

Sections 33 through 37 deal with 2-dimensional compact projective planes $\mathcal{P}$ whose collineation groups Σ have relatively large topological dimension, aiming at an explicit classification of these planes. In Section 33 we show that $\dim \Sigma \leq 8$, and that $\dim \Sigma \geq 5$ occurs only if $\mathcal{P}$ is isomorphic to the real projective plane $\mathcal{P}_2\mathbb{R}$. In Section 34 we prove that $\dim \Sigma = 4$ holds precisely if $\mathcal{P}$ is a Moulton plane. Furthermore we describe the collineation groups of the Moulton planes; these groups contain the universal covering group of $\mathrm{SL}_2\mathbb{R}$.

By Theorem (33.9), the condition $\dim \Sigma = 3$ implies that the connected component Σ^1 fixes precisely t points (and precisely t lines) where $0 \leq t \leq 2$. The case with t fixed points is treated in Section $35 + t$. For $t = 0$, one obtains the skew hyperbolic planes discovered by Salzmann [62b]. The case $t = 1$ occurs precisely for the shift planes generated by skew parabolae. Finally, $t = 2$ characterizes the planes over a special class of (explicitly described) Cartesian fields. Compared to the original papers by Salzmann from the period 1962–67, the proofs given here put slightly more emphasis on the rôle of transformation groups, see (33.11). The classification results proved in Sections 33 through 37 are summarized in (38.1).

These classification results, in particular the seminal paper Salzmann [62a], mark the beginning of topological geometry as we know it (and expound it in this book). Indeed, these results yield a paradigm for an analogous classification program for compact projective planes of higher dimensions, see Section 65 and Chapters 7, 8. Chapter 3 is in this sense a preparation for (more complicated) things to come.

We point out that Chapter 4 offers a more direct, axiomatic approach to 2-dimensional compact projective planes; indeed, in Section 31 and at the beginning of Section 32 we prove the continuity assumptions stipulated in Chapter 4.

31 $\mathbb{R}^2$-planes

In this section we study plane geometries whose point set is the surface $\mathbb{R}^2$, endowed with a system $\mathcal{L}$ of curves, called lines. We require that any two distinct points are joined by a unique line, and that every line is closed in $\mathbb{R}^2$ and homeomorphic to $\mathbb{R}$, like in the real affine plane. The topology of such a line is induced by an order of the line, hence one can define intervals on the line; the intervals (of various types) are just the connected subsets of lines. These facts lead to a convexity theory for $\mathbb{R}^2$-planes; we show that the topology of $\mathbb{R}^2$ has a basis consisting of convex triangles (or quadrangles), and that the intersection of lines is 'transversal'.

There is a natural topology for the set $\mathcal{L}$ of lines such that the geometric operation of joining two distinct points is continuous. With respect to this topology, convergence of lines is equivalent to convergence in the sense of Hausdorff. The natural topology of $\mathcal{L}$ can also be described in several other ways, see (31.19). Furthermore, it is the only topology on $\mathcal{L}$ satisfying the usual compatibility conditions of topological geometry, compare (31.22).

31.1 Definition. Let $\mathcal{L}$ be a system of subsets of $\mathbb{R}^2$. The elements of $\mathbb{R}^2$ are called points, and the elements of $\mathcal{L}$ are called lines. We say that $(\mathbb{R}^2, \mathcal{L})$ is an $\mathbb{R}^2$-*plane* if the following axioms hold:

(a) Each line is closed in the topological space $\mathbb{R}^2$ and is homeomorphic to $\mathbb{R}$.
(b) Any two distinct points p, q are contained in exactly one line $L \in \mathcal{L}$.

The joining line L in (b) is denoted by $L = p \vee q = pq$; thus we have the *join map*

$$\vee : \{ (p, q) \in \mathbb{R}^2 \times \mathbb{R}^2 \mid p \neq q \} \longrightarrow \mathcal{L}.$$

Analogously, the common point of two distinct intersecting lines K, L is denoted by $K \wedge L$; the *intersection map* $\wedge$ is defined on $\{ (K, L) \in \mathcal{L}^2 \mid K \neq L, K \cap L \neq \emptyset \}$.

As a **general assumption for Section 31**, let $(\mathbb{R}^2, \mathcal{L})$ be an $\mathbb{R}^2$-plane.

31.2 Examples. (a) The *classical* $\mathbb{R}^2$-*plane* is the real affine plane: for lines we take all sets $L_{s,t} = \{ (x, y) \in \mathbb{R}^2 \mid y = sx + t \}$ with $s, t \in \mathbb{R}$ and all 'verticals' $\{x\} \times \mathbb{R}$ with $x \in \mathbb{R}$.

The parabola model of the real affine plane is obtained as follows: as lines we take all parabolas $Q_{a,b} = \{ (x, y) \in \mathbb{R}^2 \mid y - b = (x - a)^2 \}$ with $a, b \in \mathbb{R}$ and all verticals. The map $(x, y) \mapsto (x, y - x^2)$ is an isomorphism of the parabola model onto the usual model.

(b) Fix a value $d > 1$ and and take as lines all verticals and all curves

$$\{ (x, y) \in \mathbb{R}^2 \mid y - b = |x - a|^d \} \text{ with } a, b \in \mathbb{R}.$$

These examples are affine planes, compare (31.25c), which are not desarguesian for $d \neq 2$.

(c) The *real hyperbolic plane* is the restriction of the real affine plane to the open unit disc $\{(x, y) \in \mathbb{R}^2 \mid x^2 + y^2 < 1\}$, which is homeomorphic to $\mathbb{R}^2$. The term 'restriction' indicates that now the open segments between two distinct points on the unit circle are called lines. This model is the Klein model of the real hyperbolic plane, after Klein [1871] §12. Note that the real hyperbolic plane is not an affine plane.

(d) The *real half plane* is the restriction of the real affine plane to the open set $\{(x, y) \in \mathbb{R}^2 \mid x > 0\}$.

More examples are given in (31.25).

First we derive some rather general properties of $\mathbb{R}^2$-planes.

31.3 Proposition. *For every line $L \in \mathcal{L}$, the embedding of L in $\mathbb{R}^2$ is topologically equivalent to the inclusion of $\mathbb{R} \times \{0\}$ in $\mathbb{R}^2$ (i.e., there is a homeomorphism of $\mathbb{R}^2$ mapping L onto $\mathbb{R} \times \{0\}$).*

Proof. The one-point compactification $\mathbb{R}^2 \cup \{\infty\}$ of $\mathbb{R}^2$ is the 2-sphere $\mathbb{S}_2$, and $L \cup \{\infty\}$, being compact, is homeomorphic to the circle $(\mathbb{R} \times \{0\}) \cup \{\infty\}$ via a homeomorphism h which fixes ∞. The Theorem of Schoenflies (see Moise [77] §10 Thm. 3) yields a homeomorphism $\mathbb{S}_2 \to \mathbb{S}_2$ which extends h. □

31.4 Definitions. By (31.3) the complement of every line $L \in \mathcal{L}$ has exactly two connected components. These components are the *open half-planes* defined by L, and the topological closure of such a half-plane H is the *closed half-plane* $H \cup L$. Subsets of the same open half-plane (of different open half-planes) defined by L are said to lie on the *same side* (on *different sides*) of L.

For distinct points a, b, the *closed interval* $[a, b]$ is the intersection of all connected subsets of the line $a \vee b$ that contain a and b; *open intervals* are defined by $(a, b) = [a, b] \setminus \{a, b\}$.

Note that the topology of each line is induced by an order of the line; our intervals coincide with the intervals with respect to this order or its opposite order.

A subset S of $\mathbb{R}^2$ is called *convex* (with respect to $\mathcal{L}$) if $[a, b] \subseteq S$ for any two distinct points $a, b \in S$. By (31.1a), intervals and convex sets are connected. In fact, the closed intervals contained in a line L are exactly the compact, connected subsets of L. Since intersections of convex sets are convex again, one can use the familiar definition of convex hull (with respect to $\mathcal{L}$).

Cantwell [74, 78] and Cantwell–Kay [78] develop a geometric convexity theory not only for $\mathbb{R}^2$-planes, but also for higher-dimensional geometries. See also Doignon [76], Lenz [92] and the references given in (32.24) for the interrelations between geometry and order.

31.5 Proposition.

(a) *Closed half-planes are convex.*
(b) *Let $K, L \in \mathcal{L}$ be distinct lines intersecting in the point p. Then K and L intersect 'transversally', i.e., any two distinct points $a, b \in K$ with $p \in (a, b)$ lie on different sides of L.*
(c) *A line $L \in \mathcal{L}$ and a closed interval $[a, b]$ with $a, b \notin L$ are disjoint if, and only if, a and b lie on the same side of L.*
(d) *Open half-planes are convex.*

Proof. (a) Let a, b be distinct points of a closed half-plane H defined by L, and assume that $[a, b]$ contains a point q of the opposite half-plane $\mathbb{R}^2 \setminus H$. Being connected, both $[a, q]$ and $[q, b]$ intersect L, and by (31.1b) the intersection points coincide with $c := L \wedge (a \vee b)$. Now $c \in [a, b]$ separates q from a and b, a contradiction.

(b) Assume that there are distinct points $a, b \in K$ on the same side of L with $K \wedge L = p \in (a, b)$. Then K is contained in a closed half-plane H defined by L. Let $\mathbb{R}^2 \cup \{\infty\} \approx \mathbb{S}_2$ denote the one-point compactification of $\mathbb{R}^2$. Then $K \cup \{\infty\}$ and $L \cup \{\infty\}$ are Jordan curves which touch at the two points p and ∞. Thus $H \setminus K$ decomposes into three connected components, which may be distinguished as follows: the first one contains a, but not b in its boundary, the second one has boundary K, and the last one contains b, but not a in its boundary (this follows also from the θ-curve theorem, compare Whyburn [42] VI.1). Choose a' in the first connected component.

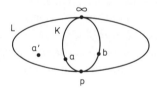

Figure 31a

The open interval (a', b) is a subset of H by part (a), and $(a', b) \cap K = \emptyset$ in view of $(a' \vee b) \wedge K = b$, hence $(a', b) \subseteq H \setminus K$. Accumulating at b, the connected set (a', b) intersects different connected components of $H \setminus K$, a contradiction.

(c) If a and b are on different sides of L, then $[a, b]$ intersects L by connectedness of $[a, b]$. If a, b lie on the same side of L, then the existence of a point $p \in (a, b) \cap L$ gives a contradiction to (b).

(d) follows from (c) using (a). □

As a consequence of (31.5d), the restriction of an $\mathbb{R}^2$-plane to any of its open half-planes is again an $\mathbb{R}^2$-plane, and the term 'open half-plane' may also denote such an $\mathbb{R}^2$-plane. Furthermore, every non-empty intersection of finitely many open halfplanes (i.e., the interior of a convex polygon) is an $\mathbb{R}^2$-plane; compare also (31.25a).

31.6 Definition. The *convex triangle* $\langle a_1, a_2, a_3 \rangle$ defined by three non-collinear points a_1, a_2, a_3 is the intersection $\langle a_1, a_2, a_3 \rangle = H_1 \cap H_2 \cap H_3$, where H_i is the closed half-plane determined by $a_j \vee a_k$ with $a_i \in H_i$, for $\{i, j, k\} = \{1, 2, 3\}$.

We claim that the interior (in the sense of the Jordan curve theorem) of the Jordan curve $J = [a_1, a_2] \cup [a_2, a_3] \cup [a_3, a_1]$ coincides with the topological interior of $\langle a_1, a_2, a_3 \rangle$. To prove this, we first argue that the union C of J with the Jordan-interior of J is the convex hull of $\{a_1, a_2, a_3\}$. Indeed, consider a line L through some point x of the Jordan-interior. Each half-line (connected component) of $L \setminus \{x\}$ is unbounded and therefore intersects J. The two intersection points u, v are distinct, and $x \in (u, v)$ since u and v belong to different components of $L \setminus \{x\}$. For the other inclusion, it suffices to show that C is convex. From $x, y \in C$ one infers that $[x, y] \subseteq C$ by an argument similar to the above. Thus we have shown that C is the convex hull of $\{a_1, a_2, a_3\}$. Our claim is now a consequence of the Schoenflies Theorem (compare Moise [77] §10 Thm. 3).

In particular, the convex triangle $\langle a_1, a_2, a_3 \rangle$ is compact and coincides with the convex hull of $\{a_1, a_2, a_3\}$.

31.7 Corollary: The Pasch Axiom. *Let a_1, a_2, a_3 be three non-collinear points.*

(a) *Every line L meeting the set $[a_1, a_2] \cup [a_2, a_3] \cup [a_3, a_1]$ intersects at least two of the sides $[a_i, a_j]$.*
(b) *A line L meets the interior of the convex triangle $\langle a_1, a_2, a_3 \rangle$ if, and only if, L meets one of the open intervals (a_i, a_j) but $L \neq a_i a_j$.*

Proof. (a) Let L be a line that meets $[a_1, a_2] \cup [a_2, a_3] \cup [a_3, a_1]$. We may assume that $L \cap \{a_1, a_2, a_3\} = \emptyset$. By (31.5b) the line L separates one of the points a_1, a_2, a_3 from the other two, so we may assume that a_1 and a_2, a_3 are on different sides of L. Using (31.5c) we obtain that L intersects the intervals $[a_1, a_2]$ and $[a_1, a_3]$.

(b) Assume now that L is a line that contains an interior point p of $\langle a_1, a_2, a_3 \rangle$. We consider the closure H of a connected component of $L \setminus \{p\}$. Since H is closed in $\mathbb{R}^2$ but not compact, it is not entirely contained in $\langle a_1, a_2, a_3 \rangle$. Being connected, the 'half-line' H meets the boundary $B = [a_1, a_2] \cup [a_2, a_3] \cup [a_3, a_1]$. Since p is an inner point, the two half-lines meet B in different points q, q', say. Now $\{q, q'\}$ is not contained in $\{a_1, a_2, a_3\}$ since the joining line $qq' = L$ contains an interior point.

Conversely, assume that a line L meets the open interval (a_1, a_2), say, and that $L \neq a_1 a_2$. Then $L \cap (a_i, a_j)$ consists of a single point q, and L meets the boundary B in another point q' by assertion (a). The closed interval $[q, q']$ is contained in the convex triangle, but not in its boundary. Hence $L = qq'$ contains interior points. □

The Pasch axiom employs the same geometric figure as the Veblen axiom; see Pickert [94] for a discussion of the differences.

31.8 Definitions. Let a, b, c, d be mutually distinct points such that $a \vee c$ and $b \vee d$ are distinct lines intersecting in a point $p \in (a, c) \cap (b, d)$. Then the pair a, b is called *opposite* to the pair c, d. We say also that the (open or closed) intervals determined by two opposite pairs are opposite. The *convex quadrangle* $\langle a, b, c, d \rangle$ is the intersection of the four closed half-planes which are defined by the lines $a \vee b$, $b \vee c$, $c \vee d$ and $d \vee a$ and contain p. We remark that this quadrangle can also be described as the convex hull of $\{a, b, c, d\}$, or as the compact 'disc' defined by the Jordan curve $[a, b] \cup [b, c] \cup [c, d] \cup [d, a]$.

31.9 Lemma. *Let $\langle a_1, a_2, a_3, a_4 \rangle$ be a convex quadrangle, and let $x_i \in (a_i, a_{i+1})$ for $1 \leq i \leq 4$, where $a_5 = a_1$. Then the intervals $[x_1, x_3]$ and $[x_2, x_4]$ intersect in an inner point of $\langle a_1, a_2, a_3, a_4 \rangle$.*

Proof. By (31.5c) the points a_2, a_3 (and, similarly, a_1, a_4) lie on different sides of $L = x_2 \vee x_4$. If a_1 and a_2 lie on different sides of L, then the open interval (a_1, a_2) intersects L. Thus L meets the boundary of the convex quadrangle in three points, a contradiction to the fact that the interior of a convex quadrangle is convex (31.5d). Hence the sets $\{a_1, a_2\}$ and $\{a_3, a_4\}$ lie on different sides of $L = x_2 \vee x_4$. Using (31.5c) again, we obtain that x_1 and x_3 lie on different sides of L, as well. Thus $[x_1, x_3] \cap (x_2 \vee x_4) \neq \emptyset$. By symmetry, we also have $(x_1 \vee x_3) \cap [x_2, x_4] \neq \emptyset$. Therefore, the intersection point of $[x_1, x_2]$ and $[x_3, x_4]$ exists; it is an inner point by (31.5d). □

31.10 Proposition. *Every point p has a neighbourhood basis consisting of finite intersections of open half-planes.*

Proof. Any neighbourhood of p contains a disc D with center p. For every point x on the boundary ∂D of D, we find a line $L_x \in \mathcal{L}$ such that p and x are on different sides of L_x, see (31.5b). Denote the corresponding half-planes by G_x and H_x in such a way that $p \in G_x$ and $x \in H_x$. By compactness, ∂D is contained in the union of finitely many half-planes H_{x_i}, $1 \leq i \leq n$. The intersection $\bigcap_{i=1}^{n} G_{x_i}$ is a neighbourhood of p which is disjoint from ∂D (by construction). This intersection is convex, hence connected, and therefore contained in D. □

31.11 Corollary. *Every point p has a neighbourhood basis consisting of convex quadrangles (or of convex triangles). In fact, one can find a neighbourhood basis consisting of convex quadrangles $\langle a, b, c, d \rangle$ which satisfy $p = ac \wedge bd$.*

Proof. Each of the neighbourhoods from (31.10) contains convex quadrangles (and triangles) with interior point p.

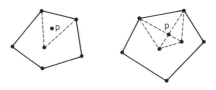

Figure 31b

□

Now we investigate systematically the compatibility of geometry and topology in $\mathbb{R}^2$-planes.

31.12 Proposition. *Collinearity and the order of (collinear) point triples are preserved under limits, in the following sense:*

(a) *If the point sequences $(a_i)_{i\in\mathbb{N}}$, $(b_i)_{i\in\mathbb{N}}$, $(c_i)_{i\in\mathbb{N}}$ have mutually distinct limits a, b, c, and if a_i, b_i, c_i are collinear for infinitely many $i \in \mathbb{N}$, then a, b, c are collinear as well.*

(b) *If, in addition, $b_i \in (a_i, c_i)$ for infinitely many $i \in \mathbb{N}$, then $b \in (a, c)$.*

Proof. (a) Passing to subsequences and adapting the notation, we may assume that $b_i \in (a_i, c_i)$ for all $i \in \mathbb{N}$. If a, b, c are not collinear, pick a line K which meets (a, b) and (b, c). Then b_i and $[a_i, c_i]$ lie on different sides of K for sufficiently large i, see (31.5), a contradiction to the fact that $b_i \in (a_i, c_i)$. Hence a, b, c are collinear.

(b) Assume that $b \notin (a, c)$. By symmetry it suffices to consider the case where $c \in (a, b)$. Let $K \neq a \vee b$ be a line which meets (c, b). Then $[a_i, c_i]$ and b_i are on different sides of K for almost all i, a contradiction to $b_i \in (a_i, c_i)$. □

Our next aim is to endow the set $\mathcal{L}$ with a topology which is compatible with the geometry $(\mathbb{R}^2, \mathcal{L})$. As a first step, we introduce a notion of convergence due to Hausdorff [14].

31.13 Definition. Let $(L_i)_{i\in\mathbb{N}}$ be a sequence of lines. Denote by $\liminf_{i\in\mathbb{N}} L_i$ the set of all limits of convergent sequences $(p_i)_{i\in\mathbb{N}}$ with $p_i \in L_i$, and denote by $\limsup_{i\in\mathbb{N}} L_i$ the set of all accumulation points of sequences $(q_i)_{i\in\mathbb{N}}$ with $q_i \in L_i$. We say that $(L_i)_{i\in\mathbb{N}}$ *converges* to $L \in \mathcal{L}$ *in the sense of Hausdorff* if $L = \liminf_{i\in\mathbb{N}} L_i = \limsup_{i\in\mathbb{N}} L_i$.

This definition has a quite intuitive interpretation: roughly speaking, it means that the points on the lines L_i approximate precisely the points on L, and that no parts of L_i can stay away from L.

It is a remarkable fact that with this notion of convergence, the join map of every $\mathbb{R}^2$-plane is sequentially continuous:

31.14 Proposition. *Let $(a_i)_{i\in\mathbb{N}}, (b_i)_{i\in\mathbb{N}}$ be convergent sequences of points with distinct limits a and b. Then the lines $L_i = a_i \vee b_i$ converge to $L = a \vee b$ in the sense of Hausdorff.*

Proof. Obviously $\liminf_{i\in\mathbb{N}} L_i \subseteq \limsup_{i\in\mathbb{N}} L_i$, hence it suffices to show that

$$\limsup\nolimits_{i\in\mathbb{N}} L_i \subseteq a \vee b \subseteq \liminf\nolimits_{i\in\mathbb{N}} L_i.$$

For the first inclusion, let x be an accumulation point of a sequence $(x_i)_{i\in\mathbb{N}}$ with $x_i \in L_i$. Then (31.12a) implies that $x \in a \vee b$.

For the second inclusion, we first consider a point $c \in (a, b)$. Let $K \ne a \vee b$ be a line through c. The points a_i, b_i lie on different sides of K for sufficiently large i, and (31.5c) shows that the intersection points $s_i = (a_i \vee b_i) \wedge K$ exist. By (31.12a), any accumulation point of $(s_i)_{i\in\mathbb{N}}$ coincides with $(a \vee b) \wedge K = c$. Consideration of a (compact) convex quadrangle with inner points a, b shows that an accumulation point exists, compare (31.11). We infer that $(s_i)_{i\in\mathbb{N}}$ converges to c, hence $c \in \liminf_{i\in\mathbb{N}} L_i$.

Now we consider a point $c \in a \vee b \setminus [a, b]$. We may assume that $b \in (a, c)$. By omitting those lines $L_i = a_i \vee b_i$ which coincide with $L = a \vee b$, we find a point $q \in L$ with $a \in (q, b)$ and $q \notin L_i$ for all $i \in \mathbb{N}$. Choose points $r_1, r_2 \notin L$ with $c \in (r_1, r_2)$.

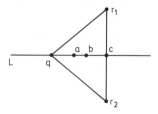

Figure 31c

If the intersection points $s_i = L_i \cap [r_1, r_2]$ exist for sufficiently large i, then (31.12a) implies that $(s_i)_{i\in\mathbb{N}}$ converges to c, hence $c \in \liminf_{i\in\mathbb{N}} L_i$. Aiming for a contradiction, we can now switch to a subsequence such that $L_i \cap [r_1, r_2] = \emptyset$ for all

$i \in \mathbb{N}$. For sufficiently large i the open interval (a_i, b_i) belongs to the interior of the convex triangle $\langle q, r_1, r_2 \rangle$. The Pasch axiom (31.7a) yields that $L_i \cap [q, r_j] \neq \emptyset$ for all $i \in \mathbb{N}$, $j \in \{1, 2\}$, hence we can choose $s_i \in L_i \cap ([q, r_1] \cup [q, r_2])$ such that $b_i \in (a_i, s_i)$; note that $s_i \neq q$, as $q \notin L_i$. The sequence $(s_i)_{i \in \mathbb{N}}$ accumulates at some point s of the compact set $[q, r_1] \cup [q, r_2]$, and (31.12a) says that s, a, b are collinear, hence $s = q$. From (31.12b) we infer that $b \in (a, s) = (a, q)$, a contradiction to the choice of q. □

31.15 Definition. The *natural topology* H on $\mathcal{L}$ is defined as follows: $A \subseteq \mathcal{L}$ is H-closed if, and only if, A contains the Hausdorff limit of any convergent sequence in A (31.13). The open sets of the topology H are just the complements of the H-closed subsets of $\mathcal{L}$.

Note that H is indeed a topology: the system of all H-closed sets is stable under intersections (by definition) and under finite unions (by a subsequence argument).

31.16 Lemma. *The join map* $\vee : \{ (p, q) \in \mathbb{R}^2 \times \mathbb{R}^2 \mid p \neq q \} \to \mathcal{L}$ *is continuous if $\mathcal{L}$ is endowed with the natural topology* H.

Proof. Let p, q be distinct points, let $U \subseteq \mathcal{L}$ be an open neighbourhood of $L := p \vee q$, and let $\{ V_i \mid i \in \mathbb{N} \}$ and $\{ W_i \mid i \in \mathbb{N} \}$ be neighbourhood bases of p and q, respectively. If $V_i \vee W_i \not\subseteq U$ for all $i \in \mathbb{N}$, then we find points $p_i \in V_i$, $q_i \in W_i$ with $p_i \vee q_i \notin U$, and by (31.14) the sequence $(p_i \vee q_i)_{i \in \mathbb{N}}$ converges to $p \vee q$ in the sense of Hausdorff. Hence $L = p \vee q$ belongs to the H-closed set $\mathcal{L} \setminus U$, a contradiction. □

31.17 Lemma. *Each pencil $\mathcal{L}_p$ of lines is homeomorphic to the circle* $\mathbb{S}_1$.

Proof. Let a, b, c be a triangle with $p \in (a, c)$. The map $[a, b] \cup [b, c] \to \mathcal{L}_p$ with $x \mapsto p \vee x$ is continuous (31.16), surjective (31.7a) and closed (compare Dugundji [66] XI.2.1 and note that $\mathcal{L}_p$ is a Hausdorff space by (31.19)), hence an identification map. Therefore $\mathcal{L}_p$ is homeomorphic to the space $([a, b] \cup [b, c])/a \sim c \approx [0, 1]/0 \sim 1 \approx \mathbb{S}_1$. □

Now we define a few more topologies on $\mathcal{L}$, and we show that all these topologies coincide with the natural topology H. The different descriptions of H obtained by this procedure turn out to be quite useful.

31.18 Definitions. The *final topology* F on $\mathcal{L}$ is the finest topology on $\mathcal{L}$ such that the join map is continuous. The *open join topology* OJ is generated by the subbasis elements $O_1 \vee O_2 = \{ p \vee q \mid p \in O_1, q \in O_2 \}$, where O_1, O_2 are disjoint open subsets of $\mathbb{R}^2$. The *interval join topology* IJ has a subbasis consisting of the sets $I_1 \vee I_2 = \{ p \vee q \mid p \in I_1, q \in I_2 \}$, where I_1, I_2 are opposite (31.8) open intervals. The *open meet topology* OM is defined by the subbasic sets $M_O = \{ L \in \mathscr{L} \mid L \cap O \neq \emptyset \}$, where O is an open set in $\mathbb{R}^2$.

31.19 Theorem. *The topologies* H, F, OJ, IJ, OM *for* $\mathscr{L}$ *coincide. With respect to this topology,* $\mathscr{L}$ *is a Hausdorff space* (T_2), *and convergence is equivalent to convergence in the sense of Hausdorff* (31.13). *Furthermore, the join map* $\vee : \{ (p, q) \in \mathbb{R}^2 \times \mathbb{R}^2 \mid p \neq q \} \to \mathscr{L}$ *is continuous and open.*

Proof. 1) We show that OJ $\subseteq$ OM $\subseteq$ H $\subseteq$ F $\subseteq$ IJ $\subseteq$ OJ.

The first inclusion follows from $O_1 \vee O_2 = M_{O_1} \cap M_{O_2}$. If O is open in $\mathbb{R}^2$, then $\mathscr{L} \setminus M_O = \{ L \in \mathscr{L} \mid L \cap O = \emptyset \}$ is H-closed; otherwise some $L \in M_O$ is the Hausdorff limit of a sequence $L_i \in \mathscr{L} \setminus M_O$. Then we find $p \in L \cap O$, and by (31.13) there exist points $p_i \in L_i$ converging to p. Almost all p_i lie in O, a contradiction to $L_i \in \mathscr{L} \setminus M_O$. Thus OM $\subseteq$ H. Now Lemma (31.16) implies that H $\subseteq$ F.

For the inclusion F $\subseteq$ IJ, let $U \subseteq \mathscr{L}$ be F-open, and let $L = p_1 \vee p_2 \in U$. Then $\{ (p, q) \in \mathbb{R}^2 \times \mathbb{R}^2 \mid p \neq q, \, p \vee q \in U \}$ is open in $\mathbb{R}^2 \times \mathbb{R}^2$, hence we find opposite open intervals I_i containing p_i such that $L \in I_1 \vee I_2 \subseteq U$ (use (31.5c) twice). This proves that F $\subseteq$ IJ.

If $\langle a_1, b_1, b_2, a_2 \rangle$ is a convex quadrangle, then $(a_1, b_1) \vee (b_2, a_2) = O_1 \vee O_2$, where O_i is the connected component of $\mathbb{R}^2 \setminus (a_1 \vee b_1 \cup b_1 \vee b_2 \cup b_2 \vee a_2 \cup a_2 \vee a_1)$ shown in Figure 31d. Since O_i is open, this shows that IJ $\subseteq$ OJ.

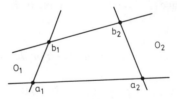

Figure 31d

2) The separation axiom (T_2) follows easily by using IJ: Given distinct lines L_1, L_2, we find lines M_1, M_2 intersecting L_1, L_2 in four mutually distinct points. For $j \in \{1, 2\}$, choose disjoint interval neighbourhoods I_{ij} of the points $L_i \wedge M_j$ in M_j. Then $I_{11} \vee I_{12}$ and $I_{21} \vee I_{22}$ are disjoint neighbourhoods of L_1 and L_2 respectively.

3) Let $(L_i)_{i \in \mathbb{N}}$ converge to $L \in \mathscr{L}$ in the sense of Hausdorff, and let U be an open H-neighbourhood of L. The H-closed set $\mathscr{L} \setminus U$ contains no sequence with Hausdorff limit L, hence it contains only finitely many L_i, and almost all L_i lie in U. This shows that Hausdorff convergence implies H-convergence.

For the converse, let $(L_i)_{i \in \mathbb{N}}$ be H-convergent to $L \in \mathscr{L}$. Choose distinct points a, b on L. For every neighbourhood O of a or b, the H-open set M_O contains almost all L_i. This proves that a and b are the limits of convergent sequences

$(a_i)_{i \in \mathbb{N}}$, $(b_i)_{i \in \mathbb{N}}$, respectively, with $a_i, b_i \in L_i$. Now (31.14) says that $L_i = a_i \vee b_i$ converges to $L = a \vee b$ in the sense of Hausdorff.

4) The join map is open (use the fact that $\mathbb{R}^2$ is a Hausdorff space, and the open join topology OJ) and continuous (31.16). □

31.20 Remarks. We mention two other descriptions of the natural topology on the set $\mathcal{L}$ of lines.

For $L, M \in \mathcal{L}$, define $d(L, M) = \sup \{ |\delta(x, L) - \delta(x, M)| e^{-|x|} \mid x \in \mathbb{R}^2 \}$, where $\delta(x, L) = \inf \{ |x - y| \mid y \in L \}$. Then d is a metric on $\mathcal{L}$ which induces the natural topology, see Busemann [55] Section 3.

Let $\mathcal{X}$ be the set of all parametrizations of lines, i.e., $\mathcal{X}$ consists of all maps $f : \mathbb{R} \to \mathbb{R}^2$ such that f is a homeomorphism onto some line $L \in \mathcal{L}$. Endow $\mathcal{X}$ with the compact-open topology (compare Dugundji [66] XII.1). Then $\mathcal{L}$ with the natural topology is the quotient of $\mathcal{X}$ obtained by identifying maps which have the same image, see Salzmann [67b] 2.8–2.12.

31.21 Proposition: Stability of intersection. *The pairs of distinct intersecting lines form an open subset of $\mathcal{L}^2$, and the intersection map $\wedge$, defined on that subset, is continuous.*

Proof. Let K, L be distinct intersecting lines. There exists a convex quadrangle $\langle a_1, a_2, a_3, a_4 \rangle$ such that $K \in I_2 \vee I_4$ and $L \in I_1 \vee I_3$, where $I_i = (a_i, a_{i+1})$ and $a_5 = a_1$. The product $(I_2 \vee I_4) \times (I_1 \vee I_3)$ is open in $\mathcal{L}^2$ (use the topology IJ) and consists of intersecting lines (31.9), which gives the first assertion. Together with (31.11), these arguments also show that $\wedge$ is continuous. □

Similar arguments in conjunction with (31.7) show that the intersection map $\wedge$ is open as well (like the join map (31.19)).

We say that the intersection map $\wedge$ is *stable* with respect to some topology on $\mathcal{L}$ if $\wedge$ is continuous and the domain of $\wedge$, consisting of the pairs of intersecting lines, is an open subset of the product space $\mathcal{L}^2$. The following theorem says that the natural topology is the unique topology satisfying the usual compatibility conditions of topological geometry.

31.22 Theorem. *Let $(\mathbb{R}^2, \mathcal{L})$ be an $\mathbb{R}^2$-plane. Then the natural topology H on $\mathcal{L}$ is the only topology such that the join map $\vee$ is continuous and the intersection map $\wedge$ is stable.*

Proof. For continuity of $\vee$ and stability of $\wedge$ see (31.16 and 21). Now let X be a topology on $\mathcal{L}$ such that $\vee$ is continuous and $\wedge$ is stable. We have $X \subseteq F = H$ by (31.19). Let $\mathcal{D} = \{ (K, L) \in \mathcal{L}^2 \mid K \neq L, K \cap L \neq \emptyset \}$ be the domain of $\wedge$. By stability the restriction $pr_1 : \mathcal{D} \to \mathcal{L}$ of the projection onto the first factor is X-open. For every open set O in $\mathbb{R}^2$, the set $M_O = pr_1(\{(K, L) \in \mathcal{D} \mid K \wedge L \in O \})$ is X-open by continuity of $\wedge$, hence OM $\subseteq$ X. Then X = H by (31.19). □

31.23 Proposition. *The set $\mathcal{L}$ of lines, endowed with the natural topology H, is a surface, i.e., a topological manifold (with countable basis) of dimension 2.*

Proof. Since $\vee$ and $\wedge$ are continuous (31.22), the topology IJ shows that $\mathcal{L}$ is locally homeomorphic to $\mathbb{R}^2$. As $\mathbb{R}^2$ has a countable basis, the OM-description of H shows that $\mathcal{L}$ has a countable basis, too. Finally, $\mathcal{L}$ is a Hausdorff space by (31.19), hence a surface. □

We mention that $\mathcal{L}$ is in fact homeomorphic to the Möbius strip, and that the space $\mathcal{L}^+$ of oriented lines is homeomorphic to the cylinder, see Salzmann [67b] 2.14, Löwen [95] 3.2; compare also (32.4).

31.24 Lemma. *Let $C \subseteq \mathbb{R}^2$ be compact. Then $\{ L \in \mathcal{L} \mid L \cap C \neq \emptyset \}$ is a compact subset of $\mathcal{L}$.*

Proof. Sequential compactness suffices, see (31.23), so we consider a sequence $(L_i)_{i \in \mathbb{N}}$ of lines meeting C. Choose points $c_i \in L_i \cap C$, an accumulation point c of $(c_i)_{i \in \mathbb{N}}$ in C, and a compact neighbourhood U of c in $\mathbb{R}^2$. Then infinitely many lines L_i contain a point of U, hence these lines (being non-compact and connected) intersect the boundary ∂U of U. The intersection points accumulate at some point $u \in \partial U$, and the lines L_i accumulate at $c \vee u$ by (31.14). □

31.25 More examples. (a) The restriction of an $\mathbb{R}^2$-plane to any of its non-empty convex open subsets is again an $\mathbb{R}^2$-plane; this generalizes (31.2c,d). In order to prove that such a subset is homeomorphic to $\mathbb{R}^2$, one can apply the results of Löwen [95], in view of (31.19 and 21).

(b) *The affine Moulton planes $\mathcal{M}_k$, $k > 1$.* We start with the real affine plane (31.2a) and replace the lines $L_{s,t}$ with negative slope s by the kinked lines

$$L^*_{s,t} = \{ (x, sx + t) \mid x \geq 0 \} \cup \{ (x, skx + t) \mid x \leq 0 \},$$

where $k > 1$ is fixed. In order to verify the axioms for affine planes, it suffices to show that every pair of points (x_1, y_1), (x_2, y_2) with $x_1 < 0 < x_2$ and $y_1 - y_2 > 0$ is joined by exactly one line $L^*_{s,t}$: the system of equations $y_1 = skx_1 + t$, $y_2 = sx_2 + t$ gives $y_2 - y_1 = s(x_2 - kx_1)$, thus s and t are uniquely determined.

It is easy to see that these planes $\mathcal{M}_k$ are not desarguesian: let $a_1 = (-1, -1)$, $b_1 = (-1, 1)$, $c_1 = (0, 0)$, $a_2 = (0, -1)$, $b_2 = (0, 1)$, $c_2 = (1, 0)$ (see Fig. 31e).

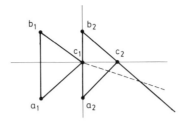

Figure 31e

Then the triangles a_1, b_1, c_1 and a_2, b_2, c_2 are perspective (the lines $a_1 \vee a_2$, $b_1 \vee b_2$, $c_1 \vee c_2$ are parallel), $a_1 \vee c_1$ is parallel to $a_2 \vee c_2$ and $a_1 \vee b_1$ is parallel to $a_2 \vee b_2$, but $b_1 \vee c_1$ and $b_2 \vee c_2$ intersect in $\mathcal{M}_k$.

The planes $\mathcal{M}_k$ have been introduced by Moulton [02]. These early examples proved to be of special importance in the classification of 2-dimensional compact projective planes, since they have rather large collineation groups, see Section 34.

(c) *Shift planes*. Apart from the vertical lines, the parabola model (31.2a) of the real affine plane consists of the curve $\{(x, y) \in \mathbb{R}^2 \mid y = x^2\}$ and its images under the shift group $\{(x, y) \mapsto (x + a, y + b) \mid a, b \in \mathbb{R}\}$. We say that an $\mathbb{R}^2$-plane is a *shift plane* if it is an affine plane which is generated in this way by a suitable curve $\{(x, y) \in \mathbb{R}^2 \mid y = f(x)\}$; see also Section 36. For higher-dimensional shift planes see Section 74.

The following criterion, due to Salzmann [65] p. 258, gives many examples: Let $f : \mathbb{R} \to \mathbb{R}$ be differentiable such that the derivative $f' : \mathbb{R} \to \mathbb{R}$ is a homeomorphism. Then the curve $\{(x, y) \in \mathbb{R}^2 \mid y = f(x)\}$ generates a shift plane as defined above.

Proof. We may assume that f' is increasing. It suffices to show that for every $a \neq 0$ the difference function f_a defined by

$$f_a(x) = f(x+a) - f(x) = \int_0^a f'(x+t)\, dt$$

is a homeomorphism of $\mathbb{R}$ onto itself, compare (74.3). Since $f_{-a}(x) = -f_a(x-a)$, we may assume that $a > 0$. The integral formula shows that f_a is strictly increasing, and that $af'(x) < f_a(x) < af'(x+a)$, whence f_a is surjective. Furthermore f_a is continuous, hence a homeomorphism. □

In fact, the differentiability of f is not necessary. According to Groh [76], a function $f : \mathbb{R} \to \mathbb{R}$ generates a shift plane if, and only if, f is strictly convex and

$$\lim_{x \to \pm\infty} (f(x) - ax) = +\infty$$

for every $a \in \mathbb{R}$, or if $-f$ satisfies these conditions.

We remark that every shift plane can be coordinatized by a Cartesian field with commutative multiplication, compare Dembowski–Ostrom [68] Thm. 6, Knarr [83], Weigand [87] Satz 5.2, Szönyi [90]; in fact, the multiplication of the Cartesian field is the map

$$(x, y) \mapsto g^{-1}(f(x+y) - f(x) - f(y)),$$

where g is the bijection defined by $g(x) = f(x+1) - f(x) - f(1)$.

(d) *Arc planes.* Groh [79, 82b] has generalized the concept of shift planes in several respects, obtaining $\mathbb{R}^2$-planes which are not necessarily affine planes; he also employs non-abelian 'shift groups'. These so-called arc planes include the following example (compare also Betten [68], Strambach [70b]): Starting with the real affine plane, we replace the lines with positive slope by the shifted arcs $\{ (x, y) \in \mathbb{R}^2 \mid y - n = e^{x-m} \}$, where $m, n \in \mathbb{R}$. It is easy to see that this gives an $\mathbb{R}^2$-plane which is not an affine plane; in fact there is no embedding into any affine $\mathbb{R}^2$-plane, compare Stroppel [94c], Löwen [81a], p. 311. Furthermore, this plane is not desarguesian: we draw a Desargues configuration in the real affine plane such that only one line L has positive slope and such that L contains the center of the Desargues configuration.

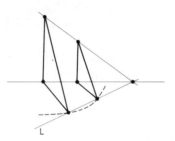

Figure 31f

Then L is replaced by some exponential function, which shows that the Desargues configuration does not close in this arc plane.

Further constructions can be found in (34.1).

31.26 Notes on Section 31. $\mathbb{R}^2$-planes are a special type of 'stable planes': A *stable plane* is an incidence structure $(P, \mathcal{L}, \mathbf{F})$, where any two points have a unique joining line, and P and $\mathcal{L}$ are endowed with non-discrete topologies such that joining is continuous and intersection is stable in the sense of (31.22). The literature on $\mathbb{R}^2$-planes and on stable planes is extensive, see Salzmann [63a, 67a,b, 69a], Strambach [67a,b, 68, 70a,b,c, 71], Betten [68], Polley [68, 72a,b], Löwen [76b, 78, 79a,b,c, 81a,b,c, 82a,b, 83a,b,c, 84b, 86a,b,c,d, 95], Groh [76, 79, 81, 82a,b], Groh–Lippert–Pohl [83], Hubig [87], Pohl [90], Seidel [90a,b, 91], Stroppel [90,

91, 92c,d, 93a,b,c,d,e, 94a,b,c,d,e]; compare also Skornjakov [57], where the results (31.5b and 16) can be found. For a recent survey see Grundhöfer–Löwen [94].

Concerning collineations of $\mathbb{R}^2$-planes, the following results have been proved by Salzmann [67b]: every collineation is continuous, only the identity fixes a non-degenerate triangle (Salzmann [67a] 1.1), and the group Σ of all collineations, endowed with the compact-open topology, is a Lie group. For affine $\mathbb{R}^2$-planes, these results will be proved in Section 32. Most of the papers quoted above contribute to the classification of stable planes with 'large' collineation groups. For the state of the art, see Stroppel [93e], and Groh–Lippert–Pohl [83] for the special case of $\mathbb{R}^2$-planes.

32 Two-dimensional compact projective planes

In this section we consider $\mathbb{R}^2$-planes which are affine planes, and we show that their projective completions are '2-dimensional compact projective planes' (32.2 and 5). These planes have the same topology as the real projective plane $\mathcal{P}_2\mathbb{R}$; their geometries, however, can be entirely different. In this section we collect basic information on collineations and on collineation groups of these planes. This will lead to first characterizations of $\mathcal{P}_2\mathbb{R}$ by its collineation group (32.20, 22 and 23).

The geometric concepts and the notation from Chapter 2 are now used systematically.

Our first aim is to give a topological version of the projective completion (21.9) of an affine $\mathbb{R}^2$-plane. This is not an entirely trivial task, as the affine Moulton planes (31.25b) show: in the point set $P_2\mathbb{R}$ of the projective completion of the real affine plane, each broken line accumulates at *two* distinct points at infinity. Thus the construction of a topological projective completion has to make essential use of the geometry.

32.1 Definitions. Let $(A, \mathcal{L})$ be an affine $\mathbb{R}^2$-plane. For $L \in \mathcal{L}$, let $\|_L$ denote the parallel class of L. Then the projective completion of $(A, \mathcal{L})$ is obtained as $(P, \hat{\mathcal{L}})$, where $P = A \cup \mathcal{L}/\|$, and $\hat{\mathcal{L}} = \{ L \cup \{\|_L\} \mid L \in \mathcal{L} \} \cup \{\mathcal{L}/\|\}$.

The natural topology H on $\mathcal{L}$ is locally compact (31.23) but not compact (e.g., consider an infinite discrete subset $D \subset L \in \mathcal{L}$, and choose a line K that meets L; then the parallels to K through points of D form an infinite closed discrete subset of $\mathcal{L}$). Identifying L with $L \cup \{\|_L\}$, we may consider $\hat{\mathcal{L}}$ as the one-point compactification of $\mathcal{L}$. We shall refer to this topology as $\hat{H}$. Note that, with respect to $\hat{H}$, every pencil $\hat{\mathcal{L}}_p$ is homeomorphic to the circle $\mathbb{S}_1$; this follows from (31.17) and from the fact that the closure of a parallel class $\|_L$ equals $\|_L \cup \{W\}$, which is the one-point compactification of $\|_L \approx \mathbb{R}$.

We topologize the point set P by a construction that is dual to that of the topology IJ on $\mathcal{L}$: for a subbase, we take the system of all sets of the form $\mathcal{J}_p \wedge \mathcal{J}_q$, where p, q are two points in P, and $\mathcal{J}_p, \mathcal{J}_q$ are disjoint open intervals in $\hat{\mathcal{L}}_p, \hat{\mathcal{L}}_q$, respectively. This 'pencil intersection' topology on P is denoted by PI.

32.2 Theorem. *Let $\mathcal{P} = (P, \hat{\mathcal{L}})$ be the projective completion of an affine $\mathbb{R}^2$-plane $(A, \mathcal{L})$. Then P and $\hat{\mathcal{L}}$, endowed with PI and $\hat{H}$, induce the natural topologies on $A = \mathbb{R}^2$ and on $\mathcal{L}$, respectively. Furthermore $\mathcal{P}$ is a topological projective plane, i.e., joining different points and intersecting different lines are continuous operations in $\mathcal{P}$. Each affine part of $\mathcal{P}$ is an $\mathbb{R}^2$-plane. The topologies PI and $\hat{H}$ are compact.*

Proof. 1) With respect to PI, every line $L \in \hat{\mathcal{L}}$ is closed. The subset A is an open subset of P, and retains its original topology.

Indeed, for any two points $p, q \in L$, the complement of L is $(\hat{\mathcal{L}}_p \setminus \{L\}) \wedge (\hat{\mathcal{L}}_q \setminus \{L\}) \in$ PI. In particular, A is open. The last assertion follows from (31.11 and 9).

2) Let u, v be distinct points in P. Then the mapping

$$c_{u,v} : (\hat{\mathcal{L}}_u \setminus \{uv\}) \times (\hat{\mathcal{L}}_v \setminus \{uv\}) \to P \setminus uv : (K, L) \mapsto K \wedge L$$

is a homeomorphism with respect to PI.

By the definition of PI, the mapping $c_{u,v}$ is open. If K and L intersect in a point of A, continuity of $c_{u,v}$ at (K, L) follows from the continuity of intersection (31.21), combined with assertion 1). Now assume that $K \wedge L \notin A$. Then $uv \neq W$, and we may assume that $u \in A$. Let $\mathcal{J}_p \wedge \mathcal{J}_q$ be a neighbourhood of $\|_K = K \wedge L$. Since $\hat{\mathcal{L}}_u \times \hat{\mathcal{L}}_v$ is first countable, it suffices to consider sequences $(K_n, L_n) \in \hat{\mathcal{L}}_u \times \hat{\mathcal{L}}_v$ that converge to (K, L). By way of contraposition, we assume that $x_n = K_n \wedge L_n$ does not accumulate at $\|_K$.

We consider the lines $X_n = x_n p$. Since $\hat{\mathcal{L}}_p$ is compact, we may assume that the sequence X_n converges to some $X \in \hat{\mathcal{L}}_p$. If $X \notin \mathcal{J}_p$, then X meets K in an affine point. Now (K_n, X_n) converges to (K, X), whence $x_n = K_n \wedge X_n$ converges to $x = K \wedge X \in A$. Since $x_n \in L \cap A$ and $\lim L_n = L$, we infer from (31.19) that $x = K \wedge L \in A$, a contradiction.

Applying the same argument to the sequence $x_n q$, we obtain that $K_n \wedge L_n$ eventually belongs to $\mathcal{J}_p \wedge \mathcal{J}_q$, and thus converges to $K \wedge L$.

3) Let $L \in \hat{\mathcal{L}}$ be an arbitrary line. With respect to the topology PI, the affine point set $P \setminus L$ is homeomorphic to $\mathbb{R}^2$, and for every $K \in \hat{\mathcal{L}} \setminus \{L\}$ the affine line $K \setminus L$ is closed in $P \setminus L$ and homeomorphic to $\mathbb{R}$. That is, every affine part of $(P, \hat{\mathcal{L}})$ is an $\mathbb{R}^2$-plane.

The assertion about $P \setminus L$ follows immediately from assertion 2), applied to $u, v \in L$. Since K is closed in P, the affine part $K \setminus L$ is closed in $P \setminus L$. For

$v = K \wedge L$ and $u \in L \setminus K$, the restriction $c_{u,v}^{-1}|_{K \setminus L}$, followed by the projection to the first factor, is a homeomorphism from $K \setminus L$ onto $\hat{\mathcal{L}}_u \setminus \{L\} \approx \mathbb{R}$.

4) For every $L \in \hat{\mathcal{L}}$, the topology $\hat{\mathrm{H}}$ coincides with the one-point compactification $\widehat{\mathrm{H}_L}$ of the natural topology H_L obtained from the topology that is induced by PI on $P \setminus L$.

Indeed, for an open subset U of P, consider $\hat{\mathcal{L}}_U = \{ X \in \mathcal{L} \mid X \cap U \neq \emptyset \}$. We have that $\hat{\mathcal{L}}_U = \hat{\mathcal{L}}_{U \setminus (L \cup W)}$, since no line has an isolated point. The OM-description for the natural topology (31.19) therefore shows that $\hat{\mathrm{H}}$ and H_L coincide on $\hat{\mathcal{L}} \setminus \{L, W\}$. Since $\widehat{\mathrm{H}_L}$ is compact, it remains to show that

$$\varphi : (\hat{\mathcal{L}}, \widehat{\mathrm{H}_L}) \to (\hat{\mathcal{L}}, \hat{\mathrm{H}}) : M \mapsto M$$

is continuous at W and at L. We choose a line $K \in \mathcal{L} \setminus \{L\}$ such that $K \wedge L \notin W$.

Assume first that W_n is a sequence in $\mathcal{L} \setminus \{L\}$ that converges to W with respect to H_L, but not with respect to $\hat{\mathrm{H}}$. Since $\hat{\mathrm{H}}$ is compact, we may assume that W_n converges to some line L' in $\mathcal{L}$ with respect to H. From the fact that the topologies $\widehat{\mathrm{H}_L}$ and $\hat{\mathrm{H}}$ coincide on $\mathcal{L} \setminus \{L\}$ we conclude that $L' = L$. The sequence of intersection points $p_n := W_n \wedge K$ converges to $W \wedge K$, since H_L is the natural topology for the line space of the affine $\mathbb{R}^2$-plane $(P \setminus L, \hat{\mathcal{L}} \setminus \{L\})$. Consequently, there is a neighbourhood of $L \wedge K$ that contains none of the points p_n, and W_n cannot converge to L with respect to H, a contradiction.

Now assume that L_n is a sequence of lines in $\mathcal{L} \setminus \{L\}$ that converges to L with respect to $\widehat{\mathrm{H}_L}$. For every compact subset C of $K \setminus L$, this implies that only a finite number of the intersection points $q_n := L_n \wedge K$ belong to C, compare (31.24). Since $K \setminus L$ is homeomorphic to $\mathbb{R}$, it is a countable union of compact sets, and we conclude that $\lim q_n = L \wedge K$. Replacing K by some $K' \in \mathcal{L} \setminus \mathcal{L}_{L \wedge K}$, and using the continuity of joining in $(A, \mathcal{L})$ we infer that L_n converges to L with respect to H.

5) With respect to the topologies PI and $\hat{\mathrm{H}}$, the projective completion of $\mathcal{P} = (P, \hat{\mathcal{L}})$ of $(A, \mathcal{L})$ is a topological projective plane.

For every pair of distinct points, there exists a line that avoids both of them. Also, for every pair of distinct lines, there exists a line that avoids their intersection. Hence continuity of join and intersection follows from the fact that every affine part is an $\mathbb{R}^2$-plane with the right topologies, see steps 1) and 4), (31.16 and 21).

6) The topology $\hat{\mathrm{H}}$ is compact by its definition. In order to prove compactness of PI, consider three lines L_1, L_2, L_3 which are not confluent. Each of the four connected components of $P \setminus (L_1 \cup L_2 \cup L_3)$ is the interior of a convex triangle in the affine $\mathbb{R}^2$-plane $\mathcal{P}^W$ for a suitable line W. Thus each of these components has a compact closure (31.6), and P is the union of these four compact sets. (Alternatively, one can use (41.7a)). □

32.3 Proposition. *Endow the projective completion $\mathcal{P} = (P, \mathcal{L})$ of an affine $\mathbb{R}^2$-plane with the topologies described in (32.2). Then each line and each pencil of $\mathcal{P}$ is homeomorphic to the circle $\mathbb{S}_1$, and P and $\mathcal{L}$ are both homeomorphic to the point space $\mathrm{P}_2\mathbb{R}$ of the real projective plane. Furthermore, the flag space $\mathbf{F} = \{(p, L) \mid p \in L\} \subseteq P \times \mathcal{L}$ is a connected topological manifold of dimension 3.*

Proof. For the assertions on lines and pencils see (32.1) and (21.6). Consider a line $L = ab$, a point $p \in P \setminus L$ and the central projection $\gamma : P \setminus \{p\} \to L$ with $\gamma(x) = xp \wedge L$. Since L decomposes as the union of two arcs A, B with common endpoints a, b, the space $P \setminus \{p\}$ is the union of the two strips $\gamma^{-1}(A), \gamma^{-1}(B) \approx [0, 1] \times \mathbb{R}$, which are glued together along the two boundary lines pa, pb. Thus, $P \setminus \{p\}$ can be represented as a quotient space

$$([0, 1] \times \mathbb{R} \cup [2, 3] \times \mathbb{R}) / \sim,$$

where $(1, t) \sim (2, f(t))$ and $(3, t) \sim (0, g(t))$ for all $t \in \mathbb{R}$ with two homeomorphisms $f, g : \mathbb{R} \to \mathbb{R}$. If $f \circ g$ is monotone, then $P \setminus \{p\}$ is (homeomorphic to) the cylinder $C \approx \mathbb{R} \times \mathbb{S}_1$, and if $f \circ g$ is antitone, then $P \setminus \{p\}$ is the Möbius strip. Furthermore, P is the one-point compactification of $P \setminus \{p\}$, see (32.2). The one-point compactification of C is $\mathbb{S}_2$ with two points identified, and the compactifying point has no neighbourhood homeomorphic to $\mathbb{R}^2$. Hence $P \setminus \{p\}$ is not homeomorphic to C, in view of $P \setminus L \approx \mathbb{R}^2$ according to (32.2). Thus the homeomorphism type of P is uniquely determined, and P is homeomorphic to $\mathrm{P}_2\mathbb{R}$ as in the classical situation. The dual arguments apply to $\mathcal{L}$ (which is locally homeomorphic to $\mathbb{R}^2$ by (31.23)).

For every line W, the map $(p, L) \mapsto (p, L \wedge W)$ is a homeomorphism of the (open) set $\{(p, L) \in \mathbf{F} \mid p \notin W, L \neq W\}$ of flags onto the product $(P \setminus W) \times W \approx \mathbb{R}^2 \times \mathbb{S}_1$. This shows that the flag space $\mathbf{F}$ is a connected topological manifold of dimension 3. □

In the situation of (32.3), one can show that the flag space $\mathbf{F}$ is homeomorphic to the flag space of the real projective plane $\mathcal{P}_2\mathbb{R}$, see Breitsprecher [72].

32.4 Corollary. *Let $(P, \mathcal{L})$ be the projective completion of an affine $\mathbb{R}^2$-plane, and let $p \in P$ and $L \in \mathcal{L}$. Then both the space $P \setminus \{p\}$ and the space $\mathcal{L} \setminus \{L\}$ of all affine lines are homeomorphic to the Möbius strip.*

Quite remarkably, a part of the topological assertions in (32.3) is strong enough to characterize the projective completions of affine $\mathbb{R}^2$-planes, see also (42.10):

32.5 Theorem. *Let $\mathcal{P} = (P, \mathcal{L})$ be a projective plane. Assume that P is a topological space homeomorphic to $\mathrm{P}_2\mathbb{R}$ such that each line is homeomorphic to $\mathbb{S}_1$. Then for every line $L \in \mathcal{L}$, the affine part $\mathcal{P}^L$ of $\mathcal{P}$ is an affine $\mathbb{R}^2$-plane, and $\mathcal{P}$*

is the projective completion of $\mathcal{P}^L$ as constructed in (32.1). In particular, $\mathcal{P}$ is a topological projective plane.

Proof. We have to show that $P\setminus L$ is homeomorphic to $\mathbb{R}^2$; then we can use (32.2). It is easy to see that $P \setminus L$ is connected: otherwise consider the line M joining two points a, b in different connected components of $P \setminus L$. Since $M \approx \mathbb{S}_1$, the two connected components of $M \setminus \{a, b\}$ intersect L in two different points, a contradiction.

Denote by $\widetilde{L}$ the full preimage of L in the two-sheeted covering $\mathbb{S}_2$ of P. The compact one-dimensional topological manifold $\widetilde{L}$ is the topological sum of one or two copies of $\mathbb{S}_1$, compare Christenson–Voxman [77] 5.A.3. The Jordan curve theorem implies that $\mathbb{S}_2 \setminus \widetilde{L}$ has at least two connected components C_1, C_2 which are homeomorphic to $\mathbb{R}^2$. Since $P \setminus L$ is connected, each component C_i gives a covering of $P \setminus L$, and these coverings are in fact homeomorphisms, because $\mathbb{S}_2 \to P$ has only two sheets. Thus $P \setminus L \approx \mathbb{R}^2$. □

In fact, (32.5) holds with even weaker hypotheses: one can drop the assumption that distinct lines intersect, see Salzmann [67b] 2.5.

32.6 Definition. The *2-dimensional compact projective planes* are the projective completions of affine $\mathbb{R}^2$-planes, endowed with the topologies described in (32.2 and 3). Theorem (32.5) gives a characterization of these planes from a projective point of view. By (32.3 and 5), the dual of a 2-dimensional compact projective plane is also a 2-dimensional compact projective plane.

The phrase '2-dimensional' refers to the topological structure of such a plane: the point space is a surface, i.e., a 2-dimensional topological manifold; compare also (42.10). In other chapters we consider topological projective planes whose point spaces are manifolds of higher dimensions. According to (32.2), every 2-dimensional compact projective plane is a topological projective plane in the sense of (41.1). Note that the continuity of joining and intersection is not assumed in (32.2) and (32.5).

32.7 Proposition. *Every (non-degenerate) quadrangle a_1, a_2, a_3, a_4 in a 2-dimensional compact projective plane $\mathcal{P} = (P, \mathcal{L})$ generates a subplane which is dense in P. In particular, that subplane is infinite, hence the three diagonal points $a_i a_j \wedge a_k a_l$ with $\{i, j, k, l\} = \{1, 2, 3, 4\}$ are never collinear.*

Proof. Let S be the point set of the subplane generated by a_1, a_2, a_3, a_4.

1) The topological closure $\overline{S}$ of S is the point set of a subplane of $\mathcal{P}$.

It suffices to prove that $ab \wedge cd \in \overline{S}$ for every non-degenerate quadrangle $a, b, c, d \in \overline{S}$. This follows readily from the continuity of joining and intersection (32.2).

We find points u, v on distinct sides of the triangle a_1, a_2, a_3 such that the connected component C of a_4 in $P \setminus (a_1a_2 \cup a_2a_3 \cup a_3a_1)$ does not accumulate at u or v. Then $W = uv$ is disjoint from $\overline{C}$, i.e., a_4 is an inner point of the convex triangle $\langle a_1, a_2, a_3 \rangle$ in the affine $\mathbb{R}^2$-plane $\mathcal{P}^W$.

2) For distinct $i, j \in \{1, 2, 3\}$, the intersection $S \cap [a_i, a_j]$ is dense in the interval $[a_i, a_j]$ of $\mathcal{P}^W$.

We may assume that $i = 1$ and $j = 2$. Proceeding indirectly, we consider a non-empty open interval $I \subseteq [a_1, a_2]$ with $I \cap \overline{S} = \emptyset$; by choosing the interval I maximal we achieve that the endpoints b_1, b_2 of I belong to $\overline{S}$. The following arguments make use of step 1). We find points $c_i \in (a_3, b_i) \cap \overline{S}$, because $a_4 \in \overline{S}$ is an inner point of $\langle a_1, a_2, a_3 \rangle$. Then $p = b_1c_2 \wedge b_2c_1 \in \overline{S}$ is an inner point of the convex quadrangle $\langle b_1, b_2, c_2, c_1 \rangle$, hence $a_3p \wedge I$ is an inner point of I belonging to $\overline{S}$, a contradiction.

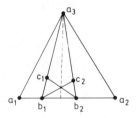

Figure 32a

Finally, we show that S is dense. Any point p of $\mathcal{P}$ may be written as $p = ab \wedge cd$ with suitable points $a, b, c, d \in [a_1, a_2] \cup [a_2, a_3] \cup [a_3, a_1]$, compare (31.7a). Now $p \in \overline{S}$ in view of steps 1) and 2). □

The following results (32.8 to 11) are of crucial importance for the theory of 2-dimensional compact projective planes.

32.8 Proposition. *The real affine plane is the only affine $\mathbb{R}^2$-plane which is a translation plane. In particular, the real projective plane $\mathcal{P}_2\mathbb{R}$ is the only 2-dimensional compact projective plane which is desarguesian.*

Proof. By (32.7) the prime field Q_0 of a quasifield Q coordinatizing such a translation plane is isomorphic to $\mathbb{Q}$ and dense in Q. Via Dedekind cuts, this field isomorphism extends to a homeomorphism of Q onto $\mathbb{R}$. That homeomorphism is additive and multiplicative on the dense subset Q_0, hence on Q, thus giving an isomorphism of Q onto $\mathbb{R}$ as a topological field. Compare also (42.6).

The topology of the affine line $Q = \mathbb{R}$ determines the topology on the affine point set $Q = \mathbb{R}^2$ and on $\mathcal{P}_2\mathbb{R}$, see (41.1). □

Now we consider collineations of 2-dimensional compact projective planes. First we prove the following basic result:

32.9 Theorem. *Every collineation between 2-dimensional compact projective planes $\mathcal{P} = (P, \mathcal{L})$ and $\mathcal{P}' = (P', \mathcal{L}')$ is a homeomorphism. Furthermore the topology of a 2-dimensional compact projective plane is uniquely determined.*

Proof. First we give a characterization of affine half-planes in purely geometric terms. Let o, u, v, e be a quadrangle in $\mathcal{P}$, let $W = uv$, and define $\varphi : P\backslash W \to P\backslash W$ by $p^\varphi = ((pu \wedge ev)o \wedge (pu \wedge oe)v)u \wedge pv$; in coordinates with respect to o, u, v, e, we have $(x, y)^\varphi = (x, y^2)$. We claim that the image $(P \backslash W)^\varphi$ of the 'folding map' φ is the closed half-plane H in $\mathcal{P}^W$ which is defined by ou and contains e.

To prove this, we observe that every line parallel to ov is left invariant by φ, and that the parallel class of ou is invariant under φ. Thus it suffices to prove that $L^\varphi = L \cap H$, where $L = ev \setminus \{v\}$. The image L^φ is connected and contains the points $(1, 1) = e$ and $(1, 0) = L \wedge ou$; furthermore L^φ accumulates at v, because the restriction $\varphi|_L$ is described by $q \mapsto (qo \wedge (qu \wedge oe)v)u \wedge ev$, and that formula gives a continuous extension of φ to ev by $v \mapsto v$. Thus $L \cap H \subseteq L^\varphi$, and it remains to prove the inclusion $L^\varphi \subseteq H$. If $q \in L \setminus H$, then $qu \wedge oe \notin H$, whence $qo \wedge (qu \wedge oe)v \in H$ and $q^\varphi \in H$, because all intersections at o are transversal (31.5), compare Figure 32b.

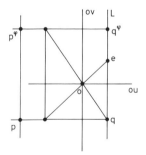

Figure 32b

Similarly we obtain $(L \cap H)^\varphi \subseteq H$, and the claim $(P \setminus W)^\varphi = H$ is proved. (Expressed in algebraic terms, these arguments exploit monotonicity properties of the squaring operation $y \mapsto y^2$, compare Salzmann [59a] p. 53, Prieß-Crampe [83]

V §3, Satz 6, p. 220; in fact, (32.9) can be derived from the fact that $\mathcal{P}$ and $\mathcal{P}'$ admit a unique order.)

Now we consider a collineation α of $\mathcal{P}$ onto $\mathcal{P}'$. The characterization of half-planes in terms of φ shows that α maps every open half-plane of $\mathcal{P}^W$ onto an open half-plane of $(\mathcal{P}')^{W^\alpha}$. The open half-planes form a subbasis for the topology (31.10), whence α is a homeomorphism. The trivial collineation $\alpha = 1\!\!1$ shows that the topology is unique. □

Theorem (32.9) is not true for $\mathcal{P}_2\mathbb{C}$, see (44.11), (55.22a). As an immediate consequence of (32.7) and (32.9), we have the following important stiffness result.

32.10 Corollary. *In every 2-dimensional compact projective plane, the identity is the only collineation that fixes all points of a non-degenerate quadrangle.* □

The following theorem is just a special case of (55.19).

32.11 Theorem. *Every collineation of a 2-dimensional compact projective plane $\mathcal{P}$ fixes some point and some line of $\mathcal{P}$.*

Proof. Since every collineation is continuous (32.9), it suffices to show that every continuous map $f : P_2\mathbb{R} \to P_2\mathbb{R}$ has a fixed point. This is well-known, compare Feigl [28] p. 373, Armstrong [83] p. 209 or Rotman [88] 9.21.

As an alternative argument, we remark that the covering $\mathbb{S}_2$ of $P_2\mathbb{R}$ is simply connected, hence f admits a lifting $\tilde{f} : \mathbb{S}_2 \to \mathbb{S}_2$, see Armstrong [83] 10.16 or Massey [67] Ch. 5, 5.1, p. 156. Thus we have $f \circ q = q \circ \tilde{f}$, where $q : \mathbb{S}_2 \to P_2\mathbb{R}$ is the covering map defined by $q(x) = \{x, -x\}$. Now f has a fixed point, because there exists an element $x \in \mathbb{S}_2$ with $\tilde{f}(x) = \pm x$, see Dugundji [66] XVI.3.4, p. 343 or Rotman [88] Thm. 6.25, p. 123. □

32.12 Lemma. *Every collineation α of order 2 of a 2-dimensional compact projective plane $\mathcal{P}$ is a reflection. Such a collineation is uniquely determined by its center and its axis.*

Proof. By (32.10), α cannot be a Baer involution, hence α is a reflection or an elation by (23.17). The latter case does not arise: pick distinct points x, y not on the axis of an involutory elation α which are not collinear with the center of α. Then the three diagonal points of the quadrangle x, y, x^α, y^α are fixed by α, hence they lie on the axis of α, a contradiction to (32.7).

Assume that $\mathcal{P}$ admits distinct reflections α_1, α_2 with the same center o and the same axis uv. We coordinatize the $\mathbb{R}^2$-plane $\mathcal{P}^{uv}$ with respect to a quadrangle o, u, v, e, compare (22.1). Using the multiplication of the corresponding ternary field, the action of α_i on $\mathcal{P}^{uv}$ is given by

$$(x, y)^{\alpha_i} = (xc_i, yc_i)$$

with suitable elements $c_i \in \mathbb{R}$ satisfying $c_i^2 = 1$, see (23.11). The homeomorphisms $x \mapsto c_i x$ and $x \mapsto xc_i$ of $\mathbb{R}$ onto itself are decreasing, because they interchange 1 and c_i. We may assume that $c_1 < c_2$. Then $1 = c_1^2 > c_1 c_2$ and $c_1 c_2 > c_2^2 = 1$, a contradiction. □

See (55.29 and 32) for generalizations of (32.12).

32.13 Lemma. *Let α, β be distinct commuting reflections of a 2-dimensional compact projective plane. Then the center c_α of α lies on the axis A_β of β, and c_β lies on A_α.*

Proof. Assume that $c_\alpha \notin A_\beta$. Since $(c_\alpha)^\beta = c_\alpha$, compare (23.12), we infer that $c_\alpha = c_\beta$, hence $c_\beta \notin A_\alpha$, thus $A_\alpha = A_\beta$ by the dual argument. Now (32.12) implies that $\alpha = \beta$, a contradiction. □

It is not known if (32.13) holds for reflections of arbitrary projective planes; compare also (55.35).

32.14 Lemma. *Let α, β be homologies of a 2-dimensional compact projective plane with common axis A and distinct centers a, b. Then the commutator $\alpha^{-1}\beta^{-1}\alpha\beta$ is an elation with axis A and center $ab \wedge A$.*

Proof. Each $\gamma \in \langle \alpha, \beta \rangle$ has axis A and fixes the line ab. If γ fixes a further line through $c = ab \wedge A$, then γ has center c, i.e., γ acts trivially on the pencil $\mathcal{L}_c$. Therefore the group Γ induced by $\langle \alpha, \beta \rangle$ on $\mathcal{L}_c \setminus \{A, ab\} \approx \mathbb{R} \setminus \{0\}$ consists of fixed-point-free homeomorphisms of $\mathbb{R} \setminus \{0\}$, together with the identity. It suffices to show that Γ is abelian, because this implies that $\alpha^{-1}\beta^{-1}\alpha\beta$ acts trivially on $\mathcal{L}_c$.

The increasing homeomorphisms in Γ form a subgroup Γ^+ of index at most 2, and $\{\gamma \in \Gamma^+ \mid \gamma(1) > 1\} = \{\gamma \in \Gamma^+ \mid \gamma(x) > x \text{ for every } x > 0\}$ is a positive cone in Γ^+, turning Γ^+ into an ordered group. If $\gamma(1) > 1$, then $\{\gamma^n(1) \mid n \in \mathbb{N}\}$ is unbounded (otherwise γ would fix the supremum of that set), hence the order on Γ^+ is archimedean. Thus Γ^+ is abelian by a well-known result of Hölder, see Fuchs [63] IV, Thm. 1, p. 45 or Prieß-Crampe [83] I §3, Satz 4, p. 8.

The elements of Γ act on Γ^+ by conjugation, inducing automorphisms of Γ^+. These automorphisms preserve the positive cone and, hence, the order of Γ^+, because the definition of the positive cone in Γ^+ is independent of the choice of an orientation of $\mathbb{R}$; indeed we have

$$\left\{\gamma \in \Gamma^+ \mid \gamma(1) < 1\right\} = \left\{\gamma \in \Gamma^+ \mid \lim_{n \to \infty} \gamma^n(x) = 0 \text{ for every } x \in \mathbb{R}\right\},$$

compare (23.11) or step 2) of the proof of (37.2). Furthermore, these order-preserving automorphisms have order at most two, hence they are trivial, and Γ is abelian. □

Now we consider topological transformation groups consisting of collineations of some 2-dimensional compact projective plane. We need the following elementary information, compare also (96.29); note that all actions are assumed to be continuous, as in (96.1).

32.15 Lemma.

(a) *If a compact topological group Δ acts on the topological space $\mathbb{R}$, then Δ induces on $\mathbb{R}$ a group of order at most 2. In particular, a compact, connected group can only act trivially on $\mathbb{R}$.*
(b) *If a compact, connected topological group Δ acts non-trivially on the circle $\mathbb{S}_1$, then Δ is transitive on $\mathbb{S}_1$.*
(c) *If a connected topological group Δ acts without fixed points on $\mathbb{R}$ or $\mathbb{S}_1$, then Δ is transitive.*
(d) *Let Δ be a compact effective transformation group on $\mathbb{S}_1$. If Δ has topological dimension 0, then Δ is finite.*

Proof. (a) Each $\delta \in \Delta$ is increasing or decreasing on $\mathbb{R}$, and every orbit X of Δ is compact, thus Δ leaves invariant $\{\min X, \max X\}$ for every (compact) orbit X of Δ. Hence $X = \{\min X, \max X\}$, and X contains at most 2 elements and every $\delta \neq 1\!\!1$ is decreasing. Connectedness of Δ carries over to X.

(b) If the orbits of Δ are proper subsets of $\mathbb{S}_1$, then they are closed intervals, being compact and connected. By homogeneity each orbit reduces to a point.

(c) If the orbits of Δ are proper subsets of $\mathbb{R}$ or $\mathbb{S}_1$, then they are open intervals, being connected and homogenous; this is a contradiction to the connectedness of $\mathbb{R}$ and $\mathbb{S}_1$.

(d) This is proved in Montgomery–Zippin [55] p. 233f; here we offer a different argument. Homeomorphisms of $\mathbb{S}_1$ preserve or reverse the orientation of $\mathbb{S}_1$, hence we may assume that Δ consists of orientation-preserving transformations only. The dimension formula (96.10) shows that every orbit B of Δ in $\mathbb{S}_1$ is a proper subset of $\mathbb{S}_1$. The connected components of $\mathbb{S}_1 \setminus B$ are open intervals (arcs), and by homogeneity each element $b \in B$ is an endpoint of one or of two of these intervals. In the latter case, the compact space B is discrete, hence finite, i.e., the stabilizer Δ_b has finite index in Δ, and Δ_b and Δ are finite by assertion (a).

It remains to consider the case where $b \in B$ is an endpoint of just one connected component I of $\mathbb{S}_1 \setminus B$. The other endpoint c of I belongs to B, hence $b = c^\delta$

for some $\delta \in \Delta$. Since I, I^δ both have endpoint b, we infer that $I = I^\delta$, hence δ exchanges b and c. But δ preserves the orientation of $\mathbb{S}_1$, a contradiction. □

32.16 Lemma. *If a compact group Δ acts effectively as a collineation group on a 2-dimensional compact projective plane, then $\dim \Delta \leq 3$.*

Proof. Let (p, L) be a flag. Then $\Delta_{p,L}$ induces a group of order at most 2 on $L \setminus \{p\} \approx \mathbb{R}$, see (32.15a). Similarly, the kernel $\Delta_{[L]}$ induces a group of order at most 2 on $\mathcal{L}_p \setminus \{L\}$. Since $\Delta_{[p,L]} = 1\!\!1$ by (32.15a and 12), $\Delta_{p,L}$ is finite (of order at most 4). The dimension formula (96.10) together with (92.4) implies that $\dim \Delta = \dim(p, L)^\Delta + \dim \Delta_{p,L} \leq 3 + 0$, since the flags form a manifold of dimension 3 by (32.3). □

Assertion (32.16) is in fact a special case of a general theorem about compact transformation groups (96.13a,b).

32.17 Lemma. *Let the compact torus group $\mathbb{T} = \mathbb{R}/\mathbb{Z} \cong \mathrm{SO}_2\mathbb{R}$ act effectively as a collineation group on a 2-dimensional compact projective plane $\mathcal{P} = (P, \mathcal{L})$. Then the involution $\sigma \in \mathbb{T}$ is a reflection, its center c is the only fixed point of $\mathbb{T}$, and the axis A of σ is the only fixed line of $\mathbb{T}$. Furthermore, $\mathbb{T}$ acts effectively and sharply transitively on each of its orbits in $P \setminus (A \cup \{c\})$, and $\mathbb{T}/\langle \sigma \rangle$ acts effectively and sharply transitively on A.*

Proof. By (32.12) the involution σ in $\mathbb{T}$ is a reflection, and the abelian group $\mathbb{T}$ fixes c and A, compare (23.12). The kernel $\mathbb{T}_{[A]} = \mathbb{T}_{[c,A]}$ of the action of $\mathbb{T}$ on A acts effectively on every line $ca \setminus \{a\} \approx \mathbb{R}$ with $a \in A$, hence $\mathbb{T}_{[A]} = \langle \sigma \rangle$ by (32.15a). By (32.15b) the group $\mathbb{T}$ is transitive on $A \approx \mathbb{S}_1$. The stabilizer $\mathbb{T}_p$ of a point $p \in P \setminus (A \cup \{c\})$ fixes $pc \wedge A$, hence every point on A. This gives $\mathbb{T}_p \leq \langle \sigma \rangle_p = 1\!\!1$, and $\mathbb{T}$ acts sharply transitively on the orbit $p^\mathbb{T}$. □

More information on the orbits of $\mathbb{T}$ in $P \setminus (A \cup \{c\})$ can be found in Salzmann [62b] Lemma 6; in fact, these orbits are ovals (in the sense of (55.7)), see Groh [71] Thm. 3, Löwen [84a].

32.18 Lemma. *The 2-Torus $\mathbb{T}^2$ cannot act effectively as a collineation group of a 2-dimensional compact projective plane.*

Proof. The three involutions in $\mathbb{T}^2$ are reflections (32.12), and their centers and axes are the vertices and sides of a (non-degenerate) triangle (32.13). The group $\mathbb{T}^2$ fixes that triangle, and $\mathbb{T}^2$, being connected, acts trivially on each side of the triangle by (32.15a). This leads to the contradiction $\mathbb{T}^2 = 1\!\!1$. □

We remark that the 2-torus $\mathbb{T}^2$ cannot act effectively on the topological space $P_2\mathbb{R}$, see Mostert [57] §6.

32.19 Lemma. *Let the topological group $\mathbb{R}$ act non-trivially on a 2-dimensional compact projective plane, inducing a group of axial collineations. Then there is a universal center and a universal axis for all elements of $\mathbb{R}$, and the action is effective.*

Proof. The first assertion is obvious for cyclic groups, for unions over chains of cyclic groups, and for closures of such groups. Since the additive group $\mathbb{R}$ contains $\mathbb{Q} = \bigcup_{n>0} \frac{1}{n!}\mathbb{Z}$ as a dense subgroup, the first assertion follows. The second assertion is a consequence of (32.17), because $\mathbb{T}$ is the only proper non-trivial quotient of $\mathbb{R}$. □

32.20 Theorem. *Let the compact group $\Delta = SO_3\mathbb{R}$ act effectively as a collineation group on a 2-dimensional compact projective plane $\mathcal{P}$. Then $\mathcal{P}$ is isomorphic to the real projective plane $\mathcal{P}_2\mathbb{R}$.*

Proof. Let (x, X) be a flag of $\mathcal{P} = (P, \mathcal{L}, \mathbf{F})$. Then $\Delta_{[x,X]} = \mathbb{1}$ by (32.15a and 12) and (23.8), whence $\Delta_{x,X}$ acts effectively on $\mathcal{L}_x \times X$, and (32.15a) implies that $|\Delta_{x,X}| \leq 4$, compare the proof of (32.16). According to the dimension formula (96.10), each Δ-orbit in the flag space $\mathbf{F}$ has dimension 3 and is therefore open in $\mathbf{F}$, see (92.14) and note that $\mathbf{F}$ is a topological manifold of dimension 3 by (32.3). The connectedness of $\mathbf{F}$ implies that Δ is transitive on $\mathbf{F}$.

The group Δ contains distinct commuting reflections σ, τ, compare (32.12). Let $A = \mathrm{Cs}_\Delta(\sigma) \cong \mathbb{T} \rtimes \mathbb{Z}_2 \cong O_2\mathbb{R}$ and $B = \mathrm{Cs}_\Delta(\tau)$. By (23.12) we have $A \leq \Delta_p$ and $B \leq \Delta_L$ where p is the center of σ and L is the axis of τ. In fact, $A = \Delta_p$ and $B = \Delta_L$, because A, B are maximal subgroups of Δ. Since σ and τ commute, the pair (p, L) is a flag (32.13).

Now we describe $\mathcal{P}$ in terms of Δ, A, B, compare (23.24): identify P with Δ/A via $p^\delta \mapsto A\delta$, and $\mathcal{L}$ with Δ/B via $L^\delta \mapsto B\delta$, for $\delta \in \Delta$. By flag-transitivity, p^γ is incident with L^δ if, and only if $(p^\varepsilon, L^\varepsilon) = (p^\gamma, L^\delta)$ for some $\varepsilon \in \Delta$. These elements ε comprise the set $A\gamma \cap B\delta$, hence p^γ is incident with L^δ if, and only if, $A\gamma \cap B\delta \neq \emptyset$. The very same description results from the usual action of $SO_3\mathbb{R}$ on the real projective plane $\mathcal{P}_2\mathbb{R}$, thus the planes $\mathcal{P}$ and $\mathcal{P}_2\mathbb{R}$ are isomorphic; every isomorphism is also a homeomorphism by (32.9). □

Remarks. (a) As a variation of the proof of (32.20), one can also describe $\mathcal{P}$ in terms of the reflections in $SO_3\mathbb{R}$, see Salzmann [62a] 4.3, 4.4. We shall use this method several times later on, compare (b) below. We remark that Theorem (32.20) is closely related to Hilbert's characterization of $\mathcal{P}_2\mathbb{R}$; in fact, the first two paragraphs of the proof above say that the pair $(\mathcal{P}, \Delta)$ satisfies Hilbert's axioms for the real projective plane and its elliptic group of motions (i.e., congruence-preserving collineations), thus one could conclude the proof by quoting the uniqueness statement in Hilbert [30] §17. Compare also the explicit axioms for the group of motions in Schur [09] No. 10.

(b) In this book, we use various methods to describe a geometry in group-theoretic terms. The general method (23.24), which applies to flag-homogeneous geometries, was used in our proof of (32.20); it will not be applied again in the same straightforward way, for lack of examples. The only flag homogeneous, compact, connected planes are the classical planes, and we want to characterize them by weaker properties. However, according to (44.9) it suffices to recognize an open dense subgeometry in order to determine a topological projective plane, and this can be done by applying (23.24) to a suitable open orbit in the flag space. See the proofs of (34.9), (64.18), (72.3) and (86.34) for instances of this method. The Hughes planes are actually constructed and shown to be projective planes by using the principle of (23.24) in an extended form, see (86.4) ff.

A related method of group-theoretical reconstruction is presented in (25.6), where certain systems of subgroups of the translation group (so-called spreads) are used to describe translation planes. Applications may be found in the proofs of (62.13), (73.19), (81.19) and (82.2 and 5), where the spread is determined from the action of some group. Also reflections may be employed in various ways for the efficient reconstruction of planes from groups. Examples appear in the proofs of (62.6) (at the end of Section 62), (63.8), (65.2), (71.10) and (86.32).

Finally we consider the full group Σ of all collineations of a 2-dimensional compact projective plane $\mathcal{P} = (P, \mathcal{L})$. We endow Σ with the compact-open topology derived from the action of Σ on P, compare (44.2) and (96.6). Note that this topology coincides with the compact-open topology derived from the action on $\mathcal{L}$. Results on locally compact (transformation) groups yield the following theorem:

32.21 Theorem. *Let $\mathcal{P} = (P, \mathcal{L}, \mathbf{F})$ be a 2-dimensional compact projective plane, and endow the group Σ of all collineations of $\mathcal{P}$ with the compact-open topology derived from the action on P.*

(a) *The group Σ is a topological transformation group on P, $\mathcal{L}$ and $\mathbf{F}$.*
(b) *The group Σ is a Lie group of dimension at most 8.*
(c) *Every analytic subgroup of Σ is closed in Σ.*

Proof. We use the fact that Σ is a locally compact transformation group (44.3). This includes assertion (a). Since P is a surface (32.3), we can infer from (96.31) that Σ is a Lie group. Here we give a proof for the slightly weaker assertion that Σ either is a Lie group or is totally disconnected. In particular, this implies that the connected component Σ^1 of Σ is a Lie group, which suffices for most of our purposes. The proof given here also serves as a preparation for the more complicated case of planes of higher dimension, see (71.2) and (87.1).

The stabilizer $\Sigma_{o,u,v,e}$ of any quadrangle o, u, v, e is trivial by the stiffness result (32.10), hence the dimension formula (96.10) gives

$$\dim \Sigma = \dim o^\Sigma + \dim u^{\Sigma_o} + \dim v^{\Sigma_{o,u}} + \dim e^{\Sigma_{o,u,v}} \leq 4 \dim P = 8.$$

By the approximation theorem (93.8), there exists an open subgroup Δ of Σ and a compact normal subgroup Θ of Δ such that Δ/Θ is a Lie group and $\dim \Theta = 0$. If Θ is finite, then Δ, and hence also Σ, is a Lie group. So assume that Θ is infinite.

We claim that the stabilizer Θ_x is finite for every point $x \in P$. Choose distinct lines $L, M \in \mathcal{L}_x$; then $\Theta_{x,L,M}$ acts effectively on $(L \setminus \{x\}) \times (M \setminus \{x\}) \approx \mathbb{R} \times \mathbb{R}$ by (32.10), whence $|\Theta_{[x]}| \leq |\Theta_{x,L,M}| \leq 4$ by (32.15a). Furthermore the group $\Theta_x/\Theta_{[x]}$ induced on $\mathcal{L}_x \approx \mathbb{S}_1$ is zero-dimensional by (93.7) and finite by (32.15d), hence Θ_x is finite as well.

Since Θ is infinite, this implies that every Θ-orbit in $P \cup \mathcal{L}$ is infinite. Therefore, an orbit x^Θ with $x \in P$ is not contained in a line (otherwise that line would be fixed by Θ). Thus x^Θ contains a triangle, hence also a quadrangle, because x^Θ contains further points and is homogeneous. As Θ is centralized by the identity component Δ^1, see (93.18), the stabilizer Δ_x^1 acts trivially on x^Θ, compare (91.1c), and we infer that $\Delta_x^1 = \mathbb{1}$ from (32.10). This holds for every $x \in P$, hence $\Delta^1 = \mathbb{1}$ by (32.11). Since Δ is open (and therefore closed) in Σ, we conclude that Σ^1 is trivial. This completes the proof of our weakened version of assertion (b).

In order to show that every analytic subgroup of Σ is closed, it suffices to prove this for every one-parameter subgroup P of Σ, see Hochschild [65] Thm. XVI.2.4, or Hilgert–Neeb [91] III.8.13. Aiming for a contradiction, we assume that P is distinct from its closure $\overline{P}$ in Σ. Then $\overline{P}$ is compact, see (93.20). Being a compact abelian Lie group (94.3a), $\overline{P}$ is isomorphic to some torus group $\mathbb{T}^a$ by (94.38). Now (32.18) implies that $a \leq 1$. But the torus $\overline{P} \cong \mathbb{T}$ contains no proper connected subgroup P, a contradiction. $\square$

32.22 Theorem. *Let Δ be a compact subgroup of the collineation group of a 2-dimensional compact projective plane $\mathcal{P}$. If $\dim \Delta \geq 2$, then $\Delta \cong \mathrm{SO}_3\mathbb{R}$ and $\mathcal{P}$ is isomorphic to the real projective plane $\mathcal{P}_2\mathbb{R}$.*

Proof. The group Δ is a Lie group by (32.21b) and (94.3a). The structure theorem (94.31c) for compact Lie groups together with (32.16 and 18) implies that the connected component Δ^1 is almost simple and 3-dimensional, hence $\Delta^1 \cong \mathrm{SO}_3\mathbb{R}$ or $\Delta^1 \cong \mathrm{SU}_2\mathbb{C}$ by (94.33).

In both cases, Δ^1 has no 2-dimensional closed subgroups, since the Lie algebra of Δ^1 is the vector space $\mathbb{R}^3$ endowed with the vector product. Therefore Δ^1 has no orbits of dimension 1, compare (96.10). If $\Delta^1 = \mathrm{SU}_2\mathbb{C}$, then there is a central involution σ in Δ^1, which is a reflection (32.12), and Δ^1 fixes the center c and the axis A of σ. Now Δ^1 acts trivially on the line A and on each line ca with $a \in A$, a contradiction. Thus $\Delta^1 = SU_2\mathbb{C}$ is not possible.

Hence $\Delta^1 = SO_3\mathbb{R}$, and (32.20) shows that $\mathcal{P} \cong \mathcal{P}_2\mathbb{R}$. Furthermore, $\Delta^1 = SO_3\mathbb{R}$ is a maximal subgroup of the collineation group $PGL_3\mathbb{R}$ of $\mathcal{P}_2\mathbb{R}$, see (94.34) or Brauer [65] or Noll [65], whence $\Delta = \Delta^1$. □

The following result is very remarkable; it says that the classical projective plane $\mathcal{P}_2\mathbb{R}$ is the only 2-dimensional compact projective plane which is homogeneous. See (63.8) for the generalization to compact projective planes of higher dimensions.

32.23 Theorem. *Let Δ be a closed subgroup of the collineation group Σ of a 2-dimensional compact projective plane $\mathcal{P} = (P, \mathcal{L})$. If Δ acts transitively on the point set P, then $\mathcal{P}$ is isomorphic to the real projective plane $\mathcal{P}_2\mathbb{R}$, and Δ is conjugate to $SO_3\mathbb{R}$ or equal to $PGL_3\mathbb{R}$.*

Proof. Result (32.3) says that P is homeomorphic to $P_2\mathbb{R}$. By (32.21b) and (94.3a), the group Δ is a Lie group, and the connected component Δ^1 of Δ is transitive on P by (96.9b).

According to a result of Montgomery (96.19), every maximal compact subgroup of Δ^1 is transitive on P; here we use the fact that the fundamental group of P is finite (of order 2). By the dimension formula (96.10), such a subgroup has dimension at least 2, and we can apply (32.22) to conclude that $\mathcal{P} \cong \mathcal{P}_2\mathbb{R}$ and that $SO_3\mathbb{R}$ is a maximal compact subgroup of Δ^1. Since $\Delta \leq PGL_3\mathbb{R}$ by (13.8), we infer the assertion about Δ from the maximality of $SO_3\mathbb{R}$ in $PGL_3\mathbb{R}$, see (94.34) or Brauer [65] or Noll [65]. □

32.24 Notes on Section 32. Most results of this section can be found in Salzmann [57, 58, 59a,b, 62a,b, 63b].

The construction of a topological projective completion in (32.2) is related to the constructions in (43.7) and in Löwen [81a] §2.

The proof of (32.2) may be regarded as a topological version of arguments that exploit the order of affine $\mathbb{R}^2$-planes; such an order extends to an order of the projective completion, which is a topological projective plane with respect to the order topology, compare Prieß-Crampe [83] V § 1, Pickert [75] 9.2, p. 227 and 10.2, p. 269; compare also (43.6). For systematic investigations of ordered planes see Prieß-Crampe [83], Karzel–Kroll [88], Kalhoff–Prieß-Crampe [90], Tecklenburg [92]. By results of Joussen and Prieß-Crampe, every finitely generated free projective plane admits an archimedean order and is therefore a (dense) subplane of a 2-dimensional compact projective plane. Result (32.7) holds in all archimedean ordered projective planes, but also in other ordered projective planes, see Prieß-Crampe [83] V §3 Satz 6 p. 220 and Kalhoff [86].

Another proof of Theorem (32.23) can be based on the following result: the groups $SO_3\mathbb{R}$ and $PGL_3\mathbb{R}$ are the only connected Lie groups with transitive effective actions on the surface $P_2\mathbb{R}$, see Mostow [50] or Betten–Forst [77]; this is related to Lie's local classification of transitive actions on surfaces.

33 Towards classification

Continuing the investigation of 2-dimensional compact projective planes $\mathcal{P}$, we obtain deeper results on the structure of collineation groups and their fixed elements. We show in (33.6b) that every plane $\mathcal{P}$ which admits a collineation group of dimension at least 5 is isomorphic to the real projective plane $\mathcal{P}_2\mathbb{R}$.

This is the first step towards the classification of all planes $\mathcal{P}$ which admit a collineation group of dimension at least 3. In fact, Theorem (33.9) yields a subdivision of this classification into four different cases, which are considered separately in Sections 34 through 37. The results of these sections are summarized in (38.1).

As a **general assumption for this section**, let $\mathcal{P}$ denote a 2-dimensional compact projective plane.

Each group Δ appearing here will be closed in the automorphism group of $\mathcal{P}$; hence Δ is a Lie group by (32.21b) and (94.3a), and the dimension of Δ as a Lie group coincides with the dimension of the topological space Δ, compare (94.5).

33.1 Proposition. *Let Δ be a closed, connected subgroup of the collineation group of $\mathcal{P}$. If the center of Δ is non-trivial, or if Δ contains a connected abelian normal subgroup $\mathsf{N} \neq \mathbb{1}$, then Δ fixes a point or a line of $\mathcal{P}$.*

Proof. 1) The assertion holds if Δ contains a normal subgroup $\Gamma \neq \mathbb{1}$ which fixes some point p of $\mathcal{P}$.

Indeed, Γ fixes every point in the orbit p^Δ, compare (91.1d), thus (32.10) implies that p^Δ contains no quadrangle. Therefore the connected set p^Δ is contained in a line, see (42.3). This line is fixed by Δ unless $p^\Delta = \{p\}$, in which case p is fixed by Δ.

2) Let $\zeta \neq \mathbb{1}$ be an element of the center of Δ. Then ζ fixes some point by (32.11), and we can apply step 1) with $\Gamma = \langle \zeta \rangle$ to conclude that Δ fixes a point or a line.

3) Let $\mathsf{N} \neq \mathbb{1}$ be a connected abelian normal subgroup of Δ. Step 2) implies that N fixes some point or some line. Thus the assertion on Δ follows from step 1) or its dual. □

33 Towards classification

We consider groups of axial collineations; see (61.19 through 28) for more comprehensive results.

33.2 Lemma. *Let Δ be a closed subgroup of the collineation group of $\mathcal{P}$. Then $\dim \Delta_{[p,L]} \leq 1$ for every point p and every line L of $\mathcal{P}$. If $p \in L$ and $\dim \Delta_{[p,L]} = 1$, then the elation group $\Delta_{[p,L]}$ is linearly transitive.*

Proof. Let M be a line through p which is distinct from L. By (23.8), the group $\Delta_{[p,L]}$ acts freely on $M' = M \setminus \{p, L \wedge M\}$, hence the dimension formula (96.10) implies that $\dim \Delta_{[p,L]} \leq 1$. Furthermore, if $\dim \Delta_{[p,L]} = 1$, then $\Delta_{[p,L]}$ is not totally disconnected, see (92.18). For $p \in L$ the space $M' = M \setminus \{p\}$ is homeomorphic to $\mathbb{R}$, and the assertion follows from (32.15c). □

33.3 Lemma. *Let Δ be a closed subgroup of the collineation group of $\mathcal{P}$, and let a, b be distinct points not on the line L. If $\Delta_{[a,L]} \neq \mathbb{1}$ and $\dim \Delta_{[b,L]} > 0$, then the elation group $\Delta_{[c,L]}$ is linearly transitive, where $c = ab \wedge L$.*

Proof. Let $\mathbb{1} \neq \alpha \in \Delta_{[a,L]}$. Then $\Xi = \{\alpha^{-1}\beta^{-1}\alpha\beta \mid \beta \in (\Delta_{[b,L]})^1\}$ is a connected subset of $\Delta_{[c,L]}$ by (32.14). By assumption, $(\Delta_{[b,L]})^1$ is not trivial, compare (93.6), hence Ξ is not reduced to $\mathbb{1}$, see (23.12). Thus $\Delta_{[c,L]}$ has a non-trivial connected component, whence $\dim \Delta_{[c,L]} > 0$, and (33.2) applies. □

We denote by L_2 the unique connected Lie group of dimension 2 which is not commutative, compare (94.40). Thus L_2 is isomorphic to the group of all bijections $x \mapsto ax + b : \mathbb{R} \to \mathbb{R}$ with $a > 0$.

33.4 Theorem. *Let W be a line of $\mathcal{P}$, and let A be a closed, connected group of axial collineations of $\mathcal{P}$ with axis W. Then A is a solvable group of dimension at most 3. More precisely, the following hold.*

(a) *If $\dim A = 1$, then $A \cong \mathbb{R}$, and all elements of A have a common center.*
(b) *If $\dim A = 2$ and if A is abelian, then $\mathcal{P} \cong \mathcal{P}_2\mathbb{R}$ and $A \cong \mathbb{R}^2$ is the group of all elations with axis W.*
(c) *If $\dim A = 2$ and if A is not abelian, then $A \cong L_2$ and A fixes precisely two lines L, W.*
(d) *If $\dim A = 3$, then $\mathcal{P}$ is isomorphic to the real projective plane $\mathcal{P}_2\mathbb{R}$ and A is the connected component of the group $(\mathrm{PGL}_3\mathbb{R})_{[W]} \cong \mathbb{R}^2 \rtimes \mathbb{R}^\times$ of all dilatations.*

Proof. We use the fact that A is a Lie group (32.21b) and (94.3a). By (32.17), A has no subgroup isomorphic to the 1-torus $\mathbb{T}$.

1) If $\dim A = 1$, then the one-parameter group A is isomorphic to $\mathbb{R}$, compare (94.39), and the assertion (a) was proved in (32.19).

2) Let $\dim A = 2$. Then $A \neq A_{[p]}$ for all points p, see (33.2). In particular, A cannot fix a point $p \notin W$. From (94.39 and 40) we infer that A is isomorphic to $\mathbb{R}^2$ or L_2. Thus A is generated by two one-parameter subgroups A_1, A_2, and $A_i \cong \mathbb{R}$ has a universal center p_i by assertion (a). We have just shown that $p_1 \neq p_2$, hence A fixes the line $L = p_1 \vee p_2$. If $L \neq W$, then A is not commutative, compare (23.12), thus $A \cong L_2$; furthermore any third line fixed by A would have to pass through $L \wedge W$, which leads to the contradiction $p_1 = L \wedge W = p_2$. If $L = W$, then A_i is (p_i, W)-transitive by (32.15c), and $A = A_1 \times A_2 \cong \mathbb{R}^2$ is the elation group of the translation plane $\mathcal{P}^W$, see (23.23). Thus $\mathcal{P} \cong \mathcal{P}_2\mathbb{R}$ by (32.8).

3) Assume that $\dim A \geq 3$. Choose $c \in W$ and lines $L, M \in \mathcal{L}_c \setminus \{W\}$. Then $A_{L,M} = A_{[c]}$ satisfies $\dim A_{L,M} = \dim A - \dim L^A - \dim M^{A_L} \geq 1$ by the dimension formula (96.10), because the orbits L^A, M^{A_L} are subsets of $\mathcal{L}_c \approx \mathbb{S}_1$. By (33.2) the elation group $A_{[c]}$ is transitive on $L \setminus \{c\}$, and this assertion holds for every $c \in W$. According to (32.8) the translation plane $\mathcal{P}$ is isomorphic to $\mathcal{P}_2\mathbb{R}$, and Δ is contained in the dilatation group

$$(\mathrm{PGL}_3\mathbb{R})_{[W]} \cong \{(x, y) \mapsto (rx + s, ry + t) \mid r, s, t \in \mathbb{R}, r \neq 0\},$$

compare (12.5 and 13). The connected component (defined by $r > 0$) of this 3-dimensional group has no proper closed subgroup of dimension 3. This proves the last assertion. □

33.5 Proposition. *Let Δ be a closed, connected subgroup of the collineation group of $\mathcal{P}$, and assume that Δ fixes a line W.*

(a) *The group Δ has a second fixed element (a point or a line) unless $\mathcal{P} \cong \mathcal{P}_2\mathbb{R}$.*
(b) *If $\dim \Delta_{[W]} \leq 1$, then Δ fixes some point of $\mathcal{P}$.*
(c) *If $\dim \Delta_{[W]} \geq 2$, then $\mathcal{P}$ is isomorphic to the real projective plane $\mathcal{P}_2\mathbb{R}$, or Δ fixes precisely one line L apart from W.*

Proof. Assertion (a) is a consequence of (b) and (c). In order to prove (b) and (c) we use the fact that $\dim \Delta_{[W]} = \dim(\Delta_{[W]})^1$, compare (93.6).

(b) If $\dim(\Delta_{[W]})^1 = 1$, then Δ fixes the (unique) universal center (32.19) of its normal subgroup $(\Delta_{[W]})^1$, compare (23.12). Thus we may assume that $(\Delta_{[W]})^1$ has dimension 0. This means that the Lie group $\Delta_{[W]}$ is discrete and therefore contained in the center of Δ, see (93.18). If $\Delta_{[W]} \neq \mathbb{1}$, then the center of any non-trivial element of $\Delta_{[W]}$ is fixed by Δ, compare (23.12), thus we may further assume that $\Delta_{[W]} = \mathbb{1}$. It suffices to consider the case where Δ has no fixed point on W. Then $\Delta \cong \Delta|_W$ is transitive on $W \approx \mathbb{S}_1$ by (32.15c), and the possibilities for Δ are given by Brouwer's theorem (96.30). In all cases Δ contains a torus subgroup $\mathbb{T}$ acting transitively on W. The involution $\sigma \in \mathbb{T}$ is a reflection (32.12).

From (32.17) we infer that W is the axis of σ, a contradiction to our assumption that $\Delta_{[W]}$ is trivial.

(c) We have $\dim(\Delta_{[W]})^1 \geq 2$, hence (33.4) implies that $\mathcal{P} \cong \mathcal{P}_2\mathbb{R}$ or $(\Delta_{[W]})^1$ fixes precisely two lines L, W. Since $(\Delta_{[W]})^1$ is normal in Δ, these lines are fixed by the connected group Δ, compare (91.1d). □

Concerning (33.5c), we point out that the 2-dimensional translation group of $(\mathcal{P}_2\mathbb{R})^W$ fixes precisely one line, namely W.

Now we can prove a chief result: we characterize the real projective plane $\mathcal{P}_2\mathbb{R}$ in terms of the dimension of its collineation group; recall that the collineation group $\mathrm{PGL}_3\mathbb{R}$ of $\mathcal{P}_2\mathbb{R}$ has dimension 8, compare (13.8).

33.6 Theorem. *Let Δ be a closed subgroup of the collineation group of a 2-dimensional compact projective plane $\mathcal{P} = (P, \mathcal{L})$.*

(a) *We have $\dim \Delta \leq 8$.*
(b) *If $\dim \Delta \geq 5$, then $\mathcal{P}$ is isomorphic to the real projective plane $\mathcal{P}_2\mathbb{R}$.*
(c) *If Δ is connected and semi-simple, then Δ is a covering group of $\mathrm{PSL}_2\mathbb{R}$ or $\mathcal{P}$ is isomorphic to $\mathcal{P}_2\mathbb{R}$.*

Proof. Assertion (a) has been deduced in (32.21b) from the dimension formula

(∗) $\qquad \dim \Delta = \dim o^\Delta + \dim u^{\Delta_o} + \dim v^{\Delta_{o,u}} + \dim e^{\Delta_{o,u,v}} \leq 4 \dim P = 8,$

where o, u, v, e is any quadrangle in $\mathcal{P}$. Now we prove (b). If $\dim \Delta = 8$, then $\dim p^\Delta = 2$ for every point $p \in P$ by (∗); thus every point orbit of Δ is open in P by (92.14), hence Δ is transitive on the connected space P, and $\mathcal{P} \cong \mathcal{P}_2\mathbb{R}$ by (32.23).

Proceeding indirectly, we now assume that $\mathcal{P} \not\cong \mathcal{P}_2\mathbb{R}$. Since $\dim \Delta^1 = \dim \Delta$ by (93.6), we may also assume that $\Delta = \Delta^1$ is a connected group satisfying $5 \leq \dim \Delta \leq 7$.

We claim that Δ is semi-simple (as defined in (94.20)) and has trivial center. Otherwise Δ has a non-trivial connected abelian normal subgroup or a non-trivial center, thus Δ fixes an element of $P \cup \mathcal{L}$ by (33.1), hence a second element of $P \cup \mathcal{L}$ by (33.5a) or its dual. If Δ fixes two points or two lines, then (∗) implies that $\dim \Delta \leq 2 \dim P = 4$, a contradiction. Similarly, if Δ fixes a non-incident point–line pair (o, uv), then

$$\dim \Delta = \dim u^\Delta + \dim v^{\Delta_u} + \dim e^{\Delta_{u,v}} \leq 1 + 1 + 2$$

for any point $e \notin ou \cup ov \cup uv$, again a contradiction. If Δ fixes just a flag (p, L), then Δ acts transitively on $L \setminus \{p\}$ by (32.15c), and the group induced on $L \setminus \{p\}$ has dimension at most 3 according to Brouwer's theorem (96.30), thus $\dim \Delta_{[L]} \geq 2$ by (93.7); this is a contradiction to (33.5c).

Thus Δ is a direct product of simple Lie groups with trivial centers, see (94.23). By the classification of simple Lie groups of small dimension, we have $\Delta \cong \mathrm{PSL}_2\mathbb{C}$, or Δ is a direct product of two factors which are isomorphic to $\mathrm{SO}_3\mathbb{R}$ or $\mathrm{PSL}_2\mathbb{R}$, compare (94.33). Now $\mathrm{PSL}_2\mathbb{C}$ contains a subgroup $\mathrm{SU}_2\mathbb{C}/\langle-1\rangle \cong \mathrm{SO}_3\mathbb{R}$, and the existence of such a subgroup implies that $\mathcal{P} \cong \mathcal{P}_2\mathbb{R}$, see (32.20). In the only remaining case $\Delta \cong \mathrm{PSL}_2\mathbb{R} \times \mathrm{PSL}_2\mathbb{R}$ we find a 2-torus $\mathrm{SO}_2\mathbb{R} \times \mathrm{SO}_2\mathbb{R}$ in Δ, a contradiction to (32.18).

In order to prove (c), we may by (b) assume that $\dim \Delta \leq 4$. Applying the classification of almost simple Lie groups of small dimension (94.33), we need to consider only the groups $\Delta \cong \mathrm{SO}_3\mathbb{R}$ and $\Delta \cong \mathrm{SU}_2\mathbb{C}$. Since both groups are compact, the assertion follows from (32.22). □

33.7 Remark. In connection with (33.6c), we mention that the semi-simple collineation groups of the real projective plane $\mathcal{P}_2\mathbb{R}$ are precisely the groups $\mathrm{PGL}_3\mathbb{R}$, $\mathrm{PSL}_2\mathbb{R} \cong \mathrm{O}'_3(\mathbb{R}, 1)$, $\mathrm{SL}_2\mathbb{R}$, $\mathrm{SO}_3\mathbb{R}$; this can be proved by using the tables (95.10). Note that the universal covering group of $\mathrm{PSL}_2\mathbb{R}$ is a collineation group of the Moulton planes (34.6 and 8).

33.8 Lemma. *Let Δ be a closed, connected subgroup of the collineation group of $\mathcal{P}$. If Δ fixes a flag (v, W) of $\mathcal{P}$, then Δ is solvable.*

Proof. We use the fact that Δ is a Lie group (32.21b). If Δ is not solvable, then Δ contains a non-trivial connected subgroup Ξ such that $\Xi = \Xi'$, see (94.16 and 17). By (33.4) and its dual, the connected components $(\Delta_{[W]})^1$, $(\Delta_{[v]})^1$ of the kernels of the actions on $W, \mathcal{L}_v$ are solvable. Thus Ξ and Δ act non-trivially on W and on $\mathcal{L}_v$. Let $p^\Xi \subseteq W \setminus \{v\}$ and $L^\Xi \subseteq \mathcal{L}_v \setminus \{W\}$ be non-trivial orbits. The orbits p^Δ and L^Δ are homeomorphic to $\mathbb{R}$, hence Brouwer's theorem (96.30) implies that Δ induces on each of these orbits the universal covering group $\widetilde{\Omega}$ of $\mathrm{PSL}_2\mathbb{R}$, which has infinite center. Choose $q \in L \setminus \{v\}$. Then the stabilizer Δ_q fixes L, hence infinitely many lines through v (the images of L under the center, compare (91.1c)). Since by (96.30) the actions of $\widetilde{\Omega}$ on L^Δ and on p^Δ are equivalent in the strong sense that the two conjugacy classes of stabilizers coincide, we infer that Δ_q fixes also infinitely many points on W. Thus $\Delta_q = \mathbb{1}$ by the stiffness result (32.10). But the dimension formula (96.10) implies that $\dim \Delta_q \geq 3 - 2$, a contradiction. □

In view of (33.6c), we shall use (33.8) mainly for covering groups Δ of $\mathrm{PSL}_2\mathbb{R}$ (to show that Δ does not fix a flag); for these almost simple groups, one could argue more directly (and avoid using Brouwer's theorem).

The following result is crucial for the classification of all planes $\mathcal{P}$ which admit a collineation group of dimension at least 3, see Sections 34 through 37.

33.9 Theorem. *Assume that $\mathcal{P} \not\cong \mathcal{P}_2\mathbb{R}$, and let Δ be a closed, connected subgroup of the collineation group of $\mathcal{P}$.*

(a) *If $\dim \Delta = 4$, then Δ fixes precisely one point a and precisely one line W; furthermore $a \notin W$.*

(b) *If $\dim \Delta = 3$, then Δ fixes precisely t points and precisely t lines with $t \in \{0, 1, 2\}$; if $t = 0$, then $\Delta \cong \mathrm{PSL}_2\mathbb{R}$.*

Proof. 0) Let $\mathcal{P} = (P, \mathcal{L})$. If Δ fixes no point and no line of $\mathcal{P}$, then (33.1) implies that Δ is semi-simple with trivial center. From (33.6c) we infer that $\Delta \cong \mathrm{PSL}_2\mathbb{R}$.

Now we assume that $\dim \Delta \geq 3$ and that Δ fixes some point or some line. By (33.5a), Δ fixes at least two elements of $P \cup \mathcal{L}$.

1) As a consequence, Δ fixes some point and some line of $\mathcal{P}$.

2) If Δ fixes two points u, v and two lines $W = uv$, $L \neq W$, then $\dim \Delta = 3$ and assertion (b) holds with $t = 2$.

Indeed, Δ acts freely on $(W \setminus \{u, v\}) \times (P \setminus (L \cup W))$ by (32.10), hence the dimension formula (96.10) gives

$$\dim \Delta = \dim x^\Delta + \dim y^{\Delta_x} \leq 1 + 2 = 3$$

for $x \in W \setminus \{u, v\}$, $y \in P \setminus (L \cup W)$. Thus $\dim \Delta = 3$, $\dim x^\Delta = 1$ and $\dim y^{\Delta_x} = 2$. The last conditions show that Δ has no further fixed elements, i.e., assertion (b) holds with $t = 2$.

3) If Δ fixes two points u, v, then $\dim \Delta = 3$, and assertion (b) holds with $t = 2$.

For the proof we may assume that $W = uv$ is the only line fixed by Δ, see step 2); this will lead to a contradiction. By (32.15c) the group Δ is transitive on $\mathcal{L}_u \setminus \{W\}$ and on $\mathcal{L}_v \setminus \{W\}$. If both kernels $\Delta_{[u]} = \Delta_{[u,W]}$, $\Delta_{[v]} = \Delta_{[v,W]}$ have positive dimension, then (33.3) implies that W is a translation line, hence $\mathcal{P} \cong \mathcal{P}_2\mathbb{R}$ by (32.8), a contradiction. Thus we may assume that $\dim \Delta_{[u]} = 0$, i.e., the Lie group $\Delta_{[u]}$ is discrete. From Brouwer's theorem (96.30) we infer that $\Delta|_{\mathcal{L}_u}$ is the simply connected covering group of $\mathrm{PSL}_2\mathbb{R}$. Hence Δ is almost simple, a contradiction to (33.8).

Assertion (b) is a consequence of 0), 1), 3), and the dual of 3).

Now we prove (a). Let $\dim \Delta = 4$. Again by 0), 1), 3) and its dual we know that Δ fixes precisely one point a and precisely one line W, and we have to show that $a \notin W$. Assume that $a \in W$ and aim for a contradiction. Then Δ acts without fixed points on $W \setminus \{a\}$, thus Δ is transitive on $W \setminus \{a\}$ by (32.15c). If the kernel of this action has dimension at least 2, then Δ fixes two lines by (33.5c), a contradiction to our assumptions. Therefore this kernel has dimension at most 1. By (93.7) and

Brouwer's theorem (96.30), $\Delta|_W$ is the universal covering group of $\mathrm{PSL}_2\mathbb{R}$, a contradiction to (33.8). □

In Sections 34 through 37, the planes satisfying (33.9) will be classified explicitly. For later use (in Sections 34, 36, and 37) we prove (33.10 and 12).

33.10 Lemma. *Let Δ be a closed, connected subgroup of the collineation group of $\mathcal{P}$ with $\dim \Delta = 2$. If Δ fixes a (non-degenerate) triangle a, b, c, then $\Delta \cong \mathbb{R}^2$; in particular, Δ is commutative.*

Proof. Δ contains no torus subgroup by (32.17), thus $\Delta \cong \mathbb{R}^2$ if Δ is commutative, compare (94.39) or use the structure theorem for locally compact, connected abelian groups in Hewitt–Ross [79] 9.14.

It suffices to derive a contradiction from the assumption that $\Delta \cong L_2$, compare (94.40). If Δ acts trivially on one of the lines ab, bc, ca, then Δ acts freely on each of the other two sides of the triangle, see (23.8), and the dimension formula (96.10) leads to the contradiction $\dim \Delta \leq 1$ (one could also use (33.4b,c)). Thus each of the lines ab, bc, ca contains a point x_i which is moved by Δ, hence $\dim \Delta_{x_i} = 1$ for $1 \leq i \leq 3$ by (96.10). Since $\Delta \cong L_2$ has only two conjugacy classes of one-dimensional closed subgroups (viz. the commutator subgroup Δ' and its complements), we infer that two of the stabilizers Δ_{x_i} are conjugate in Δ, say $\Delta_{x_1} = (\Delta_{x_2})^\delta = \Delta_{x_2^\delta}$ for some $\delta \in \Delta$. Thus Δ_{x_1} fixes a, b, c, x_1 and x_2^δ, hence a quadrangle, and this is a contradiction to (32.10). □

33.11 Methods of classification. In Sections 34 through 37 we classify 2-dimensional compact projective planes $\mathcal{P} = (P, \mathcal{L})$ with large collineation groups Σ by the following strategy. As a first step we determine the structure of the connected component $\Delta = \Sigma^1$ and the action of Δ on the point space P or on some large part of P (like an open dense orbit). Note that we may identify P with the surface $P_2\mathbb{R}$ by (32.6 and 3). Usually there are only few possibilities for the topological group Δ, and the (equivalence classes of) possible actions of a given group Δ depend only on finitely many real parameters.

As a second step in the classification, we fix such an action of Δ and determine all planes $\mathcal{P} = (P, \mathcal{L})$ admitting this action. The line L joining distinct points p, q is invariant under the stabilizer $\Delta_{p,q}$, hence the possibilities for L can be found by studying the orbits of $\Delta_{p,q}$ in P. For other methods compare the remarks b) following (32.20). In the cases of interest to us (i.e., if Δ is large enough), the planes $\mathcal{P}$ (i.e., the line systems $\mathcal{L}$) depend only on finitely many real parameters.

For example, the Moulton planes $\overline{\mathcal{M}_k}$ depend on a parameter $k > 1$ which describes the action of their collineation groups, see (34.7b), whereas the skew hyperbolic planes $\mathcal{H}_t$ depend on a parameter $t > 0$ which describes the choice of a line system $\mathcal{L}_t$ in (35.1). The parameter d of the skew parabola planes $\overline{\mathcal{E}_{c,d}}$

is determined by the isomorphism type of the corresponding collineation group (considered as a topological group), see (36.1 and 3).

As a last step in the classification, one has to look for isomorphisms between the planes found in this way. If two planes $\mathcal{P}_i = (P, \mathcal{L}_i)$ with full collineation groups Σ_i are isomorphic, i.e., if some homeomorphism $\varphi : P \to P$ induces a collineation of $\mathcal{P}_1$ onto $\mathcal{P}_2$, then $\Sigma_2 = \varphi^{-1}\Sigma_1\varphi$, hence the transformation groups (Σ_1, P) and (Σ_2, P) are equivalent (in the sense of (96.1)). If the connected components of Σ_1 and of Σ_2 coincide, with Δ say, then Δ is normalized by φ. Thus we shall usually compute the normalizer of Δ in the group of all homeomorphisms of P onto itself, in order to detect isomorphisms between the planes found in the second step. In the special case where $\mathcal{P}_1 = \mathcal{P}_2$, i.e., $\mathcal{L}_1 = \mathcal{L}_2$, this procedure allows us to compute the full collineation group Σ of the plane $(P, \mathcal{L}_1)$ if only the connected component Δ is known.

For this last step, the following is useful.

33.12 Lemma. *Denote by Γ the group of all homeomorphisms of the real line $\mathbb{R}$ onto itself. The normalizer in Γ of $\mathsf{A} = \{x \mapsto x + b \mid b \in \mathbb{R}\}$ or of $\mathsf{L}_2 = \{x \mapsto ax + b \mid a > 0\}$ is the group $\mathsf{L}_2 \rtimes \langle -\mathbb{1} \rangle = \{x \mapsto ax + b \mid a \neq 0\}$.*

Proof. Since $\mathsf{A} = \mathsf{L}'_2$ is characteristic in L_2, it suffices to prove that the normalizer of A is contained in $\mathsf{B} = \mathsf{L}_2 \rtimes \langle -\mathbb{1} \rangle$. Now B induces on A the full automorphism group of A. Moreover, the centralizer Ψ of A in Γ satisfies $\Psi = \mathsf{A}\Psi_0$, see (91.2a). By (91.1c), the stabilizer Ψ_0 is trivial. Thus the normalizer of A is contained in $\mathsf{AB} = \mathsf{B}$. □

33.13 Notes on Section 33. This section is closely related to the papers Salzmann [63b], [64] §3. Theorem (33.6) was proved in Salzmann [63b] 1.13, [62b] Satz 2. For Proposition (33.1) compare also Salzmann [67b] 5.2.

34 The Moulton planes

The main result (34.10) of this section says that the Moulton planes are the only 2-dimensional compact projective planes whose collineation groups have dimension 4. Thus the Moulton planes admit the largest collineation groups among all non-desarguesian 2-dimensional compact projective planes, see (33.6).

The line at infinity which completes the usual model (31.25b) to a projective plane is not fixed by all collineations. We describe another (radial) model of the Moulton planes, which allows a convenient description of the collineation groups Σ of the Moulton planes.

The commutator groups Σ' are isomorphic to the universal covering group $\widetilde{\Omega}$ of $\Omega = \mathrm{PSL}_2\mathbb{R}$. Thus the Moulton planes furnish natural geometries for the somewhat elusive group $\widetilde{\Omega}$. See Hilgert–Hofmann [85] for a discussion of the properties of $\widetilde{\Omega}$.

34.1 Lemma. *Let* $f : (-\frac{\pi}{2}, \frac{\pi}{2}) \to (0, \infty)$ *be a differentiable function such that* $\lim_{\varphi \to \pm \pi/2} f(\varphi) = +\infty$ *and such that the logarithmic derivative* $(\log f)' = f'/f$ *is an increasing homeomorphism of* $(-\frac{\pi}{2}, \frac{\pi}{2})$ *onto* $\mathbb{R}$. *(The 'classical' example is the function* $f(\varphi) = 1/\cos\varphi$.)

(a) *The mapping defined by* $(\varphi_1, \varphi_2) \mapsto f(\varphi_1)f(\varphi_2)^{-1}e^{i(\varphi_1 - \varphi_2)}$ *is a bijection of* $\{(\varphi_1, \varphi_2) \mid -\frac{\pi}{2} < \varphi_j < \frac{\pi}{2}, \varphi_1 \neq \varphi_2\}$ *onto* $\mathbb{C} \setminus \mathbb{R}$.

(b) *Define* $L = \{f(\varphi)e^{i\varphi} \mid -\frac{\pi}{2} < \varphi < \frac{\pi}{2}\}$ *and*

$$\mathcal{L} = \{cL \mid c \in \mathbb{C}^\times\} \cup \{\mathbb{R}e^{i\varphi} \mid -\frac{\pi}{2} < \varphi \leq \frac{\pi}{2}\}.$$

Then $(\mathbb{C}, \mathcal{L})$ *is an affine* $\mathbb{R}^2$*-plane.*

Proof. (a) Choose $re^{i\varphi} \in \mathbb{C} \setminus \mathbb{R}$ with $r > 0$, $-\pi < \varphi < \pi$, $\varphi \neq 0$. The preimage of $re^{i\varphi}$ under the map in (a) consists of the pairs (φ_1, φ_2) which satisfy $\varphi = \varphi_1 - \varphi_2$, $r = f(\varphi_1)f(\varphi_2)^{-1}$ or, equivalently,

$$\varphi_1 = \varphi_2 + \varphi, \quad \log r = \log f(\varphi_2 + \varphi) - \log f(\varphi_2) = \int_0^\varphi \frac{f'(\varphi_2 + \psi)}{f(\varphi_2 + \psi)} d\psi.$$

This integral formula shows that the continuous map $\varphi_2 \mapsto \log f(\varphi_2 + \varphi) - \log f(\varphi_2)$ is a strictly increasing map of $(-\frac{\pi}{2}, \frac{\pi}{2} - \varphi)$ into $\mathbb{R}$ or a strictly decreasing map of $(-\frac{\pi}{2} - \varphi, \frac{\pi}{2})$ into $\mathbb{R}$, depending on the sign of φ. The limit behavior of f at $\pm \frac{\pi}{2}$ implies that this map is unbounded above and below, hence is a bijection. Thus φ_2 and φ_1 are uniquely determined by r, φ.

(b) First we show that distinct points $c_1, c_2 \in \mathbb{C}$ are joined by a unique line in $\mathcal{L}$. This is obvious if $0 \in \{c_1, c_2\}$. Since the incidence structure $(\mathbb{C}, \mathcal{L})$ admits $\mathbb{C}^\times$ as a collineation group which is transitive on $\mathbb{C} \setminus \{0\}$, we may assume that $c_2 = 1$. If $c_1 \in \mathbb{R}$, then $\mathbb{R} = \mathbb{R}e^{i \cdot 0}$ is the unique line joining c_1 and c_2. Hence we may further assume that $c_1 \notin \mathbb{R}$. Then cL is a joining line if, and only if,

$$c_1 = cf(\varphi_1)e^{i\varphi_1}, \quad 1 = cf(\varphi_2)e^{i\varphi_2}$$

with suitable $\varphi_j \in (-\frac{\pi}{2}, \frac{\pi}{2})$. The equation

$$c_1 = f(\varphi_1)f(\varphi_2)^{-1}e^{i(\varphi_1 - \varphi_2)}$$

has a unique solution (φ_1, φ_2) by (a), and then the equation $c^{-1} = f(\varphi_2)e^{i\varphi_2}$ determines the joining line cL uniquely.

Every line in $\mathcal{L}$ is homeomorphic to $\mathbb{R}$, because $f(\varphi) \to \infty$ for $\varphi \to \pm\frac{\pi}{2}$. In order to verify the axiom on parallels (21.8ii) for the $\mathbb{R}^2$-plane $(\mathbb{C}, \mathcal{L})$, we use (a) again: one easily verifies that $L \cap cL = \emptyset$ if, and only if, $1 \neq c \in \mathbb{R}^\times$. Since $\{rL \mid r \in \mathbb{R}^\times\}$ is a partition of $\mathbb{C} \setminus i\mathbb{R}$, this implies that $(\mathbb{C}, \mathcal{L})$ is an affine plane. $\square$

34.2 A radial model of the Moulton planes. Fix a real number $s \geq 0$ and define the function $f : (-\frac{\pi}{2}, \frac{\pi}{2}) \to (0, \infty)$ by

$$f(\varphi) = e^{s\varphi}/\cos\varphi.$$

Then $f'(\varphi)/f(\varphi) = s + \tan\varphi$, hence f satisfies the assumptions of (34.1). By (34.1b) we obtain an affine $\mathbb{R}^2$-plane $\mathcal{M}(s)$ for every $s \geq 0$. Denote by $\overline{\mathcal{M}(s)}$ the projective completion of $\mathcal{M}(s)$; this is a 2-dimensional compact projective plane by (32.2 and 6).

We claim that $\overline{\mathcal{M}(s)}$ is isomorphic to the Moulton plane $\overline{\mathcal{M}_k}$ with $k = e^{2\pi s}$, compare (31.25b). Define a bijection $\alpha : \mathbb{C} \setminus i\mathbb{R} \to \mathbb{R}^\times \times \mathbb{R}$ by

$$(re^{i\varphi})^\alpha = \left(\frac{e^{s\varphi}}{r\cos\varphi}, \tan\varphi\right)$$

for $r > 0$, $\frac{\pi}{2} \neq \varphi \in (-\frac{\pi}{2}, \frac{3\pi}{2})$. By (44.9) it suffices to show that α is a collineation between the open dense parts $\mathbb{C} \setminus i\mathbb{R}$ of $\overline{\mathcal{M}(s)}$ and $\mathbb{R}^\times \times \mathbb{R}$ of $\overline{\mathcal{M}_k}$. This can be checked as follows: we have

$$(\mathbb{R}^\times e^{i\varphi})^\alpha = \mathbb{R}^\times \times \{\tan\varphi\},$$

and with $c = re^{i\varphi}$, $r > 0$, $0 \leq \varphi < 2\pi$ and $L = \left\{\frac{e^{s\psi}}{\cos\psi}e^{i\psi} \mid -\frac{\pi}{2} < \psi < \frac{\pi}{2}\right\}$ we obtain that

$$(cL \setminus i\mathbb{R})^\alpha = \left\{\frac{re^{s\psi}}{\cos\psi}e^{i(\varphi+\psi)} \mid -\frac{\pi}{2} < \psi < \frac{\pi}{2},\ \varphi + \psi \notin \{\frac{\pi}{2}, \frac{3\pi}{2}\}\right\}^\alpha.$$

If $\varphi \leq \pi$, then this gives

$$(cL \setminus i\mathbb{R})^\alpha = \left\{\left(\frac{e^{s\varphi}\cos\psi}{r\cos(\varphi+\psi)}, \tan(\varphi+\psi)\right) \mid -\frac{\pi}{2} < \psi < \frac{\pi}{2},\ \varphi+\psi \neq \frac{\pi}{2}\right\},$$

whereas for $\varphi > \pi$ we compute

$$(cL \setminus i\mathbb{R})^\alpha = \left\{\left(\frac{e^{s\varphi}\cos\psi}{r\cos(\varphi+\psi)}, \tan(\varphi+\psi)\right) \mid -\frac{\pi}{2} < \psi < \frac{3\pi}{2} - \varphi\right\} \cup$$
$$\cup \left\{\left(e^{-2\pi s}\frac{e^{s\varphi}\cos\psi}{r\cos(\varphi+\psi)}, \tan(\varphi+\psi)\right) \mid \frac{3\pi}{2} - \varphi < \psi < \frac{\pi}{2}\right\}.$$

The relation $\cos\psi/\cos(\varphi+\psi) = \sin\varphi\tan(\varphi+\psi) + \cos\varphi$ shows that these sets are the lines of $\mathcal{M}_k$ which are distinct from $\{0\} \times \mathbb{R}$, hence α is a collineation as asserted.

Note that $\overline{\mathcal{M}(0)} \cong \overline{\mathcal{M}_1}$ is the real projective plane $\mathcal{P}_2\mathbb{R}$.

In order to obtain a clear picture of the automorphism groups of the Moulton planes, one may profitably use the technique of vector fields and transformation groups, which we introduce next.

34.3 Vector fields and transformation groups. There are (at least) two ways of looking at a smooth vector field X defined on an open set $U \subseteq \mathbb{R}^n$. First, we may consider X as a smooth map $X : U \to \mathbb{R}^n$. This definition gives rise to the notion of *integral curves* of X. These are curves $\xi(x, t) = \xi_t(x)$ defined for each $x \in U$ on a (maximal) open interval $I_x \subseteq \mathbb{R}$ containing 0, subject to the conditions

$$\xi(x, 0) = x, \quad \tfrac{\partial}{\partial t}\xi(x, t_0) = X(\xi(x, t_0))$$

for all x and all t_0. Standard theorems on differential equations assert that these curves exist and are uniquely determined, and that the *flow* ξ associated with X is a smooth map of some open neighbourhood of $U \times \{0\}$ into U. The vector field X is said to be *complete* if ξ is defined on all of $U \times \mathbb{R}$. In this case, ξ defines an action of the group $\mathbb{R}$ on U as a smooth transformation group. Conversely, any such action defines a smooth vector field by derivation, that is, by $X(x) := \tfrac{\partial}{\partial t}\xi(x, t)|_{t=0}$. Examples of incomplete vector fields may be obtained by restricting complete ones to smaller open domains.

The second interpretation of a vector field X as above is to regard X as a differential operator on the space $\mathcal{F}$ of smooth functions $f : U \to \mathbb{R}$, defined by $X(f) := \sum_{i=1}^{n} X_i \partial_i f$, where $\partial_i f = \tfrac{\partial f}{\partial x_i}$ and $X_i \in \mathcal{F}$ is the smooth ith component function of X. In order to stress this interpretation, we write

$$X = \sum_{i=1}^{n} X_i \partial_i.$$

This leads to the notion of the *Lie bracket* or *commutator* $[X, Y]$ of two vector fields. It is defined as the operator sending f to $X(Y(f)) - Y(X(f))$. It can be shown that $[X, Y]$ is again a vector field, that is, that all second order derivatives arising in the definition cancel.

The *Theorem of Lie and Palais* asserts the following. Let $\mathfrak{x}$ be a Lie algebra of vector fields on $U \subseteq \mathbb{R}^n$, that is, $\mathfrak{x}$ is a vector space and is closed with respect to the Lie bracket. Suppose that $\mathfrak{x}$ is finite-dimensional and is generated (as a Lie algebra) by complete vector fields. Then there is a unique connected Lie transformation group Γ acting on U such that $\mathfrak{x}$ consists precisely of the vector fields associated with all one-parameter subgroups ξ in Γ. In particular, all vector fields belonging to $\mathfrak{x}$ are complete. Moreover, the Lie algebra $\mathfrak{l}\Gamma$ is isomorphic to $\mathfrak{x}$. For a proof, see Loos [69] I.3.1.

34 The Moulton planes

We shall see in (34.6) below that the transformation groups described in the sequel provide accessible models of the universal covering group $\widehat{\Omega}$ of $\Omega = \mathrm{PSL}_2\mathbb{R}$.

34.4 The Moulton groups. Let $s > 0$ be given. We construct a Lie transformation group $\Sigma = \Sigma(s)$ on $U = \mathbb{R}^2 \setminus \{0\}$ by specifying four complete vector fields H, R, E and X, which generate a 4-dimensional Lie algebra $\mathfrak{x}(s)$ of vector fields. In terms of polar coordinates, these vector fields are defined as follows.

$$H(re^{i\varphi}) = r\tfrac{\partial}{\partial r}, \qquad R(re^{i\varphi}) = sr\tfrac{\partial}{\partial r} + \tfrac{\partial}{\partial \varphi},$$

$$E(re^{i\varphi}) = r(\sin\varphi\cos\varphi + s\cos^2\varphi)\tfrac{\partial}{\partial r} + \cos^2\varphi\tfrac{\partial}{\partial \varphi},$$

$$X(re^{i\varphi}) = r(2\sin^2\varphi + 2s\sin\varphi\cos\varphi - 1)\tfrac{\partial}{\partial r} + 2\sin\varphi\cos\varphi\tfrac{\partial}{\partial \varphi}.$$

1) Elementary computation yields the commutators $[H, Y] = 0$ if Y is any of the four vector fields, $[E, X] = 2E$, $[E, R] = X$ and $[X, R] = 2R - 4E$. (The computation of $[E, X]$ may be saved, see (34.5) below.) Since the commutator is antisymmetric, this suffices to show that $\mathfrak{x}(s)$ is a Lie algebra isomorphic to the Lie algebra $\mathfrak{gl}_2\mathbb{R}$ of $\mathrm{GL}_2\mathbb{R}$, which consists of all (2×2)-matrices with the commutator $[A, B] = AB - BA$. Indeed, the following assignment defines an isomorphism $\mathfrak{x}(s) \to \mathfrak{gl}_2\mathbb{R}$:

$$H \mapsto \begin{pmatrix} 1 & 0 \\ 0 & 1 \end{pmatrix}, \quad R \mapsto \begin{pmatrix} 0 & -1 \\ 1 & 0 \end{pmatrix}, \quad E \mapsto \begin{pmatrix} 0 & 0 \\ 1 & 0 \end{pmatrix}, \quad X \mapsto \begin{pmatrix} 1 & 0 \\ 0 & -1 \end{pmatrix}.$$

2) Next we shall verify completeness of the four generating vector fields by writing down integral curves for each of them. According to the Theorem of Lie and Palais, see (34.3), this will imply that the integral curves are the orbits of one-parameter transformation groups, which generate a Lie transformation group $\Sigma(s)$ with Lie algebra $\mathfrak{l}\Sigma(s)$ isomorphic to $\mathfrak{x}(s) \cong \mathfrak{gl}_2\mathbb{R}$. As an aside, we remark that in $\mathrm{SL}_2\mathbb{R}$, the one-parameter groups associated with the three generators R, E, X of the Lie algebra represent the three conjugacy classes of one-parameter subgroups. It is easy to see that the one-parameter groups H and P associated with H and R are given respectively by

$$\eta_t(re^{i\varphi}) = e^t re^{i\varphi}$$
$$\varrho_t(re^{i\varphi}) = e^{st} re^{i(\varphi+t)};$$

they generate the ordinary action of $\mathbb{C}^\times$ on U. The elements of H are *homotheties*, and those of P will be called *spiral rotations*. The one-parameter group E associated with E will turn out to consist of the *elations* of the Moulton plane $\overline{\mathcal{M}(s)}$ having the line $i\mathbb{R}$ as axis. They are given by $\varepsilon_t|_{i\mathbb{R}} = \mathbb{1}$ and

$$\varepsilon_t(re^{i\varphi}) = \tfrac{r\cos\varphi}{u\cos\psi} e^{s(\psi-\varphi)} e^{i\psi}, \quad \text{where } \tan\psi = \tan\varphi + t.$$

and $\varphi, \psi \in (-\pi/2, \pi/2)$ or $\varphi, \psi \in (\pi/2, 3\pi/2)$. The elements ξ_u, $0 < u = e^t$, of the *hyperbolic* one-parameter group Ξ associated with X have the same form, except that $\xi_u(x) = x/u$ for $x \in i\mathbb{R}$ and that the relationship between ψ and φ is now given by

$$\tan \psi = u^2 \tan \varphi.$$

3) The relation $[E, X] = 2E$ shows that $E\Xi$ is a 2-dimensional non-abelian Lie group. An explicit isomorphism of $E\Xi$ onto the group L_2 of all maps $x \mapsto ax+b$, $a > 0$ of the real line is furnished by passing to the action $\tan \varphi \mapsto \tan \psi = u^2 \tan \varphi + t$ on the angle coordinate. This shows that $\xi_u^{-1} \varepsilon_t \xi_u = \varepsilon_{u^2 t}$.

4) The isomorphism $\mathfrak{l}\Sigma(s) \cong \mathfrak{x}(s) \cong \mathfrak{gl}_2 \mathbb{R}$ yields a local isomorphism $\Sigma \triangleq GL_2 \mathbb{R}$. Moreover, it follows that H is the connected component of the center of Σ, while the commutator group Σ' corresponds to $SL_2 \mathbb{R}$. The Lie algebra of Σ' is $\langle R, E, X \rangle$. This is why we consider P rather than the ordinary rotation group. In fact, P corresponds to $SO_2 \mathbb{R}$ under the above local isomorphism.

5) The absolute value of the radial components of the generating vector fields is bounded by some fixed multiple of r. This implies that the action of $\Sigma(s)$ on U extends to a continuous action on $\mathbb{R}^2$ if we let 0 be a universal fixed point.

6) Finally, we show that $\Sigma = \Sigma(s)$ consists of collineations of $\mathcal{M}(s)$. By the construction of $\mathcal{M}(s)$, the elements of $HP \cong \mathbb{C}^\times$ are collineations. The generating line $L = \{ e^{(s+i)\varphi} / \cos \varphi \mid -\pi/2 < \varphi < \pi/2 \}$ of $\mathcal{M}(s)$ is an orbit of the one-parameter group $E \leq \Sigma$, and the other non-trivial E-orbits are the images of this one under $H\langle -\mathbb{1}\rangle$. All conjugates of E are obtained by conjugation with elements $\varrho \in P$. Hence the lines of $\mathcal{M}(s)$ are precisely the non-trivial orbits of $H\langle -\mathbb{1}\rangle$ (augmented by 0) and of the conjugates of E. Therefore, the line system is invariant under the action of Σ. According to (32.21c), the analytic group Σ and each of its analytic subgroups are closed in $\operatorname{Aut} \mathcal{M}(s)$.

34.5 Remarks. The reader may have wondered how the vector fields E and X and their associated flows were found. The answer is easy. The isomorphism α of (34.2) carries the subgroup $\Delta = \{ (x, y) \mapsto (ux, u^2 y + t) \mid u, t \in \mathbb{R}, u > 0 \}$ of $\operatorname{Aut} \mathcal{M}_k$ acting on $\mathbb{R}^\times \times \mathbb{R}$ onto $E\Xi$ acting on the same space. The vector fields were then obtained by derivation. This provides another concrete isomorphism $E\Xi \cong L_2$. The arguments in step 6) of (34.4) could be simplified by using the fact that $\alpha^{-1} E \Xi \alpha$ is a group of automorphisms of the model $\mathcal{M}_k$ of the Moulton plane, where $k = e^{2\pi s}$.

The special value $s = 0$ of the parameter s yields the usual action of the group $\Sigma(0) = GL_2 \mathbb{R}$ on $\mathbb{R}^2$; thus this action appears as a limit of the actions of $\Sigma(s)$. Note that the isomorphism type of the groups $\Sigma(s)$ changes suddenly for $s \to 0$, as the following result shows.

34.6 Proposition: Structure of the Moulton groups. *The connected Lie group $\Sigma = \Sigma(s)$ constructed in (34.4) is locally isomorphic to $\mathrm{GL}_2\mathbb{R}$. The center of Σ is the group $\mathsf{Z} = \mathsf{H}\langle -1\!\!1\rangle \cong \mathbb{R}^\times$ of all homotheties of $\mathbb{R}^2$. The commutator group Σ' is isomorphic to the simply connected covering group $\widetilde{\Omega}$ of $\Omega = \mathrm{PSL}_2\mathbb{R}$. The intersection $\mathsf{Z} \cap \Sigma'$ is the infinite cyclic center of Σ'. These properties determine the isomorphism type of Σ as a topological group; in particular, this isomorphism type is independent of $s > 0$. The center of Σ' is generated by the spiral rotation $\zeta = \varrho_\pi : z \mapsto -e^{s\pi}z$.*

Proof. From (34.4), part 4), we know that Σ is locally isomorphic to $\mathrm{GL}_2\mathbb{R}$, and that the center of Σ has Lie algebra $\langle H\rangle$, while Σ' has Lie algebra $\langle R, E, X\rangle$ and is locally isomorphic to $\mathrm{SL}_2\mathbb{R}$. From the first two facts we infer that $\mathsf{Z} = \mathsf{H}\langle -1\!\!1\rangle$ is contained in the center. From (34.4) step 6) we know that Σ' is closed. The center of Σ' is a cyclic group $C = \langle \zeta\rangle$ since $\widetilde{\Omega}$ has infinite cyclic center (94.33). From the local isomorphism we infer that Σ is completely determined as the almost direct product obtained from $\mathsf{H} \times \Sigma'$ by identifying the copies of $\mathsf{H} \cap \Sigma' = \mathsf{H} \cap C$ lying in the two factors. Thus it suffices to compute C and $\mathsf{H} \cap C$.

The Lie algebra shows that the one-parameter group P of Σ' becomes compact if we pass to the factor group Σ'/C. Hence P intersects the center C. If C were not entirely contained in P, then $\Sigma'/\mathsf{P} \cap C$ would have a disconnected maximal compact subgroup $\mathsf{P}C/(\mathsf{P} \cap C)$, contrary to the theorem of Mal'cev and Iwasawa (93.10). Hence a generator of C has the form $\varrho_t : z \mapsto e^{(s+i)t}z$, and it fixes the ordinary lines through the origin since they are the fixed point sets of the conjugates of E. It follows that t is an integer multiple of π. On the other hand, $\zeta = \varrho_\pi$ belongs to Z and is therefore central. Thus this element is a generator of C. Using (94.2b) and the fact that $\pi_1\Omega \cong \mathbb{Z}$ has no proper infinite quotient, we conclude that $\Sigma' \cong \widetilde{\Omega}$. The group $\mathsf{H} \cap C$ is generated by $\zeta^2 = \varrho_{2\pi} = \eta_{2\pi s}$, because $\zeta^2 \in \mathsf{H}$ and $\zeta \notin \mathsf{H}$. The isomorphism type of $\Sigma \cong (\mathsf{H} \times \Sigma')/\langle(\eta_{2\pi s}, \zeta^2)\rangle$ does not depend on $s > 0$, because the pair $(\eta_{2\pi s}, \zeta^2)$ can be mapped to (η_1, ζ^2) by some automorphism of $\mathsf{H} \times \Sigma'$. □

34.7 Proposition: Moulton groups as transformation groups.

(a) *The stabilizer $\Pi = \Sigma'_p$ of $p \in U = \mathbb{R}^2 \setminus \{0\}$ is isomorphic to the semi-direct product $\mathbb{R} \rtimes \mathbb{Z}$, where a generator δ of $\mathbb{Z}$ acts on $\mathbb{R}$ via multiplication by $e^{4s\pi}$.*
(b) *Two groups $\Sigma(s)$ and $\Sigma(s')$ with $s, s' > 0$ are equivalent as transformation groups if, and only if, $s = s'$.*
(c) *The group $\Sigma(s)$ is equal to its own normalizer in the homeomorphism group of $\mathbb{R}^2$.*

Remarks. The transitive action of Σ' appears in the list of transitive actions on surfaces (Mostow [50]) as the action II.2 on p. 627, and in Betten–Forst [77] on p.42 under case b) $k = 1$, $a > 0$. Note that transfer into the model $\mathcal{M}_k$ of the Moulton plane, see (34.5), renders the action of the subgroup $\mathsf{E}\Xi$ linear.

The action of Σ appears in Mostow [50] III.6, see also Betten–Forst [77] p. 48.

Proof of (34.7). Part (b) follows immediately from (a). Indeed, if the transformation groups $\Sigma(s)$ and $\Sigma(s')$ are equivalent, then their commutator groups are equivalent, hence they have isomorphic stabilizers. Now an isomorphism γ of the two extensions of $\mathbb{R}$ by $\mathbb{Z}$ induces some map $\beta : t \mapsto bt$ on $\mathbb{R}$. Since β commutes with δ, it follows that the two extensions are actually the same.

For part (a), we choose the point $p = i = 1 \cdot e^{i\pi/2}$. Then we have $\mathsf{E} \leq \Pi$, and $\dim \Pi = \dim \Sigma - \dim U = 1$, see (96.10). Hence $\mathsf{E} = \Pi^1$ is a normal subgroup of Π. Now the normalizer of E in Σ' is $\Theta := \mathsf{E}\Xi\langle\zeta\rangle$, as can be seen from the Lie algebra. We might now pass to the model $\mathcal{M}_k$ of the Moulton plane via the isomorphism α of (34.2). Thus we consider the group $\alpha^{-1}\Theta\alpha$, which is easy to handle, compare (34.5), and compute the stabilizer of the point p^α at infinity.

We prefer to give the details for an argument within the chosen context. By (34.4), part 2), E acts trivially on the imaginary axis, and $\xi_u \in \Xi$ acts by $p \mapsto p/u$. Since ζ acts by $p \mapsto -e^{s\pi}p$, it follows that $\Pi = \mathsf{E}\langle\vartheta\rangle$, where $\vartheta = \xi_{e^{2s\pi}} \cdot \zeta^2$. The action of ϑ on E is multiplication by $e^{4s\pi}$, see part 3) of (34.4) and note that ζ is central.

For part (c), let γ be a homeomorphism of $\mathbb{R}^2$ that normalizes $\Sigma = \Sigma(s)$. In order to show that γ belongs to Σ, we shall multiply γ by elements $\sigma_n \in \Sigma$ until we obtain the identity $\mathbb{1}$. Since all maximal compact subgroups of Σ are conjugate (93.10), the first factor σ_1 may be so chosen that $\gamma_1 = \gamma\sigma_1$ normalizes any given one of them. Together with the invariance of the center Z, this means that the subgroup $\mathsf{HP} \cong \mathbb{C}^\times$ is left invariant if we please. Since Σ' is the only almost simple subgroup of Σ, the intersection $\Sigma' \cap \mathsf{HP}$ is also fixed, as well as the central one-parameter group H and the unique compact one-parameter group $\Phi \leq \mathsf{HP}$. Now γ_1 induces a linear automorphism of the universal cover $\mathbb{R}^2$ of HP, which fixes three one-dimensional subspaces and the kernel of the covering map. The only remaining possibilities are that γ_1 induces the identity or inversion on HP. Now we add a factor $\sigma_2 \in \mathsf{HP}$ such that $\gamma_2 = \gamma_1\sigma_2$ fixes some point $p \neq 0$. Then γ_2 acts on $\mathbb{R}^2 \setminus \{0\}$ in the same way as on HP. Since 0 must be fixed, the possibility of inversion disappears, and our claim follows. $\square$

34.8 Theorem: Moulton planes.

(a) *The automorphism group of the Moulton plane $\overline{\mathcal{M}(s)}$, see (34.2), is the 4-dimensional group $\Sigma(s)$ introduced in (34.4). For the structure of $\Sigma(s)$, see (34.6). The action induced on the line W at infinity is the standard action of $\Sigma/\mathsf{Z} \cong \mathrm{PSL}_2\mathbb{R}$ on the real projective line.*

(b) *The plane $\overline{\mathcal{M}(s)}$ has Lenz type III and Lenz–Barlotti type III.2. Denote the point 0 by a and the line at infinity by W. The linearly transitive elation groups are precisely the groups $\Sigma_{[a,aw]}$ with $w \in W$. They are the conjugates of the one-parameter group E, and they generate the commutator group $\Sigma' \cong \widetilde{\Omega}$, which is the universal covering group of $\Omega = \mathrm{PSL}_2\mathbb{R}$. The center $\mathsf{Z} = \Sigma_{[a,W]}$ of $\Sigma(s)$ is the only linearly transitive group of homologies.*
(c) *Two Moulton planes $\overline{\mathcal{M}(s)}$ and $\overline{\mathcal{M}(s')}$ with $s, s' > 0$ are isomorphic if, and only if, $s = s'$.*

Remark. The group $\widetilde{\Omega}$ has no faithful linear representation (95.9), hence it does not act effectively on the real projective plane.

Proof of (34.8). 1) According to (34.4) step 6), the group $\Sigma = \Sigma(s)$ is a closed subgroup of $\Gamma = \mathrm{Aut}\,\mathcal{M}(s)$.

2) It is now easy to see that E and Z are linearly transitive axial groups. By (23.19), these groups contain all collineations with the given center and axis. Hence the Lenz–Barlotti type of $\overline{\mathcal{M}(s)}$ is at least III.2, compare (24.5). By (34.2), the plane $\overline{\mathcal{M}(s)}$ is isomorphic to $\overline{\mathcal{M}_k}$, and therefore is not desarguesian (31.25b). Our assertion about the exact types follows, because type III (resp. III.2) is maximal among the Lenz (–Barlotti) types of non-desarguesian planes, see Section 24. In particular, there are no linearly transitive axial groups other than those mentioned in assertion (b). The almost simple group Σ' is generated by its one-parameter group E together with its conjugates. This completes the proof of (b).

3) It follows from the last step that the elements a and W are distinguished and are invariant under Γ. Furthermore, Σ is generated by all linearly transitive axial groups, hence Σ is a normal subgroup of Γ. Now $\Gamma = \Sigma$ follows from (34.7c). The group induced on W is (doubly) transitive; it has trivial center, because $\Sigma_{[a,W]}$ is the center of Σ. Hence, we obtain the standard action of $\mathrm{PSL}_2\mathbb{R}$, see (96.30). (This can be verified more directly.) Assertion (c) is a direct consequence of (34.7b). □

34.9 Theorem. *Let Δ be a closed, connected subgroup of the collineation group of a 2-dimensional compact projective plane $\mathcal{P} = (P, \mathcal{L})$ with $\dim \Delta \geq 3$. Assume that Δ fixes precisely one point a and precisely one line W, and that $a \notin W$. Then one of the following holds:*

(a) *$\mathcal{P}$ is isomorphic to the real projective plane $\mathcal{P}_2\mathbb{R}$, and the commutator subgroup Δ' is isomorphic to $\mathrm{SL}_2\mathbb{R}$.*
(b) *$\mathcal{P}$ is isomorphic to a Moulton plane $\overline{\mathcal{M}(s)}$ for some (uniquely determined) value $s > 0$, and Δ' is isomorphic to the universal covering group $\widetilde{\Omega}$ of $\mathrm{SL}_2\mathbb{R}$.*

In both cases we have $\dim \mathrm{Aut}\,\mathcal{P} > 3$.

Proof. Since Δ fixes no point on W, the action of Δ on W is transitive, see (32.15c). The kernel $\Delta_{[W]} = \Delta_{[a,W]}$ has dimension at most 1 by (33.2), hence $\Delta|_W = \Delta/\Delta_{[W]}$ has dimension at least 2, compare (93.7). Brouwer's theorem (96.30) implies that the action of $\Delta|_W$ on W is a (finite) covering of the usual action of $\mathrm{PSL}_2\mathbb{R}$ on the circle $\mathbb{S}_1$.

Furthermore, Δ is a Lie group, see (32.21b) and (94.3a), and the Lie algebra $\mathfrak{l}\Delta$ is isomorphic to $\mathfrak{sl}_2\mathbb{R}$ or to $\mathbb{R} \times \mathfrak{sl}_2\mathbb{R}$, compare (94.28a). The restriction $\Delta' \to \Delta'|_W = \Delta|_W$ is a covering map, and $\pi_1 W \cong \mathbb{Z}$. Thus the kernel $\Delta'_{[W]} = \Delta'_{[a,W]}$ is central (94.2b) and hence cyclic, with generator ζ, say.

1) We set $P_0 = P \setminus (W \cup \{a\})$ and $\mathcal{L}_0 = \mathcal{L} \setminus (\mathcal{L}_a \cup \{W\})$. Then Δ' is flag-transitive on the incidence structure $(P_0, \mathcal{L}_0)$.

The flag space of $(P_0, \mathcal{L}_0)$ is a connected manifold of dimension 3, see (32.3), hence it suffices to show that every flag orbit of Δ' has dimension 3, compare (96.11). In view of the dimension formula (96.10), we have to exclude the possibility that $\dim \Delta'_{p,L} > 0$ for some flag $(p, L) \in P_0 \times \mathcal{L}_0$. Now $\Delta'_{p,L}$ fixes the two points $u = L \wedge W$, $v = ap \wedge W$ on W, and $\dim \Delta'_{u,v} = 1$, hence $\dim \Delta'_{p,L} > 0$ implies that $(\Delta'_{p,L})^1 = (\Delta'_{u,v})^1$. This group fixes the points a, p, u, v. From the stiffness result (32.10) we conclude that all further fixed points of this group lie on the line av, and from (91.1d) we infer that the normalizer of $(\Delta'_{u,v})^1$ fixes the line av, hence both points v and u. This is a contradiction, because either $\Delta'|_W$ has a non-trivial center, which acts freely on W, or $\Delta'|_W \cong \mathrm{PSL}_2\mathbb{R}$ is doubly transitive on W, and in this case the normalizer of $(\Delta'_{u,v})^1$ interchanges u and v.

2) The action of $\Delta'|_W = \Delta|_W$ on W is equivalent to the usual action of $\Omega = \mathrm{PSL}_2\mathbb{R}$ on $\mathbb{S}_1$, and $\Delta'_{[W]} = \langle \zeta \rangle$ is the center of Δ'.

Indeed, assertion 1) implies that $\Delta'|_W$ is doubly transitive on W. Hence this group has trivial center, and is isomorphic to Ω. The action of Ω on W is uniquely determined, because the 2-dimensional subgroups of Ω are precisely the stabilizers of the standard action on the projective line $\mathbb{S}_1$ (for this, one can use the Lie algebra of Ω). Since Ω has trivial center, the kernel $\Delta'_{[W]} = \langle \zeta \rangle$ coincides with the center of Δ'.

3) Next, we describe the stabilizers Δ'_p, Δ'_L for a flag $(p, L) \in P_0 \times \mathcal{L}_0$.

Recall that the 2-dimensional subgroups of $\Omega = \mathrm{PSL}_2\mathbb{R}$ are precisely the stabilizers of elements of $\mathbb{S}_1$. Therefore, the maximal 2-dimensional subgroups of Δ' are precisely the groups Δ'_w with $w \in W$, and we have $\Delta'_w = (\Delta'_w)^1 \times \langle \zeta \rangle$, since $(\Delta'_w)^1 \cong L_2$ has trivial center.

We claim that the unique normal one-parameter subgroup $E_w = (\Delta'_w)'$ of $(\Delta'_w)^1$ consists of elations with axis aw and center w. Otherwise $(\Delta'_w)^1 \cong L_2$ acts effectively on $aw \setminus \{a, w\}$ or on $\mathcal{L}_w \setminus \{aw, W\}$, in fact on both sets, in view of (32.19) and the free action of E_w on $W \setminus \{w\}$. Since L_2 has only one conjugacy class of non-normal subgroups, the non-trivial stabilizer $(\Delta'_w)_{w'}$ of a point $w' \in W \setminus \{w\}$ fixes also some point in $aw \setminus \{a, w\}$ and some line in $\mathcal{L}_w \setminus \{aw, W\}$. This is a contradiction to the stiffness result (32.10).

Let $u = L \wedge W$, $v = ap \wedge W$, and let $\mathsf{M} = (\Delta'_{u,v})^1$. Then $\mathsf{E}_v \leq \Delta'_p \leq \Delta'_v = \mathsf{E}_v \mathsf{M} \times \langle \zeta \rangle$. By step 1), the group Δ'_v is transitive on $av \setminus \{a, v\}$, and the abelian group M acts sharply transitively on each half-line (connected component) of $av \setminus \{a, v\}$. Thus ζ exchanges these two half-lines, in particular $\zeta \neq \mathbb{1}$, and there is a uniquely determined element $\mu \in \mathsf{M}$ with $\mu \zeta^2 \in \Delta'_p$. Since μ fixes u and ζ belongs to $\Delta'_{[W]}$, we have $\mu \zeta^2 \in \Delta'_L$. We obtain

$$\Delta'_p = \mathsf{E}_v \langle \mu \zeta^2 \rangle, \quad \Delta'_L = \mathsf{E}_u \langle \mu \zeta^2 \rangle .$$

As Δ' is doubly transitive on W, the conjugacy class of the pair (Δ'_p, Δ'_L) can be described by choosing two arbitrary distinct groups $\mathsf{E}_v, \mathsf{E}_u$ and by specifying an element $\mu \in \mathsf{M}$, which may depend on the plane $\mathcal{P}$.

4) According to (23.24), the triple $(\Delta', \Delta'_p, \Delta'_L)$ determines the incidence structure $(P_0, \mathcal{L}_0)$, hence the projective plane $\mathcal{P}$ is also determined, compare (44.9). Since $\zeta \neq \mathbb{1}$, we infer from (32.15a) that the order of the homology ζ is either two or infinite. If $\zeta^2 = \mathbb{1}$, then $\mu = \mathbb{1}$ and $\Delta' \cong \mathrm{SL}_2\mathbb{R}$. This triple arises also in the real projective plane $\mathcal{P}_2\mathbb{R}$, whence $\mathcal{P} \cong \mathcal{P}_2\mathbb{R}$ in this case.

If $\zeta^2 \neq \mathbb{1}$, then $\mu \neq \mathbb{1}$ and $\Delta' \cong \widetilde{\Omega}$. The stabilizer Δ'_p is isomorphic to a semi-direct product $\mathbb{R} \rtimes \mathbb{Z}$, with $\mathbb{Z}$ acting non-trivially on $\mathbb{R}$ by multiplication with positive real numbers (as M is connected). The conjugacy class of the pair (Δ'_p, Δ'_L) is uniquely determined by the isomorphism type of Δ'_p, because this isomorphism type determines the automorphism group induced by $\langle \mu \rangle$ on E_v, hence also the set $\{\mu, \mu^{-1}\}$ (note that the automorphism group of E_v is abelian and that we may replace μ and ζ by their inverses). Each of these isomorphism types is realized in some Moulton plane $\overline{\mathcal{M}(s)}$ with $s > 0$, see (34.7a). Hence $\mathcal{P}$ is isomorphic to $\overline{\mathcal{M}(s)}$. The value s is uniquely determined by $\mathcal{P}$, see (34.8c).

The full collineation groups of $\mathcal{P}_2\mathbb{R}$ and $\overline{\mathcal{M}(s)}$ have dimensions 8 and 4, respectively. □

The Moulton planes are the only compact, connected projective planes of Lenz type III; this will be proved in (64.18), using (34.9) as one ingredient.

34.10 Theorem. *Let Σ be the collineation group of a 2-dimensional compact projective plane $\mathcal{P}$. If $\dim \Sigma = 4$, then $\mathcal{P}$ is isomorphic to a Moulton plane $\overline{\mathcal{M}(s)}$ for some (uniquely determined) value $s > 0$.*

Proof. The full collineation group $\mathrm{PGL}_3\mathbb{R}$ of $\mathcal{P}_2\mathbb{R}$ has dimension 8, thus $\mathcal{P} \not\cong \mathcal{P}_2\mathbb{R}$. Both Σ and its connected component Σ^1 have dimension 4, see (93.6), hence (33.9a) implies that Σ^1 fixes precisely one point a and precisely one line W, and that $a \notin W$. The assertion follows from (34.9). □

34.11 Notes on Section 34. The construction of planes described in (34.1) is due to Schellhammer [81] §7; for simplicity, we have specialized this method to differentiable generating functions f. Schellhammer proves that she obtains all 2-dimensional compact projective planes with a collineation group isomorphic to $\mathbb{C}^\times$, compare (38.5). The radial model (34.2) of the Moulton planes is due to Betten [72a], and the isomorphism α is taken from Schellhammer [81] 3.7; see Salzmann [67a] 2.3, 2.4 for yet another model. Results (34.8) and (34.10) were proved in Salzmann [63b] §2, together with further characterizations of the Moulton planes; for (34.9) compare Salzmann [64] §4.

The Moulton planes belong to the few $\mathbb{R}^2$-planes which admit a non-solvable connected collineation group, see Strambach [71]. Furthermore, they admit precisely one conjugacy class of polarities, compare Salzmann [66]. The groups of projectivities of the Moulton planes have been determined explicitly by Betten [79b]; see also Löwen [81e] 1.4, Betten–Wagner [82].

35 Skew hyperbolic planes

This section contains part of the classification of all 2-dimensional compact projective planes whose full collineation groups Σ have dimension 3. Here we consider the case where the connected component $\Sigma^1 \cong \mathrm{PSL}_2\mathbb{R}$ fixes no point and no line, compare (33.9). We prove that the planes in question are skew hyperbolic planes $\mathcal{H}_t$ as defined below, and that Σ^1 acts as the hyperbolic motion group $\mathrm{PO}_3'(\mathbb{R}, 1)$.

35.1 Skew hyperbolic planes. Let f be a quadratic form on $\mathbb{R}^3$ of Witt index 1, say $f(x, y, z) = x^2 + y^2 - z^2$. We denote by $\Omega = \mathrm{PO}_3'(\mathbb{R}, 1) = \mathrm{PO}_3(\mathbb{R}, 1)^1$ the commutator group of the corresponding hyperbolic motion group $\mathrm{PO}_3(\mathbb{R}, 1)$ of the real projective plane $\mathcal{P}_2\mathbb{R} = (P, \mathcal{L})$, compare (15.2) and (13.13). The quadric in $\mathcal{P}_2\mathbb{R}$ defined by f is the point set $Q = \{\, \mathbb{R}v \mid 0 \neq v \in \mathbb{R}^3, f(v) = 0 \,\}$. Choose $t \in \mathbb{R}$ and define a subset $L_t \subset P$ by

$$L_t = \{\, \mathbb{R}(x, 0, 1) \mid -1 \leq x \leq 1 \,\} \cup \{\, \mathbb{R}(x, y, z) \mid y^2 = t^2(x^2 - z^2) \text{ and } txy \geq 0 \,\}.$$

35 Skew hyperbolic planes

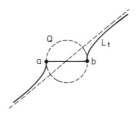

Figure 35a

For $t \neq 0$, the point set L_t consists of a straight line segment and of part of a hyperbola, including the point $\mathbb{R}(1, t, 0)$ at infinity. We define a new incidence structure $\mathcal{H}_t$ as follows: $\mathcal{H}_t = (P, \mathcal{L}_t)$ with

$$\mathcal{L}_t = \{ L \in \mathcal{L} \mid |L \cap Q| \leq 1 \} \cup \{ L_t^\omega \mid \omega \in \Omega \} \ .$$

Thus $\mathcal{H}_t$ has the point space $P = P_2\mathbb{R}$, and the lines of $\mathcal{H}_t$ are the ordinary lines which are disjoint from Q or tangent to Q, together with the images of L_t under Ω.

The action of Ω on Q is equivalent to the natural doubly transitive action of $\mathrm{PSL}_2\mathbb{R}$ on the circle $\mathbb{S}_1$, see (15.6). The points $a = \mathbb{R}(-1, 0, 1)$ and $b = \mathbb{R}(1, 0, 1)$ are the intersection points of L_t and Q, and the group $\Omega_{a,b}$ consists of (i.e., is faithfully induced by) the matrices

$$\begin{pmatrix} \sqrt{1+r^2} & 0 & r \\ 0 & 1 & 0 \\ r & 0 & \sqrt{1+r^2} \end{pmatrix}$$

with $r \in \mathbb{R}$, compare the proof of (13.17). This shows that $L_t \setminus \{a, b\}$ is the union of the two orbits of $\mathbb{R}(1, t, 0)$ and $\mathbb{R}(0, 0, 1)$ under $\Omega_{a,b}$ (or under $\Omega_{\{a,b\}} = \Omega_{a,b} \rtimes \langle \sigma \rangle$, where $\mathbb{R}(x, y, z)^\sigma = \mathbb{R}(-x, -y, z)$).

We have $\mathcal{L}_0 = \mathcal{L}$, hence $\mathcal{H}_0$ is the real projective plane $\mathcal{P}_2\mathbb{R}$. The linear transformation $(x, y, z) \mapsto (x, -y, z) : \mathbb{R}^3 \to \mathbb{R}^3$ induces a reflection $\varrho \in \mathrm{PO}_3(\mathbb{R}, 1)$, hence $Q^\varrho = Q$ and $\Omega^\varrho = \Omega$. Since $(L_t)^\varrho = L_{-t}$, we infer that ϱ is an isomorphism of $\mathcal{H}_t$ onto $\mathcal{H}_{-t}$.

By construction, Ω is a collineation group of $\mathcal{H}_t$. The isomorphism types and the full collineation groups of the *skew hyperbolic plane* $\mathcal{H}_t$ are described in (35.4).

It is a remarkable fact that each $\mathcal{H}_t$ is indeed a projective plane:

35.2 Theorem. $\mathcal{H}_t$ *is a 2-dimensional compact projective plane for every choice of* $t \in \mathbb{R}$.

Proof. Each line of $\mathcal{H}_t = (P, \mathcal{L}_t)$ is homeomorphic to the circle $\mathbb{S}_1$, hence by (32.5) it suffices to prove that $\mathcal{H}_t$ is a projective plane. The complement of each line of $\mathcal{H}_t$ is homeomorphic to $\mathbb{R}^2$ (compare the complement of L_t to that of its asymptote $a \in \mathcal{L}_0$). This implies that any two distinct lines $L, M \in \mathcal{L}_t$ intersect: otherwise

$M \subseteq P \setminus L \approx \mathbb{R}^2$, hence $P \setminus M$ is disconnected by the Jordan curve theorem, a contradiction to $P \setminus M \approx \mathbb{R}^2$.

It remains to show that distinct points $p, q \in P$ are joined by a unique line in $\mathcal{L}_t$. If $p, q \in \{ \mathbb{R}v \mid f(v) \leq 0 \}$, i.e., if neither p nor q is an exterior point of Q, then the ordinary straight line pq intersects Q in two distinct points p', q'. We have observed in (35.1) that L_t is invariant under $\Omega_{\{a,b\}}$. This means that L_t is the only element of $\mathcal{L}_t$ which contains a and b. Since Ω is doubly transitive on Q, we infer that the points p', q', hence also p, q, are contained in a unique line in $\mathcal{L}_t$.

Thus we may now assume that $p = \mathbb{R}v$ is an exterior point of Q, i.e., $f(v) > 0$. We have to prove that the pencil $(\mathcal{L}_t)_p$ yields a partition of $P \setminus \{p\}$. Since Ω is transitive on the exterior points, it suffices to consider the point $p = \mathbb{R}(0, 1, 0)$. Let $X = \{ \mathbb{R}(x, y, 1) \mid |x| < 1 \}$. The ordinary straight lines in $(\mathcal{L}_t)_p$ yield a partition of $P \setminus (X \cup \{p\})$, and it remains to show that X is partitioned by $\mathcal{X} = \{ L_t^\omega \mid \omega \in \Omega, p \in L_t^\omega \}$.

Since the exterior points of L_t form an orbit of $\Omega_{a,b}$, see (35.1), the group Ω is transitive on the flags which consist of an exterior point and of a line L_t^ω with $\omega \in \Omega$. This implies that Ω_p is transitive on $\mathcal{X}$. With a, b and σ as in (35.1), we have $\Omega_p = \Omega_{\{a,b\}} = \Omega_{a,b} \rtimes \langle \sigma \rangle$. Using (35.1), we conclude that $\Omega_{a,b}$ acts sharply transitively on $\mathcal{X}$.

The claim that X is partitioned by $\mathcal{X}$ is therefore tantamount to the assertion that L_t^α intersects each orbit of $\Omega_{a,b}$ in precisely one point, for some (hence every) line $L_t^\alpha \in \mathcal{X}$. To ensure that $L_t^\alpha \subseteq X$, we take α to be an ordinary rotation. The orbits of $\Omega_{a,b}$ in X are half-ellipses H which accumulate at a and at b, together with the open line segment between a and b, hence L_t^α meets each of these orbits.

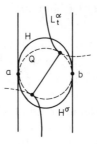

Figure 35b

Clearly $|H \cap L_t^\alpha| = 1$ if H is contained in the interior of Q. If H consists of exterior points, then the complete ellipse $H \cup H^\sigma \cup \{a, b\}$ meets the complete hyperbola determined by L_t^α in (at least) four points. The complete ellipse and the complete hyperbola are distinct conics, hence they cannot have more than four points in common. This shows that $|H \cap L_t^\alpha| = 1$. □

The following theorem characterizes the skew hyperbolic planes by their collineation group $\mathrm{PSL}_2\mathbb{R} \cong \mathrm{PO}'_3(\mathbb{R}, 1)$.

35.3 Theorem. *Let $\Delta \cong \mathrm{PSL}_2\mathbb{R}$ be a closed subgroup of the collineation group of a 2-dimensional compact projective plane $\mathcal{P} = (P, \mathcal{L})$. Then $\mathcal{P}$ is isomorphic to one of the skew hyperbolic planes $\mathcal{H}_t$ with $t \in \mathbb{R}$.*

Proof. 1) First we show that no point and no line is fixed by Δ.

Indeed, assume that Δ fixes some line L. By (32.17) each torus subgroup $\mathbb{T} \cong \mathrm{SO}_2\mathbb{R} < \Delta$ fixes precisely one line, which is the axis of the involution in $\mathbb{T}$. Thus L is the axis of each involution in Δ, hence Δ acts trivially on L, a contradiction to (33.4d).

2) The group Δ has precisely one orbit Q of dimension 1 in P. This orbit Q is a Jordan curve in P, and the action of Δ on Q is equivalent to the usual action of $\mathrm{PSL}_2\mathbb{R}$ on $\mathbb{S}_1$.

In order to prove this, we infer from (32.23) that Δ is not transitive on P. In view of step 1), the orbits of Δ in P have dimension 1 or 2. The 2-dimensional orbits are open (92.14). Since P is connected, we conclude that there exists a one-dimensional orbit Q. By Brouwer's theorem (96.30), $Q \approx \mathbb{S}_1$ is a Jordan curve and Δ acts on Q in the usual way. Note that Q is not contained in any line by step 1).

Assume that Q' is another orbit of dimension 1. The actions of Δ on Q and Q' are equivalent in the strong sense that the two conjugacy classes of stabilizers coincide (96.30), hence the stabilizer Δ_p of a point $p \in Q$ fixes also some point $p' \in Q'$ and acts transitively on $Q \setminus \{p\}$ and on $Q' \setminus \{p'\}$. Thus the line pp' is 'tangent' to both Q and Q', in the sense that $pp' \cap Q = \{p\}$ and $pp' \cap Q' = \{p'\}$. Let $q \in Q \setminus \{p\}$. Then $\Delta_{p,q}$ is a one-dimensional collineation group which fixes a non-degenerate quadrangle p, p', q, q'. This is a contradiction to the stiffness result (32.10).

3) The complement $P \setminus Q$ has precisely two connected components. One of these components, called I, is homeomorphic to $\mathbb{R}^2$, and the other component E is homeomorphic to the Möbius strip. The actions of Δ on I and on E are transitive and equivalent to the usual actions of $\Delta \cong \mathrm{PO}'_3(\mathbb{R}, 1)$ on the interior $\{\mathbb{R}v \mid f(v) < 0\}$ and on the exterior $\{\mathbb{R}v \mid f(v) > 0\}$ of the unit circle, respectively, compare (13.17).

Indeed, let σ be the involution in a torus subgroup $\mathbb{T} \cong \mathrm{SO}_2\mathbb{R}$ of Δ. By (32.17) we know that σ is a reflection and that $\mathbb{T}$ is transitive on the axis A of σ. Since by 1) the Δ-invariant point set Q cannot contain any line, we have $A \cap Q = \emptyset$. The Jordan curve theorem, applied to $Q \subseteq P \setminus A \approx \mathbb{R}^2$, yields that $(P \setminus A) \setminus Q$ has precisley two connected components, hence $P \setminus Q$ has at most two components.

Let c be the center of the reflection σ. Then $\mathbb{T} \leq \Delta_c$, hence $c \notin Q$ in view of step 2). Since $\mathbb{T}$ is a maximal subgroup of Δ (use the Lie algebra, or compare

Brauer [65] or Noll [65]), we infer that $\Delta_c = \mathbb{T}$. Thus the action of Δ on the orbit $I = c^\Delta$ is equivalent to the usual action of $\Delta = \mathrm{PO}_3'(\mathbb{R}, 1)$ on the interior of the unit circle. In particular, I is homeomorphic to $\mathbb{R}^2$.

Now we study the geometry induced on I. Consider two points $a, b \in Q$, and let $M \in \mathcal{L}$ be the line of $\mathcal{P}$ which joins a and b. Then $M \cap I$ is invariant under the group $\Omega_{\{a,b\}}$, hence is a union of orbits of this group. Apart from the straight line segment $\{ \mathbb{R}(x, 0, 1) \mid |x| < 1 \}$ between a and b, the orbits of $\Omega_{\{a,b\}} = \Omega_{a,b} \rtimes \langle \sigma \rangle$ in I are ellipses stretched between a and b, with a and b deleted.

Figure 35c

If M contains such an ellipse, then each image of this ellipse under the rotation group $\mathrm{SO}_2\mathbb{R} \subseteq \Delta$ is also contained in some line of $\mathcal{P}$, a contradiction. Thus $M \cap I$ is the straight line segment between a and b. Since Δ acts doubly transitively on Q, we conclude that M^Δ is the full line space of the real hyperbolic plane (31.2c), and the geometry induced on I is the real hyperbolic plane. The orbit M^Δ is homeomorphic to the Möbius strip; indeed, the line space of the real hyperbolic plane is homeomorphic (via intersection with the unit circle $\mathbb{S}_1$) to the quotient space $\{ (x, y) \in \mathbb{S}_1 \times \mathbb{S}_1 \mid x \neq y \}/(x, y) \sim (y, x)$, which is homeomorphic to the Möbius strip. By dualizing we obtain that P contains a Δ-orbit E which is homeomorphic to the Möbius strip. The action of Δ on E is equivalent to the action of $\mathrm{PO}_3'(\mathbb{R}, 1)$ on the exterior of the unit circle. Now E and I are distinct and open in P, compare (92.14), and Q is the only orbit that is not open, see step 2). We conclude that E and I are the connected components of $P \setminus Q$.

As suggested by 3), we identify the actions of Δ on I and on E with their counterparts in the real projective plane. Since $E \cup I$ is the topological sum of E and I, this identifies the action of Δ on $P \setminus Q = E \cup I$ with the natural action of $\Omega = \mathrm{PO}_3'(\mathbb{R}, 1)$ on the complement $\{ \mathbb{R}v \mid f(v) \neq 0 \}$ of the unit circle. (In fact, the transformation group $(\Delta, E \cup Q \cup I)$ is equivalent to (Ω, P), compare (35.4); here we avoid to discuss the question how the actions on E and on I might be glued together along Q in the sense of Betten–Ostmann [78] Satz 7.)

Now we show that $\mathcal{P} = (P, \mathcal{L})$ and $\mathcal{H}_t = (P, \mathcal{L}_t)$ coincide for some t:

4) We have $\mathcal{L} = \mathcal{L}_t$ for some $t \in \mathbb{R}$.

In $\mathcal{P}_2\mathbb{R}$, each ordinary line L which is disjoint from Q is the axis of an involution in Ω. By (32.12) this involution has an axis also in $\mathcal{P}$, hence L is a line of $\mathcal{P}$.

The Hausdorff limits (31.13) of these lines are the ordinary tangents of Q. These limits belong to $\mathcal{L}$ by (31.14).

Consider the points $a, b \in Q$ as in (35.1) and the line $M \in \mathcal{L}$ of $\mathcal{P}$ which joins a and b. As shown in step 3), $M \cap I$ is the straight line segment between a and b. In order to determine $M \cap E$, we observe that M intersects the tangents at Q in a and b transversally, see (31.5b), hence M contains some point $\mathbb{R}(x, y, 1)$ with $|x| > 1$. This point is contained in a unique orbit $\mathbb{R}(1, t, 0)^{\Omega_{a,b}}$ with $t \in \mathbb{R}$; indeed, t is determined by $y^2 = t^2(x^2 - 1)$ and $txy \geq 0$, compare (35.1). Since M is invariant under $\Omega_{a,b}$ we obtain that $L_t \subseteq M$. By (35.2) the images of L_t under Ω yield the projective plane $\mathcal{H}_t$, hence $L_t = M$ and $\mathcal{L} = \mathcal{L}_t$.

By (32.9), every collineation between $\mathcal{P}$ and $\mathcal{H}_t$ is automatically a homeomorphism. □

35.4 Proposition. *We have $\mathcal{H}_0 \cong \mathcal{P}_2\mathbb{R}$ and $\mathcal{H}_t \cong \mathcal{H}_{-t}$ for every $t \in \mathbb{R}$. The planes $\mathcal{H}_t$ with $t \geq 0$ are mutually non-isomorphic. Furthermore $\Omega = \mathrm{PO}'_3(\mathbb{R}, 1) \cong \mathrm{PSL}_2\mathbb{R}$ is the full collineation group of $\mathcal{H}_t$ for $t \neq 0$.*

Proof. For the first assertion see (35.1). The full collineation group $\Sigma = \Sigma_t$ of $\mathcal{H}_t$ contains the connected group Ω of dimension 3. If the connected component Σ^1 is larger than Ω, then $\dim \Sigma \geq \dim \Sigma^1 > 3$, compare (93.12), and $\dim \Sigma \neq 4$ by (33.9a), hence $\mathcal{H}_t \cong \mathcal{P}_2\mathbb{R}$ and $\Sigma = \mathrm{PGL}_3\mathbb{R}$ by (33.6). Thus we know that either $\Sigma^1 = \Omega$, or $\mathcal{H}_t \cong \mathcal{P}_2\mathbb{R}$ and $\Sigma^1 = \mathrm{PGL}_3\mathbb{R}$.

Let φ be an isomorphism of $\mathcal{H}_{t'} = (P, \mathcal{L}_{t'})$ onto $\mathcal{H}_t = (P, \mathcal{L}_t)$. Then Ω^φ is a collineation group of $\mathcal{H}_t$. Since $\mathrm{PGL}_3\mathbb{R}$ contains only one conjugacy class of subgroups isomorphic to Ω, compare (95.10), we have $\Omega^\varphi = \Omega^\sigma$ for some $\sigma \in (\Sigma_t)^1$. Now Ω has trivial centralizer in the homeomorphism group of P, as each point p of P is the unique fixed point of Ω_p. Hence the normalizer of Ω in the homeomorphism group of P is the group $\Omega\langle\varrho\rangle = \mathrm{PO}_3(\mathbb{R}, 1) \cong \mathrm{Aut}\,\Omega \cong \mathrm{PGL}_2\mathbb{R}$, where $\varrho : \mathbb{R}(x, y, z) \mapsto \mathbb{R}(x, -y, z)$, compare Dieudonné [71] IV §6. Thus $\varphi\sigma^{-1} \in \Omega$ or $\varphi\sigma^{-1} \in \Omega\varrho$, and this gives $t' = t$ or $t' = -t$.

Therefore, the planes $\mathcal{H}_t$ with $t \geq 0$ are mutually non-isomorphic. Let $t > 0$. Then $\mathcal{H}_t \not\cong \mathcal{H}_0 = \mathcal{P}_2\mathbb{R}$, hence the first paragraph shows that $(\Sigma_t)^1 = \Omega$. Moreover, every automorphism φ of $\mathcal{H}_t$ belongs to $(\Sigma_t)^1$. □

35.5 Notes on Section 35. The results of this section can be found in Salzmann [62b], see also [63a], [67b] 5.3. The skew hyperbolic planes $\mathcal{H}_t$ with $t \neq 0$ admit (precisely one Ω-conjugacy class of) polarities, see Bedürftig [74a] 3.4.

36 Skew parabola planes

In this section we continue the classification of all 2-dimensional compact projective planes whose full collineation groups Σ have dimension 3. Here we deal with the case where the connected component Σ^1 fixes precisely one point p and precisely one line W, compare (33.9). From (34.9) we infer that (p, W) is a flag, i.e., $p \in W$. The planes arising in this situation are shift planes which are generated by skew parabolae, see (36.1 and 2).

36.1 Skew parabola planes. Let $c, d \in \mathbb{R}$ with $c > 0$ and $d > 1$. We define the 'skew parabola' $f : \mathbb{R} \to \mathbb{R}$ by

$$f(x) = \begin{cases} x^d & \text{for } x \geq 0 \\ c(-x)^d & \text{for } x \leq 0 \end{cases}.$$

The derivative $f' : \mathbb{R} \to \mathbb{R}$ is a homeomorphism, hence f generates a shift plane, see (31.25c). We denote this (affine) shift plane by $\mathcal{E}_{c,d}$. Thus $\mathcal{E}_{c,d}$ has the point set $\mathbb{R}^2$, and the lines of $\mathcal{E}_{c,d}$ are the vertical lines $\{x\} \times \mathbb{R}$ and the images of the graph of f under the shift group $\{(x, y) \mapsto (x+a, y+b) \mid a, b \in \mathbb{R}\}$. We denote by $\overline{\mathcal{E}}_{c,d}$ the projective completion of $\mathcal{E}_{c,d}$; this is a 2-dimensional compact projective plane by (32.2).

The special case $(c, d) = (1, 2)$ gives the parabola model of the real affine plane (31.2a), hence $\overline{\mathcal{E}}_{1,2} \cong \mathcal{P}_2\mathbb{R}$.

The fixed configuration of the shift group is just the flag (v, W), where W is the line at infinity of $\mathcal{E}_{c,d}$ and v is the point at infinity of the vertical axis. The 3-dimensional collineation group $\{(x, y) \mapsto (rx + a, r^d y + b) \mid a, b, r \in \mathbb{R}, r > 0\}$ of $\mathcal{E}_{c,d}$ has the same fixed configuration. Thus the planes $\overline{\mathcal{E}}_{c,d}$ admit collineation groups of the type considered in this section.

For $c \leq 0$ or $d \leq 1$, the incidence structure $\mathcal{E}_{c,d}$ is not an affine plane; in fact, it is easily seen that in these cases suitable translates of the graph of f intersect in more than one point.

Note that $\mathcal{E}_{c,d} \cong \mathcal{E}_{c^{-1},d}$ via the map $(x, y) \mapsto (-x, c^{-1}y)$. In particular, the planes $\mathcal{E}_{c,d}$ with $0 < c \leq 1 < d$ represent all isomorphism types. In fact, this reduction yields a family of pairwise non-isomorphic representatives; see (36.3), where also the full collineation groups of the planes $\overline{\mathcal{E}}_{c,d}$ are determined. See also Section 74 for further details; in particular, 4-dimensional analogues of these skew parabola planes are described in (74.4b).

36.2 Theorem. *Let Σ be the full collineation group of a 2-dimensional compact projective plane $\mathcal{P}$. Assume that the connected component Σ^1 has dimension 3 and fixes precisely one flag (v, W) of $\mathcal{P}$. Then $\mathcal{P}$ is isomorphic to some plane $\overline{\mathcal{E}}_{c,d}$ with $0 < c \leq 1 < d$, $(c, d) \neq (1, 2)$. Conversely, all these planes satisfy our hypotheses.*

Proof. Let $\mathcal{P} = (P, \mathcal{L})$ and $\Delta = \Sigma^1$. We shall repeatedly use the dimension formula (96.10).

1) The actions of $\Delta|_W$ on $W \setminus \{v\}$ and of $\Delta|_{\mathcal{L}_v}$ on $\mathcal{L}_v \setminus \{W\}$ are both equivalent to the natural transitive action of L_2 on $\mathbb{R}$, and $\Delta_{[W]} = \Delta_{[v,W]} = \Delta_{[v]} \cong \mathbb{R}$ is a linearly transitive translation group.

Indeed, by (33.5c) the groups $\Delta_{[W]}$ and $\Delta_{[v]}$ have dimension at most 1. Since Δ acts transitively on $W \setminus \{v\}$ and on $\mathcal{L}_v \setminus \{W\}$, see (32.15c), and since Δ is solvable by (33.8), we infer from Brouwer's theorem (96.30) that these actions are equivalent to the natural action of L_2. The kernels $\Delta_{[W]}$, $\Delta_{[v]}$ are one-dimensional subgroups which are connected, because L_2 is simply connected (94.4a). As Δ fixes the universal center (32.19) of its normal subgroup $\Delta_{[W]}$, we infer that $\Delta_{[W]} = \Delta_{[v,W]} = \Delta_{[v]}$. This group of translations is isomorphic to $\mathbb{R}$ and is linearly transitive by (32.15c and 19).

2) We can identify the affine point set $P \setminus W$ with $\mathbb{R}^2$ in such a way that $\Delta_{[v,W]} = \{(x, y) \mapsto (x, y + b) \mid b \in \mathbb{R}\}$ and that $\mathsf{A} = \{(x, y) \mapsto (x + a, y + b) \mid a, b \in \mathbb{R}\}$ is a normal subgroup of Δ. Observe that A is sharply transitive on $P \setminus W$ and on $\mathcal{L} \setminus \mathcal{L}_v$.

Assertion 2) can be proved as follows. From 1) we infer that $\mathsf{A} = \Delta_{[v,W]}\Delta'$ is a 2-dimensional, connected normal subgroup which is transitive on $P \setminus W$ and on $\mathcal{L} \setminus \mathcal{L}_v$. Note that $\Delta_{[v,W]} \subseteq \Delta' = \mathsf{A}$ if $\dim \Delta' = 2$. Since Δ' acts trivially on $\Delta_{[v,W]} \cong \mathbb{R}$, we infer that A is abelian, hence A is sharply transitive on $P \setminus W$ and on $\mathcal{L} \setminus \mathcal{L}_v$, and $\mathsf{A} \cong \mathbb{R}^2$, compare (96.9a) and (94.38c).

3) Write $o = (0, 0)$. The stabilizer Δ_o fixes ov, hence it fixes also some point $u \in W \setminus \{v\}$, because L_2 has only one conjugacy class of non-normal closed subgroups of dimension 1, compare (96.30). The non-trivial orbits of $\Delta_{[v,W]}$ are the vertical lines $\{x\} \times \mathbb{R}$, and the affine line $L = ou \setminus \{u\}$ intersects each of them, hence $L = \{(x, f(x)) \mid x \in \mathbb{R}\}$ for some function $f : \mathbb{R} \to \mathbb{R}$ with $f(0) = 0$. By 2) we have $\mathcal{L} \setminus \mathcal{L}_v = L^{\mathsf{A}} = \{L_{a,b} \mid a, b \in \mathbb{R}\}$, where $L_{a,b} = \{(x, f(x - a) + b) \mid x \in \mathbb{R}\}$. This means that the affine plane $\mathcal{P}^W$ is the shift plane generated by f.

4) We claim that f or $-f$ is strictly convex, in the sense that the graph L of f intersects each ordinary affine line G in $\mathbb{R}^2$ in at most two points (compare (31.25c)). Indeed, if $|L \cap G| \geq 3$, then a suitable parallel of G meets L in three points such that the distances of the outer points from the middle one are equal. Now the vector that moves the middle point to any of the outer ones translates L to a line of $\mathcal{P}^W$ that intersects L in at least two points, a contradiction.

We have $\Delta = \mathsf{A} \rtimes \Delta_o$, and the stabilizer Δ_o fixes the affine line $Y := ov \setminus \{v\} = \{o\} \times \mathbb{R}$. Hence Δ_o is a one-parameter subgroup of the stabilizer of Y in $\mathrm{GL}_2\mathbb{R}$. In view of the action of Δ_o on $\mathcal{L}_v$, the set $L \setminus \{0\}$ is the union of the Δ_o-orbits of $(1, f(1))$ and $(-1, f(-1))$. Each of these two orbits is a continuously differentiable curve, and by convexity the values $f'(x)$ converge to a finite limit for $x \to 0$, $x > 0$.

The corresponding tangents converge to a one-dimensional subspace of $\mathbb{R}^2$, which is distinct from Y and invariant under Δ_0.

5) There exist $c, d \in \mathbb{R}$ with $0 < c \le 1 < d$ such that $\mathcal{P} \cong \overline{\mathcal{E}}_{c,d}$.

As we have seen at the end of step 4), there is a basis of $\mathbb{R}^2$ such that Δ_o is given by diagonal matrices. Since Y is one of the eigen spaces, we need only change the first vector of the basis used in step 2). Since Δ_o is a one-parameter group, there exists some $d \in \mathbb{R}$ such that these matrices have the form $\mathrm{diag}(r, r^d)$. We obtain that

$$f(x) = \begin{cases} x^d f(1) & \text{for } x \ge 0 \\ (-x)^d f(-1) & \text{for } x \le 0 \end{cases};$$

note that $f(1)$ and $f(-1)$ may have been changed by our change of basis. Up to a transformation $(x, y) \mapsto (-x, y)$ we have that $f(1) \ne 0$, because $f(1) = f(-1) = 0 = f(0)$ would imply that the lines L and $L+(1,0)$ have the two points $(0,0)$ and $(1,0)$ in common. Now the transformation $(x, y) \mapsto (x, f(1)^{-1}y)$ yields that L is described as graph of the function

$$\tilde{f}(x) = \begin{cases} x^d & \text{for } x \ge 0 \\ c(-x)^d & \text{for } x \le 0, \end{cases}$$

where $c = f(1)^{-1} f(-1)$. Note that $c > 0$ and $d > 1$ in view of the fact that $c \le 0$ or $d \le 1$ does not yield a shift plane (36.1). Finally, we may use the isomorphism $\overline{\mathcal{E}}_{c,d} \cong \overline{\mathcal{E}}_{c^{-1},d}$ in order to accomplish $c \le 1$.

We prove in (36.3) that the full collineation groups of the planes $\overline{\mathcal{E}}_{c,d}$ have the stated properties. □

36.3 Proposition. *The planes $\overline{\mathcal{E}}_{c,d}$ with $0 < c \le 1 < d$ are mutually non-isomorphic. Let $(c, d) \ne (1, 2)$. Then the full collineation group Σ of $\overline{\mathcal{E}}_{c,d}$ has dimension 3, and the connected component Σ^1, acting on the points of $\overline{\mathcal{E}}_{c,d}$, is the group*

$$\left\{ (x, y) \mapsto (rx + a, r^d y + b) \mid a, b, r \in \mathbb{R}, r > 0 \right\}.$$

Furthermore $\Sigma = \Sigma^1$ for $c \ne 1$, and $\Sigma = \Sigma^1 \langle (x, y) \mapsto (-x, y) \rangle$ for $c = 1$.

Proof. We have observed in (36.1) that $\Delta = \left\{ (x, y) \mapsto (rx + a, r^d y + b) \mid r > 0 \right\}$ is a subgroup of Σ^1. If $\Delta \ne \Sigma^1$, then $\dim \Sigma \ge \dim \Sigma^1 > \dim \Delta = 3$, compare (93.12). By (33.9a) we have $\dim \Sigma \ne 4$, and then (33.6) implies that $\overline{\mathcal{E}}_{c,d} \cong \mathcal{P}_2\mathbb{R}$. We have to show that $(c, d) = (1, 2)$. Let v and W be as in (36.1). There exists a reflection ϱ with center v and axis $L = \{ (x, f(x)) \mid x \in \mathbb{R} \}$, where $f(x) = x^d$ for $x \ge 0$ and $f(x) = c(-x)^d$ for $x \le 0$. Now ϱ normalizes the group $\mathrm{PGL}_3\mathbb{R}_{[v,W]} = \Delta_{[v,W]}$, acting on this group by inversion. Fix $(x, y) \in \mathbb{R}^2$, and write τ_b for the translation $(x', y') \mapsto (x', y' + b)$, where $b = y - f(x)$. Then

$(x, y)^\varrho = (x, f(x))^{\tau_b \varrho} = (x, f(x))^{\varrho \tau_b \varrho} = (x, f(x))^{\tau_{-b}} = (x, 2f(x) - y)$; this holds for all $x, y \in \mathbb{R}$. Two lines $L_{a,b} = \{ (x, f(x-a) + b) \mid x \in \mathbb{R} \}$ and $L_{a',b'}$ are parallel precisely if $a = a'$. The reflection ϱ inverts also the commutator group of $((\mathrm{PGL}_3\mathbb{R})_{v,W})|_W$, which is the group $\Delta'|_W$, hence $(L_{a,b})^\varrho = L_{-a,b'}$ for some $b' \in \mathbb{R}$. In particular, we have $(L_{1,0})^\varrho = L_{-1,b'}$ for some b', and this implies that $f(x+1) + f(x-1) - 2f(x) = -b'$ does not depend on x. Putting $x = 0, 1, -1$ gives $c + 1 = 2^d - 2 = c(2^d - 2)$, hence $c = 1$ and $d = 2$.

Thus we have shown that $\Sigma^1 = \Delta$ for $(c, d) \neq (1, 2)$. This implies that (v, W) is the fixed configuration of Σ.

Let φ be an isomorphism of $\overline{\mathscr{E}}_{c,d}$ onto $\overline{\mathscr{E}}_{c',d'}$, where (c, d) and (c', d') are distinct from $(1, 2)$. Since the groups $\Delta = \Sigma^1$ are not isomorphic for different values of d, we conclude that $d = d'$. We may assume that φ fixes $o = (0, 0)$. As φ normalizes the shift group $\Delta' = (\Sigma^1)'$, we have then $\varphi \in \mathrm{GL}_2\mathbb{R}$. Furthermore, φ normalizes also the group $\Delta_o = (\Sigma^1)_o$ and fixes the line ov, hence $(x, y)^\varphi = (sx, ty)$ for suitable elements $s, t \in \mathbb{R}^\times$. Now φ maps

$$L = \{ (x, x^d) \mid x \geq 0 \} \cup \{ (x, c(-x)^d) \mid x \leq 0 \}$$

onto

$$L' = \{ (x, x^d) \mid x \geq 0 \} \cup \{ (x, c'(-x)^d) \mid x \leq 0 \},$$

since W, ov and L (resp. L') are the only lines of $\overline{\mathscr{E}}_{c,d}$ (resp. $\overline{\mathscr{E}}_{c',d'}$) which are fixed by $\Delta_o = (\Sigma^1)_o$. If $s > 0$, then $t = s^d$, which implies that $\varphi \in \Delta_o$ and that $c = c'$. If $s < 0$, then $cc' = 1$, hence $c = c' = 1$, and $(x, y)^\varphi = (sx, |s|^d y)$, thus $\varphi \in \Delta_o \langle (x, y) \mapsto (-x, y) \rangle$. Note that $(x, y) \mapsto (-x, y)$ is indeed an automorphism of $\overline{\mathscr{E}}_{1,d}$.

Thus the planes $\overline{\mathscr{E}}_{c,d}$ are mutually non-isomorphic, and the automorphism groups of $\overline{\mathscr{E}}_{c,d}$ are as asserted. $\square$

36.4 Notes on Section 36. The results of this section are due to Salzmann [65], [67b] 5.16f; other proofs are given in Groh [82a] 3.1. The skew parabola planes $\overline{\mathscr{E}}_{c,d}$ admit polarities, see (38.8), Salzmann [65] p. 260, Bedürftig [74a] IIIb.

37 Planes over Cartesian fields

This section concludes the classification of all 2-dimensional compact projective planes whose full collineation group Σ has dimension 3. By (33.9) and in view of the results of Sections 35 and 36, it remains to consider the case where the connected component Σ^1 fixes precisely two points and two lines. We show that

$\Sigma^1 \cong \mathbb{R} \times L_2$ and that the planes in question are coordinatized by special Cartesian fields.

37.1 Some actions. Via $(s, t) \mapsto \begin{pmatrix} e^s & 0 \\ t & 1 \end{pmatrix}$ we identify the set $\mathbb{R}^2$ with the group L_2, see (94.40). The group operation on $\mathbb{R}^2$ becomes $(s', t')(s, t) = (s'+s, t+e^s t')$. A natural action of this group on $\mathbb{R}$ is given by $x^{(s,t)} = e^s x + t$ for $x \in \mathbb{R}$, $(s, t) \in L_2$. The direct product $\mathbb{R} \times L_2$ plays an important rôle in this section. For $\lambda \in \mathbb{R}$, the λ-action of the group $\mathbb{R} \times L_2$ on $\mathbb{R}^2$ is defined by

$$(x, y)^{(r,s,t)} = \begin{cases} (e^r x, e^s y + t) & \text{for } x \geq 0 \\ (e^{r+\lambda s} x, e^s y + t) & \text{for } x \leq 0, \end{cases}$$

where $(x, y) \in \mathbb{R}^2$, $(r, s, t) \in \mathbb{R} \times L_2$. Note that all λ-actions with $\lambda \neq 0$ are equivalent to the 1-action; indeed, conjugation of the λ-action by the homeomorphism h defined by $(x, y)^h = (\text{sign}(\lambda x)|x|^{|\lambda|}, y)$ gives

$$(x, y)^{h(r,s,t)h^{-1}} = \begin{cases} (e^{\tilde{r}} x, e^s y + t) & \text{for } x \geq 0 \\ (e^{\tilde{r}+s} x, e^s y + t) & \text{for } x \leq 0 \end{cases}$$

with $\tilde{r} = r/\lambda$ for $\lambda > 0$ and $\tilde{r} = r/|\lambda| - s$ for $\lambda < 0$, and the mapping $(r, s, t) \mapsto (\tilde{r}, s, t)$ is an automorphism of $\mathbb{R} \times L_2$. The 0-action and the 1-action are not equivalent; this may be proved by observing that groups of different dimensions are induced on the set $\{\{x\} \times \mathbb{R} \mid x \in \mathbb{R}\}$ of orbits of the commutator subgroup $\{(x, y) \mapsto (x, y + t) \mid t \in \mathbb{R}\}$, compare also (37.5). We do have a reason for introducing many actions that are equivalent to the 1-action, see (37.7).

37.2 Proposition. *Let Δ be a closed, connected subgroup of the collineation group of a 2-dimensional compact projective plane $\mathcal{P} = (P, \mathcal{L})$. Assume that Δ has dimension 3 and fixes precisely two points u, v and precisely two lines W, Y, with $W = uv$ and $v = W \wedge Y$. Then Δ is isomorphic to $\mathbb{R} \times L_2$, and the action of Δ on $P \setminus W$ is equivalent to the λ-action on $\mathbb{R}^2$ for some $\lambda \in \{0, 1\}$. Furthermore $\Delta_{[u,Y]}$ is the center of Δ, and the commutator subgroup $\Delta' = \Delta_{[v,W]}$ is linearly transitive.*

Proof. Let $Y' = Y \setminus \{v\}$.

1) The action of $\Delta|_Y$ on Y' is equivalent to the natural transitive action of L_2 on $\mathbb{R}$, and $\Delta_{[Y]} = \Delta_{[u,Y]} \cong \mathbb{R}$ is the center of Δ.

Indeed, Δ fixes no point of Y' and is therefore transitive on Y' by (32.15c). The kernel $\Delta_{[Y]} = \Delta_{[u,Y]}$ has dimension at most 1, see (33.2). By (33.8) the group Δ is solvable. From (93.7) and from Brouwer's Theorem (96.30) we infer that $\Delta|_Y \cong L_2$, acting naturally on $Y' \approx \mathbb{R}$, and that $\dim \Delta_{[Y]} = 1$. Now $\Delta/\Delta_{[Y]} \cong L_2$ is simply connected, hence $\Delta_{[Y]}$ is connected by (94.4a). This implies that $\Delta_{[Y]} \cong \mathbb{R}$, see (32.19).

The center of Δ is contained in $\Delta_{[Y]}$, in view of $\Delta/\Delta_{[Y]} \cong L_2$. By the transitivity of Δ' on Y' we have $\Delta = \Delta' \Delta_y$ for each $y \in Y'$, compare (91.2a). The group $\Delta_y = \Delta_{u,v,y}$ is 2-dimensional (96.10) and connected (94.4a), and therefore abelian by (33.10). Thus $\Delta_{[Y]} \subseteq \Delta_y$ is centralized by Δ_y. Furthermore Δ' acts trivially on the normal subgroup $\Delta_{[Y]}$, hence $\Delta_{[Y]}$ is contained in the center of $\Delta' \Delta_y = \Delta$.

2) The group $\Delta_{[Y]} = \Delta_{[u,Y]}$ has precisely two orbits $\mathcal{L}_v^+, \mathcal{L}_v^-$ on $\mathcal{L}_v \setminus \{W, Y\}$, and the action of $\Delta_{[Y]}$ on $\mathcal{L}_v$ is equivalent to the action of $\mathbb{R}$ on $\mathbb{S}_1 = \mathbb{R} \cup \{\infty\}$ defined by $x \mapsto e^s x, \infty \mapsto \infty$ for $s \in \mathbb{R}$.

This may be proved as follows. The axial group $\Delta_{[Y]} = \Delta_{[u,Y]}$ fixes no line in $\mathcal{L}_v \setminus \{W, Y\}$ and is therefore transitive on each of the two connected components $\mathcal{L}_v^{\pm} \approx \mathbb{R}$ of $\mathcal{L}_v \setminus \{W, Y\}$, see (32.15c). Since $\Delta_{[Y]} \cong \mathbb{R}$, the actions on $\mathcal{L}_v^{\pm}$ are effective and sharply transitive.

The action of $\Delta_{[Y]}$ on $\mathcal{L}_v$ is either as asserted, or equivalent to the action defined by $x \mapsto e^{\pm s} x$, where $\pm$ is the sign of x. For every homology $\delta \in \Delta_{[Y]} = \Delta_{[u,Y]}$ and for points $p, q \notin W \cup Y$, we have $q^{\delta} = (pq \wedge Y) p^{\delta} \wedge qu$, provided that p, q, u are not collinear. If δ varies in $\Delta_{[Y]}$ such that p^{δ} converges to $pu \wedge Y$, then q^{δ} converges to $qu \wedge Y$ for every point $q \notin W$. This property of groups of homologies excludes the second action mentioned above, compare also (61.3) and the proof of (44.8b).

3) Let $L \in \mathcal{L}_v \setminus \{W, Y\}$. Then $\Delta_L \cong L_2$ and $\Delta = \Delta_L \times \Delta_{[Y]} \cong L_2 \times \mathbb{R}$. Furthermore, we have that $\Delta' = \Delta_{[v,W]} \cong \mathbb{R}$ is linearly transitive.

The connected group Δ leaves $\mathcal{L}_v^+$ and $\mathcal{L}_v^-$ invariant, hence assertion 2) implies that $\Delta = \Delta_L \Delta_{[Y]}$, compare (91.2a). The axial group $\Delta_{[Y]}$ has trivial intersection with Δ_L, hence $\Delta = \Delta_L \times \Delta_{[Y]}$ and $\Delta_L \cong L_2$ by assertion 1). Now Δ_L fixes all lines in the component $L^{\Delta_{[Y]}} = \mathcal{L}_v^{\pm}$, hence $\Delta|_{\mathcal{L}_v^{\pm}} = \Delta_{[Y]}|_{\mathcal{L}_v^{\pm}}$ is abelian. This implies that Δ' acts trivially on $\mathcal{L}_v$ and (by duality) on W, thus $\Delta' = (\Delta_L)' \cong \mathbb{R}$ consists of translations with center v and axis W. By (32.15c), this group is linearly transitive.

4) The preceding steps show that we can introduce coordinates in the plane $\mathcal{P}$ with respect to a quadrangle o, e, u, v, where $o \in Y'$, in such a way that $\mathcal{L}_v^+ = \{\{x\} \times \mathbb{R} \mid x > 0\}$, $\mathcal{L}_v^- = \{\{x\} \times \mathbb{R} \mid x < 0\}$,

$$\Delta_{[u,Y]} = \{(x, y) \mapsto (e^r x, y) \mid r \in \mathbb{R}\} \quad \text{and}$$
$$\Delta' = \{(x, y) \mapsto (x, y + t) \mid t \in \mathbb{R}\}.$$

Let L be the line joining the points $(1, 0)$ and v. From $\Delta_L = \Delta' \rtimes \Delta_{(1,0)}$, compare step 3), we infer that the stabilizer $\Delta_{(1,0)}$ is isomorphic to $\mathbb{R}$. Furthermore, all points $(x, 0)$ with $x > 0$ are fixed by $\Delta_{(1,0)}$, see step 2). As shown in step 3), the group induced by Δ on $\mathcal{L}_v^-$ is abelian. Applying (33.12) to the actions of Δ' on L and of $\Delta_{[u,Y]}$ on ou, we find that $\Delta_{(1,0)}$ consists of all transformations

$$(x, y) \mapsto \begin{cases} (x, by) & \text{for } x \geq 0 \\ (b'x, by) & \text{for } x \leq 0 \end{cases}$$

with $b > 0$, where $b' = b^\lambda$ for some $\lambda \in \mathbb{R}$. Since

$$\Delta = \Delta_{[Y]} \times \Delta_L = \Delta_{[Y]} \times (\Delta' \rtimes \Delta_{(1,0)})$$

we have obtained the λ-action of $\Delta \cong \mathbb{R} \times L_2$. As explained in (37.1), we can achieve that $\lambda \in \{0, 1\}$ (without changing the actions of Δ' or of $\Delta_{[u,Y]}$). □

37.3 Construction of the planes $\mathcal{P}_{\alpha,\beta,c}$. For real parameters $\alpha, \beta, \gamma, \bar{\gamma}, c > 0$, we define a new multiplication $*$ on $\mathbb{R}$ by

$$m * x = \begin{cases} mx & \text{for } m, x \geq 0 \\ m^\alpha x & \text{for } m \geq 0, x \leq 0 \\ mx^\beta & \text{for } m \leq 0, x \geq 0 \\ |m|^\gamma c |x|^{\bar{\gamma}} & \text{for } m, x \leq 0. \end{cases}$$

It is easy to verify that $\mathbb{R}(\alpha, \beta, \gamma, \bar{\gamma}, c) = (\mathbb{R}, +, *)$ is a Cartesian field (24.4) with continuous multiplication. The affine plane coordinatized by $\mathbb{R}(\alpha, \beta, \gamma, \bar{\gamma}, c)$, compare (22.3), is an $\mathbb{R}^2$-plane, and the projective completion of this affine plane is a 2-dimensional compact projective plane, see (32.2).

Consider the special cases where $\alpha^{-1} + \beta^{-1} > 1$, $\gamma = \alpha\kappa$ and $\bar{\gamma} = \beta\kappa$ with $\kappa = (\alpha + \beta - \alpha\beta)^{-1} > 0$. We denote the projective plane coordinatized by such a Cartesian field $\mathbb{R}(\alpha, \beta, \alpha\kappa, \beta\kappa, c)$ by $\mathcal{P}_{\alpha,\beta,c}$. The conditions on $\alpha, \beta, \gamma, \bar{\gamma}$ imply that $\beta\gamma = \alpha\bar{\gamma}$ and $1 - \bar{\gamma} = (1 - \beta)\gamma$, and these equations show that the affine plane over $\mathbb{R}(\alpha, \beta, \gamma, \bar{\gamma}, c)$ admits the collineations

$$(x, y) \mapsto (a(b * x), ay + t) = \begin{cases} (abx, ay + t) & \text{for } x \geq 0 \\ (ab^\alpha x, ay + t) & \text{for } x \leq 0 \end{cases}$$

with $a, b > 0$, $t \in \mathbb{R}$. These collineations form a group isomorphic to $\mathbb{R} \times L_2$ which fixes just the points (0) and (∞) at infinity, the y-axis $\{0\} \times \mathbb{R}$ and the line W at infinity. The $(\alpha^{-1} - 1)$-action of $\mathbb{R} \times L_2$ defined in (37.1) is equivalent to (but for $\alpha \neq 1$ not identical with) the action of this collineation group; this can be seen using the homeomorphism

$$(x, y) \mapsto \begin{cases} (x, y) & \text{for } x \geq 0 \\ (-|x|^\alpha, y) & \text{for } x \leq 0. \end{cases}$$

Obviously, $\mathcal{P}_{1,1,1}$ is the real projective plane, and $\mathcal{P}_{1,1,c}$ is the projective completion of the Moulton plane $\mathcal{M}_c$, compare (31.25b), (34.2). Thus the 0-action of

$\mathbb{R} \times L_2$ occurs in the real projective plane. On the other hand, for $\alpha \neq 1$, the group induced by $\mathbb{R} \times L_2$ on the pencil $\mathcal{L}_{(\infty)}$ (or on W) in the plane $\mathcal{P}_{\alpha,\beta,c}$ is isomorphic to $\mathbb{R}^2$, which is not a subgroup of $\mathrm{PGL}_2\mathbb{R}$; indeed, the collineations $(x, y) \mapsto (a(a^{-1} * x), ay)$ fix all lines $\{x\} \times \mathbb{R}$ with $x \geq 0$.

37.4 Theorem. *Let $\mathcal{P} = (P, \mathcal{L})$ be a 2-dimensional compact projective plane. Assume that the full collineation group Σ of $\mathcal{P}$ has dimension 3 and fixes precisely two points u, v and precisely two lines W, Y. Then $\mathcal{P}$ is isomorphic to some plane $\mathcal{P}_{\alpha,\beta,c}$ with $0 < \alpha, \beta \leq 1$, $(\alpha, \beta) \neq (1, 1)$, $0 < c$, and $c \leq 1$ if $1 \in \{\alpha, \beta\}$. Conversely, all these planes satisfy our hypotheses.*

Proof. We may assume that $W = uv$ and $v = W \wedge Y$. By (37.2) the action of $\Delta = \Sigma^1$ on $P \setminus W = \mathbb{R}^2$ may be written as a λ-action with $\lambda \in \{0, 1\}$.

Since $\Delta' = \{(x, y) \mapsto (x, y + t) \mid t \in \mathbb{R}\}$ is (v, W)-transitive, the sets $\{x\} \times \mathbb{R}$ with $x \in \mathbb{R}$ are the lines through v of the affine plane $\mathcal{P}^W$. Furthermore the center $\Delta_{[u,Y]}$ of Δ consists of all transformations $(x, y) \mapsto (e^r x, y)$ with $r \in \mathbb{R}$, hence the lines of $\mathcal{P}^W$ through u are precisely the sets $\mathbb{R} \times \{y\}$ with $y \in \mathbb{R}$.

Let $L \in \mathcal{L}_{(0,0)}$ be an affine line with $L \neq X, Y$, where $X = \mathbb{R} \times \{0\}$. Then $\dim \Delta_L = \dim \Delta - \dim L^\Delta \geq 1$ by (96.10), thus Δ_L contains a one-parameter subgroup $\mathrm{P}(L) \cong \mathbb{R}$ of $\Delta_{(0,0)} = \{(r, s, 0) \mid r, s \in \mathbb{R}\} \cong \mathbb{R}^2$. Since $\mathrm{P}(L)$ is different from the center of Δ, we have $\mathrm{P}(L) = \{(\delta r, r, 0) \mid r \in \mathbb{R}\}$ for some $\delta \in \mathbb{R}^\times$, where

$$(x, y)^{(\delta r, r, 0)} = \begin{cases} (e^{\delta r} x, e^r y) & \text{for } x \geq 0 \\ (e^{\delta r + \lambda r} x, e^r y) & \text{for } x \leq 0. \end{cases}$$

Let $L = L^+ = (0, 0) \vee (1, 1)$. Then $(-1, -a) \in L^+$ for some $a > 0$, thus

$$L^+ = (1, 1)^{\mathrm{P}(L^+)} \cup \{(0, 0)\} \cup (-1, -a)^{\mathrm{P}(L^+)}$$
$$= \{(x^\delta, x) \mid x > 0\} \cup \{(0, 0)\} \cup \{(-x^{\delta+\lambda}, -ax) \mid x > 0\}.$$

Note that $\delta > 0$, as L^+ is homeomorphic to $\mathbb{R}$. Similarly, with $L = L^- = (0, 0) \vee (1, -1)$, we have $(-1, b) \in L^-$ for some $b > 0$, and with a suitable $\varepsilon > 0$ we obtain

$$L^- = \{(x^\varepsilon, -x) \mid x \geq 0\} \cup \{(-x^{\varepsilon+\lambda}, bx) \mid x \geq 0\}.$$

Now $\Delta_{(0,0)}$ contains all transformations $(x, y) \mapsto (mx, y)$ with $m > 0$. The images of L^+, L^- under these transformations together with X, Y are all the lines through $(0, 0)$, and the lines through $(0, t)$ can be obtained by applying the collineation $(x, y) \mapsto (x, y + t)$.

We have described the line pencil through $(0, 0)$ in terms of parameters $a, b, \delta, \varepsilon$. The transformation $(x, y) \mapsto (x, -y)$ exchanges δ with ε and a with b. In the special case $\lambda = 0$, the transformation $(x, y) \mapsto (-x, -y)$ replaces a, b by a^{-1}, b^{-1} without changing δ, ε. Both transformations normalize the λ-action of Δ. Thus we may assume that $\delta \leq \varepsilon$, and if $\delta = \varepsilon$ or $\lambda = 0$, then we may further assume that $b \leq a$.

One can directly verify that the bijection

$$\varphi : (x, y) \mapsto \begin{cases} (x^\delta, y) & \text{for } x \geq 0 \\ (-|x|^{\delta+\lambda}, ay) & \text{for } x \leq 0 \end{cases}$$

is an isomorphism of the affine plane over $\mathbb{R}(\alpha, \beta, \gamma, \bar\gamma, c)$ onto $\mathcal{P}^W$, where $\alpha = \delta/(\delta + \lambda) \leq 1$, $\beta = \delta/\varepsilon \leq 1$, $\gamma = \varepsilon/(\varepsilon + \lambda)$, $\bar\gamma = (\delta + \lambda)/(\varepsilon + \lambda)$ and $c = b/a$; note that $c \leq 1$ if $\alpha = 1$ or $\beta = 1$, and that $\gamma = \alpha\kappa$ and $\bar\gamma = \beta\kappa$ with $\kappa^{-1} = \alpha + \beta - \alpha\beta$. The isomorphism $\varphi : \mathcal{P}_{\alpha,\beta,c} \to \mathcal{P}$ transforms the $\lambda\delta^{-1}$-action into the λ-action of Δ, and $\lambda\delta^{-1} = \alpha^{-1} - 1$. The Moulton planes $\mathcal{P}_{1,1,c}$ and the real projective plane $\mathcal{P}_{1,1,1}$ are excluded, since they have larger collineation groups (of dimension 4 and 8, respectively).

The fact that the planes $\mathcal{P}_{\alpha,\beta,c}$ satisfy our hypotheses will be proved in (37.6). □

In order to determine the full collineation groups and the isomorphism types of the planes $\mathcal{P}_{\alpha,\beta,c}$, we need to know the normalizers of $\Delta = \mathbb{R} \times L_2$.

37.5 Lemma. *Let Γ be the group of all homeomorphisms of $\mathbb{R}^2$ onto itself, let $\alpha > 0$, and let*

$$\Delta = \left\{ (x, y) \mapsto \begin{cases} (abx, ay + t) & \text{for } x \geq 0 \\ (ab^\alpha x, ay + t) & \text{for } x \leq 0 \end{cases} \middle| a, b > 0, t \in \mathbb{R} \right\}$$

be the collineation group described in (37.3).

(a) *If $\alpha = 1$, then the normalizer of Δ in Γ is generated by Δ together with the following transformations:*

$$(x, y) \mapsto (-x, y),$$
$$(x, y) \mapsto (x, -y) \quad \text{and}$$
$$(x, y) \mapsto \begin{cases} (ax^s, y) & \text{for } x \geq 0 \\ (-b|x|^s, y) & \text{for } x \leq 0 \end{cases}$$

where $a, b, s > 0$.

(b) *If $\alpha \neq 1$, then the normalizer of Δ in Γ is generated by Δ together with the following transformations:*

$$(x, y) \mapsto (x, -y) \quad \text{and}$$
$$(x, y) \mapsto \begin{cases} (ax, y) & \text{for } x \geq 0 \\ (bx, y) & \text{for } x \leq 0 \end{cases}$$

where $a, b > 0$.

Proof. Clearly these transformations are contained in the normalizer N of Δ. The center of Δ consists of all maps $(x, y) \mapsto \begin{cases} (bx, y) & \text{for } x \geq 0 \\ (b^\alpha x, y) & \text{for } x \leq 0 \end{cases}$ with $b > 0$. As N normalizes the center of Δ, the set $\{0\} \times \mathbb{R}$ of fixed points of the center is invariant under N. Furthermore N normalizes the commutator subgroup $\Delta' = \{(x, y) \mapsto (x, y+t) \mid t \in \mathbb{R}\}$, hence by (33.12) we have $N = K\Delta\langle\varrho\rangle$, where K is the kernel of the action of N on $\{0\} \times \mathbb{R}$ and ϱ is defined by $(x, y)^\varrho = (x, -y)$. Now N permutes the orbits of the center of Δ, hence each set $\mathbb{R} \times \{y\}$ with $y \in \mathbb{R}$ is invariant under K. Lemma (33.12), applied to each orbit of the center, shows that every element of K has the form

$$(x, y) \mapsto \begin{cases} (\pm ax^s, y) & \text{for } x \geq 0 \\ (\mp b|x|^{\bar{s}}, y) & \text{for } x \leq 0 \end{cases}$$

with $a, b, s, \bar{s} > 0$. Such a transformation normalizes the center of Δ if, and only if, $s = \bar{s}$. This proves (a). In case (b), the fact that K normalizes the stabilizer $\Delta_{(0,0)}$ implies that $s = 1$ and that only the upper choice of signs is possible. □

37.6 Proposition. *Let $0 < \alpha, \beta \leq 1$, $(\alpha, \beta) \neq (1, 1)$, $0 < c$, and $c \leq 1$ if $1 \in \{\alpha, \beta\}$. Then the planes $\mathcal{P}_{\alpha,\beta,c}$ are mutually non-isomorphic, and the planes $\mathcal{P}_{\alpha,\beta,c}$ and $\mathcal{P}_{\beta,\alpha,c}$ are dual to each other. The full collineation group Σ of $\mathcal{P}_{\alpha,\beta,c}$ has dimension 3 and fixes $u = (0)$, $v = (\infty)$, $W = [\infty]$ and $Y = [0]$. The connected component Σ^1 is isomorphic to $\mathbb{R} \times L_2$, and its action on the point set of the affine plane $\mathcal{P}^W_{\alpha,\beta,c}$ is equivalent to the 0-action or to the 1-action of $\mathbb{R} \times L_2$, for $\alpha = 1$ or $\alpha \neq 1$ respectively. All these planes have Lenz type II.*

(a) *If $1 \notin \{\alpha, \beta\}$ or if $c \neq 1$, then $\Sigma = \Sigma^1$ is connected.*
(b) *If $1 \in \{\alpha, \beta\}$ and $c = 1$, then $\Sigma = \Sigma^1 \rtimes \langle\sigma\rangle$, where σ is the reflection of $\mathcal{P}_{\alpha,\beta,c}$ which is defined by $(x, y)^\sigma = (-x, -y)$ for $\alpha = 1$ and by $(x, y)^\sigma = (x, -y)$ for $\beta = 1$.*

Proof. Such a plane $\mathcal{P}_{\alpha,\beta,c}$ is not isomorphic to the real projective plane $\mathcal{P}_2\mathbb{R}$, because the corresponding Cartesian field $\mathbb{R}(\alpha, \beta, \alpha\kappa, \beta\kappa, c)$ defined in (37.3) is not isomorphic to the field $\mathbb{R}$ (e.g., no distributive law holds). Furthermore, we have observed in (37.3) that Σ contains a subgroup $\Delta \cong \mathbb{R} \times L_2$, hence dim $\Sigma \geq 3$. If dim $\Sigma > 3$, then $\mathcal{P}_{\alpha,\beta,c}$ is isomorphic to a Moulton plane $\mathcal{P}_{1,1,d}$ by (34.10), and (u, Y) is the unique fixed anti-flag of the Moulton plane (note that $d \neq 1$). By (34.8b) the plane $\mathcal{P}_{\alpha,\beta,c}$ is (u, Y)-homogeneous, hence the multiplication $*$ defined in (37.3) is associative, compare Pickert [75] 3.45 p. 102. This implies that $\alpha = \beta = 1$, a contradiction.

Thus dim $\Sigma = 3$. We infer that $\Sigma^1 = \Delta$, compare (93.12), hence Σ fixes u, v, W, Y (e.g., W is the only line carrying two fixed points of Δ). Since Σ normalizes Δ, the full group Σ is obtained using (37.5).

The planes $\mathcal{P}_{\alpha,\beta,c}$ and $\mathcal{P}_{\beta,\alpha,c}$ are dual to each other, because the corresponding Cartesian fields have opposite multiplications $*$ and $*'$; indeed, the mappings $(a,b) \mapsto [a,-b]' = \{(x, a *' x - b)' \mid x \in \mathbb{R}\}$ and $[s,t] \mapsto (s,-t)'$ yield an isomorphism of $\mathcal{P}_{\alpha,\beta,c}$ onto the dual of $\mathcal{P}_{\beta,\alpha,c}$.

Now we show that the planes $\mathcal{P}_{\alpha,\beta,c}$ are mutually non-isomorphic. The connected components Δ, $\overline{\Delta}$ of the collineation groups of $\mathcal{P}_{\alpha,\beta,c}$ and $\mathcal{P}_{\bar{\alpha},\bar{\beta},\bar{c}}$ are conjugate under some homeomorphism φ of the form

$$(x,y) \mapsto \begin{cases} (x,y) & \text{for } x \geq 0 \\ (-|x|^s, y) & \text{for } x \leq 0 \end{cases}$$

with $s > 0$. Thus every isomorphism $\mathcal{P}_{\bar{\alpha},\bar{\beta},\bar{c}} \to \mathcal{P}_{\alpha,\beta,c}$ is a product $\varphi\psi$ where ψ normalizes Δ, and (37.5) yields $(\alpha,\beta,c) = (\bar{\alpha},\bar{\beta},\bar{c})$. □

37.7 Remark on parameters. The affine plane $\mathcal{P}_{\alpha,\beta,c}^W$ admits the λ-action of $\mathbb{R} \times L_2$, where $\lambda = \alpha^{-1} - 1$, and all λ-actions with $\lambda \neq 0$ are equivalent (37.1). However, one can show that the actions of $\Sigma^1 \cong \mathbb{R} \times L_2$ on the point set of the *projective* planes $\mathcal{P}_{\alpha,\beta,c}$ are mutually inequivalent for different values of $\alpha \leq 1$; the parameter α describes how the actions of Σ^1 on the affine point set $\mathbb{R}^2$ and on the line W at infinity are glued together (in the sense of Betten–Ostmann [78] Satz 7). Since $\mathcal{P}_{\alpha,\beta,c}$ is dual to $\mathcal{P}_{\beta,\alpha,c}$, the parameter β plays an analogous rôle for the action on the line space of $\mathcal{P}_{\alpha,\beta,c}$.

The parameters α, β have also some geometric significance: the dimension of the group $\Sigma_{[v]}$ is 2 or 1, for $\alpha = 1$ or $\alpha < 1$, respectively; the same relation holds for $\dim \Sigma_{[W]}$ and β.

37.8 Notes on Section 37. This section is a variation of Salzmann [64] §2, [67b] 5.11–5.13; his parameters k, ϱ, σ for the Cartesian fields in (37.3) are our parameters $c, \alpha^{-1}, \beta^{-1}$. Compare also Groh [81] §§ 5–7, [82a] 3.2. The planes $\mathcal{P}_{\alpha,\beta,c}$ with $\alpha = \beta$ admit polarities, see Bedürftig [74a] 3.1.

38 Flexibility, rigidity and related topics

We conclude Chapter 3 by summarizing the classification results obtained in Sections 33 through 37. Furthermore, we mention some related results and concepts.

38.1 Classification Theorem. *Let Σ be the full collineation group of a 2-dimensional compact projective plane $\mathcal{P}$.*

(a) *If $\dim \Sigma \geq 5$, then $\mathcal{P}$ is isomorphic to the real projective plane $\mathcal{P}_2\mathbb{R}$.*
(b) *If $\dim \Sigma = 4$, then $\mathcal{P}$ is isomorphic to a Moulton plane $\overline{\mathcal{M}_k}$.*

(c) *If* dim $\Sigma = 3$, *then* $\mathcal{P}$ *is isomorphic to a skew hyperbolic plane* $\mathcal{H}_t$ *or to a skew parabola plane* $\overline{\mathcal{E}}_{c,d}$ *or to a plane* $\mathcal{P}_{\alpha,\beta,c}$ *over a Cartesian field.*

For proofs see (33.6 and 9), (34.2 and 10), (35.3), (36.2), (37.4) (and use the fact (93.6) that dim Σ = dim Σ^1). From the knowledge of the collineation groups of these planes one can derive the following two results, compare (94.21).

38.2 Corollary. *Let Δ be a closed, connected subgroup of the collineation group of a 2-dimensional compact projective plane $\mathcal{P}$. Then the following conditions are equivalent:*

(a) dim $\Delta \geq 3$, *and Δ fixes no point and no line of $\mathcal{P}$.*
(b) Δ *is simple as an abstract group.*

38.3 Corollary. *Let Δ be a closed, connected subgroup of the collineation group of a 2-dimensional compact projective plane $\mathcal{P}$. Assume that Δ is almost simple. Then we have one of the following cases:*

(a) $\mathcal{P}$ *is isomorphic to the real projective plane* $\mathcal{P}_2\mathbb{R}$ *and Δ is isomorphic to* $SO_3\mathbb{R}$, $SL_2\mathbb{R}$, $PSL_2\mathbb{R}$ *or* $PGL_3\mathbb{R}$.
(b) $\mathcal{P}$ *is isomorphic to a skew hyperbolic plane* $\mathcal{H}_t$ *with* $t > 0$ *and* $\Delta \cong PSL_2\mathbb{R}$.
(c) $\mathcal{P}$ *is isomorphic to a Moulton plane* $\overline{\mathcal{M}_k}$ *with* $k > 1$ *and Δ is isomorphic to the universal covering group of* $PSL_2\mathbb{R}$.

In fact, one can prove these corollaries more directly, without relying on the full classification carried out in Sections 33 through 37: disregarding the case $\mathcal{P} \cong \mathcal{P}_2\mathbb{R}$, where one can use the tables (95.10), Corollary (38.2) is a consequence of (33.6 and 9), (35.3), and Corollary (38.3) follows from (33.6, 8 and 9), (34.9), (35.3). Compare also Salzmann [63a], [67b] 5.2.

The results (33.1), (33.8) and (38.2) may be regarded as geometric analogues of the theorem of Lie which relates the solvability of linear groups with the existence of invariant subspaces.

We say that a 2-dimensional compact projective plane $\mathcal{P} = (P, \mathcal{L})$ is *flexible* if its collineation group Σ has an open orbit in the space $\{(p, L) \in P \times \mathcal{L} \mid p \in L\}$ of all flags. Since the flag space has dimension 3 according to (32.3), this implies that dim $\Sigma \geq 3$, compare (96.10). Inspection of the planes appearing in (38.1) leads to the following result:

38.4 Theorem. *A 2-dimensional compact projective plane is flexible precisely if its collineation group has dimension at least 3.*

Of course, this can be proved more directly, see Salzmann [65] Satz 2, [67b] 4.2. See (63.11) for remarks on flexible planes of higher dimensions.

As a historical remark, we mention that the concept of flexibility is related to free mobility, which was investigated by Sophus Lie in his work on the (two-dimensional case of the) Helmholtz–Lie space problem. Indeed, the results in Lie [1890] §1, [1893] §97 may be regarded as forerunners of (38.2) and (38.4).

The classification (38.1) has been extended by Groh, Lippert, Pohl, and Schellhammer, with the following result (see Pohl [90] and the references given there):

38.5 Theorem. *All 2-dimensional compact projective planes $\mathcal{P} = (P, \mathcal{L})$ whose full collineation groups have dimension at least 2 can be completely described in terms of continuous functions with special properties; in fact, most of these planes are 'pasting sums' of arc planes (31.25d) of various special types. It turns out that $\dim \operatorname{Aut} \mathcal{P} \geq 2$ holds precisely if $\operatorname{Aut} \mathcal{P}$ has an open orbit in P (and in $\mathcal{L}$).*

The arc planes appearing here are constructed from a (finite or countably infinite) family of fairly arbitrarily chosen continuous functions (arcs in $\mathbb{R}^2$ or in the half-plane), thus they form a rather large class of planes (including all shift planes), in contrast to the planes that appear in (38.1).

The proof of (38.5) has to deal with the following cases. It suffices to consider a 2-dimensional closed, connected subgroup Δ of $\operatorname{Aut} \mathcal{P}$, use (38.1), or see Groh [79] p. 172 for the existence of such a subgroup. Now Δ is isomorphic to $\mathbb{R}^2$, $\mathbb{R} \times \mathbb{S}_1 \cong \mathbb{C}^\times$, or L_2 by (32.18), (94.39 and 40). Furthermore Δ fixes some point and some line, compare Groh [76] 2.9. The case where $\Delta \cong \mathbb{C}^\times$ was settled by Schellhammer [81] and leads to planes which can be constructed by a generalization of (34.1). If $\Delta \cong \mathbb{R}^2$, then up to duality there exists an open point orbit, compare (55.42), and this point orbit is an $\mathbb{R}^2$-plane. The possible configurations of fixed elements of $\Delta \cong \mathbb{R}^2$ have been described in Groh [77]. The $\mathbb{R}^2$-planes that are induced on the open point orbit have been classified by Groh [79, 82b]; each of these $\mathbb{R}^2$-planes can be generated by a suitable family of arcs in $\mathbb{R}^2$. The case $\Delta \cong L_2$ is more complicated, but leads to similar results (one has to consider arcs in the half-plane). For these results, see Groh [79, 82b], Schellhammer [81], Lippert [86], and Pohl [90]. In most cases, the projective plane $\mathcal{P}$ is a 'pasting sum' (in the sense of Groh [81]) of arc planes.

Another type of homogeneity is assumed in the following result of Salzmann [62a] 5.5, compare also Salzmann [67b] §6:

38.6 Theorem. *If every point of a 2-dimensional compact projective plane $\mathcal{P}$ is the center of a non-trivial axial collineation, then $\mathcal{P}$ is isomorphic to the real projective plane $\mathcal{P}_2 \mathbb{R}$.*

38.7 Planes with small automorphism groups. There are several construction methods which show that the class of all (isomorphism types of) 2-dimensional compact projective planes is excessively large. Apart from the arc planes of Groh (which include the shift planes), we mention the 'Hilbert and Beltrami line systems', see Stroppel [93d] and the references given there; see also the references in Salzmann [62a] 1.2.

Some of these methods allow to construct planes whose full collineation groups are rather small, compare Anisov [92]. In fact, one can construct *rigid* planes, i.e., planes with trivial collineation group, see Steinke [85]; for this, it suffices to bend lines of $\mathcal{P}_2\mathbb{R}$ in such a way that some quadrangle is fixed by all collineations. Otte [93] constructs many examples of differentiable planes by locally disturbing the line system of $\mathcal{P}_2\mathbb{R}$ (or of the other classical planes $\mathcal{P}_2\mathbb{C}$, $\mathcal{P}_2\mathbb{H}$, $\mathcal{P}_2\mathbb{O}$).

38.8 Polarities and ovals. The Moulton planes, the skew hyperbolic planes, the skew parabola planes $\overline{\mathcal{E}}_{c,d}$, and some of the planes $\mathcal{P}_{\alpha,\beta,c}$ from Section 37 admit polarities (with absolute points), compare Bedürftig [74a]. The 2-dimensional compact projective planes which admit many polarities have been characterized by Bedürftig [74b].

If a polarity of a 2-dimensional compact projective plane $\mathcal{P}$ has absolute points, then these absolute points form an oval in $\mathcal{P}$, see Salzmann [62a] 5.8, Bedürftig [74a] 2.2. Large classes of examples are provided by planes over Cartesian fields with commutative multiplication (like $\mathcal{P}_{\alpha,\alpha,c}$), in particular, by shift planes (31.25c), like $\overline{\mathcal{E}}_{c,d}$, see also (74.5). Indeed, in the shift plane $\mathcal{P}_f$ generated by a function $f : \mathbb{R} \to \mathbb{R}$, the map

$$(a, b) \mapsto L_{-a,-b} = \{ (x, f(x + a) - b) \mid x \in \mathbb{R} \}$$

yields a polarity, and $\{ (x, y) \mid 2y = f(2x) \} \cup \{(\infty)\}$ is the corresponding oval of absolute points, compare Dembowski–Ostrom [68], Knarr [83], Szőnyi [90]. Furthermore, $\{ (x, -f(-x)) \mid x \in \mathbb{R} \} \cup \{(\infty)\}$ is also an oval in $\mathcal{P}_f$, and if $f(x) = f(-x)$ for all $x \in \mathbb{R}$, then this oval consists of the absolute points of the polarity $(a, b) \mapsto L_{a,-b}$. In addition, every ordinary straight line which is not parallel to the y-axis in $\mathbb{R}^2$ becomes an oval of $\mathcal{P}_f$ by adjoining the point (∞) at infinity of the y-axis.

Chapter 4

Compact projective planes

In Chapter 1 we studied the four classical planes over $\mathbb{R}$, $\mathbb{C}$, $\mathbb{H}$ and $\mathbb{O}$, and in Chapter 3 we considered planes which are (at least topologically) closely related to the real projective plane. Now we begin a systematic investigation of arbitrary compact projective planes.

In Section 41 we introduce the notion of a topological projective plane, and we derive some elementary topological properties which hold for arbitrary topological planes. Somewhat deeper results require stronger assumptions on the topology. We show that locally compact projective planes are metrizable and second countable. In Section 42 we add the assumption of connectedness. Locally compact, connected projective planes are in fact compact. We determine all compact, connected Moufang planes: these are precisely the planes over the classical division algebras $\mathbb{R}$, $\mathbb{C}$, $\mathbb{H}$ and $\mathbb{O}$. Thus the planes from Chapter 1 acquire a prominent position in the general theory of compact projective planes. Compact, connected projective planes have topological dimension at least two. In the two-dimensional case, the point space is homeomorphic to the 'classical' point space of the real projective plane. Hence the projective planes from Chapter 3 are precisely the compact, connected planes of minimal topological dimension, and the results from Chapter 3 hold with formally weaker assumptions, compare (42.11). Section 43 deals with the algebraic equivalent of projective planes, that is, with ternary fields. Many compact planes have been constructed by describing suitable topological ternary fields. Results of Skornjakov and Knarr–Weigand guarantee that every ternary field defined on $\mathbb{R}^n$ with continuous ternary operation coordinatizes a compact, connected projective plane. The group Σ of all automorphisms (i.e. continuous collineations) of a compact, connected projective plane is the predominant theme of the whole book. In Section 44 we show that Σ is a locally compact transformation group. Deeper results on Σ can be found in Section 55 and in Chapter 6.

41 The topology of locally compact planes

We introduce the basic notion of a topological projective plane, and we derive some elementary (topological) properties of these planes.

41.1 Definition. A *topological projective plane* is a projective plane $\mathcal{P} = (P, \mathcal{L}, \mathbf{F})$ with topologies on P and $\mathcal{L}$ such that the two geometric operations of joining distinct points and of intersecting distinct lines are continuous; furthermore we stipulate that the topologies on P and $\mathcal{L}$ are neither discrete nor indiscrete.

Our continuity assumptions say that the two mappings

$$(p, q) \mapsto pq = p \vee q : \{ (p, q) \in P^2 \mid p \neq q \} \to \mathcal{L} \quad \text{and}$$
$$(L, M) \mapsto L \wedge M : \{ (L, M) \in \mathcal{L}^2 \mid L \neq M \} \to P$$

are continuous; of course, P^2 and $\mathcal{L}^2$ are endowed with the product topologies.

We say that a topological projective plane $\mathcal{P} = (P, \mathcal{L}, \mathbf{F})$ has a topological property (like (local) compactness, connectedness etc.) if the point space P has this property.

If K is any topological (not necessarily commutative) field, then $\mathcal{P}_2 K$ is a topological projective plane, see (14.4) or Pickert [75] p. 265. By Kiltinen [73], every infinite commutative field K has very many field topologies, hence $\mathcal{P}_2 K$ carries many structures of a topological plane. On the other hand, there exist projective planes which cannot be made into topological projective planes, see Szambien [89] Prop. 1.

We point out that the concept of a topological projective plane is self-dual, and we usually give only one of a pair of dual statements.

Whenever convenient, we identify a line $L \in \mathcal{L}$ with its point row

$$\{ p \in P \mid (p, L) \in \mathbf{F} \}.$$

Occasionally we denote a projective plane just by $(P, \mathcal{L})$, without mentioning the set $\mathbf{F}$ of flags.

41.2 Proposition. *Let $\mathcal{P} = (P, \mathcal{L}, \mathbf{F})$ be a topological projective plane.*

(a) *If $(p, L) \in P \times \mathcal{L} \setminus \mathbf{F}$, then the central projection*

$$x \mapsto (x \vee p) \wedge L : P \setminus \{p\} \to L$$

is continuous. All perspectivities and projectivities, see (21.6), are homeomorphisms. In particular, any two lines, considered as point sets, are homeomorphic, and lines and pencils of lines are homeomorphic. Furthermore, each line

is triply homogeneous (i.e. the homeomorphism group of each line is triply transitive).
(b) *Each point and each line (considered as a point set) is closed in P.*
(c) *Joining $\vee$ of distinct points and intersecting $\wedge$ of distinct lines are open mappings.*
(d) *The point space of every affine part of $\mathcal{P}$ is homeomorphic to the product of two affine lines.*
(e) *P has the weak topology obtained from any three affine parts which cover P.*

Proof. (a) is an immediate consequence of Definition (41.1) and (21.6).

(b) Since P is neither discrete nor indiscrete, and since lines are doubly homogeneous, at least one (and hence every) line is a T_1-space. Since any two points are joined by a line, we infer that P is a T_1-space, i.e. points are closed. The complement of a line in P is the preimage of an affine line (i.e. a line minus a point q) under a suitable central projection, therefore this complement is open in $P \setminus \{p\}$ and in P. Thus lines are closed.

(c) Let A, B be open subsets of P, and $L = a \vee b$ with distinct points $a \in A, b \in B$. Choose distinct lines $M \in \mathcal{L}_a, N \in \mathcal{L}_b$ with $M \wedge N \notin A \cap B$ (we may assume that $A \cap B \neq P$). Then $\{ x \vee y \mid x \in A, y \in B, x \neq y \}$ contains the set

$$\{ X \in \mathcal{L} \setminus \{M, N\} \mid X \wedge M \in A \cap M, X \wedge N \in B \cap N \},$$

and the latter set contains L and is open in $\mathcal{L}$, since $A \cap M$ is open in M.

(d) For distinct points u, v on the line W, the map $p \mapsto (p \vee u, p \vee v)$ is a homeomorphism of $P \setminus W$ onto $(\mathcal{L}_u \setminus \{W\}) \times (\mathcal{L}_v \setminus \{W\})$, and $\mathcal{L}_u \setminus \{W\}$ is homeomorphic to an affine line.

(e) The affine point sets are open, hence one can apply Dugundji [66] III.9.3 and Ex. 2, p. 131. □

41.3 Definitions. A *topological affine plane* is an affine plane $(A, \mathcal{L}, \mathbf{F})$ with topologies on A and on $\mathcal{L}$ such that joining distinct points, intersecting non-parallel lines, and drawing the parallel line are continuous operations; again we stipulate that the topologies on A and on $\mathcal{L}$ are neither discrete nor indiscrete. Drawing the parallel line is the mapping $(p, L) \mapsto M : A \times \mathcal{L} \to \mathcal{L}$, where $M \in \mathcal{L}_p$ is parallel to L.

A *topological ternary field* is a ternary field (K, τ) with a topology on K which is neither discrete nor indiscrete such that $\tau : K^3 \to K$ and all the 'inverse operations' t, x, s in (22.2) are continuous (on their respective domains). Every topological ternary field (K, τ) gives rise to a topological double loop $(K, +, \cdot)$, defined by $x + y = \tau(1, x, y)$ and $x \cdot y = \tau(x, y, 0)$. In particular, the division operations

\ and / defined in (22.4) are continuous; for more details on topological double loops see Grundhöfer–Salzmann [90].

Let W be any line of a topological projective plane $\mathcal{P}$. Then $\mathcal{P}^W$ is a topological affine plane; drawing the parallel line is the mapping $(p, L) \mapsto p \vee (L \wedge W)$. Coordinatization of a topological affine plane yields a topological ternary field, because the ternary operation τ and its inverse operations have (affine) geometric interpretations, see (22.1). For example, $(x, \tau(s, x, t)) = L \wedge [x]$, where $L = [s, t] = (0, t) \vee (s)$ is the line parallel to $(0, 0) \vee (1, s)$ through $(0, t)$.

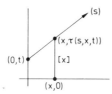

Figure 41a

The converse of this relationship is not true in general: Eisele [92b] constructs topological semifields which do not belong to topological affine planes (the trouble is caused by the lines of infinite slope). Furthermore, Eisele [90, 91a,b, 93a,b] has examples of topological affine planes which cannot be extended to *topological* projective planes. For locally compact, connected planes and ternary fields there are positive results, see (43.6 and 7).

For topological affine planes in general, a better algebraic description results if one uses two ternary operations instead of one; the second operation is obtained by exchanging the rôles of the coordinate axes. This leads to the notion of a topological biternary field, which perfectly matches the notion of a topological affine plane. See Baker–Lane–Lorimer [88], where this is carried out in the more general context of affine Hjelmslev planes.

Ternary fields (or double loops) may help to prove results about the topology of topological planes:

41.4 Proposition. *Let $\mathcal{P} = (P, \mathcal{L}, \mathbf{F})$ be a topological projective plane. Then P and $\mathcal{L}$ are regular (Hausdorff) spaces.*

Proof. First we show that each affine line $K = L \setminus \{p\}$, p a point on $L \in \mathcal{L}$, is regular. We use the topological loop $(K, +)$ obtained from a suitable ternary operation, see (22.1), (41.3). Let U be any neighbourhood of 0 in K. Since the subtraction defined by $(x - y) + y = x$ is continuous, we find a neighbourhood V of 0 with $V - V \subseteq U$. If $x \in \overline{V}$, then the neighbourhood $x + V$ of x intersects V, hence $x \in V - V \subseteq U$, which shows that $\overline{V} \subseteq U$. By homogeneity, K is regular (cp. Dugundji [66] VII.2.2(2)).

By (41.2d) the point set of each affine part of $\mathcal{P}$ is regular. Let U be a neighbourhood of a point $p \in P$. We may assume that $U \cap L_i = \emptyset$ for some triangle with lines L_1, L_2, L_3. There exist neighbourhoods V_i of p in $P \setminus L_i$ such that U contains the closure of V_i in $P \setminus L_i$. The closure of V_i in P is contained in $U \cup L_i$, hence the closure of the neighbourhood $V_1 \cap V_2 \cap V_3$ of p is contained in U. □

Proposition (41.4) shows that it does not really matter here if we include the Hausdorff property in our concept of compactness or not. As in the case of loops, it is an open problem if (41.4) can be replaced by the stronger assertion that P and $\mathcal{L}$ are completely regular.

41.5 Corollary. *Let $(P, \mathcal{L}, \mathbf{F})$ be a topological projective plane. Then the following assertions hold.*

(a) *The flag space $\mathbf{F}$ is closed in $P \times \mathcal{L}$.*
(b) *The projections $\mathbf{F} \to P$ and $\mathbf{F} \to \mathcal{L}$ are open maps.*

Proof. (a) Let $(p, L) \in (P \times \mathcal{L}) \setminus \mathbf{F}$, and choose distinct lines $L_1, L_2 \in \mathcal{L}_p$. By (41.4), the two lines L, L_1 have disjoint neighbourhoods V, V_1, and the three points $p, L \wedge L_1, L \wedge L_2$ have mutually disjoint neighbourhoods U, U_1, U_2 such that $U_1 \vee U_2 \subseteq V$ and $U \vee U_1 \subseteq V_1$. Then $U \times (U_1 \vee U_2)$ is an open neighbourhood of (p, L) by (41.2c), and this neighbourhood contains no flag, since the existence of a flag $(q, b_1 \vee b_2)$ with $q \in U$, $b_i \in U_i$ leads to the contradiction $b_1 \vee b_2 = q \vee b_1 \in V \cap V_1 = \emptyset$. Hence $(P \times \mathcal{L}) \setminus \mathbf{F}$ is open in $P \times \mathcal{L}$.

(b) Let $\mathbf{U} \subseteq \mathbf{F}$ be an open neighbourhood of the flag (p, L). By duality, it suffices to show that $\mathbf{U}$ projects onto a neighbourhood of p in P. Choose a point $q \in L \setminus \{p\}$. There exists an open neighbourhood U of p in $P \setminus \{q\}$ such that the set $\{(u, u \vee q) \mid u \in U\}$ is contained in $\mathbf{U}$. This set projects onto U. □

41.6 Corollary. *Let $(P_0, \mathcal{L}_0, \mathbf{F}_0)$ be a subplane of a topological projective plane $(P, \mathcal{L}, \mathbf{F})$. Then the topological closure $(\overline{P_0}, \overline{\mathcal{L}_0}, \overline{\mathbf{F}_0})$ is again a subplane.*

Proof. For distinct points $p_i \in \overline{P_0}$, the line $p_1 p_2$ is contained in the closure of the set $\{q_1 q_2 \mid q_i \in P_0, q_1 \neq q_2\} \subseteq \mathcal{L}_0$, hence $p_1 p_2 \in \overline{\mathcal{L}_0}$. Dually, we obtain that $L_1 \wedge L_2 \in \overline{P_0}$ for distinct lines $L_i \in \overline{\mathcal{L}_0}$. In view of (41.5a), the closure of $\mathbf{F}_0 = \mathbf{F} \cap (P_0 \times \mathcal{L}_0)$ is contained in $\mathbf{F} \cap (\overline{P_0} \times \overline{\mathcal{L}_0})$. For the converse inclusion, consider a flag $(p, L) \in \overline{P_0} \times \overline{\mathcal{L}_0}$ and neighbourhoods $U \subseteq P$ and $V \subseteq \mathcal{L}$ of p and L, respectively. We have to show that $U \times V$ contains a flag in $\mathbf{F}_0$. Choose a point $q \in P_0 \setminus L$ and a line $M \in \mathcal{L}_0$ such that $q \notin M$. If M is close enough to L, then $M \in V$ and $x := pq \wedge M \in U$. The flag (x, M) belongs to $\mathbf{F}_0 \cap (U \times V)$. □

Remark on subplanes. Let $\mathcal{P} = (P, \mathcal{L}, \mathbf{F})$ be a (topological) projective plane. If a set $X \subseteq P$ contains a non-degenerate quadrangle, then there is a smallest subplane $(P_0, \mathcal{L}_0, \mathbf{F}_0)$ of $\mathcal{P}$ such that $X \subseteq P_0$. This subplane is called *the subplane generated by* X. The point set P_0 (and, mutatis mutandis, the sets $\mathcal{L}_0$ and $\mathbf{F}_0$) may be described 'from above' as the intersection of all point sets of subplanes of $\mathcal{P}$ that contain X. There is also a description 'from below': one easily verifies that $P_0 = \bigcup_{n \geq 0} X_n$, where we set $X_0 = X$ and define X_{n+1} as the set of all points $ab \wedge cd$ such that $a, b, c, d \in \bigcup_{i \leq n} X_i$ form a non-degenerate quadrangle.

By (41.6), the topological closure $(\overline{P_0}, \overline{\mathcal{L}_0}, \overline{\mathbf{F}_0})$ is again a subplane of $\mathcal{P}$; this is the smallest *closed* subplane of $\mathcal{P}$ such that the point set contains X. We write $\langle X \rangle := \overline{P_0}$; note that $\overline{\mathcal{L}_0}$ and $\overline{\mathbf{F}_0}$ are determined by $\langle X \rangle$. We say that X *topologically generates* $(\overline{P_0}, \overline{\mathcal{L}_0}, \overline{\mathbf{F}_0})$.

41.7 Proposition. *Let $(P, \mathcal{L}, \mathbf{F})$ be a topological projective plane.*

(a) *If one of the following spaces is compact or locally compact, so are all the others: $P, \mathcal{L}, \mathbf{F}$, a line (considered as a point set), a pencil.*
(b) *If both P and $C \subseteq P$ are compact, then the set $\{ L \in \mathcal{L} \mid L \cap C \neq \emptyset \}$ of all lines meeting C is also compact.*

Proof. (a) Lines and pencils are homeomorphic (41.2a), and they inherit (local) compactness from P, see (41.2b). This gives the assertions involving $\mathbf{F}$, see (41.5) and consider the open surjective projections $\mathbf{F} \to P$ and $\mathbf{F} \to \mathcal{L}$. By duality, it now suffices to infer (local) compactness of $\mathcal{L}$ from (local) compactness of each line. In the locally compact case, this follows from (41.2b and d). In the compact case, let L_1, L_2, L_3 be lines which form a triangle. Cover each L_i by two compact point sets C_{i1}, C_{i2} such that each contains only one of the two vertices $L_i \wedge L_j$, $j \neq i$, of the triangle. Then

$$\mathcal{L} = \bigcup \{ C_{ij} \vee C_{kl} \mid C_{ij} \cap C_{kl} = \emptyset \}$$

is a union of at most 12 compact sets and therefore compact. (This argument covers the locally compact case as well, in view of (41.2c) and Dugundji [66] XI.6.5(1).)

(b) The set of lines meeting C is obtained by taking the inverse image $\mathbf{C}$ of C under the projection $\mathbf{F} \to P$ and projecting it into $\mathcal{L}$. Since both projections are continuous, we obtain that $\mathbf{C}$ is closed in $\mathbf{F}$ and that both $\mathbf{C}$ and its projection into $\mathcal{L}$ are compact. □

41.8 Theorem. *Let $\mathcal{P} = (P, \mathcal{L}, \mathbf{F})$ be a locally compact projective plane. Then P and $\mathcal{L}$ have countable bases; in particular, P and $\mathcal{L}$ are metrizable and σ-compact.*

As a consequence, sequences suffice to describe the topology, and all dimension functions mentioned in Section 92 agree on these spaces. The dimension of P coincides with the dimension of $\mathcal{L}$.

We say that $\mathcal{P}$ has dimension $\dim P = \dim \mathcal{L}$. It is an open problem whether or not each locally compact projective plane is even compact; see, however, (42.4).

Proof. Let p be a point, let M be a line through p, and let $q \in M \setminus \{p\}$. Choose a compact neighbourhood U of q in $M \setminus \{p\}$.

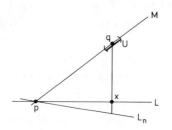

Figure 41b

As $\mathcal{L}_p$ is not discrete, all neighbourhoods in $\mathcal{L}_p$ are infinite; by local compactness, we find a sequence $L_n \in \mathcal{L}_p$ accumulating at some line L with $L \neq L_n$ for all n. For $x \in L \setminus \{p\}$ define

$$U_n(x) = (U \vee (qx \wedge L_n)) \wedge L,$$

the image of U under the projectivity $\pi(M, qx \wedge L_n, L)$, cp. (21.6). Then $U_n(x)$ is a neighbourhood of x in L.

1) $\{U_n(x) \mid n \in \mathbb{N}\}$ is a neighbourhood basis of x in L.

In order to see this, let V be an open subset of L containing x. Each $u \in U$ satisfies $(u \vee (qx \wedge L)) \wedge L = x$. Compactness of U shows that $(U \vee (qx \wedge \mathcal{W})) \wedge L \subseteq V$ for some neighbourhood $\mathcal{W}$ of L in $\mathcal{L}$. Since $\mathcal{W}$ contains infinitely many of the lines L_n, we have $U_n(x) \subseteq V$ for infinitely many $n \in \mathbb{N}$.

As a consequence of 1), all spaces occuring here are first countable. Hence by switching to a subsequence we can achieve that L_n converges to L.

2) Each compact subset C of $L \setminus \{p\}$ has a countable basis.

With $W_n(x) = C \cap U_n(x)$, we have $C = \bigcup_{c \in C} W_n(c) = \bigcup_{c \in C_n} W_n(c)$ for a finite subset C_n of C, as C is compact. Then $\{W_n(c) \mid n \in \mathbb{N}, c \in C_n\}$ is a (countable) basis, as we show now. Let V be a neighbourhood of x in C. For each $n \in \mathbb{N}$ we find $c_n \in C_n$ with $x \in W_n(c_n)$. We may assume that c_n converges to c. If y is an accumulation point of a sequence $y_n \in W_n(c_n)$, say $y_n = (u_n \vee (qc_n \wedge L_n)) \wedge L$ with $u_n \in U$, then $y = c$, since $qc_n \wedge L_n \to qc \wedge L = c$.

For the constant sequence $y_n = x$, we obtain $c = x$; furthermore, it is not possible that each $W_n(c_n)$ contains a point $y_n \notin V$. Hence $W_n(c_n) \subseteq V$ for some $n \in \mathbb{N}$.

3) Each pencil of lines has a countable basis.

According to 2), some neighbourhood $U_1 \times U_2$ of p in an affine part of P has a countable basis. Now $\mathscr{L}_p$ is a continuous open image of $U_1 \times U_2 \setminus \{p\}$ under the mapping $x \mapsto p \vee x$, and assertion 3) follows with Dugundji [66] VIII.6.2.

The point space P has a countable basis by 3) and (41.2d,e). Hence P is separable (Dugundji [66] VIII.7.3), metrizable (Dugundji [66] XI.6.4 and IX.9.2) and σ-compact (Dugundji [66] XI.6.3). For the coincidence of the dimension functions, see (92.3, 6 and 7) and the remarks preceding (92.6).

Finally, we observe that both P and $\mathscr{L}$ are locally homeomorphic to L^2, see (41.2d and a), hence they have the same (small inductive) dimension, compare (92.3). □

Because P is covered by the point sets of three affine planes, Theorem (41.8) is also a consequence of the following result:

41.9 Theorem. *Every locally compact ternary field (or double loop) is separable, metrizable, and σ-compact (but not compact, see (42.5)).*

The proof follows the same lines, see Grundhöfer–Salzmann [90] XI.1.2.

41.10 Corollary. *If $(P, \mathscr{L}, \mathbf{F})$ is a compact projective plane, and if P is not connected, then $L \in \mathscr{L}, P, \mathscr{L}$ and $\mathbf{F}$ are homeomorphic to Cantor's triadic set $\{0, 1\}^{\mathbb{N}}$.*

Proof. $P, \mathscr{L}$ and $\mathbf{F}$ are totally disconnected by (42.1) below. Now $L, P, \mathscr{L}$ and $\mathbf{F}$ are compact, totally disconnected metric spaces (41.8) without isolated points (as L is homogeneous by (41.2) and not discrete), hence these spaces are homeomorphic to the Cantor set by a result of Brouwer, see Hausdorff [27] p. 160 and p. 197, Hocking–Young [61] 2-97, 2-98, Christenson–Voxman [77] 6.C.11, or Moise [77] §12 Theorem 8. □

By (41.10), compact, disconnected projective planes cannot be distinguished topologically.

41.11 Proposition. *Let $\mathscr{P} = (P, \mathscr{L})$ be a compact projective plane, and let $\mathscr{B}$ be a closed subplane of $\mathscr{P}$ with point set B.*

(a) *For each line L of $\mathcal{B}$ and each point p on L not belonging to B, the mapping $x \mapsto x \vee p$ is a topological embedding of the affine point set of $\mathcal{B}^L$ into $\mathcal{L}_p \setminus \{L\}$, i.e. a homeomorphism of $B \setminus L$ onto its image.*
(b) *If $\mathcal{B}$ is a Baer subplane of $\mathcal{P}$, then this mapping is surjective, hence B and $\mathcal{L}_p$ are locally homeomorphic.*
(c) *Let $\mathcal{B}$ be a Baer subplane of $\mathcal{P}$, and let (K, τ) and H be ternary fields which coordinatize $\mathcal{P}$ and $\mathcal{B}$, respectively, with respect to the same quadrangle of reference. Let $k \in K \setminus H$. Then the mapping $(s, t) \mapsto \tau(s, k, t)$ is a homeomorphism of the pair $(H \times H, \{0\} \times H)$ onto the pair (K, H).*

Proof. The map $\mu : B \to \mathcal{L}_p$ with $x^\mu = x \vee p$ is continuous, and the restriction $\mu|_{B \setminus L}$ is injective, since $p \notin B$. If A is closed in $B \setminus L$, then $A \cup L$ and $(A \cup L)^\mu = A^\mu \cup \{L\}$ are compact, hence A^μ is closed in $\mathcal{L}_p \setminus \{L\}$ and in $(B \setminus L)^\mu$. This proves assertions (a) and (b). For part (c), we identify the affine line $[k]$ with K, via $(k, y) \mapsto y$. Then the dual of μ, i.e. the mapping $[s, t] \mapsto [s, t] \wedge [k]$, is just the mapping $(s, t) \mapsto \tau(s, k, t)$. □

The following lemma (which is trivial for closed subplanes) will be used in the proof of (44.3).

41.12 Lemma. *Every discrete subplane of a compact projective plane is finite.*

Proof. Let $\mathcal{P}_0 = (P_0, \mathcal{L}_0)$ be an infinite subplane of the compact projective plane $(P, \mathcal{L})$, and let $L \in \mathcal{L}_0$. Using (41.8) we find an injective sequence of points $p_n \in L \cap P_0$ such that $p_n \to p \in P$. Choose $q \in P_0 \setminus L$ and $L' \in \mathcal{L}_0 \setminus (\mathcal{L}_p \cup \mathcal{L}_q)$. Then $p'_n = p_n q \wedge L'$ converges to $p' := pq \wedge L' \neq p$. Thus $p_n p'_{n+1} \wedge p_1 q \in P_0 \setminus \{q\}$, and this sequence converges to $pp' \wedge p_1 q = q$. Hence $\mathcal{P}_0$ is not discrete. □

41.13 Notes on Section 41. Definition (41.1) is due to Skornjakov [54] and Salzmann [55]. Freudenthal [57a] stipulates that the diagonal point $ac \wedge bd$ of a quadrangle a, b, c, d should depend continuously on the quadrangle (a, b, c, d); this requires a topology only on the point set. Freudenthal's definition is equivalent to (41.1), see Arumugam [85]. Continuity of the central projections (41.2a) together with the dual condition leads to the weaker concept of Lenz-topological planes; however, for compact topologies, these two concepts are equivalent, see Hartmann [88, 89]. For a discussion of other definitions which are weaker than (41.1) see Misfeld [68] §1.

Many results from Section 41 (and 42) can be found in Salzmann [57, 81b]. The proof of Theorem (41.8) is adapted from Löwen [76b] 1.8, 1.9; see also Breitsprecher [67a] 3.6 for another proof of the metrizability of compact projective planes. Proposition (41.11) is due to Salzmann [75a] §1 (9), [79a] §1 (4) and Hähl [86b] 2.1. Lemma (41.12) is a specialized version of Löwen [76b] 1.32.

42 Compact, connected planes

As a basic dichotomy we show that a topological projective plane is either connected or totally disconnected. In the totally disconnected case only few results are known (compare (41.10)), hence we focus on the connected case. Locally compact, connected projective planes are even compact (42.4). Some preliminary information (42.6) on the vector space structure of locally compact, connected quasifields allows us to determine all compact, connected Moufang planes (42.7): we obtain precisely the four classical planes from Chapter 1. The lines of every compact, connected plane are locally and globally arcwise connected (42.8). If such a plane has topological dimension 2 (i.e. if its lines are one-dimensional), then the point space is homeomorphic to the classical point space $P_2\mathbb{R}$ of the real projective plane (42.10). These planes have been studied in Chapter 3.

42.1 Lemma. Let $(P, \mathcal{L})$ be a topological projective plane. Then $P, \mathcal{L}$, all the affine or projective lines and all pencils are simultaneously either connected or totally disconnected.

Proof. The affine lines are doubly homogeneous by (41.2), hence connected or totally disconnected. Connectedness carries over to projective lines, to pencils of lines, to affine point sets, and then to P and $\mathcal{L}$, compare (41.2).

Assume that an affine line $L' = L \setminus \{p\}$ is not connected. By duality, it suffices to show that L and P are totally disconnected. Let a, q be distinct points of L'. By double homogeneity of L' we find an open and closed subset U of L' with $a \in U, q \notin U$. The closure $\overline{U}$ of U in L satisfies $\overline{U} \subseteq U \cup \{p\}$. The same arguments for the affine line $L \setminus \{q\}$ give an open and closed subset V of $L \setminus \{q\}$ with $a \in V, p \notin V$ and $\overline{V} \subseteq V \cup \{q\}$. Then $U \cap V$ is a non-empty proper open subset of L, which is also closed in L in view of $\overline{U \cap V} \subseteq \overline{U} \cap \overline{V} = U \cap V$. Hence the projective line L is not connected and therefore totally disconnected by double homogeneity (41.2).

We have $L = A \cup B$ with non-empty disjoint open subsets A, B. Choose distinct points $p_1, p_2 \in P \setminus L$, and define central projections $\pi_i : P \setminus \{p_i\} \to L$ by $\pi_i(x) = xp_i \wedge L$. Then $A_i = \pi_i^{-1}(A)$ is open and closed in $P \setminus \{p_i\}$, and $A_1 \cap A_2$ is a non-empty proper open subset of P. The closure of A_i in P is contained in $A_i \cup \{p_i\}$, and $p_i \notin A_j$ for $\{i, j\} = \{1, 2\}$. Hence $A_1 \cap A_2$ is also closed in P, and P is not connected. If C is a (proper) connected subset of P, and if $p \in P \setminus C$, then the map $x \mapsto px \wedge M$ projects C into a given line M with $p \notin M$. As lines are totally disconnected, the set C is contained in a line, hence reduced to a point. This shows that P is totally disconnected. □

We say that $(P, \mathcal{L})$ is *connected* if P is connected. The following Corollary shows that in locally compact planes, connectedness is (equivalent to) a local property:

42.2 Corollary. *A locally compact projective plane* $(P, \mathcal{L})$ *is connected if, and only if, the topological dimension of P (or equivalently, of some line $L \in \mathcal{L}$) is positive.*

Proof. A locally compact Hausdorff space is totally disconnected if, and only if, it is zero-dimensional, see (92.18 and 5), compare also Engelking [78] 1.4.5 or Salzmann [81b] 3.0, Salzmann [57] §5. □

42.3 Lemma. *Let X be a connected set of points in a connected projective plane. Then either X is contained in a line, or X contains a quadrangle.*

Proof. Suppose otherwise. Then X contains a triangle p, q, r such that $X \subseteq pq \cup \{r\}$. Hence $\{r\} = X \setminus pq$ is open in X, a contradiction. □

42.4 Proposition. *Every locally compact, connected projective plane $\mathcal{P}$ is compact.*

Proof. Let $\mathcal{P} = (P, \mathcal{L})$, let (p, L) be a non-incident point–line pair, and let U be a compact neighbourhood of p in $P \setminus L$. Each line $X \in \mathcal{L}_p$ meets the boundary ∂U, since X is connected and $X \cap L \neq \emptyset$. Hence $\mathcal{L}_p = \{ p \vee x \mid x \in \partial U \}$ is compact, and the assertion follows with (41.7a). □

It is an open problem whether (42.4) is true without the connectedness assumption; compare Salzmann [55] §7, Grundhöfer–Salzmann [90] Section 2, Grundhöfer [87a] 5.2.

42.5 Lemma. *A (non-discrete) topological ternary field (or double loop) K is not compact.*

Proof. Let W be a proper open subset of K with $0 \in W$. We have $x \cdot 0 = 0$ for every $x \in K$, hence there exist open neighbourhoods U_x, V_x of x and 0 respectively such that $U_x \cdot V_x \subseteq W$. Compactness of K implies that K is a union of finitely many sets U_x. The intersection V of the corresponding sets V_x is a neighbourhood of 0, thus $V \neq \{0\}$ as K is not discrete. Hence $K = K \cdot V \subseteq W$, a contradiction. For the locally compact case, another argument can be found in Grundhöfer–Salzmann [90] XI.1.2. □

42.6 Theorem. *Let Q be a locally compact, connected quasifield. Then the kernel K of Q is isomorphic to $\mathbb{R}, \mathbb{C}$ or $\mathbb{H}$ as a topological field, and Q is a vector space* (25.3) *of finite dimension over K.*

Proof. The structure theorem for locally compact abelian groups (cp. Hewitt–Ross [79] 9.14) says that $(Q, +) \cong \mathbb{R}^n \times C$, where C is the largest compact subgroup of $(Q, +)$. The non-zero multiplications $x \mapsto a \cdot x$ are automorphisms of $(Q, +)$

leaving C invariant. They act transitively on $Q \setminus \{0\}$, and Q is not compact by (42.5). Therefore $C = \{0\}$, and $(Q, +)$ is a real vector group $(\mathbb{R}^n, +)$. Thus, K contains a copy $\mathbb{Q}$ of the field of rational numbers. The topological closure $\overline{\mathbb{Q}}$ of $\mathbb{Q}$ in Q is the one-dimensional subspace spanned by 1, hence $\overline{\mathbb{Q}}$ is isomorphic to $\mathbb{R}$, even as a topological field. Clearly Q has finite dimension n over $\overline{\mathbb{Q}}$. Since $\overline{\mathbb{Q}}$ is contained in the center of K, we have $K \in \{\mathbb{R}, \mathbb{C}, \mathbb{H}\}$ by a well-known algebraic result of Frobenius [1878], see e.g. Ebbinghaus et al. [90, 92] 7 §2, Palais [68]. □

As a special case ($Q = K$) of (42.6), we obtain that $\mathbb{R}, \mathbb{C}, \mathbb{H}$ are the only locally compact, connected (not necessarily commutative) fields. This is a famous result due to Pontrjagin, see Pontrjagin [32], [57] §27, Weil [67] p. 12, S. Warner [89] 27.2, Artmann [88].

Theorem (42.6) does not tell the full truth. One can show that $\dim_\mathbb{R} Q \in \{1, 2, 4, 8\}$, see Salzmann [67b] 7.12, or (52.5) for an even more general result. Furthermore $K = \mathbb{R}$ if $\dim_\mathbb{R} Q = 8$, see Buchanan–Hähl [77].

Now we determine all compact, connected Moufang planes: we obtain precisely the examples from Chapter 1; see also (63.3).

42.7 Theorem. *If $\mathcal{P}$ is a compact, connected Moufang plane, then $\mathcal{P}$ is isomorphic (as a topological plane) to one of the planes $\mathcal{P}_2 A$ with $A \in \{\mathbb{R}, \mathbb{C}, \mathbb{H}, \mathbb{O}\}$.*

Proof. $\mathcal{P}$ is coordinatized by a locally compact alternative field A (see (24.8) and (41.3)) which is connected by (42.1). By (42.6), A is an algebra over $\mathbb{R}$ of finite dimension, and $A \in \{\mathbb{R}, \mathbb{C}, \mathbb{H}\}$ if A is associative. If A is not associative, then $A \cong \mathbb{O}$ by an algebraic result of Zorn [33], compare Pickert [75] p. 177, Ebbinghaus et al. [90, 92], Zhevlakov et al. [82] 2.3. □

Every compact, disconnected Moufang plane is desarguesian (see Grundhöfer [87b]), hence results of Jacobson yield the classification of all compact, disconnected Moufang planes, see Grundhöfer–Salzmann [90], S. Warner [89] Section 27.

A topological space X is said to be *locally contractible* if for every $x \in X$, every neighbourhood U of x contains a neighbourhood V of x which is contractible in U.

42.8 Proposition. *Let $\mathcal{P}$ be a compact, connected projective plane. Then each line of $\mathcal{P}$ is locally and globally arcwise connected. Each affine line $L \setminus \{p\}$ is locally and globally contractible by a homotopy which fixes a preassigned point q.*

Proof. $K = L \setminus \{p\}$ has the structure of a ternary field with $q = 0$ as zero element, see (22.1). By (42.2 and 1), the topological dimension of K is positive, hence no neighbourhood in K is totally disconnected. This gives us a compact, connected set $C \subseteq K$ containing 0 and further elements. There exist open neighbourhoods U

of 0 such that $U \cdot C$ is arbitrarily small, and

$$U \cdot C = \bigcup \{u \cdot C \mid u \in U\} = \bigcup \{U \cdot c \mid 0 \neq c \in C\}$$

is connected and open, hence a neighbourhood of 0. Therefore K is locally connected.

By a result of Mazurkiewicz, Moore and Menger, every locally compact, locally connected metric space is locally arcwise connected, see Kuratowski [68] §50 II, p. 254, compare also Wilder [49] III, Theorem 3.9, p. 80, Whyburn [42] II.5, Whyburn–Duda [79] Section XX, and, in particular, Ball [84]. In view of (41.8), we conclude that K is locally arcwise connected. By double homogeneity, K and L are globally arcwise connected.

Let $t \mapsto e_t$, $t \in [0, 1]$, be an arc in K joining $e_0 = 1$ to $e_1 = 0$. The homotopy $F_t(x) := e_t \cdot x$ satisfies $F_0 = \mathrm{id}$, $F_1(x) = 0$ and $F_t(0) = 0$, as desired. The same homotopy also induces local contractions of K near 0, because every neighbourhood U of 0 contains a neighbourhood V of 0 such that $e_t \cdot V \subseteq U$ for every $t \in [0, 1]$, see the proof of (42.5). $\square$

42.9 Corollary. *Every compact, connected projective plane* $(P, \mathcal{L})$ *satisfies* $\dim P \geq 2$.

Proof. By (42.8), each line L contains an arc, hence a copy of the interval $[0, 1]$. Thus $\dim P = \dim L^2 \geq \dim[0, 1]^2 = 2$, compare (41.2d) and (92.4). $\square$

42.10 Proposition. *Let* $(P, \mathcal{L})$ *be a compact projective plane with* $\dim P = 2$ *or* $\dim L = 1$ *for* $L \in \mathcal{L}$. *Then* P *and* $\mathcal{L}$ *are homeomorphic to the point space* $P_2 \mathbb{R}$ *of* $\mathcal{P}_2 \mathbb{R}$, *each affine line is homeomorphic to* $\mathbb{R}$, *and the projective lines are Jordan curves in* P; *thus* $(P, \mathcal{L})$ *is a two-dimensional compact projective plane in the sense of Section* 32.

Proof. We have $\dim P = \dim L^2$, and (42.8) together with (92.10, 4 and 11) gives $2 \dim L \geq \dim L^2 \geq \dim(L \times [0, 1]) = \dim L + 1$. Hence $\dim P = 2$ is equivalent to $\dim L = 1$.

Arcwise connectedness of the affine lines (42.8) implies that L contains two arcs having only the endpoints in common; the union S of these arcs is homeomorphic to the circle $\mathbb{S}_1$. If $L \neq S$, then $S \subseteq K := L \setminus \{p\}$ for some point p on L, and the inclusion of S in K extends to a retraction of K onto S in view of $\dim K = 1$ (see (92.2) and note that K is metrizable by (41.8)). This contradicts the contractibility of K (42.8). Hence $L = S$ is homeomorphic to $\mathbb{S}_1$, and each affine line is homeomorphic to $\mathbb{R}$.

The affine point set $P \setminus L$ is homeomorphic to $\mathbb{R}^2$ by (41.2d), hence (32.3) implies that P is homeomorphic to $P_2 \mathbb{R}$. Alternatively, one could apply the classification of compact surfaces to P, see Salzmann [81b] 4.3; indeed, the surface P has genus 1 and Euler characteristic $\chi(P) = 1$. □

For another proof of (42.10) see (51.29), and for results in higher dimensions compare Sections 52 and 53.

By (42.9 and 10) the planes considered in Chapter 3 are precisely the compact, connected planes of smallest topological dimension. The results of Sections 32 and 33 together with (42.10) give characterizations of the real projective plane by rather weak properties; for example, (32.21) leads to

42.11 Theorem. *Let $\mathcal{P}$ be a compact projective plane with lines of topological dimension 1. If the collineations of $\mathcal{P}$ act transitively on the points of $\mathcal{P}$, then $\mathcal{P}$ is (geometrically and topologically) isomorphic to the real projective plane $\mathcal{P}_2\mathbb{R}$.*

The following lemma is the main step in the proof of (42.13).

42.12 Lemma. *Let $\mathcal{P} = (P, \mathcal{L})$ be a compact, connected projective plane, let $a, b \in P$ be distinct points, and let F be a compact subset of P with $ab \subseteq P \setminus F$. Then there exists a homeomorphism $\alpha : P \to P$ which fixes the points in F and maps a to b.*

Proof. Coordinatize the plane $\mathcal{P}$ by a ternary field K such that $a = (0, 0)$ and $b = (0, 1)$. As F is compact, we find a compact neighbourhood V of 0 in K with $V \times K \subseteq P \setminus F$. Since K is normal, there is an Urysohn function $f : K \to [0, 1]$ with $f(0) = 1$ and $f(K \setminus V) = \{0\}$, compare Dugundji [66] VII.4.1. By (42.8) we have a path $e_t, t \in [0, 1]$ in K with $e_0 = 0, e_1 = 1$.

Define $\alpha : P \to P$ by $(x, y)^\alpha = (x, y + e_{f(x)})$ for $(x, y) \in K^2$, and $p^\alpha = p$ for points p at infinity.

Then $a^\alpha = b$, and α fixes all points in $(K \setminus V) \times K$, hence all points in F. Furthermore, α is bijective, with inverse transformation $(x, y) \mapsto (x, y - e_{f(x)})$, where $y - z$ is defined by the equation $(y - z) + z = y$. For the continuity of α and α^{-1}, it suffices to consider convergent sequences (x_ν, y_ν) in P with $x_\nu \to \infty$ or $y_\nu \to \infty$. The points (x_ν, y_ν) with $x_\nu \notin V$ are fixed by α and α^{-1}, hence we may assume that all $x_\nu \in V$. Thus all elements $e_{f(x_\nu)}$ are contained in some compact subset of K. We infer that $y_\nu \to \infty$ and $(x_\nu, y_\nu) \to (\infty)$, and that $y_\nu \pm e_{f(x_\nu)} \to \infty$. This means that $(x_\nu, y_\nu)^{\alpha^{\pm 1}} \to (\infty)$. □

42.13 Proposition. *Let $\mathcal{P} = (P, \mathcal{L})$ be a compact projective plane. Then the homeomorphism group of P is n-transitive on P for every $n \in \mathbb{N}$.*

Proof. If P is disconnected, then P is homeomorphic to the Cantor set (41.10), and this gives the assertion (compare Moise [77] §12 Theorem 9 or Hocking–Young [61] Ex. 2-37 p. 100).

Let P be connected. Proceeding inductively, we have to show that the stabilizer of n points $p_1, \ldots, p_n$ is transitive on the complement $U = P \setminus \{p_1, \ldots, p_n\}$. Let $p, q \in U$. Since U is not covered by finitely many lines, we find an auxiliary point $r \in U$ which does not lie on any of the lines pp_i, qp_i. Now (42.12) shows that we can map p to r and then r to q by suitable homeomorphisms which fix $p_1, \ldots, p_n$. □

42.14 Notes on Section 42. Lemma (42.1) was proved by Salzmann [55], cp. also Lenz [65] p. 349, Pickert [75] p. 266. The proof of (42.6) is inspired by Jacobson–Taussky [35]. The geometric significance (42.7) of Pontrjagin's classification of all locally compact, connected (not necessarily commutative) fields was pointed out by Kolmogoroff [32]. Proposition (42.8) is contained in Freudenthal [57a], and Proposition (42.10) is due to Salzmann [59b, 81b]. The results (42.12 and 13) are taken from Buchanan–Hähl–Löwen [80].

43 Ternary fields for compact planes

Arbitrary ternary fields appear to be quite intractable. However, ternary fields are tools for proofs and for many constructions of (topological) planes. It is the aim of this section to provide criteria which characterize the ternary fields of compact projective planes. Criteria of this type have been obtained in Grundhöfer [87a] (note that this paper uses a slightly different method of coordinatization, which leads to a permutation of the arguments of the ternary operations). In general, these criteria are rather complicated. In the connected case, however, we arrive at the following very simple statement (43.6): every ternary field defined on $\mathbb{R}^n$ with continuous ternary operation coordinatizes a compact, connected projective plane. Furthermore, every locally compact, connected affine plane has a compact projective completion (43.7).

First we reduce the effort necessary for proving continuity of joining points and intersecting lines.

43.1 Theorem. *Let $\mathcal{P} = (P, \mathcal{L}, \mathbf{F})$ be a projective plane with compact Hausdorff topologies on P and $\mathcal{L}$. Then $\mathcal{P}$ is a topological projective plane if, and only if, the set $\mathbf{F}$ of flags is closed in $P \times \mathcal{L}$.*

Proof. By (41.5a), **F** is closed in $P \times \mathcal{L}$ if $\mathcal{P}$ is a topological plane. For the converse, we show that the join mapping $\vee : P^{(2)} \to \mathcal{L}$ is continuous, where $P^{(2)} = \{(p,q) \in P^2 \mid p \neq q\}$. The graph of this mapping is

$$\{(p,q,L) \in P^{(2)} \times \mathcal{L} \mid (p,L), (q,L) \in \mathbf{F}\} = (P^{(2)} \times \mathcal{L}) \cap (P \times \mathbf{F}) \cap (P \times \mathbf{F})^\pi$$

with $(p,q,L)^\pi = (q,p,L)$. Hence this graph is closed in $P^{(2)} \times \mathcal{L}$, and $\vee$ is continuous, cp. Dugundji [66] XI.2.7. Continuity of intersection follows by duality. □

In the following, we use the division operations \ and / and the inverse operations s, t, x introduced in (22.4 and 2ii).

43.2 Theorem. *Let (K, τ) be a ternary field coordinatizing the projective plane $\mathcal{P} = (P, \mathcal{L}, \mathbf{F})$. Then $\mathcal{P}$ can be made into a compact projective plane if, and only if, there exists a non-discrete locally compact, first countable Hausdorff topology on K such that the following conditions hold in the one-point compactification $K \cup \{\infty\}$ of K:*

(1) *The ternary operation $\tau : K^3 \to K$ and its inverse $t : K^3 \to K$ are continuous.*
(2) *If $\lim x_\nu = x \neq 0$, $\lim y_\nu = y \neq 0$ and $\infty \in \{x, y\}$, then*
    $\lim \tau(x_\nu, y_\nu, 0) = \infty = \lim t(x_\nu, y_\nu, 0)$.
(3) *If $\lim x_\nu = \infty$, $\lim s_\nu = s \in K$ and $\lim t_\nu = t \in K$, then*
    $\lim \tau(s_\nu, x_\nu, t_\nu)/x_\nu = s$.
(4) *If $\lim s_\nu = \infty$, $\lim x_\nu = x \in K$ and $\lim y_\nu = y \in K$, and if z_ν is defined by $\tau(s_\nu, z_\nu, t(s_\nu, x_\nu, y_\nu)) = 0$, then $\lim z_\nu = x$.*
(5) *If $\lim x_\nu = \infty = \lim s_\nu$ and $\lim r_\nu = r \in K$, then*
    $\lim \tau(s_\nu, x_\nu, t(s_\nu, r_\nu, 0))/x_\nu = \infty$.

In particular, conditions (1)–(5) imply that K with a topology as specified is a topological ternary field (41.3), and that K, considered as an affine line of $\mathcal{P}$ as in (22.1), retains its topology.

Sketch of proof. Geometric interpretation shows that conditions (1)–(5) hold in all compact projective planes, see Grundhöfer [87a] p. 93.

For the converse, we first need to define topologies on P and $\mathcal{L}$. Natural identifications of the three pencils $\mathcal{L}_{(0,0)}$, $\mathcal{L}_{(0)}$, $\mathcal{L}_{(\infty)}$ with $K \cup \{\infty\}$ give compact topologies on these three pencils, which in turn lead to locally compact 'product' topologies on the point sets of the three affine planes $\mathcal{P}^{[\infty]}$, $\mathcal{P}^{[0]}$, $\mathcal{P}^{[0,0]}$ as suggested in the proof of (41.2d). We endow P with the weak topology obtained from this covering by three affine planes, and dually for $\mathcal{L}$. Now conditions (1)–(5) imply that P and $\mathcal{L}$ are compact Hausdorff spaces, that the affine parts retain their topologies, and that **F** is closed in $P \times \mathcal{L}$, see Grundhöfer [87a] 2.2; for example, consider a sequence of points $(x_\nu, y_\nu) \in K^2$ with $x_\nu \to \infty$, $y_\nu \to y \in K$. Then $y_\nu = (y_\nu/x_\nu)x_\nu$ and (2) together imply that $y_\nu/x_\nu \to 0$, hence $(x_\nu, y_\nu) \vee (0,0) = [y_\nu/x_\nu, 0] \to [0,0]$ and $(x_\nu, y_\nu) \vee (\infty) = [x_\nu] \to [\infty]$, thus $(x_\nu, y_\nu) \to (0)$ by definition of the topology on the affine plane $\mathcal{P}^{[0]}$. If these points occur in a sequence $((x_\nu, y_\nu), [s_\nu, t_\nu])$ of flags with $s_\nu \to s \in K, t_\nu \to t \in K$, then $y_\nu/x_\nu = \tau(s_\nu, x_\nu, t_\nu)/x_\nu \to 0 = s$ by (3), hence these flags converge to $((0), [0, t])$, again a flag.

Furthermore, K, considered as an affine line of $\mathcal{P}$, retains its topology. Since $\mathcal{P}$ is a topological projective plane (43.1), we infer that the inverse operations s and x are continuous, hence K is a topological ternary field, see the remarks after (41.3). □

43.3 Remarks. (a) The assumption that K be first countable can be omitted if conditions (2)–(5) are read as conditions for nets instead of sequences (cp. (41.9)).

(b) For a Cartesian field $(K, +, \cdot)$, conditions (1)–(5) are equivalent to the following simpler conditions:

(1′) Addition and multiplication are continuous on K^2.
(2′) If $\lim x_\nu = \infty, \lim s_\nu = s \in K$ and $\lim t_\nu = t \in K$, then
 $\lim(s_\nu x_\nu + t_\nu)/x_\nu = s$.
(3′) If $\lim s_\nu = \infty, \lim x_\nu = x \in K$ and $\lim y_\nu = y \in K$, then
 $\lim s_\nu \setminus (-y_\nu + s_\nu x_\nu) = x$.
(4′) If $\lim x_\nu = \infty = \lim s_\nu$ and $\lim r_\nu = r \in K$, then $\lim(s_\nu x_\nu - s_\nu r_\nu)/x_\nu = \infty$.

See Grundhöfer [87a] 2.3. Further (algebraical or topological) simplifications for quasifields can be found in Hähl [82b], Grundhöfer [87a] Section 3.

(c) Note that the conditions in (43.2) do not stipulate, but imply continuity of the quaternary inverse functions x and s as defined in (22.2).

43.4 Corollary. *Let K be a ternary field which coordinatizes a compact projective plane. Then the mapping $x \mapsto 1/x : K^\times \to K^\times$ extends to a homeomorphism of the one-point compactification $K \cup \{\infty\}$, where $1/0 = \infty, 1/\infty = 0$. (Geometrically, this space $K \cup \{\infty\}$ appears as a projective line.)*

Proof. This is a consequence of condition (3) in (43.2). In fact, one can also argue directly: identify K with the line $[1, 0]$, as in (22.1). Then the mapping in question is just the projectivity $\pi([1, 0], (\infty), [0, 1], (0, 0), [1], (0), [1, 0])$ in the notation of (21.6), hence a homeomorphism (41.2a). □

43.5 Theorem. *Every locally compact, connected topological ternary field coordinatizes a compact, connected projective plane.*

Proof. Such a ternary field (K, τ) is first countable, see (41.9), and satisfies condition (1) of (43.2). The other conditions (2)–(5) are consequences of the following assertion, combined with the continuity of the inverse functions s, x, t:

(∗) There are no sequences $x_\nu \to \infty$, $s_\nu \to \infty$, $a_\nu \to a$, $y_\nu \to y$, $c_\nu \to c$, $d_\nu \to d$ with $a, y, c, d \in K$ and $\tau(s_\nu, x_\nu, t(s_\nu, a_\nu, y_\nu)) = \tau(c_\nu, x_\nu, d_\nu)$.

For example, we derive condition (2): if $x \neq \infty$ or $y \neq \infty$, then continuity of the inverse functions shows that $\tau(x_\nu, y_\nu, 0)$ and $t(x_\nu, y_\nu, 0)$ cannot accumulate in K; if $x = y = \infty$, then accumulation of $\tau(x_\nu, y_\nu, 0)$ or $t(x_\nu, y_\nu, 0)$ in K is a contradiction to (∗), in view of $\tau(x_\nu, y_\nu, t(x_\nu, 0, 0)) = x_\nu y_\nu = \tau(0, y_\nu, x_\nu y_\nu)$ and $\tau(x_\nu, y_\nu, t(x_\nu, 0, t_\nu)) = 0 = \tau(0, y_\nu, 0)$. Similarly, (3)–(5) follow from (∗), with constant sequences $a_\nu = y_\nu = 0$, $c_\nu = d_\nu = 0$, and $y_\nu = d_\nu = 0$, compare Grundhöfer [87a] 4.1.

Hence it suffices to prove (∗). Define $\mu : K^5 \to K$ by

$$\mu(s, x, a, c, d) = \tau(s, a, t(s, x, \tau(c, x, d))).$$

Assume that we had the situation forbidden by (∗); then $y_\nu = \mu(s_\nu, x_\nu, a_\nu, c_\nu, d_\nu)$. As K is doubly homogeneous and locally connected (by the proof of (42.8)), we find a compact, connected set $Y \subseteq K$ containing y and $\tau(c, a, d)$ as interior points. Let C be a relatively compact, open neighbourhood of c, and let $U = U_a \times U_c \times U_d$ be a compact neighbourhood of (a, c, d) such that $U_c \subseteq C$. The set

$$\{ x \in K \mid \text{there is } (a', c', d') \in U \text{ with } \mu(\partial C, x, a', c', d') \cap Y \neq \emptyset \}$$

$$= \{ x(s, t(s, a', y), c', d') \mid s \in \partial C, y \in Y, (a', c', d') \in U \}$$

is compact (note that $s \neq c'$ in view of $U_c \cap \partial C = \emptyset$). Therefore, as $x_\nu \to \infty$, the set Y is disjoint from $\mu(\partial C, x_\nu, a_\nu, c_\nu, d_\nu)$; this statement, and all the following ones, hold for sufficiently large ν. Now $\mu(\cdot; x_\nu, a_\nu, c_\nu, d_\nu)$ is a homeomorphism, because $s(x_\nu, \tau(c_\nu, x_\nu, d_\nu), a_\nu, \cdot)$ is its inverse, whence

$$\mu(\partial C, x_\nu, a_\nu, c_\nu, d_\nu) = \partial \mu(C, x_\nu, a_\nu, c_\nu, d_\nu).$$

Furthermore $\mu(C, x_\nu, a_\nu, c_\nu, d_\nu)$ contains the points

$$\mu(c_\nu, x_\nu, a_\nu, c_\nu, d_\nu) = \tau(c_\nu, a_\nu, d_\nu),$$

which converge to $\tau(c, a, d)$, an interior point of Y. Hence the connected set Y is contained in $\mu(C, x_\nu, a_\nu, c_\nu, d_\nu)$. Finally

$$y_\nu = \mu(s_\nu, x_\nu, a_\nu, c_\nu, d_\nu) \in Y \subseteq \mu(C, x_\nu, a_\nu, c_\nu, d_\nu)$$

together with bijectivity of $\mu(\cdot, x_\nu, a_\nu, c_\nu, d_\nu)$ implies that $s_\nu \in C$, a contradiction to $s_\nu \to \infty$. $\square$

Condition (∗) in the above proof characterizes those locally compact ternary fields which coordinatize compact projective planes, see Grundhöfer [87a] 4.1. The following result is most useful for the construction of compact, connected projective planes.

43.6 Theorem. *Let K be homeomorphic to $\mathbb{R}^n$ and let (K, τ) be a ternary field such that the ternary operation $\tau : K^3 \to K$ is continuous. Then (K, τ) is a topological ternary field which coordinatizes a compact, connected projective plane.*

Proof. By (43.5) it suffices to show that the inverse functions s, x, t of (K, τ) are continuous.

Now s has the domain $D = \{ (x_1, y_1, x_2, y_2) \in K^4 \mid x_1 \neq x_2 \}$, and the map

$$(x_1, a, x_2, b) \mapsto (x_1, y_1, x_2, y_2)$$

with $y_i = \tau(a, x_i, b)$ is a continuous bijection of D onto itself. By domain invariance for open subsets of $K^4 \approx \mathbb{R}^{4n}$ (cp. Dugundji [66] XVII.3.1), this bijection is also open. In particular, $s = s(x_1, y_1, x_2, y_2)$ is continuous.

Similarly, the continuous bijection $(s, x, t) \mapsto (s, x, \tau(s, x, t))$ of K^3 is open, hence $t = t(s, x, y)$ is continuous. Finally, $(s_1, t_1, s_2, x) \mapsto (s_1, t_1, s_2, t_2)$ with $t_2 = t(s_2, x, \tau(s_1, x, t_1))$ is a continuous open bijection on D, and $x = x(s_1, t_1, s_2, t_2)$ is continuous. □

The results of this section are relevant for the embedding of topological affine planes into topological projective planes, because each topological affine plane is coordinatized by a topological ternary field, as was remarked after (41.3). Indeed, (43.5) has the following consequence.

43.7 Corollary. *Every locally compact, connected affine plane has a compact projective completion.*

A topological affine plane has a compact projective completion if, and only if, its line pencils are compact, see Grundhöfer [87a] 5.1. It is an open problem whether the assumption of connectedness in (43.7) can be replaced by non-discreteness.

Sometimes it is convenient to have the following variation of (43.6).

43.8 Corollary. *Let $f : \mathbb{R}^n \times \mathbb{R}^n \times \mathbb{R}^n \to \mathbb{R}^n$ be a continuous function, and assume that the sets $L_{s,t} = \{(x, f(s, x, t)) \mid x \in \mathbb{R}^n\}$ with $s, t \in \mathbb{R}^n$ together with the sets $\{c\} \times \mathbb{R}^n, c \in \mathbb{R}^n$, are the lines of an affine plane $\mathcal{A}$ with point set $\mathbb{R}^n \times \mathbb{R}^n$. Assume further that the map $t \to f(0, x, t) : \mathbb{R}^n \to \mathbb{R}^n$ is injective for each $x \in \mathbb{R}^n$, and that*

$$L_{s,t} \| L_{s',t'} \text{ if and only if } s = s'.$$

Then $\mathcal{A}$ is a topological affine plane.

Proof. The lines $L_{0,t}$ with $t \in \mathbb{R}^n$ form a parallel class, thus each map $t \mapsto f(0, x, t)$ as above is in fact bijective. The transformation $\varphi : \mathbb{R}^{2n} \to \mathbb{R}^{2n}$: $(x, t) \mapsto (x, f(0, x, t))$ is a continuous bijection, hence a homeomorphism by domain invariance (compare Dugundji [66] XVII.3.1). Replace each line of $\mathcal{A}$ by its image under φ^{-1}. This gives a new plane, isomorphic to $\mathcal{A}$, and the new plane has a describtion analogous to $\mathcal{A}$, in terms of a new function f, which has the additional property that $f(0, x, t) = t$ for $x, t \in \mathbb{R}^n$. This means that the sets $\mathbb{R}^n \times \{t\} = L_{0,t}$ are lines of the new plane.

Let 1 denote a fixed non-zero vector of $\mathbb{R}^n$. The line $L_{1,0}$ intersects each horizontal line $\mathbb{R}^n \times \{t\}$ and each vertical line $\{c\} \times \mathbb{R}^n$ in just one point. Therefore, the map $(x, t) \mapsto (f(1, x, 0), t)$ is a continuous bijection, and even a homeomorphism, of $\mathbb{R}^{2n}$ into itself. Applying this homeomorphism, we can further achieve that the diagonal $\{(x, x) | x \in \mathbb{R}^n\} = L_{1,0}$ is a line of our plane.

Each line $L_{s,0}$ intersects the vertical line $\{1\} \times \mathbb{R}^n$ precisely in the point $(1, f(s, 1, 0))$, thus $g : s \mapsto f(s, 1, 0) : \mathbb{R}^n \to \mathbb{R}^n$ is a continuous bijection and even a homeomorphism. The line joining $o = (0, 0)$ and $(1, s)$ has slope $s' = g^{-1}(s)$.

A line $L_{s',t'}$ intersects the y-axis $\{0\} \times \mathbb{R}^n$ precisely in the point $(0, f(s', 0, t'))$. The map $h : (s', t') \mapsto (s', f(s', 0, t'))$ is a continuous bijection, and even a homeomorphism, of $\mathbb{R}^{2n}$ onto itself.

Now we coordinatize the (new) plane with respect to the quadrangle $o = (0, 0), u = (0), v = (\infty), e = (1, f(1, 1, 0)) \in L_{1,0}$. The corresponding ternary operation τ is given by $\tau(s, x, t) = f(s', x', t')$ with $s' = g^{-1}(s)$ and $(s', t') = h^{-1}(s', t)$, see (22.1). This shows that τ is continuous, and we can apply (43.6). □

43.9 Notes on Section 43. Several arguments in this section are taken from Grundhöfer [87a]. See Arumugam [85] Theorem 3 for a result related to (43.1). Theorem (43.5) can be found in Skornjakov [54], Theorem 9 and in Salzmann [67b] 7.15, and Theorem (43.6) is due to Knarr–Weigand [86]. Corollary (43.8) goes back to Betten [95]. See Salzmann [67b] 7.17 and Löwen [81a] 2.4 for other proofs of Corollary (43.7).

44 Automorphism groups

The group Σ of all continuous collineations of a compact projective plane plays a dominant rôle in this book. Here we show that Σ, equipped with the compact-open topology, is a locally compact transformation group. Furthermore we obtain some information on groups of axial collineations. Deeper results on Σ and on groups of axial collineations can be found in the next chapters, see in particular Sections 55, 61, 83 and 87.

Let $\mathcal{P} = (P, \mathcal{L})$ be a topological projective plane. A collineation of $\mathcal{P}$ is continuous with respect to the topology of P if, and only if, it is continuous with respect to the topology of $\mathcal{L}$, (use the fact that $\mathcal{L}$ carries the quotient topology with respect to $\vee$, see (41.2c)). The complex projective plane $\mathcal{P}_2\mathbb{C}$ is the only compact projective plane known to admit discontinuous collineations, compare the Notes (44.11).

An *isomorphism* between two topological projective (or affine) planes is a collineation which is also a homeomorphism; the existence of such an isomorphism is expressed by saying that the two planes are *isomorphic* (as topological planes).

44.1 Lemma. *Let $\mathcal{P} = (P, \mathcal{L})$ be a topological projective plane, and assume that $X \subseteq P$ contains a non-degenerate quadrangle. Let $\langle X \rangle$ denote the topological closure of the point set of the subplane generated by X.*

(a) *If a continuous collineation α fixes all points of X, then α fixes all points in $\langle X \rangle$.*

(b) *Let Γ, Δ be groups of continuous collineations of $\mathcal{P}$ which centralize each other. Then the stabilizer $\Gamma_{[X]}$ acts trivially on $\langle x^\delta | x \in X, \delta \in \Delta \rangle$.*

Proof. Assertion (b) is a consequence of (a), because $\Gamma_{[X]}$ acts trivially on the set $\{ x^\delta \mid x \in X, \delta \in \Delta \}$, compare (91.1c). Now we prove (a). The point set of the subplane generated by X is the union $\bigcup_{n \geq 0} X_n$, where $X_0 = X$ and X_{n+1} consists of all points $ab \wedge cd$ such that $a, b, c, d \in \bigcup_{i \leq n} X_i$ is a non-degenerate quadrangle. We conclude that α fixes each point in the union $\bigcup_{n \geq 0} X_n$, which is dense in $\langle X \rangle$. □

44.2 Automorphisms. Let $\mathcal{P} = (P, \mathcal{L})$ be a compact projective plane. Then the continuous collineations of $\mathcal{P}$ are in fact homeomorphisms of P and $\mathcal{L}$, hence they form a group Σ. They are called *automorphisms* of $\mathcal{P}$. We endow Σ with the compact-open topology derived from P, compare (96.3). By (41.8) we can choose a metric d for the topological space P. It is easy to show that the metric

$$d(\sigma, \tau) = \sup\{ d(x^\sigma, x^\tau) \mid x \in P \}$$

describes the compact-open topology on Σ, see (96.6). Convergence $\sigma_n \to \sigma$ holds in Σ if, and only if, $p_n^{\sigma_n} \to p^\sigma$ for all convergent sequences $p_n \to p$ in P, compare (96.4). It is straightforward to prove that the compact-open topology for Σ derived from P coincides with the compact-open topology derived from $\mathcal{L}$, cp. Hähl [75a] 3.6, Grundhöfer [86] Lemma 1.

44 Automorphism groups

44.3 Theorem. *Let $\mathcal{P} = (P, \mathcal{L}, \mathbf{F})$ be a compact projective plane, and denote by Σ the group of all continuous collineations of $\mathcal{P}$, endowed with the compact-open topology. Then Σ is a locally compact group with countable basis, acting as a topological transformation group on P, on $\mathcal{L}$ and on $\mathbf{F}$.*

Proof. Multiplication in Σ is continuous (cp. (96.5) or Dugundji [66] XII.2.2), and the inversion is continuous at $\mathbb{1}$ (hence everywhere), in view of $d(\sigma^{-1}, \mathbb{1}) = d(\mathbb{1}, \sigma)$. Thus Σ is a topological group. Furthermore, the action $\Sigma \times P \to P$ is continuous, and Σ has a countable basis, see (96.4 and 5). These facts are already contained in Arens [46] Theorem 3. Since the compact-open topology derived from the action on $\mathcal{L}$ coincides with the one derived from the action on P, the action $\Sigma \times \mathcal{L} \to \mathcal{L}$ is also continuous, and so is the restriction of the action on $P \times \mathcal{L}$ to $\mathbf{F}$.

We still have to show that Σ is locally compact. By (41.5a), the non-degenerate quadrangles form an open subset of P^4. Hence we can choose a non-degenerate quadrangle (e_1, e_2, e_3, e_4) in $\mathcal{P}$ and $\varepsilon > 0$ such that all quadruples (f_1, f_2, f_3, f_4) with $d(e_i, f_i) < \varepsilon$ are non-degenerate quadrangles. Define

$$d(x, L) = \inf \{ d(x, y) \mid y \in L \} \text{ for } x \in P, L \in \mathcal{L}.$$

Assume that the numbers $\sup \{ d(x, L) \mid x \in P \}$ with $L \in \mathcal{L}$ accumulate at zero. Then we find a sequence $L_n \to L$ of lines such that for every $x \notin L$ there exist points $x_n \in L_n$ with $\lim_n x_n = x$. Thus $x \in L$ by (41.5a), a contradiction. Hence we may diminish ε such that $3\varepsilon < \sup \{ d(x, L) \mid x \in P \}$ for every line $L \in \mathcal{L}$.

1) We show that $\Sigma_\varepsilon = \{ \sigma \in \Sigma \mid d(\sigma, \mathbb{1}) \leq \varepsilon \}$ has compact closure $\overline{\Sigma_\varepsilon}$ in the space P^P of all continuous maps $P \to P$, endowed with the compact-open topology.

By the Arzelà–Ascoli theorem (see Dugundji [66] XII.6.4), this follows from equicontinuity of Σ_ε on P. Assuming that Σ_ε is not equicontinuous, we find $\sigma_n \in \Sigma_\varepsilon$ and a convergent sequence $p_n \to p$ in P such that $p^{\sigma_n} \to p'$ and $p_n^{\sigma_n} \to r \neq p'$. By the choice of ε, there is a point $q \in P$ with $d(q, rp') > 3\varepsilon$. Denote by $\mathcal{P}_0 = (P_0, \mathcal{L}_0)$ a countably infinite subplane containing the points p, q, e_1, e_2, e_3, e_4. By Cantor's diagonal process (cp. Dugundji [66] p. 231) we may replace (σ_n) and (p_n) by subsequences such that (σ_n) converges on every point of $\mathcal{P}_0$. Then the map $x \mapsto x' = \lim_n x^{\sigma_n}$ is an epimorphism $\mathcal{P}_0 \to \mathcal{P}'_0$ of projective planes, since $\mathcal{P}'_0$ contains the quadrangle (e'_i); note that $d(x, x') \leq \varepsilon$ for all $x \in P_0$.

Assume that $p'q'$ is not contained in the closure of $\{ L' \mid L \in \mathcal{L}_0, p \in L \neq pq \}$, and aim for a contradiction. The pre-images of points under a non-injective epimorphism intersect each line either trivially or in infinitely many points, compare Prieß-Crampe [83] V, §4, Lemma 4, p. 246, or Hughes [60], Mortimer [75]. We conclude first that the epimorphism $\mathcal{P}_0 \to \mathcal{P}'_0$ is injective, as the restriction to

$(\mathcal{L}_0)_p$ is injective at pq. Hence $\mathcal{P}'_0$ is infinite. Furthermore $p'q'$ is isolated in the pencil $(\mathcal{L}'_0)_{p'}$, thus $\mathcal{P}'_0$ is discrete (41.2), a contradiction to (41.12).

Thus $p'q'$ is contained in the closure of $\{ L' \mid L \in \mathcal{L}_0, p \in L \neq pq \}$. Hence we find lines $L_m \in (\mathcal{L}_0)_p \setminus \{pq\}$ with $\lim_m L'_m = p'q'$. Then

$$\lim_{n \to \infty} (p_n q \wedge L_m) = pq \wedge L_m = p \quad \text{and} \quad \lim_{n \to \infty} (p_n q \wedge L_m)^{\sigma_n} = rq' \wedge L'_m ,$$

which converges to $rq' \wedge p'q' = q'$ for $m \to \infty$; note that $rq' \neq p'q'$ by the choice of q. Now $\sigma_n \in \Sigma_\varepsilon$ gives $d(p, rq' \wedge L'_m) \leq \varepsilon$ and, upon passage to the limit, $d(p, q') \leq \varepsilon$, hence $d(p', q) \leq 3\varepsilon$, a contradiction to $d(q, rp') > 3\varepsilon$. This proves 1).

2) We complete the proof by showing that $\overline{\Sigma_\varepsilon} = \Sigma_\varepsilon$.

Let $\sigma = \lim_n \sigma_n \in \overline{\Sigma_\varepsilon}$ with $\sigma_n \in \Sigma_\varepsilon$. The elements $\sigma_n^{-1} \in \Sigma_\varepsilon^{-1} = \Sigma_\varepsilon \subseteq \overline{\Sigma_\varepsilon}$ accumulate at some $\tau \in P^P$, by step 1). Since composition is a continuous operation on P^P, compare Dugundji [66] XII.2.2, we get $\mathbb{1} = \tau\sigma = \sigma\tau$. Hence σ is bijective, $\sigma \in \Sigma$ and $\sigma \in \Sigma_\varepsilon$. □

By Gleason–Palais [57] 5.6, the compact-open topology is the only topology which has a countable basis and renders Σ a locally compact transformation group on P.

The following technical lemma describes the compatibility of the compact-open topology with various restriction maps. For sharper versions of (44.4d) see Grundhöfer–Stroppel [92]. We give an example which shows that not every restriction is a quotient mapping: The group $\Gamma = (GO_2\mathbb{R})^1$ of all orientation-preserving similarities of the Euclidean vector space $\mathbb{R}^2$ is isomorphic to $\mathbb{R}/\mathbb{Z} \times \mathbb{R}$. The element $(\sqrt{2} + \mathbb{Z}, 1)$ generates a discrete subgroup $Z \cong \mathbb{Z}$. Restriction of Γ to the line at infinity of the affine plane $\mathbb{R}^2$ corresponds to the projection of Γ to its first factor. This projection maps Z onto a dense subgroup, hence this projection, restricted to Z, is not a quotient map.

44.4 Lemma. *Let $\mathcal{P} = (P, \mathcal{L})$ be a compact projective plane and let Σ be the automorphism group of $\mathcal{P}$, endowed with the compact-open topology.*

(a) *Let $M \neq \emptyset$ be an open subset of P and let $\Delta = \{ \delta \in \Sigma \mid M^\delta = M \}$. Then the restriction $\Delta \to \Delta|_M : \delta \mapsto \delta|_M$ is a homeomorphism of Δ (as a subspace of Σ) onto $\Delta|_M$, endowed with the compact-open topology derived from M.*
(b) *For every line $L \in \mathcal{L}$, the topology of the stabilizer Σ_L as a subspace of Σ coincides with the compact-open topology for Σ_L derived from the affine point set $P \setminus L$.*
(c) *Let (K, τ) be the ternary field coordinatizing $\mathcal{P}$ with respect to a quadrangle o, u, v, e. Then the stabilizer $\Sigma_{o,u,v,e}$ is isomorphic as a topological group to*

the group $\mathrm{Aut}(K, \tau)$ of all continuous automorphisms of (K, τ), endowed with the compact-open topology derived from K.

(d) Let Δ be a closed subgroup of Σ, and let $B \subseteq P$ be the point set of a closed Baer subplane which is invariant under Δ. Endow $\Delta|_B$ with the compact-open topology derived from B. Then the restriction $\delta \mapsto \delta|_B : \Delta \to \Delta|_B$ is a quotient map, hence $\Delta/\Delta_{[B]} \cong \Delta|_B$ as topological groups.

Proof. (a) The restriction is continuous, and injective, as $\mathcal{P}$ is generated by M; it remains to prove continuity of $\delta|_M \mapsto \delta$. Let $\delta_n, \delta \in \Delta$ such that $m^{\delta_n} \to m^\delta$ for every convergent sequence $m_n \to m$ in M. It suffices to show that $p_n^{\delta_n} \to p^\delta$ for every convergent sequence $p_n \to p$ in P, compare (44.2). As M is open, we find points $a, a', b, b' \in M$ such that a, b, p are not collinear and $a' \in ap \setminus \{a, p\}$, $b' \in bp \setminus \{b, p\}$. For sufficiently large n, we have $p_n = ap_n \wedge bp_n = ar_n \wedge bs_n$ with $r_n = ap_n \wedge a'b' \to a'$ and $s_n = bp_n \wedge a'b' \to b'$. Since M contains almost all the points r_n, s_n, we infer that $r_n^{\delta_n} \to a'^\delta$ and $s_n^{\delta_n} \to b'^\delta$ and therefore

$$p_n^{\delta_n} = a^{\delta_n} r_n^{\delta_n} \wedge b^{\delta_n} s_n^{\delta_n} \to a^\delta a'^\delta \wedge b^\delta b'^\delta = (aa' \wedge bb')^\delta = p^\delta.$$

(b) is the special case of (a) where $M = P \setminus L$.

(c) The mapping $\alpha \mapsto (\alpha, \alpha) : \mathrm{Aut}(K, \tau) \to \Sigma_{o,u,v,e}|_{K^2}$ is bijective (23.5), continuous and open, hence the assertion follows from (b).

(d) See Grundhöfer–Stroppel [92] (15). □

We conjecture that in a compact projective plane the stabilizer $\Sigma_{o,u,v,e}$ of a quadrangle o, u, v, e is always compact. This conjecture has been verified in various situations, e.g. for translation planes:

44.5 Theorem. *In every compact projective translation plane, the stabilizer $\Sigma_{o,u,v,e}$ of every quadrangle o, u, v, e is compact.*

For a proof, and for more information, see (81.5), Hähl [75a] 2.6 for connected planes, and Grundhöfer [86] for the general case. See also (32.10), (55.21) and the remarks after (55.22) for related results.

Another conjecture says that for compact, connected projective planes the group Σ is always a (real) Lie group, compare (32.21), (71.2), (87.1). This is easily seen to be true for translation planes:

44.6 Theorem. *Let $\mathcal{P} = (P, \mathcal{L})$ be a compact, connected projective translation plane with translation line W. Then the group Σ of all continuous collineations of $\mathcal{P}$ is a linear Lie group. The stabilizer Σ_W acts on the set $P \setminus W \approx \mathbb{R}^{2n}$ of affine* *points of $\mathcal{P}^W$ as a closed subgroup of the affine group* $\mathrm{AGL}_{2n}\mathbb{R}$; *in fact,*

$$\Sigma_W = \mathsf{T} \rtimes \Sigma_{W,o}$$

is the semidirect product of the translation group $\mathsf{T} = \Sigma_{[W,W]} \cong (\mathbb{R}^{2n}, +)$ *and of the closed linear group* $\Sigma_{W,o} \le \mathrm{GL}_{2n}\mathbb{R}$ *for any point* $o \in P \setminus W$.

Proof. If $\mathcal{P}$ is a Moufang plane, then (42.7) and (13.6), (17.8) give the assertion. Otherwise Σ fixes the unique translation line W of $\mathcal{P}$, see the Theorem of Skornjakov–San Soucie (24.9). Let Q be a quasifield coordinatizing the affine plane $\mathcal{P}^W$. We can identify Σ_W with its restriction to the affine point set $P' = P \setminus W = Q^2$, even as a topological group (44.4b), and Q is a vector space of finite dimension n over $\mathbb{R}$ by (42.6). As $\mathsf{T} \cong \mathbb{R}^{2n}$ is a normal subgroup acting sharply transitively on $P \setminus W$, we have $\Sigma_W = \mathsf{T} \rtimes \Sigma_{W,o}$, and every element of $\Sigma_{W,o}$ is semi-linear over the kernel of Q, see (25.5). Hence the elements of $\Sigma_{W,o}$ are linear over $\mathbb{R}$, in view of $\mathbb{R} = \overline{\mathbb{Q}}$. Thus Σ_W is a subgroup of the affine group $\mathrm{AGL}(Q^2, \mathbb{R}) = \mathrm{AGL}_{2n}\mathbb{R}$. Note that the compact-open topology of $\mathrm{AGL}_{2n}\mathbb{R}$ coincides with the natural topology of that group (because an affine map is uniquely determined by its values on finitely many vectors). The group Σ_W is a closed subgroup of $\mathrm{AGL}_{2n}\mathbb{R}$, because the collinear triples $\{ (p, q, r) \in P^{(3)} \mid pq = pr \}$ of distinct points form a closed subset of $P^{(3)} = \{ (p, q, r) \in P^{(3)} \mid p \ne q \ne r \ne p \}$. Every closed subgroup of a linear Lie group is also a linear Lie group, see (94.3a). □

The following two results are concerned with the topological behavior of (groups of) axial collineations. More information on axial collineations in compact, connected planes can be found in Section 61.

44.7 Lemma. *Every axial collineation α of a topological projective plane $\mathcal{P} = (P, \mathcal{L})$ is a homeomorphism of P and of $\mathcal{L}$.*

Proof. The restriction of α to any line which avoids the center of α is a perspectivity and therefore a homeomorphism (41.2a), and dually for pencils. By (41.2d,e), the topologies on P and $\mathcal{L}$ are uniquely determined by the topologies on these lines and pencils, hence α is a homeomorphism. □

44.8 Proposition. *Let $\mathcal{P} = (P, \mathcal{L})$ be a compact projective plane, and let Δ be a closed subgroup of the automorphism group Σ of $\mathcal{P}$, endowed with the compact-open topology.*

(a) *The center and the axis of a non-trivial axial collineation in Σ depend continuously on the collineation.*

(b) *Let $c \in P$ and $A \in \mathcal{L}$. Then the topological group $\Delta_{[c,A]}$ is isomorphic to a closed subgroup of the additive loop $(K, +)$ (for $c \in A$) or of the multiplicative loop $K^\times$ (for $c \notin A$) of some ternary field K coordinatizing the affine plane $\mathcal{P}^A$. Every non-trivial orbit of $\Delta_{[c,A]}$ is closed in $P \setminus (A \cup \{c\})$ and homeomorphic to $\Delta_{[c,A]}$. If Δ is (c, A)-transitive, then $\Delta_{[c,A]}$ is isomorphic to $(K, +)$ or to $K^\times$, respectively.*

(c) *Let $p \in P \setminus A$. Then the evaluation $\delta \mapsto p^\delta$ is a homeomorphism of the elation group $\Delta_{[A,A]}$ onto the orbit of p under $\Delta_{[A,A]}$, and this orbit is closed in $P \setminus A$.*

Proof. (a) Let $\sigma \in \Sigma \setminus \mathbb{1}$ be axial. There exist points $p, q \in P$ such that $p \neq p^\sigma$, $q \neq q^\sigma$ and $pp^\sigma \neq qq^\sigma$. Let τ be an axial collineation in the open neighbourhood $\{\tau \in \Sigma \mid p \neq p^\tau, q \neq q^\tau \text{ and } pp^\tau \neq qq^\tau\}$ of σ. Then the center of τ is given by $pp^\tau \wedge qq^\tau$, which depends continuously on τ. For the axis, the dual arguments apply.

(b) Coordinatize $\mathcal{P}^A$ by a ternary field K such that $c \in \{o, u\}$, where $o = (0, 0)$, $u = (0) \in A$. We regard Δ_A as a transformation group of the affine point set K^2, compare (44.4b). Then

$$\Delta_{[o,A]} = \{(x, y) \mapsto (xa, ya) \mid a \in S_1\} \quad \text{and}$$
$$\Delta_{[u,A]} = \{(x, y) \mapsto (x + a, y) \mid a \in S_2\}$$

with suitable closed subgroups S_1 of $K^\times$ and S_2 of $(K, +)$, see (23.11). The orbits of a point $p = (x, y) \notin A \cup \{c\}$ are the sets $\{(xa, ya) \mid a \in S_1\}$ and $\{(x + a, y) \mid a \in S_2\}$, respectively. Hence the assertions follow from the continuity of addition, multiplication and their inverse operations (cp. Dugundji [66] XII.3.1(1)).

(c) Clearly the evaluation is a continuous bijection. Let $q \in P \setminus A$ and $\delta_n \in \Delta_{[A,A]}$ such that $p^{\delta_n} \to q$. It suffices to prove that $\delta_n \to \delta$ for some $\delta \in \Delta_{[A,A]}$.

The elation δ, if it exists, is uniquely determined by $p^\delta = q$. Hence for the construction of δ, we may switch to a subsequence of (δ_n) such that the centers c_n of δ_n converge to some point $c \in A$. Let $X = P \setminus (A \cup pc)$ and define $\delta : X \to X$ by $x^\delta = (xc \wedge ((xp \wedge A)q)$ for $x \in X$.

Let $x_n \to x$ be a convergent sequence in X. Then $x_n^{\delta_n} = x_n c_n \wedge (x_n p)^{\delta_n} = x_n c_n \wedge (x_n p \wedge A) p^{\delta_n}$ converges to x^δ. This means that $\delta_n \to \delta$ in the compact-open topology of X, see (44.2) or (96.4). The transformation $x \mapsto xc \wedge ((xq \wedge A)p)$ is inverse to δ, hence δ is bijective and therefore extends to a collineation, in fact to an elation with center c. Now (44.4b) says that $\delta_n \to \delta$ in the compact-open topology of P. □

44.9 Proposition. *Let $\mathcal{P} = (P, \mathcal{L})$, $\mathcal{P}' = (P', \mathcal{L}')$ be topological projective planes, let $U \subseteq P$ and $U' \subseteq P'$ be open subsets, and let $\varphi : U \mapsto U'$ be a homeomorphism such that $x, y, z \in U$ are collinear if, and only if, $x^\varphi, y^\varphi, z^\varphi$ are collinear. If U*

is dense in P, then φ has a unique extension to a (geometrical and topological) isomorphism of $\mathcal{P}$ onto $\mathcal{P}'$.

Proof. The extension is unique, as U is dense in P. In order to construct an extension, choose distinct points $u, v \in U$, and let $A = P \setminus uv$, $A' = P' \setminus u^\varphi v^\varphi$. It suffices to show that the restriction $\varphi|_{U \cap A}$ extends to an isomorphism $\psi : A \to A'$ of topological affine planes (note that by (41.2) the topology of P is determined uniquely by the topology of A and the geometry of $\mathcal{P}$). The line pencils $\mathcal{L}_u, \mathcal{L}_v$ are contained in $\mathcal{L}_U = \{ L \in \mathcal{L} \mid |L \cap U| \geq 2 \}$, since U is open and P is not discrete. By assumption, φ induces a well-defined mapping on $\mathcal{L}_U$, hence we can define
$$x^\psi = (xu)^\varphi \wedge (xv)^\varphi$$
for $x \in A$. Then $\psi : A \to A'$ is a continuous extension of $\varphi|_{U \cap A}$ with continuous inverse
$$y \mapsto y^{\psi^{-1}} = (yu^\varphi)^{\varphi^{-1}} \wedge (yv^\varphi)^{\varphi^{-1}}.$$

Now we show that the collinear triples of distinct points in $A \cap U$ form a dense subset of all collinear triples in A; together with the corresponding statement for A', this implies that ψ and ψ^{-1} are collineations.

Let V, V', V'' be open neighbourhoods of three distinct collinear points p, p', p'' in A. By (41.2c) the set $\mathcal{L}_V = \{ x \vee y \mid x, y \in V, x \neq y \}$ is an open neighbourhood of $p' \vee p''$, hence we may assume that $V' \cap V'' = \emptyset$ and $V' \vee V'' \subseteq \mathcal{L}_V$. Since $V \cap U$ is open and non-empty, the set $\{ (a, b) \in V' \times V'' \mid a \vee b \in \mathcal{L}_{V \cap U} \}$ is open in A^2 and non-empty, hence it meets the dense set U^2. Thus we find a collinear triple in $(V' \times V'' \times V) \cap U^3$. □

Proposition (44.9) implies that most of the methods for the construction of finite projective planes cannot be applied to topological projective planes; the construction of Hughes planes from suitable nearfields is a noteworthy exception, see Section 86.

44.10 Corollary. *Let φ be a continuous collineation between two locally compact, connected affine planes. Then the unique extension (23.4) of φ is a continuous collineation between the projective completions.*

Proof. This special case of (44.9) may also be proved directly. One has to show only that the extension described in (23.4) is continuous, and this follows from the fact the topology on a projective plane is determined uniquely by the topology of an affine part, see (41.2). □

44.11 Notes on Section 44. Information on discontinuous automorphisms of $\mathbb{C}$ can be found in Segre [47], Kestelman [51], Baer [70], Dieudonné [74], Kallman–Simmons [85], Keller [86], Schnor [92]. In fact, discontinuous automorphisms of a topological ternary field $K \approx \mathbb{R}^n$ are highly discontinuous: the image of the sphere $\mathbb{S}_{n-1} \subset K$ is dense in K, see (55.22a). Compare also Knarr [87a].

Versions of Theorem (44.3) can be found in Löwen [76b] 2.9, Salzmann [75a] (∗); see also Bödi [93b] for an analogous result on locally compact double loops. See Burns–Spatzier [87] for a generalization to compact, metric Tits-buildings. Lemma (44.4) extends Hähl [75a] 3.5, and Proposition (44.9) is adapted from Löwen [81c] §3.

Chapter 5

Algebraic topology of compact, connected planes

All planes $\mathcal{P} = (P, \mathcal{L}, \in)$ in this chapter will be compact, connected projective planes.

Recall that this is equivalent to P being locally compact and of positive topological dimension, $\dim P > 0$ (42.2 and 4). In this chapter we consider the local and global topological properties of the topological spaces P (point space), L (a line), and $\mathbf{F}$ (flag space) associated with a plane $\mathcal{P}$. We shall see that the presence of a geometry imposes most effective restrictions on these spaces.

In all examples that are known sufficiently well, the spaces under consideration are manifolds homeomorphic to their counterparts in one of the classical planes $\mathcal{P}_2\mathbb{F}$, where $\mathbb{F} \in \{\mathbb{R}, \mathbb{C}, \mathbb{H}, \mathbb{O}\}$. In particular, L is homeomorphic to a sphere $\mathbb{S}_l$ whose dimension l divides 8, and the point space is a manifold of dimension $2l$. At present, it seems impossible to prove that this is true in general, but we shall offer an efficient substitute. For planes of finite topological dimension, many of the invariants of algebraic topology will be computed in this chapter, and indeed they show no difference between P, L, $\mathbf{F}$ and their classical counterparts. In particular, the dimension $\dim P$ is one of the numbers $2l \in \{2, 4, 8, 16\}$. The topological resemblance to classical planes has strong geometric consequences, some of which will be discussed in Section 55.

The range of validity of our results is limited only by our assumption of finite dimensionality. It is still an open problem whether or not this hypothesis is satisfied by all compact planes. Typical situations where this property can be proved are those considered in Sections 63 and 64. On the other hand, we remark that the hypothesis of (local) compactness is essential for our results. In fact, Ursul [87] constructs topological fields of inductive dimension n for each positive integer n. They define projective planes with n-dimensional lines by a result of Salzmann [57] Section 14.

The logical dependencies within this chapter are such that the line of investigation should proceed from local to global properties. The order of presentation is inverted, however, because the proofs become much easier if one is willing to grant that lines are manifolds. The organization of the chapter easily allows the reader to make this assumption. He may then skip Section 54, where the local properties are studied using methods of sheaf theory, and still see as much as possible of the arguments needed in the general case.

In order to understand the subsequent chapters, it suffices to be acquainted with the main *results* of this chapter or to refer back to them where necessary. The deeper topological *methods* used to prove them will not play a dominant rôle outside this chapter.

51 General properties

Our first aim is to show that $P, \mathcal{L}$ and $\mathbf{F}$ have many of the useful properties of topological manifolds known from dimension theory, compare Hurewicz–Wallman [48]. Moreover, we consider product and fiber bundle structures arising from the geometry of a plane and we study contraction and deformation properties as well as simple connectedness.

Note beforehand that it is not necessary to discuss the spaces $\mathcal{L}$ and $\mathcal{L}_x$ (line pencil), because they appear respectively as the point space and as a line of the dual plane. Moreover, $\mathcal{L}_x$ is homeomorphic to any line L via projectivities (21.6), hence P is locally homeomorphic to $\mathcal{L}$ (41.2d); compare (51.2) below. The line space of an affine plane is the complement of an element in the dual point set.

We start by recalling some results from Chapter 4.

51.1 Review: Topological homogeneity.

(a) *The homeomorphism group of a line L is triply transitive; we say that L is 3-homogeneous* (41.2a).
(b) *The point set P is n-homogeneous for all $n \in \mathbb{N}$ by* (42.13). □

This says that topological homogeneity is a general property of compact, connected planes. Note by contrast that geometric homogeneity as studied in Chapter 6 is rare.

Two spaces X, Y are said to be *locally homeomorphic* if for every pair of points $x \in X$, $y \in Y$ there exist open neighbourhoods $U \subseteq X$ and $V \subseteq Y$ of x and y, respectively, and a homeomorphism $U \approx V$ sending x to y. For example, a sphere $\mathbb{S}_n$ is locally homeomorphic to $\mathbb{R}^n$.

51.2 Review: Product structures. *Both P and $\mathcal{L}$ are locally homeomorphic to $L \times L = L^2$, and $\mathbf{F}$ is locally homeomorphic to L^3.*

Proof. For P and $\mathcal{L}$, use (41.2a,d) and the dual assertions to obtain a homeomorphism, e.g., of $P \setminus L$ onto the space $(L \setminus \{a\})^2$; here, a denotes any point of L.

The image space is homogeneous as a consequence of (51.1), hence the homeomorphism can be adjusted such that it maps a given point x to another given point, y.

For $\mathbf{F}$, consider the homeomorphism $(q, K) \mapsto (q, K \wedge L)$ mapping the open subset $\{(q, K) \in \mathbf{F} \mid q \notin L\}$ of $\mathbf{F}$ onto the space $(P \setminus L) \times L$, which in turn is homeomorphic to $(L \setminus \{a\})^2 \times L$ by (41.2d). Again, the last space is homogeneous by (51.1). □

51.3 Definition. Let X, Y be topological spaces. By a convenient abuse of language, a *homotopy* $F : X \times [0, 1] \to Y$ will always be written as a collection of maps $F_t : X \to Y$, $0 \le t \le 1$, and continuity of the map F will be tacitly understood. If $X = Y$ and F_t is a homeomorphism for each $t \ne 1$, then F is called a *pseudo-isotopy* on X. The space X is *pseudo-isotopically contractible* if there is a pseudo-isotopy F_t on X which contracts X, i.e., satisfies $F_0 = \mathrm{id}_X$ and $F_1(X) = \{x_0\}$.

The following proposition formulates one of the basic principles supporting this chapter. It is essential for almost everything that is done in Sections 51 to 54.

51.4 Review: Contractibility. *A punctured line $L \setminus \{x\}$ is contractible by a pseudo-isotopy that fixes a preassigned point $y \ne x$. Consequently, $P \setminus L \approx (L \setminus \{x\})^2$ also is pseudo-isotopically contractible. The spaces P, L and $\mathbf{F}$ are locally contractible. The quotient space P/L introduced in (51.6) below also has this property, see (51.27).*

Proof. For local and global contractibility of $L \setminus \{x\}$, see (42.8). The contraction $F_t(x) = e_t \cdot x$ constructed there is a pseudo-isotopy. In view of (51.2), it follows that P and $\mathbf{F}$ are locally contractible. □

51.5 Remark. In particular, the complement of every point in L has trivial homology groups. According to McCord [66], spaces X with this property resemble spheres in many respects. If moreover $H_n X \ne 0$ for $n = 1$ or 2, then X is in fact homeomorphic to an n-sphere. If such a space X is an n-dimensional polyhedron (n arbitrary), then X is a homology manifold (compare Section 54). In general, X satisfies some weak form of Alexander duality. The last results are, however, not suitable for our purposes, since we cannot directly verify any of the additional hypotheses. See (52.5) and Section 54 in this respect.

51.6 The point space modulo a line. An important rôle will be played by the quotient space

$$P/L$$

obtained from P by shrinking the line L to a point. Note that P/L may be interpreted as the one-point compactification of the affine plane $P \setminus L$. Indeed, the identification map induces a homeomorphism of $P \setminus L$ onto $P/L \setminus \{L\}$. Eventually it will turn out that P/L has the homotopy type of a sphere (52.12). In the classical plane $\mathcal{P}_2\mathbb{F}$, the space $P/L \approx \mathbb{F}^2 \cup \{\infty\}$ is in fact homeomorphic to a sphere.

51.7 Definition. A compact, connected and locally connected metric space is called a *Peano continuum*. The name reflects the fact that these spaces are precisely those Hausdorff spaces which admit a 'space-filling curve' or 'Peano curve', i.e., which are continuous images of the unit interval (theorem of Hahn and Mazurkiewicz, see Christenson–Voxman [77] 9B3).

The following result will be needed in (61.5).

51.8 Proposition: Peano continua. *The spaces P, L, P/L and $\mathbf{F}$ are Peano continua.*

Proof. By (41.7a and 8), (42.8) and (51.2), the spaces P, L and $\mathbf{F}$ are Peano continua as defined in (51.7). The quotient space P/L is easily seen to be a Hausdorff space. The defining properties of Peano continua carry over to Hausdorff quotient spaces, see Christenson–Voxman [77] 8C6 and 10C8. Alternatively, the theorem of Hahn and Mazurkiewicz quoted in (51.7) can be used to show that a Hausdorff space which is a continuous image of a Peano continuum is itself a Peano continuum. □

51.9 Problem: Finiteness of dimension. For the deeper results of this chapter, we need the hypothesis that the topological dimension

$$l = \dim L$$

of a line is finite; for the definition, see Section 92. Using the theory of locally compact groups, we shall prove this under rather weak geometric homogeneity assumptions, see (53.1 to 4). Thus, at least all the nice examples will be finite-dimensional, and it is tempting to conjecture that no infinite-dimensional compact planes exist. No promising tool for a general proof has been found to date. On the other hand, there are infinite-dimensional complete topological fields and corresponding topological projective planes.

Specifically, there are the following two problems:

(1) Is $l = \dim L$ finite for all compact planes?
(2) Is there another topological property (e.g., completeness?) to replace compactness such that interesting phenomena occur in infinite-dimensional planes with this property?

The assumption of finite dimensionality will be needed in order to obtain that lines are ANR (51.12). This fact will allow us to apply Whitehead's theorem in the proof of (52.5). It will also be an essential prerequisite for our theorem that lines are homology manifolds (54.10); this result rests on Bredon's characterization of homology manifolds, which has the hypothesis of finite dimensionality.

The framework of Section 54 seems to be the most likely context where an argument leading to a positive solution of (1) might be found. However, one can hardly expect to obtain more than finite cohomological dimension in this way, and it is now known that this does not imply finite covering dimension, see Dranishnikov [88b, 89], Dydak–Walsh [93].

We recall from (41.8) that all dimension functions considered in Section 92 agree on each of the spaces L, P, $\mathscr{L}$ and $\mathbf{F}$. This is also true for the quotient space $P/L \approx (P \setminus L) \cup \{\infty\}$ introduced in (51.6), which is a separable metric space, compare (51.8). From (51.2) we infer that $\dim P = \dim \mathscr{L}$.

51.10 Proposition. *We have* $\dim P = \dim \mathscr{L} = \dim P/L$. *If one of the spaces L, P or $\mathbf{F}$ has finite dimension then the same is also true for the other two.*

Proof. The first equation has been proved. Next, observe that $l = \dim L \leq \dim P \leq \dim \mathbf{F}$ by the subspace theorem (92.4). On the other hand, the product theorem (92.10) together with (51.2) shows that $\dim P \leq 2l$ and $\dim \mathbf{F} \leq 3l$. This proves the second statement, and it remains to verify the equation $\dim P/L = \dim P =: d$. At the points of $P \setminus L$, the inductive dimension of $(P \setminus L) \cup \{\infty\} \approx P/L$ clearly agrees with d. Now consider the point ∞. The boundary B of a neighbourhood of ∞ is a subset of $P \setminus L$ with empty interior. By (92.16) and (51.1b and 4), this implies that $\dim B < \dim P$. This proves that $\dim P/L = d$. □

51.11 Definition. A metric space X is called an *absolute neighbourhood retract*, abbreviated ANR, if for every metric space Y containing (a homeomorphic copy of) X, there is a neighbourhood U of X in Y and a *retraction* $r : U \to X$, i.e., a continuous map inducing the identity on X. These spaces are closely related to polyhedra (spaces composed of simplices) or, more generally, to CW-complexes (spaces composed of cells) and therefore are very convenient to work with in algebraic topology. Fortunately, they can be characterized by local properties: Any *finite-dimensional* locally contractible metric space is an ANR; see Dugundji [58]. Infinite-dimensional counterexamples are known, see Borsuk [67] 11.1.

According to Hurewicz–Wallman [48] V.3 p.60, every finite-dimensional separable metric space X can be embedded into some Euclidean space $\mathbb{R}^n$. If moreover X is an ANR, then this implies that X is a *Euclidean neighbourhood retract* (ENR), i.e., homeomorphic to a retract of some open subset in $\mathbb{R}^n$; see Dold [72] pp. 81–84 for general information. Together with (51.4 and 10), these remarks prove the first part of the next result.

A space Y is said to be *locally homogeneous* if for any two points $x, y \in Y$ there are open neighbourhoods U of x and V of y that are homeomorphic via a homeomorphism sending x to y. It follows from (51.1a and b) that the spaces Y considered here have this property. This establishes the last part of the following result.

51.12 Proposition. *Every open subset X of P, L, P/L or $\mathbf{F}$ is locally contractible. If $l = \dim L < \infty$, then this implies that X is an ANR and in fact an ENR. Moreover, P, L and $\mathbf{F}$ are locally homogeneous.* □

51.13 Corollary. *If $l < \infty$, then every open subset of P, L, P/L or $\mathbf{F}$ is homotopy equivalent to some CW-complex. The spaces P, L, P/L and $\mathbf{F}$ themselves are even homotopy equivalent to compact polyhedra.*

Proof. Every ANR is homotopy equivalent to some CW-complex according to a result of J.H.C. Whitehead, see Weber [68] p. 218. In the case of compact ANR, West [77] 5.4 shows that this complex may be chosen to be a compact polyhedron. For a more recent proof, see Ranicki-Yamasaki [95]. □

51.14 Note. The assertion about open subsets in (51.13) will be an important ingredient in the proof of Theorem (52.5b) on the homotopy type of doubly punctured lines. The assertion about L itself is one of the cornerstones for Theorem (54.10), which states that L is a homology manifold.

In fact, this application of West's deep result in the proof of (54.10) can be avoided. Instead, one may use the weaker and more easily accessible domination theorem, which can be found, e.g., in R.F. Brown [71] p. 41. It asserts that every compact ANR X is *dominated* by some compact polyhedron Q, i.e., there are maps $f : X \to Q$, $g : Q \to X$ such that $g \circ f \simeq \mathrm{id}_X$. This has the following consequences. The total homology $H_*Q = \bigoplus_{n \in \mathbb{N}} H_n Q$ is finitely generated, because it can be interpreted as the simplicial homology, see Spanier [66] 4.6.8. It decomposes as $f_*(H_*X) \oplus \ker g_*$, and f_* is injective, because the induced homomorphisms satisfy $g_* \circ f_* = \mathrm{id}$. Hence, H_*X is finitely generated as well (X is of *finite type*), and this is all we shall need in the proof of (54.10).

51.15 Definition. A (topological) *n-manifold* is a separable Hausdorff space locally homeomorphic to $\mathbb{R}^n$. Abstracting one of the useful properties of n-manifolds (see Hurewicz–Wallman [48] p. 48, Cor. 1), we define a *Cantor n-manifold* to be an n-dimensional, locally compact, connected, metric space X such that $X \setminus A$ is connected for every closed subset $A \subseteq X$ with $\dim A \leq n - 2$. Thus, a connected n-manifold is a Cantor n-manifold, but not conversely. Note that we deviate from the terminology of the literature, where Cantor manifolds usually are compact by definition.

51.16 Proposition: Separation. *If $l = \dim L < \infty$, then L is a Cantor l-manifold.*

A more general fact is proved in (51.21) below. We record the following proof of the special statement mainly because of its simplicity.

Proof. Let $A \subseteq L$ be a closed set of dimension $\leq l - 2$. If $L \setminus A$ is disconnected, choose points $0, y \in L \setminus A$ in different components U, V and make $L \setminus \{y\}$ into a ternary field with 0 as zero element, compare (41.3). The boundary $\partial U \subseteq A$ is at most $(l-2)$-dimensional by the subspace theorem (92.4), and $\overline{U} \subseteq L \setminus V$ is compact. Taking $a \in L$ close to 0, we obtain arbitrarily small neighbourhoods $a \cdot \overline{U} \approx \overline{U}$ of 0 with an at most $(l-2)$-dimensional boundary. Therefore, the inductive dimension ind L is at most $l-1$ at the point 0, and hence everywhere by homogeneity (51.1). By (92.5 to 7), we have the contradiction $l = \dim L = \text{ind } L \leq l - 1$. $\square$

51.17 Remark. If $l = \infty$, the same proof shows that $L \setminus A$ is connected for finite-dimensional closed sets A.

An analogous result is proved by Krupski [90] for locally compact, connected, homogeneous, l-dimensional metric spaces in general; they are Cantor l-manifolds if l is finite, with similar extensions to the case $l = \infty$.

51.18 Definition. Following Brouwer, we say that a space X has the *domain invariance property* if the following is true: If $U \subseteq X$ is open and $f : U \to V \subseteq X$ is a homeomorphism, then V is open in X as well. (This is trivially satisfied if f extends to a homeomorphism of X.) Euclidean spaces have the domain invariance property, see Dugundji [66] XVII 3.1 or Spanier [66] 4.7.16, and this carries over to all n-manifolds, see Eilenberg–Steenrod [52] X 3.11, Hocking–Young [61] 6-54. We state two useful consequences (51.19 and 20) of the domain invariance property, which are not quite as widely known as they deserve to be.

51.19 Lemma: Open mapping. *Suppose that X and Y are locally homeomorphic Hausdorff spaces, where X is locally compact and Y has the domain invariance property. Then every continuous injection $f : X \to Y$ is an open map.*

Proof. Consider a point x of an open set $U \subseteq X$. Then x has an open neighbourhood $V \subseteq U$ with compact closure and an open neighbourhood W homeomorphic to an open subset of Y. The intersection $Z = V \cap W$ has both of these properties. By compactness, f induces a homeomorphism $\overline{Z} \approx f(\overline{Z})$. Thus, $f(Z) \approx Z$ is homeomorphic to an open subset of Y. By domain invariance, $f(Z)$ is open and $f(x)$ is an interior point of $f(U)$. $\square$

The pigeon-hole principle of combinatorics asserts that a set of finite cardinality cannot be mapped injectively into a set of smaller cardinality. In other words, any injection between two sets of the same finite cardinality is bijective. The following result, which can be applied to mappings between compact, connected manifolds of equal dimensions, is a topological analogue of this fact. Combined with (51.21) below, it provides a tool that is used in a way similar to the typical counting arguments of finite geometry.

51.20 Corollary: Topological pigeon-hole principle. *In addition to the hypotheses of* (51.19), *assume that X is compact and Y is connected. Then every continuous injection* $f : X \to Y$ *is surjective (in fact, a homeomorphism).*

Proof. The image $f(X)$ is closed, and is open by the preceding lemma. □

51.21 Theorem: Open sets, domain invariance and separation. *Assume that* $l = \dim L < \infty$, *and let X be any open subset of P, L or* **F**. *Then the following assertions hold.*

(a) *A closed subset* $A \subseteq X$ *satisfies* $\dim A = \dim X$ *if, and only if, A has non-void interior.*
(b) *X has the domain invariance property.*
(c) *If X is connected, then X is a Cantor manifold.*

Proof. Properties (a), (b), (c) have been proved for finite-dimensional, locally compact, locally homogeneous, separable ANR by Seidel [85] A,B,C, generalizing a result of Łysko. The proofs use homology techniques introduced by Bing and Borsuk in their paper [65]; later on, we shall find their work most useful again, see (53.7). By (51.10), X is finite-dimensional, so that Seidel's results apply to X in view of (51.12).

Properties (a), (b), (c) also follow from the much deeper fact that X is a homology manifold (54.10 and 8e). References are given in Löwen [83b] p. 120. □

Our next result (51.23) and its consequence (51.26) are the key to the global properties of P, once those of L are known sufficiently well. First we need a definition.

51.22 Definition. Let A, E and B be topological spaces. A map $f : E \to B$ is called a *trivial fiber bundle with fibers homeomorphic to* A if there is a homeomorphism (a *bundle chart*) $\varphi : B \times A \to E$ such that the composite $f \circ \varphi : B \times A \to B$ is the projection onto the first factor. The map f is called a *locally trivial fiber bundle with fibers homeomorphic to* A if there is an open cover $\mathcal{U}$ of B such that for every $U \in \mathcal{U}$ the restriction of f to $f^{-1}(U)$ is a trivial fiber bundle with fibers homeomorphic to A. Note that Spanier [66] omits the words 'locally trivial' and simply speaks of fiber bundles.

By Spanier [66] 2.7.14, locally trivial fiber bundles over a paracompact Hausdorff base space B are *fibrations* in the sense of Spanier [66] p. 66; that is, they have the homotopy lifting property for every space. In particular, they have the homotopy lifting property with respect to all cubes $[0, 1]^n$, hence they are *weak fibrations* in the sense of Spanier [66] p. 374. Therefore, they give rise to an exact sequence of homotopy groups, see Spanier [66] 7.2.10. This is where our interest in fiber bundles lies. In compact planes, we meet with fiber bundles in two places at least:

51.23 Proposition: Geometric fiber bundles.

(a) *For $p \in P \setminus L$, the central projection $f : P \setminus \{p\} \to L$ defined by $x \mapsto xp \wedge L$ is a locally trivial fiber bundle whose fibers are the punctured lines $xp \setminus \{p\}$, see Figure 51a.*

(b) *The projection $P \times \mathcal{L} \to P$ restricts to a locally trivial fiber bundle $g : \mathbf{F} \to P$, defined by $(q, K) \mapsto q$, with fibers homeomorphic to $K \in \mathcal{L}$.*

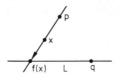

Figure 51a

Proof. 1) For $q \in L$ and $U = L \setminus \{q\}$, $\mathcal{A} = \mathcal{L}_q \setminus \{pq\}$, the map

$$\varphi : U \times \mathcal{A} \to f^{-1}(U),$$
$$(y, A) \mapsto py \wedge A$$

is a homeomorphism such that $f \circ \varphi$ is the projection onto the first factor. $\mathcal{A}$ is homeomorphic to a punctured line via a perspectivity.

2) For $L \in \mathcal{L}$ and $U = P \setminus L$, the map

$$\varphi : U \times L \to \mathbf{F},$$
$$(x, q) \mapsto (x, xq)$$

is a bundle chart for g restricted to $g^{-1}(U)$. (The inverse map φ^{-1} has been considered in the proof of (51.2).) □

51.24 Remark: Hopf fibrations. Recall from (14.11) and (16.15) that the classical Hopf fibrations are obtained from the maps f of (51.23a) associated with the classical planes $\mathcal{P}_2 \mathbb{F}$ by restriction to a sphere $\mathbb{S}_{2l-1}$ around the point p. There is no geometric way to define an analogue of these spheres in arbitrary planes, but we may regard the restriction of f to $P \setminus (L \cup \{p\})$ as a 'Hopf fibration up to homotopy', noting that $P \setminus (L \cup \{p\}) \approx \mathbb{R}^{2l} \setminus \{0\}$ has the homotopy type of a sphere if L is an l-manifold (52.3).

The map f will play an essential rôle in the determination of the topology of P, see (51.26), Section 52 and (53.11ff). For further applications of this map, see (55.44).

51.25 Definition. A subset R of a space X is called a *strong deformation retract* of X if there is a homotopy $F_t : X \to X$ such that $F_0 = \mathrm{id}_X$, $F_1(X) \subseteq R$ and F_t keeps R pointwise fixed for all t. Intuitively, this means that X is gradually deformed onto R, keeping R fixed all the time. X is then homotopy equivalent to R.

51.26 Proposition: Deformation of punctured planes. *For $p \in P \setminus L$, the line L is a strong deformation retract of $P \setminus \{p\}$. The deformation can be achieved by a pseudo-isotopy.*

Proof. The first assertion is a special case of the following well-known fact, see Dold [63] 2.8 and 3.2: If $f : E \to B$ is a locally trivial fiber bundle with contractible fibers and (para-) compact base B, then f admits a cross-section, and each cross-section is a deformation retract of E. In our situation, $f : P \setminus \{p\} \to L$ is the fiber bundle of (51.23). For this special case, we construct a deformation retraction onto the cross-section L provided by the geometric structure.

Figure 51b

We choose points such that $L = a_1 a_2$, and we define $A_i = p a_i$ and $\mathcal{W}_i = \mathcal{L}_{a_i} \setminus \{A_i\}$, see Figure 51b. The subset $\{L\}$ is a strong deformation retract of $\mathcal{W}_i$; in fact, the dual of (51.4) yields a pseudo-isotopy $F_t^i : \mathcal{W}_i \to \mathcal{W}_i$ that deforms $\mathcal{W}_i$ onto $\{L\}$ and fixes L. Then $P \setminus A_i$ is deformed onto $L \setminus \{a_i\}$ by

$$G_t^i : P \setminus A_i \to P \setminus A_i,$$
$$x \mapsto xp \wedge F_t^i(xa_i).$$

The problem is to piece G_1 and G_2 together to a single homotopy. In order to do this, choose open subsets $U_i \subseteq L$ such that $L = U_1 \cup U_2$ and $a_i \notin \overline{U_j}$ for $i \neq j$. Next, let $V_i \subseteq L \setminus \overline{U_j}$ be a closed neighbourhood of a_i, and choose Urysohn functions $\varphi_i : L \to [0, 1]$ such that $\varphi_i(V_i) = \{0\}$ and $\varphi_i(\overline{U_j}) = \{1\}$ for $i \neq j$; see

Dugundji [66] VII 4.1. Now define a pseudo-isotopy

$$H^i_t : P \setminus \{p\} \to P \setminus \{p\}$$

by $H^i_t = \mathrm{id}$ on A_i and

$$H^i_t(x) := G^i_s(x), \quad \text{where} \quad s = t\varphi_i(f(x)),$$

for $x \in P \setminus A_i$. Then H^i_t is the identity on $f^{-1}(V_i)$ and on L. Moreover, H^i_t maps each f-fiber $f^{-1}(x)$ to itself and deforms $f^{-1}(U_j)$ to U_j. Therefore, the composite pseudo-isotopy $H^1_t \circ H^2_t$ deforms $P \setminus \{p\}$ to L, as desired. □

51.27 Corollary: Local Contractibility. *The quotient space P/L defined in (51.6) is locally contractible. Every element $y \in P/L$ has a pseudo-isotopically contractible complement. In particular, (51.5) applies to P/L.*

Proof. We prove the second assertion first. If $y = L$, then $P/L \setminus \{y\} \approx P \setminus L$, and we may refer to (51.4) to obtain that this space is locally contractible and globally pseudo-isotopically contractible.

If $y \in P \setminus L$, let $U = P \setminus \{y\}$. By passage to the quotient, the strong deformation retraction $U \times [0, 1] \to U$ of U onto L given by (51.26) induces a strong deformation retraction of U/L onto $\{L\}$; see Dugundji [66] VI 4.3 and note that $U/L \times [0, 1]$ is a quotient space of $U \times [0, 1]$ by Dugundji [66] XII 4.1. The induced deformation fixes $\{L\}$, hence it may be used to prove local contractibility at this point. All global deformations may be chosen to be pseudo-isotopies. □

51.28 Proposition: Simple connectedness. *Let $L = ab$ and assume that $l = \dim L > 1$. Then the following assertions hold.*

(a) $L \setminus \{a, b\}$ *is path-connected.*
(b) *P, L and $\mathbf{F}$ are simply connected.*
(c) *If $l > 2$, then $L \setminus \{a, b\}$ is also simply connected.*
(d) *P/L is simply connected, even for $l = 1$.*

Remarks. (a) The space $L \setminus \{a, b\}$ carries the multiplicative loop of a suitable coordinatizing ternary field, compare (22.1) and (41.3).

(b) If $l = 1$, then $\pi_1 L \cong \mathbb{Z}$ and $\pi_1 P \cong \mathbb{Z}_2$, compare (32.3) and (52.14).

Proof. 1) The set $L \setminus \{a, b\}$ is connected by (51.16 and 17). Since L is locally path-connected (42.8), assertion (a) follows. In order to show that L is simply connected, write L as the union of the two contractible (51.4) and, hence, simply connected sets $L \setminus \{a\}$ and $L \setminus \{b\}$. Their intersection is path-connected by (a), so that van Kampen's theorem implies our claim; see Armstrong [83] Theorem 5.12.

Less directly, Massey [67] IV 2.1 can be applied. Assertion (d) is obtained in the same way, using (51.27).

2) For P, apply van Kampen's theorem to three affine subplanes $P \setminus L_i$, $i = 1, 2, 3$, whose union is P. An affine plane is homeomorphic to $(L \setminus \{a\})^2$ and hence is simply connected (51.4). The intersection of two of them is an affine plane minus a line and is homeomorphic to the product $(L \setminus \{a\}) \times (L \setminus \{a, b\})$, which is path-connected. By van Kampen's theorem, $P \setminus (L_1 \cap L_2) =: X$ is simply connected. Likewise, $X \cap (P \setminus L_3)$ is path-connected, hence $P = X \cup (P \setminus L_3)$ is simply connected.

3) In order to treat $\mathbf{F}$, we look at the exact homotopy sequence of the fibration $g : \mathbf{F} \to P$ of (51.23b), cf. (51.22) and Spanier [66] 7.2.10. Its last terms are

$$\ldots \to \mathbb{1} = \pi_1 L \to \pi_1 \mathbf{F} \to \pi_1 P = \mathbb{1}.$$

This shows that $\pi_1 \mathbf{F} = \mathbb{1}$.

4) Now assume that $l > 2$. Instead of considering $A = L \setminus \{a, b\}$ as the multiplicative loop, we prefer to use a ternary structure on $K = L \setminus \{b\}$ with unit $a = 1$, so that $A = K \setminus \{1\}$. Take any closed curve $\alpha : [0, 1] \to A$ with $\alpha(0) = \alpha(1) = 0$. We have to show that α is homotopic in A to a constant curve. The end points have to be fixed during the homotopy.

First we construct a homotopy of α to a curve β such that $\dim \beta([0, 1]) = 1$. This can be done by choosing β to be piecewise injective and close enough to α. More precisely, consider a neighbourhood U of 0 in K that can be contracted to 0 within the neighbourhood $\{ x \mid 1 \notin \alpha([0, 1]) + x \}$ of 0, see (51.4). We may apply a contraction of the form $x \mapsto e_t \cdot x$, where e_t is a path in K joining 1 and 0; in particular, this contraction fixes 0. Now using a compactness argument and the fact that L is locally arcwise connected (42.8), a piecewise injective closed curve β with $\beta(0) = \alpha(0)$ can be found such that $\beta(s) = \alpha(s) + \gamma(s)$, where $\gamma(s) \in U$ for all s. Then $F_t(s) := \alpha(s) + (e_t \cdot \gamma(s)) \in A$ is a homotopy $\alpha \simeq \beta$ fixing the end points.

The image set $B = \beta[0, 1]$ is one-dimensional by the sum theorem (92.9). Let $\iota : L \to L$ be the inversion homeomorphism introduced in (43.3); thus $\iota(x) \cdot x = 1$ in general, but $\iota(0) = \infty$ and $\iota(\infty) = 0$. By (51.16 and 17), the complement $L \setminus \iota(B)$ is connected and path-connected. Choose a path f_t in this set such that $f_0 = 1$ and $f_1 = 0$. Then the homotopy $G_t(s) = f_t \cdot \beta(s) \in A$ contracts the closed curve β in $A = K \setminus \{1\}$. □

The next proposition has been partly proved in (42.10), using different methods. A related result is (32.3); part of it is used in our proof here.

51.29 Proposition: Characterization of two-dimensional planes. *The following conditions are equivalent:*

(i) $\dim L = 1$ (v) $L \approx \mathbb{S}_1$
(ii) $\dim P = 2$ (vi) $P \approx \mathrm{P}_2\mathbb{R}$
(iii) *L is a 1-manifold*   (vii) *L is not simply connected*
(iv) *P is a 2-manifold*   (viii) *P is not simply connected*

Proof. 1) We show first that (i) $\Rightarrow$ (iii) $\Rightarrow$ (iv) $\Rightarrow$ (ii) $\Rightarrow$ (i). The second implication is obvious in view of the local product structure of P, see (51.2), and the third one is trivial. If $\dim L = 1$, let $B \approx [0, 1]$ be an arc in L, see (42.8). By (51.21a), the interior $\operatorname{int} B$ is non-void, and some point of L has an open neighbourhood homeomorphic to $\mathbb{R}$. By homogeneity (51.1), L is a 1-manifold. Finally, assume that $\dim P = 2$. Then $l = \dim L > 0$ by (42.2), and $l \leq \dim P$ by the subspace theorem (92.4). Moreover, $l \neq 2$ because for $l = 2$ the line L would contain an open subset of P by (51.21), and then $L \cap M = \{L \wedge M\}$ would be open in M for a line $M \neq L$, a contradiction.

2) The only compact, connected 1-manifold is the circle $\mathbb{S}_1$, see Christenson–Voxman [77] 5.A.3. Thus, (iii) is equivalent to (v). (An alternative argument for $L \approx \mathbb{S}_1$ was given in (42.10).) Next, (iii) implies that every affine part of $\mathcal{P}$ is an $\mathbb{R}^2$-plane, cf. (41.2d). Then (32.3) shows that (vi) holds, which in turn implies (iv). By (42.2) and (51.28b), (i) follows from (vii) and from (viii). Conversely, (v) implies (vii), and (vi) implies (viii), because there are non-trivial covering maps $\mathbb{R} \to \mathbb{S}_1$ and $\mathbb{S}_2 \to \mathrm{P}_2\mathbb{R}$, see Armstrong [83] Section 5.3. □

51.30 Notes on Section 51. The basic contraction principle (51.4) is due to Freudenthal [57a], who also conjectured that the lines of a compact, connected plane should be spheres. The consequences (51.12), (51.13) and (51.14) were noted jointly by Dugundji and Salzmann in 1975. They appear explicitly in Buchanan–Hähl–Löwen [80] 1.1 and Salzmann [81b] 6.7.

The argument given in (51.16) to prove that lines are Cantor manifolds can be found in Salzmann [75a] p. 220, where it is used for a different purpose, and in Salzmann [79a] (12). Theorem (51.21b and c) on domain invariance and separation is due to Buchanan, Hähl and Löwen (assertion 1.4 in their joint paper [80]). The characterization of open sets (51.21a) was obtained by Dugundji in collaboration with Salzmann, see the result C announced in Salzmann [79a]. Proofs were published by Löwen [83a] 3.3 and [83b] Theorem 11.

The essential rôle of the geometric fiber bundles (51.23) was recognized and exploited by Breitsprecher [71] and Salzmann [75a]. The deformation retraction of a punctured plane onto a line (51.26) was given by Löwen [83b] 5.6, and Proposition (51.28) on simple connectedness is due to Salzmann [75a] Section 1. The characterization of 2-dimensional planes (51.29) goes back to Salzmann [59b].

52 Assuming that lines are manifolds

The considerations of this section apply to planes whose lines are manifolds. For most of the results, it suffices to assume that the lines have the same local homology as manifolds. Under this assumption, we compute the global homology of a line L and of the point set P. In fact, we determine the homotopy type of the lines. In Section 54 we shall show that L does have the local homology of a manifold whenever $l = \dim L$ is finite. Hence, the results of the present section are far more general than they appear to be.

52.1 Proposition. *If L is an l-manifold, then the local product structure (51.2) of the spaces P, $\mathcal{L}$ and $\mathbf{F}$ implies that they, too, are manifolds, of dimension $2l$, $2l$ and $3l$, respectively; the spaces have countable bases by (41.8).* □

52.2 Warning. The converse of (52.1) is not known to be true. In other words, it is conceivable that the point set is a manifold while the lines are not. The product structure alone does not preclude this, because there exist non-manifolds X whose Cartesian squares $X \times X$ are homeomorphic to $\mathbb{R}^n$, see the survey by Daverman [80]. This is why we shall assume that L, rather than P, is a manifold.

52.3 Theorem: Spheres. *If the line L is an l-manifold, then it is homeomorphic to the sphere $\mathbb{S}_l$, and the affine point set $P \setminus L$ is homeomorphic to $\mathbb{R}^{2l}$.*

Proof. The line L is the one-point compactification of a ternary field $K = L \setminus \{p\}$, and $P \setminus L$ is homeomorphic to K^2 by (41.2d). Hence, it suffices to prove that $K \approx \mathbb{R}^l$.

In order to do this, we use a pseudo-isotopic contraction F_t of K fixing 0, see (51.4). Choose a neighbourhood U of 0 in K that is a closed l-cell, i.e., is homeomorphic to the Euclidean l-ball. For any compact set $X \subseteq K$ there is a t close to 1 such that X is mapped by F_t into the open l-cell $\operatorname{int} U \approx \mathbb{R}^l$. Thus X is contained in the interior of the closed l-cell $F_t^{-1}(U)$.

By (41.8), K is a countable union of compact sets X_k. Inductively, we can find closed l-cells $U_k \subseteq K$, $k \in \mathbb{N}$, such that $X_k \cup U_{k-1}$ is contained in the open l-cell $\operatorname{int} U_k$. Hence, K is the 'monotone' union of a sequence of open l-cells. By a theorem of M. Brown [61], the space K itself is an open l-cell, as desired. □

We remark that Brown's theorem can be obtained as a corollary of the theory of topological regular neighbourhoods developed by Siebenmann, Guillou and Hähl in their paper [73]. Indeed, an l-cell is a regular neighbourhood of each of its points, and in a σ-compact space a monotone union of regular neighbourhoods is again a regular neighbourhood; see loc. cit. 4.3. The uniqueness theorem for regular neighbourhoods, loc. cit. 2.1, implies that this monotone union is an open l-cell.

The preceding proof works for true manifolds only. Next, we prove a result (52.5) which draws a weaker conclusion from a weaker hypothesis, compared to (52.3). Its advantage is that combined with Section 54 it can be applied whenever $\dim L$ is finite.

52.4 Local homology. We use singular homology with integer coefficients as in Spanier [66], unless the contrary is stated. The *local homology* of a space X at a point $x \in X$ is the relative homology

$$H_q(X, X \setminus \{x\}).$$

If X is an n-manifold and if U is a neighbourhood of x homeomorphic to the n-ball, then by the excision theorem (Spanier [66] 4.6.5) we have $H_q(X, X\setminus\{x\}) \cong H_q(U, U\setminus\{x\})$. We write down the reduced version of the exact homology sequence of the pair $(U, U \setminus \{x\})$, see Spanier [66] p. 184f. It runs

$$\ldots \to \widetilde{H}_q U \to H_q(U, U \setminus \{x\}) \to \widetilde{H}_{q-1}(U \setminus \{x\}) \to \widetilde{H}_{q-1} U \to \ldots$$

Since U is contractible, we have that $\widetilde{H}_q U = 0$ for all q, hence exactness implies that

$$H_q(U, U \setminus \{x\}) \cong \widetilde{H}_{q-1}(U \setminus \{x\}) \cong \widetilde{H}_{q-1}(\partial U) = \widetilde{H}_{q-1}\mathbb{S}_{n-1}.$$

Thus, the local homology of an n-manifold is

$$H_q(X, X \setminus \{x\}) = \begin{cases} \mathbb{Z} & \text{if } q = n \\ 0 & \text{else.} \end{cases}$$

We shall say that an arbitrary space X *has the local homology of an n-manifold* if the above relation holds for all $x \in X$.

52.5 Theorem: Homotopy spheres and dimension. *Assume that the line $L = ab$ has the local homology of an m-manifold, $m > 0$, and that $l = \dim L < \infty$. Then the following are true.*

(a) *L is homotopy equivalent to $\mathbb{S}_m$.*
(b) *The multiplicative loop $L \setminus \{a, b\} = K^\times$ of an associated ternary field is homotopy equivalent to $\mathbb{S}_{m-1}$.*
(c) *m divides 8.*

52.6 Notes. (a) The values $m \in \{1, 2, 4, 8\}$ allowed by (52.5c) occur in the classical planes over $\mathbb{F} \in \{\mathbb{R}, \mathbb{C}, \mathbb{H}, \mathbb{O}\}$. In these planes, we even have homeomorphisms $L \approx \mathbb{S}_m$ and $K^\times \approx \mathbb{S}_{m-1} \times \mathbb{R}$, see (14.7 and 9) and (16.13).

(b) The hypotheses of (52.5) are satisfied if L is an m-manifold. In that case, we have $l = m$, see (92.12). In fact, we shall show in Section 54 that $l < \infty$ alone suffices as a hypothesis; again, we shall obtain that $l = m$.

(c) For later use, we note that $l = 1$ if, and only if $m = 1$; this is a by-product of our proof given below.

(d) In Löwen–Salzmann [82], assertion (52.5a) is proved in a stronger form. It is shown that the pair (L, x) is homotopy equivalent to $(\mathbb{S}_m, e)$ for $x \in L$ and $e \in \mathbb{S}_m$. This is then used to show that the quotient space P/L obtained from P by shrinking L to a point is homotopy equivalent to $\mathbb{S}_{2m}$; compare (52.12) for a different proof of this fact.

(e) The main step in the proof of (52.5a and b) is to establish a purely topological characterization of spheres up to homotopy, see (52.7) below.

Proof of (52.5). 1) We show first that $H_q L \cong H_q \mathbb{S}_m$ and $H_q K^\times \cong H_q \mathbb{S}_{m-1}$ for all q. For $H_q L$, this follows from the assumption about the local homology via the reduced homology sequence of the pair $(L, L\setminus\{a\})$ (Spanier [66] p. 184f), because $K = L \setminus \{a\}$ is contractible (51.4). For $K^\times$, we use the reduced Mayer–Vietoris sequence of the decomposition $L = A \cup B$, where $A = L \setminus \{a\}$, $B = L \setminus \{b\}$ and $A \cap B = K^\times$, see Spanier [66] 4.6 p. 189. The open sets A, B form an excisive couple, as required, and the sequence runs

$$ \cdots \to \widetilde{H}_{q+1} A \oplus \widetilde{H}_{q+1} B \to \widetilde{H}_{q+1}(A \cup B) \to \widetilde{H}_q(A \cap B) \to \widetilde{H}_q A \oplus \widetilde{H}_q B \to \cdots $$

Since A and B are contractible, both end terms vanish. By exactness, the middle homomorphism is an isomorphism $\widetilde{H}_{q+1} L \cong \widetilde{H}_q K^\times$.

2) We treat the cases $m = 1$ and $l = 1$ by a separate argument. If $m = 1$, then $H_1 L \cong \mathbb{Z}$ by step 1). According to a theorem of Hurewicz, $H_1 L$ is the commutator factor group of the fundamental group $\pi_1 L$, see Spanier [66] 7.5.5 and p. 391. Hence, L is not simply connected, and $l = 1$ by (51.28b) and (42.2). If $l = 1$, then $L \approx \mathbb{S}_1$ by (51.29), and $m = 1$.

From now on, assume that $\min\{m, l\} > 1$.

3) We want to prove (a), using the isomorphism $H_q L \cong H_q \mathbb{S}_m$. In general, it is far from true that spaces with isomorphic homology groups are homotopy equivalent. This does hold for simply connected CW-complexes X, Y, provided that there is a map $f : X \to Y$ inducing isomorphisms $H_q X \to H_q Y$ for all q. Indeed, such a map f is a *weak homotopy equivalence*, that is, f induces isomorphisms of all homotopy groups; see Spanier [66] 7.5.9. (The groups $H_q(X, \{x_0\})$ used by Spanier are naturally isomorphic to $\widetilde{H}_q X$, as the reduced pair sequence shows.) By a theorem of J.H.C. Whitehead [48], see the statement in Spanier [66] 7.6.24, a weak homotopy equivalence between CW-complexes is a homotopy equivalence.

To make use of this fact we had to assume that $l < \infty$; this ensures that L and $K^\times$ are ANR (51.12), hence there are CW-complexes $Q \simeq L$ and $R \simeq K^\times$ by (51.13). Note that $\mathbb{S}_m$ also is a CW-complex. The original version of Whitehead's theorem directly applies to compact ANR and to maps inducing isomorphisms of homology. This works for L, but not for $K^\times$, for lack of compactness.

4) By the preceding explanations, all we need in order to prove (a) is a map $h : \mathbb{S}_m \to Q$ inducing isomorphisms $h_q : H_q \mathbb{S}_m \to H_q Q$ for all q; recall that $m > 1$ and $l > 1$, so that $\mathbb{S}_m$ and Q are simply connected, cf. (51.28b). The theorem of Hurewicz about the first non-vanishing reduced homology group of simply connected spaces (Spanier [66] 7.5.5 and p. 391) hence yields that $\pi_m Q \cong H_m Q \cong \mathbb{Z}$. We choose a map $h : \mathbb{S}_m \to Q$ whose homotopy class $[h]$ generates $\pi_m Q$. Then the induced homomorphism

$$h_\sharp : \pi_m \mathbb{S}_m \to \pi_m Q,$$
$$[f] \mapsto [h \circ f],$$

sends the generator $[\mathrm{id}] \in \pi_m \mathbb{S}_m \cong \mathbb{Z}$ (compare Spanier [66] 7.5.6) to the generator $[h] \in \pi_m Q$. Thus, $h_\sharp$ is an isomorphism. This is not yet what we want, but close to it. By naturality of the Hurewicz isomorphism $\pi_m Q \cong H_m Q$ (Spanier [66] 7.4.3b), we have a commutative diagram

$$\begin{array}{ccc} \pi_m \mathbb{S}_m & \xrightarrow{h_\sharp} & \pi_m Q \\ \cong \downarrow & & \downarrow \cong \\ H_m \mathbb{S}_m & \xrightarrow{h_m} & H_m Q \end{array}$$

Hence, h_m is an isomorphism as well. Trivially, h_q is an isomorphism for $q \neq m$. Applying step 3), we obtain that $\mathbb{S}_m \simeq Q \simeq L$, as desired. We interrupt our proof to state the purely topological fact that we have obtained:

52.7 Lemma: Homotopy spheres. *Let $n > 1$ and let X be a simply connected ANR with $H_q X \cong H_q \mathbb{S}_n$ for all q. Then X is homotopy equivalent to $\mathbb{S}_n$.* □

Proof of (52.5), resumed. 5) If $l > 2$, then assertion (b) follows like (a) from (52.7), for then $K^\times$ is simply connected (51.28c). In the remaining case $l = 2$, the methods used so far do not work. Now a theorem of Bing and Borsuk, see Borsuk [67] VII.16.10, asserts that every 2-dimensional, connected, homogeneous ANR is a manifold. This theorem applies to L in view of (51.1) and (51.12). By (52.3), we obtain that $L \approx \mathbb{S}_2$ and hence $K^\times \approx \mathbb{C}^\times \simeq \mathbb{S}_1$ as desired. We defer a more thorough examination of the case $l = 2$ until Section 53, see (53.7).

6) It remains to prove the dimensional restriction (c). We obtain this as a consequence of the presence of a continuous multiplication $(x, y) \mapsto x \cdot y$ with unit 1

on $K^\times$. Such a multiplication is provided by (22.1) and (41.3). Let $f : K^\times \to \mathbb{S}_{m-1}$ and $g : \mathbb{S}_{m-1} \to K^\times$ be homotopy inverses of each other, and define a multiplication μ on $\mathbb{S}_{m-1}$ by $\mu(u, v) := f(g(u) \cdot g(v))$. Then μ is continuous, and the two maps $\lambda : u \mapsto \mu(u, f(1))$ and $\varrho : u \mapsto \mu(f(1), u)$ of $\mathbb{S}_{m-1}$ to itself are homotopic to the identity. Indeed, a homotopy $\lambda \simeq f \circ g \ (\simeq \mathrm{id})$ is given by $G_t(u) = f(g(u) \cdot F_t(1))$, where F_t is a homotopy $g \circ f \simeq \mathrm{id}_{K^\times}$.

Spaces with a multiplication of this kind (continuous, with two-sided unit up to homotopy) will be called H-spaces. Most authors reserve this term for multiplications with a strict unit, but all that we need is that the maps λ, ϱ of $\mathbb{S}_{m-1}$ are of degree 1, i.e., that they induce the identity on $H_{m-1}\mathbb{S}_{m-1}$. This can be expressed by saying that the *bidegree* of $\mu : \mathbb{S}_{m-1} \times \mathbb{S}_{m-1} \to \mathbb{S}_{m-1}$ is $(d_1, d_2) = (1, 1)$, compare Husemoller [66] p. 90.

Now the Hopf construction assigns to μ a map $\mathrm{H}(\mu) : \mathbb{S}_{2m-1} \to \mathbb{S}_m$, see Husemoller [66] p. 199, 3.3. For m even, the Hopf invariant of this map is $d_1 \cdot d_2 = 1$ (Husemoller [66] p. 199, 3.5). By a famous theorem of Adams [60], maps $\mathbb{S}_{2m-1} \to \mathbb{S}_m$ with Hopf invariant 1 exist for $m = 2, 4$ or 8 only. See Husemoller [66] p. 201, 4.3 for a proof using K-theory. It is easier to see that H-multiplications do not exist on $\mathbb{S}_{m-1}$ for $1 < m$ odd, so some texts hardly mention the odd case. Indeed, according to Hopf [35] Satz V, the multiplication μ is null homotopic in this case, contrary to the facts that λ, the restriction of μ to the left factor, is homotopic to the identity, and that $\mathrm{id}_{\mathbb{S}_{m-1}} \not\simeq 0$. Other proofs for the odd case can be found in Spanier [66] 5.8.14, Stöcker–Zieschang [88] 15.3.8 p. 390 and Stauffer [75]. □

Let us record the theorem of Adams.

52.8 Theorem: Spheres that are H-spaces. *The only spheres that are H-spaces are $\mathbb{S}_0$, $\mathbb{S}_1$, $\mathbb{S}_3$, and $\mathbb{S}_7$.* □

We mention here a related theorem due to Hubbuck [69] that will be used in a similar way, see (55.14) and (74.6). An H-space X with multiplication μ is said to be *homotopy commutative* if the two maps $(x, y) \mapsto \mu(x, y)$ and $(x, y) \mapsto \mu(y, x)$ from $X \times X$ to X are homotopic.

52.9 Theorem: Commutative H-spaces. *If X is a connected, homotopy commutative H-space and if X is homotopy equivalent to a compact CW-complex, then X is either contractible or homotopy equivalent to a torus $\mathbb{S}_1 \times \ldots \times \mathbb{S}_1$.* □

52.10 Corollary. *If the lines of a plane are m-manifolds and if some coordinatizing ternary field has (homotopy) commutative multiplication, then $m \in \{1, 2\}$.*

Proof. The homotopy commutative H-space structure on $K^\times$ carries over to $\mathbb{S}_{m-1} \simeq K^\times$ as in the proof of (52.5), step 6). For $m > 1$, (52.9) can then be applied to $X = \mathbb{S}_{m-1}$. For $m > 2$, this yields a contradiction, because X is simply connected but not contractible. □

52.11 Note. The assumption that lines are manifolds may be replaced by the condition that $l = \dim L < \infty$. One has to use Section 54, where it is proved that then the lines have the local homology of l-manifolds, see (54.11).

52.12 Theorem: The point set modulo a line. *Assume that $l = \dim L < \infty$ and that P has the local homology of a $2m$-manifold. Then the quotient space P/L introduced in (51.6) also has this property. Moreover, P/L is homotopy equivalent to $\mathbb{S}_{2m}$. In particular, the Euler characteristic is $\chi(P/L) = 2$.*

We shall prove later that $l = m$, compare (52.14a).

Proof. By (51.12) and (51.28d), P/L is a simply connected ANR. Thus, it suffices to compute the local and global homology of P/L and to apply (52.7). Since we have $P/L \setminus \{L\} \approx P \setminus L$, the excision theorem (Spanier [66] 4.6.5) gives isomorphisms of local homology groups

$$H_q(P, P \setminus \{p\}) \cong H_q(P \setminus L, P \setminus (L \cup \{p\})) \cong H_q(P/L, P/L \setminus \{p\})$$

for each $p \notin L$. Moreover, the reduced pair sequence of $(P/L, P/L \setminus \{x\})$ yields isomorphisms $\widetilde{H}_q(P/L) \cong H_q(P/L, P/L \setminus \{x\})$ for every element $x \in P/L$, because $P/L \setminus \{x\}$ is contractible (51.27). For a different proof, see Löwen–Salzmann [82]. □

For planes whose lines have the local homology of manifolds, we are now ready to compute the global invariants of the point space. See the note (52.14a) for comments on the hypotheses.

52.13 Theorem: Homotopy invariants of the point set. *Assume that a compact, connected projective plane $\mathcal{P} = (P, \mathcal{L})$ and one of its lines $L \in \mathcal{L}$ satisfy the following conditions:*

(i) $\dim L = l < \infty$,
(ii) *L has the local homology of an m-manifold, and*
(iii) *P has the local homology of a $2m$-manifold.*

Then the following invariants of P depend only on the value of m:

(a) *All homology groups $H_q P$ and $H_q(P, L)$,*
(b) *the Euler characteristic χP,*
(c) *the homotopy groups $\pi_q P$ for $q \leq m$.*

52.14 Notes. (a) As in (52.5), the hypotheses of (52.13) are trivially satisfied if L is a manifold. Moreover, (ii) implies (iii); this follows from the local product structure of P, see (51.2), by an application of the Künneth formula, see Spanier [66] 5.3.10. In fact, a much weaker assumption suffices. We shall show in (54.11) that the conditions (ii) and (iii) with the specific value $l = m$ follow from (i) alone; moreover, we show in (54.11) that $\dim P = 2l$.

(b) According to (52.5), there is $\mathbb{F} \in \{\mathbb{R}, \mathbb{C}, \mathbb{H}, \mathbb{O}\}$ with $m = \dim_\mathbb{R} \mathbb{F}$. Then (52.13) says that P and $P_2\mathbb{F}$ have the same invariants. Moreover, since P and $\mathcal{L}$ are locally homeomorphic (51.2), the plane $\mathcal{P}$ and its dual have the same value of m, hence P and $\mathcal{L}$ have the same invariants.

(c) Specifically, the computations in the proof will produce the following results.

$$H_q(P, L) = \begin{cases} \mathbb{Z} & \text{for } q = 2m \\ 0 & \text{else} \end{cases}$$

For $m = 2, 4$ or 8,

$$H_q P = \begin{cases} \mathbb{Z} & \text{for } q = 0, m, 2m \\ 0 & \text{else} \end{cases}$$

$$\chi P = 3$$

$$\pi_q P \cong \widetilde{H}_q P \text{ for } q \leq m$$

For $m = 1$,

$$H_q P = \begin{cases} \mathbb{Z} & \text{for } q = 0 \\ \mathbb{Z}_2 & \text{for } q = 1 \\ 0 & \text{else} \end{cases}$$

$$H_q(P; \mathbb{Z}_2) = \begin{cases} \mathbb{Z}_2 & \text{for } q = 0, 1, 2 \\ 0 & \text{else} \end{cases}$$

$$\chi P = 1$$

$$\pi_1 P = \mathbb{Z}_2$$

(d) Using a Gysin exact sequence, one can compute the homology of the flag space **F** from the data listed in (c). Details are given in Löwen [83b] Theorem 5.

Proof of (52.13). 1) In the subsequent steps, we shall compute the homology groups, and so far verify the assertions of (52.14c) above. The claims about the Euler characteristic will then follow from its definition as the alternating sum of the ranks of the homology groups with integer coefficients. Regarding $\pi_q P$, note first that the space P is path-connected by (42.8) and (41.2), whence the homotopy set $\pi_0 P$ is a singleton (Spanier [66] p. 371). By (52.6c) we have $m > 1$ if, and only if $l > 1$. In this case, P is simply connected (51.28b). Then the theorem of Hurewicz (Spanier [66] 7.5.5, Switzer [75] 10.25) says that $\pi_q P \cong \widetilde{H}_q(P; \mathbb{Z})$ for q less than or equal to the degree of the first non-vanishing reduced homology group, i.e., for $q \leq m$; Spanier uses $H_q(P, x_0)$ instead of $\widetilde{H}_q P$, but these two groups are isomorphic as shown by the pair sequence of (P, x_0).

We need a special argument for π_1 in the case $m = l = 1$: First, we refer to (51.29) to obtain that $P \approx P_2 \mathbb{R}$ in this case. Then we use the two-sheeted covering $\mathbb{S}_2 \to P_2 \mathbb{R}$ to compute that $\pi_1(P_2 \mathbb{R}) = \mathbb{Z}_2$, see Armstrong [83] Theorem 5.13.

2) Let $U = P \setminus \{p\}$, where $p \in P \setminus L$. From (51.26) we know that the inclusion map $L \to U$ is a homotopy equivalence; hence it induces isomorphisms of homology groups. By Spanier [66] 4.8.6, the inclusion map $(P, L) \to (P, U)$ then has the same property, i.e., it induces isomorphisms $H_q(P, L) \to H_q(P, U)$ for every q. The latter groups are the local homology groups of P, hence by hypothesis (iii) we obtain that $H_q(P, L)$ is infinite cyclic for $q = 2m$ and zero else.

3) The homology groups of P can be computed from the exact homology sequence of the pair (P, U), where $U = P \setminus \{p\}$ as above:

$$\ldots \to H_{q+1}(P, U) \to H_q U \xrightarrow{\alpha} H_q P \xrightarrow{\beta} H_q(P, U) \to H_{q-1} U \to \ldots$$

The evaluation of this sequence is easy because $H_q U \cong H_q L \cong H_q \mathbb{S}_m$ and $H_q(P, U) \cong \widetilde{H}_q \mathbb{S}_{2m}$ are known respectively from (52.5) and from hypothesis (iii). If $m > 1$, or if $m = 1$ and $q \geq 3$, then most of these groups are trivial, and we may conclude in each case that α or β is an isomorphism.

4) For $m = 1$, we noted in step 1) that $P \approx P_2 \mathbb{R}$. The homology of P is well known in this case, but we want to compute it using our method set up above. P is a 2-manifold and $L \approx \mathbb{S}_1$, hence the relations $H_q(P, U) \cong \widetilde{H}_q \mathbb{S}_2$ and $H_q U \cong H_q L \cong H_q \mathbb{S}_1$ hold for the coefficient rings $R = \mathbb{Z}$ and $R = \mathbb{Z}_2$; note that $H_q(\mathbb{S}_n; R) = R$ for $q = 0$ or n and $H_q(\mathbb{S}_n; R) = 0$ else. Substituting this in the exact sequence from 3), starting with $H_2 L$, we obtain

$$\ldots \to 0 \to H_2 P \xrightarrow{\beta} R \xrightarrow{\gamma} R \xrightarrow{\delta} H_1 P \to 0.$$

We need further information to evaluate this. If $R = \mathbb{Z}$, then we use the theorem of Hurewicz (Spanier [66] 7.5.5 and p. 391, Greenberg [67] 12.1), which says that $H_1(P; \mathbb{Z})$ is $\pi_1 P$ made abelian, so that $H_1 P \cong \pi_1 P \cong \mathbb{Z}_2$, compare step 1). The sequence now shows that the image of $\mathbb{Z}$ under γ corresponds to $2\mathbb{Z}$, so that γ is injective and $\beta = 0$. Hence, $H_2(P; \mathbb{Z}) = 0$.

If $R = \mathbb{Z}_2$, we work from the other end of the sequence. We have that $H_2(P; \mathbb{Z}_2) \cong \mathbb{Z}_2$ like for every compact, connected 2-manifold P, see Greenberg [67] 22.30 or Spanier [66] 6.2.9 and 6.2.18. Since β is injective by exactness, we conclude that $\gamma = 0$ and δ is injective. By exactness, δ is also surjective, and $H_1(P; \mathbb{Z}_2) \cong \mathbb{Z}_2$. □

52.15 Theorem: Homeomorphism of point and line spaces. *Assume that the affine plane $P \setminus L$ is homeomorphic to $\mathbb{R}^{2l}$ via a homeomorphism that sends all lines through some point o onto vector subspaces of $\mathbb{R}^{2l}$. (This is satisfied, e.g., in translation planes, compare (64.4).) Then the point space P is homeomorphic to the line space $\mathcal{L}$.*

Proof. We identify $P \setminus L$ with $\mathbb{R}^{2l}$ by means of the given homeomorphism. By (51.6), the quotient space P/L in this way becomes the one-point compactification $\mathbb{R}^{2l} \cup \infty$. We define a homeomorphism $\iota : P/L \to P/L$ ('inversion') by interchanging the origin o with ∞, and by setting

$$p^\iota := p \|p\|^{-2}$$

for $p \notin \{o, \infty\}$. The pencil $\mathcal{L}_o$ is homeomorphic to the one-point compactification of an affine line. By our assumption, this is an l-sphere, hence we have an 'antipodal map' $\alpha : \mathcal{L}_o \to \mathcal{L}_o$ (corresponding to $x \mapsto -x$ on $\mathbb{S}_l$), which is a fixed-point free involution. Now a bijective map $\delta : P \to \mathcal{L}$ is obtained by

$$p^\delta := \begin{cases} L & \text{if } p = o \\ (po)^\alpha & \text{if } p \in L \\ p^\iota\big((po)^\alpha \wedge L\big) & \text{else.} \end{cases}$$

Continuity of the map δ is easily verified using sequences. By compactness of P, this implies that δ is a homeomorphism. □

52.16 Further results on the topology of P and $\mathbf{F}$ may be found in Section 53 and in Breitsprecher [71] and Löwen [83b].

52.17 Notes on Section 52. The conjecture that the lines of a compact, connected projective plane should be spheres goes back to Freudenthal's paper [57a], which also introduced one of the main tools (51.4) that later made it possible to come close to a proof of the conjecture. Under the hypothesis that lines are manifolds, the conjecture was proved by Salzmann [67b] 7.12. The more direct proof given here (52.3) is taken from Breitsprecher [71] 2.1. In fact, both proofs rest on Brown's theorem on the monotone union of cells. Under the same restrictive hypothesis, Salzmann and Breitsprecher also deduced from the theorem of Adams that the possible dimensions of lines are $l \in \{1, 2, 4, 8\}$, and Breitsprecher [71] moreover computed the cohomology ring of the point and flag spaces.

The collaboration of Dugundji and Salzmann in 1975 led to the first significant progress in the general case, see the announcements in Salzmann [79a,b] and the exposition Salzmann [81b]. Several of their results and methods have been integrated into our present approach, although their considerations started from the opposite end compared to ours. The smallest hypothetical unpleasant structure that they could not exclude was a line L of dimension 7 with $L \setminus \{a, b\} \simeq \mathbb{S}_3 \times \mathbb{S}_3$.

The breakthrough happened when Löwen [83b] had the idea to deduce the global homology of L and P from the local one rather than the other way round. This was inspired by a characterization of homology spheres given by Mitchell [78]. We have divided Löwen's arguments into the determination of the local invariants on the one hand, which was his new contribution and will be presented in Section 54, and the deduction of the global invariants on the other hand, which owes much to the earlier development and is given in (52.5 and 13).

The homeomorphism of point and line spaces (52.15) is due to Eisele [92a].

53 Conditions implying that lines are manifolds

Recall first that the point space of a compact, connected plane $\mathcal{P}$ is a manifold if the lines are manifolds (52.1); the converse is not known to be true. If the automorphism group of $\mathcal{P}$ has an open orbit in the point space or in a line then the theory of locally compact groups allows us to show that this orbit is a manifold. In particular, transitivity of axial groups, as well as the related algebraic conditions, imply that lines are manifolds. In fact, the point spaces of planes of Lenz type V can be determined up to homeomorphism. We present the geometric parts of the proof of this fact.

If $l = \dim L \leq 2$, then a characterization of low-dimensional manifolds due to Bing and Borsuk can be applied, and the lines are spheres. The topology of the planes is completely known in this case. As far as the point space is concerned, we shall give a proof.

Throughout, $\mathcal{P} = (P, \mathcal{L})$ denotes a compact, connected, projective plane.

We want to collect conditions ensuring that lines are manifolds. First, we consider criteria involving group actions. The first result of this kind (53.2) will be a consequence of the following proposition, which provides some general criteria for the recognition of transitive actions.

53.1 Proposition: Open orbits. *Consider a compact, connected, projective plane $\mathcal{P} = (P, \mathcal{L})$ and a line $L \in \mathcal{L}$, and let G be a locally compact group which has a countable basis and acts on $X \in \{P, L, \mathcal{L}, \mathbf{F}\}$ as a transformation group. For example, G could be a closed subgroup of $\mathrm{Aut}\,\mathcal{P}$, see (44.3). Then the following assertions hold.*

(a) *A G-orbit $U \subseteq X$ is open if, and only if $\dim U = \dim X < \infty$.*
(b) *If this is the case, then X is a manifold and the induced group $G/G_{[U]} = G|_U$ is a Lie group.*

53 Conditions implying that lines are manifolds

(c) *Suppose that a compact group G has an open orbit U as in (a). Then G is transitive on X, that is, $X = U$.*
(d) *If G leaves a connected subset $Y \subseteq X$ invariant and if all G-orbits $U \subseteq Y$ satisfy $\dim U = \dim X < \infty$, then G is transitive on Y.*

Proof. If G is compact, then U is closed in the connected space X, hence assertion (c) follows. Similarly, (d) is a consequence of (a). Now assume first that U is open in X. The kernel $G_{[U]}$ of the action on U is a closed subgroup of G, and the factor group $G/G_{[U]} \cong G|_U$ is again a locally compact group with a countable basis and acts as an effective transformation group on U, see (96.2c). To this action we apply Szenthe's theorem (96.14), which is related to the positive solution of Hilbert's 5th problem. We know that U is locally contractible (51.4), hence Szenthe's theorem may be applied. We obtain that $G/G_{[U]}$ is a Lie group and U is a manifold. By homogeneity (51.1), the space X is a manifold, and $\dim U = \dim X < \infty$.

Conversely, suppose that $\dim U = \dim X < \infty$. The orbit U is a union of countably many compact sets U_i, because G is locally compact and second countable. By the sum theorem (92.9), some set U_i has the same dimension as U itself. According to (51.21a), the compact set U_i has non-void interior relative to X. Being homogeneous, the orbit U is open. This argument is a specialization of (96.11a). □

53.2 Corollary: Automorphism groups with open orbits. *Let $\mathcal{P}$, L and X be as in (53.1), and let Δ be a closed subgroup of $\Sigma = \mathrm{Aut}\,\mathcal{P}$ with $\Delta \leq \Sigma_L$ if $X = L$. If $U \subseteq X$ is a Δ-orbit which is open or, equivalently, satisfies $\dim U = \dim X < \infty$, then X is a manifold and Δ induces a Lie group on U. If $X = L$, then it follows that all lines are manifolds homeomorphic to $\mathbb{S}_l$, see (52.3). If $X \in \{P, \mathcal{L}, \mathbf{F}\}$, then Σ is a Lie group.*

Remark. In the case $X = L$, we cannot prove that the stabilizer Δ_L or even Δ itself is a Lie group.

Proof. The group Δ is a locally compact transformation group acting on X and has a countable basis (44.3). Hence, (53.1) may be applied. If $X \in \{P, \mathcal{L}, \mathbf{F}\}$, then Δ acts effectively on U by (44.4a) and hence coincides with the induced group $\Delta|_U$. □

The previous results may be applied to linearly transitive groups of axial collineations, see (23.21). This shows that planes with a non-trivial Lenz–Barlotti type (see Section 24) have manifold lines (also compare (64.1)):

53.3 Corollary: Transitive axial groups. *If $\mathcal{P}$ admits at least one linearly transitive group $\Delta = \Sigma_{[c,A]}$ of axial collineations (homologies or elations), then the lines of $\mathcal{P}$ are manifolds and Δ is a Lie group.*

Proof. This follows from (53.2), applied to a line $L \neq A$ containing the center c. Note that the orbit $U = L \setminus (A \cup \{c\})$ is open in L and that the kernel of the action of Δ on U is trivial (23.8). □

According to Hughes–Piper [73] VI, Sections 3 and 4, the assumption made in (53.3) that there is a linearly transitive group of axial collineations amounts to stipulating that there is a coordinatizing ternary field which is linear and has either associative addition or associative multiplication (or both); compare also (23.11). In fact, we may obtain the conclusion of (53.3) without using linearity of the ternary field. If $\dim L$ is finite, then it suffices to apply the solution of Hilbert's problem (93.2) to the additive or multiplicative loop whenever the latter is a group. In fact, applying Szenthe's theorem (96.14) to the action of that group on itself by left multiplication, we see that finite dimensionality is not needed:

53.4 Proposition: Associative loops. *If some ternary field coordinatizing $\mathcal{P}$ has associative addition or associative multiplication, then the lines of $\mathcal{P}$ are manifolds.* □

Notes. (a) For associative multiplication, it follows moreover that the multiplicative group $K^\times$ is isomorphic to $\mathbb{R}^\times, \mathbb{C}^\times$ or $\mathbb{H}^\times$, see (64.1b) or Grundhöfer–Salzmann [90] XI.8.8, where this is proved even for double loops.

(b) See (53.9) below for a result about ternary fields with commutative multiplication.

(c) The point space of a compact, connected plane of Lenz type V is homeomorphic to that of the classical plane of the same topological dimension, see (53.17).

Next, we treat special results valid for planes of dimension 2 or 4.

For the sake of completeness, we repeat a result which has been proved several times, see (32.3), (42.10) and (51.29). Its proof given in (32.3) is a very simple model for the proof of the corresponding result about 4-dimensional planes, see (53.15) below.

53.5 Review: Topology of 2-dimensional planes. *If $\dim L = 1$ or $\dim P \leq 2$, then L and P are homeomorphic to their classical counterparts $\mathbb{S}_1$ and $\mathrm{P}_2\mathbb{R}$.* □

53.6 Remark. A sharper result has been obtained by Breitsprecher [72]. He proves that for any two 2-dimensional planes, the corresponding pairs $(P \times \mathcal{L}, \mathbf{F})$ and $(P' \times \mathcal{L}', \mathbf{F}')$ are homeomorphic by a homeomorphism f that sends each fiber $\{x\} \times \mathcal{L}$ onto some fiber $\{x'\} \times \mathcal{L}'$. (Note that f gives rise to an isomorphism of planes if, and only if, f respects the fibers $P \times \{L\}$ as well.) In this sense, all 2-dimensional planes can be considered as deformations of any given one, e.g., of $\mathcal{P}_2 \mathbb{R}$.

53.7 Proposition: Small dimensions. *If $\dim L \leq 2$ or $\dim P \leq 4$, then the lines and, hence, the spaces P, $\mathcal{L}$ and $\mathbf{F}$ are manifolds.*

Proof. By (51.2), it suffices to prove the statement about lines. Every connected, homogeneous ANR of dimension ≤ 2 is a manifold; this was shown by Bing and Borsuk in their paper [65], see also Borsuk [67] VII 16.10. In fact, the property used in their proof is local contractibility, which characterizes the ANR among finite-dimensional metric spaces, see (51.11). In view of (51.1) and (51.12), we only need to show that $\dim P \leq 4$ implies $\dim L \leq 2$ and then apply that theorem to L. Now if $\dim L \geq 3$, then the local product structure of P, see (51.2), together with (92.13) yields that $\dim P \geq 5$.

An independent proof can be based on a characterization of the spheres $\mathbb{S}_l$, $l \leq 2$, due to McCord [66] Theorem 3. In order to apply it, we have to check the characteristic properties that L is a Hausdorff space, that $\widetilde{H}_q(L \setminus \{x\}) = 0$ for every $q \leq l$ and $x \in L$ (which is granted by (51.4)), and that $H_l L \neq 0$ (which holds by (52.5a) and (54.11)). $\square$

53.8 Note. Bing and Borsuk conjectured that compact homogeneous ANR of arbitrary finite dimension are manifolds. This is still an open question, but, according to Jakobsche [80], a counterexample to the Poincaré conjecture would yield a 3-dimensional counterexample to the Bing–Borsuk conjecture. Bing and Borsuk proved the conjecture for ANR of dimension 1 or 2. Some of their preparatory results remain valid, however, for all dimensions; compare Seidel [85]. Those facts have been used in Section 51. For a characterization of higher dimensional manifolds, see Quinn [83, 87]. Compare also (54.13).

53.9 Corollary: Commutative multiplication. *If $\mathcal{P}$ can be coordinatized by a ternary field with commutative multiplication, and if $l = \dim L < \infty$, then $l \in \{1, 2\}$, and each line L is homeomorphic to $\mathbb{S}_l$.*

Proof. The corollary (52.10) of Hubbuck's theorem can be generalized to our situation, as explained in (52.11); this uses (54.11) below. We obtain that $l \leq 2$, and the assertion follows from (53.7) and (52.3). $\square$

53.10 Corollary: Baer subplanes. *If* $\dim P \leq 8$, *and if* $\mathcal{P}$ *contains a closed Baer subplane, then the lines of* $\mathcal{P}$ *are manifolds homeomorphic to* $\mathbb{S}_l$; *compare also* (55.6).

Proof. The lines of $\mathcal{P}$ are locally homeomorphic to the point set Q of the Baer subplane (41.11b). By (51.2) and (92.13), it follows that $\dim P \geq 2\dim Q - 1$, hence $\dim Q \leq 4$ and Q is a manifold by (53.7). By homogeneity, the lines are manifolds and, in fact, spheres (52.3). □

The following definition prepares a tool that will enable us to compare the point spaces or the line spaces of two planes with homeomorphic lines. If lines are manifolds, then the fiber bundle $x \mapsto px \wedge L$ of (51.23a) has base $L \approx \mathbb{S}_l$. It can be covered by two trivializing charts as in the proof of (51.23a). The change from one chart to the other may be expressed using the 'characteristic map' that we are going to introduce. It follows from the theory of fiber bundles that the homotopy type of this map determines the equivalence class of the fiber bundle; see, e.g., Steenrod [51] Section 18. This background of our constructions will, however, not become explicitly apparent. See (53.13) and Breitsprecher [71] or Buchanan [79b] for more information.

53.11 The characteristic maps of a ternary field. Let K be a locally compact, connected ternary field. We shall define two maps, χ^* and χ, both of which describe the left multiplications in K, with slight differences regarding the domains involved. For $s \in K$, the map $\chi^*(s) : K \to K$ is defined by

$$\chi^*(s, x) = \chi^*(s)(x) := sx,$$

provided that K is a linear ternary field, i.e., that $\tau(s, x, t) = sx + t$ holds in K. This special case is the intuitive basis of the following general definition. First, we define the additive right inverse a' of $a \in K$ by the equation $a + a' = 0$. Now $\chi^*(s, x)$ is defined by the relation $\tau\bigl(s, x, \chi^*(s, x)'\bigr) = 0$, which expresses the condition that, in the plane coordinatized by K,

(∗)   the points $(x, 0)$, $\bigl(0, \chi^*(s, x)'\bigr)$, and (s) are collinear.

More precisely, we consider χ^* as a map from K into the homeomorphism group of the space K, endowed with the compact-open topology, see (96.3)ff. Continuity of χ^* follows from continuity of τ and its inverses, by (96.4). Hence, the image of K is contained in the connected component $\mathcal{H}^*$ of this group containing the identity $\chi^*(1) = \mathbb{1}$. Thus, by restriction to $s \neq 0$ we obtain the *dual characteristic map* of K,

$$\chi^* : K^\times = K \setminus \{0\} \to \mathcal{H}^*.$$

The definition of χ is as follows. Deleting the zero element from the one-point compactification of K, we obtain a space

$$K^{-1} := (K \cup \{\infty\}) \setminus \{0\}.$$

We denote by $\mathcal{H}$ the identity component of the homeomorphism group of this space, and we define the *characteristic map* of K,

$$\chi : K^\times \to \mathcal{H},$$

by

$$\chi(s)(x) = \chi(s, x) := sx, \quad \chi(s, \infty) = \infty.$$

This map is continuous by (96.4) together with (43.2(2)).

53.12 Theorem: Homeomorphism of planes. *Let K and $\overline{K}$ be two topological ternary fields, both defined on the topological space $\mathbb{R}^l$ and with zero element 0, coordinatizing planes $\mathcal{P} = (P, \mathcal{L})$ and $\overline{\mathcal{P}} = (\overline{P}, \overline{\mathcal{L}})$.*

(a) *If the dual characteristic maps χ^* and $\overline{\chi^*} : \mathbb{R}^l \setminus \{0\} \to \mathcal{H}^*$ are homotopic, then the line spaces $\mathcal{L}$ and $\overline{\mathcal{L}}$ are homeomorphic.*
(b) *Similarly, if the characteristic maps χ and $\overline{\chi} : \mathbb{R}^l \setminus \{0\} \to \mathcal{H}$ are homotopic, then the point spaces P and $\overline{P}$ are homeomorphic.*

Proof. (a) It suffices to establish a homeomorphism $\alpha : \mathcal{L} \setminus \{L\} \to \overline{\mathcal{L}} \setminus \{\overline{L}\}$, where L and $\overline{L}$ are the lines at infinity. Indeed, the assertion then follows by extending α to the one-point compactifications. We write $\vee$ and $\cup$, respectively, for the join operations in $\mathcal{P}$ and in $\overline{\mathcal{P}}$. The additive right inverses of y in K and in $\overline{K}$ will be denoted respectively by y' and by y''. We use the same notation $(x, 0)$ or $(0, y)$ for points on the coordinate axes in either plane, and similarly for points (s) at infinity; here, $x, y \in \mathbb{R}^l$ and $s \in \mathbb{R}^l \cup \{\infty\}$.

Deviating from the usual range of parameters, we denote the given homotopy by χ_t^*, $1 \le t \le 2$, where $\chi_1^* = \chi^*$ and $\chi_2^* = \overline{\chi^*}$. The desired homeomorphism will be written down in three pieces α_1, α_2 and α_3. The first and third parts are very natural, and the second one is designed to bridge the gap between them. Observe that α maps every line to a line with the same slope s.

$$\alpha_1\big((0, y') \vee (s)\big) := (0, y'') \cup (s) \quad \text{for} \quad \|s\| \le 1,$$
$$\alpha_2\big((x, 0) \vee (s)\big) := \big(0, \chi_{\|s\|}^*(s, x)''\big) \cup (s) \quad \text{for} \quad 1 \le \|s\| \le 2,$$
$$\alpha_3\big((x, 0) \vee (s)\big) := (x, 0) \cup (s) \quad \text{for} \quad 2 \le \|s\| \le \infty.$$

The point is that α_1 and α_3 would not fit together, even if we extended their domains of definition, but α_2 matches with both of them. Indeed, using the defining property (∗) of dual characteristic maps, see (53.11), we obtain that, for $\|s\| = 1$,

$$\alpha_2\big((x, 0) \vee (s)\big) = \big(0, \chi^*(s, x)''\big) \cup (s)$$
$$= \alpha_1\big((0, \chi^*(s, x)') \vee (s)\big) = \alpha_1\big((x, 0) \vee (s)\big).$$

If $\|s\| = 2$, then similarly

$$\alpha_2\big((x, 0) \vee (s)\big) = \big(0, \overline{\chi}^*(s, x)''\big) \cup (s)$$
$$= (x, 0) \cup (s) = \alpha_3\big((x, 0) \vee (s)\big).$$

The map α is thus well defined. It is continuous, because the partial maps α_i are continuous. Note that continuity of α_2 follows from continuity of the evaluation map $\varepsilon : \mathcal{H}^* \times \mathbb{R}^l \to \mathbb{R}^l$, defined by $\varepsilon(f, x) = f(x)$; continuity of ε is a basic property of the compact-open topology, see (96.4). Moreover, α is bijective, and has an inverse which may be written down in a similar way. Thus, α is in fact a homeomorphism.

(b) The proof is similar in case of the point sets. However, continuity is not quite as obvious in this case. We shall omit the necessary sequence arguments, which are easily formulated using (43.2(2)). We write $x \cdot y$ and $x \circ y$, respectively, for multiplication in K and in $\overline{K}$, and we define continuous divisions from the left, $x \setminus y$ and $x \setminus\!\setminus y$, by the equations

$$x \cdot (x \setminus y) = y,$$
$$x \circ (x \setminus\!\setminus y) = y,$$
$$\infty \setminus y = \infty \setminus\!\setminus y = 0.$$

We now construct a homeomorphism $\alpha : P \setminus \{o\} \to \overline{P} \setminus \{\overline{o}\}$, where o and $\overline{o}$ denote the origins, i.e., $o = (0, 0)$. We use a homotopy $\chi_t : K^\times \to \mathcal{H}$ satisfying $\chi_1 = \overline{\chi}$ and $\chi_2 = \chi$. The homeomorphism α will map the point row of a line through o to the point row of the line through $\overline{o}$ of the same slope s. The definition is given by $\alpha(s) = (s)$, together with the following equations, which apply to $x, y \in K^\times = K \setminus \{0\}$.

$$\alpha_1(x, s \cdot x) = (x, s \circ x) \quad \text{for} \quad \|s\| \leq 1,$$
$$\alpha_2(x, s \cdot x) = \big(s \setminus\!\setminus \chi_{\|s\|}(s, x), \chi_{\|s\|}(s, x)\big) \quad \text{for} \quad 1 \leq \|s\| \leq 2,$$
$$\alpha_3(s \setminus y, y) = (s \setminus\!\setminus y, y) \quad \text{for} \quad 2 \leq \|s\| \leq \infty.$$

The proof of compatibility uses the following computations. For $\|s\| = 1$, we have that

$$\alpha_2(x, s \cdot x) = \big(s \setminus\!\setminus (s \circ x), s \circ x\big) = (x, s \circ x) = \alpha_1(x, s \cdot x),$$

and for $\|s\| = 2$, we compute that

$$\alpha_2(x, s \cdot x) = \big(s \setminus\!\setminus (s \cdot x), s \cdot x\big) = \alpha_3(x, s \cdot x).$$

This ends the proof. □

53.13 Note. The proof of (53.12b) actually shows that the fiber bundle

$$f : P \setminus \{o\} \to L$$

of (51.23a), defined by $x \mapsto ox \wedge L$, is *bundle equivalent* to the analogous bundle $\overline{f}$ derived from $\overline{\mathcal{P}}$ by an equivalence sending the cross section L to $\overline{L}$. Here, L denotes the line at infinity, and 'bundle equivalent' means that there are homeomorphisms $\alpha : P \setminus \{o\} \to \overline{P} \setminus \{\overline{o}\}$ and $\beta : L \to \overline{L}$ such that the following diagram commutes:

$$\begin{array}{ccc} P \setminus \{o\} & \xrightarrow{\alpha} & \overline{P} \setminus \{\overline{o}\} \\ {\scriptstyle f}\downarrow & & \downarrow{\scriptstyle \overline{f}} \\ L & \xrightarrow{\beta} & \overline{L} \end{array}$$

Commutativity of the diagram expresses that α maps the fiber $f^{-1}(x) = (o \vee x) \setminus \{o\}$ of $x \in L$ homeomorphically onto $\left(\overline{f}\right)^{-1}(\beta(x)) = (\overline{o} \vee \beta(x)) \setminus \{\overline{o}\}$.

Theorem (53.12) entails the following question.

53.14 Problem. What are the homotopy types of the characteristic maps of all ternary fields $K \approx \mathbb{R}^l$?

In the sequel, we present the known partial solutions of this problem. The case $l = 1$ is omitted, because it is trivial. Note that, in fact, the proof of Proposition (32.3) on the homeomorphism type of 2-dimensional planes may be taken as a simple and instructive model of our present reasoning.

53.15 Theorem: Topology of 4-dimensional planes. *Assume that* $\dim L = 2$ *or, equivalently, that* $2 < \dim P \leq 4$. *Then the following statements hold.*

(a) L *is homeomorphic to the complex line* $\mathbb{S}_2 = \mathbb{C} \cup \{\infty\}$.
(b) P *is homeomorphic to the complex point space* $P_2\mathbb{C}$.
(c) **F** *is homeomorphic to the complex flag space* $\mathbf{F}_{\mathbb{C}}$.

Notes. 1) The proof of (b) uses (53.12). Hence, without extra effort, it produces a homeomorphism with the additional property that the elements of one pencil $\mathcal{L}_o$ and one line $L \notin \mathcal{L}_o$ are mapped to their complex counterparts, see (53.13).

2) It is by no means an easy matter to deduce (b) from (a). If (a) holds, then we know that P is obtained from the open cell $P \setminus L \approx \mathbb{R}^4$ by gluing it to $L \approx \mathbb{S}_2$. A priori, this could be done in many ways. By analogy, observe that the open 2-cell

can be glued to a circle to produce either the closed disc or the real projective plane or various non-manifolds. Even the fact that P may be covered by three cells $P \setminus L_i$, $i = 1, 2, 3$, is of no immediate use unless we find out exactly how the cells overlap. Basically, this is what we do in the proof.

Proof of (53.15). Part (a) has been proved, see (53.7) and (52.3). Consequently, we know that P is a 4-manifold (51.2). Part (b) will be proved by determining the homotopy type of a characteristic map. The fiber bundle method used in the original versions of this proof may still be recognized, see the remarks preceding (53.11). By a refinement of those methods, Buchanan [79b] proves (c); that proof will not be reproduced here.

We shall determine the homotopy class of the characteristic map χ of a ternary field K coordinatizing $\mathcal{P}$, and then apply (53.12). Recall that χ maps $K^\times$ to the identity component $\mathcal{H}$ of the homeomorphism group of $K^{-1} \approx \mathbb{R}^2$, sending s to the map $x \mapsto sx$. The image of $K^\times$ is contained in the stabilizer $\mathcal{H}_\infty$. We compose χ with the evaluation map $\varepsilon : \mathcal{H}_\infty \to K^\times$, which sends a homeomorphism $h \in \mathcal{H}_\infty$ to its value $h(1)$ at the multiplicative unit. We see that

$$(*) \qquad \varepsilon \circ \chi = \mathbb{1},$$

the identity map of $K^\times$.

Next, we shall use a theorem of Kneser [26] to show that ε has a (left) homotopy inverse δ, which means that $\delta \circ \varepsilon \simeq \mathbb{1}$. Using $(*)$, we infer that then $\chi \simeq \delta$, which ends the proof because it shows that both δ and χ are determined up to homotopy.

First note that we may replace the pair (K^{-1}, ∞) by the homeomorphic pair $(\mathbb{R}^2, 0)$. The homeomorphism group $\mathcal{G}$ of $\mathbb{R}^2$ retracts by deformation onto the stabilizer $\mathcal{G}_0$, via $F_t(g)(x) = g(x) - tg(0)$. Kneser's result as stated by Friberg [73] asserts that the orthogonal group $O_2\mathbb{R}$ is also a deformation retract of $\mathcal{G}$. Consequently, $SO_2\mathbb{R} \subseteq \mathcal{G}^1$ is a deformation retract of the identity component $\mathcal{G}^1$, and the inclusion $SO_2\mathbb{R} \to \mathcal{G}_0^1$ is a homotopy equivalence. Since the restricted evaluation map $\bar{\varepsilon} : SO_2\mathbb{R} \to \mathbb{R}^2 \setminus \{0\}$ clearly is another homotopy equivalence, it follows that $\varepsilon : \mathcal{G}_0^1 \to \mathbb{R}^2 \setminus \{0\}$ has the same property and, in fact, admits a two-sided homotopy inverse δ. □

53.16 Translation planes. For the background on translation planes that is needed here, see Section 25 and (44.6), (64.4). In particular, note that a compact, connected translation plane is coordinatized by a quasifield $Q \approx \mathbb{R}^l$, compare also (53.4). This allows us to apply (53.12). Recall furthermore that the line space of such a translation plane is homeomorphic to the point space by (52.15). Hence, we may use the slightly more convenient dual characteristic map χ^* of Q in order to determine the topology of both P and $\mathcal{L}$. Since quasifields are linear ternary fields, χ^* takes the simple form $\chi^*(s, x) = sx$. This agrees with $\chi(s, x)$, but $\chi(s)$ and $\chi^*(s)$ have different domains of definition.

In the special case of semifields, which coordinatize the planes of Lenz type V, Buchanan [79a] proved that the map χ^* is homotopic to the corresponding map arising in the classical plane $\mathcal{P}_2\mathbb{F}$ of the same dimension. His proof uses Hermitian interpolation, which allows him to construct a deformation retraction that retracts the space of $l \times l$-matrices without real eigenvalues onto the space of orthogonal skew-symmetric matrices. By (53.12) and the preceding remarks, his result has the following consequence.

53.17 Theorem: Topology of planes of Lenz type V. *The point space and the line space of a compact, connected plane $\mathcal{P}$ of Lenz type V are both homeomorphic to $P_2\mathbb{F}$ for some $\mathbb{F} \in \{\mathbb{R}, \mathbb{C}, \mathbb{H}, \mathbb{O}\}$.* □

Hähl [87d] 3.5 proves that a translation plane coordinatized by a differentiable quasifield $Q = (\mathbb{R}^l, \cdot)$ has a dual characteristic map χ^* homotopic to the dual characteristic map of some plane of Lenz type V. That plane is defined by a semifield obtained from Q by differentiating. The precise definition of a *differentiable quasifield* requires that the map sending $(x, y) \in \mathbb{R}^l \times \mathbb{R}^l$ with $x \neq 0$ to the left multiplication $\lambda_a \in \text{End}\,\mathbb{R}^l$ satisfying $\lambda_a(x) = ax = y$ is differentiable. Combining this with (53.12a) and (53.17), we obtain

53.18 Theorem: Planes over differentiable quasifields. *The point space and the line space of a translation plane coordinatized by a differentiable quasifield are both homeomorphic to the point space of some classical plane $\mathcal{P}_2\mathbb{F}$, where $\mathbb{F} \in \{\mathbb{R}, \mathbb{C}, \mathbb{H}, \mathbb{O}\}$.* □

53.19 Further results. Using different methods (involving characteristic cohomology classes), Kramer [94a] proves that a projective plane carrying a smooth structure in the sense of (75.2) has a point space homeomorphic to the point space of some classical plane $\mathcal{P}_2\mathbb{F}$, where $\mathbb{F} \in \{\mathbb{R}, \mathbb{C}, \mathbb{H}, \mathbb{O}\}$. More results on smooth planes will be recorded in (75.5).

53.20 Notes on Section 53. The results on open orbits (53.1 and 2) are due to Salzmann [75a]. The special case (53.3) concerning axial groups was proved much earlier by Hofmann [62]. The algebraic version (53.4) is given in Salzmann [67b] Section 7. The assertion that 4-dimensional planes have manifold lines (53.7) is a result of Salzmann [69b] 1.2. The topology of the point space of a 4-dimensional plane (53.15) was determined by Breitsprecher [71]. The version of the proof presented here is inspired by Buchanan [79b]. As we mentioned in the text, (53.16 and 17) are taken from Buchanan [79a].

54 Lines are homology manifolds

We show that the main theorems (52.5 and 13) of Section 52 hold under the only assumption that L (or P) is of finite topological dimension. To prove this, we have to show that then L, P and $\mathbf{F}$ have the local homology of manifolds. In fact, we show more than this, namely, that these spaces are 'homology manifolds' (also called 'generalized manifolds'). There are many variations of this notion. Besides specifying the local homology groups, one usually requires that the local homology groups at nearby points can be compared in a natural way ('local orientability'), a property that can be deduced from the other defining conditions under certain mild assumptions, as shown by Bredon [69]. We shall directly prove global orientability and use this to determine the local homology.

It is natural to study generalized manifolds even if one is primarily interested in true manifolds. For example, if a product $X \times Y$ is a manifold, then the factors X and Y are generalized manifolds (Bredon [67] V 15.8), but as we remarked in (52.1), they may not be manifolds. Generalized manifolds also appear as fixed point sets of compact groups acting on manifolds, see Borel et al. [60]. There is a characterization of manifolds based on the notion of homology manifold, due to Quinn [83, 86, 87]. The present state of knowledge does not permit to verify Quinn's conditions in order to apply his result to projective planes. See (54.13) for further comments on this question.

A technical tool needed in order to understand and work with local orientability is the notion of sheaf. Without using the term, Greenberg [67] Section 22 gives an introduction to the classical applications of the local homology sheaf of a manifold. For more information, see Bredon [67].

54.1 (Co-)homological local connectivity. The definition of homology manifolds and several theorems that we are going to use require some kind of local connectivity property, such as the following.

HLC (*homology locally connected*): A topological space X is HLC if for every open neighbourhood U of a point $x \in X$ there is an open neighbourhood $V \subseteq U$ of x such that the inclusion $V \hookrightarrow U$ maps $\widetilde{H}_q V$ to 0 for all q.

clc (*cohomology locally connected*): as before, but $\widetilde{H}^q U$, instead of $\widetilde{H}_q V$, is mapped to 0.

There are several variations due to the use of different (co-)homology theories and coefficients, and weaker conditions such as clc^∞ (i.e., V depends on q) or clc^n (i.e., $q \leq n$). Since we are dealing with locally contractible spaces, we can always choose V such that $V \to U$ is null homotopic, hence the strongest conditions HLC and clc are satisfied for any reasonable (co-)homology theory and any coefficient domain.

54.2 Presheaves and sheaves. For general reference, see Bredon [67]. Let R be a principal ideal domain; in fact, only $R = \mathbb{Z}$ will be used. A *presheaf* of R-modules on a Hausdorff space X is a contravariant functor $U \mapsto A(U)$ from the category of all open subsets $U \subseteq X$ and all inclusion maps to the category of R-modules; in other words, the inclusion $V \to U$ induces a homomorphism of the associated modules in the reverse direction. The elements of $A(U)$ may be imagined to be some kind of functions on U, and the induced homomorphism $A(U) \to A(V)$ is then thought of as 'restriction'. Of interest to us is the *local homology presheaf* of $\mathbb{Z}$-modules

$$U \mapsto H_q(X, X \setminus U).$$

As always, we use singular homology with integer coefficients unless stated otherwise.

The *sheaf* $\mathcal{A}$ associated with a presheaf A on X is constructed as follows. The *stalk* $\mathcal{A}_x$ at $x \in X$ is defined as the direct limit (as in Greenberg [67] Section 25),

$$\mathcal{A}_x := \varinjlim_{x \in U} A(U),$$

and $\mathcal{A}$ is defined as the disjoint union of all stalks $\mathcal{A}_x$. For $a \in A(U)$ and $x \in U$ let $a_x \in \mathcal{A}_x$ be the image of a under the canonical map $A(U) \to \mathcal{A}_x$ (the *germ* of a at x). By definition of the limit, $a_x = b_x$ for $b \in A(V)$ if, and only if, there is an open subset $W \subseteq U \cap V$ containing x such that the restrictions of a and b to W are the same. The sets $\{ a_x \mid x \in U \}$ for all open sets U and all $a \in A(U)$ form a basis for a topology on $\mathcal{A}$. This topology makes the projection

$$\pi : \mathcal{A} \to X \quad \text{defined by} \quad a_x \mapsto x$$

a local homeomorphism. Moreover, it renders $\mathcal{A}_x$ discrete and makes the 'module' operations in $\mathcal{A}$ continuous. More precisely, for

$$\mathcal{A} \times_\pi \mathcal{A} := \{ (a, b) \in \mathcal{A} \times \mathcal{A} \mid \pi(a) = \pi(b) \},$$

the map

$$R \times (\mathcal{A} \times_\pi \mathcal{A}) \to \mathcal{A},$$
$$(r, a, b) \mapsto ra - b$$

is continuous; here, R is discrete.

54.3 Locally constant sheaves. A sheaf $\mathcal{A}$ on X is said to be *constant* if there is an R-module B (with discrete topology) and a homeomorphism $f : X \times B \xrightarrow{f} \mathcal{A}$ inducing module isomorphisms $B \to \{x\} \times B \xrightarrow{f} \mathcal{A}_x$ for every $x \in X$. The sheaf $\mathcal{A}$ is said to be *locally constant* if X can be covered by open sets U such that $\mathcal{A}$ is *constant over* each U, i.e., there is a homeomorphism $f_U : U \times B_U \to \pi^{-1}(U)$

inducing module isomorphisms on the fibers as before. Note that a constant sheaf is also locally constant.

54.4 Lemma: Constant sheaves. *Assume that the canonical projection*

$$A(U) \to \mathcal{A}_x$$

is an isomorphism for every x in the open set U. Then $\mathcal{A}$ is constant over U, with stalks isomorphic to $A(U)$.

Proof. Endow $A(U)$ with the discrete topology. A map $f_U : U \times A(U) \to \pi^{-1}(U) \subseteq \mathcal{A}$ is defined by sending (x, a) to a_x. By our hypothesis, f_U sends $\{x\} \times A(U)$ isomorphically onto $\mathcal{A}_x$. In particular, f_U is bijective. By definition of the topology on $\mathcal{A}$, any open set of the form $U \times \{a\}$ is mapped homeomorphically onto an open set of $\mathcal{A}$. Since these sets cover $U \times A(U)$, the map f_U is a homeomorphism, and $\mathcal{A}$ is constant over U. □

Remark. Let $\mathcal{A}$ be constant. We may replace $A(X)$ by 0 without changing $\mathcal{A}$, unless some point has no neighbourhoods other than X itself. This shows that the converse of (54.4) is not true.

54.5 The (q-dimensional) local homology sheaf of X. This is the sheaf $\mathcal{H}_q(X)$ associated with the presheaf $U \mapsto H_q(X, X \setminus U)$. It is customary to take the direct sum over all q, but we shall avoid this. Let $x \in X$ be fixed. Inclusion induces a compatible family $(\beta_U)_{x \in U}$ of homomorphisms from the directed system $\bigl(H_q(X, X \setminus U)\bigr)_{x \in U}$ into $H_q(X, X \setminus \{x\})$, hence the universal property of the limit yields a commutative diagram

$$\begin{array}{ccc}
 & H_q(X, X \setminus U) & \\
 & {\scriptstyle \alpha_U} \downarrow \quad \searrow {\scriptstyle \beta_U} & \\
\mathcal{H}_q(X)_x = \varinjlim H_q(X, X \setminus U) & \xrightarrow{\gamma} & H_q(X, X \setminus \{x\})
\end{array}$$

where α_U is the canonical map. We shall show that γ is an isomorphism. Every singular chain u in $X \setminus \{x\}$ is carried by some compact set $K \subseteq X \setminus \{x\}$, so that u is a chain in $X \setminus U$ for $U = X \setminus K$. Now any class $\sigma \in H_q(X, X \setminus \{x\})$ can be represented by a chain v in X such that ∂v lies in $X \setminus \{x\}$. By the last remark, the same v represents a class in some $H_q(X, X \setminus U)$, hence $\sigma = \beta_U[v] = \gamma \alpha_U[v]$. On the other hand, if $\tau = \alpha_U[u]$ is mapped to 0 by γ, then $\beta_U[u] = 0$, and there is a chain w in X such that $\partial w - u$ lies in $X \setminus \{x\}$. Again, $\partial w - u$ lies in some compact set $X \setminus V$. We may assume that $X \setminus V \supseteq X \setminus U$. Then u represents 0 in $H_q(X, X \setminus V)$, so that $\tau = 0$. Thus, we have proved the following proposition.

54.6 Proposition: Local homology at a point. *The stalk $\mathcal{H}_q(X)_x$ may be identified with the local homology group $H_q(X, X \setminus \{x\})$ in such a way that the canonical map $H_q(X, X \setminus U) \to \mathcal{H}_q(X)_x$ defined by $u \mapsto u_x$ becomes the homomorphism $H_q(X, X \setminus U) \to H_q(X, X \setminus \{x\})$ induced by inclusion.* □

54.7 Homology manifolds. A locally compact space X is called a *homology n-manifold* (over $\mathbb{Z}$) if the following conditions are satisfied, where homology is singular homology with coefficients in $\mathbb{Z}$.

(i) $\dim X < \infty$.
(ii) X is HLC.
(iii) The stalks of the local homology sheaf satisfy $\mathcal{H}_q(X)_x = 0$ for $q \neq n$ and $\mathcal{H}_n(X)_x \cong \mathbb{Z}$.
(iv) *Local orientability*: $\mathcal{H}_q(X)$ is a locally constant sheaf for all q.

We shall have to consider (iii) and (iv) independently of each other, therefore we do not restrict (iv) to the only non-trivial case $q = n$. By (54.6), condition (iii) can be rephrased as (iii') below, and by (54.4), condition (iv) is implied by (iv'):

(iii') X has the local homology of an n-manifold.
(iv') Every point $x \in X$ has an open neighbourhood U such that $H_q(X, X \setminus U) \to H_q(X, X \setminus \{u\})$ is an isomorphism for every $u \in U$.

54.8 Comments. (a) It is a classical fact that topological n-manifolds have these properties; compare (52.4) or Greenberg [67] Section 22. Condition (iv') is Greenberg's 22.4. His exposition of Poincaré and Lefschetz duality relies on these facts.

(b) In (i) the covering dimension $\dim X$ is sometimes replaced by the (sheaf theoretic) *cohomology dimension* $\dim_{\mathbb{Z}} X$ defined in Bredon [67] Section II.15. If $\dim X < \infty$ and X is a paracompact Hausdorff space, then $\dim_{\mathbb{Z}} X = \dim X$ by Deo–Singh [82]; compare Löwen [83b] Section 4.

For some applications, it is required that the number $\dim_{\mathbb{Z}_p} X$ is finite, which is defined over $\mathbb{Z}_p$ in the analogous way. Now if X is separable metric and $\dim X < \infty$, then X is homeomorphic to a subset of some Euclidean space $\mathbb{R}^n$, see Hurewicz–Wallman [48] V.3 p. 60. Since $\dim_{\mathbb{Z}_p} \mathbb{R}^n = n$ (Bredon [67] IV.7.6) and $\mathbb{Z}_p$-dimension satisfies the subspace theorem for metric spaces (Bredon [67] II.15.8 and I.6.2), it follows that finiteness of $\dim X$ implies finiteness of $\dim_{\mathbb{Z}_p} X$.

(c) A striking theorem of Bredon [67] V 15.2 asserts that for connected spaces X, condition (iii), which appears to be the only specific one among the four conditions defining homology n-manifolds, is redundant if all stalks $\oplus_q \mathcal{H}_q(X)_x$ of the 'total' local homology sheaf are finitely generated. The value of n is then given by $n = \dim_{\mathbb{Z}} X$. This will be the basis of our proof that lines are homology manifolds. Note, however, that Bredon uses a different homology theory.

(d) Some authors drop the HLC condition from the definition of homology manifolds and reserve the term 'generalized n-manifold' for those that are HLC.

(e) Using the excision theorem (Spanier [66] 4.6.5), one can prove without difficulty that, for a regular space X, the restriction of the local homology sheaf $\mathcal{H}_q(X)$ to any open set $U \subseteq X$ is isomorphic to $\mathcal{H}_q(U)$. Consequently, a non-empty open subset of a homology n-manifold is a homology n-manifold. Conversely, if every point of X has an open neighbourhood which is a homology n-manifold and if $\dim X < \infty$, then X itself is a homology n-manifold.

We return to compact, connected planes $\mathcal{P} = (P, \mathcal{L})$.

54.9 Lemma: Orientability of lines. *The local homology sheaf $\mathcal{H}_q(L)$ of a line L is constant with stalk $\mathcal{H}_q(L)_x \cong \widetilde{H}_q(L)$. The analogous statement is true for the quotient space P/L defined in (51.6).*

Proof. The reduced homology sequence (compare Spanier [66] p. 184) of the pair $(L, L \setminus \{x\})$ shows that $H_0(L, L \setminus \{x\}) = 0$ and that the inclusion map $(L, \emptyset) \to (L, L \setminus \{x\})$ induces isomorphisms $H_q L \to H_q(L, L \setminus \{x\})$ for $q > 0$; remember that $L \setminus \{x\}$ is contractible (51.4). Substituting $U = L$ in (54.4 and 6) and observing that $H_q(L, L \setminus L) = H_q L$, we obtain that $\mathcal{H}_q(L)$ is constant. We used this argument in the reverse direction in the proof of (52.5). For P/L, the same proof works with (51.27) in place of (51.4). □

54.10 Theorem: Lines are homology manifolds. *Let $\mathcal{P} = (P, \mathcal{L})$ be a compact, connected projective plane. Assume that $l = \dim L < \infty$. Then the following assertions hold.*

(a) *L is a homology l-manifold.*
(b) *P, P/L and $\mathcal{L}$ are homology $2l$-manifolds, and $\dim P = \dim \mathcal{L} = 2l$.*
(c) *$\mathbf{F}$ is a homology $3l$-manifold, and $\dim \mathbf{F} = 3l$.*

54.11 Corollary: Extending the range of Section 52. *If $l = \dim L < \infty$, then L has the local homology of an l-manifold and P has the local homology of a $2l$-manifold. In particular, Theorems (52.5, 12, and 13) hold for $\mathcal{P}$, with the specific value $m = l$. Thus, $\dim P = 2l$ divides 16. Moreover, we have $L \simeq \mathbb{S}_l$ and $P/L \simeq \mathbb{S}_{2l}$.* □

Proof of (54.10). By (54.1 and 9), the space L satisfies all conditions defining a homology manifold except perhaps (iii). It only remains to determine the stalks $\mathcal{H}_q(X)_x$. In order to do this, we shall translate what we have proved in (54.9) into the setting of *Borel–Moore homology* as defined in Bredon [67], obtain our result there, and then translate back. We write $\overline{H}_q$ for Borel–Moore homology with (constant) coefficients in $\mathbb{Z}$ and with compact supports. The resulting local homology sheaf will be denoted by $\overline{\mathcal{H}}_q$.

A preparatory step is to replace the presheaf $U \mapsto H_q(L, L \setminus U)$ generating the local homology sheaf by $U \mapsto H_q(L, L \setminus \overline{U})$. This does not change the associated sheaf. Indeed, since L is a regular space, every open neighbourhood of $x \in L$ contains a closed neighbourhood and conversely, hence the limit over open or closed neighbourhoods of x alone is the same as the limit over both kinds of neighbourhoods together; compare Greenberg [67] 25.11. Moreover, the topology on $\mathcal{H}_q(L)$ remains the same.

Now since both L and $L \setminus \overline{U}$ are locally contractible, the groups $H_q(L, L \setminus \overline{U})$ are naturally isomorphic to $\overline{H}_q(L, L \setminus \overline{U})$. This follows from Bredon [67] V 12.6 p. 231, together with p. 181; note that in the compact space L, locally finite singular chains are actually finite. Thus, $H_q(L, L \setminus \overline{U})$ and $\overline{H}_q(L, L \setminus \overline{U})$ are isomorphic presheaves. Replacing $\overline{H}_q(L, L \setminus \overline{U})$ by $\overline{H}_q(L, L \setminus U)$ as we did for the singular presheaf, we obtain that $\mathcal{H}_q(L)$ and $\overline{\mathcal{H}}_q(L)$ are isomorphic sheaves.

We know that $\dim_{\mathbb{Z}} L = l$, cf. (54.8b), and that L is $\mathrm{clc}_{\mathbb{Z}}^{\infty}$, see (54.1). Moreover, $\overline{\mathcal{H}}_q(L)$ is constant with stalks $\overline{\mathcal{H}}_q(L)_x \cong \widetilde{H}_q(L)$ by (54.9). The reduced singular homology $\widetilde{H}_*(L)$ is finitely generated by (51.13 or 14). This is all that is needed in order to apply Bredon's characterization of homology n-manifolds (Bredon [67] V 15.2 p. 240, compare (54.8c)). The theorem asserts that $\overline{\mathcal{H}}_q(L)_x = 0$ for $q \neq l$ and $\overline{\mathcal{H}}_l(L)_x \cong \mathbb{Z}$. Note in passing that this argument shows that $\dim X = n$ for every homology n-manifold X. Using the isomorphism $\mathcal{H}_q(L) \cong \overline{\mathcal{H}}_q(L)$ we obtain that L is a homology l-manifold with respect to singular homology as well.

Exactly the same proof works for P/L. Products of connected, locally contractible homology manifolds are homology manifolds of the appropriate dimension by Bredon [67] V 15.8 p. 242, combined with V 15.7 p. 241. Using the above translation techniques again, we may transfer this result from Borel–Moore homology to singular homology. In view of the local product structures of P and F (51.2), this yields assertions (b) and (c); note that being a homology manifold is a local property (54.8e). □

54.12 Remark on alternative proofs. Using the topological homogeneity of L, we could have obtained (54.10) by a mere quotation of any of the following works: Bredon [70], Bryant [87], Dydak–Walsh [87, 88]. These papers have in common that the main amount of work is invested to deduce local orientability (which comes for free in our situation) from homogeneity. Therefore, we preferred to give a proof that concentrates on the heart of the matter, which is, of course, Bredon's characterization of homology manifolds. A simplified proof of the latter is contained in Dydak–Walsh [87]. Note also that the proof of (54.10) given in Löwen [83b] uses a slightly different approach.

54.13 Characterization of manifolds. Quinn [83] announced a theorem characterizing topological n-manifolds for $n \geq 5$ as those homology n-manifolds that in addition satisfy the so-called *disjoint disks property*. Based on this announcement, it was concluded in Löwen [83b] that the point space of a finite-dimensional affine plane is a manifold, because it could be proved that the affine point set $P \setminus L$ satisfies the disjoint disks property. In the meantime, Quinn discovered a gap in his proof, see Quinn [86, 87]; compare also Bryant–Ferry–Mio–Weinberger [93]. As a result, the manifold characterization was restated in Quinn [87] with an additional hypothesis which appears to be hard to check for planes. Thus, the following problem remains open:

Problem. Are the lines or the point space of a finite-dimensional compact projective plane necessarily topological manifolds?

The homogeneity of the lines seems to be of little use regarding this problem, compare (53.8). A survey of the problem of characterizing manifolds is given by Cannon [78].

We do not know of any major geometric problem that could be solved only if the above question had a positive answer (see, however, (84.4)). In fact, most of the benefit that can be drawn from the manifold property is available to us on the basis of Sections 51 and 54. For example, homology manifolds satisfy Poincaré duality, compare Lacher [84], Skljarenko [71a,b] and Bredon [67] V.9.2. Moreover, Jordan–Brouwer type separation theorems can be proved, see Raymond [60]. Transformation groups on homology manifolds are investigated in Borel et al. [60].

54.14 Other notions of (co-)homology manifolds. From our proof of (54.10) it is clear that each space $X \in \{P, L, P/L, \mathbf{F}\}$ is a homology manifold with respect to both singular and Borel–Moore homology with integer coefficients. For applications in 'local' Smith theory (e.g., if one wants to show that fixed point sets of $\mathbb{Z}_p$-actions on X are again homology manifolds), one needs to replace the coefficient ring $\mathbb{Z}$ by $\mathbb{Z}_p$. Of course, condition (54.7iii) has to be changed accordingly. First, recall that $\dim_{\mathbb{Z}_p} X$ is finite by (54.8b). Since the homotopy equivalence $L \simeq \mathbb{S}_l$ is now known thanks to (54.11), we can use the proof of (54.9) to obtain conditions (iii) and (iv) of (54.7) with respect to $\mathbb{Z}_p$-coefficients for L. This carries over to the other spaces X as in the proof of (54.10).

Since X is locally contractible, X is also a cohomology manifold according to Bredon [67] V 15.7 p. 241. Moreover, theorems 1 and 2 of Raymond [65a] allow us to change freely between the notions of cohomology manifolds or homology manifolds with respect to singular or Borel–Moore homology and between the use of different rings $R \in \{\mathbb{Z}, \mathbb{Z}_p, \mathbb{Q}\}$ of coefficients. Compare also Raymond [65b] Section 3.

54.15 Notes on Section 54. This section is an adaptation of Löwen [83b] to the special case of projective planes; Löwen considers stable planes. More on the history is said in the notes on Section 52.

55 Geometric consequences

In this section, $\mathcal{P} = (P, \mathcal{L})$ always denotes a compact, connected and finite-dimensional plane, and L is a line; recall that $\dim P = 2 \dim L = 2l$ divides 16 by (54.11).

It is not hard to imagine that the results of the preceding four sections are fundamental for all deeper investigations into the structure of compact, connected planes and their automorphisms. The consequences that we are going to point out in the present section are the more immediate ones, that have become standard tools of the theory. The main topics treated are subplanes, ovals, special properties of ternary fields and automorphisms of planes. In particular, fixed points and finite groups of automorphisms are considered.

At this point, we emphasise a fact which is now obvious, and which can hardly be overstressed. According to (54.11), each of the planes $\mathcal{P}$ under consideration has dimension $\dim P = 2 \dim L = 2l$ dividing 16. By (42.7), there is a unique Moufang plane of the same topological dimension $2l$, namely the classical plane $\mathcal{P}_2\mathbb{F}$ with $\dim_\mathbb{R} \mathbb{F} = l$; compare (14.9) and (16.14). We shall refer to this fact as *domination by classical planes*. The dominating Moufang plane sets the standard by which all properties of a given plane $\mathcal{P}$ are assessed. By (54.11) and (52.5 and 13), the *topology* of these two planes is very similar, but the *geometry* may of course be totally different.

The domain invariance theorem (51.21b) together with the topological pigeon-hole principle (51.20) forms a tool that can be used to mimic some of the typical counting arguments of finite geometry in the topological context. For examples demonstrating its potential, see (43.6), (55.1, 5, 10, 11 and 14), (61.4 and 17), (64.8 and 9), and (65.2). The results (96.11, 13b, 19 and 25) and their applications in Chapter 6 are also related. Thus, there is a strong analogy between the rôles played by the dimension of a compact, connected plane and the (logarithm of the) order of a finite plane.

In the light of this comparison, consider the analogue of domination by classical planes in the finite case. It is still unknown whether or not there are projective planes of finite orders other than prime powers; those are the orders of the desarguesian planes. For a given prime power p^r, there is precisely one desarguesian plane of order p^r. Needless to say, the analogue of classical domination does not make much sense for infinite planes in general.

Subplanes

In a classical projective plane $\mathcal{P}_2\mathbb{F}$, where $\mathbb{F} \in \{\mathbb{R}, \mathbb{C}, \mathbb{H}, \mathbb{O}\}$, every proper closed subplane $\mathcal{Q}$ is isomorphic to some other classical plane of smaller dimension; in particular, $\mathcal{Q}$ is connected. This follows from the results of Chapter 1. Indeed, Aut $\mathcal{P}_2\mathbb{F}$ is transitive on quadrangles, see (13.5) and (17.6), hence $\mathcal{Q}$ can be coordinatized by some substructure $A \leq \mathbb{F}$ containing the unit element. Then A contains the closure $\mathbb{R} = \overline{\mathbb{Q}}$ of the subfield generated by 1. Using Hamilton and Cayley triples, see (11.9, 10 and 16), it is easy to show that A is a classical division algebra.

In an arbitrary compact, connected plane $\mathcal{P} = (P, \mathcal{L})$ of finite dimension $\dim P = 2l = 2^r$, cf. (54.11), one expects to find a similar pattern of subplanes. Indeed, by (54.11) only the numbers $2^s \neq 1$ dividing 2^r can occur as the dimension of a *connected* closed subplane $\mathcal{Q} \leq \mathcal{P}$. By (51.21a), equality $s = r$ means that the point set of $\mathcal{Q}$ contains a non-empty open subset of P and thus implies that $\mathcal{Q} = \mathcal{P}$. This proves the following.

55.1 Proposition: Dimension of subplanes. *In a plane $\mathcal{P}$ of dimension 2^r, every closed, connected, proper subplane has dimension 2^s, where $0 < s < r$. In particular, the length of a nested sequence of distinct, connected, closed subplanes is at most r.* □

Note. It is not at all clear that, in a given plane $\mathcal{P}$, each of the numbers allowed by (55.1) does occur as the dimension of some subplane of $\mathcal{P}$. It can happen that a 4-dimensional plane $\mathcal{P}$ is the closure of the subplane generated by one of its quadrangles, see Knarr [87b] for examples. This indicates that even planes of dimension ≥ 4 without any proper closed subplanes might exist.

55.2 Problem. Is every closed subplane of a plane $\mathcal{P}$ connected?

This is true in the classical planes by the introductory remarks above, and in planes of dimension 2 or 4, see (55.3 and 4) below. However, it is still conceivable that some plane $\mathcal{P}$ of dimension 8 or 16 might contain a seven-point plane or an infinite, 0-dimensional, compact plane. This constitutes one of the major unsolved problems of the theory. Apart from its independent interest, it is also one of the stumbling blocks on the way towards a proof of another pivotal conjecture. That conjecture says that the automorphism group of a coordinatizing ternary field is always compact; compare (44.5). This group is isomorphic to the stabilizer of a quadrangle in Aut $\mathcal{P}$, see (44.4c).

55.3 Review: 2-dimensional planes. *A 2-dimensional plane $\mathcal{P}$ contains no proper, closed subplanes* (32.7). □

55.4 Theorem: 4-dimensional planes.

(a) *If $\mathcal{Q} = (Q, \mathcal{M})$ is a proper, closed subplane of a 4-dimensional plane $\mathcal{P}$, then $\mathcal{Q}$ is connected and 2-dimensional. Moreover, $\mathcal{Q}$ is a Baer subplane of $\mathcal{P}$, as defined in (21.7).*
(b) *If K_1 is a closed, proper ternary subfield of a 2-dimensional ternary field K, then the pair (K, K_1) is homeomorphic to $(\mathbb{R}^2, \mathbb{R})$.*

Proof. We treat (a) and (b) simultaneously, assuming that $\mathcal{P}$ and $\mathcal{Q}$ are coordinatized respectively by K and K_1; cf. (43.2 and 5). We shall show that $\dim K_1 > 0$. Then Q is connected (42.2), and $\dim Q = 2$ by (55.1), hence $K_1 \approx \mathbb{R}$ by (51.29). Moreover, (55.5) below will show that $\mathcal{Q}$ is a Baer subplane of $\mathcal{P}$, and then (b) follows from (41.11c).

We shall apply Brouwer's theorem on plane translations (Brouwer [12], compare E. Sperner [34], M. Brown [84]). It asserts in particular that for any fixed-point free, orientation preserving homeomorphism $h : \mathbb{R}^2 \to \mathbb{R}^2$, the iterated images $h^n(x)$ of each point x converge to ∞ as $n \to \infty$.

By (53.15a) we have that $K \approx \mathbb{R}^2$. Assume that $\dim K_1 = 0$. Our homeomorphism $h : K \to K$ will be given by ternary addition $h(x) = x + q$, where $0 \neq q \in K_1$. Joining q to 0 by a path q_t in K, we obtain an isotopy $h_t : x \to x + q_t$ between h and id_K, hence h preserves the orientation. Clearly, h fixes no point. The points $h^n(0)$ belong to K_1 and converge to ∞, hence K_1 is infinite. By (41.10), the line $K_1 \cup \{\infty\}$ of $\mathcal{Q}$ is homeomorphic to Cantor's triadic set, so that K_1 contains a set X which is both compact and open in K_1. Continuity of addition implies that $X + U \subseteq X$ for some neighbourhood U of 0 in K_1. Choosing $q \in U \setminus \{0\} \neq \emptyset$, we obtain that $h(X) \subseteq X$ contrary to Brouwer's theorem. □

Note. There are easy counterexamples to Brouwer's theorem in dimensions ≥ 3, hence a general solution of (55.2) would have to be based on different grounds.

It follows from the definition given in (21.7) that a Baer subplane of a projective plane $\mathcal{P}$ is a maximal projective subplane. On the other hand, scarcity of subplanes may cause existence of maximal subplanes which are nevertheless not Baer subplanes. The same can happen if we consider closed subplanes of compact planes. Baer subplanes of compact planes may be characterized as follows.

55.5 Proposition: Baer subplanes. *Let $\mathcal{Q} = (Q, \mathcal{M})$ be a compact subplane of $\mathcal{P}$. Then $\mathcal{Q}$ is a Baer subplane if, and only if $\dim P = 2 \dim Q$.*

Proof. Let l and m be the respective dimensions of the lines of $\mathcal{P}$ and $\mathcal{Q}$. By (54.10b), we have $\dim P = 2l$ and $\dim Q = 2m$. If $\mathcal{Q}$ is a Baer subplane, then (41.11b) asserts that Q is locally homeomorphic to a line L of $\mathcal{P}$, and we infer that $2 \dim Q = 2 \dim L = 2l = \dim P$.

Conversely, let $l = 2m = \dim Q$, and assume that $p \in P \setminus Q$ is not incident with any line of $\mathcal{Q}$. Then the map $\varphi : Q \to \mathcal{L}_p$ defined by $q \mapsto pq$ is continuous and injective, hence it is an embedding by compactness. We have $\dim \varphi(Q) = 2m = l = \dim \mathcal{L}_p$, hence (51.21a) shows that the image $\varphi(Q)$ contains a non-empty open set U of $\mathcal{L}_p$. Then $\varphi^{-1}(U) \approx U$ is open in Q, and Q is locally homeomorphic to $\mathcal{L}_p$ because both spaces are homogeneous. By the topological pigeon-hole principle (51.20 and 21b), φ is a homeomorphism. However, according to (54.11) the homology groups of $\mathcal{L}_p \approx L$ and of Q are given respectively by (52.5) and (52.13c), with contradictory results. The dual arguments show that every line of $\mathcal{P}$ is incident with some point of $\mathcal{Q}$, hence $\mathcal{Q}$ is indeed a Baer subplane. □

55.6 Corollary: Nested sequences. *If a compact plane $\mathcal{P}$ of dimension 2^r contains a nested sequence of compact, connected subplanes of extremal length r, then the lines of $\mathcal{P}$ are spheres.*

Proof. It suffices to show that lines are manifolds (52.3). The given subplanes $\mathcal{Q}_i$, $1 \leq i \leq r$, have point sets satisfying $\dim Q_i = 2^i$. By (55.5), each $\mathcal{Q}_i$ is a Baer subplane of $\mathcal{Q}_{i+1}$. We use induction along the sequence, starting with (53.5) for $\mathcal{Q}_1$. If the lines of $\mathcal{Q}_i$ are manifolds, then the same is true for the point set of $\mathcal{Q}_i$ by (51.2) and, hence, for the lines of $\mathcal{Q}_{i+1}$, see (41.11b). This furnishes the inductive step. □

Note. A similar result holds for nested sequences of length $r - 1$ starting with a 4-dimensional plane. The proof uses the same arguments; induction is started using (53.7).

Ovals

Ovals in projective planes may be used to construct circle planes, such as Möbius, Laguerre, or Minkowski planes. For a survey of this topic, see Steinke [95]. Within the theory of projective planes, we shall use ovals in connection with shift planes; see Section 74 and, in particular, the proofs of (74.6 and 13); compare also (74.17).

55.7 Definition. An *oval* in a projective plane is a set O of points such that the following conditions are satisfied.

(i) Convexity: Each line L meets O in at most 2 points.
(ii) Smoothness: For each point $x \in O$ there is a unique line, denoted T_x, such that $T_x \cap O = \{x\}$.

We denote by $\mathscr{L}_i$ the set of all lines meeting O in precisely i points. For $i = 2, 1, 0$, these lines are respectively called *secants, tangents*, and *exterior lines*. The smoothness condition requires that each $x \in O$ lies on a unique tangent.

Mazurkiewicz [14, 69] and Barlotti [67] give a transfinite construction producing ovals in any infinite plane. In compact, connected planes, this construction can be modified so as to produce ovals that are at the same time dense in the point set and connected; the necessary ideas can be found in Freudenthal [55]. Thus, some further condition is needed in order to come closer to the intuitive idea of ovals as represented by convex simple closed curves in $\mathbb{R}^2$ or by conics in $\mathscr{P}_2\mathbb{C}$. To prepare this condition, we introduce

55.8 Symmetric squares. The *symmetric square* $X * X$ of a Hausdorff space X is defined as the quotient space of $X \times X$ modulo the equivalence relation $a \sim \sigma(a)$, where $\sigma(x, y) = (y, x)$. Denote the class of (x, y) by $\{x, y\}$. A net $\{a_i, b_i\}$ converges to $\{x, y\}$ if, and only if $\{a_i, b_i\} = \{x_i, y_i\}$ with $x_i \to x$ and $y_i \to y$. For $x, y \in X$, consider the pair $(\{x, y\}, x) \in (X * X) \times X$. From the description of convergence one obtains:

(a) *The map* $(\{x, y\}, x) \mapsto y$ *is continuous.*

The quotient map $s_X : X \times X \to X * X$ is closed and open because the saturated hull $A \cup \sigma(A)$ of a closed or open set $A \subseteq X \times X$ is respectively closed or open. This implies that the restriction of s_X to A is again a quotient map. Hence, s_X maps the diagonal $\{(x, x) \mid x \in X\} \approx X$ in $X \times X$ homeomorphically onto the diagonal of $X * X$:

(b) $$X \approx D_X := \{\{x, x\} \mid x \in X\} \subseteq X * X$$

Similarly, s_X induces the second homeomorphism in

(c) $$X \approx X \times \{x_0\} \approx s_X(X \times \{x_0\}).$$

The restriction of s_X to the complement of the diagonal is continuous, open, and locally injective. Thus, if X is locally homogeneous, then

(d) $(X * X) \setminus D_X$ *is locally homeomorphic to* $X \times X$.

55.9 Definition. Let O be an oval in a compact, connected plane $\mathscr{P}$ of finite dimension and endow O with the topology induced from P. We have a bijective

map
$$\varphi : O * O \to \mathcal{L}_1 \cup \mathcal{L}_2$$
defined by
$$\{x, y\} \mapsto \begin{cases} xy & \text{if } x \neq y \\ T_x & \text{if } x = y. \end{cases}$$

We say that O is a *topological oval* if φ is a homeomorphism. The definition given in Buchanan–Hähl–Löwen [80] looks weaker than this one, but by 2.5 of that paper, the two definitions are equivalent. Non-trivial examples of topological ovals are given in the same article and in (74.5).

If O is a topological oval, then φ maps the diagonal D_O homeomorphically onto the set $\mathcal{L}_1$ of tangents, hence by (55.8b) we have $O \approx \mathcal{L}_1$ via $x \mapsto T_x$. Similarly, (55.8c) implies that φ induces a homeomorphism $O \approx \mathcal{L}_x$ for each $x \in O$. In particular, a topological oval is compact and hence closed in P. Since we assume that P is finite-dimensional, we can prove the converse. The proof rests on the following purely topological fact taken from Buchanan [79c] Section 4. Buchanan only considers the case where Y is a sphere, but his proof, an ingenious application of domain invariance (51.18), carries over to the general situation verbatim. We, too, are mainly interested in the special case; this explains the caption that we have chosen. Since Buchanan's paper is in German, we mention Löwen [94] as another source.

55.10 Lemma: Characterization of spheres. *Let X and Y be compact metric spaces and assume that Y has the domain invariance property. If for every $x \in X$ there is a point $y \in Y$ and a continuous bijection $X \setminus \{x\} \to Y \setminus \{y\}$, then X is homeomorphic to Y.* □

This result can be applied to a compact oval O as follows: X is the oval O, and Y is a line pencil $\mathcal{L}_x$, where $x \in O$; note that these pencils are pairwise homeomorphic. A continuous bijection $O \setminus \{x\} \to \mathcal{L}_x \setminus \{T_x\}$ is given by the join map. Hence, (55.10) implies that $O \approx \mathcal{L}_x \approx L$. Therefore, (55.8d) together with (51.2) shows that $U := O * O \setminus D_O$ is locally homeomorphic to $\mathcal{L}$. The restriction of φ to U is continuous by definition, hence the open mapping theorem (51.19 and 21b) shows that φ induces a homeomorphism $U \approx \mathcal{L}_2$, and moreover that $\mathcal{L}_2$ is open in $\mathcal{L}$. Now an easy compactness argument shows that φ^{-1} is continuous at every tangent, so that φ^{-1} is in fact a globally continuous map. Moreover, its domain of definition, the set $\mathcal{L}_1 \cup \mathcal{L}_2$, is compact because O is compact (41.7b). Thus, φ^{-1} is a homeomorphism, and we have proved:

55.11 Theorem: Compact ovals. *An oval O in $\mathcal{P}$ is a topological oval if, and only if, O is a closed subset of P. In this case, both O and the set $\mathcal{L}_1$ of tangents* are homeomorphic to a line, and we have that $O * O \approx \mathcal{L}_1 \cup \mathcal{L}_2$. Moreover, $\mathcal{L}_2$ is open in $\mathcal{L}$. □

55.12 Corollary. *There is a homeomorphism* $P_2\mathbb{C} \approx \mathbb{S}_2 * \mathbb{S}_2$.

Proof. This follows from (55.11), applied to a conic in $\mathcal{P}_2\mathbb{C}$. Note that $\mathcal{L}_0 = \emptyset$ in this case, because $\mathbb{C}$ is algebraically closed. □

Incidentally, Massey [73] uses this homeomorphism to show that the quotient space of $P_2\mathbb{C}$ modulo the Baer involution induced by complex conjugation is a 4-sphere. Compare also Marin [86]. We remark that the space $\mathbb{S}_1 * \mathbb{S}_1$ associated with a conic in $\mathcal{P}_2\mathbb{R}$ is a Möbius strip with boundary.

Closely related to our notion of topological ovals is the concept of 'generalized Mazurkiewicz sets' $X \subseteq \mathbb{R}^2$, defined by the property that a line separating two points in X meets X in exactly two points. For such sets, a result similar to (55.11) has been proved by Loveland–Loveland [95].

A famous theorem of Segre asserts that in a finite, desarguesian plane of odd order, every oval is a conic; see Hughes–Piper [73] Section XII.6. The following analogous result was proved by Buchanan [79c]; note the contrast to the real case.

55.13 Theorem: Complex conics. *Every closed oval O in $\mathcal{P}_2\mathbb{C}$ is a conic.*

Proof. Introduce coordinates in an affine subplane such that the line at infinity is a tangent of O, touching O in the point at infinity of the y-axis of $A_2\mathbb{C} = \mathbb{C} \times \mathbb{C}$. Then the affine part of O is the graph of a function $f : \mathbb{C} \to \mathbb{C}$, and (55.11) implies that this function is complex differentiable. Indeed, the slope of the secant through $(x, f(x))$ and $(y, f(y))$ converges to the slope s of the tangent through $(x, f(x))$ as $y \to x$; in other words, we have

$$\frac{f(y) - f(x)}{y - x} \to s.$$

Since f can attain a given value at most twice by convexity, the theorem of Casorati–Weierstraß implies that the singularity of f at infinity is inessential. Thus, f is a polynomial function. Using the polynomial derivative, it is easily seen that the graph of a complex polynomial function is an oval if, and only if, the polynomial is of degree 2. □

A 'quadratic polynomial' over $\mathbb{H}$ does not define an oval. For example, consider the graph of $f(x) = x^2$ in the affine plane $\mathbb{H} \times \mathbb{H}$. Its intersection with the line $y = -1$ is described by the equation $x^2 + 1 = 0$. The set of solutions of this equation is the unit sphere $\{x \in \mathrm{Pu}\,\mathbb{H} \mid \|x\| = 1\} \approx \mathbb{S}_2$, see (11.6(1)). The next theorem gives a topological 'explanation' of this fact.

55.14 Theorem: Dimension of planes containing ovals. *If the plane $\mathcal{P}$ contains a closed (and hence topological) oval O, then the lines have dimension $l \le 2$. The oval admits exterior lines if, and only if $l = 1$.*

We first give a short proof and then add a more elementary one, valid for planes with manifold lines. For further variations, see Buchanan–Hähl–Löwen [80].

First Proof. Choose points $0, 1, \infty \in O$ and let $L = 0 \vee \infty$. By (55.11), we have $O \setminus \{0, \infty\} \approx L \setminus \{0, \infty\}$. Following Artzy [66], we introduce a commutative, continuous multiplication on this space; Figure 55a shows how this can be done. Continuity of this multiplication follows from (55.8a). By (54.11) and (52.5b), the space $L \setminus \{0, \infty\}$ is homotopy equivalent to $\mathbb{S}_{l-1}$. As in the proof of (52.10), we infer that $l \le 2$. The argument this far is due to Knarr.

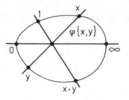

Figure 55a

For $l \in \{1, 2\}$, we have $O \approx L \approx \mathbb{S}_l$ by (53.5 and 15a). Hence (55.11) implies that the space $\mathcal{A} := \mathcal{L}_1 \cup \mathcal{L}_2$ is homeomorphic to $\mathbb{S}_l * \mathbb{S}_l$. The same is true for a conic $\widehat{O}$ in the classical plane $\widehat{\mathcal{P}}$ dominating $\mathcal{P}$, i.e., in the plane $\mathcal{P}_2 \mathbb{F}$ of the same dimension, and we have $\mathcal{A} \approx \widehat{\mathcal{A}}$. Applying the topological pigeon-hole principle (51.20 and 21b) to the inclusion map $\mathcal{A} \to \mathcal{L}$, we see that the compact set $\mathcal{A}$ is a manifold if, and only if, exterior lines do not exist. The homeomorphism $\mathcal{A} \approx \widehat{\mathcal{A}}$ now implies that O behaves like the conic $\widehat{O}$ in this respect.

Second proof. Assume that the lines of $\mathcal{P}$ are manifolds of dimension $l > 1$. We want to show that $\mathcal{L}_0 = \emptyset$ and that $l = 2$. Assume first that $\mathcal{L}_0 \ne \emptyset$. The set $\mathcal{A} = \mathcal{L}_1 \cup \mathcal{L}_2 \subseteq \mathcal{L}$ is closed, so that $\mathcal{L}_0$ is open. Moreover, $\mathcal{L}_2$ is open in $\mathcal{L}$ by (55.11). This proves that $\mathcal{L} \setminus \mathcal{L}_1$ is disconnected. Since $\mathcal{L}$ is a manifold and, hence, a Cantor manifold, it follows that $2l - 1 = \dim \mathcal{L} - \dim \mathcal{L}_1 \le 1$, a contradiction.

Thus, $\mathcal{L} = \mathcal{A} \approx O * O$, where $O \approx L \approx \mathbb{S}_l$. For $x \in O$, we obtain an induced homeomorphism

$$\mathcal{L} \setminus \mathcal{L}_x \approx (O \setminus \{x\}) * (O \setminus \{x\}).$$

Since $\mathcal{L} \setminus \mathcal{L}_x$ is the point set of an affine plane, this yields a homeomorphism $\mathbb{R}^{2l} \approx \mathbb{R}^l * \mathbb{R}^l$, and the desired equality $l = 2$ follows from Lemma (55.15) below:

55.15 Lemma: Euclidean symmetric squares. *The symmetric square $\mathbb{R}^l * \mathbb{R}^l$ is homeomorphic to $\mathbb{R}^{2l}$ if, and only if $l = 2$.*

Given a space W and a map $f : W \to W$, we shall denote by W/f the quotient space obtained from W by the identification $w = f(w)$.

Proof of (55.15). The linear involution $\sigma(x, y) = (y, x)$ of $\mathbb{R}^l \times \mathbb{R}^l$ is conjugate to the involution $\tau(x, y) = (x, -y)$. Thus, $\mathbb{R}^l * \mathbb{R}^l = (\mathbb{R}^l \times \mathbb{R}^l)/\sigma$ is homeomorphic to $X := (\mathbb{R}^l \times \mathbb{R}^l)/\tau = \mathbb{R}^l \times (\mathbb{R}^l/\iota)$, where $\iota = -\mathbb{1}$. Assume that X is homeomorphic to $\mathbb{R}^{2l}$. Then the complement $X \setminus R$ of the closed subset $R = \mathbb{R}^l \times \{0\}$ can be written as $(X \cup \{\infty\}) \setminus (R \cup \{\infty\})$, the two spaces being respectively homeomorphic to $\mathbb{S}_{2l}$ and to $\mathbb{S}_l$. By Alexander duality, this implies that $H_1(X \setminus R) = 0$ for $l > 2$; for the special case at hand, Spanier [66] 4.7.14 is a good reference. On the other hand, using $A = \mathbb{R}^l \setminus \{0\}$, we may describe the space in question as $X \setminus R = \mathbb{R}^l \times (A/\iota)$. This product can be homotopically deformed onto the right factor, which in turn can be radially deformed onto the projective space $P_{l-1}\mathbb{R} = \mathbb{S}_{l-1}/\iota$. The first homology group of the projective space is non-zero; see, e.g., Greenberg [67] 19.23. This contradiction shows that $\mathbb{R}^l * \mathbb{R}^l$ is not homeomorphic to $\mathbb{R}^{2l}$ if $l > 2$.

Finally, the right factor $\mathbb{R}^l/\iota$ of X is homeomorphic to $\mathbb{R}^2$ for $l = 2$ and to $[0, \infty[$ for $l = 1$; homeomorphisms are induced by the maps $z \mapsto z^2$ from $\mathbb{C}$ to $\mathbb{C}$ or $t \mapsto t^2$ from $\mathbb{R}$ to $[0, \infty[$. Thus, $\mathbb{R}^l * \mathbb{R}^l$ is homeomorphic to $\mathbb{R}^4$ in the first case and to a closed half-plane in the second case. □

55.16 Note. As a consequence of (55.14), there are no locally compact, connected circle planes (Laguerre, Minkowski, or Möbius planes) with point spaces of finite dimension $d > 4$; see Buchanan–Hähl–Löwen [80] Section 3 and Steinke [89] Cor. C. In fact, the point set of such a Möbius plane is a 2-sphere (see also Groh [73]), whereas Laguerre and Minkowski planes exist in dimensions 2 and 4.

55.17 Proposition: Duality. *If O is a topological oval in $\mathcal{P}$, then the set $\mathcal{L}_1$ of tangents is a topological oval in the dual plane $\mathcal{P}^*$.*

Remark. This is analogous to the situation in finite planes of odd order; see Dembowski [68] 3.2.23, p. 148.

Proof. $\mathcal{L}_1$ is compact by (55.11). Hence, it suffices to verify that $\mathcal{L}_1$ satisfies the conditions dual to (55.7) (i) and (ii). This means, in terms of O, that

(i*) every $x \in P$ lies on at most 2 tangents, and
(ii*) every $T \in \mathcal{L}_1$ contains precisely one point that lies on no other tangent.

Both conditions follow from

(*) The number of tangents through a point $x \in P \setminus O$ is 0 or 2.

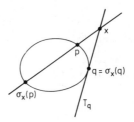

Figure 55b

We prove (*), employing the *bundle involutions* $\sigma_x : O \to O$, where $x \in P \setminus O$. They are defined by requiring that $\{p, \sigma_x(p)\} = O \cap px$; see Figure 55b. It follows from (55.8a) that σ_x is a homeomorphism of O. The map $\alpha_x : O \to \mathcal{L}_x$ defined by $p \mapsto px$ sends the set of fixed points of σ_x bijectively onto $\mathcal{L}_x \cap \mathcal{L}_1$. If $\sigma_x = \mathrm{id}$, then $\mathcal{L}_1 \subseteq \mathcal{L}_x$, and some points $y \in P \setminus O$ are on precisely one tangent. Hence, it suffices to prove (*) for all points x with $\sigma_x \neq \mathrm{id}$; then (*) holds in general.

Now by (55.11 and 14) together with (53.5 and 15), the oval O is homeomorphic to a sphere $\mathbb{S}_l$, where $l \in \{1, 2\}$. According to Brouwer [19] or Kerékjártó [19], every involutory homeomorphism of $\mathbb{S}_l$, $l \leq 2$, is topologically equivalent to a linear involution of $\mathbb{S}_l$. Hence, the number of fixed points of σ_x is indeed 0 or 2, unless $l = 2$ and σ_x is equivalent to a reflection of $\mathbb{S}_2$ in a great circle. In that case, α_x induces an embedding of the quotient space O/σ_x, the closed disc, into $\mathcal{L}_x$. Then, α_x is not surjective, and $\mathcal{L}_x$ contains exterior lines contrary to (55.14). □

The following variation of (55.11) will be used in (74.5). As the caption suggests, it will be applied in the situation where O' is contained in an affine plane and p is a point at infinity.

55.18 Proposition: Affine test of topological ovals. *In a compact plane $\mathcal{P}$ whose lines are l-manifolds, let O be an oval containing some point p such that $O' = O \setminus \{p\}$ is an l-manifold. Then O is a topological oval.*

Proof. We shall formulate the proof for the case $l > 1$; the ideas needed for $l = 1$ are very similar. The continuous bijection $g : x \mapsto x \vee p$ of O' onto $\mathcal{L}_p \setminus \{T_p\}$ is a homeomorphism by domain invariance (51.19). According to (52.3), this yields a homeomorphism $h : \mathbb{R}^l \to O'$. It suffices to show that O is closed in the point set, see (55.11). Thus, we have to show that every sequence $x_k \to \infty$ in O' converges to p when considered as a sequence in P. Indeed, this condition means that h extends to a homeomorphism $\mathbb{S}_l \to O$. The homeomorphism g shows that the lines $x_k \vee p$ converge to T_p, therefore every accumulation point of x_k lies on T_p.

Now consider a fixed point $q \in O'$ and its tangent T_q. The map $f_q : O' \setminus \{q\} \to \mathcal{L}'_q = \mathcal{L}_q \setminus \{pq, T_q\}$ sending x to xq is again a continuous bijection and hence a homeomorphism (51.19). Let $S \subseteq O'$ correspond to a sphere $\mathbb{S}_{l-1}$ centered at $h^{-1}(q)$ under the homeomorphism $h^{-1} : O' \to \mathbb{R}^l$. Then S separates $O' \setminus \{q\}$ into two domains U and V, where $q \in \overline{U}$ and the closure $\overline{U}$ of U in O' is compact. Similarly, $\mathcal{S} = f_q(S)$ separates $\mathcal{L}'_q$ into $\mathcal{U} = f_q(U)$ and $\mathcal{V} = f_q(V)$, and moreover both $\mathcal{U} \cup \mathcal{S}$ and $\mathcal{V} \cup \mathcal{S}$ are non-compact. According to the theorem of Jordan–Brouwer, see Spanier [66] 4.7.15, the complement $\mathcal{L}_q \setminus \mathcal{S}$ also has precisely two connected components, and it follows that pq and T_q belong to different complementary components of $\mathcal{S}$. Hence, either pq belongs only to the closure of $\mathcal{U}$ and T_q belongs only to that of $\mathcal{V}$, or vice versa. We claim that the second possibility holds. Then $f_q(x_k) = x_k \vee q \in \mathcal{V}$ converges to pq, hence $x_k \to T_p \wedge pq = p$, and the proof is complete.

In the other case, the same reasoning shows that $x_k \to T_p \wedge T_q = r$. Since $\lim x_k$ does not depend on the choice of q, this would mean that r lies on all tangents of O. Then no tangent other than T_p passes through an arbitrary third point $a \in T_p \setminus \{p, r\}$. The absence of tangents allows us to construct a continuous bundle involution $\sigma_a : O' \to O'$, using the same definition as in the proof of (55.17). We simply observe that the join map induces a continuous bijection of $O' * O' \setminus D_{O'}$ onto the set of secants of O', and that this map is open by domain invariance (51.19), combined with (55.8d). Now continuity of σ_a follows from (55.8a), as in the proof of (55.17). The absence of tangents means that the involution σ_a of $O' \approx \mathbb{R}^l$ is fixed-point free, contrary to a theorem of P.A. Smith, see Zieschang [81] 36.1; compare also (55.23) below. □

Automorphisms

We consider automorphisms of planes and of their coordinatizing ternary fields. The two objects are connected via (44.4c). An elementary proof of the following result has been given in the case $l = 1$, see (32.11).

55.19 Theorem: Fixed elements. *Every automorphism φ of a plane $\mathcal{P}$ fixes some point and some line.*

Proof. We shall apply the fixed point theorem of Lefschetz in order to show that in fact every homeomorphism $f : P \to P$ fixes some point. By duality, the assertion about lines will follow. It will be convenient to use the cohomology of P with coefficients in the field $\mathbb{Z}_2$. This can be computed from the integral homology given in (52.14c); cf. (54.11). We use the universal-coefficient theorem, Spanier [66] 5.5.3, applying it to the singular chain complex C of P; this is possible because C is free. We obtain that

$$H^q(P; \mathbb{Z}_2) \cong \mathrm{Hom}(H_q C; \mathbb{Z}_2) \oplus \mathrm{Ext}(H_{q-1} C; \mathbb{Z}_2) = \begin{cases} \mathbb{Z}_2 & \text{if } q \in \{0, l, 2l\} \\ 0 & \text{else.} \end{cases}$$

The trace of the isomorphism $f^q : H^q(P; \mathbb{Z}_2) \to H^q(P; \mathbb{Z}_2)$ is 1 or 0, according to whether or not $q \in \{0, l, 2l\}$. The *Lefschetz number* of f,

$$\Lambda(f) = \sum_{q \geq 0} (-1)^q \mathrm{\,trace\,} f^q$$

therefore never vanishes. In view of (51.12), the theorem of Lefschetz for compact ANR (R.F. Brown [71] p.42) yields that f has a fixed point. $\square$

Remark. The proof can also be obtained using the integral homology version of the theorem of Lefschetz (Spanier [66] 4.7.7). This version carries over from compact polyhedra to ANR just as in the proof by Brown quoted above. There is another proof of (55.19) by Breitsprecher [71] 2.3.2, see the Notes (55.45).

The next result exhibits a topological background of algebraic closure. Moreover, in every ternary field of dimension ≥ 2 it provides us with a compact set that is invariant under all automorphisms.

55.20 Proposition: Square roots. *Let K be a locally compact, connected ternary field of finite dimension l. Then the set*

$$\sqrt{-1} := \{\, x \in K \mid (x \cdot x) + 1 = 0 \,\}$$

is compact. It is non-empty if, and only if $l > 1$.

Note. The proof shows that $\sqrt{b}$ is non-empty for $l > 1$ and $b \in K$ arbitrary.

Proof. First note that K coordinatizes a plane with lines of dimension l, see (43.5). Let -1 be the left additive inverse of 1, satisfying $(-1) + 1 = 0$. We consider the squaring map $\sigma : K \to K$, defined by $x \mapsto x \cdot x$, and we extend it to $L = K \cup \{\infty\}$ by $\sigma(\infty) = \infty$. This map is continuous (43.2(2)), so that $\sqrt{-1} = \sigma^{-1}(-1)$ is closed in L and hence is compact.

For $l = 1$, we shall exploit the order properties of K, as we did in the proof of (32.9). Let $<$ be the unique order on $K \approx \mathbb{R}$ inducing the topology and satisfying

$0 < 1$. Right addition $\alpha_a : x \mapsto x + a$ is a homeomorphism, hence it is either monotone or antitone. Since antitone maps of $\mathbb{R}$ have fixed points, α_a is in fact monotone. It follows that $-1 = \alpha_1^{-1}(0) < \alpha_1^{-1}(1) = 0$.

Left multiplication $\mu_a : x \mapsto a \cdot x$ by $a \neq 0$ is a homeomorphism as well, and it fixes 0. Since $\mu_a(1) = a$, it follows that μ_a is monotone or antitone according to whether $a > 0$ or $a < 0$. Thus, $a \cdot a > 0$ in both cases, and $a \cdot a \neq -1$.

Now let $l > 1$. If $\sqrt{-1}$ is empty, then $\sigma(L)$ is contained in the contractible set $L \setminus \{-1\}$, see (51.4), and then the induced homomorphism $\sigma_l : H_l L \to H_l L$ is zero. We show that this is not in fact the case. First we consider the homomorphism $\sigma_\sharp : \pi_{l-1} K^\times \to \pi_{l-1} K^\times$ induced by the restriction of σ. By (54.11) and (52.5b), we have that $K^\times \simeq \mathbb{S}_{l-1}$, hence $\pi_{l-1} K^\times \cong H_{l-1} K^\times \cong \mathbb{Z}$ under the natural Hurewicz isomorphism, see Spanier [66] 7.5.6 and p. 391. According to Spanier [66] 1.6.10, addition in $\pi_{l-1} K^\times$ can be expressed using the continuous multiplication of $K^\times$: For $\alpha, \beta : \mathbb{S}_{l-1} \to K^\times$, define $(\alpha \cdot \beta)(x) = \alpha(x) \cdot \beta(x) \in K^\times$; then $[\alpha] + [\beta] = [\alpha \cdot \beta]$. For a generator $[\alpha]$ of $\pi_{l-1} K^\times$, this implies that

$$\sigma_\sharp[\alpha] = [\sigma \circ \alpha] = [\alpha \cdot \alpha] = [\alpha] + [\alpha] \neq 0.$$

It remains to translate this result to homology. The Mayer–Vietoris sequence of the decomposition $L = (L \setminus \{0\}) \cup (L \setminus \{\infty\})$ of L into two contractible spaces gives a natural isomorphism $H_{l-1} K^\times \to H_l L$; see Greenberg [67] 17.6 and 7 or Massey [91] VIII 5.1 for the Mayer–Vietoris sequence and its naturality. Combining this with the Hurewicz isomorphism $\pi_{l-1} K^\times \cong H_{l-1} K^\times$ we obtain a commutative diagram, in which all horizontal maps are isomorphisms:

$$\begin{array}{ccccc} \pi_{l-1} K^\times & \longrightarrow & H_{l-1} K^\times & \longrightarrow & H_l L \\ \sigma_\sharp \downarrow & & \sigma_{l-1} \downarrow & & \downarrow \sigma_l \\ \pi_{l-1} K^\times & \longrightarrow & H_{l-1} K^\times & \longrightarrow & H_l L \end{array}$$

This shows that σ_l is multiplication by 2, which ends the proof. □

Recall that a 1-dimensional ternary field has no non-trivial automorphisms (32.10). The next proposition asserts that, similarly, the automorphism group of a 2-dimensional ternary field cannot be larger than that of $\mathbb{C}$, which is generated by complex conjugation. Remember that by their definition automorphisms are in particular homeomorphisms.

55.21 Theorem: Automorphisms of 2-dimensional ternary fields. *Consider a 4-dimensional plane $\mathcal{P}$ and a (2-dimensional) ternary field K coordinatizing $\mathcal{P}$, and write $\Sigma = \mathrm{Aut}\,\mathcal{P}$. Then the following statements hold.*

(a) *Every non-trivial automorphism of K reverses the orientation and fixes some 1-dimensional subternary elementwise; in particular, $\mathrm{Aut}\,K \leq \mathbb{Z}_2$.*

(b) *Every non-trivial collineation σ of $\mathcal{P}$ fixing a quadrangle $e_1, \ldots, e_4$ is a Baer involution (23.14 and 17) and reverses the orientation of each of its fixed lines. Hence, $\Sigma_{e_1,\ldots,e_4} \leq \mathbb{Z}_2$.*
(c) *A connected subgroup $\Delta \leq \Sigma$ contains no Baer involution.*

Proof. Assertion (b) is a restatement of (a) in geometric terms; see (44.4c) and note that $K \approx \mathbb{R}^2$ by (53.15a). Observe also that every fixed line L of σ may be chosen to carry the coordinatizing ternary field; this shows that the orientation of L is indeed reversed. In order to prove (a), it suffices to show that $\gamma = \mathrm{id}$ if $\gamma \in \mathrm{Aut}\, K$ is orientation preserving, for then $\alpha \circ \beta = \mathrm{id}$ whenever both α and β are non-trivial.

Every automorphism γ of K acts trivially on the closure $R = \overline{\langle 1 \rangle}$ of the ternary field generated by $1 \in K$. If $\gamma \neq \mathrm{id}$, then $R \neq K$, hence $(K, R) \approx (\mathbb{R}^2, \mathbb{R})$ by (55.4b); moreover, γ fixes no element of $K \setminus R$. Therefore, γ is orientation preserving if, and only if, γ maps each of the two connected components $H_1, H_2 \approx \mathbb{R}^2$ of $K \setminus R$ onto itself. By Brouwer's theorem on plane translations (Brouwer [12], E. Sperner [34], M. Brown [84]), the iterated images $\gamma^n(x)$ of any point $x \in H_i$ form an unbounded subset of H_i if γ is non-trivial and orientation preserving. On the other hand, γ leaves the compact set $\sqrt{-1} \neq \emptyset$ invariant, and $\sqrt{-1} \cap R = \emptyset$, see (55.20). This is a contradiction.

For the proof of (c), we may assume that Δ is closed in Σ. Hence, Δ is a Lie group, see (71.2) and (94.3a). Now suppose that there is a Baer involution $\delta \in \Delta$. By the theorem of Mal'cev–Iwasawa (93.10), this involution is contained in a compact, connected subgroup $\Phi \leq \Delta$. Moreover, δ belongs to a torus subgroup $\Psi \leq \Delta$ by (94.31a). There exists $\sigma \in \Psi$ such that $\sigma^2 = \delta$, and σ has a fixed line L by (55.19). Then $\delta = \sigma^2$ cannot reverse the orientation of L, a contradiction. $\square$

The field of complex numbers is the only known example of a locally compact, connected ternary field whose underlying algebraic structure admits discontinuous automorphisms. Examples of such automorphisms can be obtained by extending an arbitrary permutation of a maximal algebraically independent set; compare Baer [70], Dieudonné [74], Keller [86]. Part (a) of the following proposition generalizes known properties of these automorphisms, see Kestelman [51] and Kallman–Simmons [85]. The result was obtained by Salzmann [69c] for $l = 2$ and by Hähl [75a] p. 312 in general; we shall not reproduce the proof. The geometric application in part (b) is due to Knarr [83, 87a].

55.22 Proposition: Discontinuous automorphisms.

(a) *Let $K \approx \mathbb{R}^l$ be a topological ternary field and let α be a discontinuous automorphism of its underlying algebraic structure. If $X \subseteq K$ is a subset homeomorphic to the sphere $\mathbb{S}_{l-1}$, then the image X^α is dense in K.*

(b) *Let $\mathcal{P}$ be a 4-dimensional compact plane. If $\mathcal{P}$ contains a desarguesian subplane $\mathfrak{Q} = (Q, \mathcal{M})$ and admits a discontinuous collineation γ, then $\mathcal{P}$ is isomorphic to the complex plane $\mathcal{P}_2\mathbb{C}$.*

Proof of part (b). We may assume that Q is closed in P since the closure inherits the Desargues property. There are ternary fields K_1 and K_2 coordinatizing $\mathcal{P}$ such that $K_1 \cap Q = L$ is a line of $\mathfrak{Q}$, and such that K_2 contains L^γ. Moreover, we may assume that γ induces an isomorphism $K_1 \to K_2$, see (23.5). As in (a), the image of $X = L$ is dense in K_2. This shows that $\mathfrak{Q}^\gamma$ is a dense desarguesian subplane of $\mathcal{P}$. Hence, $\mathcal{P}$ is desarguesian, and the assertion $\mathcal{P} \cong \mathcal{P}_2\mathbb{C}$ follows by (42.7 or 6). □

For higher-dimensional ternary fields, we consider the group $\operatorname{Aut} K$ with the compact-open topology, which turns it into a locally compact transformation group of K, see (44.4c and 3). We compare $\operatorname{Aut} K$ with its classical counterpart $\operatorname{Aut} \mathbb{H}$ or $\operatorname{Aut} \mathbb{O}$ according as $\dim K = 4$ or 8. We know that these are compact Lie groups of dimension $d = 3$ or $d = 14$, respectively, see (11.29 and 33). In all known examples, $\operatorname{Aut} K$ is a compact Lie group of dimension not exceeding d. For ternary fields with associative addition, this can be proved. In fact, in this case $\operatorname{Aut} K$ is isomorphic to a subgroup of $\operatorname{Aut} \mathbb{H} \cong \operatorname{SO}_3 \mathbb{R}$ or of $\operatorname{Aut} \mathbb{O} = G_2$, respectively, see Grundhöfer–Salzmann [90] Section XI.9. Compare also the weaker result (83.6).

A related result of Bödi [93a] asserts that for a ternary field (or even a double loop) whose addition is differentiable of class $\mathcal{C}^2$, the group of differentiable automorphisms is compact. If K is a locally compact, connected ternary field with associative addition, then the additive group is a Lie group, compare the remarks leading to (53.4). In this case, all continuous automorphisms are differentiable (94.9), and Bödi's result implies that $\operatorname{Aut} K$ is indeed compact.

It is conjectured that $\operatorname{Aut} K$ is compact and at most d-dimensional in general, but a proof still seems out of reach. What has been proved in this direction without imposing algebraic conditions on K largely depends on (55.20) and on the following results from P.A. Smith's theory of periodic maps. We formulate them for the special situation at hand. Groups that are denoted by roman letters $G, H, \ldots$ are not assumed to consist of collineations.

55.23 Proposition: Cyclic actions on contractible spaces. *No finite group $G \neq \mathbb{1}$ can act continuously and freely on a locally compact, contractible subset $X \subseteq P$. In particular, this applies to the affine point set $X = P \setminus L$ and to an affine line $X = L \setminus \{a\}$, see* (51.4).

Proof. Consider a subgroup $\mathbb{Z}_p \leq G$, p a prime. It acts on the space X, which is finite-dimensional, locally compact and contractible (51.4). By the homotopy property of Čech homology (Eilenberg–Steenrod [52] Section IX.5 p. 240), this implies that the Čech homology groups $\check{H}_q(X;\mathbb{Z}_p)$ are isomorphic to those of a point, i.e., $\check{H}_0 = \mathbb{Z}_p$ and $\check{H}_q = 0$ else. In this situation, Floyd [52] 5.1 asserts that $\mathbb{Z}_p$ fixes a point contrary to our hypothesis. □

Note. After slightly more preparation, one could also apply the original theorem of Smith [41], which preceded the result of Floyd. For manifolds, an easy proof of a directly applicable theorem is given by Zieschang [81] 36.1; compare also Cartan–Eilenberg [56] Section XVI.9 p. 357. The same proofs work in general if we show that the homology $H_q(X/\mathbb{Z}_p;\mathbb{Z}_p)$ vanishes above a certain degree if $\mathbb{Z}_p$ acts freely. This can be done because X and, hence, $X/\mathbb{Z}_p$ are homology manifolds over both $\mathbb{Z}$ and $\mathbb{Z}_p$ (54.10 and 14), so that their homology vanishes in all degrees $q > \dim X$; compare Lacher [84] Section 3.

Actually, (55.23) is a special case of Proposition (55.24) below; indeed, the latter implies that ∞ cannot be the only fixed point of $\mathbb{Z}_p$ acting on $(L \setminus \{a\}) \cup \{\infty\} \approx L$ or $(P \setminus L) \cup \{\infty\} \approx P/L$. We formulated (55.23) mainly because it has easy proofs, which may serve as an introduction to Smith theory.

55.24 Proposition: Cyclic actions on spheres. *Let G be an abelian group of order a prime power $d = p^\alpha$. If G acts continuously on a line $X = L$ of $\mathcal{P}$ or on $X = P/L$, then the Čech homology of the fixed point set $Y = \mathrm{Fix}_X(G)$ satisfies*

(a) $\check{H}_q(Y;\mathbb{Z}_p) \cong \check{H}_q(\mathbb{S}_r;\mathbb{Z}_p)$ *for some r, $-1 \le r \le n$, where $\mathbb{S}_{-1} := \emptyset$ and $n = \dim X$.*
(b) *If p is odd, then $n - r$ is even.*

Proof. From (52.5a and 12) together with (54.11), we know that $X \simeq \mathbb{S}_n$. Using the homotopy property of Čech homology (Eilenberg–Steenrod [52] p. 240), we conclude that $\check{H}_q(X;\mathbb{Z}_p) \cong \check{H}_q(\mathbb{S}_n;\mathbb{Z}_p)$. The assertions now follow from Floyd [52] 5.2. Note that Floyd uses his hypothesis that p is odd only for the proof of (b). A different proof is given by Borel [55]; see also Bredon [72] 3.7.11. □

55.25 Notes. (a) In particular, (55.24) excludes the possibility that Y is a single point. For $n > 1$ and p odd, the possibility that $Y = \emptyset$ is also ruled out; remember that n is even by (52.5c) and (54.11). Here, we used the Čech homology of finite, discrete spaces Y, which is easily computed from the definition: the homology $\check{H}_q(Y;R)$ with coefficients in $R = \mathbb{Z}_p$ or $\mathbb{Z}$ vanishes for $Y = \emptyset$; for Y consisting of k points, it vanishes in degrees $q \ne 0$, and $\check{H}_0(Y,R) = R^k$. The sphere $\mathbb{S}_0$ consists of two points.

(b) Using sheaf-theoretic versions of Smith theory, see Borel et al. [60] III.4.5 or Bredon [67] 5.17.1, it is possible to deduce that Y is a homology manifold over $\mathbb{Z}_p$ if X has this property, see (54.14). In our situation, however, it will usually be easier to obtain this from geometry, because Y will be a line of some subplane or a subplane modulo a line.

55.26 Proposition: Codimension 2. *Assume that a finite group G acts continuously on $\mathbb{S}_n$ with fixed point set $Y \approx \mathbb{S}_{n-2}$, acting freely on the complement of Y. Then G is cyclic. (Remember that $\mathbb{S}_0$ consists of two points.)*

Proof. This is the corollary on p. 408 of Smith [60]. □

If Y is a sphere of smaller dimension, then one needs the hypothesis that G is abelian:

55.27 Proposition: Abelian actions on spheres. *Assume that a finite abelian group G acts continuously on a line L of $\mathcal{P}$, leaving a closed set $Y \subseteq L$ invariant and acting freely on $L \setminus Y$. Then each of the following conditions implies that G is cyclic:*

(i) *Y has the same Čech cohomology as $\mathbb{S}_r$, where $0 \leq r \leq l - 2$ and $l = \dim L$ (e.g., $l \geq 2$ and Y consists of two points).*
(ii) *There is a connected, closed subplane $(Q, \mathcal{M})$ with $L \in \mathcal{M}$ and $Y = L \cap Q$, and $l > 2$.*

Proof. Property (ii) implies that $Y \simeq \mathbb{S}_r$, see (54.11) and (52.5a). Using the homotopy property of Čech cohomology (Eilenberg–Steenrod [52] p. 240), we obtain (i) in each case. Thus, L and Y are what Smith [60] calls integral homology spheres, and his theorem on p. 407 asserts that G 'has cohomology of period $l - r$'. What he actually proves is that $\hat{H}^{l-r}(G; A) \cong \hat{H}^0(G; A)$ for every G-module A. In particular, $\hat{H}^{l-r}(G; \mathbb{Z})$ is cyclic of order $|G|$, compare Cartan–Eilenberg [56] p. 237. The latter property defines what Cartan and Eilenberg call a period for G, see loc.cit., p. 260f. Their characterization of the finite groups having a period > 0 implies that abelian groups with this property are in fact cyclic, see loc.cit., p. 262, XII.11.6. □

We are now ready to apply the results of Smith theory in order to obtain geometric conclusions.

55.28 Proposition: Elations. *Non-trivial elations of a plane $\mathcal{P}$ are of infinite order.*

Proof. Apply (55.23) to the free action on the complement of the axis, and use (23.8). □

By (23.17), any involution $\sigma \in \operatorname{Aut} \mathcal{P}$ either is a Baer collineation or is axial. In particular, the distinction between planar collineations and Baer collineations

(23.14) is immaterial for involutions. Therefore, we use freely either of the expressions 'σ is a Baer involution' and 'the involution σ is planar'. By (55.28), no involution can be an elation. This proves the following.

55.29 Corollary: Involutions. *Any involution $\sigma \in \operatorname{Aut} \mathcal{P}$ either is a homology (reflection) or is planar (in fact, a Baer involution).* □

55.30 Involutions of classical planes. In the classical planes $\mathcal{P}_2 \mathbb{F}$, an involution σ is uniquely determined by its fixed point set. Indeed, if σ is axial, then we may assume that its center and axis are as considered in (12.13); hence σ is given by multiplication with some involutory element $x \in \mathbb{F}^\times$, and using (11.4) we obtain that $x = -1$. A Baer involution σ is determined by its action on the algebra $\mathbb{F}$ coordinatizing the plane with respect to a given quadrangle of fixed points, see (23.5). By (11.28), the induced automorphism of $\mathbb{F}$ is an orthogonal involution, hence it is uniquely determined by its fixed point set.

It is conjectured that involutions of arbitrary compact, connected planes are determined by their fixed point sets, but only the following results (55.31 and 32) have been obtained.

55.31 Proposition: Cyclic collineation groups. *Let $\Xi \leq \Sigma = \operatorname{Aut} \mathcal{P}$ be a finite subgroup. Assume that $\dim L = l \leq 2$ or that Ξ is abelian. Then Ξ is cyclic if*

(i) *Ξ consists of axial collineations with common center and axis, or if*
(ii) *Ξ leaves some proper, closed, connected subplane $\mathcal{Q} = (Q, \mathcal{M})$ and some line $L \in \mathcal{M}$ invariant, acting freely on $L \setminus Q$.*

Proof. Suppose that $\Xi \neq \mathbb{1}$ satisfies condition (i), so that $\Xi \leq \Sigma_{[c,A]}$, say. By (55.28), we have $c \notin A$, and the action of Ξ on $L \in \mathcal{L}_c$ is effective according to (23.8). For $l = 1$, we may refer to (32.15a), so let $l > 1$. Then Ξ satisfies condition (i) of (55.27) with $Y = \{c, L \wedge A\}$, and the assertion follows.

Likewise, for $l > 2$ the present condition (ii) implies condition (ii) of (55.27), and the assertion follows. Next assume that condition (ii) holds, and that $l \leq 2$. By (55.3), we have in fact $l = 2$, and then $(L, L \cap Q) \approx (\mathbb{S}_2, \mathbb{S}_1)$ by (55.4b). There are two connected components $X_1, X_2 \approx \mathbb{R}^2$ of $L \setminus Q$, and the subgroup $\Omega \leq \Xi$ preserving the components has index at most 2. The theorem of Smith (55.23) may be applied to the free action of Ω on either component. It follows that $\Omega = \mathbb{1}$, whence $\Xi \leq \mathbb{Z}_2$. □

55.32 Theorem: Uniqueness of involutions. *Let $\sigma, \varrho \in \Sigma = \operatorname{Aut} \mathcal{P}$ be involutions with the same fixed point set. Then $\sigma = \varrho$ follows if*

(i) $\dim P \leq 4$, *or*
(ii) σ *and* ϱ *commute, or*

(iii) $\dim P = 8$, and σ and ϱ are two Baer involutions that are both contained in some compact Lie group $\Delta \leq \Sigma$.

Proof. By (32.12), we may assume that $\dim P \geq 4$. According to (55.29), the involutions σ and ϱ are either both planar or both axial. By maximality of Baer subplanes and by (23.8), the group $\Phi \leq \Sigma$ generated by σ and ϱ acts freely outside the common fixed point set of σ and ϱ. Hence, the assertion follows from (55.31) if Φ is abelian and hence finite, and also if $\dim P = 4$ and Φ is finite. If $\dim P = 4$ and Φ is planar (finite or infinite), then we refer to (55.21b). If Φ is a finite group of Baer involutions on an 8-dimensional plane, then we apply (55.26) to the action of Φ on a fixed line L. Note that $L \approx \mathbb{S}_4$ contains a fixed point set homeomorphic to $\mathbb{S}_2$; this is obtained from (53.15a), (55.5 and 6), and the subsequent Note.

The remaining possibilities are that Φ is infinite and either (1) $\dim P = 4$ and $\Phi \leq \Sigma_{[c,A]}$ or (2) $\dim P = 8$ and Φ is contained in some compact Lie group. (In the second case, we also assumed that Φ is planar rather than axial; this was needed for the finite case and will not be used again.) In both cases, Φ is a semidirect product $\Psi \rtimes \langle \sigma \rangle$ of $\langle \sigma \rangle \cong \mathbb{Z}_2$ and the cyclic normal subgroup Ψ generated by $\varrho\sigma$. It follows that Ψ is infinite. Conjugation by σ inverts the elements of Ψ because $\sigma(\varrho\sigma)\sigma = \sigma\varrho = (\varrho\sigma)^{-1}$.

The closure $\overline{\Psi}$ is a Lie group in each case. Indeed, a closed subgroup of a Lie group is a Lie group (94.3a), and in a 4-dimensional plane the group $\Sigma_{[c,A]}$ is a Lie group since it acts effectively on a 2-sphere $L \in \mathcal{L}_c$, see (44.3) and (96.31). See also (71.2) for a stronger result. If $\overline{\Psi}$ is compact, then $\dim \overline{\Psi} > 0$ because Ψ is infinite and because a 0-dimensional Lie group is discrete. By (94.31a), the group $\overline{\Psi}$ contains a torus, which in turn contains an involution τ. Since $\overline{\Psi}$ is abelian and σ inverts $\overline{\Psi}$, we see that σ, τ are distinct commuting involutions. This possibility has been excluded, hence $\overline{\Psi}$ is not compact. In particular, case (2) is now ruled out.

According to (44.8b), the group $\Sigma_{[c,A]}$ is isomorphic to some closed subgroup $H \leq K^\times$ of the multiplicative loop of a suitably chosen ternary field for $\mathcal{P}$. In the subgroup of H corresponding to Ψ, we find a sequence h_n converging to 0 or ∞. By (43.2) we may assume that $h_n \to 0$ and $h_n^{-1} \to \infty$. But h_n^{-1} is conjugate to h_n by the element $s \in H$ corresponding to σ, hence conjugation with s is discontinuous at 0 because $s^{-1} \cdot 0 \cdot s = 0$. This contradiction ends our proof. $\square$

55.33 Elementary abelian 2-groups on classical planes. We claim that the classical plane $\mathcal{P}_2\mathbb{F}$ of dimension 2^s admits a group $\mathbb{Z}_2^{s+1}$ of automorphisms. We shall describe an action of $\mathbb{Z}_2^5$ on $\mathcal{P}_2\mathbb{O}$ that induces such groups on the other classical planes $\mathcal{P}_2\mathbb{F} \leq \mathcal{P}_2\mathbb{O}$. We start with a triangle e_1, e_2, e_3 and consider 3 reflections fixing that triangle, see (12.13). It follows from (12.13) together with (11.4) that these reflections are uniquely determined by the given triangle. Using (23.12), we infer that they form a group $\mathbb{Z}_2^2$, and that this group is centralized by the stabilizer

$\wedge$ of a quadrangle $e_1, \ldots, e_4$. By (44.4c), we have $\wedge \approx \operatorname{Aut} \mathbb{O}$, hence it suffices to exhibit a subgroup $\mathbb{Z}_2^3$ in $\operatorname{Aut} \mathbb{O}$. Take a Cayley triple $u, v, z \in \mathbb{O}$, see (11.16). Seven other Cayley triples are obtained by replacing some of the elements u, v, z by their additive inverses. According to (11.16), the group $\operatorname{Aut} \mathbb{O}$ is sharply transitive on the set of Cayley triples, and the action on $\mathbb{O}$ is $\mathbb{R}$-linear (11.28). Hence we obtain a group $\mathbb{Z}_2^3$ permuting our 8 Cayley triples.

The next result says that no plane $\mathcal{P}$ can admit larger groups $\mathbb{Z}_2^r$ than its classical counterpart.

55.34 Corollary: Elementary abelian 2-groups. *The following assertions hold for each group $\Phi \leq \operatorname{Aut} \mathcal{P}$ isomorphic to $\mathbb{Z}_2^r$.*

(a) *Φ has a triangle of fixed points.*
(b) *If Φ is generated by reflections, then $r \leq 2$.*
(c) *If $\dim P = 2^s$, then $r \leq s + 1$.*
(d) *If Φ does not contain any reflections, then $r \leq s - 1$.*

Proof. If σ and ϱ are commuting reflections, then ϱ fixes the center c_σ and the axis A_σ of σ. Thus, either $c_\sigma = c_\varrho$ and $A_\sigma = A_\varrho$, or $c_\sigma \in A_\varrho$ and $c_\varrho \in A_\sigma$. In the former case, the reflections are equal by (55.32). In the latter case, the fixed points c_ϱ, c_σ and $c = A_\varrho \wedge A_\sigma$ form a triangle. We claim that $\varrho\sigma$ is a reflection with axis $c_\varrho c_\sigma$ and center c. Indeed, $\varrho\sigma$ is an involution and acts freely on the sides $A_\varrho \setminus \{c, c_\sigma\}$ and $A_\sigma \setminus \{c, c_\varrho\}$ of the triangle. Hence, $\varrho\sigma$ is not a Baer involution, and our claim follows in view of (23.17). By similar arguments, any reflection commuting with ϱ and σ belongs to the set $\{\varrho, \sigma, \varrho\sigma\}$. This proves (b).

Statement (d) will be proved in (55.39); it has been included here merely for comparison. The other assertions are obtained by induction on s, the case $s = 1$ being covered by the preceding paragraph; observe that the elements of Φ are reflections in this case (32.12). If $s > 1$ and Φ contains a Baer involution σ, then $\Phi/\langle\sigma\rangle \cong \mathbb{Z}_2^{r-1}$ acts on the associated Baer subplane, which is of dimension 2^{s-1} by (55.5). The action is effective by (55.32). □

55.35 Corollary: Commuting reflections. *Let $\sigma, \varrho \in \operatorname{Aut} \mathcal{P}$ be different reflections. Then σ commutes with ϱ if, and only if, the axis of σ contains the center of ϱ and vice versa. In this case, the product $\sigma\varrho$ is a reflection, and the centers and axes of σ, ϱ and $\sigma\varrho$ form a non-degenerate triangle.*

Proof. The last statements and the 'only if'-part of the first assertion have been shown in the last proof. The converse follows from Hughes–Piper [73] Lemma 4.22, which asserts that the product of the given reflections is an involution if their centers and axes are mutually incident. □

55.36 Torus rank. Consider the desarguesian planes $\mathcal{P}_2\mathbb{F}$, where $\mathbb{F} \in \{\mathbb{R}, \mathbb{C}, \mathbb{H}\}$. The torus rank (94.31) of $\Sigma = \operatorname{Aut} \mathcal{P}_2\mathbb{F}$ is $\operatorname{rk} \Sigma = s$ for the plane of dimension 2^s. To verify this, it suffices to look at a maximal compact subgroup (93.10). We know that $\Sigma^1 = \operatorname{PGL}_3\mathbb{F}$, see (13.6 and 8), and the elliptic motion group $\operatorname{PU}_3\mathbb{F}$ is a maximal compact subgroup, compare (94.33). Now the torus ranks of the compact almost simple Lie groups are well known, see e.g. Adams [69] or the table in the proof of (63.8).

The same formula for $\operatorname{rk} \Sigma$ works for the octonion plane $\mathcal{P}_2\mathbb{O}$ of dimension 2^4, where the elliptic motion group $F_{4(-52)}$ of rank 4 is a maximal compact subgroup of Σ by (18.18). We prove next that the automorphism groups of the classical planes have the largest possible torus ranks:

55.37 Theorem: Torus actions. *Let* $\mathsf{T} \leq \operatorname{Aut} \mathcal{P}$ *be an r-torus group of automorphisms of a plane $\mathcal{P}$ of dimension 2^s. Then we have*

(a) $r \leq s$.
(b) *If* T *fixes a non-degenerate quadrangle, then* $r \leq \max\{0, s - 2\}$.

Proof. Part (a) is proved by induction on s. For $s = 1$, apply (32.18). Suppose next that $s \geq 2$ and $r > s$. The r-torus T contains a subgroup $\mathbb{Z}_2^r$, hence it contains a Baer involution σ by (55.34b). The plane $\mathcal{F}_\sigma$ of fixed elements has dimension 2^{s-1} by (55.5), and the kernel of the action of T on the point space F_σ is at most 1-dimensional because it cannot contain two commuting involutions (55.32). Thus, we obtain a contradiction to the inductive hypothesis. The same arguments also prove (b) if we start the induction with the results (32.17) and (55.21b,c), which imply that $r = 0$ for $s \leq 2$. □

In general, it is not known whether or not two Baer subplanes of a compact, connected plane can be disjoint. This may be ruled out, however, for the planes of fixed elements associated with Baer involutions:

55.38 Proposition: Common fixed elements. *Any two Baer involutions* $\alpha, \beta \in \operatorname{Aut} \mathcal{P}$ *have a common fixed point and a common fixed line.*

Proof. Suppose that the Baer subplanes $\mathcal{F}_\alpha, \mathcal{F}_\beta$ of fixed elements have disjoint point sets F_α, F_β. Then the intersection $\mathcal{M}_\alpha \cap \mathcal{M}_\beta$ of the line sets consists of precisely one line W. Indeed, the intersection point of any two common fixed lines would be a common fixed point, and one common fixed line can be constructed as follows. A fixed point p of $\alpha\beta$ as provided by (55.19) satisfies $p^\alpha = p^\beta \neq p$. Hence, we obtain a line $W := pp^\alpha$, which belongs to $\mathcal{M}_\alpha \cap \mathcal{M}_\beta$; indeed, $W^\alpha = p^\alpha p = W$, and $W^\beta = (p^{\alpha\beta} p^\alpha)^\beta = p^\alpha p^{\alpha\beta} = W$.

Now any line $L \in \mathcal{M}_\beta \setminus \{W\}$ contains a unique point $x \in F_\alpha$, and $x \notin W$ since $L \wedge W \in F_\beta$. Conversely, each point $y \in F_\alpha \setminus W$ lies on a unique line

$M \in \mathcal{M}_\beta \setminus \{W\}$. Using compactness of the flag space (41.7a), one easily shows that the mappings $L \mapsto x$ and $y \mapsto M$ are continuous. Hence, the affine point set $F_\alpha \setminus W$ is homeomorphic to $\mathcal{M}_\beta \setminus \{W\}$. The former space is contractible, see (51.4), but the latter one is homotopy equivalent to a line pencil of $\mathcal{F}_\beta$ by the dual of (51.26); this pencil is not contractible, as shown by (54.11) together with the dual of (52.5a). This contradiction ends our proof. □

55.39 Proposition: Commuting Baer involutions.

(a) *Suppose that α, β and $\alpha\beta$ are commuting Baer involutions of a plane $\mathcal{P}$ of dimension $2l$. Then α induces a Baer involution on the plane $\mathcal{F}_\beta$ of fixed elements of β. Consequently, the intersection $\mathcal{F}_\alpha \cap \mathcal{F}_\beta$ is a plane of dimension $l/2$.*

(b) *An abelian group Φ such that $\Phi \setminus \{\mathbb{1}\}$ consists of Baer involutions has order at most l.*

Proof. By (55.32), the planes $\mathcal{F}_\alpha$ and $\mathcal{F}_\beta$ are distinct, hence α induces on $\mathcal{F}_\beta$ either a Baer involution or a homology (55.29). In the first case, $\mathcal{F}_\alpha \cap \mathcal{F}_\beta$ is the Baer subplane of $\mathcal{F}_\beta$ formed by the fixed elements of the induced Baer involution and has dimension $l/2$ by repeated application of (55.5).

If the induced map $\alpha|_{\mathcal{F}_\beta} = \alpha\beta|_{\mathcal{F}_\beta}$ is a homology, then the axis A of this reflection is a line of $\mathcal{F}_\beta$ and is contained in a line of each of the subplanes $\mathcal{F}_\alpha$ and $\mathcal{F}_{\alpha\beta}$. Now the three lines in question all have the same dimension $l/2$. By the domain invariance property (51.21b), it follows that the line A is open in the other two as well as closed. By connectedness, the three lines are equal. This implies that the non-cyclic, abelian group generated by α and β acts freely on $L \setminus A$, where L is the line of $\mathcal{P}$ containing A. This contradicts (55.31), and (a) is proved.

Part (b) is obtained by induction over t, where $l = 2^t$. Baer involutions do not exist for $t = 0$, see (32.12). For the inductive step, one passes to the group of Baer involutions induced on $\mathcal{F}_\alpha$, where $\alpha \in \Phi \setminus \{\mathbb{1}\}$. By (55.32), one obtains an effective action of $\Phi/\langle\alpha\rangle$. □

Next, we record a very typical concrete application of our general results on involutions. In spite of its seemingly rather specialized nature, this will be a most powerful tool. In fact, our observation will greatly reduce the list (94.33) of almost simple Lie groups that could possibly act on planes, because so many of these candidates contain a subgroup isomorphic to $SO_5\mathbb{R}$. For a formal treatment of the notion of an *action* appearing here, compare also (71.5 and 6).

55.40 Corollary. *The group $SO_5\mathbb{R}$ does not act non-trivially on any compact projective plane of finite dimension.*

Proof. In $SO_5\mathbb{R}$, there are two conjugacy classes of involutions; namely, those with a 1-dimensional space of fixed vectors (in the natural representation), and those where this space has dimension 3. With respect to an orthonormal basis, the subgroup of diagonal matrices is an elementary abelian group of order 2^4 which contains 5 involutions from the first conjugacy class, and 10 involutions of the second kind. In view of (55.34b), there are no reflections in this group, and we have reached a contradiction to (55.39b). □

The following results make use of the fact that the automorphism group Σ of a plane $\mathcal{P}$ is a locally compact transformation group of both P and $\mathcal{L}$, see (44.3). These properties carry over to any closed subgroup $\Delta \leq \Sigma$.

55.41 Proposition: Abelian groups. *Let Δ be a closed, connected, abelian group of automorphisms of a plane $\mathcal{P}$ of dimension $d = 2l$. Then we have one of the following (mutually exclusive) cases:*

(a) *Each point orbit under Δ is contained in a line; that is, Δ is straight on $\mathcal{P}$.*
   In this case, the stabilizer of any line moved by Δ is trivial, and $\dim \Delta \leq d$.
(b) *There is a point p whose orbit p^Δ topologically generates $\mathcal{P}$.*
   Consequently, the stabilizer Δ_p is trivial, and $\dim \Delta \leq d$.
(c) *There are points whose orbits are not contained in any line, and each of these orbits topologically generates a proper closed subplane.*
   In this case, $\dim \Delta \leq \frac{3}{2}l < d$.

Proof. By (42.3), each orbit p^Δ either is contained in a line or contains a quadrangle. In the latter case, p^Δ topologically generates a closed subplane with point set $\langle p^\Delta \rangle$, compare (41.6) and the subsequent Remarks. Hence, the three cases of the assertion are exhaustive and mutually exclusive. Since Δ is abelian, the stabilizer Δ_p acts trivially on p^Δ and, hence, on $\langle p^\Delta \rangle$, see (44.1a).

1) In case (a), the orbit of each point x moved by Δ is contained in a unique line F_x, which is fixed by Δ. Consequently, the stabilizer Δ_L of a line L moved by Δ acts trivially on L and on each of the images of L. Having two axes, the group Δ_L is trivial (23.8), and $\dim \Delta = \dim L^\Delta$ by the dimension formula (96.10). Hence, $\dim \Delta \leq \dim \mathcal{L} = d$ by the subspace theorem (92.4) together with (51.2).

2) In case (b), we have $\Delta_p = \mathbb{1}$ and $\dim \Delta = \dim p^\Delta \leq d$.

3) In case (c), we proceed by induction on l. For $l = 1$, there are no proper closed subplanes by (32.7), and case (c) cannot occur. Assume that $l > 1$, and let $\mathcal{Q} = (Q, \mathcal{M})$ be a maximal, closed, Δ-invariant subplane on which Δ acts non-trivially. By (55.1), we have that $\dim Q \leq l$. By induction, the closure of the group induced on Q is at most l-dimensional. In view of (83.3), it suffices now to show that the kernel Θ of that action has dimension at most $\frac{l}{2}$.

Choose a point $y \in P \setminus Q$. By maximality of $\mathcal{Q}$, we have $\langle Q \cup y^\Delta \rangle = P$. Since Θ_y acts trivially on $Q \cup y^\Delta$, that group is the identity, and $\dim \Theta = \dim y^\Theta$ by the dimension formula (96.10). We may assume that $\Theta \neq \mathbb{1}$ and that $\Delta \neq \Theta$. If we choose y on a line $L \in \mathcal{M}$ which is moved by Δ, then y^Θ is contained in L, but y^Δ is not. This implies that y^Δ is not contained in any line, whence $H = \langle y^\Delta \rangle$ is a closed subplane. We have $H \neq P$ by our assumption, and y^Θ is contained in $L \cap H$. Using (54.10b), we see that $\dim(L \cap H) = \frac{1}{2} \dim H \leq \frac{l}{2}$. Since $\dim \Theta = \dim y^\Theta$ and $y^\Theta \subseteq L \cap H$, this yields the desired bound for $\dim \Theta$. □

55.42 Corollary: Large abelian groups. *Let Δ be a closed, connected, abelian group of automorphisms of a plane of dimension $d = 2l$, and assume that $\dim \Delta \geq d$. Then Δ has an open orbit either in the point space or in the line space. The action of Δ on this open orbit is sharply transitive, and Δ is a Lie group of dimension d.*

Moreover, Δ fixes some point or some line, but leaves no proper, closed subplane invariant. Every orbit of points either is open or else is contained in some line.

Proof. 1) By (55.41), there is an orbit $X \subseteq P \cup \mathcal{L}$ such that $\langle X \rangle = P$, and the action of Δ on X is free. In particular, X is d-dimensional and hence open (53.1a), and Δ is a Lie group homeomorphic to X, see (53.2). Since P is connected and since Δ cannot be transitive by (63.8), there is an orbit $B \subseteq P$ such that $\langle B \rangle \neq P$. (The quotation of a result from the next chapter can be avoided by comparing the homology of P to that of an abelian Lie group, see (52.14 and 9) and (94.38c).) If B is contained in a line, then we obtain a fixed element. If not, then $\langle B \rangle$ is a proper, closed subplane, see (42.3) and the Remarks following (41.6). It only remains to be shown that there is no proper, closed subplane invariant under Δ. If $l = 1$, then this follows from (32.7).

2) Now suppose that our claim is false for $l \geq 2$. We consider a Δ-invariant, proper, closed subplane $\mathcal{Q} = (Q, \mathcal{M})$ of maximal dimension e, and the kernel Θ of the action of Δ on Q. According to (55.41), the closure of the group induced on $\mathcal{Q}$ is at most e-dimensional, and we have $e \leq l$, see (55.1). This implies that $\dim \Theta \geq d - e \geq l$; compare (83.3) for details. If we choose $y \in P \setminus Q$ on a line L of $\mathcal{Q}$, then $L = L^\Theta$, and $\Theta_y = 1\!\!1$ by maximality of e. Using the dimension formula (96.10) together with (53.1a), we infer that $e = l$, and that each Θ-orbit in $L \setminus Q$ is open. If $l = 2$, then this observation contradicts (55.21), and the corollary is proved in that case. If $l > 2$, then $L \setminus Q$ is connected (51.16), hence Θ is transitive on this set.

3) We use induction over l. The inductive hypothesis yields that, up to duality, the line L may be chosen so that $L = L^\Delta$. The stabilizer Δ_y is l-dimensional by (96.10) and acts trivially on the orbit y^Θ, which is dense in L. Thus, the group Δ_y has axis L, and each point $q \in L \cap Q$ is fixed by $\Delta = \Delta_y \Theta$. The action induced by Δ on $\mathcal{L}_q \setminus \mathcal{M}_q$ coincides with the transitive action of Θ, hence Δ acts trivially on each $\mathcal{M}_q$, a contradiction. □

55.43 Examples. (a) On the quaternion plane, the 5-dimensional abelian group

$$\left\{ \begin{pmatrix} s & 0 & x \\ 0 & s & y \\ 0 & 0 & s \end{pmatrix} \;\middle|\; s, x, y \in \mathbb{C}, \; s\bar{s} = 1 \right\}$$

acts almost effectively and leaves the complex subplane invariant. Here, the bound in (55.41c) is almost attained. Using results on the dimensions of groups that act trivially on a subplane, see (32.10), (55.21), (83.18 and 23) and the remarks (83.16 and 25), this bound may be improved.

(b) The bounds in (55.41b and a) are attained respectively by full translation groups and by their duals. However, case (b) covers more than just the dual of (a). Indeed, there exist abelian two-dimensional subgroups of $\mathrm{PSL}_3\mathbb{R}$ that are not translation groups, such as

$$\left\{ \begin{pmatrix} a & b & \\ & a & \\ & & 1 \end{pmatrix} \;\middle|\; a > 0, b \in \mathbb{R} \right\} \cong \mathbb{R}^2,$$

$$\left\{ \begin{pmatrix} a & b & \\ -b & a & \\ & & 1 \end{pmatrix} \;\middle|\; (a, b) \in \mathbb{R}^2 \setminus \{(0, 0)\} \right\} \cong \mathbb{C}^\times.$$

Another example is given in (73.9). Far more possibilities arise in non-desarguesian, 2-dimensional planes, compare (38.5). For the possible actions of 4-dimensional, abelian groups on 4-dimensional, compact, projective planes, see (71.11).

55.44 Further results. More information about topological ovals may be found in Buchanan–Hähl–Löwen [80]. Buchanan [79b] 9.24 shows that the absolute points of a polarity of a 4-dimensional plane form an oval if, and only if, the polarity reverses the orientation of the flag space; the proof uses the cohomology ring of the flag space. The notion of topological ovals has been generalized by Löwen [94] so as to make up for the lack of 4-dimensional ovoidal Laguerre planes caused by (55.13). Ovals arising as orbits of collineation groups are considered in Groh [71] and Löwen [84a].

The geometric fiber bundle (51.23a) is used in Grundhöfer–Hähl [90] in order to show that the only differentiable projective planes of Lenz type V are the classical planes $\mathcal{P}_2\mathbb{F}$, where $\mathbb{F} \in \{\mathbb{R}, \mathbb{C}, \mathbb{H}, \mathbb{O}\}$, compare 75.5. Applications of the geometric fiber bundle (51.23a) to problems related to the Blaschke conjecture in differential geometry are given by Hähl [87d].

In Buchanan–Hähl [77], it is shown that the kernel of an 8-dimensional quasifield does not contain the complex numbers and hence is isomorphic to $\mathbb{R}$, and that the left nucleus of an 8-dimensional quasifield does not contain a torus subgroup in its multiplicative group. These results are obtained using the non-existence of cross-sections for certain fibrations associated with classical groups.

55.45 Notes on Section 55. The results on subplanes of 4-dimensional planes (55.4) are due to Salzmann [69c]. The characterization of Baer subplanes (55.5) and its consequence (55.6) were also obtained by Salzmann [79a,b]. Like some of the other results in this section, they owe their present general form to the results of Löwen reproduced in Section 54. The results about ovals are taken from Buchanan–Hähl–Löwen [80], except for (55.13), which is due to Buchanan [79c], and (55.18), which was conjectured by Polster. For the case $l = 1$, the latter proposition is proved in Polster–Steinke [95].

Assuming that lines are manifolds, Breitsprecher [71] 2.3.2 gave the first proof of the fixed point property of planes. He used the cohomology rings of the spaces of points and flags. Without that assumption, the cohomology rings have been computed by Löwen [83b]. The proof of the fixed point property using the Lefschetz fixed point theorem as in (55.19) was found by Salzmann and Dugundji, see Salzmann [79a] (B). Existence of square roots (55.20) was proved by Salzmann [70] in the 4-dimensional case and by Hähl (presented by Salzmann [79a]) for planes with manifold lines. Salzmann [69b, 70] determined the stabilizer of a quadrangle in a 4-dimensional plane (55.21). The proposition on the infinite order of elations (55.28) and the ensuing classification of involutions (55.29) were given by Salzmann [75a] for the simplest (but typical) case. The uniqueness theorems for involutions (55.31 and 32) go back to Salzmann [70, 79a,b]; similar results for stable planes were obtained by Löwen [81a] and Stroppel [93c]. The rank estimate (55.37b) is taken from Grundhöfer–Salzmann [90] XI 9.6. Existence of common fixed points of two Baer involutions (55.38) was proved by Salzmann [79a]. The bound on the order of abelian groups of Baer involutions (55.39) and its consequence, the nonexistence of $SO_5\mathbb{R}$-actions (55.40), were proved by M. Lüneburg [92]. The results (55.41 and 42) on abelian groups are taken from Stroppel [93a].

Chapter 6

Homogeneity

Consider a compact, connected projective plane with point set P, and assume that $\dim P$ is finite. According to the results of the last chapter, these planes split into four classes defined by the value of $\dim P = 2l \in \{2, 4, 8, 16\}$. Each of the four classes contains a distinguished representative, namely the classical plane $\mathcal{P}_2\mathbb{F}$, where $\mathbb{F} \in \{\mathbb{R}, \mathbb{C}, \mathbb{H}, \mathbb{O}\}$ ('domination by classical planes', see the introduction to Section 55). All other planes in the same class have a topology very similar to the classical representative. We may therefore roughly think of these other planes as the point space $P_2\mathbb{F}$ endowed with a distorted version of the classical line system. In fact, most of the known examples of compact planes have been obtained by some kind of deformation of the classical planes; compare also (53.6).

Research in the past decades followed the leading idea that it should be possible to measure the amount of distortion by somehow measuring the degree of homogeneity of the resulting plane, i.e., the size of its automorphism group. This idea turned out to be very fruitful. Indeed, it could be shown that the classical plane in each class has by far the biggest group. Moreover, the most homogeneous ones among the non-classical planes can be completely classified. Usually they can be divided into a finite number of subclasses, some of which are finite while others are naturally parametrized by a finite number of real parameters; often the classical planes are obtained for particular values of those parameters. If one further relaxes the homogeneity requirements, one also obtains classes parametrized by functions of some kind. At some point after that, with very little homogeneity left, all recognizable patterns are lost. The planes in this region are practically invisible, because for lack of regularity they cannot be obtained by any manageable construction. In particular, no examples of rigid planes (having no automorphisms at all) have been constructed for $l > 1$ although experience with $l = 1$ suggests that their name must be legion.

For these ideas to make sense, the vague notion of *size of the automorphism group* must be made precise. There are many ways to do this, which will be considered individually in the different sections of this chapter. The first four sections concern notions of homogeneity that make sense also for abstract projective planes. However, they tend to work much better and give far stronger results if they are combined with topological assumptions, because then strong tools, mainly from the theory of Lie groups, become available.

This is true in particular for all kinds of transitivity assumptions concerning general (as opposed to axial) automorphisms. These are treated in Section 63, where we show, among other things, that mere transitivity of the automorphism group on the set of points suffices to characterize the classical planes $\mathcal{P}_2\mathbb{F}$, where $\mathbb{F} \in \{\mathbb{R}, \mathbb{C}, \mathbb{H}, \mathbb{O}\}$.

The situation is very much different in abstract planes. In their article [73], Kegel and Schleiermacher embed every projective plane into a plane whose group is transitive on triangles; compare also Tits [77] Satz 3. A plane with a group transitive on quadrangles may still be very far from satisfying the Moufang condition. Even in finite planes, the implications of transitivity are not completely understood. It is known that twofold transitivity of the automorphism group on the point set characterizes pappian planes in the finite case. See Hughes–Piper [73] 14.13 for this famous theorem of Ostrom and Wagner. It is not known whether or not the same is true under the hypothesis of mere transitivity. A phenomenon creating massive difficulties is the existence of Singer groups, i.e., of groups that are sharply transitive on the point set, see Hughes–Piper [73] XIII Section 5. As a consequence of the fixed point property (55.19), a compact, connected projective plane does not admit any sharply transitive, continuous group actions.

Transitivity conditions for axial collineations on the other hand are equivalent to strong algebraic assumptions about suitable coordinate structures, and therefore rather easily lead to good results. Some basic facts about this type of homogeneity are needed throughout this chapter and are therefore dealt with in Section 61, whereas the systematic analysis is given in Section 64. There, interesting classes of non-classical planes appear, which will also occupy us in the remaining two chapters.

In Section 62, we consider a plane admitting an action of one of the motion groups of its dominating classical plane, as introduced in Sections 13 and 18. Again it turns out that this characterizes the classical planes. Section 65 treats a measure of homogeneity that is meaningful only for topological planes and that has proved to be the most successful tool if one tries to single out the few pleasant specimens among the chaotic mass of all (compact, connected) planes. This measure is provided by the topological dimension of the automorphism group. Section 65 is intended to give a rough idea of how to work with this notion. A detailed analysis is given in Chapters 3, 7 and 8. This organization scheme reflects the fact that, unlike the other notions of homogeneity, this one requires very elaborate and extended arguments that have to be set up individually for every possible value of l. The brief final Section 66 merely sketches how homogeneity can be measured by the size of the group of projectivities rather than the group of automorphisms. Here, the classical planes have the smallest groups.

61 Axial collineations

61.0 General Assumptions. Throughout this section, $\mathcal{P} = (P, \mathcal{L})$ denotes a compact projective plane with l-dimensional lines, $0 < l < \infty$ (so that $l \in \{1, 2, 4, 8\}$ by (52.5) together with (54.11)), and Σ is its locally compact automorphism group (44.3). The letter Δ stands for some closed subgroup of Σ. The assumption that Δ be closed is needed in the proofs of some basic tools (61.1 and 3). Hence it is is indispensable for our main results such as (61.20), which depend on the use of those tools. Some other results, like (61.4 and 11), would also work under the weaker hypothesis that Δ is a locally compact group acting effectively on $\mathcal{P}$, compare (71.6). For example, if Σ is a Lie group, then Δ could be an analytic subgroup.

Our goal in this section is to provide techniques which can be used to ascertain the existence of sufficiently large groups of elations. Such techniques will be a key ingredient for many proofs of classification results; compare Chapters 7 and 8. We shall mostly refrain from formulating the duals of our statements, which may nevertheless be useful.

As in Sections 23 and 44, the symbol $\Delta_{[c,A]}$ denotes the group of axial collineations in Δ having center c and axis A. We recall the distinction between homologies ($c \notin A$) and elations ($c \in A$), as well as the following facts, see (44.7 and 8b).

61.1 Review.

(a) *Axial collineations are continuous.*
(b) *If $\Delta \leq \Sigma_{[c,A]}$ is a closed subgroup, then Δ is isomorphic to some closed subgroup S of the additive ($c \in A$) or multiplicative ($c \notin A$) loop of some suitable ternary field for $\mathcal{P}$. Orbits of Δ in $X = P \setminus (A \cup \{c\})$ are closed in X and are homeomorphic to Δ.*
(c) *If $\Delta \leq \Sigma_{[A,A]}$ is closed, then the orbits of Δ in $Y = P \setminus A$ are closed and are homeomorphic to Δ.* □

More precisely, in the case of homologies the connection obtained in (44.8b) between Δ and S is as follows. Take a ternary field K such that $P \setminus A$ can be identified with $K \times K$, the element c corresponding to $(0, 0)$. Then Δ corresponds to the set of all mappings

$$(x, y) \mapsto (xs, ys), \quad s \in S.$$

If $S \cong \Delta$ is not compact, then S accumulates at 0 or at ∞. By (43.4), this implies that $\overline{S} = S \cup \{0, \infty\}$ is a compactification. If moreover Δ is connected, then this means that the group $\Delta \cong S$ is two-ended, so that it satisfies the hypothesis of (93.13). This yields the next proposition.

61.2 Proposition: Two-ended homology groups. *A closed, connected subgroup Δ of a homology group $\Sigma_{[c,A]}$, $c \notin A$, is either compact or two-ended. In the latter case, Δ is a direct product $\Omega \times \mathsf{P}$ of a compact, connected group Ω and a one-dimensional vector group $\mathsf{P} \cong \mathbb{R}$, see (93.13).* □

The description of Δ given above also yields the following contraction principle if we just note that multiplication is continuous at $(x, 0) \in K \times K$.

61.3 Proposition: Contraction. *Let $\Delta \leq \Sigma_{[c,A]}$, $c \notin A$, be a closed group of homologies (not necessarily connected). If Δ is not compact, then every compact set $C \subseteq P \setminus A$ can be mapped into an arbitrarily small neighbourhood of c by some element of Δ.* □

61.4 Proposition: Dimension bounds.

(a) *Let $\Delta \leq \Sigma_{[c,A]}$ be a d-dimensional subgroup. Then $d \leq l$, and Δ is (c, A)-transitive if, and only if, equality $d = l$ holds and one of the following conditions is satisfied: $c \in A$, or $l > 1$, or Δ is not connected. In this case, we have that $\Delta = \Sigma_{[c,A]}$.*
(b) *Let $\Delta \leq \Sigma_{[A,A]}$ be d-dimensional. Then $d \leq 2l$, and equality holds precisely when Δ is transitive on $P \setminus A$. In this case, we have that $\Delta = \Sigma_{[A,A]}$.*
(c) *If $\Delta \leq \Sigma_{[A,A]}$ has two l-dimensional subgroups $\Delta_{[c_1,A]}$ and $\Delta_{[c_2,A]}$, where $c_1 \neq c_2$, then Δ is transitive on $P \setminus A$.*

Proof. (a) Every non-trivial Δ-orbit B is d-dimensional by (61.1b). (The same holds for Δ^1-orbits, see (93.6).) Moreover, B is contained in some line $L \in \mathcal{L}_c$. By the subspace theorem (92.4), we have that $d \leq l$. According to (53.1a), the orbit B is open in L if, and only if $d = l$. In this case, Δ^1 is transitive on any connected component of $X := L \setminus (A \cup \{c\})$. Now X is connected unless $l = 1$ and $c \notin A$, in which case there are two components. The last assertion follows from (23.9). The proof of part (b) is similar, and (c) follows from (a), see (23.23). □

61.5 Proposition: Connected elation groups are Lie groups.
The connected component $\Phi = \Sigma_{[A,A]}{}^1$ of the group of elations with axis A is a Lie group containing no compact subgroups other than $\mathbb{1}$.

Proof. We have $\dim \Phi \leq 2l < \infty$, and Φ acts freely on $X = P \setminus A$. In particular, the stabilizers of this action are (trivial) Lie groups. According to Löwen [76a] Theorem (d), this implies that Φ is a Lie group, provided that we also show that the one-point compactification $X \cup \{\infty\}$ is a Peano continuum with Euler characteristic $\chi \notin \{0, 1\}$. Now $X \cup \{\infty\}$ is homeomorphic to the quotient space P/A, see (51.6); the latter has the desired properties by (51.8) and (52.12). A maximal compact subgroup $\Psi \leq \Phi$ is connected by (93.10), and thus contains an involution if

$\Psi \neq \mathbb{1}$, cf. (94.31a). This is impossible as we proved in (55.28), using a fixed point theorem of Smith. □

61.6 Corollary: Structure of elation groups. *If $\Sigma_{[A]}$ contains non-trivial elations with different centers, then $\Sigma_{[A,A]}{}^1 \cong \mathbb{R}^k$.*

Proof. By our hypothesis, $\Sigma_{[A,A]}$ is commutative, see (23.13), so that $\Sigma_{[A,A]}{}^1 \cong \mathbb{R}^k \times (\mathrm{SO}_2\mathbb{R})^s$ according to (94.38). It follows from (61.5) that $s = 0$. Without using the theorem of Smith, this can also be inferred from the fact that all elements in $\Sigma_{[A,A]}$ have the same order (23.13). □

61.7 The center map. Let $\Upsilon \subseteq \Sigma$ be the set of non-trivial axial collineations. A most useful tool is provided by the continuous *center map*

$$\zeta : \Upsilon \to P$$

sending an axial collineation to its center (44.8a). We recall the easy proof of continuity: Take points p, q that are moved by $\gamma \in \Upsilon$ and satisfy $pp^\gamma \neq qq^\gamma$. For $\delta \in \Upsilon$ close to γ, the center of δ is $c = pp^\delta \wedge qq^\delta$.

The following result on the center map is a special case of a more general fact, see (61.25).

61.8 Proposition: Axial one-parameter groups. *The center map is constant on any one-parameter group Ψ consisting of axial collineations.*

Proof. Let $\alpha : \mathbb{R} \to \Psi$ be a continuous epimorphism. Then ζ is constant on $(\frac{1}{n}\mathbb{Z})^\alpha$ for any $n \in \mathbb{N}$, and hence on $\mathbb{Q}^\alpha$, which is dense in Ψ. By continuity, ζ is constant on Ψ. □

61.9 Corollary: Connectedness of elation groups. *For a connected group Δ contained in $\Sigma_{[A,A]}$, the subgroups $\Delta_{[a]}$ are connected for all $a \in A$. If $\Delta_{[a]} \neq \Delta$, then $\Delta_{[a]} \cong \mathbb{R}^t$.*

Proof. We may assume that $\Delta \neq \Delta_{[a]}$, so that $\Delta \cong \mathbb{R}^k$ by (61.6) and (94.41). A non-trivial element $\gamma \in \Delta_{[a]}$ lies on a one-parameter group $\Psi \leq \Delta$. Now (61.8) shows that $\gamma \in \Psi \leq \Delta_{[a]}{}^1$. Since $\Delta_{[a]}$ is closed in Δ and connected, the second assertion follows from (94.41). □

Let $\Delta \leq \Sigma_{[A]}$ be a Lie group, and consider the projective space $P\mathfrak{l}\Delta$ associated with its Lie algebra $\mathfrak{l}\Delta$, topologized as in (14.1); $P\mathfrak{l}\Delta$ may be considered as the space of one-parameter subgroups of Δ.

61.10 Corollary: Projective center map. *The center map induces a continuous map* $\mathrm{P}\mathfrak{l}\Delta \to P$, *which will be denoted by the same symbol ζ.* □

61.11 Proposition: Large elation groups. *Let $\Delta \leq \Sigma_{[A,A]}$ be given. For every point $c \in A$ we have*

(a) $\dim \Delta - \dim \Delta_{[c]} \leq l$.
(b) *If $\dim \Delta > l$, then every group $\Delta_{[c]}$, $c \in A$, contains a one-parameter group, hence the projective center map sends $\mathrm{P}\mathfrak{l}\Delta$ onto A.*
(c) *If $\dim \Delta = 2l - 1$, then $\dim \Delta_{[c]} = l - 1$ for all $c \in A$ with at most one exception.*

Proof. Assertion (a) can be seen from the free action of $\Delta/\Delta_{[c]}$ on $\mathcal{L}_c \setminus \{A\}$, and (b) is an immediate consequence. For (c), we note that $\dim \Delta_{[c]} \in \{l, l-1\}$ by (a) and (61.4). If $\dim \Delta_{[c]} = l$ occurs twice, then (61.4c) yields that Δ is transitive on $P \setminus A$ and $\dim \Delta = 2l$. □

Improving (61.11c), the following theorem shows that $\dim \Delta = 2l - 1$ implies the existence of an exceptional point $a \in A$ with $\dim \Delta_{[a]} = l$.

61.12 Theorem: Almost homogeneous elation groups. *Let $\Delta \leq \Sigma_{[A,A]}$ be a closed subgroup, and let $a \in A$. If $\dim \Delta_{[c]}$ has the same value $k > 0$ for every $c \in A \setminus \{a\}$, then $\dim \Delta_{[a]} = l$, and hence Δ is (a, A)-transitive.*

The proof of (61.12) will be given after that of (61.15); it shows in addition that $\Delta = \Delta_{[a]} \times \Delta_{[c]}$. The following corollary is immediate from (61.12 and 4).

61.13 Corollary: Homogeneous elation groups. *Let $\Delta \leq \Sigma_{[A,A]}$ be a closed subgroup. If $\dim \Delta_{[c]}$ has the same value $k > 0$ for every $c \in A$, then $k = l$ and $\Delta = \Sigma_{[A,A]}$ is transitive.* □

61.14 Remark. Examples of almost homogeneous elation groups are easily obtained in the classical planes. Simply take any $(2l-1)$-dimensional vector subgroup Δ of $\Sigma_{[A,A]} \cong \mathbb{R}^{2l}$; according to (61.11c), this elation group is almost homogeneous with $k = l - 1$. Examples where $\Sigma_{[A,A]}$ itself is almost homogeneous are constructed by Betten [84] and Weigand [87], see (82.30 through 33). Their planes have $l \geq 2$ and $\dim \Sigma_{[A,A]} = 2l - 1$ for some line A. As before, this elation group is almost homogeneous.

Before we can prove Theorem (61.12), we need the following lemma about fibrations of vector spaces by subspaces.

61.15 Lemma: Vector space fibrations. *Let $p : \mathbb{R}^s \setminus \{0\} \to A$ be a continuous map onto a metric space A. Assume that for every $b \in A$, the fiber $F_b = p^{-1}(b)$ gives rise to a vector subspace $V_b = F_b \cup \{0\} \leq \mathbb{R}^s$.*

If $\dim V_b$ has a constant value $k \notin \{0, s\}$ on some non-empty open set $B \subseteq A$, then p restricted to $E := p^{-1}(B)$ is a locally trivial fibre bundle in the sense of (51.22). In particular, there is an associated exact homotopy sequence, which runs

$$\ldots \to \pi_q \mathbb{S}_{k-1} \to \pi_q E \to \pi_q B \to \pi_{q-1} \mathbb{S}_{k-1} \to \ldots$$

Proof. 1) References for the homotopy sequence of a fibre bundle are given in (51.22). Above, we have written down this sequence, replacing the homotopy group $\pi_q F_b$ of $F_b \approx \mathbb{R}^k \setminus \{0\} \simeq \mathbb{S}_{k-1}$ by the isomorphic group $\pi_q \mathbb{S}_{k-1}$.

2) Let $c \in B$ and consider a decomposition $\mathbb{R}^s = W \oplus V_c$. Let U be the set of all $u \in B$ such that $\mathbb{R}^s = W \oplus V_u$ or, equivalently, such that $F_u \cap W = \emptyset$. Then $B \setminus U = p(\mathbb{S}_W) \cap B$, where $\mathbb{S}_W$ denotes the unit sphere in W. In particular, U is open in B.

3) We shall exhibit a trivialization of p over U by proving that the continuous map

$$\alpha : p^{-1}(U) \to U \times F_c,$$
$$x \mapsto (p(x), \psi(x)),$$

is a homeomorphism, where ψ denotes the projection of $\mathbb{R}^s = W \oplus V_c$ onto the first summand. Since ψ induces isomorphisms $F_u \to F_c$ for all $u \in U$, the map α is bijective. To see that α^{-1} is continuous, let $(p(x_n), \psi(x_n))$ converge to $(p(x), \psi(x)) \in U \times F_c$. If x_n accumulates at $y \in \mathbb{R}^s$, then $y \neq 0$ and $y = x$ by continuity of ψ and p, hence it suffices to show that $\|x_n\|$ cannot converge to ∞. In that case, the bounded sequence $x'_n := \|x_n\|^{-1} \cdot x_n$ would accumulate at some $x' \neq 0$ such that $p(x') = p(x) = u \in U$ and $\psi(x') = \lim \|x_n\|^{-1} \psi(x_n) = 0$. In other words, we have the contradiction $x' \in F_u \cap W$. □

Proof of (61.12). 1) The dimension of $\Delta_{[c]}$ does not change if we replace Δ by Δ^1, compare (93.6). In view of (61.6), we may assume that $\Delta \cong \mathbb{R}^s$, so that $\Delta_{[c]} \cong \mathbb{R}^k$ and $\Delta_{[a]} \cong \mathbb{R}^t$ by (61.9). We may therefore apply Lemma (61.15) to the center map $\zeta : \Delta \setminus \{\mathbb{1}\} \to A$, taking $p = \zeta$ and $B = A \setminus \{a\}$, so that $E = \Delta \setminus \Delta_{[a]}$.

2) We have $E = \mathbb{R}^s \setminus \mathbb{R}^t \approx \mathbb{R}^t \times (\mathbb{R}^{s-t} \setminus \{0\}) \simeq \mathbb{S}_{s-t-1}$, and $\pi_q B$ is trivial since $B = A \setminus \{a\}$ is contractible (42.8). The exact sequence thus shows that $\mathbb{S}_{k-1}$ and $\mathbb{S}_{s-t-1}$ have isomorphic homotopy groups, so that $s = k+t$. Since $\mathbb{R}^s$ contains two k-dimensional subspaces with trivial intersection, we have $s \geq 2k$ and $t \geq k > 0$. Moreover, we obtain that $\Delta = \Delta_{[a]} \times \Delta_{[c]}$ for every $c \in B$.

3) Choose a non-trivial element $\delta \in \Delta_{[c]}$. For $w \neq a$, both $\Delta_{[c]}$ and $\Delta_{[w]}$ are complements of $\Delta_{[a]}$, hence $\Delta_{[a]} \times \{\delta\} \subseteq E$ meets the fibre $\Delta_{[w]} \setminus \{\mathbb{1}\}$ of $p = \zeta$ exactly once. We conclude that p maps $\Delta_{[a]} \times \{\delta\}$ bijectively and continuously onto B. We shall show that this map is a homeomorphism, so that $\mathbb{R}^t \approx \Delta_{[a]} \times \{\delta\} \approx B$ and $t = \dim B = l$. Indeed, assume that $b_n \to b$ in B and consider $\{x_n\} = \Delta_{[b_n]} \cap (\Delta_{[a]} \times \{\delta\})$. If x_n accumulates at x, then $x \in \Delta_{[b]} \cap (\Delta_{[a]} \times \{\delta\})$. It remains to exclude the possibility that $x_n \to \infty$. In that case, $\|x_n\|^{-1} \cdot x_n$ accumulates at some non-zero element of $\Delta_{[b]} \cap \Delta_{[a]}$, a contradiction. □

Elations generated by homologies

61.16 Definition. For a subset $\Xi \subseteq \Sigma_A$, let

$$Z(\Xi) := \left(\Xi_{[A]} \setminus \{\mathbb{1}\}\right)^\zeta \setminus A$$

denote the set of centers of non-trivial homologies $\xi \in \Xi$ with axis A. We shall consider a closed subgroup

$$\Delta \leq \Sigma_{[A]},$$

and we write

$$\Phi := \Delta_{[A,A]}{}^1.$$

Our aim will be to draw conclusions about the size of Φ from information about the size of $Z(\Delta)$. This is in the spirit of a result of Baer stating that the product of two reflections with the same axis and distinct centers is an elation, see (23.20).

61.17 Lemma on Cosets. *With the notation of* (61.16), *let* $\delta \in \Delta_{[c,A]}$ *be a non-trivial homology. Then the coset* $\delta\Phi$ *clearly consists of homologies, and we claim that* $Z(\delta\Phi) = c^\Phi$.

61.18 Corollary on Products. *If* $c \notin A$ *and* $\Delta = \Delta_{[c]} \cdot \Phi$, *then* $Z(\Delta) = c^\Phi$, *and* $\Delta_{[A,A]}$ *is connected.*

Proof of corollary. The first assertion follows directly from the lemma since $\Delta \setminus \Phi$ decomposes into cosets $\delta\Phi$ represented by elements of $\Delta_{[c]}$. Moreover, $\delta\Phi$ contains elations only if $\delta = \mathbb{1}$. Hence, $\Delta_{[A,A]} = \Phi$ is connected. □

Proof of lemma. 1) Consider an arbitrary point $a \in A$ and let $\Psi = \Phi_{[a]}$. It suffices to prove that then $Z(\delta\Psi) = c^\Psi$, because Φ is the union of the subgroups Ψ. Note that Ψ is normal in Δ by (23.12) and that Ψ is a connected Lie group by (61.5 and 9).

2) We have $c^\Psi \subseteq Z(\delta\Psi)$ because c^ψ is the center of $\psi^{-1}\delta\psi$, which belongs to $\Psi\delta\Psi = \delta\Psi\Psi = \delta\Psi$. Now c^Ψ is a connected manifold (homeomorphic to Ψ) and is closed in $P \setminus A$ by (61.1c). We shall show in step 3) that $Z(\delta\Psi)$ also is homeomorphic to Ψ, and then the topological pigeon-hole principle will yield $c^\Psi = Z(\delta\Psi)$. More precisely, we apply (51.20) to the (closed) inclusion map $\alpha : c^\Psi \hookrightarrow Z(\delta\Psi)$, to obtain that α is open and hence surjective by connectedness.

3) Choose a point $p \notin A \cup ca$. We construct a homeomorphism $f : Z(\delta\Psi) \approx p^{\delta\Psi}$; this suffices since $p^{\delta\Psi} \approx \Psi$. The construction is described by Figure 61a. Since $\delta\Psi$ fixes ca, the center set $Z(\delta\Psi)$ is contained in that line. The orbit $p^{\delta\Psi}$ is contained in $p^\delta a$, and f maps the line ca onto $p^\delta a$ by the perspectivity with center p. □

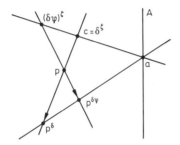

Figure 61a

The following lemma contains a central argument for the proof of the major result (61.20). It is useful by itself because unlike (61.20a) it does not need the hypothesis that Δ is an axial group. The assumption that Δ is a Lie group can be replaced by the weaker one that $\dim \Delta < \infty$; this suffices for Mostert [53] to supply the cross-section required in the proof.

61.19 Lemma. *Assume that Δ is a Lie group and that $1 \neq \gamma \in \Delta_{[c]}$. Let $\dim c^\Delta = k$. Then the following assertions hold.*

(a) *The conjugacy class γ^Δ contains a compact, connected k-dimensional subset Ξ which contains γ and consists of collineations with pairwise different centers. In particular, we have $\dim \gamma^\Delta \geq k$.*

(b) *If moreover γ is an involution and Δ fixes $A \in \mathcal{L} \setminus \mathcal{L}_c$, then $\dim \Delta_{[A,A]} = k$.*

Proof. Consider a local cross section $\Lambda \subseteq \Delta$ for the canonical map $\pi : \Delta \to \Delta/\Delta_c$. This means that π is injective on the compact, connected set Λ containing 1 and that $\dim \Lambda + \dim \Delta_c = \dim \Delta$, whence $\dim \Lambda = k$ by (96.10). Such a set Λ exists by (94.3b). Since c^λ is the center of γ^λ, the map $\lambda \mapsto \gamma^\lambda$ is a homeomorphism onto a set Ξ with the required properties. This proves (a).

If γ is an involution and Δ fixes A, then $\gamma \cdot \Xi \approx \Xi$ is contained in $\Delta_{[A,A]}$ by (23.20). Hence, $\dim \Delta_{[A,A]} \geq k$. The converse inequality follows from the inclusion $c^\Phi \subseteq c^\Delta$. □

An easy special case of the following result has been proved in (33.3).

61.20 Theorem: Elations generated by homologies. *Assume that $\Delta \leq \Sigma_A$ is a closed subgroup. Let $\Phi = \Delta_{[A,A]}^{1}$ and let $Z(\Delta)$ denote the set of centers of non-trivial homologies $\delta \in \Delta_{[A]}$.*

(a) *If Δ acts trivially on A and if the connected component of Δ contains a non-trivial homology with center c, then Δ is a semi-direct product $\Delta = \Delta_{[c]} \ltimes \Phi$. Consequently, by (61.18), we have that $Z(\Delta) = c^\Phi$, and $\Delta_{[A,A]} = \Phi$ is connected.*

(b) *If Δ is a Lie group, and if $\Delta_{[c]} \neq \mathbb{1}$ for some $c \notin A$, then $\dim Z(\Delta) = \dim \Phi = \dim c^\Delta$.*

61.21 Remark. For the sake of simplicity, we have weakened Hähl's original version of this result [81]. In fact he shows that an exhaustive description of $Z(\Delta)$ has to be more complicated, for in certain two-dimensional projective planes constructed by Salzmann [58] the set of all centers of non-trivial homologies with some fixed axis A is non-empty but is not a single orbit of $\Sigma_{[A,A]}$. In particular, assertion (a) depends on the assumption that $\Delta^1_{[c]} \neq \mathbb{1}$ and not just $\Delta_{[c]} \neq \mathbb{1}$.

Proof of theorem. 1) We begin by proving (a) in the case where Δ is connected. From (61.4), combined with the dimension formula (96.10), we infer that $\dim \Delta$ is finite. We may further assume that Δ is a Lie group. If this is not the case, then (93.9) implies that the center Θ of Δ is non-trivial. Applying (23.12) twice, we conclude that then $\Theta \leq \Delta_{[c,A]}$ and $\Delta = \Delta_{[c,A]}$, and (a) follows. The group $\Delta_c = \Delta_{[c]}$ has trivial intersection with Φ. In view of (94.3f), it suffices therefore to show that now

$$\dim \Delta_{[c]} + \dim \Phi \geq \dim \Delta.$$

2) If $\Delta_{[c]}$ contains an involution, then (61.19b) shows that $\dim c^\Delta = \dim \Phi$. On the other hand, $\dim c^\Delta = \dim \Delta - \dim \Delta_{[c]}$ by the dimension formula (96.10), and the proof of (a) is complete in this case.

3) We claim that Δ contains a one-parameter group of homologies. If not, then (61.8) implies that no exponential neighbourhood (94.7) contains any non-trivial homologies. Since Δ is generated by every open subset (93.14a), we obtain that $\Delta = \Phi$ contrary to our hypothesis. Hence, the set

$$D := \{ d \in P \setminus A \mid \dim \Delta_{[d]} > 0 \}$$

is non-empty. If there is some $d \in D$ such that the group $\Delta_{[d]}$ contains a non-trivial compact, connected subgroup, then it contains an involution, compare (94.31a). By step 2), the assertion holds for d in place of c; in particular, we have $Z(\Delta) = d^\Phi$. The point c belongs to this set, hence the assertion holds for c as well. By (61.2), the group $\Delta_{[d]}^1$ is isomorphic to $\mathbb{R}$ if it contains only the trivial compact subgroup. Only the case where this happens for all $d \in D$ is still open.

4) Consider the Lie algebras $\mathfrak{d} = \mathsf{l}\Delta$ and $\mathfrak{f} = \mathsf{l}\Phi$, and the projective center map $\zeta : \mathsf{P}\mathfrak{d} \to Z(\Delta) \cup A$ as in (61.10), which is continuous and closed because $\mathsf{P}\mathfrak{d}$ is compact. For every $d \in D$, the Lie algebra $\mathsf{l}\Delta_{[d]}$ is a point of $\mathsf{P}\mathfrak{d}$ by 3), so that ζ maps $\mathsf{P}\mathfrak{d} \setminus \mathsf{P}\mathfrak{f}$ injectively and hence homeomorphically onto the Δ-invariant manifold $D \subseteq Z(\Delta)$. According to (61.3), every loop α in D can be mapped into a Euclidean neighbourhood of $d \in D$ by a homeomorphism of D, hence α is null homotopic and $D \approx \mathsf{P}\mathfrak{d} \setminus \mathsf{P}\mathfrak{f}$ is simply connected. Now $\mathsf{P}\mathfrak{d} \setminus \mathsf{P}\mathfrak{f}$ is doubly covered by $S \setminus \mathfrak{f}$, where S is the unit sphere in $\mathfrak{d}$. Thus, $S \setminus \mathfrak{f}$ must be disconnected, which shows that $\dim \Delta - \dim \Phi = 1 = \dim \Delta_{[d]}$. This proves assertion (a) for the point d; as in step 3), the result carries over to c.

5) To complete the proof of (a), assume no longer that Δ is connected, but that $\Delta^1_{[c]} \neq \mathbb{1}$. Then $c^\Phi = Z(\Delta^1)$ is Δ-invariant, since Δ^1 is normal. Therefore, $c^\Phi = c^\Delta$ and hence $\Delta = \Delta_c \Phi = \Delta_{[c]} \Phi$, see (91.2a).

6) It suffices to prove (b) for groups $\Delta \leq \Sigma_{[A]}$, because for arbitrary groups $\Omega \leq \Sigma_A$ and $\Delta := \Omega_{[A]}$ this implies the inequalities

$$\dim Z(\Omega) = \dim Z(\Delta) = \dim \Delta_{[A,A]} = \dim c^\Delta \leq \dim c^\Omega \leq \dim Z(\Omega).$$

If Δ^1 contains non-trivial homologies, then $Z(\Delta) = c^\Phi = c^\Delta$ is homogeneous by (a) and step 5). Thus, we may now assume that $\Delta^1 = \Phi$, and then Φ is an open subgroup of the Lie group Δ. Since Δ has a countable basis, it follows that $\Delta \setminus \Delta_{[A,A]}$ is a countable union of cosets $\delta_i \Phi$, where $\delta_i \in \Delta_{[d_i]}$ and $d_i \notin A$. Then the coset lemma (61.17) shows that $Z(\Delta)$ is the union of the closed orbits d_i^Φ, and the sum theorem (92.9) implies that $\dim Z(\Delta) = \dim \Phi$. Finally, the inclusions $c^\Phi \subseteq c^\Delta \subseteq Z(\Delta)$ show that $\dim c^\Delta = \dim \Phi$. □

We use (55.42) in order to obtain another variation on the theme 'elations generated by homologies', compare (61.20). For details on the notion of *action of a group on a plane* appearing here, compare (71.5 and 6).

61.22 Proposition: A criterion for translation planes. *Let $\mathcal{P}$ be a $2l$-dimensional compact plane admitting an effective action $\Omega \to \operatorname{Aut} \mathcal{P}$ of a locally compact group $\Omega = \Delta \cdot \langle \sigma \rangle$ of the following kind: The group Δ is connected and of dimension at least $2l$, and σ is an involution inducing inversion on Δ, that is, $\delta^\sigma = \delta^{-1}$ for each $\delta \in \Delta$.*

Then, up to duality, $\mathcal{P}$ is a translation plane with Δ acting as the translation group, and σ is a reflection having the same axis as the elements of Δ.

It follows that $\Delta \cong \mathbb{R}^{2l}$, see (61.6) or (64.4). Every compact, connected translation plane admits a group Ω as described here.

Proof. 1) It suffices to prove the proposition in the case where Δ is a closed subgroup of $\mathrm{Aut}\,\mathcal{P}$. Then the general assertion follows, since we may first replace Δ by its image in $\mathrm{Aut}\,\mathcal{P}$ and then pass to the closure. This does not affect the hypotheses about the structure and the dimension of Δ (note that Δ contains compact subsets of dimension equal to $\dim \Delta$, and apply the subspace theorem (92.4)). It turns out finally that the image of the original group is a locally compact, $2l$-dimensional subgroup of the translation group $\mathbb{R}^{2l}$, hence the two groups are the same by (93.12).

2) The hypothesis implies that inversion is an automorphism of Δ, hence Δ is an abelian normal subgroup of Ω. By (55.42), Δ is a Lie group of dimension $2l$. The involution σ acts either as a reflection or as a Baer involution (55.29), and we shall first exclude the latter possibility.

3) Suppose that σ is a Baer involution with fixed point set F. By (55.42), some point $x \in F$ is moved by Δ; moreover, either the orbit x^Δ is open in P, or else it is contained in a unique line $L = L_x$ fixed by Δ. If x has an open orbit, then we claim that x is an isolated fixed point of σ; this would be a contradiction to the assumption that F is a subplane. We have $x^{\delta\sigma} = x^{\sigma\delta\sigma} = x^{\delta^{-1}}$, and Δ acts freely on the open set x^Δ. Hence every fixed point $x^{\delta^{-1}} = x^\delta$ arises from an involution δ in the Lie group Δ. Now the unit element in a Lie group is not a limit of involutions. Indeed, as a consequence of Ado's theorem (94.6), the group Δ admits a covering homomorphism onto a linear Lie group G, and every limit of a sequence of involutions in G inherits their eigenvalue -1. Our claim that x is an isolated fixed point now follows.

We conclude that each point $y \in F$ that is moved by Δ lies on a unique line L_y fixed by Δ; consequently, σ fixes L_y. The stabilizer $\Delta_{x,y}$ of two such points acts trivially on the set $x^\Delta \cup y^\Delta$, which contains a quadrangle. In other words, $\Delta_{x,y}$ is a planar group, and hence is trivial because Δ does not leave any proper closed subplane invariant (55.42). Using the dimension formula (96.10) together with (53.1a), we infer that x^Δ is l-dimensional and hence open in L_x. As before, the action of σ on x^Δ may be described as the action of σ on the factor group Δ/Δ_x by conjugation, that is, by inversion. Hence, x is an isolated fixed point of σ in L_x, contrary to the observation we made that L_x belongs to the Baer subplane of fixed elements of σ. This proves that σ is indeed a reflection.

4) By (55.42), Δ fixes some element, for example, a point p. If p is moved by σ, then Ω fixes the line $p^\sigma p$. Hence, we may assume, up to duality, that Ω fixes

some line L. If L happens to be the axis of σ, then the entire group Ω has axis L, since it is generated by the conjugates of σ; note that

(∗) $$\sigma\delta^{-1}\sigma\delta = \delta^2.$$

If Δ contains non-trivial homologies, then Δ fixes their centers, and $\dim \Delta \leq l$ by (61.4a). By (61.4b), it follows that $\Delta = \Delta_{[L,L]}$ is transitive on $P \setminus L$.

There remains the possibility that L contains the center c of σ. We claim that then c is fixed by Δ, so that we have the dual of the situation treated previously. Consider $\delta \in \Delta_c$ and $K \in \mathcal{L}_c$. Equation (∗) shows that δ^2 fixes K, so that δ^2 has center c. Therefore, the abelian group Δ fixes c. □

Groups containing axial collineations

61.23 Lemma: Abelian groups containing axial collineations. *Let Θ be an abelian group of automorphisms of a compact, connected, projective plane $\mathcal{P}$, and let E_Θ and H_Θ respectively denote the sets of all elations and of all homologies in Θ. Then the following assertions hold.*

(a) *Up to duality, all elements of E_Θ have the same axis W. Consequently, E_Θ is a subgroup of Θ. If a subgroup $\Delta \leq \mathrm{Aut}\,\mathcal{P}$ normalizes Θ, then it normalizes both E_Θ and the identity component of E_Θ, hence it fixes W (unless $E_\Theta = \mathbb{1}$). If Δ is connected, but the identity component of E_Θ is trivial, then Δ centralizes E_Θ.*
(b) *The centers and axes of elements of H_Θ form part of a non-degenerate triangle. If a connected subgroup $\Delta \leq \mathrm{Aut}\,\mathcal{P}$ normalizes Θ, then it fixes some anti-flag (unless $H_\Theta = \mathbb{1}$). If Θ contains non-trivial homologies with different centers, then Δ fixes a triangle pointwise.*

Proof. Assume that $\alpha, \beta \in \Theta$ have different centers a and b. Since the collineations commute, the center of α lies on the axis of β, and vice versa, see (23.12). If α and β are elations, then the axes coincide. The rest of assertion (a) follows easily. If α and β are homologies, then their axes are different, and we obtain that (a, b, c) is a non-degenerate triangle, where c is the intersection of the axes of α and β. For each element $\gamma \in H_\Theta$ with center $z \notin \{a, b\}$ we conclude that $z = c$. Finally, the connected group Δ acts trivially on the discrete set of centers and axes of homologies in Θ. □

61.24 Corollary: Minimal normal subgroups. *Assume that Θ is minimal among the connected abelian normal subgroups of a connected group $\Delta \leq \mathrm{Aut}\,\mathcal{P}$. If Θ contains a non-trivial elation, then either $\Theta = E_\Theta \leq \Delta_{[W,W]}$ or E_Θ is a totally disconnected subgroup of the center of Δ.*

Proof. The connected component of the group E_Θ is normal in Δ. By minimality, this component either is trivial or coincides with Θ. In the first case, E_Θ lies in the center of the connected group Δ, see (93.18). □

We add another result which is also designed to be applied to minimal connected abelian normal subgroups of Aut $\mathcal{P}$. As a by-product, this yields a second, independent proof of Proposition (61.8) on axial one-parameter groups.

61.25 Proposition: Connected, abelian groups of axial collineations. *Let $\Delta \leq$ Aut $\mathcal{P}$ be a connected abelian subgroup entirely consisting of axial collineations. Then either $\Delta = \Delta_{[x,x]}$ is an elation group with center or axis $x \in P \cup \mathcal{L}$, or $\Delta = \Delta_{[c,A]}$ is a homology group with center c and axis A.*

Proof. By (61.23a), we may assume that Δ contains non-trivial homologies with axis A and center c, say. We may also assume that Δ is closed in Aut $\mathcal{P}$ and, hence, is locally compact. We claim that there are at most 4 pairs $(p, L) \in P \times \mathcal{L}$ such that $\Delta_{[p,L]} \neq 1$. By (61.23b), we know that there are at most 3 antiflags of this kind. If (p, L) is a flag corresponding to a non-trivial elation, then $p \in A$ and $c \in L$ by (23.12). If (p', L') is another such flag, then we have $p = p'$ or $L = L'$ because the elations commute, and the conditions $p, p' \in A$ and $c \in L, L'$ imply that $(p, L) = (p', L')$. This verifies our claim. We infer that Δ is the union of at most 4 closed subgroups of the form $\Delta_{[p,L]}$. By Baire's category theorem, one of these subgroups contains an open set. This subgroup is both open and closed in the connected group Δ, and hence coincides with it. □

61.26 Lemma: Compact axial groups. *Let $\Delta \leq \Sigma_{[A]}$ be a compact group of axial collineations with a fixed axis. If Δ is not discrete, then there exists a point $c \notin A$ such that $\Delta = \Delta_{[c,A]}$. Consequently, Δ acts freely on each non-trivial orbit, and $\dim \Delta < l = \dim L$.*

Proof. The connected component $\Delta^1 \neq 1$ is compact and normal in Δ, and $\dim \Delta^1 = \dim \Delta$ by (93.6). Hence we may assume that $\Delta = \Delta^1$, compare (23.12). From (61.5) and (93.18), we infer that $\Delta_{[A,A]}$ is totally disconnected and central in Δ. Therefore, Δ contains non-trivial homologies, and we see that in fact $\Delta_{[A,A]} = 1$. If the center $C(\Delta)$ contains a homology $\delta \neq 1$, then Δ fixes the center c of δ. Hence, we may assume that $C(\Delta) = 1$, and then Δ is a Lie group by (93.9); observe that $\dim \Delta$ is finite by (61.4) and the dimension formula (96.10). Consequently, there is an involution $\varrho \in \Delta$ with center $c \notin A$, see (94.31a). If Δ does not fix c, then (23.12 and 20) together show that Δ contains non-trivial elations, a contradiction. The last assertion follows, because an l-dimensional orbit would coincide with the line containing it, see (53.1c), and then the center would be moved. □

61.27 Further results on axial collineations can be found in the following places: (55.28 to 35), (44.7 and 8) and Section 64.

Homologies of translation planes are described by the kernel and the nuclei of a coordinatizing quasifield in a manner analogous to (61.1b), see (25.4). For 8-dimensional quasifields, an investigation of the possibilities for these substructures is given in Buchanan–Hähl [77]. Among other things, they prove that in the special case of 8-dimensional semifields, each of these structures is isomorphic to $\mathbb{R}$. This generalizes (11.20). On the other hand, there exist 4-dimensional semifields and nearfields with kernels isomorphic to $\mathbb{C}$, see Rees [50], Knarr [95], (82.1 and 16), (64.19).

Planes where all groups $\Sigma_{[c]}$ are non-trivial are treated by Salzmann [58], [62a] Section 5, Strambach [70c], Löwen [81a, 84b] and Seidel [90b]. For 4-dimensional planes containing a non-empty open set of reflection centers, see Löwen [81c].

61.28 Notes on Section 61. Corollary (61.13) on homogeneous translation groups is analogous to a theorem of Gleason [56] 1.6 for finite planes. For infinite planes in general, this theorem is not true as was shown by André [58]. The first topological version was given by Salzmann [70] 4.6 for 4-dimensional planes. There, he also introduced the projective center map (61.10). Corollary (61.13) was proved directly by Salzmann [85], and Theorem (61.12) on almost homogeneous translation groups is due to Salzmann [90] (G). The analogue of that theorem for finite planes is proved in Hughes–Piper [73] 4.26.

Theorem (61.20) on homologies and elations also has an analogue in finite planes, proved by André [54a], Satz 3. For topological planes, similar results appeared first in Salzmann [58, 62a], [67b] Section 6, and in a more developed form in Salzmann [73a]. The most general form of the theorem is due to Hähl [81], who also proved (61.17). The version given here is a simplification of his comprehensive results, cf. (61.21).

62 Planes admitting a classical motion group

We shall consider three types of motion groups (named the elliptic, hyperbolic and Euclidean motion groups) for each of the four classical planes $\mathcal{P}_2\mathbb{F}$, where $\mathbb{F} \in \{\mathbb{R}, \mathbb{C}, \mathbb{H}, \mathbb{O}\}$. Our aim is, roughly speaking, to show that given any non-trivial action of one of these 'classical motion groups' on any plane of the appropriate dimension, both the plane and the action have to be the classical ones. As always in similar situations, the idea of the proof is to gather enough information about the action until the geometry can be reconstructed from the group. It may be helpful to recall a simple instance of that procedure (32.20) before reading the present section, where the actual reconstruction takes place right at the end, starting with (62.13 and 14).

A priori, it is not clear how the term 'classical motion group' should be defined. We decided to give a definition simply by enumerating three groups for each of the four classical planes. Our motive for selecting these particular groups is the fact that each of them has a unique open point orbit on which it acts in a flag transitive way. In fact, the groups that we consider are precisely the minimal subgroups having this property in the automorphism groups of the classical planes. This is a consequence of a result of Löwen [83a] 1.4, where it is also shown that the flag homogeneous stable planes are precisely the 12 flag homogeneous orbits just described. Compare also the remarks on isotropic points below.

There are other good reasons supporting our choice of groups to be considered. In Chapter 1, we defined the elliptic and hyperbolic motion groups as centralizers of certain polarities of the classical planes, and we studied their actions extensively. In fact, we determined all (continuous) polarities of the classical planes, see (13.18) and (18.29 and 31), and so it would be natural to consider all centralizers of those polarities as 'classical motion groups'. However, mainly the elliptic and hyperbolic groups among them have received much attention in the past, together with the Euclidean groups, which are not obtained from polarities.

Moreover, the remaining three centralizers of the standard planar polarities of $\mathcal{P}_2\mathbb{F}$, where $\mathbb{F} \in \{\mathbb{C}, \mathbb{H}, \mathbb{O}\}$, have much smaller dimensions. These centralizers are isomorphic respectively to $PO_4(\mathbb{R}, 1)$, $PU_4(\mathbb{C}, 1)$ and $PU_4(\mathbb{H}, 1)$, compare (18.32), and their dimensions are 6, 15 and 36, respectively. The corresponding elliptic and hyperbolic motion groups are of dimension 8, 21 and 52, and the Euclidean motion groups have dimension 7, 18 and 52. Nevertheless, results similar to those obtained in the present section have been proved for the centralizers of the planar polarities, see (72.4), Stroppel [91, 94d] and Priwitzer [94b].

As a preparation for our proof we consider what we call isotropic points, that is, points p such that there is a group of collineations acting transitively on the pencil $\mathcal{L}_p$. It turns out that p is isotropic if, and only if, $\mathrm{Spin}_{l+1}\mathbb{R}$ acts as a group of collineations fixing p, and that such an action is topologically equivalent to the one that appears within the classical motion group. Here, as always, l denotes the dimension of a line in a given compact, connected plane $\mathcal{P}$. These results are then applied in order to show that every action of a classical motion group has a flag homogeneous open orbit of points or lines.

The theorems obtained in the present section will be used in the subsequent ones, where we consider groups satisfying various transitivity assumptions. The case $l = 1$ is special in many respects (e.g., the universal covering of $SO_2\mathbb{R}$ is the non-compact group $\mathbb{R}$) and will be treated separately.

62 Planes admitting a classical motion group

62.1 Definition: Motion groups of the real plane. We shall consider the groups

$$\Phi_1^+ := \mathrm{PO}_3\mathbb{R} \cong \mathrm{SO}_3\mathbb{R} \quad \textit{elliptic motion group}$$
$$\Phi_1^- := \mathrm{PO}_3(\mathbb{R}, 1) \cong \mathrm{PGL}_2\mathbb{R} \quad \textit{hyperbolic motion group}$$
$$\Phi_1^0 := \mathbb{R}^2 \cdot \mathrm{O}_2\mathbb{R} \quad \textit{Euclidean motion group}$$
$$\Phi_1^\flat := \mathbb{R}^2 \cdot \langle -\mathbb{1} \rangle \leq \Phi_1^0.$$

The subscript 1 stands for $l = 1$; groups Φ_l for $l > 1$ will be defined in (62.4). The connected component of the hyperbolic group (the group $\mathrm{PO}_3'(\mathbb{R}, 1)$ of *direct hyperbolic motions*) is isomorphic to $\mathrm{PSL}_2\mathbb{R}$, see (15.6). The Euclidean group is a semidirect product determined by the natural action of $\mathrm{O}_2\mathbb{R}$ on $\mathbb{R}^2$. Its subgroup $\Phi_1^\flat$ is generated by all central affine reflections.

If $\Phi^\sharp$ is any one of the preceding groups, then, apart from the classical action $(\mathcal{P}_2\mathbb{R}, \Phi^\sharp)$, there is the dual action $(\mathcal{P}_2^*\mathbb{R}, \Phi^\sharp)$ of $\Phi^\sharp$ on the dual plane. In the elliptic and hyperbolic cases, the two actions are isomorphic since $\Phi^\sharp$ is the centralizer of a polarity, but the Euclidean action differs from its dual.

In the next theorem, and in the remainder of this section, we use the notion of *action of a group on a plane* in an informal way. For a formal definition and properties of actions, see (71.5 and 6).

62.2 Theorem: Real motion groups. *Assume that a group Φ isomorphic to $\Phi_1^\sharp \in \{\Phi_1^+, \Phi_1^-, \Phi_1^0, \Phi_1^\flat\}$ acts effectively as a group of automorphisms on some 2-dimensional, compact plane $\mathcal{P}$. Then $(\mathcal{P}, \Phi)$ is isomorphic to $(\mathcal{P}_2\mathbb{R}, \Phi_1^\sharp)$ or to its dual $(\mathcal{P}_2^*\mathbb{R}, \Phi_1^\sharp)$, i.e., both the plane and the action are classical.*

62.3 Remark. The theorem does not extend to groups *locally* isomorphic to $\Phi_1^\sharp$. The group of direct hyperbolic motions can also act on the skew hyperbolic planes found by Salzmann, see (35.1, 2 and 4). Its universal covering group acts effectively on the Moulton planes (34.6 and 8a), and the twofold covering group $\mathrm{SL}_2\mathbb{R}$ has its natural effective affine action on $\mathcal{P}_2\mathbb{R}$. On the other hand, no proper covering group of Φ_1^0 can act effectively on any 2-dimensional plane, see Löwen [86b] p. 122. For finite coverings, this will be clear from our proof, which uses the subgroup $\Phi_1^\flat$ only.

In contrast, for $l > 1$, it will be sufficient to assume that Φ is a group locally isomorphic to $\Phi_l^\sharp$ as defined in (62.4) in order to obtain a result analogous to (62.2), see (62.6). However, $\Phi_l^\flat$ will not be considered since it can act on any translation plane, compare (61.22), and translation planes abound for $l > 1$.

Proof of (62.2). 1) The elliptic case has been treated in (32.20), and the hyperbolic case in (35.3); note that by (35.4) the skew hyperbolic planes admit an action of the direct hyperbolic motion group $\mathrm{PO}_3'(\mathbb{R}, 1)$, but not of the full group Φ_1^-. Uniqueness of the actions is obtained in the proofs of the theorems quoted. Alternatively, one can show using (13.6) that every action of Φ on the real plane comes from a representation on $\mathbb{R}^3$. Representation theory then shows that the action of the connected component is unique (95.10). The disconnected group Φ_1^- is also determined, being the normalizer in $\operatorname{Aut} \mathcal{P}_2 \mathbb{R}$ of its connected component.

2) The group $\Phi_1^\flat$ has the structure considered in the criterion (61.22) for translation planes, i.e., it is a product $\Lambda \cdot \langle \omega \rangle$, where $\Lambda \cong \mathbb{R}^2$ and ω is an involution acting on Λ by inversion. (By (32.12), we know that ω acts on $\mathcal{P}$ as a reflection, and the proof of (61.22) may be simplified if this is known.) The criterion shows that, up to duality, $\mathcal{P}$ is a translation plane with translation group Λ. By (32.8), it follows that $\mathcal{P}$ is isomorphic to $\mathcal{P}_2 \mathbb{R}$, and Φ is uniquely determined.

3) The group Φ_1^0 contains $\Phi_1^\flat$ as a normal subgroup. This shows that any plane admitting Φ_1^0 is isomorphic to $\mathcal{P}_2 \mathbb{R}$. Up to duality, Φ_1^0 fixes the unique fixed line W of $\Phi_1^\flat$. Now let $\Sigma = \operatorname{Aut} \mathcal{P}_2 \mathbb{R}$; then $\Sigma_W = \mathbb{R}^2 \cdot \operatorname{GL}_2 \mathbb{R}$ by (12.11), and $\Phi_1^\flat$ is determined as a subgroup of Σ_W according to step 2). The subgroup $\operatorname{SO}_2 \mathbb{R} \leq \Phi_1^0$ is a maximal compact, connected subgroup of Σ_W and hence is determined up to conjugacy in Σ_W, see (93.10a). Finally, $\Phi_1^0 \leq \Sigma_W$ is determined as the only $\mathbb{Z}_2$-extension of its connected component $\mathbb{R}^2 \cdot \operatorname{SO}_2 \mathbb{R}$. □

62.4 Definition: Classical motion groups for $l \geq 2$. The *elliptic* and *hyperbolic motion groups* of the classical $2l$-dimensional planes have been defined in Chapter 1, see (13.13) and Section 18. We shall denote them by Φ_l^+ and Φ_l^-, respectively. By (13.14) and (18.8), the point stabilizer of the elliptic group is $U_2 \mathbb{C}$ or $U_2 \mathbb{H} \cdot \operatorname{Spin}_3 \mathbb{R}$ or $\operatorname{Spin}_9 \mathbb{R}$ according as $l = 2$, 4 or 8. By (11.26), the group $U_2 \mathbb{C} = \operatorname{SU}_2 \mathbb{C} \cdot \mathbb{S}_1$ is isomorphic to $\operatorname{Spin}_3 \mathbb{R} \cdot \operatorname{SO}_2 \mathbb{R}$. Similarly, $U_2 \mathbb{H} \cdot \operatorname{Spin}_3 \mathbb{R}$ is isomorphic to $\operatorname{Spin}_5 \mathbb{R} \cdot \operatorname{Spin}_3 \mathbb{R}$, see (18.9). In both cases, the products are obtained from the direct products by identifying the central involutions of the two factors. The 'interior' orbit of the hyperbolic group has the same point stabilizer by (13.14) and (18.4).

Each of the point stabilizers just described contains a unique subgroup Ψ isomorphic to $\operatorname{Spin}_{l+1} \mathbb{R}$, the universal cover of $\operatorname{SO}_{l+1} \mathbb{R}$. Its dimension is $\binom{l+1}{2}$, and Ψ fixes some point and its polar W. We use this group Ψ to define the *Euclidean motion group* as the semi-direct product $\mathbb{F}^2 \cdot \Psi$ of the affine translation group $\mathbb{F}^2 = \Sigma_{[W,W]}$ by Ψ.

Our definitions are summarized by the following table.

	elliptic Φ_l^+	hyperbolic Φ_l^-	Euclidean Φ_l^0
$l = 2$	$\operatorname{PU}_3 \mathbb{C}$	$\operatorname{PU}_3(\mathbb{C}, 1)$	$\mathbb{C}^2 \cdot \operatorname{SU}_2 \mathbb{C}$
$l = 4$	$\operatorname{PU}_3 \mathbb{H}$	$\operatorname{PU}_3(\mathbb{H}, 1)$	$\mathbb{H}^2 \cdot \operatorname{U}_2 \mathbb{H}$
$l = 8$	$F_4(-52)$	$F_4(-20)$	$\mathbb{O}^2 \cdot \operatorname{Spin}_9 \mathbb{R}$

62.5 Remarks. We assume throughout that $l \geq 2$.

(a) The classical motion groups are all connected. For the Euclidean groups, this follows from the connectedness of $\mathrm{Spin}_{l+1}\mathbb{R}$. In the other cases, it is proved in (18.15 and 26) for $l = 8$, and it can be found in Porteous [81] 20.83 for $l < 8$.

(b) All point stabilizers of the classical motion groups are also connected. To see this, it suffices to show that all orbits are simply connected (96.9c). This can be inferred from the results of Chapters 1 and 5. We shall not need this fact, except for one particular case, see (62.16).

(c) Each classical motion group has a special point orbit X_l such that the stabilizer of each point $x \in X_l$ contains a group $\Psi \cong \mathrm{Spin}_{l+1}\mathbb{R}$. The special orbits are the following: the entire point set $P_2\mathbb{F}$ in the elliptic case, the affine point set in the Euclidean case and the interior point orbit in the hyperbolic case. The orbit X_l and the group Ψ will be key objects in the proofs in this section. In fact this motivates our particular definition of the Euclidean groups. The point stabilizer on X_l has been described in (62.4).

(d) The classical motion groups are all center-free, and the Euclidean groups are homeomorphic to $\mathbb{R}^{2l} \times \mathrm{Spin}_{l+1}\mathbb{R}$ and hence are simply connected. Thus, any group Φ locally isomorphic but not isomorphic to one of the classical motion groups is a covering group of $\Phi_l^\pm$. The center of the universal covering is $\mathbb{Z}_3$, $\mathbb{Z}$, or $\mathbb{Z}_2$, respectively, for Φ_2^+, Φ_2^- or $\Phi_4^\pm$, while $\Phi_8^\pm$ is itself simply connected. Compare (94.33 and 2) and (18.15 and 26); see also Tits [67] p. 28.

(e) As in the real case (62.1), the elliptic and hyperbolic actions are isomorphic to the corresponding dual actions, whereas the Euclidean groups have two different actions.

62.6 Theorem: Classical motion groups. *For $l \in \{2, 4, 8\}$, assume that a group Φ locally isomorphic to $\Phi_l^\sharp \in \{\Phi_l^+, \Phi_l^-, \Phi_l^0\}$ acts effectively as a group of automorphisms on some compact projective plane $\mathcal{P}$ of dimension $\leq 2l$. Then $(\mathcal{P}, \Phi)$ is isomorphic to $(\mathcal{P}_l, \Phi_l^\sharp)$ or to its dual $(\mathcal{P}_l^*, \Phi_l^\sharp)$, where $\mathcal{P}_l$ denotes the classical $2l$-dimensional plane. In other words, both the plane and the action are classical.*

For the notion of an action, compare also (71.5 and 6). The remainder of the present section will be devoted to the proof of Theorem (62.6).

Actions of $\mathrm{Spin}_{l+1}\mathbb{R}$ and isotropic points

62.7 Definition. Let $\mathcal{P}$ be a plane with full automorphism group Σ. A point x is said to be *isotropic* if the stabilizer Σ_x is transitive on the line pencil $\mathcal{L}_x$. This should not be confused with the notion of an isotropic point with respect to a polarity, called an absolute point in this book.

Roughly speaking, a point in a $2l$-dimensional plane is isotropic if, and only if, it is fixed by some group isomorphic to $\mathrm{Spin}_{l+1}\mathbb{R}$. More precisely, we have

62.8 Theorem: Isotropic points. *Let x be a point of a $2l$-dimensional, compact, projective plane.*

(a) *If $2 \leq l \leq \infty$ and x is isotropic, then $l < \infty$ and Σ_x contains a subgroup $\Psi \cong \mathrm{Spin}_{l+1}\mathbb{R}$. In fact, every closed subgroup $\Delta \leq \Sigma_x$ that is transitive on $\mathcal{L}_x$ contains such a subgroup Ψ.*

(b) *Conversely, if $1 \leq l < \infty$ and Σ_x contains a compact, connected subgroup Ψ locally isomorphic to $\mathrm{SO}_{s+1}\mathbb{R}$, where $s \geq l$, then Ψ is transitive on $\mathcal{L}_x$, hence x is isotropic. Moreover, we have $s = l$, and Ψ is isomorphic to $\mathrm{SO}_2\mathbb{R}$ if $l = 1$ and to $\mathrm{Spin}_{l+1}\mathbb{R}$ in all other cases.*

(c) *Likewise, if $\Psi \leq \Sigma_x$ is a compact, connected, almost simple subgroup and $\dim \Psi > \binom{l}{2}$, then $\Psi \cong \mathrm{Spin}_{l+1}\mathbb{R}$ and Ψ is transitive on $\mathcal{L}_x$.*

Proof. 1) Suppose first that $\infty \neq l \geq 2$ and that a group $\Psi \cong \mathrm{SO}_{l+1}\mathbb{R}$ of automorphisms acts on $\mathcal{L}_x \approx \mathbb{S}_l$ by the standard orthogonal action. We show that this is impossible. Indeed, there would be commuting involutions $\sigma, \tau \in \Psi$ such that σ fixes precisely two lines $K, L \in \mathcal{L}_x$ and τ interchanges those lines. Moreover, σ cannot be planar, or else there would be more fixed lines in $\mathcal{L}_x$. By (23.17), the involution σ has axis K or L, and τ has to fix the axis (23.12), a contradiction.

2) Assume that $l \geq 2$ and that Δ is transitive on $\mathcal{L}_x$. By (44.3) and (96.2), the effective quotient group $\Omega = \Delta / \Delta_{[x]}$ is a locally compact transformation group on $\mathcal{L}_x$ and has a countable basis. Since $\mathcal{L}_x$ is a locally contractible space, see (51.4), Szenthe's theorem (96.14) asserts that Ω is a Lie group and $\mathcal{L}_x$ is a manifold. In particular, we have $l \in \{2, 4, 8\}$ and $\mathcal{L}_x \approx \mathbb{S}_l$ by (52.5 and 3). Moreover, the connected component Ω^1 is transitive on $\mathcal{L}_x$ by (96.9b). Any maximal compact subgroup $\Lambda \leq \Omega^1$ is transitive (96.19) and connected (93.10). Moreover, Λ is simple since l is even (96.21), and $\Lambda \cong \mathrm{SO}_{l+1}\mathbb{R}$ acts on $\mathcal{L}_x \approx \mathbb{S}_l$ in the standard way according to (96.22 or 23). By (94.27), there is a connected subgroup $\Psi \leq \Sigma_x$ that is locally isomorphic to Λ (and maps onto Λ). We have $\pi_1(\mathrm{SO}_{l+1}\mathbb{R}) = \mathbb{Z}_2$, cf. (94.33), hence the only possibilities are $\Psi = \Lambda \cong \mathrm{SO}_{l+1}\mathbb{R}$ and $\Psi \cong \mathrm{Spin}_{l+1}\mathbb{R}$. The first of these possibilities has been ruled out in step 1).

3) We assume conversely that $\Psi \leq \Sigma_x$ satisfies the hypotheses of (b) or of (c), and we show that Ψ acts transitively on $\mathcal{L}_x$ and has a 0-dimensional kernel $\Psi_{[x]}$. At first suppose that Ψ is not locally isomorphic to $\mathrm{SO}_4\mathbb{R}$. Thus, Ψ is a compact group without any proper, connected, closed, normal subgroups, and $\dim \Psi > \binom{l}{2} = \dim \mathrm{SO}_l\mathbb{R}$. Note that this is also true if $l = 1$ and $\Psi \cong \mathrm{SO}_2\mathbb{R}$. If Ψ has a non-trivial orbit $\mathcal{B} \subseteq \mathcal{L}_x$, then the kernel of the action on $\mathcal{B}$ must be 0-dimen-

sional. Using (96.13a), we infer that $\dim \mathcal{B} > l - 1$, so that Ψ is transitive on $\mathcal{L}_x$ by (53.1c). On the other hand, if $\Psi \leq \Sigma_{[x]}$, then Ψ acts trivially on at most one line $L \in \mathcal{L}_x$, and is transitive on all other ones by the argument used before. This is impossible since x is fixed. (We could have used the general result (61.26) here.)

If Ψ is locally isomorphic to $SO_4\mathbb{R}$, then $\Psi = \Psi_1 \cdot \Psi_2$ with commuting factors Ψ_i locally isomorphic to $SO_3\mathbb{R}$, see (11.23). If $l \leq 2$, then our previous arguments show that neither factor can act trivially on $\mathcal{L}_x$, and that both are in fact transitive. This contradicts the last statement in (96.21).

4) Let Λ be the group induced on $\mathcal{L}_x$ by Ψ. As in step 2), we see that $\Lambda \cong SO_{l+1}\mathbb{R}$, with standard action on $\mathcal{L}_x \approx \mathbb{S}_l$. This is also true for $l = 1$, see (96.13a) or (96.30a). We claim now that Ψ is a Lie group. This will imply that Ψ is a covering group of Λ, so that $\Psi \cong \mathrm{Spin}_{l+1}\mathbb{R}$ by step 1).

In part (b) of the assertion, Ψ is assumed to be locally isomorphic to $SO_{l+1}\mathbb{R}$, and then there is nothing to prove. Regarding part (c), we show that Ψ cannot be an almost simple non-Lie group. If this were the case, then $\Psi_{[x]}$ would be an infinite 0-dimensional group contained in the center of Ψ, compare (93.8b). There would be subgroups N of this group having a finite index $k > 2$. Then the factor group Ψ/N is a Lie group locally isomorphic to Λ and has a center of order k, contrary to the fact that the fundamental group of Λ has order 2. (In fact, we shall show in (62.11) that existence of an isotropic point forces Σ to be a Lie group; the proof of that result will be independent of (62.8c), so that it provides an alternative to our last arguments.) □

62.9 Theorem: $\mathrm{Spin}_{l+1}\mathbb{R}$-actions. *Assume that $l \geq 2$ and that $\Psi \cong \mathrm{Spin}_{l+1}\mathbb{R}$ acts effectively on a compact, projective plane $\mathcal{P} = (P, \mathcal{L})$ of dimension $2r \leq 2l$. Then we have $r = l$, and the action of Ψ on the point set P is equivalent to the action of Ψ on the point set $P_2\mathbb{F}$ of the classical $2l$-dimensional plane described in Chapter 1; compare (62.4).*

We list some properties of the action of Ψ explicitly. Technically, their proof will precede that of (62.9).

62.10 Corollary. *In the situation of (62.9), the group $\Psi \cong \mathrm{Spin}_{l+1}\mathbb{R}$ fixes some antiflag $x \notin W$ in the given plane $\mathcal{P}$. Moreover, the following are true:*

(a) *The center $C(\Psi) \cong \mathbb{Z}_2$ is generated by a reflection $\sigma_x \in \Psi_{[x,W]}$.*
(b) *$\Psi/C(\Psi) \cong SO_{l+1}\mathbb{R}$ acts on the l-spheres W and $\mathcal{L}_x$ in the standard way.*
(c) *If $L = x \vee y$ and $y \notin W$, then $y^\Psi \approx \mathbb{S}_{2l-1}$ and $y^{\Psi_L} \approx \mathbb{S}_{l-1}$; moreover, Ψ_L fixes a triangle with sides K, L, W such that $x = K \wedge L$.*
(d) *If $l > 2$, then $\Psi_y \cong \mathrm{Spin}_{l-1}\mathbb{R}$ and $\Psi_L \cong \mathrm{Spin}_l\mathbb{R}$; moreover, there is a unique reflection $\sigma_L \in \Psi_{[L]}$, and $C(\Psi_L) = \{1, \sigma_x, \sigma_L, \sigma_K = \sigma_x\sigma_L\} \cong \mathbb{Z}_2{}^2$.*

(e) *If $l > 2$, then an involution $\tau \in \Psi$ is conjugate to σ_L if, and only if $\mathrm{Cs}_\Psi(\tau) \cong \mathrm{Spin}_l \mathbb{R}$.*
(f) *If $l > 2$, then Ψ_y is determined up to conjugacy in $\mathrm{Ns}_\Psi \Psi_L$ by the properties that $\sigma_x \notin \Psi_y$ and $\Psi_y \cong \mathrm{Spin}_{l-1}\mathbb{R}$; on the other hand, $\Psi_y = \mathbb{1}$ for $l = 2$.*
(g) *The space P/Ψ of orbits is homeomorphic to the closed interval $[0, 1]$.*

Remark. In 2-dimensional planes, the theorem is false for $\Psi = \widetilde{\mathrm{SO}_2\mathbb{R}} = \mathbb{R}$, but it holds for $\Psi = \mathrm{SO}_2\mathbb{R}$ by a similar, but simpler argument. For $\mathrm{SO}_3\mathbb{R}$ acting on 4-dimensional planes, an analogous result will be proved (71.10); these actions are fixed-point free.

Proof of (62.9 and 10). 1) If an involution $\sigma \in C(\Psi)$ is axial, then Ψ fixes its center x and axis W, which form an antiflag by (55.28). We claim that this is the only possibility for σ, which in view of (23.17) amounts to claiming that σ cannot be a Baer involution. Indeed, if this happens, then we have $r \geq 4$ by (55.21c), so that Ψ is almost simple. Ψ acts on the fixed point set F_σ, which is an r-dimensional plane by (55.5). The action is non-trivial (and hence almost effective), or else we obtain non-isotropic fixed elements of Ψ contrary to (62.8). We apply (96.13a) to the action on a non-trivial orbit in F_σ, like in step 3) of the last proof; this shows that $r = l$ and Ψ is transitive on F_σ. Since F_σ is simply connected (51.28b), a point $p \in F_\sigma$ has a connected stabilizer Ψ_p by (96.9c). The dimension formula (96.10) shows that $\dim \Psi_p = \binom{l+1}{2} - l = \binom{l}{2}$. According to Montgomery–Samelson [43] Lemma 7, it follows that $\Psi_p/\langle\sigma\rangle$ is conjugate to $\mathrm{SO}_l\mathbb{R}$ in $\Psi/\langle\sigma\rangle = \mathrm{SO}_{l+1}\mathbb{R}$. Thus, F_σ is an l-sphere contrary to our results on the topology of projective planes; see (52.14), or (51.26) together with (52.5) and (54.11). This proves our claim.

2) By (62.8b), we obtain that $l = r$ and that Ψ is transitive on W and on $\mathcal{L}_x$. Thus, both spaces are manifolds, see (96.9a) and (94.3a), and are homeomorphic to $\mathbb{S}_l$ by (52.3) and its dual. By construction, $\sigma = \sigma_x$ acts trivially, and the action of $\Psi/\langle\sigma_x\rangle \cong \mathrm{SO}_{l+1}\mathbb{R}$ on these spheres is the standard action according to (96.22). We have $C(\Psi) = \langle\sigma_x\rangle$ because $l+1$ is odd, see (94.33). This completes the proof of assertions (a) and (b) of the corollary.

3) For $L \in \mathcal{L}_x$, we wish to determine the stabilizer Ψ_L. The exact homotopy sequence associated with the transitive action of Ψ on $\mathcal{L}_x \approx \mathbb{S}_l$, see (96.12), contains the section

$$\ldots \to \pi_2 \mathbb{S}_l \to \pi_1 \Psi_L \to \pi_1 \Psi \to \pi_1 \mathbb{S}_l \to \pi_0 \Psi_L \to \pi_0 \Psi.$$

Since both Ψ and $\mathbb{S}_l$ are connected and simply connected, the sequence shows that $\pi_0 \Psi_L$ is trivial, i.e., Ψ_L is connected; compare (96.9c). Moreover, since $\pi_2 \mathbb{S}_l = 0$ for $l > 2$, we obtain that Ψ_L is simply connected in this case. We know that $\Psi_L/\Psi_{[x]}$ is the stabilizer $\mathrm{SO}_l\mathbb{R}$ of the standard action of $\mathrm{SO}_{l+1}\mathbb{R}$ on $\mathbb{S}_l$, and we conclude that $\Psi_L \cong SO_2\mathbb{R}$ if $l = 2$ and $\Psi_L \cong Spin_l\mathbb{R}$ if $l > 2$. It follows from (b) that Ψ_L fixes a triangle as stated in (c). As in step 1) of this proof, Ψ_L is actually determined up to conjugacy in Ψ.

4) A point $y \in L$ as considered in (c), (d) and (f) is moved by σ_x, whence y^{Ψ_L} is a non-trivial orbit and has dimension $d \geq l - 1$ by (96.13a). The orbit is compact and is a proper subset of $L \approx \mathbb{S}_l$, therefore it is not open. Using (53.1a or c), we infer that $d < l$ and in fact $d = l - 1$. The dimension formula (96.10) shows that $\Psi_y = \Psi_{L,y}$ has codimension $l - 1$ in Ψ_L and Ψ_L has codimension l in Ψ, hence $\dim y^\Psi = 2l - 1$. Therefore, Théorème a of Poncet [59] can be applied to Ψ acting on $P \setminus W \approx \mathbb{R}^{2l}$ and to Ψ_L acting on $L \setminus W \approx \mathbb{R}^l$. It asserts that both actions are linear, so that in particular the orbits are spheres, and (c) is proved.

5) For $l > 2$ we have $C(\Psi_L) = C(Spin_l\mathbb{R}) = \mathbb{Z}_2^2$, see (94.33). Since $y^{\sigma_x} \neq y$, the stabilizer Ψ_y contains at most one of the central involutions of Ψ_L. We shall determine Ψ_y in the subsequent steps in order to prove (62.9), and it will turn out that there is an involution $\sigma_L \in \Psi_y \cap C(\Psi_L)$. Since Ψ_L is linear on L and transitive on the sphere $\mathbb{S}_{l-1} \subseteq L$, this implies that $\sigma_L \in \Psi_{[L]}$, and (d) will follow. Moreover, we obtain one direction of the equivalence (e), since $Cs_\Psi(\sigma_L) = \Psi_L$ is isomorphic to $Spin_l\mathbb{R}$. Conversely, by Montgomery–Samelson [43] Lemma 7, any subgroup isomorphic to $Spin_l\mathbb{R}$ in Ψ is conjugate to Ψ_L, and the two involutions σ_K, σ_L whose centralizer equals Ψ_L are conjugate by any element $\psi \in \Psi$ that interchanges K and L.

6) For $l = 2$ we claim that Ψ_L acts freely on $L \setminus W$; in other words, we claim that $\Psi_y = \mathbb{1}$. Indeed, Ψ_L is normalized by an element τ of order 4 which interchanges K and L and induces inversion on Ψ_L. Hence, $\Psi_y = \Psi_y{}^\tau$ fixes $y \in L$ and $y^\tau \in K$, and acts trivially on their orbits under the abelian group Ψ_L. By step 4), these orbits are 1-spheres, and the actions on K and L are linear. This implies our claim that $\Psi_y = \mathbb{1}$ and establishes (f) for $l = 2$.

7) For $l \geq 4$, the homotopy sequence of the transitive action of Ψ_L on $y^{\Psi_L} \approx \mathbb{S}_{l-1}$ shows as in 3) that Ψ_y is connected and simply connected. Moreover, we know that Ψ_y has codimension $l - 1$ in Ψ_L.

For $l = 4$, where $\Psi_L \cong Spin_4\mathbb{R} = Spin_3\mathbb{R} \times Spin_3\mathbb{R}$, the fact that $Spin_3\mathbb{R}$ has no 2-dimensional subgroups implies that the 3-dimensional group Ψ_y either is one of the factors or projects isomorphically onto both factors. The 'diagonal' groups of the latter kind contain an involution σ with factor group $\Psi_L/\langle\sigma\rangle \cong SO_4\mathbb{R}$. This property characterizes σ_x, compare 3), but σ_x does not fix y, hence Ψ_y must be a factor of Ψ_L. Then we have even $\Psi_y = \Psi_{[L]}$ and by symmetry the other factor of $\Psi_L = \Psi_K$ is $\Psi_{[K]}$. Since the normalizer of Ψ_L contains elements exchanging K and L, this proves (f) in the case under consideration.

If $l = 8$, then there are 3 conjugacy classes of subgroups of codimension $l - 1$ in $\Psi_L \cong Spin_8\mathbb{R}$, see (17.16). Each of the classes can be characterized as the set of those subgroups isomorphic to $\mathrm{Spin}_7\mathbb{R}$ that contain a particular one of the 3 central involutions of Ψ_L. To see this, consider the action of Ψ_L on the octonion plane and observe that the triality automorphism permutes both the central involutions and the classes of $\mathrm{Spin}_7\mathbb{R}$-subgroups cyclically. Hence, the proof of (f) can be completed in this case as in the previous one.

8) Finally we apply a theorem of Mostert [57] Section 5 on actions of compact groups possessing an orbit of codimension 1. According to that result, the action of Ψ on P can be reconstructed from its point stabilizers, as follows. There are two kinds of orbits: the 'non-singular' ones with minimal stabilizers, and at most two singular orbits. Here the singular orbits are $\{x\}$ and W. Since the non-singular orbits are of codimension one, there is a cross-section for all orbits; more precisely, since there are two singular orbits, we obtain an arc $A \approx [0, 1]$ in P meeting every orbit exactly once. Its end points are in the singular orbits. Moreover, the arc can be adjusted so that all its interior points have the same stabilizer Θ_1. The action of Ψ on P is determined up to isomorphism by the quadruple of groups $(\Psi, \Theta_1, \Theta_2, \Theta_3)$, where Θ_2 and Θ_3 are the stabilizers of the end points. In our case, one end point is x, so that $\Theta_2 = \Psi$. The arc is contained in one line $L \in \mathcal{L}_x$ since the stabilizer of an interior point $y \in A$ fixes precisely two such lines. Therefore, the other end point is $u = L \wedge W$, with stabilizer $\Theta_3 = \Psi_u = \Psi_L$. Above, we determined Ψ_L to within conjugacy in Ψ, and Ψ_y to within conjugacy in the normalizer of Ψ_L. This ends the proof of (62.9).

9) If an action of a compact group on a connected manifold has generic orbits of codimension 1 and two singular orbits, then the orbit space is homeomorphic to $[0, 1]$. This is the first major step in the proof of Mostert's theorem quoted above, see also Montgomery–Zippin [55] p. 252. Thus assertion (g) of (62.10) is proved. □

Suppose that $\mathcal{P}$ contains an isotropic point. Then it follows directly from (62.10c) and (96.24) that the connected component of $\mathrm{Aut}\,\mathcal{P}$ is a Lie group (observe that (32.23) settles the case $l = 1$, and that the point set is a $2l$-manifold by (62.10b) and (52.1)). We can, however, improve this result.

62.11 Corollary. *If a compact, connected plane $\mathcal{P}$ contains an isotropic point, then the automorphism group $\Sigma = \mathrm{Aut}\,\mathcal{P}$ is a Lie group.*

Proof. The plane has finite dimension $2l$ by (62.8a), and we may assume that $l \geq 4$, see (32.21) and (71.2). The dimension of Σ is also finite; this will be proved later for planes of finite dimension in general, compare (83.2). In our special situation, finiteness of $\dim \Sigma$ also follows from the fact that Σ^1 is a Lie group, see the remarks preceding (62.11), and the observation (93.6) that $\dim \Sigma = \dim \Sigma^1$. Yet another proof may be obtained by examining the stabilizer of an isotropic point, compare the proof of (62.8), and using the dimension formula (96.10).

According to (93.8), there exist an open subgroup Δ of Σ and a compact, 0-dimensional, normal subgroup $\Theta \triangleleft \Delta$ such that Δ/Θ is a Lie group. Our aim is to show that Θ is discrete; then (94.3d) implies that Δ and, hence, Σ are Lie groups. By (62.8a), there is a subgroup $\Psi \cong \mathrm{Spin}_{l+1}\mathbb{R}$ of Σ; in fact, $\Psi \leq \Delta$, since Ψ is connected. The connected group Ψ centralizes Θ, see (93.18). Hence Θ acts on the orbit space P/Ψ, which is homeomorphic to $[0, 1]$ by (62.10g). The kernel Ω of the action of Θ on P/Ψ is a Lie group, because we may apply Szenthe's theorem (96.14) to the action of $\Omega\Psi$ on any $(2l-1)$-dimensional Ψ-orbit B as in (62.10c); observe that this action is effective, because $\langle B \rangle = \mathcal{P}$, compare (44.1a). On the other hand, the effective factor group Θ/Ω induced on the interval is finite, see (32.15a). Now Θ is a Lie group of dimension 0, hence it is discrete. □

62.12 Proposition: Primitivity of flag homogeneous orbits. *Let Δ be a closed subgroup of the automorphism group of a $2l$-dimensional, compact plane $\mathcal{P}$ with $l \geq 2$. Assume that Δ has a non-trivial point orbit X such that the stabilizer Δ_x of a point $x \in X$ is transitive on the pencil $\mathcal{L}_x$ or, equivalently, such that Δ_x contains a subgroup $\Psi \cong \mathrm{Spin}_{l+1}\mathbb{R}$. Then the following assertions hold.*

(a) *X is open,*
(b) *the action of Δ on X is topologically primitive, i.e., Δ_x is maximal among closed subgroups of Δ,*
(c) *Δ has trivial center.*

Proof. By (62.10c), every Ψ-orbit near x except $\{x\}$ itself is $(2l-1)$-dimensional. Since $X \neq \{x\}$ by assumption and P is a manifold by (62.10b), it follows from (96.25) that X is open. The same argument shows that every closed, connected subgroup $\Omega \leq \Delta$ properly containing Δ_x is transitive on X, so that $\Omega = \Delta$ by (91.3). On the other hand, a disconnected, closed subgroup $\Omega \leq \Delta$ with $\Omega^1 = \Delta_x^{\,1}$ normalizes $\Delta_x^{\,1}$ and permutes its orbits. Since x is the only fixed point of $\Delta_x^{\,1}$, we obtain that $\Omega \leq \Delta_x$ in this case. This proves (b). For assertion (c) observe that every $x \in X$ is the center of a reflection (62.10a); hence by (23.12) the center of Δ acts trivially on X and hence everywhere. □

62.13 Theorem: Translation planes admitting $\mathrm{Spin}_{l+1}\mathbb{R}$. *Let $l \geq 2$, and assume that $\Psi \cong \mathrm{Spin}_{l+1}\mathbb{R}$ acts effectively on a compact translation plane $\mathcal{P}$ of dimension $2l$. Then $\mathcal{P}$ is a classical plane $\mathcal{P}_2\mathbb{F}$, where $\mathbb{F} \in \{\mathbb{R}, \mathbb{C}, \mathbb{H}, \mathbb{O}\}$, and Ψ acts in the usual way.*

Remarks. (a) Together with the dual of (62.8a), this theorem implies that a compact, connected translation plane is classical if its automorphism group is transitive on the translation axis.

(b) If we just assume a non-trivial action of Ψ (rather than an effective one), then the theorem of Skornyakov and San Soucie (24.9) implies that either $\mathcal{P}$ is a Moufang plane (and hence is classical, see (42.7), (63.3)), or Ψ fixes the translation axis. In the latter case, the action *is* effective by the dual of (62.8b).

Proof of (62.13). Consider fixed elements $x \notin A$ of Ψ as provided by (62.10). Then Ψ is transitive on A according to the dual of (62.8b). We claim that A is a translation axis. Indeed, if $W \neq A$ is a translation axis, then Ψ moves $W \wedge A$ and hence does not fix W. In this case, every line (including A) is a translation axis by the theorem of Skornjakov and San Soucie (24.9).

If we identify $P \setminus A$ with the sharply transitive group $\Sigma_{[A,A]}$ via $x^\sigma \mapsto \sigma$, then the action of Ψ on $P \setminus A$ becomes the action on $\Sigma_{[A,A]} \cong \mathbb{R}^{2l}$ by conjugation; see (44.6). Because Ψ is compact, this linear action may be considered as an orthogonal one (95.3), and Ψ is transitive on the sphere $\mathbb{S}_{2l-1}$ by (62.10c). In particular, the representation of Ψ on $\mathbb{R}^{2l}$ is irreducible, and therefore it is uniquely determined by the dimension $2l$, see (95.10).

Let $l > 2$. By (62.10d and e), Ψ contains a conjugacy class of axial involutions σ_L with axes $L \in \mathcal{L}_x$, and this class is uniquely determined by the property that $\mathrm{Cs}_\Psi(\sigma_L) \cong \mathrm{Spin}_l \mathbb{R}$. The line L in turn is determined as the fixed point set of σ_L. Using homogeneity of $P \setminus A$, we see that the affine plane $\mathcal{P}^A$ is completely determined.

If $l = 2$, then we may identify $\mathbb{R}^4$ with $\mathbb{H}$ and describe the group induced by Ψ as

$$\{ w \mapsto aw \mid a \in \mathbb{H}, \|a\| = 1 \}.$$

Indeed, this group is irreducible and isomorphic to $\mathrm{Spin}_3 \mathbb{R} \cong \mathbb{S}_3 \subseteq \mathbb{H}^\times$, see (11.26). It is a normal subgroup of

$$\mathrm{SO}_4 \mathbb{R} = \{ w \mapsto awb \mid a, b \in \mathbb{H}, \|a\| = \|b\| = 1 \},$$

see (11.23).

Transitivity of the translation group implies that the plane is completely determined by the line pencil $\mathcal{L}_x$. Under the identification $P \setminus A \cong \mathbb{R}^4$ via $\Sigma_{[A,A]}$ as described above, lines through x become vector subspaces, see (61.9) or (64.4). Thus, the pencil $\mathcal{L}_x$ corresponds to a Ψ-orbit in the Grassmann manifold $G_{4,2}$ of all 2-dimensional subspaces in $\mathbb{R}^4$. Since Ψ is normal in the group $\mathrm{SO}_4\mathbb{R}$, which is transitive on $G_{4,2}$, any two Ψ-orbits $\mathcal{G}, \mathcal{G}'$ in $G_{4,2}$ are equivalent. In other words, there is a vector space automorphism α of $\mathbb{R}^4$ (an element of $\mathrm{SO}_4\mathbb{R}$) that sends $\mathcal{G}$ onto $\mathcal{G}'$ and normalizes Ψ. Thus α is a Ψ-equivariant isomorphism of the translation plane defined by $\mathcal{L}_x = \mathcal{G}$ onto a plane having line pencil $\mathcal{G}'$. Since the classical plane is one of the planes satisfying our hypotheses, the given plane $\mathcal{P}$ is isomorphic to it. $\square$

The action of a group Φ locally isomorphic to $\Phi_l^\sharp$

Here we begin to consider the situation of (62.6). We are given a group Φ acting effectively on a compact plane $\mathcal{P}$ of dimension $2r$, and Φ is locally isomorphic to $\Phi_l^\sharp \in \{\Phi_l^+, \Phi_l^-, \Phi_l^0\}$ with $l \geq \max\{r, 2\}$. According to (62.5d), Φ is in fact a covering group of $\Phi_l^\sharp$, and the kernel of the covering map $\pi : \Phi \to \Phi_l^\sharp$ is $C(\Phi)$, the center of Φ. This implies that in addition to the given action on $\mathcal{P}$, there is an action of Φ on the classical plane $\mathcal{P}_l$. Our aim is to compare these two actions.

Observe that Φ contains a closed, connected group Ψ covering a subgroup $\mathrm{Spin}_{l+1}\mathbb{R}$ contained in $\Phi_l^\sharp$. Since $\mathrm{Spin}_{l+1}\mathbb{R}$ is simply connected, we have $\Psi \cong \mathrm{Spin}_{l+1}\mathbb{R}$. According to (62.10), we have that $r = l$, and Ψ fixes a point and a line in $\mathcal{P}$, which are therefore isotropic. First we deal with the case where one of these elements is fixed by all of Φ. By duality, it suffices to consider the following case.

62.14 Lemma: Actions with fixed element. *Let $l \geq \max\{r, 2\}$, and let Φ be a group locally isomorphic to a classical motion group $\Phi_l^\sharp$. If Φ acts effectively on a compact plane $\mathcal{P}$ of dimension $2r$ and fixes a line A, then $r = l$ and Φ contains a transitive elation group with axis A. Moreover, $\mathcal{P}$ is classical and Φ acts as the Euclidean motion group.*

Proof. By the preceding remarks, we have $r = l$, and Φ contains a subgroup $\Psi \cong \mathrm{Spin}_{l+1}\mathbb{R}$ acting transitively on some fixed line W. It follows that $W = A$. Let σ be the central involution in Ψ. By (62.10), this is a reflection with axis A and center p, say. Since the covering map $\pi : \Phi \to \Phi_l^\sharp$ induces an isomorphism on Ψ, the involution σ^π is a reflection of the classical plane. Either its center or its axis has an open orbit under $\Phi_l^\sharp$; to see this, inspect the classical actions or use (62.10c) and (96.25). By (61.19a) the conjugacy classes of σ^π and hence of σ are at least $2l$-dimensional. If Φ fixes p, then the closure $\overline{\Phi}$ in $\mathrm{Aut}\,\mathcal{P}$ does the same, and we obtain that $\dim \overline{\Phi}_{[p,A]} \geq \dim \sigma^\Phi \geq 2l$, which is absurd (61.4). Since all Ψ-orbits near the fixed point p are $(2l-1)$-dimensional, (96.25) shows that $\dim p^\Phi = 2l$. Thus $\dim \Phi_{[A,A]} = 2l$ according to (61.19b), and all assertions follow readily from (61.4) and (62.13). $\square$

The proof of (62.6) is finished by the last lemma in all cases where Φ fixes one of the fixed elements p, A of $\Psi \cong \mathrm{Spin}_{l+1}\mathbb{R}$ in $\mathcal{P}$. In the remaining cases we may consider the orbit $X = p^\Phi$ and conclude from (62.12) that Φ is center-free. This implies that $\Phi \cong \Phi_l^\sharp$, hence Φ acts on the two $2l$-dimensional planes $\mathcal{P}$ and $\mathcal{P}_l$ (the classical plane). Next, we look at a special orbit X_l of Φ in $\mathcal{P}_l$ as considered in (62.5c). The stabilizer $(\Phi_l^\sharp)_x = \Phi_x$ of $x \in X_l$ contains $\mathrm{Spin}_{l+1}\mathbb{R}$, hence X_l is flag homogeneous, e.g., by (62.8). The orbit X_l is uniquely determined by those properties, but we shall not need this fact.

62.15 Lemma: Recognition of the flag homogeneous orbit. *Let $l \geq \max\{r, 2\}$ and let Φ be locally isomorphic to a classical motion group $\Phi_l^\sharp$. If Φ acts effectively on a compact plane $\mathcal{P}$ of dimension $2r$, then $r = l$ and Φ is globally isomorphic to $\Phi_l^\sharp$. Moreover, up to duality there is a point orbit X equivalent to the flag homogeneous classical orbit X_l with stabilizer containing $\mathrm{Spin}_{l+1}\mathbb{R}$.*

Proof. The assertion is true in the situation treated in (62.14). Hence, we may assume that Φ fixes no point and no line. By the remarks following (62.14), we have $\Phi = \Phi_l^\sharp$ and $r = l$. Choose a point $x \in X_l$ and consider a subgroup $\Psi \cong \mathrm{Spin}_{l+1}\mathbb{R}$ contained in Φ_x. By (62.10), Ψ fixes a point p and a line W in $\mathcal{P}$, and by our assumption we have $p^\Psi = p \neq p^\Phi$. We define $X = p^\Phi$, and we have to show that $\Phi_x = \Phi_p$.

By (62.10), the central involution $\sigma \in \Psi$ is a reflection both at p and at x. We claim that σ is central in Φ_x. No proof is needed in those cases where $\Phi_x = \Psi$, i.e., in the Euclidean and octonion cases (18.8). For the remaining elliptic and hyperbolic groups over $\mathbb{C}$ and $\mathbb{H}$, we know that Φ_x is respectively isomorphic to $\mathrm{U}_2\mathbb{C}$ and to $\mathrm{U}_2\mathbb{H} \cdot \mathrm{U}_1\mathbb{H}$, see (62.4), and that σ corresponds to the element $-1\!\!1$ in the unitary group U_2. This proves our claim, and it follows that Φ_x fixes p, the center of σ, so that $\Phi_x \leq \Phi_p$. Now the maximality of Φ_x, see (62.12), implies that either $\Phi_p = \Phi_x$ or $\Phi_p = \Phi$. We have ascertained that $p \neq p^\Phi$, hence the latter possibility is excluded and the lemma is proved. □

62.16 Lemma. *Let $l \geq 2$ and consider the action of the hyperbolic group $\Phi = \Phi_l^-$ on the classical plane. For a line L meeting the flag homogeneous orbit X_l, the stabilizer Φ_L is connected and is equal to its own normalizer in Φ. It contains a unique reflection σ_L in the line L.*

Proof. 1) In order to prove that Φ_L is connected, it suffices to show that the set L^Φ of all lines meeting X is simply connected and to apply (96.9c). Now, by (13.17c) and (18.24), the hyperbolic polarity π^- maps L^Φ homeomorphically onto the orbit E of exterior points with respect to the hyperbolic quadric, which can be described as the complement of the affine unit ball in the projective point set P. Clearly, E is homotopy equivalent to the complement $P \setminus \{p\}$ of a point; by (51.26), this implies that $E \simeq L \simeq \mathbb{S}_l$. Compare also (18.24) for a different proof in the case $l = 8$.

2) The assertion about the normalizer follows if we prove that Φ_L fixes no other line (meeting X_l), and for that it is enough to show that Φ_L fixes no point on L. For $x \in L \cap X_l$, the stabilizer $\Phi_{L,x}$ fixes both x and the intersection point $a_x = A_x \wedge L$ with the axis $A_x = x^{\pi^-}$ of σ_x. By (62.10c), these are the only fixed points of $\Phi_{L,x}$ on L. They are not fixed by Φ_L since a_x changes if we replace x with a different point of $L \cap X_l$.

3) In the desarguesian cases, the last assertion is easily obtained by linear algebra. In the octonion case, we refer to the theorem in (18.22). □

Proof of (62.6). 1) We resume the considerations taken up after (62.13). So let Φ be a group locally isomorphic to $\Phi_l^\sharp$ that acts effectively on a $2r$-dimensional plane $\mathcal{P}$, where $r \leq l$ and $l \geq 2$. Until now we have seen that we may assume that $\Phi = \Phi_l^\sharp$ and that Φ fixes no point and no line in $\mathcal{P}$. Moreover, we know that $r = l$ and that the point set P of $\mathcal{P}$ contains a Φ-orbit X equivalent to the flag homogeneous classical Φ-orbit X_l. Recall that the stabilizer Φ_p of $p \in X$ contains a subgroup $\Psi \cong \mathrm{Spin}_{l+1}\mathbb{R}$.

2) Consider the Euclidean group Φ_2^0, which is not simple. We have to use a fact here that will be proved in Chapter 7 using the fixed point property of collineations (55.19), namely: a connected group of automorphisms of a 4-dimensional plane either is simple or fixes a point or a line (71.4 and 8). By (62.14), this completes the proof for Φ_2^0.

3) For all remaining cases, our next aim is to show that every line L meeting X is the axis of a reflection $\sigma_L \in \Phi$. Moreover, we claim that these involutions form a conjugacy class Ξ that can be identified by means of group theoretical properties. This will imply that the geometry induced by $\mathcal{P}$ on X (i.e., the family of all sets $L \cap X$, where $L \in \mathcal{L}$) is entirely determined by the action of Φ and therefore is classical. Note that the axis and the center can be distinguished within the fixed point set of σ since the center is isolated.

4) For $l > 2$, the claim of 3) follows from (62.10d and e). Indeed, Φ_p is isomorphic to $\mathrm{Spin}_9\mathbb{R}$ if $l = 8$ and to $\mathrm{U}_2\mathbb{H}$ or $\mathrm{U}_2\mathbb{H} \cdot \mathrm{U}_1\mathbb{H} \cong \mathrm{Spin}_5\mathbb{R} \cdot \mathrm{Spin}_3\mathbb{R}$ if $l = 4$, see (62.4); hence, Φ_p contains a unique group $\Psi \cong \mathrm{Spin}_{l+1}\mathbb{R}$, which in turn contains a reflection σ_L in a given line $L \in \mathcal{L}_p$. Now (62.10e) characterizes the conjugacy class $\sigma_L{}^\Phi = \Xi$ by a group theoretical property.

If $l = 2$ and Φ is not the Euclidean group, then $\Phi_p \cong \mathrm{U}_2\mathbb{C}$ acts on some neighbourhood of p in X like on the unit ball of $\mathbb{C}^2$, see the proof of (13.17). Linear algebra shows that apart from the central involution, Φ_p contains one other conjugacy class of involutions. Its elements σ are homologies of $\mathcal{P}$ by (55.21c). The action on $\mathbb{C}^2$ shows that the axis of σ passes through p.

5) In the elliptic case, the proof is complete since the compact orbit X must be the whole point set, e.g., by the pigeon-hole principle (51.20 and 21b), applied to the inclusion map $X \to P$. In the Euclidean cases, X is an affine plane, and we show that $\mathcal{P}$ is the projective completion of that plane. Indeed, every line L meeting X belongs to the subplane because X is open. In view of the parallel axiom, this implies that $L \setminus X$ is a single point. If W is a line joining two points in $P \setminus X$, then W must therefore be contained in $P \setminus X$, and finally $P \setminus X = W$.

6) In the hyperbolic case we continue by reconstructing the exterior point orbit of the classical action within the point set P of $\mathcal{P}$. Consider the reflection σ_p at $p \in X$, which is central in $\Psi \leq \Phi_p$ and hence in Φ_p, compare the proof of (62.15). The axis A of σ_p is an orbit of Φ_p, because Φ_p acts on A and on $\mathcal{L}_p$ in the same way. It follows that A does not meet X, or else X would contain A; being flag homogeneus, X would then be the entire point set. The last assertion can also be seen by inspection of the classical action.

We have $\Phi_p \leq \Phi_A$, hence maximality of Φ_p, see (62.12), together with (62.14) and (96.10) implies that $\Phi_A = \Phi_p$ and $\dim A^\Phi = \dim p^\Phi = 2l$, so that A^Φ is open in $\mathcal{L}$. Now for $a \in A$, the orbit a^Φ is equal to the set $\bigcup A^\Phi$ of all points incident with some image A^φ; remember that $A = a^{\Phi_p}$. In particular, $E = a^\Phi$ is open since the join operation is continuous, and this is the orbit we have been looking for.

Our next aim is to show that E is equivalent to the orbit of all lines meeting X. For $L \in \mathcal{L}_p$, we have constructed a reflection $\sigma_L \in \Psi$. Since Φ_p fixes A, the center of σ_L is a point $a \in A$. We want to show that $\Phi_L = \Phi_a$. First we observe that Φ_L centralizes σ_L by (62.16) and hence fixes the center a. Thus we have $\Phi_L \leq \Phi_a$. For the converse inclusion, note first that $\dim \Phi_L = \dim \Phi_a$ by (96.10); we use the fact that the corresponding orbits are open in $\mathcal{L}$ or P, and hence have the same dimension as these spaces. Therefore, Φ_a normalizes the connected component $\Phi_L{}^1 = \Phi_a{}^1$, and $\Phi_a = \Phi_L$ by (62.16).

7) The result of 6) may be rephrased by stating that the action of Φ on $U = X \cup E$ is equivalent to the classical action of Φ_l^- on the complement of the hyperbolic quadric. The axial reflections $\sigma_p = \sigma_A$ and σ_L have been singled out, and A and $L \cap U$ are determined as their fixed point sets (minus the isolated centers). This gives us two out of the three classical line orbits, the missing one being the orbit of absolute lines. Using (13.17a) and (18.21) and applying the polarity, we see that this orbit is homeomorphic to $\mathbb{S}_{2l-1}$, hence it does not contain a non-empty open subset of $\mathcal{L}$. Therefore, the missing absolute lines can be represented as limits of sequences of known lines, and we have completely described the geometry induced on U, i.e., the pair $\bigl(U, \{\, G \cap U \mid G \in \mathcal{L} \,\}\bigr)$. This geometry is isomorphic to its classical counterpart, the complement of the hyperbolic quadric, which is dense in the classical projective plane. Now (44.9) asserts that such an isomorphism can be extended to an isomorphism of the entire projective planes. □

62.17 Notes on Section 62. Theorems (62.2 and 6) have a long history, and various special cases have been proved separately. The first result of this kind was given by Salzmann [62b], concerning the case $l = 1$; compare also Salzmann [63a], Strambach [71]. The case $l = 2$ is treated by Salzmann [70] and Betten [73c]. For $l = 4$ or 8, see Salzmann [79a, 82b]. The present form of (62.6) is due to Löwen [86b] where, more generally, stable planes are considered. There the proof is given by reduction to results of Löwen [83a] on stable planes containing more than one isotropic point, and the latter paper heavily uses methods of differential geometry: Theorem (62.10) is used to impose the structure of an affine symmetric space on any flag homogeneous orbit X, the symmetry at p being the reflection σ_p. Then the same theorem is used again to show that the point stabilizer in the motion group of this symmetric space is transitive on the directions in the tangent space, and that the space is therefore Riemannian and two-point homogeneous. Finally, the classification of such spaces due to Tits [55] p. 183 is used; compare also Wolf [84] 8.12.2. The proof of (62.6) presented here would not work for stable planes. The crucial result (62.10) is essentially contained in Salzmann [75b] and, in an extended form, in Löwen [86b].

63 Transitive groups

Homogeneity of geometries can be measured in many different ways. A particularly simple criterion is existence of large orbits of the automorphism group. For projective planes in general, however, homogeneity in this sense is of no consequence whatsoever for the structure of the plane: in their article [73], Kegel and Schleiermacher show that *every* projective plane can be embedded into a plane with an automorphism group transitive on triangles. Tits [77] Satz 3 proves a similar result for generalized n-gons; a projective plane is a generalized 3-gon. See also the remarks on finite, homogeneous planes in the introduction to Chapter 6. In contrast, we show in this section that in the realm of compact, connected planes the four classical planes may be characterized by rather weak homogeneity conditions of this kind. Even the hypothesis of finite dimensionality, otherwise a habitual rite in this book, is unnecessary at this point.

The first step in this direction was taken in the previous section when we considered isotropic points or, by duality, isotropic lines; see (62.7)ff. The following result, as well as Theorems (63.5) and (64.18), may be viewed as an extension of that investigation. Together they show that a compact, connected plane with a doubly homogeneous line is either classical or a Moulton plane.

63.1 Theorem: Line homogeneous affine planes. *Let $\mathcal{A} = (A, \mathcal{H})$ be a locally compact, connected, affine plane. If $\mathcal{A}$ admits a group Δ of automorphisms that acts transitively on the line W at infinity and fixes no point in A, then $\mathcal{A}$ is one of the four classical planes $\mathcal{A}_2\mathbb{F}$, where $\mathbb{F} \in \{\mathbb{R}, \mathbb{C}, \mathbb{H}, \mathbb{O}\}$. In particular, this holds if Δ is transitive on $\mathcal{H}$.*

Proof. 0) If $\mathcal{A}$ is an affine part of a compact translation plane $\mathcal{P}$, and if Δ is a closed subgroup of $\operatorname{Aut} \mathcal{P}$, then the assertion follows from (62.13) together with the dual of (62.8a). The general assertion will be proved by reduction to this special case.

1) The given group Δ may be considered as a subgroup of $\Sigma = \operatorname{Aut} \overline{\mathcal{A}}$, cf. (23.4) and (44.4b). We may assume that Δ is closed in Σ and, hence, is locally compact. By (42.10) and (62.8 and 10b), the spaces W, A and $\mathcal{H}$ are manifolds, compare (52.1). More precisely, we have $W \approx \mathbb{S}_l$, where $l \in \{1, 2, 4, 8\}$. From (62.11) we know that Δ is a Lie group. Moreover, Δ^1 is transitive on W, see (96.9b). The set of fixed points of the normal subgroup $\Delta^1 \trianglelefteq \Delta$ in A is invariant under Δ and thus cannot consist of a single point. Hence, transitivity of Δ^1 on W implies that there are no fixed points, and we may assume that Δ is connected.

2) We treat the case $l = 1$ separately. Then Δ cannot be effective on W. Indeed, if Δ is effective, then we obtain from (96.30a) that either $\Delta \cong \operatorname{SO}_2\mathbb{R}$ or Δ is a covering group of $\operatorname{PSL}_2\mathbb{R}$ with finite center. In both cases, Δ contains a reflection σ, see (32.12); its axis is W, because $\operatorname{Cs}(\sigma)$ is transitive on W (or because the connected group Δ preserves the orientation of A, whereas a reflection with affine axis reverses the orientation). This shows that Δ is not effective on W, after all.

If Δ contains non-trivial elations, then the compatible actions of Δ on $\Lambda := \Delta_{[W,W]}$ and on W (compare (23.12)) show that $\dim \Lambda \geq 2$, and Λ is transitive on A by (61.4b). Hence $\mathcal{A}$ is classical, cf. (24.6) and (32.8). If Δ contains homologies with axis W, then their centers have orbits of positive dimension, and $\dim \Lambda > 0$ by (61.20b). Again, $\mathcal{A}$ is classical.

3) Now let $l \geq 2$. Since W is isotropic, the dual of (62.8) provides a subgroup $\Psi \cong \operatorname{Spin}_{l+1}\mathbb{R}$ contained in Δ and transitive on W. By (62.10), the group Ψ fixes some point $a \in A$. Since all Ψ-orbits in $A \setminus \{a\}$ are $(2l-1)$-dimensional (62.10c), we obtain from (96.25) that a^Δ is open. Now Ψ contains a reflection $\sigma \in \Psi_{[a,W]}$ by (62.10a), and from (61.19b) or (61.20b) we see that $\dim \Lambda = \dim a^\Delta = 2l$. Hence, $\mathcal{A}$ is a translation plane (61.4b), and $\mathcal{A}$ is classical by (62.13). □

63.2 Corollary: Doubly homogeneous projective planes. *Let $\mathcal{P}$ be a compact, connected projective plane. If $\operatorname{Aut} \mathcal{P}$ acts doubly transitively on the point set, then $\mathcal{P}$ is classical.*

Proof. Apply (63.1) to any affine part of the dual plane $\mathcal{P}^*$. □

Note that we are going to prove a stronger theorem (63.8). Compare also the arguments in Section 65, where implicitly we give another independent proof of (63.2).

63.3 Corollary: Moufang planes. *The four classical planes $\mathcal{P}_2\mathbb{F}$, where $\mathbb{F}$ belongs to $\{\mathbb{R}, \mathbb{C}, \mathbb{H}, \mathbb{O}\}$, are the only compact, connected Moufang planes. The algebras $\mathbb{R}, \mathbb{C}, \mathbb{H}$, and $\mathbb{O}$ are the only locally compact, connected alternative fields.*

Proof. As an immediate consequence of the definition, Moufang planes are doubly homogeneous, hence the first assertion follows from (63.2). For the second one, note that a locally compact, connected alternative field coordinatizes a compact, connected Moufang plane $\mathcal{P}$, and that $\mathcal{P} \cong \mathcal{P}_2\mathbb{F}$ by what we just proved. Now Aut $\mathcal{P}_2\mathbb{F}$ is transitive on quadrangles, see (13.5) and (17.6), hence all ternary fields coordinatizing $\mathcal{P}_2\mathbb{F}$ are isomorphic (23.5). □

63.4 Note. The last corollary also follows from Theorem (63.8), which will be proved independently. It should be noted here that both these proofs are anything but direct. Our intention merely is to point out that the classification of compact, connected Moufang planes is a simple special case of our stronger results. For a straight proof of that classification, see the references given in (42.7).

The following result complements Theorem (63.1). For the definition of Lenz type III, see (24.5).

63.5 Theorem: Doubly homogeneous lines. *Let $\mathcal{P} = (P, \mathcal{L})$ be a compact, connected projective plane and let Δ be a closed subgroup of $\Sigma = \text{Aut}\,\mathcal{P}$. Then Δ is a group of Lenz type III if, and only if, Δ fixes some antiflag $o \notin W$ and is doubly transitive on W.*

63.6 Remark. The planes admitting such groups will be determined in (64.18). They are precisely the Moulton planes and the classical planes. There is a shortcut leading to this result if one assumes that $l = 1$. In that situation, one may leave the proof of (63.5) as soon as it becomes clear that the group Δ, which fixes an antiflag, is at least 3-dimensional, and refer directly to (34.9).

Proof of (63.5). 1) As an easy consequence of the definition, a group of Lenz type III is doubly transitive on its fixed line W and fixes a point $o \notin W$. Conversely, let $\Delta \leq \Sigma$ be a closed subgroup fixing $o \notin W$ that is doubly transitive on W. By (62.11), we know that Δ is a Lie group. We apply (96.9b) to the action of Δ on the connected set of all pairs of distinct elements of $W \approx \mathbb{S}_l$; this shows that we may assume that Δ is connected. Considering the values $l \leq 2$ and $l \geq 4$ separately, we shall show in the subsequent steps that Δ is of Lenz type III.

2) Let $l \leq 2$. By (96.16 and 17), the action induced on W is the standard action of $\text{PSL}_2\mathbb{F}$, where $\mathbb{F} = \mathbb{R}$ or $\mathbb{C}$. Since $\dim \Delta_{[o]} \leq 2$ by (61.4a), a Levi subgroup $\Lambda \leq \Delta$ is a covering group of $\text{PSL}_2\mathbb{F}$. In fact, for $l = 2$ we have $\Lambda \cong \text{SL}_2\mathbb{C}$ because $\pi_1(\text{PSL}_2\mathbb{C}) = \mathbb{Z}_2$, see (94.33), and because $\text{PSL}_2\mathbb{C}$ itself contains $\text{SO}_3\mathbb{R}$, which cannot act with a fixed point (62.8). We shall not need to know if Λ is closed in Δ; it is a Lie group with a topology possibly finer than its subgroup topology, and this Lie group still acts continuously on $\mathcal{P}$.

For $uv = W$, consider the group Λ_u^1. It is a covering group of the affine group $\mathbb{F} \rtimes \mathbb{F}^*$ induced on $W \setminus \{u\}$, where $\mathbb{F}^*$ denotes the connected component of $\mathbb{F}^\times$. More precisely, $\Lambda_u^1 = \Pi\Omega$, where $\Pi = (\Lambda_u^1)'$ induces the left factor $\mathbb{F}$ and $\Omega = \Lambda_{u,v}^1$ induces some conjugate of $\mathbb{F}^*$. We claim that $\Pi \leq \Lambda_{[u,ou]}$; it will follow that Π is a (u, ou)-transitive group of elations (61.4a), and that Δ is of Lenz type III.

There is an element $\lambda \in \Lambda$ which interchanges u and v and normalizes the group Ω, inducing inversion on it, compare (15.8). Therefore, Ω acts freely on $X = ou \setminus \{o, u\}$, or we would obtain a non-trivial collineation fixing a quadrangle, contrary to (32.10) or (55.21); note that Ω is connected. Since $\Lambda_u^1 \setminus \Pi$ is partitioned by the conjugates of Ω, it follows that $\Lambda_{u,x}^1 \leq \Pi$ for every $x \in X$. Now $\dim \Lambda_{u,x} \geq l$ by the dimension formula (96.10), hence Π acts trivially on ou. Since Π is abelian and transitive on $W \setminus \{u\}$, the elements of Π have center u.

3) If $l > 2$, then Δ contains reflections with axis $K = ob$ and center a, where $a, b \in W$, see (62.10d). By double transitivity, (61.19b) implies that the elation group $\Delta_{[b,K]}$ is l-dimensional, hence (b, K)-transitive (61.4a). Therefore, Δ is of type III. □

63.7 Remark. Theorem (63.5) becomes false if the hypothesis of double transitivity is replaced simply by transitivity. Indeed, Schellhammer [81] constructs planes with $l = 1$ and with automorphism group $\Sigma^1 \cong \mathbb{C}^\times$ transitive on some punctured affine part, see (34.1). Similarly, P. Sperner [90] produces planes with $l = 2$ and with group $\Sigma^1 \cong GU_2\mathbb{C}$, again acting transitively on some punctured affine part.

63.8 Theorem: Point homogeneous planes. *Let $\mathcal{P}$ be a compact, connected projective plane. If $\Delta \leq \Sigma = \mathrm{Aut}\,\mathcal{P}$ is a closed subgroup acting transitively on the point set P, then $\mathcal{P}$ is a classical plane $\mathcal{P}_2\mathbb{F}$, where $\mathbb{F} \in \{\mathbb{R}, \mathbb{C}, \mathbb{H}, \mathbb{O}\}$.*

Moreover, either $\Delta^1 = \Sigma^1$ or Δ^1 is conjugate in Σ to the elliptic motion group of $\mathcal{P}$. In particular, $\dim P$ is finite and Δ^1 is even flag transitive.

Organization of Proof. For information on the elliptic motion groups, see Sections 13, 18 and 62. In Lemma (63.9), we shall prove that the elliptic motion group is maximal among connected, closed subgroups of $\mathrm{Aut}\,\mathcal{P}_2\mathbb{F}$. Using (94.34), this can also be deduced from its maximality among compact, connected subgroups, cf. (18.18) and (94.33). Thus the essential task is to prove that a transitive group $\Delta \leq \Sigma$ forces $\mathcal{P}$ to be classical and that Δ has to contain the elliptic group.

Planes with $l = 1$ have very special topological features. They are not simply connected, and their Euler characteristic $\chi P = 1$ is not a prime. This makes it necessary to give a separate treatment for the case $l = 1$. That has been done in (32.23). To deal with the case $l \geq 2$, the following instruments are available:

(i) The topological results from Chapter 5, in particular the fact that $\chi P = 3$.
(ii) The classification of homogeneous spaces of prime Euler characteristic due to Borel and De Siebenthal, see their joint article [49] and Borel [49].
(iii) The results of Section 62 on planes admitting an action of the elliptic motion group.

We give three versions of a proof of (63.8), relying on these sources of information to different degrees. In the first version, the instruments from (i) and (ii) are used freely, and a very quick proof results. In this case, Section 62 is used only marginally. The second version incorporates a simplified proof (partly following Wolf [84]) of those small parts of Borel's and De Siebenthal's classification that are needed if one allows full use of Section 62. The third version is independent of the deeper topological results of Chapter 5, replacing them by methods from the theory of transformation groups. Chapter 5 is used several times in the second version, hence it would not make sense to combine versions 2 and 3.

Note that another independent proof can be obtained by combining the original proof of Salzmann [75a], who assumed that $\dim P \leq 16$, with (54.11).

Proof of Theorem (63.8): *First Version.* The group Δ is locally compact and has a countable basis (44.3) and acts transitively on the compact and locally contractible space P, see (51.4). Therefore, Δ is a Lie group and P is a manifold according to Szenthe's theorem (96.14). By (96.9b), we may assume that Δ is connected. As P is simply connected (51.28b), a theorem of Montgomery (96.19) asserts that a maximal compact subgroup $\Phi \leq \Delta$ is also transitive on P. Hence, (96.9a) shows that P is a homogeneous space Φ/Φ_p. Note that Φ is connected (93.10a).

Now $\chi P = 3$ is a prime (52.14c), and among all simply connected, homogeneous spaces of compact Lie groups, those having prime characteristic have been completely classified by Borel [49]. Borel used the classification he had obtained with De Siebenthal (Borel–De Siebenthal [49]) of maximal subgroups of maximal rank in compact Lie groups. A convenient reference is Wolf [84] 8.10.16. For the prime $p = 3$, every homogeneous space of this kind is equivalent either to one of the classical point sets $P_2\mathbb{F}$ of dimension $2l \geq 4$ with the standard action of the elliptic motion group Φ_l, or to one exceptional space $X = G_2/SO_4\mathbb{R}$. It remains to exclude the space X, which by (11.31b) may be considered as the space of all quaternion subfields of $\mathbb{O}$, and to reconstruct the classical projective plane $\mathcal{P}_2\mathbb{F}$ from the action of the elliptic group on its point set.

We use the shorthand G_2 for the compact form $G_2(-14)$ of the exceptional simple Lie group of dimension 14, which is simply connected, see (94.33). We write down the exact homotopy sequence (96.12) associated with the transitive action of G_2 on X,

$$\ldots \to \pi_2 X \to \pi_1 SO_4\mathbb{R} \to \pi_1 G_2 = \mathbb{1}.$$

From (11.23) we may infer the well-known fact that $\pi_1 SO_4\mathbb{R} = \mathbb{Z}_2$, and the exact sequence yields that $\pi_2 X \neq 0$. As we know from (52.14c) that $\pi_2 P = 0$, we conclude that $P = X$ is impossible.

We are left with the standard actions of the elliptic groups, and we may apply Theorem (62.6) on classical motion groups. In fact, since we know the action and not just the group, we can omit all the difficult parts of Section 62 and only use the reconstruction of the geometry from the elliptic action given at the end of that section; see steps 3) and 4) in the proof of (62.6) or step 3) in the proof of (65.2). Briefly, the lines are described as the fixed point sets of the elliptic reflections (disregarding the isolated centers), and these reflections may be singled out in Φ by group-theoretical properties. For this reconstruction, compare also Salzmann [75a] pp. 228, 233.

Proof of Theorem (63.8): *Second Version.* 1) As in the first version, theorems of Szenthe and Montgomery show that Δ is a Lie group containing a compact, connected, transitive subgroup Φ. The stabilizer Φ_p is connected since P is simply connected, see (51.28b) and (96.9c).

2) Let $\mathsf{T} \leq \Phi$ be a maximal torus subgroup (94.31). There is a generator $\tau \in \mathsf{T}$, i.e., an element whose powers form a dense subset; see Adams [69] 4.3. By (55.19), the collineation τ fixes some point $p \in P$. Then Φ_p contains T, hence Φ_p is a subgroup of *maximal rank* in Φ.

3) Since T is its own centralizer according to Adams [69] 4.26, it follows that Φ_p contains the center of Φ. Hence, the center is trivial. This fact may also be obtained by geometrical arguments, using transitivity together with the fixed point property (55.19). We infer that Φ is a direct product

$$\Phi = \Phi_1 \times \ldots \times \Phi_k$$

of center-free compact simple Lie groups, see (94.31c). The stabilizer Φ_p is of the form

$$\Phi_p = \Psi_1 \times \ldots \times \Psi_k,$$

where $\Psi_i \leq \Phi_i$. This is true for maximal subgroups of maximal rank in general, see Wolf [84] 8.10.1; the proof uses the way in which T acts on the Lie algebra of Φ.

4) The *Weyl group* of Φ is the group

$$W = \mathrm{Ns}_\Phi(\mathsf{T})/\mathsf{T}$$

of automorphisms of T induced by Φ. It is a compact subgroup of the group $\mathrm{Aut}\,\mathsf{T}$, which is discrete, see Hewitt–Ross [79] 26.18 (h). Hence W is a finite group. A theorem of Hopf and Samelson, proved in their paper [41], asserts that the Euler characteristic χP equals the quotient $\sharp W/\sharp W_p$, where W_p denotes the Weyl group of Φ_p and $\sharp$ denotes cardinality. According to (52.14c) we obtain in our case that

$$\sharp W/\sharp W_p = 3.$$

5) The stabilizer Φ_p is maximal among connected subgroups of Φ. Indeed, suppose that $\Phi_p \leq \Psi \leq \Phi$ and let W' be the Weyl group of Ψ. Then

$$3 = \sharp W/\sharp W' \cdot \sharp W'/\sharp W_p = \chi(\Phi/\Psi) \cdot \chi(\Psi/\Phi_p),$$

hence either Φ_p and Ψ have the same Weyl group, or the same is true for Ψ and Φ; note that all Weyl groups are contained in Aut T. Now a center-free, compact, connected Lie group is completely determined by a maximal torus and the Weyl group, hence Ψ coincides with Φ_p or with Φ; details are given by H.-C. Wang [49, proof of Theorem 1]. Wang's Theorem 1 says that a simply connected homogeneous space of a compact Lie group, with Euler characteristic equal to one, is a point; we could have used this directly.

6) Combining the maximality statement 5) with the description $\Phi_p = \Psi_1 \times ... \times \Psi_k$ obtained in 3), we infer that $\Psi_i = \Phi_i$ except for one value $i = i_0$. On the other hand, Φ_p does not contain any non-trivial, normal subgroup of Φ since Φ acts effectively. This implies that $k = 1$, and Φ is a simple Lie group. From (55.37a) we know, moreover, that the rank $r = \dim T$ of Φ is bounded by s, where $2l = 2^s$ is the dimension of the plane. In particular, we have $r \leq 4$.

7) In order to determine the group Φ, we look at the compact simple Lie groups of rank $r \leq 4$ one by one. We use the following table, which lists these groups together with their Weyl group orders. The groups and their dimensions can be found in (94.33). For the Weyl group orders, see Humphreys [90] 2.11 or Bourbaki [68] chap. vi, §4.

type	Φ	$\sharp W$	r	$\dim \Phi$ for $r \leq 4$
A_r	$PSU_{r+1}\mathbb{C}$	$(r+1)!$	$r \geq 1$	3, 8, 15, 24
B_r	$SO_{2r+1}\mathbb{R}$	$2^r r!$	$r \geq 2$	$-$, 10, 21, 36
C_r	$PU_r\mathbb{H}$	$2^r r!$	$r \geq 3$	$-$, $-$, 21, 36
D_r	$PSO_{2r}\mathbb{R}$	$2^{r-1} r!$	$r = 4$	$-$, $-$, $-$, 28
G_2		12	$r = 2$	14
F_4		$2^7 3^2$	$r = 4$	52

The group Φ_p is a product of some of the groups of the list and, possibly, a torus, the total rank being equal to that of Φ. It follows directly from the definition that the Weyl group of Φ_p is the direct product of the Weyl groups of the simple factors, hence the quotient of $\sharp W$ divided by the product of their orders must be 3, compare step 4). Moreover, we shall use the fact that $\dim \Phi - \dim \Phi_p = \dim P = 2l$, see (96.10). Now consider the possibilities for the type of Φ.

(a) A_4: Then $\sharp W = 5!$, and $\sharp W_p$ must be divisible by 5, which is impossible.

(b) A_3: Then $\dim \Phi = 15$ and $2l = 8$, so that $\dim \Phi_p = 7$. Thus Φ_p is locally isomorphic to $A_1{}^2 \cdot SO_2\mathbb{R}$, and $\sharp W/\sharp W_p = 4!/4 \neq 3$.

(c) A_2: This is the complex elliptic motion group. If $l \leq 2$, then Theorem (62.6) says that $\mathcal{P}$ is classical. On the other hand, $l > 2$ implies that $P \approx \Phi$, which is impossible because $\pi_1 \Phi = \mathbb{Z}_3$, see (94.33). Alternatively, we may obtain the contradiction from the fact that $\chi \Phi = 0$; indeed, this holds for every Lie group.

(d) Types A_1 and B_2 cannot occur because 3 does not divide their Weyl group orders.

(e) B_4 or C_4: We obtain that $\sharp W_p = 2^4 4!/3 = 2^7$, which is obviously impossible.

(f) B_3: Then $\dim \Phi = 21$ and $\dim P = 8$ or 16, hence $\dim \Phi_p = 13$ or 5. We have $\sharp W_p = 2^4$, hence the only possibility is $\Phi_p = B_2 \cdot A_1$, $\dim \Phi_p = 13$ and $\dim P = 8$. Such a group is not contained in $\Phi = SO_7\mathbb{R}$, because the smallest representation of B_2 with centralizer $\mathbb{H}$ is 8-dimensional, see (95.10). Alternatively, we obtain a contradiction by considering an involution $\beta \in \Phi$ corresponding to the reflection of $\mathbb{R}^7$ in some one-dimensional subspace. We claim that $\Omega = \mathrm{Cs}_\Phi(\beta)^1 = SO_6\mathbb{R}$ fixes a point of $\mathcal{P}$, contrary to (62.8b). Indeed, if β is axial, then the center is fixed. By (23.17), the remaining possibility is that β is a Baer involution. The plane F of fixed points is 4-dimensional by (55.5). Now Ω cannot act effectively on F because $\mathrm{rk}\,\Omega = 3$. Thus, the simple group Ω acts trivially on F, confirming our claim.

(g) The center-free group of type C_3 is the quaternion elliptic motion group. By Theorem (62.6), we only have to exclude the possibility that $l = 8$. In that case, we have $\dim \Phi_p = 5$, which is impossible because of $\sharp W_p = 2^4$.

(h) The group of type D_4 is 28-dimensional, and $l = 8$ implies that $\dim \Phi_p = 12$. We have $\sharp W_p = 2^6$, a contradiction.

(i) G_2 is of dimension 14, hence $\dim P = 4$ or 8, and accordingly $\dim \Phi_p = 10$ or 6. Moreover, $\sharp W_p = 4$, so that Φ_p is locally isomorphic to $SO_4\mathbb{R}$. The argument used to exclude $G_2/SO_4\mathbb{R}$ in the first version of the proof shows that Φ_p is simply connected, whence $\Phi_p \cong \mathrm{Spin}_4\mathbb{R}$ contains a central involution. But by (11.33 and 31d), all involutions in G_2 are conjugate, and their centralizers are isomorphic to $SO_4\mathbb{R}$, a contradiction.

(k) F_4 has been treated in (62.6).

Proof of Theorem (63.8): *Third Version*. 1) The starting point this time is to consider the compact, connected transitive Lie group Φ obtained as in the first version, and to show geometrically that Φ has a stabilizer of maximal rank. However, we shall avoid using the Lefschetz fixed point theorem as we did in the second version when we applied (55.19). What has to be shown is that a maximal torus $\mathsf{T} \leq \Phi$ fixes some point. This is clear if T contains an axial involution, because then T fixes the center. Suppose therefore that all involutions $\sigma \in \mathsf{T}$ are Baer involutions (23.17). Then T acts on the corresponding planes F_σ of fixed points. By (55.5), we have $\dim F_\sigma = \frac{1}{2} \dim P$; note that $\dim P$ is finite by Szenthe's theorem. Applying the same argument to the group induced by T on F_σ and using downward induction we obtain a reflection on some subplane, and the center is fixed. By duality we have now proved:

2) Every maximal torus $\mathsf{T} \leq \Phi$ fixes some point and some line. In particular, Φ_p has maximal rank.

3) Let F_T be the non-empty set of T-fixed points. Then F_T is left invariant by the normalizer $\mathrm{Ns}_\Phi(\mathsf{T})$. In fact, we claim that F_T is an orbit of $\mathrm{Ns}_\Phi(\mathsf{T})$. Consider $p, q \in F_\mathsf{T}$ and let $q = p^\varphi$, where $\varphi \in \Phi$. Both T and T^φ are maximal tori in Φ_q, hence they are conjugate in this group (94.31b). Thus, there is $\psi \in \Phi_q$ such that $\varphi\psi$ normalizes T and satisfies $q = p^{\varphi\psi}$.

Using (96.9a), we conclude that F_T is homeomorphic to the coset space

$$\mathrm{Ns}_\Phi(\mathsf{T})/(\mathrm{Ns}_\Phi(\mathsf{T}) \cap \Phi_p) = \mathrm{Ns}_\Phi(\mathsf{T})/\mathrm{Ns}_{\Phi_p}(\mathsf{T})$$
$$= \mathrm{Ns}_\Phi(\mathsf{T})/\mathsf{T} \Big/ \mathrm{Ns}_{\Phi_p}(\mathsf{T})/\mathsf{T} = W/W_p.$$

Therefore, using Hopf–Samelson [41], we obtain that

$$\sharp F_\mathsf{T} = \sharp W/\sharp W_p = \chi P.$$

4) In order to compute $\sharp F_\mathsf{T} = \chi P$, we note that $\chi P \neq 1$ by Wang [49] as P is simply connected (51.28b), and we shall prove presently that given any T-fixed line L as in 2), the torus T fixes precisely one point in the affine plane $P \setminus L$. This will imply that F_T is a triangle, and hence that

$$\chi P = 3.$$

We have seen that P is a manifold by Szenthe's theorem, hence we have $P \setminus L \cong \mathbb{R}^{2l}$ according to (52.3). The action of T on this space extends to an action on the one-point compactification $\mathbb{S}_{2l}$. According to Borel et al. [60] IV, 5.7, the non-void finite fixed point set of the extended action is a homology sphere with respect to rational coefficients. Hence, it must have precisely two elements because $H_0(X; \mathbb{Q})$ counts the connected components of a space X.

5) Having computed the Euler characteristic of P, we are now in position to use the classification of homogeneous spaces of prime characteristic just as in the first version of our proof. In the course of that argument, we used the fact that $\pi_2 P = 0$

for $l = 4$ in order to exclude the possibility $P = G_2/SO_4\mathbb{R}$. This fact can be obtained without using the deep results of Chapter 5, as follows: Steps 2) and 3) in the proof of (52.13) show that $H_2P \cong H_2L$ for a line L. Using a Mayer–Vietoris sequence as in the proof of (52.5), step 1), one sees that $H_2L \cong H_1(K^\times)$, where $K^\times = L \setminus \{a, b\}$. Finally, $K^\times$ is simply connected by (51.28c), hence $H_1(K^\times) = 0$ by the Hurewicz isomorphism theorem, see Spanier [66] 7.5.5. Thus, $H_2P = 0$, and $\pi_2 P = 0$ again by (51.28b) and the Hurewicz theorem. □

63.9 Proposition: Maximality of elliptic motion groups. *For each of the classical planes $\mathcal{P}$, the elliptic motion group Φ of $\mathcal{P}$ is a maximal subgroup of $\Sigma^1 = (\mathrm{Aut}\,\mathcal{P})^1$. In view of (63.8), this implies that Φ and Σ^1 are the only closed, connected, point transitive groups of automorphisms.*

Proof. This can be seen from (94.34) because Φ is a maximal compact subgroup of Σ^1, cf. (94.33). Here we give another proof of the weaker assertion that Φ is maximal among closed, connected proper subgroups of Σ. We use some facts from Lie group theory without further reference.

The adjoint representation $\mathrm{Ad}\,\Phi$ on the Lie algebra $\mathfrak{l}\Sigma$ preserves the non-degenerate Killing form β, hence a β-orthogonal decomposition $\mathfrak{l}\Sigma = \mathfrak{l}\Phi \oplus \mathfrak{d}$ remains fixed; note that β is negative definite on the first summand. If Δ is a connected Lie group in between Φ and Σ, then $\mathfrak{l}\Delta \cap \mathfrak{d}$ is also invariant. Thus, it suffices to show that $\mathrm{Ad}\,\Phi$ acts irreducibly on $\mathfrak{d}$. We have $\mathrm{Cs}_\Sigma \Phi = \mathbb{1}$ since Φ is transitive and every collineation of $\mathcal{P}$ has a fixed point (55.19), hence $\mathrm{Ad}\,\Phi$ fixes no elements of $\mathfrak{d}$, and $\mathfrak{d}$ is a sum of irreducible subspaces. It suffices to show that there can be no irreducible subspace of dimension at most $\frac{1}{2} \dim \mathfrak{d}$. Now $\dim \mathfrak{d} = \dim \Sigma - \dim \Phi \in \{5, 8, 14, 26\}$, depending on the value of $l \in \{1, 2, 4, 8\}$, and by (95.10) the dimensions of the smallest irreducible representations of Φ are respectively 3, 6, 12, 26, which is too large in each case. □

We note the following result of Löwen [83a] without giving the proof, which combines the results of Section 62 with methods from differential geometry.

63.10 Theorem: Planes with isotropic points. *If $\mathcal{P}$ is a compact, connected projective plane containing more than one isotropic point (62.7), then $\mathcal{P}$ is isomorphic to a classical plane $\mathcal{P}_2\mathbb{F}$, where $\mathbb{F} \in \{\mathbb{R}, \mathbb{C}, \mathbb{H}, \mathbb{O}\}$, or to a skew hyperbolic plane, see Section 35.* □

Examples of planes containing exactly one isotropic point have been given for $l = 1$ (Schellhammer [81], see (34.1)) and $l = 2$ (P. Sperner [90]).

63.11 Other transitivity conditions. In particular, (63.10) shows that a plane $\mathcal{P}$ is classical or skew hyperbolic if Aut $\mathcal{P}$ has an open orbit of points on which it acts flag transitively. This last condition is indispensable. If we just require an open orbit of points then we obtain an overwhelming multitude of examples, including, e.g., all translation planes; compare also (38.5).

If one wishes to capture 'all the interesting examples', a good transitivity condition to use is *flexibility*, which is traditionally defined as existence of an open orbit in the flag space. For $l = 1$, all flexible planes are known, see (38.1 and 4). It turns out that the flexible planes are precisely those with an automorphism group of dimension at least three. For partial results in the case $l = 2$, see (73.3) and (74.21). The flexible planes of Lenz–Barlotti type II.2 with $l = 2$ have been determined by Knarr [87b]. See (82.34) for a description of these planes. A complete classification of the flexible planes with $l = 2$ has not yet been obtained, but there is some hope, compare Betten [90]. In contrast, for $l \geq 4$, there is an abundance of planes of Lenz type V, i.e., translation planes whose duals also are translation planes, see Section 82; they are all flexible in the above sense. Therefore, it seems preferable to redefine flexibility as finiteness of the set of all flag orbits. This implies existence of an open orbit; indeed, by the sum theorem (92.9), there is an orbit of dimension $3l$, and this orbit is open by (53.1a).

Four-dimensional planes admitting a group transitive on some oval are treated by Löwen [84a].

63.12 Notes on Section 63. Theorem (63.8) on point homogeneous planes was proved by Salzmann first for $l = 1$ [62a] and $l = 2$ [70], and then for $l \leq 16$ [75a]. It was not known at the time that $l < \infty$ implies $l \leq 16$, as is proved in Chapter 5. The first published proof of (63.8) for arbitrary l is due to Löwen [81d]. The proofs given here follow his ideas. Theorem (63.1) on line homogeneous affine planes was obtained by Salzmann [75b].

64 Transitive axial groups

A very special kind of homogeneity is (p, L)-homogeneity, as considered in (23.21). A (p, L)-transitive axial group $\Sigma_{[p,L]}$ is homeomorphic to a line minus one or two points. This imposes severe restrictions on the structure of such groups. Moreover, (p, L)-transitivity translates into strong algebraic properties of a suitably chosen ternary field, where $\Sigma_{[p,L]}$ reappears as the additive or multiplicative structure. Therefore, (p, L)-homogeneous planes are very convenient to work with, and the proofs in this section will be more elementary than in most other parts of this book. In particular, we rely on Chapter 5 to a lesser degree. A survey stressing the algebraic aspect of our topic can be found in Grundhöfer–Salzmann [90].

The Lenz type of a plane $\mathcal{P}$ is determined by the set of flags (p, L) such that $\mathcal{P}$ is (p, L)-homogeneous, compare Section 24. The Lenz–Barlotti type is a refinement obtained by considering transitive homology groups $(p \notin L)$ as well. The regularity implied by assumptions about the type allows us to completely classify the planes of some special types, such as III and IVa.2. We present some results of this kind. Other types, such as IV.1 and V, are very interesting (although they are far too large for complete classification), because they provide a wealth of manageable examples that are particularly homogeneous in more than one respect. In the present section, we shall consider the basic properties of those types. They will be examined more closely in Sections 73 and 82. It will turn out that under suitable additional assumptions, e.g. on the dimension of the automorphism group, a complete classification becomes possible.

Throughout this section, $\mathcal{P} = (P, \mathcal{L})$ is a compact, connected projective plane with lines of dimension l and automorphism group Σ. As in the preceding section, we shall see that our homogeneity assumptions ensure finiteness of dimension.

64.1 Proposition: (p, L)-Transitivity. *Assume that Σ is (p, L)-transitive for some pair $(p, L) \in P \times \mathcal{L}$. Then the lines of $\mathcal{P}$ are manifolds (in fact, spheres of dimension $l \in \{1, 2, 4, 8\}$). Moreover, we have the following.*

(a) *If $p \in L$, then $\Sigma_{[p,L]}$ is a Lie group homeomorphic to $\mathbb{R}^l$, containing no compact subgroup $\neq 1\!\!1$.*

(b) *If $p \notin L$, then $l \leq 4$ and $\Sigma_{[p,L]}$ is isomorphic to the multiplicative group $\mathbb{F}^\times$ of the l-dimensional classical field $\mathbb{F} \in \{\mathbb{R}, \mathbb{C}, \mathbb{H}\}$.*

Proof. For any line $K \in \mathcal{L}_p \setminus \{L\}$, the group $\Delta := \Sigma_{[p,L]}$ is homeomorphic to its domain of transitivity $K \setminus \{p, K \wedge L\}$, see (44.8b), which is locally compact and locally contractible (42.8) and has a countable basis (41.8). By Szenthe's theorem (96.14), Δ is a Lie group and, in particular, Δ and K are l-manifolds. If one assumes that $l < \infty$ then one may use the positive solution of Hilbert's 5th problem (93.2) instead of Szenthe's theorem. By (52.3), we have $K \approx \mathbb{S}_l$, whence $\Delta \approx \mathbb{R}^l$ or $\Delta \approx \mathbb{R}^l \setminus \{0\}$ according to whether $p \in L$ or $p \notin L$. In the first case, absence of compact subgroups was proved in (61.5). Here is an alternative proof: a maximal compact subgroup $\Phi \neq 1\!\!1$ would be contractible, because the theorem of Mal'cev–Iwasawa (93.10) asserts that $\Delta \approx \Phi \times \mathbb{R}^k$; on the other hand, Φ is a compact, connected $(l - k)$-manifold, hence Φ has a non-trivial homology group $H_{l-k}(\Phi; \mathbb{Z}_2)$, see Greenberg [67] 22.30.

If $p \notin L$ and $l = 2$, then Δ is a Lie group homeomorphic to $\mathbb{C}^\times$, and the classification of 2-dimensional Lie algebras and of the corresponding Lie groups shows that $\Delta \cong \mathbb{C}^\times$, compare (94.39 and 40).

If $l > 2$, then we have $\Delta \cong \mathbb{R} \times \Phi$, where Φ is a maximal compact subgroup, see (61.2). We obtain homotopy equivalences $\Phi \simeq \Delta \approx \mathbb{R}^l \setminus \{0\} \simeq \mathbb{S}_{l-1}$. Thus, Φ

is simply connected, so that Φ does not contain a direct factor which is a torus. Hence, the structure theorem for compact Lie groups (94.31c) implies that Φ is a direct product of simply connected, almost simple, compact Lie groups. Since $\dim \Phi = l - 1 \leq 7$, the classification (94.33) implies that each factor is isomorphic to the subgroup $\mathbb{S}_3 \cong \mathrm{Spin}_3 \mathbb{R}$ of $\mathbb{H}^\times$, cf. (11.25). The only possibility is that $\Delta \cong \mathbb{R} \times \mathrm{Spin}_3 \mathbb{R} \cong \mathbb{H}^\times$ and $l = 4$.

Finally, if $l = 1$, then $\Delta^1 \approx \mathbb{R}$ and hence $\Delta^1 \cong \mathbb{R}$. Furthermore, Δ has precisely two connected components, hence Δ is generated by Δ^1 together with one element $\alpha \in \Delta \setminus \Delta^1$ satisfying $\alpha^2 = \delta \in \Delta^1$. Since $\mathrm{Aut}\,\mathbb{R}$ consists of the multiplication maps $r \mapsto ar$, $a \neq 0$, it follows that conjugation by α induces either the identity or inversion on Δ^1. Now recall that Δ is isomorphic to the multiplicative group of some ternary field for $\mathcal{P}$, see (44.8b). This shows that, unlike inversion, conjugation by α maps null sequences in Δ^1 to null sequences. Hence, α centralizes Δ^1. Using a square root $\gamma \in \Delta^1$ of δ, we obtain an element $\beta = \gamma^{-1} \cdot \alpha$ such that $\beta^2 = 1$ and $\Delta = \Delta^1 \times \langle \beta \rangle \cong \mathbb{R}^\times$. □

64.2 Problem. Is every linearly transitive elation group $\Sigma_{[p,L]}$ isomorphic to $\mathbb{R}^l$?

By (64.1a) and (94.38), the answer is 'yes' if $\Sigma_{[p,L]}$ is abelian. This is known to be true if $\Sigma_{[L,L]}$ contains non-trivial elations with different centers, see (61.6).

Our next aim is a closer study of translation planes. According to (25.6), a translation plane can be described by a spread in $\mathbb{R}^{2l}$. In order to understand the conditions on the spread ensuring that a topological plane is obtained, we have to consider the spread as a subset of the Grassmann manifold of all l-dimensional subspaces of $\mathbb{R}^{2l}$, and we have to study this Grassmann manifold first.

64.3 Grassmann manifolds. (a) The set $G_{n,k}$ of all k-dimensional vector subspaces of $\mathbb{R}^n$ admits transitive actions of $\mathrm{GL}_n\mathbb{R}$ and $\mathrm{SO}_n\mathbb{R}$ and, hence, can be identified with a coset space of either group, compare (93.16). Since $\mathrm{SO}_n\mathbb{R}$ is compact, both groups induce the same coset space topology on $G_{n,k}$. To see this, restrict the quotient map $\mathrm{GL}_n\mathbb{R} \to G_{n,k}$ to the subgroup $\mathrm{SO}_n\mathbb{R}$. The same topology may also be obtained as a quotient topology from the set of all ordered bases of k-dimensional subspaces, viewed as a subset of $(\mathbb{R}^n)^k$. This follows, e.g., from (96.9a). In this way, one obtains a compact, connected manifold (94.3a), called a *Grassmann manifold*. Compare Bourbaki [66b] VI, §3.5, where the topology is introduced via bases. Note that $G_{n,1}$ is the projective space $P_{n-1}\mathbb{R}$ with the topology introduced in (14.1).

We say that two subspaces of $\mathbb{R}^n$ are *incident* if one of them is contained in the other. For $j < k < n$ the *incidence relation*

$$I_{j,k} \subseteq G_{n,j} \times G_{n,k}$$

defined in this way is a closed set of pairs. Indeed, $I_{j,k}$ is an orbit of the compact group $SO_n\mathbb{R}$ acting on the product of the Grassmann manifolds.

(b) Let $c = n - k$. For $Y \in G_{n,c}$ let $U \subseteq G_{n,k}$ be the set of all complements of Y, and fix $X \in U$. Then every element of U can be represented in $\mathbb{R}^n = X \times Y$ as the graph $\{(x, f(x)) \mid x \in X\}$ of a unique linear map $f : X \to Y$. In this way, U is mapped bijectively onto the space $L(X, Y) \cong \mathbb{R}^{c \times k}$ of linear maps. In fact, U is an open set, and the map $U \to \mathbb{R}^{c \times k}$ onto the space of matrices is a homeomorphism, compare Bourbaki [66b] VI, §3.5. We shall refer to this homeomorphism as the *Grassmann coordinates* of $G_{n,k}$ with respect to the decomposition $\mathbb{R}^n = X \times Y$.

(c) Let $U \subseteq G_{n,k}$ be the set of subspaces complementary to $Y \in G_{n,c}$ as in (b) and choose $v \notin Y$. Let $Z = Y + v$ and consider the *intersection map*

$$h : U \to Z \quad \text{defined by} \quad V \mapsto V \cap Z.$$

Then h is well defined and continuous. Indeed, suppose that $V_n \to V$ in U and that $z_n = h(V_n)$ converges to $z \in Z$. As incidence is closed by (a), the one-dimensional subspaces $\mathbb{R}z_n = \langle z_n \rangle \leq V_n$ converge to $\langle z \rangle \leq V$, whence $z = h(V)$. On the other hand, if $z_n = y_n + v$ converges to infinity, then $z_n \|z_n\|^{-1}$ accumulates at some non-zero vector $y \in Y \cap V$, a contradiction.

Some of the proofs in this section could be simplified by making more systematic use of the continuity of geometric operations in real projective spaces. The same remark applies to the proof of (61.12), for instance. We decided not to deviate from the main theme of this book by including a full treatment of that topic. The interested reader is referred to Kühne–Löwen [92] for a convenient approach and for historical remarks.

64.4 Theorem: Translation planes and spreads. *Assume that the elation group $\Delta = \Sigma_{[A,A]}$ of $\mathcal{P}$ with axis A is transitive on $P \setminus A$. Then the following conditions (a), (b) and (c) hold.*

(a) $\Delta \cong \mathbb{R}^{2l}$, *and $P \setminus A$ may be identified with Δ via the map $p^\delta \mapsto \delta$, where $p \notin A$ is an arbitrary point. Under this identification, the pencil $\mathcal{L}_p$ becomes a spread (25.6) in $\mathbb{R}^{2l}$ consisting of l-dimensional vector subspaces. This spread $\mathcal{S}$ determines $\mathcal{P}$ since $\mathcal{L} \setminus \{A\} = \bigcup \mathcal{S}^\Delta = \{L^\delta \mid L \in \mathcal{S}, \delta \in \Delta\}$.*

(b) *The spread $\mathcal{S}$ is closed in the Grassmann manifold $G_{2l,l}$. The two spaces $\mathcal{L}$ and $G_{2l,l}$ induce the same topology on $\mathcal{S}$, and $\mathcal{S}$ is homeomorphic to $\mathbb{S}_l$ in this topology.*

(c) *Either the line A is fixed by $\Sigma = \text{Aut } \mathcal{P}$, or $\mathcal{P}$ is a classical plane $\mathcal{P}_2\mathbb{F}$.*

(d) *Conversely, every spread $\mathcal{S} \subseteq G_{2l,l}$ which is closed in the Grassmann manifold gives rise to a topological projective translation plane with affine part $(\mathbb{R}^{2l}, \bigcup \mathcal{S}^\Delta)$.*

Note. In other words, (b) and (d) together say that the spread $\mathcal{S}$ defines a topological translation plane if, and only if, $\mathcal{S}$ is compact in the topology induced by the Grassmann manifold.

Proof. 1) The proof of (a) may be obtained from the general facts about translation planes outlined in Section 25, using the additional information that the kernel of a corresponding quasifield contains the real numbers (42.6). Alternatively, we may use the results of the present chapter, as follows.

As in the proof of (64.1a), we obtain that $\Delta \approx P \setminus A \approx \mathbb{R}^{2l}$, a homeomorphism being given by $\delta \mapsto p^\delta$ for $p \notin A$. By (61.6), Δ is abelian and isomorphic to $\mathbb{R}^{2l}$. Clearly, p^δ belongs to the line $L \in \mathcal{L}_p$ if, and only if $\delta \in \Delta_{[A \wedge L]}$. By (64.1a), the latter is a connected, closed subgroup, hence a vector subspace of $\Delta \cong \mathbb{R}^{2l}$, compare (94.41). The condition that every $q \in P \setminus (A \cup \{p\})$ lies on exactly one line $L \in \mathcal{L}_p$ translates to the condition that the set $\mathcal{S} = \{\Delta_{[A \wedge L]} \mid L \in \mathcal{L}_p\}$ is a spread.

2) We identify $P \setminus A$ with $\mathbb{R}^{2l}$ and accordingly write 0 instead of p. Since $\mathcal{L}_0$ is homeomorphic to $\mathbb{S}_l$ by (52.3), assertion (b) will be proved if we show that the identity map of $\mathcal{L}_0$ is continuous as a map from $\mathcal{L}_0 \subseteq \mathcal{L}$ into $G_{2l,l}$. So let $L_n \to L$ in $\mathcal{L}_0$. Since $G_{2l,l}$ is also compact, we may assume that $L_n \to U \in G_{2l,l}$ and we have to show that $L = U$; in fact, $L \subseteq U$ suffices as both vector spaces are l-dimensional. Now continuity of the geometric operations in the given plane shows that every point $x \in L \setminus \{0\}$ is the limit of some sequence $x_n \in L_n$. This implies that $\langle x_n \rangle \to \langle x \rangle$ in $G_{2l,1} = P_{2l-1}\mathbb{R}$. Hence, $\langle x \rangle \leq U$ because incidence is closed with respect to the Grassmann topology. This proves (b). Assertion (c) follows from (24.9) together with (42.7) or (63.3), or with (64.24) below, and we are left with the proof of (d).

3) From the assumption that $\mathcal{S}$ is a spread it follows easily that $(\mathbb{R}^{2l}, \bigcup \mathcal{S}^\Delta)$ is an affine translation plane, compare (25.6). Let $\mathcal{P} = (P, \mathcal{L})$ be its projective completion. We prove using (43.1) that $\mathcal{P}$ is a topological projective plane if $\mathcal{S}$ is compact. To this end, we exhibit compact Hausdorff topologies on P and on $\mathcal{L}$ and show that incidence is closed. The topology on P will extend the natural one on $P \setminus A = \mathbb{R}^{2l}$.

We consider the sets $X_j = G_{2l+1,j+1}$ of j-flats in the $2l$-dimensional real projective space, where $0 \leq j \leq 2l - 1$. For $x \in X_k$ and $j \leq k$ we define $X_j(x) = \{y \in X_j \mid y \leq x\}$. The spread $\mathcal{S}$ may be considered as a subset of $X_{l-1}(h)$ for some hyperplane $h \in X_{2l-1}$. Following Bruck–Bose [64], we describe $\mathcal{P}$ as follows: points are

(i) the points of $X_0 \setminus X_0(h)$ and
(ii) the elements of $\mathcal{S}$,

and lines are

(i) the hyperplane h and
(ii) the l-flats $x \in X_l$ such that $x \wedge h \in \mathcal{S}$.

Incidence is inclusion. In order to see that indeed the affine translation plane $(\mathbb{R}^{2l}, \bigcup \mathcal{S}^\Delta)$ is obtained from this geometry by removing the line h and its points, the essential step is to observe that parallelity of affine subspaces of $\mathbb{R}^{2l}$ appears in the projective closure of $\mathbb{R}^{2l}$ as coincidence of intersections with h.

Next we reformulate the description of $\mathcal{P}$, introducing topologies at the same time. We construct quotient spaces with identification maps

$$\pi : X_0 \to P, \quad \lambda : \mathcal{K} \to \mathcal{L},$$

defined as follows. The map π sends all of $X_0(s)$ to $s \in \mathcal{S}$ and maps $x \in X_0 \setminus X_0(h)$ to itself; the set $\mathcal{K} \subseteq X_l$ consists of all $x \in X_l$ containing some $s \in \mathcal{S}$, and λ sends x to itself unless $x \leq h$, in which case $\lambda(x) = h$. Then incidence in $\mathcal{P}$ is $(\pi \times \lambda)(I)$, where $I = I_{0,l} \cap (X_0 \times \mathcal{K})$ and the incidence relation $I_{0,l}$ is defined as in (64.3a). By (64.3a), $I_{k,l}$ is always compact. As $\mathcal{S}$ is compact, it follows that $\mathcal{K}$ is closed in X_l, and hence P, $\mathcal{K}$, $\mathcal{L}$ and I are compact. Moreover, the affine point set $P \setminus \pi(h)$ is homeomorphic to $X_0 \setminus X_0(h) \approx \mathbb{R}^{2l}$, compare (14.3).

Thus, all that remains to be done is to show that P and $\mathcal{L}$ are Hausdorff spaces. It is enough to verify that the equivalence relations defining the quotient spaces P and $\mathcal{L}$ are closed subsets of $X_0 \times X_0$ or $\mathcal{K} \times \mathcal{K}$, respectively, see Bourbaki [66a] I § 10.4, Prop. 8. This verification is easily carried out using sequences and applying the fact that both the spread $\mathcal{S}$ and the incidence relations $I_{k,l}$ are compact (64.3a). □

64.5 Remarks. (a) The description of the topology of a compact translation plane as a quotient of $P_{2l}\mathbb{R}$ given in the last proof is of interest even in the case of the classical planes.

(b) Compact spreads $\mathcal{S} \subseteq G_{2l,l}$ give rise to topological translation planes as we have just seen. Therefore they exist only when $l \in \{1, 2, 4, 8\}$, see (52.5). In contrast, arbitrary spreads can be constructed for any l using transfinite induction, see Bernardi [73]. Many constructions of new spreads from old ones, familiar from the abstract theory of translation planes, also produce non-compact spreads. For example, this happens when one constructs a Hall plane from $\mathcal{P}_2\mathbb{C}$. One starts from the regular spread in $G_{4,2}$ defining the complex plane and replaces one of its reguli by the opposite regulus, see André [54b] p. 183. This is a special instance of the process of derivation (Hughes–Piper [73] chap. X). As a consequence of (44.9), this process can never work within the realm of compact, connected planes.

64.6 Corollary. *Consider a locally compact, affine translation plane $\mathcal{A}$ with point set $\mathbb{R}^{2l}$, defined by a spread $\mathcal{S}$. Then the scalar multiplications $\varrho : x \mapsto rx$, $r \in \mathbb{R} \setminus \{0\}$, are homologies of $\mathcal{A}$ with center 0. In particular, the kernel of $\mathcal{A}$*

contains a subfield which is isomorphic to $\mathbb{R}$ *and, hence, coincides with the closure of the prime field.*

Proof. This is obvious since $\mathscr{S}$ is fixed elementwise by ϱ and the kernel describes homologies with improper axis (25.4). For a direct proof of (64.6), see (42.6). □

64.7 Corollary. *A 2-dimensional, locally compact translation plane is isomorphic to* $\mathscr{P}_2\mathbb{R}$.

Proof. A spread of $\mathbb{R}^2$ necessarily consists of all one-dimensional subspaces. Compare also (32.8). □

What is needed according to Theorem (64.4) in order to construct a topological translation plane is a compact spread $\mathscr{S} \subseteq G_{2l,l}$. A candidate might be represented in Grassmann coordinates (64.3b). Then sometimes the spread condition is easily tested. In such cases, the verification of the topological conditions may be simplified by the following Proposition (64.8a). See step 7) in the proof of (73.10) for a typical application. In a similar way, one may also use the result (43.8).

On the other hand, Knarr proved in [95] that the spread condition can be weakened if one knows the topology of $\mathscr{S}$, see (64.9). He uses this in order to describe all four-dimensional topological translation planes in a very simple way. The same device (64.11) works for most of the 'nice' 8- and 16-dimensional examples. In particular, his method gives a clear idea of the abundance of topological translation planes.

64.8 Proposition: Spreads represented in Grassmann coordinates.
Let $\mathscr{S}$ be a spread in $\mathbb{R}^{2l} = \mathbb{R}^l \times \mathbb{R}^l$, and assume that $Y = \{0\} \times \mathbb{R}^l \in \mathscr{S}$. Using Grassmann coordinates (64.3b), consider $\mathscr{S}' := \mathscr{S} \setminus \{Y\}$ as a subset of the space $\mathbb{R}^{l \times l}$ of $(l \times l)$-matrices.

(a) $\mathscr{S}$ *is compact (and hence defines a topological translation plane* $\mathcal{A}$) *if, and only if* $\mathscr{S}' \approx \mathbb{R}^l$.
(b) *For a translation plane* $\mathcal{A}$ *as in* (a) *with the line at infinity* A, *the affine shear group with axis Y is linearly transitive if, and only if,* $\mathscr{S}'$ *is a real affine subspace of* $\mathbb{R}^{l \times l}$. (*In this case,* $\mathcal{A}$ *is of Lenz type at least* V, *see* (24.7).)

Proof. (a) The assertion follows from (43.8), but we also give a direct proof. Note first that Grassmann coordinates are relevant only for the application of (a), not for the proof. If $\mathscr{S}$ is compact, then $\mathscr{S} = \mathscr{L}_0$ is a line pencil in a $2l$-dimensional topological translation plane and hence is an l-sphere (64.4d and b). Conversely, assume that $\mathscr{S}' \approx \mathbb{R}^l$. In order to see that $\mathscr{S}$ is compact, we shall show that every sequence $L_n \in \mathscr{S}'$ converging to ∞' in the one-point compactification $\mathscr{S}' \cup \infty'$ converges to Y in $G_{2l,l}$. It suffices to show that each $x \in Y \setminus \{0\}$ is the limit of some sequence $x_n \in L_n$. Since incidence is closed in the Grassmann topology, this will imply that $\langle x \rangle \leq X$ whenever the sequence L_n accumulates at $X \in G_{2l,l}$. Consequently, $Y \leq X$, and $X = Y$ is the only possibility.

In order to construct a sequence $x_n \to x$, choose $W \in \mathscr{S}'$ and let $W_1 = W + x$. Consider the perspectivity $g : \mathscr{S}' \setminus \{W\} \to W_1 \setminus \{x\}$ sending L to $L \wedge W_1$. As incidence is closed, g is continuous (64.3c). Being a bijection between spaces homeomorphic to $\mathbb{R}^l \setminus \{0\}$, the map g is a homeomorphism by domain invariance (51.19). For $K_n \to W$ in $\mathscr{S}'$, the images $g(K_n)$ cannot accumulate at any $w \in W_1$, or we would obtain that $\langle w \rangle \leq W$. Thus $g(K_n)$ converges to the point ∞ of the one-point compactification $W_1^+ = W_1 \cup \{\infty\}$, and g extends to a continuous map h between the spaces $\mathscr{S}'$ and $W_1^+ \setminus \{x\}$, which are both homeomorphic to $\mathbb{R}^l$. As before, this map is a homeomorphism, hence it extends to a homeomorphism of the one-point compactifications W_1^+ and $\mathscr{S}' \cup \{\infty'\}$ of those spaces. Hence, $x_n := h(L_n) = g(L_n) \in L_n$ converges to the compactifying point x of $W_1^+ \setminus \{x\}$.

(b) Let $K, L \in \mathscr{S}'$ be represented by the linear maps $\varphi_K : \mathbb{R}^l \to \mathbb{R}^l$ and φ_L. Then

$$\alpha : (x, y) \mapsto \bigl(x, y + (\varphi_K - \varphi_L)(x)\bigr)$$

is the unique linear map that is the identity on Y and on $\mathbb{R}^{2l}/Y$ and sends L to K. Since a shear sending L to K has the same properties, all shears are obtained in this way. It follows that the shear group is transitive if, and only if, $\varphi_M + \varphi_K - \varphi_L$ represents a line whenever φ_M, φ_K and φ_L do so, in other words if, and only if, $\mathscr{S}' - \varphi_K \leq \mathrm{L}(\mathbb{R}^l, \mathbb{R}^l)$ is an additive subgroup. Since $\mathscr{S}'$ is assumed to be closed and connected, $\mathscr{S}' - \varphi_K$ is even a real subspace if it is a subgroup, compare (94.12). □

64.9 Proposition: Spread recognition. *Let $\mathscr{S} \subseteq G_{2l,l}$ be a set of pairwise complementary l-dimensional subspaces of $\mathbb{R}^{2l}$. If $\mathscr{S}$ is homeomorphic to the sphere $\mathbb{S}_l$ in the Grassmann topology, then $\mathscr{S}$ is a spread and hence defines a topological translation plane.*

Proof. The missing part of the spread condition is that every $x \in \mathbb{R}^{2l} \setminus \{0\}$ lies on some element of $\mathscr{S}$. To verify this, choose $W \in \mathscr{S}$ such that $x \notin W$, and consider the map

$$f : \mathscr{S}' = \mathscr{S} \setminus \{W\} \to W + x,$$
$$L \mapsto L \cap (W + x),$$

which is well defined and injective since the elements of $\mathscr{S}$ are pairwise complementary. We prove that the image of f is both open and closed in $W + x$; it will follow that x lies on some $L \in \mathscr{S}$. If $y_n = f(L_n) \to y$ in $W + x$, then $\langle y \rangle$ is contained in any $L \in \mathscr{S}$ such that the sequence L_n accumulates at L, because incidence is closed (64.3a). In this case, $y = f(L)$. Openness of $f(\mathscr{S}')$ follows from the domain invariance theorem, see (51.19), because f is continuous (64.3c) and injective. □

For a typical application of the last proposition, see the proof of (73.19), step 2). Part (a) of the following result does not hold for abstract translation planes. Counterexamples are given by Bruen–Fisher [69] and by Bernardi [73] Teorema 2.2.

64.10 Theorem: Transposition.

(a) *Let $\mathcal{S} \subseteq G_{2l,l}$ be a compact spread on $V = \mathbb{R}^{2l}$. Then a compact spread $\mathcal{S}^\circ$ on the dual vector space V^* is obtained as the set of all annihilators $L^\circ = \{f \in V^* \mid f(L) = \{0\}\}$, where $L \in \mathcal{S}$. In other words, every hyperplane of V contains precisely one element $L \in \mathcal{S}$. The spread $\mathcal{S}^\circ$ is called the transpose of $\mathcal{S}$.*
(b) *From Grassmann coordinates of $\mathcal{S}$ with respect to a decomposition $V = X \oplus Y$, $Y \in \mathcal{S}$, one obtains Grassmann coordinates for $\mathcal{S}^\circ$ with respect to $V^* = X^\circ \oplus Y^\circ$ by matrix transposition.*
(c) *The affine translation plane defined by $\mathcal{S}$ admits a linearly transitive shear group with axis Y if, and only if, the plane defined by $\mathcal{S}^\circ$ admits a linearly transitive shear group with axis Y°.*
(d) *A linear transformation $\lambda \in \mathrm{GL}(V)$ maps $\mathcal{S}$ onto itself if, and only if, its contragredient $(\lambda^{-1})^t$ maps $\mathcal{S}^\circ$ onto itself.*

Proof. The elements of $\mathcal{S}^\circ$ are mutually complementary. Moreover, $L \mapsto L^\circ$ is a homeomorphism $\mathcal{S} \to \mathcal{S}^\circ$, cf. Kühne–Löwen [92] 2.8. By (64.8a), this implies that $\mathcal{S}^\circ \approx \mathbb{S}_l$, so that (64.9) applies to $\mathcal{S}^\circ$. This proves assertion (a). In order to prove (b), observe that $(g, h) \in X^\circ \oplus Y^\circ$ annihilates $L = \{(x, \varphi(x)) \mid x \in X\}$ if, and only if, $h|_X = -g \circ \varphi$. The minus sign can be compensated by an appropriate choice of a basis in Y°. Assertion (c) follows from (b) in view of (64.8b). Finally, the proof of (d) uses the fact that the condition $u \in L^\circ$ is equivalent to $u \circ \lambda^{-1} \in (\lambda(L))^\circ$, whence $(\lambda(\mathcal{S}))^\circ = (\lambda^{-1})^t(\mathcal{S}^\circ)$. □

64.11 Proposition: Contractions and spreads. *Let $\mathbb{F} \in \{\mathbb{R}, \mathbb{C}, \mathbb{H}, \mathbb{O}\}$ and let $\varphi : \mathbb{F} \to \mathbb{F}$ be a contraction, that is, a map satisfying the condition*

(i) *for $z \neq w \in \mathbb{F}$, $|\varphi(z) - \varphi(w)| < |z - w|$.*

If moreover

(ii) *$||z| - |\varphi(z)|| \to \infty$ for $z \to \infty$,*

then a compact spread on $\mathbb{R}^{2l} = \mathbb{F}^2$ is obtained by defining

$$\mathcal{S} = \{L_z \mid z \in \mathbb{F} \cup \{\infty\}\},$$

where $L_\infty = \{0\} \times \mathbb{F}$ and $L_z = \{(x, zx + \varphi(z)\bar{x}) \mid x \in \mathbb{F}\}$.

64.12 Remarks. 1) Knarr [95] shows that this construction, applied to $\mathbb{F} = \mathbb{C}$, in fact yields all compact spreads in $\mathbb{R}^4$, up to isomorphism. In higher dimensions, the analogous statement does not hold.

2) Condition (ii) follows from each of the less general conditions

(ii') φ is bounded,
(i') there is $c < 1$ such that $|\varphi(z) - \varphi(w)| < c|z - w|$.

Proof of (64.11). We want to use (64.9), hence we verify that the elements of $\mathscr{S}$ are pairwise complementary, which is immediate from (i) and (11.14), and that the bijection $g : \mathbb{S}_l = \mathbb{F} \cup \{\infty\} \to \mathscr{S}$ sending z to L_z is continuous, so that g is a homeomorphism by compactness. In Grassmann coordinates with respect to the decomposition $\mathbb{R}^{2l} = \mathbb{F} \times \mathbb{F}$, the image $g(z)$ of $z \in \mathbb{F}$ is represented by the linear selfmap $x \mapsto zx + \varphi(z)\bar{x}$ of $\mathbb{F}$. This shows continuity of g on $\mathbb{F}$.

It remains to be shown that $L_{z_n} \to L_\infty$ for $z_n \to \infty$. Choose $w, y \in \mathbb{F}$ arbitrarily and let $\{a_n\} = L_{z_n} \cap \big(L_w + (0, y)\big)$. Then

$$a_n = (x_n, z_n x_n + \varphi(z_n)\overline{x_n}) = (x_n, w x_n + \varphi(w)\overline{x_n} + y),$$

hence

$$|y| = |(z_n - w)x_n + (\varphi(z_n) - \varphi(w))\overline{x_n}|$$
$$\geq |x_n| \cdot \big||z_n - w| - |\varphi(z_n) - \varphi(w)|\big|$$
$$\geq |x_n| \cdot \big(\big||z_n| - |\varphi(z_n)|\big| - |w| - |\varphi(w)|\big).$$

In view of (ii), this implies that $x_n \to 0$, whence $a_n \to (0, y)$. Now $(0, y)$ is an arbitrary point of L_∞, hence we have shown that L_∞ is contained in Z whenever the sequence L_{z_n} accumulates at $Z \in G_{2l,l}$. Thus $L_{z_n} \to L_\infty$. □

In (64.21c) and in Sections 73 and 82, concrete examples of translation planes will be given. The methods used to describe these planes vary; sometimes we specify the Grassmann coordinates of the spread elements, or we write down a contraction as in (64.11), see (73.24) and the note (c) following (73.13). In other cases, such as (64.20 and 21c), (82.1, 4, 16, 18 and 21) we exhibit a coordinatizing quasifield Q, compare Section 25. By (25.2), the additive group $(Q, +)$ is isomorphic to any of the groups $\Sigma_{[p,A]}$ for p on the translation axis A. By the comments following (64.2), this means that $(Q, +) \cong \mathbb{R}^l$; compare also (44.8b) and (42.6). The following result shows that it is not necessary to verify the planarity condition (25.2v) in order to construct a quasifield of this kind; compare also (25.3). For minimal characterizations of locally compact, connected quasifields, see Hähl [82b].

64.13 Proposition: Locally compact, connected quasifields. *Let $(x, y) \mapsto x \cdot y$ be a continuous multiplication with unit on $\mathbb{R}^l$ satisfying the following conditions:*

(i) *For every $x \in \mathbb{R}^l$, left multiplication $y \mapsto x \cdot y$ is an $\mathbb{R}$-linear map $\lambda_x : \mathbb{R}^l \to \mathbb{R}^l$, and $\lambda_0 = 0$.*
(ii) *For every $a \neq 0$ and $b \in \mathbb{R}^l$, the equation $x \cdot a = b$ has a unique solution $x \in \mathbb{R}^l$.*

Then $(\mathbb{R}^l, +, \cdot)$ is a topological quasifield and coordinatizes a compact translation plane. All locally compact, connected (planar) quasifields arise in this way.

Proof. It suffices to show that $(\mathbb{R}^l, +, \cdot)$ is a quasifield; then all topological conditions follow by (43.6). Only the planarity condition is missing, which requires that for all $b, c, d \in \mathbb{R}^l$ with $b \neq c$ there is a unique $y \in \mathbb{R}^l$ such that $b \cdot y - c \cdot y = d$. In other words, we have to show that the $\mathbb{R}$-linear map $\lambda_b - \lambda_c$ is bijective for $b \neq c$. It suffices to prove that the map has trivial kernel, and this follows from (ii).
The converse is given by (42.6); compare also (64.4 and 6). □

64.14 Lemma: Center of semifields. *If $Q = (\mathbb{R}^l, +, \circ)$ is a topological semifield, that is, a left and right distributive quasifield, then the subfield $\mathbb{R}$ of the kernel obtained in (64.6) is central in $(Q, \circ)$. In particular, Q is a possibly non-associative $\mathbb{R}$-algebra.*

Proof. By (25.8), there is a (closed) central subfield F of Q. The assertion follows because $\mathbb{R}$ is a minimal closed subfield, hence $F \cap \mathbb{R} = \mathbb{R}$. (It is also easy to verify directly, using the two distributive laws, that $\mathbb{N}$, $\mathbb{Q}$ and hence $\mathbb{R}$ are contained in the center.) □

64.15 Corollary: Lenz type V. *If $\mathcal{P}$ is a non-desarguesian compact, connected projective plane of Lenz type V, that is, if both $\mathcal{P}$ and its dual are non-desarguesian translation planes, then $\mathcal{P}$ is 8- or 16-dimensional.*

Proof. In view of (64.7), it suffices to exclude the possibility that $l = 2$. In this case, $\mathcal{P}$ is coordinatized by a 2-dimensional semifield Q, see (24.7), and we show that $Q \cong \mathbb{C}$. Let $\{1, j\}$ be a basis of the $\mathbb{R}$-vector space Q. By (64.14), it suffices to show that we can choose j such that $j \circ j = -1$. This is possible by (55.20). Alternatively, we may show that Q is a commutative field extension of $\mathbb{R}$ by observing that the associative law holds for the elements of a basis $\{1, b\}$, and hence is generally satisfied. □

64.16 Examples of topological semifields and hence of compact planes of Lenz type V can be obtained as *mutations* of the non-commutative classical semifields $\mathbb{F} = \mathbb{H}$ or $\mathbb{O}$: for a real $t \neq \frac{1}{2}$, define $\mathbb{F}_t$ as the set $\mathbb{F}$ with the usual addition and with the multiplication

$$x \circ_t y := txy + (1-t)yx.$$

It is easily seen that $\mathbb{F}_t$ is a semifield; it defines a compact plane by (64.13). For $t = 0$ or 1, the semifield $\mathbb{F}_t$ is isomorphic to $\mathbb{F}$. More about $\mathbb{F}_t$ and the corresponding plane is said in (82.21 and 22). For further examples of topological semifields see Grundhöfer–Salzmann [90] or (82.1 and 16).

64.17 General remarks on Lenz–Barlotti types. So far, the present section was concerned with the basic methods of construction of planes of types IV and V. In the remainder of the section, we shall treat the Lenz–Barlotti types III, IVa.2 and VII. The compact, connected planes of these types can and will be completely classified.

At this point, we shall briefly discuss the remaining types. First recall that transitive homology groups do not occur for $l = 8$ by (64.1b). Next, there are no topological planes of type I.6, see Pickert [75] p. 342. The reason for this is that some coordinatizing ternary field would have to satisfy the identity $x^{-1} + x = 1$ for all $x \neq 0$. If $x \to 0$, then this implies that $x^{-1} \to 1$, whence $x \to 1$. This is a contradiction. In general, existence of planes of type I.6 is an open problem.

Presumably, rigid planes (admitting only the trivial automorphism) abound. Yet they are rarely observed, just because they are not 'nice'. Only for $l = 1$ there are many known examples; see, e.g., (38.7), Schellhammer [81] and Steinke [85]. Steinke's examples are obtained by gluing together two desarguesian half planes. Rigid planes obviously are of type I.1. For $l > 2$, examples of type I.1 are the Hughes planes of dimension 8 and 16, see Section 86.

Planes of type I.2 have been constructed by Schellhammer [81] for $l = 1$, see (34.1), and by P. Sperner [90] for $l = 2$. Type I.3 does not exist in any dimension except perhaps $l = 4$. Type I.4 is known to exist for $l = 1$. Examples are given by Naumann [54], compare also Salzmann [57] Section 13, Bedürftig [74b]. See the note following (71.11) for comments on the case $l = 2$.

Planes of types II.1 and II.2 with $l = 1$ were found by Salzmann [64], see (37.6). Other planes of type II.1 are the 4-dimensional shift planes of Betten and Knarr; see Section 74, in particular (74.9). Further examples are obtained by gluing together two halves of different translation planes, see Betten [84] ($l = 2$) and Weigand [87] ($l \geq 4$); compare (82.30)ff. More planes of types II.1 and II.2 can be found in Knarr [87b] and Plaumann–Strambach [74], compare (82.34).

Each of the classical planes $\mathcal{P}_2\mathbb{F}$, where $\mathbb{F} \in \{\mathbb{R}, \mathbb{C}, \mathbb{H}, \mathbb{O}\}$, admits a group of Lenz type III, e.g., the full stabilizer $\Sigma_{o,W}$ of an antiflag $o \notin W$. The Moulton planes also admit such a group, but the antiflag determined by the centers and axes of elations is unique in this case; see (34.8b). In fact, we have the following result.

64.18 Theorem: Lenz type III. *The Moulton planes (see (31.25b) and (34.2)) and the classical planes $\mathcal{P}_2\mathbb{F}$ for $\mathbb{F} \in \{\mathbb{R}, \mathbb{C}, \mathbb{H}, \mathbb{O}\}$ are the only compact, connected* *projective planes of Lenz type* (*at least*) III. *More precisely, the Moulton planes are of type* III.2.

Proof. 1) Let $\mathcal{P}$ be of Lenz type III. Then there is an antiflag $o \notin W$ such that for each $a \in W$ the elation group $\Theta(a) := \Sigma_{[a,ao]}$ is (a, ao)-transitive. By (64.1), W is a manifold of dimension $l \in \{1, 2, 4, 8\}$ and $\Theta(a)$ is a Lie group homeomorphic to $\mathbb{R}^l$. In particular, the closed subgroup $\Delta \leq \Sigma_{o,W}$ generated by these elation groups is connected. Δ acts doubly transitively on W and acts transitively on the set of 'generic' flags, i.e., flags (p, G) such that $p \notin W$ and $o \notin G$. By (96.14), this implies that Δ is a Lie group; compare also (62.11).

2) For $l \geq 2$, the method of our proof will be to identify the group Δ and to show that the unordered pair of stabilizers $\{\Delta_p, \Delta_G\}$ for a generic flag (p, G) is determined to within conjugacy. According to (23.24), this will imply that the geometry $(P', \mathcal{L}')$ of generic flags is determined up to duality for any given value of $l \geq 2$; here $P' = P \setminus (W \cup \{o\})$ and $\mathcal{L}' = \mathcal{L} \setminus (\mathcal{L}_o \cup \{W\})$. By a density argument, the pencil $\mathcal{L}_o$ is also determined, and finally the plane $\mathcal{P}$ is determined as the projective closure of the affine plane $\mathcal{P}^W$. Since the $2l$-dimensional classical plane $\mathcal{P}_2\mathbb{F}$ is of Lenz type III, it follows that $\mathcal{P} \cong \mathcal{P}_2\mathbb{F}$. A different way to complete the proof would be to reconstruct the generic lines from the action on the generic point set as the orbits of the elation groups $\Theta(a)$.

3) For l arbitrary, the group $\Omega := \Delta|_W$ induced by Δ on W is simple (96.16). Hence, (96.17) asserts that Ω is equivalent to the group of projectivities of the projective line over $\mathbb{F}$, where $\mathbb{F} \in \{\mathbb{R}, \mathbb{C}, \mathbb{H}, \mathbb{O}\}$ and $\dim_\mathbb{R} \mathbb{F} = l$. These groups have been treated in Section 15, see (15.6). Another proof of the stated equivalence can be found in the original paper by Salzmann [74]; it makes stronger use of the geometric situation. For $l = 1$, where $\Omega \cong \mathrm{PSL}_2\mathbb{R}$, we can now refer to (34.9) for the proof that $\mathcal{P}$ is either classical or a Moulton plane. In the Moulton planes, $\Sigma_{[o,W]}$ is (o, W)-transitive, hence their type is indeed III.2, compare (34.8b).

Henceforth, let $l \geq 2$.

4) The groups $\Theta(a)$ and $\Delta_{[o]}$ have trivial intersection and normalize each other, hence they commute. Since the groups $\Theta(a)$ generate Δ, this implies that $\Xi := \Delta_{[o]}^1$ satisfies

$$\Xi \leq C(\Delta)^1 \leq \sqrt{\Delta} \leq \Xi;$$

here $\sqrt{\Delta}$ denotes the radical, which is connected by definition (94.18), and the last inclusion holds because $\Delta/\Delta_{[o]} = \Omega$ is simple. Consequently, a Levi subgroup Λ of Δ is a covering group of Ω, and we obtain a covering epimorphism $\Xi \times \Lambda \to \Delta$, which sends (ξ, λ) to $\xi\lambda$ and has a discrete kernel isomorphic to a subgroup of $\Lambda_{[o]}$.

5) We have $\Omega \cong \mathrm{PSO}_{l+2}(\mathbb{R}, 1)$ by (15.5 and 6), hence a maximal compact subgroup of Ω is (locally) isomorphic to $\mathrm{SO}_{l+1}\mathbb{R}$, see (94.33). By (62.8 and 9), Λ has a maximal compact subgroup $\Psi \cong \mathrm{Spin}_{l+1}\mathbb{R}$, and Ψ acts on P in the same way as on the classical point set. By (93.10b), we have $\pi_1(\Lambda) = \pi_1(\Psi) = \mathbb{1}$, hence it follows that Λ is the simply connected covering group of Ω,

$$\Lambda = \mathrm{Spin}_{l+2}(\mathbb{R}, 1).$$

The kernel of the covering map has order 2, see (94.33), hence it coincides with $\Psi_{[o]} \cong \mathbb{Z}_2$, compare (62.10a and b).

Let $a, b \in W$. Since $\Theta(a) \leq \Delta_a$ is sharply transitive on $W \setminus \{a\}$, the action of $\Psi_{a,b} \cong \mathrm{Spin}_l\mathbb{R}$ on $\Theta(a)$ is equivalent to the action on $W \setminus \{a\}$, which is the standard action of $\mathrm{SO}_l\mathbb{R}$ on $\mathbb{R}^l$, see (62.10b). Thus, $\Psi_{a,b}$ acts irreducibly on $\Theta(a)$, and in particular $\Theta(a)$ is abelian; note that $\Theta(a) \approx \mathbb{R}^l$ cannot be simple. Moreover, we obtain that either $\Theta(a) \cap \Lambda = \mathbb{1}$ or $\Theta(a) \leq \Lambda$. The former possibility implies that $\Theta(a) \cong \mathbb{R}^l$ is isomorphic to a subgroup of $\Xi/(\Xi \cap \Lambda)$, contrary to (64.1b). Since Δ is generated by the elation groups, we obtain that

$$\Delta = \Lambda.$$

6) From (15.6) and (15.7), it can be seen that Ω_a acts on $W \setminus \{a\} \approx \mathbb{R}^l$ as $\mathbb{R}^l \cdot \mathrm{GO}_l^+\mathbb{R}$, with standard action. Note that the quoted results also apply to $l = 8$. The normal subgroup induced by $\Theta(a)$ meets the factor $\Pi \cong \mathbb{R}^l$ non-trivially, because the factor group $\mathrm{GO}_l^+\mathbb{R} \cong \mathbb{R} \times \mathrm{SO}_l\mathbb{R}$ has no l-dimensional normal subgroup. Thus in fact $\Theta(a)$ induces Π, and $\Theta(a)$ itself is uniquely determined by the action (Δ, W) together with the point a; indeed, $\Theta(a)$ is the unique connected subgroup of Δ which induces Π.

For $a, b \in W$, the group $\Omega_{a,b}$ has a unique maximal compact subgroup $\mathsf{P} = \mathrm{SO}_l\mathbb{R}$. Replacing the maximal compact subgroup $\Psi \leq \Delta$ by a suitable conjugate, we may arrange that P is induced by $\Psi_{a,b}$, compare (93.10a). We have to select Ψ such that it contains the compact inverse image of P; this has the side-effect that then $\Psi_a = \Psi_{a,b}$. This stabilizer is isomorphic to $\mathrm{Spin}_l\mathbb{R}$ for $l \geq 2$ and to $\mathrm{SO}_2\mathbb{R}$ for $l = 1$, compare (62.10b and d). The group $\Psi_{a,b}$ is completely determined by a, b and the action of Δ on W.

7) Now consider a point $p \in A = a \vee o \setminus \{a, o\}$. Its stabilizer $\Psi_p = \Psi_{a,b,p}$ has been described in (62.10f). For $l = 2$, we have $\Psi_p = \mathbb{1}$, and for $l \geq 4$ the group $\Psi_p \cong \mathrm{Spin}_{l-1}\mathbb{R}$ is determined up to conjugacy in $\mathrm{Ns}_\Psi \Psi_{a,b}$. This conjugation could possibly exchange $\Theta(a)$ with $\Theta(b)$.

8) Consider the line $G = p \vee b$. We show finally that $\{\Delta_p, \Delta_G\}$ can be constructed using $\Theta(a)$, $\Theta(b)$ and Ψ_p, which completes the program outlined in step 2). Indeed,

$$\{\Delta_p, \Delta_G\} = \{\Psi_p \Theta(a), \Psi_p \Theta(b)\}$$

holds because the groups on the right hand side are contained in those on the left hand side and because the dimensions are correct, as an easy calculation using the dimension formula (96.10) shows. One also needs to know that Δ_p and Δ_G are connected. For Δ_p, this follows from (96.9c) because $p^\Delta = P \setminus (W \cup \{o\}) \simeq \mathbb{S}_{2l-1}$ is simply connected, and the dual argument works for Δ_G. □

64.19 Definition. Recall that a nearfield is defined as a quasifield with associative multiplication (25.7). Kalscheuer [40] constructs nearfields $N_r = (\mathbb{R}^4, +, \circ)$ for $0 \leq r \in \mathbb{R}$ by deforming the quaternion skew field $\mathbb{H}$ as follows. Addition is the usual one, and multiplication $\circ$ is defined, using quaternion multiplication $\cdot$, by setting $0 \circ x := 0$ and

$$a \circ x := a \cdot |a|^{-ir} \cdot x \cdot |a|^{ir} \quad \text{for} \quad a \neq 0.$$

Here, $i = \sqrt{-1} \in \mathbb{C}$. Using the homomorphism

$$\varphi_r :]0, \infty[\to \mathbb{C}^\times \leq \mathbb{H}^\times : t \mapsto t^{ir} = e^{ir \ln t},$$

the product in N_r may also be written as

$$a \circ x = a \cdot x^{\varphi_r(|a|)}.$$

This presents N_r as a Dickson nearfield, compare Wähling [87] Kap. 3. Note that $N_0 = \mathbb{H}$.

64.20 Theorem: Kalscheuer nearfields.

(a) *For every real $r \geq 0$, the construction above produces a topological nearfield N_r, and every locally compact, connected proper nearfield is isomorphic to N_r for a unique $r > 0$.*
(b) *Only $N_0 = \mathbb{H}$ is a (skew) field. In all other cases, the kernel of N_r is the subfield $\mathbb{C}$ with its usual field structure, and the multiplicative group $N_r^\times$ is isomorphic to $\mathbb{H}^\times$. The center of $N_r^\times$ is $\langle -1 \rangle \times C_r$, where*

$$C_r = \{t \cdot \varphi_r(t) \mid 0 < t \in \mathbb{R}\}.$$

We shall prove this theorem together with its corollary (64.22) after the statement of the latter.

64.21 Remarks. (a) The second assertion of part (a) is true without the planarity assumption, which is inherent in our definition of nearfields. The proof is virtually the same.

(b) Let F be one of the nearfields $\mathbb{R}, \mathbb{C}$ and N_r, $r \geq 0$, and consider the group $A(F)$ of all affine maps $F \to F$, defined by

$$x \mapsto a \circ x + b, \quad \text{where} \quad a, b \in F.$$

Tits [52, 56] shows that up to isomorphism, these groups with their natural actions are the only sharply 2-transitive Lie transformation groups on manifolds. Actually, his topological assumptions are weaker than stated here. The proof of the classification of nearfields presented in the sequel is essentially his proof given in [56].

(c) Suppose we replace the homomorphism φ_r in the construction of N_r by an arbitrary continuous map φ sending the positive real numbers into $\mathrm{Spin}_3 \mathbb{R} \leq \mathbb{H}^\times$ and satisfying $\varphi(1) = 1$. Then we obtain 4-dimensional topological quasifields, and the associated topological translation planes are precisely the 8-dimensional translation planes admitting an effective action of $\mathrm{Spin}_4 \mathbb{R}$. This was proved by Hähl [80a]. The planes are examples for Lenz–Barlotti type IVa.1, unless φ is a homomorphism.

(d) The Kalscheuer nearfields are closely related to the Hughes planes of dimension 8, see Section 86, Salzmann [81a] Satz 2, and Maier-Stroppel [95].

64.22 Corollary: Lenz–Barlotti type IVa.2 (nearfield planes). *The locally compact, connected projective planes of Lenz–Barlotti type (at least) IVa.2 are precisely the planes over the Kalscheuer nearfields N_r, $r > 0$, as described in (64.19), and the classical desarguesian planes $\mathcal{P}_2 \mathbb{F}$, where $\mathbb{F} \in \{\mathbb{R}, \mathbb{C}, \mathbb{H}\}$. These planes are pairwise non-isomorphic. Their automorphism groups will be determined in (82.24).*

Proof of Theorem (64.20) *and Corollary* (64.22). 1) We prove that N_r is a nearfield with continuous addition and multiplication. Then (64.13) will show that N_r is in fact a topological nearfield coordinatizing a compact plane. Continuity of multiplication near 0 follows from the observation that $|a \circ x| = |a| \cdot |x|$. The remaining continuity properties are trivial. We verify the algebraic axioms, writing $\varphi = \varphi_r$. Distributivity is an easy consequence of the fact that conjugation by $\varphi(t)$ is additive. Associativity is proved by the following computation, which uses the relation $|x \circ y| = |x| \cdot |y|$:

$$x \circ (y \circ z) = x \cdot (y \cdot z^{\varphi(|y|)})^{\varphi(|x|)} = x \cdot y^{\varphi(|x|)} \cdot z^{\varphi(|y|) \cdot \varphi(|x|)}$$
$$= x \cdot y^{\varphi(|x|)} \cdot z^{\varphi(|x \circ y|)} = (x \circ y) \circ z.$$

In order to solve $y \circ b = c$ for y, note that $|y| = |b|^{-1} \cdot |c| = t$ and compute y from $y \cdot b^{\varphi(t)} = c$.

2) It is easily seen that the elements $t \cdot \varphi_r(t)$ form a subgroup of the center $C(N_r^\times)$. Conversely, if a is central, then $a \cdot x^{\varphi_r(|a|)} = x \cdot a$ for all $x \in \mathbb{S}_3 = \{x \in \mathbb{H} \mid |x| = 1\}$. Consequently, $a \cdot \varphi_r(|a|)^{-1}$ belongs to $C(\mathbb{H}^\times)$, hence this must be a real number t. We infer that $|t| = |a|$, whence $a = t \cdot \varphi_r(|t|)$. This verifies the statement made about the center.

It follows that N_r is a (skew) field only if $r = 0$; indeed, if both distributive laws hold in N_r, then $\mathbb{R}$ is central in N_r by (64.14), and therefore $r = 0$. Observe that this also follows from our assertion that N_r and N_q are non-isomorphic for $r \neq q$.

The multiplication in N_r agrees with quaternion multiplication whenever the right factor belongs to $\mathbb{C}$, because $\varphi_r(|a|) \in \mathbb{C}$ centralizes $\mathbb{C}$. Therefore, the right distributive law holds in this situation, and $\mathbb{C}$ is contained in the kernel K of N_r. If K is a proper (skew) field extension of $\mathbb{C}$, then $\dim_\mathbb{R} K = 4$ and $N_r = K$ is a skew field, hence $r = 0$.

From what we have seen it follows that

$$N_r^\times = C_r \times \mathbb{S}_3 \cong \mathbb{R} \times \mathrm{Spin}_3 \mathbb{R} \cong \mathbb{H}^\times;$$

note that in $\mathbb{S}_3$ the multiplication of N_r agrees with quaternion multiplication. This completes the proof of (b).

3) By (25.7), a projective plane is of type at least IVa.2 if, and only if, it can be coordinatized by a nearfield. The improper locally compact, connected nearfields are the fields $\mathbb{R}$, $\mathbb{C}$ and $\mathbb{H}$, compare (42.6). The planes defined by N_r and N_q are isomorphic if, and only if $N_r \cong N_q$, see André [55] Satz 7 or Pickert [75] p. 219. This shows that (64.22) is indeed a corollary of (64.20). What remains to be proved is that every locally compact, connected proper nearfield is isomorphic to N_r for a unique $r > 0$.

In order to do this, we need the following auxiliary result.

64.23 Lemma: Isomorphism of nearfields. *Let $(N, +, \circ)$ and $(N, +, *)$ be nearfields with the same additive group. The two nearfields are isomorphic if, and only if, the subgroup $L_\circ = \{x \mapsto a \circ x \mid a \in N^\times\} \leq \mathrm{Aut}(N, +)$ is conjugate to the corresponding subgroup L_*.*

Proof. Let $\alpha : (N, +, \circ) \to (N, +, *)$ be an isomorphism. Then the relation

$$\alpha^{-1}(\alpha(a) * \alpha(x)) = a \circ x$$

shows that $\alpha^{-1} L_* \alpha = L_\circ$. Conversely, let $\alpha \in \mathrm{Aut}(N, +)$ with the latter property be given. Let 1 and e be the respective units of $(N, \circ)$ and $(N, *)$. There is a unique map $\beta \in L_*$ sending $\alpha(1)$ to e. The composite $\tilde{\alpha} := \beta\alpha \in \mathrm{Aut}(N, +)$ maps 1 to e and also satisfies $\tilde{\alpha}^{-1} L_* \tilde{\alpha} = L_\circ$. We rewrite this as $L_* \tilde{\alpha} = \tilde{\alpha} L_\circ$ and obtain a bijection $\gamma : N \to N$ satisfying $\gamma(0) = 0$ and

$$\gamma(a) * \tilde{\alpha}(x) = \tilde{\alpha}(a \circ x)$$

for all $a, x \in N$. Taking $x = 1$ we see that $\gamma(a) = \tilde{\alpha}(a)$, and $\tilde{\alpha}$ is an isomorphism of nearfields. □

Note that we have explicitly constructed an isomorphism from the given map α.

Proof of (64.20), *resumed.* 4) Let $N = (N, +, \circ)$ be a proper, locally compact, connected nearfield. By (25.2 and 4), the additive or multiplicative group of N is respectively isomorphic to a linearly transitive elation or homology group of the corresponding plane; see also (44.8b) for the fact that this is an isomorphism of topological groups. By (64.7), we have $l \geq 2$. If $l = 2$, then (64.1b) implies that $N^\times$ is commutative. Thus, both distributive laws hold in N, and N is a commutative field. Therefore, $l \geq 4$, and in fact $l = 4$ and $N^\times \cong \mathbb{H}^\times \cong \mathbb{R} \times \mathrm{Spin}_3 \mathbb{R}$ by (64.1b). Moreover, we may identify $(N, +)$ with $(\mathbb{H}, +)$ by (42.6), compare (64.2). Only this part of the proof requires a few additional arguments if planarity is not assumed.

5) Consider the subgroup $L_\circ \leq \mathrm{GL}_4 \mathbb{R}$ as defined in the lemma. We have $L_\circ \cong N^\times$, hence $L_\circ = C \times S$, where $S \cong \mathrm{Spin}_3 \mathbb{R}$ is the unique maximal compact subgroup of $L_\circ$ and C is the connected component of the center. Now $\mathrm{Spin}_3 \mathbb{R}$ has only one faithful 4-dimensional representation (95.10), so S is conjugate in $\mathrm{GL}_4 \mathbb{R}$ to $S_\mathbb{H} = \{ x \mapsto a \cdot x \mid a \in \mathbb{S}_3 \}$. According to the lemma, we may even arrange that $S = S_\mathbb{H}$ if we replace N by a suitable isomorphic nearfield with the same addition.

By (95.10), the centralizer of S in $\mathrm{GL}_4 \mathbb{R}$ is $R = \{ x \mapsto x \cdot b \mid b \in \mathbb{H}^\times \}$. The one-parameter group $C \leq R$ has elements of the form $x \mapsto x \cdot b_t$, where $t \mapsto b_t$ is a multiplicative homomorphism $[0, \infty[\to \mathbb{H}^\times$. We may change C using conjugation by $\varrho \in R$; this fixes S and replaces b_t by a conjugate $b^{-1} \cdot b_t \cdot b$ in $\mathbb{H}^\times$. The elements b_t generate a commutative subfield $F \leq \mathbb{H}$, and using (11.24) we see that $F = \mathbb{C}$ up to conjugacy. This implies that $b_t = t^{c+ir}$ for some $c, r \in \mathbb{R}$. If $c = 0$, then $L_\circ = S \times C$ is compact, a contradiction. Hence, we may reparametrize C so that $c = 1$. Finally, we may conjugate by the map $\varrho : x \mapsto x \cdot j$, which sends i to $-i$. This replaces r by $-r$, hence we may assume that $r > 0$ and $C = \{ x \mapsto x \cdot t^{1+ir} \mid 0 < t \in \mathbb{R} \}$. On the other hand, in N_r the corresponding group C_r contains the mappings $x \mapsto (t \cdot \varphi_r(t)) \circ x = x \cdot t^{1+ir}$, hence we obtain the same result there.

By several steps of conjugation, we have now transformed $L_\circ$ into L_r, the group of left multiplications of N_r. By the lemma, this means that $N \cong N_r$. We have $r > 0$ since we started with a proper nearfield.

6) It remains to be proved that $N_r \cong N_q$ only if $r = q$, where r and q are assumed to be positive. Any isomorphism $\lambda : N_r \to N_q$ maps the connected component C_r of the center of $N_r^\times$ to C_q and preserves the subfield $\mathbb{C}$, which contains the centers

and is the common kernel of N_r and N_q by step 1). Now λ induces the identity or conjugation on $\mathbb{C}$, hence $q \in \{r, -r\}$ and in fact $q = r$ as both numbers are positive. — For variations of step 6), see Kalscheuer [40], Gröger [92]. □

64.24 Theorem: Lenz type VII (Moufang planes). *The only compact, connected projective Moufang planes are the four classical planes* $\mathcal{P}_2\mathbb{F}$, *where* $\mathbb{F} \in \{\mathbb{R}, \mathbb{C}, \mathbb{H}, \mathbb{O}\}$.

For *proofs*, see (42.7) and (63.3 and 4). □

64.25 Further results on translation planes. Characterizations of translation planes are given by (61.22) and (71.11) (the latter for dimension 4), and we mention Theorem (66.2) on groups of projectivities of translation planes. Systematic treatments of translation planes are given separately for the possible dimensions 4, 8 and 16 in Sections 73 (see also (72.10 and 11)) and 81, 82. More results on spreads and partial spreads can be found in Plaumann–Strambach [90] and Löwen [94]. The Kalscheuer nearfields (64.20) have been characterized by H. Löwe [94] as the only locally compact, connected quasifields besides the classical division algebras $\mathbb{R}, \mathbb{C}, \mathbb{H}, \mathbb{O}$ which satisfy the first of the Moufang identities (12.15). The duals of the nearfield planes contain the only known non-classical symmetric planes, see H. Löwe [94, 95].

64.26 Notes on Section 64. The description of translation planes by spreads is due to André [54b] and to Bruck and Bose in their joint papers [64, 66]. Theorem (64.4), saying that precisely the compact spreads lead to *topological* translation planes, is implicit in Breuning [70]. The proof given here is taken from Löwen [89]. The results (64.8, 9 and 11) on the recognition of topological spreads have developed out of Betten's method of 'transversal homeomorphisms' for the case $l = 2$, see Betten [72b]. The characterization of spreads (64.9) as well as the method of representing translation planes by contractions (64.11) were obtained by Knarr [95]. Corollary (64.10) on transposition was proved by Buchanan and Hähl in their article [78], using different methods. The characterization of the Moulton planes as the only compact, connected planes of Lenz type exactly III is due to Salzmann [64, 72a, 74]. We follow his proof closely. The classification of locally compact, connected nearfields was given by Kalscheuer [40] and Tits [52, 56]. Grundhöfer [90a] shows that also every disconnected, locally compact nearfield of characteristic zero is a Dickson nearfield.

65 Groups of large dimension

In this brief section, we introduce the basic ideas connected with that measure of homogeneity which is best adapted to topological planes, namely the dimension of the automorphism group. This concept is meaningful because for a compact plane $\mathcal{P}$, the group $\Sigma = \operatorname{Aut} \mathcal{P}$ is a locally compact topological group with a countable basis (44.3), so that it makes sense to define $d = \dim \Sigma$, compare (93.5 and 6). Assumptions about the size of d relative to the dimension $l = \dim L$ of a line can be compared to assumptions about the order of the automorphism group of a finite plane relative to the order of the plane. To exploit these assumptions, we rely on finiteness of the invariants d and l together with topological facts like the topological pigeon-hole principle (51.20 and 21b) and the dimension formula (96.10). These powerful tools allow us to mimic the counting arguments known from finite geometry.

Yet assumptions about d would be quite useless if they were not complemented by some information about the general *stiffness* of the given class of geometries, that is, by upper bounds on the dimensions of the stabilizers of suitable configurations in $\mathcal{P}$. Such bounds depend on the dimension l and have to be established for each possible value of l individually. For example, bounds are known for the dimension of the stabilizer Δ of a non-degenerate quadrangle. According to (44.4c), the group Δ may be interpreted as the automorphism group of a coordinatizing ternary field. Hence, Δ is trivial for $l = 1$ (32.10) and is of order at most 2 for $l = 2$ (55.21). The general conjecture that the dimension of the stabilizer of a non-degenerate quadrangle never exceeds that of its counterpart in the l-dimensional classical plane is still unproved in the case $l = 4$. For $l \geq 4$ it is therefore more convenient to look at the stabilizer of a triangle, for which the analogous assertion will be proved in Chapter 8. In order to explain with as little technicalities as possible how to work with this result, we shall quote it here (65.1) and then apply it to prove a rudimentary form of the characterization of the classical planes by the dimension of their automorphism groups (65.2).

To describe the situation somewhat more closely, let C_l denote the dimension of the automorphism group of the $2l$-dimensional classical plane $\mathcal{P}_2\mathbb{F}$. We have $C_1 = 8$, $C_2 = 16$, $C_4 = 35$ and $C_8 = 78$ by the results of Chapter 1; see (13.6 and 8) and (17.10) for the groups and refer to (94.33) for their dimensions. In the present section we shall be content to prove that a plane $\mathcal{P}$ is classical if it satisfies $d \geq C_l - 1$. In particular, this shows that there is a *critical dimension* c_l such that non-classical planes with $d = c_l$ exist but $d > c_l$ forces a plane to be classical. In fact the values of c_l are known, see (33.6), (34.8a), (38.1), (73.1), (84.28), (85.16) and (82.27). We have

$$c_1 = 4 = \tfrac{1}{2}C_1 \qquad\qquad c_2 = 8 = \tfrac{1}{2}C_2$$
$$c_4 = 18 = \tfrac{1}{2}(C_4 + 1) \qquad\qquad c_8 = 40 = \tfrac{1}{2}(C_8 + 2).$$

In all cases, c_l may be described as $c_l = 2(3l - m - 1)$, where $l = 2^m$. All planes with $d = c_l$ are known. For $l = 1$, they are precisely the Moulton planes (38.1); in all other cases, they are translation planes. We refer to Chapters 7 and 8 for a detailed discussion.

By a *triangle* we mean three non-collinear points.

65.1 Theorem: Stiffness. *In a compact, connected projective plane of finite dimension $2l$, the dimension of the stabilizer $\Delta \leq \operatorname{Aut} \mathcal{P}$ of a triangle is bounded above by $C_l - 6l$, that is, by* 2, 4, 11, 30 *for $l = 1, 2, 4, 8$, respectively. The bounds are attained in the classical planes.*

Proof. For $l \in \{4, 8\}$, we refer to Chapter 8, see (83.26). For $l \leq 2$, choose a point x not on any side of the given triangle. Observe that the group Δ_x fixes a non-degenerate quadrangle. Hence, (32.10) and (55.21b) assert that the order of Δ_x is at most 2. Using the dimension formula (96.10) and the subspace theorem (92.4) (applied to $x^\Delta \subseteq P$), we infer that $\dim \Delta = \dim x^\Delta + \dim \Delta_x \leq 2l + 0$.

In the classical desarguesian planes, the group Δ is described in (12.19). In the octonion case, Δ is given by (12.17); compare also (17.13). For its dimension, refer to (94.33). □

65.2 Corollary: Groups of almost maximal dimension. *If $\dim \Sigma \geq C_l - 1$ holds for the automorphism group Σ of a compact, connected projective plane $\mathcal{P} = (P, \mathcal{L})$ of dimension $2l < \infty$, then $\mathcal{P}$ is isomorphic to the $2l$-dimensional classical plane $\mathcal{P}_2\mathbb{F}$, where $\mathbb{F} \in \{\mathbb{R}, \mathbb{C}, \mathbb{H}, \mathbb{O}\}$, and then $\dim \Sigma = C_l$.*

Proof. 1) We show that $\dim \Sigma \geq C_l - 1$ implies that Σ is doubly transitive on P. Suppose that this is not the case; then some stabilizer Σ_x is not transitive on $P \setminus \{x\}$. This set is connected, compare (51.26), hence there must be a Σ_x-orbit $B \subseteq P \setminus \{x\}$ with $\dim B < \dim P = 2l$, because orbits of dimension $2l$ are open (53.1a). Either $B \cup \{x\}$ contains a triangle, or B is contained in a unique line $L \in \mathcal{L}_x$, which is fixed by Σ_x because B is invariant. Applying the dimension formula (96.10) and leaving the case $l = 1$ aside, which was treated in (33.6), it is now easy to find a triangle with stabilizer Δ such that $\dim \Delta \geq \dim \Sigma - 6l + 2 > C_l - 6l$. This lower bound is larger than what the stiffness theorem (65.1) allows.

2) By (96.15), the connected component Σ^1 is a doubly transitive Lie group acting on the compact space P. Thus Σ^1 is simple (96.16), and we can use the list (96.17) of possibilities for this action. The actions (96.17b) are easily ruled out. Indeed, if P were homeomorphic to a sphere, then the complement of a point would be contractible, but in fact it has the homotopy type of a line according to (51.26); note that lines are l-spheres (and hence are not contractible), because every point is isotropic (53.2). Similarly, if P were homeomorphic to a projective space $P_n\mathbb{F}$

with $n \geq 3$ and $\mathbb{F} \neq \mathbb{O}$, then the complement of a point could be retracted onto a hyperplane h not containing that point by a projection map. (In fact, the projection would be a homotopy equivalence, compare the proof of (51.26).) However, a space which is homotopy equivalent to a sphere cannot be retracted onto a projective space, because the projective space has more non-zero homology groups than the sphere, see Spanier [66] 5.8.4 and 5.8.5. Hence from the list (96.17) of actions, only the actions of the maximal connected automorphism groups of the classical planes can occur in our situation.

3) It remains to reconstruct the lines of $\mathcal{P}_2\mathbb{F}$ from one of these actions. Consider any non-trivial homology $\sigma \in \Sigma^1$ of the classical plane. The fixed point set F of σ is disconnected. In fact $F = L \cup \{c\}$, where $L \approx \mathbb{S}_l$ is the classical axis of σ and $c \notin L$ is the center. In particular, F cannot be a projective subplane, since compact projective planes are locally homogeneous (51.1). Therefore, L has to be contained in some line L' of $\mathcal{P}$, or we could generate additional fixed points contrary to the definition of F. Now the topological pigeon-hole principle (51.20) implies that $L = L'$ and $L \in \mathcal{L}$. Since Σ^1 is transitive on the classical line set, every classical line belongs to $\mathcal{L}$. These must be all the lines of $\mathcal{P}$ because they form a projective plane. □

65.3 Note. Using (63.8), the last result could have been improved slightly. Other proofs of similar results can be based on (63.1 or 2) or (63.5).

66 Remarks on von Staudt's point of view

This chapter would not be complete without at least a few remarks on one more measure of homogeneity. Whereas all measures considered previously concern the size of the group of continuous collineations (= automorphisms), which is never larger than it is in the classical cases and is probably trivial in 'most' cases, this one deals with the group of projectivities (21.6) or *von Staudt group* Π of a plane $\mathcal{P}$, and behaves rather in the opposite way. Namely, the group Π acting on a line is triply transitive in every projective plane (21.6), and it is sharply triply transitive if, and only if, the plane is pappian. For this well-known theorem of von Staudt, see Pickert [75] 5.1.4, p. 139 or Dembowski [68] 3.4.9 and 3.4.10, p. 160f. A fascinating application of von Staudt's theorem will be given in (75.1). In general, Π tends to be excessively large. Various relaxations of the condition of sharp triple transitivity have been successfully investigated both in abstract and in compact, connected projective planes. See the conference report Plaumann–Strambach [81] for a survey; the topological case is treated in the chapter Löwen [81e] of that book. Some of the major results, due to Strambach [77] and Löwen [77], are summarized by the following theorem; compare Löwen [81e] 5.1. A transformation group is said to act ω-*regularly* if the stabilizer of some finite set of points is discrete.

66.1 Theorem: Projectivities of projective planes. *Let* Π *be the group of projectivities of a line L in a compact, connected projective plane* 𝒫, *endowed with the compact-open topology* (96.3). *The following conditions are equivalent.*

(a) Π *is locally compact.*
(b) *The closure of* Π *in the homeomorphism group of L is locally compact.*
(c) $\dim P$ *is finite and* Π *acts ω-regularly.*
(d) 𝒫 *is a classical plane* $\mathcal{P}_2 \mathbb{F}$, *where* $\mathbb{F} \in \{\mathbb{R}, \mathbb{C}, \mathbb{H}, \mathbb{O}\}$.

The proof of (d) from condition (a) or (b) uses the classification of triply transitive transformation groups (96.18) and von Staudt's theorem. In the proof that (c) implies (a), one first establishes the fact that Π has at most two path-connected components and then applies results of Gleason–Palais [57].

Similarly, consider a locally compact, connected affine plane 𝒜 and its group A of affine projectivities. These are the projectivities of the projective closure $\overline{\mathcal{A}}$ obtained by using projection centers on the line at infinity only; in other words, affine projectivities are products of parallel projections. Part (a) of the following theorem is proved in the articles quoted above. Part (d), which characterizes Moufang planes among locally compact, connected translation planes, is obtained in Grundhöfer–Strambach [86]. This has been used by Otte [93] in his proof that differentiable projective translation planes are Moufang planes, compare (75.5).

66.2 Theorem: Affine projectivities. *Let* 𝒜 *be a locally compact, connected affine plane of finite dimension, and let* A *be its group of affine projectivities with respect to an affine line K. Then the following assertions hold.*

(a) *If* A *is locally compact or acts ω-regularly, then* 𝒜 *is an affine translation plane.*
(b) *If* 𝒜 *is the desarguesian plane over* $\mathbb{F} \in \{\mathbb{R}, \mathbb{C}, \mathbb{H}\}$, *then* A *acts on K as the group of all affine maps* $x \mapsto ax + b$.
(c) *If* 𝒜 *is the octonion affine plane* $\mathcal{A}_2 \mathbb{O}$, *then* A *acts on K as the group of direct similitudes of* $\mathbb{R}^8$.
(d) *For all other translation planes,* A *is the connected component of the group of affine transformations of K considered as a vector space over the kernel of* 𝒜.

For the Moulton planes, the group Π has been computed by Betten [79b], compare also Löwen [81e] 1.4, and its structure has been investigated by Betten and Wagner in their paper [82]. For further examples, see Betten [83] and Betten–Weigand [85].

Chapter 7

Four-dimensional planes

Planes of dimension 4 have many special features that make a separate treatment desirable. For example, in Section 74 we treat the important class of shift planes and its intimate relationship with the class of translation planes. This attractive phenomenon is without parallel in higher dimensions, as we show in (74.6). We shall describe the state of knowledge on 4-dimensional planes; it may be summed up by saying that the critical dimension in the sense of Section 65 is $c_2 = 8$, and that all planes with an automorphism group of dimension $d \geq 7$ and all shift planes with $d \geq 6$ are known. Almost all planes with $d \geq 7$ are translation planes or dual translation planes, the only exception being one single shift plane found by Knarr (74.24, 26 and 27).

Space does not permit us to give full proofs of these statements. What we omit are mainly the contents of Salzmann [73b], Betten [90], Löwen [90], Betten [75, 76, and 77] and Knarr [86]. The first three of these articles show that $d \geq 7$ is possible only in shift planes and (dual) translation planes, and the last four classify translation planes with $d = 7$ and shift planes with $d = 6$. Especially Salzmann's contribution is beyond the scope of this book. Nevertheless, we hope to convey a clear picture of the theory. We present the basic facts about the action of special types of groups (compact, semi-simple, abelian) in Section 71, various group-theoretical characterizations of the complex plane in Section 72, and a systematic treatment of translation planes and shift planes in Sections 73 and 74. We shall prove that 8 is the critical dimension (73.1), and we shall completely classify translation planes with $d \geq 8$. Moreover, we shall prove that there is a unique shift plane with $d \geq 7$. The classification of translation planes and shift planes for the next lower values of d will be outlined. The final Section 75 shows that only the complex plane admits a complex analytic structure.

71 Automorphism groups

Throughout this section, $\mathcal{P} = (P, \mathcal{L})$ denotes a 4-dimensional, compact, projective plane.

We consider the action of the automorphism group Σ of a 4-dimensional plane and establish the fundamental results that will be used throughout this chapter. In particular, we obtain that Σ is a Lie group and that the action of Σ has strong stiffness properties. In this respect, the case of 4-dimensional planes is almost as convenient as the 2-dimensional case. Moreover, for subgroups of Σ belonging to some special classes of groups (compact groups, semi-simple groups, abelian groups of dimension 4), we determine all possibilities for the groups and for their actions.

71.1 Review. For the convenience of the reader, we collect some of the facts that have emerged so far concerning the topology of 4-dimensional planes and concerning their automorphism groups.

Every line L is homeomorphic to the 2-sphere $\mathbb{S}_2$, and the point set P is homeomorphic to the point set $\mathrm{P}_2\mathbb{C}$ of the complex plane (53.15). The automorphism group $\Sigma = \mathrm{Aut}\,\mathcal{P}$ is a locally compact transformation group of P and of $\mathcal{L}$ and has a countable basis (44.3). Every automorphism fixes some point and some line (55.19). The stabilizer of a non-degenerate quadrangle either is trivial or is generated by a Baer involution, and a connected group does not contain any Baer involution (55.21b and c). Hence, every involution contained in the identity component Σ^1 is a reflection (55.29). A reflection is uniquely determined by its center and axis (55.32i). Together with (55.34b), the non-existence of Baer involutions in Σ^1 implies that Σ^1 does not contain a subgroup isomorphic to $\mathbb{Z}_2^3$; in particular, Σ does not contain a 3-torus (55.37a). A 2-torus $\mathsf{T} \le \Sigma$ fixes the triangle formed by the centers of the reflections contained in T (55.35). Abelian subgroups of Σ are at most 4-dimensional (55.41).

71.2 Theorem: Lie group. *The automorphism group Σ of a 4-dimensional, compact plane $\mathcal{P}$ is a Lie group of dimension at most 16.*

Proof. 1) We prove the dimension bound first. A point orbit of a closed subgroup of Σ is at most 4-dimensional by the subspace theorem (92.4). Since the stabilizer of a non-degenerate quadrangle is discrete by (55.21b), repeated application of the dimension formula (96.10) shows that $\dim \Sigma \le 16$.

2) Next, we show that the stabilizer $\Sigma_{p,L}$ of any pair $(p, L) \in P \times \mathcal{L}$ is a Lie group. We use the facts that a locally compact effective transformation group of a 2-manifold is a Lie group (96.31), and that a locally compact group is a Lie group if it contains a closed normal Lie subgroup with Lie factor group (94.3d). Since

the axial group $\Sigma_{[p,L]}$ acts effectively on each line through p other than L, it is a Lie group. It is the kernel of the action of $\Sigma_{[p],L}$ on L, and the effective quotient group (96.2c) is a Lie group. This shows that $\Sigma_{[p],L}$ is a Lie group, and repeating the argument we obtain that $\Sigma_{p,L}$ has the same property.

3) According to (93.8a), Σ has an open subgroup Δ such that the factor group Δ/Δ^1 is compact. It suffices to show that Δ is a Lie group. By the approximation theorem (93.8c), we know that Δ has arbitrarily small 0-dimensional, compact, normal subgroups Θ such that Δ/Θ is a Lie group. Assume that Δ is not a Lie group; then $\Theta \neq 1$. Since the topology of Σ is the topology of uniform convergence on P, we may assume that all point orbits of Θ have diameter less than a given number $\varepsilon > 0$. Every element of Θ fixes some point and some line by (55.19), hence there is some pair (p, L) such that the Lie group $\Delta_{p,L}$ intersects Θ in a compact Lie group $\Theta_{p,L} \neq 1$. The latter group is 0-dimensional, hence discrete and even finite. Now a theorem of Newman (96.27) asserts that every non-trivial, finite group of homeomorphisms of a compact manifold has an orbit of diameter at least δ, where $\delta > 0$ is a constant depending only on the manifold and its metric. If we choose $\varepsilon = \delta$, then we obtain a contradiction. □

The next proposition essentially says that an axial group $\Sigma_{[A]}$ is always solvable. For an application in (71.8) we need in fact a somewhat stronger statement.

71.3 Proposition: Axial groups are solvable. *If a closed, connected subgroup $\Delta \leq \Sigma$ fixes all points of some non-empty, open subset U of a line A, then Δ is solvable.*

Proof. 1) By (71.2), Δ is a Lie group. Suppose that Δ is not solvable. Then Levi's theorem (94.28b) together with (94.23) implies that there is an almost simple, analytic subgroup $\Pi \leq \Delta$. In its Lie topology (94.14), the group Π is at least 3-dimensional by (94.39 and 40).

2) Assume first that $\Pi \leq \Delta_{[A]}$ is axial. We claim that Π contains non-trivial homologies. Suppose that Π consists of elations. There are elations with different centers in Π, because a group $\Delta_{[c,A]}$ is at most 2-dimensional (61.4a). Then the closure $\overline{\Pi} =: \Lambda$ is abelian by (23.13), a contradiction.

Now (61.20a) asserts that Λ is a semi-direct product of a connected, normal subgroup consisting of elations and a group $\Lambda_{[c]}$ of homologies. Being almost simple, Π is locally isomorphic to some analytic subgroup of the factor $\Lambda_{[c]} \leq \Sigma_{[c,A]}$. Hence Π is at most 2-dimensional, a contradiction.

3) If Π contains an involution π, then π is a reflection (55.21 and 29). If π belongs to the center of Π, then Π fixes the center $c \notin A$ of π and each point $u \in U$ and, hence, acts freely on the remainder of the line cu (as in the case of axial collineations). This is impossible by the dimension formula (96.10). If π

does not belong to the center, then the conjugacy class of π generates the almost simple group Π. In this case, Π is strictly axial, and we may refer to step 2).

4) The only almost simple Lie group containing no involution is the universal covering group Ω of $SL_2\mathbb{R}$, compare (94.37). It has an infinite cyclic center Z. The adjoint representation of Ω leaves the Killing form invariant, which is non-degenerate and has index 1. Using this fact one may show that Ω contains precisely 3 conjugacy classes of one-parameter subgroups, compare Hilgert–Hofmann [85]. Suppose that $\Pi \cong \Omega$. By the dimension formula (96.10), every line L satisfying $L \wedge A = u \in U$ is fixed by some one-parameter group P. Every group P' conjugate to P also fixes some line in $\mathcal{L}_u$. Some conjugacy class has this property with respect to two points u_1 and u_2. A group P belonging to this class fixes a triangle $\{u_1, u_2, p\}$, and moreover P fixes all lines pu for $u \in U$. The point p is uniquely determined by this condition, or else P would act trivially. Hence Z fixes p, and fixes all elements of an open subset of $\mathcal{L}_p$. By an argument symmetric to the preceding one, p must be fixed by Π. This is impossible since $\dim \Pi = 3$, compare step 3). □

71.4 Lemma: Fixed elements. *Let $\Delta \leq \operatorname{Aut}\mathcal{P}$ be a closed, connected group of automorphisms of a 4-dimensional, compact plane. Then either Δ fixes some point or some line, or Δ is a direct product of (center-free) simple Lie groups.*

Proof. If Δ is not a direct product of center-free simple groups, then it follows from (94.23 and 26) that Δ either has a non-trivial center or contains a connected abelian normal subgroup. In both cases, we have to show that Δ fixes some element. As every single automorphism has this property (55.19), it suffices to prove the following assertion: If subgroups $\mathbb{1} \neq \Theta \trianglelefteq \Lambda \leq \Delta$ are given such that Λ is connected and normalizes Θ, and if Θ fixes some element, then also Λ fixes some (possibly different) element.

Hence assume, for example, that Θ fixes a point x. Then Θ acts trivially on the connected orbit x^Λ. If this orbit is not reduced to x, but is contained in a line L, then Λ fixes L. In the remaining case, the orbit contains a quadrangle according to (42.3), and Θ fixes this quadrangle, contradicting (55.21b and c). □

71.5 Definition. We say that a topological group Δ *acts on a plane* $\mathcal{P}$ if a continuous homomorphism $\Delta \to \operatorname{Aut}\mathcal{P}$ is given. If this homomorphism is injective, then we say that Δ acts *effectively* on $\mathcal{P}$.

71.6 Lemma. *Assume that we have a locally compact, connected group Δ and a 4-dimensional, compact, projective plane $\mathcal{P}$, as well as a monomorphism $\varphi : \Delta \to \Sigma = \operatorname{Aut}\mathcal{P}$. Then the following conditions are equivalent.*

(a) Δ *acts on $\mathcal{P}$ via φ, in other words, φ is continuous.*
(b) *Via φ, the group Δ becomes a topological transformation group of the point space P.*
(c) *Up to isomorphism, Δ is an analytic subgroup of Σ, endowed with its Lie topology, and φ is the inclusion homomorphism.*

Proof. Assertion (a) implies (b), because Σ is a transformation group. Since Σ carries the compact-open topology, the converse implication follows from the fact that the compact-open topology on Δ^φ is the coarsest one that makes this group into a topological transformation group, see Arens [46]. If $\varphi : \Delta \to \Sigma$ is a continuous monomorphism, then Δ is a Lie group, see (94.3e). According to (94.14), this means that Δ^φ is an analytic subgroup. Conversely, if Δ is an analytic subgroup, then its Lie topology is finer than the subgroup topology, see (94.14), hence Δ acts on $\mathcal{P}$ via the inclusion homomorphism. □

The equivalence of (a) and (b) does not depend on the fact that Σ is a Lie group, and thus remains true if we pass to higher-dimensional planes.

The next lemma adapts the stiffness result (55.21) to the case of actions of connected groups, and formulates some consequences that will be used frequently.

71.7 Lemma: Stiffness. *Let Δ be a locally compact, connected group acting effectively on a 4-dimensional, compact, projective plane $\mathcal{P}$. Let $\mathfrak{e} = (a, b, c, d)$ be a quadrangle in P, that is, a quadruple of distinct, non-collinear points.*

(a) *If $\mathfrak{e}$ is non-degenerate, that is, if no three of the four points a, b, c, d are collinear, then the stabilizer $\Delta_\mathfrak{e}$ of those four points is trivial.*
(b) *If a, b, c are collinear points and $d \notin ab$, then $\dim \Delta_\mathfrak{e} \leq 2$.*
(c) *In the situation of (b), assume that $\dim \Delta_\mathfrak{e} = 2$. Then the group $\Delta_\mathfrak{e}$ is isomorphic to $\mathbb{C}^\times$, and the 1-torus $\mathsf{T} \leq \Delta_\mathfrak{e}$ is axial, $\mathsf{T} \leq \Delta_{[d,ab]}$. In particular, T contains a reflection with center d and axis ab.*

Proof. Assertion (a) is immediate from (55.21b and c). This implies that the group $\Delta_\mathfrak{e}$ considered in (b) and (c) acts freely on the punctured line $X := da \setminus \{d, a\}$. The dimension formula (96.10) shows that $\dim \Delta_\mathfrak{e} \leq 2$ in this case. If the bound is attained, then each orbit of $\Delta_\mathfrak{e}$ in X is open by (53.1a). Hence, $\Delta_\mathfrak{e}$ is transitive on the connected space $X \approx \mathbb{C}^\times$. By (96.8), the group $\Delta_\mathfrak{e}$ is homeomorphic and then even isomorphic (94.39 and 40) to the group $\mathbb{C}^\times$. If $\mathsf{T} \cong \mathrm{SO}_2\mathbb{R}$ acts non-trivially on the 2-sphere ab, then the action is equivalent to the standard one, consisting of the rotations of $\mathbb{S}_2$ about a fixed axis; see Mostert [57] Theorems 4 and 5. This action does not fix three points. □

71.8 Theorem: Semi-simple groups. *Let Δ be a locally compact, connected group acting effectively on a 4-dimensional, compact, projective plane $\mathcal{P}$. If Δ is semi-simple, then Δ is even almost simple. In fact, Δ is either*

(a) *locally isomorphic to $\mathrm{SL}_2\mathbb{R}$, or*
(b) *isomorphic to one of the groups*

$$\mathrm{SO}_3\mathbb{R}, \quad \mathrm{Spin}_3\mathbb{R}, \quad \mathrm{SL}_2\mathbb{C}, \quad \mathrm{PSL}_2\mathbb{C},$$
$$\mathrm{PSU}_3(\mathbb{C},0), \quad \mathrm{PSU}_3(\mathbb{C},1), \quad \mathrm{PSL}_3\mathbb{R}, \quad \mathrm{PSL}_3\mathbb{C}.$$

Note. All groups in the list (b) act on $\mathcal{P}_2\mathbb{C}$. This is obvious except for $\mathrm{Spin}_3\mathbb{R}$, which acts as $\mathrm{SU}_2\mathbb{C}$, see (11.26), and for $\mathrm{PSL}_2\mathbb{C}$, which acts as $\mathrm{SO}_3\mathbb{C}$, compare (18.32) and (94.33). It is also clear that $\mathrm{SL}_2\mathbb{R}$ and $\mathrm{PSL}_2\mathbb{R}$ act on $\mathcal{P}_2\mathbb{C}$, the latter group being isomorphic to the real hyperbolic motion group. The more than two-sheeted covering groups of $\mathrm{PSL}_2\mathbb{R}$ do not act on the complex plane $\mathcal{P}_2\mathbb{C}$; indeed, the Fundamental Theorem (13.6) together with (94.27) shows that such an action would be induced by a faithful linear representation on $\mathbb{C}^3$, contrary to (95.10). It is not known whether or not they act on any other 4-dimensional plane. The universal cover does act, however, on the (2-dimensional) Moulton planes, see Section 34.

Proof of (71.8). 1) By (71.6), Δ is a Lie group. First, suppose that $\Delta = \Pi\Lambda$ is a product of two almost simple groups that centralize each other. By (94.39 and 40), both factors are at least 3-dimensional. In steps 2) to 6), we show that this is impossible.

2) We claim that the center Z of Δ is non-trivial. Otherwise, each factor contains infinitely many involutions (94.37), which must be reflections by (55.21 and 29). The reflections in Π commute with those in Λ, and hence all the centers of the former are incident with all the axes of the latter (55.35). Distinct reflections have either distinct axes or distinct centers (55.32). If x and y are different points that occur as centers of reflections in Π, for example, then the reflections in Λ all have the same axis $L = xy$. Since the involutions generate a normal subgroup of Λ, this means that Λ acts trivially on L, contrary to (71.3).

3) We claim that, up to duality, the center Z is an axial group, $\mathsf{Z} \leq \Delta_{[W]}$. By (71.4) and step 2) we may assume that Δ fixes some line W. The group Z is straight in the sense of (23.15). Indeed, if the orbit x^Z of $x \notin W$ contains a triangle, then the group $\Delta_x \neq \mathbb{1}$ fixes a quadrangle, contrary to (71.7a). By (23.15 and 16) together with (55.21c), this implies that Z consists of axial collineations. Hence, Δ fixes the centers of all non-trivial elements of Z, and the previous arguments work in the dual plane, as well. Using (23.16) again, we infer that Z is an axial group.

4) We consider the action of Δ on W. If there is no 2-dimensional Δ-orbit in W, then the classification of 1-dimensional actions (96.30) shows that on each orbit

one of the factors acts trivially. Hence, W is the union of the two closed sets formed by all fixed points of Π and of Λ. At least one of these sets has non-void interior, contrary to (71.3).

5) Let $U \subseteq W$ be a 2-dimensional orbit of Δ, as provided by step 4). According to (53.1a), the set U is open in W. None of the factors of Δ can act transitively on U. Indeed, if $u^\Lambda = U$, for example, then the kernel of the action on U contains the stabilizer Π_u, with $\dim \Pi_u \geq 3 - 2$; it follows that the almost simple group Π acts trivially, contradicting (71.3). Now Π permutes the Λ-orbits in U transitively, hence those orbits are all 1-dimensional, and the same is true for the Π-orbits.

6) By step 3), both factors of $\Delta/\Delta_{[W]}$ are simple groups. Step 5) together with the list (96.30) shows that they are in fact isomorphic to $PSL_2\mathbb{R}$. By step 5) again, the 4-dimensional stabilizer of $u \in U$ intersects each factor in a 2-dimensional subgroup. Now $PSL_2\mathbb{R}$ contains only one conjugacy class of 2-dimensional subgroups. Hence, Δ_u is determined up to conjugacy, and U is homeomorphic to the torus $\mathbb{S}_1 \times \mathbb{S}_1$, with the Cartesian square of the natural $PSL_2\mathbb{R}$-action. Using the topological pigeon-hole principle (51.20), we conclude that the inclusion map $U \to W$ is a homeomorphism between a torus and a sphere. This is a contradiction.

7) We know now that Δ is an almost simple Lie group of dimension $f \leq 16$. Among the list (94.33) of almost simple Lie groups, those satisfying $10 \leq f \leq 15$ can be excluded because each of them contains some almost direct product of two 3-dimensional, almost simple groups. For example, the compact group of type G_2 contains $SO_4\mathbb{R}$, which is locally isomorphic to $SO_3\mathbb{R} \times SO_3\mathbb{R}$, see (11.31b). The symplectic group $Sp_4\mathbb{R}$ contains $Sp_2\mathbb{R} \times Sp_2\mathbb{R} = SL_2\mathbb{R} \times SL_2\mathbb{R}$, and the unitary group $SU_4(\mathbb{C}, 1)$ contains $SU_2(\mathbb{C}, 1) \times SU_2\mathbb{C} \cong SL_2\mathbb{R} \times Spin_3\mathbb{R}$; for the isomorphism, see Dieudonné [71] Chap. II, §4c and §5 no. 1.

8) If Δ is locally isomorphic to the elliptic or hyperbolic motion group $PSU_3(\mathbb{C}, i)$ of the complex plane, then (62.6) asserts that Δ is in fact isomorphic to that group. The only proper covering group of $PSL_3\mathbb{C}$ is the simply connected, threefold cover $SL_3\mathbb{C}$, compare (94.2a and 33). It contains the threefold cover $SU_3\mathbb{C}$ of $PSU_3\mathbb{C}$ by (94.30). Since $SU_3\mathbb{C}$ cannot act effectively, the same is true for $SL_3\mathbb{C}$.

9) The only group that remains to be excluded is the universal covering group Ω of $PSL_3\mathbb{R}$. Its maximal compact subgroup is $K \cong Spin_3\mathbb{R}$, compare (94.30). If $\Delta \cong \Omega$, then the center $Z \cong \mathbb{Z}_2$ of Δ contains a reflection $\zeta \in \Delta_{[c,A]}$, and Δ fixes c and A. The action of K forces c to be isotropic (62.8b). The classification of transitive actions on surfaces (Mostow [50], compare (96.32)) now shows that the action induced by Δ on A is the standard action of $SL_3\mathbb{R}$ on the 2-sphere formed by the oriented lines through $0 \in \mathbb{R}^3$. Hence, Δ contains a 3-dimensional subgroup Ξ fixing all points of some simple closed curve $S \subseteq A$; such a subgroup of Δ/Z

is formed by the matrices

$$\begin{pmatrix} a^{-2} & b & c \\ & a & \\ & & a \end{pmatrix},$$

where $a, b, c \in \mathbb{R}$ and $a > 0$. If $\xi \in \Xi$ fixes an additional point on one of the fixed lines sc, where $s \in S$, then ξ fixes a quadrangle, and $\xi = 1\!\!1$ by (71.7a). Hence, $\Xi_s = 1\!\!1$, contrary to the dimension formula (96.10). □

71.9 Proposition: Compact groups. *Let Δ be a compact, connected group acting effectively on a 4-dimensional, compact, projective plane $\mathcal{P}$. Then Δ is isomorphic to some subgroup of the classical elliptic motion group $\mathrm{PSU}_3\mathbb{C}$, that is, to a torus of dimension at most 2, or to $\mathrm{U}_2\mathbb{C}$, $\mathrm{SU}_2\mathbb{C}$, $\mathrm{SO}_3\mathbb{R}$, or to $\mathrm{PSU}_3\mathbb{C}$ itself.*

Proof. According to (94.31c), Δ is a locally isomorphic image of some direct product, where each factor is either a torus or a compact, simply connected, almost simple group. By (55.37a), the ranks of the factors add up to 2 at most. By (71.8), there is at most one almost simple factor, and the corresponding subgroup of Δ can only be $\mathrm{SO}_3\mathbb{R}$, or $\mathrm{SU}_2\mathbb{C} \cong \mathrm{Spin}_3\mathbb{R}$, or $\mathrm{PSU}_3\mathbb{C}$. The rank condition allows to multiply the first two by a torus, but this is not in fact possible for $\mathrm{SO}_3\mathbb{R}$. Indeed, the involution contained in the torus would be axial by (55.21c and 29), and $\mathrm{SO}_3\mathbb{R}$ would fix the center, contradicting (62.8b). Alternatively, we may argue that $\mathrm{SO}_3\mathbb{R} \times \mathrm{SO}_2\mathbb{R}$ contains a subgroup $\mathbb{Z}_2{}^3$, which would consist of reflections contrary to (55.34b). The group $\mathrm{U}_2\mathbb{C}$ is a product of $\mathrm{SU}_2\mathbb{C}$ and a 1-dimensional torus. Suppose that Δ is such a product. Each factor contains a unique involution, and these are amalgamated in $\mathrm{U}_2\mathbb{C}$. This is in fact the only possibility for Δ. Indeed, the reflection τ contained in the torus centralizes the involution $\sigma \in \mathrm{SU}_2\mathbb{C}$, hence τ fixes the center c and the axis A of σ. Since $\mathrm{SU}_2\mathbb{C}$ is transitive on A by (62.8b), the center of τ cannot belong to A, whence it coincides with c. Similarly, A is the axis of τ, and (55.32) shows that $\sigma = \tau$. The intersection of the two factors is contained in the center $\mathbb{Z}_2$ of $\mathrm{SU}_2\mathbb{C}$, hence no further amalgamation can take place. □

71.10 Theorem: Actions of compact groups. *If $\Delta \in \{\mathrm{SO}_3\mathbb{R}, \mathrm{SU}_2\mathbb{C}\}$ acts effectively on a 4-dimensional, compact plane $\mathcal{P}$, then the action on the point set P is equivalent to the corresponding standard action on the point set $P_2\mathbb{C}$ of the complex plane.*

In particular, $\mathrm{SU}_2\mathbb{C}$ fixes an antiflag $p \notin L$, and p and L are

isotropic in the sense of (62.7). Every action of $\Delta = \mathrm{SO}_3\mathbb{R}$ preserves a desarguesian Baer subplane with point set Q, on which Δ acts in the standard way, and has an orbit O equivalent to $\mathbb{S}_2$ with the standard action; the points of all other Δ-orbits have stabilizers of order two.

Addendum. It is not too difficult to prove that the exceptional $SO_3\mathbb{R}$-orbit O is a topological oval in $\mathcal{P}$, see Löwen [84a].

Proof of (71.10). 1) For $SU_2\mathbb{C} \cong Spin_3\mathbb{R}$, we may refer to (62.9 and 10). If the real, elliptic motion group $\Delta = SO_3\mathbb{R}$ acts on $\mathcal{P}$, we consider the conjugacy class $\mathcal{I} \subseteq \Delta$ formed by all reflections of $\mathcal{P}_2\mathbb{R}$. By (55.21c and 29), they act as reflections on $\mathcal{P}$, as well. In $\mathcal{P}_2\mathbb{R}$, the correspondence between elliptic reflections and their centers is bijective, because the axis of the reflection σ_p at p is the polar p^{π^+} with respect to the elliptic polarity π^+ introduced in (13.12). Hence, there is a continuous map $\alpha : P_2\mathbb{R} \to P$ sending a point p to the center in $\mathcal{P}$ of σ_p. There is a similar map β for lines.

2) By (23.12), these maps are equivariant, i.e., compatible with the actions of Δ. In particular, the stabilizer $\Delta_p \cong O_2\mathbb{R}$ of a point $p \in P_2\mathbb{R}$ fixes the image $q = p^\alpha$. Now $O_2\mathbb{R}$ is a maximal subgroup of Δ, hence it coincides with Δ_q; note that $\Delta_q \neq \Delta$ by (62.8b). Thus, α is a homeomorphism onto its image set $Q \subseteq \mathcal{P}$, and Q is an orbit of Δ equivalent to $P_2\mathbb{R}$. Similarly, β is a homeomorphism onto a set $\mathcal{M} \subseteq \mathcal{L}$.

3) Moreover, we claim that the pair (α, β) of mappings preserves incidence. The relation $p \in L$ in $\mathcal{P}_2\mathbb{R}$ implies that the two elliptic reflections σ_p with center p and σ_L with axis L commute. To see this, note that the axis of σ_p is the polar p^{π^+}, which contains L^{π^+}, and apply (55.35). Using (55.35) in the converse direction, one obtains that $p^\alpha \in L^\beta$.

4) In view of (55.5), our assertion that we have a desarguesian Baer subplane $\mathcal{Q} = (Q, \mathcal{M})$ with standard action of Δ will be proved if we show that the α-image of a triangle in $\mathcal{P}_2\mathbb{R}$ is never contained in a line. If this did happen, then the sides of the triangle would be mapped into that line, and the same would happen to all lines, since each line meets the triangle in at least two points, and since α is injective. Then Δ would fix a line, contrary to the dual of (62.8c).

5) For $q \in Q$, we determine the action of $\Delta_q \cong O_2\mathbb{R}$ on the pencil $\mathcal{L}_q$. First, we note that the kernel of the action on both $\mathcal{M}_q$ and $\mathcal{L}_q$ is generated by the reflection σ_q at q. From Mostert [57] Theorems 4 and 5, it follows that the induced action $\Xi|_{\mathcal{L}_q}$ of the identity component $\Xi = \Delta_q^{\,1}$ on $\mathcal{L}_q$ is the standard action of $SO_2\mathbb{R}$ on the 2-sphere. In particular, Ξ fixes precisely two lines K, L in this pencil (the 'poles'), which do not belong to the orbit $\mathcal{M}_q \approx \mathbb{S}_1$ (the 'equator'). Each element $\delta \in \Delta_q \setminus \Xi$ is an involution and acts as a reflection on both $\mathcal{P}$ and $\mathcal{Q}$. Since $\delta \neq \sigma_q$, we infer that δ has an axis $A \in \mathcal{M}_q$ and one more fixed line in $\mathcal{M}_q$, and fixes no other line in $\mathcal{L}_q$. In particular, δ interchanges K and L. Now $\mathcal{Q}$ is a Baer subplane, hence every line $G \in \mathcal{L} \setminus \mathcal{M}$ meets Q in a unique point q, and Δ_G fixes this point. By our previous considerations, we have thus determined all stabilizers of lines. There is one Δ-orbit, $\mathcal{M}$, of lines with stabilizer $O_2\mathbb{R}$, and another orbit, K^Δ, with

stabilizer $SO_2\mathbb{R}$; hence, the latter orbit is equivalent to the 2-sphere with standard action. All remaining lines have stabilizers of order 2. By duality, the orbit pattern in P is the same.

6) Finally, we shall show that the equivalence class of the action of Δ on $\mathcal{L}$ is uniquely determined. By duality, this will complete the proof of (71.10). We use Section 5 of Mostert [57], as we did in the proof of the corresponding result for $Spin_3\mathbb{R}$, see (62.9), step 8). Consider an arc X in $\mathcal{L}_q$ which is a cross-section for the Δ_q-orbits and joins the two lines A and K introduced above (e.g., a 'quarter meridian'). The stabilizers of these lines are $\Delta_A \cong O_2\mathbb{R}$ and $\Xi = \Delta_q^1 \cong SO_2\mathbb{R}$, and all other elements of the arc have stabilizer $\langle\sigma_q\rangle$. The same arc X is also a cross-section for the Δ-orbits in $\mathcal{L}$. The corresponding triple of stabilizers has been described in terms of the standard action on $\mathfrak{Q} \cong \mathcal{P}_2\mathbb{R}$, hence we may apply Mostert's result.

7) If we had proved the Addendum, then we could use the homeomorphism $\mathcal{L} \approx O * O$ provided by (55.11 and 14) in order to show very easily that the action of Δ on $\mathcal{L}$ is uniquely determined. □

71.11 Theorem: Abelian groups. *Let Δ be an abelian, closed, connected subgroup of the automorphism group of a 4-dimensional, compact plane $\mathcal{P}$. Then Δ is at most 4-dimensional, and if $\dim \Delta = 4$, then precisely one of the following statements is true.*

(a) $\Delta \cong (\mathbb{C}^\times)^2$. *The group Δ fixes a triangle $\{p, q, r\}$, and is (x, yz)-transitive in the sense of (23.21) whenever $\{x, y, z\} = \{p, q, r\}$.*
(b) $\Delta \cong \mathbb{R}^3 \times SO_2\mathbb{R}$. *The group Δ fixes two points p, q and two lines $K = pq$ and $L \in \mathcal{L}_q$ and is both (q, K)-transitive and (p, L)-transitive.*
(c) $\Delta \cong \mathbb{R}^4$. *The group Δ is (q, K)-transitive for precisely one flag (q, K) and is sharply transitive both on $P \setminus K$ and on $\mathcal{L} \setminus \mathcal{L}_q$.*
(d) $\Delta \cong \mathbb{R}^4$, *and Δ is a transitive elation group, $\Delta = \Delta_{[x,x]}$ for some $x \in P \cup \mathcal{L}$.*

Note. Some comments are in order before we prove this result. First we note that examples for the four possibilities are easily found in the complex plane. Cases (b), (c), and (d) are known to occur in non-desarguesian planes. There are groups of type (b) acting on the planes over the so-called Hurwitz ternaries, see Plaumann–Strambach [74]. Cases (c) and (d) characterize shift planes (74.8) and translation planes, respectively; see Sections 74 and 73 for examples.

Non-desarguesian planes admitting a group of type (a) must be of Lenz–Barlotti type I.4. Finding examples requires the construction of 2-dimensional, planar neofields. The only known examples of 2-dimensional neofields are due to Hofmann [58, 61]. It is an open question whether or not they are planar.

We shall see in (72.12) that the complex plane is the only one whose automorphism group contains more than one subgroup Δ as considered in the theorem. In particular, Δ is normal in $\operatorname{Aut}\mathcal{P}$ unless $\mathcal{P}$ is the complex plane. The planes having an at least 7-dimensional automorphism group with a 4-dimensional, abelian subgroup are known, see (74.26).

For the analogous problem in 2-dimensional planes, compare (38.5). That situation is distinctly more complicated. For example, there is the extra possibility described in (34.1). Moreover, the fact that the multiplicative group of positive real numbers is isomorphic to the additive group $\mathbb{R}$ makes it impossible to obtain a neat case distinction as presented above for the 4-dimensional case.

Proof of (71.11). Assume that $\dim \Delta \geq 4$. Then $\dim \Delta = 4$ by (55.41).

1) We claim that all Δ-orbits of points and lines are of even dimension. It suffices to treat a point orbit $B = p^\Delta$. Assume that $\dim B < 4$. The stabilizer $\Delta_p \neq \mathbb{1}$ acts trivially on B, hence B contains no quadrangle (71.7a), and B is contained in some line (42.3). In particular, only the case $\dim B = 1$ has to be considered.

If Δ_p moves a line $L \in \mathcal{L}_p$, then the stabilizer $\Delta_{p,L}$ has dimension at least $4-1-2=1$ and fixes each element of the orbit L^Δ. This set contains a quadrangle of the dual plane, contrary to (71.7). Indeed, we may apply the dual of (42.3); the orbit L^Δ is not contained in any pencil because it has more than one element in common with the pencil of each point $b \in B$.

2) Up to duality, there is a fixed line $K = K^\Delta$, see (71.4). Any 0-dimensional orbit of the connected group Δ is a fixed element. According to step 1), all non-trivial orbits of Δ in K are 2-dimensional and hence open, see (53.1a). If the number of fixed points on K is finite, then this implies that Δ is transitive on the connected set of non-fixed points on K. In the sequel, this argument will be applied several times to the actions of Δ on various point rows or line pencils.

3) By (62.8), the abelian group Δ is not transitive on K. In view of step 2), it follows that Δ fixes some flag (q, K).

4) If Δ fixes a triangle $\{p, q, r\}$, then Δ fixes no further points by (71.7b). By step 2), Δ is transitive on the side $S = pq \setminus \{p, q\}$ of the triangle. The 2-dimensional kernel of the action on S is a linearly transitive group of (r, pq)-homologies (61.4). This establishes the situation (a).

5) We may now assume that all fixed points of Δ are incident with one line K and that all fixed lines pass through one fixed point $q \in K$. If there are additional fixed elements $p \in K$ and $L \in \mathcal{L}_q$, then Δ is transitive on $L \setminus \{q\}$ by step 2). The 2-dimensional kernel of this action is a linearly transitive group of (p, L)-homologies. In particular, Δ is transitive on $K \setminus \{p, q\}$ as well, and the kernel of this action is a linearly transitive group of (K, q)-elations. Therefore, Δ satisfies assertion (b).

6) If Δ fixes $p, q \in K$ and no line other than K, then Δ is transitive on $\mathcal{L}_p \setminus \{K\}$ and on $\mathcal{L}_q \setminus \{K\}$ by step 2), and the corresponding kernels are linearly transitive elation groups with axis K. Thus we have obtained the situation (d), see (23.23).

7) The only remaining possibility is that only the flag (q, K) is fixed. By step 2) we have transitive actions on $K \setminus \{q\}$ and on $\mathcal{L}_q \setminus \{K\}$. The kernel of either action is a linearly transitive group of (q, K)-elations. We infer that Δ is transitive on $P \setminus K$ and on $\mathcal{L} \setminus \mathcal{L}_q$. This yields the situation (c) and ends the proof. $\square$

71.12 Notes on Section 71. Salzmann [70] proved that the automorphism group Σ is a Lie group (71.2). A somewhat simpler method due to Löwen [78] Theorem A can be used to show that the connected component of Σ is a Lie group. Solvability of axial groups (71.3) is a strengthened version of a result of Salzmann [70]. The stiffness results (71.4 and 7) are taken from Salzmann [69b, 70]. The semi-simple groups acting on 4-dimensional planes (71.8) were determined by Löwen [78], even in the general situation of stable planes. For further results in this context, see Löwen [86d]. The result on $SO_3\mathbb{R}$-actions (71.10) is due to Löwen [84a], and the classification of large abelian actions (71.11) was obtained by Knarr [88a].

72 Characterizing $\mathcal{P}_2\mathbb{C}$

Throughout this section, $\mathcal{P} = (P, \mathcal{L})$ denotes a 4-dimensional, compact, projective plane.

We shall present several characterizations of the complex projective plane that are based on assumptions concerning the automorphism group. In particular, we consider non-solvable groups, compact groups, groups of dimension at least 9 and abelian groups. Moreover, we characterize the complex plane among translation planes of dimension 4 by various properties of the automorphism group.

We recall that there are no 4-dimensional, compact, projective planes of Lenz type exactly III or V, or of Lenz–Barlotti type exactly IVa.2. This has been proved in (64.15, 18, and 22).

Planes admitting a non-solvable group

72.1 Review: Motion groups. *If one of the motion groups $PSU_3\mathbb{C}$, $PSU_3(\mathbb{C}, 1)$, or $\mathbb{C}^2 \rtimes SU_2\mathbb{C}$ acts effectively on a 4-dimensional, compact plane $\mathcal{P}$, then the plane is isomorphic to the classical plane $\mathcal{P}_2\mathbb{C}$, and the action is the standard action.*

Proof. This has been proved in (62.6). For the compact, 8-dimensional elliptic motion group $\Delta = \mathrm{PSU}_3\mathbb{C}$, a different proof is obtained if we apply (63.8). Indeed, (71.9) shows that each stabilizer $\Delta_x \neq \Delta$ satisfies $\dim \Delta_x \leq 4$. Using (96.10) and (53.1c or d), we infer that Δ is transitive, so that (63.8) applies. □

This result implies that $\mathcal{P}$ is isomorphic to the complex plane if $\mathrm{PSL}_3\mathbb{C}$ acts effectively on $\mathcal{P}$, a fact which also follows from (65.2).

72.2 Proposition: Groups containing $\mathrm{SU}_2\mathbb{C}$**.** *Let Δ be a locally compact, connected group acting effectively on a 4-dimensional, compact, projective plane $\mathcal{P}$. If Δ contains a subgroup isomorphic to $\mathrm{SU}_2\mathbb{C} \cong \mathrm{Spin}_3\mathbb{R}$, then either $\mathcal{P}$ is isomorphic to the complex plane $\mathcal{P}_2\mathbb{C}$, or Δ is a subgroup of the 5-dimensional group $\mathrm{GU}_2\mathbb{C} \cong \mathrm{SU}_2\mathbb{C} \cdot \mathbb{C}^\times$. In particular, $\mathcal{P}$ is isomorphic to the complex plane if $\Delta \cong \mathrm{SL}_2\mathbb{C}$.*

Note. Planes $\mathcal{P}$ with $\mathrm{Aut}\,\mathcal{P} \cong \mathrm{GU}_2\mathbb{C}$ have been constructed by P. Sperner [90].

Proof. We may assume that Δ is a closed subgroup of $\mathrm{Aut}\,\mathcal{P}$, so that Δ is a Lie group (71.2).

1) By (62.9 and 10), the action of $\Phi = \mathrm{SU}_2\mathbb{C}$ is equivalent to the standard action. In particular, Φ fixes some antiflag (c, A), and the unique involution $\sigma \in \Phi$ belongs to $\Delta_{[c,A]}$. The action induced by Φ on the pencil $\mathcal{L}_c$ is the standard action of $\mathrm{SO}_3\mathbb{R}$ on the 2-sphere.

2) First we consider the case where Δ fixes (c, A). If the transitive group induced by Δ on $\mathcal{L}_c$ is larger than $\mathrm{SO}_3\mathbb{R}$, then according to Mostow [50] the only possibilities for this group are $\mathrm{PSL}_2\mathbb{C}$ or $\mathrm{SL}_3\mathbb{R}$, both with standard action. In the first case, the action is triply transitive, and (63.5) together with (64.18) implies that $\mathcal{P} \cong \mathcal{P}_2\mathbb{C}$. In the second case, we infer from (71.8) that a Levi subgroup $\Psi \leq \Delta$ containing Φ is isomorphic to $\mathrm{PSL}_3\mathbb{R}$, see (94.28). Then Ψ would be a simple group, contrary to the fact that $\sigma \in \Phi_{[c,A]} \trianglelefteq \Psi$.

If $\Delta \neq \Phi$, then $\dim \Delta > \dim \Phi$, see (94.3f), and the only remaining possibility is that the kernel $\Xi = \Delta_{[c,A]}$ of the action on $\mathcal{L}_c$ has positive dimension e. By (61.4), we know that $e \leq 2$. The identity component Ξ^1 is isomorphic to $\mathbb{R}$, $\mathrm{SO}_2\mathbb{R}$, or $\mathbb{C}^\times$; this follows from (94.39) if $e = 1$, and from (64.1) if $e = 2$. The group $\Phi \cong \mathrm{SU}_2\mathbb{C}$ centralizes Ξ^1, whence Δ is isomorphic to the factor group of $\Phi \times \Xi^1$ modulo the intersection $\Phi \cap \Xi^1 \leq \Phi_{[c]} \cong \mathbb{Z}_2$. By (55.32(i)), there is at most one involution in $\Delta_{[c,A]}$. Thus, Δ is isomorphic to some subgroup of $\mathrm{GU}_2\mathbb{C}$.

3) We return to the general case. If a Levi subgroup Ψ of Δ is larger than Φ, then (71.8) tells us that either Ψ contains a classical motion group, or $\Psi \cong \mathrm{SL}_2\mathbb{C}$. Note that neither $\mathrm{PSL}_2\mathbb{C}$ nor $\mathrm{PSL}_3\mathbb{R}$ contains a subgroup isomorphic to $\mathrm{SU}_2\mathbb{C}$; indeed, both contain $\mathrm{SO}_3\mathbb{R}$ as a maximal compact subgroup, compare (94.30 and 33). The

possibility $\Psi \cong SL_2\mathbb{C}$ leads to the case considered in step 2), because σ is central. The classical motion groups act on $\mathcal{P}_2\mathbb{C}$ only, see (63.6).

By (94.26), the remaining possibility is that Δ contains a minimal connected, abelian, normal subgroup $\Lambda \neq 1\!\!1$; note that $\dim \Lambda \leq 4$ by (55.41) or by (71.11). If σ centralizes Λ, then step 2) applied to $\Lambda\Phi$ shows that $\Lambda \leq \Delta_{[c,A]}$. Since Λ is normal in Δ, it follows that Δ fixes (c, A), and we obtain the assertion by applying step 2) to Δ. Now assume that σ acts non-trivially on Λ. By (95.5), the representation of Φ on Λ then is the standard representation of $SU_2\mathbb{C}$ on $\mathbb{C}^2$, and Δ contains the complex Euclidean motion group. By (62.6), this implies that $\mathcal{P} \cong \mathcal{P}_2\mathbb{C}$.

4) We indicate a different argument for the case where Δ is not simple and does not fix (c, A). By (71.4), there is some fixed element, and step 1) shows that this must be either c or A. By (96.25), the other of the two elements has an open orbit. Now (61.19b) shows that we have a translation plane, and (62.13) completes the proof. □

According to Dembowski [71b], the finite Hughes planes are characterized among finite planes by the following set of properties: they are non-desarguesian, projective planes admitting a group Δ of collineations that leaves some desarguesian Baer subplane $\mathcal{Q} \cong \mathcal{P}_2 F$ invariant and induces on $\mathcal{Q}$ the full projective linear group $PGL_3 F$. We shall refer to this as the *Hughes situation*. Suppose that the Hughes situation occurs for a 4-dimensional, compact plane $\mathcal{P}$ with a closed subplane $\mathcal{Q}$, and that Δ is a closed, connected group of continuous collineations. By (55.21c), Δ acts effectively on $\mathcal{Q}$. Since $\mathcal{Q}$ is 2-dimensional (55.4 or 5), we have that $\mathcal{Q} \cong \mathcal{P}_2\mathbb{R}$ and $\Delta \cong PSL_3\mathbb{R}$; compare also (71.8). The next result shows that in fact this cannot happen. A systematic treatment of Hughes planes will be given in the context of higher dimensional planes, where examples do exist; see Section 86.

72.3 Theorem: The Hughes situation. *Let Δ be a locally compact, connected group containing a connected subgroup Φ locally isomorphic to $PSL_3\mathbb{R}$. If Δ acts effectively on a 4-dimensional, compact, projective plane $\mathcal{P}$, then $\mathcal{P}$ is isomorphic to the complex plane $\mathcal{P}_2\mathbb{C}$, and Δ is isomorphic to $PSL_3\mathbb{R}$ or to $PSL_3\mathbb{C}$, and acts in the standard way.*

Proof. 1) According to (71.6c), we may consider Δ as an analytic subgroup of the Lie group $\Sigma = \text{Aut}\,\mathcal{P}$. Using (94.14 or 15), we infer that Φ is an analytic subgroup as well. By (71.8), we know that Φ is even globally isomorphic to $PSL_3\mathbb{R}$. We shall prove that both $\mathcal{P}$ and the action of Φ are classical. This suffices, because $PSL_3\mathbb{R}$ is a maximal connected subgroup of $PSL_3\mathbb{C}$. Maximality follows, for instance, from the list of possible semi-simple groups (71.8) together with Lemma (71.4) on fixed elements of groups that are not semi-simple.

2) The stabilizer of Φ at a point of $\mathcal{P}_2\mathbb{R}$ is not semi-simple, hence it fixes some point q of $\mathcal{P}$, see (71.4). In view of the action of $\Psi = SO_3\mathbb{R} \leq \Phi$, see (71.10), it follows that the orbit $Q = q^\Psi = q^\Phi$ is the point set of a Baer subplane $\mathcal{Q} = (Q, \mathcal{M})$, with standard action of Φ. Hence we are, indeed, in the Hughes situation.

3) We consider the outer geometry of $\mathcal{Q}$, that is, the geometry $(P \setminus Q, \mathcal{L} \setminus \mathcal{M})$. We shall show that the unordered pair of the stabilizers of a point p and a line L incident with p in this geometry is determined to within conjugacy in Φ. According to (23.24), this implies that the outer geometry is isomorphic or dual to the (self-dual) outer geometry of $\mathcal{P}_2\mathbb{R}$ in $\mathcal{P}_2\mathbb{C}$, and that the action of Φ on it is the standard action. Our assertion then follows by a density argument, compare (44.9).

4) We claim that Φ is sharply transitive on the open set $\mathcal{A}$ of all pairs (L, M) of outer lines meeting in an outer point $p = L \wedge M$. Observe that $P \setminus Q$ is path connected because $L \setminus Q \approx \mathbb{R}^2$ for any outer line L; dually, $\mathcal{L}_p \setminus \mathcal{M} \approx \mathbb{R}^2$ for $p \in P \setminus Q$, and $\mathcal{A} \cap (\mathcal{L}_p)^2$ is path connected. It follows that $\mathcal{A}$ is connected (and 8-dimensional), like Φ.

Using the dimension formula (96.10) together with (53.1d), we see that it suffices to show that $\Phi_{L,M} = \mathbb{1}$. Now if an element $\varphi \in \Phi$ fixes L and M, then it fixes the unique line $G \in \mathcal{M}$ containing p, as well as the unique points $a \in Q$ and $b \in Q$ incident with L and M, respectively, and the line $H = ab \in \mathcal{M}$; see Figure 72a. It is easily seen that a, b and $q = G \wedge H$ are three distinct points on $H \cap Q$, so that the automorphism induced by φ on the desarguesian plane $\mathcal{Q}$ has axis $H \cap Q$, compare (12.12), Addendum (a). The center $c \in Q$ must be incident with the fixed line G. By intersection with the lines in $\mathcal{M}_c$, we obtain additional fixed points in $\mathcal{P}$ lying on L and M, so that $\varphi = \mathbb{1}$ by (71.7a).

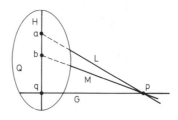

Figure 72a

5) The stabilizer of an outer flag (p, L) fixes the corresponding 'inner' elements $G \in \mathcal{L}_p$ and $a \in L \setminus G$, compare 4). By step 4), the stabilizer $\Phi_{p,L}$ is sharply transitive on the punctured affine plane $Q \setminus (G \cup \{a\}) \approx \mathbb{C}^\times$. This implies that $\Phi_{p,L}$ is homeomorphic and, hence, even isomorphic to $\mathbb{C}^\times$, see (96.9a) and (94.39 and 40). The 1-dimensional torus T contained in this group is determined up to conjugation in Φ, see (93.10) and (94.31b). Hence, $\Phi_{p,L}$ is also determined in the same sense, being the centralizer of T.

6) With the notation of step 5), the stabilizer of p is sharply 2-transitive on the real affine plane $Q \setminus G$. Therefore, the conjugates in Φ_p of the reflection in T generate a $\mathbb{Z}_2$-extension of the translation group Λ of $\mathfrak{Q}$ with axis $G \cap Q$. Hence, we have that $\Phi_p = \Lambda \cdot \Phi_{p,L}$. There are precisely two elation groups isomorphic to $\mathbb{R}^2$ in Φ that are normalized by $\Phi_{p,L}$, namely Λ and M, where M is the elation group with center a. This shows that the unordered pair $\{\Phi_p, \Phi_L\}$ is determined up to conjugacy. □

72.4 Proposition: The complex orthogonal group. *If $\Delta \cong \mathrm{PSL}_2\mathbb{C}$ (which is isomorphic to $\mathrm{SO}_3\mathbb{C}$) acts effectively on a 4-dimensional, compact, projective plane $\mathcal{P}$, then $\mathcal{P}$ is isomorphic to the complex plane $\mathcal{P}_2\mathbb{C}$, and the action of Δ is the standard action of $\mathrm{SO}_3\mathbb{C}$.*

Proof. Let Q be a desarguesian Baer subplane that is invariant under some subgroup $\Phi \cong \mathrm{SO}_3\mathbb{R}$ of Δ, see (71.10). This subplane is moved by Δ, since Δ is simple and does not act effectively on $\mathcal{P}_2\mathbb{R}$, see (32.20 and 23). Moreover, (71.10) shows that each point $p \notin Q$ near Q has a 3-dimensional orbit p^Φ. According to (96.25), this implies that $\dim q^\Delta = 4$ for $q \in Q$. We shall show below that the action of Δ on q^Δ is equivalent to the transitive action of $\mathrm{SO}_3\mathbb{C}$ on the complement X of the conic $C = \{(z_1, z_2, z_3) \mid z_1^2 + z_2^2 + z_3^2 = 0\}$ in $P_2\mathbb{C}$. Now $\mathrm{SO}_3\mathbb{C}$ contains a reflection in each secant of the conic, hence the secants are determined by the action. Since tangents may be described as limits of secants, and since exterior lines do not exist, we then obtain an isomorphism between the geometries induced on the sets X and q^Δ respectively by $\mathcal{P}_2\mathbb{C}$ and $\mathcal{P}$. This isomorphism extends to the entire planes by density (44.9).

It remains to prove the claim about the equivalence of actions. The action of $\mathrm{SO}_3\mathbb{C}$ on the conic C is equivalent to the triply transitive standard action of $\mathrm{PSL}_2\mathbb{C}$ on $P_1\mathbb{C} \approx \mathbb{S}_2$, see Berger [77, 87] 16.2 and 16.3. Each unordered pair $\{a, b\}$ of distinct points of C determines a point of X, namely, the intersection of the tangents at these points. This defines a bijection of the set of unordered pairs onto X, compare the dual of (55.11), and it follows that Δ is transitive on X.

We compare the stabilizers Δ_x (of $x \in X$) and Δ_q; they are both 2-dimensional (96.10). Since Φ acts on Q in the ordinary way, Φ_q is isomorphic to $O_2\mathbb{R}$. Moreover, Δ does not contain a 2-torus. Hence $\Delta_q^1 \cong \mathbb{C}^\times$, see (94.39 and 40), and Δ_q normalizes Φ_q^1 and preserves the set $\{a, b\}$ of fixed points of Φ_q^1 in $P_1\mathbb{C}$. This set corresponds to an element $x \in X$, and we see that $\Delta_q \leq \Delta_x$. Moreover, $(\Delta_q)^1 = \Delta_{a,b}$, compare (15.7). Since $\Phi_q \leq \Delta_q$ contains elements interchanging the points a and b, it follows that Δ_q is in fact equal to Δ_x. □

72.5 Proposition: Groups containing $\mathrm{SO}_3\mathbb{R}$. *Let Δ be a locally compact, connected group acting on a 4-dimensional, compact, projective plane $\mathcal{P}$. If Δ contains* a subgroup $\Phi \cong SO_3\mathbb{R}$ and $\dim \Delta \geq 4$, then $\mathcal{P}$ is isomorphic to the complex plane $\mathcal{P}_2\mathbb{C}$.

Proof. By (71.10), the group Φ does not fix any elements of $\mathcal{P}$. Hence, (71.4) shows that Δ is a product of simple groups. If $\Delta \neq \Phi$, then the assertion follows from (71.8) together with (72.1, 3, and 4). □

Combining (71.9) with (72.2 and 5), we obtain the following theorem.

72.6 Theorem: Groups containing compact subgroups. *Let Δ be a locally compact group acting on a 4-dimensional, compact, projective plane $\mathcal{P}$. If Δ contains a compact subgroup of dimension at least 3, then either $\dim \Delta \leq 5$, or $\mathcal{P}$ is isomorphic to the complex plane $\mathcal{P}_2\mathbb{C}$.* □

72.7 Note. Let Δ be a locally compact, connected group acting effectively on a 4-dimensional plane $\mathcal{P}$, and assume that Δ contains a subgroup locally isomorphic to $SL_2\mathbb{R}$. Using (61.22), it can be shown that $\mathcal{P}$ is a translation plane if $\dim \Delta \geq 7$. If $\dim \Delta = 6$, then Δ is isomorphic to the real affine group $\mathbb{R}^2 \cdot GL_2\mathbb{R}$, see Löwen [90], and a list of all possible actions of Δ can be given (unpublished result of Löwen). The translation planes admitting an action of $SL_2\mathbb{R}$ have been determined by Betten, see (73.22). Together with the preceding results, this provides complete insight into planes admitting a non-solvable automorphism group of dimension at least 7; compare (74.27).

Planes admitting a group of large dimension

72.8 Theorem: Groups of large dimension. *If $\mathcal{P}$ is a 4-dimensional, compact, projective plane, and if Δ is an at least 9-dimensional, locally compact group acting effectively on $\mathcal{P}$, then $\mathcal{P}$ is isomorphic to the complex plane $\mathcal{P}_2\mathbb{C}$.*

Proof. The aim of this proof is to show that $\mathcal{P}$ is a translation plane, up to duality. The theorem will then follow from (72.11g) below. It will be easier to formulate the proof if we argue by contradiction; that is, we assume that $\mathcal{P}$ is neither a translation plane nor a dual translation plane, and we show that this is impossible.

1) We may assume that Δ is connected (93.6) and closed in $\text{Aut}\,\mathcal{P}$. Indeed, consider $\varphi : \Delta \to \text{Aut}\,\mathcal{P}$; the image of this homomorphism is an analytic subgroup of its closure Θ, hence $\dim \Theta \geq 9$ and we may replace Δ by Θ if necessary. By (72.6), we may also assume that Δ contains no compact subgroups of dimension greater than 2. Together with (71.8), this implies that Δ contains no semi-simple subgroups of dimension greater than 3. From (71.4), we know that, up to duality,

Δ fixes some line L. For lack of compact subgroups, Δ cannot be transitive on L, see (96.19 and 20).

2) We claim that Δ is transitive on $P \setminus L$. Here and in subsequent steps, we shall repeatedly (and tacitly) use the dimension formula (96.10) and the fact that an intransitive action on a connected k-manifold must have an orbit of dimension less than k, compare (53.1d). We may choose points $u, v \in L$ such that $\dim \Delta_{u,v} \geq 7$. If some point $a \notin L$ has an orbit of dimension less than four, then the group $\Xi := \Delta_{a,u,v}$ is at least 4-dimensional. This group acts freely on the complement of the set $T \subseteq P$ formed by the sides of the triangle a, u, v, see (71.7a), hence Ξ is sharply transitive on $P \setminus T$ and in particular is 4-dimensional. In view of $\dim a^\Delta < 4$, this implies that a^Δ is contained in T and, hence, is at most 2-dimensional. But then Ξ is at least 5-dimensional, a contradiction.

3) The center $C(\Delta)$ is trivial. Indeed, the usual dimension considerations show that the stabilizer of an arbitrary point $a \in P \setminus L$ fixes no other point in this affine plane. Hence, a is fixed by $C(\Delta)$.

4) We show that Δ fixes some point $v \in L$ and that the stabilizer Δ_a of $a \in P \setminus L$ is transitive on $L \setminus \{v\}$. Since we know that Δ is not transitive on L, it suffices to exclude the possibilities that L contains either a 1-dimensional Δ_a-orbit or two fixed points of Δ_a. If x, y, z are points of a 1-dimensional orbit, then the group $\Omega := \Delta_{a,x,y,z}$ is at least 2-dimensional. This implies that Ω contains a reflection with axis L, see (71.7c), and we obtain a transitive translation group using step 2) and (61.19b or 20b). This contradicts our assumptions. Similar arguments work if $x, y \in L$ are fixed by Δ_a and z is an arbitrary third point on L. In this case, $\dim \Omega > 2$, contrary to (71.7b).

5) Repeating the last arguments once more, we may improve 4) and obtain that Δ_a is doubly transitive on $L \setminus \{v\}$.

6) Suppose that Δ contains an involution σ. By (55.21 and 29), σ is a reflection, $\sigma \in \Delta_{[c,A]}$. If $A = L$ or $c = v$, then (61.19b or 20b) together with step 2) and its dual implies that either $\mathcal{P}$ or the dual plane is a translation plane. Hence we may assume that $v \in A$ and $c \in L$; then a similar argument using 4) shows that $\Delta_{[v,A]}$ is linearly transitive. By step 2), the same is true for every line $A' \in \mathcal{L}_v \setminus \{L\}$, and it follows that $\mathcal{P}$ is a dual translation plane.

7) By (61.20b) and step 2), we obtain a translation plane unless $\Delta_{[a]} = \mathbb{1}$. In that last case, Δ_a acts effectively and, by step 5), doubly transitively on $L \setminus \{v\} \cong \mathbb{R}^2$. For $u \in L \setminus \{v\}$, consider the transitive action of $\Delta_{a,u}$ on the projective line $P_1\mathbb{R}$, compare (96.16b). Since the kernel of that action is at most 1-dimensional and $SL_2\mathbb{R}$ contains no 2-dimensional transitive subgroups, we infer that either $\Delta_{a,u} \cong \mathbb{C}^\times$ or $\Delta_{a,u}$ contains $SL_2\mathbb{R}$. In each case, Δ contains an involution; this ends the proof, by step 6). $\square$

72.9 Note. Salzmann [73b] proves that every 4-dimensional, compact, projective plane with an 8-dimensional automorphism group is either isomorphic or dual to a translation plane. These translation planes have been determined by Betten, see Section 73. The planes with a 7-dimensional group are also known, see (74.27).

Translation planes

We restate the following crucial result, see (64.4c):

72.10 Proposition: Unique translation axis. *Let $\mathcal{P}$ be a compact, connected projective translation plane. Then either $\mathcal{P}$ is isomorphic to a classical plane $\mathcal{P}_2\mathbb{F}$, where $\mathbb{F} \in \{\mathbb{R}, \mathbb{C}, \mathbb{H}, \mathbb{O}\}$, or $\mathcal{P}$ has a unique translation axis W. In the latter case, W is fixed by all automorphisms.*

Recall from (25.6) or (64.4) that a 4-dimensional affine translation plane $\mathcal{A}$ may be represented as $\mathbb{R}^4$ endowed with a spread $\mathcal{S}$ consisting of 2-dimensional vector subspaces; the lines of the affine plane are the translates of the elements of $\mathcal{S}$. Assume that the plane is non-desarguesian. Then by (72.10) the automorphism group Σ of the projective completion $\mathcal{P} = \overline{\mathcal{A}}$ is the extension of the translation group $\mathbb{R}^4$ by the stabilizer Σ_o of the origin $o = 0$, and Σ_o may be described as the set of all linear automorphisms of $\mathbb{R}^4$ that preserve $\mathcal{S}$, compare (25.5) and (44.6).

72.11 Theorem: Homogeneous translation planes. *Let $\mathcal{P}$ be a 4-dimensional, projective translation plane with axis W and with automorphism group Σ. Each one of the following conditions implies that $\mathcal{P}$ is isomorphic to the complex plane $\mathcal{P}_2\mathbb{C}$.*

(a) *The homology group $\Sigma_{[o]}$ contains elements other than the homotheties $x \mapsto rx$ with real factors $r \in \mathbb{R}^\times$.*
(b) *Some affine line L is the axis of a 2-dimensional group $\Sigma_{[c,L]}$ of homologies.*
(c) *The translation axis W contains three points u, v, w such that the stabilizer of the degenerate quadrangle o, u, v, w is at least 2-dimensional.*
(d) *For some affine line L, the elation group $\Sigma_{[L \wedge W, L]}$ is 2-dimensional.*
(e) *The automorphism group of $\mathcal{P}$ contains a 2-dimensional torus subgroup.*
(f) *The action of Σ on W is transitive.*
(g) $\dim \Sigma \geq 9$.

Note. Compare also (64.8b), where condition (d) is expressed in terms of a matrix description of the spread $\mathcal{S}$.

Proof. By (25.4), the group $\Sigma_{[o]}$ consists of the homotheties with factors taken from the kernel K of $\mathcal{P}$. The kernel is a locally compact field of dimension 1 or 2, hence it is isomorphic to one of the fields $\mathbb{R}$ or $\mathbb{C}$. If K is isomorphic to $\mathbb{C}$, then the plane is isomorphic to $\mathcal{P}_2\mathbb{C}$. This proves the assertion under condition (a).

Condition (b) implies that the group $\Sigma_{[c,L]}$ is linearly transitive, see (61.4a). By (64.1b), this group is isomorphic to $\mathbb{C}^\times$ and, in particular, is abelian. Note that $c \in W$ since W is invariant (72.10). By (44.8b), the multiplicative loop of some coordinatizing quasifield Q is isomorphic to $\mathbb{C}^\times$, hence Q satisfies the second distributive law and is a field.

By (71.7c), condition (c) implies (a). Property (d) leads to Lenz type at least V, see (61.4a) and (24.7), and hence implies that $\mathcal{P}$ is the complex plane (64.15).

If $\Pi \le \Sigma$ is a 2-dimensional torus group, then Π fixes the triangle formed by the centers of the reflections contained in Π, see (55.34a) or (61.23). Moreover, the action of Π on W has a 1-dimensional kernel; this follows from the fact that there is no effective action of a 2-torus on $\mathbb{S}_2$, see Mostert [57] or (96.28). Thus, condition (e) reduces to (a).

By (63.1), condition (f) characterizes the complex plane. Finally, assume condition (g), that is, let $\dim \Sigma \ge 9$. Consider the at least 4-dimensional subgroup $\Omega \le \Sigma_o$ consisting of all elements of determinant one, and define $\Theta = \Omega^1$. If (a) does not hold, then Ω acts on W with kernel $\mathbb{Z}_2$, and we may also assume that this action is not transitive. Hence, there is a Θ-orbit of dimension 0 (that is, a fixed point), or one of dimension 1, see (53.1d). If there is either a 1-dimensional orbit or a pair of fixed points, then the dimension formula (96.10) leads to condition (c); remember that Θ contains at most two of the real homotheties. Hence, we may assume that Θ fixes one point $v \in W$ and acts transitively on the remainder of W. In fact, the action on $W \setminus \{v\}$ is even doubly transitive, or else we may use (c) again. As in the proof of (72.8), step 7), we infer from this fact that the group induced by Θ contains the Euclidean motion group Φ of $\mathbb{R}^2$. It follows that Θ contains a connected subgroup $\widetilde{\Phi}$ that is locally isomorphic to Φ. This group has no faithful linear representation on $\mathbb{R}^2$, hence the kernel $\widetilde{\Phi}_{[ov]}$ of the action on the vector space ov has positive dimension. The structure of $\widetilde{\Phi}$ shows that $\widetilde{\Phi}_{[ov]}$ contains the commutator group Λ of $\widetilde{\Phi}$, which is isomorphic to $\mathbb{R}^2$ and is sharply transitive on $W \setminus \{v\}$. Hence, Λ is a linearly transitive elation group, and condition (d) is satisfied. □

Abelian groups

72.12 Theorem: Abelian groups. *If the automorphism group of a 4-dimensional, compact, projective plane $\mathcal{P}$ contains two distinct 4-dimensional, closed, connected, abelian subgroups Δ_1 and Δ_2, then $\mathcal{P}$ is isomorphic to the complex plane $\mathcal{P}_2\mathbb{C}$.*

Proof. We use Theorem (71.11), which classifies the possible actions of 4-dimensional abelian groups. By Δ we denote the smallest closed subgroup of $\Sigma = \mathrm{Aut}\,\mathcal{P}$ containing Δ_1 and Δ_2.

1) It follows from (71.11) that an abelian group Δ_i fixing a triangle (71.11a) or two points and two lines (71.11b) is uniquely determined by its fixed configuration, because Δ_i is the product of two linearly transitive axial groups. If Δ_1, say, is of one of these types, then we infer that the fixed configuration is not invariant under Δ. Hence, the Lenz–Barlotti type of Δ is strictly larger than that of Δ_1, which is either I.4 or II.2. In the first case, it follows directly from Pickert [75] Anhang 6 that $\mathcal{P}$ is desarguesian. In the other case, Pickert shows that the Lenz–Barlotti type of Δ is at least III.2 or IVa.2 (up to duality). Then we obtain from (64.18 and 22) that $\mathcal{P}$ is in fact desarguesian. From now on, we exclude groups Δ_i satisfying (71.11a or b).

2) If Δ_1 is a transitive elation group with axis W as in (71.11d), and if we assume that $\mathcal{P}$ is not desarguesian, then (72.10 and 11d) say that Δ_1 contains all linearly transitive groups of elations of $\mathcal{P}$. In particular, Δ_1 is uniquely determined, and the linearly transitive elation group contained in the group Δ_2, which must be of the kind (71.11c), coincides with $\Theta = (\Delta_1)_{[p]}$ for some $p \in W$. Then (p, W) is the unique flag fixed by Δ_2. As Δ is generated by the abelian groups Δ_1 and Δ_2, we observe that $\Theta \subseteq \Delta_1 \cap \Delta_2$ is contained in the center of Δ. In fact, we have $\Theta = \Delta_1 \cap \Delta_2$, because Δ_2 is transitive on $W \setminus \{p\}$ by (71.11) and hence cannot centralize any elations in $\Delta_1 \setminus \Theta$. Applying the dimension formula for sums and intersections of vector spaces, we infer that (the Lie algebra of) Δ is 6-dimensional.

Since $\Delta_1 \trianglelefteq \Delta$ is sharply transitive on $P \setminus W$, the group Δ is a semi-direct product $\Delta = \Delta_1 \rtimes \Delta_x$, the right factor being the stabilizer of a point $x \notin W$. The factor Δ_x is abelian since it is isomorphic to a subgroup of the quotient group Δ/Δ_1, which also contains a dense subgroup isomorphic to Δ_2/Θ. Thus we obtain another 4-dimensional abelian group $\Delta_3 = \Theta\Delta_x$. This group is closed, or else its closure would be more than 4-dimensional contrary to (71.11). Now Δ_3 fixes both W and xp, hence Δ_3 cannot be of the type (71.11c), and we are led to cases that have been excluded.

3) Finally assume that both Δ_1 and Δ_2 satisfy (71.11c). Thus each of them contains a unique linearly transitive elation group. If these groups are different, then the hierarchy of Lenz types (Section 24) implies that $\mathcal{P}$ is of Lenz type at least III or IV. In the first case, $\mathcal{P}$ is desarguesian by (64.18), and in the second case we have a group Δ_3 satisfying (71.11d), and we are in a situation that has been treated. We may therefore assume that Δ_1 and Δ_2 fix the same flag (p, L).

The linearly transitive group $\Delta_{[p,L]}$ is contained in $\Delta_1 \cap \Delta_2$, which is a central subgroup of Δ. If $\dim \Delta = 5$, then the 3-dimensional intersection $\Delta_1 \cap \Delta_2$ together with any one-parameter group fixing a point $x \notin L$ generates an abelian group Δ_3 that does not act freely on $P \setminus L$ and hence is not of the type (71.11c). Again, this situation has been treated.

It remains to consider the case that $\dim \Delta \geq 6$ and $\dim \Delta_x \geq 2$. Now Δ_x coincides with the axial group $\Delta_{[px]}$ because $\Delta_{[p,L]}$ centralizes Δ_x. Our aim is to show that $\Delta_{[p,px]}$ is linearly transitive; then $\mathcal{P}$ is a dual translation plane and the proof is finished. By (61.4a), it suffices to show that Δ_x consists of elations.

As a preliminary step, we show that $\dim \Delta_{[p,px]} \geq 1$. Suppose that all one-parameter groups of Δ_x consist of homologies, compare (61.8). Either the center q is always the same, and then $\Delta_{[q,px]} \cong \mathbb{C}^\times$ together with $\Delta_{[p,L]}$ generates a group Δ_3 of type (71.11b), or we obtain a one-parameter group of elations by the theorem on elations generated by homologies (61.20a).

By a similar argument, we may asssume that $\dim \Delta_x < 3$, so that $\dim \Delta = 6$: if not, then $\Delta_{q,x} = \Delta_{[q,px]}$ is non-trivial for every $q \in L \setminus \{p\}$, and then the elation group $\Delta_{[p,px]}$ is linearly transitive by (61.20a).

Now consider the group $\Xi = \Delta/\Delta_{[p]}$, which is at most 3-dimensional. It is generated by two abelian subgroups $\Delta_1/\Delta_{[p]}$ and $\Delta_2/\Delta_{[p]}$, which are sharply transitive on $\mathcal{L}_p \setminus \{L\}$ and, hence, 2-dimensional. Their intersection Θ is central in Ξ, hence $\Delta_x = \Delta_{[px]}$ fixes all lines $(px)^\vartheta \in \mathcal{L}_p$ where $\vartheta \in \Theta$. Thus, Δ_x consists of elations. □

72.13 Corollary. *Let $\mathcal{P}$ be a non-desarguesian, 4-dimensional, compact plane. Then every 4-dimensional, closed, connected, abelian subgroup Δ of $\Sigma = \operatorname{Aut} \mathcal{P}$ is normal in Σ.* □

72.14 Further characterizations of the complex plane may be found in Sections 66, 74 and 75.

72.15 Notes on Section 72. The result on $SU_2\mathbb{C}$-actions (72.2) is due to Löwen [86d]; his version includes actions on stable planes, and he constructs a non-classical stable plane admitting an action of $SL_2\mathbb{C}$. Proposition (72.3) on the Hughes situation is taken from Salzmann [73c]. Löwen [84a] obtained the results (72.4 and 5) on groups containing $SO_3\mathbb{R}$; in fact, he proves more than what we have stated here. The characterization (72.8) of $\mathcal{P}_2\mathbb{C}$ by the dimension of the automorphism group was given by Salzmann [71]. The results (72.10 and 11) on translation planes are due to Betten [72b]. The uniqueness of 4-dimensional abelian groups (72.12 and 13) was proved by Knarr [88a].

73 Four-dimensional translation planes

Recall from Section 65 that the critical dimension c_l is defined as the largest dimension that the automorphism group of a non-desarguesian, $2l$-dimensional, compact plane $\mathcal{P}$ can have. We already know (72.8) that $c_2 < 9$, and we shall see in this section that in fact $c_2 = 8$. Moreover, we shall explicitly describe all planes $\mathcal{P}$ with an 8-dimensional automorphism group. A result of Salzmann [73b] (which will not be proved here) asserts that such a plane must be a translation plane or a dual translation plane. Using this, we shall prove that there are exactly two single planes and a one-parameter family of planes of this kind. The automorphism group of one of the single planes is solvable, and in the other cases the group is not solvable.

The 4-dimensional translation planes $\mathcal{P}$ with $\dim \operatorname{Aut} \mathcal{P} = 7$ have also been classified by Betten. In this case, $\operatorname{Aut} \mathcal{P}$ is solvable, and we shall roughly describe this classification by listing some characteristic dimension parameters that distinguish between the families of planes appearing here. For details, we refer to Betten [75, 76, 77]. We remark that a 4-dimensional translation plane is flexible in the sense of (63.11) if, and only if, its automorphism group is at least 7-dimensional.

More generally, all 4-dimensional compact planes with an automorphism group of dimension at least 7 have been determined; in addition to (dual) translation planes, there is exactly one special plane with this property, namely, a shift plane found by Knarr. We shall come back to this point in Section 74.

We remark that the first examples of non-desarguesian 4-dimensional planes that were discovered were translation planes, see Betten [70] and Plaumann–Strambach [70].

73.1 Theorem: Critical dimension. *The critical dimension for the automorphism group Σ of a 4-dimensional, compact, projective plane $\mathcal{P}$ is $c_2 = 8$, and each plane $\mathcal{P}$ with $\dim \Sigma = 8$ is either isomorphic or dual to a translation plane, but not both.*

The first statement follows from (72.8) combined with the presentation of explicit examples to be given in this section, see (73.10, 13 and 19). A plane which is both isomorphic and dual to a translation plane is of Lenz type V and hence is desarguesian (64.15). The core of Theorem (73.1) is the assertion that the property $\dim \Sigma = 8$ forces the plane to be a translation plane up to duality. This is proved by Salzmann [73b] using a considerable refinement of the methods applied in (72.8). In particular, he makes extensive use of the classification due to Lie and Mostow of transitive actions of Lie groups on surfaces. Compare M. Lüneburg [88] for a proof that partly avoids the use of that classification.

73.2 Notation and review of results. (a) Recall from (64.4) that a 4-dimensional, affine translation plane $\mathcal{A}$ may be represented as the translation group $\mathsf{T} = \mathbb{R}^4$ endowed with a spread $\mathcal{S}$ consisting of 2-dimensional vector subspaces; the points of the affine plane are the elements of T, and the lines are the translates of the elements of $\mathcal{S}$.

Assume that the plane is not desarguesian. Then it is not even a Moufang plane (42.7), and the theorem of Skornjakov and San Soucie (24.9) implies that the line W at infinity is fixed by every automorphism of the projective completion

$\mathcal{P} = \overline{\mathcal{A}}$; compare (72.10). (This argument applies to a hypothetical discontinuous collineation as well.) Hence, the automorphism group $\Sigma = \operatorname{Aut} \mathcal{P}$ is a semi-direct product

$$\Sigma = \mathsf{T} \rtimes \Sigma_o$$

of the translation group T and the stabilizer Σ_o of the origin $o = 0$. If we identify the affine point set with $\mathsf{T} = \mathbb{R}^4$ as indicated above, then the action of T on itself is given by left multiplication, and the factor Σ_o acts on T by conjugation. In this way, Σ_o appears as the set of all continuous group automorphisms of $\mathbb{R}^4$ that preserve $\mathcal{S} = \mathcal{L}_o$, and thus becomes a subgroup of $\operatorname{GL}_4\mathbb{R}$. More generally, any collineation fixing o corresponds to an automorphism of T which is (semi-)linear over the kernel $\mathbb{R}$ of a coordinatizing quasifield, compare (25.5) and (64.6). This shows that in fact every collineation of $\mathcal{P}$ is continuous and hence is an automorphism, compare (44.6). This is not true for the complex plane, where the kernel $\mathbb{C}$ admits discontinuous automorphisms; see the remarks preceding (55.22).

(b) Observe that $\dim \Sigma \geq 5$ as Σ_o contains at least the real homotheties. There need not be any other elements, as the examples of Riesinger [91] show. They are obtained by a generalization of the construction described in (73.13). In the non-desarguesian case, (72.11a) implies that the real homotheties form the group $\Sigma_{[o]}$. In addition, (72.11) asserts that the following subgroups of Σ are at most 1-dimensional in the non-desarguesian case: the stabilizer of three lines in $\mathcal{L}_o$, the group of elations with a given affine axis L, and every group of homologies $\Sigma_{[c,L]}$. Moreover, we have $\dim \Sigma \leq 8$, and Σ is not transitive on W. According to (96.33 and 11a), this means that the connected component $\Delta = \Sigma^1$ has a closed orbit B of dimension 0 or 1 in W; in fact, B is a fixed point or $B \approx \mathbb{S}_1$, compare (96.9a).

(c) In order to describe the spread $\mathcal{S}$, we shall use the Grassmann coordinates introduced in (64.3b and 8). That is, we write T as $\mathsf{T} = X \oplus Y$, where $X, Y \in \mathcal{S}$, and we represent $\mathcal{S}' = \mathcal{S} \setminus \{Y\}$ by a set of 2×2-matrices A; after a choice of bases, a matrix defines a map $x \mapsto Ax$ from X to Y, and A stands for the subspace

$$\{(x, Ax) \mid x \in X\} \leq X \oplus Y.$$

As we have seen in (64.8), the plane is of Lenz type V if, and only if, this set of matrices forms a vector space; note that $X \in \mathcal{S}$ is represented by the zero matrix. In the 4-dimensional case considered here this implies, in view of (64.15), that only the complex plane $\mathcal{P}_2\mathbb{C}$ is represented by a vector space $\mathcal{S}'$.

A collineation fixing the flag (o, Y) may be represented by a 4×4 block matrix

$$\begin{pmatrix} B & \\ C & D \end{pmatrix},$$

and an easy calculation shows that this map sends the line given by A to the one defined by $(C + DA)B^{-1}$.

(d) We fix some notation for subgroups of Σ that will be used frequently. The connected component will be denoted by $\Sigma^1 = \Delta$. Moreover, we introduce

$$\Omega = \{\, \omega \in \Sigma_o \mid \det \omega = 1 \,\}$$

and

$$\Theta = \Omega^1.$$

We shall refer to either group as the *reduced stabilizer*. If the plane is not desarguesian, then (72.11a) asserts that Ω acts on W with a kernel of order 2. Note that Σ is a semi-direct product of Ω and the connected, normal subgroup H that is generated by the translations and the real homotheties $x \mapsto rx$ where $r > 0$. This implies that $\Delta = H \rtimes \Theta$, whence $\Theta = \Omega \cap \Delta$.

73.3 Proposition: Flexible planes. *A non-desarguesian, 4-dimensional translation plane is flexible in the sense of (63.11) if, and only if, its automorphism group has dimension 7 or 8.*

Proof. We prove only the easy part, concerning the automorphism group Σ of a flexible translation plane. The action of Σ on the line W at infinity has a kernel of dimension at least 5, containing the translations and the real homotheties. An open Σ-orbit in the line space yields an open orbit in W via intersection, and thus enforces $\dim \Sigma \geq 7$. In view of (72.11g), this implies the necessity of the dimension condition.

The converse statement may be verified by inspection as soon as the 4-dimensional translation planes with a group of dimension $d \in \{7, 8\}$ have been classified. The classification is obtained in the present section for $d = 8$, and references for $d = 7$ will be given in (73.23), together with a rough description of the results. □

Translation planes with solvable automorphism groups

73.4 Proposition: Fixed flag. *Let Λ be a connected, solvable group of automorphisms of a 4-dimensional, affine translation plane. Then Λ fixes some point ∞ at infinity.*

Proof. Let H be the solvable group generated by all translations and all homotheties $x \mapsto rx$, where $0 < r \in \mathbb{R}$. As a first step, we replace Λ by the solvable Lie group $\overline{H\Lambda}$, compare the remarks preceding (94.16). This is no loss of generality.

It suffices to consider the action of the stabilizer $\Lambda_o \cong \Lambda/T$, which is connected and solvable. The linear representation of Λ_o on the translation group $\mathbb{R}^4$ leaves some subspace V of dimension 1 or 2 invariant (Lie's theorem, see Bourbaki [89] I.5.3 Cor.4). If V is contained in a line L, then Λ_o fixes both L and the point

$\infty = L \wedge W$. If a 2-dimensional Λ_o-irreducible space V is not contained in any line, then the action of Λ_o on V is faithful by (55.21), and Λ_o is isomorphic to $\mathbb{C}^\times$; indeed, this is the only connected, solvable subgroup of $GL_2\mathbb{R}$ that contains the real homotheties and does not fix any 1-dimensional subspace, compare (95.6b). The subgroup $SO_2\mathbb{R} \le \Lambda_o$ acts non-trivially on $\mathcal{L}_o \approx \mathbb{S}_2$, because it is transitive on the set of 1-dimensional subspaces of V. From Mostert [57] Theorems 4 and 5, it follows that the induced action on $\mathcal{L}_o$ is the standard action, which fixes precisely 2 elements. These must be fixed by the connected, abelian group Λ_o. □

73.5 Strategy. We shall now assume that a translation plane $\mathcal{P}$ with translation axis W and with 8-dimensional, solvable automorphism group $\Sigma = \mathrm{Aut}\,\mathcal{P}$ is given. Then the plane is not desarguesian; we shall frequently use this fact without explicit mention. Let ∞ be a point at infinity that is fixed by the connected component $\Delta = \Sigma^1$, see (73.4). We choose the right summand of $\mathsf{T} = X \oplus Y$ to be the line $Y = o \vee \infty$.

In order to obtain a firm grip on the action of Δ on $\mathbb{R}^4$, we shall determine three actions on $\mathbb{R}^2$ that are either subactions or quotient actions of that action. These auxiliary actions are

1) the action induced on $W \setminus \{\infty\} \approx \mathbb{R}^2$, called the *action at infinity*,
2) the action induced on the pencil $\mathcal{L}_\infty \setminus \{W\}$, called the *action on the basis*,
3) the action of Δ_Y on the line $Y = o \vee \infty$, called the *action on the fiber*.

The terminology refers to the geometric fiber bundle $p \mapsto p \vee \infty$, compare (51.23a). The action at infinity is also induced by the reduced stabilizer Θ, because $\Delta = \mathrm{H}\Theta$, see (73.2d). In order to describe the action on the basis, we identify the orbit $\mathcal{L}_\infty \setminus \{W\}$ of T with the factor group T/Y, see (96.9a). Then T acts on the factor group by translation, and Δ_o acts by conjugation. We shall see presently that each of the above actions is a transitive, affine action on $\mathbb{R}^2$ containing all $\mathbb{R}^2$-translations.

73.6 Lemma. *If Σ is solvable and 8-dimensional, then the action at infinity is a transitive affine action. More precisely, the reduced stabilizer Θ contains a closed, normal subgroup isomorphic to $\mathbb{R}^2$ that is sharply transitive on $W \setminus \{\infty\}$.*

Proof. The reduced stabilizer Θ is a 3-dimensional, connected, solvable Lie group and acts on W with a kernel K of order at most 2, see (73.2d). The stabilizer $\Theta_{p,q}$ of any two points $p \ne q \in W \setminus \{\infty\}$ is 0-dimensional; this is obtained by applying (72.11c) to the group generated by $\Theta_{p,q}$ together with the real homotheties, which fixes a degenerate quadrangle. Using the dimension formula (96.10), we infer that Θ fixes no point on W other than ∞. Moreover, the center of Θ fixes each point $p \in W \setminus \{\infty\}$ because it preserves the set of fixed points of Θ_p. Hence $C(\Theta) \le \mathsf{K}$ has order at most 2. Now the structure of 3-dimensional, solvable Lie algebras (and Lie groups) is known, see, e.g., Jacobson [62] I.4. Since Θ has finite center, we obtain that the commutator group Θ' is abelian and 2-dimensional. The group Θ' cannot have two fixed points in $W \setminus \{\infty\}$, and a single fixed point would also be fixed by Θ. The action of Θ' on a 1-dimensional orbit would have a kernel of dimension 1. It follows that all Θ'-orbits in $W \setminus \{\infty\}$ are 2-dimensional and, hence, open, see (53.1a). This implies that the induced group $\Theta'/(\Theta' \cap K)$ is sharply transitive and isomorphic to $\mathbb{R}^2$. Now Θ' covers this group (94.2a), but $\mathbb{R}^2$ has no proper covering group, hence Θ' acts effectively. □

73.7 Lemma. *If Σ is solvable and 8-dimensional, then the following statements hold.*

(a) *The kernel $\Delta_{[\infty]}$ of the action of Δ on the basis $\mathcal{L}_\infty \setminus \{W\}$ is 3-dimensional and consists of elations:*

$$\Delta_{[\infty]} = \Delta_{[\infty,\infty]}.$$

(b) *The kernel $\Delta_{[Y]}$ of the action of Δ_Y on the fiber Y is 1-dimensional and consists of elations:*

$$\Delta_{[Y]} = \Delta_{[\infty,Y]}.$$

(c) *Both actions are equivalent to the standard action of the 5-dimensional affine group $\mathbb{R}^2 \rtimes \mathsf{P}$, where P denotes the identity component of the group of lower triangular 2×2-matrices.*

Proof. 1) Both actions are transitive affine actions, since there are sharply transitive, normal subgroups isomorphic to $\mathbb{R}^2$, namely, the groups $X = \mathsf{T}_X$ and $Y = \mathsf{T}_Y$, respectively. We shall show that the induced groups are 5-dimensional in both cases. This will imply assertion (c), since any 3-dimensional, connected, solvable subgroup $G \leq \mathrm{GL}_2\mathbb{R}$ is equivalent to P. Indeed, using the Lie algebra one sees that $\mathrm{SL}_2\mathbb{R}$ contains precisely two conjugacy classes of 2-dimensional, closed subgroups, namely, the stabilizers of points in the projective line $\mathsf{P}_1\mathbb{R}$ and their identity components; applying this to the intersection $G \cap \mathrm{SL}_2\mathbb{R}$, which is a 2-dimensional, normal subgroup of G, one obtains that G can be represented by triangular matrices.

2) The kernel $\Delta_{[\infty]}$ is at least 3-dimensional, or else the induced group $\Delta|_{\mathcal{L}_\infty}$ is the connected component of the full affine group, which is not solvable. Suppose that $\Delta_{[\infty]}$ contains homologies with an axis $L \notin \mathcal{L}_\infty$. Since Δ acts transitively on the set of these lines (73.6), the dual of (61.20b) then implies that $\dim \Delta_{[\infty,\infty]} = 4$, contrary to (64.15). This shows that $\Delta_{[\infty]} = \Delta_{[\infty,\infty]}$, and that this group is 3-dimensional.

3) Similar arguments show that $\Delta_{[Y]}$ contains no homologies, and that $\Delta_{[Y]} = \Delta_{[\infty,Y]}$ is at most 1-dimensional. On the other hand, this kernel has positive dimension by step 2), combined with the dual of (61.11).

4) Since Δ is transitive on $\mathcal{L}_\infty \setminus \{W\}$, the stabilizer Δ_Y is 6-dimensional. Finally, we obtain that $\dim \Delta - \dim \Delta_{[\infty]} = 5 = \dim \Delta_Y - \dim \Delta_{[Y]}$. □

In contrast to the 5-dimensional affine actions considered in (73.7), the action at infinity is only 3-dimensional. The following lemma provides a useful description of that action in a somewhat more general situation.

73.8 Lemma: Action at infinity. *Suppose that the automorphism group Σ of a 4-dimensional, affine translation plane fixes the point $\infty \in W$ belonging to the line Y, and that Δ_o normalizes some 3-dimensional subspace $\Xi < T$ that contains $Y = T_Y$.*

Then the action of Δ_o on $W \setminus \{\infty\}$ is isomorphic to the affine action on $\mathbb{R}^2$ that is obtained if we let Δ_o act on the real projective plane $P(\Xi)$ and restrict this action to an affine plane by deleting the Δ_o-invariant line $P(Y)$.

Proof. According to (61.11c), the group Ξ is an 'almost homogeneous translation group'; in the case at hand, this means that for each $p \in W \setminus \{\infty\}$ the group $\Xi_{[p]}$ is a 1-dimensional vector space, whereas $\dim \Xi_{[\infty]} = 2$. The map that sends $\Xi_{[p]}$ to p is a Δ_o-equivariant homeomorphism of $P(\Xi) \setminus P(Y)$ onto $W \setminus \{\infty\}$. □

If Δ is 8-dimensional and solvable, then the last lemma applies, in view of (73.7c). Indeed, since the action of Δ_o on the basis is triangular, T/Y contains a 1-dimensional invariant subspace. We want more information on the affine action at infinity. We know from (73.6) that Θ contains a closed, normal subgroup $A \cong \mathbb{R}^2$ that is sharply transitive on $W \setminus \{\infty\}$. The induced group may or may not be the group of translations of $\mathbb{R}^2$. The possibilities will be clarified by the following lemma. The key step in the investigation of translation planes with an 8-dimensional, solvable group will be to apply (73.8) together with this lemma in order to give an explicit description of A in terms of 4×4-matrices.

In order to formulate the lemma, we regard the real affine plane as the set of 1-dimensional subspaces of $\mathbb{R}^3$ that are not contained in $\{0\} \times \mathbb{R}^2$. We may represent the affine group $\text{AGL}_2\mathbb{R}$ as the group of 3×3-matrices A whose first row is $(1, 0, 0)$. This is analogous to (13.7), but, for technical reasons, the coordinates are permuted cyclically. The corresponding Lie algebra consists of the matrices with first row equal to zero.

73.9 Lemma: Sharply transitive groups on $\mathbb{R}^2$. *Let A be a transitive, abelian group of affine maps of $\mathbb{R}^2$, and assume that A fixes the direction of the y-axis; that is, the transformations in A are represented by lower triangular 3×3-matrices. Then A is a closed subgroup of $\text{AGL}_2\mathbb{R}$ and hence is a Lie group. Moreover, there* is a real number a such that the Lie algebra $\mathfrak{l}A$ is generated by the matrices

$$M = \begin{pmatrix} 0 & & \\ 1 & 0 & \\ 0 & a & 0 \end{pmatrix}, \quad N = \begin{pmatrix} 0 & & \\ 0 & 0 & \\ 1 & 0 & 0 \end{pmatrix}.$$

Notes. (a) We emphasize that the lemma determines the possible Lie algebras $\mathfrak{l}A$ as sets of matrices, and not only up to some sort of equivalence.

(b) The group A is isomorphic to the additive group of its Lie algebra via the exponential map

$$\exp(sM + tN) = \begin{pmatrix} 1 & & \\ s & 1 & \\ \frac{a}{2}s^2 + t & as & 1 \end{pmatrix}.$$

In particular, the group A is also determined; it consists of the mappings $(x, y) \mapsto (x + s, y + asx + \frac{a}{2}s^2 + t)$. In fact, it is easy to twist the proof given below such that one first determines the group and then its Lie algebra. However, we are more interested in the latter.

Proof of (73.9). 1) The closure of A is abelian. Hence, it is again sharply transitive, and coincides with A.

2) We claim that the matrices $A = (a_{ij})_{i,j} \in A$ satisfy $a_{22} = 1$. Indeed, on the set of all parallels of the y-axis, A induces a transitive abelian subgroup of the group of affine maps $x \mapsto rx + t$. The only such group is the one defined by $r = 1$, because all elements with $r \neq 1$ have fixed points. Now r corresponds to the matrix element a_{22}. (Remember that $a_{11} = 1$ by definition.)

3) We claim further that $a_{33} = 1$. Indeed, if A contains a matrix A with $a_{33} \neq 1$, then the generalized eigenspace $\ker(A-\mathbb{1})^2 = \mathbb{R}^2 \times \{0\}$ is invariant under the abelian group A, a contradiction to the transitivity of A. It follows that the elements of $\mathfrak{l}A$ are lower triangular matrices with zero diagonal, because they are obtained by differentiating one-parameter groups in A at the parameter 0.

4) The stabilizer of $0 \in \mathbb{R}^2$ in the affine group is the subgroup defined by the equations $a_{21} = a_{31} = 0$, and the corresponding Lie algebra is the subalgebra defined by the same equations. By the hypotheses of commutativity and transitivity, $\mathfrak{l}A$ has trivial intersection with this subalgebra and, hence, is mapped isomorphically onto $\mathbb{R}^2$ by $A \mapsto (a_{21}, a_{31})$. We infer that there is a basis of $\mathfrak{l}A$ consisting of two matrices

$$M = \begin{pmatrix} 0 & & \\ 1 & 0 & \\ 0 & a & 0 \end{pmatrix}, \quad N = \begin{pmatrix} 0 & & \\ 0 & 0 & \\ 1 & c & 0 \end{pmatrix}.$$

Since the Lie bracket $[M, N] = MN - NM$ is zero, we obtain that $c = 0$. $\square$

After these preparations we are able to classify the 4-dimensional translation planes having an 8-dimensional, solvable collineation group. We shall actually use a weaker hypothesis and thus include a certain single plane with a 7-dimensional group. The characterization of this plane was originally proved in Betten [73a] Satz 2a using lengthy calculations, whereas we obtain it as a by-product of our treatment of the 8-dimensional case at virtually no extra cost.

73.10 Theorem: Solvable group. *Let $(P, \mathcal{L})$ be a 4-dimensional translation plane with a solvable automorphism group Σ and let $\Delta = \Sigma^1$ be the connected component. We know that there is a fixed point ∞ on the translation axis W (73.4).*

Moreover, we assume that the stabilizer Δ_o acts as the lower triangular group P both on the invariant line $o \vee \infty = Y < T$ and on T/Y, and that Δ_o contains a commutative, normal subgroup that is sharply transitive on $W \setminus \{\infty\}$.

Then the plane is generated by one of two spreads, which are represented in Grassmann coordinates as follows:

$$\mathscr{S}_h = \{Y\} \cup \left\{ \begin{pmatrix} t & s \\ -\frac{s^3}{3} + hs & s^2 + t \end{pmatrix} \,\Big|\, s, t \in \mathbb{R} \right\}, \quad \text{where} \quad h \in \{0, -1\}.$$

We define the following groups:

$$\mathsf{B} = \left\{ \operatorname{diag}(1, b, b^2, b^3) \,\big|\, b > 0 \right\},$$

$$\mathsf{\Pi} = \mathbb{R}^4 \rtimes \left\{ \begin{pmatrix} r & & & \\ 0 & r & & \\ d & 0 & r & \\ 0 & d & 0 & r \end{pmatrix} \,\Bigg|\, d \in \mathbb{R}, r > 0 \right\},$$

$$\mathsf{M}_h = \left\{ \begin{pmatrix} 1 & & & \\ c & 1 & & \\ \frac{c^2}{2} & c & 1 & \\ \frac{c^3}{6} + hc & \frac{c^2}{2} & c & 1 \end{pmatrix} \,\Bigg|\, c \in \mathbb{R} \right\}$$

and consider the transformations $\sigma = -\mathbb{1}$ and $\kappa = \operatorname{diag}(1, -1, 1, -1)$. The automorphism groups of the planes may be described as follows:

(a) *If $h = -1$, then $\Delta = \langle \mathsf{\Pi}, \mathsf{M}_{-1} \rangle$, and Δ is 7-dimensional.*
(b) *If $h = 0$, then $\Delta = \langle \mathsf{\Pi}, \mathsf{M}_0, \mathsf{B} \rangle$, and Δ is 8-dimensional.*
(c) *In both cases, the full automorphism group is $\Sigma = \langle \Delta, \sigma, \kappa \rangle$, whence $\Sigma/\Delta \cong \mathbb{Z}_2 \times \mathbb{Z}_2$.*
(d) *σ is a reflection, and κ is a Baer involution.*

We postpone the proof for a moment. It follows from our results (73.4, 6, and 7) that a plane with an 8-dimensional solvable group Δ satisfies the assumptions of this theorem, so that we obtain the following corollary.

73.11 Corollary: *There exists a unique 4-dimensional translation plane with an 8-dimensional, solvable automorphism group. It is defined by the spread $\mathcal{S}_0$.* □

Note. (a) The proof of (73.10) will show that the planes admit a 1-dimensional elation group $\Delta_{[\infty,Y]}$, see step 2). According to Knarr [95] Prop. 5.3, this implies that $\mathcal{S}_h$ may be covered by a set R of reguli containing Y such that R induces a partition of the complement $\mathcal{S}_h \setminus \{Y\}$.

(b) The groups $\langle \Pi, M_{-1} \rangle$ and $\langle \Pi, M_0 \rangle$ are isomorphic. This becomes evident if one conjugates the Lie algebra of M_h by a matrix that differs from $\mathbb{1}$ in position $(3, 1)$ only. (That Lie algebra will be computed in the proof given below.) However, this conjugation does not respect our convention that the factor $\mathbb{R}^2 \times \{0\}$ of T should be a line. This is why we have chosen the parameter-dependent description of the group. The same situation will reappear in (74.23), but without the need to preserve the first factor.

(c) The planes described in (73.10) have finite analogues, as was discovered by Walker [76]. A characterization of Walker's planes has been given by Bartolone [83]. For generalizations, see Spera [86]; compare also the survey Jha–Johnson [85].

Proof of (73.10). 1) By our hypothesis, there is a basis of T such that the line X is generated by the first two basis elements and Y is generated by the remaining two, and such that the stabilizer Δ_o is represented by lower triangular matrices with respect to this basis. (Since Δ is connected, all diagonal entries are positive.) Remove the first row and the first column of these matrices. According to (73.8), the affine maps of $\mathbb{R}^2$ induced by the truncated matrices describe the action of Δ_o on the line W at infinity. In other words, the truncation homomorphism $E \mapsto E^\flat$ corresponds to the restriction map $\Delta_o \to \Delta_o|_W$. Our plane is non-desarguesian, because its automorphism group is solvable. Therefore, the kernel of the restriction map consists precisely of the real homotheties with positive factors (72.11a). Hence, the subgroup E of all 4×4-matrices in Δ_o having an entry 1 in position $(2,2)$ is mapped isomorphically via truncation onto the matrix group describing the action at infinity. According to our hypothesis, E contains an abelian, normal subgroup A that is transitive on $W \setminus \{\infty\}$.

2) From (73.9) we obtain that the Lie algebra of $A^\flat$ is generated by two 3×3-matrices,

$$M^\flat = \begin{pmatrix} 0 & & \\ 1 & 0 & \\ 0 & a & 0 \end{pmatrix} \quad \text{and} \quad N^\flat = \begin{pmatrix} 0 & & \\ 0 & 0 & \\ 1 & 0 & 0 \end{pmatrix}.$$

If the parameter a is zero, then $A \leq \Delta_{[Y]}$ because the action on Y is given by the bottom right 2×2-blocks of the matrices $\exp H^\flat \in A^\flat$, which are all equal to $\mathbb{1}$

in this case. Since A acts freely on $W \setminus \{\infty\}$, we have in fact $\mathsf{A} \leq \Delta_{[\infty,Y]}$, and the plane is desarguesian and has a non-solvable group Σ, see (72.11d). Hence, $a \neq 0$, and the same argument shows that the one-parameter group $\mathsf{N} = \exp \mathbb{R}N$ is contained in the elation group $\Delta_{[\infty,Y]}$ and forms its identity component.

3) Our next aim is to reconstruct the missing first columns of the generators of $\mathfrak{l}\mathsf{A}^{\flat}$. Note that we may normalize using conjugation $H \mapsto Q^{-1}HQ$ by block matrices

$$Q = \begin{pmatrix} A & \\ & B \end{pmatrix},$$

where A and B are lower triangular 2×2-matrices; this does not disturb any of our previous arrangements. We may think of this process as a revision of our choice of basis. The formula $\exp(Q^{-1}HQ) = Q^{-1}(\exp H)Q$ shows that it does not matter whether we conjugate elements H of the Lie algebra or the corresponding elements $\exp H$ of the group.

A non-trivial elation $\exp tN$ has diagonal blocks equal to $\mathbb{1}$ because it fixes every parallel of Y and acts trivially on Y. The bottom left block D is a regular matrix because all fixed points must belong to the axis Y, and a look at the generating matrix N tells us that D is lower triangular. We may conjugate with a block matrix defined by $A = \mathbb{1}$ and $B = D$ to obtain that the bottom left block of $\exp tN$ becomes $t\mathbb{1}$ and the completed matrix N is given by

$$N = \begin{pmatrix} 0 & 0 \\ \mathbb{1} & 0 \end{pmatrix}.$$

The Lie algebra of the new matrix group A obtained by this conjugation again has a generator having the shape of M; this will replace our previous M.

4) We are still allowed to conjugate by block matrices as above provided that we choose B to be a scalar multiple of A. This has the harmless side-effect of multiplying N by the same scalar. By conjugating with $\mathrm{diag}(a, 1, 1, a^{-1})$, we obtain $a = 1$. The Lie bracket $[M, N]$ is zero, hence we know that the completed matrix M is of the form

$$\begin{pmatrix} 0 & & & \\ 1 & 0 & & \\ g & 1 & 0 & \\ h & 0 & 1 & 0 \end{pmatrix}.$$

The parameter g may be reduced to zero by first conjugating with

$$A = B = \begin{pmatrix} 1 & \\ -\frac{g}{2} & 1 \end{pmatrix},$$

and replacing M by $M - \frac{g}{2}N$ afterwards. Finally, conjugation by $\mathrm{diag}(1, b, b^2, b^3)$ and subsequent multiplication by b transforms h into $b^{-2}h \in \{1, 0, -1\}$. The resulting matrix

$$M_h = \begin{pmatrix} 0 & & & \\ 1 & 0 & & \\ 0 & 1 & 0 & \\ h & 0 & 1 & 0 \end{pmatrix},$$

where $h \in \{1, 0, -1\}$, generates the one-parameter group $\exp \mathbb{R} M_h = \mathsf{M} = \mathsf{M}_h$ appearing in the statement of the theorem; at the moment, however, there is an additional possibility, $h = 1$, for the parameter h.

5) The spread $\mathscr{S}$ is entirely determined by the group $\mathsf{A} = \mathsf{M} \times \mathsf{N}$, which acts transitively on $W \setminus \{\infty\}$ and hence also on $\mathscr{S} \setminus \{Y\}$. Applying a lower triangular matrix

$$\begin{pmatrix} A & \\ B & C \end{pmatrix}$$

to $X \in \mathscr{S}$, the left factor of $T = X \oplus Y$, we obtain the space $\{(Ax, Bx) \mid x \in X\}$. Via Grassmann coordinates, this space is represented by the matrix BA^{-1}. Applying first an element of M and then an element of N we produce the representative matrices

$$\begin{pmatrix} -\frac{c^2}{2} + d & c \\ -\frac{c^3}{3} + hc & \frac{c^2}{2} + d \end{pmatrix}, \quad \text{where} \quad c, d \in \mathbb{R}.$$

After substituting $t = -\frac{c^2}{2} + d$ and $s = c$ we obtain the system $\mathscr{S} = \mathscr{S}_h$ of lines as written down in the theorem, still with the additional possibility $h = 1$.

6) Next we prove that the system $\mathscr{S}_h$ is a spread precisely in the cases $h = 0$ and $h = -1$. We have to check whether the origin o is joined to each point $(x, y, u, v) \in \mathbb{R}^4$ by a unique line. This is true for $(0, 0, u, v) \in Y$. The other cases lead to the following system of equations for s and t:

$$tx + sy = u$$
$$(-\tfrac{s^3}{3} + hs)x + (s^2 + t)y = v.$$

The solution is unique if $x = 0$. If $x \neq 0$, we eliminate t:

$$-\tfrac{x}{3}s^3 + ys^2 + (hx - \tfrac{y^2}{x})s + \tfrac{u}{x}y = v.$$

The solution s is uniquely determined if, and only if, the polynomial function on the left hand side is either increasing or decreasing, that is, if and only if its derivative never changes its sign. This is the case if, and only if, $h = 0$ or $h = -1$.

7) The map $\mathbb{R}^2 \to \mathscr{S} \setminus \{Y\}$ sending the pair (s, t) to the corresponding line is a homeomorphism; remember that the correspondence between subspaces and their Grassmann coordinates is a homeomorphism (64.3). By (64.8a), this implies that

$\mathcal{S}$ is compact and defines a topological translation plane. As an alternative, we may apply (43.8) directly in order to show that a topological affine plane is obtained.

8) Consider the plane $\mathcal{P}_h$ defined by $\mathcal{S}_h$ and its full automorphism group $\Sigma = \Sigma_h$. To conclude the proof, we determine this group in the cases $h = 0$ and $h = -1$. As in step 5), we see that $\mathcal{S} \setminus \{Y\}$ is an orbit of the group $\mathsf{A} = \mathsf{A}_h = \mathsf{M}_h \mathsf{N}$, whence $\mathsf{A} \leq \Sigma$.

We claim that none of the planes $\mathcal{P}_h$ is desarguesian. Assume to the contrary that $\mathcal{P}_h^W \cong \mathcal{A}_2\mathbb{C}$; then the stabilizer of Y acts on $W \setminus \{\infty\}$ as the group $\mathsf{L}_2\mathbb{C}$ of affine transformations of $\mathbb{C}$. The only sharply transitive subgroup of this group is induced by the shears with axis Y. Therefore, A must induce real homotheties on Y, but this is not in fact the case.

By (72.11g), we may conclude that $\dim \Sigma \leq 8$. Thus, Σ_o is an at most 4-dimensional group containing a central one-parameter group E (of homotheties) and a 2-dimensional abelian subgroup A such that $\mathsf{E} \cap \mathsf{A} = \mathbb{1}$. This implies that the commutator subalgebra of $\mathfrak{l}\Sigma_o$ is at most 2-dimensional and, hence, solvable. As a consequence, $\Gamma := (\Sigma_o)^1$ and Σ^1 are also solvable.

We have either $\dim \Sigma = 7$ and $\mathsf{EA} = \Gamma$ or $\dim \Sigma = 8$. In the first case, it is clear that $\mathsf{A} \triangleleft \Gamma$, and in the second case the same follows from the previous steps of this proof together with the remarks preceding (73.11).

As A acts transitively on $W \setminus \{\infty\}$, it suffices to determine the form of a collineation $\delta \in \Sigma_o$ that leaves the lines X (and Y) invariant, compare (91.3). Now $\mathsf{A} \triangleleft \Gamma$ fixes a unique maximal flag of $\mathbb{R}$-subspaces of T. By normality, the same is true for Γ. As δ in turn normalizes Γ, it follows that δ is given by a lower triangular matrix which we may write as a block matrix $\text{diag}(A, D)$ with (2×2)-matrices A and D. Invariance of the shear group N, see step 2), implies that $D = kA$, where $k \neq 0$. We infer that, up to real homotheties, δ is given by a matrix

$$\begin{pmatrix} 1 & & & \\ p & b & & \\ & & k & \\ & & kp & kb \end{pmatrix}.$$

Applying this map to the spreads, we obtain first that $p = 0$ and $k = b^2$, that is, $\delta = \text{diag}(1, b, b^2, b^3)$, where $b \neq 0$. Moreover, we see that for $h = 0$ all values $b \neq 0$ are admissible and that for $h = -1$ only $b = -1$ (and $b = 1$) give collineations.

The remaining assertion (d) is obvious. □

Translation planes with non-solvable automorphism groups

The 4-dimensional translation planes with a non-solvable automorphism group Σ will be completely determined, see (73.12, 13, and 19). First we determine a maximal semi-simple, connected, analytic subgroup (= Levi subgroup) of the automorphism group. Note incidentally that such a Levi subgroup is closed in Σ by (94.28e); this result applies because the affine group of $\mathbb{R}^4$ contains Σ and has a faithful linear representation on $\mathbb{R}^5$.

73.12 Levi subgroups. *If the automorphism group Σ of a non-desarguesian, 4-dimensional translation plane is not solvable, then* $\dim \Sigma = 8$ *and the stabilizer Σ_o of the origin contains a Levi subgroup $\Phi \cong \mathrm{SL}_2\mathbb{R}$ of Σ.*

The action of Φ on T by conjugation is equivalent either to the standard action of $\mathrm{SL}_2\mathbb{R}$ on $\mathbb{C}^2$ or to the unique irreducible, 4-dimensional, real representation of $\mathrm{SL}_2\mathbb{R}$.

An explicit description of the irreducible representation will be given below, see (73.17). The $\mathbb{C}$-linear action is the direct sum of two copies of the standard representation of $\mathrm{SL}_2\mathbb{R}$ on $\mathbb{R}^2$. Remember that the action of Φ on T by conjugation corresponds to the action on the affine point set (73.2).

Proof of (73.12). The semi-direct decomposition $\Delta = \Sigma^1 = \mathsf{T} \rtimes \Delta_o$ shows that Δ_o maps onto Δ/T under the canonical homomorphism. Hence we infer from (94.27) that there is a Levi subgroup $\Phi \leq \Sigma$ that is contained in Δ_o. Since $\dim \Delta_o \leq 4$ by (72.11g), Φ must be locally isomorphic either to $\mathrm{SO}_3\mathbb{R}$ or to $\mathrm{SL}_2\mathbb{R}$, compare the list (94.33). The first possibility leads to a transitive action on the line W at infinity (62.8b) and to a contradiction (72.11f).

The space $\mathsf{T} \cong \mathbb{R}^4$ is a direct sum of Φ-irreducible subspaces T_i, see (95.2). Suppose that there is a 3-dimensional component T_1. Then there is a 1-dimensional invariant complement, determining an invariant line L. The intersection $\mathsf{T}_1 \cap L$ is invariant and 1-dimensional, and T_1 is not irreducible, a contradiction.

There is only one irreducible representation of a group locally isomorphic to $\mathrm{SL}_2\mathbb{R}$ in dimension 2 and one in dimension 4, see the tables (95.10). Thus, it only remains to exclude the possibility that Φ acts trivially on some 2-dimensional subspace T_1. In this case, an element τ contained in an invariant complement of T_1 has a 1-dimensional stabilizer $\Phi_\tau = \mathrm{Cs}_\Phi(\tau)$ that acts trivially on a 3-dimensional subspace $\mathsf{T}_1 \oplus \mathbb{R}\tau$, a contradiction to stiffness. □

73.13 Theorem: Reducible $\mathrm{SL}_2\mathbb{R}$-action. *Let $\mathcal{P} = (P, \mathcal{L})$ be a 4-dimensional translation plane such that the automorphism group Σ contains a subgroup $\Phi \cong \mathrm{SL}_2\mathbb{R}$ which acts on the translation group $\mathsf{T} \cong \mathbb{C}^2$ in the standard $\mathbb{C}$-linear way. Then the plane is generated by one of the spreads*

$$\mathcal{S}_w = \{Y\} \cup \left\{ \begin{pmatrix} s & -t \\ t & s \end{pmatrix} \,\bigg|\, s, t \in \mathbb{R}, t \geq 0 \right\} \cup \left\{ \begin{pmatrix} s & -t \\ wt & s \end{pmatrix} \,\bigg|\, s, t \in \mathbb{R}, t \leq 0 \right\},$$

where $w \geq 1$. The plane $\mathcal{P}_w$ defined by $\mathcal{S}_w$ is desarguesian if, and only if $w = 1$, and different parameters w and w' define non-isomorphic planes $\mathcal{P}_w$ and $\mathcal{P}_{w'}$. For

$w > 1$, *the full collineation group Σ of $\mathcal{P}_w$ is 8-dimensional and has 4 connected components. More precisely, Σ is the group $\mathbb{C}^2 \rtimes \mathrm{GL}_2\mathbb{R}$ (with standard action on $\mathbb{C}^2$), extended by a Baer involution β that will be described in Note (a) below.*

The action of Φ on the line at infinity has 3 orbits, namely, a circle $\mathcal{R} \approx \mathbb{S}_1$ with the standard doubly transitive action of Φ, and the two complementary open disks, on which Φ acts like the direct real hyperbolic motion group (acting on interior points).

Notes. (a) We describe the situation in more detail. The spread $\mathcal{S}_w$ is composed of two partial spreads $\mathcal{S}_w^+$ and $\mathcal{S}_w^-$ consisting of Y together with the elements satisfying $\pm t \geq 0$. They are both isomorphic to some part of the desarguesian complex spread. $\mathcal{S}_w^+$ is in fact equal to the set of complex lines with slope $z = s+it$ such that $t \geq 0$, including $z = \infty$, and $\mathcal{S}_w^-$ is obtained from $\mathcal{S}_w^+$ by applying an $\mathbb{R}$-linear map β (the Baer involution appearing in the theorem). Using the number $g > 0$ such that $g^4 = w$, we may describe β by the matrix $\mathrm{diag}(B_g, B_g)$, where

$$B_g = \begin{pmatrix} & g^{-1} \\ g & \end{pmatrix}.$$

The intersection $\mathcal{R} = \mathcal{S}_w^+ \cap \mathcal{S}_w^-$ consists of the complex lines of real slope and is homeomorphic to $\mathrm{P}_1\mathbb{R} \approx \mathbb{S}_1$. The group Φ acts transitively on $\mathcal{R}$ and on the two open disks $\mathcal{B}_w^\pm = \mathcal{S}_w^\pm \setminus \mathcal{R}$. The Baer involution β interchanges the halves of $\mathcal{S}_w$ and fixes the elements of $\mathcal{R}$.

(b) We remark without proof that the dual plane $\mathcal{P}_w^*$ may also be obtained by gluing together two copies of the complex cylinder plane $\{(x, y, z) \in \mathrm{P}_2\mathbb{C} \mid x\bar{x} \geq y\bar{y}\}$ along their boundaries.

(c) Knarr [95] Prop. 5.7 describes the plane $\mathcal{P}_w$ using a contracting map $\varphi : \mathbb{C} \to \mathbb{C}$ as in (64.11). The contraction is given by

$$\varphi(a + ib) = \begin{cases} 0 & \text{if } b \geq 0 \\ i\mu b & \text{if } b < 0, \end{cases}$$

where μ satisfies $(w+1)\mu = w - 1$.

Proof of (73.13). 1) We keep the notation introduced in Note (a) above. Given the standard action of Φ on $\mathbb{C}^2$, we may reconstruct the complex lines $R \in \mathcal{R}$ of real slope as the fixed point sets of those shears of the complex plane that belong to Φ. Indeed, an element $\varphi \in \Phi$ having eigenvalue 1 on $\mathbb{C}^2$ has the same eigenvalue on $\mathbb{R}^2$, because the two characteristic polynomials are the same; hence, the axis of φ belongs to $\mathcal{R}$. In the given plane $\mathcal{P}$, either the elements of $\mathcal{R}$ are lines, or they contain quadrangles (42.3). Theorem (55.21) excludes the latter possibility, whence $\mathcal{R}$ is contained in the spread $\mathcal{S}$ defining $\mathcal{P}$.

2) Φ is transitive on $\mathcal{R}$ because the action is the ordinary one. We claim that Φ acts transitively on the two open disks $\mathcal{C}^+$ and $\mathcal{C}^-$ that form the complement $\mathcal{S} \setminus \mathcal{R}$. If this is not the case, then there are additional Φ-orbits of dimension less than 2, see (53.1d). Locally (that is, on the level of the Lie algebra), all 1-dimensional Φ-actions have the same set of stabilizers, compare (96.30), hence the stabilizer of a pair of real lines would fix at least 3 lines in all. Adding the real homotheties we obtain a 2-dimensional group that fixes a degenerate quadrangle. This forces $\mathcal{P}$ to be desarguesian (72.11c); our claim and the theorem are true in this case. From now on we shall exclude the desarguesian plane.

3) We claim that the action of Φ on $\mathcal{C}^\pm$ is isomorphic to the action of $\Phi/\mathbb{Z}_2 \cong \mathrm{PO}'_3(\mathbb{R}, 1)$ as the direct hyperbolic motion group on the set of interior points with respect to the real hyperbolic polarity (13.17). As $\mathrm{SO}_2\mathbb{R}$ is a maximal subgroup (94.34), it suffices to show that $\mathrm{SO}_2\mathbb{R} \leq \Phi$ fixes an element in $\mathcal{C}^\pm$. This is indeed the case because $\mathrm{SO}_2\mathbb{R}$ acts on $\mathcal{S} \approx \mathbb{S}_2$ in the standard way, see Mostert [57] Theorems 4 and 5.

To continue the proof, we need to know the normalizer of Φ in $\mathrm{GL}_4\mathbb{R}$. The necessary information is provided by the following lemma.

73.14 Lemma. *The normalizer* N *of* $\Phi = \mathrm{SL}_2\mathbb{R} \leq \mathrm{GL}_2\mathbb{C}$ *in* $\mathrm{GL}_4\mathbb{R}$ *is a product of two commuting copies of* $\mathrm{GL}_2\mathbb{R}$. *The first factor* A *is the subgroup* $\mathrm{GL}_2\mathbb{R} \leq \mathrm{GL}_2\mathbb{C}$, *and the second factor* B *acts on* $\mathbb{C} \times \mathbb{C} = \mathbb{R}^2 \times \mathbb{R}^2$ *as the group of all transformations* $(u, v) \mapsto (Bu, Bv)$, *where* B *is an invertible real* 2×2-*matrix.*

Proof. In $\mathbb{C}^2$, consider the set $\mathcal{R}$ of all complex lines through the origin having real slope. We show that the normalizer N acts on $\mathcal{R}$. Each line $R \in \mathcal{R}$ is the axis of some shear $\varphi \in \Phi$. A conjugate φ^ν, where $\nu \in \mathsf{N}$, belongs to Φ and hence is $\mathbb{C}$-linear. Therefore, the eigenspace $L = R^\nu$ of φ^ν is a complex line. Now the two characteristic polynomials of φ^ν considered as a 2×2-matrix over $\mathbb{R}$ or $\mathbb{C}$ are the same. This implies that L belongs to $\mathcal{R}$.

The subgroup $\mathsf{A} \leq \mathsf{N}$ induces on $\mathcal{R}$ the triply transitive group $\mathrm{PGL}_2\mathbb{R}$ of projectivities of the real projective line. The Frattini argument (91.2a) shows that N is generated by A together with the subgroup of N fixing three elements of $\mathcal{R}$, e.g., those having slopes 0, 1, and ∞. That subgroup is equal to B. □

Note. The subgroup $\mathrm{SL}_2\mathbb{R} \cdot \mathrm{SL}_2\mathbb{R} \leq \mathsf{N}$ coincides with the connected component of the group $\mathrm{SO}_4(\mathbb{R}, 2)$. Upon passage to the projective space $\mathrm{P}_3\mathbb{R}$, the set $\mathcal{R}$ of lines of real slope becomes one of the reguli on the ruled quadric associated with this group.

The stated equality of groups may be seen as follows. Consider the isomorphism of $\mathbb{C}^2$ onto the space of real 2×2-matrices which sends $(x + iy, p + iq)$ to the matrix

$$\begin{pmatrix} x & y \\ p & q \end{pmatrix}.$$

This transforms the action of N into the action given by matrix multiplication, $X \mapsto AXB^t$, where $A, B \in \mathrm{GL}_2\mathbb{R}$. If A and B have determinant 1, then this action preserves the quadratic form f of index 2 given by $f(X) = \det X$.

Proof of (73.13), *resumed.* 4) We may use elements of N in order to normalize; note that this may also change the complex structure on $\mathbb{R}^4$ that we are using. An element of B maps the line $\{(u, Mu) \mid u \in \mathbb{C}\}$ defined by a real 2×2-matrix M to the subspace defined by the matrix BMB^{-1}.

5) The subgroup $\mathrm{SO}_2\mathbb{R} \leq \Phi$ fixes precisely two lines in $\mathcal{S}$, one in each component $\mathcal{C}^\pm$, see step 3). Being complementary to $\mathbb{C} \times \{0\} \in \mathcal{R}$, these lines are defined by two matrices $I, J \in \mathrm{GL}_2\mathbb{R}$. We shall identify the pair (I, J) up to the action of B, that is, up to simultaneous conjugation in $\mathrm{GL}_2\mathbb{R}$. The plane $\mathcal{P}$ is completely determined by this pair of matrices since Φ is transitive on $\mathcal{C}^\pm$; moreover, equivalent pairs yield isomorphic planes.

6) The lines fixed by $\mathrm{SO}_2\mathbb{R}$ are mapped into themselves by any endomorphism generating the Lie algebra of this one-parameter group. Such generators are obtained by derivation of matrices at the parameter value 0; one of them maps $(u, v) \in \mathbb{C}^2$ to $(-v, u)$. In particular, (x, Mx) is taken to $(-Mx, x) = (z, -M^{-1}z)$. It follows that M and the line defined by M are invariant if, and only if $M^2 = -\mathbb{1}$. Thus we are looking for the conjugacy classes of pairs of square roots of $-\mathbb{1}$ in $\mathrm{GL}_2\mathbb{R}$. Instead of conjugation, we may think of changing our choice of an $\mathbb{R}$-basis in $\mathbb{C}$.

In the complex plane $\mathcal{A}_2\mathbb{C}$, the corresponding matrices I_1 and J_1 represent complex slopes. Our general arguments show that these slopes are the square roots i and $-i$ of -1, so that

$$I_1 = \begin{pmatrix} 0 & -1 \\ 1 & 0 \end{pmatrix}, \quad J_1 = \begin{pmatrix} 0 & 1 \\ -1 & 0 \end{pmatrix}.$$

7) For each $v \in \mathbb{C} \setminus \{0\}$, passage to the basis $\{v, Iv\}$ transforms the matrix I to I_1. We may rotate this basis without changing I, and we claim that we may rotate so that both I and J interchange the 1-dimensional subspaces $\mathbb{R}v$ and $\mathbb{R}Iv$. That is, we want a rotation ϱ such that $\varrho(v) \perp J\varrho(v)$. If the scalar product $\langle v, Jv \rangle$ is negative, then $\langle Jv, J^2v \rangle = \langle Jv, -v \rangle$ is positive and vice versa. Using a family of rotations ϱ_t such that $\varrho_0 = \mathbb{1}$ and $\varrho_1(v) = Jv$, we find that $\langle \varrho_t(v), J\varrho_t(v) \rangle = 0$ for some t. By passing to the rotated basis, the pair (I, J) is transformed to

$$I = \begin{pmatrix} 0 & -1 \\ 1 & 0 \end{pmatrix}, \quad J = J_h = \begin{pmatrix} 0 & h^{-1} \\ -h & 0 \end{pmatrix},$$

where $0 \neq h \in \mathbb{R}$.

8) We shall write $\mathcal{C}_h^+$ and $\mathcal{C}_h^-$ for the half pencils obtained respectively from these matrices I and J by applying Φ to the corresponding lines; it will turn out that $\mathcal{C}_h^\pm = \mathcal{B}_w^\pm$ are the half pencils constituting the spread $\mathcal{S}_w$ for the parameter $w = h^2$. If $h = 1$, then we obtain the halves $\mathcal{C}_1^\pm$ of the complex line pencil, see step 6). We denote by R and $A^\pm$ the sets of points $(u, v) \in \mathbb{C}^2 \setminus \{0\}$ that are covered by $\mathcal{R}$ and by $\mathcal{C}_1^\pm$, respectively. Thus, $(u, v) \in A^\pm$ if, and only if, $u \neq 0$ and $\pm \operatorname{Im}(u^{-1}v) > 0$. We give an equivalent condition in terms of the function $f(u, v) = \operatorname{Im}(\bar{u}v)$, which is expressed in real coordinates by $f(a, b, c, d) = ad - bc$. We see that $p \in \mathbb{C}^2 \setminus \{0\}$ belongs to R or to $A^\pm$ if, and only if, $f(p) = 0$ or $\pm f(p) > 0$, respectively. Incidentally, f is the quadratic form considered in (73.14).

Clearly, the closure $\overline{\mathcal{C}_h^+} = \overline{\mathcal{C}_1^+} = \mathcal{C}_1^+ \cup \mathcal{R}$ covers precisely the set $A^+ \cup R$. For an admissible choice of h, the resulting set $\mathcal{C}_h^-$ of lines must cover precisely the remaining set A^- of points. In particular, we want the line defined by J to be contained in this set. This is the case if, and only if $h > 0$. Applying the quarter rotation defined by $B = J_1$, we may transform h to h^{-1}, whence $h > 1$ gives all isomorphism types.

9) We shall now exhibit an involution $\beta \in \mathsf{B}$ that interchanges $\mathcal{C}_1^+$ with $\mathcal{C}_h^-$ and fixes the elements of $\mathcal{R}$. Furthermore, β satisfies $f(p^\beta) = -f(p)$, hence β interchanges A^+ with A^-. This implies that $\mathcal{C}_h^-$ covers A^-, whence $\mathcal{T}_h = \mathcal{C}_1^- \cup \mathcal{C}_h^+ \cup \mathcal{R}$ is a spread for each $h > 1$. (We remark that it is not difficult to verify the spread property by a more direct computation.) Moreover, $\mathcal{T}_h = \overline{\mathcal{C}_1^+} \cup (\overline{\mathcal{C}_1^+})^\beta$ is compact and hence defines a compact translation plane according to (64.4d).

As in Note (a) following (73.13), the map β is defined by the matrix $\operatorname{diag}(B, B)$, where

$$B = \begin{pmatrix} 0 & g^{-1} \\ g & 0 \end{pmatrix}$$

and $g^2 = h$. In real coordinates, β sends $p = (a, b, c, d)$ to $(g^{-1}b, ga, g^{-1}d, gc)$, and $f(p^\beta) = -f(p)$. As B commutes with the matrices $s\mathbb{1}$, all lines of real slope are fixed by β, see step 4). Conjugation by B transforms I to $J = J_h$, hence β interchanges the corresponding Φ-orbits $\mathcal{C}_1^+$ and $\mathcal{C}_h^-$; note that β normalizes Φ. On the other hand, direct computation shows that $(\mathcal{C}_1^+)^\beta$ is the half pencil $\mathcal{B}_w^-$ of $\mathcal{S}_w$ as described in the theorem, where $w = h^2 = g^4$. The compact translation plane now obtained will be denoted by $\mathcal{P}_w$.

10) If the affine plane $\mathcal{P}_w \setminus W$ is desarguesian, then all automorphisms fixing the origin o are either $\mathbb{C}$-linear or $\mathbb{C}$-antilinear maps, see (12.10). If $w > 1$, then the Baer involution β satisfies neither condition, and the plane $\mathcal{P}_w$ is non-desarguesian. This implies that its automorphism group Σ is 8-dimensional (72.11g), whence $\Delta = \mathsf{T} \rtimes \mathrm{GL}_2^+\mathbb{R}$ is the connected component of Σ. Other components are represented by the Baer involutions β and $\gamma = \operatorname{diag}(1, -1, -1, 1)$. These involutions do not belong to Δ, see (55.21c), and they belong to different components because

β interchanges the half pencils $\mathcal{B}_w^{\pm}$ and γ preserves them. The group generated by all these collineations answers to the description given in the statement of the theorem.

11) Suppose that there is some further automorphism α of a non-desarguesian plane $\mathcal{P}_w$. By (72.10) we know that α fixes W; moreover, we may assume that $\alpha \in \Sigma_o$. Since Φ is the only Levi subgroup of $\Delta_o \trianglelefteq \Sigma_o$, we infer that α belongs to $\mathsf{N} = \mathrm{Ns}\,\Phi$. In particular, the set $\mathcal{R}$ of all lines of real slope is invariant under α. Now we may use the transitivity properties of the known automorphisms in order to arrange that α also preserves the components $\mathcal{B}_w^{\pm}$ and fixes the line defined by the matrix I. Then α normalizes the stabilizer $\mathrm{SO}_2\mathbb{R} = \Psi \leq \Phi$ of that line. Hence, α also preserves the other fixed line of Ψ, which is represented by the matrix J.

With the notation of (73.14), we have that $\mathrm{Ns}_\mathsf{N}(\Psi) = \mathrm{Ns}_\mathsf{A}(\Psi) \cdot \Gamma$, where $\Gamma \leq \mathsf{B}$ is the subgroup defined by $|\det B| = 1$. Now the normalizer $\mathrm{Ns}_\mathsf{A}(\Psi)$ of $\mathrm{SO}_2\mathbb{R}$ is $\mathrm{GO}_2\mathbb{R}$, and its identity component $\mathrm{GO}_2^+\mathbb{R}$ is contained in Δ_o. Hence, we may assume that α belongs to the group $\mathbb{Z}_2 \cdot \Gamma$, where the generator of $\mathbb{Z}_2 \leq \mathsf{A}$ is represented by the matrix $\mathrm{diag}(1,-1)$. This group acts on $\mathbb{R}^4$ by block matrices $\mathrm{diag}(B, \varepsilon B)$, where $B \in \mathrm{GL}_2\mathbb{R}$ has determinant ± 1 and $\varepsilon \in \{1, -1\}$. Using the collineation γ obtained in step 10) we may reduce this to $\varepsilon = 1$.

Now we use the condition that conjugation by B fixes the matrices I and J. This implies that B centralizes the two one-parameter groups $\exp \mathbb{R}I$ and $\exp \mathbb{R}J$, which generate Φ. Finally, this shows that $B = \pm 1\!\!1$, which completes the computation of the group Σ. By a variation of step 11), we see that an isomorphism of the planes defined by J_h and J_k enforces that $h \in \{k, k^{-1}\}$. Indeed, the same reductions as before lead to a matrix B which centralizes I (and hence belongs to $\mathrm{GO}_2^+\mathbb{R}$) and conjugates J_h into J_k. $\square$

73.15 Note. Our construction of planes involved the ordering of the real numbers. Hence, it is not surprising that there appear to be no finite analogues of the planes constructed here. At least we can say that by classifying finite planes under analogous group-theoretical hypotheses, one obtains only the desarguesian planes and the Hall planes, see H. Lüneburg [80] Theorem 49.6, p. 254. Our planes here are not Hall planes; in fact, the Hall construction (replacing a regulus by the opposite regulus) cannot produce a topological plane, because the complement of the regulus is dense, and thus entirely determines the compact spread; see also (44.9).

73.16. Next, we consider the case that the Levi subgroup $\Phi \cong \mathrm{SL}_2\mathbb{R}$ of Δ acts irreducibly on $\mathsf{T} \cong \mathbb{R}^4$. According to (95.10), this representation is uniquely determined. We shall first describe it explicitly in terms of matrices (73.17). (As usual, matrices act from the left on column vectors, but the same vectors are also written as rows whenever this is more convenient.) Then (73.18) we derive some properties of the action of Φ on the sets of 1- or 2-dimensional subspaces of T; as a by-product, (73.18) will imply that the representation written down in (73.17) is indeed irreducible.

Of course, this irreducible representation is well known, see e.g. Bourbaki [75] Chap. VIII, §1. A deeper understanding may be gained by looking at the associated sharply transitive action of $\mathrm{SL}_2\mathbb{R}$ on the twisted cubic

$$\{ (a^3, 3a^2b, 3ab^2, b^3) \mid a, b \in \mathbb{R} \} \subset \mathbb{R}^4,$$

see H. Lüneburg [80] Section 43.

73.17 Lemma. *Let ϱ be the unique irreducible representation $\mathrm{SL}_2\mathbb{R} \to \mathrm{GL}_4\mathbb{R}$, see (95.10). This representation is faithful, and it may be described as follows.*

$$\begin{pmatrix} a & b \\ c & d \end{pmatrix}^\varrho = \begin{pmatrix} a^3 & a^2b & ab^2 & b^3 \\ 3a^2c & 2abc + a^2d & 2abd + b^2c & 3b^2d \\ 3ac^2 & 2acd + bc^2 & 2bcd + ad^2 & 3bd^2 \\ c^3 & c^2d & cd^2 & d^3 \end{pmatrix}.$$

Proof. Consider the space $\mathscr{F}$ of polynomial functions $f : \mathbb{R}^2 \to \mathbb{R}$ that is generated by the functions x^3, x^2y, xy^2 and y^3. The usual action of $\mathrm{SL}_2\mathbb{R}$ on $\mathbb{R}^2$ induces an action on $\mathscr{F}$ by $(Af)(v) = f(A^t v)$ for $A \in \mathrm{SL}_2\mathbb{R}$ and $v \in \mathbb{R}^2$; here, A^t denotes the transpose of A. With respect to the given basis, this representation is described by the matrices written down in the lemma, hence these matrices do define a representation ϱ of $\mathrm{SL}_2\mathbb{R}$. A glance at the first and last rows of the matrices shows that this representation is faithful. It remains to be shown that ϱ is irreducible, and this will follow from (73.18a) below, since a reducible 4-dimensional representation of the reductive group $\mathrm{SL}_2\mathbb{R}$ has an invariant subspace of dimension 1 or 2, see (95.2). □

Via the representation ϱ, we let $\Phi = \mathrm{SL}_2\mathbb{R}$ act on the Grassmann manifolds $G_{4,i}$, $i \in \{1, 2\}$, of all i-dimensional subspaces of $\mathbb{R}^4$, see (64.3). We consider the special elements $X, Y \in G_{4,2}$ given respectively by the equations $x_3 = x_4 = 0$ and $x_1 = x_2 = 0$. Elements of $G_{4,2}$ skew to Y will be given by their Grassmann coordinates (64.3c) with respect to the decomposition $\mathbb{R}^4 = X \oplus Y$. In other words, a (2×2)-matrix L represents the subspace $\{ (v, Lv) \mid v \in \mathbb{R}^2 \}$ of $\mathbb{R}^4$. Moreover, we consider the special element $U \in G_{4,1}$ generated by the vector $(0, 0, 0, 1)$.

73.18 Lemma. *Let $\varrho : \Phi = \mathrm{SL}_2\mathbb{R} \to \mathrm{GL}_4\mathbb{R}$ be the representation introduced in (73.17). With the above notation, we have:*

(a) *The stabilizers Φ_U and Φ_Y are both equal to the lower triangular subgroup*

$$\Lambda = \left\{ \begin{pmatrix} a & 0 \\ c & a^{-1} \end{pmatrix} \;\middle|\; a, c \in \mathbb{R}, \; a \neq 0 \right\}$$

of Φ, and Λ fixes no other 1- or 2-dimensional subspaces.
(b) *The actions of Φ on the orbits U^Φ and Y^Φ are both equivalent to the standard action of Φ on the circle $P_1\mathbb{R}$.*
(c) *The orbit Y^Φ contains X. Moreover, any two elements of Y^Φ are complementary.*
(d) *The subgroup $\Psi = SO_2\mathbb{R} \le \Phi$ fixes precisely two elements $L_+, L_- \in G_{4,2}$. In Grassmann coordinates, these subspaces are given by*

$$L_+ = \begin{pmatrix} 1 & 0 \\ 0 & 1 \end{pmatrix} \quad \text{and} \quad L_- = \begin{pmatrix} -3 & 0 \\ 0 & -\frac{1}{3} \end{pmatrix}.$$

(e) Ψ *acts effectively on L_+ but has a kernel K of order 3 on L_-.*

Proof. 1) It is evident that Λ is the stabilizer of U and Y. Conversely, consider the following matrix B, which is obtained as the derivative (at $c = 0$) of the one-parameter subgroup P of Λ^ϱ defined by $a = 1$, and which therefore generates the Lie algebra $\mathfrak{l}P$:

$$B = \begin{pmatrix} 0 & & & \\ 3 & 0 & & \\ & 2 & 0 & \\ & & 1 & 0 \end{pmatrix}.$$

It is easily seen that no 1-dimensional subspace other than U and no 2-dimensional subspace other than Y is mapped into itself by this endomorphism, and this proves (a). Assertion (b) now follows from (96.9a) because Λ is a stabilizer for the action on $P_1\mathbb{R}$.

2) The transformation

$$\begin{pmatrix} & & & -1 \\ & & 1 & \\ & 1 & & \\ 1 & & & \end{pmatrix}^\varrho = \begin{pmatrix} & & & -1 \\ & & 1 & \\ & -1 & & \\ 1 & & & \end{pmatrix}$$

clearly interchanges X with Y. This gives the first half of (c). For the other half, note that Φ is doubly transitive on Y^Φ by (b), and that the elements X and Y of this orbit are complementary.

3) We verify first that L_+ and L_- are indeed invariant under Ψ or, equivalently, under the following matrix Q, which generates the Lie algebra of Ψ^ϱ:

$$Q = \begin{pmatrix} 0 & 1 & & \\ -3 & 0 & 2 & \\ & -2 & 0 & 3 \\ & & -1 & 0 \end{pmatrix}.$$

More precisely, we find that Q induces endomorphisms of L_+ and L_- that are given respectively by the matrices

$$\begin{pmatrix} 0 & 1 \\ -1 & 0 \end{pmatrix} \quad \text{and} \quad \begin{pmatrix} 0 & 3 \\ -3 & 0 \end{pmatrix},$$

with respect to the bases $\{(1, 0, 1, 0), (0, 1, 0, 1)\}$ and $\{(\frac{1}{3}, 0, -1, 0), (0, 1, 0, -\frac{1}{3})\}$ of L_+ and L_-. Applying the exponential function to these matrices, we see that Ψ rotates L_+ with normal speed and L_- with triple speed, which implies assertion (e).

Now let $V \neq L_-$ be any 2-dimensional subspace of $\mathbb{R}^4$ that is invariant under Q. We claim that $V = L_+$. Indeed, there are vectors $w_\pm \in L_\pm$ such that $w_+ \neq 0$ and $v = w_+ + w_- \in V$. Then $0 \neq v - v^\kappa \in V \cap L_+$ whenever $\kappa \neq \mathbb{1}$ belongs to the kernel K of Ψ acting on L_-. As Ψ acts irreducibly on L_+, our claim is verified. □

73.19 Theorem: Irreducible $\mathrm{SL}_2\mathbb{R}$-action. *Let $\mathcal{P} = (P, \mathcal{L})$ be a 4-dimensional translation plane such that the automorphism group Σ contains a subgroup Φ^ϱ, where $\Phi = \mathrm{SL}_2\mathbb{R}$ acts irreducibly via ϱ on the translation group $\mathsf{T} \cong \mathbb{R}^4$. Then $\mathcal{P}$ is isomorphic to the plane constructed as follows.*

(a) *The generating spread $\mathcal{S}$ is given by*

$$\mathcal{S} = Y^\Phi \cup L_+{}^\Phi \cup L_-{}^\Phi = \mathcal{S}_0 \cup \mathcal{S}_+ \cup \mathcal{S}_-;$$

here, $Y \leq \mathbb{R}^4$ is defined by $x_1 = x_2 = 0$, and the Grassmann coordinates of L_+, L_- are

$$L_+ = \begin{pmatrix} 1 & 0 \\ 0 & 1 \end{pmatrix}, \quad L_- = \begin{pmatrix} -3 & 0 \\ 0 & -\frac{1}{3} \end{pmatrix};$$

the action of Φ on $\mathsf{T} = \mathbb{R}^4$ is given by the representation ϱ of (73.17).

(b) *The full automorphism group of this plane is $\Sigma = \mathsf{T} \rtimes \mathrm{GL}_2\mathbb{R}$, where the action of $\mathrm{GL}_2\mathbb{R}$ is given by the formula in (73.17) defining the representation of Φ. We have $\dim \Sigma = 8$, and the stabilizer $\Sigma_o = \mathrm{GL}_2\mathbb{R}$ preserves the three Φ-orbits constituting $\mathcal{S}$. The action of Φ on $\mathcal{S}_0$ is equivalent to the action on the real projective line. On $\mathcal{S}_\pm$ an action equivalent to the direct real hyperbolic motion group (acting on interior points) is induced.*

73.20 Remark. Explicitly, the elements of the spread $\mathcal{S}$ are as follows:

$$\mathcal{S} = \{Y\} \cup \left\{ \begin{pmatrix} -3s^2 + t & 2s \\ -2s^3 - 2st & s^2 + t \end{pmatrix} \,\middle|\, s, t \in \mathbb{R},\ t \geq 0 \right\}$$
$$\cup \left\{ \begin{pmatrix} -3s^2 + 3t & 2s \\ -2s^3 + 2st & s^2 + \frac{t}{3} \end{pmatrix} \,\middle|\, s, t \in \mathbb{R},\ t \leq 0 \right\}.$$

The orbits $\mathcal{S}_0$, $\mathcal{S}_+$, and $\mathcal{S}_-$ are given respectively by $t = 0$, $t > 0$, and $t < 0$.

73 Four-dimensional translation planes

Proof of (73.19). 1) By (95.10), we may assume that the representation ϱ of Φ occuring in the given plane $\mathcal{P}$ coincides with the one described in (73.17). According to (73.18a), the 2-dimensional subgroup $\Lambda \leq \Phi$ fixes the 1-dimensional subspace $U \leq \mathbb{R}^4$. Hence Λ fixes the line $L \in \mathcal{S}$ determined by U, as well. Now (73.18a) implies that $L = Y$, whence the orbit $\mathcal{S}_0 = Y^\Phi \approx \mathbb{S}_1$ is contained in $\mathcal{S}$. From Mostert [57] Theorems 4 and 5, it follows that the subgroup $\Psi = \mathrm{SO}_2\mathbb{R}$ of Φ acts in the ordinary way on the 2-sphere $\mathcal{L}_o$. Hence, Ψ fixes a line in each of the complementary open disks of the orbit $\mathcal{S}_0$, and (73.18d) shows that these lines must be L_+ and L_-. We infer that $\mathcal{S}$ is contained in the line pencil $\mathcal{L}_o$. Since we shall show in step 2) below that $\mathcal{S}$ is a spread, it follows that in fact $\mathcal{S} = \mathcal{L}_o$.

Before going on, we check that $\mathcal{S}$ is described correctly in (73.20). Applying the Frattini argument (91.2a) to the action on $\mathcal{S}_0$, which is equivalent to the action on $\mathrm{P}_1\mathbb{R}$ by (73.18b), we see that $\Phi = \Psi\Lambda$. Thus, we obtain two of the orbits as $\mathcal{S}_\pm = L_\pm{}^\Lambda$. Moreover, Φ is doubly transitive on $\mathcal{S}_0$, whence $\mathcal{S}_0 \setminus \{Y\} = X^\Lambda$; note that X belongs to Y^Φ by (73.18c). Now the elements of Λ^ϱ are block matrices

$$\begin{pmatrix} A & 0 \\ B & C \end{pmatrix},$$

and a block matrix like this sends the subspace given by the matrix L to that given by $(B+CL)A^{-1}$. These remarks reduce the verification of (73.20) to simple matrix computation.

2) In order to show that the description given in the theorem actually defines a compact translation plane, we shall verify the following two conditions: (i) $\mathcal{S}$ is a partial spread, that is, any two elements of $\mathcal{S}$ are complementary, and (ii) $\mathcal{S} \approx \mathbb{S}_2$. These conditions together imply that $\mathcal{S}$ is a compact spread and defines a topological translation plane, see (64.9). See (73.21c) for a sketch of a completely different proof.

Clearly, the map sending $(s, t) \in \mathbb{R}^2$ to the line $L_{s,t}$ defined by these parameters as in (73.20) is a homeomorphism of $\mathbb{R}^2$ onto $\mathcal{S}\setminus\{Y\}$. As $\mathcal{S} \subseteq G_{4,2}$ is a Hausdorff space and Y is not invariant under the action of Φ, it follows that $\mathcal{S}$ is a connected 2-manifold which can be obtained from $\mathbb{R}^2$ by adding a point ∞. Such a manifold is compact, being the union of a disk neighbourhood D of ∞ and the bounded domain of $\mathbb{R}^2$ defined by the Jordan curve ∂D. It follows that $\mathcal{S}$ is homeomorphic to the one-point compactification of $\mathbb{R}^2$, or in other words, that $\mathcal{S} \approx \mathbb{S}_2$.

Having thus verified condition (ii), we turn to (i). First, Y^Φ is a partial spread (73.18c). Clearly, $L_\pm$ is complementary to Y, and hence to every element of $\mathcal{S}_0 = Y^\Psi$. The action of Φ shows that $L_\pm$ may be replaced by any member of $\mathcal{S}_\pm$ in this statement. Again in view of the action of Φ, it suffices to prove the additional conditions that L_+ is complementary to every other line in $\mathcal{S}_+ \cup \mathcal{S}_-$ and that L_- is complementary to every other line in $\mathcal{S}_-$. To prove that two lines are complementary, one checks that the determinant of the difference of the defining matrices is non-zero; this amounts to some computations involving nothing more than (bi-)quadratic equations.

3) The plane $\mathcal{P}$ is not isomorphic to the complex plane because the representation ϱ is not equivalent to a complex linear one, see (95.10) and (12.10). It remains to determine the full automorphism group Σ. Since $\mathsf{E}^\varrho = (\mathrm{GL}_2\mathbb{R})^\varrho$ normalizes the group Φ^ϱ, which entirely determines the spread $\mathcal{S}$ according to step 1), we see that E^ϱ is contained in Σ_o. As an alternative, one may use the fact that E^ϱ is generated by Φ^ϱ together with the real homotheties and the map $\varepsilon = \mathrm{diag}(1, -1, 1, -1) = \mathrm{diag}(1, -1)^\varrho$, and that these generators are collineations. The unique 1-dimensional orbit $\mathcal{S}_0$ is preserved by E^ϱ, and the action on it is equivalent to the action on the conjugacy class of $\Lambda = \Phi_Y$, that is, to the triply transitive standard action of $\mathrm{PGL}_2\mathbb{R}$ on $\mathrm{P}_1\mathbb{R}$.

Both E^ϱ and Σ_o are 4-dimensional groups (72.11g); hence, their connected components coincide (94.3f), and Σ_o normalizes the almost simple factor Φ^ϱ. Now the centralizer of Φ^ϱ in $\mathrm{GL}_4\mathbb{R}$ consists of the real homotheties (95.10), and all automorphisms of Φ are induced by (inner automorphisms of) $\mathrm{GL}_2\mathbb{R}$, see Dieudonné [71] Chap. IV §2. We conclude from these facts that E^ϱ is the normalizer of Φ^ϱ, whence $\Sigma_o = \mathsf{E}^\varrho$ and $\Sigma = \mathsf{T} \rtimes \mathsf{E}^\varrho$.

Finally we remark that E does not interchange the orbits $\mathcal{S}_+$ and $\mathcal{S}_-$. This may be seen by computing the action of the element $\varepsilon = \mathrm{diag}(1, -1, 1, -1) \in (\mathsf{E}\setminus \mathsf{E}^1)^\varrho$. As an alternative, one may use the fact that Ψ acts effectively on L_+ but not on L_-, see (73.18e). □

73.21 Notes. (a) The plane constructed in the last theorem has analogues over all finite fields, discovered by Hering [70], Ott [75] and Schaeffer, see H. Lüneburg [80] Section 45.

(b) Seidel [91] discovered a 3-symmetric plane contained in the dual of the translation plane considered in (73.19). Essentially, this means that each line in the orbit $\mathcal{S}_-$ of L_- is the axis of a homology σ of order 3 (this follows from 73.18e), and that σ acts as the point symmetry of a generalized symmetric space with point set $\mathcal{S}_-$. This example is of special interest for at least two reasons. It is the only generalized symmetric plane of dimension ≤ 4 admitting no 2-symmetric structure, and the only one which is neither an affine translation plane nor an open (stable) subplane of a classical plane. Moreover, no global geometric description of the underlying 3-symmetric space had been found previously.

(c) We sketch a different proof of the fact that $\mathcal{S}$ is a topological spread, compare step 2) of the last proof. The plan is to verify the spread condition directly and then to apply (43.8). Using the group action, the spread property is reduced to the condition that the intersection map sends the set $\mathcal{S} \setminus \{Y\}$ bijectively onto each of the lines $L_1 = \{(0, 1)\} \times \mathbb{C}$ and $L_2 = \{(1, 0)\} \times \mathbb{C}$. Bijectivity is easily verified in the case of L_1. For L_2, what one has to show is that the following self-map α of $\mathbb{R}^2$ is bijective:

$$\alpha(s, t) = \begin{cases} (-3s^2 + t, -2s^3 - 2st) & \text{if } t \geq 0 \\ (-3s^2 + 3t, -2s^3 + 2st) & \text{if } t \leq 0. \end{cases}$$

The map appearing here for $t \geq 0$ is known in the theory of singularities as a Whitney pleat, and the other half of α appears in algebraic geometry in connection with the operation of blowing up a point of $\mathbb{R}^2$. The verification of bijectivity involves cubic equations.

Outline of the classification

Combining theorems (73.11, 12, 13 and 19), we obtain a complete description of all 4-dimensional translation planes with an automorphism group of dimension $d = 8$. According to (73.1), these planes and their duals are in fact the only 4-dimensional, compact, projective planes for which d is equal to the critical dimension 8. We summarize these results as follows.

73.22 Theorem. *Let $\mathcal{P}$ be a 4-dimensional, compact, projective plane and $d = \dim \operatorname{Aut} \mathcal{P}$. If $d > 8$, then $\mathcal{P}$ is isomorphic to the complex plane $\mathcal{P}_2\mathbb{C}$, and in fact $d = 16$. If $d = 8$, then $\mathcal{P}$ is isomorphic or dual to one of the following translation planes:*

(i) *A unique plane with a group $\operatorname{SL}_2\mathbb{R} \leq \operatorname{Aut} \mathcal{P}$ acting irreducibly on the translation group, see (73.19).*
(ii) *A one-parameter family of planes with a group $\operatorname{SL}_2\mathbb{R} \leq \operatorname{Aut} \mathcal{P}$ acting reducibly on the translation group, see (73.13).*
(iii) *A unique plane with a solvable automorphism group, see (73.11).*   □

73.23 Flexible translation planes. The above planes with $d = 8$ were all discovered and classified by Betten, see Betten [73b, 72b, 73a] for the cases (i), (ii) and (iii), respectively. He also classified the compact, 4-dimensional translation planes with $d = 7$, see Betten [73a, 75, 76, 77], so that in fact all flexible, 4-dimensional, compact translation planes are known, compare (73.3). We shall roughly describe the results. The automorphism group Σ of such a plane (with $d = 7$) is solvable, because the reduced stabilizer Θ is 2-dimensional and hence solvable. By (73.4), the connected component $\Delta = \Sigma^1$ fixes some point $\infty \in W$.

It is a general experience that classification problems in topological geometry like this one lead to an enumeration of several families of planes with distinct geometric features, and that the elements of each individual family may be represented in a uniform way using some free parameters; these parameters may be integers or real numbers or functions, compare (82.21), and each choice of parameters defines an isomorphism type of planes. Sometimes, members of other families are obtained if the parameters are chosen in the boundary of the admissible domain. For an example, take the value $w = 1$ in (73.13).

In order to distinguish the families of planes in our situation, we introduce a set of invariants. We let ε denote the dimension of the group of elations having center ∞,

$$\varepsilon = \dim \Delta_{[\infty,\infty]}.$$

Next, we look at the orbits of Δ on $W \setminus \{\infty\}$, and we let μ denote the minimal dimension of such an orbit. Thus, $\mu = 2$ means that Δ acts transitively on $W \setminus \{\infty\}$, compare (53.1d). If Δ has a 1-dimensional orbit but no fixed point on $W \setminus \{\infty\}$, then we have $\mu = 1$. Existence of a second fixed point on W is expressed by $\mu = 0$. Two more invariants are provided by the number ν of 1-dimensional normal subgroups of Δ and by $d = \dim \Sigma$.

Now Betten's classification of 4-dimensional, compact translation planes with a solvable group $\mathrm{Aut}\,\mathcal{P}$ of dimension $d \geq 7$ is summarized by the following table; for comparison the plane with an 8-dimensional solvable group is included here once more. The entry 'number of planes' is meant to indicate the number of parameters determining the isomorphism types of planes; the expression $\omega_0 \omega^2$ indicates, for example, that the types depend on one integer and 2 real numbers as parameters. These numbers have been corrected in our table according to the results of Knarr [95]. The entry 'article, theorem' refers to the original papers by Betten.

$d\ \varepsilon\ \mu\ \nu$	number of planes	article, theorem	this section
8 3 2 1	1	[73a], 3	(73.11)
7 3 2 1	1	[73a], 2a	(73.10b)
7 3 2 0	ω	[77]	(73.24)
7 3 1 2	ω^5	[75], 1	
7 3 1 1	ω^3	[75], 2	
7 3 1 0	ω^3	[75], 3	(73.24)
7 2 1 1	ω^4	[75], 5	
7 2 0 2	ω^2	[76], 1	
7 2 0 1	ω	[76], 2	
7 2 0 0	$\omega_0 \omega^2$	[76], 3 and 4	(73.24)

4-dimensional translation planes with a solvable group

73.24 Contractions. For some of these planes, Knarr [95] gives particularly simple descriptions in terms of contractions $\varphi : \mathbb{C} \to \mathbb{C}$, compare (64.11). For example, the planes of Betten [77] are obtained using $\varphi(a+ib) = ce^{ia}$, where $c \in \mathbb{C}$ and $0 < |c| \leq 1$.

The planes of Betten [75] Satz 3 are obtained as follows. Consider a pair $(u, v) \in \mathbb{C}^2 \setminus \{(0,0)\}$ such that $5|u|^2 \leq 1$ and $5|v|^2 \leq 1$, and define φ by

$$\varphi(a+ib) = \begin{cases} ube^{2i \log |b|} & \text{if } b < 0 \\ 0 & \text{if } b = 0 \\ vbe^{2i \log |b|} & \text{if } b > 0. \end{cases}$$

The planes defined by (u, v) and (cu, cv) or (cv, cu) are isomorphic for $c \in \mathbb{C}$, $|c| = 1$.

Finally, the planes of Betten [76] Theorems 3 and 4 arise from the following contractions. The parameters are $k \in \mathbb{N} \cup \{0\}$ and $u, v \in \mathbb{R}$, where $0 < u < 1$ and $0 \leq v$, and $u^2 v^2 \leq (1-u^2)(1-k^2 u^2)$. The map φ is defined by

$$\varphi(z) = u \frac{z^k}{|z|^{k-1}} e^{iv \log |z|}.$$

73.25 Problem: The 'space' of isomorphism types. Our description of the classification of flexible translation planes, see (73.23), suggests that there is such a thing as the *space of all isomorphism types* of compact, connected translation planes or even of compact, connected projective planes in general. This idea has not been made precise. Several approaches suggest themselves in the case of translation planes. It seems necessary to start by considering a space of planes and then pass to the quotient space consisting of the isomorphism types. The space of translation planes with point set $\mathbb{R}^{2l}$ could be endowed with a topology by embedding it in the space of all compact l-dimensional submanifolds of the Grassmannian $G_{2l,l}$, using the description of a plane by its generating spread (64.4). A topology might also be obtained from the description of planes by means of contractions, see (64.11 and 12); this has been carried out by Gluck–Warner [83]. Another approach would be to use quasifields in order to introduce a topology. In the case of division algebras, where the multiplication is bilinear, a finite number of structure constants suffices to define a plane. This idea has been pursued by Kuzmin [66]. For projective planes in general, one might consider all planes having flag manifolds homeomorphic to a given one, and introduce a topology on the space of all pairs of projection maps $\mathbf{F} \to \mathcal{P}$ and $\mathbf{F} \to \mathcal{L}$.

One important part of the problem is to compare the topologies introduced in these ways. As soon as a reasonable topology is available, there arise numerous questions regarding contractibility properties of the space of isomorphism types or regarding the position of the families of relatively homogeneous planes within this space. For an interesting application of these ideas, see Eisele [93b], where passage to the 'limit' of a family of planes is used as a method of construction, resulting in a pathological example.

74 Four-dimensional shift planes

Shift planes generalize the *parabola model* of the classical affine plane $\mathcal{A}_2\mathbb{F}$ over $\mathbb{F} \in \{\mathbb{R}, \mathbb{C}\}$, which consists of all vertical lines of $\mathbb{F}^2$ together with the images of the parabola with equation $y = x^2$ under all *shifts* $u \mapsto u+v$ of $\mathbb{F}^2$. The generalization is obtained by replacing x^2 with some suitable function $f(x)$. We have already met 2-dimensional shift planes in (31.25c) and in Section 36. The dimension of a locally compact shift plane is at most 4, see (74.6), and 4-dimensional shift planes are the most interesting ones.

According to (71.11), there are exactly two possibilities for an effective action of $\mathbb{R}^4$ on a 4-dimensional compact plane: $\mathbb{R}^4$ can act as the translation group of a translation plane (this situation was studied in the previous section), and in one other fashion. It turns out (74.8) that the second type of action characterizes the shift groups $\Lambda \cong \mathbb{R}^4$ of shift planes obtained from $\mathbb{F} = \mathbb{C}$.

The relationship between translation planes and shift planes is in fact much deeper. It turns out that in all reasonably homogeneous affine shift planes the lines are differentiable submanifolds of $\mathbb{R}^{2l}$, and the collection of their affine tangent spaces is the line set of some translation plane. Reversing this operation of 'derivation', one may obtain examples of shift planes by 'integration' of known translation planes. In connection with this construction there are several intriguing questions that we cannot answer. However, in (74.19) we shall present a new sufficient condition for integrability of translation planes which makes it easy to handle most of the practical problems.

Like the translation group of a proper (that is, non-desarguesian) translation plane, the shift group of a proper shift plane is unique (72.12), hence the full automorphism group splits as a semi-direct product of the normal shift group and a point stabilizer. This accounts for many similarities in the treatment of translation planes and shift planes.

The only 2-dimensional shift planes with an at least 3-dimensional automorphism group are the skew parabola planes (including the desarguesian plane), see Section 36.

In the classification of 4-dimensional projective planes with an automorphism group of large dimension d, shift planes are placed third after the complex plane and after some translation planes. The translation planes with $d = 8$ are the most homogeneous non-desarguesian planes, see Section 73. The planes with $d = 7$ are precisely the translation planes described in (73.23) and a single shift plane found by Knarr [83]. This result of Betten [90] and Löwen [90] will be stated in (74.27), but will not be proved here. However, we shall construct Knarr's plane and show that it is the only *shift* plane with $d = 7$, see (74.26).

Furthermore, Betten and Knarr classified all flexible 4-dimensional shift planes, see Knarr [83, 86] and Betten–Knarr [87]. All planes appearing in the classification are obtained from translation planes by integration. Apart from the single plane with $d = 7$, they have groups of dimension $d = 6$. There are a single plane with point stabilizer $\Sigma_o^1 \cong \mathbb{R}^2$, a 2-parameter family with $\Sigma_o^1 \cong \mathbb{C}^\times$, and four more families with $\Sigma_o^1 \cong L_2$, see (74.23). We shall describe some of these examples, stressing the connection with translation planes via integration.

A striking feature of shift planes is that each of them admits a polarity whose absolute points form an oval (74.5). In particular, the dual of a shift plane is a shift plane and is isomorphic to the given plane. The existence of ovals is responsible for the dimensional restriction mentioned in the first paragraph; another reason is the existence of a ternary field with commutative multiplication (74.6). The ovals are used again in the proof of the stiffness theorem (74.13).

74.1 Shift planes. For $v \in \mathbb{R}^{2l}$, we consider the *shift* $\lambda_v : \mathbb{R}^{2l} \to \mathbb{R}^{2l}$ defined by $\lambda_v(u) = u + v$. We use it to define an incidence structure on the point set $\mathbb{R}^{2l}$ by applying the *shift group* $\Lambda = \{ \lambda_v \mid v \in \mathbb{R}^{2l} \}$ to the graph $\{ (x, f(x)) \mid x \in \mathbb{R}^l \}$ of some *generating function* $f : \mathbb{R}^l \to \mathbb{R}^l$. For $v = (a, b)$, the image of the graph is the line

$$L(a, b) = \{ (x, f(x - a) + b) \mid x \in \mathbb{R}^l \}.$$

We add the *verticals* $L(c) = \{c\} \times \mathbb{R}^l$ to our system of lines and thus obtain an incidence structure $\mathcal{E}_f$, which clearly admits the shift group Λ as a group of collineations.

We say that f is a *planar function* if f generates an affine plane $\mathcal{E}_f$. When this happens, we call $\mathcal{E}_f$ the *shift plane generated by* f and denote its projective completion by $\mathcal{P}_f$. A necessary and sufficient condition for planarity is given in (74.3) below.

74.2 First examples. (a) The function $f(x) = x^2$ on $\mathbb{F} \in \{\mathbb{R}, \mathbb{C}\}$ defines an affine shift plane, which is isomorphic to $\mathcal{A}_2\mathbb{F}$; this is the so-called *parabola model* of the plane $\mathcal{A}_2\mathbb{F}$. Indeed, the homeomorphism $(x, y) \mapsto (x, y - x^2)$ of $\mathbb{F}^2$ sends the line $L(a, b) = \{ (x, x^2 - 2ax + a^2 + b) \mid x \in \mathbb{F} \}$ onto the line $[-2a, a^2 + b]$ of $\mathcal{A}_2\mathbb{F}$. The resulting isomorphism of planes transforms the shift group into the group of all collineations $(x, y) \mapsto (x + a, y - 2ax + c)$ of $\mathcal{A}_2\mathbb{F}$. This comparison may help to clarify the distinction between shifts and translations; compare also (74.4).

(b) Let $w > 1$ and $\mathbb{F} = \mathbb{R}$. If we modify example (a) by setting $f(x) = wx^2$ whenever $x > 0$, then we obtain the *real skew parabola planes* found by Salzmann [65], see Section 36. Betten [79a] discovered complex analogues of these planes, see (74.4b); these were the first examples of shift planes with $l = 2$.

(c) We have encountered a different, but very similar, situation in (34.1). There, we have two distinct abelian groups, $SO_2\mathbb{R}$ and the multiplicative group $\mathbb{R}_{\text{pos}}$ of positive real numbers, and a generating function f from a subset of $SO_2\mathbb{R}$ to $\mathbb{R}_{\text{pos}}$. The shift condition of (74.3) below is satisfied, but the domain of f_d is smaller than that of f. The resulting plane is a punctured affine plane.

74.3 Proposition: Planarity.

(a) *A function $f : \mathbb{R}^l \to \mathbb{R}^l$ is planar if, and only if, f satisfies the following 'shift condition': For each $d \in \mathbb{R}^l \setminus \{0\}$, the map $f_d : \mathbb{R}^l \to \mathbb{R}^l$ defined by*

$$f_d(x) = f(x+d) - f(x)$$

is a bijection.

(b) *The affine shift plane $\mathcal{E}_f$ defined by a planar function f is a topological affine plane (with respect to the usual topology on the point set $\mathbb{R}^{2l}$) if, and only if, f is continuous. In this case, $\mathcal{P}_f$ is a compact, projective plane.*

Proof. Suppose that f satisfies the shift condition. Two points (x_1, y_1) and (x_2, y_2) with $x_1 = x_2$ are uniquely joined by the vertical $L(x_1)$. If $x_1 \neq x_2$, then the shift condition provides a unique line joining those two points, namely $L(x_1 - x_0, y_1 - f(x_0))$, where x_0 is defined by $f_{x_2-x_1}(x_0) = y_2 - y_1$.

The parallel axiom is clearly satisfied for the verticals. The lines $L(a, b)$ for a fixed a are also mutually disjoint and cover the whole point set. Hence, it suffices to show that $L(a_1, b_1)$ intersects $L(a_2, b_2)$ whenever $a_1 \neq a_2$. This follows again from the shift condition; the intersection point is $(x_0, f(x_0 - a_1) + b_1)$, where x_0 is defined by $f_{a_2-a_1}(x_0 - a_2) = b_2 - b_1$.

Conversely, assume that every point (x, y) of $\mathcal{E}_f$ with $x \neq 0$ is joined to $(0, 0)$ by a unique line $L(a, b)$. Then a and b are determined by the equations $y = f(x - a) + b$ and $b = -f(0 - a)$, hence the equation $f_x(z) = y$ has the unique solution $z = -a$.

Suppose that $\mathcal{E}_f$ is a topological plane with respect to the usual topology of $\mathbb{R}^2$. We use the point ∞ at infinity defined by the verticals and the continuous maps $i : x \mapsto (x, 0)$ and $p : (x, y) \mapsto y$ to write $f(x)$ as $p(L(0, 0) \wedge (i(x) \vee \infty))$. This shows that f is continuous. Conversely, if f is continuous, then our description of $\mathcal{E}_f$ satisfies the conditions of (43.8), and hence we have a topological plane. □

Notation. The line at infinity of the projective shift plane $\mathcal{P}_f$ will be denoted by W. The point at infinity incident with all vertical lines will be written (∞) or simply ∞, and the point at infinity corresponding to the parallel class $\{ L(a, b) \mid b \in \mathbb{R}^l \}$ will be denoted by (a).

74.4 More examples. (a) Continuous planar functions $f : \mathbb{R} \to \mathbb{R}$ are not at all rare. According to a result of Groh [76], see (31.25c), a function f is planar if, and only if, f is strictly convex (or concave) and for each $a \in \mathbb{R}$, the function $f(x) - ax$ converges to $+\infty$ (to $-\infty$ if f is concave) whenever $|x| \to \infty$. Note that f need not even be differentiable: for example, take $f(x) = x^2$ for $x \geq 0$ and $f(x) = (x-1)^2 - 1$ for $x < 0$. If the automorphism group of $\mathcal{E}_f$ is at least 3-dimensional, then f is a (differentiable) skew parabola, see Section 36.

(b) Given two planar functions f and g on $\mathbb{R}$ which are both convex, Polster [93] constructs a planar function $f * g$ on $\mathbb{R}^2$, called the *product* of f and g, as follows:

$$(f * g)(x_1, x_2) = (f(x_1) - g(x_2), x_1 x_2) \quad \text{for } (x_1, x_2) \in \mathbb{R}^2.$$

Consider the special case $f = g = q$, where $q(x) = x^2$. The resulting product $q * q$ clearly is equivalent to the complex squaring map, which generates the parabola model of $\mathcal{A}_2\mathbb{C}$. If we choose $f = q$ and let g be a real skew parabola as in (74.2b), then we obtain the complex skew parabolae of Betten [79a] as the products $f * g$. They form part of a larger family that will be characterized in (74.30). If the planar functions f and g are not differentiable, then $f * g$ is not differentiable either.

The plane $\mathcal{E}_{f*g}$ contains isomorphic copies of the given planes $\mathcal{E}_f$ and $\mathcal{E}_g$ as Baer subplanes; if $f = g$, then also $\mathcal{E}_q \cong \mathcal{A}_2\mathbb{R}$ occurs as a subplane. The point sets of those subplanes consist respectively of all points of the form $(x_1, 0, y_1, 0)$, $(0, x_2, y_1, 0)$ and $(x_1, x_1, 0, y_2)$.

We prove planarity of $f * g$: if $(r, s) \in \mathbb{R}^2 \setminus \{(0, 0)\}$, then

$$(f * g)_{(r,s)}(x_1, x_2) = (f_r(x_1) - g_s(x_2), x_1 s + x_2 r + rs).$$

Bijectivity of this function is trivial if $rs = 0$. For $rs \neq 0$, we have to show that the following system of two equations has a unique solution $(x_1, x_2) \in \mathbb{R}^2$:

$$f_r(x_1) - g_s(x_2) = c$$
$$x_1 s + x_2 r + rs = d$$

Eliminating x_2 we reduce this to the condition that for each $d \in \mathbb{R}$ the function $h_d(x_1) = f_r(x_1) - g_s(d/r - x_1 s/r - s) = f_r(x_1) + g_{-s}(d/r - x_1 s/r)$ is bijective. As f and g are convex, the functions f_t and g_t are increasing for $t > 0$ and decreasing for $t < 0$. It follows that the functions f_r and $x_1 \mapsto g_{-s}(d/r - x_1 s/r)$ are either both increasing or both decreasing, and then h_d shares this property and is bijective. □

74.5 Proposition: Ovals and Polarity.

(a) *The projective shift plane $\mathcal{P}_f$ admits a polarity π, given by*

$$(a, b) \leftrightarrow L(-a, -b), \quad \infty \leftrightarrow W \quad \text{and} \quad (c) \leftrightarrow L(-c).$$

In particular, the plane $\mathcal{P}_f$ is self-dual.

(b) *If $\mathcal{P}_f$ is a topological plane, then π is continuous, and the absolute points of π form a topological oval*

$$O_\pi = \left\{ \left(x, \tfrac{1}{2}f(2x)\right) \mid x \in \mathbb{R}^l \right\} \cup \{\infty\}.$$

(c) *There is another oval in $\mathcal{P}_f$, which again is a topological oval if the shift plane is a topological plane, namely:*

$$O = \left\{ (x, -f(-x)) \mid x \in \mathbb{R}^l \right\} \cup \{\infty\}.$$

Proof. 1) We begin by proving (c). This will suffice to derive the corollary (74.6), which in turn will be used in the proof of (b), see step 2). The affine part of O is homeomorphic to $\mathbb{R}^l$, hence we may apply (55.18) to obtain the desired topological properties once we have checked that O has the geometric properties of an oval.

We have to determine the number of intersection points of O with a line $L(a, b)$. These are the points (x, y) satisfying $(x, y) = (x, -f(-x)) = (x, f(x - a) + b)$. Suppose we have two solutions x_1, x_2 of this equation. Subtracting the equations, we obtain $f(x_1 - a) - f(-x_2) = f(x_2 - a) - f(-x_1)$. This is equivalent to $f_c(-x_2) = f_c(-x_1)$, where $c = x_1 + x_2 - a$. The shift condition says that this can happen only if $c = 0$, that is, if $x_2 = a - x_1$. Conversely, if x is a solution of $f(x - a) + b = -f(-x)$, then $a - x$ is another solution. Thus, there are at most two intersection points, and $L(a, b)$ is a tangent if, and only if, $x = \tfrac{a}{2}$ defines an intersection point. In this case, $b = -f(-x) - f(x - a) = -2f(-x)$, hence the tangent of O at the point $(x, -f(-x))$ is the line $L(2x, -2f(-x))$. Clearly, W is the unique tangent at the point ∞.

2) We turn now to the proof that π is a polarity. It follows from the definitions that $(x, y) \in L(a, b)$ if, and only if, $(-a, -b) \in L(-x, -y)$; this is the essential step. By (44.9), continuity of π only has to be checked on the affine part, where it is obvious. Therefore, the absolute points form a closed set O_π. The correctness of the given description of this set is easily verified. In view of (55.11), it only remains to be shown that O_π is an oval. Since this fact will not be needed in the sequel, we will be content to prove it essentially by a quotation; our claim follows from Bedürftig [74a] for $l = 1$ and from Buchanan–Hähl–Löwen [80] Satz 4.3 for the remaining case $l = 2$ (compare (74.6) below). In order to apply the latter result, we verify the condition that there is a non-absolute line containing exactly 2 absolute points. Indeed, the absolute points incident with a line $L(c)$ are precisely ∞ and $\left(c, \tfrac{1}{2}f(2c)\right)$. □

Many further examples of ovals and polarities of shift planes are presented by Polster [93].

74.6 Corollary: Dimension of shift planes. *A topological shift plane has dimension* $2l \in \{2, 4\}$.

Proof. This follows from (74.5c) together with (55.14). In his original proof, Weigand applied (52.10 and 11), see Weigand [87] 5.2 and 5.4. He had to show that a shift plane may be coordinatized by a ternary field with commutative multiplication, compare Dembowski–Ostrom [68] and (31.25c). □

The oval O will be used once more in the proof of the stiffness theorem (74.13).

Automorphisms

74.7 The shift group. The shift group Λ of a shift plane $\mathcal{E}_f$ is a direct product $\Lambda = \Xi \times \Upsilon$, where

$$\Upsilon = \{ \lambda_{(0,b)} \mid b \in \mathbb{R}^l \} = \Lambda_{[\infty, W]}$$

is a linearly transitive elation group, whereas

$$\Xi = \{ \lambda_{(a,0)} \mid a \in \mathbb{R}^l \}$$

does not contain any non-trivial elations. Indeed, a line $L(c, d)$ is parallel to its image $L(c + a, d + b)$ under the shift $\lambda_{(a,b)}$ if, and only if $a = 0$, compare the proof of (74.3).

As an immediate consequence of the definitions, the abelian group Λ is sharply transitive both on the affine point set $P \setminus W$ and on $\mathcal{L} \setminus \mathcal{L}_\infty$. In other words, Λ acts as described in (71.11c), a property which characterizes shift groups, as we shall show next (74.8). The subgroup Ξ is sharply transitive both on $W \setminus \{\infty\}$ and on the set $\mathcal{L}_\infty \setminus \{W\}$ of verticals.

The shift group Λ is closed in the automorphism group of $\mathcal{P}_f$, because the closure of Λ is abelian and hence is sharply transitive on $P \setminus W$, like Λ.

74.8 Proposition: Characterization of shift planes.

(a) *A compact, projective plane* $\mathcal{P} = (P, \mathcal{L})$ *of dimension* $2l$ *is a shift plane if, and only if, it admits an abelian group* Λ *of automorphisms which fixes some flag* (W, ∞) *and acts transitively both on* $P \setminus W$ *and on* $\mathcal{L} \setminus \mathcal{L}_\infty$.

(b) *A 4-dimensional, compact, projective plane is a shift plane if, and only if, it admits a group* $\Lambda \cong \mathbb{R}^4$ *of automorphisms which does not entirely consist of elations.*

(c) *A* $2l$-*dimensional topological shift plane* $\mathcal{P}$ *is non-desarguesian if, and only if, its shift group* Λ *is the only* $2l$-*dimensional, closed, connected, abelian group of automorphisms of* $\mathcal{P}$.

Proof. We have seen in (74.7) that a shift group has the properties stated in assertion (a). Conversely, assume that Λ has those properties.

Then the abelian group $\overline{\Lambda}$ is sharply transitive on $P \setminus W$, whence $\overline{\Lambda} = \Lambda$ is a Lie group (53.1b) homeomorphic to $P \setminus W \approx \mathbb{R}^{2l}$, see (96.9a). It follows that $\Lambda \cong \mathbb{R}^{2l}$, compare (94.38c). We claim that the axial group $\Upsilon = \Lambda_{[\infty,W]}$ is linearly transitive. Indeed, the restriction $\Lambda|_{\mathcal{L}_\infty}$ is sharply transitive on $\mathcal{L}_\infty \setminus \{W\}$, hence this group and the kernel $\Lambda_{[\infty]}$ are both l-dimensional (96.9a). Each element of the kernel has an axis, which is fixed by the abelian group Λ and hence coincides with W. By (61.4a), the group $\Lambda_{[\infty]} = \Upsilon$ is linearly transitive and isomorphic to $\mathbb{R}^l$.

We identify $P \setminus W$ with Λ via the action (96.9a), and we choose an arbitrary complement $\Xi \cong \mathbb{R}^l$ to obtain a decomposition $\Lambda = \Xi \times \Upsilon$. The fibers $\{\xi\} \times \Upsilon$ are orbits of the elation group Υ, hence they are affine lines. We infer that each line $L \in \mathcal{L} \setminus \mathcal{L}_\infty$ intersects each of these fibers exactly once. Thus, L is the graph of a function $f : \Xi \to \Upsilon$. It follows that $\mathcal{P}$ is a shift plane with shift group Λ and generating function f.

Suppose that $l = 2$. Then assertion (b) follows from (a) in view of (71.11), and (c) is a consequence of (72.12). If a non-desarguesian shift plane with $l = 1$ admits another 2-dimensional abelian group besides the shift group, then the automorphism group of that plane is at least 3-dimensional. Assertion (c) states that this does not occur, and we may confirm this claim by an an inspection of the classification (38.1). The structure of the hyperbolic motion group and the fixed configurations of the other automorphism groups show that the skew parabola planes are the only shift planes in this list, and they do not admit two distinct 2-dimensional abelian groups, as can be seen from (36.3). □

Note. Assertion (74.8b) has no analogue for $l = 1$. Even the real affine plane admits actions of $\mathbb{R}^2$ that are not transitive on the point set and, hence, induce neither a shift group nor a translation group. An example is given by the action $(x, y) \mapsto (e^a x, y + b)$. Compare also the note after (71.11), concerning abelian groups on 2-dimensional planes in general.

A non-desarguesian shift plane will briefly be called a *proper shift plane*.

74.9 Corollary: Lenz–Barlotti type. *Let $\mathcal{P}$ be a compact, proper shift plane with shift group Λ and automorphism group Σ. Then we have the following.*

(a) *$\mathcal{P}$ is of Lenz type II. Furthermore, $\Sigma_{[o,W]} = \mathbb{1} = \Sigma_{[\infty,L(a,b)]}$, so that $\mathcal{P}$ is of Lenz–Barlotti type II.1.*
(b) *The fixed flag (W, ∞) of Λ is fixed by all collineations.*
(c) *The shift group Λ is normal in $\Sigma = \operatorname{Aut} \mathcal{P}$.*
(d) *Each reflection of $\mathcal{P}$ has an axis $L(c) \in \mathcal{L}_\infty \setminus \{W\}$.*

Proof. Note first that (c) implies (b), and that (c) follows from (74.8c) directly. The Lenz type of $\mathcal{P}$ is at least II by (74.7). It cannot be III since there is no fixed antiflag, and type IV is excluded by (74.8c). In view of the hierarchy of Lenz types, see Section 24, this proves the first part of (a). If $\Sigma_{[W]}$ contains a non-trivial homology, then its conjugates generate a transitive elation group (61.20b), contrary to (74.8c). The dual statement about homologies with center ∞ follows by (74.5). This proves (a); finally, (a) and (b) together imply (d). □

74.10 Corollary: Automorphism group. *Let $\mathcal{P}$ be a $2l$-dimensional, proper shift plane with shift group $\Lambda = \Xi \times \Upsilon$, compare (74.7). Then the following assertions hold.*

(a) *The automorphism group $\Sigma = \operatorname{Aut}\mathcal{P}$ is a semi-direct product $\Sigma = \Lambda \rtimes \Sigma_o$ with closed factors, and the stabilizer Σ_o of the origin $o = (0,0)$ acts linearly on the affine point set $P \setminus W = \mathbb{R}^l \times \mathbb{R}^l$, leaving the second factor invariant. The dual statement is also true.*
(b) *The polarity π of (74.5) normalizes the shift group Λ and induces inversion on it.*
(c) *The action of π on the factor group Σ/Λ is trivial. In particular, every subgroup of Σ that contains Λ is invariant under π.*

Proof. 1) The shift group $\Lambda \le \Sigma$ is a sharply transitive, closed, normal subgroup (74.7 and 9c), hence the Frattini argument (91.2a) yields a semidirect decomposition as stated in the assertion. We identify $P \setminus W$ with Λ via $o^\lambda \mapsto \lambda$, compare (96.9a), and we observe that then Σ_o acts on $P \setminus W = \Lambda$ by conjugation. From (74.7), we have a decomposition $\Lambda = \Xi \times \Upsilon$, where both factors are isomorphic to $\mathbb{R}^l$, and (74.9a) implies that the right factor is invariant. This proves (a).

2) In order to prove (b), we compute the effect of π on the shift λ_v, where $v \in \mathbb{R}^{2l}$. We simplify notation by writing $L(a,b) = L(u)$ for $u = (a,b) \in \mathbb{R}^{2l}$. We have

$$u^{\lambda_v \pi} = (u+v)^\pi = L(-u-v) = L(-u)^{\lambda_{-v}} = u^{\pi \lambda_v^{-1}}.$$

3) As to assertion (c), we observe that Λ is equal to its own centralizer in Σ by (a). Hence, the two elements σ and σ^π define the same coset $\Lambda\sigma = \Lambda\sigma^\pi$ if, and only if, they induce the same automorphism of Λ by conjugation. Using (b), we obtain

$$\lambda^{(\sigma^\pi)} = \pi^{-1}\sigma^{-1}\pi\lambda\pi^{-1}\sigma\pi$$
$$= \pi^{-1}(\sigma^{-1}\lambda^{-1}\sigma)\pi$$
$$= \lambda^\sigma.$$
□

74.11 Corollary: Isomorphism. *Let $f, g : \mathbb{R}^l \to \mathbb{R}^l$ be planar functions. The shift planes $\mathcal{P}_f$ and $\mathcal{P}_g$ are isomorphic if, and only if, there are vectors $a, b \in \mathbb{R}^l$ and real $(l \times l)$-matrices A, C and B, where A and B are invertible, such that the* *following identity holds:*

(∗) $$g(x) = Bf\left(A^{-1}(x-a)\right) + C'x + b.$$

Note. Every isomorphism of proper shift planes is the projective extension of an isomorphism of the form $\varphi : (x, y) \mapsto (Ax, By + Cx)$, with matrices as above, followed by a shift $\lambda_{(u,v)}$. Such a product $\lambda_{(u,v)} \circ \varphi$ is an isomorphism of $\mathcal{E}_f$ onto $\mathcal{E}_g$ if, and only if, there are $a, b \in \mathbb{R}^l$ such that (∗) is satisfied with $C' = CA^{-1}$.

Proof. If the two isomorphic planes are desarguesian, then we may use their homogeneity to arrange that we have an isomorphism ψ of the *affine* shift planes and that the shift groups correspond under ψ; compare (73.9) for $l = 1$. If the planes are proper shift planes, then this is automatically so by (74.8c). By composition with a shift of $\mathcal{E}_g$, we obtain an isomorphism $\varphi = \lambda \circ \psi$ that sends the origin of $\mathcal{E}_f$ onto that of $\mathcal{E}_g$. Identifying the point sets with the shift groups as in the proof of (74.10a), we see that φ is an isomorphism of the shift groups sending the elation group of $\mathcal{E}_f$ onto that of $\mathcal{E}_g$. Thus, φ has the form described in the note above. Now φ sends some line $L(c, d)$ of $\mathcal{E}_f$ onto the graph of g. Using $a = Ac$, $b = Bd$ and $C' = CA^{-1}$, we obtain the identity (∗). Reading the same arguments backwards, we obtain the converse assertion. □

74.12 Proposition: Classical planes. *Let $\mathcal{E}_f$ be a $2l$-dimensional shift plane. The following conditions are necessary and sufficient for $\mathcal{E}_f$ to be isomorphic to the classical plane $\mathcal{A}_2\mathbb{F}$ over $\mathbb{F} \in \{\mathbb{R}, \mathbb{C}\}$. If $l = 1$, then f has to be a quadratic polynomial. If $l = 2$, then f has to be of the form $f = (f_1, f_2)$, where f_1 and f_2 are real quadratic polynomials in 2 variables.*

The proof will show that not all quadratic functions $f = (f_1, f_2)$ are planar.

Proof. We consider the case $l = 2$ only. If $\mathcal{E}_f$ is classical, then f may be obtained from the complex squaring map by a transformation (∗) as in (74.11), hence both components of f are quadratic polynomials. Conversely, if $f = (f_1, f_2)$ is a planar map composed of quadratic polynomials f_1 and f_2, then we may use a transformation (∗) to obtain that the components of f are homogeneous quadratic forms $f_\nu(x_1, x_2) = r_\nu x_1^2 + s_\nu x_1 x_2 + t_\nu x_2^2$, $\nu = 1, 2$. Replacing $f(x)$ by $f(A^{-1}x)$, we may transform f_1 to diagonal form, that is, we obtain $s_1 = 0$. Then $r_1 \neq 0$, or else the first component of $f_{(1,0)}$ is identically zero and f is not planar. Passing to an equivalent map Bf, we may now enforce that $r_2 = 0$, and then $s_2 \neq 0$ by the previous argument, so that f_2 is indefinite. Transformation by a suitable matrix A_1 now yields $f_2(x_1, x_2) = x_1 x_2$, and multiplication by some matrix B_1 restores the property $s_1 = 0$, which we may have lost on our way. Planarity implies that the determinant $q = 2r_1 d_1^2 - 2t_1 d_2^2$ of the linear part of f_d is non-zero for each $d \neq (0,0)$. Thus, q is a definite quadratic form in (d_1, d_2); consequently, $r_1 t_1 < 0$, and f is equivalent to the squaring map $(x_1^2 - x_2^2, 2x_1 x_2)$. □

74.13 Theorem: Stiffness. *Let $\mathcal{E}_f$ be a compact, connected proper shift plane, and let Δ be the identity component of $\operatorname{Aut} \mathcal{E}_f$. If K and L are two non-vertical lines, and if $K \wedge L \notin W$, then $\Delta_{K,L} = \mathbb{1}$.*

Proof. This very remarkable proof is due to Knarr [83]. It uses the oval O that was constructed in (74.5c) as the graph of the function g, defined by $g(x) = -f(-x)$. The function g satisfies the shift condition, like f. Moreover, the automorphisms of $\mathcal{E}_f$ and of $\mathcal{E}_g$ are the same. Indeed, consider the identity $(*)$ of (74.11); we have $f(x) = Bf(A^{-1}(x - a)) + C'x + b$ for all x if, and only if $g(x) = Bg(A^{-1}(x + a)) + C'x - b$ for all x.

By (74.5), we may replace the assertion by the dual statement, concerning the stabilizer of two points $p, q \notin W$ such that $\infty \notin p \vee q$. Now $\Delta_{p,q}$ fixes the line of $\mathcal{E}_g$ joining these points. This line arises from O by application of an automorphism of $\mathcal{E}_f$, hence it is an oval O' in $\mathcal{E}_f$. The group $\Delta_{p,q}$ fixes the tangents of O' at the points p and q, and we have seen in the proof of (74.5) that these tangents are not parallel. Thus we obtain another fixed point r for $\Delta_{p,q}$ which together with the given fixed points (including the point $\infty \in O$) forms a quadrangle. Now (32.10) and (55.21b and c) yield the assertion for $l = 1$ and $l = 2$, respectively. By (74.6), the proof is complete.

We add another, more pedestrian proof of the fact that $\Delta_{p,q} = \mathbb{1}$, concentrating on the case $l = 2$. According to (74.10), the group $\Delta_{p,q}$ acts on the point space $\mathbb{R}^4$ by affine maps, whence it fixes all points on the line of the affine space $\mathbb{R}^4$ that joins p and q. That line is a connected set which is not contained in any line of $\mathcal{E}_f$; this follows from the shift condition. Using (42.3) and (44.1), we infer that $\Delta_{p,q}$ acts trivially on some connected subplane of $\mathcal{E}_f$, and (55.21b and c) imply that $\Delta_{p,q} = \mathbb{1}$. □

Existence of Baer involutions is not prohibited by the above result, and in fact many examples are given by Polster [93]. Let us just remark that each of the three Baer subplanes $\mathcal{E}_f$, $\mathcal{E}_g$ and $\mathcal{E}_q$ of $\mathcal{E}_{f*g}$ mentioned in (74.4b) is the fixed point set of a Baer involution, sending (x_1, x_2, y_1, y_2) respectively to the point $(x_1, -x_2, y_1, -y_2)$, $(-x_1, x_2, y_1, -y_2)$ or $(-x_2, -x_1, -y_1, y_2)$, provided that the condition $g(x) = g(-x)$, $f(x) = f(-x)$ or $f = g$ is satisfied.

Differentiable planar functions and spreads

As the examples (74.4) show, planar maps need not be differentiable. Nevertheless, the classification shows that all reasonably homogeneous shift planes are generated by differentiable planar maps, and there is a very powerful construction method that makes use of differentiability. The starting point is the observation that the differentials $D_{(s,t)}q$ of the map $q(s, t) = (s^2 - t^2, 2st)$ generating the parabola model of $\mathcal{A}_2\mathbb{C}$ are given by the matrices

$$\begin{pmatrix} 2s & -2t \\ 2t & 2s \end{pmatrix},$$

which, via Grassmann coordinates (64.3b), describe the elements of the complex spread. This leads to the question which other spreads may be obtained in this way, and whether, conversely, planar maps may be constructed by 'integrating' spreads. We concentrate on the case $l = 2$, compare (74.6).

74.14 Definition. Let $f = (g, h) : \mathbb{R}^2 \to \mathbb{R}^2$ be a differentiable map (of class $\mathcal{C}^1$). Consider the differential

$$Df : (s, t) \mapsto D_{(s,t)}f = \begin{pmatrix} g_s & g_t \\ h_s & h_t \end{pmatrix},$$

where (s, t) denotes a point in $\mathbb{R}^2$. Let also a spread $\mathcal{S}$ in $\mathbb{R}^4$ be given which contains the vertical $Y = \{0\} \times \mathbb{R}^2$, and consider a homeomorphism w of $\mathbb{R}^2$ onto $\mathcal{S}' = \mathcal{S} \setminus \{Y\}$. (According to (64.8a), such a homeomorphism exists if, and only if, $\mathcal{S}$ defines a topological translation plane $\mathcal{A}_\mathcal{S}$. On the other hand, if we know that $\mathcal{S}$ is compact, then in view of (64.4d and b) and (51.19) it suffices to require that $w : \mathbb{R}^2 \to \mathcal{S}'$ is a continuous bijection.)

Now assume that $Df = w$. Then we say that f is an *integral function* of w or that $\mathcal{S}$ is obtained from f by derivation. If an integral function f of $\mathcal{S}$ is planar, then we say that the shift plane $\mathcal{E}_f$ is an *integral plane* of the translation plane $\mathcal{A}_\mathcal{S}$ or that $\mathcal{A}_\mathcal{S}$ is the *derivative* of $\mathcal{E}_f$. As we have seen, the derivative of the parabola model of the complex affine plane $\mathcal{A}_2\mathbb{C}$ is the standard model of $\mathcal{A}_2\mathbb{C}$.

74.15 Note. Sometimes we say, by abuse of language, that f is an integral function of $\mathcal{S}$ in the above situation. It should be kept in mind, however, that the choice of Y is essential and that the parametrization w of the remainder $\mathcal{S}'$ is the given object that is being integrated. This becomes obvious if we consider the analogous situation in one variable ($l = 1$). Via derivation, all differentiable, planar functions yield the unique spread in $\mathbb{R}^2$, but with different parametrizations.

Similarly, the classical complex spread has all quadratic, complex polynomials as integral functions; however, they all define the same plane $\mathcal{A}_2\mathbb{C}$, up to isomorphism. The quadratic polynomials are the only integral functions of the complex spread $\mathcal{S}_\mathbb{C}'$. Indeed, the condition $D_{(s,t)}f \in \mathcal{S}_\mathbb{C}'$ implies that the differential is a complex linear map, whence f is complex differentiable. Since we require that Df is a bijection of $\mathbb{C}$ onto $\mathcal{S}_\mathbb{C}'$, the complex derivative $f' : \mathbb{C} \to \mathbb{C}$ is bijective. Hence f' is an affine map of $\mathbb{C}$, and f is quadratic.

In general, the search for an integrable parametrization of a given spread is part of the problem. Again, the (isomorphism type of the) integral plane is not uniquely determined by the (isomorphism type of the) spread; see (74.31c) for examples of isomorphic spreads $\mathscr{S}_1$, $\mathscr{S}_2$ having non-isomorphic integral planes $\mathscr{E}_{f_1}$ and $\mathscr{E}_{f_2}$. This example may be interpreted as different parametrizations of the *same* spread leading to non-isomorphic shift planes via integration. Indeed, in this example the isomorphism sending $\mathscr{S}_1$ onto $\mathscr{S}_2$ is of the form $(x, y) \mapsto (Bx, By)$ for some $B \in \mathrm{GL}_2\mathbb{R}$, and the same transformation sends the integral plane $\mathscr{E}_{f_1}$ onto an isomorphic integral plane $\mathscr{E}_{Bf_1B^{-1}}$ of $\mathscr{S}_2$. Then this plane is not isomorphic to the given integral plane $\mathscr{E}_{f_2}$ of the same spread.

Recall that a map $u : \mathbb{R}^2 \to \mathbb{R}^2$ is called *proper* if $\lim_{x \to \infty} u(x) = \infty$, or in other words, if $u(\infty) = \infty$ defines a continuous extension of u to the 2-sphere.

74.16 Theorem: Planarity of integral functions. *An integral function f of a compact spread $\mathscr{S}$ in $\mathbb{R}^4$ is planar (and hence generates a topological, affine shift plane) if, and only if, it satisfies the following 'proper shift condition': for each $d \in \mathbb{R}^2 \setminus \{0\}$, the map $f_d : \mathbb{R}^2 \to \mathbb{R}^2$ defined by $f_d(x) = f(x + d) - f(x)$ is proper.*

Proof. We know that f is planar if, and only if, it satisfies the shift condition of (74.2), which requires that each map f_d is bijective. A continuous bijection of $\mathbb{R}^2$ is a homeomorphism (51.19) and hence is proper. Conversely, assume that f_d is a proper map. Since the differential $D_x f_d = D_{x+d} f - D_x f$ is the difference of two distinct spread matrices, it is invertible, and f_d is a local homeomorphism. Observe that we have used the injectivity of $Df = w : \mathbb{R}^2 \to \mathscr{S}$. Now a proper, local homeomorphism u of $\mathbb{R}^n$ is bijective and hence is a homeomorphism. Indeed, F.E. Browder [54] Theorem 4 shows that u is a covering map (note that a proper map is closed, as required by Browder), and this covering is one-sheeted because $\mathbb{R}^n$ is simply connected. Applying this to $u = f_d$, we obtain the theorem. □

74.17 Problems. (a) Is every integral function f of a compact spread $\mathscr{S}$ in $\mathbb{R}^4$ a planar map? (For instance, can one prove the proper shift condition for integral functions?)

(b) Conversely, does every differentiable planar map f on $\mathbb{R}^2$ define a spread by derivation?

(c) If f is a planar integral function of $\mathscr{S}$, is it true that the lines $L = L(a, b)$ of $\mathscr{E}_f$ define topological ovals $L \cup \{\infty\}$ in $\mathscr{A}_\mathscr{S}$?

(d) Conversely, let O be a topological oval in a projective translation plane, such that the translation axis is a tangent. Do the translates of the affine part of O form an affine plane? In other words, is the affine part of O the graph of a planar map?

(e) Does every 4-dimensional translation plane admit an integral plane?

Positive answers to these intriguing questions would greatly facilitate the investigation of shift planes. For example, (a) would help to construct planes, and (b) could be used to rule out undesired candidates, compare (74.29) and step 3) in the proof of (74.24).

For $l = 1$, the question analogous to (a) has a positive answer, see (31.25c). The proof does not carry over to $l = 2$ for lack of a sufficiently strong mean value theorem. The difficulty is that one has to derive global information from local assumptions. Almost trivially, questions (b) through (e) have positive answers for $l = 1$.

The answer to (d) is 'yes' in the complex plane, because then O is a conic, see (55.13), so that we obtain the parabola model of the complex plane (74.2a). For some other examples, (c) has been verified directly; see Polster [93] 3.5.2 and Betten–Polster [94]. In general, the difficulties are similar to those with problem (a). The answer to problem (e) might well depend on the choice of the vertical line. For further comments on this problem, see (74.25a).

At present, an answer to problem (a) is not available as a tool for the construction of shift planes. However, we shall prove a new planarity criterion (74.19) that is almost equally useful for this practical purpose. It is based on the following characterization of diffeomorphisms of $\mathbb{R}^n$.

74.18 Theorem of Hadamard. *Let $u : \mathbb{R}^n \to \mathbb{R}^n$ be a local diffeomorphism of class $\mathcal{C}^1$ and suppose that the matrix norm of the inverse differential $(D_x u)^{-1}$ is bounded above by some constant k. Then u is bijective and hence is a global diffeomorphism.*

Proof. The following argument is adapted from Ortega–Rheinboldt [70] 5.3.10.

1) We show that every linear map $a : \mathbb{R} \to \mathbb{R}^n$ lifts over u to a unique map $b : \mathbb{R} \to \mathbb{R}^n$ with any prescribed initial value $b(0) = b_0$, provided that $u(b_0) = a(0)$. This will imply surjectivity of u. Like in the theory of covering maps, b may be constructed on an open interval $]p, q[$ containing 0, and the difficulty is to define $b(p)$ and $b(q)$. Now let $q_n < q$ be a sequence converging to q. Since we have $b'(t) = (D_{b(t)}u)^{-1}a'(t)$, we see that $\|b(q_m) - b(q_n)\| \leq k\|a'\| \, |q_m - q_n|$. Thus $b(q_n)$ is a Cauchy sequence, and we may define $b(q) = \lim b(q_n)$.

2) The proof of injectivity again resembles standard arguments from the theory of coverings. If $u(x_1) = u(x_2) = y$, consider the path $r(s) = u(sx_1 + (1-s)x_2)$ and the homotopy $F(s, t) = ty + (1-t)r(s)$. This homotopy has a unique lift G satisfying $G(s, 0) = sx_1 + (1-s)x_2$, which is obtained by lifting each of the linear paths $t \mapsto F(s_0, t)$ with the given initial value $G(s_0, 0)$. Now F is constant on the

connected set $(\{0, 1\} \times [0, 1]) \cup ([0, 1] \times \{1\})$ and u is a local homeomorphism, hence G is also constant on that set and $x_1 = G(1, 0) = G(0, 0) = x_2$. □

74.19 Corollary: Hadamard criterion for planarity. *Let $f : \mathbb{R}^2 \to \mathbb{R}^2$ be of class $\mathcal{C}^1$. For each fixed $d \in \mathbb{R}^2 \setminus \{0\}$ we suppose that the differentials $D_x f_d$ are all invertible, and that the matrix norms of their inverses are bounded above. Then f satisfies the shift condition and hence defines a topological shift plane.* □

74.20 Notes. (a) We do not explicitly require that f is an integral function of some spread. However, invertibility of all differentials $D_x f_d$ is equivalent to the condition that Df defines at least a partial spread.

(b) It does not matter which matrix norm we use to make the criterion explicit. We shall write $f_d = (g_d, h_d)$ and $x = (s, t)$ and take the norm

$$\left\|(D_x f_d)^{-1}\right\|^2 = \frac{\left(\partial_s g_d(x)\right)^2 + \left(\partial_t g_d(x)\right)^2 + \left(\partial_s h_d(x)\right)^2 + \left(\partial_t h_d(x)\right)^2}{\left(\partial_s g_d(x)\partial_t h_d(x) - \partial_t g_d(x)\partial_s h_d(x)\right)^2}.$$

Flexible shift planes

74.21 Proposition: Flexible shift planes. *A $2l$-dimensional, compact, connected shift plane is flexible in the sense of (63.11) if, and only if, its automorphism group Σ is at least $3l$-dimensional.*

Proof. An orbit is open in the flag space $\mathbf{F}$ if, and only if, its dimension is equal to $\dim \mathbf{F} = 3l$, see (52.1) and (53.1a). Existence of such an orbit implies that $d = \dim \Sigma \geq 3l$ by the dimension formula (96.10). Conversely, assume that $d \geq 3l$ and that the plane is non-desarguesian (otherwise, our assertion is trivial). We want to show that some orbit $(o, K)^\Sigma \subseteq \mathbf{F}$ is $3l$-dimensional. The dimension formula implies that $\dim(o, K)^\Sigma = \dim o^\Sigma + \dim K^{\Sigma_o}$. Therefore, it suffices to show that the identity component $\Omega = \Sigma_o^1$ of the stabilizer has an l-dimensional orbit $K^\Omega \subseteq \mathcal{L}_o$. By (74.13), a non-trivial element of Ω does not fix two lines in $\mathcal{L}_o \setminus \mathcal{L}_\infty$. Since $\dim \Omega = d - 2l \geq l$ by (74.10a), an l-dimensional orbit exists for $l = 1$.

If $l = 2$ and the maximal dimension of Ω-orbits in $\mathcal{L}_o$ is 1, then the vertical line $Y = o \vee \infty$ forms the only trivial orbit, or else we obtain a contradiction to (74.13) by fixing one additional line. For the same reason, Ω acts effectively on each of the 1-dimensional orbits L^Ω, and Ω is 2-dimensional and non-abelian, whence $\Omega \cong L_2$, see (94.40). The stabilizer Ω_L belongs to the unique conjugacy class of non-normal one-parameter subgroups in Ω. Hence Ω fixes a line in every orbit. Again, this contradicts (74.13). □

74.22 Proposition: Structure of the stabilizer. *Let $\mathcal{E}_f$ be a 4-dimensional, proper shift plane whose automorphism group Σ has dimension $d \geq 6$. Then Σ is a solvable group of dimension 6 or 7, and the identity component $\Omega = \Sigma_o^1$ of the stabilizer answers to one of the following descriptions:*

(a) Ω *contains a closed, normal subgroup* $\Lambda \cong \mathbb{R}^2$ *which is sharply transitive on* $\mathcal{L}_o \setminus \{Y\}$, *where* $Y = o \vee \infty$.
(b) Ω *is isomorphic to* $\mathbb{C}^\times$ *and fixes precisely one line* $L \in \mathcal{L}_o$ *other than* Y. *On the remainder of* $\mathcal{L}_o$, *the group* Ω *is sharply transitive.*
(c) $\Omega \cong L_2$ *is 2-dimensional and non-abelian.*

In particular, we have $d = 6$ in cases (b) and (c).

Proof. 1) Recall that the action of Ω on the affine point set $\mathbb{R}^4$ is linear (74.10a). By the stiffness theorem (74.13) and its dual, the stabilizer Ω_L of $L \in \mathcal{L}_o \setminus \mathcal{L}_\infty$ acts freely on $L \setminus \{o\}$ and on $\mathcal{L}_o \setminus \{L, Y\}$. For $d \geq 8$, we use the dimension formula (96.10) and (53.1d) to infer from this fact that Ω_L is sharply transitive on both spaces. It follows that Ω_L is homeomorphic and hence isomorphic to $\mathbb{C}^\times$, see (96.9a) and (94.39 and 40).

2) Conversely, any subgroup $\Pi \cong \mathbb{C}^\times$ of Ω contains a reflection σ, see (55.21c and 29). By (74.9d), the axis of σ is Y, and σ fixes some other line $L \in \mathcal{L}_o$. By the arguments of step 1), it follows that Π acts on the 2-sphere $\mathcal{L}_o$ in the standard, effective way. If L is not fixed by Ω (in particular, if $d \geq 7$), then Ω is even transitive on $W \setminus \{\infty\}$, and we obtain a linearly transitive elation group $\Sigma_{[\infty, Y]}$ using (61.19b). Then $\mathcal{E}_f$ is a dual translation plane because $\Sigma_{[\infty, W]}$ is also transitive (74.7), and (74.8c or 9a) yields a contradiction since $\mathcal{E}_f$ is a proper shift plane by assumption. Thus, $d \geq 8$ is impossible, and $\dim \Omega \leq 3$; if $\mathbb{C}^\times \leq \Omega$, then assertion (b) holds.

3) If Ω is not solvable, then (94.28 and 40) show that Ω is 3-dimensional and almost simple. By (95.2), there is an Ω-invariant complement X for the invariant subspace $Y \leq \Lambda$, see (74.10a), and Ω acts freely on $X \setminus \{0\}$ by the dual of (74.13), a contradiction to $\dim \Omega = 3$. Thus, Ω and $\Sigma^1 = \Lambda \rtimes \Omega$ are both solvable. Suppose that Ω is 3-dimensional (and solvable). Such a group contains a 2-dimensional, abelian, normal, analytic subgroup Λ; this follows from the corresponding fact for Lie algebras, see Jacobson [62] I.4. By (94.38), there are three possibilities for the structure of Λ, two of which are ruled out easily: $\Lambda \cong \mathbb{C}^\times$ contradicts step 2), and a 2-torus Λ would contain reflections with different axes (55.35), contrary to (74.9d). The remaining possibility is that $\Lambda \cong \mathbb{R}^2$, and then Λ acts freely on $\mathcal{L}_o \setminus \{Y\}$. Indeed, suppose that Λ_L is non-trivial for $L \in \mathcal{L}_o \setminus \{Y\}$. Then Λ_L acts trivially on L^Λ, hence $L^\Lambda = \{L\}$ by stiffness (74.13), and Λ is sharply transitive on $\mathcal{L}_o \setminus \{Y, L\} \approx \mathbb{R}^2 \setminus \{0\}$ (again by stiffness). This is a contradiction. The same

reasoning applied to the closure $\overline{\mathsf{A}}$ shows that A is in fact closed. Finally, if Ω is 2-dimensional and non-abelian, then we have case (c), see (94.40). □

74.23 Outline of the classification. All 4-dimensional flexible shift planes have been determined by Betten and Knarr, see Knarr [83, 86] and Betten–Knarr [87]. It turns out a posteriori that they are all obtained from translation planes by integration. The planes are naturally divided into three classes according to the isomorphism type of $\Omega = \Sigma_o^1$, see (74.22).

The case $\mathbb{R}^2 \leq \Omega$ leads to two single planes, obtained by integrating the translation planes of (73.10), see (74.24). One of them has $d = \dim \Sigma = 7$, the other has $d = 6$. The former plane is in fact the only 4-dimensional projective plane with $d \geq 7$ that is neither isomorphic nor dual to a translation plane. This was proved by Betten [90] and Löwen [90]. Under a rather strong additional assumption, we prove this in (74.26). The plane with $d = 6$ is the only flexible shift plane with a nilpotent automorphism group. In fact, it is the only 4-dimensional, compact, projective plane that is flexible and has a nilpotent group, see Betten [91a].

There is a 2-parameter family of shift planes with $\Omega \cong \mathbb{C}^\times$, see (74.28). Finally, the case $\Omega \cong L_2$ leads to four families of planes that are obtained by integrating the translation planes of (73.13) and those of Betten [75] Theorems 1, 2 and 5. We shall construct some of these shift planes (74.30 and 32) in order to demonstrate that one translation plane may have different integral planes (74.31c).

74.24 Theorem: Two single shift planes. *Let $\mathscr{E}_f$ be a 4-dimensional, proper shift plane such that the stabilizer Σ_o of $\Sigma = \operatorname{Aut} \mathscr{E}_f$ contains a subgroup $\mathsf{A} \cong \mathbb{R}^2$. Up to isomorphism this implies that f is one of the planar maps $f_{[0]}$ or $f_{[-1]}$ of $\mathbb{R}^2$ defined by*

$$f_{[k]}(s,t) = (st - \tfrac{1}{3}s^3 + ks, \tfrac{1}{2}(t^2 + ks^2) - \tfrac{1}{12}s^4).$$

The derivative of the plane $\mathscr{E}_{f_{[k]}}$ exists and is isomorphic to the translation plane generated by the spread $\mathscr{S}_k$ of (73.10).

In order to describe the automorphism groups, we define $\beta = \operatorname{diag}(-1, 1, -1, 1)$, and we introduce one-parameter groups P, M *and* N *as follows:*

$$\mathsf{P} = \{ \operatorname{diag}(r, r^2, r^3, r^4) \mid r > 0 \}, \text{ and}$$

$$\mathsf{M} = \left\{ \begin{pmatrix} 1 & & & \\ b & 1 & & \\ \tfrac{b^2}{2} & b & 1 & \\ \tfrac{b^3}{6} & \tfrac{b^2}{2} & b & 1 \end{pmatrix} \,\middle|\, b \in \mathbb{R} \right\}, \quad \mathsf{N} = \left\{ \begin{pmatrix} 1 & & & \\ & 1 & & \\ c & & 1 & \\ & c & & 1 \end{pmatrix} \,\middle|\, c \in \mathbb{R} \right\}.$$

We have $\Sigma = \Lambda \rtimes \Sigma_o$, where $\Sigma_o = \langle \mathsf{M}, \mathsf{N}, \mathsf{P}, \beta \rangle$ for $k = 0$ and $\Sigma_o = \langle \mathsf{M}, \mathsf{N}, \beta \rangle$ for $k = -1$. The dimension of Σ is 7 for $k = 0$ and 6 for $k = -1$. The *automorphism β is a Baer involution defining a desarguesian subplane $\mathcal{F}_\beta$. As a consequence, all collineations of $\mathcal{E}_{f_{[k]}}$ are continuous.*

Proof. 1) We show first that $f_{[k]}$ is an integral function of some spread. The differential

$$D_{(s,t)} f_{[k]} = \begin{pmatrix} t - s^2 + k & s \\ -\frac{1}{3} s^3 + ks & t \end{pmatrix}$$

is almost equal to the spread element $w(s, t) \in \mathcal{S}_k$ defined by the parameters $(\tilde{s}, \tilde{t}) = (s, t - s^2)$. The only difference is the summand k in the top left position. It is removed by subtracting a constant matrix from all spread elements, which results in an isomorphic spread.

2) We may now prove planarity of the maps $f = f_{[k]}$ using the Hadamard criterion (74.19 and 20); it is not necessary to check the invertibility of differentials, see (74.20a). For $d = (a, b) \neq (0, 0)$, we obtain

$$\left\| (D_{(s,t)} f_{(a,b)})^{-1} \right\|^2 = \frac{(b - 2sa - a^2)^2 + a^2 + \left(\frac{1}{3}(3s^2 a + 3sa^2 + a^3) + ka \right)^2 + b^2}{\left((b - 2sa - a^2)b + \frac{1}{3}a(3s^2 a + 3sa^2 + a^3) - a^2 k \right)^2}.$$

This continuous function of s either is constant (if $a = 0$) or converges to a^{-2} as $s \to \infty$, whence it is bounded in each case. This proves existence of the planes $\mathcal{E}_{f_{[k]}}$.

3) Conversely, consider a shift plane $\mathcal{E}_f$ satisfying the assumption that the stabilizer Σ_o contains a subgroup $\mathsf{A} \cong \mathbb{R}^2$. In order to avoid case distinctions, we shall work with the 6-dimensional group $\Delta = \Lambda \mathsf{A}$ instead of Σ. If we knew that we obtain a translation plane $\mathcal{A}_\mathcal{F}$ by derivation (compare Problem (74.17b)), then Δ would act on that plane, satisfying the hypotheses of (73.10), and we would obtain complete information about the action of Δ on the affine point set. Without this knowledge, we can at least borrow most of our arguments from the proof of (73.10).

4) We claim that the translation group $\mathsf{T} = \Delta_{[W,W]}$ and the shear group $\Delta_{[\infty, \infty]}$ dual to it are both 3-dimensional. Since the action of Δ is self-dual by (74.10c), it suffices to compute $\dim \mathsf{T}$. First recall from (74.9a) that $\mathsf{T} = \Delta_{[W]}$ and that $\dim \mathsf{T} \leq 3$; compare also (74.8c).

We use the action of Δ on $W' = W \setminus \{\infty\}$. The restricted group $\Delta|_{W'}$ has a sharply transitive, normal subgroup $\Lambda|_{W'} \cong \mathbb{R}^2$, because $\lambda_{(a,b)}$ sends the point $(x) = L(x, 0) \wedge W$ at infinity to $(x + a)$. Hence $\Delta|_{W'}$ is an affine group, and it contains another sharply transitive, abelian subgroup $\mathsf{A}|_{W'}$ by our hypothesis, compare (74.22a). Since $\Delta = \Lambda \mathsf{A}$, we see that $\Delta|_{W'}$ is an extension of the group $\Lambda|_{W'}$ by an abelian subgroup B of $\mathrm{GL}_2 \mathbb{R}$. If B fixes no direction in $\mathbb{R}^2$, then $\mathsf{B} \leq \mathrm{GO}_2 \mathbb{R}$ up to conjugacy, compare (95.6b), and then $\mathsf{A}|_{W'} = \Lambda|_{W'} = \Delta|_{W'}$, because the translations are the only similarities of $\mathbb{R}^2$ without fixed points. It follows that B does fix a direction, whence we may apply (73.9) to $\mathsf{A}|_{W'}$. We find that $\mathsf{A}|_{W'}$ has a one-parameter group in common with $\mathsf{\Lambda}|_{W'}$. Therefore, we have $\dim \Delta|_{W'} \leq 3$, whence $\dim \mathsf{T} \geq 3$ by (93.7). This proves our claim; as a by-product, we obtain that $\dim \Delta|_{W'} = 3$ and that $\mathsf{A}|_{W'}$ is one of the affine groups described in (73.9).

5) Like for Ξ in (73.8), we may argue that T is an almost homogeneous translation group and that the action $\Delta|_{W'}$ is therefore isomorphic to the action of Δ on the real projective plane PT associated with $\mathsf{T} = \mathbb{R}^3$, minus the line $\mathsf{P\Upsilon}$ defined by the group $\Upsilon = \mathsf{T} \cap \Lambda$ of vertical translations, compare (74.7). However, we shall not be able to use this fact as directly as in the corresponding situation in (73.10) since we are identifying the point space $\mathbb{R}^4$ with Λ, and T is not a subgroup of Λ. Nevertheless, we keep in mind that the Lie subalgebra $\mathfrak{a} \leq \operatorname{End} \mathfrak{l}\mathsf{T}$ defined by the action of A on $\mathsf{T} = \mathfrak{l}\mathsf{T}$ is determined, modulo the Lie algebra $\mathbb{R}\mathbb{1}$ of the real homothety group, according to (73.9). We obtain that $\mathfrak{a}$ is generated by two matrices

$$M^{\flat} = \begin{pmatrix} m & & \\ 1 & m & \\ 0 & a & m \end{pmatrix} \quad \text{and} \quad N^{\flat} = \begin{pmatrix} n & & \\ 0 & n & \\ 1 & 0 & n \end{pmatrix}.$$

Observe that the action of $\mathfrak{a}$ on $\mathfrak{l}\mathsf{T}$ is given by the adjoint representation in $\mathfrak{l}\Delta$, i.e., by the Lie bracket.

6) Let B, X_3, X_4 be the basis of $\mathfrak{l}\mathsf{T}$ used for the above matrix representation. Then $\mathfrak{l}\Upsilon = \mathfrak{l}\mathsf{T} \cap \mathfrak{l}\Lambda$ is generated by X_3 and X_4. Since $\mathfrak{l}\Delta = \mathfrak{l}\Lambda \oplus \mathfrak{l}\mathsf{A}$ as a vector space, we may write $B = X_2 - A$, where $A \in \mathfrak{l}\mathsf{A}$ and $X_2 \in \mathfrak{l}\Lambda$. If X_2, X_3, X_4 are linearly dependent, then $A \in \mathfrak{l}\mathsf{T}$ acts trivially on T and, hence, on W, a contradiction. It follows that we may choose X_1 such that $X_1, \ldots, X_4$ form a basis of $\mathfrak{l}\Lambda$.

Now let M, N be basis elements of $\mathfrak{l}\mathsf{A}$ acting as $M^{\flat}$ and $N^{\flat}$ on $\mathfrak{l}\mathsf{T}$. Since A is abelian, we have

$$M^{\flat}B = [M, B] = [M, A + B] = [M, X_2] \in \mathfrak{l}\mathsf{T} \cap \mathfrak{l}\Lambda = \mathfrak{l}\Upsilon.$$

This implies two things: first, that $m = 0$ and similarly $n = 0$, and secondly, that M and N act on $\mathfrak{l}\Lambda = \langle X_1, \ldots, X_4 \rangle$ by the following matrices, which we shall denote by the same symbols M and N:

$$M = \begin{pmatrix} 0 & & & \\ b & 0 & & \\ c & 1 & 0 & \\ d & 0 & a & 0 \end{pmatrix}, \quad N = \begin{pmatrix} 0 & & & \\ * & 0 & & \\ * & 0 & 0 & \\ * & 1 & 0 & 0 \end{pmatrix}.$$

Now we may argue almost exactly as in steps 2) to 4) in the proof of (73.10) to compute or normalize the unknown constants in these matrices. As in step 2) of that proof we obtain first that N generates the 1-dimensional shear group $\exp \mathbb{R}N$

with axis $L(0) = o^\tau$ and that $a \neq 0$. This implies that we can proceed as in the former step 3) to arrange that the first column of N reads $(0, 0, 1, 0)$. As we did in step 4) of the earlier proof, we may obtain that $a = b = 1$ and that $c = 0$. We have even more freedom now, because there is no reason to keep $\langle X_1, X_2 \rangle$ invariant. Therefore, we may reduce d to zero, using conjugation by a $(3, 1)$-elementary matrix, that is, by a matrix which differs from $\mathbb{1}$ in position $(3, 1)$ only. The final form of M and N is as follows.

$$M = \begin{pmatrix} 0 & & & \\ 1 & 0 & & \\ 0 & 1 & 0 & \\ 0 & 0 & 1 & 0 \end{pmatrix}, \quad N = \begin{pmatrix} 0 & & & \\ 0 & 0 & & \\ 1 & 0 & 0 & \\ 0 & 1 & 0 & 0 \end{pmatrix}.$$

(The symbols a, b, c, d have now done their duty and will be used with fresh meanings in the next step.)

7) We have thus determined the group Δ; computing exponentials, we see that it is equal to the group $\mathbb{R}^4 \rtimes \langle \mathsf{M}, \mathsf{N} \rangle$ appearing in the theorem. The next step will be to determine the stabilizer Δ_L of the line $L = o^\pi = L(0, 0)$, where π is the polarity introduced in (74.5). According to (74.10c), the generators of $\mathfrak{l}\Delta_L$ differ from those of $\mathfrak{l}\mathsf{A} = \mathfrak{l}\Delta_o$ by some elements of $\mathfrak{l}\Lambda$. This follows perhaps more easily from the fact that $\Delta = \Lambda \rtimes \mathsf{A} = \Lambda \rtimes \Delta_L$. Thinking of $\Delta = \Lambda \rtimes \mathsf{A}$ as a subgroup of the affine group of Λ, we represent the generators of $\mathfrak{l}\Delta_L$ by (5×5)-matrices S, T with first rows equal to zero; the first column represents the component from $\mathfrak{l}\Lambda$, and the remaining (4×4)-blocks of S and T are the matrices M and N, respectively. Conjugating by a shift, represented by a (5×5)-matrix which differs from $\mathbb{1}$ only in the first column, we may reduce the first column of T to $(0, a, b, 0, 0)$; this involves the passage to a different line L. The fact that $[S, T] = 0$ now implies that $a = 0$ and that the first column of S is $(0, b, 0, c, d)$. If $b = 0$, then $T = N$ belongs to $\mathfrak{l}\mathsf{A} \cap \mathfrak{l}\Delta_L$, contrary to the fact that A acts freely on W'. Thus we may obtain $b = 1$ if we replace X_1 by $b^{-1}X_1$.

8) There is a point $(0, 0, u, v) \in L$, and we obtain all points of L if we apply the transformations $(\exp xS)(\exp yT)$ to this point. The resulting image point is

$$\left(x, y + \tfrac{x^2}{2}, xy + cx + \tfrac{x^3}{6} + u, \tfrac{y^2}{2} + \tfrac{x^2y}{2} + dx + \tfrac{cx^2}{2} + \tfrac{x^4}{24} + ux + v\right).$$

Substituting $x = s$ and $y = t - \tfrac{s^2}{2}$, we obtain

$$\left(s, t, st - \tfrac{s^3}{3} + cs + u, \tfrac{t^2}{2} - \tfrac{s^4}{12} + \tfrac{cs^2}{2} + (d+u)s + v\right).$$

Applying the shift $\lambda_{(0,0,-u,-v)}$ and replacing d by $(d+u)$ we remove the constants u and v without changing the plane or the group. The constant d disappears if we apply a suitable $(4, 1)$-elementary (4×4)-matrix. This matrix commutes with

M and N, hence Δ remains unaltered. We have now arrived at the generating function $f_{[c]}$. For $c = k$, this is the function that appears in the assertion. We shall see in steps 9) and 10) below that the planes for $c < 0$ are mutually isomorphic, and that $f_{[c]}$ is not planar for $c > 0$.

9) By (74.12), the planes generated by our maps are non-desarguesian. Using (74.11), it is now an easy matter to determine all automorphisms and isomorphisms between them. At this point, we may also use the fact that the automorphism group of the derivative of $\mathcal{E}_{f_{[k]}}$ is known (73.10) and contains $\mathrm{Aut}\,\mathcal{E}_{f_{[k]}}$. One finds that the group P permutes the planes with $c < 0$ transitively and acts on the plane with $c = 0$. Moreover, all planes admit the Baer involution β.

The set of fixed points of β is given by $x_1 = x_3 = 0$. The intersection of this set with the graph of $f_{[c]}$ consists of the points $(0, t, 0, \frac{t^2}{2})$, so that the Baer subplane $\mathcal{F}_\beta$ is the parabola model of the real affine plane. By (55.22b), this implies that all automorphisms of $\mathcal{E}_{f_{[k]}}$ are continuous.

10) Suppose that $c > 0$. An easy computation shows that there is some $d \in \mathbb{R}^2 \setminus \{0\}$ such that the determinants $j_d(x) = \det D_x f_d$ are not all of the same sign; these determinants form the denominators of the matrix norms appearing in step 2). If f_d were a homeomorphism, then it would be either orientation preserving or orientation reversing, and the numbers $j_d(x)$ would be all ≥ 0 or all ≤ 0 accordingly. □

74.25 Notes. (a) We observe that the automorphism groups of these shift planes, extended by the real homotheties, yield precisely the automorphism groups of the derived translation planes; remember that a linear transformation is needed for this comparison in the case $k = -1$, see step 1) of the last proof. Surprisingly, this appears to be a widespread phenomenon. In general, one can only expect to compare the stabilizers of the point ∞. This makes no difference in the present example, but it does in others, see (74.31d). The call for a general theorem might be added to the list (74.17) of problems.

At least one of the following bold conjectures is false: (i) 'every translation plane admits an integral plane with respect to an arbitrarily chosen vertical line Y', and (ii) 'the dimensions of the stabilizers of ∞ in the automorphism groups of these two planes differ by 1'. If both statements were true, then each of the translation planes with 7-dimensional groups (73.23) would lead to some integral plane with a 6-dimensional group, in view of (73.4). The classification of shift planes surveyed in (74.23) shows that this is not the case. In fact, several of the families of translation planes of Betten [75, 76, 77] do not occur as the derivatives of shift planes with 6-dimensional groups.

(b) The derivative of $\mathcal{E}_{f_{[0]}}$ is a translation plane which has finite analogues, see the note (c) following (73.10). Knarr [83] 4.6 showed that the shift plane itself does not share this distinction. □

74.26 Corollary: Large abelian groups. *Let $\mathcal{P}$ be a 4-dimensional, compact, projective plane such that $\Sigma = \operatorname{Aut} \mathcal{P}$ is at least 7-dimensional and contains a 4-dimensional, abelian subgroup Π. Then, up to duality or isomorphism, $\mathcal{P}$ is either a translation plane (possibly desarguesian) or the unique proper shift plane $\mathcal{E}_{f_{[0]}}$ with a 7-dimensional automorphism group, see (74.24).*

Proof. The possibilities for Π have been described exhaustively in (71.11). We know that the cases (c) and (d) appearing there characterize respectively shift planes and translation planes, see (74.8a). If $\mathcal{P}$ is a proper shift plane, then the stabilizer Σ_o contains a subgroup isomorphic to $\mathbb{R}^2$ by (74.22), and the plane is uniquely determined, see (74.24). If $\mathcal{P}$ is any non-desarguesian plane satisfying our hypothesis, then the 4-dimensional, connected, abelian subgroup $\Pi \leq \Sigma$ is unique according to (72.12). Therefore, the remaining cases (a) and (b) of (71.11) cannot occur. If they do, then Σ preserves the fixed configuration of Π, which either is a triangle or consists of two points and two lines; in both cases, we obtain a contradiction to stiffness (71.7). □

The same result can be proved without assuming the presence of a 4-dimensional abelian group, see Betten [90] and Löwen [90]. Note that the translation planes appearing in (74.26) are known, see (73.22 and 23). We summarize these results as follows.

74.27 Classification Theorem. *Let $\mathcal{P}$ be a 4-dimensional, compact, projective plane such that $\Sigma = \operatorname{Aut} \mathcal{P}$ is at least 7-dimensional. If $d = \dim \Sigma > 8$, then $\mathcal{P}$ is isomorphic to the complex plane. If $d = 8$, then $\mathcal{P}$ is isomorphic or dual to one of the translation planes (one family and 2 single planes) of Theorems (73.11, 13 and 19). If $d = 7$, then $\mathcal{P}$ is either isomorphic to the distinguished shift plane $\mathcal{E}_{f_{[0]}}$ of Theorem (74.24), or else $\mathcal{P}$ is isomorphic or dual to one of the translation planes of (73.23).* □

We continue the investigation of shift planes with a 6-dimensional automorphism group.

74.28 Theorem. *Let $\mathcal{E}$ be a 4-dimensional shift plane such that $\Omega = \Sigma_o^1 \cong \mathbb{C}^\times$, as in (74.22b). Then there exists $w = (u + iv) \in \mathbb{C}$ such that $\mathcal{E}$ is isomorphic to the plane $\mathcal{E}_w$ generated by the following multiplicative homomorphism $f_{[w]} : \mathbb{C} \to \mathbb{C}$:*

$$f_{[w]}(z) = z^2 |z|^{w-2}.$$

A sufficient condition for planarity of the map $f_{[w]}$ is that the parameters satisfy the conditions $1 < u \le 4$ and

$$v^6 + 2(u^2+4)v^4 + (u^4 + 16u^3 - 40u^2 + 16u + 16)v^2 + 16u(u-1)^3(u-4) \le 0.$$

The plane $\mathcal{E}_2$ is the complex plane, and the other planes $\mathcal{E}_w$ satisfy the hypothesis $\Omega \cong \mathbb{C}^\times$; isomorphic planes result precisely for parameter pairs w and $\overline{w}$. The derivatives of these proper shift planes are among the translation planes of Betten [76] Theorems 3 and 4. The group Ω consists of the maps

$$(x, y) \mapsto \bigl(cx, f_{[w]}(c)y\bigr)$$

of $\mathbb{C}^2$, where $c \in \mathbb{C}^\times$. We have $\Sigma_o = \Omega$ if $w \ne \overline{w}$, while for $w = \overline{w} \ne 2$ there is a Baer involution β, given by $\beta(x, y) = (\overline{x}, \overline{y})$, such that $\Sigma_o = \Omega\langle\beta\rangle$.

Proof. We shall only show that, up to isomorphism, all shift planes satisfying $\Omega \cong \mathbb{C}^\times$ are generated by some map $f_{[w]}$. The comparison with the spreads of Betten [76] and the verification of planarity are necessarily somewhat technical, and we refer to Betten–Knarr [87] for this part of the proof.

The involution $\sigma \in \Omega$ is a reflection (55.29 and 21c), and its axis is the line $Y = L(0)$, see (74.9d). If $U \le \Lambda$ denotes the (-1)-eigenspace of σ, then $\Lambda = U \oplus Y$ is an Ω-invariant decomposition. The action of Ω on U is effective by the dual of (74.13). Hence, we may identify U with $\mathbb{C}$ in such a way that Ω acts by sending $(x, y) \in U \oplus Y$ to $(cx, y^{\varphi(c)})$, where $c \in \mathbb{C}^\times$ and $\varphi : \mathbb{C}^\times \to \mathrm{GL}_2\mathbb{R}$ is a homomorphism satisfying $\varphi(-1) = \mathbb{1}$. Now σ fixes a unique line $L \in \mathcal{L}_o \setminus \{Y\}$, and hence $L = L^\Omega$; compare also (74.22b). We infer that $L \setminus \{o\}$ may be obtained as the orbit $(1, a)^\Omega = \bigl\{ (c, a^{\varphi(c)}) \bigm| c \in \mathbb{C}^\times \bigr\}$ for a suitable $a \in \mathbb{C}^\times$. Indeed, such a set must be contained in L, and it exhausts $L \setminus \{o\}$, because it meets every parallel of Y. Thus we have represented L as the graph of the function $f(c) = a^{\varphi(c)}$, extended by $f(0) = 0$.

Let ψ be the restriction of φ to the 1-torus $\mathbb{T} \le \mathbb{C}^\times$, and suppose that the kernel of ψ is non-trivial. Then the kernel contains all $2k$th roots of unity for some integer $k \ge 2$, and f takes the same value a on all these numbers. A regular $2k$-gon has a pair of parallel sides, hence there are two pairs (u_j, v_j) of such roots with the same difference $u_j - v_j = d$, $j \in \{1, 2\}$. Then $f_d(v_j) = 0$ for $j \in \{1, 2\}$, and f does not satisfy the shift condition. This shows that $\ker \psi = \langle -1 \rangle$.

In particular, we may also identify Y with $\mathbb{C}$ in such a way that $c \in \mathbb{T}$ acts by $y^{\psi(c)} = c^2 y$. Since $\varphi(\mathbb{C}^\times)$ is abelian, it follows that there is a homomorphism $\alpha : \mathbb{C}^\times \to \mathbb{C}^\times$ such that $y^{\varphi(c)} = \alpha(c)y$ for each $c \in \mathbb{C}^\times$. Moreover, we obtain that $\alpha(re^{i\vartheta}) = r^w e^{2i\vartheta}$ for some $w \in \mathbb{C}$, that is, $f(c) = \alpha(c)a = c^2|c|^{w-2}a$. Applying the transformation $(x, y) \mapsto (x, ya^{-1})$, which centralizes Ω, we may now reduce the constant a to 1. □

74.29 Note. The proof of (74.28) given in Betten–Knarr [87] yields more precise information. Namely, the planes $\mathcal{E}_w$, where w belongs to the range specified in (74.28), are precisely those shift planes with $\Omega \cong \mathbb{C}^\times$ that can be obtained by integration of spreads. Whether or not there are other shift planes with $\Omega \cong \mathbb{C}^\times$ is thus a question connected with the problem (74.17b). In any case, such planes would be of the form $\mathcal{E}_w$ with w outside the specified range.

74.30 Theorem: Skew parabola planes. *Let $\mathcal{E}$ be a 4-dimensional shift plane such that $\Omega = \Sigma_o^1 \cong L_2$, as in (74.22c). Suppose that Ω acts on the vertical line $Y = L(0)$ inducing only real homotheties. Then $\mathcal{P}$ is isomorphic to precisely one of the skew parabola planes $\mathcal{E}_f = \mathcal{E}_{g,k}$ defined by the following planar maps:*

$$f(s,t) = \left(\tfrac{1}{2}s^2 + a(t)t^2,\ st + b(t)t^2\right),$$

where

$$(a(t), b(t)) = \begin{cases} (-\tfrac{1}{2}, 0) & \text{if } t \geq 0 \\ (g, k) & \text{if } t < 0 \end{cases}$$

with parameters satisfying $k \geq 0$, $2g + k^2 \leq -1$ and $(g,k) \neq (-\tfrac{1}{2}, 0)$. All collineations of these planes are continuous; in order to describe the automorphism groups $\Sigma = \Lambda \rtimes \Sigma_o$, we define the shear group N as in (74.24), and we let

$$\mathsf{P} = \{\operatorname{diag}(r, r, r^2, r^2) \mid 0 < r \in \mathbb{R}\},$$

$$\beta(x_1, x_2, x_3, x_4) = (-x_1, x_2, x_3, -x_4),$$

$$\mu_k(x_1, x_2, x_3, x_4) = (x_1 - kx_2, -x_2, x_3 - kx_4, -x_4).$$

We have

$$\Sigma_o = \begin{cases} \langle \mathsf{P}, \mathsf{N}, \beta \rangle & \text{if } g \neq -\tfrac{1}{2},\ k = 0 \\ \langle \mathsf{P}, \mathsf{N}, \mu_k \rangle & \text{if } 2g + k^2 = -1 \\ \langle \mathsf{P}, \mathsf{N} \rangle & \text{else.} \end{cases}$$

The derivatives of these shift planes are the translation planes with reducible action of $\operatorname{SL}_2\mathbb{R}$, see (73.13).

74.31 Notes. (a) We shall not prove this theorem of Knarr [86] 4.1, but we shall demonstrate the existence of the skew-parabola planes and determine their derivatives, see (74.32) below.

(b) The excluded parameters $(g, k) = (-\tfrac{1}{2}, 0)$ define the complex plane.

(c) The arguments in (74.32) will show that there are non-isomorphic skew-parabola planes having isomorphic derivatives. In fact, the isomorphism type of a skew-parabola plane depends on a pair (g, k) of parameters, and the isomorphism type of the derivative depends on the single parameter $w = w(g, k)$. The proof shows that

$$w(g, k) = \tfrac{-1}{2g+k^2} \left(\tfrac{1}{2} - g + \sqrt{(g + \tfrac{1}{2})^2 + k^2} \right)^2.$$

This function is not injective. In fact, the values $(g, k) = (-1, 0)$ and $(g, k) = (-\tfrac{3}{4}, \tfrac{1}{3})$ both lead to $w = 2$, but the shift plane has a disconnected automorphism group in the first case and a connected one in the second case. This shows that the integral plane of a translation plane is not uniquely determined; the parametrization of the spread enters into the integration process, as we remarked earlier (74.15).

(d) In these examples, the difference between the dimensions of the automorphism groups of the derivatives and of the shift planes is 2. This is due to the fact that only the stabilizer of the point ∞ survives the integration process, compare (74.25a).

(e) Betten [79a] found the first one-parameter family of 4-dimensional shift planes, and called them 'complex skew parabola planes'. We have extended this name to the full two-parameter family considered here, because $2f$ coincides with the complex squaring map on pairs (s, t) such that $t \geq 0$; compare also (74.2a).

74.32 Existence of the skew parabola planes. 1) We consider the differentials

$$D_{(s,t)}f = \begin{pmatrix} s & 2a(t)t \\ t & s + 2b(t)t \end{pmatrix},$$

and compare with the spreads $\mathscr{S}_w$ constructed in (73.13), leaving the axis Y aside. The differentials with $t \geq 0$ define a partial spread $\mathscr{S}_w^+ = \mathscr{S}^+$, which does not depend on w and is contained in the complex spread. Let $\mathscr{B}$ be the set of subspaces defined by the differentials with $t < 0$. This set is invariant under the one-parameter groups N and $\Theta \leq \Phi = \mathrm{SL}_2\mathbb{R}$ which act on $\mathbb{C}^2$ by $(u, v) \mapsto (u, v + qu)$ and by $(u, v) \mapsto (ru, r^{-1}v)$, respectively, where $q, r \in \mathbb{R}$. Under the action of $\mathsf{N}\Theta \cong \mathrm{L}_2$, the differentials are mapped according to $(s, t) \mapsto (r^{-2}s + q, r^{-2}t)$, hence the action on $\mathscr{B}$ is transitive. Moreover, if $2g + k^2 < 0$, then $\mathscr{B}$ contains a matrix J such that $J^2 = -\mathbb{1}$, namely,

$$J = \sqrt{\tfrac{-1}{2g+k^2}} \cdot \begin{pmatrix} k & -2g \\ -1 & -k \end{pmatrix}.$$

In the proof of (73.13), step 6), we have seen that this condition means that J is fixed by the subgroup $\Psi = \mathrm{SO}_2\mathbb{R} \leq \Phi$. Since $\Phi = \Psi\mathsf{N}\Theta$, we infer that $\mathscr{B} = J^\Phi$.

2) We claim that the elements of $\mathscr{B}$ are contained in the set $A^- \cup \{0\} \subseteq \mathbb{C}^2$ considered in step 8) of the proof of (73.13); the set A^- consists of the points not covered by $\mathscr{S}^+$. Since A^- is preserved by Φ, it suffices to apply the test function $f(a, b, c, d) = ad - bc$ defined in the earlier proof to a single element of $\mathscr{B}$, for example, to the element defined by $(s, t) = (0, -1)$. One finds that f induces on this 2-dimensional vector space a quadratic form whose matrix has determinant

$-2g - k^2 > 0$ and trace $-1 + 2g \le -2$; hence the restriction of f is negative definite, and this observation confirms our claim.

3) Step 7) of the proof of (73.13) shows that J is conjugate by a rotation matrix B to one of the matrices J_h introduced there. We may show this by direct computation and obtain the value $h = \sqrt{w}$ as a by-product. Let (x, y) and $(-y, x)$ be the column vectors of B, that is, the rotated standard basis vectors. We are looking for $h > 1$ such that the equation

$$J \begin{pmatrix} x \\ y \end{pmatrix} = -h \begin{pmatrix} -y \\ x \end{pmatrix}$$

has a non-trivial solution (x, y). This leads to the value $w = w(g, k) = h^2$ given in (74.31c).

As in steps 8) and 9) of the proof of (73.13) we may conclude that the linear transformation $\beta : (u, v) \mapsto (Bu, Bv)$ of $\mathbb{R}^4$ maps $\mathcal{S}_w^+$ onto itself and sends the orbit J^Φ onto $\mathcal{S}_w^-$. Thus, our whole set of differentials yields precisely the spread $\mathcal{S}_w \setminus \{Y\}$. (We remark without proof that the condition $2g + k^2 < 0$ is sufficient to produce a plane, but all isomorphism types are already obtained for $2g + k^2 \le -1$.)

4) The existence proof is completed by applying the Hadamard criterion (74.19); compare also (74.20a). For a fixed $d = (c, e) \ne (0, 0)$, the differential

$$D_{(s,t)} f_{(c,e)} = \begin{pmatrix} c & 2a(t+e)(t+e) - 2a(t)t \\ e & c + 2b(t+e)(t+e) - 2b(t)t \end{pmatrix}$$

is independent of s and takes constant values for $\{t, t+e\} < 0$ and for $0 < \{t, t+e\}$. Hence, the norm of the inverse matrix is bounded. □

74.33 Notes on Section 74. The notion of shift plane was introduced in Dembowski–Ostrom [68], where (74.3a) is proved. The theory of 4-dimensional shift planes was developed in the papers Betten [79a], Knarr [83, 86, 88b], Betten–Knarr [87], Weigand [87], Polster [93], and Betten–Polster [94]. Betten's article introduces the concept of integration of spreads and the idea of the proper shift condition (74.14 and 16); moreover, Betten constructs a subfamily of the family of skew-parabolae given in (74.30). In Knarr's first thesis, he constructs the polarity with an absolute oval (74.5) (following Dembowski–Ostrom loc. cit., who treat the finite case), and he proves the characterizations of shift planes in general (74.8) and of flexible shift planes in particular (74.21), the results about automorphisms (74.10 and 11) (the case $l = 1$ is due to Groh [82b] 3.5), and the stiffness theorem (74.13); the paper culminates in the construction and characterization of the two singular shift planes (74.24 and 26). Knarr's second thesis classifies the shift planes with point stabilizer L_2. Knarr [88b] gives a special case of Polster's product construction (74.4b). The joint article of Betten and Knarr contains the classification of shift planes with point stabilizer $\mathbb{C}^\times$. Weigand proves the dimension result

(74.6). Polster and Betten–Polster are mainly concerned with the construction of examples and the interrelations with translation planes, Laguerre planes, and ovals. In particular, the oval in (74.5c) is due to Polster, who was led by ideas of Jungnickel [87].

75 Analytic planes

The purpose of this section is the proof of the following theorem of Breitsprecher [67b, 71].

75.1 Theorem: Analytic planes. *Every complex analytic, projective plane $\mathcal{P} = (P, \mathcal{L})$ is isomorphic as an analytic plane to the complex plane $\mathcal{P}_2\mathbb{C}$ with its standard analytic structure, as defined in (75.4).*

In particular, there are no 8- or 16-dimensional, complex analytic planes. Before we can prove this result, we need to define the relevant notions and to observe some simple facts.

75.2 Definition. Let $\mathscr{C}$ denote a category of manifolds and continuous morphisms, for example, the category $\mathscr{C}^0$ of topological manifolds and continuous maps or the category $\mathscr{C}^1$ of differentiable manifolds and continuously differentiable maps. The main example in this section will be the class $\mathscr{C}^\omega$ of complex analytic manifolds and complex analytic maps.

A $\mathscr{C}$-*structure* on a projective plane $\mathcal{P} = (P, \mathcal{L})$ is a pair of $\mathscr{C}$-manifold structures on P and $\mathcal{L}$ such that the operations of joining two points and of intersecting two lines are $\mathscr{C}$-morphisms. Clearly, this implies that $\mathcal{P}$ is a compact, connected projective plane with respect to the topologies underlying the two $\mathscr{C}$-structures; see (42.4) for the compactness. A plane with a $\mathscr{C}$-structure will briefly be called a $\mathscr{C}$-*plane*, or an *analytic plane* in the case $\mathscr{C} = \mathscr{C}^\omega$.

Earlier definitions of $\mathscr{C}$-planes required, in addition to our conditions, that all lines are $\mathscr{C}$-submanifolds of P and the dual property (Breitsprecher [67b, 71]) or that the flag space **F** has a $\mathscr{C}$-structure such that the projections from **F** to P and to $\mathcal{L}$ are $\mathscr{C}$-morphisms. For $\mathscr{C}^0$, there is a problem, see (52.2), but for differentiable structures, we shall prove those additional conditions. Different proofs are given by Kramer [94a] and Otte [95].

75.3 Proposition. *Let $\mathscr{C}$ be a category of manifolds and differentiable maps, and let $\mathcal{P} = (P, \mathcal{L})$ be a $\mathscr{C}$-plane. Then we have the following.*

(a) *The lines are $\mathcal{C}$-submanifolds of the point manifold P, and projectivities are $\mathcal{C}$-isomorphisms. The dual statement holds for line pencils.*
(b) *There is a $\mathcal{C}$-structure on the flag space* **F** *such that the projections* **F** $\to$ **P** *and* **F** $\to \mathcal{L}$ *are $\mathcal{C}$-fiber bundles.*
(c) *The $\mathcal{C}$-structure on a single line or line pencil entirely determines the $\mathcal{C}$-structure on $\mathcal{P}$.*

Proof. 1) Choose a triangle $\{x, p, q\}$, and let π_p and π_q be the differentiable maps of the open submanifold $P \setminus pq$ into $\mathcal{L}$ given by $\pi_p(y) = yp$ and $\pi_q(y) = yq$. We claim that the kernels U_x and V_x of the differentials $D_x \pi_p$ and $D_x \pi_q$ are l-dimensional if $\dim P = 2l$. Since $y = y^{\pi_p} \wedge y^{\pi_q}$ holds in a neighbourhood of x, the kernels have trivial intersection. Thus, after exchanging p and q if necessary, we have that U_x is at most l-dimensional. Any small $\mathcal{C}$-submanifold $X \subseteq P$ satisfying $T_x P = T_x X \oplus U_x$ is mapped $\mathcal{C}$-isomorphically by π_p onto some $\mathcal{C}$-submanifold $\mathcal{M} \subseteq \mathcal{L}$ which is contained in $\mathcal{L}_p$. Hence $l \leq 2l - \dim U_x = \dim \mathcal{M} \leq l$, and equality follows. In particular, the point q enjoys the same properties as p. By the open mapping theorem (51.19 and 21), $\mathcal{M}$ is an open subset of $\mathcal{L}_p$. Thus, we have a $\mathcal{C}$-submanifold chart at every element of every line pencil. By duality, the same holds for lines. Projectivities are obtained from join and intersection by restriction to these submanifolds, hence they are $\mathcal{C}$-isomorphisms.

2) In (51.2) and (51.23), assertions analogous to (b) and (c) were proved for the category $\mathcal{C}^0$. The same maps that were used as local charts in those proofs may now serve to define the desired $\mathcal{C}$-structures and $\mathcal{C}$-isomorphisms. □

75.4 Examples: The classical planes. In (14.3 and 8) and (16.12 and 14), we have shown that the point and line spaces P and $\mathcal{L}$ of the classical planes $\mathcal{P}_2 \mathbb{F}$ over $\mathbb{F} \in \{\mathbb{R}, \mathbb{C}, \mathbb{H}, \mathbb{O}\}$ are topological manifolds. The charts used in those proofs may also serve to introduce $\mathcal{C}$-structures on these planes; here, $\mathcal{C}$ is any reasonable category like those considered above, but $\mathcal{C}^\omega$ is allowed only for $\mathbb{F} = \mathbb{C}$. One has to show that change of charts, as well as join and intersection, are $\mathcal{C}$-morphisms.

For simplicity, we restrict ourselves to $\mathbb{F} \neq \mathbb{O}$, so that we may use homogeneous coordinates. Similar arguments work for Veronese coordinates. The point space and the line space are treated in the same way. It suffices to use three charts

$$\varphi_1 : (x_1, x_2) \mapsto (1, x_1, x_2)\mathbb{F},$$
$$\varphi_2 : (x_1, x_2) \mapsto (x_1, 1, x_2)\mathbb{F},$$
$$\varphi_3 : (x_1, x_2) \mapsto (x_1, x_2, 1)\mathbb{F},$$

where $x_i \in \mathbb{F}$. Change of charts from φ_1 to φ_2 is given by the $\mathcal{C}$-morphism $(x_1, x_2) \mapsto (x_1^{-1}, x_2 x_1^{-1})$ for $x_1 \neq 0$; this is where we use that $\mathbb{F} = \mathbb{C}$ in the analytic case. Now one obtains easily that the actions of $\mathrm{PGL}_3 \mathbb{F}$ on P and on $\mathcal{L}$

are $\mathscr{C}$-actions. Using this fact, the *first* proof of (14.3) may be reinterpreted as a proof that the join and intersection operations are $\mathscr{C}$-morphisms.

A different approach is worked out by Otte [93]; it applies to the octonions without complication. He defines two charts sending a pair $(x, y) \in \mathbb{F}^2$ to the affine point (x, y) and to the affine line $[x, y]$, respectively, and adds the charts obtained from these by composition with the triality collineation τ of (16.6) and with its square τ^2. The proof of differentiability of join and intersection is again based on ideas from the first proof of (14.3). Instead of using the action of the collineation group in order to reduce the verification of differentiability to the case where two affine points are joined, Otte employs only the triality collineation. (One easily checks that this collineation is a $\mathscr{C}$-morphism; in fact, its description in affine coordinates looks exactly like our change of charts in the first approach.) Therefore, he needs also to use the implicit function theorem in order to prove differentiability of the join operation near a pair consisting of an affine point and a point at infinity. Incidentally, Otte's method offers a third alternative way of proving Theorem (14.3) about continuity of the geometric operations.

Proof of (75.1). By (52.3 and 5), we know that a line L is homeomorphic to a sphere $\mathbb{S}_l$, where l divides 8. In a complex analytic plane, L carries a complex analytic structure, which obviously excludes the possibility $l = 1$. It is known that the spheres $\mathbb{S}_4$ and $\mathbb{S}_8$ do not carry complex structures either; see Hsiung [86] Corollary 3 for this deep fact. Thus, only the case $l = 2$ remains, and Riemann's mapping theorem, see Griffiths–Harris [78] p. 222, implies that L is isomorphic to $\mathrm{P}_1\mathbb{C}$ with the standard complex structure. The group Ω of complex analytic automorphisms of L corresponds to $\mathrm{PGL}_2\mathbb{C}$ under this isomorphism, see Griffiths–Harris [78] p. 64. In particular, this group is sharply triply transitive. By (75.3a), we know that Ω contains the group Π of projectivities of L, which is triply transitive, and hence coincides with Ω. In particular, Π is sharply triply transitive (21.6b), and von Staudt's theorem (see Section 66) implies that $\mathscr{P}$ is isomorphic to $\mathscr{P}_2\mathbb{C}$ as a topological projective plane; compare also (66.1).

It remains to be shown that the given complex analytic structure of $\mathscr{P}$ is isomorphic to the standard structure. By what we have seen, both structures lead to the same group of analytic automorphisms of L, namely, to the group $\Pi = \mathrm{PGL}_2\mathbb{C}$ of projectivities. Now Riemann's mapping theorem provides an isomorphism $f : L \to L$ of the two structures on L, and we infer that f normalizes Π. The group Π has trivial centralizer in the homeomorphism group of L, because Π contains elements with precisely one fixed point. It follows that f is determined by the automorphism of Π induced by f. According to Dieudonné [71] Chap. IV §§1, 6, all automorphisms of $\mathrm{PGL}_2\mathbb{C}$ are induced by $\mathrm{P\Gamma L}_2\mathbb{C}$, hence f belongs to that group. Since f is continuous, we infer that the two analytic structures on L either coincide or differ by complex conjugation. In view of (75.3c), the same is then true for the structures on $\mathscr{P}$. □

75.5 Further results. The first indication that differentiable planes are more special than topological ones was found by Betten [71], who showed that the Moulton planes and the planes over Cartesian fields (Section 37) do not admit differentiable structures. Kramer [94a] shows that, for fixed $l > 0$, the point spaces of all $2l$-dimensional projective $\mathcal{C}^\infty$-planes are homeomorphic; see also (53.19). Here, $\mathcal{C}^\infty$ denotes the category of smooth (infinitely differentiable) manifolds and smooth maps. Otte [93, 95] proves a variation of Breitsprecher's theorem (75.1), namely, he shows that every projective $\mathcal{C}^1$-translation plane is classical. His proof uses the theorem of Grundhöfer and Strambach on projectivities of affine translation planes (66.2). In contrast, Otte [93] constructs large families of non-classical, projective $\mathcal{C}^\infty$-planes of arbitrary dimension $2l$. He obtains them by disturbing the classical ternary operation of $\mathbb{F} \in \{\mathbb{R}, \mathbb{C}, \mathbb{H}, \mathbb{O}\}$ near 0 in a smooth way.

The situation is different in differentiable *affine* planes, which may be defined in a similar way. The differentiability conditions are less stringent in this case, since nothing is required when elements at infinity are involved. Nevertheless, Otte's characterization of classical planes has an affine analogue, which is older than his theorem and is due to Grundhöfer and Hähl, see their paper [90]. They prove that the only $\mathcal{C}^1$-affine planes of Lenz type V are the classical planes $\mathcal{P}_2 \mathbb{F}$ over $\mathbb{F} \in \{\mathbb{R}, \mathbb{C}, \mathbb{H}, \mathbb{O}\}$. (Actually, the result is stated for projective planes, but the proof, which uses the geometric fiber bundle (51.23a), works in the affine case.) In contrast to this, Otte [93] constructs many non-classical affine $\mathcal{C}^\infty$-translation planes of arbitrary dimension $2l \geq 4$.

75.6 Notes on Section 75. Theorem (75.1) is due to Breitsprecher [67b, 71], as was mentioned in the text. Strambach [75] obtained a related result. He considers *algebraic planes* (defined as $\mathcal{C}$-planes for a suitable category $\mathcal{C}$ of algebraic varieties over an algebraically closed field), and he shows that such planes are necessarily Pappian.

We add here a few comments on two papers of Kuiper [57] and Segre [56], who were the first to consider questions of differentiability in non-desarguesian planes. Answering a prize question posed by the Dutch 'Wiskunding Genootschap' in 1955, both authors independently constructed non-desarguesian planes on the point set $P_2\mathbb{R}$ of the real projective plane, such that the lines are real algebraic (and, in particular, real analytic) subvarieties of the point variety. They do not consider the question whether the planes are, for example, real analytic or algebraic planes, that is, whether the geometric operations belong to one of these categories.

Chapter 8

Eight- and sixteen-dimensional planes

Homogeneity conditions on compact, connected planes of finite dimension have been discussed in Chapter 6. They are mainly of two sorts: either certain orbits are required to be large, or some lower bound for the dimension of the automorphism group Σ is specified. The choice of such bounds will depend on the dimension $2l$, $l \in \{1, 2, 4, 8\}$, of the plane. Of particular interest in this respect is the critical dimension c_l introduced at the beginning of Section 65. Its defining properties are as follows. (i) A plane of dimension $2l$ satisfying $\dim \Sigma > c_l$ is classical, that is, isomorphic to the unique compact, connected Moufang plane $\mathcal{P}_2\mathbb{F}$, $\mathbb{F} \in \{\mathbb{R}, \mathbb{C}, \mathbb{H}, \mathbb{O}\}$ of the same dimension. (ii) There are non-classical planes of dimension $2l$ satisfying $\dim \Sigma = c_l$. This critical dimension is approximately half the dimension of the group of all automorphisms of the corresponding classical plane $\mathcal{P}_2\mathbb{F}$. Planes for which $\dim \Sigma$ is close to c_l are the nearest relatives of $\mathcal{P}_2\mathbb{F}$. For $l \leq 4$, the planes satisfying $\dim \Sigma \geq c_l - 1$ can be determined explicitly, and under additional assumptions regarding the structure of the group Σ and its action (e.g., existence of fixed elements), the classification can be extended to planes where $\dim \Sigma$ is slightly smaller. For $l = 8$, work is still in progress; the planes satisfying $\dim \Sigma \geq c_l$ are known.

For $l = 1$ and $l = 2$, pertaining results can be found in Sections 33 ff and 72 ff. The present chapter is devoted to a discussion of this classification program for the higher-dimensional planes, i.e., for $l = 4$ (with $c_4 = 18$) and $l = 8$ (with $c_8 = 40$). There are three types of results:

1) *General structural results*, offering insights into the structure of certain subgroups of Σ.
2) *Classification results* in which all planes satisfying a certain hypothesis on Σ are explicitly determined. Typical hypotheses require that $\dim \Sigma > d$ for some bound d, or that Σ contains a subgroup of a certain kind. In most cases, the hypotheses are satisfied in the classical plane $\mathcal{P}_2\mathbb{F}$, $\mathbb{F} \in \{\mathbb{H}, \mathbb{O}\}$, which then appears as an extremal case.
3) *Characterization results for the classical planes* $\mathcal{P}_2\mathbb{F}$, $\mathbb{F} \in \{\mathbb{H}, \mathbb{O}\}$, or, in other words, classification results which lead to the classical plane $\mathcal{P}_2\mathbb{F}$ only.

Within the scope of this book, only a selection of the known results can be proved. They are chosen in such a way as to illustrate typical methods of proof while avoiding technicalities. Therefore, many results will not be proved in the strongest form known.

Work on the classification program splits into two clearly separated phases:

Phase 1: The reduction to the case of translation planes. For instance, one proves that each plane with $\dim \Sigma \geq c_l$ is a translation plane (up to duality).

Phase 2: The classification of translation planes. Here, more information and special techniques are available.

The *prevalence of translation planes* indicated by the result of phase 1 reaches even further than stated; this is a consequence of results of the classification program which are not proved in this book. Namely, among 4-dimensional planes, there is exactly one plane satisfying $\dim \Sigma \geq c_l - 1$ that is not a (dual) translation plane, see (74.27). Among 8-dimensional planes, it is known that the only ones satisfying the same conditions are the Hughes planes of Section 86, which form a one-parameter family. Their analogues in dimension 16 have $\dim \Sigma = 37 = c_l - 3$. The only known non-translation planes of dimension 16 having a larger group are the glued planes of (82.33) with $\dim \Sigma = 38 = c_l - 2$, and it is conjectured that $\dim \Sigma \geq c_8 - 1$ occurs only in (dual) translation planes.

Section 87 presents an outline of the reduction to the case of translation planes (phase 1). Groups slightly below the critical dimension are included in the considerations, and this means that, for $l = 8$, work in progress is touched upon. For this reason, that section has been placed at the end. Readers who would like to begin directly at this crucial point of the classification program are invited to do so; however, they should take notice of Section 83 first, whose stiffness results serve as a technical basis for the later sections. Under much stronger assumptions than in Section 87, a selection of results supporting the claim of phase 1 will be given with full proofs in Sections 84 and 85, see the more detailed overview below. The chapter begins, however, with those parts of the classification program which take place within the domain of translation planes (Sections 81 and 82). Thus, our presentation more or less inverts the order of phases 1 and 2. We have done so for several reasons. Translation planes are more accessible due to their inherent vector space structure, implying in particular that their automorphism group Σ is a linear group, and their study can be used as a gentle introduction to higher-dimensional planes. Before being confronted with the problems involved in the reduction to the case of translation planes, the reader should have the opportunity of seeing what can be gained from it. The classification program for translation planes is independent of the other parts of the general program and a stepping stone for the latter; indeed, many general results are obtained by reducing the problem to translation planes and then applying a result proved for these, which only afterwards appears merely as a special case of that more general result. Also, there are properties which are conjectured to hold for higher-dimensional planes in general, but which

8 Eight- and sixteen-dimensional planes

up to now have been established for translation planes only, so that the theory of translation planes may serve as a model and an incentive for the general theory. For the same reasons, the subject evolved in this order in the original papers rather than in the logical order in which phases 1 and 2 were presented above.

In Section 81, we prove fundamental theorems on the structure of higher-dimensional translation planes and of their automorphism groups, and we deduce characterizations of the classical planes among translation planes. It is shown that an 8-dimensional translation plane with $\dim \Sigma > 21$ and a 16-dimensional translation plane with $\dim \Sigma > 48$ is classical. These bounds are still far from the critical dimensions. In Section 82, then, a complete classification of 8-dimensional translation planes with $\dim \Sigma \geq 17 = c_4 - 1$ and of 16-dimensional translation planes with $\dim \Sigma \geq 38 = c_8 - 2$ is presented. In contrast to Section 81, we have refrained in Section 82 from giving complete proofs, except for some paradigmatic results; we have rather concentrated on conveying a general impression, with partial proofs, outlines of proof or mere citations. These results will not be used in later sections. In accordance with the assertion about the prevalence of translation planes mentioned above, the 8-dimensional translation planes emerging from this classification, together with their duals and with the 8-dimensional Hughes planes of Section 86, exhaust the list of all 8-dimensional compact planes with $\dim \Sigma \geq 17$. (This is one of the main results of the classification program for compact planes in general, which will not be fully proved in this book.) It is conjectured that similarly the 16-dimensional translation planes obtained in the classification described in Section 82 comprise, up to duality, all possibilities for 16-dimensional compact planes with $\dim \Sigma \geq 39$; for $\dim \Sigma \geq 40$ this is known.

Planes of higher dimension (8 or 16) are fundamentally different from those of low dimension (2 or 4) in several respects. One distinctive feature is the fact that the stabilizer Λ in Σ of a non-degenerate quadrangle is not generally discrete. Section 83 deals with this phenomenon separately, as a preparation for Sections 84–87. General results and dimension estimates for Λ are proved, resulting in first dimension estimates for Σ. Further differences become apparent in Section 83, as well: in low-dimensional planes, lines are always manifolds and Σ is a Lie group, whereas for higher-dimensional planes it is not known whether these properties hold in general. Instead, one is forced to establish such properties in various special situations, if this is necessary for the further course of the argument.

Sections 84 and 85 contain the proofs for a selection of characterization results for 8- and 16-dimensional compact planes in general, without assuming translation planes. Again, the dimension bounds for Σ are chosen much higher than the critical dimensions, in order to permit complete proofs without exceeding the scope of this book. We briefly describe the global result in terms of a closed, connected subgroup Δ of Σ having large dimension. In the case of an 8-dimensional plane, it is proved that if Δ is semi-simple and $\dim \Delta > 19$, then the plane is classical. If Δ is not semi-simple and $\dim \Delta > 22$, then the plane is shown to be a translation plane up to duality; by the results of Section 81, the plane then turns out to be classical again. In the case of 16-dimensional planes, analogous results are obtained for semi-simple subgroups Δ satisfying $\dim \Delta > 45$ and for subgroups Δ which are not semi-simple and satisfy $\dim \Delta > 56$.

Thus, Sections 84 and 85 together with Section 81 and the stiffness results of Section 83 contain a complete set of proofs for characterizations of $\mathcal{P}_2\mathbb{H}$ and $\mathcal{P}_2\mathbb{O}$ as the only compact planes of dimension 8 or 16 satisfying $\dim \Sigma > 22$ or $\dim \Sigma > 56$, respectively.

Section 86 is devoted to the construction and a characterization of the Hughes planes. These are planes of dimension 8 or 16 which are not translation planes and satisfy $\dim \Sigma = 17$ or $\dim \Sigma = 37$, respectively. The significance of these planes for the classification program has already been mentioned. The 8-dimensional Hughes planes finally turn out to be the only 8-dimensional compact planes satisfying $\dim \Sigma \geq 17$ which are neither translation planes nor dual translation planes, cf. (84.28).

The final Section 87, whose central rôle for the subject of this chapter has already been emphasized, shows how the classification can be carried on, beyond the results which are expounded in the previous sections. Within a systematic outline, the present state of affairs is described, references to the literature are given, and a program for further investigation is laid down. Also, it is shown that Σ is a Lie group at least if $\dim \Sigma$ is sufficiently large.

81 Translation planes

The prominent rôle played by translation planes in the classification of higher-dimensional compact projective planes has already been indicated above. Whenever feasible, one tries to reduce general problems to the case of translation planes, where stronger tools are available.

Here, some of these tools shall be introduced and used for proving characterizations of the classical planes $\mathcal{P}_2\mathbb{H}$ and $\mathcal{P}_2\mathbb{O}$ among translation planes; these characterizations are the main subject of the present section. The overall result will be that, up to isomorphism, $\mathcal{P}_2\mathbb{H}$ is the only 8-dimensional compact translation plane whose automorphism group Σ satisfies $\dim \Sigma > 21$, and analogously for $\mathcal{P}_2\mathbb{O}$ with $\dim \Sigma > 48$. In Sections 84 and 85, these results will be used to obtain similar characterizations of the classical planes among compact 8- and 16-dimensional projective planes in general.

The bounds $\dim \Sigma > 21$ and $\dim \Sigma > 48$ are far from being best possible; they have been chosen so as to allow proofs at reduced cost which nevertheless convey a true impression of the methods. Since we follow the same principle in Sections 84 and 85, the results of the present section suffice for those final classification results which will be given in this book with complete proofs. Sharp characterization and classification results concerning 8- and 16-dimensional translation planes will be presented in Section 82, often without proof.

The basis for the classification of higher-dimensional translation planes are stiffness properties restricting the structure and the size of stabilizers in Σ of non-degenerate quadrangles and even of triangles. For instance, the stabilizer of a non-degenerate quadrangle is compact (81.5), and, what is more, even the stabilizer of a non-degenerate triangle is determined to a large extent by a compact normal subgroup (81.8). This is proved with the help of a module function which can be defined for locally compact quasifields in a similar way as the classical module function of locally compact fields. The stiffness results for compact, connected planes in general, see Section 83, do not have the same strength.

The theory of 8- and 16-dimensional translation planes is quite different in flavour from that of 4-dimensional translation planes, cf. Section 73. We mention a few reasons for this. For 4-dimensional planes, stiffness questions are settled once and for all by the general result (55.21b) that the stabilizer of a non-degenerate quadrangle has at most 2 elements. The world of 4-dimensional compact translation planes is neatly separated into two compartments, as the commutator subgroup of a point stabilizer Σ_o is either solvable or semi-simple (and then almost simple); the organisation of Section 73 reflects this division. For higher-dimensional translation planes, and in particular for the more homogeneous ones, the mixed case is the rule. Also, the number of possible semi-simple subgroups of Σ rapidly increases with the dimension. One of the main problems with non-classical translation planes consists in mastering the action of Σ on the translation line W (which is fixed by Σ). Unlike the action of Σ on the affine part, the action on W is not a priori supported by a linear structure. In 4-dimensional translation planes, W is a 2-sphere, and the actions of Lie groups on surfaces, starting with transitive ones, are well understood. For 8- or 16-dimensional planes, with W homeomorphic to $\mathbb{S}_4$ or $\mathbb{S}_8$, respectively, this problem becomes increasingly hard.

81.0 General assumptions and basic information. (a) In this section, we always consider a compact projective translation plane $\mathcal{P} = (P, \mathcal{L})$ of dimension $2l \in \{8, 16\}$; here, as usual, $l \in \{4, 8\}$ denotes the dimension of a line (considered as a subspace of P). By definition of a translation plane, there is a *translation line* W such that the *translation group* T consisting of all elations with axis W is (sharply) transitive on $P \setminus W$, see (24.6) and (25.1). Very often, we regard the given plane as the affine plane $\mathcal{P}^W$ with point set $P \setminus W$; then W is considered as the line at infinity, see (21.9). First, we present basic information, which for the most part has already been obtained earlier.

(b) Let o, u, v, e be a non-degenerate quadrangle such that $u, v \in W$. Then, with respect to this quadrangle of reference, the plane is coordinatized by a locally compact, connected quasifield Q, see (22.1 and 3), (24.6), (25.2), (41.3 and 7a), and (42.1). The additive group of Q is isomorphic to $\mathbb{R}^l$, see the remarks preceding (64.13) and also (42.6). What is more, Q is made into a right vector space over its kernel $\mathbb{K}$ by restriction of its multiplication to $Q \times \mathbb{K}$, and $\mathbb{K}$ is isomorphic to $\mathbb{R}$, $\mathbb{C}$, or $\mathbb{H}$, cf. (25.3) and (42.6).

In particular, a projective line of $\mathcal{P}$ is homeomorphic to the one-point compactification $\mathbb{R}^l \cup \{\infty\} \approx \mathbb{S}_l$, compare also (52.3).

By our coordinatization, the affine point set $P \setminus W$ is identified with $Q \times Q \cong \mathbb{R}^{2l}$. Thus, it has a natural vector space structure over the kernel $\mathbb{K}$, and, hence, over the closed subfield $\mathbb{R}$ of $\mathbb{K}$. The translation group T is just the group of vector space translations, see (25.4), and the group $\Sigma_{[o,W]}$ of homologies with axis W and center o consists precisely of the homotheties of the right $\mathbb{K}$-vector space $P \setminus W$, cf. (25.4). Thus,

$$\mathsf{T} \cong \mathbb{R}^{2l} \quad \text{and} \quad \Sigma_{[o,W]} \cong \mathbb{K}^\times .$$

The second isomorphism shows in particular that the isomorphism type of $\mathbb{K}$ as a topological field is independent of the choice of the quadrangle of reference (as long as $u, v \in W$). The (affine) lines through the origin o are $\mathbb{K}$-linear subspaces of $P \setminus W$, which form a spread of this vector space; compare (25.6), where also the independence of $\mathbb{K}$ from the quadrangle is explained in a still more conceptual way. The other affine lines are obtained as the images of the lines through o under the translations. Thus, the geometry of the plane is entirely determined once the system $\mathcal{L}_o$ of lines through o is known.

(c) If $\mathcal{P}$ is not isomorphic to the classical plane $\mathcal{P}_2\mathbb{F}$, $\mathbb{F} \in \{\mathbb{H}, \mathbb{O}\}$, then the group Σ of continuous collineations leaves the translation line W invariant (64.4c), and

$$\Sigma = \Sigma_o \ltimes \mathsf{T} .$$

For connected components, the same is true, since T is connected:

$$\Sigma^1 = (\Sigma_o)^1 \ltimes \mathsf{T} .$$

The kernel of the action of Σ_o on W is just the group $\Sigma_{[o,W]}$ of homologies. By (25.5), the elements of Σ_o are semilinear maps of the $\mathbb{K}$-vector space $P \setminus W \cong Q \times Q$. Moreover, by continuity, the elements of Σ_o are linear over the closure $\mathbb{R}$ of the prime field $\mathbb{Q}$ of $\mathbb{K}$, and Σ_o is a closed subgroup of $\mathrm{GL}_{2l}\mathbb{R}$, see (44.6). Hence, closed subgroups of Σ are Lie groups (94.3a).

The same argument shows that the group $\mathrm{Aut}\, Q$ of continuous automorphisms of the quasifield $Q \cong \mathbb{R}^l$ is a closed subgroup of $\mathrm{GL}_l\mathbb{R}$. This may also be viewed as a special case of the above, since $\mathrm{Aut}\, Q$ can be identified with the stabilizer $\Sigma_{o,u,v,e}$ according to (44.4c).

Conversely, in order to check that a given $\mathbb{R}$-linear transformation σ of $P \setminus W$ yields an element of Σ_o, it suffices to verify that the images of the elements of the spread $\mathcal{L}_o$ under σ belong to $\mathcal{L}_o$ again. Indeed, σ normalizes T, so that then every affine line, being the image of a line from $\mathcal{L}_o$ under a translation, is mapped to a line by σ. Via an obvious extension to the line at infinity, σ thus yields a continuous automorphism of the projective plane $\mathcal{P}$, see (44.10).

In many instances, we need to know that large parts of Σ_o are even $\mathbb{K}$-linear. For this, the possibilities for the kernel $\mathbb{K}$ have to be inspected more closely. If $l = 4$, and if $\mathcal{P}$ is not isomorphic to the classical plane $\mathcal{P}_2\mathbb{H}$, we have that $\mathbb{K} \in \{\mathbb{R}, \mathbb{C}\}$. If $l = 8$, a result by Buchanan–Hähl [77] says that $\mathbb{K}$ is always equal to $\mathbb{R}$. Hence, if $\mathcal{P}$ is not classical, then the kernel $\mathbb{K} \in \{\mathbb{R}, \mathbb{C}\}$ has at most two continuous automorphisms, so that the connected component $(\Sigma_o)^1$ is even $\mathbb{K}$-linear. Now we consider the closed subgroup

$$\Omega = \{\sigma \in \Sigma_o \mid \sigma \text{ is } \mathbb{K}\text{-linear, } \det_{\mathbb{K}} \sigma = 1\}.$$

As in (73.2d), we call this subgroup the *reduced stabilizer* of o in Σ. It satisfies the relations

(1) $\qquad (\Sigma_o)^1 = \Omega^1 \cdot (\Sigma_{[o,W]})^1 \subseteq \Omega \cdot \Sigma_{[o,W]} \subseteq \Sigma_o,$

of which only the inclusion $(\Sigma_o)^1 \subseteq \Omega^1 \cdot (\Sigma_{[o,W]})^1$ needs to be explained. For this, it suffices to show that $\det_{\mathbb{K}}$ takes the same values on $(\Sigma_o)^1$ and on $(\Sigma_{[o,W]})^1$. Now recall that $\Sigma_{[o,W]} \cong \mathbb{K}^\times$ is the group of homotheties of the $\mathbb{K}$-vector space $P \setminus W$. If $\mathbb{K} \cong \mathbb{C}$, then $\Sigma_{[o,W]}$ is connected, and every element of $\mathbb{K}^\times$ is the determinant of a homothety. If $\mathbb{K} = \mathbb{R}$, then an element of the connected component $(\Sigma_o)^1$ has positive determinant, which again is also the determinant of a homothety obtained by multiplication with a suitable positive scalar, that is, of an element of the connected component $(\Sigma_{[o,W]})^1$.

Concerning the action on the translation line W, we note that by (1) the group of transformations $(\Sigma_o)^1|_W$ of W induced by $(\Sigma_o)^1$ is the same as $\Omega^1|_W$. Studying Ω^1 instead of $(\Sigma_o)^1$ has the advantage that Ω^1 and Ω act almost effectively on W, since by definition Ω has finite intersection with $\Sigma_{[o,W]}$. Combined with (1), this shows in particular that

$$\dim \Sigma_o = \dim(\Sigma_o)^1 = \dim \Omega \cdot \Sigma_{[o,W]} = \dim \Omega + \dim \mathbb{K}.$$

(d) In the classical planes $\mathcal{P}_2\mathbb{H}$ and $\mathcal{P}_2\mathbb{O}$, analogous considerations apply to the stabilizer Σ_W instead of Σ. In the classical plane over $\mathbb{H}$, with $P \setminus W = \mathbb{H}^2$, the subgroup Ω has to be replaced by

$$\mathrm{SL}_2\mathbb{H} = \{\sigma \in \mathrm{GL}_2\mathbb{H} \mid \det_{\mathbb{R}} \sigma = 1\}.$$

Indeed, this group has finite intersection with the group $\Sigma_{[o,W]}$ of homotheties of $\mathbb{H}^2$, and from the description (12.11) of the affine collineation group one derives that $\Sigma_{W,o} = \mathrm{SL}_2\mathbb{H} \cdot \Sigma_{[o,W]}$. For this, one may use the fact (11.29) that every automorphism α of $\mathbb{H}$ is of the form $x^\alpha = a^{-1}xa$ for a suitable $a \in \mathbb{H}$, $\|a\|^2 = 1$, so that the collineation $\mathbb{H}^2 \to \mathbb{H}^2 : (x, y) \mapsto (x^\alpha, y^\alpha)$ appearing in (12.11) can be written as the product of the homothety $(x, y) \mapsto (xa, ya)$ by the transformation $(x, y) \mapsto (a^{-1}x, a^{-1}y)$, which belongs to $\mathrm{SL}_2\mathbb{H}$.

(e) Now assume in addition that our translation plane has Lenz type at least V, which means that the dual plane is a translation plane as well, cf. (24.7). Let v be the point of the original plane which has the rôle of a translation line in the dual plane. Dually to what has been stated at the beginning of (c), it follows as in (64.4c) that v is fixed under every collineation unless the plane is classical; in particular, v is incident with W. For every affine line L through v, the group $\Sigma_{[v,L]}$ of shears with axis L is linearly transitive; this is expressed by saying that v is a *shear point*.

Conversely, among the translation planes discussed here, those which have Lenz type at least V can be recognized by the property that for a point $v \in W$ the group $\Sigma_{[v,ov]}$ of shears with axis ov satisfies

$$\dim \Sigma_{[v,ov]} \geq l.$$

Indeed, we then have $\dim \Sigma_{[v,ov]} = l$, which is equivalent to linear transitivity of $\Sigma_{[v,ov]}$, see (61.4). This already suffices to conclude that all shear groups with center v and arbitrary axis are linearly transitive, so that the plane has indeed Lenz type at least V, see (24.7).

(f) *Different notational conventions.* In this and the following section, we frequently refer to the literature, in particular to papers by Hähl. Therefore, the reader should be aware that the notation used in this book systematically deviates from Hähl's. In particular, the symbol Σ, which here always denotes the full automorphism group of the plane, is used in those papers for certain compact subgroups. The translation line, which here is denoted by W, is called L_∞ there, whereas W often is used for the first coordinate axis.

The following observations will be employed frequently.

81.1 Lemma. *Let L_1, L_2 be two distinct affine lines through o, and let Δ be a subgroup of Σ_W fixing L_1, L_2, and a point q not incident with L_2. (Note that $\Sigma_W = \Sigma$ if the plane is not classical.)*

If q is not incident with L_1, either, then Δ acts effectively on L_1 and L_2, and the $\mathbb{R}$-linear representations of Δ induced on L_1 and L_2 (as linear subspaces of $P \setminus W \cong \mathbb{R}^{2l}$) are equivalent.

If q is not incident with W, then the action of Δ on W is equivalent to the action on L_2; in particular, it is effective.

Proof. First, assume that q is not incident with L_1. The line $L' = oq$ is a fixed line of Δ. The actions of Δ on L_1, L', and L_2 are equivalent; equivariant bijections are the projectivities $\pi' = \pi(L_1, L_2 \wedge W, L')$ and $\pi_2 = \pi(L', L_1 \wedge W, L_2)$, see (21.6). We show that these projectivities are linear isomorphisms $L_1 \to L'$ and $L' \to L_2$. By symmetry, it suffices to deal with $\pi_2 : L' \to L_2$. Now, for affine points $p' \in L'$, $p_2 \in L_2$, the relation $p_2 = \pi_2(p')$ holds if, and only if, p' and p_2 belong to the same parallel of L_1, or equivalently, if $p' - p_2 \in L_1$. The vector space $P \setminus W$ is the direct sum of the subspaces L_1 and L_2, so that every $p' \in L'$ decomposes uniquely as $p' = p_1 + p_2$, where $p_1 \in L_1, p_2 \in L_2$, and the map $p' \mapsto p_2$ is linear; but by the above this map is just π_2, since $p' - p_2 = p_1 \in L_1$. The equivalent actions of Δ on L_1 and L_2 are effective, since a collineation with two axes is the identity (23.8).

If $q \notin W$, then the actions of Δ on the fixed lines L_2, $L'_2 = q \vee (L_2 \wedge W)$, and W are equivalent via the equivariant bijections $\pi(L_2, L_1 \wedge W, L'_2)$ and $\pi(L'_2, o, W)$. As above, it follows that Δ acts effectively on L_2 and on W. □

The following lemma on homomorphisms of topological groups will often be applied to a closed subgroup Δ of the reduced stabilizer Ω and to the restriction homomorphism $\Delta \to \Delta|_W$.

81.2 Lemma. *Let Δ be a Lie group, and let $\varrho : \Delta \to \Lambda$ be a continuous surjective homomorphism with discrete kernel onto a topological group Λ whose topology is the quotient topology defined by ϱ.*

Then, for a closed, connected subgroup $\Gamma \leq \Lambda$, the connected component $\widetilde{\Gamma} = (\Gamma^{\varrho^{-1}})^1$ satisfies $\widetilde{\Gamma}^\varrho = \Gamma$ and is a covering group of Γ via ϱ. In particular, if Γ is simply connected, then $\varrho : \widetilde{\Gamma} \to \Gamma$ is an isomorphism.

Remark. The assumption that Λ carries the quotient topology is automatically satisfied for any continuous, surjective homomorphism $\Delta \to \Lambda$ if Δ is compact, or if Λ is locally compact, since ϱ is obviously closed in the first case, and open in the second by (96.9d); in both cases, hence, ϱ is identifying, cf. (92.20b).

Proof of (81.2). By (93.16), the assumptions imply that ϱ is open, and the same then is true for the restriction $\Gamma^{\varrho^{-1}} \to \Gamma$. Now, $\Gamma^{\varrho^{-1}}$ is a Lie group (94.3a), in which the connected component $\widetilde{\Gamma}$ is open. Hence, $\widetilde{\Gamma}^\varrho$ is open in Γ, and $\widetilde{\Gamma}^\varrho = \Gamma$ by connectedness (93.14a). The open map $\varrho : \widetilde{\Gamma} \to \Gamma$ with discrete kernel is a covering homomorphism (94.2). A covering of a simply connected group by a connected group is an isomorphism, see for instance Hochschild [65] Theorem IV.1.4, p. 46. □

Stiffness

One of the open problems for higher-dimensional compact planes in general is the question whether the stabilizers of non-degenerate quadrangles, or, equivalently, the automorphism groups of coordinatizing ternary fields, are always compact, see Section 83. For translation planes and quasifields, this can be proved. The main instrument is the module function of a locally compact quasifield. This concept has been in use for locally compact fields in order to construct absolute values, see Braconnier [45], Weiss–Zierler [58], Weil [67] Chap. I §2. It can be adapted to the more general context of locally compact quasifields, where it looses its algebraic qualities in most cases, but retains its topological usefulness. For 4-dimensional translation planes, which were treated in Section 73, the following results are valid, as well, but are of less relevance; in fact, stabilizers of non-degenerate quadrangles in the automorphism groups of these planes have at most two elements (55.21b).

81.3 Definition: The module function of a quasifield. In a locally compact, connected quasifield $Q = (\mathbb{R}^l, +, \cdot)$, the left multiplication by a fixed element $a \in Q$,

$$\lambda_a : Q \to Q : x \mapsto a \cdot x,$$

is linear over the kernel $\mathbb{K}$, see (25.3), and, in particular, over $\mathbb{R} \subseteq \mathbb{K}$. For $a \neq 0$, this map is bijective, so that for arbitrary a we have

$$\lambda_a \in \mathrm{GL}_l \mathbb{R} \cup \{0\}.$$

Define

$$\mathrm{mod}\, a = |\det \lambda_a|.$$

Remark. The determinant is meant to be the determinant over $\mathbb{R}$. It is related to the Haar measure μ on $\mathbb{R}^l$; for every measurable subset U of $\mathbb{R}^l$ and for $\varphi \in \mathrm{End}\,\mathbb{R}^l$ one has $\mu(\varphi(U)) = |\det \varphi| \cdot \mu(U)$, hence $\mu(\lambda_a(U)) = \mathrm{mod}\, a \cdot \mu(U)$. In this form, the definition of mod is usually given for locally compact fields.

In the non-associative case, the module function does not in general retain the algebraic properties which it has for locally compact fields. (See, however, the references on this question in (81.22)). Notwithstanding, the following topological properties survive and will be extremely useful.

81.4 Lemma. *The module function* $\mathrm{mod} : Q \to [0, \infty) \subseteq \mathbb{R} : a \mapsto \mathrm{mod}\, a$ *is continuous and describes the topology of Q; more precisely, the balls*

$$B_r = \{\, a \in Q \mid \mathrm{mod}\, a \leq r \,\},$$

where $0 < r \in \mathbb{R}$, form a neighbourhood base at 0 in Q. Moreover, each ball B_r is compact.

Proof. Continuity is clear from the continuity of multiplication. Thus, the balls B_r are closed neighbourhoods of 0.

For compactness, it suffices to prove that the balls B_r are bounded. This follows immediately if we show that, for a sequence $a_\nu \in Q$ converging to ∞, the values $\mod a_\nu$ also converge to ∞. For every $c \in Q$, we obtain from (43.3)(3′) that $(\lambda_{a_\nu})^{-1}(c) = a_\nu \backslash c \xrightarrow[\nu]{} 0$; the use of this result is justified by (43.5). Hence, $(\lambda_{a_\nu})^{-1} \xrightarrow[\nu]{} 0$ in $\operatorname{End} \mathbb{R}^l$, so that $\det(\lambda_{a_\nu})^{-1} \xrightarrow[\nu]{} 0$ and consequently $\mod a_\nu = |\det \lambda_{a_\nu}| \xrightarrow[\nu]{} \infty$.

The assertion that the balls B_r form a neighbourhood base at 0 now follows by a standard compactness argument from the fact that $\bigcap_{r>0} B_r = \{0\}$, see for instance Dugundji [66] XI.1.6(b), p. 226. □

81.5 Theorem: Compact automorphism groups. *The group* $\operatorname{Aut} Q$ *of continuous automorphisms of a locally compact, connected quasifield* $Q = (\mathbb{R}^l, +, \cdot)$ *is a compact subgroup of* $\operatorname{GL}_l \mathbb{R}$.

Corollary. *The stabilizer of a non-degenerate quadrangle in the automorphism group* Σ *of a compact, connected translation plane is compact.*

Proof of theorem. The fact that $\operatorname{Aut} Q \subseteq \operatorname{GL}_l \mathbb{R}$ has already been observed in (81.0c). For $\alpha \in \operatorname{Aut} Q$ and $a \in Q$, we have $\alpha \circ \lambda_a \circ \alpha^{-1} = \lambda_{\alpha(a)}$ and hence $\mod \alpha(a) = |\det \lambda_{\alpha(a)}| = |\det \lambda_a| = \mod a$. Thus, $\operatorname{Aut} Q$ leaves the compact balls B_r of (81.4) invariant. In particular, the orbits of $\operatorname{Aut} Q$ are relatively compact. Moreover, since those balls form a neighbourhood base at 0, it is immediate that $\operatorname{Aut} Q$ is equicontinuous. Hence, by the Arzelà–Ascoli theorem, see Dugundji [66] XII.6.4, p. 267, we obtain that $\operatorname{Aut} Q$ is relatively compact in the space of continuous mappings of $\mathbb{R}^l$ into itself, endowed with the compact-open topology. Now, the space $\operatorname{End} \mathbb{R}^l$ of linear endomorphisms of $\mathbb{R}^l$ is a closed subspace of this space and hence contains the closure $\overline{\operatorname{Aut} Q}$; moreover, the topology induced on $\operatorname{End} \mathbb{R}^l$ by the compact-open topology is the usual topology. Since $\operatorname{Aut} Q$ is a closed subgroup of $\operatorname{GL}_l \mathbb{R}$, we may now conclude using the following standard argument that relative compactness of $\operatorname{Aut} Q$ in $\operatorname{End} \mathbb{R}^l$ implies compactness.

For $\alpha \in \overline{\operatorname{Aut} Q}$ and a sequence $\alpha_\nu \in \operatorname{Aut} Q$ converging to α, the sequence α_ν^{-1} accumulates at some $\alpha' \in \operatorname{End} \mathbb{R}^l$ by relative compactness, and the constant sequence $\alpha_\nu \alpha_\nu^{-1} = \operatorname{id}$ accumulates at $\alpha \alpha'$, so that $\alpha \alpha' = \operatorname{id}$ and likewise $\alpha' \alpha = \operatorname{id}$. Hence, $\overline{\operatorname{Aut} Q}$ is contained in $\operatorname{GL}_l \mathbb{R}$; consequently, $\operatorname{Aut} Q = \overline{\operatorname{Aut} Q}$ is compact. □

Proof of corollary. If the plane is not classical, then Σ leaves W invariant (81.0c), and the stabilizer of any non-degenerate quadrangle is contained in the stabilizer of a non-degenerate quadrangle o, u, v, e with the additional property that the points u and v lie on the translation line W. Hence, it suffices to consider such a quadrangle.

This is also true in the classical planes, where every line is a translation line; one could equally well use the transitivity properties of the automorphism groups of these planes, cf. (13.5) and (17.6).

With respect to such a quadrangle o, u, v, e, the plane is coordinatized by a locally compact, connected quasifield $Q = (\mathbb{R}^l, +, \cdot)$, see (81.0b), and the stabilizer $\Sigma_{o,u,v,e}$ is isomorphic to $\operatorname{Aut} Q$ according to (23.5) and (44.4c). Now the assertion follows from the preceding theorem. □

81.6 Note. With Theorem (81.5) as starting point, the list of all possible subgroups of $\operatorname{GL}_l \mathbb{R}$ which may occur as the automorphism group of a locally compact, connected quasifield $Q = (\mathbb{R}^l, +, \cdot)$ can be narrowed down closely. For $l = 4$ one obtains that either $\operatorname{Aut} Q \cong \operatorname{SO}_3 \mathbb{R}$ or $\dim \operatorname{Aut} Q \leq 1$. For $l = 8$, $\operatorname{Aut} Q$ either is isomorphic to one of the groups $G_2 = \operatorname{Aut} \mathbb{O}$, $\operatorname{SU}_3 \mathbb{C} \rtimes \mathbb{Z}_2$, $\operatorname{SU}_3 \mathbb{C}$, or $\operatorname{SO}_4 \mathbb{R}$, or $\dim \operatorname{Aut} Q \leq 4$. For these and further results, see Hähl [75b, 76]. Moreover, for the easier special case of division algebras instead of general quasifields, it is shown there that each of the groups mentioned above is actually the full automorphism group of a suitable 4- or 8-dimensional real division algebra, and all such division algebras admitting one of these groups as automorphism group are explicitly determined. The same subject, for real division algebras, was taken up again later by Benkart–Osborn [81a,b], with almost corresponding results, but in a different language (for derivation algebras instead of automorphism groups), and with different methods. Some differences derive from the fact that the division algebras considered there need not have a multiplicative unit. As to generalizations for arbitrary locally compact, connected ternary fields, see (83.16 and 25) and the references given there.

For classification results about compact, connected translation planes, it is of great value that the method of the module function can be extended from stabilizers of non-degenerate quadrangles to stabilizers of triangles in order to prove that there is a compact subgroup of small codimension.

81.7 The stabilizer of the standard triangle. In the plane over a locally compact, connected quasifield $Q = (\mathbb{R}^l, +, \cdot)$, we consider the standard triangle $o = (0, 0), u = (0)$ and $v = (\infty)$. In affine coordinates over Q, a collineation $\gamma \in \nabla = \Sigma_{o,u,v}$ is an $\mathbb{R}$-linear transformation of $Q \times Q$, see (81.0c), which has the form

$$\gamma : (x, y) \mapsto (A_\gamma x, B_\gamma y),$$

where $A_\gamma, B_\gamma \in \operatorname{GL}_l \mathbb{R}$. On the line W at infinity, γ induces a homeomorphism

$$\gamma|_W : (s) \mapsto (P_\gamma s), \ (\infty) \mapsto (\infty),$$

where P_γ is a homeomorphism of Q; as $u = (0)$ is a fixed point of γ, we have $P_\gamma 0 = 0$. The line through o of slope s is mapped to the line of slope $P_\gamma s$; this is expressed algebraically by

(∗) $$B_\gamma(s \cdot x) = P_\gamma s \cdot A_\gamma x.$$

One of the technical difficulties which one encounters here is that P_γ is in general not linear. For compact subgroups of ∇, this problem is less serious. Indeed, there exist general theorems about actions of compact groups on a sphere $\mathbb{S}_l$, which often imply that the action is equivalent to a linear one, like Richardson's theorem (96.34) for $l = 4$. This is one of the reasons why we are interested in showing that geometrically relevant subgroups of Σ_o like ∇ have relatively large compact subgroups.

For the following theorem, recall the definition of the reduced stabilizer Ω from (81.0c). We use the notation introduced in (81.7) for the collineations in ∇.

81.8 Theorem: Large compact subgroups of triangle stabilizers. *In the plane over the locally compact, connected quasifield* $Q = (\mathbb{R}^l, +, \cdot)$, *consider the stabilizer* ∇ *of the standard triangle and the normal subgroup*

$$\mathsf{M} = \{ \gamma \in \nabla \mid |\det A_\gamma| = 1 = |\det B_\gamma| \} \trianglelefteq \nabla$$

(where det *denotes the real determinant). The following assertions hold.*

(a) *The group* M *is the largest compact subgroup of* ∇.
(b) *Let* Δ *be a closed, connected subgroup of* ∇ *contained in* Ω, *and consider the subgroup*

$$\mathsf{K} = \Delta \cap \mathsf{M}.$$

According to (a), K *is the largest compact subgroup of* Δ. *In particular, if* Δ *is compact, then* $\Delta = \mathsf{K}$. *If* Δ *is not compact, then* Δ *is the direct product*

$$\Delta = \mathsf{K} \times \Upsilon$$

of K *by a closed one-parameter subgroup* Υ *isomorphic to* $\mathbb{R}$. *(We remark that* K *is connected, and that* K *acts almost effectively on* W *by definition of* Ω.)
(c) *Assume that* Δ *contains a closed one-parameter subgroup* Υ *isomorphic to* $\mathbb{R}$. *Then, there is an isomorphism* $\eta : \mathbb{R} \to \Upsilon : t \mapsto \eta(t)$ *having the following property:*

For $t \to -\infty$, *the maps* $\eta(t)|_{W \setminus \{v\}}$ *converge uniformly to the constant map* $W \setminus \{v\} \to \{u\}$ *on each compact subset of* $W \setminus \{v\}$. *For* $t \to +\infty$, *an analogous convergence property holds with the rôles of* u *and* v *interchanged.*

Because of the convergence properties just stated, a one-parameter subgroup Υ as in (c) will sometimes be referred to as a *compression subgroup*.

Proof. 1) It is clear that M contains every compact subgroup of ∇, since the maps $\gamma \mapsto \det A_\gamma$ and $\gamma \mapsto \det B_\gamma$ of ∇ to $\mathbb{R}^\times$ are continuous homomorphisms, and since $\{1, -1\}$ is the only non-trivial compact subgroup of $\mathbb{R}^\times$. In order to prove that M itself is compact, we use the module function introduced in (81.3). For $\gamma \in \nabla$ and $s \in Q$ we infer from (81.7(∗)) that

(1) $$\operatorname{mod} P_\gamma s = |\det A_\gamma|^{-1} \cdot |\det B_\gamma| \cdot \operatorname{mod} s \,.$$

Therefore, if $\gamma \in M$, then the corresponding transformation P_γ is contained in the following group of homeomorphisms of Q:

$$\mathcal{M} = \{ P_\gamma \mid \gamma \in \nabla, \ \forall_{s \in Q} \ \operatorname{mod} P_\gamma s = \operatorname{mod} s \} \,.$$

We prove that the obvious homomorphism $P : M \to \mathcal{M} : \mu \mapsto P_\mu$ is surjective. Indeed, let $\gamma \in \nabla$ such that $P_\gamma \in \mathcal{M}$, which means that $|\det A_\gamma| = |\det B_\gamma|$ according to (1). By composing γ with a suitable real homothety, we obtain a collineation $\gamma' \in \nabla$ such that $|\det A_{\gamma'}| = 1 = |\det B_{\gamma'}|$. Then $\gamma' \in M$; moreover, γ' and γ induce the same transformation $P_{\gamma'} = P_\gamma$ on the line W at infinity, since composition with a homothety does not change the action on W.

The kernel of the homomorphism $P : M \to \mathcal{M}$ consists of $\mathbb{K}$-homotheties having real determinant 1, see (81.0b). The determinant condition is satisfied if, and only if, the homothety in question is multiplication by a scalar of norm 1. Hence, the kernel of $M \to \mathcal{M}$ is compact. The main effort of the proof of assertion (a) will consist in showing that $\mathcal{M}$ is a compact group of homeomorphisms of $Q = \mathbb{R}^l$ (in the compact-open topology), see steps 2) and 3). Before attacking this problem, we verify that compactness of M itself follows easily, once $\mathcal{M}$ is known to be compact. Indeed, by (96.9) it then follows that the surjective homomorphism $P : M \to \mathcal{M}$ is open. If C is a compact neighbourhood of id in M, then for $\gamma \in M$ the image $P(C\gamma)$ is a neighbourhood of P_γ. By compactness of $\mathcal{M}$, there are finitely many $\gamma_1, \ldots, \gamma_n \in M$ such that the sets $P(C\gamma_\nu)$ cover $\mathcal{M}$. Then the sets $P^{-1}(P(C\gamma_\nu)) = C\gamma_\nu \cdot \ker P$ cover M; these are compact since $\ker P$ is. Thus, M is shown to be compact. In order to prove that $\mathcal{M}$ is compact, we establish the following assertion.

2) $\mathcal{M}$ *is equicontinuous.*

Indeed, if this were not the case, then we could find $s \in Q$, $\varepsilon > 0$, a sequence $s_\nu \in Q$ converging to 0, and collineations $\gamma_\nu \in \nabla$ such that the corresponding transformations $P_\nu := P_{\gamma_\nu}$ belong to $\mathcal{M}$ and satisfy

(2) $$\|P_\nu(s + s_\nu) - P_\nu(s)\| > \varepsilon$$

for all $\nu \in \mathbb{N}$. For each ν, consider the $\mathbb{R}$-linear transformations $A_\nu := A_{\gamma_\nu}$, $B_\nu := B_{\gamma_\nu}$ of Q describing γ_ν on the coordinate axes according to (81.7).

By definition of $\mathcal{M}$, the images $P_\nu(s)$ are contained in the compact ball $\{x \in Q \mid \bmod x \leq \bmod s\}$, see (81.4). If we put $r_\nu = 1 \cdot \|A_\nu(1)\|^{-1} \in Q$, then $\|A_\nu(r_\nu)\| = 1$. By considering suitable subsequences, we may assume that the sequences $A_\nu(r_\nu)$ and $P_\nu(s)$ converge. Put

$$c := \lim A_\nu(r_\nu) \neq 0 \quad \text{and} \quad z := \lim P_\nu(s).$$

As the real multiples r_ν of 1 are contained in the kernel of Q, the following identity is obtained from (81.7($*$)).

(3)
$$\begin{aligned} P_\nu(s + s_\nu) \cdot A_\nu(r_\nu) &= B_\nu\big((s + s_\nu) \cdot r_\nu\big) = B_\nu(sr_\nu + s_\nu r_\nu) \\ &= P_\nu(s) \cdot A_\nu(r_\nu) + P_\nu(s_\nu) \cdot A_\nu(r_\nu). \end{aligned}$$

Now, we have $\bmod P_\nu(s_\nu) = \bmod s_\nu$ by definition of $\mathcal{M}$. Since according to (81.4) the balls defined by the module function form a neighbourhood base at 0 and since the sequence s_ν converges to 0, we infer that $P_\nu(s_\nu)$ converges to 0, as well. Hence, the term in the second line of (3) converges to $z \cdot c$. Using $\lim A_\nu(r_\nu) = c \neq 0$, we conclude from (3) that $P_\nu(s + s_\nu)$ converges to $(z \cdot c)/c = z = \lim P_\nu(s)$. This contradicts assumption (2), so that claim 2) is proved.

3) Moreover, by definition of $\mathcal{M}$, every orbit of $\mathcal{M}$ is contained in a ball defined by the module function and hence is relatively compact (81.4). Again by the Arzelà–Ascoli theorem (see Dugundji [66] XII.6.4, p. 267) we conclude that the homeomorphism group $\mathcal{M}$ has compact closure $\overline{\mathcal{M}}$ in the space Q^Q of continuous self-mappings of Q. Since $\mathcal{M}$ is a group, we infer that $\overline{\mathcal{M}}$ is contained in the group $\mathcal{H}(Q)$ of homeomorphisms of Q, essentially by the same standard argument as at the end of the proof of Theorem (81.5). In order to prove that $\mathcal{M}$ is compact, it thus remains to be shown that, in fact, $\mathcal{M}$ is closed in $\mathcal{H}(Q)$. This is a technical argument concerning automorphisms of compact projective planes which is independent of the other parts of the proof and will therefore be relegated to step 8). Modulo that step, the compactness of $\mathcal{M}$ may be taken for granted. As explained at the end of step 1), it follows that M is compact, and assertion (a) is proved.

4) Now let Δ be a closed, connected subgroup of ∇ contained in Ω, and $K = \Delta \cap M$, as in assertion (b). Again, we consider $\delta \in \Delta$ as an $\mathbb{R}$-linear transformation $\delta : (x, y) \mapsto (A_\delta x, B_\delta y)$. Belonging to Ω, this transformation is $\mathbb{K}$-linear and has $\mathbb{K}$-determinant 1; hence the same is true for the real determinant, $\det A_\delta \cdot \det B_\delta = 1$. By definition of M, the kernel of the multiplicative homomorphism

$$\psi : \Delta \to (0, \infty) \subseteq \mathbb{R}^\times : \delta \mapsto |\det A_\delta|^{-1} \cdot |\det B_\delta|$$

is

$$\ker \psi = \Delta \cap M = K.$$

Consider the transformation P_δ of Q representing the action of $\delta \in \Delta$ on W. By (1), the homomorphism ψ describes how P_δ affects the values of the module function:

(4) $$\mod P_\delta s = \psi(\delta) \cdot \mod s$$

for all $s \in Q$. We now prove the following assertion.

5) *A one-parameter subgroup Υ of Δ is closed and isomorphic to $\mathbb{R}$ if, and only if, $\psi(\Upsilon) \neq \{1\}$, and then ψ induces an isomorphism of Υ onto $(0, \infty)$.*

Indeed, Υ is closed and isomorphic to $\mathbb{R}$ if, and only if, $\overline{\Upsilon}$ is not compact (93.20). By assertion (a), this is equivalent to $\Upsilon \not\subseteq M$, that is, to $\Upsilon \not\subseteq K = \ker \psi$ and hence to $\psi(\Upsilon) \neq \{1\}$. Thus, $\psi|_\Upsilon$ is a non-trivial homomorphism between two groups isomorphic to $\mathbb{R}$ and consequently is an isomorphism $\Upsilon \to (0, \infty)$.

6) Now we show that the isomorphism

(5) $$\eta : \mathbb{R} \to \Upsilon : t \mapsto (\psi|_\Upsilon)^{-1}(e^t)$$

satisfies the compression properties stated in assertion (c). For $t \to -\infty$, we have that $\psi(\eta(t)) = e^t$ converges to 0. From (4), it is immediate that the corresponding transformations $P_{\eta(t)}$ converge uniformly to the constant zero map on the compact balls defined by mod. (Here, we use again that these balls form a neighbourhood base at 0, see (81.4)). Likewise, if we consider $P_{\eta(t)}$ as a transformation of $(Q \cup \{\infty\}) \setminus \{0\}$ by putting $P_{\eta(t)}(\infty) = \infty$, we see that for $t \to \infty$, with $\psi(\eta(t)) = e^t \to \infty$, these transformations converge uniformly to the constant map $x \mapsto \infty$ $(x \neq 0)$ on compact subsets. As $P_{\eta(t)}$ describes the action of $\eta(t)$ on W, this proves assertion (c).

7) By assertion (a), Δ is compact if, and only if $\Delta \subseteq M$, that is, if $\Delta = \Delta \cap M = K = \ker \psi$. Being connected, Δ is generated by its one-parameter subgroups, see (94.7) and (93.14). Hence, if Δ is not compact, then there is a one-parameter subgroup Υ such that $\psi(\Upsilon) \neq \{1\}$, i.e., having the properties stated in step 5). As $\psi(\Upsilon) = (0, \infty) = \psi(\Delta)$, we see that Δ is the semidirect product of $\ker \psi = K$ by Υ. In particular, K is connected since Δ is. (Indeed, the semidirect product above is a product also topologically; the map $K \times \Upsilon \to \Delta : (\gamma, \upsilon) \mapsto \gamma \upsilon$ is a homeomorphism with inverse map $\Delta \to K \times \Upsilon : \delta \mapsto (\delta \cdot \upsilon(\delta)^{-1}, \upsilon(\delta))$, where $\upsilon(\delta) = (\psi|_\Upsilon)^{-1}(\psi(\delta))$.) For the proof of assertion (b), all that remains to be shown is that Υ may be chosen so as to centralize K.

In the Lie algebra $\mathfrak{l}\Delta$, the Lie algebra $\mathfrak{l}K$ is a subalgebra which is invariant under the adjoint representation of K on $\mathfrak{l}\Delta$, see e.g. Hochschild [65] IX.2, p. 100. From the above, we know that $\dim \mathfrak{l}K = \dim \mathfrak{l}\Delta - 1$. Hence, as a vector space, $\mathfrak{l}\Delta$ is the direct sum of $\mathfrak{l}K$ and a 1-dimensional subspace R, and R can be chosen to be invariant under the adjoint representation of K, as well, by complete reducibility of compact groups (95.3). The compact, connected group K then acts trivially on R. Consequently, the one-parameter subgroup $\Upsilon = \{\exp r \mid r \in R\} \leq \Delta$ corresponding to R, see (94.7), is centralized by K (this follows from the second part of (94.9), applied to inner automorphisms of Δ). Moreover, Υ is not contained in $K = \ker \psi$, so that this one-parameter subgroup has all the properties discussed above, beginning with step 5).

8) It remains to prove that $\mathcal{M}$ is closed in the homeomorphism group $\mathcal{H}(Q)$; this was postponed in step 3). Obviously, $\mathcal{M}$ is closed in $\{P_\gamma \mid \gamma \in \nabla\}$. The latter group describes the action of $\nabla = \Sigma_{o,u,v}$ on $W = uv$, hence the proof is complete if we show that $\nabla|_W$ is closed in the group $\mathcal{H}(W)$ of homeomorphisms of W. Note that the following proof of this fact does not need a translation plane; the argument is valid for any compact projective plane admitting sufficiently many homologies with axis W and center o.

Let σ be a homeomorphism of W such that there is a sequence of collineations $\sigma_\nu \in \nabla$ whose restrictions $\sigma_\nu|_W$ converge to σ. In order to show that $\sigma \in \nabla|_W$, we have to construct a collineation $\gamma \in \nabla$ inducing σ on W. Choose a point $a \in ou \setminus \{o, u\}$. There are homologies ϱ_ν with center o and axis W such that all the points $a^{\sigma_\nu \varrho_\nu}$ belong to some fixed compact subset of the punctured affine plane $P \setminus (W \cup \{o\})$. Hence, putting $\sigma'_\nu = \sigma_\nu \varrho_\nu$, we may assume that the sequence $a^{\sigma'_\nu}$ converges to a point $a' \in ou \setminus \{o, u\}$. Note that the sequence $\sigma'_\nu|_W$ converges to $\sigma = \lim \sigma_\nu|_W$, since $\varrho_\nu|_W = \mathrm{id}$.

For any homeomorphism φ of W fixing u and v, and for any $c \in ou \setminus \{o, u\}$, the map

$$\gamma(\varphi, c) : P \setminus ou \to P \setminus ou : p \mapsto \left(o \vee (op \wedge W)^\varphi\right) \wedge \left(c \vee (ap \wedge W)^\varphi\right)$$

is a homeomorphism extending $\varphi|_{W \setminus \{u\}}$. A direct verification using the properties of collineations shows that $\gamma(\sigma'_\nu|_W, a^{\sigma'_\nu})$ is just the restriction of the collineation σ'_ν to $P \setminus ou$. Hence, these restrictions converge pointwise to $\gamma_1 := \gamma(\sigma, a')$. Likewise, the restrictions of the collineations σ'_ν to $P \setminus ov$ converge pointwise to a homeomorphism γ_2. The two homeomorphisms γ_1 and γ_2 fit together to yield a homeomorphism γ of $P \setminus \{o\}$. By one-point compactification, γ extends to a homeomorphism of P, which is the (pointwise) limit of the collineations σ'_ν and hence is a collineation itself. Now we note that $\gamma \in \nabla$, and that $\gamma|_W = \sigma$ by construction, hence we obtain that $\sigma \in \nabla|_W$. □

First characterization results

81.9 Theorem. *An 8-dimensional compact translation plane whose automorphism group Σ has dimension at least 22 is isomorphic to the classical quaternion plane $\mathcal{P}_2 \mathbb{H}$.*

81.10 Theorem. *A 16-dimensional compact translation plane whose automorphism group Σ has dimension at least 49 is isomorphic to the classical octonion plane $\mathcal{P}_2\mathbb{O}$.*

Remarks. 1) The automorphism group of $\mathcal{P}_2\mathbb{H}$ is $\mathrm{PGL}_3\mathbb{H}$, see (13.8), and has dimension 35. The automorphism group of $\mathcal{P}_2\mathbb{O}$ has dimension 78, see (17.10).

2) The preceding theorems are far from being best possible. In fact, their conclusions are valid under the weaker assumptions $\dim \Sigma \geq 19$ for 8-dimensional planes and $\dim \Sigma \geq 41$ for 16-dimensional planes. There are non-classical 8-dimensional translation planes with $\dim \Sigma \in \{17, 18\}$ and 16-dimensional translation planes with $\dim \Sigma \in \{38, 39, 40\}$, and all such translation planes can be determined explicitly. For detailed results, with outlines of proof and precise references, see Section 82, in particular (82.25, 27, and 28).

3) Analogous characterization results can be obtained for 8- and 16-dimensional compact projective planes in general, cf. Section 87 and (84.28), (85.16), by employing a reduction to the case of translation planes.

Theorems (81.9 and 10) will be proved in (81.13 and 21). This needs some preparations. The following situation will be frequently encountered.

81.11 Proposition. *Assume that $\Sigma_{o,u,v}$ contains a compact, connected subgroup Φ whose orbits on $W \setminus \{u, v\}$ all have dimension $l - 1$, and an involutory axial collineation ι whose axis is not W. If u is not a fixed point of Σ_v, then the shear group $\Sigma_{[v,ov]}$ is transitive on $W \setminus \{v\}$, so that v is a shear point and the plane has Lenz type at least V.*

Proof. If u were a fixed point of the connected component $\Delta = (\Sigma_{o,v})^1$, then it would be the only fixed point on $W \setminus \{v\}$ and hence would be a fixed point of $\Sigma_{o,v}$ and of Σ_v, as well. Thus, u is not a fixed point of Δ. In view of the $(l-1)$-dimensional orbits of Φ, then, the orbit u^Δ has dimension l, by the general principle (96.25).

By (55.28), the involution ι is a homology. We now generate elations with axis ov and center v from this homology applying the method described in (61.19). The points u and v are the center and the point at infinity of the axis of ι, or vice versa, cf. (23.8). First assume that v is the center; then the axis is ou. By the above, the orbit of the line ou under Δ, which is contained in the line pencil $\mathcal{L}_o$, has dimension l. Applying (61.19b) in the dual plane, we infer that the elation group $\Delta_{[v,v]}$ has dimension l. Its elements all have axis ov, since o is a fixed point of Δ. By (81.0e), the plane has Lenz type (at least) V, with shear point v.

The case that u is the center of ι is treated analogously; in this case, (61.19) need not be dualized. □

Next, we discuss a general scheme for estimating the dimension of Σ; this will be useful not only for the proofs of (81.9 and 10), but also in Section 82.

81.12 Dimension estimates. 1) The dimension of a Lie group Δ equals the dimension of its connected component Δ^1, see (93.6); this shall be frequently used.

Assume that the translation plane under consideration is not classical. We recall the following facts, see (81.0c). The translation line W is invariant under Σ, and $\Sigma = \Sigma_o \ltimes \mathsf{T}$, where $\mathsf{T} \cong \mathbb{R}^{2l}$ is the translation group. In particular, $\Sigma|_W = \Sigma_o|_W$, and $\dim \Sigma = \dim \Sigma_o + 2l$. Moreover, $\dim \Sigma_o = \dim \Omega + \dim \mathbb{K}$, where $\Omega \subseteq \Sigma_o$ is the reduced stabilizer defined in (81.0c), and $\mathbb{K}$ is the kernel of a coordinatizing quasifield.

In the sequel, we write $\Theta = \Omega^1$. For points $u \neq v$ on the translation line W, let $\Delta = (\Omega_{u,v})^1 = (\Theta_{u,v})^1$. From (81.8) we recall that Δ has a largest compact subgroup $\mathsf{K} = \Delta \cap \mathsf{M}$, which is connected and satisfies $\dim \Delta \leq \dim \mathsf{K} + 1$; moreover, equality holds if, and only if Δ is not compact. Combining the preceding estimates with the dimension formula (96.10) applied to the action of Θ on W, we obtain

(1) $$\dim \Sigma \leq 2l + \dim \mathbb{K} + \dim v^\Theta + \dim u^{\Theta_v} + \dim \mathsf{K} + 1,$$

and this inequality is strict if $\Delta = (\Theta_{u,v})^1$ is compact.

2) We now may use this inequality, together with a suitable choice of the points u and v, for a first concrete estimate of $\dim \Sigma$, which is rather rough and will be refined later in various situations. From (63.1), we know that Σ is not transitive on W, since the plane is not classical. As W is connected, there is an orbit in W which is not open in W and therefore has dimension $< l$, see (53.1a). Choose v from this orbit. If this orbit is not a fixed point, then choose a second point u in it; otherwise, choose $u \in W \setminus \{v\}$ arbitrarily. Then, $\dim v^\Theta + \dim u^{\Theta_v} \leq 2(l-1)$, and we infer from (1) that

(2) $$\dim \Sigma \leq 4l - 1 + \dim \mathbb{K} + \dim \mathsf{K}.$$

81.13 Proof of Theorem (81.9). 1) We consider an 8-dimensional translation plane ($l = 4$) which is not isomorphic to the classical quaternion plane. Then $\mathbb{K} \in \{\mathbb{R}, \mathbb{C}\}$. We want to show that $\dim \Sigma \leq 21$. We choose points $u, v \in W$ as above (81.12), and consider the largest compact subgroup K of $(\Theta_{u,v})^1$. Then

$$\dim \Sigma \leq 17 + \dim \mathsf{K}.$$

2) The action of the compact, connected group K on the 4-sphere W is not transitive, and it is almost effective because of $\mathsf{K} \subseteq \Omega$ (81.0c). By Richardson's results (96.34) on such actions, either we have $\dim \mathsf{K} \leq 3$, in which case $\dim \Sigma \leq 20$, or K acts on $W \approx \mathbb{R}^4 \cup \{\infty\}$ as $\mathrm{SO}_4\mathbb{R}$ or as $\mathrm{U}_2\mathbb{C}$ in their natural actions.

We now deal with the latter possibilities, i.e., $\mathsf{K}|_W \cong \mathrm{SO}_4\mathbb{R}$ or $\mathsf{K}|_W \cong \mathrm{U}_2\mathbb{C}$. Then $\dim \mathsf{K} \leq 6$, and $\mathsf{K}|_W$ contains a subgroup $\mathrm{Spin}_3\mathbb{R} \cong \mathrm{SU}_2\mathbb{C}$, see (11.26), which acts freely on $W \setminus \{u, v\} \approx \mathbb{R}^4 \setminus \{0\}$. The connected component Ψ of the preimage of this subgroup under the restriction homomorphism $\mathsf{K} \to \mathsf{K}|_W$ is mapped isomorphically onto $\mathrm{Spin}_3\mathbb{R}$, see (81.2). The central involution ι of Ψ is not a Baer collineation (23.14), because on W it has just the fixed points u and v. Hence, it is an axial collineation (23.17); the axis must be one of the fixed lines ov or ou, see (23.8).

3) If $u^{\Sigma_v} = \{u\}$, then the dimension estimate (81.12(1)) gives $\dim \Sigma \leq 8+2+4+0+6+1 = 21$. If $u^{\Sigma_v} \neq \{u\}$, then (81.11) applies, since the orbits of $\Psi \cong \mathrm{SU}_2\mathbb{C}$ on $W \setminus \{u, v\}$ are 3-spheres, so that the plane has Lenz type V, with shear point v, which is then a fixed point of Σ, see (81.0e). The dimension estimate (1) of (81.12) now says

(1) $\qquad \dim \Sigma \leq 8 + 2 + 0 + 4 + \dim \mathsf{K} + 1 ,$

so that again $\dim \Sigma \leq 21$. $\qquad \square$

The preceding results suffice as a basis for Section 84, where characterizations of $\mathcal{P}_2\mathbb{H}$ among 8-dimensional compact planes in general are proved. Readers who are mainly interested in these characterizations may now turn to Section 83, where further prerequisites for Section 84 are discussed.

81.14 Eight-dimensional planes of Lenz type V. For this special case, the preceding considerations will be pursued a little further.

1) Let $\Psi \cong \mathrm{Spin}_3\mathbb{R}$ be a subgroup of $\Sigma_{o,u,v}$ as above, acting effectively on $W = uv$. The central involution $\iota \in \Psi$ has axis ou or ov; hence, on one of these two lines, Ψ does not act effectively. We show that Ψ cannot act trivially there, either, if the plane has Lenz type V and is not isomorphic to the classical quaternion plane.

Assume to the contrary that Ψ acts trivially on ov. In other words, the collineations in Ψ have axis ov; the center must then be the fixed point u, so that $\Psi \leq \Sigma_{[u,ov]}$. Now the latter group of homologies can be described in coordinates using the multiplicative group of the middle nucleus $N_m(Q)$ of a coordinatizing quasifield, $N_m(Q) = \{ c \in Q \mid (xc)y = x(cy)$ for all $x, y \in Q \}$, see (25.4). Since we are assuming a plane of Lenz type V, the quasifield Q is in fact a division algebra (semifield) containing $\mathbb{R}$ in its center, see (24.7) and (64.14). It is straightforward that then $N_m(Q)$ is a closed subfield of Q containing $\mathbb{R}$, and that Q is a vector space over this subfield, cf. (25.8). Since Q has dimension 4 over $\mathbb{R}$, we conclude that either $\dim N_m(Q) \leq 2$ or $N_m(Q) = Q$. In our situation, the latter is true, since by assumption the multiplicative group of $N_m(Q)$ contains a subgroup isomorphic to $\mathrm{Spin}_3\mathbb{R}$. Thus, Q is a field, so that $Q \cong \mathbb{H}$, see (42.6).

If Ψ acts trivially on ou, then one argues in exactly the same way using the description of $\Sigma_{[v,ou]}$ by the left nucleus $N_l(Q)$, see (25.4).

2) Thus, the group $\Psi \cong \mathrm{Spin}_3\mathbb{R}$ acts non-trivially but not effectively on one of the two coordinate axes (provided the plane under consideration is not the quaternion plane). This action is linear; hence the coordinate axis in question may be identified with $\mathbb{R}^4 = \mathbb{R}^3 \times \mathbb{R}$ in such a way that Ψ acts as $\mathrm{SO}_3\mathbb{R} = \mathrm{Spin}_3\mathbb{R}/\langle \iota \rangle$ in the usual way on the factor $\mathbb{R}^3$ and acts trivially on the other factor $\mathbb{R}$. (Up to equivalence, this is the only representation satisfying the given conditions, as can be obtained, for instance, from the tables in (95.10).) The space of fixed vectors of this action is $\{0\} \times \mathbb{R}$; it follows that the kernel $\mathbb{K}$ of the plane cannot be larger than $\mathbb{R}$, since the connected group Ψ is $\mathbb{K}$-linear (81.0c).

3) We now recall that in a non-classical plane of Lenz type V the shear point v is fixed under Σ, see (81.0e). We consider once more the connected component $\Theta = \Omega^1$ of the reduced stabilizer and the largest compact subgroup K of $(\Theta_{u,v})^1$. Using Richardson's theorem (96.34) in a more detailed fashion than above (81.13), we obtain that one of the following cases holds.

(a) $\dim \mathsf{K} \leq 2$.
(b) $\mathsf{K}|_W \cong \mathrm{SO}_3\mathbb{R}$.
(c) $\mathsf{K}|_W$ is isomorphic to $\mathrm{SU}_2\mathbb{C}$, $\mathrm{U}_2\mathbb{C}$ or $\mathrm{SO}_4\mathbb{R}$.

(We remark that in a plane of Lenz type V the use of Richardson's theorem could be avoided, because then the action of $\Sigma_{o,u,v}$ on $W \approx \mathbb{R}^4 \cup \{\infty\}$ is known to be linear, see step 1) of the proof of (82.2).)

In case (a), the dimension estimate (1) of (81.13) immediately gives $\dim \Sigma \leq 17$.

In case (b), we consider the natural restriction map $\mathsf{K} \to \mathsf{K}|_W \cong \mathrm{SO}_3\mathbb{R}$. It is a covering homomorphism, since $\mathsf{K} \subseteq \Theta = \Omega^1$ acts almost effectively on W, see (81.0c and 2). Thus, there are two subcases. Either the restriction map is an isomorphism, or K is the two-fold universal covering group $\mathrm{Spin}_3\mathbb{R} \cong \mathrm{SU}_2\mathbb{C}$ of $\mathrm{SO}_3\mathbb{R}$, cf. (11.25 and 26).

If $\mathsf{K} \cong \mathrm{SO}_3\mathbb{R}$, then one may conclude exactly as in step 2) that the kernel of the plane is $\mathbb{R}$. The dimension estimate (1) of (81.12) then yields $\dim \Sigma \leq 8 + 1 + 0 + 4 + 3 + 1 = 17$.

If $\mathsf{K} \cong \mathrm{Spin}_3\mathbb{R}$, $\mathsf{K}|_W \cong \mathrm{SO}_3\mathbb{R}$, then K might (and will) act effectively on the coordinate axes ou and ov, hence we cannot draw the same conclusion about the kernel of the plane, and we only obtain the estimate $\dim \Sigma \leq 8 + 2 + 0 + 4 + 3 + 1 = 18$. In fact, planes of this type with $\dim \Sigma = 18$ exist, see (82.2).

In case (c), it can be seen as in step 2) of the previous proof (81.13) that K contains a subgroup Ψ isomorphic to $\mathrm{Spin}_3\mathbb{R}$ which acts effectively on W. As a result of the analysis of this type of subgroup in steps 1) and 2), we obtain that the kernel $\mathbb{K}$ is $\mathbb{R}$. Hence, for the cases $\mathsf{K}|_W \cong \mathrm{SU}_2\mathbb{C}$ and $\mathsf{K}|_W \cong \mathrm{U}_2\mathbb{C}$, in which $\dim \mathsf{K} \leq 4$, we obtain the dimension estimate $\dim \Sigma \leq 8 + 1 + 0 + 4 + 4 + 1 = 18$. Planes of this type with $\dim \Sigma = 18$ also exist, see (82.2).

4) We finally deal with the case $K|_W \cong SO_4\mathbb{R}$ and show that in this case our plane of Lenz type V would have to be classical. The group $SO_4\mathbb{R}$ contains two distinct normal subgroups isomorphic to $Spin_3\mathbb{R}$ and centralizing each other (see the beginning of the proof of (11.32)). Accordingly, K has two distinct subgroups Ψ_1 and Ψ_2 which are isomorphic to $Spin_3\mathbb{R}$, centralize each other and act effectively on W, see (81.2). By step 1), on one of the coordinate axes ou or ov the action of Ψ_1 is not effective; this coordinate axis will be denoted by Y. We show that either Ψ_1 or Ψ_2 acts trivially on Y; then, we conclude by step 1) that the plane is the classical quaternion plane. If Ψ_1 acts non-trivially on Y, then it induces the linear group $SO_3\mathbb{R}$ on $Y = \mathbb{R}^4 = \mathbb{R}^3 \times \mathbb{R}$ according to step 2). This action is centralized by Ψ_2. On the other hand, its centralizer in $GL_4\mathbb{R}$ consists of the maps $\mathbb{R}^3 \times \mathbb{R} \to \mathbb{R}^3 \times \mathbb{R} : (z, r) \mapsto (t_1 z, t_2 r)$ for $t_1, t_2 \in \mathbb{R} \setminus \{0\}$, and does not contain any non-trivial compact, connected subgroup. Hence, the action of Ψ_2 on Y is trivial.

At this point we leave this subject until later (82.2), where all planes of Lenz type V with $\dim \Sigma = 18$ will be explicitly determined. As a last preparation, we record here that the automorphism group of such a plane always contains a subgroup of Σ_o isomorphic to $Spin_3\mathbb{R}$. This is a by-product of the dimension estimates of the preceding discussion, which also yield the following result.

81.15 Theorem. *An 8-dimensional compact plane of Lenz type at least V whose automorphism group has dimension at least* 19 *is isomorphic to the classical quaternion plane* $\mathcal{P}_2\mathbb{H}$. □

We turn now to 16-dimensional planes. The proof of Theorem (81.10) characterizing the octonion plane among 16-dimensional translation planes will be prepared by a few results concerning subgroups of Σ that are locally isomorphic to $SO_7\mathbb{R}$. A description of the simply connected, two-fold covering group $Spin_7\mathbb{R}$ of $SO_7\mathbb{R}$ as a group acting on the octonion plane can be found in (17.14).

81.16 Lemma. *If $\Sigma_{o,u,v}$ contains a closed, connected subgroup Ψ locally isomorphic to $SO_7\mathbb{R}$, then Ψ is isomorphic to $Spin_7\mathbb{R}$. On at least one of the coordinate axes ou and ov, the linear action of Ψ is effective and equivalent to the irreducible representation of $Spin_7\mathbb{R}$ on $\mathbb{R}^8$. Up to an exchange of u and v, we may assume that this is the case for the line ov. Then there are the following two possibilities for the action of Ψ on ou.*

(i) *The linear action of Ψ on ou is effective and equivalent to the action on ov.*
(ii) *The group $\Psi|_{ou}$ induced by Ψ on ou is equivalent to $SO_7\mathbb{R}$ in its usual linear representation on $\mathbb{R}^8 = \mathbb{R}^7 \times \mathbb{R}$. Then, the central involution ι of $\Psi \cong Spin_7\mathbb{R}$ is a homology with axis ou and center v. Furthermore, the action of Ψ on W is equivalent to the action on ov.*

The orbits of Ψ on $ov \setminus \{o, v\} \approx \mathbb{R}^8 \setminus \{0\}$ are 7-spheres, and the same is true in case (ii) *for the orbits of Ψ on $W \setminus \{u, v\}$.*

Proof. By dimension reasons, Ψ cannot be a group of homologies (61.4a), hence Ψ acts non-trivially on both coordinate axes. Up to equivalence, the representations mentioned in our assertion are the only non-trivial representations of connected Lie groups locally isomorphic to $\mathrm{SO}_7\mathbb{R}$ on $\mathbb{R}^8$, see (95.2 and 10).

We show that Ψ cannot induce $\mathrm{SO}_7\mathbb{R}$ on both coordinate axes. Assume to the contrary that this is the case. Then, on both axes, Ψ has fixed points other than o, hence Ψ fixes a non-degenerate quadrangle. By (81.1), the actions of Ψ on both axes are effective and equivalent. We find an involution β in Ψ which on both axes fixes the points of a 6-dimensional subspace, so that the set of its affine fixed points is a subspace of dimension 12. But β is a Baer involution (23.14 and 17), in other words, its fixed points constitute the point set of a Baer subplane, which has dimension 8 by (41.11b). This is a contradiction.

Hence, on at least one of the coordinate axes, ov say, Ψ induces $\mathrm{Spin}_7\mathbb{R}$. The restriction homomorphism $\Psi \to \Psi|_{ov} \cong \mathrm{Spin}_7\mathbb{R}$ is an isomorphism by (81.2), since $\mathrm{Spin}_7\mathbb{R}$ is simply connected. The assertion about the orbits of Ψ on ov may be obtained from (17.14). There, such a group isomorphic to $\mathrm{Spin}_7\mathbb{R}$ is found in the automorphism group of the octonion plane, namely the group $\mathrm{S}\nabla_{(1)}$; see also the beginning of the next proof, where that group is described again. (It induces the effective, irreducible representation on both coordinate axes.)

Now assume that Ψ induces the group $\mathrm{SO}_7\mathbb{R}$ on ou. Recall from (17.14) that the only non-trivial, proper normal subgroup of Ψ is $\{\mathrm{id}, \iota\}$; this normal subgroup must then be the kernel of the action of Ψ on ou. Furthermore, Ψ has fixed points on ou other than o and u; therefore the action on W is effective and equivalent to the action on ov by the general argument (81.1). □

81.17 Proposition. *Let $\mathcal{P}$ be a 16-dimensional compact translation plane such that $\Sigma_{o,u,v}$ contains a subgroup Ψ isomorphic to $\mathrm{Spin}_7\mathbb{R}$ which acts effectively on both coordinate axes ou and ov. Then $\mathcal{P}$ is isomorphic to the classical octonion plane.*

Proof. 1) The result will be obtained via a comparison to the octonion plane, guided by the group Ψ. We therefore recall the description of such a group of collineations of the octonion plane, the group $\mathrm{S}\nabla_{(1)}$ discussed in (17.14). It consists of the transformations

$$(A, A \,|\, C) : (x, y) \mapsto (Ax, Ay), \ (s) \mapsto (Cs),$$

where $A \in \mathrm{SO}_8\mathbb{R}$ and $C \in \mathrm{SO}_7\mathbb{R}$ satisfy

(∗) $$A(s \cdot x) = Cs \cdot Ax$$

for all $s, x \in \mathbb{O}$. Here, $C \in SO_7\mathbb{R}$ means that C belongs to $SO_8\mathbb{R}$ and fixes $1 \in \mathbb{O} = \mathbb{R}^8$. On each of the coordinate axes $\mathbb{O} \times \{0\}$ and $\{0\} \times \mathbb{O}$, the group $S\nabla_{(1)} \cong \mathrm{Spin}_7\mathbb{R}$ induces the effective, irreducible 8-dimensional representation (which is unique up to equivalence). On the line W at infinity, $S\nabla_{(1)}$ induces the group $SO_7\mathbb{R}$; the action there is described by the homomorphism

$$\mathrm{pr}_3 : S\nabla_{(1)} \to SO_7\mathbb{R} : (A, A\,|\,C) \mapsto C,$$

whose kernel is generated by the reflection $\iota_o : (x, y) \mapsto (-x, -y), \ (s) \mapsto (s)$.

2) Now we study — still in the classical octonion plane — the stabilizer

$$\Xi = S\nabla_{(1),(i)}$$

of the points (1) and (i) at infinity. This stabilizer fixes every point (c) on W where c belongs to the subfield $\mathbb{C} = \mathbb{R} + \mathbb{R}i$ of $\mathbb{O}$. The group induced by Ξ on $W \cong \mathbb{O} \cup \{\infty\}$ is

$$\mathrm{pr}_3(\Xi) = \Xi|_W \cong SO_6\mathbb{R}.$$

The kernel of the homomorphism $\mathrm{pr}_3 : \Xi \to SO_6\mathbb{R}$ is $\langle \iota_o \rangle$; in particular, $\Xi/\langle \iota_o \rangle \cong SO_6\mathbb{R}$, and Ξ is almost simple since $SO_6\mathbb{R}$ is. We now shall show that the group induced by Ξ on $\mathbb{O} \times \{0\}$ and on $\{0\} \times \mathbb{O}$ is $SU_4\mathbb{C}$; here, we consider $\mathbb{O}$ as a $\mathbb{C}$-vector space by scalar multiplication with the elements of the subfield $\mathbb{C} \leq \mathbb{O}$ from the *left*. Indeed, if $(A, A\,|\,C) \in \Xi$, then for $c \in \mathbb{C}$ we have $Cc = c$, and for $x \in \mathbb{O}$ we obtain from $(*)$ that $A(c \cdot x) = c \cdot Ax$, hence A is $\mathbb{C}$-linear. By orthogonality, it follows that $A \in U_4\mathbb{C}$; this is to be understood with respect to a suitable identification $\mathbb{R}^8 = \mathbb{O} = \mathbb{C}^4$, giving $O_8\mathbb{R} \cap GL_4\mathbb{C} = U_4\mathbb{C}$, compare the discussion in (11.34). The homomorphism $\delta : \Xi \to \mathbb{C}^\times : (A, A\,|\,C) \mapsto \det_\mathbb{C} A$ induces an isomorphism $\Xi/\ker\delta \to \Xi^\delta$. Since Ξ is almost simple, either $\ker\delta$ is discrete, or $\ker\delta = \Xi$. In the former case, $\Xi/\ker\delta \cong \Xi^\delta \leq \mathbb{C}^\times$ would be almost simple (94.24) and commutative at the same time, a contradiction. Thus, we have $\Xi^\delta = \{1\}$, and $\Xi \cong \Xi|_{\mathbb{O}\times\{0\}} \leq SU_4\mathbb{C}$. Finally, these two groups have the same dimension and hence coincide by (94.3f); indeed, since $\Xi/\langle \iota_o \rangle \cong SO_6\mathbb{R}$, we know that $\dim \Xi = \dim SO_6\mathbb{R} = 15 = \dim SU_4\mathbb{C}$, see (93.7). We thus have obtained that

$$\Xi|_{\mathbb{O}\times\{0\}} = SU_4\mathbb{C} = \Xi|_{\{0\}\times\mathbb{O}}.$$

(By the way, this is a pleasant method for showing that $SU_4\mathbb{C}$ is isomorphic to the universal covering group $\mathrm{Spin}_6\mathbb{R}$ of $SO_6\mathbb{R}$. In the same way, it can be proved that $SU_2\mathbb{H} \cong \mathrm{Spin}_5\mathbb{R}$ by fixing the additional point $(j) \in W$.)

The following step is easy but important.

3) *Every 8-dimensional $\mathbb{R}$-linear subspace of $\mathbb{O} \times \mathbb{O}$ which is invariant under the action of Ξ is a line of the classical affine octonion plane.*

Indeed, let L be such a subspace. Since $\Xi \cong \mathrm{SU}_4\mathbb{C}$ acts irreducibly on the coordinate axes, we may assume that L intersects the axes trivially. Then L is of the form $L = \{ (x, Sx) \mid x \in \mathbb{O} = \mathbb{R}^8 \}$, where $S \in \mathrm{GL}_8\mathbb{R}$. The invariance of L under an element $(A, A \mid C) \in \Xi$ translates into $AS = SA$. Hence S belongs to the centralizer of $\mathrm{SU}_4\mathbb{C}$ in $\mathrm{GL}_8\mathbb{R}$, that is, S is given by multiplication with a scalar $c \in \mathbb{C} \setminus \{0\}$. As this scalar multiplication is the same as multiplication in $\mathbb{O}$ from the left, L is just the octonion line of slope c.

4) Now we turn to an arbitrary 16-dimensional translation plane $\mathcal{P}$ with the property that $\Sigma_{o,u,v}$ has a subgroup Ψ isomorphic to $\mathrm{Spin}_7\mathbb{R}$ which acts effectively on the coordinate axes ou and ov. As the same is true in the octonion plane, and since $\mathrm{Spin}_7\mathbb{R}$ has only one effective representation on $\mathbb{R}^8$, see (95.10), we may identify the affine point set of this plane with the affine point set $\mathbb{O} \times \mathbb{O}$ of the classical octonion plane in such a way that Ψ coincides with the group $S\nabla_{(1)} \leq \mathrm{Aut}\,\mathcal{P}_2\mathbb{O}$ described above.

Our main task is to compare the action of Ψ on the line at infinity W of the plane $\mathcal{P}$ to the situation in the octonion plane. For $\mathcal{P}$, it is not obvious that this action is supported by a linear structure. There is a general result, however, that every action of $\mathrm{Spin}_7\mathbb{R}$ on $\mathbb{S}_8$ is equivalent to a linear one, see (96.36). In the sequel, we shall give a more elementary argument which suffices for our special situation.

The reflection $\iota_o \in S\nabla_{(1)} = \Psi$ is a homothety of the $\mathbb{R}$-vector space $\mathbb{O} \times \mathbb{O}$ and hence leaves all lines through the origin invariant, in $\mathcal{P}$ as well as in $\mathcal{P}_2\mathbb{O}$. Hence, ι_o acts trivially on W. The factor group $\Psi/\langle\iota_o\rangle$ is isomorphic to $\mathrm{SO}_7\mathbb{R}$, as it describes the action of $S\nabla_{(1)}$ on W in $\mathcal{P}_2\mathbb{O}$ according to step 1); in particular, this factor group is simple. Consequently, the effective transformation group induced by Ψ on the translation line W of $\mathcal{P}$ is $\mathrm{SO}_7\mathbb{R}$, as well.

By simplicity, $\Psi|_W \cong \mathrm{SO}_7\mathbb{R}$ acts effectively on every non-trivial orbit in W. From (96.13) we infer that these orbits have dimension at least 6 (and at most 8, of course). Existence of an orbit of dimension 8 would imply transitivity on W by (53.1c), which is impossible in view of the fixed points u and v.

Next, orbits of dimension 7 will be excluded. Assume to the contrary that there is such an orbit $w^\Psi \subseteq W$. Then, all orbits on $W \setminus \{u, v\}$ are 7-dimensional, and there is a global cross section, so that $W \setminus \{u, v\}$ is homeomorphic to $w^\Psi \times \mathbb{R}$; this is a fundamental result of Montgomery–Zippin [55] p. 252. In particular, w^Ψ is simply connected, so that the stabilizer Ψ_w is connected, by the exact homotopy sequence (96.12). According to the dimension formula (96.10), the stabilizer Ψ_w has codimension 7 in Ψ. Now, a closed, connected subgroup of $\mathrm{SO}_7\mathbb{R}$ having codimension 7 is isomorphic to G_2, see (95.12); in particular, it is simply connected (17.15c). Hence, we would obtain an exact homotopy sequence $0 = \pi_1(\mathrm{G}_2) \to \pi_1(\mathrm{SO}_7\mathbb{R}) \to \pi_1(w^\Psi) = 0$, contradicting the fact that $\mathrm{SO}_7\mathbb{R}$ is not simply connected.

Thus, all non-trivial orbits of Ψ on W have dimension 6, and the corresponding stabilizers have codimension 6, that is, dimension 15 by the dimension formula (96.10). We shall now see that all connected, closed subgroups of dimension 15 of $\Psi|_W \cong SO_7\mathbb{R}$ are conjugate; this will show in particular that every such subgroup has a fixed point in every orbit of Ψ on W.

A compact, connected subgroup $\Theta \leq SO_7\mathbb{R}$ of dimension 15 can be found in the list (95.12) of large compact subgroups of $GL_8\mathbb{R} \geq GL_7\mathbb{R}$. According to that list, Θ is a conjugate of $SO_6\mathbb{R} \leq SO_7\mathbb{R}$ in $GL_8\mathbb{R}$. In particular, in the natural action on $\mathbb{R}^7$ as a subgroup of $SO_7\mathbb{R}$, the subspace of fixed vectors of Θ has dimension 1. The stabilizer in $SO_7\mathbb{R}$ of these fixed vectors is also isomorphic to $SO_6\mathbb{R}$ and hence coincides with Θ. Now, since $SO_7\mathbb{R}$ is transitive on the set of 1-dimensional subspaces in $\mathbb{R}^7$, all these stabilizers are conjugate in $SO_7\mathbb{R}$.

We now apply this to the subgroup Ξ of Ψ described in step 2); recall that $\Xi|_W \cong \Xi/\langle\iota_o\rangle \cong SO_6\mathbb{R}$. By the preceding argument, Ξ has a fixed point in every Ψ-orbit on W; put in a different way, this means that every line through o is the image of a fixed line of Ξ under an element of $\Psi = S\nabla_{(1)}$. Now, according to step 3), a fixed line of Ξ in $\mathcal{P}$ containing o is also a line of the octonion plane, and since $S\nabla_{(1)}$ is a group of collineations of the octonion plane, we infer that every line of $\mathcal{P}$ containing o is an octonion line. As these lines cover the affine plane, we obtain that, conversely, every octonion line through o is a line of $\mathcal{P}$, hence $\mathcal{P}$ is the classical octonion plane. □

81.18 Corollary. *Let $\mathcal{P}$ be a 16-dimensional plane of Lenz type at least V, with v as shear point, and assume that $\Sigma_{o,u,v}$ contains a subgroup Ψ isomorphic to $Spin_7\mathbb{R}$. If Ψ acts effectively on the second coordinate axis ov, then $\mathcal{P}$ is isomorphic to the classical octonion plane.*

Proof. Let Ψ be a subgroup of this kind. If Ψ acts effectively on the first coordinate axis ou, as well, then the assertion follows immediately from the preceding result (81.17). If Ψ does not act effectively on ou, then $\Psi|_{ou} \cong SO_7\mathbb{R}$, see (81.16), and Ψ acts effectively on W. In the dual plane, which in case of Lenz type V is also a translation plane, the rôle of the line at infinity is played by the line pencil $\mathcal{L}_v$, and the line pencils $\mathcal{L}_o$ and $\mathcal{L}_u$ represent two lines of the dual plane which we may use as coordinate axes. The actions of Ψ on $\mathcal{L}_o$ and on $\mathcal{L}_u$ are equivalent to the actions on W and on ov. In particular, these actions are effective, and we infer from the preceding result (81.17) again that the dual plane is isomorphic to $\mathcal{P}_2\mathbb{O}$. Now, $\mathcal{P}_2\mathbb{O}$ is isomorphic to its dual via any of its polarities, e.g., the standard elliptic polarity (16.7). □

Remarks. 1) The corollary is in fact true without the assumption of effectivity of Ψ on ov. Indeed, if this assumption is not satisfied in the original plane, then it can be shown to hold in the transposed plane, see (64.10). The proof of (82.2), step 2) deals with an analogous situation for 8-dimensional planes.

2) There are non-classical 16-dimensional translation planes that are not of Lenz type V but satisfy the other conditions of the corollary. In fact, these planes will be determined explicitly, see (82.5).

81.19 Corollary. *Up to isomorphism, the classical octonion plane is the only 16-dimensional compact translation plane such that $\Sigma_{o,u,v}$ contains a subgroup locally isomorphic to $SO_8\mathbb{R}$.*

Proof. Let Δ be such a subgroup. Its orbits in W have dimension at most 7, since existence of an orbit of larger dimension would imply transitivity on W by (53.1c), and hence is inconsistent with the presence of the fixed points u and v. On the other hand, Δ cannot act trivially on W, that is, as a group of homologies, by its dimension (61.4a). Hence there is a point $w \in W \setminus \{u, v\}$ whose orbit is non-trivial. According to the dimension formula (96.10), the connected component Ψ of Δ_w has codimension at most 7 in Δ. Now, the only proper closed, connected subgroups of $SO_8\mathbb{R}$ of such codimension are locally isomorphic to $SO_7\mathbb{R}$, see (95.12). By (81.16), the group Ψ is isomorphic to $Spin_7\mathbb{R}$, and it acts effectively on ou and ov, since its action on $W \setminus \{v\} \cong \mathbb{R}^8$ cannot be equivalent to the effective, irreducible representation of $Spin_7\mathbb{R}$ because of the fixed point w. The assertion now follows by (81.17). □

The next lemma will considerably facilitate the subsequent arguments, though it could be dispensed with to a certain extent. It rests on a result by Buchanan–Hähl [77] on the kernel and the nuclei of 8-dimensional locally compact quasifields, according to which the kernel is always isomorphic to $\mathbb{R}$, and the connected components of the nuclei (25.4) are either trivial or isomorphic to $\mathbb{R}$. (This is a specific feature of that particular dimension.)

81.20 Lemma. *For a 16-dimensional translation plane, let Φ be a compact subgroup of $\Sigma_{o,u,v}$. Then the kernels $\Phi_{[ou]}$, $\Phi_{[ov]}$, and $\Phi_{[uv]}$ of the actions of Φ on the lines ou, ov, and $W = uv$ are finite.*

Proof. In terms of a coordinatizing quasifield, these kernels may be interpreted as compact subgroups of the multiplicative groups of the nuclei and of the kernel, respectively, see (25.4). By the above-mentioned result, the groups $\Phi_{[ou]}$, $\Phi_{[ov]}$, and $\Phi_{[uv]}$ have trivial connected components and hence are discrete, being Lie groups. By compactness, they are finite. □

81.21 *Proof of Theorem* (81.10). We consider a 16-dimensional translation plane which is not isomorphic to the classical octonion plane. We want to show that $\dim \Sigma \leq 48$.

From the above-mentioned result of Buchanan–Hähl [77], we know that the kernel $\mathbb{K}$ of a coordinatizing quasifield is isomorphic to $\mathbb{R}$. As in step 2) of (81.12), let $\Theta = \Omega^1$, let $u, v \in W$ be points at infinity such that $\dim v^\Theta + \dim u^{\Theta_v} \leq 14$, and let K be the largest compact (connected) subgroup of $(\Theta_{u,v})^1$. Equation (2) of (81.12) says that $\dim \Sigma \leq 32 + \dim \mathsf{K}$. If $\dim \mathsf{K} \leq 16$, our claim is proved. From now on, we assume that

$$\dim \mathsf{K} \geq 17.$$

By (81.20 and 2), K is locally isomorphic to the linear group $\mathsf{K}|_{ov} \leq \mathrm{GL}_8\mathbb{R}$ induced on the line ov. By (95.12), there are only few possibilities: $\mathsf{K}|_{ov}$ may be isomorphic to $\mathrm{SO}_7\mathbb{R}$, to $\mathrm{Spin}_7\mathbb{R}$ or to $\mathrm{SO}_8\mathbb{R}$ itself. Moreover, in a non-classical plane, $\mathrm{SO}_8\mathbb{R}$ is excluded from this list by (81.19). Hence K is isomorphic to $\mathrm{Spin}_7\mathbb{R}$ by (81.16). Only in the classical octonion plane it can happen that the action of such a subgroup is effective on both coordinate axes ou and ov, see (81.17). Thus, according to (81.16) again, and up to an exchange of u and v, we may assume that the action of K on ov is effective and equivalent to the irreducible representation of $\mathrm{Spin}_7\mathbb{R}$ on $\mathbb{R}^8$, and that K acts ineffectively on ou, inducing the group $\mathrm{SO}_7\mathbb{R}$. Since the estimate $\dim v^\Theta + \dim u^{\Theta_v} \leq 14$ possibly has been invalidated by an exchange of u and v, fresh estimates will now be made.

If u is fixed under Θ_v, then $\dim v^\Theta + \dim u^{\Theta_v} \leq 8$, and from equation (1) of (81.12) we obtain that $\dim \Sigma \leq 16 + 1 + 8 + \dim \mathsf{K} + 1 = 47$.

Now assume that u is not a fixed point of Θ_v. The central involution ι of $\mathsf{K} \cong \mathrm{Spin}_7\mathbb{R}$ is a homology with axis ou and center v, see (81.16). Furthermore, the orbits of K on $W \setminus \{u, v\} \approx \mathbb{R}^8 \setminus \{0\}$ are 7-spheres. Thus, by (81.11), the plane has Lenz type at least V; hence, by (81.18), it is classical. □

81.22 Notes on Section 81. The module function and its applications in the theory of locally compact fields were devised by Braconnier [45], see also Weiss–Zierler [58] (who say that the idea was suggested to them by the 1950 thesis of J.T. Tate), and Weil [67] Chap. I § 2. Outside the area of associative structures, the utilization of the module function for the structure theory of the automorphism groups of locally compact, connected quasifields and of compact, connected translation planes as expounded in (81.4 through 8) is due to Hähl [75a]. Connectedness is not essential for these results; Grundhöfer [86] has given versions of (81.4 and 5) which are valid more generally, namely for locally compact quasifields which have non-discrete kernels or which coordinatize compact projective planes. (In the latter case, the proofs given here may essentially be used as they stand.) Concerning the connection of the module function with places of the quasifield and homomorphisms of the corresponding translation plane, see

Grundhöfer [88]. In the case of locally compact nearfields, whose multiplication is associative, the module function essentially retains the algebraic qualities which it has for locally compact fields, yielding a valuation in the disconnected case, see Grundhöfer [88] and Wähling [88]. Taking up a method used by Hähl [75b] in a special case, Hefendehl-Hebeker [83] systematically employed the module function for the investigation of isomorphism and isotopy problems in her classification of 4-dimensional quadratic division algebras over Hilbert fields. The characterization results (81.9 through 19) are extracted from the classification of 8- and 16-dimensional compact translation planes having large automorphism groups by Hähl, in particular from Hähl [75c, 79, 86a, 88], where much stronger results are obtained.

82 Classification of translation planes

The preceding section was mainly concerned with the basic structure theory of compact, connected translation planes and with first characterizations of the classical planes among these. Here we present the entire classification of 8-dimensional compact translation planes with an automorphism group of dimension at least 17, and of 16-dimensional compact translation planes with an automorphism group of dimension at least 38. All these planes can be determined explicitly. Many of them emerged in the process of the classification program on which we report here and had not been known before. The section closes, despite its heading, with a look at certain constructions producing higher-dimensional planes which are neither translation planes nor dual translation planes and have fairly large automorphism groups all the same; often, they are hybrids of translation planes.

The results of the present section are at the heart of the classification of higher-dimensional compact projective planes in general, if this classification is pursued beyond the stage that is aimed at in this book, compare the introduction of this chapter and Sections 84, 85 and 87, in particular (84.28) and (87.5 through 7).

In later parts of this section, there are fewer and fewer proofs, and for large segments of the classification, besides precise statements of the results and references to the original papers, only an outline of the central ideas will be given. The results about translation planes which are needed for proofs in later sections, and which are weaker versions of the results presented here, have been proved completely in Section 81 for that purpose. The first classification results of the present section will also be proved throughout, in order to demonstrate again, in a few cases which are typical for higher-dimensional translation planes, how whole families of planes can be explicitly determined from assumptions about collineation groups. For this purpose, we have chosen results which are both typical and within easy reach. The planes appearing here are of different types and dimensions, and they finally play a rôle in the classification of all 8- and 16-dimensional translation planes having automorphism groups of particularly large dimension. The subsequent outline of the entire classification, see (82.15 ff and 26 ff), will be preceded by a few general and far-reaching results about orbits of automorphism groups on the translation line (82.7 through 14); these results play a decisive rôle in the classification. Parts of the proofs of these orbit theorems have been included.

Throughout this section, we continue to use the terminology and notation introduced in (81.0). As we shall often refer to papers by Hähl, we point out again that notation here deviates from his, see (81.0f).

Special classification results

First, we determine all 8-dimensional planes of Lenz type V having automorphism groups of dimension 18. The automorphism groups of such planes have already been analyzed in (81.14). The planes in question will turn out to be coordinatized by the division algebras which we now describe.

82.1 A family of 4-dimensional real division algebras. On the quaternion skew field $\mathbb{H}$, with its usual addition $+$ and multiplication $\cdot$, we introduce a new multiplication $\circ$ in the following way.

1) Consider a fixed quaternion of the special form

$$h = h_1 + h_2 i \quad \text{such that} \quad h_1, h_2 \in \mathbb{R}, \ h_1 > 0, h_2 \geq 0 \quad \text{and} \quad \|h\|^2 = h_1^2 + h_2^2 = 1.$$

Recall from (11.6) that the space $\text{Pu}\,\mathbb{H}$ of pure quaternions is a 3-dimensional subspace of the $\mathbb{R}$-vector space $\mathbb{H}$. The subspaces $\mathbb{R} = \mathbb{R} \cdot 1$ and $(\text{Pu}\,\mathbb{H}) \cdot h$ are complementary, since $h^{-1} = \overline{h} = h_1 - h_2 i$ is not pure by assumption. Hence, every $a \in \mathbb{H}$ can be decomposed uniquely in the form

(1) $\qquad a = r(a) + p(a) \cdot h \quad \text{with } r(a) \in \mathbb{R}, \ p(a) \in \text{Pu}\,\mathbb{H},$

and the maps

$$r : \mathbb{H} \to \mathbb{R} : a \mapsto r(a) \quad \text{and} \quad p : \mathbb{H} \to \text{Pu}\,\mathbb{H} : a \mapsto p(a)$$

are clearly $\mathbb{R}$-linear.

2) We now introduce a new multiplication on $\mathbb{H}$ by

$$a \circ x = r(a) \cdot x + p(a) \cdot x \cdot h,$$

and we show that it defines a real division algebra $(\mathbb{H}, +, \circ)$. (We remark that this division algebra is non-associative except for $h = 1$, in which case it is just the quaternion algebra itself.)

Indeed, the multiplication is clearly bilinear over $\mathbb{R}$. Hence, by (64.13), it suffices to show that, for given $c, d \in \mathbb{H}$ with $c \neq 0$, the equation $y \circ c = d$ always has a unique solution y. Such a solution satisfies $r(y) \cdot c + p(y) \cdot c \cdot h = d$, hence $r(y) + c^{-1}p(y)ch = c^{-1}d$. Since $c^{-1}p(y)c$ is pure, see (11.24 or 28), the latter equation describes the decomposition of $c^{-1}d$ according to (1), so that $r(y) = r(c^{-1}d)$ and $c^{-1}p(y)c = p(c^{-1}d)$, hence $p(y) = cp(c^{-1}d)c^{-1}$. This determines y uniqely; conversely, an immediate verification shows that this element y satisfies the equation $y \circ c = d$.

The idea behind the construction of these division algebras will appear more clearly in the proof of the following theorem, cf. step 4).

82.2 Theorem: 8-dimensional planes of Lenz type V with 18-dimensional collineation groups. *The automorphism group Σ of an 8-dimensional compact plane of Lenz type V has dimension at least 18 if, and only if, Σ_o contains a subgroup isomorphic to $\mathrm{Spin}_3\mathbb{R}$. A plane of this kind is isomorphic to one of the following planes.*

(o) *The classical quaternion plane.*
(i) *The planes over the division algebras introduced in (82.1), with $h \neq 1$.*
(ii) *The dual planes of the planes in (i).*
(iii) *The planes obtained from the planes in (ii) by spread transposition.*

In cases (i)–(iii), Σ has dimension 18.

Remarks. 1) For the technique of spread transposition, see (64.10) in connection with (64.4); cf. also Buchanan–Hähl [78].

2) The automorphism group $\mathrm{PGL}_3\mathbb{H}$ of the classical quaternion plane, see (13.8), has dimension 35.

3) In case (i), the automorphism group will be described explicitly in the proof, see step 5). This description carries over to the other cases by dualization and transposition. The effect of transposition on the automorphism group is indicated at the end of step 2) of the proof.

4) It can be shown that the planes of the assertion are mutually non-isomorphic.

5) More information, including explicit descriptions of the planes mentioned in (ii) and (iii) and of their automorphism groups, can be found in Hähl [75c] 5.2–5.6 p. 348 ff and Hähl [86a] p. 329 ff (in particular, Section 5).

6) Under the hypothesis $\dim \Sigma \geq 18$, the assumption of Lenz type V is redundant; in fact, every 8-dimensional compact projective plane whose automorphism group has dimension at least 18 is of this type. This is one of the final results of the classification of 8-dimensional compact planes. It will not be proved in this book; see, however, (82.25) and (84.28) for further information.

Proof of (82.2). 1) Before we go into the specific situation of the theorem, a few preliminary considerations on planes of Lenz type V in general are in order. Consider an 8-dimensional plane of Lenz type V which is not isomorphic to the classical quaternion plane, with translation line W and shear point v, see (81.0e). For points $o \notin W$, $u \in W \setminus \{v\}$ and $e \notin ou \cup ov \cup W$, let $D = (\mathbb{R}^4, +, \circ)$ be the real division algebra coordinatizing the plane with respect to the quadrangle o, u, v, e, see (24.7), (25.8) and (64.14). The multiplication of D is denoted by $\circ$ in order to distinguish it from classical quaternion multiplication $\cdot$, which will also be used later. In affine coordinates over D, the shear group $\Sigma_{[v,ov]}$ consists precisely of the transformations

(1) $\qquad (x, y) \mapsto (x, y + d \circ x) \quad \text{for } d \in D = \mathbb{H}$,

see (24.7) and (25.4). This shear group is sharply transitive on $W \setminus \{v\}$ and isomorphic to $(D, +) = \mathbb{R}^4$. Hence, the stabilizer Σ_o of the automorphism group Σ, which fixes the flag (v, W), see (81.0e), is the semidirect product of $\Sigma_{[v,ov]}$ by $\Sigma_{o,u} = \Sigma_{o,u,v} = \nabla$.

In particular, a maximal compact subgroup M of ∇ is a maximal compact subgroup of Σ_o. (Indeed, the canonical epimorphism $\Sigma_o \to \nabla \cong \Sigma/\Sigma_{[v,ov]}$ induces a monomorphism on a maximal compact subgroup N of Σ_o containing M, since $\Sigma_{[v,ov]} \cong \mathbb{R}^4$ has no compact subgroup except $\{id\}$, and N is mapped onto a compact subgroup of ∇ containing M; hence N = M by maximality of M among compact subgroups of ∇.) Since maximal compact subgroups are conjugate (93.10), it follows that every compact subgroup of Σ_o fixes a point in $W \setminus \{v\}$.

The map

$$\Sigma_{[v,ov]} \to W \setminus \{v\} : \sigma \mapsto u^\sigma$$

is a homeomorphism by sharp transitivity (96.9a). If we let $\nabla = \Sigma_{o,u,v}$ act on the normal subgroup $\Sigma_{[v,ov]} \cong \mathbb{R}^4$ by conjugation, then that map is an isomorphism of this action onto the action of ∇ on $W \setminus \{v\}$. In particular, the action is $\mathbb{R}$-linear. This special feature of planes of Lenz type V, which is not a priori shared by more general translation planes, allows to analyze the action of Σ on W using representation theory, and makes it unnecessary to use stronger results from the theory of transformation groups (like Richardson's theorem (96.34)).

2) We now add the further hypothesis $\dim \Sigma \geq 18$. In our analysis of planes of Lenz type V in (81.14), we have concluded after step 4) that $\dim \Sigma \geq 18$ is only possible if Σ_o contains a subgroup Ψ isomorphic to $\mathrm{Spin}_3 \mathbb{R}$.

Henceforth, we shall not directly use the assumption $\dim \Sigma \geq 18$ any more, but only the existence of such a subgroup Ψ. It fixes a further point $u \in W \setminus \{v\}$ at infinity according to step 1). The only non-trivial normal subgroup of Ψ is the group of order 2 generated by the central involution ι. Hence, Ψ acts effectively on at least one of the fixed lines ou, ov, and $W = uv$, and the linear action there is equivalent to the standard action of $\mathrm{SU}_2\mathbb{C}$ on $\mathbb{C}^2 \cup \{\infty\}$, see (95.10). In particular, ι has precisely two fixed points on this line, so it cannot be a Baer collineation, and hence is an axial collineation (23.17). The axis of ι is another one of the fixed lines ou, ov, and $W = uv$, and on the third one, the action of Ψ is again effective.

We now show that Ψ cannot act trivially on the axis of ι. If the action of Ψ on W were trivial, Ψ would be contained in the homology group $\Sigma_{[o,W]}$, which is isomorphic to the multiplicative group $\mathbb{K}^\times$ of the kernel $\mathbb{K} \in \{\mathbb{R}, \mathbb{C}\}$, see (81.0b) and (25.4); this is impossible. Much in the same way, it has already been shown in step 1) of (81.14) that Ψ cannot act trivially on ou or ov, either (for this argument, the assumption of Lenz type V is crucial). Hence, Ψ acts on the axis of ι as the factor group $\mathrm{SO}_3\mathbb{R} = \mathrm{Spin}_3\mathbb{R}/\langle\iota\rangle$. Again, this action is linear on the punctured line and hence equivalent to the natural action of $\mathrm{SO}_3\mathbb{R}$ on $\mathbb{R} \times \mathbb{R}^3$, use (95.3 and 10). We now distinguish the following cases.

Case (i) : The axis of ι is W.
Case (ii) : The axis of ι is ou.
Case (iii) : The axis of ι is ov.

These three cases will lead us to the planes having the corresponding number in the assertion of our theorem. Indeed, by transposition and dualization, these cases will be reduced to case (i), which shall then be treated explicitly.

First, we deal with case (iii). We shall see that spread transposition transforms the plane into a plane with a group $\Psi^* \cong \mathrm{Spin}_3\mathbb{R}$ fitting into case (ii). The affine point set of the transposed plane is the dual space $(\mathbb{R}^{2l})^*$; we denote the group of automorphisms of the transposed plane by Σ^*. According to (64.10), the stabilizer Σ_o^* is the image of the contragredient representation of Σ_o on $(\mathbb{R}^{2l})^*$. In particular, Σ^* has the same dimension as Σ. Moreover, the transposed plane has Lenz type V again; in fact, the annihilator $Y^\circ \subseteq (\mathbb{R}^{2l})^*$ of the line $Y = ov$ is the axis of a linearly transitive group of shears, so that the shear point is the point at infinity of Y°. As the vector space $P \setminus W$ is the direct sum $X \oplus Y$, where $X = ou$, it is clear that, in the contragredient representation, ι induces the identity on the polar X°. This explains the transition from case (iii) to case (ii) and back. Cf. also Buchanan–Hähl [78] Propositions 2 and 3, where the same facts are proved.

We now consider a plane corresponding to case (ii) and show that its dual falls under case (i). In the dual plane, which is also a plane of Lenz type V, the rôle of the line at infinity is played by the line pencil $\mathscr{L}_v$, and the line pencils $\mathscr{L}_o$ and $\mathscr{L}_u$ represent two lines of the dual plane which we may use as the first and second coordinate axis. The actions of Ψ on $\mathscr{L}_o$, $\mathscr{L}_u$, and $\mathscr{L}_v$ are equivalent to the actions on W, ov, and ou, respectively. Hence, the situation in the dual plane corresponds to case (i).

3) Thus, up to transposition and dualization, it suffices to settle case (i). Then, as explained in step 2), the group induced by Ψ on $W \setminus \{v\} \approx \mathbb{R}^4$ is equivalent to $\mathrm{SO}_3\mathbb{R}$ in its usual action on $\mathbb{R} \times \mathbb{R}^3$. In particular, Ψ has fixed points in $W \setminus \{u, v\}$

and hence leaves a line $E \in \mathcal{L}_o \setminus \{ou, ov\}$ invariant. Let $e \neq o$ be an affine point on E. Using the quadrangle o, u, v, e as reference quadrangle, we coordinatize the plane by a real division algebra $D = (\mathbb{R}^4, +, \circ)$ as in step 1). By (81.7), the collineations from Ψ may be written using affine coordinates over D in the form $\psi : (x, y) \mapsto (A_\psi x, B_\psi y)$ with $A_\psi, B_\psi \in \mathrm{GL}_4\mathbb{R}$, and since they fix the line $E = \{(x, x) \mid x \in D\}$, one obtains $A_\psi = B_\psi$. The representation $\psi \mapsto A_\psi$ of $\Psi \cong \mathrm{Spin}_3\mathbb{R}$ is faithful, since we are considering case (i). Hence we may identify D and $\mathbb{H}$ as $\mathbb{R}$-vector spaces in such a way that Ψ consists precisely of the transformations of $\mathbb{R}^8 = \mathbb{H} \times \mathbb{H}$ of the form

$$(2) \qquad (x, y) \mapsto (ax, ay) \quad \text{for } a \in \mathbb{S}_3 ;$$

here, $\mathbb{S}_3$ denotes the unit sphere $\mathbb{S}_3 = \{ a \in \mathbb{H} \mid \|a\|^2 = 1 \}$ of $\mathbb{H}$, and multiplication is the classical multiplication within $\mathbb{H}$. Indeed, the $\mathbb{R}$-linear transformations

$$\lambda_a^{\mathbb{H}} : \mathbb{H} \to \mathbb{H} : x \mapsto ax$$

belong to the faithful representation $a \mapsto \lambda_a^{\mathbb{H}}$ of $\mathbb{S}_3 \cong \mathrm{Spin}_3\mathbb{R}$ (11.25) on $\mathbb{R}^4 = \mathbb{H}$, and up to equivalence this is the only faithful 4-dimensional representation, see (95.10).

4) Now we recall the coordinate description of the shear group $\Sigma_{[v,ov]}$, see (1) in step 1). Since this is a normal subgroup of $\Sigma_o = \Sigma_{o,v}$, we obtain from the coordinate description (2) of Ψ that the following set $\mathcal{D}$ of $\mathbb{R}$-linear transformations of $D \cong \mathbb{H} \cong \mathbb{R}^4$ is invariant under conjugation by $\lambda_a^{\mathbb{H}}$ for $a \in \mathbb{S}_3$:

$$\mathcal{D} = \{ x \mapsto d \circ x \mid d \in D \} \subseteq \mathrm{End}\,\mathbb{R}^4 .$$

Now, as D is a 4-dimensional real division algebra, an easy verification shows that $\mathcal{D}$ is a 4-dimensional $\mathbb{R}$-linear subspace of $\mathrm{End}\,\mathbb{R}^4$. (This corresponds to (64.8b); indeed, $\mathcal{D}$ represents the spread of the plane in Grassmann coordinates.) Moreover, we have a representation of the group $\mathbb{S}_3$ on $\mathcal{D}$ by conjugation (which, by the way, corresponds to the action of Ψ on $W \setminus \{v\}$ via the free action of $\Sigma_{[v,ov]}$ there). This representation is non-trivial, for else Ψ would act trivially on the line at infinity (indeed, the transformations in $\mathcal{D}$ would commute with the transformations $\lambda_a^{\mathbb{H}}$, and all lines $\{(x, d \circ x) \mid d \in D\}$ for $d \in D$ through the origin would be fixed under the transformations (2)).

The map $x \mapsto 1 \circ x$ is the identity; hence $\mathbb{R} \cdot \mathrm{id}$ is a subspace of $\mathcal{D}$, which is fixed under the representation of $\mathbb{S}_3$. As $\mathbb{S}_3$ is compact, there is a complementary 3-dimensional subspace $\mathcal{C}$ of $\mathcal{D}$ which is also $\mathbb{S}_3$-invariant, see (95.3), and on which $\mathbb{S}_3$ acts non-trivially and hence induces the group $\mathbb{S}_3/\{1, -1\} \cong \mathrm{SO}_3\mathbb{R}$ in its ordinary action on $\mathbb{R}^3$, cf. (95.10). We now note that the fixed vectors of an involution $\varepsilon \in \mathrm{SO}_3\mathbb{R}$ form a 1-dimensional subspace of $\mathbb{R}^3$, which is contained in the eigenspace corresponding to the eigenvalue -1 of every other involution in $\mathrm{SO}_3\mathbb{R}$

commuting with ε. For the three pairwise commuting involutory transformations of $\mathscr{C}$ given by conjugation with $\lambda_i^{\mathbb{H}}$, $\lambda_j^{\mathbb{H}}$, and $\lambda_k^{\mathbb{H}}$, this means the existence of an endomorphism $\delta \in \mathscr{C} \setminus \{0\}$ such that $(\lambda_i^{\mathbb{H}})^{-1} \circ \delta \circ \lambda_i^{\mathbb{H}} = \delta$, $(\lambda_j^{\mathbb{H}})^{-1} \circ \delta \circ \lambda_j^{\mathbb{H}} = -\delta$, and $(\lambda_k^{\mathbb{H}})^{-1} \circ \delta \circ \lambda_k^{\mathbb{H}} = -\delta$. In particular, δ satisfies $\delta(i) = i \cdot \delta(1)$, $\delta(j) = -j \cdot \delta(1)$, and $\delta(k) = -k \cdot \delta(1)$. For general $x = x_1 \cdot 1 + x_2 i + x_3 j + x_4 k \in \mathbb{H} \cong D$, where $x_\nu \in \mathbb{R}$, we then obtain $\delta(x) = (x_1 \cdot 1 + x_2 i - x_3 j - x_4 k) \cdot \delta(1) = -i(x_1 \cdot 1 + x_2 i + x_3 j + x_4 k)i \cdot \delta(1)$. In short, we find that $\delta(x) = ixh$, where $h = -i \cdot \delta(1)$. For $a \in \mathbb{S}_3$ and $r \in \mathbb{R}$, the endomorphism $r(\lambda_a^{\mathbb{H}})^{-1} \circ \delta \circ \lambda_a^{\mathbb{H}} : x \mapsto ra^{-1}iaxh$ is an element of $\mathscr{C}$, as well. Here, $ra^{-1}ia$ is a pure quaternion, and every pure quaternion can be written in this way. (This is immediate from the fact that the action of $SO_3\mathbb{R}$ on $\text{Pu}\,\mathbb{H}$ can be described by conjugation in $\mathbb{H}$, see (11.24 and 25).) The endomorphisms $x \mapsto pxh$ for $p \in \text{Pu}\,\mathbb{H}$ form a 3-dimensional subspace of $\text{End}\,\mathbb{R}^4$, whence

$$\mathscr{C} = \{ x \mapsto pxh \mid p \in \text{Pu}\,\mathbb{H} \},$$

and $\mathscr{D} = \mathscr{C} + \mathbb{R} \cdot \text{id}$. Now initially, $\mathscr{D}$ was defined by the multiplication of D, so that for $r \in \mathbb{R}$ and $p \in \text{Pu}\,\mathbb{H}$ the endomorphism $x \mapsto rx + pxh$ equals left multiplication $x \mapsto d \circ x$ for a suitable $d \in D$, namely for $d = d \circ 1 = r + ph$. Thus

$$(r + ph) \circ x = rx + pxh \quad \text{for } r \in \mathbb{R},\ p \in \text{Pu}\,\mathbb{H} \text{ and } x \in \mathbb{H},$$

and the division algebra answers to the description given in (82.1), provided that h can be made to satisfy the restrictions imposed there.

Since D is a division algebra, the map $D \to D : d \mapsto d \circ h$, and equivalently, the map $\varrho_h : \mathscr{D} \to D : \delta \mapsto \delta(h)$ are injective; in particular, since $\mathscr{C}$ is complementary to the space of homotheties, the image $\varrho_h(\mathscr{C}) = \{ ph^2 \mid p \in \text{Pu}\,\mathbb{H} \}$ does not contain a real multiple of h, whence $h \notin \text{Pu}\,\mathbb{H}$. Multiplying h by a suitable scalar, which does not change the description of the space $\mathscr{C}$, we may assume that $\|h\|^2 = 1$ and that the real part of h is positive. In order to see finally that, up to isomorphism, h may be chosen within the quadrant $(0, \infty) \cdot 1 + [0, \infty) \cdot i$, we use an automorphism $\alpha \in \text{Aut}\,\mathbb{H} \cong SO_3\mathbb{R}$, see (11.25), mapping h to an element $\tilde{h}$ of this quadrant, and we modify the identification of D with $\mathbb{H}$ by α. In the description of $\mathscr{C}$ and $\circ$, then, h is replaced by $\tilde{h}$.

Thus, we have proved that in case (i) the plane may be coordinatized by a division algebra of the type described in (82.1).

5) Now, conversely, let $D = (\mathbb{H}, +, \circ)$ be any one of these division algebras, with $h \neq 1$, so that D is not associative. We convince ourselves that the plane over D has the properties discussed here. In particular, we show that $\dim \Sigma = 18$; we shall in fact determine the connected component of Σ. As D is not associative, the plane is not the classical quaternion plane, compare (24.8). The map (2) is a collineation: for $r \in \mathbb{R}$ and $p \in \text{Pu}\,\mathbb{H}$ it maps the line $\{ (x, rx + pxh) \mid x \in \mathbb{H} \}$ through the origin o of slope $r + ph$ to $\{ (ax, rax + apxh) \mid x \in \mathbb{H} \}$, which is a

line of slope $r + apa^{-1}h$, since $apxh = (apa^{-1})(ax)h$, and since $apa^{-1} \in \operatorname{Pu}\mathbb{H}$, see (11.24). The group Ψ of these collineations is isomorphic to $\operatorname{Spin}_3\mathbb{R}$. Also, we find that

(3) $$(x, y) \mapsto (x, ry) \quad \text{for } 0 < r \in \mathbb{R}$$

and

(4) $$(x, y) \mapsto (xc, yc) \quad \text{for } 0 \neq c \in \mathbb{C} = \mathbb{R} + \mathbb{R} \cdot i$$

are collineations. For (3), this is obvious since D is a real division algebra. Concerning the transformations (4), it turns out, more precisely, that they leave every line through the origin invariant, just because c and h both belong to $\mathbb{R} + \mathbb{R} \cdot i = \mathbb{C}$ and therefore commute. Hence, the collineations (4) are homologies with center o and axis W.

Furthermore, we have the 4-dimensional group of shears (1). The subgroup of Σ_o generated by all these collineations (1)–(4) is closed and connected and has dimension $4 + 3 + 1 + 2 = 10$. Together with the 8-dimensional translation group we obtain a closed, connected subgroup of Σ having dimension 18. From (81.15), we already know that $\dim \Sigma \leq 18$; we conclude that $\dim \Sigma = 18$ and that the subgroup described above is in fact the connected component of Σ, cf. (93.6 and 12). More precisely, it is now easy to show that Σ is generated by Σ^1 and the reflection $(x, y) \mapsto (x, -y)$, see the end of the proof of Theorem 4.4, p. 329 in Hähl [86a]. □

82.3 Remarks. The planes of Theorem (82.2), that is, all 8-dimensional compact planes of Lenz type V whose automorphism groups have dimension (at least) 18, can be obtained alternatively from a result of Knarr [95] Theorems 6.5 and 6.6. There, all 8-dimensional compact planes of Lenz type V whose kernel (in the sense of (81.0b)) is isomorphic to $\mathbb{C}$ are determined and described by Knarr's contraction method, see (64.11). Moreover, these planes are grouped according to the dimensions of their automorphism groups, so that those planes having an automorphism group of dimension at least 18 can be singled out. Now with the arguments of the beginning of the preceding proof on the basis of (81.14), it can be seen that an 8-dimensional plane of Lenz type V having an automorphism group of dimension 18 either has complex kernel, or is the dual or the transpose of the dual of such a plane with complex kernel. In other words, up to duality and transposition, it can be found in Knarr's list. Indeed, step 2) of the preceding proof shows for such planes that, up to transposition and duality, Σ_o contains a subgroup $\Psi \cong \operatorname{Spin}_3\mathbb{R}$ such that $\Psi|_W$ is isomorphic to $\operatorname{SO}_3\mathbb{R}$. From the discussion in (81.14) one then infers that in fact $\Psi = K$, the largest compact subgroup of $(\Theta_{u,v})^1$, and that therefore the automorphism group Σ can attain dimension 18 only if the kernel of the plane is isomorphic to $\mathbb{C}$.

One may ask if there are 16-dimensional planes which are analogous to the planes of Theorem (82.2), described as 8-dimensional planes of Lenz type V with the property that $\Sigma_{o,u,v}$ contains a subgroup isomorphic to $\mathrm{Spin}_3 \mathbb{R}$. These would be 16-dimensional planes of Lenz type V such that $\Sigma_{o,u,v}$ contains a subgroup isomorphic to $\mathrm{Spin}_7 \mathbb{R}$. Now, by (81.18) and the subsequent remark 1), non-classical 16-dimensional planes of this type do not exist.

It turns out that the 4-dimensional division algebras constructed in (82.1) are intimately related to the 4-dimensional locally compact, connected nearfields, which have been described in (64.19 and 20). This relationship is discussed in Hähl [86a] Section 3. The division algebras (82.1) are infinitesimal algebras of the Kalscheuer nearfields; conversely, the 4-dimensional subspace $\mathscr{D} \subseteq \mathrm{End}\,\mathbb{R}^4$ considered in step 4) of the preceding proof is in fact a Lie subalgebra; the corresponding subgroup of $\mathrm{GL}_4 \mathbb{R}$ is the group of left multiplications in one of the Kalscheuer nearfields, compare (64.23).

These phenomena do not carry over to 16-dimensional planes; the lack of associativity of the octonions may be considered to be the profound reason. Indeed, there are no 8-dimensional locally compact, connected nearfields (64.20). Real division algebras of dimension 8 analogous to (82.1) may be constructed all the same; but the 16-dimensional planes coordinatized by them are far less nice.

Thus, the planes discussed in (82.2) are a special trait of the 8-dimensional case. This combines with the fact that they are the only 8-dimensional compact translation planes whose automorphism groups have dimension 18, as was already mentioned, see (82.25).

However, if the assumption of Lenz type V is left aside, there are non-classical translation planes of dimension 16 such that $\Sigma_{o,u,v}$ contains a subgroup isomorphic to $\mathrm{Spin}_7 \mathbb{R}$. As an example of a classification result about 16-dimensional translation planes, these shall now be determined explicitly. They will be obtained as the planes over the quasifields which we now construct. We use the unique decomposition (11.6) of an octonion $x \in \mathbb{O}$ into its real part and its pure part,

$$x = \mathrm{Re}\,x + \mathrm{Pu}\,x, \quad \text{where } \mathrm{Re}\,x \in \mathbb{R} = \mathbb{R} \cdot 1 \text{ and } \mathrm{Pu}\,x \in \mathrm{Pu}\,\mathbb{O}.$$

82.4 Construction. *Choose any homeomorphism $\varrho \neq \mathrm{id}$ of the interval $[0, \infty)$ with $\varrho(0) = 0$ and $\varrho(1) = 1$, and define a new multiplication $\circ$ on $\mathbb{O}$, using the classical octonion multiplication $\cdot$, by $0 \circ x = 0$ and*

$$a \circ x = a \cdot \left(\mathrm{Re}\,x + \frac{\varrho(\|a\|)}{\|a\|} \mathrm{Pu}\,x \right) \quad \text{for } a \neq 0.$$

Then $\mathbb{O}^{(\varrho)} := (\mathbb{O}, +, \circ)$ is a topological quasifield.

Proof. The map $x \mapsto a \circ x$ is $\mathbb{R}$-linear. For $c, d \in \mathbb{O}$ such that $c \neq 0$, we show that the equation $y \circ c = d$ has a unique solution y. Indeed, $\|y \circ c\|^2 = \|y\|^2 (\text{Re } c)^2 + \varrho(\|y\|)^2 \|\text{Pu } c\|^2$ is a monotone and unbounded continuous function of $\|y\|^2$, so that the equation $y \circ c = d$ uniquely determines $\|y\|^2$ and then y itself. Now the assertion follows from (64.13). □

Remarks. 1) For $\varrho = \text{id}$, the multiplication $\circ$ is the classical octonion multiplication.

2) Each automorphism of $\mathbb{O}$ is an automorphism of $\mathbb{O}^{(\varrho)}$, as well, since by (11.28) it is an $\mathbb{R}$-linear orthogonal map and respects the decomposition into real and pure parts.

3) The same construction can be applied to $\mathbb{H}$ instead of $\mathbb{O}$, leading to 4-dimensional locally compact quasifields $\mathbb{H}^{(\varrho)}$. See also (82.6).

82.5 Theorem: 16-dimensional translation planes admitting $\text{Spin}_7 \mathbb{R}$.

(a) *The 16-dimensional compact translation planes with the property that $\Sigma_{o,u,v}$ contains a subgroup Ψ isomorphic to $\text{Spin}_7 \mathbb{R}$ are precisely the planes over the quasifields $\mathbb{O}^{(\varrho)}$ of (82.4).*

(b) *For $\varrho \neq \text{id}$, the automorphism group of the plane over $\mathbb{O}^{(\varrho)}$ satisfies $\dim \Sigma \in \{38, 39\}$, and $\dim \Sigma = 39$ if, and only if, ϱ is a multiplicative homomorphism.*

(c) *For two homeomorphisms ϱ and ϱ' as in (82.4), the planes over the quasifields $\mathbb{O}^{(\varrho)}$ and $\mathbb{O}^{(\varrho')}$ are isomorphic if, and only if, there exists $0 < r_0 \in \mathbb{R}$ such that $\varrho'(t) = \varrho(r_0)^{-1} \cdot \varrho(r_0 t)$ for all $t \in [0, \infty)$.*

Remark. The connected component of the automorphism group Σ of these planes will be determined entirely, see step 9) of the proof.

Proof. 1) By (81.16 and 17) we may assume, up to an exchange of u and v, that

$$\Psi|_{ou} \cong \text{SO}_7\mathbb{R} \quad \text{and} \quad \Psi|_{ov} \cong \text{Spin}_7\mathbb{R}.$$

The actions of Ψ on the affine lines ou and ov are then linear representations which are uniquely determined by this assumption.

In order to describe the given plane as a perturbation of the classical octonion plane, we convince ourselves that the automorphism group of the octonion plane also has a subgroup isomorphic to $\text{Spin}_7\mathbb{R}$ and acting in this way, namely the stabilizer $S\nabla_{(1,0)}$ of the affine point $(1, 0)$, where $S\nabla \cong \text{Spin}_8\mathbb{R}$ is the group described in (12.17) and (17.13). The required properties follow from similar properties of the stabilizer $S\nabla_{(1)}$, which has been described in detail in (17.14). Indeed, $S\nabla_{(1,0)} = \tau S\nabla_{(1)} \tau^{-1}$ where τ is the triality collineation (16.6), which normalizes $S\nabla$, see (17.12), and maps $(1, 0)$ to (1) and the line ou to the axis W of the central involution ι_o of $S\nabla_{(1)}$.

Exactly as at the beginning of the proof of (17.14), one infers directly from the description of $S\nabla$ in (12.17) that $S\nabla_{(1,0)}$ consists of the collineations which in affine coordinates are of the form

$$(A, B \mid B) : (x, y) \mapsto (Ax, By), \quad (s) \mapsto (Bs)$$

with $A \in \mathrm{SO}_7\mathbb{R}$ and $B \in \mathrm{SO}_8\mathbb{R}$ satisfying the triality condition

(∗) $$B(s \cdot x) = Bs \cdot Ax$$

for all $s, x \in \mathbb{O}$. Recall that $\mathrm{SO}_8\mathbb{R}$ acts in the usual way on $\mathbb{O} = \mathbb{R}^8$, and $\mathrm{SO}_7\mathbb{R}$ denotes the stabilizer of 1 in $\mathrm{SO}_8\mathbb{R}$. This stabilizer fixes all elements of the subfield $\mathbb{R}$ of $\mathbb{O}$ and preserves the orthogonal complement, the space $\mathrm{Pu}\,\mathbb{O} = \mathbb{R}^7$ of pure octonions. In particular, concerning the decomposition of an octonion into its real part and its pure part, we have for $A \in \mathrm{SO}_7\mathbb{R}$ and $x \in \mathbb{O}$ that

(1) $$\mathrm{Re}\,Ax = \mathrm{Re}\,x \quad \text{and} \quad \mathrm{Pu}\,Ax = A(\mathrm{Pu}\,x)\,.$$

Because of $\mathrm{Aut}\,\mathbb{O} \subseteq \mathrm{SO}_7\mathbb{R}$ (11.28), the collineations

(2) $$(x, y) \mapsto (x^\alpha, y^\alpha) \quad \text{for } \alpha \in \mathrm{Aut}\,\mathbb{O}$$

are contained in $\mathrm{S}\nabla_{(1,0)}$.

2) By uniqueness of the representations of $\mathrm{Spin}_7\mathbb{R}$ involved, we now may identify the affine point set of the yet unknown translation plane with $\mathbb{O} \times \mathbb{O}$ in such a way that the collineation group Ψ is the group $\mathrm{S}\nabla_{(1,0)}$ of the classical octonion plane described above.

We consider the affine line L_r of the unknown plane joining $(0, 0)$ to $(1, r)$ for $0 < r \in \mathbb{R} \subseteq \mathbb{O}$. This line has the form $L_r = \{(x, C_r x) \mid x \in \mathbb{O}\} \subseteq \mathbb{O} \times \mathbb{O}$, where C_r is an $\mathbb{R}$-linear bijection of $\mathbb{O}$ satisfying $C_r(1) = r$. Since the collineations (2) fix $(1, r)$ and hence leave L_r invariant, C_r centralizes $\mathrm{Aut}\,\mathbb{O}$. Now $\mathrm{Aut}\,\mathbb{O}$ is transitive on the unit sphere in $\mathrm{Pu}\,\mathbb{O}$, see (11.30). In particular, $\mathbb{R}$ and $\mathrm{Pu}\,\mathbb{O}$ are the only $\mathbb{R}$-linear subspaces which are invariant under $\mathrm{Aut}\,\mathbb{O}$, hence they are invariant under C_r, as well. Because the dimension is odd, C_r has an eigenvector of norm 1 in $\mathrm{Pu}\,\mathbb{O}$. If $\varrho(r) \in \mathbb{R} \setminus \{0\}$ is the corresponding eigenvalue, then $C_r p = \varrho(r) \cdot p$ holds in fact for all $p \in \mathrm{Pu}\,\mathbb{O}$ by transitivity of $\mathrm{Aut}\,\mathbb{O}$. For arbitrary $x \in \mathbb{O}$, we then have $C_r x = r \cdot \mathrm{Re}\,x + \varrho(r) \cdot \mathrm{Pu}\,x$, so that

$$L_r = \{(x, r \cdot \mathrm{Re}\,x + \varrho(r) \cdot \mathrm{Pu}\,x) \mid x \in \mathbb{O}\}\,.$$

In this description, we may assume that

$$\varrho(1) = 1,$$

just by modifying the identification of the given affine plane with $\mathbb{O} \times \mathbb{O}$ by an $\mathbb{R}$-linear bijection of the form $(x, y) \mapsto (\mathrm{Re}\,x + r_0 \cdot \mathrm{Pu}\,x, y)$ for a suitable $r_0 \in \mathbb{R} \setminus \{0\}$; such a modification does not affect the description of Ψ by $\mathrm{S}\nabla_{(1,0)}$.

3) For arbitrary $a \in \mathbb{O} \setminus \{0\}$, we now obtain the line L_a joining $(0,0)$ to $(1,a)$ in the unknown plane by applying a suitable collineation $\psi \in \Psi$ to the line L_r, where $r = \|a\|$. From the description of $\mathrm{Spin}_7\mathbb{R} \cong S\nabla_{(1)}$ in (17.14), we know that the orbits of the effective representation of $\mathrm{Spin}_7\mathbb{R}$ on $\mathbb{R}^8$ are the spheres centered at 0. Hence, Ψ contains a collineation $(A, B \mid B) \in S\nabla_{(1,0)} = \Psi$ such that $B(r \cdot 1) = a$. Using the triality condition $(*)$ and (1), we find that this collineation maps the point $\left(x, r \cdot \mathrm{Re}\, x + \varrho(r) \cdot \mathrm{Pu}\, x\right)$ of the line L_r to the point

$$\left(Ax, B\left(r \cdot (\mathrm{Re}\, x + \tfrac{\varrho(r)}{r} \mathrm{Pu}\, x)\right)\right) = \left(Ax, Br \cdot A\left(\mathrm{Re}\, x + \tfrac{\varrho(r)}{r} \mathrm{Pu}\, x\right)\right)$$
$$= \left(Ax, a \cdot \left(\mathrm{Re}\, Ax + \tfrac{\varrho(r)}{r} \mathrm{Pu}\, Ax\right)\right).$$

Hence, the line L_a is

$$L_a = \left\{ \left(x, a \cdot \left(\mathrm{Re}\, x + \tfrac{\varrho(\|a\|)}{\|a\|} \mathrm{Pu}\, x\right)\right) \,\Big|\, x \in \mathbb{O} \right\},$$

and all lines through the origin are determined.

4) The map which to $a \in \mathbb{O}$ associates the second coordinate of the point on L_a with fixed first coordinate $x \in \mathbb{O}\setminus\{0\}$ is a homeomorphism, since it may be obtained within the given plane by a projectivity. If x is chosen from $\mathrm{Pu}\, \mathbb{O}$, then this implies that the induced map $\mathbb{R} \to \mathbb{O} : t \mapsto \mathrm{sign}(t) \cdot \varrho(|t|) \cdot x$, $0 \mapsto 0$ is a homeomorphism of $\mathbb{R}$ onto a closed subset of $\mathbb{O}$ and hence onto $\mathbb{R}x$. By connectedness and because of $\varrho(1) = 1$, we infer that $\varrho(r) > 0$ for $r > 0$ and that the map $r \mapsto \varrho(r), 0 \mapsto 0$ is a homeomorphism of the interval $[0, \infty)$ onto itself. Thus we have proved that our plane is the plane over the quasifield $\mathbb{O}^{(\varrho)}$ as described in (82.4).

Conversely, for $A \in \mathrm{SO}_7\mathbb{R}$ and $B \in \mathrm{SO}_8\mathbb{R}$ satisfying $(*)$, a similar computation as in step 3) shows that the transformations $(A, B \mid B)$ are indeed collineations of the plane over $\mathbb{O}^{(\varrho)}$. Thus, this plane admits a group of collineations isomorphic to $\mathrm{Spin}_7\mathbb{R}$, and assertion (a) is proved.

5) We now prove assertions (b) and (c) concerning $\dim \Sigma$ and isomorphisms. We assume $\varrho \neq \mathrm{id}$. Then the multiplication of $\mathbb{O}^{(\varrho)}$ is clearly non-linear in the first argument; in particular, $\mathbb{O}^{(\varrho)}$ is not an alternative field, so that the plane over $\mathbb{O}^{(\varrho)}$ is not isomorphic to the classical octonion plane (24.3 and 8). By (81.18), it therefore cannot have Lenz type V, either.

6) Consider the action of $\Psi \cong \mathrm{Spin}_7\mathbb{R}$ on $W \setminus \{v\}$. By (81.1), this action is equivalent to the action on $ov \setminus \{v\}$ and hence to the effective representation of $\mathrm{Spin}_7\mathbb{R}$ on $\mathbb{R}^8$. Thus, as was already stated in step 3), the orbits of Ψ on W are 7-spheres except for the fixed points u and v. Such a situation occurs frequently; we shall deal with it later in full generality (82.13) and anticipate the result here. Step 4) of the proof there deals exactly with the present situation and shows that u and v are fixed points of Σ^1, since the plane does not have Lenz type V.

7) On the basis of this fact we now examine relevant parts of Σ. In 16-dimensional translation planes, the kernel $\mathbb{K}$ is always just $\mathbb{R}$ by a result of Buchanan–Hähl [77]. Let $\Omega = \mathrm{SL}_{16}\mathbb{R} \cap \Sigma_o$ be the reduced stabilizer introduced in (81.0c). The compact, connected group Ψ is contained in the connected component $\Theta = \Omega^1$. By the conclusion of step 6), Θ fixes the triangle o, u, v. Hence, by applying (81.8b) to $\Delta = \Theta$, we see that Θ is the direct product of its largest compact subgroup K by a closed subgroup Υ which is either trivial or isomorphic to $\mathbb{R}$, and K contains Ψ and is connected. Now, $\mathsf{K}|_{ou}$ is a compact, connected subgroup of $\mathrm{GL}_8\mathbb{R}$ containing $\Psi|_{ou} \cong \mathrm{SO}_7\mathbb{R}$, hence from the list (95.12) of large compact subgroups of $\mathrm{GL}_8\mathbb{R}$ we obtain that either $\mathsf{K}|_{ou} \cong \mathrm{SO}_8\mathbb{R}$ or $\mathsf{K}|_{ou} = \Psi|_{ou}$. We infer from (81.20 and 2) that K is locally isomorphic to $\mathsf{K}|_{ou}$, hence in the first case the plane would be isomorphic to the classical octonion plane (81.19), contrary to our assumptions. Thus, $\dim \mathsf{K} = \dim \Psi$ (in fact, $\mathsf{K} = \Psi$). As explained in (81.0c), we thus obtain that $\dim \Sigma = 16 + \dim \Sigma_o = 16 + \dim \mathbb{K} + \dim \Omega = 17 + \dim \Psi + \dim \Upsilon = 38 + \dim \Upsilon$.

8) In order to prove assertion (b), it remains to be shown that $\Upsilon \cong \mathbb{R}$ if, and only if, ϱ is multiplicative. The induced group $\Upsilon|_{ov}$ centralizes $\Psi|_{ov} \cong \mathrm{Spin}_7\mathbb{R}$ and hence consists of real homotheties (95.10). By composition with elements of $\Sigma_{[o,W]}$, one thus obtains that $\Upsilon \cong \mathbb{R}$ if, and only if, $\Sigma_{[u,ov]}$ contains a one-parameter subgroup P centralizing Ψ. With the arguments of step 2), we see that the elements of $\mathsf{P} \setminus \{\mathrm{id}\}$ have the form

(3) $\qquad (x, y) \mapsto (r \cdot \mathrm{Re}\, x + r' \cdot \mathrm{Pu}\, x, y)$,

where r, r' are nonzero real numbers, and even positive, since P is connected. Such a collineation maps the point $(1, r) \in L_r$ to $(r, r) \in L_1$, where L_r is the line through the origin of slope r, as described in step 2). Hence L_r is mapped to L_1, and from the explicit description of L_r it follows immediately that $r' = \varrho(r)$. A (non-trivial) one-parameter group P of such homologies arises only if the parameter r can take all positive values, and an immediate verification shows that the transformations (3) form a group precisely if ϱ is multiplicative.

9) We thus have determined the connected component of Σ; it is generated by Ψ together with the following subgroups: (i) the one-parameter subgroup $\mathsf{P} \leq \Sigma_{[u,ov]}$ discussed in step 8), if $\Sigma_{[u,ov]}$ has such a subgroup, (ii) the connected component of $\Sigma_{[o,W]}$, i.e., the group of positive real homotheties, and (iii) the translation group. In fact, it is easy to determine all collineations, and, more generally, all possible isomorphisms φ between two such planes over the quasifields $\mathbb{O}^{(\varrho)}$ and $\mathbb{O}^{(\varrho')}$. The affine point set of both planes is $\mathbb{O} \times \mathbb{O}$, and Ψ is a group of $\mathbb{R}$-linear transformations of $\mathbb{O} \times \mathbb{O}$ which in both planes is a characteristic subgroup of Σ_o, namely the largest compact, connected subgroup. Up to a translation, an isomorphism φ is a linear bijection of $\mathbb{O} \times \mathbb{O}$ normalizing the group $\Psi \cong \mathrm{Spin}_7\mathbb{R}$, whose action is uniquely determined. Now the group $\mathrm{Spin}_7\mathbb{R}$, which is of type B_3, has only inner automorphisms; see, e.g., Loos [69] part II, Chap. V Theorem 4.5b), p. 45

and Chap. VII Table 3, p. 99. Hence we may modify φ by composition with an element of Ψ so as to make it centralize Ψ. Then, φ maps the points u and v of the first plane to the corresponding points of the second plane, since in both planes v is the center of the central involution ι of Ψ, see (81.16ii), and u is the other fixed point of Ψ on W. Now the same reasoning as in step 8) shows that φ is of the form (3) with $r' = \varrho(r)$, for some $0 < r \in \mathbb{R}$. A straightforward computation yields that this is an isomorphism between the two planes if, and only if $\varrho'(r^{-1}s)\varrho(r) = \varrho(s)$ for all $s \in [0, \infty)$. By the substitution $t = r^{-1}s$, one obtains the condition given in assertion (c).

The precise structure of the entire automorphism group which results of these considerations is stated in Hähl [87a] 5.2. More details may be found in Hähl [80b], a paper dealing with analogous 8-dimensional planes, see the following remark. □

82.6 Remark. Analogously to (82.4), one may construct quasifields $\mathbb{H}^{(\varrho)}$ by perturbation of the quaternions. The planes over these quasifields can be characterized in a similar way as in (82.5), but with Ψ isomorphic to $SO_4\mathbb{R}$ instead of $Spin_7\mathbb{R}$; the automorphism group Σ of such a plane satisfies $\dim \Sigma \in \{15, 16\}$ if $\varrho \neq \text{id}$. These planes are studied in Hähl [80b]. In that paper, details on collineations and isomorphisms can be found which equally well apply to the planes over $\mathbb{O}^{(\varrho)}$. While the latter planes hold top places in the list of 16-dimensional translation planes ranked by the dimensions of their automorphism groups, see (82.28), the 8-dimensional analogues are displaced by others on the corresponding list for 8-dimensional planes, for example by the planes discussed in (82.1 and 2), see (82.25).

Orbit theorems

We now present general results on the orbit pattern of groups of automorphisms on the line at infinity W, and on interrelations of this pattern with the structure of the automorphism group. As in Section 81, we always consider a compact projective translation plane $\mathcal{P} = (P, \mathcal{L})$ of dimension $2l$, $l \in \{4, 8\}$, and we keep the notation introduced in (81.0). In particular, $\mathbb{K}$ will be the kernel of the plane, and $\Omega \subseteq \Sigma_{o,W}$ the reduced stabilizer consisting of the $\mathbb{K}$-linear collineations of determinant 1.

82.7 Theorem: Compression subgroups and sphere orbits. *Consider a closed, connected subgroup Ξ of Ω, and a point $v \in W$ whose orbit v^Ξ is at least 2-dimensional. Assume that there is a point $u \in v^\Xi \setminus \{v\}$ such that $(\Xi_{u,v})^1$ is not compact. Then the orbit v^Ξ is homeomorphic to a sphere.*

Proof. 1) We obtain this result basically as a consequence of our insight into the structure of triangle stabilizers. In particular, (81.8) asserts that a non-compact group $\Delta = (\Xi_{u,v})^1$ contains a compression subgroup, and, hence, a collineation η

such that the restrictions $\eta^\nu|_{W\setminus\{v\}}$ converge uniformly on compact subsets of $W\setminus\{v\}$ to the constant map $W \setminus \{v\} \to \{u\}$, for $\nu \in \mathbb{N}$, $\nu \to \infty$. This fact will serve to describe $v^{\equiv} \setminus \{v\}$ as the monotone union of images of a ball around u which is blown up bigger and bigger by the inverses $\eta^{-\nu}$; then, the pertaining theorem of M. Brown will be used.

2) There are a few snags in this argument, though. First of all, we do not yet know if $v^{\equiv}$ is a manifold. We therefore provide it with a manifold topology by considering the continuous bijective map

$$\alpha : \Xi/\Xi_v \to v^{\equiv} : \Xi_v \xi \mapsto v^\xi .$$

According to (94.3a) and (96.10), the coset space Ξ/Ξ_v is a manifold of dimension $d = \dim v^{\equiv} \geq 2$. We shall use the fact that α is equivariant with respect to the action of Ξ on Ξ/Ξ_v by right multiplication. The technical difficulty is that, a priori, α might not be a homeomorphism, so that the action of the powers η^ν on Ξ/Ξ_v might not have the same compression properties as on $v^{\equiv} \subseteq W$. This problem is overcome by a connectedness argument which we now shall introduce, using the following notation. Set

$$\tilde{u} = \alpha^{-1}(u), \quad \tilde{v} = \alpha^{-1}(v) = \Xi_v, \quad \text{and} \quad M = \Xi/\Xi_v \setminus \{\tilde{v}\} .$$

Note that

$$\tilde{u}\eta = \tilde{u} \quad \text{and} \quad M\eta = M ,$$

since $u^\eta = u$ and $v^\eta = v$. We prove the following.

3) *If D is a neighbourhood of $\tilde{u}$ in M homeomorphic to a d-dimensional disk with boundary ∂D, then for every compact, connected subset C of M containing $\tilde{u}$ there exists $\nu \in \mathbb{N}$ such that $C\eta^\nu \subseteq D \setminus \partial D$.*

In order to prove this, note that ∂D separates M. If assertion 3) were not true, then for every $\nu \in \mathbb{N}$ we could find an element $x_\nu \in C$ such that $x_\nu \eta^\nu \in \partial D$. Since the points $\alpha(x_\nu)$ belong to the compact subset $\alpha(C) \subseteq v^{\equiv} \setminus \{v\}$, on which the powers η^ν converge uniformly to the constant map onto u, we conclude that $u = \lim_{\nu \to \infty} \alpha(x_\nu)^{\eta^\nu} = \lim_{\nu \to \infty} \alpha(x_\nu \eta^\nu)$, and by compactness of $\alpha(\partial D)$ this limit belongs to $\alpha(\partial D)$, contradicting $\tilde{u} \notin \partial D$.

4) M is an open submanifold of the manifold Ξ/Ξ_v of dimension $d \geq 2$. In particular, M is the union of countably many compact, connected subsets containing $\tilde{u}$. By assertion 3), M may thus be written as the union of an increasing sequence of open subsets homeomorphic to $\mathbb{R}^d$. A well-known theorem of M. Brown [61] says that, consequently, M itself is homeomorphic to $\mathbb{R}^d$. Now one easily concludes that the manifold $\Xi/\Xi_v = M \cup \{\tilde{v}\}$ is compact, see below. It is therefore homeomorphic to the one-point compactification of $\mathbb{R}^d$, that is, to $\mathbb{S}_d$. By compactness, finally, α is a homeomorphism of $\Xi/\Xi_v \approx \mathbb{S}_d$ onto $v^{\equiv}$.

The compactness of the manifold $M \cup \{\tilde{v}\}$ may be derived from the single fact that $M \approx \mathbb{R}^d$ in the following way. There is a disk neighbourhood B of $\tilde{v}$ in $M \cup \{\tilde{v}\}$ such that $M \setminus B$ is connected. Let ∂B be the boundary of B; it separates M. Since $M \approx \mathbb{R}^d$, the complement $M \setminus \partial B$ has precisely one unbounded component, see Dugundji [66] Theorem XVII.1.2, p. 356; one could also use the Jordan–Brouwer separation theorem, see Spanier [66] 4.7.15. The set $B \setminus \{\tilde{v}\}$ is closed in $M \setminus \{\tilde{v}\}$ and not compact and hence is contained in the unbounded component, which therefore cannot contain $M \setminus B$ as well. Thus, $M \setminus B$ lies in a bounded component, i.e., it has compact closure in M, and $M \cup \{\tilde{v}\} = (M \setminus B) \cup B$ is compact. □

By the preceding result, sphere orbits in W occur systematically. If the automorphism group of the plane is large, there is a good chance to encounter sphere orbits of large dimension. The following results are concerned with sphere orbits of dimension $l - 1$. First, we deal with a case which presents particularly strong homogeneity.

82.8 Theorem: Doubly homogeneous sphere orbits of codimension one. *Let $\mathcal{P}$ be a translation plane of dimension $2l$, $l \in \{4, 8\}$ admitting a closed, connected subgroup Δ of $\Sigma_{o,W}$ which is doubly transitive on some orbit w^Δ homeomorphic to $\mathbb{S}_{l-1}$ in W. Then $\mathcal{P}$ is isomorphic to the classical quaternion or octonion plane.*

Remark. For translation planes with $l = 2$, the analogous statement is not true. Counterexamples (in fact the only ones) are the two planes in (73.13 and 19). Both planes admit a group $\Phi \leq \Sigma_{o,W}$ isomorphic to $\mathrm{SL}_2\mathbb{R}$, which is triply transitive on an orbit homeomorphic to $\mathbb{S}_1$ in W. We add that Φ is transitive on the two connected components of the complement of this orbit in W.

The proof will be preceded by a few preparations.

82.9 Lemma. *Let Ψ be a semi-simple analytic subgroup of $\Sigma_{o,W}$ such that $2l$ is not the sum of dimensions of non-trivial irreducible representations of Ψ. Then the affine point space $\mathbb{R}^{2l}$ is the direct sum of a Ψ-invariant line L through o and of a Ψ-invariant linear subspace V of dimension l. In particular, an irreducible Ψ-invariant linear subspace has dimension at most l.*

Proof. Because of complete reducibility (95.2), $\mathbb{R}^{2l}$ is the direct sum of irreducible Ψ-invariant subspaces. By assumption, at least one of them is 1-dimensional. It is contained in a unique line L through o which then is Ψ-invariant. By complete reducibility again, there is a Ψ-invariant linear subspace V of dimension l complementary to L. Now L and V are direct sums of irreducible subspaces; thus we obtain a decomposition of $\mathbb{R}^{2l}$ into irreducible subspaces, all of which have dimension $\leq l$. The last assertion of the lemma now follows since the dimensions of the subspaces in such a decomposition are independent of the specific decomposition (95.2). □

82.10 Corollary. *For $l = 8$, the stabilizer $\Sigma_{o,W}$ cannot contain analytic subgroups which are locally isomorphic to* $\mathrm{SU}_5(\mathbb{C}, i)$ *or to* $\mathrm{U}_3(\mathbb{H}, i)$, *for $i = 0, 1, 2$.*

Proof. This follows immediately since the dimension of a non-trivial irreducible representation of such a group is greater than 8 and is not equal to 16, see (95.10). □

82.11 *Proof of* (82.8). We shall give a complete proof for $l = 8$. In the case $l = 4$, the proof is more involved; therefore, for some parts of it, we shall only give indications and references, see in particular step 3).

1) The proof relies heavily on the classification of doubly transitive Lie groups by Tits, see (96.16 and 17), which implies that the restriction of Δ to the orbit $w^\Delta \cong \mathbb{S}_{l-1}$ is equivalent to the standard action of one of the groups $\mathrm{PO}_5'(\mathbb{R}, 1)$, $\mathrm{PSU}_3(\mathbb{C}, 1)$, $\mathrm{PU}_2(\mathbb{H}, 1)$ for $l = 4$ and $\mathrm{PO}_9'(\mathbb{R}, 1)$, $\mathrm{PSU}_5(\mathbb{C}, 1)$, $\mathrm{PU}_3(\mathbb{H}, 1)$ for $l = 8$. By (94.27), the group Δ itself contains a semi-simple analytic subgroup Ψ locally isomorphic to one of these groups.

2) For $l = 8$, groups locally isomorphic to $\mathrm{PSU}_5(\mathbb{C}, 1)$ or to $\mathrm{PU}_3(\mathbb{H}, 1)$ are ruled out by (82.10). There remains the case that Ψ is locally isomorphic to $\mathrm{PO}_9'(\mathbb{R}, 1)$. Then, there is a compact, connected subgroup Φ locally isomorphic to $\mathrm{SO}_8\mathbb{R}$. We shall now show that Φ has two fixed points on W; then it follows by (81.19) that our plane is the classical octonion plane.

Φ is quasi-simple and hence acts almost effectively on each non-trivial orbit, which therefore must be at least 7-dimensional by (96.13a). On the other hand, Φ cannot act transitively on $W \cong \mathbb{S}_8$, see (96.22), so that there is no orbit of dimension 8 on W at all (96.11). Thus, the orbits of Φ on W are either 7-dimensional or fixed points. It is now a general fact that there must be two exceptional orbits of lower dimension, that is, fixed points, see Richardson [61] Theorem 1.1. This finishes the proof for $l = 8$.

3) For $l = 4$, we merely indicate the further steps which have to be taken and give references. One shows that $\Sigma_{o,W}$ cannot contain an analytic subgroup locally isomorphic to $\mathrm{PSU}_3(\mathbb{C}, 1)$, by analyzing the linear action of a hypothetical subgroup of this kind. The classification of representations (95.10) leaves only one possibility for this action, which can be ruled out for geometrical reasons, see Hähl [79] 3.3.

Since $\mathrm{PO}_5'(\mathbb{R}, 1)$ and $\mathrm{PU}_2(\mathbb{H}, 1)$ are isomorphic, there remains the case that Ψ is locally isomorphic to $\mathrm{U}_2(\mathbb{H}, 1)$. This case is dealt with in Hähl [79] 3.4. From the classification of representations together with (82.9) one obtains that Ψ acts on the affine point space $\mathbb{R}^8$ as $\mathrm{U}_2(\mathbb{H}, 1)$ in its natural action on $\mathbb{H}^2$. Moreover, one finds that the lines through the origin can be reconstructed from the group in a unique way. Hence, the plane is isomorphic to the classical quaternion plane, in whose automorphism group such a subgroup occurs. See also Stroppel [93a], where a related situation is treated in the more general context of stable planes. □

For the next result, we need the following criterion for doubly homogeneous orbits of Lie group actions, which is also of general use.

82.12 Lemma. *Let Δ be a connected Lie group acting continuously on a regular space with countable basis, and let A be an orbit of dimension at least 2. Then Δ is doubly transitive on A if, and only if, for some $a \in A$ and every $b \in A \setminus \{a\}$ one has $\dim b^{\Delta_a} = \dim A$.*

Proof. We endow A with the coset space topology of Δ/Δ_a via the bijection $\Delta_a \delta \mapsto a^\delta$; this topology may be different from the original one. The coset topology makes A into a connected manifold (94.3a). For every closed subgroup Γ of Δ and every $b \in A$, the dimension of b^Γ is not affected by this change of topology; indeed, this dimension is $\dim \Gamma - \dim \Gamma_b$ according to the formula (96.10) in both cases. In particular, A has the same dimension in both topologies. Now our assumption implies that Δ_a is transitive on the manifold $A \setminus \{a\}$, see (96.11); indeed, $A \setminus \{a\}$ is connected since $\dim A \geq 2$. □

82.13 Theorem: Sphere orbits of codimension 1. *Let $\mathcal{P}$ be a translation plane of dimension $2l$, $l \in \{4, 8\}$, such that some subgroup Δ of Σ_W has an orbit homeomorphic to $\mathbb{S}_{l-1}$ on the line at infinity W. Then one of the following assertions holds.*

(i) *The plane is isomorphic to the classical quaternion or octonion plane.*
(ii) *The plane has Lenz type V.*
(iii) *The connected component Σ^1 has exactly two fixed points u and v on W.*

Remarks. 1) In Hähl [79] Theorem 1.2, more information is provided in case (iii): either Σ^1 is transitive on $W \setminus \{u, v\}$, or *all* orbits of Σ^1 on $W \setminus \{u, v\}$ are homeomorphic to $\mathbb{S}_{l-1}$.

2) The theorem rests on (82.8), see step 6) of the following proof; hence for $l = 4$ it is not completely proved in this book.

3) The analogous statement for $l = 2$ is not true. The counterexamples mentioned in the remark following (82.2) are relevant here, as well.

Proof of (82.13). 1) We assume that $\mathcal{P}$ is not isomorphic to the classical quaternion or octonion plane and show that asssertion (ii) or (iii) holds. The line W at infinity is Σ-invariant, and the reduced stabilizer $\Omega \le \Sigma_o$ induces the same group of transformations of W as Σ, see (81.0c). Let $Z \subseteq W$ be an orbit of Δ homeomorphic to $\mathbb{S}_{l-1}$, and consider the subgroup Z consisting of all elements $\omega \in \Omega$ such that ω and ω^{-1} map Z into itself. This is a closed subgroup, which contains Δ, so that its connected component Z^1 and a maximal compact subgroup Π of Z^1 are both transitive on Z, by (96.9b and 19). Let Φ be a maximal compact subgroup of $\Theta := \Omega^1$ containing Π, cf. (93.10a). We show that $Z \approx \mathbb{S}_{l-1}$ is an orbit of Φ, as well.

The group Φ is connected (93.10). It cannot have orbits in W of dimension l, or else it would be transitive on W, see (53.1c), and the plane would be isomorphic to $\mathcal{P}_2\mathbb{H}$ or to $\mathcal{P}_2\mathbb{O}$ by (63.1). The Π-orbit $Z \approx \mathbb{S}_{l-1}$ is therefore contained in a Φ-orbit of the same dimension $l-1$, which is a compact, connected manifold, see (96.9a) and (94.3a), hence the two orbits coincide (96.11); compare also the general principle (96.25).

2) On the basis of Mostert [57] Theorem 1, it is shown in Richardson [61] 1.1 and 1.6 that there are two exceptional Φ-orbits of dimension less than $l-1$, and that all other orbits of Φ in W are homeomorphic to $\mathbb{S}_{l-1}$. Moreover, the complement of the two exceptional orbits is homeomorphic to $\mathbb{S}_{l-1} \times \mathbb{R}$. In particular, this shows that the orbit Z is a flatly embedded $(l-1)$-sphere in $W \approx \mathbb{S}_l$. By the generalized Schönflies theorem (M. Brown [60]), Z is the common boundary of two otherwise disjoint l-balls B_1 and B_2 whose union is W. From general theorems about compact transformation groups on $\mathbb{R}^l$, one could now deduce that Φ has two fixed points, one in each of the balls B_1 and B_2; that is, the exceptional orbits are just these fixed points. However, we prefer to obtain this fact by geometric arguments, which we need anyhow.

3) Our next step is to show that Φ contains a large almost simple normal subgroup Ψ of a certain type. For this, we use the fact that the compact, connected groups acting effectively and transitively on $\mathbb{S}_{l-1}$ are all known. By (96.21), there is an almost simple closed normal subgroup $\Psi \le \Phi$ which also acts transitively on Z. The kernel $\Psi_{[Z]}$ of Ψ on Z is finite; the factor group $\Psi/\Psi_{[Z]}$ acts effectively on $\mathbb{S}_{l-1}$ and therefore is isomorphic to $\mathrm{SU}_2\mathbb{C} \cong \mathrm{U}_1\mathbb{H}$ for $l=4$ and to one of the groups $\mathrm{SO}_8\mathbb{R}$, $\mathrm{Spin}_7\mathbb{R}$, $\mathrm{SU}_4\mathbb{C}$, or $\mathrm{U}_2\mathbb{H}$ for $l=8$; the action on Z is equivalent to the standard action of the respective group on $\mathbb{S}_{l-1}$, see (96.22). The group Ψ is a covering group of $\Psi/\Psi_{[Z]}$, cf. (94.2a). If $l=8$ and if $\Psi/\Psi_{[Z]} \cong \mathrm{SO}_8\mathbb{R}$, then we can conclude as in step 2) of (82.11) that the plane is isomorphic to $\mathcal{P}_2\mathbb{O}$, contrary to our assumptions. In the remaining cases, the groups in question are simply connected and hence have no proper connected covering groups, so that $\Psi_{[Z]} = \{\mathrm{id}\}$, and Ψ itself is isomorphic to $\mathrm{SU}_2\mathbb{C} \cong \mathrm{U}_1\mathbb{H}$ for $l=4$ and to one of the groups $\mathrm{Spin}_7\mathbb{R}$, $\mathrm{SU}_4\mathbb{C}$, or $\mathrm{U}_2\mathbb{H}$ for $l=8$. In particular, the action on W is effective.

4) The center of Ψ contains a unique involution ι. It induces the antipodal involution on the orbit $Z \approx \mathbb{S}_{l-1}$; in particular, ι has no fixed points on Z. By Brouwer's fixed point theorem, ι has a fixed point in (the interior of) each of the balls B_1 and B_2 bounded by Z, see step 2). Since the interiors of these balls are the connected components of $W \setminus Z$, the set of fixed points of ι on W is not connected. Hence, ι cannot be a Baer involution, for then, the fixed points of ι on W would constitute a line of a Baer subplane of dimension l, cf. (55.5).

Thus, ι is an axial collineation (23.17). Since ι acts non-trivially on the fixed line W, the center u of ι lies on W, and the only other fixed point of ι on W is the point at infinity v of the axis of ι, cf. (23.8). As ι is the only involution belonging to the center of the normal subgroup Ψ of Φ, it follows that u and v are fixed points of Φ and in fact are the two exceptional orbits of Φ on W, see step 2).

5) We now assume that the plane does not have Lenz type V. Then, in view of the $(l-1)$-dimensional sphere orbits of Φ on $W \setminus \{u, v\}$, Proposition (81.11) says that u is a fixed point of Σ_v, and that, vice versa, v is a fixed point of Σ_u.

6) We shall show that either u or v is a fixed point of $\Theta = \Omega^1$ and hence of Σ^1; then assertion (iii) follows by step 5). By the general principle (96.25), the Θ-orbits of dimension at most $l-1$ on W coincide with the Φ-orbits. Therefore, it suffices to prove that $\min\{\dim u^\Theta, \dim v^\Theta\} \leq l - 1$. At least, there exists $q \in W$ such that $\dim q^\Theta \leq l - 1$; this follows from (63.1) and (53.1d), since the plane is not classical. If $q \in \{u, v\}$, then q is a fixed point of Φ and, hence, of Θ.

Now assume $q \in W \setminus \{u, v\}$. Then $q^\Theta = q^\Phi \approx \mathbb{S}_{l-1}$. By (82.8), the group Θ is not doubly transitive on this orbit, hence there is $p \in q^\Phi \setminus \{q\} = q^\Theta \setminus \{q\}$ such that $\dim p^{\Theta_q} \leq l - 2$, see (82.12). Let Λ be the largest compact subgroup of $(\Theta_{p,q})^1$ according to (81.8); it satisfies $\dim \Theta_{p,q} = \dim(\Theta_{p,q})^1 \leq \dim \Lambda + 1$. Applying the dimension formula (96.10) twice, we obtain

(1)
$$\begin{aligned} \dim \Theta &= \dim q^\Theta + \dim \Theta_q \\ &= \dim q^\Theta + \dim p^{\Theta_q} + \dim \Theta_{p,q} \\ &\leq l - 1 + l - 2 + \dim \Lambda + 1 \, . \end{aligned}$$

Since every compact subgroup of Θ is contained in a maximal compact subgroup and since maximal compact subgroups are conjugate to each other (93.10a), the points p and q may be modified within their orbits in such a way that Λ is contained in Φ. Then, because of $\Lambda \subseteq \Phi_q$, we have $\dim \Phi = \dim q^\Phi + \dim \Phi_q \geq l - 1 + \dim \Lambda$. In view of $\Phi \subseteq \Theta_v$, this yields

$$\begin{aligned} \dim \Theta &= \dim v^\Theta + \dim \Theta_v \\ &\geq \dim v^\Theta + l - 1 + \dim \Lambda \, . \end{aligned}$$

Comparing this to (1), we infer that $\dim v^\Theta \leq l - 1$. □

For estimating the dimension of the automorphism group Σ by the approach explained in (81.12), every information that restricts the dimension of some orbit of Σ in W for a non-classical plane is important. It is clear from (63.1) and (53.1d) that there are orbits of dimension at most $l-1$. The following result improves this bound merely by 1, yet this is essential for the classification results presented here.

82.14 Theorem: Orbits of codimension at least two exist. *If $\mathcal{P}$ is a locally compact translation plane of dimension $2l$, $l \in \{4, 8\}$, and if $\mathcal{P}$ is not isomorphic to the classical quaternion or octonion plane, then the connected component Σ^1 of $\operatorname{Aut}\mathcal{P}$ has an orbit of dimension at most $l-2$ on the line at infinity W.*

Remarks. 1) For $l = 4$, planes are known for which this bound is met. In some of the planes constructed in Hähl [82a] 3.3, the orbits of Σ^1 on $W \approx \mathbb{S}_4$ are a 2-sphere Z and its complement $W \setminus Z$, compare loc. cit. 2.5β. For $l = 8$ it is an open question whether the theorem could be improved.

2) For $l = 2$, an analogous result does not hold. Counterexamples are again the planes which were mentioned in the remark following (82.8).

First steps of the proof of (82.14). The proof is hard; only the very first steps will be made here, in order to indicate the further route. The entire proof may be found in Hähl [78].

1) Assume that there is a non-classical translation plane $\mathcal{P}$ with $l \in \{4, 8\}$ such that the orbits of Σ^1 in W all have dimension l or $l-1$. The proof consists in deriving a contradiction. Here, we shall deduce from this assumption that the connected component $\Theta = \Omega^1$ of the reduced stabilizer $\Omega \subseteq \Sigma_o$, see (81.0c), is semi-simple. The structure theorems on large compact subgroups of triangle stabilizers (81.8), on compression subgroups and sphere orbits (82.7), and on sphere orbits of codimension one (82.13) play a crucial rôle for the argument.

2) By (81.0c), the groups Σ^1, $(\Sigma_o)^1$, and $\Theta = \Omega^1$ all have the same orbits on W. There is an orbit of dimension at most $l-1$, since the plane is not classical.

3) None of the orbits of Θ in $W \approx \mathbb{S}_l$ can be homeomorphic to a sphere. Indeed, by domain invariance (51.18), an orbit homeomorphic to $\mathbb{S}_l$ would be open in W as well as closed, which would mean transitivity on W. An orbit homeomorphic to $\mathbb{S}_{l-1}$ would imply the existence of fixed points of Θ on W by (82.13) and (81.0e), contrary to our assumptions.

4) In particular, for two points $u \neq v \in W$ belonging to the same Θ-orbit, we infer from (82.7) that $(\Theta_{u,v})^1$ is compact.

5) Now assume that Θ is not semi-simple. Then it contains a non-trivial commutative closed, connected normal subgroup Ξ. A one-parameter subgroup P of Ξ

has a fixed point $v \in W$ by the Lefschetz fixed point theorem, see Spanier [66] Theorem 4.7.12, p. 197, since W is a sphere of even dimension and hence has non-vanishing Euler characteristic. Every point of the orbit v^Ξ is now a fixed point of P by commutativity. Hence, if Ξ has no fixed point on W, then by step 4) every closed one-parameter subgroup is compact, and so then is Ξ by (94.38). On the other hand, if $v \in W$ is a fixed point of the normal subgroup Ξ, then the orbit v^Θ, which is non-trivial by our assumptions, consists of fixed points of Ξ, and it follows from step 4) again that Ξ is compact.

Thus, in all cases, Ξ turns out to be compact, and hence is a torus contained in the center of Θ, see (94.38) and (93.19). Considering again a fixed point $v \in W$ of a one-dimensional subtorus $P \leq \Xi$, we now infer that the orbit v^Θ consists of fixed points of P. But by our assumptions, this orbit has at least dimension $l - 1$, contradicting the fact that the fixed point set of a compact, connected group acting on a manifold has codimension at least 2 by Montgomery [57].

6) This contradiction shows that Θ is semi-simple. Now, Θ cannot be compact, or else Θ would have exceptional orbits of dimension at most $l - 2$, see Richardson [61] Theorem 1.1.

The further course of the proof cannot be included here for reasons of space. It is carried out in Hähl [78], pp. 280–293. There, it is shown next that the assumptions about the orbits on W imply irreducibility of the linear action of Θ on the affine point space $\mathbb{R}^{2l}$. By representation theory of semi-simple Lie groups, see Section 95, the conceivable groups and actions are explicitly known and can be inspected case by case. Dimension estimates in the manner of (81.12) help to reduce the number of cases involved. In all cases, a contradiction has to be produced; very often, this is achieved by the structure theorem on triangle stabilizers (81.8), in particular by the consequence noted in step 4) above. □

The classification

In the sequel, we shall indicate how all 8- and 16-dimensional compact translation planes with large automorphism groups can be determined explicitly.

82.15 Eight-dimensional translation planes with large groups.
We resume the arguments of (81.14), which will be generalized (from planes of Lenz type V to translation planes in general) and refined using the preceding orbit theorems. Let $\mathcal{P}$ be an 8-dimensional compact translation plane whose automorphism group Σ has dimension at least 17. We assume that $\mathcal{P}$ is not isomorphic to the classical quaternion plane.

1) In particular, the kernel $\mathbb{K}$ is isomorphic to $\mathbb{R}$ or $\mathbb{C}$, see (81.0b). Let $\Theta = \Omega^1$ be the connected component of the reduced stabilizer $\Omega \subseteq \Sigma_o$ defined in (81.0c).

Let $v \in W$ be a point at infinity whose orbit v^Θ has dimension at most 2; such an orbit exists by (82.14). We consider three cases, according to the dimension of this orbit. It will turn out that v is in fact a fixed point, see the theorem stated below. We shall frequently use the dimension estimates (81.12) and the classification of effective actions of compact, connected Lie groups on $W \approx \mathbb{S}_4$ by Richardson, see (96.34).

2) First assume that

$$\dim v^\Theta = 2.$$

This possibility will be excluded. We distinguish three subcases, of which the following is the hardest.

2.1) If the orbit v^Θ is homeomorphic to the 2-sphere $\mathbb{S}_2$, and if Θ acts doubly transitively on it, then this action is equivalent to the standard action of $\mathrm{PSL}_2\mathbb{C}$ on the complex projective line. This can be obtained from the classification of doubly transitive actions of Lie groups by Tits, see (96.16 and 17), or from the classification of all transitive Lie group actions on surfaces by Mostow [50], see also Betten–Forst [77]. Thus, Σ_o contains a closed, connected subgroup Λ locally isomorphic to $\mathrm{SL}_2\mathbb{C}$, see (94.27 and 28e). This situation is examined in Hähl [82a]; the action of Λ and the possibilities for Σ_o can be determined completely. As one of the first results, we note the following (loc.cit., 2.5 p. 82).

Proposition. *If Σ_o contains a closed, connected subgroup Λ locally isomorphic to $\mathrm{SL}_2\mathbb{C}$, then either $\Theta = \Lambda$, or Θ is an almost direct product of Λ and a 1-dimensional torus. In the latter case, the kernel of the plane equals $\mathbb{R}$. Hence, in both cases, $15 \leq \dim \Sigma \leq 16$.*

As a consequence, this situation is irrelevant for the classification of planes with $\dim \Sigma \geq 17$. Nevertheless, all the planes satisfying 2.1) are explicitly determined in the last part of the paper mentioned above.

The alternative to 2.1) will be split into the following two subcases.

2.2) If Θ is not doubly transitive on v^Θ, then by (82.12) there is a point $u \in v^\Theta \setminus \{v\}$ such that $\dim u^{\Theta_v} \leq 1$. Let K be the largest compact subgroup of $(\Theta_{u,v})^1$ according to (81.8). The dimension estimate (1) of (81.12) then gives $\dim \Sigma \leq 8 + \dim \mathbb{K} + 2 + 1 + \dim \mathsf{K} + 1$, i.e.,

$$\dim \Sigma \leq 12 + \dim \mathbb{K} + \dim \mathsf{K}.$$

Since $\dim \mathbb{K} \leq 2$, the assumption $\dim \Sigma \geq 17$ implies $\dim \mathsf{K} \geq 3$. By Richardson's theorem (96.34), the set of fixed points of K on W is at most 1-dimensional, and all other orbits of K on W have dimension at least 2. In particular, the 2-dimensional orbit v^Θ contains some 2-dimensional K-orbit. Now this is impossible because K

would have to be transitive on v^Θ by dimension reasons (96.25), but K has the fixed point $u \in v^\Theta$.

2.3) If v^Θ is not homeomorphic to the 2-sphere, then for $u \in v^\Theta \setminus \{v\}$ the stabilizer $(\Theta_{u,v})^1$ is compact (82.7). Hence, the inequality of the dimension estimate (1) of (81.12) is strict, $\dim \Sigma < 8 + \dim \mathbb{K} + 2 + 2 + \dim \mathsf{K} + 1$, so that we arrive at the same estimate and the same conclusion as in 2.2).

3) The case
$$\dim v^\Theta = 1$$
can be excluded much in the same way. For $u \in v^\Theta \setminus \{v\}$ and the largest compact subgroup K of $(\Theta_{u,v})^1$, the dimension estimate (1) of (81.12) gives $\dim \Sigma \leq 8 + \dim \mathbb{K} + 1 + 1 + \dim \mathsf{K} + 1 = 11 + \dim \mathbb{K} + \dim \mathsf{K}$. If $\dim \Sigma \geq 17$, then $\dim \mathsf{K} \geq 4$, and by Richardson's theorem each orbit of K on $W \setminus \{u,v\}$ has dimension 3, which contradicts $\dim v^\Theta = 1$.

Hence, the only remaining case is $\dim v^\Theta = 0$, in which v is a fixed point of Θ and hence of Σ^1. We pause to note this consequence explicitly.

Theorem. *If the automorphism group Σ of a non-classical 8-dimensional compact translation plane has dimension at least 17, then Σ^1 has a fixed point on the translation line W.*

4) In the sequel, $v \in W$ will denote a fixed point of Θ. Let K be a maximal compact subgroup of Θ. By Richardson's theorem, K has another fixed point $u \neq v$ on W. The dimension estimate (1) of (81.12) now says
$$\dim \Sigma \leq 8 + \dim \mathbb{K} + 0 + \dim u^\Theta + \dim \mathsf{K} + 1$$
$$\leq 13 + \dim \mathbb{K} + \dim \mathsf{K}.$$

Since $\mathbb{K} \in \{\mathbb{R}, \mathbb{C}\}$, our assumption $\dim \Sigma \geq 17$ implies $\dim \mathsf{K} \geq 2$. Recall that the kernel of the action of $\mathsf{K} \leq \Omega$ on W is finite, cf. (81.0c). By Richardson's theorem, we may distinguish the following cases.

Case 1): $\dim \mathsf{K} = 2$
Case 2): $\mathsf{K}|_W \cong \mathrm{SO}_3\mathbb{R}$
Case 3): $\mathsf{K}|_W$ is isomorphic to $\mathrm{SU}_2\mathbb{C}$, $\mathrm{U}_2\mathbb{C}$, or to $\mathrm{SO}_4\mathbb{R}$.

In cases 2) and 3), the action of $\mathsf{K}|_W$ on W is equivalent to the one-point compactification of the obvious linear action of the corresponding group on $\mathbb{R} \times \mathbb{R}^3 = \mathbb{C}^2 = \mathbb{R}^4$. All these cases actually occur in 8-dimensional translation planes with $\dim \Sigma \geq 17$.

4.1) If $\dim \mathsf{K} = 2$, then K is a 2-dimensional torus (94.39). By the dimension estimate, $\dim \Sigma \leq 17$, and this bound can be attained only if $\mathbb{K} = \mathbb{C}$ and $\dim u^\Theta = 4$.

As the elements of Ω are $\mathbb{K}$-linear by definition, the homology group $\Sigma_{[o,W]} \cong \mathbb{C}^\times$ centralizes $\Theta = \Omega^1$; it contains a 1-dimensional torus, which together with K generates a 3-dimensional torus Φ. The linear actions of Φ on the lines ou and ov have kernels $\Phi_{[ou]}$ and $\Phi_{[ov]}$ of dimension at least 1, since torus subgroups of $\mathrm{GL}_4\mathbb{R}$ have dimension at most 2. In particular, $\Phi_{[ov]}$ contains an involution ι, cf. (94.31), whose center is the fixed point u. Generating elations from the homologies conjugate to ι according to (61.19b), we obtain that $\dim \Theta_{[ov,ov]} = \dim u^\Theta = 4$. Thus, the plane has Lenz type V (see (81.0e), and note that the elements of $\Theta_{[ov,ov]}$ have center v, since W is fixed).

Let D be a 4-dimensional real division algebra coordinatizing the plane with respect to the quadrangle o, u, v, e where e is an arbitrary point not incident with ou, ov, or W, cf. (24.7) and (64.14). According to (25.4), the homology groups $\Phi_{[ou]}$ and $\Phi_{[ov]}$ are described by subgroups of the multiplicative groups of the nuclei $N_l(D)$ and $N_m(D)$. In the case of a division algebra, these nuclei are closed (not necessarily commutative) subfields containing $\mathbb{R}$, see (25.8). The presence of the compact homology groups $\Phi_{[ou]}$ and $\Phi_{[ov]}$ of positive dimension implies that the nuclei do not coincide with $\mathbb{R}$. Now, Rees [50] has determined all 4-dimensional real division algebras in which at least two of the three subalgebras $N_l(D)$, $N_m(D)$, and $\mathbb{K}$ are strictly larger than $\mathbb{R}$. They are all isotopic to the division algebras described below (82.16); since isotopic division algebras coordinatize isomorphic planes, see Hughes–Piper [73] Theorem 8.11, p. 177, the plane is isomorphic to the plane over one of these division algebras. For more information about the planes obtained in this way, see (82.17) and Hähl [75c] p. 358 ff.

4.2) The case $\mathsf{K}|_W \cong \mathrm{SO}_3\mathbb{R}$ is the most complicated and can only be sketched roughly here. In this case, we have $\dim \Sigma \leq 18$. As K is a covering group of $\mathsf{K}|_W$, see (81.2), K is isomorphic either to $\mathrm{Spin}_3\mathbb{R}$ or to $\mathrm{SO}_3\mathbb{R}$. By a thorough analysis of the structure of $\Theta = \Omega^1$ as a Lie group, of its maximal compact subgroups, its Levi complements and its radical, one finds within the latter a normal 3-dimensional vector subgroup N of Θ which is not centralized by Θ and acts freely on $W \setminus \{v\}$. The next step is to prove that N consists of shears with axis ov. In the case $\mathsf{K} \cong \mathrm{SO}_3\mathbb{R}$, the only known way to obtain this result is to try to construct a plane representing a counterexample; this construction has to go a long way until a contradiction arises. The paper Hähl [83] is devoted to these results.

Based on the description of the shear group N in terms of the distributor of a coordinatizing quasifield (25.4), it is finally possible to construct all such planes explicitly and to determine their automorphism groups. This is the subject of Hähl [84] for $\mathsf{K} \cong \mathrm{SO}_3\mathbb{R}$ and of Hähl [86a] for $\mathsf{K} \cong \mathrm{Spin}_3\mathbb{R}$. For the latter case, the principle according to which these planes are constructed is illustrated here in (82.1 and 2), where the (relatively simple) special case of planes of Lenz type V is studied. The overall result of the two papers mentioned above is stated in (82.18 through 23).

4.3) If $\mathsf{K}|_W$ is isomorphic to $\mathrm{SU}_2\mathbb{C}$, $\mathrm{U}_2\mathbb{C}$, or to $\mathrm{SO}_4\mathbb{R}$, then, by Richardson's results, K contains a subgroup $\Psi \cong \mathrm{Spin}_3\mathbb{R} \cong \mathrm{SU}_2\mathbb{C}$ acting freely on $W \setminus \{u, v\}$. In particular, the central involution ι of Ψ cannot be a Baer collineation (23.14). Hence, ι is an axial collineation (23.17); its axis is different from W.

If u is not a fixed point of Θ, then by (81.11) the plane has Lenz type V. This situation is covered by cases (ii) and (iii) of (82.2), compare step 2) of the proof there.

We now assume that u is fixed by Θ. Then our initial dimension estimate shows that $\dim \Sigma \leq 17$ since $\dim \mathsf{K} \leq 6$, and that the bound $\dim \Sigma = 17$ is only attained if $\dim \mathsf{K} = 6$ and $\mathbb{K} = \mathbb{C}$. Thus, $\mathsf{K}|_W \cong \mathrm{SO}_4\mathbb{R}$, and K is isomorphic to $\mathrm{SO}_4\mathbb{R}$ or to the two-sheeted universal covering group $\mathrm{Spin}_3\mathbb{R} \times \mathrm{Spin}_3\mathbb{R}$ of $\mathrm{SO}_4\mathbb{R}$, compare the description of $\mathrm{SO}_4\mathbb{R}$ in terms of quaternion multiplication (11.23). Now all planes admitting such a group of collineations have been explicitly determined in Hähl [80a] for $\mathsf{K} \cong \mathrm{Spin}_3\mathbb{R} \times \mathrm{Spin}_3\mathbb{R}$ and in Hähl [80b] for $\mathsf{K} \cong \mathrm{SO}_4\mathbb{R}$. The planes with $\mathsf{K} \cong \mathrm{Spin}_3\mathbb{R} \times \mathrm{Spin}_3\mathbb{R}$ are generalizations of the planes over the Kalscheuer nearfields (64.20). Our dimension estimates show that $\dim \Sigma = 17$ is only obtained if $(\Theta_{u,v})^1$ contains a compression subgroup in the sense of (81.8); in Hähl [80a] it is proved that this condition characterizes the Kalscheuer nearfield planes among those more general planes. If $\mathsf{K} \cong \mathrm{SO}_4\mathbb{R}$, it is not difficult to see that the action of K on the lines ou and ov is not $\mathbb{C}$-linear, so that the kernel $\mathbb{K}$ can only be $\mathbb{R}$. Hence, in this case, it is not possible that $\dim \Sigma \geq 17$. The planes in question are coordinatized by the 4-dimensional perturbations $\mathbb{H}^{(\varrho)}$ of the quaternion field presented in (82.6 and 4). $\square$

Several types of planes which have been found in the classification reported above will now be presented one by one. For each type, we first describe the quasifields coordinatizing the corresponding planes. Then we state a characterization result for the planes over these quasifields. Finally, we single out those planes whose automorphism groups have dimension at least 17.

82.16 The Rees algebras. Consider the following multiplication on $\mathbb{C}^2$:

$$(a, b) \circ (c, d) = (ac + \varepsilon \bar{d}b, da + b\bar{c}),$$

where $\varepsilon = e^{i\vartheta} \in \mathbb{C}$, $0 < \vartheta \leq \pi$ is a constant parameter, and $(a, c) \mapsto ac$ denotes ordinary multiplication in $\mathbb{C}$. It is easy to see that the multiplication $\circ$ defines a 4-dimensional real division algebra. For $\varepsilon = -1$, one obtains the quaternion field; the above multiplication is in fact a modification of the Cayley–Dickson process (11.1). If $\varepsilon \neq -1$, then direct verification shows that the subfield $\mathbb{C} \times \{0\}$ is at the same time the kernel $\mathbb{K}$, the middle nucleus N_m and the left nucleus N_l as defined in (25.3 and 4). By Rees [50], every finite-dimensional real division algebra in which at least two of the three subalgebras $\mathbb{K}, N_m$, and N_l are strictly larger than $\mathbb{R}$ is either isomorphic to $\mathbb{C}$ or isotopic to one of the 4-dimensional division algebras described here. Grundhöfer–Salzmann [90] XI.13.9, p. 352 further generalize the above construction.

The planes over the Rees algebras can now be characterized as follows.

82.17 Theorem. *Up to isomorphism, the planes over the Rees algebras (82.16) for $\varepsilon \neq -1$ are precisely the 8-dimensional compact translation planes with the following properties: the automorphism group Σ has dimension (at least) 17, and a maximal compact subgroup of Σ^1 is a 3-dimensional torus.*

This situation corresponds to step 4.1) of the classification report (82.15), according to which a plane of this kind may be coordinatized by one of the Rees algebras.

Conversely, consider the plane over a Rees algebra $D = (\mathbb{C}^2, +, \circ)$; it has Lenz type V by (24.7). According to (25.3 and 4), the subfield $\mathbb{C} \times \{0\} = \mathbb{K} = N_m = N_l$ gives rise to certain groups of homologies. In affine coordinates over $D = \mathbb{C}^2$, the group generated by these homologies consists of the collineations

$$(1) \quad \left(\begin{pmatrix} x_1 \\ x_2 \end{pmatrix}, \begin{pmatrix} y_1 \\ y_2 \end{pmatrix} \right) \mapsto \left(\begin{pmatrix} ax_1 c \\ ax_2 c \end{pmatrix}, \begin{pmatrix} by_1 c \\ by_2 c \end{pmatrix} \right) \quad \text{for } a, b, c \in \mathbb{C}^\times, |c| = 1.$$

This group has dimension 5 and fixes the coordinate axes. Together with the 4-dimensional group of shears

$$(x, y) \mapsto (x, y + a \circ x), \quad a \in D,$$

see (24.7) and (25.4), and with the translations, these collineations generate a closed, connected, 17-dimensional subgroup of the automorphism group Σ. We show next that $\dim \Sigma = 17$, so that in fact the group generated by these collineations is the connected component Σ^1, cf. (93.12). Indeed, the planes of Lenz type V whose automorphism groups have dimension at least 18 are known by (82.2). For these planes, Σ^1 has been determined in step 5) of the proof there, see also the end of step 2). A maximal compact subgroup of Σ^1 is isomorphic to the group of transformations $\mathbb{H} \to \mathbb{H} : x \mapsto axc$ for $a \in \mathbb{H}$, $c \in \mathbb{C} \subseteq \mathbb{H}$, $\|a\|^2 = 1 = \|c\|^2$, that is, to $U_2\mathbb{C}$, compare (11.26). It does not contain a 3-dimensional torus subgroup, whereas here the group of the collineations (1) does. Hence, the planes discussed here satisfy $\dim \Sigma = 17$.

More detailed information about these planes may be found in Hähl [75c] Theorem 6.1, p. 358 ff. There, the automorphism group is determined entirely. Also, it is shown that the planes over two Rees algebras with different parameters ε are non-isomorphic. This also follows from the fact proved by Rees [50] 3.2 that the corresponding Rees algebras are non-isotopic, cf. Hughes–Piper [73] 8.11, p. 177.

82.18 A family of 4-dimensional locally compact quasifields.
We now present a construction of quasifields which generalizes the construction of 4-dimensional real division algebras described in (82.1). Choose $h \in \mathbb{C} \subseteq \mathbb{H}$,

$h = h_1 + h_2 i$ such that $h_1, h_2 \in \mathbb{R}$, $h_1 > 0, h_2 \geq 0$ and $\|h\|^2 = h_1^2 + h_2^2 = 1$,

and let $R \subseteq \mathbb{H}$ be a closed subset which is a cross section for the cosets of the $\mathbb{R}$-linear subspace $(\operatorname{Pu} \mathbb{H}) \cdot h$ of $\mathbb{H}$, so that every $a \in \mathbb{H}$ can be uniquely written as $a = r + ph$ with $r \in R$ and $p \in \operatorname{Pu} \mathbb{H}$; moreover, assume that $0, 1 \in R$. Define a new multiplication $\circ$ on $\mathbb{H}$ by

$$(r + ph) \circ x = xr + pxh \quad \text{for } r \in R, \ p \in \operatorname{Pu} \mathbb{H}.$$

In (82.1), the special case $R = \mathbb{R}$ was discussed. Much in the same way as there, one readily proves that $(\mathbb{H}, +, \circ)$ is a topological quasifield, cf. also Hähl [86a] 2.2, p. 304.

The following theorem characterizes the planes over these quasifields. For planes of Lenz type V, the theorem was proved in (82.2). The proof in general runs along the same lines; it may be found in Hähl [86a] 2.3, p. 305 ff.

82.19 Theorem. *Up to isomorphism, the translation planes over the quasifields constructed in (82.18) are precisely the 8-dimensional compact translation planes having the following properties: for a suitable $v \in W$, the shear group $\Sigma_{[v,ov]}$ has dimension at least 3, and $\Sigma_{o,v}$ contains a subgroup $\mathsf{K} \cong \operatorname{Spin}_3 \mathbb{R}$ inducing the group $\mathsf{K}|_W \cong \operatorname{SO}_3 \mathbb{R}$ on the line at infinity.*

It is easy to see that the plane over a quasifield $(\mathbb{H}, +, \circ)$ as in (82.18) indeed admits groups of collineations of the required kind. Written in affine coordinates over $(\mathbb{H}, +, \circ)$, the transformations

$$(x, y) \mapsto (x, y + (ph) \circ x) = (x, y + pxh) \quad \text{for } p \in \operatorname{Pu} \mathbb{H}$$

constitute a 3-dimensional group of shears with axis $\{0\} \times \mathbb{H}$. Moreover, let K be the group consisting of the transformations

(1) $\qquad (x, y) \mapsto (ax, ay) \quad \text{for } a \in \mathbb{H}, \|a\|^2 = 1 \,;$

this group is isomorphic to $\operatorname{Spin}_3 \mathbb{R}$, cf. (11.25). One proves by direct computation exactly as in step 5) of the proof of (82.2) that each map (1) is a collineation which, for $r \in R$ and $p \in \operatorname{Pu} \mathbb{H}$, maps the line of slope $r + ph$ through the origin o to the line of slope $r + apa^{-1}h$. It follows that the group $\mathsf{K}|_W$ induced on W is isomorphic to $\operatorname{SO}_3 \mathbb{R}$, cf. (11.29).

The collineations above together with the real homotheties and the translations generate a 15-dimensional closed subgroup of Σ. In Hähl [86a] 4.2, p. 316 ff and 4.4, p. 321 ff it is determined precisely for which choices of R and h in the construction of the underlying quasifield the dimension of Σ is 15, 16, 17, or 18. We only report the result for the cases with $\dim \Sigma \geq 17$.

82.20 Theorem. *Let $\mathcal{P}$ be a translation plane as in Theorem (82.19), in other words, the plane over one of the quasifields (82.18), constructed from an element $h \in \mathbb{H}$ and a cross section R as specified there. Then, the following assertions hold.*

(i) *$\mathcal{P}$ is isomorphic to the classical quaternion plane if, and only if $R = \mathbb{R}$ and $h = 1$. In all other cases,*
$$15 \leq \dim \Sigma \leq 18 \, .$$
(ii) *The bound $\dim \Sigma = 18$ is attained if, and only if, the plane has Lenz type V. This is equivalent to $R = \mathbb{R}$ and $h \neq 1$, and the corresponding planes are precisely those of (82.2i).*
(iii) *One has $\dim \Sigma = 17$ if, and only if, the plane can be coordinatized by a quasifield constructed from a cross section R of the form*
$$R = \{ r \cdot 1 \mid r \geq 0 \} \cup \{ rg \mid r \geq 0 \}$$
where $g \in \mathbb{C} \setminus \{-1\}$ is such that $|g| = 1$, $\operatorname{Re} gh^{-1} < 0$.

In case (ii), the underlying quasifield is one of the division algebras of (82.1). It can be obtained by allowing the extremal value $g = -1$ in case (iii). The corresponding planes have been studied in (82.2 ff).

In case (iii), the connected component Σ^1 is generated by the collineations described in (82.19) together with the following collineations: a group of homologies with axis ov,

(1) $\qquad (x, y) \mapsto (rx, y) \quad$ for $0 < r \in \mathbb{R}$,

and the group of homologies with axis W,

(2) $\qquad (x, y) \mapsto (xc, yc) \quad$ for $c \in \mathbb{C}^\times$.

Note that in this case the cross section R consists of two linear pieces; this circumstance is responsible for the maps (1) being collineations. Moreover, it is easily verified that, under the conditions given in (iii), the kernel of the coordinatizing quasifield is the subfield $\mathbb{C} \subseteq \mathbb{H}$. This accounts for the homologies (2).

Up to isomorphism, the parameters g and h in (iii) may be chosen as $g = e^{i\gamma}$, $h = e^{i\delta}$ with $0 \leq \delta < \frac{\pi}{2}$ and $\pi < \gamma \leq \pi + 2\delta$. This is shown in Hähl [86a] 4.4, p. 321 ff. It is also proved that then the resulting planes are pairwise non-isomorphic. Moreover, the full automorphism group Σ is determined. $\square$

We now describe another family of quasifields obtained by modifying the classical multiplication. If $\alpha > 0$, and if ϱ is a homeomorphism of $\mathbb{R}$ fixing 0 and 1, then $(r + si) \circ i = -\alpha s + \varrho(r)i$ defines a quasifield multiplication on the complex numbers. This multiplication can be extended to $\mathbb{H}$ and to $\mathbb{O}$ in such a way that the full automorphism groups of $\mathbb{H}$ and $\mathbb{O}$ are retained.

82.21 Construction: Generalized mutations. With α and ϱ as above, a new multiplication $\circ$ is constructed on $\mathbb{F} \in \{\mathbb{H}, \mathbb{O}\}$ by means of the classical multiplication $\cdot$ in the following way:

$$a \circ x = a \cdot \operatorname{Re} x + \bigl(\varrho(\operatorname{Re} a) + \alpha \cdot \operatorname{Pu} a\bigr) \cdot \operatorname{Pu} x \,.$$

The resulting structure $\mathbb{F}_{\alpha,\varrho} = (\mathbb{F}, +, \circ)$ is a topological quasifield, see below.

Special case: Mutations. With $\varrho = \operatorname{id}$ and arbitrary $\alpha > 0$, one clearly obtains a division algebra. This division algebra is isomorphic to the division algebra obtained from $\mathbb{F}$ by the multiplication

$$a \circ_\tau x = \tau a x + (1 - \tau) x a \,,$$

where $\tau \in \mathbb{R} \setminus \{\tfrac{1}{2}\}$ is such that $\alpha = (2\tau - 1)^2$. Indeed, the map $x \mapsto \operatorname{Re} x + (2\tau - 1) \operatorname{Pu} x$ is an isomorphism of $\mathbb{F}_{\alpha,\operatorname{id}}$ onto this latter division algebra, which is sometimes called a mutation of $\mathbb{F}$. For verifications of this sort, it is useful to write the factors a and x of a product in the form $a = \operatorname{Re} a + a' + a''$ and $x = \operatorname{Re} x + \operatorname{Pu} x$ where $a', a'' \in \operatorname{Pu} \mathbb{F}$ and a' is a real multiple of $\operatorname{Pu} x$, whereas a'' and $\operatorname{Pu} x$ are orthogonal. Recall that then $a' \cdot \operatorname{Pu} x \in \mathbb{R}$ and $a'' \cdot \operatorname{Pu} x = - \operatorname{Pu} x \cdot a'' \in \operatorname{Pu} \mathbb{F}$, see (11.6), and note that the above multiplications are distributive with respect to such decompositions.

These quasifields and division algebras appear in the literature in still other shapes. We need to compare our description of these structures with the one obtained in Hähl [84, 90]. The map $x \mapsto \operatorname{Re} x + \alpha \cdot \operatorname{Pu} x$ is an isomorphism of $\mathbb{F}_{\alpha,\varrho}$ onto the quasifield $D^{\mathbb{F}}_{\alpha^{-1},\varrho}$ used in those papers. In the special case of the division algebras $\mathbb{F}_{\alpha,\operatorname{id}}$, this isomorphism turns the multiplication $\circ$ into a multiplication $*$ with the following properties: for $p, q \in \operatorname{Pu} \mathbb{F}$, one has $p * q = pq$ if p and q are orthogonal, and $p * p = \alpha^{-1} p^2$. This determines a multiplication table of $*$ on the standard basis of $\mathbb{F}$, compare (11.16), and is a familiar description of these division algebras.

Obviously, $\mathbb{F}_{1,\operatorname{id}} = \mathbb{F}$. Every automorphism of the classical algebra $\mathbb{F}$ is clearly an automorphism of $\mathbb{F}_{\alpha,\varrho}$, as well, since by (11.28) it fixes the elements of the subfield $\mathbb{R}$ and respects the decomposition $\mathbb{F} = \mathbb{R} \oplus \operatorname{Pu} \mathbb{F}$. Regarding the special case of the division algebras $\mathbb{F}_{\alpha,\operatorname{id}}$, it can be shown that these are precisely the real division algebras of the given dimension admitting an automorphism group isomorphic to $\operatorname{Aut} \mathbb{F}$. Because of this property and for other reasons, these division algebras have received attention from the algebraic point of view, and have been characterized in various ways; see Bannow [40] p. 244 and p. 248, Becker [72], Lex [73] Satz 29,

p. 57, Hähl [75b], [76] 4.4, p. 215, Benkart–Osborn [81b] Theorem 2.2, p. 271, Grundhöfer [90b] 4.5, p. 594 ff.

In order to prove that $\mathbb{F}_{\alpha,\varrho}$ is a topological quasifield, it suffices by (64.13) to show that the equation $y \circ c = d$ has a unique solution y for $c \neq 0$. Using decompositions as above, this equation translates into three equations

$$\operatorname{Re} y \cdot \operatorname{Re} c + y' \cdot \alpha \cdot \operatorname{Pu} c = \operatorname{Re} d$$
$$\varrho(\operatorname{Re} y) \cdot \operatorname{Pu} c + y' \cdot \operatorname{Re} c = d'$$
$$y''(\operatorname{Re} c + \alpha \cdot \operatorname{Pu} c) = d'',$$

where y', y'', d', d'' are pure. The third equation immediately yields y''. Multiplying the first equation by $\operatorname{Re} c$ and the second one by $\alpha \cdot \operatorname{Pu} c$ and subtracting, one obtains an equation with left-hand side $(\operatorname{Re} c)^2 \cdot \operatorname{Re} y - \alpha \cdot (\operatorname{Pu} c)^2 \cdot \varrho(\operatorname{Re} y)$ and real right-hand side. As $0 \geq (\operatorname{Pu} c)^2$ and $\alpha > 0$, the left-hand side is a monotone and unbounded function of $\operatorname{Re} y$, so that this equation determines $\operatorname{Re} y$ uniquely. Then, y' is obtained from one of the other equations.

The following theorem, characterizing the planes over the quaternion versions of these quasifields, is proved in Hähl [84] 3.3, p. 35 ff.

82.22 Theorem. *Up to isomorphism, the translation planes over the generalized mutations $\mathbb{H}_{\alpha,\varrho}$ are precisely the 8-dimensional compact translation planes having the following properties: For a suitable $v \in W$, the shear group $\Sigma_{[v,ov]}$ has dimension at least 3, and $\Sigma_{o,v}$ contains a subgroup $\mathsf{K} \cong \mathrm{SO}_3\mathbb{R}$.*

In the plane over $\mathbb{H}_{\alpha,\varrho}$, such groups of collineations are readily found. One verifies immediately that $\operatorname{Pu}\mathbb{H}$ is contained in the distributor (25.4) of $\mathbb{H}_{\alpha,\varrho}$; accordingly, $\Sigma_{[v,ov]}$ contains the maps

$$(x, y) \mapsto (x, y + p \circ x) \quad \text{for } p \in \operatorname{Pu}\mathbb{H},$$

which form a subgroup isomorphic to $\mathbb{R}^3$. Since $\operatorname{Aut}\mathbb{H}$ consists of automorphisms of $\mathbb{H}_{\alpha,\varrho}$, the transformations

$$(x, y) \mapsto (x^\sigma, y^\sigma) \quad \text{for } \sigma \in \operatorname{Aut}\mathbb{H}$$

constitute a group K of collineations isomorphic to $\operatorname{Aut}\mathbb{H} \cong \mathrm{SO}_3\mathbb{R}$, see (23.5) and (11.29).

These collineations together with the real homotheties and the translations generate a 15-dimensional subgroup of Σ. In Hähl [84] 3.3, p. 35 ff it is determined precisely for which parameters $\alpha > 0$ and homeomorphisms ϱ the dimension of Σ is 15, 16, or ≥ 17. We present the result for the last case.

82.23 Proposition. Let $\mathcal{P}$ be a translation plane as in (82.22), in other words the plane over one of the generalized mutations $\mathbb{H}_{\alpha,\varrho}$ of (82.21). Then $\mathcal{P}$ is isomorphic to the classical quaternion plane if, and only if, $\alpha = 1$ and $\varrho = \mathrm{id}$. In all other cases, one has $15 \leq \dim \Sigma \leq 17$.

We have $\dim \Sigma = 17$ precisely for the planes of Lenz type V, that is, for the planes over the mutations $\mathbb{F}_{\alpha,\mathrm{id}}$ with $\alpha \neq 1$.

In the latter case, Σ^1 is generated by the following five subgroups: the group K described above, the one-parameter group of homologies

$$(x, y) \mapsto (rx, y) \quad \text{for } 0 < r \in \mathbb{R},$$

the group of positive real homotheties, the group of shears with axis ov (which in this case is 4-dimensional because of Lenz type V), and the group of translations. For distinct parameters α, the resulting planes are non-isomorphic.

82.24 The automorphism group of a nearfield plane.
Here, we consider the translation plane over one of the Kalscheuer nearfields N_r, $r > 0$, see (64.20 and 22). We shall verify that the automorphism group Σ of this plane has dimension 17.

Since multiplication of N_r is associative, there are the following groups of homologies, (25.4):

$$\Sigma_{[u,ov]} = \{ (x, y) \mapsto (a \circ x, y) \mid a \in N_r \setminus \{0\} \}$$
$$\Sigma_{[v,ou]} = \{ (x, y) \mapsto (x, b \circ y) \mid b \in N_r \setminus \{0\} \} .$$

These groups centralize each other, are isomorphic to $N_r^\times \cong \mathbb{H}^\times$, and contain subgroups isomorphic to $\mathrm{Spin}_3\mathbb{R}$, so that Σ_o has a subgroup $\Phi \cong \mathrm{Spin}_3\mathbb{R} \times \mathrm{Spin}_3\mathbb{R} = \mathrm{Spin}_4\mathbb{R}$. The two homology groups are linearly transitive.

Now, the set $\{u, v\}$ of their centers is invariant under every collineation, according to a fundamental result by André [55] Satz 4, p. 143 and Satz 6, p. 148, who has shown that all nearfield planes except one plane of order 9 have this property. It follows that the connected component Σ^1 fixes u and v. Instead of André's result, we could also use Theorem (82.13) about sphere orbits of codimension 1.

According to (64.20), the kernel of N_r is $\mathbb{C}$, so that by (25.4) we obtain the homology group

$$\Sigma_{[o,W]} = \{ (x, y) \mapsto (x \circ c, y \circ c) \mid c \in \mathbb{C}^\times \} .$$

Restriction of c to $\mathbb{S}_1 = \{ c \in \mathbb{C} \mid |c| = 1 \}$ yields a one-dimensional torus subgroup Π, which intersects the product $\Sigma_{[u,ov]}\Sigma_{[v,ou]}$ in $\{\mathrm{id}, -\mathrm{id}\}$ only. Indeed, it is clear that the homothety $(x, y) \mapsto (x \circ c, y \circ c)$ for $c \in \mathbb{C}^\times$ only belongs to this product if c is an element of the center C_r of N_r, and we infer from (64.20) that $C_r \cap \mathbb{S}_1 = \{1, -1\}$. Thus, the product $\Sigma_{[u,ov]}\Sigma_{[v,ou]} \sqcap \mathsf{T}$, where T is the translation group, has dimension 17. We now show that, on the other hand, $\dim \Sigma \leq 17$; it then follows that the above product is the connected component Σ^1, cf. (93.12).

Consider the reduced stabilizer $\Omega \subseteq \Sigma_o$ defined in (81.0c). Since u and v are fixed points of Σ^1, we have $(\Omega_{u,v})^1 = \Omega^1$. The largest compact subgroup K of the latter group satisfies $\dim \mathsf{K} \leq 6$, see the beginning of (81.13). (In fact, it is easy to see that here $\mathsf{K} = \Phi \cong \mathrm{Spin}_4\mathbb{R}$.) The assertion $\dim \Sigma \leq 17$ now follows directly from the dimension estimate (1) in (81.12).

The whole group Σ is generated by Σ^1 together with the collineation $(x, y) \mapsto (y, x)$. This is proved in Hähl [80a] 4.2, p. 238 ff, where the Kalscheuer nearfields are treated within a larger family of locally compact 4-dimensional quasifields.

The existence of the reflection $(x, y) \mapsto (y, x)$ implies that the dual of the given nearfield plane contains a symmetric (stable) plane as an open subplane, see H. Löwe [94, 95] for an investigation of this remarkable geometry.

We now can state the overall result of the classification for 8-dimensional translation planes.

82.25 Theorem. *The 8-dimensional compact translation planes with an automorphism group of dimension at least 17 are precisely the following planes, up to isomorphism.*

The only plane with $\dim \Sigma \geq 19$ *is*

(i) *the classical quaternion plane, with* $\dim \Sigma = 35$, *see* (13.8).

The planes with $\dim \Sigma = 18$ *are*

(ii) *the three one-parameter families of planes of Lenz type V described in* (82.2).

The other planes satisfy $\dim \Sigma = 17$. *They are*

(iii) *the two-parameter family of those planes described in* (82.20) *which do not have Lenz type V, see case* (iii) *there.*
(iv) *the planes over the Kalscheuer nearfields* (64.20 and 22)
(v) *the planes of Lenz type V over the division algebras* $\mathbb{H}_{\alpha,\mathrm{id}}$ (*mutations of* $\mathbb{H}$) *described in* (82.21 and 23)
(vi) *the planes over the Rees algebras* (82.16 and 17).

This results from the classification scheme (82.15), see part 4) in particular, together with (82.2, 17, 19, 20, 22, 23, and 24). Note that most of the latter theorems have not been proved here. A complete proof is formed by Section 6 of Hähl [86a] p. 334 ff together with the references given there. □

82.26 Sixteen-dimensional translation planes with large groups.

Now, we give an account how all 16-dimensional compact translation planes with an automorphism group Σ of dimension at least 38 can be determined explicitly. For most cases, this cannot be more than a guide to the literature, where the classification program has been carried through for $\dim \Sigma \geq 40$. The planes with $\dim \Sigma \geq 38$ are all known, but for $\dim \Sigma \in \{38, 39\}$ the proofs of the classification are not yet completely published. We shall indicate what remains to be done.

1) In comparison with the 8-dimensional case, there are two extra difficulties. First, there are no results available about the actions of compact Lie groups on $\mathbb{S}_8$ which would be comparable to Richardson's results (96.34) for $\mathbb{S}_4$. Thus, the action of a compact subgroup Φ of $\Sigma_{o,W}$ on $W \approx \mathbb{S}_8$, which is not a priori supported by a linear structure, cannot be controlled directly. Instead, one has to resort to the linear action on the affine plane, but, as long as the plane is unknown, it is not always easy to draw conclusions about the action on W from the properties of this linear action. Related to this, there is a second difficulty. For 8-dimensional non-classical planes, it was possible to prove that Σ has a fixed point on W if $\dim \Sigma$ is large, and this result could be obtained before the hard cases of the classification had to be tackled, see the theorem at the end of step 3) in (82.15). Until now, a similar result has not been achieved for 16-dimensional planes, although at the end of the classification the fact in itself turns out to be true.

2) Let $\mathcal{P}$ be a 16-dimensional compact translation plane whose automorphism group Σ satisfies $\dim \Sigma \geq 38$. By Buchanan–Hähl [77], the kernel of a 16-dimensional compact translation plane is always isomorphic to $\mathbb{R}$. We assume that $\mathcal{P}$ is not isomorphic to the classical octonion plane. Let $\Theta = \Omega^1$ be the connected component of the reduced stabilizer $\Omega \subseteq \Sigma_o$ defined in (81.0c).

One may try to find a large compact subgroup in Θ which fixes two points $u, v \in W$ at infinity by choosing u and v from relatively small orbits of Σ on W, on the basis of the orbit theorem (82.14) and of Theorem (81.8) about the largest compact subgroup K of $(\Omega_{u,v})^1$. The linear action of K on the lines ou and ov is almost effective by (81.20); thus, K is locally isomorphic to a compact, connected subgroup of $\mathrm{GL}_8\mathbb{R}$. Using information about representations of compact Lie groups in low dimensions (as put down in Section 95) it is not difficult to establish lists showing all compact, connected subgroups $\Lambda \leq \mathrm{GL}_8\mathbb{R}$ of sufficiently high dimension, e.g., for $\dim \Lambda \geq 10$. In (95.12), we have included such a list for $\dim \Lambda \geq 14$. Along these lines, it is proved in Hähl [78] 1.7, p. 263 that, under the assumption $\dim \Sigma \geq 38$, the points $u, v \in W$ may be chosen in such a way that one of the following four cases applies.

(i) K is locally isomorphic to $\mathrm{Spin}_7\mathbb{R}$.
(ii) K contains a normal subgroup locally isomorphic to $\mathrm{Spin}_6\mathbb{R}$.
(iii) K is isomorphic to $\mathrm{G}_2 = \mathrm{Aut}\,\mathbb{O}$.

(iv) Either v is a fixed point of Θ and u is not, or $1 \leq \dim v^\Theta \leq 6$ and $u \in v^\Theta$ satisfies $1 \leq \dim u^{\Theta_v} \leq 5$. Moreover, K answers to one of the following descriptions:
   (iv1) K contains a normal closed subgroup locally isomorphic to $\mathrm{Spin}_5 \mathbb{R}$.
   (iv2) K is locally isomorphic to $\mathrm{SO}_4\mathbb{R} \times \mathrm{SO}_4\mathbb{R}$ or to $\mathrm{SO}_4\mathbb{R} \times \mathrm{U}_2\mathbb{C}$.
   (iv3) K is locally isomorphic to $\mathrm{U}_3\mathbb{C} \times \mathrm{SO}_2\mathbb{R}$.

The orbit conditions of case (iv) are very important. Later, they permit to eliminate case (iv) from the list altogether by showing, in a detailed study of the different subcases, that these additional conditions imply $\dim \Sigma \leq 37$, contrary to the original assumption. In order to ensure these orbit conditions, it is necessary to study the case of a 6-dimensional orbit on which Θ is doubly transitive, and to show that then, for a different choice of u and v, one of the cases (i)–(iii) applies. Once this list of possible cases is established, the classification proceeds case by case.

3) If K is locally isomorphic to $\mathrm{Spin}_7\mathbb{R}$, then we already know the possible planes explicitly, see (82.5); they satisfy $\dim \Sigma \in \{38, 39\}$.

4) The case that K contains a connected normal subgroup Ψ locally isomorphic to $\mathrm{Spin}_6\mathbb{R} = \mathrm{SU}_4\mathbb{C}$ is difficult and leads to an astonishing result. It is the subject of Hähl [87c]. The action of Ψ on $W \setminus \{v\} \cong \mathbb{R}^8$ turns out to be equivalent to the usual action of $\mathrm{SU}_4\mathbb{C}$ or of $\mathrm{SO}_6\mathbb{R}$. The situation that $\Psi|_W \cong \mathrm{SU}_4\mathbb{C}$ can be quickly disposed of. One uses Theorem (82.13) about sphere orbits of codimension 1, according to which either u and v are fixed points of Σ^1, or the plane has Lenz type V. In the first case, $\dim \Sigma$ would be too small by the dimension estimate (81.12). In the second case, dualization and/or transposition turns the plane into a plane with $\Psi|_W \cong \mathrm{SO}_6\mathbb{R}$, exactly in the same way as in the analogous situation for 8-dimensional planes, see step 2) of the proof of (82.2). The group Ψ itself is isomorphic to $\mathrm{SU}_4\mathbb{C}$.

The subcase with $\Psi|_W \cong \mathrm{SO}_6\mathbb{R}$ now is the crucial one. A major step is to prove that Ψ is then an almost simple factor of a Levi complement H in Θ, see Hähl [87c] 5.1, p. 334 ff. One uses the classification of almost simple Lie groups, see (94.32 and 33), together with dimension estimates as in (81.12), on the basis of Theorem (81.8) about large compact subgroups of triangle stabilizers, in order to restrict the possibilities for almost simple factors of H. It turns out that there is an almost simple factor H_1 containing a subgroup Ψ_1 isomorphic to Ψ, and a crucial step is the conclusion that H_1 cannot be larger than Ψ_1 because of the following characterization of the classical octonion plane, proved in Hähl [87b].

Theorem. *A 16-dimensional compact translation plane in which Σ_o contains a closed subgroup locally isomorphic to $O'_7(\mathbb{R}, 1)$ is isomorphic to the classical octonion plane.* □

As a consequence, if Ψ centralizes the radical $\sqrt{\Theta}$, then Ψ is a normal subgroup of Θ. This situation is covered by Section 6 of Hähl [87c]. The normalizer of Ψ contains the group Θ, which by our assumption $\dim \Sigma \geq 38$ has dimension at least 21. At the beginning of the proof of 6.1, loc. cit. p. 337, it is deduced that a Levi complement in Θ is the product of $\Psi \cong \mathrm{SU}_4\mathbb{C}$ and of a normal subgroup locally isomorphic to $\mathrm{SL}_2\mathbb{C}$, and in the subsequent theorem 6.2 it is shown that, again, such a group is only possible if the plane is classical.

There remains the case that, still under the assumption $\Psi|_W \cong \mathrm{SO}_6\mathbb{R}$, the radical $\sqrt{\Theta}$ is not centralized by Ψ. Then one finds a 6-dimensional closed vector subgroup N of $\sqrt{\Theta}$ which is normalized, but not centralized by Ψ, and which can be shown to consist of shears with axis ov, see loc. cit. Section 4. In Section 3 there, it is finally shown that this situation only leads to the classical octonion plane; this is accomplished by a direct comparison of the action of ΨN with the action of a group of the same sort in $\mathcal{P}_2\mathbb{O}$. (In this report on the contents of Hähl [87c], we have read that paper almost backwards.) The last result is somewhat surprising, if one compares this situation, where ΨN $\cong \mathrm{SU}_4\mathbb{C} \cdot \mathbb{R}^6$ and $\Psi|_W \cong \mathrm{SO}_6\mathbb{R}$, with an analogous situation for 8-dimensional planes. There, $\Psi \cong \mathrm{SU}_2\mathbb{C}$ acts on W as $\mathrm{SO}_3\mathbb{R}$, and Ψ normalizes a 3-dimensional group of shears. The investigation of the latter situation resulted in the discovery of one of the most important families of homogeneous planes, see (82.19 and 20), containing a subfamily of Lenz type V planes from which all 8-dimensional planes with $\dim \Sigma \geq 18$ can be derived, see (82.25 and 2). This contrasts with the 16-dimensional case, where only the classical plane is obtained.

5) The case $\mathsf{K} \cong \mathrm{G}_2$ leads to the most homogeneous planes. An argument which is analogous to the preceding case, but somewhat easier, shows that a large automorphism group is only obtained if Θ also contains a closed subgroup $\mathsf{N} \cong \mathbb{R}^7$, see Hähl [88] 2.2, p. 271. In Hähl [90] it is then shown that N consists of shears whose center is a fixed point of Θ, and that the planes of this kind are precisely the planes over the generalized mutations $\mathbb{O}_{\alpha,\varrho}$ constructed in (82.21), with $\dim \Sigma \in \{38, 39, 40\}$.

6) The case (iv1), in which K contains a normal closed, connected subgroup locally isomorphic to $\mathrm{Spin}_5\mathbb{R} = \mathrm{U}_2\mathbb{H}$, may be treated in the same way, see Hähl [88] Section 3, p. 277 ff. It is shown there that most subcases result in $\dim \Sigma \leq 37$; in one critical case, however, only a weaker result is obtained.

In that case, $\Psi \cong \mathrm{Spin}_5\mathbb{R} = \mathrm{U}_2\mathbb{H}$ acts on W as $\Psi|_W \cong \mathrm{SO}_5\mathbb{R}$ and normalizes a closed subgroup $\mathsf{N} \cong \mathbb{R}^5$ consisting of shears with fixed center (loc. cit., 3.9 and 3.10). This shear center is then shown to be a fixed point of Θ by some sort of 'topological combinatorics' applied to the system of N-orbits in W (loc. cit., 3.13). The usual dimension estimate (81.12) then shows that $\dim \Sigma \leq 39$. In fact it is possible to improve this estimate and to obtain $\dim \Sigma \leq 37$ in this case, as well, so that case (iv1) will be eliminated from the present classification program altogether. This step is still unpublished. Among other things, one has to consider a plane of Lenz type at least V with the property that Σ contains a subgroup locally isomorphic to $U_2\mathbb{H} \cdot \mathrm{Spin}_3\mathbb{R}$, and to show that such a plane is necessarily classical. We mention that there are non-classical planes of Lenz type V such that Σ contains a subgroup isomorphic to $U_2\mathbb{H} \cdot SO_2\mathbb{R}$ and has dimension 37.

7) In case (iv2), with K locally isomorphic to $SO_4\mathbb{R} \times SO_4\mathbb{R}$ or to $SO_4\mathbb{R} \times U_2\mathbb{C}$, an analysis of the action of K on W leads to the conclusion that $\dim \Sigma \leq 38$, see Hähl [88] Section 4, p. 289 ff. More precisely, it is shown that $\dim \Sigma \leq 37$ except if K is locally isomorphic to $SO_4\mathbb{R} \times SO_4\mathbb{R}$ and is contained in a closed subgroup Δ of Θ acting on W in one of the following ways: either it fixes v and acts transitively on $W \setminus \{v\}$, or v^Δ is homeomorphic to $\mathbb{S}_4$ and Δ is doubly transitive on this orbit. These two cases need a special argument, which is still unpublished. In fact, both situations characterize the classical octonion plane.

8) Finally, in case (iv3), the usual dimension estimate (81.12) immediately gives $\dim \Sigma \leq 38$. It is relatively easy to produce the sharper dimension estimate $\dim \Sigma \leq 36$ (not yet published), which contradicts the assumptions. □

We now formulate the result of this classification program, as published presently.

82.27 Theorem. *The 16-dimensional compact translation planes with an automorphism group of dimension at least* 40 *are precisely the following planes, up to isomorphism.*

(i) *The classical octonion plane, with* $\dim \Sigma = 78$, *see* (17.10).
(ii) *The planes of Lenz type* V *over the division algebras* $\mathbb{O}_{\alpha,\mathrm{id}}$ *(mutations of* $\mathbb{O}$*) described in* (82.21), *with* $\dim \Sigma = 40$.

This is proved in Hähl [90] Theorem 1.2, p. 182 on the basis of Section 5 of Hähl [88], p. 295 ff, where the general plan of the classification is also outlined, and of the other papers mentioned above. For the planes appearing in (ii), the group Σ will be described in (82.29). It can be shown that, up to isomorphism, the parameter α may be chosen from the interval $(0, 1)$, and that for different parameters from this interval the resulting planes are non-isomorphic. □

The proof of the following more comprehensive classification result is not yet published completely.

82.28 Theorem. *The 16-dimensional compact translation planes with an automorphism group of dimension at least 38 are precisely the following planes, up to isomorphism.*

(i) *The classical octonion plane, with* $\dim \Sigma = 78$.
(ii) *The planes over the generalized mutations* $\mathbb{O}_{\alpha,\varrho}$ *of* (82.21), $(\alpha, \varrho) \neq (1, \mathrm{id})$, *with* $\dim \Sigma \in \{38, 39, 40\}$.
(iii) *The planes over the perturbations* $\mathbb{O}^{(\varrho)}$, $\varrho \neq \mathrm{id}$, *of* $\mathbb{O}$ *described in* (82.4), *with* $\dim \Sigma \in \{38, 39\}$.

The planes of type (iii) have been discussed fully in (82.4). The planes of type (ii) are analogous to the 8-dimensional planes over the generalized mutations $\mathbb{H}_{\alpha,\varrho}$ of the quaternions, see (82.22 and 23). This analogy also extends to the automorphism group, as is expressed in the following theorem.

82.29 Theorem. *Up to isomorphism, the planes over the generalized mutations* $\mathbb{O}_{\alpha,\varrho}$ *are precisely the 16-dimensional compact translation planes having the following properties: for a suitable* $v \in W$, *the shear group* $\Sigma_{[v,ov]}$ *has dimension at least 7, and* $\Sigma_{o,v}$ *contains a subgroup isomorphic to* $G_2 \cong \mathrm{Aut}\,\mathbb{O}$.

The plane $\mathcal{P}$ *over* $\mathbb{O}_{\alpha,\varrho}$ *is isomorphic to the classical octonion plane only if* $\varrho = \mathrm{id}$ *and* $\alpha = 1$. *In all other cases, the automorphism group* Σ *of* $\mathcal{P}$ *satisfies* $38 \leq \dim \Sigma \leq 40$.

The bound $\dim \Sigma = 40$ *is attained if, and only if, the plane has Lenz type* V; *this is equivalent to* $\varrho = \mathrm{id}$ *and* $\alpha \neq 1$.

The planes of this kind with $\dim \Sigma = 39$ *can be coordinatized over a generalized mutation* $\mathbb{O}_{\alpha,\varrho}$ *defined by means of a homeomorphism* $\varrho = \varrho_g$ *of the following special form:*

$$\varrho_g(r) = \begin{cases} r & \text{for } r \geq 0 \\ gr & \text{for } r \leq 0, \end{cases}$$

where $g > 1$ *is a constant parameter.*

This is proved in Hähl [90] 3.1, p. 187 ff. We describe the connected component of the automorphism group Σ of the plane over $\mathbb{O}_{\alpha,\varrho}$ in affine coordinates. One verifies immediately that $\mathrm{Pu}\,\mathbb{O}$ is contained in the distributor (25.4) of $\mathbb{O}_{\alpha,\varrho}$; accordingly, $\Sigma_{[v,ov]}$ contains the shears

$$(x, y) \mapsto (x, y + p \circ x) \quad \text{for } p \in \mathrm{Pu}\,\mathbb{O},$$

which form a closed subgroup isomorphic to $\mathbb{R}^7$. Since the automorphisms of $\mathbb{O}$ are automorphisms of $\mathbb{O}_{\alpha,\varrho}$ as well (82.21), the maps

$$(x, y) \mapsto (x^\sigma, y^\sigma) \quad \text{for } \sigma \in \mathrm{Aut}\,\mathbb{O}$$

constitute a group Φ of collineations isomorphic to $\mathrm{Aut}\,\mathbb{O} = G_2$. These collineations together with the positive real homotheties and the translations generate a 38-dimensional closed, connected subgroup of Σ, which in most cases is the connected component Σ^1.

If $\varrho = \varrho_g$, then Σ^1 is generated by these collineations together with the one-parameter group of the homologies

$$(x, y) \mapsto (rx, y) \quad \text{for } 0 < r \in \mathbb{R}.$$

For $\varrho = \mathrm{id}$, $\alpha \neq 1$ we obtain the division algebra $\mathbb{O}_{\alpha,\mathrm{id}}$, and the corresponding plane has Lenz type V by (24.7), so that the shear group $\Sigma_{[v,ov]}$ is 8-dimensional. Together with the other collineations described above, it generates a 40-dimensional closed, connected subgroup of Σ, which is the connected component Σ^1. In fact, as a consequence of the whole classification program, we know in general that $\dim \Sigma \leq 40$. For the plane over $\mathbb{O}_{\alpha,\mathrm{id}}$, this can be seen more directly as follows.

Assume to the contrary that $\dim \Sigma > 40$. The shear point v is a fixed point of Σ, see (81.0e). By a dimension estimate as in (81.12), our assumption implies that $\Sigma_{o,u,v}$ has a compact, connected subgroup K containing the subgroup $\Phi \cong G_2$ and satisfying $\dim K > 14$. By (81.20), the linear action of K on the line ou is almost effective. Now, a compact, connected subgroup of $\mathrm{GL}_8\mathbb{R}$ of dimension > 14 and containing G_2 has a subgroup locally isomorphic to $\mathrm{Spin}_7\mathbb{R}$, see (95.12). Hence, the plane is classical by the remark 1) following (81.18).

Planes which are not translation planes

Here, we report briefly on certain constructions producing higher-dimensional compact planes which are neither translation planes nor dual translation planes and have fairly large automorphism groups. The Hughes planes will not be included here; they are the subject of Section 86.

82.30 Glued planes obtained from translation planes.

One construction consists in gluing together two halves of translation planes. Consider the point set $\mathbb{R}^{2l}$ of a locally compact affine translation plane of dimension $2l$, $l \in \{2, 4, 8\}$. A hyperplane $H \cong \mathbb{R}^{2l-1}$ of $\mathbb{R}^{2l}$ separates the point set into two halves. One can now try to glue together two such halves of (different) translation planes of dimension $2l$ along the separating hyperplanes in order to obtain a new topological affine plane. Of course, certain conditions have to be fulfilled; in the first place, the systems of subspaces of the separating hyperplanes obtained by intersection with the lines of the original planes have to match. Projective completion by a line at infinity W then produces a new compact projective plane (43.7).

The resulting plane admits a $(2l-1)$-dimensional group of translations with axis W, corresponding to certain translations of each of the original translation planes, namely those that leave the separating hyperplane invariant. Hence, one has the situation of an almost homogeneous translation group, see (61.11c and 12). It follows that there is a point $v \in W$ such that the group of translations with axis W and center v is linearly transitive in the sense of (23.21); in other words, the plane has Lenz type (at least) II and can be coordinatized by a Cartesian field, see (24.2 through 4).

This construction was conceived by Betten [84] in the context of 4-dimensional planes, and fully worked out for arbitrary dimension in the thesis of Weigand [87]. For glued planes of dimension 2 see Groh [81], Steinke [85].

From Weigand's thesis, we extract a characterization of the planes obtained in this way and two families of examples of such planes, one in dimension 8 and one in dimension 16. Weigand's thesis contains more examples of dimension 4 and 8. We shall not reproduce proofs here.

82.31 Theorem (Weigand [87] 3.4). *A given compact projective plane $\mathcal{P}$ of dimension $2l \in \{4, 8, 16\}$ is isomorphic to a plane obtained by gluing together two halves of translation planes if, and only if, there is a line W and a closed, connected subgroup T of the group of translations with axis W such that the following conditions hold:*

(i) $\dim \mathsf{T} = 2l - 1$, *that is*, $\mathsf{T} \cong \mathbb{R}^{2l-1}$.
(ii) *There is a non-trivial T-orbit H such that the positive real homotheties of $H \cong \mathsf{T} \cong \mathbb{R}^{2l-1}$ are induced by automorphisms of $\mathcal{P}$.*

Remarks. 1) The assumption $\dim \mathsf{T} = 2l - 1$ implies that $\mathsf{T} \cong \mathbb{R}^{2l-1}$ by (61.4 and 9). By (44.8b), the translation group T is homeomorphic to every non-trivial orbit.

2) In Weigand's formulation there are other conditions. As explained in (82.30), the assumption (i) implies by (61.11c and 12) that for some point $v \in W$ the subgroup $\mathsf{T}_{[v,W]}$ of translations with center v is linearly transitive. According to (53.3), the lines of $\mathcal{P}$ are manifolds. The assumptions stated above and these consequences now suffice for Weigand's proof.

82.32 Example (Weigand [87] 4.6): *Eight-dimensional planes obtained by gluing from translation planes and having automorphism groups of dimension* 14 *or* 15. Let ϱ_+ and ϱ_- be two homeomorphisms of the interval $[0, \infty)$ fixing 0 and 1. Define a continuous map $\varrho : [0, \infty) \times \mathbb{R} \to \mathbb{R}$ by

$$\varrho(s, r) = \begin{cases} \varrho_+(s) \cdot r & \text{if } r \geq 0 \\ \varrho_-(s) \cdot r & \text{if } r \leq 0 \, . \end{cases}$$

The incidence structure with point set $\mathbb{H}^2$ whose lines are the subsets $\mathbb{H} \times \{c\}$ and $\{c\} \times \mathbb{H}$ for $c \in \mathbb{H}$ and the subsets of the form

$$\left\{ \left(x + p, a \left(\tfrac{\varrho(\|a\|, \operatorname{Re} x)}{\|a\|} + \operatorname{Pu} x \right) + q \right) \;\middle|\; x \in \mathbb{H} \right\}$$

for
$$a \in \mathbb{H} \setminus \{0\}, \quad p \in \operatorname{Pu} \mathbb{H}, \quad q \in \mathbb{H}$$
is a locally compact topological affine plane. It can be obtained by applying the gluing process (82.30) to the translation planes over the quasifields $\mathbb{H}^{(\varrho_+^{-1})}$ and $\mathbb{H}^{(\varrho_-^{-1})}$ described in (82.6 and 4). Indeed, the line of slope $b \neq 0$ through the origin in the translation plane over $\mathbb{H}^{(\varrho_+^{-1})}$ is

$$\left\{ \left(x, b \left(\operatorname{Re} x + \frac{\varrho_+^{-1}(\|b\|)}{\|b\|} \operatorname{Pu} x \right) \right) \,\middle|\, x \in \mathbb{H} \right\}.$$

With $a = \varrho_+^{-1}(\|b\|) \frac{b}{\|b\|}$, this line can also be written in the form

$$\left\{ \left(x, a \left(\frac{\varrho_+(\|a\|) \cdot \operatorname{Re} x}{\|a\|} + \operatorname{Pu} x \right) \right) \,\middle|\, x \in \mathbb{H} \right\},$$

which is better suited for gluing along the hyperplane $\{ (x, y) \in \mathbb{H}^2 \mid \operatorname{Re} x = 0 \}$, and analogously for ϱ_- instead of ϱ_+.

The automorphism group Σ of such a plane contains a 7-dimensional translation group. The stabilizer of the lines $\mathbb{H} \times \{0\}$ and $\{0\} \times \mathbb{H}$ contains the positive real homotheties and a subgroup isomorphic to $\operatorname{SO}_4 \mathbb{R}$ which is inherited from the parental translation planes. Hence, $\dim \Sigma \geq 14$. If, as is generically true, the plane is neither a translation plane nor a dual translation plane, then $\dim \Sigma \leq 15$. If both ϱ_+ and ϱ_- are multiplicative, then the bound $\dim \Sigma = 15$ is attained.

82.33 Example (Weigand [87] 4.10): *Sixteen-dimensional planes obtained by gluing from translation planes and having automorphism groups of dimension* 37 *or* 38. Among the known 16-dimensional compact planes which are neither translation planes nor dual translation planes, these are the ones with the largest automorphism groups. Let $\alpha_+, \alpha_- > 0$ and let ϱ_+ and ϱ_- be two homeomorphisms of $\mathbb{R}$ fixing 0 and 1 which can be written in the form $\varrho_\pm(t) = \sigma_\pm(t) \cdot t$ where σ_+ and σ_- are continuous functions. Define continuous maps $\varrho : \mathbb{R}^2 \to \mathbb{R}$ and $\alpha : \mathbb{R} \to \mathbb{R}$ by

$$\varrho(s, r) = \begin{cases} \varrho_+(s) \cdot r & \text{if } r \geq 0, \\ \varrho_-(s) \cdot r & \text{if } r \leq 0, \end{cases}$$

$$\alpha(r) = \begin{cases} \alpha_+ r & \text{if } r \geq 0, \\ \alpha_- r & \text{if } r \leq 0. \end{cases}$$

The incidence structure with point set $\mathbb{O}^2$ whose lines are the subsets $\{c\} \times \mathbb{O}$ for $c \in \mathbb{O}$ and the subsets of the form

$$\{ (x + p, \varrho(\operatorname{Re} a, \operatorname{Re} x) + \alpha(\operatorname{Re} x) \cdot \operatorname{Pu} a + a \cdot \operatorname{Pu} x + q) \mid x \in \mathbb{O} \}$$

for
$$a \in \mathbb{O}, \ p \in \operatorname{Pu}\mathbb{O}, \ q \in \mathbb{O}$$
is a locally compact topological affine plane. It can be obtained by applying the gluing process (82.30) to the translation planes over the generalized mutations $\mathbb{O}_{\alpha_+^{-1},\varrho_+^{-1}}$ and $\mathbb{O}_{\alpha_-^{-1},\varrho_-^{-1}}$ described in (82.21). Indeed, the line of slope b through the origin of the plane over $\mathbb{O}_{\alpha_+^{-1},\varrho_+^{-1}}$ is

$$\left\{ \left(x, b \cdot \operatorname{Re} x + \left(\varrho_+^{-1} (\operatorname{Re} b) + \alpha_+^{-1} \operatorname{Pu} b \right) \operatorname{Pu} x \right) \,\middle|\, x \in \mathbb{O} \right\}.$$

With $a = \varrho_+^{-1}(\operatorname{Re} b) + \alpha_+^{-1} \operatorname{Pu} b$, this line can also be written in the form

$$\left\{ \left(x, \varrho_+(\operatorname{Re} a) \cdot \operatorname{Re} x + \alpha_+ \operatorname{Re} x \cdot \operatorname{Pu} a + a \cdot \operatorname{Pu} x \right) \,\middle|\, x \in \mathbb{O} \right\},$$

which is better suited for gluing along the hyperplane $\{\, (x, y) \in \mathbb{O}^2 \mid \operatorname{Re} x = 0 \,\}$, and analogously for ϱ_- and α_-.

The automorphism group Σ of such a plane contains a 15-dimensional translation group. In the stabilizer of $\{0\} \times \mathbb{O}$, we find a 7-dimensional closed subgroup consisting of shears with this axis, and a subgroup isomorphic to $\operatorname{Aut} \mathbb{O}$ which also fixes $\mathbb{O} \times \{0\}$; these two subgroups are inherited from the parental translation planes, see (82.29). Moreover, Σ contains the positive real homotheties. Thus, $\dim \Sigma \geq 37$. If, as is generically true, the plane is neither a translation plane nor a dual translation plane, then $\dim \Sigma \leq 38$. If both ϱ_+ and ϱ_- are of the form $\varrho_\pm(r) = r$ for $r \geq 0$ and $\varrho_\pm(r) = g_\pm r$ for $r \leq 0$ with constants $g_\pm > 0$, then Σ also contains the homotheties $(x, y) \mapsto (rx, y)$ for $0 < r \in \mathbb{R}$, so that the bound $\dim \Sigma = 38$ is attained.

82.34 More planes which are not translation planes.
There is an abundance of 2-dimensional compact projective planes, see Chapter 3 in general and (38.7); apart from the classical plane over $\mathbb{R}$, these are not translation planes, cf. (32.8) and (64.7). Certainly, there is a similar wealth of compact planes of dimension $2l \in \{4, 8, 16\}$ which are neither translation planes nor dual translation planes, but only few types of such planes are explicitly known. We mention again the shift planes (which exist for $l = 1, 2$ only), see (31.25c), Section 36 and Section 74, the planes obtained by gluing from translation planes described above, the Hughes planes treated in Section 86, the Hurwitz planes (for $l \in \{2, 4, 8\}$) constructed by Plaumann–Strambach [74], the planes of P. Sperner [90] mentioned in the note following (72.2), and the differentiable planes constructed by Otte [90], see (75.5).

We finally describe a further construction by Knarr [87b]. With a fixed parameter $r \in \mathbb{R}$ such that $|r| \leq 2$, he defines a new multiplication $*$ on $\mathbb{F} \in \{\mathbb{C}, \mathbb{H}\}$ by

$0 * w = 0 = z * 0$ and

$$z * w = z e^{ir \log |z| \log |w|} w \quad \text{for } z, w \in \mathbb{F} \setminus \{0\}.$$

Then $(\mathbb{F}, +, *)$ is a Cartesian field which coordinatizes a compact projective plane of dimension 4 or 8. If $r \neq 0$, then these planes have Lenz–Barlotti type II.2 or II.1 for $\mathbb{F} = \mathbb{C}$ or $\mathbb{F} = \mathbb{H}$, respectively. The automorphism group of the plane has dimension 6 for $\mathbb{F} = \mathbb{C}$ and 13 for $\mathbb{F} = \mathbb{H}$.

83 Stiffness

Stiffness properties are restrictions on the size of the stabilizers of certain configurations in the automorphism group Σ of the plane. Here, we deal mostly with the stabilizer Λ of a non-degenerate quadrangle. For compact planes of dimension 2 or 4, it has been shown that $\Lambda = \mathbb{1}$ or $|\Lambda| \leq 2$, respectively, see (32.10) and (55.21b). For higher-dimensional planes, Λ may be non-discrete; in $\mathcal{P}_2\mathbb{O}$, for instance, Λ is isomorphic to $\text{Aut}\,\mathbb{O} = G_2$, see (12.8) and (11.33). The type of stiffness result one can hope for in higher dimensions is that Λ is a compact group whose dimension is subject to the restrictions suggested by the classical planes. For translation planes, this has been proved, cf. (81.5 and 6), and it is conjectured that corresponding results hold in general. The main difficulty is that, up to now, it has not been possible to establish compactness of Λ in the general situation.

In this section, we shall prove stiffness properties of higher-dimensional compact projective planes in general or under various hypotheses. Moreover, we use these stiffness results in order to derive first dimension estimates concerning the full automorphism group Σ of the plane, in preparation of the following sections. In particular, we prove the results announced in (65.1 and 2). According to our general policy, the results are often not proved in the strongest form known; instead, they have been adapted to the actual needs of the final sections.

The theory of higher-dimensional planes as opposed to planes of dimension 2 or 4 presents a few fundamental difficulties: for instance, it is not known in general whether lines are manifolds, and whether the automorphism group Σ is a Lie group. These uncertainties have a direct influence on the difficulty of stiffness problems. In those favourable cases where a positive answer to these basic questions has been given, stiffness results are much easier to obtain, as we shall see in various instances.

A typical method is a kind of induction over chains of closed, connected subplanes, if there are any; in this way, one can hope to profit from the more incisive stiffness properties of lower-dimensional planes. We therefore begin by collecting basic facts and techniques concerning subplanes.

In this section, $\mathcal{P} = (P, \mathcal{L})$ will always denote a compact projective plane of dimension $2l$, where $l \in \{4, 8\}$, and $\Sigma = \operatorname{Aut} \mathcal{P}$ is the group of continuous collineations, endowed with the compact-open topology. Γ will always be a closed subgroup of Σ.

The compact-open topology on Σ is in fact induced by a metric (44.2); moreover, Σ is locally compact and has a countable basis (44.3). In particular, for subsets of Σ, the notions of topological dimension used in this book agree (92.6 and 7).

83.1 Closed, connected subplanes and dimensions of stabilizers.

(a) From (55.1), we recall that the dimension of a closed, connected subplane $\mathcal{Q}$ of $\mathcal{P}$ is $2m$ with $m \in \{1, 2, 4, 8\}$ and $m \leq l$, and that a closed subplane of $\mathcal{Q}$ having the same dimension is equal to $\mathcal{Q}$. Note that a closed subplane which is not connected is in fact totally disconnected (42.1) and hence has dimension 0, cf. (92.18). Consequently, $\dim \mathcal{Q}' \leq \frac{1}{2} \cdot \dim \mathcal{Q}$ for a proper closed subplane $\mathcal{Q}'$ of $\mathcal{Q}$.

(b) For a subset $S \subseteq P$ which generates a subplane of $\mathcal{P}$, i.e., which contains a non-degenerate quadrangle, $\langle S \rangle$ will denote the topological closure of the point set of this subplane. $\langle S \rangle$ is the point set of the smallest closed subplane containing S, see (41.6). Let $\Xi \leq \Sigma_{[S]}$ be a closed subgroup of Σ fixing S pointwise. By (44.1), then, Ξ also fixes every point of $\langle S \rangle$.

(c) Now assume that S is not totally disconnected. Then, $\langle S \rangle$ is connected, see (a), and one can find three points z_1, z_2, z_3 such that $\langle S, z_1, z_2, z_3 \rangle = P$. (In the case $l = 4$, two points suffice.) Again by (44.1), we have $\Xi_{z_1, z_2, z_3} = \mathbb{1}$. Since orbits of the points z_1, z_2, z_3 are at most $2l$-dimensional, repeated application of the dimension formula (96.10) yields $\dim \Xi \leq 6l$.

(d) A standard application of this kind of argument is concerned with the situation in which Γ centralizes a non-trivial connected subgroup Δ of Σ. The stabilizer Γ_x of a point x then fixes every point of x^Δ. If $S \subseteq P$ generates a subplane, and if $x^\Delta \neq \{x\}$, then $\langle x^\Delta, S \rangle$ is a closed, connected subplane all of whose points are fixed points of $\Gamma_{[\{x\} \cup S]}$. Of particular interest is the case where $S \subseteq x^\Delta$; then $\Gamma_{[\{x\} \cup S]} = \Gamma_x$.

(e) Now choose two non-trivial orbits x^Δ, y^Δ whose union is not contained in a line. Then $x^\Delta \cup y^\Delta$ generates a subplane, and $\langle x^\Delta \cup y^\Delta \rangle$ is connected and consists of fixed points of $\Gamma_{x,y}$. Hence, according to (c), we have $\dim \Gamma_{x,y} \leq 6l$, and by the dimension formula we conclude that $\dim \Gamma \leq 10l$.

This type of argument will be used very often. Of course, the above estimates are rather crude and will be refined considerably. Our first application is

83.2 Proposition. $\Sigma = \operatorname{Aut} \mathcal{P}$ *is a locally compact group of finite dimension.*

Remark. It will be proved later that in fact $\dim \Sigma$ is never larger than in the corresponding classical plane (83.26).

Proof of (83.2). Local compactness is known (44.3). The dimension of a locally compact group equals that of its connected component (93.6). By the Mal'cev–Iwasawa theorem (93.10) it therefore suffices to show that a compact, connected subgroup $\Phi \neq \mathbb{1}$ of Σ has finite dimension. The preceding considerations immediately tell us that this is the case if the connected component A of the center of Φ is non-trivial. If A is trivial, then, by the structure theorem for compact groups (93.11), we have that $\Phi = (\Phi_1 \times \Phi_2)/N$, where Φ_1 is a compact, almost simple Lie group (and hence finite-dimensional), Φ_2 is a product of such groups or trivial, and N is a closed, central subgroup of $\Phi_1 \times \Phi_2$ satisfying $\dim N = 0$. By (83.1e), the subgroup $\Phi_2 N/N \cong \Phi_2/\Phi_2 \cap N$ of Φ is finite-dimensional, hence the same is true for Φ_2 (93.7), for the product $\Phi_1 \times \Phi_2$ (92.10), and finally for Φ itself (93.7). This finishes the proof.

It is interesting to add that in the case $A = \mathbb{1}$ discussed last, the group Φ itself turns out to be a Lie group because it has finite dimension. □

83.3 Lemma. *Let Δ be a locally compact group having a countable basis, and assume that Δ acts continuously as a group of automorphisms of $\mathcal{P}$ and leaves a closed, connected subplane $\mathcal{Q} = (Q, \mathcal{M})$ invariant. Let Θ be a closed subgroup of $\mathrm{Aut}\,\mathcal{Q}$ such that $\Delta|_Q \leq \Theta$, and let $\Phi = \Delta_{[Q]}$. Then $\dim \Delta \leq \dim \Theta + \dim \Phi$.*

Proof. Consider the natural continuous bijective homomorphism $\varrho : \Delta/\Phi \to \Delta|_Q$. Since Δ/Φ is locally compact, and since dimension, defined as inductive dimension ind, cf. (92.3), is a local invariant, there is a compact subset C of Δ/Φ such that $\dim C = \dim \Delta/\Phi$. By compactness, $\varrho(C)$ is a subset of Θ which is homeomorphic to C. By (92.4), we have that $\dim \Theta \geq \dim \varrho(C) = \dim C = \dim \Delta/\Phi$. From (93.7), we know that $\dim \Delta/\Phi = \dim \Delta - \dim \Phi$, and the assertion follows. □

We now begin to study the stabilizer in $\Sigma = \mathrm{Aut}\,\mathcal{P}$ of a non-degenerate quadrangle. This stabilizer is isomorphic in a natural way to the group $\mathrm{Aut}\,K$ of (continuous) automorphisms of a coordinatizing locally compact, connected topological ternary field (K, τ), see (44.4c), (23.5), and (41.3). On $\mathrm{Aut}\,K$, the compact-open topologies with respect to the actions on P respectively on K coincide. The space K is (homeomorphic to) an affine line of $\mathcal{P}$, in particular we have $\dim K = l$.

83.4 Automorphisms and subternaries.

(a) For $\Xi \subseteq \mathrm{Aut}\,K$, the set F_Ξ of fixed elements of Ξ is a closed subternary of K. A closed subternary H is either connected, or totally disconnected; the latter property is equivalent to $\dim H = 0$, cf. (92.5 and 18). Closed, connected subternaries of

K coordinatize closed, connected subplanes of $\mathcal{P}$. Much of what has been said in (83.1) about subplanes may thus be restated for subternaries.

A closed, connected subternary $H \leq K$ has dimension $m \in \{1, 2, 4, 8\}$. Moreover, for closed subternaries $H' \leq H \leq K$ such that $H' \neq H$, one obtains $\dim H' \leq \frac{1}{2} \cdot \dim H$. By (55.5), equality holds if, and only if, H' is a *Baer subternary* of H, i.e., if the subplane coordinatized by H' is a Baer subplane of the subplane over H.

(b) For a subset $S \subseteq K$, the closure of the subternary generated by 0, 1, and S will be denoted by $\langle S \rangle$; we prefer to write $\langle 1 \rangle$ instead of $\langle \emptyset \rangle$. There is no danger of confusion despite of our using the same notation as for closed subplanes generated by sets of points. Indeed, the subternary $\langle S \rangle$ is the coordinatizing ternary field of the closure of the subplane generated by S together with the reference quadrangle of the coordinatization. If S is not totally disconnected, then $\langle S \rangle$ is connected.

(c) If $\alpha \in \operatorname{Aut} K$ fixes every element of S, then the same is true for the elements of $\langle S \rangle$. In particular, if $H \leq K$ is a maximal closed, connected subternary, e.g., a Baer subternary (21.7), and if α fixes every element of H and an element $x \in K \setminus H$, then $\alpha = \mathbb{1}$, since by maximality $\langle H, x \rangle = K$.

(d) If a connected subgroup Δ of $\operatorname{Aut} K$ leaves a 2-dimensional closed subternary C invariant, then Δ acts trivially on C by (55.21c and a).

(e) A 1-dimensional closed subternary R coincides with $\langle 1 \rangle$; this is a coordinate version of (32.7). The following corollary corresponds to (32.10).

(f) If $\alpha \in \operatorname{Aut} K$ leaves a 1-dimensional closed subternary R invariant, then α induces the identity on R.

The following statement is analogous to (83.1d).

(g) If a subgroup $\Gamma \leq \operatorname{Aut} K$ centralizes a connected subgroup $\Delta \neq \mathbb{1}$, then, for $c \in K$ such that $c^\Delta \neq \{c\}$, the stabilizer Γ_c fixes every element of the connected orbit c^Δ and hence acts trivially on the closed, connected subternary $\langle c^\Delta \rangle$.

83.5 Involutory automorphisms. By (23.17), an involution $\beta \in \operatorname{Aut} K$ corresponds to an involutory collineation of $\mathcal{P}$ which is a Baer involution, i.e., whose fixed elements form a Baer subplane. The set $F_\beta \leq K$ of fixed elements of β is a coordinatizing ternary field of this subplane; in particular, F_β is a Baer subternary of K. The results (55.32) about planar involutory collineations may now be rephrased for involutions in $\operatorname{Aut} K$ as follows.

Two involutions $\beta, \gamma \in \operatorname{Aut} K$ with $F_\beta = F_\gamma$ coincide if they commute, or if $l \leq 4$ and both β and γ belong to a compact Lie subgroup of $\operatorname{Aut} K$.

If K has associative addition (e.g., if K is a quasifield), then $\operatorname{Aut} K$ is known to be a compact Lie group, see Salzmann [81a] 1.1, p. 347, Grundhöfer–Salzmann

[90] XI.9.5, p. 336, and, for quasifields, (81.5). It is an open problem, however, whether Aut K is compact in general or not. This causes the main difficulty in proving stiffness results. However, the compactness criteria proved in the sequel for a closed subgroup

$$\Lambda \leq \operatorname{Aut} K$$

will suffice for our purposes.

For the proofs, recall that K is (homeomorphic to) an affine line of $\mathcal{P}$; hence the one-point compactification $\widehat{K} = K \cup \{\infty\}$ is homeomorphic to a projective line. In particular, $\widehat{K}$ is metrizable and has a countable basis (41.8), and so also has Aut K, see (44.3 and 4c). Thus, convergence of sequences suffices to describe the topologies of $\widehat{K}$ and of Aut K.

83.6 Theorem: Baer groups. *If a closed subgroup $\Lambda \leq \operatorname{Aut} K$ fixes every element of a Baer subternary B of K, then Λ is compact, and $\dim \Lambda < l$.*

Proof. Postponing the proof of compactness of Λ, we first show how $\dim \Lambda < l$ follows from it. Since the action of Λ on $K \setminus B$ is free, see (83.4c), the orbits of Λ on $K \setminus B$ are homeomorphic to Λ, and $\dim \Lambda \leq l$. Moreover, if $\dim \Lambda$ were equal to l, then every orbit in $K \setminus B$ would be open (53.1a) as well as compact, hence equal to K by connectedness. This contradicts invariance of B.

We now turn to compactness of Λ. We may also consider Aut K as a group of homeomorphisms of $\widehat{K}$. Since K is locally connected (42.8), the compact-open topologies with respect to K and to $\widehat{K}$ coincide, see Arens [46] Theorems 1,3,4, p. 596–598. In order to prove compactness of Λ, it suffices as in the proof of (44.3) to show that Λ is equicontinuous on $\widehat{K}$. In terms of sequences, this means that, for every convergent sequence $x_\nu \xrightarrow[\nu]{} x$ in $\widehat{K}$, and for every sequence λ_ν in Aut K such that x^{λ_ν} converges to some $x' \in \widehat{K}$, the sequence $x_\nu^{\lambda_\nu}$ converges to x', as well. In a locally compact topological ternary field, this may be reduced to the special case $x = 0$, with $x' = 0$. Indeed, the case of an arbitrary $x \in K$ is implied by this special case because of continuity of addition and of the inverse function t of K, cf. (22.4 and 2). The case $x = \infty$ follows via the homeomorphism $\widehat{K} \to \widehat{K} : x \mapsto 1/x$ with $1/0 = \infty$ and $1/\infty = 0$, see (43.4).

From here on, the proof of (83.6) will be integrated into a more general argument. The special case which remains to be settled gives rise to the following notion. A subternary C of K will be called Λ-*even* if $c_\nu^{\lambda_\nu} \xrightarrow[\nu]{} 0$ whenever $c_\nu \in C$, $\lambda_\nu \in \Lambda$ and $c_\nu \xrightarrow[\nu]{} 0$. What has been proved up to now may be rephrased using this notion as follows.

83.7 Lemma. *If K is Λ-even, then Λ is compact.*

Obviously, a subternary C is Λ-even if Λ acts trivially on C. Hence, the proof of (83.6) is achieved by the following lemma.

83.8 Lemma. *If $C \leq H \leq K$ are closed subternaries such that C is a Baer subternary of H, and if C is Λ-even, then H is also Λ-even.*

Proof. For fixed $h \in H \setminus C$, the map $C \times C \to H : (s,t) \mapsto \tau(s,h,t)$ is a homeomorphism, see (41.11c). Let x_ν be a sequence in H converging to 0, and write $x_\nu = \tau(s_\nu, h, t_\nu)$ with $s_\nu, t_\nu \in C$; then $s_\nu \xrightarrow[\nu]{} 0$ and $t_\nu \xrightarrow[\nu]{} 0$. As C is Λ-even, we have $s_\nu^{\lambda_\nu} \xrightarrow[\nu]{} 0$ and $t_\nu^{\lambda_\nu} \xrightarrow[\nu]{} 0$. For arbitrary $\lambda_\nu \in \Lambda$, we must show that the sequence $x_\nu^{\lambda_\nu} = \tau(s_\nu^{\lambda_\nu}, h^{\lambda_\nu}, t_\nu^{\lambda_\nu})$ converges to 0. If this is not true, then there is a subsequence $x_\mu^{\lambda_\mu}$ which does not even accumulate at 0. It follows that $h^{\lambda_\mu} \xrightarrow[\mu]{} \infty$, hence $(1/h)^{\lambda_\mu} \xrightarrow[\mu]{} 0$. Rewriting x_μ as $x_\mu = \tau(s'_\mu, 1/h, t'_\mu)$ with $s'_\mu, t'_\mu \in C$ converging to 0, we see that $x_\mu^{\lambda_\mu} = \tau\left((s'_\mu)^{\lambda_\mu}, (1/h)^{\lambda_\mu}, (t'_\mu)^{\lambda_\mu}\right)$ converges to 0, a contradiction. □

83.9 Corollary. *Assume that Λ is connected. If there are closed subternaries $C \leq H \leq K$ such that $\dim C = 2$ and $\dim H = 4$, then Λ is compact, and $\widehat{K} \approx \mathbb{S}_l$.*

Proof. Recall, as in (83.4a), that C is a Baer subternary of H, and that either $H = K$ (in case $l = 4$), or H is a Baer subternary of K. The note following (55.6), applied to the nested sequence of closed subplanes of $\mathcal{P}$ corresponding to these subternaries and beginning with a subplane of dimension 4, says that $\widehat{K}$ is homeomorphic to a sphere. We now claim that C is Λ-even; then, by repeated application of (83.8) and by (83.7), it follows that Λ is compact.

The subternary $\langle 1 \rangle$ is Λ-even since Λ acts trivially on it. If $C = \langle 1 \rangle$, our claim is trivially true. If $C \neq \langle 1 \rangle$, then $\langle 1 \rangle$ is 1-dimensional by (55.4b); in particular, $\langle 1 \rangle$ is a Baer subternary of the 2-dimensional subternary C, so that C is Λ-even according to (83.8). □

83.10 Corollary. *If $\dim K = 4$ and Λ contains an involution, or if $\dim K = 8$, and Λ contains a pair of commuting involutions, then Λ is compact, and $\widehat{K} \approx \mathbb{S}_l$.*

Proof. An involution $\beta \in \Lambda$ gives rise to the Baer subternary F_β of K. If $\gamma \in \Lambda \setminus \{\beta\}$ is a second involution commuting with β, then $F_\beta \cap F_\gamma$ is a Baer subternary of F_β by (55.39a). The dimensions of these subternaries are $\dim F_\beta = 2$ for $l = 4$ and $\dim F_\beta = 4$, $\dim F_\beta \cap F_\gamma = 2$ for $l = 8$. By (83.9), the assertion follows. □

The subsequent stiffness theorems are not quite analogous for $l=4$ and $l=8$. The two cases will therefore be treated separately. The next result is a considerable improvement of the dimension bound in (83.6).

83.11 Theorem. *If* $\dim K = 4$ *and if* Λ *leaves a Baer subternary of K invariant, then Λ is compact and* $\dim \Lambda \leq 1$.

Proof. We may assume that Λ is connected (93.6). Then Λ acts trivially on the given invariant Baer subternary B, see (83.4d). By (83.9), Λ is compact, and $\widehat{K} \approx \mathbb{S}_4$. Richardson's classification (96.34) of compact groups acting on $\mathbb{S}_4$ now shows that the Λ-orbits are at most 1-dimensional. The assertion follows, as the action of Λ on $K \setminus B$ is free, see (83.4c). □

83.12 Theorem. *If* $\dim K = 4$, *and if K contains a proper closed, connected subternary, then* $\dim \Lambda \leq 3$. *In particular, this holds if* $\mathrm{Aut}\, K$ *contains an involution.*

Proof. First suppose that K has no closed subternary of dimension 2. By assumption, there is a closed subternary of dimension 1, which then coincides with $\langle 1 \rangle$, see (83.4e), and is a maximal closed subternary. As $\Lambda|_{\langle 1 \rangle} = \mathbb{1}$, maximality implies that $\Lambda_x = \mathbb{1}$ for all $x \in K \setminus \langle 1 \rangle$, see (83.4f and c). Hence, $\dim \Lambda \leq 3$, or else all orbits of Λ in $K \setminus \langle 1 \rangle$ would be 4-dimensional (96.10). In the latter case, Λ would have to be transitive on $K \setminus \langle 1 \rangle$ according to (53.1d) since, by reasons of dimension, $K \setminus \langle 1 \rangle$ is connected (51.21c). But this is impossible, because the set $\sqrt{-1}$ of 'square roots' of -1 is a Λ-invariant non-empty compact subset of $K \setminus \langle 1 \rangle$, see (55.20).

Now assume that K has a closed 2-dimensional subternary C. This is the case, for instance, if $\mathrm{Aut}\, K$ contains an involution. If $C = \langle 1 \rangle$, then $\Lambda|_C = \mathbb{1}$, and $\dim \Lambda \leq 1$ according to (83.11). If $\langle 1 \rangle$ is a proper subternary of C, then $\langle 1 \rangle$ is 1-dimensional (55.4b), hence a closed maximal subternary of C. In particular, for $x \in C \setminus \langle 1 \rangle$ we have $\langle x \rangle = C$ and hence $\Lambda_x|_C = \mathbb{1}$, $\dim \Lambda_x \leq 1$, see (83.11). If $\dim x^\Lambda \leq 1$, the assertion follows by the dimension formula (96.10). If $\dim x^\Lambda \geq 2$, we may use Richardson's classification (96.34), which tells us that then the compact group Λ^1, acting on $\widehat{K} \approx \mathbb{S}_4$ (83.9) with many fixed points, namely the elements of $\langle 1 \rangle$, is $\mathrm{SO}_3\mathbb{R}$. □

83.13 Corollary. *If* $\dim K = 4$, *and if Λ is compact, then* $\dim \Lambda \leq 3$.

Proof. By (93.6), one may assume that Λ is connected. First suppose that the connected component A of the center of Λ is non-trivial, and choose $c \in K$ such that $c^\mathrm{A} \neq \{c\}$. If $\langle c^\mathrm{A} \rangle \neq K$, then the assertion follows from (83.12). If $\langle c^\mathrm{A} \rangle = K$, then $\Lambda_c = \mathbb{1}$, see (83.4g), and $c^\Lambda \approx \Lambda$, cf. (96.9a). If $\dim \Lambda > 3$, then c^Λ would be open (53.1a) as well as closed, hence equal to K by connectedness, but K is not compact (42.5).

526　　8 Eight- and sixteen-dimensional planes

If, on the other hand, $A = 1\!\!1$, then the group Λ is a Lie group because its dimension is finite, see the end of the proof of (83.2). In particular, Λ contains an involution (94.31a), and the assertion follows again from (83.12). □

Until now, we have only been concerned with the stabilizer of a *non-degenerate* quadrangle in the automorphism group of $\mathcal{P}$, under the disguise of Aut K. A refined analysis of the stabilizers of both non-degenerate and degenerate quadrangles will be based on the following technical lemma.

83.14 Lemma. *For $l = 4$, let $\mathfrak{f}$ be a finite set of points containing three distinct points u, v, w on a line W and a point $a \notin W$, and consider a closed, connected subgroup Ω of $\Sigma_{\mathfrak{f}}$. Assume that, for every point $c \in M = aw \setminus \mathfrak{f}$, the subset $\{c\} \cup \mathfrak{f}$ is not contained in a proper closed, connected subplane. Then, $\dim \Omega \leq 6$; equality is possible only if Ω is semi-simple.*

Proof. 1) The group Ω fixes $\mathfrak{f}$ pointwise and acts effectively on M. The assumption implies that no involution in Ω is a Baer involution; hence, an involution in Ω is a reflection with axis W and center a (23.17 and 8). In particular, Ω_c does not contain any involution, and by (55.32ii) there is no pair of commuting involutions in Ω.

2) For a connected non-trivial subgroup Δ of Ω, there is $c \in M$ such that $c^\Delta \neq \{c\}$. The set $c^\Delta \cup \mathfrak{f}$ contains the non-degenerate quadrangle $\{a, c, u, v\}$, and c^Δ is connected. Thus, the subplane $\langle c^\Delta, \mathfrak{f} \rangle$ is connected, so that, by assumption, $\langle c^\Delta, \mathfrak{f} \rangle = \mathcal{P}$. For a closed subgroup Γ of Ω centralizing Δ, the argument of (83.1d) says that $\Gamma_c = 1\!\!1$. Consequently, $\dim \Gamma \leq 4$ by the dimension formula (96.10).

3) Now assume that Ω is not semi-simple and that $\dim \Omega > 4$. We shall show that then $\dim \Omega \leq 5$. By step 2), Ω does not contain a non-trivial compact, connected subgroup in its center, and hence has a minimal closed, connected commutative normal subgroup $\Xi \cong \mathbb{R}^s$, see (94.26). Choose $c \in M$ such that $c^\Xi \neq \{c\}$. Then $\Xi_c = 1\!\!1$, and $s \leq 4$ by the dimension formula. The action of Ω_c on $\Xi \cong \mathbb{R}^s$ by conjugation is linear. Therefore, the centralizer in Ω_c of an element $\xi \in \Xi \setminus 1\!\!1$ centralizes the 1-dimensional subspace generated by ξ, and hence is trivial by step 2). In other words, the linear action of Ω_c on Ξ is free; in particular, Ω_c is a Lie group (94.3e) of dimension at most 4. Considering the faithful irreducible action of $(\Omega_c)^1$ on a minimal invariant subspace H of Ξ, we obtain from (95.6) that $(\Omega_c)^1 = \Psi Z$, where Ψ is semi-simple (in fact, the commutator subgroup), and Z is the center; moreover, Ψ is a closed subgroup of GL(H), and Z is isomorphic to a closed subgroup of $\mathbb{C}^\times$. If the linear semi-simple group Ψ were non-trivial, it would contain an involution (94.37 and 31a), in contradiction to step 1). For the same reason, $Z \cong \mathbb{C}^\times$ is impossible. Therefore, $(\Omega_c)^1 = Z$ is isomorphic to

a proper closed subgroup of $\mathbb{C}^\times$. In particular, $\dim \Omega_c \leq 1$, see (94.3f), whence $\dim \Omega \leq 5$ by the dimension formula.

4) Almost simple Lie groups have dimension either 3 or at least 6. If Ω is semi-simple, but not almost simple, we apply step 2) to products of almost simple factors of Ω, see (94.25), and obtain that these factors are of dimension 3, and that there are just two of them. Thus, in this case we have $\dim \Omega \leq 6$.

5) Finally, we assume that Ω is almost simple and that $\dim \Omega > 6$, and we derive a contradiction from this. Let Θ be a compact 0-dimensional subgroup such that Ω/Θ is an almost simple Lie group (94.24 and 25). By (94.27), there is a (possibly non-closed) subgroup $\widehat{\Omega} \leq \Omega$ which is the monomorphic image of a Lie group G covering Ω/Θ; in particular, G is almost simple. According to step 1), a torus subgroup of G is at most 1-dimensional. Hence, by the classification of almost simple Lie groups (94.32e and 33), G is locally isomorphic to $SL_3\mathbb{R}$ or isomorphic to the universal covering group of $SU_3(\mathbb{C}, 1)$ or of $Sp_4\mathbb{R}$. Now $\widehat{\Omega}$ cannot act trivially on W, since $\Omega_{[W]} = \Omega_{[a,W]}$ has dimension at most 4, see (61.4a). Thus the kernel of the action of G on W is discrete and is contained in the center of G; recall that G is almost simple and use (93.18). In particular, by step 1), every involution of $\widehat{\Omega}$ belongs to the center. But the universal covering group $SU_3(\mathbb{C}, 1)^\sim$ contains a subgroup isomorphic to $SU_2\mathbb{C}$ whose involution is not central in $SU_3(\mathbb{C}, 1)^\sim$. In the remaining cases $SL_3\mathbb{R}$ and $Sp_4\mathbb{R}$ we shall obtain subgroups which contradict the preceding arguments. In $SL_3\mathbb{R}$, the stabilizer of a 1-dimensional subspace of $\mathbb{R}^3$ is a 6-dimensional subgroup which is not semi-simple. In $Sp_4\mathbb{R}$, the stabilizer of a 1-dimensional subspace of $\mathbb{R}^4$ has dimension 7. Therefore, Ω/Θ and hence Ω contain closed subgroups of dimension at least 6 which are not semi-simple. This contradicts step 3). □

83.15 Corollary. *A closed subgroup Λ of the automorphism group of a 4-dimensional locally compact ternary field K satisfies* $\dim \Lambda \leq 5$.

Proof. The subgroup $\Lambda \leq \mathrm{Aut}\, K$ corresponds to a closed subgroup Ω of Σ fixing a non-degenerate quadrangle $\{a, u, v, e\}$. The point $w = ae \wedge uv$ is fixed, as well. If K has proper closed, connected subternaries, in particular, if Λ contains involutions, then $\dim \Lambda \leq 3$ by (83.12). We now assume that this is not the case. Then, for $\mathfrak{f} := \{a, u, v, w, e\}$ and $c \in aw \setminus \mathfrak{f}$, the subset $\{c\} \cup \mathfrak{f}$ is not contained in a proper closed, connected subplane, and we infer from (83.14) that either $\dim \Lambda \leq 5$ or else Λ is semi-simple and $\dim \Lambda = 6$.

We now assume the latter and derive a contradiction. As Λ does not contain involutions, its almost simple factors have dimension 3, see (94.25 and 37), and Λ is the product $A \cdot B$ of two such factors. Let P be a closed, connected commutative subgroup of A, e.g., the closure of a one-parameter subgroup. By our assumption about subternaries, the set F of fixed elements of P is totally disconnected; in particular, $\dim F = 0$, see (92.18). The elements of F are also fixed under the connected, closed subgroup $\Gamma = \mathsf{PB}$, which centralizes P. For $c \in K \setminus F$, the closed, connected subternary $\langle c^\mathsf{P} \rangle$ is equal to K by assumption, hence we have $\Gamma_c = 1\!\!1$ by the argument of (83.4g). As Γ is at least 4-dimensional, the dimension formula (96.10) yields $\dim c^\Gamma = 4$, so that c^Γ is open in K and $K \approx \mathbb{R}^4$ by (53.2); moreover, Γ is a Lie group. By the Mal'cev–Iwasawa theorem (93.10), as there are no involutions, Γ has no compact subgroup except $1\!\!1$, cf. (94.31), and $c^\Gamma \approx \Gamma \approx \mathbb{R}^4$. Now $K \setminus F$ is connected, see (51.15); consequently $c^\Gamma = K \setminus F$, so that finally $K \setminus F \approx \mathbb{R}^4 \approx K$. But this is forbidden by Alexander duality $H_3(K \setminus F) \cong \check{H}_c^0(F)$, see Dold [82] Chap. VIII 8.15, p. 301; accordingly, $\check{H}_c^0(F)$ would have to be 0, which is not the case, as F is totally disconnected. □

83.16 Remark. The bound given in (83.15) is not sharp. In fact, Bödi [94a] has proved $\dim \Lambda \leq 4$ even more generally for the automorphism group Λ of a 4-dimensional locally compact double loop; the proof is rather involved. Yet we conjecture that even this bound is not sharp and that (83.12) holds without any assumption on subternaries. If the closed subternary $\langle 1 \rangle$ of K is connected, then either $\langle 1 \rangle = K$, in which case $\mathrm{Aut}\, K = 1\!\!1$, or the assumptions of (83.12) are satisfied. But $\langle 1 \rangle$ might be totally disconnected, although no such examples are known.

However, the following dimension bound for the stabilizer of a *degenerate* quadrangle in the automorphism group of the plane $\mathcal{P}$ is sharp.

83.17 Theorem. *For $l = 4$, the stabilizer of a degenerate quadrangle*

$$\mathfrak{q} = \{a, u, v, w\} \quad \text{with} \quad a \notin uv = vw = uw =: W$$

satisfies $\dim \Sigma_\mathfrak{q} \leq 7$.

Proof. By (93.6), it suffices to show that $\dim \Omega \leq 7$ for every closed, connected subgroup Ω of $\Sigma_\mathfrak{q}$. The connected group Ω fixes $\mathfrak{q}$ pointwise. For $e \in aw \setminus \{a, w\}$, the stabilizer Ω_e corresponds to a subgroup of the automorphism group of the ternary field K coordinatizing the plane with respect to the non-degenerate quadrangle $\{a, u, v, e\}$. If there is $e \in aw \setminus \{a, w\}$ such that $\{e\} \cup \mathfrak{q}$ is contained in a proper closed, connected subplane, then K has proper closed, connected subternaries, and $\dim \Omega_e \leq 3$ by (83.12). Because of $e^\Omega \subseteq aw$, the dimension formula (96.10) then says that $\dim \Omega \leq 3+4 = 7$. If for $e \in aw \setminus \{a, w\}$ the set $\{e\} \cup \mathfrak{q}$ is never contained in a proper closed, connected subplane, then the assertion follows from (83.14). □

Most of the preceding results for $l = 4$ refer to automorphism groups of coordinatizing ternary fields. Before going on, we reformulate some of these results in

83.18 Corollary. *Let Φ be a closed subgroup of the group of automorphisms of a plane of dimension $2l = 8$, and assume that there is a closed subplane $\mathcal{Q}$ of dimension d which is fixed pointwise by Φ. Then, the following assertions hold.*

(a) *If $d = 4$, i.e., if $\mathcal{Q}$ is a Baer subplane, then Φ is compact and $\dim \Phi \leq 1$.*
(b) *If $d = 2$, then $\dim \Phi \leq 3$.*
(c) *If $d = 0$, then $\dim \Phi \leq 5$; if, moreover, Φ is compact, then $\dim \Phi \leq 3$.* □

83.19 Corollary. *Let Δ be a closed subgroup of the group of automorphisms of a plane of dimension $2l = 8$, and assume that Δ leaves a closed subplane $\mathcal{Q}$ of dimension d invariant. Furthermore, let Ψ be a compact subgroup of Δ. Then, the following assertions hold.*

(a) *If $d = 4$, i.e., if $\mathcal{Q}$ is a Baer subplane, then $\dim \Delta \leq 17$, and $\dim \Psi \leq 9$.*
(b) *If $d = 2$, then $\dim \Delta \leq 11$, and $\dim \Psi \leq 6$.*
(c) *If $d = 0$, then $\dim \Delta \leq 5$, and $\dim \Psi \leq 3$.*

Proof. For $d \in \{2, 4\}$ we have $\dim \mathrm{Aut}\, \mathcal{Q} \leq 4d$ by (32.21b) and (71.2). The same also holds for $d = 0$, for then $\mathcal{Q}$, and hence $\mathrm{Aut}\, \mathcal{Q}$, are totally disconnected (92.5). The assertions about $\dim \Delta$ now follow immediately from (83.18) by (83.3).

As to $\dim \Psi$, let $\Theta = \Psi|_{\mathcal{Q}}$ be the compact group of automorphisms induced by Ψ on $\mathcal{Q}$. Again, $\dim \Theta = 0$ if $d = 0$; furthermore, we know that $\dim \Theta \leq 3$ for $d = 2$ and $\dim \Theta \leq 8$ for $d = 4$, see (32.22) and (71.9). The estimates for $\dim \Psi$ then are obtained again by (83.18 and 3). □

83.20 Corollary. *In a plane $\mathcal{P}$ of dimension $2l = 8$, a semi-simple, closed, connected subgroup Δ of $\mathrm{Aut}\, \mathcal{P}$ which is not almost simple cannot leave a Baer subplane invariant. In particular, an involution of $\mathrm{Aut}\, \mathcal{P}$ centralizing Δ is a reflection.*

Proof. Assume that Δ is semi-simple and leaves a Baer subplane $\mathcal{B}$ invariant. By (83.18a), the kernel Φ of the action of Δ on $\mathcal{B}$ is at most 1-dimensional. According to (94.39), the connected component Φ^1 is abelian. From the definition of semi-simplicity, one concludes that Φ^1 is trivial, whence $\dim \Phi = 0$, cf. (92.18). The semi-simple group Δ/Φ acts effectively as a group of automorphisms on the 4-dimensional plane $\mathcal{B}$. Hence Δ/Φ is almost simple (71.8), and so is Δ.

In particular, if Δ is not almost simple, then an involutory collineation ι centralizing Δ cannot be a Baer involution, since Δ leaves the set of fixed elements of ι invariant. Thus, ι is a reflection, see (23.17) and (55.28). □

83.21 Corollary. *In a plane $\mathcal{P}$ of dimension $2l = 8$, let Δ be a connected subgroup of $\operatorname{Aut} \mathcal{P}$ which is centralized by a closed, connected subgroup Γ, and assume that the orbit x^Δ of a point x contains a non-degenerate quadrangle (this is equivalent to x^Δ not being contained in a line). Then, the following assertions hold.*

(a) *The stabilizer Γ_x satisfies $\dim \Gamma_x \leq 3$.*
(b) *If $\langle x^\Delta \rangle$ is a Baer subplane, then we even have $\dim \Gamma_x \leq 1$.*
(c) *If $\langle x^\Delta \rangle = P$, then $\Gamma_x = \mathbb{1}$.*

If $\dim x^\Delta > 2$, one of the two latter cases occurs.

Proof. The equivalence of the assumptions about x^Δ has been observed in (42.3). The stabilizer Γ_x fixes $\langle x^\Delta \rangle$ pointwise (83.1d). So, if $\langle x^\Delta \rangle = P$, then $\Gamma_x = \mathbb{1}$. In the other cases, the estimate for $\dim \Gamma_x$ is immediate from (83.18). If $\dim x^\Delta > 2$, then the only possible dimensions for the subplane $\langle x^\Delta \rangle$ are 4 and 8. □

We now turn to 16-dimensional planes (with $l = 8$).

83.22 Lemma. *A closed subgroup Φ of $\operatorname{Aut} K$ which acts trivially on a closed Baer subternary is compact and does not contain two distinct commuting involutions. In particular, Φ does not contain a 2-torus. Hence, if Φ is a Lie group, then Φ^1 is either trivial or isomorphic to $\operatorname{SO}_2 \mathbb{R}$ or to $\operatorname{Spin}_3 \mathbb{R}$.*

Proof. Compactness has already been proved in (83.6), and non-existence of commuting involutions is clear by (55.32), compare (83.5). The compact, connected Lie groups of torus rank at most 1 are known, see the remark following (94.31). The group $\operatorname{SO}_3 \mathbb{R}$, which belongs to these groups, cannot occur here because it contains pairs of distinct commuting involutions. □

Our results for $l = 4$ did not comprise a sharp bound for $\dim \operatorname{Aut} K$, see (83.16). Astonishingly, they nevertheless suffice to obtain a sharp bound in the case $l = 8$.

83.23 Theorem. *If $l = 8$, then either $\dim \operatorname{Aut} K \leq 13$, or the connected component of $\operatorname{Aut} K$ is the compact exceptional simple Lie group $G_2 = G_2(-14)$.*

Proof. Suppose that $\operatorname{Aut} K$ has dimension at least 14; then the same is true for $\Lambda = (\operatorname{Aut} K)^1$, cf. (93.6). We show that $\Lambda \cong G_2$.

1) *If Δ is a connected subgroup of Λ and $c \in K$ is not fixed under Δ, then either $\langle c^\Delta \rangle$ is a Baer subternary of K, or $\langle c^\Delta \rangle = K$.*

As Δ acts non-trivially on $\langle c^\Delta \rangle$, this follows from (83.4d and f).

2) *If $c \in K$ is not fixed under Λ, then $\langle c^\Lambda \rangle = K$.*

If $B := \langle c^\Lambda \rangle \neq K$, then B is a Baer subternary by step 1). According to (83.6), the kernel $\Lambda_{[B]}$ is compact and has dimension at most 7. Using (83.3), we conclude from (83.15) that $\dim \Lambda \leq 5 + 7$, which contradicts our assumption.

3) *If $c \in K$ is not fixed under Λ, then Λ_c is a Lie group.*

Indeed, Λ has a compact central subgroup Θ such that Λ/Θ is a Lie group (93.8c and 18). Now $\Lambda_c \cap \Theta = \Theta_c = 1\!\!1$ since $\langle c^\Lambda \rangle = K$ by step 2), compare the argument of (83.4g). Thus, Λ_c admits a continuous monomorphism into the Lie group Λ/Θ, and assertion 3) follows from (94.3e).

4) *If a closed subgroup Φ of Λ induces the identity on some closed Baer subternary B of K, then Φ is a compact Lie group, and $\dim \Phi \leq 3$.*

There is $c \in B$ such that $c^\Lambda \neq \{c\}$, since otherwise Λ would have to act freely on $K \setminus B$, see (83.4c), which is impossible by reasons of dimension (96.10). From steps 2) and 3), we infer that $\Phi \leq \Lambda_c$ is a Lie group, cf. (94.3a). The other assertions now are immediate from (83.22).

5) *If a closed subgroup Γ of Λ centralizes a connected subgroup $\Delta \neq 1\!\!1$, and if $c^\Delta \neq \{c\}$, then $\dim \Gamma_c \leq 3$, so that $\dim \Gamma \leq 11$.*

Indeed, Γ_c induces the identity on $\langle c^\Delta \rangle$, see (83.4g), so that $\dim \Gamma_c \leq 3$ by steps 1) and 4), and hence $\dim \Gamma \leq 3 + 8 = 11$ by the dimension formula (96.10).

6) In this step, we shall assume that Λ is not semi-simple, and derive a contradiction. By (94.26), then, Λ has a minimal non-trivial closed, connected commutative normal subgroup Ξ, and $\Xi \cong \mathbb{R}^s$, because Ξ cannot be a central torus by step 5). Choose $c \in K$ such that $c^\Xi \neq \{c\}$. Then Ξ_c acts trivially on $\langle c^\Xi \rangle$ by (83.4g), so that, according to steps 1) and 4), the stabilizer Ξ_c is compact and hence trivial. In particular, $s \leq 8$ by the dimension formula (96.10).

The action of Λ on $\Xi \cong \mathbb{R}^s$ by conjugation is linear. The group

$$\Gamma := (\Lambda_c)^1$$

is a connected Lie group, compare step 3). Consider a minimal Γ-invariant subspace $\Theta \cong \mathbb{R}^t$ of $\Xi \cong \mathbb{R}^s$. The centralizer $\mathrm{Cs}_\Gamma \vartheta$ of $\vartheta \in \Theta \setminus 1\!\!1$ also centralizes the 1-dimensional subspace of $\Xi \cong \mathbb{R}^s$ generated by ϑ. As $\Xi_c = 1\!\!1$, we know $c^\vartheta \neq c$, and from step 5) we infer that $\dim \mathrm{Cs}_\Gamma \vartheta \leq 3$. Applying the dimension formula (96.10) to the actions of Λ on K and of Γ on Θ, and using $\dim \Gamma = \dim(\Lambda_c)^1 = \dim \Lambda_c$, see (93.6), we obtain that $6 \leq \dim \Gamma \leq t + 3 \leq 11$, whence $t \geq 3$ and $\dim \Gamma + t \geq 9$.

We now show that $\langle c^\Theta \rangle = K$. Assume that this is not the case. According to step 1), as $\Theta_c = 1\!\!1$, the only other possibility is that $\langle c^\Theta \rangle$ is a Baer subternary. The semidirect product $\Gamma\Theta$ leaves $\langle c^\Theta \rangle$ invariant. Using (83.3) we obtain from (83.15) and step 4) that $\dim \Gamma + t = \dim \Gamma\Theta \leq 5 + 3$, which contradicts the results of the previous paragraph. Thus, indeed, $\langle c^\Theta \rangle = K$.

Hence, $\mathrm{Cs}_\Gamma \Theta = 1\!\!1$ by (83.4g); in other words, the linear representation of Γ on Θ is faithful. By minimality of Θ, this representation is irreducible. Hence, according to (95.6), the commutator subgroup Γ' is closed and semi-simple and has dimension $\dim \Gamma' \geq \dim \Gamma - 2$. Thus, our estimates of $\dim \Gamma$ imply $4 \leq \dim \Gamma' \leq 11$.

An almost simple factor of Γ' cannot be isomorphic to the universal covering group of $\mathrm{SL}_2\mathbb{R}$, since that group does not have a faithful linear representation, see (95.9 and 10). Hence, by (94.37 and 31), every almost simple factor contains an involution. On the other hand, as Λ is not compact, it does not contain a pair of commuting involutions (83.10), hence Γ' is almost simple, and the tables in (95.10) of almost simple groups admitting faithful irreducible representations in low dimensions show that $\Gamma' \cong \mathrm{SL}_2\mathbb{C}$. There is a central involution $\alpha \in \Gamma'$. On the corresponding Baer subternary F_α, see (83.5), the action of Γ' has a kernel of dimension at least 1 by (83.15 and 3). Hence the almost simple group Γ' acts trivially on F_α and is compact by (83.6), a contradiction. Thus we have proved:

7) Λ *is semi-simple.*

Now let G be a semi-simple Lie group admitting a continuous monomorphism into Λ and satisfying $\dim G = \dim \Lambda$; such a Lie group exists by (93.8) and (94.27). We prove the following assertion.

8) *Every almost simple factor of G is at least 6-dimensional and contains a torus, and Λ contains a 2-dimensional torus.*

If G is almost simple itself, this is immediate, since every almost simple Lie group of dimension at least 14 contains a 2-dimensional torus (94.32e and 33). Now assume that G is not almost simple, and let A be an almost simple factor of G. Then $G = A \cdot Z$, where Z is the product of the other almost simple factors of G, which centralizes A, see (94.23). We now use a closed one-parameter subgroup R of A to conclude from step 5) that $\dim Z < \dim RZ \leq 11$; here, step 5) is applied to the closure of the image of RZ in Λ. Since $\dim A + \dim Z = \dim G \geq 14$, it follows that $\dim A \geq 4$. Hence, $\dim A \geq 6$, by the classification of almost simple Lie groups (94.32 and 33), and A contains a torus, cf. (94.37). The same is true for a second almost simple factor, so that G and hence Λ contain a 2-torus.

9) Λ *is a compact, semi-simple Lie group, and $\dim \Lambda \leq 28$.*

Using (83.10) we infer from step 8) that Λ is compact and that $\widehat{K} \approx \mathbb{S}_8$. The dimension bound may then be obtained from (96.13b). A finite-dimensional semi-simple compact group is always a Lie group, see the end of the proof of (83.2).

10) *If an almost simple closed subgroup Ψ of Λ centralizes an involution $\alpha \in \Lambda$, then* dim $\Psi = 3$.

Indeed, Ψ leaves the Baer subternary F_α invariant, and because of simplicity, Ψ acts either trivially or almost effectively on F_α. In the first case, one may use step 4). In the second case, the compact group $\Psi|_{F_\alpha}$ contains an involution (94.27), and the assertion follows from (83.12) applied to F_α.

11) By steps 8), 9), and 10) we now conclude that Λ is almost simple. From the list of almost simple compact Lie groups of dimensions between 14 and 28, see (94.33), all groups other than $\Lambda \cong G_2$ are excluded by step 10). □

83.24 Addendum. *If $\Lambda = (\operatorname{Aut} K)^1$ is isomorphic to G_2, then $\widehat{K} \approx \mathbb{S}_8$, and the subternary of fixed elements of Λ is homeomorphic to $\mathbb{R}$.*

Proof. The group G_2 contains two commuting involutions (11.34), so that the first part follows from (83.10). The action of $\Lambda \cong G_2$ on $\widehat{K} \approx \mathbb{O} \cup \{\infty\}$ is classical (96.35); in particular, the set of fixed points on K is homeomorphic to $\mathbb{R}$, as with the classical action of $G_2 = \operatorname{Aut} \mathbb{O}$, cf. (11.28 and 30). □

83.25 Note. For the dimension of a closed, connected subgroup Λ of $\operatorname{Aut} K$, better bounds than the one given in (83.23) are known. For a compact group Λ, Salzmann [79b] has shown that it is either isomorphic to G_2, $\operatorname{SU}_3\mathbb{C}$ or to $\operatorname{SO}_4\mathbb{R}$, or satisfies dim $\Lambda \leq 5$. Bödi [95] shows in general that if $\Lambda \not\cong G_2$ then dim $\Lambda \leq 11$. If the set of fixed elements of Λ is connected and if Λ is a Lie group, then he even proves that Λ is either isomorphic to G_2 or to $\operatorname{SU}_3\mathbb{C}$, or satisfies dim $\Lambda \leq 7$, see Bödi [94c]. If K is a quasifield, the list of possible automorphism groups can be narrowed down further, see (81.6) and the references given there.

For automorphism groups of the plane $\mathcal{P}$, the preceding results, notably (83.17) and (83.23), have the following consequences.

83.26 Corollary: Stabilizers of subsets of triangles. *Upper bounds for the dimensions of the stabilizers of points and lines in $\Sigma = \operatorname{Aut} \mathcal{P}$ are given by the following table, for a line W, two points $u \neq v$ on W, and a point $a \notin W$.*

Stabilizer of		dim $\mathcal{P} = 8$	dim $\mathcal{P} = 16$
triangle	dim $\Sigma_{a,u,v} \leq$	11	30
two points	dim $\Sigma_{u,v} \leq$	19	46
flag	dim $\Sigma_{u,W} \leq$	23	54
antiflag	dim $\Sigma_{a,W} \leq$	19*	46*
point	dim $\Sigma_a \leq$	27*	62*
line	dim $\Sigma_W \leq$	27*	62*
	dim $\Sigma \leq$	35*	78*

*The bounds marked with an asterisk * may be lowered by 1 if the plane is not isomorphic to the classical plane $\mathcal{P}_2 \mathbb{F}$, $\mathbb{F} \in \{\mathbb{H}, \mathbb{O}\}$.*

Remark. In the classical planes $\mathcal{P}_2 \mathbb{F}$, the bounds are attained for $\Sigma = \text{Aut}\,\mathcal{P}_2 \mathbb{F}$. Several of the corresponding stabilizers have been studied explicitly in Chapter 1, see for instance (12.10 and 17), (13.6), (17.11 and 14). The bounds for non-classical planes will be used in Sections 84 and 85, and will eventually be refined considerably, see (84.27, 9, 14, 16 and 19) and (85.15, 5, 6, and 10); compare also (84.28) and (85.16). The bounds for dim Σ have already been announced in (65.2).

Proof of (83.26). First, we treat an 8-dimensional plane $\mathcal{P}$. Let w be a third point on W. By (83.17), we have the estimate dim $\Sigma_{a,u,v,w} \leq 7$. The dimension bounds for $\Sigma_{a,u,v}$, for $\Sigma_{u,v}$, and for $\Sigma_{u,W}$ follow from the dimension formula (96.10) applied to the actions of these groups on W and on P.

Now let $\Delta = \Sigma_{a,W}$. Again by repeated application of the dimension formula on W, we obtain that dim $\Delta \leq 2 \cdot 4 + \text{dim}\,\Sigma_{a,u,v} \leq 2 \cdot 4 + 11$. It even follows that dim $\Delta \leq 18$ unless, for every choice of different points u, v on W, the orbits u^Δ and v^{Δ_u} have dimension $4 = \text{dim}\,W$. In the latter case, Δ is doubly transitive on W by (53.1d). Then, $\mathcal{P}$ is classical according to (63.5) and (64.18).

From these bounds for dim $\Sigma_{a,W}$, the dimension bounds for Σ_a, for Σ_W, and for Σ now are obtained again by the dimension formula applied in the point space P and in the line space $\mathcal{L}$.

In the case dim $\mathcal{P} = 16$, the dimension bounds for $\Sigma_{a,u,v}$, for $\Sigma_{u,v}$, and for $\Sigma_{u,W}$ follow directly from the sharp bound (83.23) available in this case for the stabilizer of a *non*-degenerate quadrangle a, u, v, e. The rest of the argument is analogous to the 8-dimensional case.

The group of all automorphisms of the classical plane over the quaternions or the octonions has dimension 35 or 78, respectively, see (13.8) and (17.10), and is transitive on the set of non-degenerate quadrangles, cf. (13.5) and (17.6). From this, it follows readily that the given bounds are sharp, by multiple use of the dimension formula as above. □

83.27 Notes on Section 83. The results of this section are contained in Salzmann [79a] for 8-dimensional planes and in Salzmann [79b] for 16-dimensional planes; there, additional information may be found. The technique of condensing a large part of the analysis of stabilizers of both non-degenerate and degenerate quadrangles (83.15 and 17) into the proof of a unified statement (83.14) is a specific feature of the presentation here. In the case of equality in the estimate (83.17), the stabilizer of a degenerate quadrangle can be shown to be the same as in the classical plane over the quaternions, see Salzmann [79a]. The proof of (83.23) has been simplified. Further results on these topics may also be found in Salzmann [81a] for 8-dimensional planes and in Salzmann [83, 84 and 87] for 16-dimensional planes. For locally compact, connected quasifields (the coordinate domains of compact, connected translation planes) it was possible to obtain sharper results than in general, see Hähl [75a,b, 76] and also (81.5 and 6). On the other hand, many results about automorphism groups of locally compact, connected ternary fields can be proved more generally for locally compact, connected double loops, see Grundhöfer–Salzmann [90]. Further progress in these matters was recently made by Bödi [93a,b], [94a–c], [95], see also (83.16 and 25). Concerning stiffness properties in the more general context of stable planes, see Stroppel [92a]; there, pointwise stabilizers of subplanes are studied.

84 Characterizing $\mathcal{P}_2\mathbb{H}$

Throughout this section, we consider an 8-dimensional compact projective plane $\mathcal{P} = (P, \mathcal{L})$.

We show that $\mathcal{P}$ is isomorphic to $\mathcal{P}_2\mathbb{H}$, if $\dim \operatorname{Aut} \mathcal{P} \geq 23$. For the proof, we distinguish the different possibilities for the structure of the connected component of $\operatorname{Aut} \mathcal{P}$; namely the cases where the connected component is compact, semi-simple, almost simple, or not semi-simple. In the last case, our aim will be to show that — up to duality — every sufficiently large group contains a transitive translation group. Thus our problem will be reduced to the classification of translation planes, as presented in Section 82. A refinement of the methods that are used here (in particular, for the case where Δ is not semi-simple) has led to much stronger classification results, cf. (84.28). Within the scope of this book, we merely attempt to introduce the methods, but refrain from proving the best results that are presently known.

We start with the case where the structure theory is most powerful.

Compact groups

For the following collection of lemmas, let Φ be a compact connected group of automorphisms of $\mathcal{P}$.

We shall repeatedly use Weil's structure theorem for connected compact groups, see (93.11): such a group Φ is an almost direct product of a compact, connected abelian group Z and the commutator group Φ', and Φ' is an almost direct product of a number of almost simple compact Lie groups. Note that Z equals the connected component of the center of Φ.

84.1 Lemma. *If Φ has an orbit of dimension 8 in $P \cup \mathcal{L}$, then Φ is isomorphic to $\mathrm{PU}_3\mathbb{H}$, and $\mathcal{P}$ is isomorphic to $\mathcal{P}_2\mathbb{H}$.*

Proof. Every Φ-orbit is compact. From (53.2) we infer that the existence of an orbit of full dimension implies transitivity. Now (63.8) yields the assertion. □

84.2 Lemma. *If Φ leaves a proper closed subplane invariant, then $\dim \Phi \leq 9$.*

Proof. The group Φ induces an effective group of dimension at most 8 on the invariant subplane $\mathcal{E}$, cf. (71.9), and this bound reduces to 3 if $\mathcal{E}$ is not a Baer subplane (i.e., has dimension 0 or 2), cf. (32.16). The kernel of the restriction has dimension at most 3 by (83.18). If $\mathcal{E}$ is a Baer subplane, this bound reduces to 1. In each of the cases, the assertion follows from (83.3). □

84.3 Lemma. *If Φ centralizes a non-trivial automorphism ζ, then $\dim \Phi \leq 9$, or Φ fixes a point or a line.*

Proof. We consider the action of Φ on the set of fixed elements of ζ. Recall that this set is non-empty by (55.19). We may assume that there is a point p which is fixed by ζ but moved by Φ. If p^Φ generates a subplane, then ζ acts trivially on the closure $\mathcal{E}$ of this subplane. Hence, $\mathcal{E}$ is a proper subplane, and (84.2) yields that $\dim \Phi \leq 9$. If the orbit p^Φ does not generate a subplane, then p^Φ is contained in some line by (42.3), and this line is fixed by Φ. □

Much more information is available if the lines of $\mathcal{P}$ are manifolds.

84.4 Lemma. *Assume that Φ fixes a line L, and that L is a manifold. Then the following hold.*

(a) *Either Φ is transitive on L, or the induced group $\Phi|_L = \Phi/(\Phi_{[L]})$ has dimension at most 6.*

(b) *The kernel $\Phi_{[L]}$ has dimension at most 3. Consequently, $\dim \Phi \leq 9$ unless Φ is transitive on L.*
(c) *If Φ is transitive on L, then Φ is isomorphic either to $\mathrm{Spin}_5\mathbb{R}$ or to the quotient $(\mathrm{Spin}_5\mathbb{R} \times \mathsf{K})/\mathsf{Z}$, where $\mathsf{K} \in \{\mathrm{Spin}_3\mathbb{R}, \mathbb{T}\}$, and Z is generated by the pair of central involutions in $\mathrm{Spin}_5\mathbb{R}$ and K.*

Proof. Since L is a manifold, we know from (96.13b) that either $\dim \Phi/(\Phi_{[L]}) \leq 6$, or Φ acts transitively on L. This proves (a). The dimension bound in (b) has been proved in (61.26).

Now assume that Φ is transitive. From the dual of (62.8) we infer that Φ contains a subgroup Δ isomorphic to $\mathrm{Spin}_5\mathbb{R}$. On $L \approx \mathbb{S}_4$, the group Φ induces the usual action of $\mathrm{SO}_5\mathbb{R}$, see (96.23). This implies that we may choose Δ as one of the almost simple factors of Φ. Then the central involution δ of Δ even belongs to the center of Φ, and we know from (55.32ii) that $\Phi_{[L]}$ contains no involution except δ. Since the connected component Ω of the kernel $\Phi_{[L]}$ is a connected Lie group (62.11), this implies that it is either isomorphic to $\mathrm{Spin}_3\mathbb{R}$, or isomorphic to $\mathrm{SO}_2\mathbb{R}$, or trivial; recall that these three are the only compact, connected Lie groups that do not contain commuting involutions. Since $\Phi = \Delta\Omega$, the proof of assertion (c) is now completed by the observation that $\Delta \cap \Omega = \langle\delta\rangle$. □

If L is not a manifold, then (96.13b) does not apply. Restriction to the orbits on L (in order to apply (96.13a)) does not help much, since it is conceivable that Φ acts with different kernels on different orbits in L. However, this phenomenon is excluded in the case of almost simple groups.

84.5 Lemma. *If Φ is almost simple of dimension at least 10, then either $\Phi \cong \mathrm{Spin}_5\mathbb{R}$, or Φ is simple and $\dim \Phi \geq 14$.*

Proof. Assume that Φ is almost simple, but not simple, and that $\dim \Phi \geq 10$. Up to duality, there exists a point p that is fixed by Φ, cf. (84.3). Now (62.8c) yields that $\Phi \cong \mathrm{Spin}_5\mathbb{R}$. From the classification (94.33), one infers that each compact simple group of dimension at least 10 either is isomorphic to $\mathrm{SO}_5\mathbb{R}$, or has dimension at least 14. The group $\mathrm{SO}_5\mathbb{R}$ does not act by (55.40). □

84.6 Lemma. *If Φ is not semi-simple, then $\dim \Phi \leq 11$.*

Proof. Assume that Φ is not semi-simple. According to the structure theorem (93.11), the connected component Z of the center of Φ is non-trivial. If there exists some $x \in P \cup \mathcal{L}$ such that x^{Z} topologically generates a subplane $\mathcal{E}$, then Φ_x acts trivially on $\mathcal{E}$, and $\dim \Phi \leq 7 + 3 = 10$ by (84.1) and (83.18). If there exists no such point or line, we infer from (23.16) that Z acts trivially on a Baer subplane, or Z has a center and an axis. In the first case, $\dim \Phi \leq 9$ by (84.2).

In the latter case, we infer from (61.5) that $Z \leq \Phi_{[c,A]}$ is a group of homologies. Note that this implies that Φ fixes both c and A. We claim that Z equals the connected component of $\Phi_{[c,A]}$. In fact, the latter is a normal subgroup of Φ of dimension at most 3, see (61.26). Since it is not semi-simple, it is contained in the center of Φ by the structure theorem.

We study the action of Φ on A. For every point $x \in A$ we know from (53.2) that $\dim x^\Phi = 4$ implies that x^Φ is open in A, and that A is homeomorphic to the sphere $\mathbb{S}_4$. Then (96.13a) says that Φ induces a group of dimension at most 10 on A. If $\dim \Phi_{[c,A]} > 1$, then — in view of the fact that Z is a normal subgroup of Φ — Richardson's classification (96.34) implies that $\Phi^1_{[c,A]}$ contains a subgroup isomorphic to $(SO_2\mathbb{R})^2$, in contradiction to (55.32ii). We conclude that $\dim \Phi \leq 11$.

If there is a point $x \in A$ such that $1 \leq \dim x^\Phi \leq 3$, we consider the kernel Δ of the action of Φ on x^Φ. For every point $y \in cx \setminus \{c, x\}$, the set $y^Z \cup x^\Phi$ topologically generates a subplane $\mathcal{E}$. Since Z induces a non-trivial compact, connected group of homologies on $\mathcal{E}$, this subplane has dimension at least 4 by (61.26). The stabilizer Δ_y acts trivially on $\mathcal{E}$, hence $\dim \Delta_y \leq 1$ by (83.11). Since Δ cannot act transitively on the line cx, we infer that no Δ-orbit on cx has full dimension. Hence $\dim y^\Delta \leq 3$, and $\dim \Delta \leq 4$. On the other hand, $\dim \Phi/\Delta \leq 6$ by (96.13a). We conclude that $\dim \Phi \leq 10$.

Finally, there remains the case where $\Phi = \Phi_{[c,A]}$. In this case, $\dim \Phi \leq 3$ by (61.26). $\square$

84.7 Lemma. *If Φ is the product of at least two almost simple factors, then Φ has an almost simple factor of dimension 3, and $\dim \Phi \leq 13$. Equality holds only in the case where Φ is isomorphic to the stabilizer of a point, taken in the elliptic motion group $PU_3\mathbb{H}$; i.e., Φ is an almost direct product of $Spin_3\mathbb{R}$ and $Spin_5\mathbb{R}$ (with amalgamated centers). In this case, the group Φ fixes a non-incident point–line pair, and the 10-dimensional factor acts transitively on the line, while the 3-dimensional factor consists of homologies.*

Proof. Let A be an almost simple factor of Φ, and let Ψ denote the identity component of the centralizer of A in Φ. We may assume that $\dim A \leq \dim \Psi$. If $\dim A > 3$, then $\dim A \geq 8$ since A is compact, see (94.33). We claim that in this case, there exists a point x whose orbit x^A topologically generates a subplane $\mathcal{E}$ of dimension at least 4. In fact, the almost simple group A does not act straight, since (96.13a) and (53.1c) would then imply that A acts trivially (or transitively) on every fixed line. On the other hand, the group A cannot leave invariant a subplane of dimension 2, cf. (32.16) and (83.18b). Consequently, $\dim \Psi_x \leq 1$ by (83.11), and $\dim \Psi \leq 8$ by (84.1). From our assumption $8 \leq \dim A \leq \dim \Psi$ we infer that $\dim x^\Psi \geq 7$. This implies that x^Ψ topologically generates $\mathcal{P}$, whence A_x is trivial. We obtain that $\dim x^A = 8$, contradicting (84.1). Thus $\dim A = 3$.

If some orbit x^A is not contained in a line, then x^A topologically generates a subplane $\mathcal{E}$ by (42.3). We obtain that either $\dim \Psi_x \leq 1$, or $\dim \mathcal{E} = 2$ and $\dim \Psi_x \leq 3$, see (83.18). In the first case, we use that $\dim x^\Psi \leq 7$ by (84.1). In the second case, the stabilizer A_x is not trivial. Then x^Ψ cannot generate a dense subplane, whence $\dim x^\Psi \leq 4$. In both cases, we obtain that $\dim \Psi \leq 8$ and $\dim \Phi \leq 11$.

Now assume that every point orbit of A is contained in a line and, dually, every line orbit is contained in a pencil. In this case, we infer from (23.16) and (83.18a) that $A = A_{[c,L]}$. This implies that Ψ fixes both c and L. According to (55.32ii), there are no commuting involutions in $\Phi_{[c,L]}$. Consequently, A equals the connected component of $\Phi_{[c,L]}$, and the group A is isomorphic to $\mathrm{Spin}_3\mathbb{R}$. For a torus $T < A$, the group $T\Psi$ has dimension at most 11 by (84.6), whence $\dim \Psi \leq 10$. Every compact semi-simple group of dimension 10 is a covering group of $\mathrm{SO}_5\mathbb{R}$. Now the observations that $\mathrm{SO}_5\mathbb{R}$ does not act (55.40) and that the central involutions of A and Ψ have to be amalgamated (cf. (55.32ii)) complete the proof. □

84.8 Proposition. *If Φ is almost simple, then either $\Phi \cong \mathrm{PU}_3\mathbb{H}$, or $\Phi \cong \mathrm{Spin}_5\mathbb{R}$, or $\dim \Phi \leq 8$.*

Proof. There are no almost simple groups of dimension 9, cf. the list (94.33). In view of (84.5), we may assume that the group Φ is simple, and that $\dim \Phi \geq 14$. If Φ is not isomorphic to G_2, then even $\dim \Phi \geq 15$, and Φ contains a closed subgroup Λ whose simply connected covering is either $\mathrm{SU}_4\mathbb{C}$ or $\mathrm{U}_3\mathbb{H}$, see (94.35). From (84.5), we know that Λ is simple. The case that $\Lambda = \mathrm{PSU}_4\mathbb{C} \cong \mathrm{PSO}_6\mathbb{R}$ cannot occur by (55.40). It remains to exclude the case where $\Lambda = G_2$. In this case, $\dim \Lambda_p \geq 7$ for every point p by (84.1). The identity component of Λ_p is a compact, connected Lie group of rank at most 2. By dimension reasons, it is almost simple, and, therefore, a covering of $\mathrm{PSU}_3\mathbb{C}$, $\mathrm{SO}_5\mathbb{R}$, or G_2. From (11.31) we infer that $G_2 = \mathrm{Aut}\,\mathbb{O}$ has no subgroup $\Omega \cong \mathrm{Spin}_5\mathbb{R}$; in fact the central involution of Ω would imply that Ω leaves a quaternion subfield of $\mathbb{O}$ invariant. The remaining cases are excluded by (55.40) and (62.8c). □

84.9 Theorem: Compact groups on 8-dimensional planes. *Let Φ be a compact, connected group of automorphisms of a compact 8-dimensional plane.*

(a) *If $\dim \Phi > 13$, then Φ is isomorphic to the elliptic motion group $\mathrm{PU}_3\mathbb{H}$, and the plane is isomorphic to $\mathcal{P}_2\mathbb{H}$.*

(b) *If $\dim \Phi = 13$, then Φ is isomorphic to a point stabilizer in $\mathrm{PU}_3\mathbb{H}$ (i.e., to the product of $\mathrm{Spin}_3\mathbb{R}$ and $\mathrm{Spin}_5\mathbb{R}$ with amalgamated centers), and Φ acts in the usual way, fixing a non-incident point–line pair.*

Proof. According to the structure theorem (93.11), either the group Φ is almost simple, or it is a product of at least two almost simple factors, or it has a center of positive dimension. These cases have been treated in (84.8), (84.7), and (84.6), respectively. □

Large groups of automorphisms

For the remainder of Section 84, let $\mathcal{P}$ be an 8-dimensional compact projective plane, and let Δ be a closed, connected subgroup of $\operatorname{Aut}\mathcal{P}$. Recall from (83.26) that $\dim \Delta \leq 35$.

84.10 Proposition. *Assume that Δ centralizes a non-trivial automorphism ζ. Then we have one of the following (mutually exclusive) possibilities:*

(a) *The element ζ is a homology or an elation. In this case, $\dim \Delta \leq 19$.*
(b) *The fixed elements of ζ form a subplane $\mathcal{E}$, of dimension e, say. Then $\dim \Delta$ is bounded by 17, 11, or 5, according as $e = 4, 2$, or 0.*
(c) *The fixed elements of ζ do not form a subplane, and there exists a point whose orbit under $\langle \zeta \rangle$ contains a triangle. In this case, $\dim \Delta \leq 13$.*

Proof. That the situations described in (a), (b), and (c) exhaust all possibilities follows from (23.15 and 16): If there is no point whose orbit under $\langle \zeta \rangle$ contains a triangle, then ζ is a homology, or an elation, or ζ acts trivially on a Baer subplane. In the first two cases, the center c and the axis A of ζ are fixed by Δ. In any case, the stabilizer Δ_x acts trivially on $x^{\langle \zeta \rangle}$.

If ζ is a homology or an elation, we choose a point x that is moved by ζ. Then x, x^ζ, c are three distinct collinear points. For every point $a \in A \setminus xc$ we obtain from (83.17) that $\dim \Delta_{x,a} \leq 7$, hence $\dim \Delta \leq 8 + 4 + 7 = 19$. The dimension bounds in (b) follow immediately from (83.19). Finally, assume the situation of (c). Recall from (55.19) that there exists at least one point that is fixed by ζ. Since every Δ-orbit is connected, we infer from (42.3) that the following holds for every fixed point p of ζ: either p is fixed by Δ, or p^Δ is contained in a line, and this line is fixed by $\langle \Delta, \zeta \rangle$. Thus, up to duality, the group $\langle \Delta, \zeta \rangle$ fixes a line L. Since the situation assumed in (c) is self-dual, we infer that there exists a point x outside L such that the orbit $x^{\langle \zeta \rangle}$ contains a triangle. Now the stabilizer Δ_x fixes both L and the sides of this triangle, hence $\dim \Delta_x \leq 5$ by (83.18), and $\dim \Delta \leq 5 + 8 = 13$. □

84.11 Corollary. *If $\dim \Delta \geq 20$, then Δ has trivial center and, in particular, is a Lie group, cf. (93.9).*

See (87.1) for a much stronger criterion for Δ to be a Lie group.

84.12 Examples. The bounds in (84.10a) and in the first two cases ($e \in \{4, 2\}$) of (84.10b) are attained, while the remaining results might be improved, cf. also (83.16). In order to see this, consider the automorphisms of $\mathcal{P}_2\mathbb{H}$ that are induced by the following elements of $GL_3\mathbb{H}$:

$$(a_1) \begin{pmatrix} 1 & & 1 \\ & 1 & \\ & & 1 \end{pmatrix}, \quad (a_2) \begin{pmatrix} r & & \\ & 1 & \\ & & 1 \end{pmatrix}; \quad (b) \begin{pmatrix} i & & \\ & i & \\ & & i \end{pmatrix}; \quad (c) \begin{pmatrix} \varepsilon & & \\ & \varepsilon & \\ & & 1 \end{pmatrix},$$

where $r \in \mathbb{R}^\times$ and $\varepsilon^3 = 1 \neq \varepsilon \in \mathbb{C}$. In cases ($a_1$) and ($a_2$), the induced collineation is an elation resp. a homology, and the centralizer in $PGL_3\mathbb{H}$ has dimension 19. The matrix under (b) induces a planar involution, with 17-dimensional centralizer. In the last case, the set of fixed points consists of the intersection of a line and a Baer subplane, plus one point of the subplane; the centralizer has dimension 11.

84.13 Lemma. *Assume that Δ has a closed normal subgroup $\Theta \cong \mathbb{R}^t$. Then $t \leq 8$. Moreover, we have the following.*

(a) *The group Θ contains no Baer collineation.*
(b) *If Θ contains a collineation that is not axial, then $\dim \Delta \leq 11 + t \leq 19$.*
(c) *If Θ contains a non-trivial homology, then Δ fixes a non-incident point–line pair. Consequently, $\dim \Delta \leq 19$, and $\dim \Delta \leq 18$ if the plane is non-desarguesian.*
(d) *If $\dim \Delta \geq 20$ and $t > 0$, then (up to duality) Θ consists of elations with common axis W. In particular, Δ fixes W. In this case, $\dim \Delta \leq 19 + t \leq 27$. If $\dim \Delta > 26$, then the plane is isomorphic to $\mathcal{P}_2\mathbb{H}$.*

Proof. Assertion (a) follows from the observation that the subgroup generated by any non-trivial element of Θ is a non-compact closed subgroup of $\operatorname{Aut}\mathcal{P}$, and cannot act trivially on a Baer subplane by (83.6). The proof of the remaining assertions uses the fact that Δ acts linearly on Θ via conjugation. Hence the centralizer Ψ of $\vartheta \in \Theta \setminus \{\mathbb{1}\}$ centralizes the one-parameter group P of Θ that contains ϑ. Note that $\dim \Delta \leq \dim \Psi + t$, and that $t \leq 8$ by (55.41). If ϑ is not axial, then there exists a point x whose orbit x^P is not contained in a line, cf. (23.16). From (83.21) we infer that $\dim \Psi_x \leq 3$, whence $\dim \Psi \leq 8 + 3 = 11$. Thus (b) is established. Assertion (c) follows from (61.23b) and (83.26). The bound for $\dim \Delta$ in (d) is a consequence of (84.10). The rest of the assertion follows from (a)–(c), (61.23a), and (83.26). □

84.14 Lemma. *Assume that Δ has a compact, connected abelian normal subgroup $N \neq \mathbb{1}$. Then Δ centralizes N, and $\dim \Delta \leq 17$.*

Proof. According to (93.19), a compact abelian normal subgroup of a connected group is central. Assume first that N acts straight both on $\mathcal{P}$ and on the dual plane

$\mathcal{P}^*$. If N acts trivially on a Baer subplane, then $\dim \Delta \leq 17$ by (83.19). According to (23.16), there remains the case where $\mathsf{N} = \mathsf{N}_{[c,A]}$. Choosing points x, y outside $A \cup \{c\}$ such that $y \notin cx$ we obtain that $x^\mathsf{N} \cup y^\mathsf{N}$ topologically generates a connected subplane $\mathcal{E}$, and $\Delta_{x,y}$ acts trivially on $\mathcal{E}$. Since N induces a non-trivial compact, connected group of homologies on $\mathcal{E}$, this subplane cannot be two-dimensional (61.26). Hence $\dim \Delta_{x,y} \leq 1$ by (83.18a), and $\dim \Delta \leq 17$.

Up to duality, we may assume that N is not straight. Then there exists a point x such that x^N is not contained in a line. From (83.21) we infer that $\dim \Delta_x \leq 3$, and $\dim \Delta \leq 11$. □

84.15 Remark. Inspection of the classical group $\mathrm{PSL}_3\mathbb{H}$ shows that the bounds in (84.13c and d) and (84.14) are sharp. For the situation of (84.14), compare also the Hughes planes (see Section 86), and the centralizer of the diagonal matrix $\mathrm{diag}(i, 1, 1)$ in $\mathrm{SL}_3\mathbb{H}$.

84.16 Theorem: Groups that are properly semi-simple.
Assume that $\mathcal{P}$ is an 8-dimensional compact projective plane, and that Δ is a closed subgroup of $\mathrm{Aut}\,\mathcal{P}$. If Δ is an almost direct product of at least two almost simple factors, then $\dim \Delta \leq 19$.

Proof. Let A be an almost simple factor of minimal dimension, and let B be the identity component of its centralizer. If there is a point x such that the (connected) orbit x^A is not contained in a line, then x^A topologically generates a connected subplane $\mathcal{E}$ by (42.3), and the stabilizer B_x acts trivially on this subplane. Consequently, $\dim \mathsf{B}_x \leq 3$ and $\dim \mathsf{B} \leq 11$ by (83.18). Since A is almost simple, the kernel of the restriction of A to $\mathcal{E}$ has dimension 0. If $\mathcal{E}$ is two-dimensional, then $\dim \mathsf{A} \leq 8$ by (33.6), compare (83.3), whence $\dim \Delta = \dim \mathsf{A} + \dim \mathsf{B} \leq 19$. If $\mathcal{E}$ is not two-dimensional, then $\dim \mathsf{B}_x \leq 1$ by (83.18), and $9 \geq \dim \mathsf{B} \geq \dim \mathsf{A}$, whence $\dim \Delta \leq 18$. The dual reasoning applies if there exists a line L such that L^A is not contained in a pencil.

There remains the case that every point orbit under A is contained in some line, and dually. From (23.16) and (83.11) we infer that $\mathsf{A} = \mathsf{A}_{[a,L]}$. In particular, the group A acts freely on each non-trivial orbit. Since there are no simple groups of dimension 4 or ≤ 2, we obtain that $\dim \mathsf{A} = 3$. Choosing points x, y that are moved by A and such that $y \notin ax$ we infer that $\mathsf{B}_{x,y}$ acts trivially on the subplane topologically generated by $x^\mathsf{A} \cup y^\mathsf{A}$. This subplane has lines of dimension not less than $\dim x^\mathsf{A} = 3$. Consequently, these lines have dimension 4, and $x^\mathsf{A} \cup y^\mathsf{A}$ topologically generates the whole plane, cf. (55.1). We conclude that $\mathsf{B}_{x,y} = \mathbb{1}$ and $\dim \mathsf{B} \leq 16$. Thus $\dim \Delta \leq 19$. □

84.17 Remark. The bound in (84.16) is almost attained by the commutator group Ψ' of the centralizer Ψ of a reflection in $\mathrm{PGL}_3\mathbb{H}$, which is a semi-simple group of dimension 18. We have that $\dim \Psi = 19$. This indicates that more sophisticated methods are needed in order to obtain a sharp bound in the situation of (84.16). In fact, one can show that every group that satisfies the assumptions of (84.16) has dimension ≤ 15, or is isomorphic to the group Ψ' mentioned above, see Salzmann [81a] Satz 2.

84.18 Lemma. *The group* $\mathrm{PSp}_6\mathbb{R}$ *does not act on any 8-dimensional plane.*

Proof. In $\mathrm{PSp}_6\mathbb{R}$ there exist three commuting (diagonal) involutions. For each of these, the centralizer is a product of $\mathrm{Sp}_2\mathbb{R} = \mathrm{SL}_2\mathbb{R}$ and $\mathrm{Sp}_4\mathbb{R}$. From (83.20) we infer that these involutions are reflections. Their common centralizer Ψ is a product of three factors $A_\nu \cong \mathrm{SL}_2\mathbb{R}$, and it fixes the triangle (c_1, c_2, c_3) that is formed by the centers of the three reflections, cf. (55.34a). A maximal compact subgroup Ω of Ψ is isomorphic to $\mathbb{T}^3$ and contains further involutions. These must be planar by (55.35), and $\mathcal{L}_{c_1} \approx \mathbb{S}_4$ by (53.10). Applying Richardson's Theorem (96.34) to the action of Ω on $\mathcal{L}_{c_1}$, we infer that the normal subgroup $\Psi_{[c_1]} = \Psi_{[c_1,c_2c_3]}$ of Ψ has positive dimension. Consequently, it contains one of the factors A_ν, in contradiction to (61.2). $\square$

84.19 Theorem: Almost simple groups. *If* Δ *is almost simple, then either* $\Delta \in \{\mathrm{PSL}_3\mathbb{H}, \mathrm{PU}_3\mathbb{H}, \mathrm{PU}_3(\mathbb{H}, 1)\}$ *and the plane is isomorphic to* $\mathcal{P}_2\mathbb{H}$, *or* Δ *has dimension at most* 16.

Proof. Assume that $\dim \Delta > 16$. There are no almost simple groups of dimension $s \in \{17, 18, 19\}$, see (94.33). Hence $\dim \Delta \geq 20$, and Δ is a simple Lie group by (84.11). On the other hand, $\dim \Delta \leq 35$ by (83.26). Combined with the classification of simple Lie groups (94.33), our result (84.9) yields that Δ is isomorphic to one of the following groups:

$$\mathrm{PSL}_3\mathbb{H}, \ \mathrm{PU}_3\mathbb{H}, \ \mathrm{PU}_3(\mathbb{H}, 1), \ \mathrm{PSp}_6\mathbb{R}, \ \mathrm{SO}_5\mathbb{C} \cong \mathrm{PSp}_4\mathbb{C},$$

$$O'_7(\mathbb{R}, 2), \ O'_7(\mathbb{R}, 3), \ O'_8(\mathbb{R}, 4), \ \mathrm{PSL}_5\mathbb{R}, \ \text{or } \mathrm{PSU}_5(\mathbb{C}, 2).$$

Each of the groups $\mathrm{SO}_5\mathbb{C}$, $O'_7(\mathbb{R}, 2)$, and $\mathrm{PSL}_5\mathbb{R}$ contains $\mathrm{SO}_5\mathbb{R}$, and does not act by (55.40). The group $\mathrm{PSp}_6\mathbb{R}$ does not act by (84.18). There exists a subgroup isomorphic to $\mathbb{Z}_2^5$ in $\mathrm{SO}_4\mathbb{R} \times \mathrm{SO}_3\mathbb{R} < O'_7(\mathbb{R}, 3) < O'_8(\mathbb{R}, 4)$. Thus these two possibilities are ruled out by (55.34c). Finally, the group $\mathrm{PSU}_5(\mathbb{C}, 2)$ contains five commuting (diagonal) involutions such that each of them is centralized by a group that is isomorphic to $\mathrm{SU}_4(\mathbb{C}, s)$ for some s. According to (83.19a), these involutions have to be reflections, a contradiction to (55.34).

If $\Delta \in \{\mathrm{PSL}_3\mathbb{H}, \mathrm{PU}_3\mathbb{H}, \mathrm{PU}_3(\mathbb{H}, 1)\}$, then the plane is isomorphic to $\mathcal{P}_2\mathbb{H}$ by (62.6); observe that $\mathrm{PU}_3\mathbb{H}$ and $\mathrm{PU}_3(\mathbb{H}, 1)$ are subgroups of $\mathrm{PSL}_3\mathbb{H}$. $\square$

84.20 Remark. The bound in (84.19) is sharp. Indeed, the Hughes planes admit the 16-dimensional almost simple group $SL_3\mathbb{C}$, cf. Section 86.

84.21 Theorem: Groups that are not semi-simple. *Assume that Δ is not semi-simple, and let Θ be a minimal connected abelian normal subgroup.*

(a) *If $\dim \Delta \geq 18$, then Θ is isomorphic to a vector group $\mathbb{R}^t$ for some t with $1 \leq t \leq 8$. Moreover, $\dim \Delta \leq 19 + t \leq 27$.*
(b) *If $\dim \Delta \geq 20$, then, up to duality, Θ consists of elations with common axis, say W. In particular, Δ fixes W.*

Proof. A minimal connected abelian normal subgroup is either compact or isomorphic to $\mathbb{R}^t$, see (94.26). Thus (84.14 and 13) yield the assertion. □

84.22 Remark. The bound in (84.21a) is attained by the stabilizer of a line in $PGL_3\mathbb{H}$, i.e., by the full group of automorphisms of the affine plane over $\mathbb{H}$.

In the situation of (84.21b), we wish to obtain further information about the structure of the plane. This requires a close look at the group Δ, its normal subgroup Θ and their possible actions. The following steps give an idea of the methods that were used for a classification of all 8-dimensional planes admitting a group Σ with $\dim \Sigma \geq 17$, as completed by H. Salzmann [90]. In order to avoid technical complications, we shall prove only a weaker version here.

For the remainder of this section, assume that Δ is a closed subgroup of $\operatorname{Aut} \mathcal{P}$ such that $\dim \Delta \geq 20$, and that Δ is not semi-simple. Let $\Theta = \Theta_{[W,W]} \cong \mathbb{R}^t$ be a minimal connected abelian normal subgroup, as in (84.21b). Recall from (84.11) that Δ is a Lie group.

84.23 Proposition. *The group Δ fixes at most one point. If a fixed point exists, then it lies on W.*

Proof. Since Θ acts freely outside W, every fixed point of Δ belongs to W. If Δ fixes two points on W, we find points $w \in W$ and $a \in P \setminus W$ such that $\dim \Delta_{w,a} \leq 7$ by (83.17), and $\dim \Delta \leq 19$. □

We are going to discuss the two different possibilities for the configuration of fixed elements separately. Our aim is to show that $\mathcal{P}$ is a translation plane. The following general lemma will be needed.

84.24 Lemma. *Let Σ be the group of all automorphisms of an 8-dimensional compact projective plane $\mathcal{P}$, and let $\mathsf{T} \cong \mathbb{R}^d$ be a subgroup of $\Sigma_{[W,W]}$ for some line W of $\mathcal{P}$. Assume that there exists a point a outside W, and a closed subgroup*

Γ *of the normalizer of* T *in* Σ_a *such that* $\dim \tau^\Gamma = \dim T$ *for every* $\tau \in T \setminus \{1\}$. *Then either* $d = 8$, *or* $d \leq 5$.

Proof. From (61.4b), we know that $d \leq 8$. Assume that $d \in \{6, 7\}$. Since τ^Γ is open in the connected space $T \setminus \{1\}$, we infer that Γ acts transitively on $T \setminus \{1\}$. The space of all rays from 1 through τ, where $\tau \in T \setminus \{1\}$, is homeomorphic to $\mathbb{S}_{d-1}$. Since the group Γ is transitive on $T \setminus \{1\}$, it is also transitive on this sphere, and we deduce from (96.19 to 22) that Γ contains a compact almost simple group Φ that covers one of the groups $SO_6\mathbb{R}$, $SU_3\mathbb{C}$, or G_2. According to (62.8c), the group Φ cannot act on a line of $\mathcal{P}$, in contradiction to the fact that $W = W^\Gamma$. □

84.25 Theorem. *Let* $\mathcal{P}$ *be an 8-dimensional compact projective plane, let* Δ *be a closed subgroup of* $\mathrm{Aut}\,\mathcal{P}$ *of dimension at least* 20 *such that* Δ *is not semi-simple, and let* $\Theta \leq \Delta_{[W,W]}$ *be a minimal closed, connected abelian normal subgroup, as provided by* (84.21b). *If* Δ *fixes no point, and if* $\dim \Delta \geq 22$, *then* $\mathcal{P}$ *is a translation plane.*

Proof. We assume that $\mathcal{P}$ is not a translation plane, and aim for a contradiction. By our assumptions, $\dim \Theta_{[w]} \leq 3$ for every $w \in W$, cf. (61.4c). Moreover, there exists a point $z \in W$ such that $\dim \Theta_{[z]} \leq 2$ by (61.13). From (61.11a) we know that $\dim \Theta \leq 6$.

We investigate the stabilizer $\Gamma = \Delta_a$ of a point $a \in P \setminus W$. According to (84.24), there exists $\vartheta \in \Theta \setminus \{1\}$ such that $\dim \vartheta^\Gamma \leq 5$. Since Δ fixes no point, there exists $\tau \in \Theta \setminus \{1\}$ such that ϑ and τ have different centers. If $\dim \vartheta^\Gamma = 5$, we may even choose $\tau \in \vartheta^\Gamma$. In any case, we obtain that the centralizer Λ of ϑ and τ in Γ satisfies $\dim \Lambda \geq \dim \Gamma - 10$. Now Λ centralizes the vector subspace Ω that is spanned by ϑ and τ in $\Theta \cong \mathbb{R}^t$. Since a^Ω is not contained in a line, we infer from (83.21) that $\dim \Lambda \leq 3$, and that $\dim \Delta \leq 8 + 10 + 3 = 21$. □

We turn to the case where Δ fixes a point on W. From (83.26) we know that $\dim \Delta \leq 23$. This bound is attained by the stabilizer Λ of a flag, taken in $\mathrm{Aut}\,\mathcal{P}_2\mathbb{H} = \mathrm{PGL}_3\mathbb{H}$. If Δ fixes a flag, then our assumptions that Δ is not semi-simple and that $\dim \Delta \geq 20$ entail that the action of Δ resembles the action of Λ.

84.26 Theorem. *Let* $\mathcal{P}$ *be an 8-dimensional compact projective plane, let* Δ *be a closed subgroup of* $\mathrm{Aut}\,\mathcal{P}$ *of dimension at least* 20 *such that* Δ *is not semi-simple, and let* $\Theta \leq \Delta_{[W,W]}$ *be a minimal closed, connected abelian normal subgroup, as provided by* (84.21b). *If* Δ *fixes a point* z, *then* $\Theta \leq \Delta_{[z,W]}$.

Proof. From (84.23) we know that $z \in W$. Assume that Θ is not contained in $\Delta_{[z,W]}$. By minimality of Θ, we infer that $\Theta_{[z]} = \{1\}$; recall that $\Theta_{[z]}$ is connected (61.9). Using (61.11a) we infer that $\dim \Theta \leq 4$. We choose a point a outside W.

The centralizer of a non-trivial element $\vartheta \in \Theta$, taken in Δ_a, fixes infinitely many points on the line that joins a and the center of ϑ. Since z does not belong to this line, we infer from (83.17) that $\dim \Delta_a \leq \dim \Theta + 7$. Hence $\dim \Delta \leq 19$. □

84.27 Theorem. *Let $\mathcal{P}$ be an 8-dimensional compact projective plane, and let Δ be a closed subgroup of $\operatorname{Aut} \mathcal{P}$. If $\dim \Delta \geq 23$, then $\mathcal{P} \cong \mathcal{P}_2\mathbb{H}$. In this case, either the group Δ is equal to $\operatorname{Aut} \mathcal{P} \cong \operatorname{PGL}_3\mathbb{H}$, or Δ fixes a point or a line.*

Proof. According to (81.9), it suffices to show that $\mathcal{P}$ is a translation plane or a dual translation plane. In view of (84.16, 19, 21, 25, and 26), we may assume that Δ has a minimal normal subgroup $\Theta \cong \mathbb{R}^t$, and that $\Theta \leq \Delta_{[z,W]}$. In particular, Δ fixes the flag (z, W). Combining (84.21a) and (61.4a), we infer that $\dim \Theta = 4$. Choose a point a outside W, and let $\Gamma = \Delta_a$. For every $\vartheta \in \Theta \setminus \mathbb{1}$, the centralizer Ψ of ϑ in Γ acts trivially on an infinite subset of the line az. Therefore $\dim \Psi_w \leq 7$ for every point $w \in W \setminus \{z\}$ by (83.17). We conclude that

$$(*) \qquad \dim \Gamma \leq \dim \vartheta^\Gamma + \dim w^\Psi + \dim \Psi_w \leq 4 + 4 + 7 = 15,$$

and $\dim a^\Delta = 8$. Since a was chosen arbitrarily in $P \setminus W$, we infer from (53.1d) that Δ acts transitively on $P \setminus W$. Moreover, the inequality $(*)$ is an equality. In particular, we have that $\dim w^\Gamma = \dim w^\Psi = 4$ for every $w \in W \setminus \{z\}$, and that $\dim \vartheta^\Gamma = 4$ for every $\vartheta \in \Theta \setminus \mathbb{1}$. This implies that $W \approx \mathbb{S}_4$ by (53.2), and that Γ acts transitively on $\Theta \setminus \mathbb{1}$. From (96.19 to 22), we know that Γ contains a subgroup $\Omega \cong \operatorname{SU}_2\mathbb{C} \cong \operatorname{U}_1\mathbb{H}$ which acts freely on $\Theta \setminus \mathbb{1}$.

We investigate the action of the central involution $\iota \in \Omega$. If ι has axis W, then ι has center a. Since Δ acts transitively on $P \setminus W$, this implies that for every point p outside W there exists a reflection with axis W and center p. Consequently, $\mathcal{P}$ is a translation plane (61.20b); recall that Δ is a Lie group (84.11).

If $\iota \notin \Delta_{[W]}$, then Ω acts effectively on $W \approx \mathbb{S}_4$. From Richardson's classification (96.34), we infer that Ω acts in the standard way on $W \approx \mathbb{C}^2 \cup \{\infty\}$. In particular, the involution ι fixes exactly two points z, w on W. Therefore, the involution ι is not planar, but has center z or w. Choose $\vartheta \in \Theta \setminus \mathbb{1}$. We have that $a^{\vartheta\varphi} = a^{\varphi^{-1}\vartheta\varphi}$ for every $\varphi \in \Omega$. Since Ω acts freely on $\Theta \setminus \mathbb{1}$, and Θ acts freely on az, we conclude that az is not the axis of ι. Thus $\iota \in \Delta_{[z,aw]}$. The centralizer of ι in Γ fixes w. Therefore, $\dim \iota^\Gamma \geq \dim w^\Gamma = 4$. This implies that $\iota^\Gamma \iota$ is a 4-dimensional set of elations with center z and axis az, see (23.20), compare also (61.19a). Now $\dim \Delta_{[z,z]} = 8$ by (61.19b), and $\mathcal{P}$ is a dual translation plane. □

We conclude this section by stating the classification of 8-dimensional compact projective planes with at least 17-dimensional groups of automorphisms, as obtained in Salzmann [90]. Note that this classification depends on the corresponding classification result for translation planes, which was proved in Hähl [86a].

84.28 Theorem. *Assume that $\mathcal{P}$ is an 8-dimensional compact projective plane. If* $\dim \operatorname{Aut} \mathcal{P} \geq 17$*, then one of the following cases occurs.*

(a) *The plane $\mathcal{P}$ is isomorphic to $\mathcal{P}_2\mathbb{H}$, and $\operatorname{Aut} \mathcal{P}$ is the 35-dimensional simple group* $\operatorname{PGL}_3\mathbb{H}$.
(b) *The plane $\mathcal{P}$ is a Hughes plane, see Section 86, and $\operatorname{Aut} \mathcal{P}$ is isomorphic to the 17-dimensional group* $\operatorname{GL}_3\mathbb{C}/(\mathbb{R}^\times \cdot \mathbb{1})$.
(c) *Up to duality, the plane $\mathcal{P}$ is a translation plane. These planes have been classified, see* (82.25). *In particular, it turns out that* $\dim \operatorname{Aut} \mathcal{P} \leq 18$.

84.29 Notes on Section 84. This section presents a weakened version of the results of Salzmann [79a, 81a]. We refrained from proving the full results, thus ideas of Stroppel [93e] became available to simplify the proofs. Theorem (84.9) has first been proved in Salzmann [79a] Section 3. The result (84.19) has been generalized to the case of stable planes in Stroppel [94e]. For the case of projective planes, a considerably stronger result is proved in Stroppel [94f]: The only 8-dimensional compact projective planes that admit an almost simple group of dimension greater than 10 are the Hughes planes (including $\mathcal{P}_2\mathbb{H}$).

85 Characterizing $\mathcal{P}_2\mathbb{O}$

Throughout this section, we consider a 16-dimensional compact projective plane $\mathcal{P} = (P, \mathcal{L})$, and a closed, connected subgroup Δ of $\operatorname{Aut} \mathcal{P}$.

Our aim is to show that $\mathcal{P}$ is isomorphic to $\mathcal{P}_2\mathbb{O}$, if $\dim \operatorname{Aut} \mathcal{P} \geq 57$. Note that much better results may be found in the existing literature, compare Salzmann [87], and see Section 87 for an outline.

Within the scope of this book, we merely attempt to introduce the methods, but refrain from proving the best results that are presently known.

As in Section 84, we distinguish the different possibilities for the structure of the connected component of $\operatorname{Aut} \mathcal{P}$; namely, the cases where the connected component is semi-simple, almost simple, or not semi-simple. In the last case, we shall reduce our problem to the classification of translation planes, as presented in Section 82. In fact, we shall show that — up to duality — every sufficiently large group contains a transitive translation group. In contrast with Section 84, we shall not prove a bound for the dimension of compact subgroups of $\operatorname{Aut} \mathcal{P}$. However, such a bound is known, see Stroppel [94a]: namely, a compact subgroup of $\operatorname{Aut} \mathcal{P}$ either is isomorphic to the (52-dimensional) elliptic motion group (whence $\mathcal{P} \cong \mathcal{P}_2\mathbb{O}$), or has dimension at most 36.

85.1 Proposition. *Assume that Δ commutes with a non-trivial automorphism ζ. Then we have one of the following (mutually exclusive) possibilities:*

(a) *The element ζ is a homology or an elation. In this case, $\dim \Delta \leq 46$.*
(b) *The fixed elements of ζ form a subplane $\mathcal{E}$, of dimension e, say. Then $\dim \Delta$ is bounded by $42, 29, 22$, or 13, according as e equals $8, 4, 2$, or 0.*
(c) *The fixed elements of ζ do not form a subplane, and there exists a point whose orbit under $\langle \zeta \rangle$ contains a triangle. In this case, $\dim \Delta \leq 30$.*

Proof. The proof is quite similar to the proof of the analogous result (84.10) for 8-dimensional planes. That the situations described in (a), (b), and (c) exhaust all possibilities follows from (23.15 and 16): If there is no point whose orbit under $\langle \zeta \rangle$ contains a triangle, then either ζ is a homology, or an elation, or ζ acts trivially on a Baer subplane. In the first two cases, the center c and the axis A of ζ are fixed by Δ. In any case, the stabilizer Δ_x acts trivially on $x^{\langle \zeta \rangle}$.

If ζ is a homology or an elation, we choose a point x that is moved by ζ. Then x, x^ζ, c are three distinct collinear points. For every choice of two points $a, a' \in A \setminus xc$, we obtain from (83.23) that $\dim \Delta_{x,a,a'} \leq 14$, hence $\dim \Delta \leq 16 + 8 + 8 + 14 = 46$.

In the situation of case (b), we consider the restriction of Δ to $\mathcal{E}$. The group Δ induces a group of dimension at most d on $\mathcal{E}$, where d equals $35, 16, 8,$ or 0, according as $\dim \mathcal{E}$ equals $8, 4, 2$, or 0, compare (84.28), (71.9), (33.6). If $\dim \mathcal{E} = 8$, then the kernel of the restriction has dimension at most 7 by (83.6). Thus $\dim \Delta \leq 42$ in this case. If $\dim \mathcal{E} \in \{0, 4\}$, we have that the kernel of the restriction has dimension at most 13 by (83.24), while the kernel may be 14-dimensional if $\dim \mathcal{E} = 2$. This completes the proof of assertion (b).

Finally, assume the situation of (c). Recall from (55.19) that there exists at least one point that is fixed by ζ. Since every Δ-orbit is connected, we infer from (42.3) that the following holds for every fixed point p of ζ: either p is fixed by Δ, or p^Δ is contained in a line, and this line is fixed by $\langle \Delta, \zeta \rangle$. Thus, up to duality, the group $\langle \Delta, \zeta \rangle$ fixes a line L. Since the situation assumed in (c) is self-dual, we infer that there exists a point x outside L such that the orbit $x^{\langle \zeta \rangle}$ contains a triangle. Now the stabilizer Δ_x fixes both L and the sides of this triangle, hence $\dim \Delta_x \leq 14$ by (83.23), and $\dim \Delta \leq 14 + 16 = 30$. □

85.2 Corollary. *If $\dim \Delta \geq 47$, then Δ has trivial center and, in particular, is a Lie group, cf. (93.8).*

See the beginning of Section 87 for a much stronger criterion for Δ to be a Lie group.

85.3 Remark. The bound in (85.1a) is attained in $\operatorname{Aut} \mathcal{P}_2\mathbb{O}$. Using better estimates for the dimension of stabilizers of quadrangles, some of the bounds in (85.1b) can be improved.

85.4 Lemma. *Assume that Δ has a closed normal subgroup $\Theta \cong \mathbb{R}^t$. Then $t \leq 16$. Moreover, we have the following.*

(a) *The group Θ contains no Baer collineation.*
(b) *If Θ contains a collineation that is not axial, then $\dim \Delta \leq 30 + t \leq 46$.*
(c) *If Θ contains a non-trivial homology then Δ fixes a non-incident point–line pair. Consequently, $\dim \Delta \leq 46$, and $\dim \Delta \leq 45$ if the plane is not a Moufang plane.*
(d) *If $\dim \Delta \geq 47$, then (up to duality) Θ consists of elations with common axis W. In particular, Δ fixes W. In this case, $\dim \Delta \leq 46+t \leq 62$. If $\dim \Delta = 62$, then the plane is isomorphic to $\mathcal{P}_2\mathbb{O}$.*

Proof. The proof is similar to the proof of (84.13). Assertion (a) follows from the observation that the subgroup generated by any non-trivial element of Θ is a non-compact closed subgroup of $\operatorname{Aut} \mathcal{P}$, and cannot act trivially on a Baer subplane by (83.6). The proof of the remaining assertions uses the fact that Δ acts linearly on Θ via conjugation. Hence the centralizer Ψ of $\vartheta \in \Theta \setminus \{\mathbb{1}\}$ centralizes the one-parameter group P of Θ that contains ϑ. Note that $\dim \Delta \leq \dim \Psi + t$, and that $t \leq 16$ by (55.41). If ϑ is not axial, then there exists a point x whose orbit x^P is not contained in a line, cf. (23.16). In this case, the orbit x^P generates a subplane, cf. (42.3). From (83.23) we infer that $\dim \Psi_x \leq 14$, whence $\dim \Psi \leq 16 + 14 = 30$. Thus (b) is established. Assertion (c) follows from (61.23b) and (83.26). The bound for $\dim \Delta$ in (d) is a consequence of (85.1). The rest of the assertion follows from (a)–(c), (61.23a), and (83.26). □

The following estimate is far from being sharp, but it will be sufficient for the purposes of this section.

85.5 Lemma. *Assume that Δ has a compact, connected abelian normal subgroup $\mathsf{N} \neq \mathbb{1}$. Then Δ centralizes N, and $\dim \Delta \leq 45$.*

Proof. According to (93.19), a compact abelian normal subgroup of a connected group is central. Assume first that N acts straight both on $\mathcal{P}$ and the dual plane $\mathcal{P}^*$. If N acts trivially on a Baer subplane, then $\dim \Delta \leq 42$ by (85.1b). According to (23.16), there remains the case where $\mathsf{N} = \mathsf{N}_{[c,A]}$. Choosing points x, y outside $A \cup \{c\}$ such that $y \notin cx$ we obtain that $x^\mathsf{N} \cup y^\mathsf{N}$ topologically generates a connected subplane $\mathcal{E}$, and $\Delta_{x,y}$ acts trivially on $\mathcal{E}$. Since N induces a non-trivial compact, connected group of homologies on $\mathcal{E}$, this subplane cannot be two-dimensional (61.26). Hence $\dim \Delta_{x,y} \leq 13$ by (83.24), and $\dim \Delta \leq 45$.

Up to duality, we may assume that N is not straight. Then there exists a point x such that x^N is not contained in a line. From (83.23) we infer that $\dim \Delta_x \leq 14$, and $\dim \Delta \leq 30$. □

85.6 Theorem: Groups that are properly semi-simple. *Assume that $\mathcal{P}$ is a 16-dimensional compact projective plane, and that Δ is a closed subgroup of $\operatorname{Aut}\mathcal{P}$. If Δ is an almost direct product of at least two almost simple factors, then $\dim \Delta \leq 38$.*

Proof. Let A be an almost simple factor of minimal dimension, and let B be the identity component of its centralizer.

1) If there is a point x such that the (connected) orbit x^A is not contained in a line, then x^A generates a subplane, and the stabilizer B_x acts trivially on the closure $\mathcal{E}$ of this subplane. Consequently, $\dim \mathsf{B}_x \leq 14$ by (83.23), and $\dim \mathsf{B} \leq 30$.

Assume first that $\mathcal{E}$ is a proper subplane. If $\dim x^\mathsf{B} > 8$, then x^B cannot be contained in a line or in a proper closed subplane. Hence A_x is trivial, and $\dim \mathsf{A} \leq \dim \mathcal{E} \leq 8$. This yields that $\dim \Delta \leq 38$. If $\dim x^\mathsf{B} \leq 8$, then $\dim \mathsf{B} \leq 22$. Since A is almost simple, the kernel of the restriction of A to $\mathcal{E}$ has dimension 0. If $\dim \mathcal{E} \leq 4$, then $\dim \mathsf{A} \leq 16$ by (33.6) and (71.9), compare (83.3), whence $\dim \Delta \leq 38$. If $\mathcal{E}$ is a Baer subplane, then $\dim \mathsf{B}_x \leq 7$ by (83.6), and $15 \geq \dim \mathsf{B} \geq \dim \mathsf{A}$. Thus $\dim \Delta \leq 30$ in this case.

If x^A topologically generates $\mathcal{P}$, then B_x is trivial, and $16 \geq \dim \mathsf{B} \geq \dim \mathsf{A}$. This yields that $\dim \Delta \leq 32$.

2) There remains the case that every point orbit under A is contained in some line, and every line orbit is contained in some pencil. From (23.16) we infer that either A acts trivially on some Baer subplane $\mathcal{B}$, or $\mathsf{A} = \mathsf{A}_{[a,L]}$. In the first case, we obtain that $\mathcal{B}$ is Δ-invariant. Then the connected component K of the kernel of the restriction to $\mathcal{B}$ is a semi-simple group, which is compact by (83.6). This implies that K is a Lie group (93.11), and $\dim \mathsf{K} = 3$ by (83.22). In particular, we have that $\mathsf{K} = \mathsf{A}$, and B acts with a kernel of dimension 0 on $\mathcal{B}$. According to (84.28), we have that $\dim \mathsf{B} \leq 35$, and $\dim \Delta \leq 38$.

Finally, assume that $\mathsf{A} = \mathsf{A}_{[a,L]}$. Since A acts freely on each non-trivial orbit, we have that $\dim \mathsf{A} \leq 8$, and equality is possible only if A acts transitively on $K \setminus \{a, K \wedge L\}$ for every line K through a, see (61.4a). This is impossible by (61.2). Thus $\dim \mathsf{A} \leq 7$, and in fact $\dim \mathsf{A} \in \{3, 6\}$ since A is almost simple, see (94.33). For suitable points x, y, we have that $x^\mathsf{A} \cup y^\mathsf{A}$ topologically generates a subplane $\mathcal{E}$, and $\dim \mathcal{E} \geq 2 \dim \mathsf{A}$. If $\dim \mathsf{A} = 6$, then $\mathsf{B}_{x,y}$ is trivial, and we obtain that $\dim \Delta \leq 6 + 16 + 16 = 38$. There remains the case where $\dim \mathsf{A} = 3$. In this case, we have that $\dim \mathcal{E} \geq 8$, and $\dim \mathsf{B}_{x,y} \leq 7$. Let Z denote the center of B. If $\mathsf{Z}_{x,y}$ is not trivial, then $x^\mathsf{B} \cup y^\mathsf{B}$ does not generate a dense subplane of $\mathcal{P}$. This implies that $\dim \mathsf{B} \leq 8 + 8 + \dim \mathsf{B}_{x,y}$, and $\dim \Delta \leq 26$. If $\mathsf{Z}_{x,y}$ is trivial, then the connected component of the stabilizer $B_{x,y}$ injects into the Lie group B/Z, and is therefore a Lie group. According to (83.22), we have that $\dim B_{x,y} \leq 3$, and $\dim \Delta \leq 3 + 16 + 16 + 3 = 38$. □

85.7 Remark. In $\operatorname{Aut} \mathcal{P}_2\mathbb{O}$, the centralizer of a planar involution is a properly semi-simple group of dimension 38. Thus the bound in (85.6) is sharp.

85.8 Proposition. *Let $\mathcal{P}$ be a 16-dimensional compact projective plane, and assume that Δ is a closed, connected, almost simple subgroup of $\operatorname{Aut} \mathcal{P}$. Then the factor group of Δ modulo its center is not the split real form $F_4(4)$ of the exceptional simple Lie group $F_4^{\mathbb{C}}$.*

Proof. Assume that the center factor group of Δ is the split real form $F_4(4)$ of $F_4^{\mathbb{C}}$. From (85.2) we infer that Δ is a simple Lie group. Each maximal compact subgroup of Δ is an almost direct product of a subgroup $\Theta \cong \operatorname{SU}_2\mathbb{C}$ by a subgroup $\Phi \cong \operatorname{U}_3\mathbb{H}$, see (94.33).

1) From (84.9), we infer that Φ has no subgroup of codimension less than 8. By (53.1c), this implies that the group Φ would act transitively on any fixed line or line pencil, in contradiction to (62.8). Therefore, it fixes no point and no line of $\mathcal{P}$, and the set B of fixed points of the central involution φ of Φ carries a Baer subplane $\mathcal{B}$. According to (62.6), the subplane $\mathcal{B}$ is isomorphic to the quaternion plane $\mathcal{P}_2\mathbb{H}$, and the group Φ induces the elliptic motion group. In particular, the involutions $\alpha = \operatorname{diag}(-1, -1, 1)$ and $\beta = \alpha\varphi$ in $\operatorname{U}_2\mathbb{H}$ induce a reflection on $\mathcal{B}$. We denote its axis by A, and its center by c. Thus, the line $A \in \mathcal{L}$ is fixed by φ and has the property that $A \cap B$ is fixed pointwise by both α and β. According to (55.27ii), the group $\langle \alpha, \varphi \rangle$ does not act freely on $A \setminus B$. Hence either α or $\beta = \alpha\varphi$ is a reflection of $\mathcal{P}$. Let τ denote an element of Φ that permutes the coordinates cyclically. Then $\alpha = \alpha^\tau \alpha^{\tau^2} = \beta^\tau \beta^{\tau^2}$. From (55.35) it follows that α is a reflection on $\mathcal{P}$.

2) Investigation of the action of Φ on B shows that the centralizer Ψ of α in Φ acts transitively on $A \cap B$, cf. (13.14). Since φ moves every point in $A \setminus B$, we obtain that Ψ does not fix any point on A.

3) We claim that the stabilizer $\Lambda = \Delta_A$ is not contained in the stabilizer of c. Indeed, if this were the case, we could proceed as follows. Since $\Phi \cong \operatorname{U}_3\mathbb{H}$ acts on $\mathcal{B} \cong \mathcal{P}_2\mathbb{H}$ in the usual way, we have that $c \in A^\tau$, and the group $\operatorname{M} := \Lambda \cap \Lambda^\tau = \Lambda_{A^\tau}$ fixes the triangle c, c^τ, c^{τ^2} pointwise. For any point $b \in B$ that is not contained in any of the sides of this triangle, we have that the stabilizer M_b is compact by (83.9); recall that every quadrangle in $\mathcal{B} \cong \mathcal{P}_2\mathbb{H}$ is contained in some 4-dimensional subplane. Since $c \in A^\tau$, the orbit $(A^\tau)^\Lambda$ is contained in the pencil $\mathcal{L}_c$. Therefore $\dim \operatorname{M}_b \geq \dim \Delta - \dim A^\Delta - \dim(A^\tau)^\Lambda - 16 \geq 12$, and from the structure theory of compact, connected Lie groups we infer that either $\operatorname{M} \cong G_2$, or M has

rank at least 3. But the group G_2 is not contained in Φ by (84.8), and the stabilizer of a quadrangle does not contain $\mathbb{T}^3$ by (55.37b).

4) The set $\mathscr{E} = \{ \alpha^\lambda \alpha \mid \lambda \in \Lambda \setminus \Lambda_c \}$ generates a non-trivial closed subgroup Ξ of $\Delta_{[A]}$. Since the group Ψ centralizes α, it fixes the axis A of α, and therefore normalizes Ξ. Conjugation by α induces inversion on $\mathscr{E}$. Since $\mathscr{E}$ has non-empty interior in Ξ, cf. (61.20b), this implies that Ξ^1 is abelian. From (23.20) we know that Ξ consists of elations with axis A. According to (61.6), the connected component Ξ^1 is isomorphic to $\mathbb{R}^t$, where $t = \dim \Xi > 0$. Since Ψ commutes also with β, we obtain that Ψ leaves the β-eigenspaces in Ξ^1 invariant; these are $\Xi^+ = \Xi^1 \cap \mathrm{Cs}\,\beta$ and $\Xi^- = \{ \xi \in \Xi^1 \mid \xi^\beta = \xi^{-1} \}$. Since $\varphi = \alpha\beta$ and α induces inversion on Ξ, we obtain that $\Xi^- = \Xi^1 \cap \mathrm{Cs}\,\varphi$. We conclude that the orbits c^{Ξ^+} and c^{Ξ^-} are contained in the set of fixed points of β and φ, respectively. Therefore, each of the eigenspaces has dimension at most 8. Since α does not commute with any non-trivial element of Ξ, we obtain that Ψ acts effectively on each non-trivial β-eigenspace. The fact that Ψ contains $U_2\mathbb{H} \cong \mathrm{Spin}_5\mathbb{R}$ entails that every non-trivial eigenspace has dimension 8, see (95.10).

5) If both β-eigenspaces are non-trivial, we obtain that Λ contains a transitive elation group, and that $\mathscr{P} \cong \mathscr{P}_2\mathbb{O}$ by (24.9) and (64.24), since $\Lambda \neq \Delta$ by step 1). This implies that Δ is contained in the full automorphism group $\Sigma = E_6(-26)$. From step 2) we infer that the group Δ fixes no point and no line. Since Δ contains a transitive elation group, this yields that it acts transitively on P. By (63.8), this implies that Δ is either isomorphic to Σ or to the compact elliptic motion group. This is a contradiction.

6) There remains the case where $\Xi^1 \cong \mathbb{R}^8$ coincides with one of the β-eigenspaces. In this case, the orbit $C = c^{\Xi^1}$ generates a Baer subplane $\mathscr{C}$; in fact, this orbit carries an affine part either of $\mathscr{B}$ or of the subplane of fixed elements of β, depending on whether or not β acts non-trivially on it (recall that α induces inversion on Ξ). We infer that C consists of all centers of reflections in Λ with axis A, whence C is Λ-invariant. Since Λ fixes the line A of $\mathscr{C}$, we infer from (83.26) that Λ induces a group of dimension at most 27 on $\mathscr{C}$. But the kernel of this restriction has dimension at most 3 by (83.22), a contradiction to the fact that $\dim \Lambda \geq \dim \Delta - 16 = 36$. □

85.9 Remark. The proof of (85.8) is essentially due to Salzmann [82] (3.1d), p. 452. An alternative proof of (85.8) may be based on the fact that $F_4(4)$ contains a solvable subgroup of dimension 32. Such a group cannot act by M. Lüneburg [92] II, Satz 1.

85.10 Theorem: Almost simple groups. *Assume that $\mathscr{P}$ is a 16-dimensional compact projective plane, and that Δ is a closed subgroup of $\mathrm{Aut}\,\mathscr{P}$. If Δ is almost* simple and dim $\Delta > 45$, then the plane $\mathcal{P}$ is isomorphic to $\mathcal{P}_2\mathbb{O}$, and Δ is either isomorphic to $\operatorname{Aut}\mathcal{P}_2\mathbb{O}$ or to the elliptic or hyperbolic motion group.

Proof. If Δ is almost simple and dim $\Delta > 45$, then even dim $\Delta \geq 48$ by the classification (94.33), and Δ is a simple Lie group by (85.2). If dim $\Delta \geq 77$, then $\Delta = \operatorname{Aut}\mathcal{P}_2\mathbb{O}$ by (65.2). Therefore, we need only consider groups of dimension at most 76. From the classification of simple Lie groups, see (94.32), we infer that Δ is isomorphic to one of the following groups:

$$\operatorname{PSL}_5\mathbb{C},\ \operatorname{PSL}_6\mathbb{C};\ \operatorname{PSO}_8\mathbb{C},\ \operatorname{PSO}_9\mathbb{C};\ \operatorname{PSp}_8\mathbb{C};$$
$$\operatorname{PSU}_7(\mathbb{C},r) \text{ for } 0 \leq r \leq 3,\ \operatorname{PSU}_8(\mathbb{C},r) \text{ for } 0 \leq r \leq 4;$$
$$\operatorname{PSL}_7\mathbb{R},\ \operatorname{PSL}_8\mathbb{R};\ \operatorname{PSL}_4\mathbb{H};$$
$$O'_{11}(\mathbb{R},r) \text{ for } 0 \leq r \leq 5,\ O'_{12}(\mathbb{R},r) \text{ for } 0 \leq r \leq 6;$$
$$\operatorname{PU}_5(\mathbb{H},r) \text{ for } 0 \leq r \leq 2;\ \operatorname{PU}_6^\alpha\mathbb{H};\ \operatorname{PSp}_{10}\mathbb{R};$$
$$F_4(c) \text{ for } c \in \{-52, -20, 4\}.$$

In view of (62.6), we may assume that Δ is not isomorphic to one of the motion groups $F_4(-52)$ and $F_4(-20)$ of $\mathcal{P}_2\mathbb{O}$. The case $\Delta \cong F_4(4)$ is excluded by (85.8).

Obviously, each of the groups $\operatorname{PSL}_5\mathbb{C}$, $\operatorname{PSO}_8\mathbb{C}$, $\operatorname{PSL}_7\mathbb{R}$ and $O'_{11}(\mathbb{R},r)$ contains a subgroup isomorphic to $\operatorname{SO}_5\mathbb{R}$, which cannot act by (55.40). The groups $\operatorname{PSp}_8\mathbb{C}$ and $\operatorname{PSL}_4\mathbb{H}$ both contain a subgroup isomorphic to $\operatorname{PU}_4\mathbb{H}$, which contains $\operatorname{PU}_2\mathbb{H} \cong \operatorname{SO}_5\mathbb{R}$. The groups $\operatorname{PSp}_{10}\mathbb{R}$ and $\operatorname{PU}_6^\alpha\mathbb{H}$ as well as each of the groups $\operatorname{PSU}_7(\mathbb{C},r)$ and each of the groups $\operatorname{PU}_5(\mathbb{H},r)$ contain a 5-dimensional torus group, and cannot act by (55.37). □

85.11 Theorem: Groups that are not semi-simple. *Assume that $\mathcal{P}$ is a 16-dimensional compact projective plane, and that Δ is a closed, connected subgroup of $\operatorname{Aut}\mathcal{P}$. Assume that Δ is not semi-simple, and let Θ be a minimal connected abelian normal subgroup. Then the following hold.*

(a) *If dim $\Delta \geq 46$, then Θ is isomorphic to a vector group $\mathbb{R}^t$ for some t with $1 \leq t \leq 16$. Moreover, dim $\Delta \leq 46 + t \leq 62$.*
(b) *If dim $\Delta \geq 47$, then, up to duality, Θ consists of elations with common axis, say W. In particular, Δ fixes W.*

Proof. A minimal connected abelian normal subgroup is either compact or isomorphic to $\mathbb{R}^t$, see (94.26). Thus (85.5 and 4) yield the assertion. □

85.12 Remark. The bound in (85.11a) is attained by the stabilizer of a line in $\operatorname{Aut}\mathcal{P}_2\mathbb{O}$, i.e., the full group of automorphisms of the affine plane over $\mathbb{O}$.

In the situation of (85.11b), we wish to obtain further information about the structure of the plane. This requires a close look at the group Δ, its normal subgroup Θ and their possible actions.

For the remainder of this section, assume that Δ is a closed subgroup of $\operatorname{Aut}\mathcal{P}$ such that $\dim\Delta \geq 47$, and that Δ is not semi-simple. Let $\Theta = \Theta_{[W,W]} \cong \mathbb{R}^t$ be a minimal connected abelian normal subgroup, as in (85.11b). Recall from (85.2) that Δ is a Lie group.

85.13 Proposition. *The group Δ fixes at most one point. If a fixed point exists, then it belongs to W.*

Proof. Since Θ acts freely outside W, every fixed point of Δ belongs to W. If Δ fixes two points on W, we find points $a, b \in P \setminus W$ such that $\dim \Delta_{a,b} \leq 14$ by (83.23), and $\dim \Delta \leq 46$. □

85.14 Lemma. *Let Σ be the group of all automorphisms of a 16-dimensional compact projective plane $\mathcal{P}$, and let $\mathsf{T} \cong \mathbb{R}^t$ be a subgroup of $\Sigma_{[W,W]}$ for some line W of $\mathcal{P}$. Assume that there exists a point a outside W, and a closed subgroup Γ of the normalizer of T in Σ_a such that $\dim \tau^\Gamma = \dim \mathsf{T}$ for every $\tau \in \mathsf{T} \setminus \{\mathbb{1}\}$. Then either $t = 16$, or $t \leq 12$.*

Proof. From (61.4b), we know that $t \leq 16$. Assume that $13 \leq t \leq 15$. Since τ^Γ is open in the connected space $\mathsf{T} \setminus \{\mathbb{1}\}$, we infer that Γ acts transitively on $\mathsf{T} \setminus \{\mathbb{1}\}$. The space of all rays $\{\tau^r \mid r > 0\}$, where $\tau \in \mathsf{T} \setminus \{\mathbb{1}\}$, is homeomorphic to $\mathbb{S}_{t-1}$. Since the group Γ is transitive on $\mathsf{T} \setminus \{\mathbb{1}\}$, it is also transitive on this sphere, and we deduce from (96.19 to 22) that Γ contains a compact almost simple group Ω that covers one of the groups $\mathrm{SO}_{13}\mathbb{R}$, $\mathrm{PSU}_7\mathbb{C}$, or $\mathrm{SO}_{15}\mathbb{R}$. According to (62.8c), the group Ω cannot act on a line of $\mathcal{P}$, in contradiction to the fact that $W = W^\Gamma$. □

85.15 Theorem. *Let $\mathcal{P}$ be a 16-dimensional compact projective plane, and assume that Δ is a closed subgroup of $\operatorname{Aut}\mathcal{P}$ such that $\dim \Delta \geq 57$. Then $\mathcal{P} \cong \mathcal{P}_2\mathbb{O}$.*

Proof. By (85.6 and 10), it suffices to consider the case where Δ is not semi-simple. According to (85.11b), there exists (up to duality) a minimal connected abelian normal subgroup $\Theta = \Theta_{[W,W]} \cong \mathbb{R}^t$. In particular, the group Δ fixes the line W. From (83.26) we infer that Δ fixes no point. If $\mathcal{P}$ is not a translation plane, then $\dim \Theta_{[w]} \leq 7$ for every $w \in W$, cf. (61.4c). Moreover, there exists a point $z \in W$ such that $\dim \Theta_{[z]} \leq 6$ by (61.13). From (61.11a) we know that $\dim \Theta \leq 14$.

We investigate the stabilizer $\Gamma := \Delta_a$. According to (85.14), there exists $\vartheta \in \Theta \setminus \{\mathbb{1}\}$ such that $\dim \vartheta^\Gamma \leq 13$. Since Δ fixes no point, we find $\tau \in \Theta \setminus \{\mathbb{1}\}$ such that ϑ and τ have different centers. If $\dim \vartheta^\Gamma \geq 9$, we may even choose $\tau \in \vartheta^\Gamma$. In any case, we obtain that the centralizer Λ of ϑ and τ in Γ satisfies $\dim \Lambda \geq \dim \Gamma - 26$. Since Λ acts trivially on the vector subspace that is spanned by ϑ and τ in $\Theta \cong \mathbb{R}^t$, we infer from (83.23) that $\dim \Lambda \leq 14$, and that $\dim \Delta \leq 16 + 26 + 14 = 56$. □

We conclude this section by stating a characterization of $\mathcal{P}_2\mathbb{O}$ among the 16-dimensional compact projective planes. This characterization was accomplished in Salzmann [87], its proof relies on the corresponding results about 16-dimensional compact, connected translation planes, see Hähl [88].

85.16 Theorem. *Assume that $\mathcal{P}$ is a 16-dimensional compact projective plane. If $\dim \mathrm{Aut}\, \mathcal{P} > 40$, then $\mathcal{P}$ is isomorphic to $\mathcal{P}_2\mathbb{O}$, and $\mathrm{Aut}\, \mathcal{P}$ is the 78-dimensional simple exceptional Lie group* $\mathrm{E}_6(-26)$.

Recall from (82.27) that there exist translation planes with 40-dimensional automorphism group. Note also that the proof of (85.16) uses the characterization of $\mathcal{P}_2\mathbb{O}$ among the 16-dimensional translation planes, as given in Section 81.

85.17 Notes on Section 85. The material presented in this section is taken from Salzmann [82b, 83, 84 and 87]. Some of the assertions have been weakened, thus ideas of Stroppel [93e] became available to simplify the proofs. Section 87 sketches the proof of (85.16), and indicates directions for further refinements of the classification.

86 Hughes planes

We construct the 8- and 16-dimensional Hughes planes. These are non-Moufang planes containing a desarguesian Baer subplane all of whose linear automorphisms extend to the larger plane. In contrast to the existing literature, our construction as well as the proofs of all relevant properties (regarding both incidence and topology) are almost identical in the two cases; hence, the cases will be treated simultaneously. In dimension 4, there are no planes with the above properties, see (72.3).

There is a different approach to the 8-dimensional Hughes planes, which shows the connection of these planes with the Kalscheuer nearfields (64.19) more clearly and makes the topological properties more easily accessible; however, this approach does not generalize to the 16-dimensional case. We remark that only the 8-dimensional planes are (generalized) Hughes planes in the usual sense. Our extended use of the name is suggested by the characteristic properties of the automorphism groups stated above. For distinction, our 16-dimensional planes were called 'Moufang-Hughes planes' by Hähl [86b]. The classical planes $\mathcal{P}_2\mathbb{H}$ and

$\mathcal{P}_2\mathbb{O}$ will appear as special cases of our construction, hence we shall not always insist that a 'Hughes plane' be non-classical.

The first two subsections deal with the construction of the Hughes planes as projective planes and as topological projective planes. In the final, third subsection, we shall characterize the compact, connected Hughes planes as the only compact, non-Moufang planes of dimension $4d$ admitting an action of a group locally isomorphic to the linear collineation group of the desarguesian plane of dimension $2d$.

86.1 The Hughes groups of the classical planes.
(a) The inclusion of $\Gamma = \Gamma^{\mathbb{C}} = SL_3\mathbb{C}$ in $GL_3\mathbb{H}$ defines an effective action of Γ on the quaternion plane $\mathcal{P}_2\mathbb{H}$, which leaves the complex plane $\mathcal{P}_2\mathbb{C}$ invariant and induces its group of linear automorphisms, compare (13.6). The kernel of the restricted action is the center $Z \cong \mathbb{Z}_3$ of Γ. Moreover, each elation of $\mathcal{P}_2\mathbb{C}$ is induced by an elation of $\mathcal{P}_2\mathbb{H}$ that belongs to Γ.

It is not difficult to compute the stabilizer $\Pi = \Gamma_p$ of a point of $\mathcal{P}_2\mathbb{H}$ that does not belong to the complex plane. We may assume that p is a point of the line W at infinity, and then W is the only line of the complex plane that contains p. Hence, Π fixes W and may be represented by matrices as in (13.7). Clearly, all translations and all homotheties belong to Π. If we select p as the point with slope $k \in \mathbb{H}$, where k belongs to a Hamilton triple i, j, k with $i \in \mathbb{C}$, see (11.10), then an easy computation shows that the complex 2×2-matrices fixing p are precisely the unitary matrices. Thus, Π coincides with the group Π_0 that will be defined in the sequel (86.2).

(b) Next, we consider the group $\Gamma = \Gamma^{\mathbb{H}} = SL_3\mathbb{H}$, compare the 'conventions' below. We construct an action of Γ on the octonion plane $\mathcal{P}_2\mathbb{O}$ with the following properties, similar to those of the action of $\Gamma^{\mathbb{C}}$ described in (a). The group Γ leaves the quaternion plane $\mathcal{P}_2\mathbb{H}$ invariant and induces its full automorphism group $\Omega = PGL_3\mathbb{H} = PSL_3\mathbb{H} = \Gamma/\mathbb{Z}_2$, see (13.8). Moreover, each elation of $\mathcal{P}_2\mathbb{H}$ is induced by an elation of $\mathcal{P}_2\mathbb{O}$ that belongs to Γ.

Consider the affine planes $\mathcal{A}_2\mathbb{H} \leq \mathcal{A}_2\mathbb{O}$ as described in Section 12, with their obvious inclusion. Let $\{o, u, v\}$ be the triangle in the quaternion plane whose sides are the x-axis X, the y-axis Y and the line W at infinity. The description of elations given in (12.5c) shows that every elation of $\mathcal{P}_2\mathbb{H}$ with axis X and center $u = X \wedge W$ extends to a unique elation of the octonion plane. Now the triality collineation (16.6) of $\mathcal{P}_2\mathbb{O}$ leaves the quaternion plane and the triangle invariant, and conjugation by this collineation shows that the elation groups $\Omega_{[W,v]}$ and $\Omega_{[Y,o]}$ also extend. Together, these elation groups generate the simple group Ω; see (17.2) for a proof of the analogous statement in the octonion plane. Hence, there is a closed subgroup $\Delta \leq \text{Aut}\,\mathcal{P}_2\mathbb{O}$ which induces (a group containing) Ω. By (94.27), we may assume that Δ is a covering group of Ω, which leaves only the possibilities $\Delta \cong \Omega$ and $\Delta \cong \Gamma$. It will turn out in (86.36) that in fact $\Delta \cong \Gamma$. The same theorem will also show that the pair of stabilizers of an incident pair of elements of $\mathcal{P}_2\mathbb{O}$ not belonging to $\mathcal{P}_2\mathbb{H}$ is conjugate to the pair (Π_0, Λ_0) constructed in (86.2) below.

Hughes planes as projective planes

Conventions. Let $\mathbb{F} \in \{\mathbb{C}, \mathbb{H}\}$. We shall write $d = \dim \mathbb{F}$; observe that this is the dimension of the lines of $\mathcal{P}_2\mathbb{F}$, and that we are going to construct a plane with lines of dimension $2d$.

We deviate from the conventions in Section 13 in our representation of the plane $\mathcal{P}_2\mathbb{F}$ and of the action of $\mathrm{GL}_3\mathbb{F}$ on that plane. Points will be generated by elements of the *left* vector space $\mathbb{F}^3$; they will be considered as *row* vectors, and will be multiplied by matrices from the *right*. Dually, lines are represented by column vectors, with matrices acting from the left.

We shall work with the group $\mathrm{SL}_3\mathbb{H}$, which may be defined either as the subgroup of $\mathrm{GL}_3\mathbb{H}$ generated by all elations of the affine space $\mathbb{H}^3$, or as the subgroup consisting of the elements of determinant 1, compare (94.32f). The latter description requires the non-commutative determinant (Artin [57] Chap. IV Section 1), which may be obtained in the following way. Each quaternion x may be represented by a matrix $M_x \in \mathrm{GL}_4\mathbb{R}$, namely, the matrix of the right multiplication $y \mapsto yx$ with respect to the standard basis $1, i, j, k$ of $\mathbb{H}$. Replacing every entry a of $A \in \mathrm{GL}_n\mathbb{H}$ by the corresponding 4×4-block M_a, we obtain an embedding $f : \mathrm{GL}_n\mathbb{H} \to \mathrm{GL}_{4n}\mathbb{R}$, and we define $\det A = \sqrt[4]{\det_\mathbb{R} f(A)}$. This has many of the usual properties; in particular, $\det AB = \det A \det B$, and the determinant of a block matrix $\begin{pmatrix} A & 0 \\ X & B \end{pmatrix}$ may be computed as $\det A \det B$. Moreover, for a 1×1-matrix (a) we have $\det(a) = \sqrt[4]{\det_\mathbb{R} M_a} = |a|$. Observe that $\det A$ is always a non-negative real number.

86.2 Some subgroups of $\mathrm{SL}_3\mathbb{F}$. Let $\mathbb{F} \in \{\mathbb{C}, \mathbb{H}\}$. The vector $(1, 0, 0)$ represents a point o of $\mathcal{P}_2\mathbb{F}$, whose stabilizer in the group $\Gamma = \Gamma^\mathbb{F} = \mathrm{SL}_3\mathbb{F}$ is

$$\Gamma_o = \left\{ \begin{pmatrix} m & r \\ s^t & R \end{pmatrix} \,\middle|\, 0 \neq m \in \mathbb{F},\ s \in \mathbb{F}^2,\ R \in \mathrm{GL}_2\mathbb{F} \right\} \cap \Gamma.$$

By transposition, we obtain a column vector $(1, 0, 0)^t$ representing a line L of $\mathcal{P}_2\mathbb{F}$, whose stabilizer is

$$\Gamma_L = \Gamma_o^{\,t}.$$

We have the elation groups

$$\Theta = \Gamma_{[L,L]} = \left\{ \begin{pmatrix} 1 & u \\ & \mathbb{1} \end{pmatrix} \,\middle|\, u \in \mathbb{F}^2 \right\} \cong \mathbb{R}^{2d}$$

and, dually,
$$P = \Gamma_{[o,o]} = \Theta^t.$$

In order to verify this statement, observe that $\vartheta \in \Theta \setminus \{\mathbb{1}\}$ fixes precisely those points whose first coordinate is zero, and that no other linear collineations have this property.

Next, we consider a 'spiral subgroup' $H_\alpha = \{e^{(1+i\alpha)r} \mid r \in \mathbb{R}\}$ of the multiplicative group of complex numbers, where α denotes a fixed real parameter. We use H_α to define two subgroups of Γ, as follows.

$$\Pi = \Pi_\alpha = \left\{ \begin{pmatrix} h^{-2} & z \\ & |h|C \end{pmatrix} \;\middle|\; h \in H_\alpha,\; z \in \mathbb{F}^2,\; C \in U_2\mathbb{F} \right\} \cap \Gamma,$$

$$\Lambda = \Lambda_\alpha = \Pi_\alpha{}^t.$$

The intersection
$$\Delta = \Pi \cap \Lambda$$

is the subgroup of Π defined by the condition $z = 0$. In the quaternion case, the matrices written down in the definition of Π have determinant 1, hence it is not necessary to intersect with Γ. However, in the complex case, the (complex) determinants of these matrices merely have absolute value 1, and all such determinants occur. In this case, by taking the intersection with Γ we decrease the dimension of the group by 1.

86.3 Lemma. *The groups introduced above have the following properties.*

(a) $\Gamma_o \cap \Lambda\Pi = \Lambda$ *and* $\Gamma_L \cap \Lambda\Pi = \Pi$.
(b) $\Gamma_L \cap \Pi\Lambda\Pi = \Pi$.
(c) $\Pi = \Delta\Theta = \Theta\Delta$ *and* $\Lambda = \Delta P = P\Delta$.
(d) $\Pi\Lambda\Pi = \Delta\Theta P\Theta = \Pi P\Theta$.

Proof. For the first equation in (a), we observe that $\Lambda \subseteq \Gamma_o$ and $\Gamma_o \cap \Pi \subseteq \Lambda$ by definition, and we multiply the last relation by Λ from the left. The second equation follows by matrix transposition. Multiplying this by Π from the left yields (b). Assertion (c) is immediate from the definitions, and (d) follows from (c). □

86.4 Definition: The Hughes planes. Let $\mathbb{F} \in \{\mathbb{C}, \mathbb{H}\}$. Fix a real number α and consider the subgroups $\Pi = \Pi_\alpha$ etc. of $\Gamma = SL_3\mathbb{F}$ introduced in (86.2). We add a set N of 'outer' points and a set $\mathcal{N}$ of outer lines to the classical plane $\mathcal{P}_2\mathbb{F} = (D, \mathcal{D})$, as follows. We define

$$N = \Gamma/\Pi \quad \text{and} \quad \mathcal{N} = \Gamma/\Lambda,$$

so that the outer elements are right cosets of Π or Λ. The elements of $\mathcal{P}_2\mathbb{F}$ will be represented by their stabilizers in Γ, which form the conjugacy classes

$$D = \Gamma_o{}^\Gamma \quad \text{and} \quad \mathcal{D} = \Gamma_L{}^\Gamma.$$

We define the *Hughes plane* $\mathcal{H}_\alpha = \mathcal{H}_\alpha^\mathbb{F}$ as $(P, \mathcal{L})$, where $P = D \cup N$ and $\mathcal{L} = \mathcal{D} \cup \mathcal{N}$, endowed with the following incidence relation I. Incidence of 'inner' elements $d \in D$ and $M \in \mathcal{D}$ is the usual one. For $\sigma, \tau \in \Gamma$, the inner point $o^\tau = \Gamma_o{}^\tau$ is incident with the outer line $\Lambda\sigma$ if, and only if, $\Lambda^\sigma \subseteq \Gamma_o{}^\tau$ (which means inclusion of stabilizers with respect to the action of Γ introduced below). Incidence of inner lines and outer points is defined dually. Finally, for two outer elements we let $\Pi\sigma$ I $\Lambda\tau$ if, and only if, $\Pi\sigma \cap \Lambda\tau \neq \emptyset$. The incidences between the special elements o, L, $p = \Pi$ and $A = \Lambda$ are shown by Figure 86a. (We shall often give an element two different names; this should help to avoid confusion caused by the double rôle played by some subgroups of Γ, which appear both as elements of the geometry and as groups of automorphisms. When we use Latin symbols for lines, we even feel free to consider lines as subsets of the point set.)

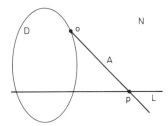

Figure 86a

The group Γ acts on $\mathcal{H}_\alpha$ via the usual action on inner elements (which corresponds to conjugation of stabilizers) and via right translation of cosets on the outer elements. This action has orbits D, $\mathcal{D}$, N and $\mathcal{N}$, and it preserves incidence. Note that the coset $\Pi\mathbb{1} = p$ coincides with the stabilizer $\Gamma_p = \Pi$, while in general the stabilizer of $\Pi\tau$ is Π^τ.

We remark that the groups Π_0 and Λ_0 are the 'outer' stabilizers of Γ acting on the $4d$-dimensional classical plane. This was shown to be true in the complex case, see (86.1), and it will later turn out to hold for $\mathbb{F} = \mathbb{H}$. This assertion about stabilizers is closely related to the fact that $\mathcal{H}_0$ is isomorphic to the classical quaternion or octonion plane. (Of course, this is not a Hughes plane in the proper sense, because it is a Moufang plane.)

For $\mathbb{F} = \mathbb{C}$, the isomorphism $\mathcal{H}_0^\mathbb{C} \cong \mathcal{P}_2\mathbb{H}$ is already proved by what we have said. For $\mathbb{F} = \mathbb{H}$, we shall proceed the other way round. We know from (86.1) that Γ acts on the octonion plane, and we shall deduce (86.34) that the octonion plane is isomorphic to $\mathcal{H}_\alpha$ for some α. Finally, we shall exclude the possibility that

$\alpha \neq 0$ and infer that $\mathcal{H}_0 \cong \mathcal{P}_2\mathbb{O}$, see (86.36). Moreover, the action of Γ is uniquely determined (86.35), and $\Gamma_p = \Pi_0$ holds. We emphasize that this is independent of the fact, to be proved later, that $\mathcal{H}_\alpha$ is a projective plane for $\alpha \neq 0$; this is important because the equality $\Gamma_p = \Pi_0$ will be used in that proof, see (86.9).

Our task is to show in general that $\mathcal{H}_\alpha$ is a projective plane, to introduce topologies on the point and line sets such that a compact plane of dimension $4d \in \{8, 16\}$ results, and finally to verify that the properties of the action of Γ on $\mathcal{H}_\alpha$ are similar to those encountered in the quaternion or octonion situation (86.1). This task will be accomplished in (86.12 and 31). We remark here that the definition of $\mathcal{H}_\alpha$ is self-dual. Hence we may often omit one half of the proof of a self-dual assertion about $\mathcal{H}_\alpha$.

Notes. Hughes planes in the original sense are finite. The construction of these planes from finite nearfields, due to Hughes, can be found in Hughes–Piper [73] IX.6. Various successful attempts to generalize this construction reached a first culmination point with Dembowski's paper [71a]; he allows infinite nearfields whose kernel is not necessarily contained in the center. A correction to that paper was given by Biliotti [79].

In Salzmann's article [81a], Hughes planes made their first appearance on the topological stage. The 8-dimensional Hughes planes constructed in that paper are more closely related to the Kalscheuer nearfields than the 16-dimensional ones discovered later. In fact they can be described by 'homogeneous coordinates' taken from a nearfield in a way very similar to the desarguesian case. For an exposition of this construction, which is not limited to the topological situation, see Maier–Stroppel [95].

Various group-theoretical characterizations of abstract Hughes planes have been given. They can be found in Dembowski [71b] and H. Lüneburg [76]. Very similar characterizations were obtained in the topological context by Salzmann [81a, 82a] and Hähl [86b]; they are reproduced here in (86.34) and in the remark following (86.33). This similarity of group-theoretical properties prompted us to call the planes constructed above Hughes planes, although the 16-dimensional ones are not even 'generalized Hughes planes' in the sense of Dembowski [71a]. As we know, there are no 8-dimensional proper nearfields (64.20a), hence there is no chance of constructing 16-dimensional 'generalized Hughes planes' in this sense.

The following is an easy consequence of the definition of incidence.

86.5 Lemma.

(a) *The action of Γ on the outer incidence structure $(N, \mathcal{N})$ is flag transitive.*
(b) *The substructure $(D, \mathcal{D})$ of $\mathcal{H}_\alpha$ is a projective plane isomorphic to $\mathcal{P}_2\mathbb{F}$. It satisfies the Baer condition, that is, each outer element is incident with a unique inner element.*

(c) *The stabilizer Γ_L is transitive on the set $L \cap N$ of outer points of the inner line L.*

Proof. Let $(\Pi\sigma, \Lambda\tau)$ be an outer flag, and select an element $\varrho \in \Pi\sigma \cap \Lambda\tau$. Then $(\Pi\sigma, \Lambda\tau) = (\Pi\varrho, \Lambda\varrho) = (p, L)^\varrho$. The isomorphism of the inner structure with $\mathcal{P}_2 \mathbb{F}$ was almost part of the definition, and the Baer condition follows from the observation that the elation group $\Theta \leq \Pi$ acts transitively on $D \setminus L$. Indeed, this shows that Π does not fix any inner line other than L, whence, by definition, p is not incident with any such line. Finally, a point $q = p^\sigma \in N = p^\Gamma$ belongs to a unique inner line, namely to L^σ, hence $q \in L$ if, and only if, $\sigma \in \Gamma_L$. □

86.6 Lemma. *The elation group Θ of $\mathcal{P}_2 \mathbb{F}$ is sharply transitive both on $D \setminus L$ and on the set $\mathcal{N}_p$ of outer lines through p. The join map $q \mapsto q \vee p$ induces a well-defined, bijective map $D \setminus L \to \mathcal{N}_p$. In particular, no outer line $\Lambda\tau$ other than Λ is incident with both p and o. Moreover, the join map is an isomorphism between the actions of Θ (and even of Π) on the two spaces.*

Proof. Flag transitivity on the outer geometry implies that Π is transitive on $\mathcal{N}_p$. We know that the group Θ is sharply transitive on $D \setminus L$ and fixes p. Moreover, $\Pi \leq \Gamma_L$ leaves $D \setminus L$ invariant. Hence $\Pi = \Gamma_p$ is transitive on this set. Since o is the unique inner point on Λ, it follows that the points of $D \setminus L$ are precisely the inner points of the outer lines through p.

All that remains to be shown is that the point o is not incident with any outer line $\Lambda\tau \neq \Lambda$ through p. Assume by contradiction that this happens. Since o^τ is the unique inner point of $\Lambda\tau$, we have that $\tau \in \Gamma_o$. Incidence of p and $\Lambda\tau$ implies that $\tau \in \Lambda\Pi$. Then (86.3a) shows that even $\tau \in \Lambda$, so that $\Lambda\tau = \Lambda$. □

By (86.5b) and (86.6) we know that two points of $\mathcal{H}_\alpha$ are joined by a unique line whenever at least one of them is an inner point. Next, we deal with the problem of joining two outer points. Their join may be an inner or outer line. In the first case, the solution follows easily from (86.5b and c):

86.7 Lemma. *Two distinct outer points p^σ and p^γ are incident with the same inner line if, and only if $\gamma\sigma^{-1} \in \Gamma_L$. In this case, $p^\sigma \vee p^\gamma = L^\sigma$ is uniquely determined.* □

We still have to show that two points p^σ and p^γ are joined by a (unique) outer line if, and only if, they are not joined by an inner one. A first step in this direction is the following.

86.8 Lemma. *Two distinct outer points p^σ and p^γ can be joined by an outer line if, and only if $\gamma\sigma^{-1} \in \Pi\Lambda\Pi$. In this case, $\gamma\sigma^{-1} \notin \Gamma_L$; hence, by (86.7), the same points cannot be joined by an inner line.*

Proof. We may assume that $\sigma = 1\!\!1$. Suppose that $\Lambda\tau$ is an outer line incident with both $p = \Pi$ and $p^\gamma = \Pi\gamma$. By the definition of incidence, we obtain elements $\pi, \pi' \in \Pi$ and $\lambda, \lambda' \in \Lambda$ such that $\pi = \lambda\tau$ and $\pi'\gamma = \lambda'\tau$. Then $\gamma = \pi'^{-1}\lambda'\lambda^{-1}\pi \in \Pi\Lambda\Pi$.

Conversely, suppose that $\gamma \in \Pi\Lambda\Pi$ and write $\gamma = \pi\lambda\pi'$ accordingly. Then the relations $\pi' \in \Pi \cap \Lambda\pi'$ and $\lambda\pi' \in \Pi\gamma \cap \Lambda\pi'$ show that $\Lambda\pi'$ is an outer line joining p to p^γ. We have $\gamma \notin \Pi$ since we are considering distinct points. Hence, the equation $\Gamma_L \cap \Pi\Lambda\Pi = \Pi$ of (86.3b) shows that $\gamma \notin \Gamma_L$. □

86.9 Proposition. *For two distinct outer points p^σ and p^γ, the following conditions are equivalent.*

(i) p^σ and p^γ can be joined by an outer line.
(ii) $\gamma\sigma^{-1} \in \Pi\Lambda\Pi = \Pi P\Theta$, compare (86.3d).
(iii) $\gamma\sigma^{-1} \notin \Gamma_L$.
(iv) p^σ and p^γ cannot be joined by an inner line.

Proof. The implications (i) ⇔ (ii) ⇒ (iii) ⇔ (iv) follow from (86.8 and 7). We shall complete the proof by showing that (iii) ⇒ (ii). It suffices to do this for $\sigma = 1\!\!1$.

The proof will be based on a comparison with the case $\alpha = 0$, which yields the classical plane $\mathcal{P}_2\mathbb{H}$ or $\mathcal{P}_2\mathbb{O}$, see (86.4 and 36). In that case, conditions (i) and (iv) are equivalent, hence the proposition is true for $\alpha = 0$. Using (ii) for $\alpha = 0$, the given element $\gamma = \begin{pmatrix} m & w \\ s^t & R \end{pmatrix} \in \Gamma \setminus \Gamma_L$ can be written as

$$\gamma = \begin{pmatrix} h^{-2} & z \\ & |h|C \end{pmatrix} \begin{pmatrix} 1 & \\ v^t & 1\!\!1 \end{pmatrix} \begin{pmatrix} 1 & u \\ & 1\!\!1 \end{pmatrix} \in \Pi_0 P\Theta,$$

where in particular $h \in H_0 = \mathbb{R}$. We have $s \neq 0$ and $v \neq 0$ since $\gamma \notin \Gamma_L$. Conjugation by a unitary matrix $\mu = \begin{pmatrix} 1 & \\ & U \end{pmatrix} \in \Pi \cap \Pi_0$ has the following effect: γ is replaced with $\mu\gamma\mu^{-1}$, the points p and p^γ are replaced with p and $p^{\gamma\mu^{-1}}$, and condition (i) remains unaffected. Moreover, the decomposition of γ is transformed into another decomposition in $\Pi_0 P\Theta$, because μ normalizes those factors, but the element v^t is changed to Uv^t. Using a suitable matrix U, we may change v to $(r, 0) \in \mathbb{F}^2$, where r is a positive real number; here we use the transitivity of the unitary group on the unit sphere in $\mathbb{F}^2$. After these changes have been made, we return to the original notation.

Now we choose a complex number $f = e^{ia}$ such that $fh^2 \in H_\alpha$, and we define $b \in \mathbb{F}$ by $rb + 1 = f$. Using $c = (b, 0) \in \mathbb{F}^2$ and $F = \text{diag}(f, 1) \in U_2\mathbb{F}$, we compute

$$\begin{pmatrix} 1 & \\ v^t & 1\!\!1 \end{pmatrix} = \begin{pmatrix} f^{-1} & c \\ & F \end{pmatrix} \begin{pmatrix} 1 & \\ v^t f^{-1} & 1\!\!1 \end{pmatrix} \begin{pmatrix} 1 & -c \\ & 1\!\!1 \end{pmatrix}.$$

Substituting this in the decomposition of γ, we obtain

$$\gamma = \begin{pmatrix} h^{-2}f^{-1} & h^{-2}c + zF \\ & |h|CF \end{pmatrix} \begin{pmatrix} 1 & \\ v^t f^{-1} & \mathbb{1} \end{pmatrix} \begin{pmatrix} 1 & u-c \\ & \mathbb{1} \end{pmatrix}.$$

This decomposition shows that $\gamma \in \Pi_\alpha P\Theta$. □

The following is a general observation on flag homogeneous incidence structures, due to Freudenthal [85] 6.3.3.

86.10 Lemma. *The condition that each pair of outer points can be joined by at most one outer line may be translated into the condition* $\Pi\Lambda \cap \Lambda\Pi = \Pi \cup \Lambda$.

Proof. Suppose that this condition holds. By flag-transitivity (86.5a), it suffices to show that a point q I A distinct from $p = \Pi\mathbb{1}$ cannot be joined to p by any outer line other than $A = \Lambda\mathbb{1}$. By the definition of incidence, we have $q = \Pi\lambda$ for some $\lambda \in \Lambda \setminus \Pi$. Any other outer line B through p has the form $B = \Lambda\pi$ with $\pi \in \Pi$. If q is incident with B, then we find $\pi' \in \Pi$ and $\lambda' \in \Lambda$ such that $\pi'\lambda = \lambda'\pi$. By our hypothesis, this element must belong to Π or to Λ. Since λ does not belong to Π, we infer that $\pi'\lambda \in \Lambda$, which implies that $\pi \in \Lambda$ and $B = \Lambda\pi = \Lambda = A$. Each step of this argument can be reversed, and one obtains the converse implication. In fact, we shall only need the statement that we proved in detail. □

86.11 Proposition. *Two distinct outer points are joined by at most one outer line.*

Proof. We show that the condition of (86.10) is satisfied. So let $\gamma \in \Pi\Lambda \cap \Lambda\Pi = \Pi P \cap \Lambda\Theta$ be given, compare (86.3c). This means that γ may be written in the following two ways.

$$\gamma = \begin{pmatrix} h^{-2} & z \\ & |h|C \end{pmatrix} \begin{pmatrix} 1 & \\ s^t & \mathbb{1} \end{pmatrix} = \begin{pmatrix} f^{-2} & \\ v^t & |f|F \end{pmatrix} \begin{pmatrix} 1 & u \\ & \mathbb{1} \end{pmatrix},$$

with the usual conventions; in particular, $h, f \in H_\alpha$. Our aim is to show that either $s = 0$ or $u = 0$. As in the proof of (86.9), we may conjugate the equation for γ by a suitable unitary matrix μ which normalizes each of the groups Δ, Θ and P, and we obtain that $s = (r, 0)$ for some non-negative real number r. Computing the above matrix products and comparing coefficients, we obtain the equations

$$f^2 h^{-2} + us^t = 1 \quad \text{and} \quad s^t u + |fh^{-1}|C^{-1}F = \mathbb{1}.$$

Abbreviating $g = fh^{-1}$ and $G = C^{-1}F$ and using $s = (r, 0)$ and $u = (u_1, u_2)$, we rewrite this as $g^2 + ru_1 = 1$ and

(1) $$\begin{pmatrix} ru_1 & ru_2 \\ 0 & 0 \end{pmatrix} = \mathbb{1} - |g|G.$$

Thus, $|g|^{-1}$ is an eigenvalue of the unitary matrix G, whence $|g| = 1$. In view of $g \in H_\alpha$, it follows that $g = 1$, and we have $ru_1 = 0$. Using (1) and the fact that G is unitary again, we infer that also $ru_2 = 0$. To sum up, we have either $r = 0$ and hence $s = 0$, or $u = 0$. In other words, $\gamma \in \Pi \cup \Lambda$. □

Combining the results (86.6, 7, 9 and 11) and their duals, we may formulate our first main result (86.12). For the assertions regarding the group, see (86.4 and 5) and (13.8).

86.12 Theorem. *Let α be a real number, and let $\mathbb{F} \in \{\mathbb{C}, \mathbb{H}\}$. The Hughes plane $\mathcal{H}_\alpha = \mathcal{H}_\alpha^\mathbb{F}$ as defined in (86.4) is a projective plane. There is an action of $\Gamma = \mathrm{SL}_3\mathbb{F}$ on $\mathcal{H}_\alpha$ which leaves the 'inner' Baer subplane $\mathcal{P}_2\mathbb{F} \leq \mathcal{H}_\alpha$ invariant and induces its linear automorphism group $\mathrm{PGL}_3\mathbb{F}$. The action of Γ on the 'outer' structure is flag transitive.* □

A practical description of the join operation is given by the following result.

86.13 Lemma.

(a) *Two distinct outer points $\Pi\sigma$ and $\Pi\gamma$ such that $\gamma\sigma^{-1} \notin \Gamma_L$ are joined by the outer line $\Lambda\tau$, where τ is any element of $\Lambda\Pi\sigma \cap \Lambda\Pi\gamma$.*
(b) *The outer point $\Pi\sigma$ is joined to the inner point $q \notin L^\sigma$ by the outer line $\Lambda\tau$, where τ is any element of $\Pi\sigma$ satisfying $o^\tau = q$.*

Proof. (a) We know from (86.10) that there exists a unique outer line $\Lambda\tau$ joining the two points. The required incidences mean that $\Lambda\tau$ meets both $\Pi\sigma$ and $\Pi\gamma$, or in other words, that τ belongs to both $\Lambda\Pi\sigma$ and $\Lambda\Pi\gamma$.

(b) This follows by applying the collineation τ to the relation $o \vee p = A = \Lambda$. □

86.14 Example. Let $\Pi\sigma$ be an outer point on the line L. According to (86.5c), this means that $\sigma \in \Gamma_L$, whence $\sigma = \begin{pmatrix} m & u \\ & R \end{pmatrix}$. The outer line joining $\Pi\sigma$ to o is $\Pi\sigma \vee o = \Lambda\tau$, where $\tau = \begin{pmatrix} m & \\ & R \end{pmatrix}$. This follows from (86.13b), because τ belongs to $\Pi\sigma \cap \Gamma_o$.

86.15 Proposition: Further transitivity properties.

(a) *The group Δ is transitive on the punctured inner affine plane $D \setminus (L \cup \{o\})$ and, dually, on $\mathcal{D} \setminus (\mathcal{D}_o \cup \{L\})$.*
(b) *Δ is also transitive on $A \setminus \{o, p\}$ and, dually, on $\mathcal{N}_p \setminus \{A\}$.*
(c) *Γ is transitive on the set Q of pairs of distinct outer points that are joined by an outer line and, dually, on the set $\mathcal{Q}$ of pairs of outer lines with outer intersection point.*

Proof. Observe first that Δ fixes o, p, A and L. As in (13.7), the action of Δ on the inner affine point set $\mathbb{F}^2$ is given by $(x, y) \mapsto h^2|h|(x, y)C$, where $h \in H_\alpha$ and $C \in U_2\mathbb{F}$; in the complex case, these elements have to satisfy $h^2 = |h|^2 \det C$. We have $o = (0, 0)$, and the special unitary group is transitive on the unit sphere $\mathbb{S}_{2d-1} \subseteq \mathbb{F}^2$. Moreover, elements with $h \neq 1$ may be used to permute the spheres $r \cdot \mathbb{S}_{2d-1}$, $r > 0$, transitively. This proves the first half of (a), and the second half is obtained by transposition. Statement (b) is easily deduced from (a), using the Baer condition; compare, e.g., (86.6). Assertion (c) follows from the flag transitivity of Γ on the outer geometry together with (b). □

Hughes planes as topological projective planes

86.16 Lemma. *The groups Γ_o, Γ_L, Π, Λ and $\Delta = \Pi \cap \Lambda$ defined in (86.2) are closed subgroups of the Lie group $\Gamma = \mathrm{SL}_3\mathbb{F}$. The dimensions of these groups are shown by the following table.*

$\mathbb{F}$	Γ	Γ_o	Γ_L	Π	Λ	Δ
$\mathbb{C}$	16	12	12	8	8	4
$\mathbb{H}$	35	27	27	19	19	11

Proof. For Γ itself, see (94.33). The standard action on $\mathcal{P}_2\mathbb{F}$ shows that the stabilizers Γ_o and Γ_L are closed subgroups. Their dimensions can be computed using transitivity of this action and the dimension formula (96.10). Alternatively, one may consider the matrix entries m, s and R, whose contribution to the dimension is d, $2d$ and $4d$, respectively, and subtract 2 or 1 from the total to account for the passage to the subgroup of determinant 1. A similar computation yields the values of $\dim \Pi = \dim \Lambda$ and $\dim \Delta$; one uses that $\dim U_2\mathbb{C} = 4$ and $\dim U_2\mathbb{H} = 10$. The groups Π and Δ are closed because H_α is closed in $\mathbb{C}^\times$ and $U_2\mathbb{F}$ is compact. □

86.17 Lemma. *The products $\Pi\Lambda$ and $\Lambda\Pi$ are closed subsets of Γ.*

Proof. It suffices to give the proof for $\Lambda\Pi = \Delta P \Theta$, compare (86.3c). The elements of this set have the form

$$\gamma = \begin{pmatrix} h^{-2} & \\ & |h|C \end{pmatrix} \begin{pmatrix} 1 & \\ v' & \mathbb{1} \end{pmatrix} \begin{pmatrix} 1 & u \\ & \mathbb{1} \end{pmatrix} = \begin{pmatrix} h^{-2} & h^{-2}u \\ |h|Cv' & |h|C(v'u + \mathbb{1}) \end{pmatrix}.$$

If a sequence γ_ν of such matrices converges to a matrix $\sigma \in \Gamma$, then the entries h_ν belonging to the spiral group H_α converge to some $h \in H_\alpha$. Moreover, we may assume that the entries C_ν belonging to the compact unitary group $U_2\mathbb{F}$ converge to $C \in U_2\mathbb{F}$. Now convergence of the remaining elements u_ν and v_ν follows immediately, and we infer that $\sigma \in \Lambda\Pi$. □

86.18 Introducing topologies on $\mathcal{H}_\alpha$. The sets $N = \Gamma/\Pi$ and $\mathcal{N} = \Gamma/\Lambda$ of outer points and outer lines are endowed with the natural coset space topologies, that is, with the quotient topologies with respect to the canonical maps $\Gamma \to N$ and $\Gamma \to \mathcal{N}$, respectively. According to (86.16) and (94.3a), this makes both N and $\mathcal{N}$ a manifold of dimension $4d \in \{8, 16\}$, and Γ acts as a topological transformation group on either manifold (96.2).

Now consider the space

$$N^{(2)} = \{ (q, r) \in N^2 \mid q \neq r \}$$

with the topology inherited from N^2. Define the topology of $\mathcal{L}$ as the quotient topology with respect to the surjective join map

$$\vee : N^{(2)} \to \mathcal{L}.$$

Note that this induces a new topology on $\mathcal{N}$. It will turn out that the two topologies of $\mathcal{N}$ agree (86.20c); until that moment, only the coset space topology on $\mathcal{N}$ will be used. Dually, the topology of P is defined as the quotient topology with respect to $\wedge : \mathcal{N}^{(2)} \to P$.

We shall repeatedly use the following spaces, which were first introduced in (86.15c).

$$Q = \{ (q, r) \in N^{(2)} \mid q \vee r \in \mathcal{N} \} \quad \text{and} \quad \mathcal{Q} = \{ (A, B) \in \mathcal{N}^{(2)} \mid A \wedge B \in N \}.$$

According to (86.15), the group Γ is transitive on each of these sets.

Observe that the transitive actions of Γ on the sets D and $\mathcal{D}$ of inner elements also yield natural coset space topologies, which make these spaces compact $2d$-manifolds. These are the usual topologies on the point and line spaces of the classical plane $\mathcal{P}_2 \mathbb{F}$ as introduced in Section 14; compare also (64.3a).

86.19 Lemma.

(a) *The space $\mathcal{N}_p$ of outer lines through p is closed in $\mathcal{N}$ and is homeomorphic to $\Theta \cong \mathbb{R}^{2d}$ under the map $\vartheta \mapsto A^\vartheta$. Dually, the P-orbit $A \cap N \approx \mathbb{R}^{2d}$ is closed in N.*
(b) *The space $L \cap N$ of outer points on the line L is a $2d$-dimensional manifold.*

Proof. (a) At the moment, the topology of $\mathcal{N}$ is still defined as the coset space topology. Consider $\mathcal{N}_p$ as a subset of $\mathcal{N} = \Gamma/\Lambda$. By (86.5a), the union of this set of cosets is the subset $\Lambda \Pi \subseteq \Gamma$, which is closed (86.17). Hence $\mathcal{N}_p$ is closed in $\mathcal{N}$. Now the sharply transitive action of $\Theta \cong \mathbb{R}^{2d}$ on $\mathcal{N}_p$, see (86.6), yields our assertion in view of (96.9a).

(b) By (86.5c), the set $L \cap N$ is an orbit of the closed subgroup $\Gamma_L \leq \Gamma$. Hence $L \cap N = \Gamma_L/\Pi \subseteq N = \Gamma/\Pi$ is closed. As above, $L \cap N$ is a manifold, and its dimension is $\dim \Gamma_L - \dim \Pi = 2d$, compare (94.3a) and (86.16). □

86.20 Proposition.

(a) *The Γ-orbits $Q \subseteq N^{(2)}$ and $\mathcal{Q} \subseteq \mathcal{N}^{(2)}$ are both open.*
(b) *The group Γ acts on both P and $\mathcal{L}$ as a topological transformation group.*
(c) *The topologies induced by P and $\mathcal{L}$ on the spaces D, $\mathcal{D}$, N and $\mathcal{N}$ of inner and outer elements agree with their standard coset space topologies, compare (86.18).*
(d) *The subsets $N \subseteq P$ and $\mathcal{N} \subseteq \mathcal{L}$ are open.*

Proof. We prove only one of each pair of dual statements.

(a) Combining (86.15b, 16 and 19a) with the dimension formula (96.10), we may compute the dimension of the stabilizer Δ_q of a point $q \in A \setminus \{o, p\}$. Remember that N still has the coset space topology. We find that $\dim \Delta_q = 4 - 4 = 0$ if $\mathbb{F} = \mathbb{C}$, and $\dim \Delta_q = 11 - 8 = 3$ for $\mathbb{F} = \mathbb{H}$. Now Δ_q is also the stabilizer of Γ fixing the element $(p, q) \in Q$. The dimension formula (94.3a) yields that Q with its coset space topology is a manifold of dimension $\dim Q = 16 - 0$ or $\dim Q = 35 - 3 = 32$, respectively. Hence, Q and $N^{(2)}$ are manifolds of the same dimension, and openness follows by domain invariance (51.19).

(b) Since N carries the coset space topology, the actions of Γ on N and on $N^{(2)}$ are continuous. Let $f : N^{(2)} \to \mathcal{L}$ be the identification map (i.e., $f = \vee$). Then the product map $\mathbb{1} \times f : \Gamma \times N^{(2)} \to \Gamma \times \mathcal{L}$ is an identification, as well, see Dugundji [66] XII 4.1. By passage to the quotients, we obtain that the action $\Gamma \times \mathcal{L} \to \mathcal{L}$ is continuous, see (92.20a).

(c) The topology induced by $N^{(2)}$ on the open subset Q is the coset space topology of the orbit Q of Γ, see (96.9a). Since Q is a saturated, open subset of $N^{(2)}$, the restriction $f : Q \to \mathcal{N}$ is an identification; this is the first time that we use the induced topology on $\mathcal{N} \subseteq \mathcal{L}$. However, the same map is also an identification if we consider $\mathcal{N}$ with its coset space topology, because the equivariant map f is just the natural map between two coset spaces of the same group that is induced by an inclusion of stabilizers. It follows that the two quotient topologies on $\mathcal{N}$ coincide, hence assertion (c) is proved.

Let $\mathcal{D}$ and $\mathcal{D}'$ denote the space of inner lines, endowed with the topology induced by $\mathcal{L}$ and with the coset space topology, respectively. The continuous, transitive action of Γ on $\mathcal{D}$ shows that the identity map $\mathcal{D}' \to \mathcal{D}$ is continuous; this follows from the universal property of the quotient topology (92.19).

In order to show that the identity map in the converse direction is also continuous, we observe first that the set U of all outer points incident with some element of an open set $\mathcal{U} \subseteq \mathcal{D}'$ is open in N. Indeed, we may write $\mathcal{U} = L^\Omega$ for some open subset $\Omega \subseteq \Gamma$, and U is the union of the open sets q^Ω for all points $q \in L \setminus D$.

Now we show that $\mathcal{U}$ is open in $\mathcal{D}$. It suffices to verify that $\mathcal{V} = \mathcal{U} \cup \mathcal{N}$ is open in $\mathcal{L}$, which means that the inverse image $f^{-1}(\mathcal{V})$ under the quotient map $f : N^{(2)} \to \mathcal{L}$ is open in $N^{(2)}$. Now it is easy to check that $f^{-1}(\mathcal{V}) = Q \cup U^{(2)}$. This set is indeed open by (a) together with the last remarks.

(d) This assertion follows directly from assertion (a) together with the definition of the quotient topologies on P and $\mathcal{L}$. □

86.21 Corollary. *The operations* $\vee : N^{(2)} \to \mathcal{L}$ *and* $\wedge : \mathcal{N}^{(2)} \to P$ *of joining and intersecting outer elements are continuous.*

Proof. According to (86.20c), the topology of P is the quotient topology with respect to the map $\wedge : \mathcal{N}^{(2)} \to P$. □

86.22 Corollary. *The map sending an outer line to its unique inner point is continuous. Dually, the map sending an outer point q to the inner line incident with q is continuous.*

Proof. In the proof of (86.20c), we have seen that $\mathcal{D} = \mathcal{D}'$, and that an open subset $\mathcal{U} \subseteq \mathcal{D}'$ defines an open set $U \subseteq N$ of points incident with elements of $\mathcal{U}$. Both statements together prove the second assertion of the corollary. □

At the present stage, we do not know that P and $\mathcal{L}$ are Hausdorff spaces. Thus the following result is by no means trivial.

86.23 Lemma. *Each compact subset $K \subseteq N$ is closed in P.*

Proof. 1) We show that K has a closed inverse image in $\mathcal{N}^{(2)}$. First we treat the case where $K = \{q\}$ is a single point. Then we have to examine the set $(\mathcal{N}_q \times \mathcal{N}_q) \cap \mathcal{N}^{(2)}$, which is indeed closed by (86.19a).

2) In the general case, we have to prove that for each convergent sequence (B_ν, C_ν) in $\mathcal{N}^{(2)}$ such that $q_\nu = B_\nu \wedge C_\nu \in K$, the limit (B, C) also satisfies $B \wedge C \in K$. We may assume that the sequence q_ν, $\nu \in \mathbb{N}$, has a limit point $q \in K$, and we claim that $q = B \wedge C$. By step 1), we know that this is true if $q_\nu = q$ is a constant sequence.

Now N carries the coset space topology (86.20c). Using local cross sections of the canonical map $\Gamma \to N$, see (94.3b), we express q_ν as $q_\nu = q^{\gamma_\nu}$, where $\gamma_\nu \to 1\!\!1$ is a convergent sequence in Γ. Let $\sigma_\nu = \gamma_\nu^{-1}$. Then $\sigma_\nu \to 1\!\!1$, and the sequence $B_\nu^{\sigma_\nu} \wedge C_\nu^{\sigma_\nu} = q_\nu^{\sigma_\nu} = q$ is constant. Moreover, $(B_\nu^{\sigma_\nu}, C_\nu^{\sigma_\nu})$ converges to (B, C). Using step 1), we conclude that $B \wedge C = q$, as desired. □

Since N is open in P and each point of the manifold N has a compact neighbourhood, (86.23) has the following consequence.

86.24 Corollary. *Each pair consisting of an inner point and an outer point can be separated by open sets of P. The dual statement holds for pairs of lines.* □

86.25 Proposition. *Both the line pencil $\mathcal{L}_p$ of an outer point and the point row $A \subseteq P$ of an outer line are compact Hausdorff spaces homeomorphic to the sphere $\mathbb{S}_{2d}$.*

Proof. In view of the Baer condition (86.5b), the Hausdorff property follows immediately from (86.24). By (86.19a), it suffices to show that $\mathcal{L}_p$ and A are compact. Let B be the boundary of a compact neighbourhood of p in N, and observe that this neighbourhood is closed in P by (86.23). Then the image of $B \times \{p\}$ under the join map is a compact subset of $\mathcal{L}_p$, see (86.21). It is enough to show that this set contains all outer lines through p. Now such an outer line intersects N in a closed, connected, non-compact set (86.19a), hence it meets B. □

86.26 Lemma. *The restriction of the join map to $D \times \{p\}$ is continuous.*

Proof. The restriction $f : (D \setminus L) \times \{p\} \to \mathcal{N}_p$ of the join map is a homeomorphism; this follows from (86.6, 22 and 25) (by domain invariance) or from the fact that the action of Θ defines homeomorphisms $\Theta \approx D \setminus L$ and $\Theta \approx \mathcal{N}_p$, see (86.19a). Hence f is a proper map, and the extension sending all points of $D \cap L$ to L is continuous. □

The following result is crucial for the proof that the Hughes planes are topological planes, especially because it overcomes the trouble caused by hypothetical non-Hausdorff topologies. This is a remarkable contribution by Niemann [90], who also proved (86.28). Generally, our exposition in this section up to that point owes him a good deal. We thank him for permitting us to use his unpublished results in this way.

86.27 Proposition. *The affine point set $P \setminus L$ is open in P and is homeomorphic to $\mathbb{R}^{4d}$. Consequently, P and $\mathcal{L}$ are Hausdorff spaces and, in fact, manifolds of dimension $4d \in \{8, 16\}$.*

Proof. 1) Consider the affine line $X = A \setminus \{p\} \approx \mathbb{R}^{2d}$ and the group $\Theta \cong \mathbb{R}^{2d}$, see (86.25 and 2). Our aim is to prove that the map

$$\psi : X \times \Theta \to P \setminus L, \quad \text{defined by} \quad \psi(x, \vartheta) = x^\vartheta,$$

is a homeomorphism, and that ψ is open, even as a map into P. The fact that Θ is sharply transitive on $\mathcal{N}_p$, see (86.6), implies that ψ is bijective. From (86.20b) it follows that ψ is continuous. Since we do not know in advance that P is a manifold, we cannot apply domain invariance in order to conclude that ψ is also open. Nevertheless, it will be possible to shift the whole problem to a situation where this can be done. At least we can use domain invariance to show that the restriction $(X \setminus \{o\}) \times \Theta \to N \setminus L$ of ψ is open; it is open also as a map into N or even into P, see (86.20d). The only problem is to show that every open neighbourhood of (o, ϑ) in $X \times \Theta$ has an image that is open in P. Using the action, we may reduce this problem to the special case where $\vartheta = 1$.

2) For an open neighbourhood U of o in A not containing p and an open neighbourhood $\Omega \subseteq \Theta$ of 1, we have to show that U^Ω is open in P, which means by definition that the inverse image $f^{-1}(U^\Omega)$ with respect to the identification map $f = \wedge : \mathcal{N}^{(2)} \to P$ is open.

3) We use an auxiliary map in order to prove this. Let $W = L \cap N$ and define

$$\varphi : W^{(2)} \times U \times \Omega \to \mathcal{N}^{(2)} \quad \text{by} \quad (b, c, a, \vartheta) \mapsto (ba, ca)^\vartheta.$$

The image set of φ contains $f^{-1}(o^\Omega)$. Therefore, we obtain a decomposition

$$f^{-1}(U^\Omega) = f^{-1}((U \setminus \{o\})^\Omega) \cup \varphi(W^{(2)} \times U \times \Omega).$$

Since only open sets containing o present a problem, this decomposition reduces our problem to the task of verifying that φ is an open map.

4) At this point, domain invariance will come in again. By (86.19 and 25), both the domain and the range of φ are manifolds of dimension $8d$. The fact that Θ acts freely on $\mathcal{N}_p$ implies that φ is injective. Thus, it suffices to show that φ is continuous. In fact, is is enough to show that the 'first half' $(b, a, \vartheta) \mapsto (ba)^\vartheta$ of φ is continuous. Remember that Θ acts continuously, see (86.20b); hence continuity of the restricted join map $\eta : (b, a) \mapsto ba$ on $W \times U$ suffices. We know that joining of outer elements is continuous (86.21); this implies that continuity of η has to be verified only at (b, o).

5) Now let $b_\nu \in W$ converge to b and let $a_\nu \in A$ converge to o. We have to show that $b_\nu a_\nu \to bo$. We write $b_\nu = p^{\sigma_\nu}$ and $b = p^\sigma$, where

$$\sigma_\nu = \begin{pmatrix} m_\nu & v_\nu \\ & R_\nu \end{pmatrix} \quad \text{and} \quad \sigma = \begin{pmatrix} m & v \\ & R \end{pmatrix},$$

are suitable elements of Γ_L, see (86.5c). We may arrange that $\sigma_\nu \to \sigma$, using a local cross-section for the action of Γ_L on L, see (94.3b). By (86.14), we have $bo = A^\delta$, where $\delta = \begin{pmatrix} m & \\ & R \end{pmatrix}$. A similar formula holds for $b_\nu o$, hence we may assume that a_ν is always different from o, which means that a_ν is an outer point. By (86.19a), it follows that we may write $a_\nu = p^{\varrho_\nu}$, where

$$\varrho_\nu = \begin{pmatrix} 1 & \\ w_\nu^t & \mathbb{1} \end{pmatrix} \in P$$

and $w_\nu \to \infty$, because $a_\nu \to o$.

6) The fact that $b_\nu a_\nu = p^{\sigma_\nu} p^{\varrho_\nu}$ is an outer line means that $\gamma_\nu = \varrho_\nu \sigma_\nu^{-1} \in \Pi P\Theta$, see (86.9), and we obtain a decomposition

(1) $\quad \gamma_\nu = \begin{pmatrix} 1 & \\ w_\nu^t & \mathbb{1} \end{pmatrix} \begin{pmatrix} m_\nu & v_\nu \\ & R_\nu \end{pmatrix}^{-1} = \begin{pmatrix} h_\nu^{-2} & z_\nu \\ & |h_\nu|C_\nu \end{pmatrix} \begin{pmatrix} 1 & \\ s_\nu^t & \mathbb{1} \end{pmatrix} \begin{pmatrix} 1 & u_\nu \\ & \mathbb{1} \end{pmatrix}.$

The proof of (86.8) shows that $p \vee p^{\gamma_\nu} = A^{\zeta_\nu}$, where ζ_ν is the last factor in the three term decomposition of γ_ν. It follows that $b_\nu a_\nu = A^{\zeta_\nu \sigma_\nu}$, and we have

$$\zeta_\nu \sigma_\nu = \begin{pmatrix} m_\nu & v_\nu + u_\nu R_\nu \\ & R_\nu \end{pmatrix}.$$

If we can show that $u_\nu \to -vR^{-1}$, then $\zeta_\nu \sigma_\nu \to \delta$, and our proof is complete.

7) Multiplying equation (1) by σ_ν from the right and comparing coefficients, we obtain the following equations.

(2) $\quad 1 = (h_\nu^{-2} + z_\nu s_\nu^t) m_\nu$
(3) $\quad 0 = (h_\nu^{-2} + z_\nu s_\nu^t) v_\nu + ((h_\nu^{-2} + z_\nu s_\nu^t) u_\nu + z_\nu) R_\nu$
(4) $\quad w_\nu^t = |h_\nu| C_\nu s_\nu^t m_\nu$
(5) $\quad \mathbb{1} = |h_\nu| C_\nu s_\nu^t v_\nu + (|h_\nu| C_\nu s_\nu^t u_\nu + |h_\nu| C_\nu) R_\nu.$

Substituting (2) in (3), we obtain

(6) $\quad v_\nu = -(u_\nu + m_\nu z_\nu) R_\nu.$

Substituting (6) in (5) and using equation (4), we arrive at

(7) $\quad R_\nu^{-1} = |h_\nu| C_\nu - w_\nu^t z_\nu.$

8) We may assume that C_ν converges to $C \in U_2 \mathbb{F}$. The sequence h_ν might converge to ∞, but this leads to a contradiction. Indeed, we would infer from equation (5) that

$$s_\nu^t v_\nu R_\nu^{-1} + s_\nu^t u_\nu + \mathbb{1} = |h_\nu|^{-1} C_\nu^{-1} R_\nu^{-1} \to 0,$$

and together with equation (6) this would yield that $s_\nu^t m_\nu z_\nu \to \mathbb{1}$. Now a matrix of the form $s^t z$ is always singular, and a sequence of singular matrices cannot converge to $\mathbb{1}$.

There remain the cases where h_ν converges to $h \in H_\alpha \cup \{0\}$. Then equation (7) implies that $w_\nu^t z_\nu \to |h|C - R^{-1}$. As w_ν tends to infinity, it follows that $z_\nu \to 0$ and $m_\nu z_\nu \to 0$, whence $u_\nu \to -vR^{-1}$ by equation (6). This ends the proof. □

86.28 Proposition. *The join operation* $\vee : P^{(2)} \to \mathcal{L}$ *of* $\mathcal{H}_\alpha$ *is continuous except, perhaps, at pairs* $(q, r) \in D^{(2)}$ *of inner points. Similarly, intersection is continuous except at pairs of inner lines.*

Proof. 1) Suppose that a sequence of pairs $(q_\nu, p_\nu) \in \mathcal{P}^{(2)}$ converges to (q, p) in $\mathcal{P}^{(2)}$, where $p \in N$. Because of the group action, it does not matter which point p we consider. We have to show that $q_\nu \vee p_\nu \to q \vee p$ as $\nu \to \infty$. Using a local cross-section (94.3b) for the canonical map $\Gamma \to N$, we may arrange that the sequence p_ν is constant, $p_\nu = p$.

2) If $q \in P \setminus L$, where L is the inner line through p, then the inverse β of the homeomorphism ψ constructed in (86.27) may be applied to all points w in the neighbourhood $P \setminus L$ of q. We have $w^\beta = (x, \vartheta)$, where $x \in A$ and $x^\vartheta = w$. Thus, $wp = A^\vartheta$ depends continuously on w.

3) In view of (86.21), only the case that $q \in D \cap L$ remains to be considered. We use the general fact that a sequence converges to x if each of its subsequences contains another subsequence that converges to x. By (86.25), we may thus assume that the lines $q_\nu p$ converge to some line $B \in \mathcal{L}_p$. We have to show that $B = L$. If this is not the case, then B is an outer line. Then select an outer line $C \in \mathcal{L}_p$ that contains none of the points q_ν. There is a point $p' \in C \setminus \{p\}$ such that qp' is an outer line. The point $r = B \wedge qp'$ is well defined and distinct from q since $q \notin B$. Now step 1) shows that $q_\nu p' \to qp'$. Since $\mathcal{N}$ is open (86.20d) and intersection is continuous on $\mathcal{N}^{(2)}$, see (86.21), we infer that $q_\nu = q_\nu p \wedge q_\nu p' \to B \wedge qp' = r$, a contradiction. □

86.29 Proposition. *The spaces P and $\mathcal{L}$ are compact.*

Proof. Let A and B be two outer lines intersecting in an outer point p. Write $X = A \setminus \{p\}$ and $Y = B \setminus \{p\}$. According to (86.28), the bijective map $\mathcal{L} \setminus \mathcal{L}_p \to X \times Y$ sending G to $(G \wedge A, G \wedge B)$ is continuous. By (86.27), domain and range of this map are manifolds of dimension $4d$. By domain invariance, the inverse map $X \times Y \to \mathcal{L}$ sending (x, y) to $x \vee y$ is continuous. Now the proof of compactness can be given exactly as for (41.7a). □

86.30 Proposition. *The join map is continuous.*

Proof. Consider a sequence of pairs $(a_\nu, b_\nu) \in \mathcal{P}^{(2)}$ converging to $(a, b) \in \mathcal{P}^{(2)}$. We use the same test of convergence as in step 3) of (86.28). By (86.29), we may thus assume that the sequence $G_\nu = a_\nu \vee b_\nu$ converges to some line G, and the problem is to show that $G = a \vee b$. There is an outer line B meeting G in an outer point $r \notin \{a, b\}$. We have $G_\nu \wedge B = r_\nu \to r$ by (86.28), and $G_\nu = a_\nu \vee r_\nu \to a \vee r$ for the same reason. Similarly, $G_\nu \to b \vee r$, and $a \vee b = a \vee r = b \vee r = G$. □

Summing up what we have proved, we formulate the following main result.

86.31 Theorem. *For each $\alpha \in \mathbb{R}$, the Hughes plane $\mathcal{H}_\alpha^{\mathbb{F}}$ over $\mathbb{F} \in \{\mathbb{C}, \mathbb{H}\}$ is a compact, topological projective plane of dimension $4d$, where $d = \dim \mathbb{F}$. It contains the classical projective plane $\mathcal{P}_2\mathbb{F}$ as a closed Baer subplane, and it admits an action of the group $\mathrm{SL}_3\mathbb{F}$ that induces the group $\mathrm{PGL}_3\mathbb{F}$ on this Baer subplane.* □

Further information on these planes, concerning, for example, their automorphisms and isomorphisms, will be given towards the end of this section.

Besides the desarguesian subplane mentioned in (83.31), every 16-dimensional Hughes plane $\mathcal{H}_\alpha^{\mathbb{H}}$ also contains the 8-dimensional Hughes plane $\mathcal{H}_\alpha^{\mathbb{C}}$ as a Baer subplane. Indeed, the inclusion map $\mathrm{SL}_3\mathbb{C} \to \mathrm{SL}_3\mathbb{H}$ is compatible with our construction of the Hughes planes and thus induces an inclusion of the planes. By (55.5), we obtain Baer subplanes in this way.

Characterization of the Hughes planes

The following result has an analogue in the 4-dimensional case, see (71.10); in fact, a stronger assertion was proved in that case.

86.32 Theorem: Actions of unitary groups. *Let $\Delta = \mathrm{SU}_3\mathbb{F}$, where $\mathbb{F} = \mathbb{C}$ or $\mathbb{F} = \mathbb{H}$, act non-trivially on a compact plane $\mathcal{P}$ of dimension 8 or 16, respectively. Then Δ leaves a desarguesian subplane $\mathcal{E} \cong \mathcal{P}_2\mathbb{F}$ of $\mathcal{P}$ invariant and acts on $\mathcal{E}$ as the elliptic motion group. Moreover, the reflections of $\mathcal{E}$ are induced by reflections of $\mathcal{P}$.*

Proof. We will closely follow steps 1) to 4) of the proof of (71.10). It will be sufficient to comment on those arguments that do not carry over directly. First of all, the action of Δ may have a non-trivial kernel; it is contained in the center, which has order 3 or 2 for $\mathbb{F} = \mathbb{C}$ or $\mathbb{F} = \mathbb{H}$, respectively. In step 1), we need a different argument to ensure that each elliptic reflection $\sigma \in \mathcal{I}$ of $\mathcal{P}_2\mathbb{F}$ also acts as a reflection on $\mathcal{P}$. First observe that σ does not act trivially. Now the centralizer $\mathrm{Cs}_\Delta(\sigma)$ is equal to the stabilizer Δ_c of the center c of σ in $\mathcal{P}_2\mathbb{F}$, and we have $\Delta_c \cong \mathrm{SU}_1\mathbb{F} \cdot \mathrm{SU}_2\mathbb{F}$, compare (13.14). If $\mathbb{F} = \mathbb{H}$, then σ is the central involution of $\mathrm{SU}_2\mathbb{H} \cong \mathrm{Spin}_5\mathbb{R}$. If this involution is planar, then an action of $\mathrm{SO}_5\mathbb{R}$ on the plane $\mathcal{F}_\sigma$ of fixed elements of σ results, contrary to (55.40). By (55.29), it follows that σ acts as a reflection. If $\mathbb{F} = \mathbb{C}$, then $\mathrm{Cs}_\Delta(\sigma)$ contains a 2-torus, which in turn contains another involution τ belonging to $\mathcal{J}$. If these involutions are both planar, then (55.32ii) asserts that $\mathcal{F}_\sigma \neq \mathcal{F}_\tau$. From (55.39), we infer that τ induces a Baer involution on $\mathcal{F}_\sigma$. This is impossible, because τ is contained in a connected group acting on $\mathcal{F}_\sigma$, see (55.21c).

The remainder of step 1) works, and we obtain continuous maps α and β by sending the center and axis of σ in $\mathcal{P}_2\mathbb{F}$ respectively to the center and axis of σ in $\mathcal{P}$. As in step 2), these maps are Δ-equivariant. In order to show that a point p of $\mathcal{P}_2\mathbb{F}$ has the same stabilizer as $q = p^\alpha$, we need to know that Δ_p is a maximal subgroup of Δ and that Δ cannot fix q. The first assertion follows from (62.12b), applied to the action of Δ on $\mathcal{P}_2\mathbb{F}$, and the second one follows from (62.8c). No changes are needed in steps 3) and 4). □

It is now easy to deduce the following result, which should be compared with Theorem (72.3) on 4-dimensional planes. See also Theorem (86.34).

86.33 Proposition: The Hughes situation. *Let* $\Gamma = \mathrm{SL}_3\mathbb{F}$, *where* $\mathbb{F} = \mathbb{C}$ *or* $\mathbb{F} = \mathbb{H}$, *act non-trivially on a compact plane* $\mathcal{P}$ *of dimension* 8 *or* 16, *respectively. Then the following assertions hold.*

(a) Γ *leaves a desarguesian subplane* $\mathcal{E} = (D, \mathcal{D}) \cong \mathcal{P}_2\mathbb{F}$ *of* $\mathcal{P}$ *invariant and acts on* $\mathcal{E}$ *in the ordinary way.*
(b) *Each reflection or elation of* $\mathcal{E}$ *is induced by a reflection or elation of* $\mathcal{P}$, *respectively.*

Remark. The following converse statement is true: If $\mathcal{P}$ is a compact plane of dimension 8 or 16 which contains a Baer subplane $\mathcal{E} \cong \mathcal{P}_2\mathbb{F}$ such that each linear collineation of $\mathcal{E}$ extends to a collineation of $\mathcal{P}$, then there is an action of $\mathrm{SL}_3\mathbb{F}$ on $\mathcal{P}$ of the kind described in the proposition. This follows from (94.27), applied to the subgroup of $\mathrm{Aut}\,\mathcal{P}$ consisting of all automorphisms that preserve $\mathcal{E}$. Compare again the 4-dimensional case (72.3).

Proof of (86.33). 1) Applying (86.32) to $\Delta = \mathrm{SU}_3\mathbb{F} \leq \Gamma$, we obtain that the elliptic reflections in Δ are reflections of $\mathcal{P}$, and that their centers and axes form a Δ-invariant, desarguesian Baer subplane $\mathcal{E} = (D, \mathcal{D})$ of $\mathcal{P}$. In order to prove (a), we show that $\mathcal{E}$ is invariant under Γ. We consider the conjugacy class $\mathcal{J}$ of Γ which contains the elliptic reflections as a proper subset. It suffices to prove that, up to duality, the set of centers of this larger class of reflections of $\mathcal{P}$ is again equal to D.

2) Let $c(\sigma) \in \mathrm{P}_2\mathbb{F}$ and $c'(\sigma) \in P$ denote the two centers of $\sigma \in \mathcal{J}$, and similarly for the axes $A(\sigma)$ and $A'(\sigma)$. For each pair $\sigma_1, \sigma_2 \in \mathcal{J}$ satisfying $c(\sigma_1) = c(\sigma_2)$

and $A(\sigma_1) \neq A(\sigma_2)$, there is a third involution $\sigma_3 \in \mathcal{J}$ that commutes with both σ_1 and σ_2. This is obtained by applying (55.35) to the action of Γ on $\mathcal{P}_2\mathbb{F}$; it suffices to choose any reflection $\sigma_3 \in \mathcal{J}$ with center $c = A(\sigma_1) \wedge A(\sigma_2)$ and axis A such that $c \notin A$ and $c(\sigma_1) \in A$.

3) All pairs $\sigma_1 \neq \sigma_2$ with $c(\sigma_1) = c(\sigma_2)$ are conjugate, because Γ is transitive on quadrangles in $\mathcal{P}_2\mathbb{F}$, see (13.5), and because a reflection of a classical plane is determined by its center and axis, see (12.13a). Hence if $c'(\sigma_1) \neq c'(\sigma_2)$ holds for one such pair, then the same happens for all such pairs. We claim that up to dualizing either $\mathcal{P}_2\mathbb{F}$ or $\mathcal{P}$, the implication

$$c(\sigma_1) = c(\sigma_2) \Rightarrow c'(\sigma_1) = c'(\sigma_2)$$

holds. Suppose, on the contrary, that each pair of involutions having precisely one of the two elements c and A in common disagrees with respect to both c' and A'. Given such a pair $\sigma_1, \sigma_2 \in \mathcal{J}$ with $A(\sigma_1) \neq A(\sigma_2)$, we choose σ_3 as in step 2), and we obtain that $A'(\sigma_3) = c'(\sigma_1) \vee c'(\sigma_2)$ and $c'(\sigma_3) = A'(\sigma_1) \wedge A'(\sigma_2)$ are both uniquely determined. However, there are several choices of σ_3, and all these have $c(\sigma_3) = A(\sigma_1) \wedge A(\sigma_2)$. This contradicts our assumption, and the proof of (a) is complete.

4) The proof so far has also shown that all reflections of $\mathcal{E}$ are induced by reflections of $\mathcal{P}$. Now each elation of $\mathcal{E}$ can be written as a product of two reflections with the same axis; indeed, a reflection σ with axis L acts on $\tau \in \Gamma_{[L,L]} \cong \mathbb{R}^d$ by inversion, whence $\tau^2 = \sigma\tau^{-1}\sigma\tau$. Since reflections of $\mathcal{E}$ extend to reflections of $\mathcal{P}$, it follows from (23.20) that the same is also true for elations. □

We have seen in (86.31) that, in contrast to the 4-dimensional case (72.3), there are 8- and 16-dimensional examples of the Hughes situation. We shall now prove that the planes obtained there are the only examples. The idea of the proof is the same as in (72.3). However, we cannot show directly that Γ is transitive on the set of pairs of outer lines with outer intersection, as in step 4) of that proof. This accounts for some extra difficulties. Nevertheless, this transitivity statement is true; it follows from (86.15) together with the next result.

86.34 Theorem: Characterization of the Hughes planes.
Let $\Gamma = \mathrm{SL}_3\mathbb{F}$, where $\mathbb{F} = \mathbb{C}$ or $\mathbb{F} = \mathbb{H}$, act non-trivially on a compact plane $\mathcal{P}$ of dimension 8 or 16, respectively. Then the plane $\mathcal{P}$ is isomorphic to one of the Hughes planes $\mathcal{H}_\alpha^\mathbb{F}$ constructed in the preceding subsections, and the action of Γ is the one described there. In particular, Γ acts effectively on $\mathcal{P}$.

For other characterizations of the Hughes planes, see (87.2 and 3).

Proof. 1) We use the invariant 'inner' Baer subplane $\mathcal{E} = (D, \mathcal{D})$ obtained in (86.33). Let $N = P \setminus D$ and $\mathcal{N} = \mathcal{L} \setminus \mathcal{D}$ respectively denote the set of outer points and outer lines. Choose an outer flag (p, A) and let $\Pi = \Gamma_p$ and $\Lambda = \Gamma_A$ be the corresponding stabilizers. We shall show that the pair (Π, Λ) is conjugate to a pair $(\Pi_\alpha, \Lambda_\alpha)$ as defined in (86.2). As a consequence, we shall obtain that the outer geometry $(N, \mathcal{N})$ is isomorphic to the outer geometry of some Hughes plane $\mathcal{H}_\alpha$, see (23.24). By density of $\mathcal{N} \subseteq \mathcal{L}$, the isomorphism also respects inner lines, and by (44.9) it extends to an isomorphism $\mathcal{P} \cong \mathcal{H}_\alpha$.

2) We have $\Pi \leq \Gamma_L$, where L is the unique inner line containing p, and dually $\Lambda \leq \Gamma_o$, where o is the inner point on A. The intersection $\Delta = \Pi \cap \Lambda = \Pi_o = \Lambda_L$ is the stabilizer of the outer flag (p, A). Moreover, the elation groups $\Theta = \Gamma_{[L,L]}$ and $\mathsf{P} = \Gamma_{[o,o]}$ of $\mathcal{E}$ satisfy $\Theta \leq \Pi$ and $\mathsf{P} \leq \Lambda$ by (86.33b). Now Θ is transitive on $D \setminus L$ and P is transitive on $\mathcal{D} \setminus \mathcal{L}_o$, so that we have semi-direct decompositions $\Pi = \Delta\Theta$ and $\Lambda = \Delta\mathsf{P}$ as in (86.3c). Thus, it suffices to determine Δ up to conjugation as a subgroup of the stabilizer $\Omega = \Gamma_{L,o}$ of an antiflag in $\mathcal{P}_2\mathbb{F}$.

3) The transitivity of Θ on $D \setminus L$ moreover implies that Π is transitive on the set $\mathcal{N}_p$ of outer lines containing the (arbitrary) outer point p. From this it is easily deduced that Γ is transitive on $\mathcal{N}$, and by duality it follows that Γ is even transitive on the space of outer flags. This space has dimension $6d$ if $\dim \mathbb{F} = d$, hence we obtain that $\dim \Delta = \dim \Gamma - 6d$ equals $16 - 12 = 4$ or $35 - 24 = 11$ for $\mathbb{F} = \mathbb{C}$ or $\mathbb{F} = \mathbb{H}$, respectively.

4) By (14.9) and (41.11b), the lines of $\mathcal{E}$ and $\mathcal{P}$ are spheres of dimension d and $2d$, respectively. By step 3), the flag (p, A) can be mapped to every flag (q, qo) with $q \in L \setminus D$. It follows that

$$\Omega = \Gamma_{L,o}$$

is transitive on the complement $L \setminus D \approx \mathbb{S}_{2d} \setminus \mathbb{S}_d$. This space can also be described as $\mathbb{R}^{2d} \setminus \mathbb{R}^d = \mathbb{R}^d \times (\mathbb{R}^d \setminus \{0\})$, hence it is homotopy equivalent to $\mathbb{S}_{d-1}$. The stabilizer of that action is $\Pi_o = \Delta$, and we may use the homotopy sequence (96.12) in order to obtain results on the homotopy type of Δ.

5) First we need information on the homotopy type of Ω. If $\mathbb{F} = \mathbb{C}$, then Ω consists of block matrices $\mathrm{diag}(z, R)$, where $R \in \mathrm{GL}_2\mathbb{C}$ and $z \det R = 1$. By (93.10), the group $\Omega \cong \mathrm{GL}_2\mathbb{C}$ is homotopy equivalent to its maximal compact subgroup $\Xi = \mathrm{U}_2\mathbb{C}$, which has $\mathrm{SU}_2\mathbb{C} \times \mathrm{SO}_2\mathbb{R} \approx \mathbb{S}_3 \times \mathbb{S}_1$ as a double cover. Thus $\pi_3\Omega = \mathbb{Z}$, $\pi_2\Omega = 0$ and $\pi_1\Omega = \mathbb{Z}$, compare (94.36). If $\mathbb{F} = \mathbb{H}$, then Ω consists of block matrices $\mathrm{diag}(z, |z|^{-1/2}R)$, where $R \in \mathrm{SL}_2\mathbb{H}$ and $z \in \mathbb{H}^\times$. This shows that Ω is a direct product $\Omega = \mathrm{SL}_2\mathbb{H} \times \mathbb{H}^\times$, which is homotopy equivalent to its maximal compact subgroup $\Xi = \mathrm{SU}_2\mathbb{H} \times \mathrm{Spin}_3\mathbb{R}$. This group is simply connected, and (94.36) shows that $\pi_3\Omega = \mathbb{Z} \times \mathbb{Z}$ and $\pi_2\Omega = 0$.

6) We obtain the following homotopy sequences, where (1) represents the general form of the sequence and (2) and (3) show the concrete shape that this sequence assumes for $\mathbb{F} = \mathbb{C}$ or $\mathbb{F} = \mathbb{H}$, respectively. We use that $\pi_4 \mathbb{S}_3 = \mathbb{Z}_2$, see Spanier [66] 8.5.12.

(1) $\quad \ldots \pi_4 \mathbb{S}_{d-1} \to \pi_3 \Delta \to \pi_3 \Omega \to \pi_3 \mathbb{S}_{d-1} \to \pi_2 \Delta \ldots$
$\quad \ldots \pi_2 \mathbb{S}_{d-1} \to \pi_1 \Delta \to \pi_1 \Omega \to \pi_1 \mathbb{S}_{d-1} \to \pi_0 \Delta \to \pi_0 \Omega$

(2) $\quad \ldots 0 \to \pi_3 \Delta \to \mathbb{Z} \to 0 \ldots 0 \to \pi_1 \Delta \to \mathbb{Z} \to \mathbb{Z} \to \pi_0 \Delta \to 0$

(3) $\quad \ldots \mathbb{Z}_2 \to \pi_3 \Delta \to \mathbb{Z} \times \mathbb{Z} \to \mathbb{Z} \to 0 \ldots 0 \to \pi_1 \Delta \to 0 \to 0 \to \pi_0 \Delta \to 0$

Evaluation of the sequences shows first that $\pi_3 \Delta$ has rank 1, whence a maximal compact subgroup $\Phi \leq \Delta$ has precisely one almost simple factor Ψ, see (94.36). For $\mathbb{F} = \mathbb{H}$, we also see that Δ is connected and simply connected. In fact, connectedness of Δ can be shown for $\mathbb{F} = \mathbb{C}$ and $\mathbb{F} = \mathbb{H}$ by a uniform argument, see below. Then the sequence (2) shows that the group $\pi_1 \Delta$ is trivial for $\mathbb{F} = \mathbb{C}$, as well. It follows that Φ is connected (93.10a) and has no torus factors, whence $\Phi = \Psi$ is almost simple, compare (94.31c).

We show that Δ is always connected. We may obtain Δ from Γ by taking first the stabilizer $\Pi = \Gamma_p$ of the transitive action on N, and then the stabilizer Π_o of the transitive action of Π on $D \setminus L$. By (96.9c), it suffices to show that both of these homogeneous spaces are simply connected. This is clear for $D \setminus L \approx \mathbb{F}^2$, and for N we shall prove it using the fact that P and D are simply connected (52.14c). Being homogeneous spaces, both P and D are manifolds, and it follows from (41.11c) that the embedding $D \subseteq P$ is locally flat; that is, the pair (P, D) is locally homeomorphic to $(\mathbb{R}^{2d}, \mathbb{R}^d)$. Therefore, (P, D) satisfies the conditions of Eilenberg–Wilder [42] Theorem 9, which shows that N is simply connected.

7) We have thus obtained that the maximal compact subgroup $\Phi \leq \Delta$ is almost simple and connected. We claim that Φ is isomorphic to $\mathrm{SU}_2 \mathbb{F}$. This is clear for $\mathbb{F} = \mathbb{C}$, where $\Xi = \mathrm{U}_2 \mathbb{C}$ is a maximal compact subgroup of Ω. For $\mathbb{F} = \mathbb{H}$, we have $\Xi = \mathrm{SU}_2 \mathbb{H} \times \mathrm{Spin}_3 \mathbb{R}$. We may assume that $\Phi = \Xi \cap \Delta$, which shows that $\dim \Delta - \dim \Phi \leq \dim \Omega - \dim \Xi = 6$ and $\dim \Phi \geq 5$ by step 3). Being almost simple, Φ projects trivially into the small factor of Ξ and, hence, is contained in $\mathrm{SU}_2 \mathbb{H}$. Moreover, $\dim \Phi \geq 8$ by (94.33), but $\dim \Phi < 10$ is impossible by (96.13a), applied to the homogeneous space $\mathrm{SU}_2 \mathbb{H}/\Phi$, which has dimension $10 - \dim \Phi$ by (94.3a). This leaves only the possibility that $\Phi = \mathrm{SU}_2 \mathbb{H}$, see (93.12) or (94.12).

8) Since $\dim \Delta - \dim \Phi = 1$ by steps 3) and 7), there must be a one-parameter subgroup A of Δ which is centralized by Φ. Such a one-parameter group is obtained by decomposing the Lie algebra of Δ into Φ-irreducible subspaces; details are given in the proof of (81.8), step 7). As $\Delta \leq \Omega$, it follows that A is contained in

the centralizer $C_{s_\Omega}(\Phi)$. By Schur's lemma, this centralizer is equal to the factor $\mathbb{F}^\times$ of Ω, compare (95.4). Moreover, Λ is not compact, hence Λ is conjugate to some spiral subgroup H_α as introduced in (86.2). It follows that $\Delta = \Phi H_\alpha = \Pi_\alpha \cap \Lambda_\alpha$ as desired.

9) We see finally that Γ acts effectively on $\mathcal{P}$, because Π_α does not contain the center of Γ. □

86.35 Proposition: Automorphisms of Hughes planes. *Let B_α denote the group of all automorphisms of the Hughes plane $\mathcal{H}_\alpha^\mathbb{F}$ that preserve the desarguesian subplane $\mathcal{P}_2\mathbb{F}$. Write $\Phi_0 = \mathrm{Spin}_3\mathbb{R}$ for $\mathbb{F} = \mathbb{H}$, and $\Phi_\alpha = \mathrm{SO}_2\mathbb{R}$ in all other cases.*

Then the connected component B_α^1 is obtained from the direct product $\Gamma \times \Phi_\alpha$ with $\Gamma = \mathrm{SL}_3\mathbb{F}$ by amalgamating the center $Z \in \{\mathbb{Z}_2, \mathbb{Z}_3\}$ of Γ with the unique subgroup of Φ_α that is isomorphic to Z. In particular, for $\mathbb{F} = \mathbb{H}$ we have $\dim B_\alpha = 36$ if $\alpha \neq 0$ and $\dim B_0 = 38$, and we obtain $\dim B_\alpha = 17$ if $\mathbb{F} = \mathbb{C}$.

The group B_α is disconnected only if $\mathbb{F} = \mathbb{C}$ and $\alpha = 0$. In this case, we have $B_\alpha = B_\alpha^1 \rtimes \langle \tau_0 \rangle$, where τ_0 is the collineation of $\mathcal{H}_0^\mathbb{C} = \mathcal{P}_2\mathbb{H}$ defined by the automorphism $x \mapsto jxj^{-1}$ of $\mathbb{H}$.

Remarks. (a) In the case $\mathbb{F} = \mathbb{C}$, the group B_α^1 and its action may be described more directly. We have

$$B_\alpha^1 = \{ A \in \mathrm{GL}_3\mathbb{C} \mid |\det A| = 1 \},$$

and the action is obtained if one simply replaces Γ with B_α^1 in the construction of $\mathcal{H}_\alpha^\mathbb{C}$, see (86.4).

(b) Our description of Φ_α given below differs from the one used by Hähl [86b]. He writes $\gamma_c = c\mathbb{1}$, which has the same effect as our γ_c for $\mathbb{F} = \mathbb{H}$, but does not carry over to $\mathbb{F} = \mathbb{C}$.

Proof of (86.35). 1) We know that Γ induces all linear collineations of $\mathcal{P}_2\mathbb{F}$, see (13.4 and 6). According to (13.6 and 8), further automorphisms of $\mathcal{P}_2\mathbb{F}$ exist for $\mathbb{F} = \mathbb{C}$ only, and $\mathrm{Aut}\,\mathcal{P}_2\mathbb{C}$ is the extension of its connected component (which is induced by Γ) by the Baer involution τ defined by complex conjugation. If $\alpha = 0$, then τ is induced by τ_0. In all other cases, we shall show in step 5) that τ does not extend to an automorphism of $\mathcal{H}_\alpha^\mathbb{C}$.

Thus, it only remains to determine all Baer collineations of $\mathcal{H}_\alpha^\mathbb{F}$ that fix $\mathcal{P}_2\mathbb{F}$ elementwise. We shall exhibit a group $\Phi \cong \Phi_\alpha$ of automorphisms of $\mathcal{H}_\alpha^\mathbb{F}$ which contains all those Baer collineations. This completes the proof of the proposition.

2) The group Φ may be described as follows. Let $c \in \mathbb{H}$ satisfy $|c| = 1$. We also require that c belongs to the subfield $\mathbb{C} \leq \mathbb{H}$ unless $\alpha = 0$ and $\mathbb{F} = \mathbb{H}$. Define a

diagonal matrix γ_c by
$$\gamma_c = \mathrm{diag}(c^{-2}, c, c).$$
We claim that there is an automorphism $\varphi_c \in \Phi$ which acts on the sets Γ/Π and Γ/Λ of outer elements by
$$\varphi_c(\Pi\sigma) = \Pi\gamma_c\sigma, \quad \varphi_c(\Lambda\sigma) = \Lambda\gamma_c\sigma,$$
and that these automorphisms form a group Φ having the properties mentioned above. Amalgamation takes place as stated, because φ_c belongs to Γ if, and only if, c is a second or third root of unity for $\mathbb{F} = \mathbb{H}$ or $\mathbb{F} = \mathbb{C}$, respectively.

3) The matrix γ_c normalizes both Π and Λ, hence φ_c is well defined and preserves incidence of outer elements, as defined in (86.4). By a density argument, compare (44.9), one obtains that φ_c extends to an automorphism of the plane $\mathcal{H}_\alpha^{\mathbb{F}}$.

4) Let φ be an arbitrary automorphism of $\mathcal{H}_\alpha$ that fixes every element of $\mathcal{P}_2\mathbb{F}$. This property of φ implies that it commutes with each elation of $\mathcal{P}_2\mathbb{F}$. Since the elations generate Γ, we infer that φ centralizes Γ. In particular, φ maps the point $p = \Pi$ to another point $q = p^\varphi$ lying on the same inner line L and having the same stabilizer $\Gamma_q = \Pi$. It follows that $q = p^\beta$ for some $\beta \in \Gamma_L$ normalizing Π. We may multiply β by an arbitrary elation with axis L, hence we may assume that β is a block matrix $\mathrm{diag}(a, Q)$. Now Q has to normalize $\mathrm{SU}_2\mathbb{F}$, hence we may write $Q = rU$, where $U \in \mathrm{U}_2\mathbb{F}$ and $r > 0$, compare the remark following this proof. Since $\det \beta = 1$, we obtain that $|a| = r^{-2}$. Moreover, a has to normalize the spiral group H_α, hence a is a complex number if $\alpha \neq 0$.

There is a quaternion c of norm 1 such that $ac^2 \in H_\alpha$; if $\alpha \neq 0$, then even $c \in \mathbb{C}$. The element $h \in H_\alpha$ satisfying $h^{-2} = ac^2$ has norm $|h| = r$. As $\det \gamma_c = 1$, it follows that the matrix π defined by $\beta = \pi\gamma_c$ belongs to Π. Since φ centralizes Γ, we obtain that $\varphi(\Pi\sigma) = p^{\sigma\varphi} = q^\sigma = p^{\pi\gamma_c\sigma} = \Pi\gamma_c\sigma$, whence $\varphi \in \Phi$ as we have claimed.

5) Now let $\mathbb{F} = \mathbb{C}$ and $\alpha \neq 0$, and suppose that the Baer involution $\tau \in \mathrm{Aut}\,\mathcal{P}_2\mathbb{C}$ extends to an automorphism (again denoted τ) of $\mathcal{H}_\alpha^{\mathbb{C}}$. We shall show that this is impossible. Our results so far suffice to prove the assertion about B_α^1; in particular, we know that Γ is the commutator group of B_α^1, whence Γ is normalized by τ. Since Γ acts almost effectively on $\mathcal{P}_2\mathbb{C}$ and since the automorphism σ of Γ induced by τ is determined by its derivative $\mathfrak{l}\sigma$, see (94.9), we infer that τ acts on Γ by complex conjugation of matrix entries.

The image Π^τ is the stabilizer of p^τ in Γ. We know the action of τ on $\mathcal{P}_2\mathbb{C}$; in particular, we know that τ fixes L, whence $p^\tau \in L$; compare (86.2). Recall that L is the unique inner line containing p. Together with transitivity of Γ on the set N of outer points, this implies that $p^\tau = p^\gamma$ and $\Pi^\tau = \Pi^\gamma$ for some $\gamma \in \Gamma_L$. Now the first column of γ contains only one non-zero entry (in position $(1, 1)$), hence the $(1, 1)$-entries of the elements of Π are not affected by the action of γ. These entries belong to the spiral group H_α, see (86.2), and we infer that H_α is invariant under complex conjugation. This is true only in the case where $\alpha = 0$, a contradiction. □

In the last proof, we needed to know the normalizer of $\mathrm{SU}_2 \mathbb{F}$ in $\mathrm{GL}_2 \mathbb{F}$. Any element α of this normalizer maps the standard scalar product to another one which is invariant under the same unitary group and, hence, defines the same set of spheres around 0. It follows that the two scalar products differ by a factor $r^2 > 0$, whence $r^{-1}\alpha \in \mathrm{U}_2 \mathbb{F}$.

86.36 Corollary. *A compact Hughes plane $\mathcal{H}_\alpha^{\mathbb{F}}$ is a classical (Moufang) plane if, and only if, $\alpha = 0$. In all other cases, $\operatorname{Aut} \mathcal{H}_\alpha^{\mathbb{F}}$ preserves the subplane $\mathcal{P}_2 \mathbb{F}$, and hence coincides with the group B_α considered in (86.35).*

Proof. 1) We know from (86.4) that $\mathcal{H}_0^{\mathbb{C}}$ is isomorphic to the quaternion plane as an abstract plane. The isomorphism was obtained from the equivalence of outer actions and, hence, is continuous on the outer elements. By a density argument, we have an isomorphism of topological planes, compare (44.9). Conversely, all embeddings of $\mathcal{P}_2 \mathbb{C}$ in $\mathcal{P}_2 \mathbb{H}$ are equivalent, because the automorphism group of the latter plane is transitive on quadrangles (13.5) and because the stabilizer $\operatorname{Aut} \mathbb{H}$ of a quadrangle is transitive on the set of all subfields of $\mathbb{H}$ that are isomorphic to $\mathbb{C}$, see (11.29 and 10). The action of Γ on $\mathcal{P}_2 \mathbb{H}$ preserving such a subplane is uniquely determined (86.35), hence these actions are mutually equivalent. Now the embeddings of Π_α in Γ_L, see (86.2), are inequivalent for $\alpha = 0$ and $\alpha \neq 0$. This implies that only $\mathcal{H}_0^{\mathbb{C}}$ is isomorphic to the quaternion plane.

2) The case $\mathbb{F} = \mathbb{H}$ is made easier by our previous results. We have observed in (86.1) that $\mathcal{P}_2 \mathbb{O}$ admits an action of $\Gamma = \mathrm{SL}_3 \mathbb{H}$. Hence (86.34) implies that $\mathcal{P}_2 \mathbb{O}$ is isomorphic to some plane $\mathcal{H}_\alpha$. Now $\mathbb{O}$ admits a group $\Psi \cong \mathrm{Spin}_3 \mathbb{R}$ of automorphisms fixing the subfield $\mathbb{H} \leq \mathbb{O}$ elementwise, see (11.31c). This gives rise to a 3-dimensional group Φ as considered in (86.35). By that result, only $\mathcal{H}_0$ admits such a group, hence only $\mathcal{H}_0$ can be isomorphic to $\mathcal{P}_2 \mathbb{O}$.

3) The invariance of $\mathcal{P}_2 \mathbb{F}$ follows from the remark that Γ is always transitive on the outer points; if the inner point orbit were not invariant, then $\operatorname{Aut} \mathcal{H}_\alpha^{\mathbb{F}}$ would be transitive on the whole point set, contrary to Theorem (63.8). □

86.37 Proposition: Isomorphism. *For different values $\alpha, \beta \geq 0$, the Hughes planes $\mathcal{H}_\alpha^{\mathbb{F}}$ and $\mathcal{H}_\beta^{\mathbb{F}}$ over $\mathbb{F} \in \{\mathbb{C}, \mathbb{H}\}$ are not isomorphic.*

For a proof using ideas similar to (86.36), we refer to Hähl [86b] 4.6. It is formulated for $\mathbb{F} = \mathbb{H}$, but carries over to $\mathbb{F} = \mathbb{C}$.

86.38 Notes on Section 86. For references and comments concerning the general theory of Hughes planes and its relationship to our subject, see the note following (86.4). Salzmann [81a] characterized the 8-dimensional Hughes planes and proved their existence. Stroppel [94b] extended the characterization to stable planes. The proof of the existence of the 16-dimensional Hughes planes was sketched by Salzmann [82a]. The exposition presented here is a condensed and supplemented version of the work of Niemann [90]. The characterization of the 16-dimensional Hughes planes and the results on their automorphisms and isomorphisms are due to Hähl [86b]. Salzmann [82a] had already shown that only the octonion plane admits a 38-dimensional group containing $SL_3\mathbb{H}$.

87 Principles of classification

So far, we have discussed several classes of $2l$-dimensional compact planes with $l \in \{4, 8\}$ such that each plane has an automorphism group Σ of dimension at least $5l - 4$, and we have obtained some characterizations of the classical planes. In fact, there are non-classical planes with a group of maximal ('critical') dimension c_l, and $\dim \Sigma > c_l$ implies that the plane is classical, and then $\dim \Sigma = C_l$, compare the beginning of Section 65. As stated without proof in (84.28), any 8-dimensional plane with $\dim \Sigma \geq 17$ is a (dual) translation plane or a Hughes plane, and $c_4 = 18$. Similarly, $c_8 = 40$ according to (85.16). Note that $5l - 3 < \frac{1}{2}C_l < c_l$. Translation planes are also prominent among the 16-dimensional planes with a group of large dimension. It is not known, however, if all planes with a group of dimension 39 are translation planes. All translation planes with $\dim \Sigma \geq \frac{1}{2}C_l - 1$ are known explicitly; their classification is outlined in Section 82.

A natural question arises: how can an analogous classification be achieved without assuming that Σ contains a transitive translation group? Translation planes have the following nice properties (among others): (1) lines are manifolds, (2) the automorphism group Σ is a linear Lie group, and (3) the stabilizer ∇ of a triangle has a compact subgroup Φ of codimension $\dim \nabla/\Phi$ at most 2. At present, there seems to be little hope that similar properties can be obtained for all compact, connected planes. This accounts for some of the difficulties in the general classification program. Instead of (1), the results of Section 52 are available by Corollary (54.11). The stiffness results of Section 83 serve as a substitute for (3). From (83.2) and (93.8c) it follows that some open subgroup of Σ can be approximated by Lie groups of the same dimension. In all known examples, Σ itself is a Lie group. This is true in general if Σ has an open point orbit (53.2) or if the stabilizer of some line is transitive on that line (62.11). Moreover, the connected component Δ of Σ is a Lie group if $\dim \Sigma \geq 4(l-1)$ (Priwitzer [94a,b]). For the classification of planes with $\dim \Sigma$ close to c_l, this theorem is quite convenient but not actually necessary; it becomes indispensable, however, if one wants to study groups of a smaller dimension. We shall prove a weaker result in (87.1) below.

Since the connected component of a locally compact group has the same dimension as the group itself (93.6), it suffices to consider the connected component of Σ or a suitable connected, closed subgroup Δ of Σ of dimension $\dim \Delta = h \geq 5l-3$. For $l = 4$ and $h \geq 17$, the classification has been completed, see (82.25) together with statement (84.28). For $l = 8$, results are still less complete, and we shall concentrate on this case. The methods are similar in both cases, and (84.28) can be proved in much the same way as (87.3, 4 and 5). The second half of this section (after the proof of (87.1)) may be regarded as a program for further investigation.

87.1 Proposition. *If Δ is a connected, closed subgroup of Σ and if $\dim \Delta \geq 5(l-1)$, then Δ is a Lie group.*

We remark without proof that under the same assumptions Σ itself is a Lie group.

Proof of (87.1). For $l < 4$, the group Σ is always a Lie group, see (32.21) and (71.2). For $l = 8$ the proposition improves a weaker result of Löwen–Salzmann [82]. Suppose that Δ is not a Lie group. Then $l \in \{4, 8\}$ and there is a compact central subgroup $\Theta \neq \mathbb{1}$ such that Δ/Θ is a Lie group, but Θ is not (93.8, 9 and 18). First we show:

1) *There is no Δ-invariant closed proper subplane.*

Assume that $\mathcal{2}$ is such a subplane. Let Q be its point set and put $\Phi = \Delta_{[Q]}$. From $\dim \Delta \geq 5(l-1)$ it follows that $\mathcal{2}$ is a Baer subplane: for $l = 4$, this has been shown in (83.19); if $l = 8$ and $\dim Q \leq 4$, then $\dim \Delta|_Q \leq 16$ by (71.2), and, as $\dim \Phi \leq 14$ by (83.23), we would obtain $\dim \Delta \leq 30$, see (83.3).

For $l = 4$, we know from (83.18) that $\dim \Phi \leq 1$; for $l = 8$, Theorem (83.6) gives $\dim \Phi \leq 7$. Applying (83.3) to $\Delta|_Q \leq \text{Aut}\,\mathcal{2}$, we infer that $\dim \text{Aut}\,\mathcal{2} \geq 14$ for $l = 4$ and $\dim \text{Aut}\,\mathcal{2} \geq 28$ for $l = 8$. By (72.8) and (84.27), it follows that $\mathcal{2} \cong \mathcal{P}_2\mathbb{F}$ with $\mathbb{F} \in \{\mathbb{C}, \mathbb{H}\}$ is classical.

Now Δ/Φ injects into the Lie group $\text{Aut}\,\mathcal{2}$ and therefore is a Lie group itself (94.3e), so we may assume that $\Theta \leq \Phi$. The action of Φ on $P \setminus Q$ is free (23.14), and since Θ is not a Lie group, the orbit x^Θ of any point $x \notin Q$ is infinite. Θ fixes the unique line L_x of $\mathcal{2}$ through x; consequently, $x^\Theta \subseteq L_x$. For a second point $y \notin Q \cup L_x$, the set $x^\Theta \cup y^\Theta$ contains a non-degenerate quadrangle and consists of fixed points of $\Lambda = \Delta_{x,y}$; hence, the fixed elements of Λ constitute a closed subplane $\mathcal{F} \neq \mathcal{2}$.

Next, we show that $\mathcal{F}$ is connected. The lines $x^\vartheta \vee y$ for $\vartheta \in \Theta$ are all different and they do not belong to $\mathcal{2}$. Their unique points in $\mathcal{2}$ are fixed under Λ, i.e., they belong to $\mathcal{2} \cap \mathcal{F}$. Hence $\mathcal{2} \cap \mathcal{F}$ is infinite and contains at least 3 collinear points, say (0), (1) and (∞). By the Fundamental Theorem (12.12) Addendum (a), the plane $\mathcal{F}$ contains the connected set of all points (s) with $s \in \mathbb{R}$.

Now $\dim x^\Delta < 2l$, $\dim y^\Delta < 2l$, or else Δ would be a Lie group (53.2). By the dimension formula (96.10), we obtain $\dim \Lambda = \dim \Delta - \dim x^\Delta - \dim y^{\Delta_x} \geq 5(l-1) - 2(2l-1) = l - 3$. In particular, $\Lambda \neq 1$, so that $\mathcal{F}$ is a proper subplane.

First assume that $\dim \mathcal{F} = 8$. In this case, $l = 8$ and $\dim \Lambda \geq 5$. For a point p not belonging to $\mathcal{F}$, the stabilizer $\Lambda_p = \Delta_{x,y,p}$ is trivial as $\mathcal{F}$ is a Baer subplane (55.5), so that $\dim \Lambda = \dim p^\Lambda$. Choose $p \in Q$, consider the unique line L through p belonging to $\mathcal{F}$, and note that $p^\Lambda \subseteq L$. If L is not a line of $\mathcal{Q}$, then $p = L \cap Q$ is fixed by Λ. If L belongs to $\mathcal{Q}$, then p^Λ is contained in the 4-dimensional line $L \cap Q$ of $\mathcal{Q}$. Thus, in all cases, $\dim \Lambda = \dim p^\Lambda \leq 4$, contradicting $\dim \Lambda \geq l - 3 = 5$.

There remain the cases that $\dim \mathcal{F} \in \{2, 4\}$. As $\Theta \leq \Phi$ acts freely outside $\mathcal{Q}$, its action on $\mathcal{F}$ is effective. In other words, Θ injects into the Lie group $\mathrm{Aut}\,\mathcal{F}$, see (32.21) or (96.31), and (71.2), and so is a Lie group itself (94.3e), contradicting our assumption.

Up to duality, we now may assume the following:

2) Δ *fixes a line*.

To see this, let a be a fixed point (55.19) of some element $\vartheta \in \Theta \setminus 1$. The orbit a^Δ is pointwise fixed under ϑ. Hence, it is contained in some line L, or else it would contain a non-degenerate quadrangle (42.3), so that the fixed elements of ϑ would constitute a closed subplane invariant under Δ, contradicting 1). If a is not a fixed point of Δ, then L is a fixed line.

3) *The group Θ is straight on $\mathcal{P}$, see* (23.15).

Assume, to the contrary, that the orbit c^Θ of a point c outside the fixed line L of Δ contains three non-collinear points. These are fixed points of Δ_c, hence, with the fixed line L taken into consideration, Δ_c fixes a non-degenerate quadrangle. By (83.15 and 23), $\dim \Delta_c \leq 2l - 2$, and $\dim \Delta \leq 2l + 2l - 2 < 5(l-1)$, in contradiction to our assumption.

4) Θ *is contained in a group $\Delta_{[a,W]}$ of homologies. In particular, Δ fixes the elements a and W.*

In fact, no element in Θ can be planar by 1). Because of Lemma (23.16) and the remark preceding it, Θ consists of axial collineations. The center of any $\vartheta \in \Theta \setminus 1$ is a fixed point of Δ. Hence the assertions dual to 2) and 3) are also true, and (23.16) shows that $\Theta = \Theta_{[p,W]}$ for some p and W. If $p \in W$, then any stabilizer Δ_x with $x \notin W$ injects into the Lie group Δ/Θ and therefore is a Lie group itself. As in the proof of (61.5) it would follow that the connected group Δ is a Lie group.

5) *A closed, connected, Θ-invariant proper subplane $\mathcal{E}$ has dimension 8.*

Indeed, Θ acts effectively on $\mathcal{E}$ because of 4). If $\dim \mathcal{E} < 8$, then Θ would be a Lie group as at the end of the proof of 1).

6) *There is a point $c \notin W$, $c \neq a$ such that the connected component Γ of Δ_c satisfies $3(l-1) \leq \dim \Gamma \leq 4(l-1)$. Moreover, Γ is a Lie group and Γ acts effectively on W.*

The group Δ cannot be transitive on W by (62.11), and (53.1d) implies that $\dim v^\Delta < l$ for some point $v \in W$. Let $c \in Y = av \setminus \{a, v\}$. If $\dim c^{\Delta_v} = l$, then $\Theta \cong \Theta|_Y \leq \Delta_v|_Y$ would be a Lie group (53.2). Hence $\dim \Delta/\Gamma \leq 2(l-1)$. This gives the lower bound for $\dim \Gamma$. Again by (53.2), there is some point $x \notin av \cup W$ such that $\dim x^\Gamma < 2l$. The stiffness results of Section 83 may now be applied to the connected component K of Γ_x. If $\dim \mathsf{K} \geq 14$, then $\mathsf{K} \cong \mathsf{G}_2$ fixes a 2-dimensional subplane by (83.24). This contradicts step 5). With (83.18 and 23), we obtain $\dim \mathsf{K} \leq 2l-3$, hence the upper bound for $\dim \Gamma$. Finally, Γ is a Lie group since $\Gamma \cap \Theta = 1\!\!1$ and Γ injects into Δ/Θ, and $\Gamma_{[W]} = 1\!\!1$ because $\Gamma \leq \Delta_{a,c}$.

7) *If $u \in W \setminus \{v\}$, then $u^\Gamma \neq u$, and no proper closed subplane contains a, c and the orbit u^Γ.*

Assume $u^\Gamma = u$, and let $x \in au \setminus \{a, u\}$. Then $\dim x^\Gamma < l$ as in the proof of 6). Again, $\dim \mathsf{K} \leq 2l-3$. This gives $\dim \Gamma \leq 3l-4$ contrary to step 6). If there were a proper closed subplane $\mathcal{F} = \langle u^\Gamma, a, c \rangle$, then the stiffness results (83.18) and (83.22 and 23) would imply $\dim \Gamma \leq \dim \Gamma|_\mathcal{F} + \dim \Gamma_{[\mathcal{F}]} < 3l-6$.

8) *Consider connected, closed subgroups A and B of Γ which centralize each other. If $u \neq u^\mathsf{A} \subseteq W$, then B_u is compact. Moreover, $\mathsf{B}_u = 1\!\!1$, or $l = 8$ and $\dim \mathsf{B}_u \leq 3$.*

Indeed, B_u fixes each element of the connected subplane $\mathcal{B} = \langle u^\mathsf{A}, a, c \rangle$. If $\mathsf{B}_u \neq 1\!\!1$, then $l = 8$ and $\mathcal{B}$ is a Baer subplane by step 5). Because Γ and, hence, B_u are Lie groups, assertion 8) follows from (83.22).

As immediate consequence of 6) and 8) we get

9) *The connected component A of the center of Γ is trivial.*

We can now distinguish three cases according to the structure of Γ, see (94.26 and 25). Each of the cases 10), 11) and 12) will lead to a contradiction.

10) If Γ is not semi-simple, then Γ has a minimal closed normal vector subgroup $\Xi \cong \mathbb{R}^t$, $t > 0$, by (94.26) and step 9). From $u^\Xi = u \in W \setminus \{v\}$ it follows that Ξ fixes u^Γ pointwise. By step 7), this would imply $\Xi = 1\!\!1$. Hence $u^\Xi \neq u$, and 8) with $\mathsf{A} = \mathsf{B} = \Xi$ shows that Ξ_u is compact. Consequently, Ξ acts freely on $W \setminus \{v\}$, and $t \leq l$. Consider a one-parameter subgroup Π in Ξ and the connected component Ψ of its centralizer in Γ. Since Γ acts linearly on $\Xi \cong \mathbb{R}^t$, we have $\dim \Gamma - \dim \Psi \leq t$ and $\dim \Psi \leq l + \dim \Psi_u$. With 8) we obtain $\dim \Psi_u \leq 3$ and $\dim \Gamma < 3(l-1)$. This contradicts step 6).

11) Assume that Γ is semi-simple but not almost simple. Then $\Gamma = AB$, where A is an almost simple factor of minimal dimension, and B is the product of the other almost simple factors. There is some $u \in W \setminus \{v\}$ such that $u^A \neq u$, and we may apply step 8). If $l = 4$, then $\dim A \leq \dim B \leq 4$. If $l = 8$, then $\dim B \leq 11$. The possible dimensions of almost simple Lie groups are $3, 6, 8, 10$, or at least 14, see (94.33). From $\dim B = 11$ it follows that B has a 3-dimensional factor. Thus, $\dim A = 3$ by minimality, or else $\dim A \leq \dim B \leq 10$. In any case, $\dim \Gamma < 3(l-1)$; this contradicts 6) again.

12) There remains only the possibility that Γ is almost simple. If Γ would contain a reflection σ, then σ^Γ would generate a connected subgroup of $\Gamma_{[av]}$; by (61.20a), the elation group $\Gamma_{[v,av]}$ would then be a connected, proper normal subgroup. Hence all involutions in Γ are planar (55.29), and Γ contains some involution α by (94.37). Let $\mathcal{F}_\alpha$ be the corresponding Baer subplane and note that $\mathcal{F}_\alpha$ is Θ-invariant. From 5) we know that $\dim \mathcal{F}_\alpha = l = 8$, and by 6) we have $21 \leq \dim \Gamma \leq 28$. Inspection of the list (94.33) shows that Γ has torus rank at least 2, and there are commuting Baer involutions α, β and $\alpha\beta$. By (55.39), this implies that $\mathcal{F}_\alpha \cap \mathcal{F}_\beta$ is a 4-dimensional Θ-invariant subplane in contradiction to step 5). $\square$

We are now in a position to describe in general terms how one may proceed in order to determine all compact planes which admit a group of sufficiently large dimension. As mentioned in the introduction, we shall do this in the case $l = 8$. So let Δ denote a connected, closed subgroup of the automorphism group of a 16-dimensional compact plane $\mathcal{P}$, and assume first that $\dim \Delta = h \geq 35 = 5(l-1)$. Then we know from the preceding proposition that Δ is in fact a Lie group, and either Δ is semi-simple or Δ has a minimal connected abelian normal subgroup Θ, compare (94.26). If Δ is semi-simple, then $\mathcal{P}$ is known to be classical if $h \geq 37$, see Salzmann [82b], and a good deal can be said for $h \geq 35$. If Δ is not semi-simple, however, most arguments require $h > 35$, and a complete classification has been obtained only for $h > 39$. In any case, a key to the classification is the existence of involutions and an analysis of their centralizers. The smaller the dimension of Δ, the more important becomes Proposition (87.1) because a locally compact group may have fewer involutions than an approximating Lie group of the same dimension. (Note that there are (compact, connected) solenoids of arbitrary dimension without any involution.) In a semi-simple group there are many involutions (94.37). If Δ is not semi-simple, and if Θ is not a torus, then Θ is a vector group by (94.26), and Δ induces an irreducible linear representation on Θ. In this case, Δ has a semi-simple subgroup of dimension at least $\dim(\Delta/\operatorname{Cs} \Theta) - 2$, see (95.6).

If the group Δ is semi-simple, its center Z may either be trivial or contain a planar element or consist of homologies or elations. In non-classical planes, Δ

turns out to be even almost simple. For topological reasons, Δ has torus rank at most 4, see (55.37) or Chang–Skjelbred [76]. Moreover, there are at most 3 pairwise commuting reflections (55.34b), and Δ has no subgroup isomorphic to $SO_5\mathbb{R}$, see (55.40). On the other hand, $h \geq 35$ implies that Δ contains a 3-torus and hence some planar involution β. An almost simple subgroup $\Psi \not\cong SU_2\mathbb{C}$ in $Cs\,\beta$ must act on $\mathcal{F}_\beta$ by (83.22). This leads to further restrictions for Δ. Only the simple group $PSp_6\mathbb{C}$ and the almost simple groups of dimension $h \leq 36$ have to be inspected more closely. In fact, the group $PSp_6\mathbb{C}$ cannot act on any compact plane, compare Salzmann [82b] (3.2f). The existence of a planar element in Z is equivalent to $\Delta \cong SL_3\mathbb{H}$ and leads to the Hughes planes (the only planes admitting a group $\Delta \stackrel{\circ}{=} SL_3\mathbb{H}$, see (86.34)). If Z contains a homology with axis W, a study of the action of Δ on W shows that $\Delta \cong Spin_9(\mathbb{R}, r)$ with $r \leq 2$ if $\mathcal{P}$ is not classical. In the case $r = 2$ it follows that $u^\Delta = W \setminus \{v\}$, and this leads to a contradiction, see Priwitzer [94b]. If Z consists of elations, then Z is torsion free by (55.28), and Δ covers a group $O'_9(\mathbb{R}, 2)$, $PSp_8\mathbb{R}$, or $PSU_6(\mathbb{C}, r)$ where $r > 0$. In the first two cases, Δ would contain a group $SO_5\mathbb{R}$. Similarly, the group $PSU_6(\mathbb{C}, 1)$ has a subgroup $SO_5\mathbb{R} < SU_5\mathbb{C}$. The remaining two unitary groups can be excluded by a study of the actions of their maximal compact subgroups. Finally, if $Z = \mathbb{1}$, only the groups $PU_4(\mathbb{H}, r)$ with $r \leq 2$ require a more detailed study. It turns out that the scalar matrices $i\mathbb{1}$ and $j\mathbb{1}$ induce planar involutions with a 4-dimensional subplane $\mathcal{C}$ as common fixed point set. Moreover, $\Omega = PO'_4(\mathbb{R}, r)$ acts effectively on $\mathcal{C}$, and Ω is simple by (71.8). This leaves only the possibility $r = 1$. The group $PU_4(\mathbb{H}, 1)$ is the centralizer of a planar polarity of $\mathcal{P}_2\mathbb{O}$, see (18.32). According to a theorem of Priwitzer [94b], this planar motion group can act only on the classical Moufang plane. Eventually, we get

87.2 Proposition. *If Δ is a semi-simple group of automorphisms of a non-classical compact plane $\mathcal{P}$ and if $\dim \Delta \geq 35$, then Δ is even almost simple, and the center Z of Δ is generated by an involution ι. There are the following two possibilities.*

(a) *If ι is planar, then $\Delta \cong SL_3\mathbb{H}$ and $\mathcal{P}$ is a Hughes plane (86.34).*
(b) *If ι is a reflection, then $\Delta \cong Spin_9(\mathbb{R}, r)$. Either $r = 0$ and Δ is transitive on the axis W of ι, or $r = 1$ and Δ has 3 orbits on W; one of these is a 7-sphere.*

Remark. The same results have been obtained by Priwitzer [94b] under the weaker hypothesis that $\dim \Delta > 30$.

Next, we turn to the case where Δ contains a normal (and hence central) torus $\Theta \cong \mathbb{T}$. Assume again that $h \geq 35$. If the involution $\iota \in \Theta$ is a reflection, then Θ is in fact contained in a group $\Delta_{[a,W]}$ of homologies. Whenever $\Lambda < \Delta$ fixes a quadrangle, then Θ acts faithfully on $\mathcal{F}_\Lambda$, and $\dim \mathcal{F}_\Lambda \geq 4$. A theorem of Bödi (83.25) implies that $\dim \Lambda \leq 8$, and each orbit of Δ on W has dimension at least 7. If Δ were transitive on W, then Δ would contain a group $\mathrm{Spin}_9\mathbb{R}$ by (62.8), and Δ would have torus rank > 4. Hence $\dim v^\Delta = 7$ for some $v \in W$. Either Δ is doubly transitive on v^Δ, or Δ_v is doubly transitive on a 6-dimensional orbit. Using the classification of doubly transitive groups (96.16 and 17), both possibilities can be excluded. Therefore, ι is planar, and Δ induces on $\mathscr{F}_\iota = \mathscr{F}_\Theta$ the full group of the quaternion plane, cf. (83.22), (84.16, 19 and 21). This implies

87.3 Proposition. *If Δ has a normal torus subgroup $\Theta \neq \mathbb{1}$, and if $\dim \Delta \geq 35$, then $\mathcal{P}$ is a Hughes plane, $\Delta' \cong \mathrm{SL}_3\mathbb{H}$, and $\sqrt{\Delta} = \Theta \cong \mathbb{T}$.*

By (94.26), there remains only the possibility that Δ has a minimal normal subgroup $\Theta \cong \mathbb{R}^t$. This case is more difficult, and we will have to suppose $h \geq 36$ throughout the remainder of this section. There are several subcases. In some of these we can show that Θ is contained in a transitive translation group. Typically, each one-parameter subgroup Π of Θ is straight and hence (83.6) consists of axial collineations. By commutativity, Θ is then a group of elations with common axis (or with common center), or Θ is a group of homologies (61.25). In fact, if Π is not straight, and if Ψ denotes the connected component of $\mathrm{Cs}\,\Pi$, then $\dim \Delta/\Psi \leq t$, some orbit a^Π generates a connected subplane $\mathscr{F}$, and $\Psi_a = \Lambda$ induces the identity on $\mathscr{F}$. Consequently,

$$36 \leq h \leq t + \dim a^\Psi + \dim \Lambda \leq t + 30,$$

and $\Lambda^1 \cong G_2$ or $t \geq 12$ by Bödi's theorem (83.25). A detailed analysis of the situation and of the irreducible representation of Δ on Θ in particular reveals that $\Theta \cong \mathbb{R}^8$, that $\Delta = \Phi\sqrt{\Delta}$ with $\Phi \cong \mathrm{Spin}_7\mathbb{R}$, and that $\mathrm{Cs}\,\Theta$ contains a minimal normal subgroup $\Xi \cong \mathbb{R}^7$ of Δ which consists of elations. Thus, substituting Ξ for Θ if feasible, we may always assume that $\Theta \leq \Delta_{[W]}$ for some axis W.

If $t > 8$, then $\dim \Theta_{[z]} > 0$ for each center $z \in W$ by (61.11b). Consequently, $\mathrm{Cs}\,\Theta \leq \Delta_{[W]}$ and then $\mathrm{Cs}\,\Theta = \mathsf{T} = \Delta_{[W,W]}$. Minimality of Θ implies that Δ fixes no point on W. If T is not transitive, it can be shown that a maximal semi-simple subgroup Ψ of Δ acts faithfully and irreducibly on Θ. Hence $\sqrt{\Delta}/\mathsf{T} \leq \mathbb{C}^\times$ and $\Delta \circ \sqrt{\Delta} \leq \mathsf{T}$. Moreover, $\dim \mathsf{T} < 15$ by (61.11c and 12), the radical $\sqrt{\Delta}$ is contained in $\Delta_{[W]}$ and $\dim \sqrt{\Delta} \leq 2 + \dim \mathsf{T} \leq 16$. Therefore, $h \geq 36$ implies $\dim \Psi \geq 20$. As above, Ψ has torus rank at most 4 by (55.37), and Ψ does not contain a subgroup $\mathrm{SO}_5\mathbb{R}$, see (55.40). Proposition (87.2), applied to Ψ instead of Δ, yields $\dim \Psi \leq 36$ and $\Psi \not\cong \mathrm{SL}_3\mathbb{H}$. If Ψ is the complexification of the compact group G_2, then Ψ has fixed points on W by (96.35), (96.10) and (94.34), hence Ψ cannot be irreducible on Θ. For an almost simple group Ψ inspection of the irreducible representations in dimension $t \leq 14$ together with the last remarks shows that either Ψ is a real form of type A_4 and $t = 10$ or Ψ is a real form of type C_3 and $t = 14$. In the latter case, the representation is absolutely irreducible, and $\dim \sqrt{\Delta} \leq t + 1$. Hence $h = 36$ whenever Ψ is almost simple. Using Lemma

(95.5), the same can be proved for a properly semi-simple group Ψ. In other words, $h \geq 37$ and $t > 8$ imply that T is transitive. This is another indication that the planes with the largest groups are to be expected among the translation planes. For these, stronger methods are available as we have seen in Sections 81 and 82. Here, we shall be content to reduce the general classification problem to the case of translation planes. In (82.28), all translation planes with $h \geq 38$ have been described; in particular, the following is true: either $\mathcal{P}$ is classical or Δ fixes some point on W and $t \leq 8$.

In the last case, where $t \leq 8$ and T is not transitive, it can be proved that Θ also has a unique center. If Θ consists of homologies, then $t = 1$ by (61.2), and either Δ has a fixed point on W (and $h \leq 38$), or $\Delta = \Theta \Delta'$ with $\Delta' \cong \mathrm{Spin}_9(\mathbb{R}, r)$, where $r \leq 1$, and $h = 37$.

With the notation of this section, the different possibilities may be summarized as follows.

87.4 Proposition. *Let Δ be a connected group of automorphisms of a non-classical compact plane. If Δ has a minimal normal vector subgroup $\Theta \cong \mathbb{R}^t$, and if $\dim \Delta = h \geq 36$, then there are only the following possibilities.*

$$\begin{cases} \Theta \leq \Delta_{[a,W]}, \ a \notin W \begin{cases} \Delta' \cong \mathrm{Spin}_9(\mathbb{R}, r), \ r \leq 1, \text{ and } h = 37 \\ v^\Delta = v \in W \text{ and } h \leq 38 \end{cases} \\ \Theta \leq \mathsf{T} = \Delta_{[W,W]} \begin{cases} t > 8 \begin{cases} h = 36 \\ \mathsf{T} \cong \mathbb{R}^{16} \text{ and } h = 37 \end{cases} \\ t = 8 \Rightarrow \Theta = \Delta_{[v,W]} \text{ or } \mathsf{T} \cong \mathbb{R}^{16} \\ t < 8 \Rightarrow \Theta \leq \Delta_{[v,W]}, \ v \in W \end{cases} \end{cases}$$

Least can be said if $\Theta \leq \Delta_{[v,W]}$ with $v \in W$. For smaller values of $s = \dim \mathsf{T}$, and if $\mathsf{T} \leq \Delta_{[v,W]}$ in particular, the representation of Δ on T does not tell much. In this case, it is useful (M. Lüneburg [92]) to consider a connected solvable subgroup M of Δ of maximal dimension. If N/T is a minimal normal subgroup of M/T, then $\dim \mathsf{N} \leq s + 2$, and there is a point $b \in a^\mathsf{N}, b \notin av$ such that $\dim \mathsf{M}/\mathsf{M}_{a,b} \leq s + 18$. This argument, together with suitable stiffness results, shows that $\dim \mathsf{M} \leq 30$ and $\dim \mathsf{M} \leq 23 + s$. Let $\kappa : \Delta \to \Delta/\sqrt{\Delta}$ be the canonical map and note that $\dim \mathsf{M}^\kappa = \mathrm{rk}\,\Delta^\kappa$ if M^κ is compact (and much larger otherwise, namely at least equal to the signature $\mathrm{sg}\lfloor \Delta^\kappa$ as defined in (94.32c); this follows from the Iwasawa decomposition, see Hilgert–Neeb [91] III.6.27 or Onishchik–Vinberg [90] Chap. 5 § 4.5° p. 275). From $h > 36$ it follows that Δ contains a semi-simple subgroup Ψ such that either $\Psi \cong \mathrm{U}_2\mathbb{H}$ or $\dim \Psi > 10$, compare (82.26) case (iv) for translation planes with a group $\mathrm{U}_2\mathbb{H}$. Better bounds for the dimension of a Levi complement in Δ can be obtained if $h \geq 39$ or if s is small. A detailed analysis (M. Lüneburg [92] III Satz 6) leads to

87.5 Theorem. *If* $\dim \Delta \geq 40$, *then* Δ *contains a transitive translation group, or Δ is almost simple and $\mathcal{P}$ is classical, that is, isomorphic to the octonion plane.*

This was first proved in a different way by Hubig [90] in his dissertation.

From Propositions (87.2, 3, and 4) we obtain

87.6 Corollary. *If $\mathcal{P}$ is not classical and* $\dim \Delta > 36$, *then $\mathcal{P}$ is a translation plane, or Δ fixes a point and a line.*

In view of (87.5), we can restate Theorem (82.27) in the following final form.

87.7 Theorem. *If $\mathcal{P}$ is not classical, and if* $\dim \Sigma \geq 40$, *then $\mathcal{P}$ can be coordinatized by a mutation of the octonions (82.21), and* $\dim \Sigma = 40$. *In particular, $\mathcal{P}$ has Lenz type* V.

Chapter 9

Appendix: Tools from topology and Lie theory

In this book, we use several results from topology, Lie theory, linear representations and transformation groups. Here, we survey some aspects of these theories which have turned out to be relevant for our purposes, and we collect a number of theorems in a suitable form. We have tried to give convenient references and to guide the reader to the literature.

The first section contains a few easy but rather useful facts on (infinite) permutation groups. Prerequisites from topology are easily accessible, and so we discuss only the notion of topological dimension, which is rarely treated in texts on topology. Theorem (92.16) is particularly significant. A few remarks on quotients conclude this section. In Section 93 we deal with locally compact groups and their approximation by Lie groups (Hilbert's fifth problem). Section 94 is concerned with fundamentals of Lie theory and the structure of Lie groups. For each simple Lie group of dimension at most 52, one (finite) covering group is given in the list (94.33). The only larger simple Lie group which plays a rôle in the theory of compact projective planes is the full automorphism group of the classical Moufang plane (18.19).

Often, a collineation group acts on some normal vector group Θ, and Θ consists of translations, whence $\dim \Theta \leq 16$. Therefore, irreducible real representations in dimension at most 16 play a crucial rôle. Section 95 lists all such representations of almost simple Lie groups and gives the necessary background. The final section collects various theorems on locally compact transformation groups. It includes a description of all compact transitive groups on spheres and of Tits' classification of doubly transitive groups.

91 Permutation groups

In dealing with collineation groups, some basic arguments concern permutation groups of the point set. Generalities on finite and infinite permutation groups can be found in Aschbacher [86] Sect. 5, see also Suzuki [82] Chap. 1, §7. We say that the group Γ acts on the set M if $x^{\mathbb{1}} = x$ and $(x^\alpha)^\beta = x^{\alpha\beta} \in M$ for all $x \in M$

and $\alpha, \beta \in \Gamma$. The orbits x^Γ form a partition of M. For $a \in M$, the subgroup $\Gamma_a = \{\gamma \in \Gamma \mid a^\gamma = a\}$ is called the stabilizer of a in Γ. If $A^\Gamma = A \subseteq M$, then $\mathrm{Fix}_A \Gamma$ is the set of fixed points of Γ in A and $\gamma \mapsto \gamma|_A$ is the restriction homomorphism of Γ into the group of all permutations of A. Its kernel is $\Gamma_{[A]} = \bigcap_{a \in A} \Gamma_a$, and its image $\Gamma|_A$ is isomorphic to $\Gamma/\Gamma_{[A]}$. In particular, $\Gamma/\Gamma_{[M]}$ is the effective permutation group induced by Γ on M. Permutation groups (Γ, M) and (Δ, N) are said to be *equivalent* if there is a bijection $\eta : M \to N$ and an isomorphism $\sigma : \Gamma \cong \Delta$ such that $x^{\eta \gamma^\sigma} = x^{\gamma \eta}$ for all $x \in M$ and $\gamma \in \Gamma$. If this is true with $\sigma = \mathbb{1}$, the actions of Γ on M and on N are called *isomorphic*. The assertions of the next proposition follow from the trivial identity $(x^\gamma)^\tau = (x^\tau)^{\gamma^\tau}$, which expresses the observation that mappings of M are related to conjugations of Γ.

91.1 Proposition.

(a) *Conjugate stabilizers:* $(\Gamma_a)^\delta = \Gamma_{a^\delta}$.
(b) *A commutative, transitive effective group is sharply transitive.*
(c) *If Δ is in the centralizer of Γ_a, then Γ_a acts trivially on the orbit a^Δ.*
(d) *If N is normal in Γ, then Γ permutes the orbits of N and the fixed points of N. Moreover, Γ_a leaves the orbit a^N invariant.*

The following easy facts are quite useful:

91.2 Proposition.

(a) *If T is a transitive subset of Γ, then $\Gamma = \Gamma_a \mathsf{T}$ (Frattini argument).*
(b) *Assume that T is a normal subgroup of Γ and that T is sharply transitive on M. Identify T with M via the map $\gamma \mapsto a^\gamma$. Then the action of Γ_a on M is given by conjugation on T.*
(c) *If $\Delta \leq \Gamma$, then Γ acts transitively on the coset space Γ/Δ by $\Delta \xi \mapsto \Delta \xi \gamma$, and any transitive action (Γ, M) is isomorphic to $(\Gamma, \Gamma/\Gamma_a)$ for each $a \in M$.*

91.3 Corollary. *Let Γ and Δ be transitive permutation groups on M. If $\Gamma \leq \Delta$ and $\Delta_a = \Gamma_a$, then $\Gamma = \Delta$.*

The group Γ is *doubly transitive* on M if $a^\Gamma \neq a$ and Γ_a is transitive on $M \setminus \{a\}$, for some $a \in M$, and Γ is said to be *primitive* if Γ is transitive and if each stabilizer Γ_a is a maximal subgroup of Γ, see Aschbacher [86] (5.19).

91.4 Proposition.

(a) *A doubly transitive group is primitive.*
(b) *A normal subgroup $\mathsf{N} \neq \mathbb{1}$ of a transitive and primitive effective permutation group acts transitively.*

Proof. (a) If $\Gamma_a < \Delta \leq \Gamma$, then $a^\Delta = M$ and $\Gamma = \Gamma_a \Delta = \Delta$ by (91.2a).
(b) $a^\mathsf{N} \neq a$ implies $\Gamma_a < \Gamma_a \mathsf{N} = \Gamma$ and $a^\mathsf{N} = a^\Gamma = M$. □

92 Topological dimension and remarks on general topology

There are several notions of dimension for normal (or more general) spaces. Most of the dimension functions agree on nicer spaces such as separable metric spaces or locally compact groups. The dimension theory for separable metric spaces is developed in Hurewicz–Wallman [48]. This suffices for most of our purposes. Widely used are the inductive dimensions ind and Ind and the covering dimension dim (the latter is related to the dimensions defined by Čech homology or cohomology). Here we collect a few basic facts. Details can be found in any of the books Nagami [70], Pears [75], Engelking [78], Nagata [83]. A comprehensive survey of dimension theory is given by Fedorchuk [90].

A good dimension function d should satisfy $d(\mathbb{R}^n) = n$. For formal reasons, let $d(\emptyset) = -1$. For a space X, it is sufficient to define the relation $d(X) \leq n$. If there is no such $n \in \mathbb{N}$, put $d(X) = \infty$. Otherwise $d(X)$ is the smallest number n with $d(X) \leq n$. Only regular spaces will be considered in this section.

92.1 Definition.

(a) A covering $\mathfrak{v}$ of X is a (precise) refinement of a covering $\mathfrak{u}$ if there is a (bijective) map $\varphi : \mathfrak{v} \to \mathfrak{u}$ such that $V \subseteq \varphi(V)$ for each $V \in \mathfrak{v}$.
(b) ord $\mathfrak{v} = k$ if there are k but not $k + 1$ distinct sets $V \in \mathfrak{v}$ with non-empty intersection.
(c) dim $X \leq n$ if, and only if, every finite open covering $\mathfrak{u}$ of X has an open refinement $\mathfrak{v}$ of order ord $\mathfrak{v} \leq n + 1$. (Then $\mathfrak{u}$ has also a precise open refinement $\mathfrak{w}$ with ord $\mathfrak{w} \leq n + 1$. In fact, the sets $W \in \mathfrak{w}$ can be choosen as $\bigcup \{ V \in \mathfrak{v} \mid \varphi(V) = U \}$.)

For normal spaces, there is a very useful equivalent characterization:

92.2 Theorem. dim $X \leq n$ if, and only if, for every closed subspace $A \subset X$ each continuous map $\varphi : A \to \mathbb{S}_n$ has a continuous extension $\psi : X \to \mathbb{S}_n$.

For a *proof* see Engelking [78] 3.2.10 or Pears [75] 3.2.2. □

Recall that a set $S \subseteq X$ has the boundary $\partial S = \overline{S} \cap \overline{X \setminus S}$.

92.3 Inductive definition.

(a) ind $X \leq n$ if, and only if, every point has arbitrarily small open neighbourhoods U with ind $\partial U \leq n-1$.
(b) Ind $X \leq n$ if, and only if, every closed subset A of X has arbitrarily small open neighbourhoods U with Ind $\partial U \leq n-1$.

By an obvious induction, ind $X \leq$ Ind X. Easy verification shows the following.

92.4 Proposition. ind $A \leq$ ind X for any subspace $A \subset X$. If A is closed in X, then also dim $A \leq$ dim X.

In contrast to Ind and dim, the dimension ind is a local invariant by its very definition: ind $X \leq n$ if each point in X has a neighbourhood U with ind $U \leq n$. Therefore, it is useful to know relations between the various dimensions. According to the remarks at the beginning of this section, most counter-examples necessarily are fairly complicated and will not be discussed here.

92.5 Proposition. *If X is normal, then dim $X \leq$ Ind X. Moreover, dim $X = 0$ implies Ind $X = 0$ (but there is a normal space Z with ind $Z = 0$, dim $Z = 1$, and Ind $Z = 2$ and, for any $n \geq 2$, there is a compact space X_n with dim $X_n = 1$ and ind $X_n = n$). Any space X with ind $X = 0$ is totally disconnected.*

A *proof* of the inequality is given in Pears [75] 4.2.10. The first example is due to Nagami [70] 19-2. The spaces X_n have been constructed by Chatyrko [90]. □

The next two results together imply that the dimensions dim, ind, and Ind coincide on all regular spaces with a countable basis (i.e., on separable metric spaces, cp. Dugundji [66] IX, Th. 5.6 and Cor. 9.2). A proof of the coincidence is given in Nagata [83] Th. IV 1. Recall that a space with a countable basis has the Lindelöf property (Dugundji [66] VIII, Th. 6.3).

92.6 Theorem.

(a) *Any regular Lindelöf space is hypocompact (= strongly paracompact: any open covering has a star-finite open refinement).*
(b) *If X is hypocompact, then dim $X \leq$ ind X.*

Proof. (a) Pears [75] 2.2.9, (b) Engelking [78] 3.1.29, cp. also Nagami [70] 19-1. □

92.7 Theorem (Katetov–Morita). *If X is metrizable, then Ind $X =$ dim X.*

For *proofs* see Nagata [83] Th. II.7, Engelking [78] 4.1.3, or Pears [75] 4.5.4. □

For normal spaces in general, the covering dimension behaves better than the inductive dimensions. Hence the following results will be formulated for the dimension dim mainly.

92.8 Proposition. *If $A \cup B = X$ is normal, then $\dim X \leq \dim A + \dim B + 1$.*

This is proved in Pears [75] 3.5.11. The example $\mathbb{Q} \cup (\mathbb{R} \setminus \mathbb{Q})$ shows that the bound is sharp in general. A better inequality holds if A is closed in X.

92.9 The sum theorem. *If $X = \bigcup_{\nu \in \mathbb{N}} F_\nu$ is normal, and if each F_ν is closed in X and satisfies $\dim F_\nu \leq n$, then $\dim X \leq n$.*

Proofs can be found in Engelking [78] 3.1.8, Pears [75] 3.2.5, or Nagami [70] 9-10, cp. also Nagata [83] Th. VII.2. □

Many applications of dimension theory to topological planes are related to the fact that an affine plane is homeomorphic to the product of a line with itself. For statements on the dimension of $X \times Y$, the case $X = Y = \emptyset$ has to be excluded.

92.10 Proposition.

(a) *If $X \times Y$ is normal, then $\mathrm{ind}(X \times Y) \leq \mathrm{ind}\, X + \mathrm{ind}\, Y$.*
(b) *If X and Y are normal and $X \times Y$ is hypocompact, then we have the inequality $\dim(X \times Y) \leq \dim X + \dim Y$.*

Proof. Part (a) is a rather immediate consequence of (92.9) and the definition, cp. Engelking [78] 1.5.16 and 2.4.5. Part (b) is proved in Pears [75] 9.3.3 and in Engelking [78] 3.2.14. □

Even for compact spaces, equality need not hold: there are compact, 2-dimensional spaces $A, B \subseteq \mathbb{R}^4$ with $\dim(A \times B) = 3$, see Pontrjagin [30]. The following is known, however.

92.11 Theorem. *Assume that X and Y are paracompact. If $Y = [0, 1]$, or if X is locally compact and $\dim Y = 1$, then $\dim(X \times Y) = \dim X + 1$.*

The *proof* uses Theorem (92.2) (Morita [53]) or cohomology, cp. Nagami [70] 42-3. □

Without some compactness condition, the assertion is no longer true: the space Q of rational points in Hilbert space is separable metric (and hence paracompact); it has dimension 1 by Engelking [78] 1.5.17 or Pears [75] 4.1.8. Obviously, $Q \approx Q^2$ and therefore $\dim Q^n = 1$ for all $n \in \mathbb{N}$.

92.12 Corollary. *Euclidean spaces satisfy* $\dim \mathbb{R}^n = n$. *More generally,* $\operatorname{ind} X = n$ *if X is locally homeomorphic to $\mathbb{R}^n$.*

The non-trivial part $\dim \mathbb{R}^n \geq n$ follows also directly from Theorem (92.2): let B denote the unit ball in $\mathbb{R}^n$ and assume $\dim B \leq n - 1$. The antipodal map α of $\partial B = \mathbb{S}_{n-1}$ then has a continuous extension $\beta : B \to \mathbb{S}_{n-1}$. By construction β is fixed point free. This contradicts Brouwer's fixed point theorem. □

92.13 Theorem: Defect. *If X is separable metric and locally compact, then the defect $2 \dim X - \dim X^2$ is at most 1.*

This is due to Fáry [61]. The *proof* depends again on cohomology, cp. Nagami [70] 41-5. For examples with $\dim X^2 < 2 \dim X$ see Karno–Krasinkiewicz [90]. □

The next results are of particular importance in this book.

92.14 Theorem. *Let M be an n-dimensional topological manifold. Any set $X \subseteq M$ with $\operatorname{ind} X = n$ contains a non-empty open subset of M.*

Equivalently, $\operatorname{ind}(\mathbb{R}^n \setminus A) < n$ whenever A is a countable dense subset. The following lemma reduces the last assertion to the easy case $A = \mathbb{Q}^n$, cp. Hurewicz–Wallman [48] Th. IV 3:

92.15. *If A and C are countable dense subsets of $\mathbb{R}^n$, then there exists a homeomorphism σ of $\mathbb{R}^n$ with $A^\sigma = C$.*

Sketch of proof. Let π_ν denote the projection to the νth coordinate. By a suitable rotation, we may assume that each π_ν is injective on A and on C. Then a bijection $\varrho : A \to C$ can be constructed inductively such that $a\pi_\nu < b\pi_\nu$ if, and only if, $a^\varrho \pi_\nu < b^\varrho \pi_\nu$ for all $a, b \in A$ and every ν. By Dedekind cuts in each coordinate, ϱ extends to a homeomorphism σ of $\mathbb{R}^n$. See also Morayne [87]. □

In more general spaces than manifolds, (92.14) is known only for closed subsets. A topological space M is *locally homogeneous* if any two points have homeomorphic open neighbourhoods, and M is *locally contractible* if for any $a \in M$ and for any open neighbourhood V of a there is an open open neighbourhood U and a continuous map $\eta : U \times [0, 1] \to V$ such that $\eta(u, 1) = u$ and $\eta(u, 0) = a$ for all $u \in U$. The following is proved in Seidel [85]:

92.16 Theorem. *Let M be a locally compact separable metric space which is locally homogeneous and locally contractible. If X is closed in M and $\dim X = \dim M < \infty$, then X contains a non-empty open subset of M.*

Finally, we need

92.17 Theorem. *If X is normal and Y is paracompact, and if $f : X \to Y$ is a closed continuous surjection, then $\dim X \leq \operatorname{Ind} Y + \sup_{y \in Y} \dim f^{-1}(y)$.*

For a *proof* see Nagami [70] 21-1 or Nagata [83] Th. VII.12. Weaker versions are given in Pears [75] 9.2.6 and Engelking [78] 4.3.6. □

92.18 Proposition. *If X is locally compact and totally disconnected, then $\operatorname{ind} X = 0$.*

The *proof* consists of the following easy steps:

1) Since ind is defined locally, we may assume that X is compact.

2) Let C be the intersection of all open and closed subsets containing a. By compactness, C has arbitrarily small open and closed neighbourhoods.

3) Normality is applied to show that C is connected, and then $C = \{a\}$ by assumption. The assertion follows now from 2), cp. also Engelking [78] 1.4.5, Fedorchuk [90] Chap. 1, Th. 5. □

Notions and facts from general topology which are used in this book can be found in most texts, e.g. in Dugundji [66]. A few remarks on quotient spaces may be helpful, however.

92.19 Definition. Let X be a topological space and $f : X \to Y$ a map of X onto the set Y. The *quotient topology* on Y is the finest topology for which f is continuous: V is open in Y if, and only if, the preimage $f^{-1}(V)$ is open in X. This is the unique topology on Y such that for any map $g : Y \to Z$ into a topological space Z continuity of g and of $g \circ f$ are equivalent (*universal property*, verification is easy). A surjective map $f : X \to Y$ is called *identifying* (or an *identification map*) if the topology on Y is the quotient topology.

92.20 Remarks. (a) Often the universal property is applied in the following special form: consider the commutative diagram

$$\begin{array}{ccccc} h: & X & \longrightarrow & X' \\ & f \downarrow & & \downarrow f' \\ g: & Y & \longrightarrow & Y' \end{array}$$

of continuous maps between topological spaces, and assume that h is a homeomorphism, that f and f' are identifying, and that g is a bijection. Then g is also a homeomorphism. (The map g will be called the map obtained from h by passing to the quotient.)

(b) If $f : X \to Y$ is a continuous open or closed surjection, then Y carries the quotient topology.

Note that the quotient topology need not satisfy any separation properties, even if X is compact (map $[0, 1]$ onto $\{0,1\}$).

92.21 Proposition. *If there exists a Hausdorff topology on Y such that the surjection $f : X \to Y$ is continuous, then Y is a Hausdorff space in the quotient topology with respect to f.*

93 Locally compact groups and Lie groups

Because of Theorem (44.3), we are primarily concerned with locally compact groups having a countable basis for the topology, usually even with Lie groups. The book of Husain [66] or Chapter II of Hewitt–Ross [79] may serve as a first introduction to topological groups in general; better suited to our purposes is Pontrjagin's classic [66, 86]. More on Lie groups can be found in Hochschild [65] and Varadarajan [74] and in the next section. Note that the connected component Γ^1 of the identity in a topological group Γ is a closed normal subgroup of Γ.

93.1 Definition. A topological group Γ is called a *Lie group* if Γ is a topological manifold admitting local coordinates near the identity that render the group operations real analytic (i.e., expressible by power series).

Hilbert's 5th problem asks for a characterization of Lie groups without differentiability assumptions. The following solution is contained in Montgomery–Zippin [55] Chap. IV, cp. also Kaplansky [71], Yang [76] and primarily (93.4).

93.2 Theorem. *Let Γ be a locally compact topological group. If Γ is locally connected and $\dim \Gamma < \infty$, then Γ is a Lie group.*

93.3 Corollary. *Let Γ be a topological group such that Γ^1 is open in Γ. If Γ^1 is a topological manifold, then Γ is a Lie group.*

Using non-standard analysis, a simplified proof of (93.3) has been given by Hirschfeld [90]. Theorem (93.2) is a direct consequence of case (b) of the following

93.4 Local structure theorem. *Any locally compact group Γ has arbitrarily small open neighbourhoods Ω of the identity which split into a direct product of a connected local Lie group Λ and a compact group Φ. Either* (a) *Λ can be chosen of arbitrarily large dimension, or* (b) *there is some Ω such that Φ is totally disconnected, and* $\operatorname{ind}\Gamma = \operatorname{ind}\Omega = \operatorname{ind}\Lambda < \infty$.

Proof. Gluškov [57, 60] Th. A and B. The last assertion follows with (92.18) and (92.10). For a precise definition of a local Lie group or 'germ' see Bourbaki [89] III §1 no. 10 p. 288. □

93.5 Corollary. *If Γ is a locally compact group, then* $\dim\Gamma = \operatorname{ind}\Gamma$.

Proof. Let $A = A^{-1}$ be a compact neighbourhood of the identity. Then $\Delta = \bigcup_{n\in\mathbb{N}} A^n$ is an open subgroup of Γ, and Δ is a Lindelöf space because each A^n is compact. By (92.6a) the group Δ is hypocompact, and so is Γ, being a union of the open cosets of Δ. From (92.6b) it follows that $\dim\Gamma \leq \operatorname{ind}\Gamma$, and (93.4b) implies $\dim\Lambda \leq \dim\Gamma \leq \operatorname{ind}\Gamma = \operatorname{ind}\Lambda = \dim\Lambda$. For generalizations see Shakhmatov [89] and Tkachenko [90]. □

93.6 Corollary. *If Γ is a locally compact group, then $\dim\Gamma$ is the supremum of all numbers n such that Γ contains an n-ball; in particular,* $\dim\Gamma = \dim\Gamma^1$.

This follows from (93.4 and 5).

Corollary (93.5) has been extended to homogeneous spaces (= coset spaces, see (93.16)):

93.7 Theorem. *If Δ is a closed subgroup of the locally compact group Γ, then* $\dim\Gamma = \dim\Delta + \dim\Gamma/\Delta$. *Moreover, $X = \Gamma/\Delta$ is hypocompact, and* $\dim X = \operatorname{ind} X = \operatorname{Ind} X$ *whenever one of these numbers is finite.*

The first part is *proved* in Nagami [62] Th. 2.1. An analogous result holds for the inductive dimension ind, see Mostert [62] Sect. 5, Cor. 2. If Γ has finite dimension, then (93.5) gives $\dim X = \operatorname{ind} X$. The general case is treated in Pasynkov [62], the proof uses the representation of an open subgroup of Γ as a projective limit of Lie groups as described in

93.8 Approximation theorem.

(a) *Every locally compact group Γ has an open subgroup Δ which is an extension of its connected component Δ^1 by a compact group.*
(b) *If Δ is locally compact and Δ/Δ^1 is compact, then Δ has arbitrarily small compact normal subgroups Θ such that Δ/Θ is a Lie group.*
(c) *If, moreover, $\dim\Delta$ is finite, then $\dim\Theta = 0$ for each sufficiently small subgroup $\Theta \leq \Delta$.*

Proof. Montgomery–Zippin [55] Chap. IV and 2.3.1, Gluškov [60] Th. 9 and the arguments of (c) below. The approximation theorem is the key to many results on locally compact groups. It is used in the proof of the local structure theorem (93.4), and it can easily be deduced from the latter:

(a) If $\Omega = \Lambda \times \Phi$ is a neighbourhood as in (93.4), then Ω generates an open subgroup $\Delta \leq \Gamma$ such that $\Lambda \subseteq \Delta^1$, and Δ/Δ^1 is an epimorphic image of the compact group Φ.

(b) Now choose $\Omega = \Lambda \times \Phi$ in a given neighbourhood of $\mathbb{1}$ in Δ. Then $\langle \Omega \rangle = \Sigma$ is an open (and hence closed) subgroup of Δ, and $\Delta^1 \leq \Sigma$. The compact group Δ/Δ^1 is mapped continuously onto the discrete coset space Δ/Σ, hence Σ has finite index in Δ. Therefore, $\Delta/\bigcap_{\delta \in \Delta} \Sigma^\delta$ is finite, and we may assume that Σ is normal in Δ. By construction, $\Phi < \Sigma$, and Σ/Φ is a Lie group. Since Δ/Σ is a finite group, there are only finitely many conjugates of Φ in Δ. Put $\Theta = \bigcap_{\delta \in \Delta} \Phi^\delta$. Then Σ/Θ is also a Lie group, and so is Δ/Θ, see (94.3d).

(c) follows from (93.4b) together with (92.6 and 18) because Lie groups do not have arbitrarily small subgroups (use the exponential map (94.7)). □

93.9 Corollary. *Let Γ be a locally compact, connected group of finite dimension. If each compact subgroup in the center of Γ is a Lie group, then Γ itself is a Lie group.*

Proof. Γ has a totally disconnected, compact normal subgroup Θ such that Γ/Θ is a Lie group, and Θ is in the center of Γ by (93.18). If Θ is a Lie group, then $\Theta^1 = \mathbb{1}$ is open in Θ. Being compact, Θ is finite, and Γ is locally isomorphic to Γ/Θ. □

The next important fact has been proved for Lie groups first; by means of (93.8b) it extends to locally compact groups (Iwasawa [49] Th. 13, p. 549, see also Hofmann–Terp [94]):

93.10 Mal'cev–Iwasawa theorem. *Let Γ be a locally compact, connected group. Then the following hold.*

(a) *There exists a maximal compact subgroup Φ in Γ, and each maximal compact subgroup is connected and conjugate to Φ. Moreover, each compact subgroup of Γ is contained in a maximal one.*

(b) *There are one-parameter subgroups $\mathsf{P}_\kappa \cong \mathbb{R}$ such that the multiplication map*

$$(\varrho_1, \ldots, \varrho_k, \varphi) \mapsto \varrho_1 \ldots \varrho_k \varphi : \mathsf{P}_1 \times \ldots \times \mathsf{P}_k \times \Phi \to \Gamma$$

is a homeomorphism. In particular, $\Gamma \approx \mathbb{R}^k \times \Phi$. If Γ is commutative, then this map is even an isomorphism.

For the purely topological aspect cp. Cleary–Morris [88], see also Hofmann [93]. The group Φ is a product of a commutative group and some almost simple Lie groups (see (94.20 and 24) and the next remark), more precisely:

93.11 Theorem. *Any compact, connected group Φ is of the form $(A \times \Lambda)/N$, where A is the connected component of the center of Φ and Λ is a direct product of (possibly infinitely many) compact, simply connected, almost simple Lie groups; the kernel N is totally disconnected and closed in $A \times \Lambda$. Hence N is in the center of $A \times \Lambda$, see (93.18), and $\dim N = 0$ by (92.18). If $\dim \Lambda < \infty$, then Φ is a Lie group if, and only if, A is a Lie group.*

Proof. Bourbaki [82] Chap. 9 App. I no. 3 Prop. 2 p. 101, Weil [51], see the text (and not only the result) of §25. □

Remark. The compact almost simple Lie groups are known explicitly, see, e.g., Pontrjagin [66], Helgason [78] Chap. X, and (94.33) or the table in the proof of (63.8), Second Version, step 7). Any compact, connected commutative group is the dual (character group) D^* of a discrete torsion free commutative group D, and $\dim D^*$ is the rank of D. A *torus* $\mathbb{T}^n = (\mathbb{R}/\mathbb{Z})^n$ is dual to $\mathbb{Z}^n$. More on the structure of D^* can be found in Hewitt–Ross [79] §25. The dual of a non-cyclic subgroup D of the discrete additive group $\mathbb{Q}$ is compact and connected but is not a Lie group; it is called a 1-dimensional *solenoid*.

93.12 Proposition. *If Δ is a closed subgroup of the locally compact, connected group Γ, and if $\dim \Delta = \dim \Gamma < \infty$, then $\Delta = \Gamma$.*

Proof. As a continuous image of Γ, the coset space Γ/Δ is connected, and it is totally disconnected by (93.7). Hence Γ/Δ is a singleton. □

93.13 Two-ended groups. *If a locally compact, connected group Γ can be densely embedded into a compact topological space K such that $K \setminus \Gamma$ consists of two distinct points, then Γ is a direct product of a subgroup $P \cong \mathbb{R}$ and a compact (connected) group.*

Proof. Freudenthal [51], Iwasawa [51a], Hofmann–Mostert [63] 13.3, compare also Zippin [50]. □

Having thus surveyed some relevant aspects of the structure theory of locally compact groups, we conclude this section with a few elementary facts. Where nothing else is said, the easy proofs can be found in Hewitt–Ross [79] §5 and §7.

93.14 Proposition.

(a) *Any open subgroup of a topological group Γ is closed in Γ. A connected group has no proper open subgroup. If Γ is a locally connected group, then its connected component Γ^1 is open.*
(b) *A subgroup Δ of a locally compact group Γ is closed in Γ if, and only if, Δ is locally compact.*

93.15 Proposition. *A locally compact group is strongly paracompact ($=$ hypocompact) and hence normal.*

Proof. The first part is contained in the proof of (93.5). Any paracompact space is normal (Dugundji [66] p. 163). □

93.16 Quotients. *If Δ is a subgroup of the topological group Γ, and if κ denotes the canonical map $\gamma \mapsto \Delta\gamma$ from Γ onto the (homogeneous) coset space Γ/Δ provided with the quotient topology as described in (92.19), then κ is an open continuous surjection. A continuous epimorphism $\eta : \Gamma \to \Phi$ of topological groups is open if, and only if, η is identifying.*

Proof. If Ω is open in Γ, then $\Omega^{\kappa\kappa^{-1}} = \Delta\Omega = \bigcup_{\delta \in \Delta} \delta \cdot \Omega$ is also open. The converse is clear by (92.20b). The last part can be proved analogously. □

93.17 Proposition. *Assume, moreover, that Γ is locally compact and that Δ is closed in Γ. Then*

(a) *Γ/Δ is a locally compact regular Hausdorff space.*
(b) *If Δ is compact, then κ is a closed map.*
(c) *If Δ is open, then Γ/Δ is discrete.*

93.18 Proposition. *A totally disconnected normal subgroup Ξ of a topological group Γ centralizes each connected subgroup Δ of Γ.*

Proof. Each set $\xi^\Delta \subseteq \Xi$ is connected and hence consists of ξ only. □

93.19 Theorem. *A compact commutative normal subgroup Θ of a connected group Γ is contained in the center of Γ.*

Proof. Hewitt–Ross [79] Th. (26.10). If Θ is a connected Lie group and therefore a torus, the proof follows immediately from (93.18) and the fact that the dense torsion subgroup Ξ of Θ is normal in Γ and totally disconnected. □

93.20 One-parameter groups. *The image* Π *of a continuous homomorphism* $\varphi : \mathbb{R} \to \Gamma$ *is called a one-parameter subgroup of* Γ. *If* Γ *is locally compact, then either* $\overline{\Pi} = \Pi \cong \mathbb{R}$, *or* $\overline{\Pi}$ *is a compact subgroup of* Γ.

Proof. Hewitt–Ross [79] Th. (9.1) or Hochschild [65] XVI Prop. 2.3. □

94 Lie groups and their structure

Lie groups have been defined in (93.1). A concise introduction to the theory of Lie groups is given in the first two chapters of Adams [69]. For basic notions see also Tondeur [65]. More on differentiable manifolds can be found in F. W. Warner [83]. A survey on the topology of Lie groups is given in Samelson [52]. Most aspects of Lie theory which are relevant in this book are presented clearly in Onishchik–Vinberg [90].

Typical Lie groups are the general linear group $\mathrm{GL}_n\mathbb{R}$ (which is clearly an open subset of the Euclidean space $\mathbb{R}^{n \times n}$ of all real $n \times n$ matrices) and its closed subgroups (94.3a). These linear Lie groups are treated in Freudenthal–de Vries [69]. The Lie algebra $\mathfrak{gl}_n\mathbb{R}$ of $\mathrm{GL}_n\mathbb{R}$ is the vector space $\mathbb{R}^{n \times n}$ with the commutator $[A, B] = AB - BA$ as Lie product. For Lie algebras in general see Humphreys [72], Samelson [69], Jacobson [62], Wan [75] and several books on Lie groups, in particular Bourbaki [89].

If $C \in \mathbb{R}^{n \times n}$, let $\exp C = \sum_{k=0}^{\infty} \frac{1}{k!} C^k$. Then $\exp$ maps $\mathfrak{gl}_n\mathbb{R}$ analytically into $\mathrm{GL}_n\mathbb{R}$ and, e.g., sends the subalgebra $\mathfrak{o}_n\mathbb{R}$ of the skew-symmetric matrices onto the orthogonal group $\mathrm{SO}_n\mathbb{R}$.

The general situation is quite similar. In fact, any Lie group is locally isomorphic to a linear one.

94.1 Definition. Two topological groups Γ and Δ are *locally isomorphic*, $\Gamma \stackrel{\circ}{=} \Delta$, if there is a neighbourhood Ω of $\mathbb{1}$ in Γ and a homeomorphism ι of Ω onto an open subset of Δ such that $\alpha\beta = \gamma$ is equivalent to $\alpha^\iota \beta^\iota = \gamma^\iota$ for all $\alpha, \beta, \gamma \in \Omega$.

94.2 Coverings.

(a) *A homomorphism* $\varphi : \Gamma \to \Delta$ *of topological groups is a covering map if each element in* Δ *has a neighbourhood* Ω *which is 'evenly covered', i.e., there exist a discrete space* Ξ *and a homeomorphism* $\sigma : \Omega \times \Xi \approx \varphi^{-1}(\Omega)$ *such that* $\varphi \circ \sigma$ *is the projection map* $\Omega \times \Xi \to \Omega$. *Equivalently,* φ *is an open epimorphism with discrete kernel* $\mathsf{N}(\approx \Xi)$. *If such a map* $\varphi : \Gamma \to \Delta$ *exists,* Γ *is said to be*

a covering group of Δ. A homomorphism of a Lie group Γ onto a Lie group Δ is a covering if, and only if, it has discrete kernel.
(b) Any connected Lie group Λ has a simply connected (= universal) covering group $\widetilde{\Lambda}$ (unique up to isomorphism). The kernel Θ of the covering map $\widetilde{\Lambda} \to \Lambda$ is a discrete central subgroup of $\widetilde{\Lambda}$, and $\widetilde{\Lambda} \to \Lambda$ is identifying. Moreover, $\Theta \cong \pi_1 \Lambda$, the fundamental group of Λ.
(c) Any analytic homomorphism $\varphi : \Gamma \to \Delta$ between connected Lie groups is induced by a unique homomorphism $\widetilde{\varphi} : \widetilde{\Gamma} \to \widetilde{\Delta}$, and $\Gamma \stackrel{\circ}{=} \Delta$ if, and only if, $\widetilde{\Gamma} \cong \widetilde{\Delta}$, compare also (94.9).

Proof. Hochschild [65] IV, Pontrjagin [66] IX, F. W. Warner [83] Th. 3.25, see also Spanier [66] Chap. 2. The last part of (a) depends on (94.9), compare Onishchik–Vinberg [93] Chap. 1, §4.3. □

94.3 Theorem. *Assume that Δ is a closed subgroup of the Lie group Γ.*

(a) *Δ is also a Lie group. Moreover, Γ/Δ is an (analytic) manifold (of dimension $\dim \Gamma - \dim \Delta$), and the canonical action of Γ on Γ/Δ is analytic.*
(b) *The canonical projection $\varrho : \Gamma \to \Gamma/\Delta$ admits local cross-sections: there exists an open set $\Omega \subseteq \Gamma/\Delta$ and a continuous map $\sigma : \Omega \to \Gamma$ such that $\varrho \circ \sigma = \mathbb{1}$.*
(c) *If Δ is normal in Γ, then Γ/Δ is a Lie group.*
(d) *Conversely, a topological group Γ is a Lie group if Δ and Γ/Δ are Lie groups.*

If Λ is a locally compact group and if $\mu : \Lambda \to \Gamma$ is a continuous monomorphism, then

(e) *Λ is a Lie group.*
(f) *If Γ is connected, then μ is surjective or $\dim \Lambda < \dim \Gamma$.*

Proof. (a) Hochschild [65] VIII Sect. 1 and 2, F. W. Warner [83] 3.42, 58 and 63, Varadarajan [74] Th. 2.12.6 and 2.9.4, Bourbaki [89] III §8 no.2 p. 340 and §1 no. 6 p. 219.

(b,c) Varadarajan [74] Th. 2.9.5 and 6, F. W. Warner [83] 3.58 and 64.

(d) is used in the proof of the approximation theorem (93.8), see Gluškov [60] Th. 7 and 6. On the other hand, (93.8) implies (d) since Γ cannot have arbitrarily small compact subgroups.

(e) follows from (93.8) because all compact subgroups of Δ are Lie groups by (a), see also Hochschild [65] VIII Th.1.1.

(f) Note that $\dim \Lambda = \mathrm{ind}\, \Lambda \leq \mathrm{ind}\, \Lambda^\mu \leq \mathrm{ind}\, \Gamma = n$. If $\dim \Lambda = n$, then also $\mathrm{ind}\, \Lambda^\mu = n$, and Λ^μ is open in Γ by (92.14). Since Γ is connected, (93.14a) implies $\Lambda^\mu = \Gamma$. □

If Γ is not a Lie group, then (f) may fail: any solenoid (cp. the remark after (93.11)) contains a dense one-parameter subgroup, see Hewitt–Ross [79] Th. (10.13).

94.4 Proposition. *Let Δ be a closed subgroup of the connected Lie group Γ.*

(a) *If the coset space Γ/Δ is simply connected, then Δ is connected.*
(b) *If Γ is simply connected and Δ is connected, then Γ/Δ is simply connected.*

Proof. (a) The connected component Δ^1 is open in Δ, and there is a symmetric neighbourhood Ω in Γ such that $\Delta \cap \Omega^2 \subseteq \Delta^1$. It is easy to verify that the canonical map $\Gamma/\Delta^1 \to \Gamma/\Delta$ is a covering (with evenly covered neighbourhoods of the form $\{\Delta\omega\gamma \mid \omega \in \Omega\}$ for $\gamma \in \Gamma$). Since Γ/Δ is simply connected and hence has no proper connected covering, $\Delta = \Delta^1$.

(b) The manifold $M = \Gamma/\Delta$ has a universal covering $\pi : \widetilde{M} \to M$, and Γ induces a transitive transformation group on $\widetilde{M}$ with $(x^\gamma)^\pi = (x^\pi)^\gamma$. Choose $c \in \widetilde{M}$ such that c^π corresponds to the coset Δ. The stabilizer Γ_c is an open subgroup of $\Delta = \Gamma_{c^\pi}$. If Δ is connected, then $\Gamma_c = \Delta$ by (93.14a) and π is bijective and hence a homeomorphism. □

The proposition is also a direct consequence of the exact homotopy sequence (96.12).

According to Definition (93.1), Γ is a Lie group if there exists an analytic group whose underlying topological group is Γ.

From now on, every Lie group will be considered together with a given analytic structure. By (94.9) this analytic structure is in fact unique.

The theory of Lie group germs is equivalent to that of (real) Lie algebras:

94.5 Theorem. *There is a functor $\mathfrak{l}$ mapping each Lie group Γ to its Lie algebra $\mathfrak{l}\Gamma = \mathfrak{g}$. As a vector space, $\mathfrak{g}$ is the tangent space of the manifold Γ at $\mathbb{1}$. In particular, $\dim \Gamma = \dim \mathfrak{g}$. If $\varphi : \Gamma \to \Delta$ is an analytic homomorphism, then $\mathfrak{l}\varphi$ is the differential $D\varphi_{\mathbb{1}}$. For any Lie algebra $\mathfrak{g}$ there is a unique simply connected Lie group Γ with $\mathfrak{l}\Gamma \cong \mathfrak{g}$. Hence $\Gamma \stackrel{\circ}{=} \Delta$ if, and only if, $\mathfrak{l}\Gamma \cong \mathfrak{l}\Delta$.*

Proof. Most books on Lie groups. For the existence of a Lie group with a given Lie algebra, in particular, see Hochschild [65] XII Sect. 1, Varadarajan [74] Th. 3.15.1, or Bourbaki [89] III §6 no.3 p. 310. □

The last-mentioned proof depends on

94.6 Ado's theorem. *Any (real) Lie algebra is isomorphic to a subalgebra of some linear Lie algebra* $\mathfrak{gl}_n\mathbb{R}$.

See Varadarajan [74] Th. 3.17.7 or Bourbaki [89] I §7 no. 3 p. 72. □

94.7 Exponential mapping. *If* Γ *is a Lie group and* $\mathfrak{g} = \mathfrak{l}\Gamma$ *is its Lie algebra, then there exists a unique analytic mapping* $\exp : \mathfrak{g} \to \Gamma$ *which induces a homomorphism on each additive subgroup* $\mathbb{R}a$ *of* $\mathfrak{g}$ *and has differential* $\mathrm{D}\exp_0 = \mathbb{1}_\mathfrak{g}$. *There are neighbourhoods* W *of* 0 *in* $\mathfrak{g}$ *and* Ω *of* $\mathbb{1}$ *in* Γ *such that* $\exp$ *induces a homeomorphism* $W \approx \Omega$ *(but in general* $\exp$ *is neither injective nor surjective). The images of the 1-dimensional subspaces of* $\mathfrak{g}$ *are the one-parameter groups* (93.20) *in* Γ.

Proof. Bourbaki [89] III §4 no. 3 p. 284 and §6 no. 4 Th. 4 p. 311; for the existence of exp see also Adams [69] Chap. 2 and most other books on Lie groups, e.g., Onishchik–Vinberg [93] Chap. 2, §3. The uniqueness is not always explicitly stated. □

The relation between the Lie product [,] and the group multiplication can be expressed as follows:

94.8 Hausdorff product. *Let* $\mathfrak{g} = \mathfrak{l}\Gamma$. *There is a neighbourhood* W *of* 0 *in* $\mathfrak{g}$ *and a multiplication* $\mathbin{\vdash}\!\!\mathbin{\dashv} : W \times W \to \mathfrak{g}$ *such that* $\exp$ *induces an isomorphism of* $(W, \mathbin{\vdash}\!\!\mathbin{\dashv})$ *onto a local subgroup of* Γ, *and* $\mathbin{\vdash}\!\!\mathbin{\dashv}$ *is of the form* $x \mathbin{\vdash}\!\!\mathbin{\dashv} y = x + y + \frac{1}{2}[x, y] + \eta(x, y)$, *where* $\eta(x, y)$ *is a convergent series of Lie products in* x *and* y *of degrees* > 2.

Proof. Hochschild [65] X Th. 3.1, Varadarajan [74] Th. 2.15.4, compare also Bourbaki [89] III §4 no. 2 p. 281. □

A typical application of the exponential is

94.9 Proposition. *If* Γ *and* Δ *are Lie groups and* $\varphi : \Gamma \to \Delta$ *is a continuous homomorphism, then* φ *is an analytic map. Moreover,* $\varphi \circ \exp = \exp \circ \mathfrak{l}\varphi$.

Proof. Hochschild [65] VII Th. 4.2 and Th. 3.3, Varadarajan [74] Sect. 2.11 and Th. 2.10.3, Bourbaki [89] III §8 no. 1 p. 337 and §6 no. 4 Prop. 10(i) p. 312. □

94.10 Theorem. *Let* Γ, Δ *be Lie groups and* $\lambda : \mathfrak{l}\Gamma \to \mathfrak{l}\Delta$ *a homomorphism. If* Γ *is simply connected, then there exists a unique continuous homomorphism* $\varphi : \Gamma \to \Delta$ *with* $\mathfrak{l}\varphi = \lambda$.

Proof. Varadarajan [74] Th. 2.7.5, Bourbaki [89] III §6 no. 1 p. 305. □

94.11 Corollary. *If Γ is a simply connected Lie group with Lie algebra $\mathfrak{g}$, and if $\operatorname{Aut}\Gamma$ is taken with the compact-open topology, see (96.3), then $\mathfrak{l}$ induces an isomorphism $\operatorname{Aut}\Gamma \cong \operatorname{Aut}\mathfrak{g}$ of topological groups, in fact even of Lie groups.*

Proof. Hochschild [65] IX Th. 1.2, Bourbaki [89] III §10, no. 2 Th. 1 p. 363. □

94.12 Proposition. *Let Δ be a closed, connected subgroup of the Lie group Γ. Then $\mathfrak{d} = \mathfrak{l}\Delta$ is a subalgebra of $\mathfrak{l}\Gamma$, and $\langle \exp \mathfrak{d} \rangle = \Delta$. If Γ is connected, then Δ is normal in Γ if, and only if, $\mathfrak{d}$ is an ideal of $\mathfrak{l}\Gamma$.*

Proof. The first part is an immediate consequence of (94.7 and 9) and the fact that a connected group is generated by any open subset, compare (93.14a). For the second part see Bourbaki [89] III §6 no. 6 Prop. 14 p. 316 or Onishchik–Vinberg [93] Chap. 2 Th. 2.7 Cor. □

Note, however, that for a given subalgebra $\mathfrak{a} < \mathfrak{l}\Gamma$ the group $\langle \exp \mathfrak{a} \rangle$ is not necessarily closed in Γ, not even if Γ is simply connected (consider, e.g., the one-parameter groups in $\mathrm{SU}_3\mathbb{C}$).

94.13 Proposition. *Let Γ be a simply connected Lie group with Lie algebra $\mathfrak{g}$. If $\mathfrak{a}$ is an ideal of $\mathfrak{g}$, then $\langle \exp \mathfrak{a} \rangle$ is a closed, simply connected normal subgroup of Γ.*

Proof. Varadarajan [74] Sect. 3.18, Bourbaki [89] III §6 no. 6 p. 316. □

The assertion of (94.13) need not be true if Γ is not simply connected. A very typical counter-example is given by a dense one-parameter subgroup in a torus $\mathbb{T}^2$. This phenomenon leads to the following

94.14 Definition. Let Γ be a Lie group with Lie algebra $\mathfrak{g}$. Then A is said to be an *analytic subgroup* of Γ if $\mathrm{A} = \langle \exp \mathfrak{a} \rangle$ for some subalgebra $\mathfrak{a} \leq \mathfrak{g}$. By (94.5, 9 and 10), this is equivalent to the existence of a connected Lie group Δ and a continuous homomorphism $\varphi : \Delta \to \Gamma$ such that $\Delta^\varphi = \mathrm{A}$. In the quotient topology, A is a Lie group; this *Lie topology* is finer than the subgroup topology of A in Γ.

For details compare Godement [82] §6.13, compare also F. W. Warner [83] 3.16–21. In the terminology of Bourbaki [89] III §6 no.2 p. 306, the group A is called an integral subgroup. □

94.15 Analytic subgroups. *A subgroup A of a Lie group Γ is an analytic subgroup if, and only if, A is arcwise connected.*

For a *proof* see Goto [69]. □

For subsets A, B in a topological group Γ, let [A, B] be the subgroup generated by all commutators $\alpha^{-1}\beta^{-1}\alpha\beta$ with $\alpha \in A, \beta \in B$. Put $\Gamma' = [\Gamma, \Gamma]$ and $\Gamma^{(\nu)} = \Gamma^{(\nu-1)'}$ as usual. The *derived group* $D\Gamma = \overline{\Gamma'}$ is the closure of Γ' in Γ. Note that $\Gamma^{(\nu)} \triangleleft \Gamma$, and that $D^\nu \Gamma = \overline{\Gamma^{(\nu)}}$ since $[\overline{A}, \overline{B}] \leq \overline{[A, B]}$. The group Γ is solvable as an abstract group if, and only if, $D^k \Gamma = \mathbb{1}$ for some $k \in \mathbb{N}$. In a Lie algebra $\mathfrak{g}$, the commutator ideal $[\mathfrak{g}, \mathfrak{g}]$ is denoted by $\mathfrak{g}'$, and $\mathfrak{g}$ is called *solvable* if $\mathfrak{g}^{(k)} = 0$ for some k.

94.16 Proposition. *Let Γ be a connected Lie group and $\mathfrak{g} = \mathfrak{l}\Gamma$. Then $\Gamma^{(\nu)} = \langle \exp \mathfrak{g}^{(\nu)} \rangle$. If Γ is simply connected, then $\Gamma^{(\nu)} = D^\nu \Gamma$ is closed and simply connected.*

Proof. Varadarajan [74] Th. 3.18.8, Bourbaki [89] III §9 no. 2 p. 344. □

94.17 Corollary. *A connected Lie group is solvable if, and only if, its Lie algebra is solvable.*

94.18 Definition. The *radical* $\sqrt{\mathfrak{g}}$ of a Lie algebra $\mathfrak{g}$ is the largest solvable ideal of $\mathfrak{g}$. The radical $\sqrt{\Gamma}$ of a Lie group Γ is the largest connected solvable normal subgroup of Γ.

94.19 Proposition. *If Γ is a Lie group, then $\sqrt{\Gamma}$ is closed in Γ, and $\mathfrak{l}\sqrt{\Gamma} = \sqrt{\mathfrak{l}\Gamma}$.*

See e.g. Bourbaki [89] III §9 no. 7 p. 354. □

94.20 Definition.

(a) A (non-trivial) locally compact, connected group Γ is *semi-simple* if, and only if, each solvable (or equivalently, each commutative) connected normal subgroup of Γ is trivial.
(b) A (non-trivial) Lie algebra $\mathfrak{g}$ is *semi-simple* if, and only if, $\sqrt{\mathfrak{g}} = 0$. Note that $\mathfrak{g}/\sqrt{\mathfrak{g}}$ is always semi-simple (or trivial). As usual, $\mathfrak{g}$ is called *simple* if $\mathfrak{g}$ is not commutative and if $\mathfrak{g}$ and $\{0\}$ are the only ideals of $\mathfrak{g}$.
(c) A (non-commutative) connected Lie group Γ is called *almost simple* (also quasi-simple, often just simple) if $\mathfrak{l}\Gamma$ is a simple Lie algebra, cp. also (94.24).

94.21 Proposition. *An almost simple Lie group Γ with center $Z = \mathbb{1}$ is simple as an abstract group.*

Proof. If $\mathbb{1} \neq \Theta \triangleleft \Gamma$, then the arc component A of $\mathbb{1}$ in Θ is a proper normal subgroup of Γ, and $Z = \mathbb{1}$ implies $A \neq \mathbb{1}$ since each Γ-orbit in Θ is arcwise connected. From (94.12 and 15) it follows that $A = \langle \exp \mathfrak{a} \rangle$ for some ideal $\mathfrak{a} \triangleleft \mathfrak{l}\Gamma$. Hence $\mathfrak{l}\Gamma$ is not simple. □

94.22 Proposition. *A semi-simple Lie algebra* $\mathfrak{g}$ *is a direct product of simple Lie algebras, viz. of the minimal ideals of* $\mathfrak{g}$. *Each ideal of* $\mathfrak{g}$ *is then a product of some of the minimal ideals.*

For a *proof* and for more details see Bourbaki [89] I §6. □

94.23 Semi-simple Lie groups. *A connected Lie group* Γ *is semi-simple if, and only if,* $\sqrt{\Gamma} = \mathbb{1}$ *or, equivalently, if the Lie algebra* $\mathfrak{l}\Gamma$ *is semi-simple. This implies that the center* Z *of* Γ *is discrete and that* Γ/Z *is the direct product of a finite number of simple Lie groups, moreover* $\Gamma' = \Gamma$.

This can be *proved* using successively (94.19), (94.22) together with (94.12 and 13), and (94.16). □

94.24 Almost simple groups. *A non-commutative, connected, locally compact group* Γ *is called almost simple 'if* Γ *has no connected proper closed normal subgroup other than* $\mathbb{1}$. *Such a group* Γ *is semi-simple. This definition generalizes* (94.20c); *use* (94.23 *and* 22).

The following example shows that an almost simple group need not be a Lie group.

Example. The projective limit Γ of an inverse sequence of proper finite coverings of the group $\Omega = \mathrm{PSL}_2\mathbb{R}$ is a locally compact group of the form $\Gamma = \Sigma\Lambda$, where Σ is a 1-dimensional solenoid and $\Lambda \approx \mathbb{R}^2$. In particular, Γ is not a Lie group. The center Z of Γ is compact and 0-dimensional and $\Gamma/Z = \Omega$ is a simple Lie group. Moreover, there is a dense embedding of some covering group of Ω into Γ.

94.25 Proposition. *Consider a locally compact, connected, semi-simple group* Γ *of finite dimension and a compact, 0-dimensional normal (hence central) subgroup* Θ *such that* Γ/Θ *is a Lie group* (93.8c *and* 18). *Then* Γ/Θ *is also semi-simple, and* (94.23) *implies that* Γ *is a product of a finite number of minimal connected, closed normal subgroups, the almost simple factors of* Γ. *These factors centralize one another. Each connected closed normal subgroup of* Γ *is a product of some of the factors.*

Proof. Apply (93.12) in order to show that the product of the almost simple factors is indeed all of Γ. □

94.26 Proposition. *A locally compact, connected group* Γ *of finite dimension is not semi-simple if, and only if,* Γ *has a minimal connected, commutative, closed normal subgroup* $\Xi \neq \mathbb{1}$. *Either* Ξ *is a vector group or* Ξ *is compact and central.*

If Γ is a Lie group and if Ξ is compact, then Ξ is in fact a 1-dimensional torus, see (94.38).

For a *proof* use (93.10, 12 and 19). □

94.27 Proposition. *Consider a locally compact, connected group Γ with countable basis, a homomorphism λ of Γ onto a Lie group Δ, and the inclusion map ι of an analytic subgroup Σ into Δ. Assume either that Σ is semi-simple or $\dim \Sigma = 1$, or that $\dim \ker \lambda = 0$. Then there exist a covering $\kappa : \Psi \to \Sigma$ and a monomorphism $\mu : \Psi \to \Gamma$ such that $\mu\lambda = \kappa\iota$.*

Proof. By the approximation theorem, Γ is the projective limit of a sequence of Lie groups Γ_ν. In each of the three cases, the universal covering group $\widetilde{\Sigma}$ of Σ maps into each group Γ_ν and hence into Γ. Choose Ψ as the image of $\widetilde{\Sigma}$ in Γ, see Löwen [83a] Lemma (3.9) for more details. □

A clue for many results on the structure of Lie groups is

94.28 Levi's theorem. *Consider a connected Lie group Γ and its Lie algebra $\mathfrak{g} = \mathfrak{l}\Gamma$.*

(a) *The vector space $\mathfrak{g}$ is the direct sum of the radical $\sqrt{\mathfrak{g}}$ and any maximal semi-simple subalgebra $\mathfrak{s}$.*
(b) *$\Sigma = \langle \exp \mathfrak{s} \rangle$ is a maximal semi-simple analytic subgroup of Γ, called a Levi complement (of the radical in Γ), and $\Gamma = \Sigma \sqrt{\Gamma}$.*
(c) *Any two Levi complements are conjugate by an element in $\sqrt{\Gamma}'$.*
(d) *If Γ is simply connected, then Σ is closed in Γ and $\Sigma \cap \sqrt{\Gamma} = 1\!\!1$, that is, Γ is a semi-direct product $\Sigma \ltimes \sqrt{\Gamma}$.*
(e) *If the center $Z(\Sigma)$ is finite, then Σ is closed in Γ. Note that $Z(\Sigma)$ is finite if Σ has a faithful linear representation (95.2 and 6b).*

Proof. (a) Jacobson [62] III Sect. 9, Hochschild [65] XI Sect. 3, Varadarajan [74] Sect. 3.14, Bourbaki [89] I §6 no. 8 p. 63. Proofs of (b), (c) and (d) are based on (a), see Varadarajan [74] Th. 3.18.13, or, for simply connected groups, Bourbaki [89] III §9 no. 8 Prop. 31 p. 356. Assertion (e) is proved in Djoković [76] Prop. 4; for the linear case compare also Hochschild [65] XVIII. 4. □

The proof for the existence of a Lie group with given Lie algebra (94.5) can also be based on the Lie algebra version (a) of Levi's theorem.

94.29 Weyl's theorem. *The simply connected covering group of a compact simple Lie group Γ is compact. Equivalently, the fundamental group $\pi_1 \Gamma$ is finite.*

Proof. Pontrjagin [66] Th. 110, Bredon [72] 0.6.10, Varadarajan [74] Th. 4.11.6 or Bourbaki [82] IX §1 no. 4 Th. 1 p. 5. □

94.30 Coverings of Lie groups. *If $\lambda : \Delta \to \Gamma$ is a covering of Γ by a connected Lie group Δ, and if Φ is a maximal compact subgroup of Γ, then Δ has a connected, closed subgroup Ψ such that $\lambda|_\Psi = \kappa : \Psi \to \Phi$ is a covering and $\ker \kappa = \ker \lambda$. Thus, the connected coverings of Γ are in a one-to-one correspondence with the connected coverings of Φ.*

Proof. 1) If the kernel of λ is finite, then $\Phi^{\lambda^{-1}}$ is compact and connected (93.10a).

2) Let Δ be simply connected, and note that $\Gamma \approx \Phi \times \mathbb{R}^k$ by (93.10b). The universal covering space $\widetilde{\Gamma}$ of Γ is homeomorphic to $\widetilde{\Phi} \times \mathbb{R}^k$, and the local group structure of Γ extends in a unique way to $\widetilde{\Gamma}$. Hence $\Delta = \widetilde{\Gamma}$ up to isomorphism, and we may put $\Psi = \widetilde{\Phi}$.

3) The general result follows from 2) by taking quotients. □

94.31 Theorem. *Let Γ be a compact, connected Lie group and $\mathfrak{g}$ its Lie algebra. Then*

(a) *the mapping $\exp : \mathfrak{g} \to \Gamma$ is surjective, and each element of Γ belongs to a torus, hence Γ contains an involution,*

(b) *any two maximal torus subgroups of Γ are conjugate, their dimension is called the rank $\mathrm{rk}\,\Gamma$,*

(c) *Γ is covered by a direct product of a torus and some compact, simply connected, almost simple Lie groups, and Γ is semi-simple if, and only if, the fundamental group $\pi_1 \Gamma$ is finite, or, equivalently, if the torus factor is trivial.*

Proof. Adams [69] 4.21–23 or Hochschild [65] XIII, Th. 3.2 and §4; assertion (c) follows from (93.11) together with (94.29 and 38). □

Remark. The rank of an arbitrary Lie group Γ, more precisely, the *torus rank* $\mathrm{rk}\,\Gamma$ is defined as the rank of a maximal compact connected subgroup of Γ. The only compact, connected Lie groups of rank 1 are the torus $\mathbb{T} \cong \mathrm{SO}_2\mathbb{R}$, the simple group $\mathrm{SO}_3\mathbb{R}$ and its universal covering group $\mathrm{Spin}_3\mathbb{R}$, compare Bourbaki [82] IX §3 no. 6 p. 22.

94.32 Classification. (a) The classification of the complex simple Lie algebras is presented in several texts, see e.g., Humphreys [72] or Varadarajan [74]. The corresponding simply connected complex almost simple Lie groups are the special linear groups $\mathrm{SL}_{n+1}\mathbb{C}$, the double coverings $\mathrm{Spin}_n\mathbb{C}$ of the orthogonal groups $\mathrm{SO}_n\mathbb{C}$, the symplectic groups $\mathrm{Sp}_{2n}\mathbb{C}$ and 5 exceptional groups. These groups may also be considered as almost simple real Lie groups. If $\mathfrak{g}$ is the Lie algebra of any other almost simple Lie group, then the tensor product $\mathfrak{g} \otimes \mathbb{C}$ is a complex simple Lie algebra, and $\mathfrak{g}$ is called a *real form* of its complexification $\mathfrak{g} \otimes \mathbb{C}$.

(b) The complex simple Lie algebra $\mathfrak{sl}_{n+1}\mathbb{C}$ is also denoted by A_n. Analogously, $B_n = \mathfrak{o}_{2n+1}\mathbb{C}$, $C_n = \mathfrak{sp}_{2n}\mathbb{C}$, and $D_n = \mathfrak{o}_{2n}\mathbb{C}$. A complex or real almost simple Lie group Γ is said to be of type X_n if $\mathfrak{l}\Gamma = X_n$ or $\mathfrak{l}\Gamma \otimes \mathbb{C} = X_n$ respectively. In particular, $SU_{n+1}(\mathbb{C}, r)$ and $SL_{n+1}\mathbb{R}$ are of type A_n, $SL_n\mathbb{H}$ is of type A_{2n-1}, $O'_{2n+1}(\mathbb{R}, r)$ is of type B_n, $U_n(\mathbb{H}, r)$ and $Sp_{2n}\mathbb{R}$ are of type C_n, and $O'_{2n}(\mathbb{R}, r)$ is of type D_n, see also (f) below.

(c) All real forms are determined in Hausner–Schwartz [68] and in Onishchik–Vinberg [90, 93], cp. also Koornwinder [82], Magyar [89] or Freudenthal–de Vries [69]. In this classification, a simple real Lie algebra $\mathfrak{g}$ is characterized by the isomorphism type of $\mathfrak{g} \otimes \mathbb{C}$ and the *signature* sg $\mathfrak{g}$, the maximal dimension of a subspace of $\mathfrak{g}$ on which the Killing form $(g, h) \mapsto \text{trace}\,(\text{ad}\,g\,\text{ad}\,h)$ is positive definite.

(d) For our purposes, we need to know the almost simple Lie groups of dimension up to 52. Each of them is locally isomorphic to one group in the following list (94.33). There, Z is the center of the almost simple Lie group Γ, and K is a maximal compact subgroup. Absence of an entry for K indicates that Γ itself is compact. In all cases, K is indeed a compact subgroup of Γ of the largest possible dimension $\dim K = \dim \Gamma - \text{sg}\,\mathfrak{l}\Gamma$. The fundamental groups π_1 of Γ and of K coincide by (93.10). For a calculation of the homotopy groups see Husemoller [75] §7.12 or Steenrod [51] §§22, 25, compare also Freudenthal–de Vries [69] §62. Essentially, our list is adopted from Tits [67].

(e) As a rule, a simple group Γ of type X_n has torus rank n, and the same is true for covering groups with finite center. Noteworthy exceptions are $\text{rk}\,SL_{2n}\mathbb{R} = \text{rk}\,SL_{2n+1}\mathbb{R} = \text{rk}\,SL_n\mathbb{H} = n$ and, if r is odd, $\text{rk}\,O'_{2n}(\mathbb{R}, r) = n - 1$. If the center Z of Γ is infinite, then $\text{rk}\,\Gamma = \text{rk}\,\Gamma/Z - 1$. An almost simple Lie group Γ with $\text{rk}\,\Gamma \leq 3$ has dimension at most 48, thus Γ is locally isomorphic to a group in the list.

(f) Notation is mostly standard. As customary, S indicates the subgroup of determinant 1. The connected component of the real orthogonal group $SO_n(\mathbb{R}, r)$ of a form of Witt index r coincides with the commutator group $O'_n(\mathbb{R}, r)$. The group $\text{Spin}_n(\mathbb{R}, r)$ is a double cover of $O'_n(\mathbb{R}, r)$; note that it is not simply connected for $r \geq 2$. The *anti-unitary group* is $U_n^a\mathbb{H} = \left\{U \in GL_n\mathbb{H} \mid \overline{U}^t iU = i\mathbb{1}\right\}$, see (13.18); this is a group of type D_n. Note that $U_n(\mathbb{H}, r) = SU_n(\mathbb{H}, r)$ and $U_n^a\mathbb{H} = SU_n^a\mathbb{H}$ in contrast to the complex case (cp. Dieudonné [73] Th. 6 for the case $r > 0$, or use the fact that $\begin{pmatrix} a & c \\ b & d \end{pmatrix} \in SL_2\mathbb{H}$ if, and only if, $da - cb^a$ or $ad^c - bc$ has norm 1). The group $U_2^a\mathbb{H}$ is doubly covered by $SU_2\mathbb{C} \times SL_2\mathbb{R}$. For details on the elliptic motion group $\text{Ell}(\mathbb{F})$ and the hyperbolic motion group $\text{Hyp}(\mathbb{F})$ see (13.13) and (18.15 and 26).

94.33 Almost simple Lie groups and their maximal compact subgroups.

Notation is as in (94.32).

dim Γ	Γ	Z	dim K	K	π_1	Remarks
3	$SU_2\mathbb{C}$	$\mathbb{Z}_2$	3		$\mathbb{1}$	$\Gamma/Z \cong SO_3\mathbb{R}$, $\Gamma \cong U_1\mathbb{H}$
	$SL_2\mathbb{R}$	$\mathbb{Z}_2$	1	$SO_2\mathbb{R}$	$\mathbb{Z}$	$\Gamma/Z \cong O'_3(\mathbb{R}, 1)$
6	$SL_2\mathbb{C}$	$\mathbb{Z}_2$	3	$SU_2\mathbb{C}$	$\mathbb{1}$	$\Gamma/Z \cong SO_3\mathbb{C} \cong O'_4(\mathbb{R}, 1)$
8	$SU_3\mathbb{C}$	$\mathbb{Z}_3$	8		$\mathbb{1}$	$\Gamma/Z = \mathrm{Ell}(\mathbb{C})$
	$SU_3(\mathbb{C}, 1)$	$\mathbb{Z}_3$	4	$U_2\mathbb{C}$	$\mathbb{Z}$	$\Gamma/Z = \mathrm{Hyp}(\mathbb{C})$
	$SL_3\mathbb{R}$	$\mathbb{1}$	3	$SO_3\mathbb{R}$	$\mathbb{Z}_2$	
10	$U_2\mathbb{H}$	$\mathbb{Z}_2$	10		$\mathbb{1}$	$\Gamma/Z \cong SO_5\mathbb{R}$
	$U_2(\mathbb{H}, 1)$	$\mathbb{Z}_2$	6	$\mathrm{Spin}_4\mathbb{R}$	$\mathbb{1}$	$\Gamma/Z \cong O'_5(\mathbb{R}, 1)$
	$Sp_4\mathbb{R}$	$\mathbb{Z}_2$	4	$U_2\mathbb{C}$	$\mathbb{Z}$	$\Gamma/Z \cong O'_5(\mathbb{R}, 2)$
14	$G_2(-14)$	$\mathbb{1}$	14		$\mathbb{1}$	$\Gamma = G_2 \cong \mathrm{Aut}\,\mathbb{O}$
	$G_2(2)$	$\mathbb{1}$	6	$SO_4\mathbb{R}$	$\mathbb{Z}_2$	
15	$SU_4\mathbb{C}$	$\mathbb{Z}_4$	15		$\mathbb{1}$	$\Gamma/Z^2 \cong SO_6\mathbb{R}$
	$SL_2\mathbb{H}$	$\mathbb{Z}_2$	10	$U_2\mathbb{H}$	$\mathbb{1}$	$\Gamma/Z \cong O'_6(\mathbb{R}, 1)$
	$SU_4(\mathbb{C}, 1)$	$\mathbb{Z}_4$	9	$U_3\mathbb{C}$	$\mathbb{Z}$	$\Gamma/Z^2 \cong U_3^\alpha\mathbb{H}$
	$SU_4(\mathbb{C}, 2)$	$\mathbb{Z}_4$	7	$S(U_2\mathbb{C} \times U_2\mathbb{C})$	$\mathbb{Z}$	$\Gamma/Z^2 \cong O'_6(\mathbb{R}, 2)$
	$SL_4\mathbb{R}$	$\mathbb{Z}_2$	6	$SO_4\mathbb{R}$	$\mathbb{Z}_2$	$\Gamma/Z \cong O'_6(\mathbb{R}, 3)$
16	$SL_3\mathbb{C}$	$\mathbb{Z}_3$	8	$SU_3\mathbb{C}$	$\mathbb{1}$	
20	$Sp_4\mathbb{C}$	$\mathbb{Z}_2$	10	$U_2\mathbb{H}$	$\mathbb{1}$	$\Gamma/Z \cong SO_5\mathbb{C}$
21	$SO_7\mathbb{R}$	$\mathbb{1}$	21		$\mathbb{Z}_2$	
	$O'_7(\mathbb{R}, 1)$	$\mathbb{1}$	15	$SO_6\mathbb{R}$	$\mathbb{Z}_2$	
	$O'_7(\mathbb{R}, 2)$	$\mathbb{1}$	11	$SO_5\mathbb{R} \times SO_2\mathbb{R}$	$\mathbb{Z}_2 \times \mathbb{Z}$	
	$O'_7(\mathbb{R}, 3)$	$\mathbb{1}$	9	$SO_4\mathbb{R} \times SO_3\mathbb{R}$	$\mathbb{Z}_2 \times \mathbb{Z}_2$	
21	$U_3\mathbb{H}$	$\mathbb{Z}_2$	21		$\mathbb{1}$	$\Gamma/Z = \mathrm{Ell}(\mathbb{H})$
	$U_3(\mathbb{H}, 1)$	$\mathbb{Z}_2$	13	$U_2\mathbb{H} \times SU_2\mathbb{C}$	$\mathbb{1}$	$\Gamma/Z = \mathrm{Hyp}(\mathbb{H})$
	$Sp_6\mathbb{R}$	$\mathbb{Z}_2$	9	$U_3\mathbb{C}$	$\mathbb{Z}$	
24	$SU_5\mathbb{C}$	$\mathbb{Z}_5$	24		$\mathbb{1}$	
	$SU_5(\mathbb{C}, 1)$	$\mathbb{Z}_5$	16	$U_4\mathbb{C}$	$\mathbb{Z}$	
	$SU_5(\mathbb{C}, 2)$	$\mathbb{Z}_5$	12	$S(U_3\mathbb{C} \times U_2\mathbb{C})$	$\mathbb{Z}$	
	$SL_5\mathbb{R}$	$\mathbb{1}$	10	$SO_5\mathbb{R}$	$\mathbb{Z}_2$	

28	$SO_8\mathbb{R}$	$\mathbb{Z}_2$	28		$\mathbb{Z}_2$	$Z(\mathrm{Spin}_8\mathbb{R}) \cong \mathbb{Z}_2 \times \mathbb{Z}_2$
	$O'_8(\mathbb{R},1)$	$\mathbb{1}$	21	$SO_7\mathbb{R}$	$\mathbb{Z}_2$	
	$O'_8(\mathbb{R},2)$	$\mathbb{Z}_2$	16	$SO_6\mathbb{R} \times SO_2\mathbb{R}$	$\mathbb{Z}_2 \times \mathbb{Z}$	$\Gamma \stackrel{\circ}{=} U_4^\alpha\mathbb{H} > U_4\mathbb{C} \stackrel{\circ}{=} K$
	$O'_8(\mathbb{R},3)$	$\mathbb{1}$	13	$SO_5\mathbb{R} \times SO_3\mathbb{R}$	$\mathbb{Z}_2 \times \mathbb{Z}_2$	
	$O'_8(\mathbb{R},4)$	$\mathbb{Z}_2$	12	$SO_4\mathbb{R} \times SO_4\mathbb{R}$	$\mathbb{Z}_2 \times \mathbb{Z}_2$	
28	$G_2^{\mathbb{C}}$	$\mathbb{1}$	14	G_2	$\mathbb{1}$	
30	$SL_4\mathbb{C}$	$\mathbb{Z}_4$	15	$SU_4\mathbb{C}$	$\mathbb{1}$	$\Gamma/Z^2 \cong SO_6\mathbb{C}$
35	$SU_6\mathbb{C}$	$\mathbb{Z}_6$	35		$\mathbb{1}$	
	$SU_6(\mathbb{C},1)$	$\mathbb{Z}_6$	25	$U_5\mathbb{C}$	$\mathbb{Z}$	
	$SL_3\mathbb{H}$	$\mathbb{Z}_2$	21	$U_3\mathbb{H}$	$\mathbb{1}$	
	$SU_6(\mathbb{C},2)$	$\mathbb{Z}_6$	19	$S(U_4\mathbb{C} \times U_2\mathbb{C})$	$\mathbb{Z}$	
	$SU_6(\mathbb{C},3)$	$\mathbb{Z}_6$	17	$S(U_3\mathbb{C} \times U_3\mathbb{C})$	$\mathbb{Z}$	
	$SL_6\mathbb{R}$	$\mathbb{Z}_2$	15	$SO_6\mathbb{R}$	$\mathbb{Z}_2$	
36	$SO_9\mathbb{R}$	$\mathbb{1}$	36		$\mathbb{Z}_2$	
	$O'_9(\mathbb{R},1)$	$\mathbb{1}$	28	$SO_8\mathbb{R}$	$\mathbb{Z}_2$	
	$O'_9(\mathbb{R},2)$	$\mathbb{1}$	22	$SO_7\mathbb{R} \times SO_2\mathbb{R}$	$\mathbb{Z}_2 \times \mathbb{Z}$	
	$O'_9(\mathbb{R},3)$	$\mathbb{1}$	18	$SO_6\mathbb{R} \times SO_3\mathbb{R}$	$\mathbb{Z}_2 \times \mathbb{Z}_2$	
	$O'_9(\mathbb{R},4)$	$\mathbb{1}$	16	$SO_5\mathbb{R} \times SO_4\mathbb{R}$	$\mathbb{Z}_2 \times \mathbb{Z}_2$	
36	$U_4\mathbb{H}$	$\mathbb{Z}_2$	36		$\mathbb{1}$	
	$U_4(\mathbb{H},1)$	$\mathbb{Z}_2$	24	$U_3\mathbb{H} \times SU_2\mathbb{C}$	$\mathbb{1}$	
	$U_4(\mathbb{H},2)$	$\mathbb{Z}_2$	20	$U_2\mathbb{H} \times U_2\mathbb{H}$	$\mathbb{1}$	
	$Sp_8\mathbb{R}$	$\mathbb{Z}_2$	16	$U_4\mathbb{C}$	$\mathbb{Z}$	
42	$SO_7\mathbb{C}$	$\mathbb{1}$	21	$SO_7\mathbb{R}$	$\mathbb{Z}_2$	
42	$Sp_6\mathbb{C}$	$\mathbb{Z}_2$	21	$U_3\mathbb{H}$	$\mathbb{1}$	
45	$O'_{10}(\mathbb{R},r)$	$\langle(-\mathbb{1})^{r+1}\rangle$		$SO_{10-r}\mathbb{R} \times SO_r\mathbb{R}$		$0 \leq r \leq 5$
	$U_5^\alpha\mathbb{H}$	$\mathbb{Z}_2$	25	$U_5\mathbb{C}$	$\mathbb{Z}$	
48	$SU_7(\mathbb{C},r)$	$\mathbb{Z}_7$		$S(U_{7-r}\mathbb{C} \times U_r\mathbb{C})$		$0 \leq r \leq 3$
	$SL_7\mathbb{R}$	$\mathbb{1}$	21	$SO_7\mathbb{R}$	$\mathbb{Z}_2$	
48	$SL_5\mathbb{C}$	$\mathbb{Z}_5$	24	$SU_5\mathbb{C}$	$\mathbb{1}$	
52	$F_4(-52)$	$\mathbb{1}$	52		$\mathbb{1}$	$\Gamma = F_4 = \mathrm{Ell}(\mathbb{O})$
	$F_4(-20)$	$\mathbb{1}$	36	$\mathrm{Spin}_9\mathbb{R}$	$\mathbb{1}$	$\Gamma = \mathrm{Hyp}(\mathbb{O})$
	$F_4(4)$	$\mathbb{1}$	24	$\dfrac{SU_2\mathbb{C} \times U_3\mathbb{H}}{\langle(-1,-1)\rangle}$	$\mathbb{Z}_2$	

Remark. From the given list it is obvious how to continue the infinite series of types $A_n, \ldots, D_n$. The only other groups appearing in the classification are the exceptional groups of types E_6, E_7, E_8. Their real forms have dimensions 78, 133, and 248 respectively.

The following is noteworthy.

94.34 Theorem. *A maximal compact subgroup of a simple Lie group Γ is even a maximal subgroup of Γ considered as an abstract group.*

See Helgason [78] Chap. VI Ex. A3(iv) pp. 276, 567.

94.35 Proposition. *Any simply connected, compact, almost simple Lie group $\Gamma \not\cong U_3 \mathbb{H}$ with $\dim \Gamma \geq 15$ contains the group $\mathrm{Spin}_6 \mathbb{R} \cong \mathrm{SU}_4 \mathbb{C}$.*

The compact groups can easily be recognized in (94.33) by the absence of an entry for K, see also the table in the Second Version of the proof of Theorem (63.8).

For a *proof* of (94.35) note that $\mathrm{SO}_n \mathbb{R} < \mathrm{SO}_{n+1} \mathbb{R}$ and that analogous relations hold for the unitary groups; use also $\mathrm{SU}_4 \mathbb{C} < \mathrm{U}_4 \mathbb{H}$ and $\mathrm{Spin}_9 \mathbb{R} < F_4$, see (18.15 and 8). From (18.15, 18 and 19) it follows that $F_4 < E_6$ via complexification and passage to maximal compact subgroups. $\square$

Because of the Mal'cev–Iwasawa theorem (93.10), the homotopy of a Lie group is given by the homotopy of a maximal compact subgroup. The first homotopy groups of the compact simple Lie groups are known. Up to π_9 they can be found in Bott [70], [58], Mimura–Toda [64], Mimura [67], Kervaire [60], and Bott–Samelson [58], see also Husemoller [66] Section 7.12. In particular, the following is shown in W. Browder [61] Th. 6.11 and Bott [56] p. 253:

94.36 Theorem. *If Γ is a compact almost simple Lie group, then $\pi_2 \Gamma = 0$ and $\pi_3 \Gamma \cong \mathbb{Z}$.*

Using the classification, we obtain

94.37 Corollary. *Let Γ be an almost simple Lie group other than the universal covering group of $\mathrm{SL}_2 \mathbb{R}$. Then Γ has a maximal compact, connected subgroup $K \neq \mathbb{1}$. Hence Γ contains a torus (94.31a), and Γ is not contractible.*

Note that a proper covering group of $\mathrm{SL}_2 \mathbb{R}$ has no faithful linear representation, see Hochschild [65] XVIII Prop.4.1 or (95.10).

The next results are easy consequences of the general theory.

94.38 Commutative Lie groups. *For a connected Lie group Δ and its Lie algebra $\mathfrak{d} = \mathfrak{l}\Delta$ the following are equivalent.*

(a) $\mathfrak{d}' = [\mathfrak{d}, \mathfrak{d}] = 0$.
(b) Δ *is commutative.*
(c) Δ *is a direct product of a torus and a vector group.*

Proof. If $\mathfrak{d}' = 0$, then (94.8) implies that Δ has a covering group $\widetilde{\Delta} \cong (\mathfrak{d}, +)$, cp. Adams [69] Th. 2.19; see also Pontrjagin [34] p. 380, Bourbaki [89] III §6 no. 4 Prop. 11(ii) p. 314, or Fegan [91] Th. 4.9 and 12. □

Note that $\mathfrak{d}' = 0$ if $\dim \mathfrak{d} = 1$. Applying (93.8c) to a 1-dimensional connected group P, one obtains $P' \leq \Theta$ and hence

94.39 Corollary. *Each 1-dimensional, locally compact, connected group P is commutative; if P is a Lie group, then $P \cong \mathbb{R}$ or $P \cong \mathbb{T} = \mathbb{R}/\mathbb{Z}$. The only commutative 2-dimensional, connected Lie groups are $\mathbb{R}^2$, $\mathbb{R} \times \mathbb{T}$, and $\mathbb{T}^2$.*

94.40 Proposition. *Any non-commutative 2-dimensional, connected Lie group Γ is isomorphic to the linear group $\Lambda = \left\{ \begin{pmatrix} a & \\ b & 1 \end{pmatrix} \,\middle|\, a > 0,\ b \in \mathbb{R} \right\} = L_2$.*

Proof. There is exactly one real Lie algebra $\mathfrak{l}$ such that $\dim \mathfrak{l} = 2$ and $\mathfrak{l}' \neq 0$. Hence $\Gamma \stackrel{\circ}{=} \Lambda$ by (94.5). Since Λ is simply connected and has trivial center, (94.2b) implies that $\Gamma \cong \Lambda$. □

94.41 Proposition. *A closed subgroup of $\mathbb{R}^n$ is isomorphic to a group $\mathbb{R}^m \times \mathbb{Z}^k$ with $m + k \leq n$, and it is embedded into $\mathbb{R}^{m+k}$ in a canonical way.*

For a *proof* see Hewitt–Ross [79] Th. (9.11). □

Finally, we note

94.42 Proposition. *Any non-discrete Lie group contains a closed one-parameter group.*

The *proof* follows from (93.20) and (94.31). □

95 Linear representations

A linear representation of dimension $n \in \mathbb{N}$ of a topological group Γ is a continuous homomorphism of Γ into the group $\mathrm{GL}_n\mathbb{R}$ (all representations considered in this section are finite-dimensional). The structure theory of Lie groups is intimately related to their representations, and the (finite-dimensional) linear representations of semi-simple Lie groups are well understood; the special case of complex linear representations is treated in many texts, e.g., in Varadarajan [74], Humphreys [72], Bourbaki [89, 75], Samelson [90]. See Iwahori [59] for an account of E. Cartan's theory of real linear representations, cp. also Tits [71], Freudenthal–de Vries [69].

In the study of compact, connected projective planes, linear representations frequently arise in the following way: some collineation group Γ has a minimal normal subgroup $\Theta \cong \mathbb{R}^n$, thus inducing on Θ an irreducible subgroup $\Gamma/\mathrm{Cs}_\Gamma \Theta$ of $\mathrm{GL}_n\mathbb{R}$; usually Θ is a group of translations, and $n \leq 16$, compare (61.4b and 6) and (55.41).

In this section we mention a few basic results on linear representations of (Lie) groups, and then we give a short summary of the representation theory of complex or real simple Lie algebras. We use this theoretical basis for an enumeration of all pairs (Γ, ϱ) where Γ is an almost simple (linear) Lie group and $\varrho : \Gamma \to \mathrm{GL}_n\mathbb{R}$ is an irreducible representation of dimension $n \leq 16$.

As usual, two linear representations on a vector space V are said to be *equivalent* if they differ only by an inner automorphism of $\mathrm{GL}(V)$. Note that quasi-equivalence as defined in (95.11) is a slightly more general concept, compare also the beginning of Section 91 and (96.1).

95.1 Lie groups and Lie algebras. Let Γ be a connected Lie group with universal covering group $\widetilde{\Gamma}$, cp. (94.2b). Each continuous linear representation $\varrho : \Gamma \to \mathrm{GL}_n\mathbb{R}$ yields a representation $\widetilde{\varrho} : \widetilde{\Gamma} \to \mathrm{GL}_n\mathbb{R}$. Conversely, a representation $\widetilde{\varrho}$ of $\widetilde{\Gamma}$ induces a representation of $\Gamma = \widetilde{\Gamma}/N$ if, and only if, $\widetilde{\varrho}$ is trivial on the kernel N of the covering map $\widetilde{\Gamma} \to \Gamma$.

The continuous linear representations $\widetilde{\varrho} : \widetilde{\Gamma} \to \mathrm{GL}_n\mathbb{R}$ are analytic (94.9) and correspond bijectively to the Lie algebra representations $\varphi : \mathfrak{g} \to \mathfrak{gl}_n\mathbb{R}$ of the Lie algebra $\mathfrak{g} = \mathfrak{l}\widetilde{\Gamma} = \mathfrak{l}\Gamma$ via $\varphi = \mathfrak{l}\widetilde{\varrho}$, compare (94.10). Furthermore $\widetilde{\varrho}$ is irreducible, or *completely reducible* (i.e., every invariant subspace has an invariant complement, or equivalently, $\mathbb{R}^n$ is a direct sum of irreducible subspaces), if, and only if, $\mathfrak{l}\widetilde{\varrho}$ has the same property, see Bourbaki [89] III.6.5 Cor. 2 p. 315.

95.2 Reductivity. A Lie algebra $\mathfrak{g}$ is called *reductive* if, and only if, the nilpotent radical (as defined in Bourbaki [89] I.5.3 p. 44) of $\mathfrak{g}$ is zero; equivalently, the commutator subalgebra $\mathfrak{g}'$ is semi-simple, and $\mathfrak{g}$ is the direct sum of its center and $\mathfrak{g}'$, cp. Bourbaki [89] I.6.4 Prop. 5 p. 56.

A linear Lie algebra $\mathfrak{g} \leq \mathfrak{gl}_n\mathbb{R}$ is completely reducible if, and only if, $\mathfrak{g}$ is reductive and the center of $\mathfrak{g}$ consists of semi-simple (i.e., diagonalizable) endomorphisms, see Bourbaki [89] I.6.5 Thm. 4 p. 58. In particular, the Lie algebra of every connected completely reducible linear Lie group is reductive. Furthermore, every linear representation of a semi-simple Lie algebra (or of a connected semi-simple Lie group) is completely reducible, compare Bourbaki [89] I.6.2 Theorem 2 p. 52.

In any two decompositions of a given representation into irreducible subspaces, the isomorphism types of irreducible components and their multiplicities are the same. This is just a special case of the general Krull–Schmidt Theorem. For our purposes, it suffices to apply the version for modules given in Jacobson [80] 3.4 to the associative algebra generated by the image of the representation.

95.3 Compact groups. *Every compact subgroup Γ of $\mathrm{GL}_n\mathbb{R}$ is contained in a conjugate of the orthogonal group $\mathrm{O}_n\mathbb{R}$. Thus Γ is completely reducible.*

Proof. Integration of the usual scalar product over Γ (with respect to the Haar measure of Γ) yields a Γ-invariant scalar product which is positive definite, thus $\Gamma \leq \mathrm{O}_n\mathbb{R}$ up to conjugation, cp. Hewitt–Ross [79] 22.23, Freudenthal–de Vries [69] 35.1, Hein [90] Satz 30 p. 174. The arguments in Danzer–Laugwitz–Lenz [57] 2.1 or in Onishchik–Vinberg [90] Chap. 3 §4 p. 130 f do not need the Haar measure of Γ. Furthermore, Γ is completely reducible, because the orthogonal complement of an invariant subspace is invariant as well. □

95.4 Centralizers. *Let $\varrho : \Gamma \to \mathrm{GL}_n\mathbb{R}$ be an irreducible representation. Then the centralizer C of the image Γ^ϱ in the endomorphism ring $\mathrm{End}_\mathbb{R}\mathbb{R}^n$ is isomorphic to $\mathbb{R}$, $\mathbb{C}$ or $\mathbb{H}$; furthermore ϱ may be considered as a representation $\varrho : \Gamma \to \mathrm{GL}_m C$ over C with $m = n/\dim_\mathbb{R} C$.*

Proof. Indeed, C is a (not necessarily commutative) field by Schur's lemma (compare Aschbacher [86] 12.4). Furthermore C contains $\mathbb{R}$ in its center, hence $C \in \{\mathbb{R}, \mathbb{C}, \mathbb{H}\}$ by a result of Frobenius [1878], cp. also Ebbinghaus et al. [90, 92], or Palais [68]. □

95.5 Lemma. *Let V be a vector space of finite dimension, let N be a normal subgroup of an irreducible group $\Gamma \leq \mathrm{GL}(V)$, and let $U \neq \{0\}$ be an N-irreducible subspace of V. Then $V = \bigoplus_{i=1}^t U^{\gamma_i}$ is a direct sum of N-irreducible subspaces U^{γ_i} with suitable $\gamma_i \in \Gamma$; in particular, $\dim U = t^{-1} \dim V$ divides $\dim V$.*

Proof. All subspaces U^γ with $\gamma \in \Gamma$ are N-irreducible, and V is generated by these subspaces (by irreducibility of Γ). If a direct sum $V_0 = \bigoplus U^{\gamma_i}$ is a proper subspace of V, then $U^\gamma \not\subseteq V_0$ for some $\gamma \in \Gamma$, hence $U^\gamma \cap V_0 = \{0\}$ by N-irreducibility of U^γ, and $U^\gamma \oplus V_0$ is a larger direct sum of the required shape. This gives the assertion. □

95.6 Structure of irreducible groups.

(a) *Every semi-simple analytic subgroup of* $\mathrm{GL}_n\mathbb{R}$ *is closed in* $\mathrm{GL}_n\mathbb{R}$.

(b) *Let* $\Gamma \leq \mathrm{GL}_n\mathbb{R}$ *be an analytic connected subgroup which acts irreducibly on* $\mathbb{R}^n$. *Then* Γ *and its commutator subgroup* Γ' *are closed subgroups of* $\mathrm{GL}_n\mathbb{R}$, *and* Γ' *is semi-simple. Furthermore* $\Gamma = \Gamma'Z$ *is the product of* Γ' *and the center* Z *of* Γ. *The center* Z *is isomorphic to a subgroup of* $\mathbb{C}^\times$, *and* $\Gamma' \cap Z$ *is cyclic of order dividing* n. *If* Γ' *is not irreducible on* $\mathbb{R}^n$, *then every non-trivial* Γ'-*irreducible subspace of* $\mathbb{R}^n$ *has dimension* $n/2$, *and* $\mathbb{R}^n$ *is the direct sum of any two of these subspaces, which are* Γ'-*equivalent.*

(c) *Let* Γ *be a locally compact, connected group, and let* $\varrho : \Gamma \to \mathrm{GL}_n\mathbb{R}$ *be an irreducible linear representation. Then* $\Gamma/\ker\varrho$ *and* Γ^ϱ *are isomorphic as topological groups. Furthermore,* Γ^ϱ *is a closed subgroup of* $\mathrm{GL}_n\mathbb{R}$, *and* (b) *applies to* Γ^ϱ.

Proof. (a) See Djoković [76] Proposition 3 or Hilgert–Neeb [91] III.8.15.

(b) According to Djoković [77], Γ is closed in $\mathrm{GL}_n\mathbb{R}$, and Γ' is closed by Hochschild [65] XVIII.4.5, see also Djoković [76] Proposition 2, Hilgert–Neeb [91] III.10.9. The Lie algebra of Γ is reductive, and $\Gamma = \Gamma'Z$, see (95.2). The center Z generates a subfield F of the centralizer of Γ, hence $F \cong \mathbb{R}$ or $F \cong \mathbb{C}$ by (95.4), and $Z \leq F^\times$. Since $\det_F(\Gamma') = 1$, we infer that $\Gamma' \cap Z \subseteq \{\zeta \in F \mid \zeta^n = 1\}$. If $F \cong \mathbb{R}$, then Γ' is irreducible on $\mathbb{R}^n$. Otherwise $F \cong \mathbb{C}$, and $\Gamma'\langle -1\rangle$ has index 2 in the irreducible group $\Gamma'\langle i\rangle \leq \Gamma\mathbb{C}^\times$. Because Γ' and $\Gamma'\langle -1\rangle$ have the same irreducible subspaces, the last assertions follow from (95.5).

(c) As ϱ is continuous, the factor group $\Gamma/\ker\varrho$ is locally compact by (93.17a). Applying (94.3e) to the inclusion $\Gamma/\ker\varrho \to \mathrm{GL}_n\mathbb{R}$, we infer that $\Gamma/\ker\varrho$ is a Lie group. Now ϱ induces an analytic bijection of $\Gamma/\ker\varrho$ onto Γ^ϱ, see (94.9), hence Γ^ϱ is an analytic subgroup of $\mathrm{GL}_n\mathbb{R}$, see (94.14). Thus (b) applies to Γ^ϱ. In particular, Γ^ϱ is closed in $\mathrm{GL}_n\mathbb{R}$, hence locally compact, and by (96.9d) the bijection between $\Gamma/\ker\varrho$ and Γ^ϱ induced by ϱ is an isomorphism of topological groups. □

95.7 Complex Lie algebras. Let $\mathfrak{l}$ be a semi-simple Lie algebra over $\mathbb{C}$. We need the concept of a *dominant weight* of $\mathfrak{l}$, see e.g., Humphreys [72] 13.1, Samelson [90] 3.1, or the beginning of Section 4.6 in Varadarajan [74]. For our purposes, it suffices to know that the dominant weights of $\mathfrak{l}$ form a free semi-group (a semi-lattice)

$$\Lambda_+(\mathfrak{l}) = \bigoplus_{i=1}^{l} \mathbb{N}_0 \lambda_i \subset \mathbb{R}^l,$$

which is generated by the *fundamental weights* λ_i, $1 \leq i \leq l$; here l is the *rank* of $\mathfrak{l}$, i.e., the dimension of the Cartan subalgebras of $\mathfrak{l}$.

Theorem. *There exists a natural bijection between the elements of $\Lambda_+(\mathfrak{l})$ and the equivalence classes of irreducible $\mathbb{C}$-linear representations of $\mathfrak{l}$ (on vector spaces of finite positive dimension over $\mathbb{C}$).*

See Humphreys [72] 21.2, Samelson [90] 3.2, Bourbaki [75] VIII.7.2, Varadarajan [74] Thm. 4.7.1, Jacobson [62] VII.3, Freudenthal–de Vries [69] 44.1.

The *Weyl formula* can be used to express the dimension dim(λ) of the representation corresponding to a given $\lambda = \sum_{i=1}^{l} m_i \lambda_i \in \Lambda_+(\mathfrak{l})$ as a polynomial in $m_1, m_2, \ldots, m_l$; this polynomial is an increasing function of each m_i, compare Humphreys [72] 24.3, Samelson [90] 3.8, Jacobson [62] p. 257, Bourbaki [75] VIII.9.2, Varadarajan [74] Thm. 4.14.6, Freudenthal–de Vries [69] 47.8. For example, if $\mathfrak{l} = \mathfrak{sl}_{l+1}\mathbb{C}$, then that dimension is given by

$$\dim(\lambda) = \prod \frac{\sum_{i=a}^{b}(m_i+1)}{b+1-a},$$

where the product is taken over all pairs (a, b) satisfying $1 \leq a \leq b \leq l$, compare Dynkin [57b] p. 358. Extensive tables of evaluations of the Weyl formula can be found in McKay–Patera [81], cp. also Onishchik–Vinberg [90] Table 5 p. 299ff.

If $\mathfrak{l} = \mathfrak{l}_1 \oplus \mathfrak{l}_2$ is a direct sum of two ideals $\mathfrak{l}_1, \mathfrak{l}_2$, then $\Lambda_+(\mathfrak{l}) = \Lambda_+(\mathfrak{l}_1) \oplus \Lambda_+(\mathfrak{l}_2)$; the irreducible representations of $\mathfrak{l}$ are precisely the 'tensor sums' $\varrho_1 \otimes \mathbb{1} + \mathbb{1} \otimes \varrho_2$ of the irreducible representations ϱ_i of $\mathfrak{l}_i$, see Samelson [90] 3.4 Thm. E p. 104, Iwahori [59] Lemma 3, p. 71.

95.8 Real Lie algebras. Let $\mathfrak{l}$ be a semi-simple Lie algebra over $\mathbb{R}$, and denote by σ the generator of the Galois group of $\mathbb{C}$ over $\mathbb{R}$ (i.e., σ is complex conjugation). Then σ acts on $\mathfrak{l}^{\mathbb{C}} = \mathfrak{l} \otimes_{\mathbb{R}} \mathbb{C}$ (on the second factor) and on $\Lambda_+(\mathfrak{l}^{\mathbb{C}})$, cp. Tits [71] 3.1, Iwahori [59] §4; both actions depend on $\mathfrak{l}$ and not just on $\mathfrak{l}^{\mathbb{C}}$ (see the remarks on split types and compact types below). If we identify the elements of $\Lambda_+(\mathfrak{l}^{\mathbb{C}})$ with complex irreducible representations ϱ as in (95.7), then ϱ^{σ} is obtained from ϱ by applying σ to every matrix entry. Note that the equivalence type of ϱ^{σ} is determined uniquely by the equivalence type of ϱ.

Theorem. *The equivalence classes of irreducible $\mathbb{R}$-linear representations of $\mathfrak{l}$ (on vector spaces of finite positive dimension over $\mathbb{R}$) correspond bijectively to the orbits of $\langle \sigma \rangle$ on $\Lambda_+(\mathfrak{l}^{\mathbb{C}})$.*

Write $\varrho_\lambda = \varrho_{\lambda^\sigma}$ for the representation of $\mathfrak{l}$ corresponding to $\{\lambda, \lambda^\sigma\}$, and let $\varrho_\lambda \otimes \mathbb{1}_\mathbb{C}$ denote the complex representation of $\mathfrak{l}^\mathbb{C}$ obtained from ϱ_λ. There exists an additive map

$$\alpha_\mathfrak{l} : \{\, \lambda \in \Lambda_+(\mathfrak{l}^\mathbb{C}) \mid \lambda^\sigma = \lambda \,\} \to \mathbb{Z}_2 = \{0, 1\}$$

such that the following holds for $\lambda \in \Lambda_+(\mathfrak{l}^\mathbb{C})$:

(a) If $\lambda^\sigma = \lambda$ and $\alpha_\mathfrak{l}(\lambda) = 0$, then ϱ_λ has centralizer $\mathbb{R}$. In this case, ϱ_λ is absolutely irreducible (i.e., $\varrho_\lambda \otimes \mathbb{1}_\mathbb{C}$ is irreducible), and the real dimension of ϱ_λ coincides with the complex dimension $\dim(\lambda)$ of the representation $\varrho_\lambda \otimes \mathbb{1}_\mathbb{C}$ determined by λ.
(b) If $\lambda^\sigma = \lambda$ and $\alpha_\mathfrak{l}(\lambda) \ne 0$, then ϱ_λ has centralizer $\mathbb{H}$. In this case, $\varrho_\lambda \otimes \mathbb{1}_\mathbb{C}$ is a direct sum of two equivalent complex representations corresponding to λ, and the real dimension of ϱ_λ is $2 \cdot \dim(\lambda)$.
(c) If $\lambda^\sigma \ne \lambda$, then ϱ_λ has centralizer $\mathbb{C}$. In this case, $\varrho_\lambda \otimes \mathbb{1}_\mathbb{C}$ is a direct sum of two inequivalent complex representations corresponding to λ, λ^σ, and the real dimension of ϱ_λ is $2 \cdot \dim(\lambda)$.

For proofs cp. Iwahori [59], Tits [71] 3.5, 7.2, 8.2, Freudenthal–de Vries [69] Section 55; see also Onishchik–Vinberg [90] p. 290ff and Bröcker–tom Dieck [85] II.6, VI.4.

This result allows one to derive the real irreducible representations of $\mathfrak{l}$ from the complex irreducible representations of $\mathfrak{l}^\mathbb{C}$, provided that $\alpha_\mathfrak{l}$ and the action of σ on $\Lambda_+(\mathfrak{l}^\mathbb{C})$ are known explicitly. The action of σ on $\Lambda_+(\mathfrak{l}^\mathbb{C})$ is additive, and the permutation induced by σ on the fundamental weights is a graph automorphism of the Dynkin diagram of $\mathfrak{l}^\mathbb{C}$, cp. Iwahori [59] Lemma 7. In Tits [67], Bödi–Joswig [93], this permutation $\lambda_i \mapsto \overline{\lambda_i}$ is described for each simple Lie algebra $\mathfrak{l}$ (these permutations can also be deduced from the Satake diagrams in Onishchik–Vinberg [90] table 9 p. 312ff). For the computation of $\alpha_\mathfrak{l}(\lambda)$ (which is sometimes called the index and denoted by ε) see Onishchik–Vinberg [90] p. 291f, Freudenthal–de Vries [69] Sections 57, 58, Tits [71] 3.5, 5.5, 6.1, Bödi–Joswig [93]. In Tits [67], the (so-called real) fundamental representations λ_i with $\alpha_\mathfrak{l}(\lambda_i) = 0$ are listed for each simple Lie algebra $\mathfrak{l}$; see also Tits [67] p. 21.

Many special cases are covered by the following remarks.

If $\mathfrak{l}$ is of split type (i.e., if some Cartan subalgebra is diagonalizable in its adjoint action on $\mathfrak{l}$), then σ acts trivially, and $\alpha_\mathfrak{l} = 0$; in other words, all irreducible representations of $\mathfrak{l}$ are absolutely irreducible, and the representation theories of $\mathfrak{l}$ and of $\mathfrak{l}^\mathbb{C}$ are very much the same, see Jacobson [62] p. 223, Bourbaki [75] VIII.7, Tits [71] 2.5. The Lie algebras $\mathfrak{l} = \mathfrak{sl}_n\mathbb{R}, \mathfrak{sp}_{2n}\mathbb{R}, \mathfrak{o}_n(\mathbb{R}, r)$ with $r = [n/2]$ are of split type (they are also called anticompact).

If $\mathfrak{l}$ is compact (i.e., if the Killing form of $\mathfrak{l}$ is negative definite), then λ^σ corresponds to the dual (or contragredient) of the representation determined by λ;

hence $\lambda = \lambda^\sigma$ if, and only if, this representation is self-dual, compare Iwahori [59] §12, and then $\alpha_\mathfrak{l}(\lambda) = 0$ holds if, and only if, that representation is orthogonal (i.e., it preserves a non-trivial symmetric bilinear form), see Bourbaki [82] IX.7.2 and Appendice II, Onishchik–Vinberg [90] p. 291. Thus λ^σ and $\alpha_\mathfrak{l}(\lambda)$ can be computed using the results in Samelson [90] 3.11, Onishchik–Vinberg [90] Chap. 4 §3 Ex. 5-14, Bourbaki [75] VIII.7.5 or the tables in McKay–Patera [81]. In this way the representation theory of $\mathfrak{l}$ is reduced to that of $\mathfrak{l}^\mathbb{C}$; cp. also Varadarajan [74] 4.14.3, Bröcker–tom Dieck [85] Chap. VI, Adams [69]. Examples of compact Lie algebras are $\mathfrak{l} = \mathfrak{o}_n\mathbb{R}$, $\mathfrak{su}_n\mathbb{C}$, $\mathfrak{u}_n\mathbb{H}$.

If $\mathfrak{l}$ is a complex semi-simple Lie algebra considered as a real Lie algebra, then one can write $\mathfrak{l}^\mathbb{C} = \mathfrak{l} \oplus \bar{\mathfrak{l}} \cong \mathfrak{l} \oplus \mathfrak{l}$ as complex Lie algebras, hence $\Lambda_+(\mathfrak{l}^\mathbb{C}) \cong \Lambda_+(\mathfrak{l}) \oplus \Lambda_+(\mathfrak{l})$ by (95.7). In this case, the representation (λ, μ) of $\mathfrak{l}^\mathbb{C}$ corresponds to the representation $\lambda \otimes \mathbb{1} + \mathbb{1} \otimes \mu^{\text{dual}}$ of $\mathfrak{l} \oplus \mathfrak{l}$, the action of σ is described by $(\lambda, \mu)^\sigma = (\mu, \lambda)$, and $\alpha_\mathfrak{l} = 0$, see Bödi–Joswig [93] p. 247, Iwahori [59] §11, Onishchik–Vinberg [90] p. 292. In (95.10) this is applied to the Lie algebras $\mathfrak{l} = \mathfrak{sl}_n\mathbb{C}$, $\mathfrak{sp}_{2n}\mathbb{C}$, $\mathfrak{o}_n\mathbb{C}$.

We point out that many (real or complex) representations of non-exceptional Lie algebras $\mathfrak{l}$ and $\mathfrak{l}^\mathbb{C}$ can be obtained by Weyl's method: starting with the natural module V, one decomposes tensor powers $V^{\otimes k}$ (or exterior powers or the matrix algebra $\text{End}(V) \cong V \otimes V^{\text{dual}}$) into irreducible submodules, cp. Hein [90] III.2, IV.3, Samelson [90] 3.6, Jacobson [62] VII.6, Varadarajan [74] end of Section 4.7, Onishchik–Vinberg [90] table 5, p. 299ff.

95.9 Linear Lie groups and kernels. A *linear Lie group* is a Lie group admitting a faithful linear representation. Let Γ be a connected semi-simple Lie group, and denote by Θ the intersection of the kernels of all linear representations of Γ. Then Θ is a discrete subgroup of the center of Γ, and Γ/Θ is a linear Lie group, see Onishchik–Vinberg [90] 5.3.6 Thm. 8 p. 264, Hilgert–Neeb [91] III.9.18, cp. also Tits [67] 10.2 p. 17.

As far as linear representations of almost simple Lie groups are concerned, it clearly suffices to consider the linearized versions Γ/Θ of simply connected almost simple Lie groups Γ; every linear Lie group locally isomorphic to Γ is a quotient of Γ/Θ, cp. (95.1).

If Γ is compact, or if the Lie algebra of Γ is the real restriction of a complex semi-simple Lie algebra, then $\Theta = \{\mathbb{1}\}$, see Onishchik–Vinberg [90] 5.2.4 Thm. 10 p. 245 and 4.3.4 Thm. 6 p. 190, Bourbaki [82] IX.7.2. The linearized versions Γ/Θ of all other simply connected Lie groups Γ are given in table 10 in Onishchik–Vinberg [90] p. 318f, cp. also Tits [67]. Apart from real orthogonal groups and from $U_5^\alpha\mathbb{H}$, all the groups Γ appearing in the table (94.33) are linearized versions of simply connected almost simple Lie groups.

Some linear Lie groups Γ/Θ, like $\mathrm{Spin}_8\mathbb{C}$ and $\mathrm{Spin}_8(\mathbb{R},r)$ with $r = 0, 2, 4$, admit no faithful *irreducible* linear representation (as a consequence of (95.4), this is true in all cases where the center of Γ/Θ is not a subgroup of $\mathbb{C}^\times$).

Let Γ be the linearized version of a simply connected almost simple Lie group. The kernels of the linear representations of Γ can be found as follows. Γ embeds into the corresponding simply connected complex Lie group $\Gamma^\mathbb{C}$ (with Lie algebra $\mathfrak{l}\Gamma \otimes \mathbb{C}$), and each representation of Γ extends to a $\mathbb{C}$-linear representation of $\Gamma^\mathbb{C}$, see Onishchik–Vinberg [90] 5.3.6, proof of Thm. 8, Tits [67] 10.2. The embedding of the center Z of Γ into the center $Z^\mathbb{C}$ of $\Gamma^\mathbb{C}$ is described in the tables in Tits [67], see also Bödi–Joswig [93]; if Γ is a complex Lie group, then the center Z of Γ is embedded diagonally into the center $Z^\mathbb{C} = Z \times Z$ of $\Gamma^\mathbb{C} = \Gamma \times \Gamma$, see Bödi–Joswig [93] p. 247. The restriction to $Z^\mathbb{C}$ of an irreducible $\mathbb{C}$-linear representation of $\Gamma^\mathbb{C}$ depends multiplicatively on the dominant weight λ of the representation (95.7), in the sense that $\varrho(\xi) = \varrho'(\xi)\varrho''(\xi)$ for every $\xi \in Z^\mathbb{C}$ if the representations ϱ', ϱ'', ϱ have dominant weights λ, μ, $\lambda + \mu$, respectively, cp. Tits [67] 7.9. Hence it suffices to know the kernels of the fundamental representations λ_i of $\Gamma^\mathbb{C}$; these kernels can be found in the tables in Tits [67].

95.10 Irreducible representations up to dimension 16.
We enumerate (up to equivalence of representations) all pairs (Γ, ϱ) where Γ is an almost simple linear Lie group and $\varrho : \Gamma \to \mathrm{GL}_n\mathbb{R}$ is an irreducible representation of dimension $n \leq 16$ (in other words, we enumerate the conjugacy classes of irreducible almost simple analytic subgroups of the Lie groups $\mathrm{GL}_n\mathbb{R}$ for $n \leq 16$). See Bödi–Joswig [93] for an algorithm which produces similar tables for an arbitrary range of dimensions; there is also a computer implementation of this algorithm.

Denoting by $\mathfrak{g} = \mathfrak{l}\Gamma$ the (real) Lie algebra of Γ, we identify an irreducible representation ϱ by a representative dominant weight $\lambda \in \Lambda_+(\mathfrak{g}^\mathbb{C})$ of the corresponding σ-orbit (95.8); we express λ as a linear combination of the fundamental weights λ_i, which are indexed as in Tits [67].

As indicated in (95.9), it suffices to consider the linearized versions Γ of simply connected almost simple Lie groups: if Υ is a subgroup of the center of Γ, then all representations of Γ/Υ are obtained from representations of Γ containing Υ in their kernel (95.1).

The tables below are ordered first by the (dimensions of) the groups Γ, and then by the dimensions n of the representations. The tables also contain the centralizer (95.4) and the kernel of each representation, cp. (95.8 and 9).

The trivial representations on vector spaces of dimension 0 or 1 are omitted (the latter corresponds to the weight $\lambda = 0$).

Several representations arise naturally in the study of classical planes, see the table at the end of the introduction to Chapter 1.

$\Gamma = \mathrm{SU}_2\mathbb{C} \cong \mathrm{Spin}_3\mathbb{R}$ has center $Z = \langle -\mathbb{1} \rangle \cong \mathbb{Z}_2$, and $\Gamma/Z \cong \mathrm{SO}_3\mathbb{R}$.

weight λ	$2\lambda_1$	λ_1	$4\lambda_1$	$6\lambda_1$	$3\lambda_1$	$8\lambda_1$	$10\lambda_1$	$5\lambda_1$	$12\lambda_1$	$14\lambda_1$	$7\lambda_1$
dimension n	3	4	5	7	8	9	11	12	13	15	16
centralizer	$\mathbb{R}$	$\mathbb{H}$	$\mathbb{R}$	$\mathbb{R}$	$\mathbb{H}$	$\mathbb{R}$	$\mathbb{R}$	$\mathbb{H}$	$\mathbb{R}$	$\mathbb{R}$	$\mathbb{H}$
kernel	Z	$\mathbb{1}$	Z	Z	$\mathbb{1}$	Z	Z	$\mathbb{1}$	Z	Z	$\mathbb{1}$

See also Itzkowitz–Rothman–Strassberg [91].

$\Gamma = \mathrm{SL}_2\mathbb{R}$ has center $Z = \langle -\mathbb{1} \rangle \cong \mathbb{Z}_2$, and $\Gamma/Z \cong \mathrm{PSL}_2\mathbb{R}$. This group Γ has precisely one irreducible $\mathbb{R}$-linear representation in each dimension $n \in \mathbb{N}$, corresponding to the weight $\lambda = (n-1)\lambda_1$; the centralizer is always $\mathbb{R}$, and the kernel is $\langle (-\mathbb{1})^n \rangle$ for $n > 1$, cp. Bourbaki [75] VIII.1.

$\Gamma = \mathrm{SL}_2\mathbb{C}$ has center $Z = \langle -\mathbb{1} \rangle \cong \mathbb{Z}_2$, and $\Gamma/Z \cong \mathrm{PSL}_2\mathbb{C} \cong \mathrm{O}'_4(\mathbb{R}, 1) \cong \mathrm{SO}_3\mathbb{C}$.

weight λ	$(\lambda_1, 0)$	(λ_1, λ_1)	$(2\lambda_1, 0)$	$(3\lambda_1, 0)$	$(2\lambda_1, 2\lambda_1)$	$(4\lambda_1, 0)$
dimension n	4	4	6	8	9	10
centralizer	$\mathbb{C}$	$\mathbb{R}$	$\mathbb{C}$	$\mathbb{C}$	$\mathbb{R}$	$\mathbb{C}$
kernel	$\mathbb{1}$	Z	Z	$\mathbb{1}$	Z	Z

weight λ	$(5\lambda_1, 0)$	$(2\lambda_1, \lambda_1)$	$(6\lambda_1, 0)$	$(7\lambda_1, 0)$	$(3\lambda_1, \lambda_1)$	$(3\lambda_1, 3\lambda_1)$
dimension n	12	12	14	16	16	16
centralizer	$\mathbb{C}$	$\mathbb{C}$	$\mathbb{C}$	$\mathbb{C}$	$\mathbb{C}$	$\mathbb{R}$
kernel	$\mathbb{1}$	$\mathbb{1}$	Z	$\mathbb{1}$	Z	Z

$\Gamma = \mathrm{SU}_3(\mathbb{C}, r)$ with $r = 0, 1$ has center $Z \cong \mathbb{Z}_3$, and $\Gamma/Z \cong \mathrm{PSU}_3(\mathbb{C}, r)$.

weight λ	λ_1	$\lambda_1 + \lambda_2$	$2\lambda_1$
dimension n	6	8	12
centralizer	$\mathbb{C}$	$\mathbb{R}$	$\mathbb{C}$
kernel	$\mathbb{1}$	Z	$\mathbb{1}$

$\Gamma = \mathrm{SL}_3\mathbb{R}$ has trivial center; the centralizer is always $\mathbb{R}$, and all kernels are trivial.

weight λ	λ_1	λ_2	$2\lambda_1$	$2\lambda_2$	$\lambda_1 + \lambda_2$	$3\lambda_1$	$3\lambda_2$	$4\lambda_1$	$4\lambda_2$	$\lambda_1 + 2\lambda_2$	$2\lambda_1 + \lambda_2$
dimension n	3	3	6	6	8	10	10	15	15	15	15

$\Gamma = U_2(\mathbb{H}, r) = \mathrm{Spin}_5(\mathbb{R}, r)$ with $r = 0, 1$ has center $Z = \langle -\mathbb{1} \rangle \cong \mathbb{Z}_2$, and $\Gamma/Z \cong O'_5(\mathbb{R}, r)$.

weight λ	λ_1	λ_2	$2\lambda_2$	$2\lambda_1$
dimension n	5	8	10	14
centralizer	$\mathbb{R}$	$\mathbb{H}$	$\mathbb{R}$	$\mathbb{R}$
kernel	Z	$\mathbb{1}$	Z	Z

$\Gamma = \mathrm{Sp}_4\mathbb{R} = \mathrm{Spin}_5(\mathbb{R}, 2)$ has center $Z = \langle -\mathbb{1} \rangle \cong \mathbb{Z}_2$, and $\Gamma/Z \cong O'_5(\mathbb{R}, 2)$.

weight λ	λ_2	λ_1	$2\lambda_2$	$2\lambda_1$	$\lambda_1 + \lambda_2$
dimension n	4	5	10	14	16
centralizer	$\mathbb{R}$	$\mathbb{R}$	$\mathbb{R}$	$\mathbb{R}$	$\mathbb{R}$
kernel	$\mathbb{1}$	Z	Z	Z	$\mathbb{1}$

$\Gamma = G_2(-14)$ and $\Gamma = G_2(2)$ have trivial center. All irreducible representations have centralizer $\mathbb{R}$ and trivial kernel. There are only two non-trivial irreducible representations of dimension $n \leq 16$, viz. the natural representation of dimension 7 and the adjoint representation of dimension 14 (corresponding to the fundamental weights λ_1 and λ_2).

$\Gamma = \mathrm{SU}_4(\mathbb{C}, r)$ with $r = 0, 1, 2$ has center $Z \cong \mathbb{Z}_4$, and $\Gamma/\langle -\mathbb{1} \rangle \cong O'_6(\mathbb{R}, r)$ for $r = 0, 2$; furthermore $\mathrm{SU}_4(\mathbb{C}, 1)/\langle -\mathbb{1} \rangle \cong U_3^\alpha \mathbb{H}$.

	$r = 0, 2$			$r = 1$		
weight λ	λ_2	λ_1	$\lambda_1 + \lambda_3$	λ_1	λ_2	$\lambda_1 + \lambda_3$
dimension n	6	8	15	8	12	15
centralizer	$\mathbb{R}$	$\mathbb{C}$	$\mathbb{R}$	$\mathbb{C}$	$\mathbb{H}$	$\mathbb{R}$
kernel	$\langle -\mathbb{1} \rangle$	$\mathbb{1}$	Z	$\mathbb{1}$	$\langle -\mathbb{1} \rangle$	Z

$\Gamma = \mathrm{SL}_2\mathbb{H}$ has center $Z = \langle -\mathbb{1} \rangle \cong \mathbb{Z}_2$, and $\Gamma/Z \cong \mathrm{PSL}_2\mathbb{H} \cong O'_6(\mathbb{R}, 1)$.

weight λ	λ_2	λ_1	λ_3	$2\lambda_1$	$2\lambda_3$	$\lambda_1 + \lambda_3$
dimension n	6	8	8	10	10	15
centralizer	$\mathbb{R}$	$\mathbb{H}$	$\mathbb{H}$	$\mathbb{R}$	$\mathbb{R}$	$\mathbb{R}$
kernel	Z	$\mathbb{1}$	$\mathbb{1}$	Z	Z	Z

$\Gamma = \mathrm{SL}_4\mathbb{R}$ has center $Z = \langle -\mathbb{1} \rangle \cong \mathbb{Z}_2$, and $\Gamma/Z \cong \mathrm{PSL}_4\mathbb{R} \cong \mathrm{O}'_6(\mathbb{R}, 3)$.

weight λ	λ_1	λ_3	λ_2	$2\lambda_1$	$2\lambda_3$	$\lambda_1 + \lambda_3$
dimension n	4	4	6	10	10	15
centralizer	$\mathbb{R}$	$\mathbb{R}$	$\mathbb{R}$	$\mathbb{R}$	$\mathbb{R}$	$\mathbb{R}$
kernel	$\mathbb{1}$	$\mathbb{1}$	Z	Z	Z	Z

$\Gamma = \mathrm{SL}_3\mathbb{C}$ has center $Z \cong \mathbb{Z}_3$, and $\Gamma/Z \cong \mathrm{PSL}_3\mathbb{C}$.

weight λ	$(\lambda_1, 0)$	$(\lambda_2, 0)$	(λ_1, λ_1)	(λ_2, λ_2)	$(2\lambda_1, 0)$	$(2\lambda_2, 0)$	$(\lambda_1 + \lambda_2, 0)$
dimension n	6	6	9	9	12	12	16
centralizer	$\mathbb{C}$	$\mathbb{C}$	$\mathbb{R}$	$\mathbb{R}$	$\mathbb{C}$	$\mathbb{C}$	$\mathbb{C}$
kernel	$\mathbb{1}$	$\mathbb{1}$	Z	Z	$\mathbb{1}$	$\mathbb{1}$	Z

$\Gamma = \mathrm{Sp}_4\mathbb{C} \cong \mathrm{Spin}_5\mathbb{C}$ has center $Z \cong \mathbb{Z}_2$, and $\Gamma/Z \cong \mathrm{PSp}_4\mathbb{C} \cong \mathrm{SO}_5\mathbb{C}$.

weight λ	$(\lambda_2, 0)$	$(\lambda_1, 0)$	(λ_2, λ_2)
dimension n	8	10	16
centralizer	$\mathbb{C}$	$\mathbb{C}$	$\mathbb{R}$
kernel	$\mathbb{1}$	Z	Z

$\Gamma = \mathrm{Spin}_7(\mathbb{R}, r)$ with $r = 0, 1, 2, 3$ has center $Z \cong \mathbb{Z}_2$, and $\Gamma/Z \cong \mathrm{O}'_7(\mathbb{R}, r)$.

	$r = 0, 3$		$r = 1, 2$	
weight λ	λ_1	λ_3	λ_1	λ_3
dimension n	7	8	7	16
centralizer	$\mathbb{R}$	$\mathbb{R}$	$\mathbb{R}$	$\mathbb{H}$
kernel	Z	$\mathbb{1}$	Z	$\mathbb{1}$

$\Gamma = \mathrm{U}_3(\mathbb{H}, r)$ with $r = 0, 1$ has center $Z \cong \mathbb{Z}_2$, and $\Gamma/Z \cong \mathrm{PU}_3(\mathbb{H}, r)$.

weight λ	λ_1	λ_2
dimension n	12	14
centralizer	$\mathbb{H}$	$\mathbb{R}$
kernel	$\mathbb{1}$	Z

$\Gamma = \mathrm{Sp}_6\mathbb{R}$ has center $Z = \langle -\mathbb{1} \rangle \cong \mathbb{Z}_2$, and $\Gamma/Z \cong \mathrm{PSp}_6\mathbb{R}$.

weight λ	λ_1	λ_2	λ_3
dimension n	6	14	14
centralizer	$\mathbb{R}$	$\mathbb{R}$	$\mathbb{R}$
kernel	$\mathbb{1}$	Z	$\mathbb{1}$

$\Gamma = \mathrm{SU}_5(\mathbb{C}, r)$ has only one non-trivial irreducible representation of dimension at most 16, viz. the natural representation on $\mathbb{R}^{10} = \mathbb{C}^5$, corresponding to the weight $\lambda = \lambda_1$, with centralizer $\mathbb{C}$ and trivial kernel.

$\Gamma = \mathrm{SL}_5\mathbb{R}$ has trivial center; the centralizer is always $\mathbb{R}$, and all kernels are trivial.

weight λ	λ_1	λ_4	λ_2	λ_3	$2\lambda_1$	$2\lambda_4$
dimension n	5	5	10	10	15	15

$\Gamma = \mathrm{Spin}_8(\mathbb{R}, r)$ with $r = 0, 1, 2, 3, 4$ has center $Z = \langle z, z' \rangle \cong \mathbb{Z}_2 \times \mathbb{Z}_2$ if r is even, and $Z \cong \mathbb{Z}_2$ if r is odd. The image of the representation determined by λ_1 is the group $\mathrm{O}'_8(\mathbb{R}, r)$, and $\mathrm{Spin}_8(\mathbb{R}, 2)/\langle z \rangle \cong \mathrm{Spin}_8(\mathbb{R}, 2)/\langle z' \rangle \cong \mathrm{U}_4^\alpha \mathbb{H}$.

	$r = 0, 4$			$r = 1, 3$		$r = 2$		
weight λ	λ_1	λ_3	λ_4	λ_1	λ_3	λ_1	λ_3	λ_4
dimension n	8	8	8	8	16	8	16	16
centralizer	$\mathbb{R}$	$\mathbb{R}$	$\mathbb{R}$	$\mathbb{R}$	$\mathbb{C}$	$\mathbb{R}$	$\mathbb{H}$	$\mathbb{H}$
kernel	$\langle zz' \rangle$	$\langle z \rangle$	$\langle z' \rangle$	Z	$\mathbb{1}$	$\langle zz' \rangle$	$\langle z \rangle$	$\langle z' \rangle$

$\Gamma = \mathrm{G}_2^\mathbb{C}$ has only one non-trivial irreducible representation of dimension at most 16, viz. the natural representation on $\mathbb{R}^{14} = \mathbb{C}^7$, corresponding to the weight $\lambda = \lambda_1$, with centralizer $\mathbb{C}$ and trivial kernel.

$\Gamma = \mathrm{SL}_4\mathbb{C}$ has center $Z \cong \mathbb{Z}_4$, and $\Gamma/Z \cong \mathrm{PSL}_4\mathbb{C}$, $\Gamma/\langle -\mathbb{1} \rangle \cong \mathrm{SO}_6\mathbb{C}$.

weight λ	$(\lambda_1, 0)$	$(\lambda_3, 0)$	$(\lambda_2, 0)$	(λ_1, λ_1)	(λ_3, λ_3)
dimension n	8	8	12	16	16
centralizer	$\mathbb{C}$	$\mathbb{C}$	$\mathbb{C}$	$\mathbb{R}$	$\mathbb{R}$
kernel	$\mathbb{1}$	$\mathbb{1}$	$\langle -\mathbb{1} \rangle$	Z	Z

$\Gamma = \mathrm{SU}_6(\mathbb{C}, r)$ with $r = 0, 1, 2, 3$ has only one non-trivial irreducible representation of dimension at most 16, viz. the natural representation on $\mathbb{R}^{12} = \mathbb{C}^6$, corresponding to the weight $\lambda = \lambda_1$, with centralizer $\mathbb{C}$ and trivial kernel.

$\Gamma = \mathrm{SL}_3\mathbb{H}$ has center $Z \cong \mathbb{Z}_2$, and $\Gamma/Z \cong \mathrm{PSL}_3\mathbb{H}$.

weight λ	λ_1	λ_5	λ_2	λ_4
dimension n	12	12	15	15
centralizer	$\mathbb{H}$	$\mathbb{H}$	$\mathbb{R}$	$\mathbb{R}$
kernel	$\mathbb{1}$	$\mathbb{1}$	Z	Z

$\Gamma = \mathrm{SL}_6\mathbb{R}$ has center $Z \cong \mathbb{Z}_2$, and $\Gamma/Z \cong \mathrm{PSL}_6\mathbb{R}$.

weight λ	λ_1	λ_5	λ_2	λ_4
dimension n	6	6	15	15
centralizer	$\mathbb{R}$	$\mathbb{R}$	$\mathbb{R}$	$\mathbb{R}$
kernel	$\mathbb{1}$	$\mathbb{1}$	Z	Z

$\Gamma = \mathrm{Spin}_9(\mathbb{R}, r)$ with $r = 0, 1, 2, 3, 4$ has center $Z \cong \mathbb{Z}_2$, and $\Gamma/Z \cong \mathrm{O}'_9(\mathbb{R}, r)$. For $r = 2, 3$, the natural representation in dimension $n = 9$ is the only non-trivial irreducible representation of dimension $n \leq 16$; it has centralizer $\mathbb{R}$ and kernel Z.

For $r = 0, 1, 4$:

weight λ	λ_1	λ_4
dimension n	9	16
centralizer	$\mathbb{R}$	$\mathbb{R}$
kernel	Z	$\mathbb{1}$

See (18.8) for the representation of $\mathrm{Spin}_9(\mathbb{R}, 0)$ on $\mathbb{R}^{16} = \mathbb{O}^2$.

$\Gamma = \mathrm{U}_4(\mathbb{H}, r)$ with $r = 0, 1, 2$ has only one non-trivial irreducible representation of dimension $n \leq 16$, viz. the natural representation on $\mathbb{R}^{16} = \mathbb{H}^4$, corresponding to the weight $\lambda = \lambda_1$, with centralizer $\mathbb{H}$ and trivial kernel.

The groups $\Gamma = \mathrm{Sp}_n\mathbb{R}$ with $n = 8, 10, 12, 14, 16$ have only one non-trivial irreducible representation of dimension at most 16, viz. the natural representation of dimension n, corresponding to the weight $\lambda = \lambda_1$, with centralizer $\mathbb{R}$ and trivial kernel.

$\Gamma = \mathrm{Spin}_7\mathbb{C}$ has center $Z \cong \mathbb{Z}_2$, and $\Gamma/Z \cong \mathrm{SO}_7\mathbb{C}$.

weight λ	$(\lambda_1, 0)$	$(\lambda_3, 0)$
dimension n	14	16
centralizer	$\mathbb{C}$	$\mathbb{C}$
kernel	Z	$\mathbb{1}$

$\Gamma = \mathrm{Sp}_6\mathbb{C}$ has only one non-trivial irreducible representation of dimension at most 16, viz. the natural representation on $\mathbb{R}^{12} = \mathbb{C}^6$, corresponding to the weight $\lambda = \lambda_1$, with centralizer $\mathbb{C}$ and trivial kernel.

$\Gamma = \mathrm{Spin}_{10}(\mathbb{R}, r)$ with $r = 0, 1, 2, 3, 4, 5$ has center $Z \cong \mathbb{Z}_4$ if r is even, and $Z \cong \mathbb{Z}_2$ if r is odd.
For $r = 0, 2, 3, 4$ the group Γ has only one non-trivial irreducible representation of dimension $n \leq 16$, viz. the natural representation of dimension 10 as $\mathrm{O}'_{10}(\mathbb{R}, r)$, corresponding to the weight $\lambda = \lambda_1$, with centralizer $\mathbb{R}$ and kernel Z.

For $r = 1, 5$:

weight λ	λ_1	λ_4	λ_5
dimension n	10	16	16
centralizer	$\mathbb{R}$	$\mathbb{R}$	$\mathbb{R}$
kernel	Z	$\mathbb{1}$	$\mathbb{1}$

The representation of $\mathrm{Spin}_{10}(\mathbb{R}, 1)$ on $\mathbb{R}^{16} = \mathbb{O}^2$ is given by the stabilizer of o in the group $A_\mathbb{F}$ in (15.6).

The groups $\Gamma = \mathrm{SU}_7(\mathbb{C}, r)$ with $r = 0, 1, 2, 3$ and $\Gamma = \mathrm{SU}_8(\mathbb{C}, r)$ with $r = 0, 1, 2, 3, 4$ have only one non-trivial irreducible representation of dimension at most 16, viz. the natural representation on $\mathbb{R}^{14} = \mathbb{C}^7$ and $\mathbb{R}^{16} = \mathbb{C}^8$, respectively, corresponding to the weight $\lambda = \lambda_1$, with centralizer $\mathbb{C}$ and trivial kernel.

The groups $\Gamma = \mathrm{SL}_n\mathbb{R}$ with $7 \leq n \leq 16$ have only two non-trivial irreducible representations of dimension at most 16, viz. the natural representation of dimension n and its dual representation, corresponding to the weights $\lambda = \lambda_1, \lambda_{n-1}$, with centralizer $\mathbb{R}$ and trivial kernel.

The groups $\Gamma = \mathrm{SL}_n\mathbb{C}$ with $5 \leq n \leq 8$ have only two non-trivial irreducible representations of dimension at most 16, viz. the natural representation on $\mathbb{R}^{2n} = \mathbb{C}^n$ and its dual representation, corresponding to the weights $\lambda = (\lambda_1, 0), (\lambda_{n-1}, 0)$, with centralizer $\mathbb{C}$ and trivial kernel.

The groups $\Gamma = \mathrm{Spin}_n(\mathbb{R}, r)$ with $11 \leq n \leq 16$ and $0 \leq 2r \leq n$ have only one non-trivial irreducible representation of dimension at most 16, viz. the natural representation of dimension n as $\mathrm{O}'_n(\mathbb{R}, r)$, corresponding to the weight $\lambda = \lambda_1$, with centralizer $\mathbb{R}$ and kernel of order 2.

$\Gamma = \mathrm{Spin}_8\mathbb{C}$ has center $Z = \langle z, z' \rangle \cong \mathbb{Z}_2 \times \mathbb{Z}_2$, and $\Gamma/\langle zz' \rangle \cong \mathrm{SO}_8\mathbb{C}$.

weight λ	$(\lambda_1, 0)$	$(\lambda_3, 0)$	$(\lambda_4, 0)$
dimension n	16	16	16
centralizer	$\mathbb{C}$	$\mathbb{C}$	$\mathbb{C}$
kernel	$\langle zz' \rangle$	$\langle z \rangle$	$\langle z' \rangle$

$\Gamma = \mathrm{SL}_4\mathbb{H}$ has only two non-trivial irreducible representations of dimension at most 16, viz. the natural representation on $\mathbb{R}^{16} = \mathbb{H}^4$ and its dual representation, corresponding to the weights $\lambda = \lambda_1, \lambda_7$, with centralizer $\mathbb{H}$ and trivial kernel.

$\Gamma = \mathrm{Sp}_8\mathbb{C}$ has only one non-trivial irreducible representation of dimension at most 16, viz. the natural representation on $\mathbb{R}^{16} = \mathbb{C}^8$, corresponding to the weight $\lambda = \lambda_1$, with centralizer $\mathbb{C}$ and trivial kernel.

95.11 Quasi-equivalence. Two representations $\varrho_i : \Gamma \to \mathrm{GL}_n\mathbb{R}$, $i = 1, 2$, are said to be quasi-equivalent if ϱ_1 is equivalent to $\alpha\varrho_2$ for some automorphism α of the Lie group Γ. Then the two transformation groups $(\Gamma, \mathbb{R}^n)$ defined by ϱ_1 and ϱ_2 are equivalent in the sense of (96.1).

For example, the natural representations of the groups $\mathrm{SL}_n\mathbb{R}, \mathrm{SL}_n\mathbb{C}, \mathrm{SL}_n\mathbb{H}$ are quasi-equivalent to their dual representations. The triality automorphism, compare (17.16), shows that various representations of $\mathrm{Spin}_8\mathbb{C}$ and of $\mathrm{Spin}_8(\mathbb{R}, r)$, with $r \in \{0, 4\}$, are quasi-equivalent.

Note that quasi-equivalent representations ϱ_i define isomorphic semidirect products $\mathbb{R}^n \rtimes \Gamma$.

95.12 Large compact subgroups of $\mathrm{GL}_8\mathbb{R}$. *Let Γ be a compact, connected subgroup of* $\mathrm{GL}_8\mathbb{R}$ *with* $\dim \Gamma \geq 14$. *Then Γ is conjugate in* $\mathrm{GL}_8\mathbb{R}$ *to one of the following groups (each acting naturally as indicated):*

(a) $\mathrm{SO}_8\mathbb{R}$ *of dimension* 28.
(b) $\mathrm{SO}_7\mathbb{R}$ *acting on* $\mathbb{R}^7 \oplus \mathbb{R}$ *or* $\mathrm{Spin}_7\mathbb{R}$ *acting irreducibly on* $\mathbb{R}^8$, *compare the action of* $\mathrm{S}\nabla_{(1)}$ *on Y in* (17.14); *these groups have dimension* 21.
(c) $\mathrm{SU}_4\mathbb{C}$ *or* $\mathrm{U}_4\mathbb{C}$, *acting irreducibly on* $\mathbb{C}^4$; *these groups have dimensions* 15 *and* 16, *respectively.*
(d) $\mathrm{SO}_6\mathbb{R}$ *or* $\mathrm{SO}_6\mathbb{R} \times \mathrm{SO}_2\mathbb{R}$, *acting on* $\mathbb{R}^6 \oplus \mathbb{R}^2$; *these groups have dimensions* 15 *and* 16, *respectively.*
(e) G_2 *of dimension* 14, *acting on* $\mathbb{R}^7 \oplus \mathbb{R}$, *compare* (11.33).

This result can be proved using the structure theory of compact Lie groups (94.31c), together with the information provided by (95.3, 4 and 6) and the tables (95.10). Compare also Hähl [78] 2.8 for a similar enumeration of all groups Γ with $\dim \Gamma \geq 10$.

Note that inclusions and non-inclusions in the list above can be inferred from the invariant subspaces.

96 Transformation groups

Several crucial arguments in the theory of topological geometries use the fact that the automorphism group acts as a topological transformation group on the point set. An introduction to the theory of transformation groups is given in Kawakubo [91]. Most of the deeper results on topological transformation groups concern compact groups, and the books of Palais [60], Bredon [72], tom Dieck [87] and Onishchik [94] concentrate on these. Many papers on non-compact groups deal with smooth actions, as did Lie's original treatise. All transitive actions on surfaces are known (Mostow [50]), and so are the locally compact doubly transitive groups on manifolds (Tits [55, 57]) and the compact transitive groups on spheres. In the following, we present a few rather unrelated results which have been useful in topological geometry.

96.1 Definition. (Γ, M) is a *(topological) transformation group* if Γ is a topological group, M is a topological space, and $(x, \gamma) \mapsto x^\gamma : M \times \Gamma \to M$ is a continuous action, i.e., a continuous map satisfying $x^{(\alpha\beta)} = (x^\alpha)^\beta$ and $x^\mathbb{1} = x$. In particular, each $\gamma \in \Gamma$ induces a homeomorphism of M. The kernel $\Gamma_{[M]}$ is defined as in Section 91. If $\dim \Gamma_{[M]} = 0$, then Γ is said to be almost effective. Two transformation groups (Γ, M) and (Δ, N) are *equivalent* if there is a homeomorphism $\eta : M \approx N$ and an isomorphism $\sigma : \Gamma \cong \Delta$ of topological groups such that $y^{\gamma^{\sigma}} = y^{\eta^{-1}\gamma\eta}$ for all $y \in N$ and all $\gamma \in \Gamma$.

96.2 Remarks. (a) If Δ is a closed subgroup of the topological group Γ, then Γ acts on the coset space Γ/Δ (taken with the quotient topology) by translation $\Delta\xi \mapsto \Delta\xi\gamma$, and $(\Gamma, \Gamma/\Delta)$ is a topological transformation group.

(b) Let (Γ, M) be a topological transformation group. The *orbit space* M/Γ is by definition the set $\{ x^{\Gamma} \mid x \in M \}$ taken with the quotient topology. Note that the canonical map $M \to M/\Gamma$ is open. If Θ is a normal subgroup of Γ, then $(\Gamma, M/\Theta)$ is again a topological transformation group. (Use the fact that $M \to M/\Theta$ is open.)

(c) If $A^{\Gamma} = A \subseteq M$, then $(\Gamma/\Gamma_{[A]}, A)$ is also a topological transformation group. (Note that the canonical map $\Gamma \to \Gamma/\Gamma_{[A]}$ is open (93.16), and apply the universal property of quotient maps.)

96.3 Definition. Let Γ be a group acting on the topological space M so that each $\gamma \in \Gamma$ induces a homeomorphism of M. The *compact-open topology* (also called c-topology) on Γ is generated by the subbasis consisting of all sets $\Omega_{C,U} = \{ \gamma \in \Gamma \mid C^{\gamma} \subseteq U \}$, where C is compact and U is open in M.

96.4 Proposition. *Let M be a locally compact space with a countable basis, and let Γ be endowed with the compact-open topology. Then Γ has a countable basis. Moreover, a sequence γ_{ν} converges to γ if, and only if, $x_{\nu}^{\gamma_{\nu}} \to x^{\gamma}$ whenever $x_{\nu} \to x$. Consequently, the compact-open topology is the coarsest topology on Γ for which the action $M \times \Gamma \to M$ is continuous.*

Proof. Dugundji [66] XII Th. 5.2 and Section 7. □

96.5 Proposition. *If M is locally compact, and if Γ carries the compact-open topology, then the action of Γ on M and the group multiplication in Γ are continuous. Hence (Γ, M) is a topological transformation group if, and only if, the inversion map $\gamma \mapsto \gamma^{-1}$ of Γ is continuous.*

96.6 Proposition. *If M is a compact metric space, then the compact-open topology on Γ coincides with the topology of uniform convergence defined by the supremum metric on Γ. Hence Γ is a topological transformation group by (96.5).*

The *proof* is a straightforward application of compactness, cp. Dugundji [66] XII Sect. 8. □

Less immediate is the criterion of Arens [46] (see also Bourbaki [66b] X §3 no. 5 and Exerc. 17a):

96.7 Theorem. *If Γ is a group of homeomorphisms of a locally compact, locally connected Hausdorff space M, and if Γ is given the compact-open topology, then inversion is continuous, and (Γ, M) is a topological transformation group (96.5).*

The following useful proposition is due to Freudenthal [36] (see also Hohti [87]):

96.8 Open action. *Let (Γ, M) be a transitive transformation group and $a \in M$. If Γ and M are locally compact and Γ is a Lindelöf space (in particular if Γ is connected or if Γ has a countable basis), then $\gamma \mapsto a^\gamma : \Gamma \to M$ is an open map.*

Sketch of proof. If Φ is any compact symmetric neighbourhood of the identity in Γ, then Γ is a countable union of sets $\Phi\gamma$. Let A° denote the interior of A in M. By Baire's theorem (Dugundji [66] XI 10.1-3 or Bourbaki [66b] IX §5 no. 3 Th. 1(i) p. 193), some $(a^{\Phi\gamma})^\circ \neq \emptyset$. Hence there is $\varphi \in \Phi$ with $a^\varphi \in (a^\Phi)^\circ$ and consequently $a \in (a^{\Phi^2})^\circ$. For an open subset Ω of Γ and $\omega \in \Omega$, choose Φ so that $\Phi^2\omega \subseteq \Omega$; then $a^\omega \in (a^{\Phi^2\omega})^\circ \subseteq a^\Omega$, and a^Ω is open in M. □

96.9 Corollary.

(a) *Under the same hypotheses, $\gamma \mapsto a^\gamma$ induces a homeomorphism $\Gamma/\Gamma_a \approx M$, and (Γ, M) is equivalent to $(\Gamma, \Gamma/\Gamma_a)$ as a topological transformation group.*
(b) *If, moreover, M is connected and if the connected component Γ^1 is open in Γ (in particular, if Γ is a Lie group), then Γ^1 is also transitive on M.*
(c) *If Γ is connected and M is simply connected, then Γ_a is connected (94.4a).*
(d) *If $\varphi : \Gamma \to \Delta$ is a continuous surjective homomorphism between locally compact groups, and if Γ is a Lindelöf space, then φ is open.*

Proof of (d). Apply (96.8) to the action $(\delta, \gamma) \mapsto \delta\gamma^\varphi$ and put $\delta = 1$. □

Part (a) has the obvious consequence $\dim a^\Gamma = \dim \Gamma/\Gamma_a$. This holds even without a^Γ being locally compact (Halder [71]):

96.10 Dimension formula. *If the locally compact Lindelöf group Γ acts continuously on the regular space M and if M has a countable basis (i.e., if M is separable metric), then $\dim \Gamma = \dim \Gamma_a + \dim a^\Gamma$ whenever the orbit $a^\Gamma \subseteq M$ has finite dimension.*

Proof. Again $\Gamma = \bigcup_{\nu \in \mathbb{N}} \Phi\gamma_\nu$ with a compact neighbourhood Φ of the identity. The map $\gamma \mapsto a^\gamma$ induces a continuous bijection $\Gamma/\Gamma_a \to a^\Gamma$ and a homeomorphism $\Phi/\Gamma_a \approx a^\Phi$. The sum theorem (92.9) applied to Γ/Γ_a and to a^Γ gives $\dim \Gamma/\Gamma_a = \dim a^\Phi = \dim a^\Gamma$. Now use (93.7). □

96.11 Corollary. *Assume, moreover, that M is locally compact, locally homogeneous, and locally contractible as in* (92.16).

(a) *If* $\dim a^\Gamma = \dim M < \infty$, *then* a^Γ *is open in M. If, in addition, Γ is compact and M is connected, then* $a^\Gamma = M$.

(b) *If Γ is transitive on M, and if $\dim M < \infty$ and M is connected, then Γ^1 is transitive.*

Proof. (a) Let Φ be a compact neighbourhood in Γ. By (92.16) and the last proof, the interior of a^Φ is non-empty. Because Γ is transitive on the orbit, a^Γ has only interior points.

(b) The dimension formula together with (93.6) shows $\dim a^{\Gamma^1} = \dim a^\Gamma$. Hence each orbit of Γ^1 is open by (a). □

96.12 Theorem. *Let (Γ, M) be a transitive transformation group, and assume that Γ and M are locally compact. If $\dim \Gamma < \infty$, then there is an exact homotopy sequence*

$$\pi_k \Gamma_a \to \pi_k \Gamma \to \pi_k M \to \pi_{k-1} \Gamma_a \to \ldots \to \pi_0 \Gamma \to \pi_0 M.$$

The *proof* consists of 4 parts: (a) The canonical map $\kappa : \Gamma \to \Gamma/\Gamma_a \approx M$ admits local cross sections, see Nagami [63] theorem and footnote, cp. also Skljarenko [63] Th. 13.

(b) If $\varrho : \Gamma \to M$ has local cross sections, then ϱ determines a locally trivial bundle with fiber Γ_a. This follows directly from the definitions, cp. Steenrod [51] 7.4 Cor.

(c) A locally trivial bundle is a Serre fibration by Gray [75] Th.11.4, cp. also Hu [59] III Th.4.1 or Rotman [88] Th.11.52.

(d) If $\varrho : \Gamma \to M$ is a Serre fibration with fiber $F = a\varrho^{-1}$, then the corresponding homotopy sequence is exact, Hu [59] V Sect. 6, Spanier [66] 7, Sect. 2, Th. 10, Gray [75] Cor. 11.9, or Rotman [88] Th. 11.51. □

A torus $\mathbb{T}^k$ of arbitrarily large dimension k can act effectively on a compact, connected 2-dimensional non-manifold or on a disconnected 1-manifold (see Mann [67]). The following is true, however.

96.13 Dimensions of compact transformation groups.
Let (Γ, M) be an effective transformation group. Assume that Γ is a compact group of dimension k and that M is a connected space of covering dimension n.

(a) *If M and Γ have countable bases, and if Γ is transitive on M, then $k \leq \binom{n+1}{2}$.*

(b) *If M is a manifold and Γ is not transitive, then $k \leq \binom{n}{2}$.*
(c) *If M is a manifold, and if the connected component of Γ is not a Lie group, then $n \geq 4$ and $k \leq \binom{n-3}{2} + 1$.*

Note that $\binom{n+1}{2}$ is the dimension of the orthogonal group $\mathrm{SO}_{n+1}\mathbb{R}$ acting on $\mathbb{S}_n$.

Proof. For (a) and (c) see Mann [67] Cor. of Th. 1 and Th. 7 together with Th. 8 and Bredon [58] Th. 8. In case (b), we may assume by (93.6) and (c) that Γ is a connected Lie group. Since M is connected, (96.11) implies $\dim x^\Gamma \leq n - 1$ for each orbit. The assertion follows now from Montgomery–Zippin [55] Sect. 6.3 Th. 2 on p.246. □

A locally compact group which acts effectively on a manifold is known to be a Lie group only under additional assumptions. An important case is

96.14 Szenthe's theorem. *Assume that (Γ, M) is a transitive effective transformation group, and that both Γ and M are locally compact. If Γ has a countable basis and if M is connected and locally contractible, then Γ is a Lie group (and hence M is a manifold).*

See Szenthe [74] Th. 4. Actually, Szenthe states his theorem under the hypotheses that Γ is σ-compact (equivalently, a Lindelöf space) and that Γ/Γ^1 is compact. The second condition is dispensable by (93.8a) and (96.8). As he uses approximation by a well-ordered chain of Lie groups, it is not quite clear that σ-compactness suffices. If Γ has a countable basis, however, then some open subgroup can be approximated even by a *sequence* of Lie groups. □

96.15 Doubly transitive groups. *Let (Γ, M) be a doubly transitive effective transformation group, and assume that Γ is a locally compact Lindelöf space. If $\dim \Gamma/\Gamma_a > 1$ for $a \in M$, then Γ is a Lie group, and Γ^1 is also doubly transitive on M. Hence M is connected.*

Proof. 1) The locally compact coset space $X = \Gamma/\Gamma_a$ contains a connected subset consisting of more than one point by (92.18), and X is connected because Γ is doubly transitive on X. Moreover, $\operatorname{ind} X > 1$ by (93.7). Since $\operatorname{ind} X$ is determined locally, there is also a compact subset C of X with $\operatorname{ind} C > 1$.

2) The set Y of pairs of distinct points of X is also connected: otherwise there would exist a continuous bijection $\iota : X \to K$ onto a linearly ordered space K, see Eilenberg [41] Th. I. Obviously, $\operatorname{ind} K \leq 1 < \operatorname{ind} C^\iota$. This contradicts (92.4).

3) The action of Γ on Y is open by (96.8), and each open subgroup Δ of Γ is transitive on Y and hence doubly transitive on X and on M.

4) The approximation theorem (93.8) gives an open subgroup Δ with an arbitrarily small normal subgroup Θ such that Δ/Θ is a Lie group. If Θ is sufficiently small, then $a^\Theta \neq M$, and (91.4) shows $\Theta = \mathbb{1}$, so Γ is also a Lie group (93.3). □

Remark. In the case $\dim \Gamma/\Gamma_a = 1$ the space Y may have two connected components (Eilenberg [41]). Hence Γ is a Lie group by the same arguments, but Γ^1 is not doubly transitive if $\Gamma/\Gamma_a \approx \mathbb{R}$, see also (96.30).

(96.15) is the basis for a complete classification of all locally compact doubly transitive transformation groups given by Tits [55] p. 222 ff, see also Tits [57].

96.16 Theorem. *If (Γ, M) satisfies the assumptions of (96.15), and if Γ is connected, then either*

(a) Γ *is simple, and M is compact, see (96.17), or*
(b) Γ *has a sharply transitive normal subgroup $\Theta \cong \mathbb{R}^n$, and Γ_a acts transitively on $\mathbb{R}^n \setminus \{0\}$ and, hence, on the sphere $\mathbb{S}_{n-1}$ consisting of the rays in $\mathbb{R}^n$, see (96.21 and 22).*
(c) *If Γ is even triply transitive on M, then M is homeomorphic to a sphere, see (96.18).*

For the case $\dim M = 1$ see (96.30).

Sketch of Proof. Note that Γ has center $Z = \mathbb{1}$ since x is the only fixed point of Γ_x in M.

Case (a) Γ is semi-simple. Then Γ is a direct product of simple Lie groups (94.23). If there is more than one factor, then each factor A is sharply transitive by (91.4 and 1c). In fact, Γ is then equivalent to the group of maps $\xi \mapsto \alpha^{-1}\xi\beta$ of A, but this group is not doubly transitive. Therefore Γ is simple. The stabilizer Γ_a is a maximal subgroup of Γ by (91.4a), and double transitivity implies $2 \dim \Gamma_a \geq \dim \Gamma$. All such subgroups are known, see Dynkin [52ab, 57ab] and Tits [55] p. 149/50 for the complex case, Tits [55] p. 160/61 for the classical real Lie groups, Chen [73] for the hyperbolic motion group of type F_4, Seitz [87, 91] for algebraic groups over algebraically closed commutative fields, and Komrakov [90] for maximal real subgroups with non-maximal complexification. Inspection shows that either Γ/Γ_a is compact or Γ cannot be doubly transitive, compare also Mostow [61].

Case (b) Γ is not semi-simple. (94.26) gives a minimal commutative normal subgroup Θ which is transitive on M by (91.4). Moreover, Θ is a vector group. Because Γ_a is transitive on $\Theta \setminus \mathbb{1}$, a maximal compact subgroup of Γ_a is a transitive subgroup of $SO_n\mathbb{R}$ up to conjugation, see (96.19). All possibilities for Γ'_a are listed in Völklein [81].

(c) is an immediate consequence of (a) and (b) since Γ_a is doubly transitive on $M \setminus \{a\}$, which is not compact. □

The groups of (96.16a) are exactly the projective groups in their natural actions and the hyperbolic groups acting on the absolute points.

96.17 Theorem. *If (Γ, M) is a doubly transitive transformation group, and if Γ is a simple Lie group, then*

(a) (Γ, M) *is equivalent to* $\mathrm{PSL}_n \mathbb{F}$ *on* $\mathrm{P}_{n-1}\mathbb{F}$ *with $n \geq 2$ and $\mathbb{F} \in \{\mathbb{R}, \mathbb{C}, \mathbb{H}\}$, or Γ is the automorphism group* $\mathrm{E}_6(-26)$ *of the projective octonion plane, acting on the point space, or*

(b) (Γ, M) *is equivalent to* $\mathrm{PO}'_{n+1}(\mathbb{R}, 1)$ *on* $\mathbb{S}_{n-1}$, $\mathrm{PSU}_{n+1}(\mathbb{C}, 1)$ *on* $\mathbb{S}_{2n-1}$, *or* $\mathrm{PU}_{n+1}(\mathbb{H}, 1)$ *on* $\mathbb{S}_{4n-1}$ *with $n \geq 2$, or to* $\mathrm{F}_4(-20)$ *on* $\mathbb{S}_{15}$.

Note that the groups $\mathrm{PSL}_2\mathbb{F}$ are equivalent to $\mathrm{PO}'_{n+2}(\mathbb{R}, 1)$, where $n = \dim \mathbb{F}$, cp. (15.6). The proof of the theorem is due to Tits [55] IV.F, see also the resumé in Tits [57]. The group $\mathrm{F}_4(-20)$ is discussed in more detail by Chen [73], its action is described in (18.23 and 26). □

96.18 Corollary. *The real hyperbolic groups* $\mathrm{PO}'_{n+1}(\mathbb{R}, 1)$ *with $n > 2$ in their standard actions on* $\mathbb{S}_{n-1}$ *are the only triply transitive locally compact, connected groups.*

96.19 Theorem. *Let (Γ, M) be a transitive transformation group. If Γ is a connected Lie group and M is a compact, connected manifold with finite fundamental group, then any maximal compact subgroup Φ of Γ is transitive on M.*

Proof. 1) Assume first that M is simply connected. By (93.10a), for some $a \in M$ the stabilizer Φ_a is a maximal compact subgroup of Γ_a. Moreover, since Γ_a is connected by (96.9c), we have homotopy equivalences $\Phi \simeq \Gamma$ and $\Phi_a \simeq \Gamma_a$, see (93.10b). From (96.12) and the 5-lemma (Spanier [66] 4.5.11) it follows that the inclusion map η of $X = a^\Phi$ into M induces isomorphisms $\eta_\sharp : \pi_q X \cong \pi_q M$ between the homotopy groups. The space $X \approx \Phi/\Phi_a$ is also a compact manifold (94.3a). Now the Whitehead theorem (Spanier [66] p. 399) shows that the simply connected, compact manifolds M and X have the same integral singular homology. Any simply connected manifold is orientable, see Greenberg [67] (22.15) or Spanier [66] 6.2.10, p. 294. If M is a compact, connected, orientable n-manifold, then $H_q(M) = 0$ for $q > n$ and $H_n(M) \cong \mathbb{Z}$, see Greenberg [67] Th. (22.24 and 28). Hence $\dim X = \dim M$, and X is open in M by (96.11). Since X is compact and M is connected, we have $X = M$, and the theorem is proved in the case $\pi_1 M = \mathbb{1}$.

2) We may replace Γ by its universal covering group, cf. Bredon [72] Chap. I, 9.1. Let $\Delta = (\Gamma_a)^1$ and consider the canonical covering map $\Gamma/\Delta \to \Gamma/\Gamma_a$. We have $\Gamma/\Gamma_a \approx M$ by (96.9a). Because of (96.12) and the hypothesis, $\Gamma_a/\Delta \cong \pi_1 M$ is finite, whence Γ/Δ is compact, and Γ/Δ is simply connected by (94.4b). Step 1) shows that Φ is transitive on Γ/Δ. Hence $\Gamma = \Delta\Phi = \Gamma_a\Phi$, and Φ is also transitive on M. □

For the original proof see Montgomery [50] or Montgomery–Zippin [55] p. 226.

96.20 Theorem. *If a compact, connected group Γ acts effectively and transitively on a sphere $\mathbb{S}_{n-1}$, then Γ is equivalent to a subgroup of $SO_n\mathbb{R}$.*

In other words, the action of Γ is equivalent to a linear one. Based on the following theorem, this is proved in Poncet [59] Th. a. □

96.21 Theorem. *Let Γ be a compact, connected group which acts effectively and transitively on $\mathbb{S}_{n-1}$. If $n \geq 3$, then $\Gamma = \Lambda\Phi$ is a product of two compact, connected subgroups such that Φ is almost simple and transitive on $\mathbb{S}_{n-1}$, and $\Lambda \leq \mathrm{Cs}\,\Phi$. Moreover, Λ is a subgroup of $U_1\mathbb{H}$. If n is odd, then Γ is simple.*

This has been *proved* in Montgomery–Samelson [43]. In the situation of (96.16b), linearity of the action is already known. If Γ is assumed to act linearly on $\mathbb{S}_{n-1}$, the assertions of (96.21) can be obtained in the following way: either Γ itself is almost simple, or Γ is a product of compact, connected subgroups Λ and Φ such that $\Lambda \cap \Phi$ is discrete and $[\Lambda, \Phi] = \mathbb{1}$, see (94.31c). Transitivity of Γ on $\mathbb{S}_{n-1}$ implies that one of the factors, say Φ, acts irreducibly on $\mathbb{R}^n$: in fact, let $U^\Lambda = U$ and $V^\Phi = V$ be proper minimal invariant subspaces of $\mathbb{R}^n$. Each subspace U^γ is also Λ-invariant, thus we may assume $U \cap V \neq 0$. From (95.5) it follows that $\dim U \leq \frac{n}{2}$ and $\dim V \leq \frac{n}{2}$. Now let $0 \neq s \in U \cap V$; then $n - 1 = \dim \Gamma/\Gamma_s \leq \dim \Lambda/\Lambda_s + \dim \Phi/\Phi_s \leq 2(\frac{n}{2} - 1)$, a contradiction. By Schur's lemma (95.4), $\mathbb{R}^n$ may be identified with $\mathbb{F}^k$ where $\mathbb{F} \in \{\mathbb{R}, \mathbb{C}, \mathbb{H}\}$ such that $\Phi \leq U_k\mathbb{F}$ and $n = k \cdot \dim \mathbb{F}$, and Λ is a compact subgroup of $\mathbb{F}^\times$. In particular, $\mathrm{rk}\,\Lambda \leq 1$. The same is true for any possible decomposition of Γ. It follows easily that Φ is almost simple. Moreover, Φ is transitive on the projective space $P_{k-1}\mathbb{F}$. One can show that $\dim s^\Phi = n-1$ for $0 \neq s \in \mathbb{F}^k$. Hence Φ is transitive on $\mathbb{S}_{n-1}$ by (96.11). Finally, if n is odd, then $\mathbb{F} = \mathbb{R}$, and any element ζ in the center of $\Gamma = \Phi$ has an eigenvalue ± 1. Transitivity of Γ implies $\zeta = \pm \mathbb{1}$, and $\det \zeta = 1$ shows that $\zeta = \mathbb{1}$. Hence Γ is simple. □

The groups of (96.21) are known explicitly:

96.22 Theorem. *The almost simple, compact, connected transitive transformation groups on spheres are exactly*
(a) *the groups* $SO_n\mathbb{R}$ *with* $n \neq 2, 4$, *and the groups* $SU_m\mathbb{C}$ *and* $U_k\mathbb{H}$, *all in their natural actions, and*
(b) *three groups contained in the elliptic motion group* F_4 *of the octonion plane, namely the point stabilizer* $Spin_9$ *on* $\mathbb{S}_{15}$ *and its subgroups* $Spin_7$ *on* $\mathbb{S}_7$ *and* G_2 *on* $\mathbb{S}_6$, *see* (18.12), (17.14), *and* (11.30 *and* 33).

This is due to Borel [49], [50]. Linearity being known, it can be proved by methods of representation theory (Tits [55] IV.C, Freudenthal [56], cp. also Freudenthal [64] Sect. 12). For a proof mainly using homotopy see Onishchik [63] Th. 6a. In the cases which are relevant in this book the proof is easy:

If a compact almost simple group Γ is transitive on $\mathbb{S}_{n-1}$, then the exact homotopy sequence (96.12) shows $\pi_k \Gamma_x \cong \pi_k \Gamma$ for $k \leq n-3$. The cases $n < 5$ are covered by (96.34) below. For $n \geq 5$ the stabilizer Γ_x is semi-simple by (94.31c), for $n > 5$ even almost simple by (94.36). Note that Γ is simple if n is odd (96.21). Up to $n \leq 11$, the condition $\pi_1 \Gamma \cong \pi_1 \Gamma_x$ excludes all proper subgroups of $SO_n\mathbb{R}$ in the list (94.33) other than those of the theorem. In each case, Γ_x is in fact uniquely determined up to conjugation. This suffices for our purposes. Making use also of π_4 and π_5 in a few cases (see Bott [70], Mimura [67]), the argument can be extended up to and beyond $n = 16$. □

96.23 Corollary. *The only compact, connected, effective, and transitive transformation groups on even-dimensional spheres are* $SO_{2k+1}\mathbb{R}$ *on* $\mathbb{S}_{2k}$ *and* G_2 *on* $\mathbb{S}_6$.

96.24 Theorem. *If a locally compact, connected group* Γ *acts effectively on a connected n-dimensional manifold* M, *and if* $\dim a^\Phi \geq n - 2$ *for a compact subgroup* Φ *of* Γ *and some* $a \in M$, *then* Γ *is a Lie group.*

Proof. By (93.10 and 3) it suffices that Φ is a Lie group. This follows from Bredon [58] Th. 10 and 11. □

96.25 Proposition. *Assume that the connected Lie group* Γ *acts on a space* M. *If* Φ *is a compact subgroup of* Γ *and if* $c \in a^\Gamma \setminus a^\Phi$ *for some point* $a \in M$, *then* $\dim c^\Phi < \dim c^\Gamma$. *If, in particular,* M *is an n-manifold and* $\dim c^\Phi = n - 1$, *then* a^Γ *is open in* M.

Proof. Consider the canonical projection κ from Γ onto the connected manifold $B = \Gamma/\Gamma_c$ and the injective continuous mapping μ from B onto $c^\Gamma \subseteq M$. Compactness of Φ implies $\Phi^\kappa \approx c^\Phi$, and $\dim B \leq \dim c^\Gamma$ because B has compact subsets of the full dimension. If $\dim \Phi^\kappa = \dim B$, then the compact orbit Φ^κ is open in the connected manifold B by (92.14 or 16), and $B = \Phi^\kappa$, see also (96.11). Applying μ, we obtain that $a \in c^\Gamma = c^\Phi$ contradicting the assumption $c \notin a^\Phi$. □

The following two useful results have first been proved for groups of order 2 by Newman [31].

96.26 Theorem. *If a compact Lie group acts effectively on a connected manifold, then the set of fixed points has empty interior.*

96.27 Theorem. *Let M be a connected metric manifold. Then there exists some $\varepsilon > 0$ such that any compact Lie group (in particular, any finite group) which acts non-trivially on M has an orbit of diameter $> \varepsilon$.*

A *proof* of these two theorems is given in Bredon [72] Th. 9.5 and Cor. 9.6, cp. Dress [69]. □

96.28 Torus actions. *If a torus group Γ acts effectively on a compact n-manifold M, then there is a point $a \in M$ with $\Gamma_a = 1\!\!1$.*

Proof. Suppose the theorem is false. According to Mostow [57], there are only finitely many different stabilizers Γ_x. Hence there is a finite set $\Delta \subseteq \Gamma \setminus 1\!\!1$ such that M is covered by the fixed point sets F_δ with $\delta \in \Delta$. Each F_δ has empty interior by (96.26). This contradicts Baire's Theorem, cf. Dugundji [66] XI 10.1-3 or Bourbaki [66b] IX §5 no. 3 Th. 1(i) p. 193. □

A few facts of a more special nature will conclude this section.

96.29 Connected groups on 1-manifolds. *Consider a connected group Γ acting effectively on $\mathbb{R}$ or $\mathbb{S}_1$ (the only connected 1-manifolds).*

(a) *If Γ has no fixed point, then Γ is transitive.*
(b) *Any non-trivial compact subgroup of Γ acts freely on $\mathbb{S}_1$; it cannot act on $\mathbb{R}$.*
(c) *If Γ is compact, then $\Gamma = 1\!\!1$, or $\Gamma \cong \mathrm{SO}_2\mathbb{R}$ and Γ is sharply transitive on $\mathbb{S}_1$.*

Proof. (a) Each orbit is connected and homogeneous, hence open.

(b) Being connected, Γ preserves the orientation of the manifold. Any compact orbit in $\mathbb{R}$ coincides with its endpoints.

(c) If $\Gamma \neq 1\!\!1$, then $\Gamma \approx \mathbb{S}_1$ by (b) and (a), and (94.39) gives the assertion. □

96.30 Brouwer's theorem. *Let Γ be a locally compact, connected, effective and transitive transformation group on a connected 1-manifold M. Then $\dim \Gamma \leq 3$.*

(a) *If $M \approx \mathbb{S}_1$, then $\Gamma \cong \mathrm{SO}_2\mathbb{R}$, or Γ is a finite covering group of $\Omega = \mathrm{PSL}_2\mathbb{R}$.*
(b) *If $M \approx \mathbb{R}$, then $\Gamma \cong \mathbb{R}$, or Γ is equivalent to the connected component L_2 of the affine group of $\mathbb{R}$, or Γ is the simply connected covering group $\widetilde{\Omega}$ of Ω.*

In each case, the action of Γ is equivalent to the ordinary one in the strong sense that the conjugacy class of the stabilizers is uniquely determined.

This theorem has first been *proved* by Brouwer [09]. It follows easily from (96.29) and some general results:

1) Γ is a Lie group by (93.10) and (96.29c), compare also Montgomery–Zippin [55] p. 233.

2) If Γ is not semi-simple, then Γ has a minimal commutative connected normal subgroup N by (94.26). Because of (91.1d) and (96.29a), the group N is transitive on M, and (91.1c) shows that the centralizer of N is sharply transitive. Hence $\operatorname{Cs} \mathsf{N} = \mathsf{N}$ and $\Gamma/\mathsf{N} \leq \operatorname{Aut} \mathsf{N}$.

3) If Γ is semi-simple, the same arguments show that any almost simple factor of Γ is transitive and that its centralizer acts freely on M. Hence Γ is almost simple. A maximal compact subgroup of Γ is isomorphic to $\mathrm{SO}_2\mathbb{R}$, or $\Gamma \cong \widetilde{\Omega}$.

A different proof determines all Lie algebras with a subalgebra of codimension 1 (Hofmann [65], [90], Hofmann–Mostert [68], Poguntke [92]). □

Special features of the topology of *surfaces* (= connected 2-manifolds) imply

96.31 Theorem. *If (Γ, M) is a locally compact effective transformation group and M is a surface, then Γ is a Lie group.*

Proof. For connected groups this follows from (96.24). In the general case, $\dim \Gamma$ is finite by (93.6), (93.10b), and (96.13). Hence the approximation theorem (93.8) gives an open subgroup Δ of Γ and a compact group $\Theta \trianglelefteq \Delta$ such that Δ/Θ is a Lie group and $\dim \Theta = 0$. A theorem of Montgomery–Zippin [55] p. 249 states that any compact 0-dimensional effective transformation group of a surface is finite. Consequently, Δ is a Lie group, and so is Γ. □

96.32 Remark. All transitive effective transformation groups on surfaces have been determined, see Mostow [50], Betten–Forst [77]. In contrast to (96.30), there are transitive groups on $\mathbb{R}^2$ of any dimension $k \geq 2$.

The locally compact group $\mathbb{R}$ can act on the torus $\mathbb{T}^2$ without any closed orbit. By a result of Halder [73], the situation is different on the 2-sphere or the point space $\mathrm{P}_2\mathbb{R}$ of the real projective plane:

96.33 Theorem. *If (Γ, M) is a locally compact transformation group where Γ/Γ^1 is compact and M is a subsurface of $\mathbb{S}_2$ or of $\mathrm{P}_2\mathbb{R}$, then Γ has a closed orbit.* □

All compact, connected groups on the 4-sphere can be determined. The result is very useful for investigating 8-dimensional planes:

96.34 Theorem (Richardson). *For a compact, connected, effective transformation group $(\Gamma, \mathbb{S}_4)$ with $\dim \Gamma > 1$ the following conditions are equivalent:*

(i) $\dim a^\Gamma > 1$ *for some point a,*
(ii) Γ *is a Lie group,*
(iii) Γ *is equivalent to a subgroup of $\mathrm{SO}_5 \mathbb{R}$ in its linear action.*

There are exactly 8 actions satisfying these conditions:

(a) Γ *has no fixed point, and Γ is equivalent to $\mathrm{SO}_5\mathbb{R}$, to $\mathrm{SO}_3\mathbb{R} \times \mathrm{SO}_2\mathbb{R}$ on $\mathbb{R}^3 \times \mathbb{R}^2$, or to $\mathrm{SO}_3\mathbb{R}$ in its irreducible representation on $\mathbb{R}^5$, restricted to $\mathbb{S}_4$.*
(b) Γ *has exactly two fixed points, and Γ is equivalent to $\mathrm{SO}_4\mathbb{R}$ or to one of the subgroups $\mathrm{U}_2\mathbb{C}$, $\mathrm{SU}_2\mathbb{C}$, or to $(\mathrm{SO}_2\mathbb{R})^2$.*
(c) *The fixed points of Γ on $\mathbb{S}_4$ form a Jordan curve, and Γ is equivalent to $\mathrm{SO}_3\mathbb{R}$ on $\mathbb{R}^3 \times \mathbb{R}^2$.*

In particular, Γ does not contain $\mathbb{T}^3$.

This is actually *proved* but not fully stated in Richardson [61]. The implication (i $\Rightarrow$ ii) is a consequence of (96.24). Now assume (ii) and use (94.31c). If $\Gamma' \neq \mathbb{1}$, then Γ' is semi-simple by (94.31c), and (96.29c) shows that $\dim \Gamma/\Gamma_x \neq 1$. Any torus group acts faithfully on some of its orbits (96.28). Hence (iii $\Rightarrow$ ii $\Rightarrow$ i). At the end of his paper, Richardson shows that Γ cannot contain a 3-torus. For each of the possible groups of rank at most 2 he proves that any action on $\mathbb{S}_4$ is equivalent to a linear one. □

96.35 Proposition. *Any non-trivial action of the compact group $\Gamma = \mathrm{G}_2$ on $\mathbb{S}_8$ is equivalent to the action of G_2 on $\mathbb{O} \cup \{\infty\}$ as group of automorphisms of $\mathbb{O}$.*

Proof. If $c^\Gamma \neq c \in \mathbb{S}_8$, then $5 \leq \dim c^\Gamma < 8$ by (96.13a, 11 and 22). The dimension formula (96.10) gives $7 \leq \dim \Gamma_c \leq 9$. Hence the connected component Δ of Γ_c has rank 2, and (94.31c) together with the classification (94.33) shows that Δ is locally isomorphic to $\mathrm{SU}_3\mathbb{C}$. In fact, Δ is even conjugate to the stabilizer Γ_i in the usual action of Γ on $\mathbb{O}$, see (11.35). Now Γ_i has index 2 in the global stabilizer $\Gamma_\mathbb{C}$. Let $\Gamma_\mathbb{C} < \Lambda \leq \Gamma$. From (11.30a) and (96.25), applied to the action of Λ on $\mathbb{S}_6 \subset \mathrm{Pu}\,\mathbb{O}$, it follows that Λ is transitive on $\mathbb{S}_6$. Consequently, $\dim \Lambda/\Gamma_\mathbb{C} = 6$ and $\dim \Lambda = 14$. Thus $\Lambda = \Gamma$, see (93.12). This proves that $\Gamma_\mathbb{C}$ is maximal in Γ. Hence $c^\Gamma \approx \Gamma/\Delta \approx \Gamma/\Gamma_i \approx \mathbb{S}_6$ or, possibly, $c^\Gamma \approx \Gamma/\Gamma_\mathbb{C} \approx \mathrm{P}_6\mathbb{R}$. The second alternative will be excluded by showing $H_1(c^\Gamma) = 0$. According to Richardson [61] 1.2, 1.3, and 1.6, the fixed point set F of Γ is homeomorphic to $\mathbb{S}_1$, and $\mathbb{S}_8 \setminus F \simeq c^\Gamma$. By Alexander duality (see, e.g., Dold [72] VIII.8.15), we get $H_1(c^\Gamma) \cong H_1(\mathbb{S}_8 \setminus F) \cong$

$H^6(\mathbb{S}_1) = 0$. The same conclusion follows from Borel [60] Chap. XV (Bredon), Cor. 1.6 (1 and 4) together with Bredon [61] Th. Therefore, $x^\Gamma \approx \mathbb{S}_6$ for each $x \notin F$, and Poncet [59] Th. b shows that Γ acts linearly on $\mathbb{S}_8 \setminus \{a\}$ whenever a is a fixed point. Finally, (95.10) implies that the action of Γ is uniquely determined up to equivalence. □

Similar arguments show

96.36 Proposition. *Any non-trivial action of the group* $\Sigma = \mathrm{Spin}_7\mathbb{R}$ *on* $\mathbb{S}_8$ *has a fixed point a and is linear on* $\mathbb{R}^8 = \mathbb{S}_8 \setminus \{a\}$.

Sketch of proof. Since Σ is not transitive, $D = \{\dim x^\Sigma \mid x \in \mathbb{S}_8\}$ is a subset of $\{0, 6, 7\}$, and D contains at least two numbers by Bredon [61] Cor. 3 or Richardson [61] 1.1. If $\dim x^\Sigma = 7$, then Richardson [61] 1.6 implies that the union M of the principal orbits is homeomorphic to $\mathbb{R} \times x^\Sigma$. By Alexander duality, $H^6(\mathbb{S}_8 \setminus M) \cong H_1(M) \leq \mathbb{Z}_2$. Hence the exceptional orbits are points and not 6-spheres. Linearity on the complement of a fixed point follows again from Poncet [59]. □

Bibliography

Adams, J. F.
[60] *On the non-existence of elements of Hopf invariant one*,
 Ann. of Math. 72, 20–104.
[69] *Lectures on Lie groups*, New York: Benjamin.

André, J.
[54a] *Über Perspektivitäten in endlichen projektiven Ebenen*, Arch. Math. 6, 29–32.
[54b] *Über nicht-Desarguessche Ebenen mit transitiver Translationsgruppe*,
 Math. Z. 60, 156–186.
[55] *Projektive Ebenen über Fastkörpern*, Math. Z. 62, 137–160.
[58] *Affine Ebenen mit genügend vielen Translationen*, Math. Nachr. 19, 203–210.
[62] *Über verallgemeinerte Moulton-Ebenen*, Arch. Math. 13, 290–301.
[63] *Bemerkung zu meiner Arbeit "Über verallgemeinerte Moulton-Ebenen"*,
 Arch. Math. 14, 359–360.

Anisov, S. S.
[92] Uspekhi Mat. Nauk 47, no. 3, 147–148, (Russian) translated as
 The collineation group of Hilbert's example of a projective plane,
 Russian Math. Surveys 47, no. 3, 163-164.

Arens, R. F.
[46] *Topologies for homeomorphism groups*, Amer. J. Math. 68, 593–610.

Arkhangel'skiĭ, A. V. – Pontryagin, L. S.
[90] *General topology I*, Encycl. Math. Sci. 17, Berlin etc.: Springer.

Armstrong, M. A.
[83] *Basic topology*, Berlin etc.: Springer
 (first ed. Maidenhead (UK): McGraw–Hill 1979).

Artin, E.
[57] *Geometric algebra*, New York: Interscience.

Artmann, B.
[88] *The concept of number: from quaternions to monads and topological fields*,
 Chichester: Ellis Horwood.
 (Translation of *"Der Zahlbegriff"*, Göttingen: Vandenhoek & Ruprecht 1983).

Artzy, R.
[66] *Non-euclidean incidence planes*, Israel J. Math. 4, 43–53.

Arumugam, S.
[85] *On topological projective planes I*, Publ. Math. Debrecen 32, 251–253.

Aschbacher, M.
[86] *Finite group theory*, Cambridge Univ. Press.

Aslaksen, H.
[91] *Restricted homogeneous coordinates for the Cayley projective plane*, Geom. Dedicata 40, 245–250.

Baer, R.
[42] *Homogeneity of projective planes*, Amer. J. Math. 64, 137–152.
[46] *Projectivities with fixed points on every line of the plane*, Bull. Amer. Math. Soc. 52, 273–286.
[52] *Linear algebra and projective geometry*, New York: Academic Press.
[70] *Die Automorphismengruppe eines algebraisch abgeschlossenen Körpers der Charakteristik 0*, Math. Z. 117, 7–17.

Baker, C. A. – Lane, N. D. – Lorimer, J. M.
[88] *A construction for topological non-desarguesian affine Hjelmslev planes*, Arch. Math. 50, 83–92.

Ball, B. J.
[84] *Arcwise connectedness and the persistence of errors*, Amer. Math. Monthly 91, 431–433.

Bannow, E.
[40] *Die Automorphismengruppen der Cayley-Zahlen*, Abh. Math. Sem. Hansische Univ. (Hamburg) 13, 240–256.

Barlotti, A.
[67] *Sulle 2-curve nei piani grafici*, Rend. Sem. Mat. Univ. Padova 37, 91–97.

Bartolone, C.
[83] *On some translation planes admitting a Frobenius group of collineations*, in: A. Barlotti et al. (eds), Combinatorics '81, Proc. Roma 1981, Ann. Discrete Math. 18, pp. 37–53, Amsterdam: North-Holland.

Beardon, A. F.
[84] *A Primer on Riemann surfaces*, London Math. Soc. Lecture Note Series 78, Cambridge University Press.

Becker, E.
[72] *Über eine Klasse flexibler quadratischer Divisionsalgebren*, J. Reine Angew. Math. 256, 25–57.

Bedürftig, T.
[74a] *Polaritäten ebener projektiver Ebenen*, J. Geom. 5, 39–66.
[74b] *Ebene projektive Ebenen über Neokörpern*, Geom. Dedicata 3, 21–34.

Benkart, G. M. – Osborn, J. M.
[81a] *The derivation algebra of a real divison algebra*, Amer. J. Math. 103, 1135–1150.

[81b] *An investigation of real division algebras using derivations*,
Pacific J. Math. 96, 265–300.

Berger, M.
[77] *Géometrie 1–5*, Paris: CEDIC, translated as:
[87] *Geometry I–II*, Berlin etc.: Springer.

Bernardi, M.
[73] *Esistenza di fibrazioni in uno spazio proiettivo infinito*,
Ist. Lombardo Accad. Sci. Lett. Rend. A 107, 528–542.

Besse, A. L.
[78] *Manifolds all of whose geodesics are closed*, Berlin etc.: Springer.

Betten, D.
[68] *Topologische Geometrien auf dem Möbiusband*, Math. Z. 107, 363–379.
[70] *Nicht-desarguessche 4-dimensionale Ebenen*, Arch. Math. 21, 100–102.
[71] *2-dimensionale differenzierbare projektive Ebenen*, Arch. Math. 22, 304–309.
[72a] *Projektive Darstellung der Moulton-Ebenen*, J. Geom. 2, 107–114.
[72b] *4-dimensionale Translationsebenen*, Math. Z. 128, 129–151.
[73a] *4-dimensionale Translationsebenen mit 8-dimensionaler Kollineationsgruppe*,
Geom. Dedicata 2, 327–339.
[73b] *4-dimensionale Translationsebenen mit irreduzibler Kollineationsgruppe*,
Arch. Math. 24, 552–560.
[73c] *Die komplex-hyperbolische Ebene*, Math. Z. 132, 249–259.
[75] *4-dimensionale Translationsebenen mit genau einer Fixrichtung*,
Geom. Dedicata 3, 405–440.
[76] *4-dimensionale Translationsebenen mit 7-dimensionaler Kollineationsgruppe*,
J. Reine Angew. Math. 285, 126–148.
[77] *4-dimensionale Translationsebenen mit kommutativer Standgruppe*,
Math. Z. 154, 125–141.
[79a] *Komplexe Schiefparabel-Ebenen*, Abh. Math. Sem. Univ. Hamburg 48, 76–88.
[79b] *Die Projektivitätengruppe der Moulton-Ebenen*, J. Geom. 13, 197–209.
[80] *Zur Klassifikation 4-dimensionaler projektiver Ebenen*, Arch. Math. 35, 187–192.
[81] *Topologische Geometrien auf 3-Mannigfaltigkeiten*, Simon Stevin 55, 221–235.
[83] *Die Projektivitätengruppe einer Klasse 4-dimensionaler Translationsebenen*,
J. Geom. 21, 19–32.
[84] *4-dimensionale projektive Ebenen mit 3-dimensionaler Translationsgruppe*,
Geom. Dedicata 16, 179–193.
[87] *Einige Klassen topologischer 3-Räume*, Results in Math. 12, 37–61.
[90] *4-dimensional compact projective planes with a 7-dimensional collineation group*,
Geom. Dedicata 36, 151–170.
[91a] *4-dimensional compact projective planes with a nilpotent collineation group*,
Mitt. Math. Gesellsch. Hamburg 12, 741–747.
[91b] *Orbits in 4-dimensional compact projective planes*, J. Geom. 42, 30–40.
[95] *4-dimensional compact projective planes with a 5-dimensional nilradical*, Geom.
Dedicata, to appear .

Betten, D. – Forst, M.
[77] *Transitive Wirkungen auf Flächen / Effektive Lie-Algebren-Paare der Codimension* 2, Math. Seminar d. Univ. Kiel.

Betten, D. – Horstmann, C.
[83] *Einbettung von topologischen Raumgeometrien auf $\mathbb{R}^3$ in den reellen affinen Raum*, Resultate Math. 6, 27–35.

Betten, D. – Klein, H.
[96] *4-dimensional compact projective planes with two fixed points*, J. Geom., to appear.

Betten, D. – Knarr, N.
[87] *Rotationsflächen-Ebenen*, Abh. Math. Sem. Univ. Hamburg 57, 227–234.

Betten, D. – Ostmann, A.
[78] *Wirkungen und Geometrien der Gruppe* $L_2 \times \mathbb{R}$, Geom. Dedicata 7, 141–162.

Betten, D. – Polster, B.
[94] *Topological ovals in 4-dimensional translation planes*, Arch. Math. 63, 284–288.

Betten, D. – Wagner, A.
[82] *Eine stückweise projektive topologische Gruppe im Zusammenhang mit den Moulton-Ebenen*, Arch. Math. 38, 280–285.

Betten, D. – Weigand, C.
[85] *Groups of projectivities of topological planes*,
C. R. Math. Rep. Acad. Sci. Canada 7, 73–78.

Biliotti, M.
[79] *Su una generalizzazione di Dembowski dei piani di Hughes*,
Boll. Un. Mat. Ital. (5) 16-B, 674–693.

Bing, R. H. – Borsuk, K.
[65] *Some remarks concerning topologically homogeneous spaces*,
Ann. of Math. 81, 100–111.

van der Blij, F. – Springer, T. A.
[60] *Octaves and triality*, Nieuw Arch. Wisk. (3) 8, 158–169.

Bödi, R.
[92] *Automorphismengruppen lokalkompakter Doppelloops*, Dissertation, Tübingen.
[93a] *Automorphism groups of differentiable double loops*, Geom. Dedicata 46, 61–72.
[93b] *Automorphism groups of locally compact connected double loops are locally compact*, Arch. Math. 61, 291–294.
[94a] *On the dimensions of automorphism groups of four-dimensional double loops*, Math. Z. 215, 89–97.
[94b] *On the dimensions of automorphism groups of eight-dimensional double loops*, Monatsh. Math. 117, 1–16.
[94c] *On the dimensions of automorphism groups of eight-dimensional ternary fields, II*, Geom. Dedicata 53, 201–216
[95] *On the dimensions of automorphism groups of eight-dimensional ternary fields I*, J. Geom. 52, 30–40

Bödi, R. – Joswig, M.
[93] *Tables for an effective enumeration of real representations of quasi-simple Lie groups*, Seminar Sophus Lie 3, 239–253.

Borel, A.
[49] *Some remarks about Lie groups transitive on spheres and tori*,
      Bull. Amer. Math. Soc. 55, 580–587.
[50] *Le plan projectif des octaves et les sphères comme espaces homogènes*,
      C. R. Acad. Sci. Paris 230, 1378–1380.
[55] *Nouvelle démonstration d'un théorème de P. A. Smith*,
      Comment. Math. Helv. 29, 27–39.

Borel, A. – De Siebenthal, J.
[49] *Les sous-groupes fermés de rang maximum des groupes de Lie clos*,
      Comment. Math. Helv. 23, 200–221.

Borel, A. et al.
[60] *Seminar on transformation groups*, Ann. of Math. Stud. 46, Princeton Univ. Press.

Borsuk, K.
[67] *Theory of retracts*, Warsaw: Polish Scientific Publishers.

Bott, R.
[56] *An application of the Morse Theory to the topology of Lie-groups*,
      Bull. Soc. Math. France 84, 251–281.
[58] *The space of loops on a Lie group*, Michigan Math. J. 5, 35–61.
[70] *The periodicity theorem for the classical groups and some of its applications*,
      Adv. in Math. 4, 353–411.

Bott, R. – Milnor, J.
[58] *On the parallelizability of the spheres*, Bull. Amer. Math. Soc. 64, 87–89.

Bott, R. – Samelson, H.
[58] *Applications of the theory of Morse to symmetric spaces*,
      Amer. J. Math. 80, 964–1029.

Bourbaki, N.
[66a] *General topology, part 1*, Paris: Hermann.
[66b] *General topology, part 2*, Paris: Hermann.
[68]   *Groupes et algèbres de Lie, Chap. 4, 5, 6*, Paris: Hermann.
[75]   *Groupes et algèbres de Lie, Chap. 7, 8*, Paris: Hermann.
[82]   *Groupes et algèbres de Lie, Chap. 9, Groupes de Lie réels compacts*,
       Paris: Masson.
[89]   *Lie groups and Lie algebras, Chap. 1–3*, Berlin etc.: Springer.

Braconnier, J.
[45] *Groupes d'automorphismes d'un groupe localement compact*,
      C. R. Acad. Sci. Paris 220, 383–384.

Brada, C. – Pecaut-Tison, F.
[87] *Géométrie du plan projectif des octaves de Cayley*, Geom. Dedicata 23, 131–154.

Brauer, R.
[65] *On the relation between the orthogonal group and the unimodular group*,
Arch. Rational Mech. Anal. 18, 97–99.

Bredon, G. E.
[58] *Some theorems on transformation groups*, Ann. of Math. 67, 104–118.
[61] *Transformation groups with orbits of uniform dimension*,
Michigan Math. J. 8, 139–147.
[67] *Sheaf theory*, New York etc.: McGraw–Hill.
[69] *Wilder manifolds are locally orientable*,
Proc. Nat. Acad. Sci. USA 63, 1079–1081.
[70] *Generalized manifolds, revisited*, Proc. Univ. of Georgia topology of manifolds institute 1969, Athens (Georgia), pp. 461–469, Chicago: Markham.
[72] *Introduction to compact transformation groups*, New York: Academic Press.

Breitsprecher, S.
[67a] *Uniforme projektive Ebenen*, Math. Z. 95, 139–168.
[67b] *Einzigkeit der reellen und der komplexen projektiven Ebene*,
Math. Z. 99, 429–432.
[71] *Projektive Ebenen, die Mannigfaltigkeiten sind*, Math. Z. 121, 157–174.
[72] *Zur topologischen Struktur zweidimensionaler projektiver Ebenen*,
Geom. Dedicata 1, 21–32.

Breuning, P.
[70] *Translationsebenen und Vektorraumbündel*, Mitt. Math. Sem. Gießen 86.

Bröcker, Th. – tom Dieck, T.
[85] *Representations of compact Lie groups*, New York etc.: Springer.

Brouwer, L. E. J.
[09] *Die Theorie der endlichen kontinuierlichen Gruppen, unabhängig von den Axiomen von Lie*, Math. Ann. 67, 246–267.
[12] *Beweis des ebenen Translationssatzes*, Math. Ann. 72, 37–54.
[19] *Über die periodischen Transformationen der Kugel*, Math. Ann. 80, 39–41.

Browder, F. E.
[54] *Covering spaces, fibre spaces, and local homeomorphisms*,
Duke Math. J. 21, 329–336.

Browder, W.
[61] *Torsion in H-spaces*, Ann. of Math. 74, 24–51.

Brown, M.
[60] *A proof of the generalized Schoenflies theorem*, Bull. Amer. Math. Soc. 66, 74–76.
[61] *The monotone union of open n-cells is an open n-cell*,
Proc. Amer. Math. Soc. 12, 812–814.
[84] *A new proof of Brouwer's lemma on translation arcs*, Houston J. Math. 10, 35–41.

Brown, R. B.
[67] *On generalized Cayley–Dickson algebras*, Pacific J. Math. 20, 415–422.

Brown, R. F.
[71]   *The Lefschetz fixed point theorem*, Glenview (Ill.): Scott, Foresman and Co.

Bruck, R. H. – Bose, R. C.
[64]   *The construction of translation planes from projective spaces*,
       J. Algebra 1, 85–102.
[66]   *Linear representations of projective planes in projective spaces*,
       J. Algebra 4, 117–172.

Bruen, A. – Fisher, J. C.
[69]   *Spreads which are not dual spreads*, Canad. Math. Bull. 12, 801–803.

Bryant, J. L.
[87]   *Homogeneous ENR's*, Topology Appl. 27, 301–306.

Bryant, J. – Ferry, S. – Mio, W. – Weinberger, S.
[93]   *Topology of homology manifolds*, Bull. Amer. Math. Soc. 28, 324–328.

Buchanan, T.
[79a] *Zur Topologie der projektiven Ebenen über reellen Divisionsalgebren*,
       Geom. Dedicata 8, 383–393.
[79b] *The topology of the flag space of a topological projective plane with 2-spheres as
       point rows*, Dissertation, Erlangen.
[79c] *Ovale und Kegelschnitte in der komplexen projektiven Ebene*,
       Math.-Phys. Semesterber. 26, 244–260.

Buchanan, T. – Hähl, H.
[77]   *On the kernel and the nuclei of 8-dimensional locally compact quasifields*,
       Arch. Math. 29, 472–480.
[78]   *The transposition of locally compact, connected translation planes*,
       J. Geom. 11, 84–92.

Buchanan, T. – Hähl, H. – Löwen, R.
[80]   *Topologische Ovale*, Geom. Dedicata 9, 401–424.

Buekenhout, F.
[95]   (editor) *Handbook of incidence geometry*, Amsterdam: North Holland.

Burns, K. – Spatzier, R.
[87]   *On topological Tits buildings and their classification*, Publ. I.H.E.S. 65, 5–34.

Busemann, H.
[55]   *The geometry of geodesics*, New York: Academic Press.

Cannon, J. W.
[78]   *The recognition problem: what is a topological manifold?*,
       Bull. Amer. Math. Soc. 84, 832–866.

Cantwell, J.
[74]   *Geometric convexity. I*, Bull. Inst. Math. Acad. Sinica 2, 289–307.
[78]   *Geometric convexity. II: Topology*, Bull. Inst. Math. Acad. Sinica 6, 303–311.

Cantwell, J. – Kay, D. C.
[78]   *Geometric convexity. III: Embedding*, Trans. Amer. Math. Soc. 246, 211–230.

Cartan, H. – Eilenberg, S.
[56]   *Homological algebra*, Princeton Univ. Press.

Cecil, T. E. – Ryan, P. J.
[85]   *Tight and taut immersions of manifolds*, Marshfield (Mass.): Pitman.

Chatyrko, V. A.
[90]   *Compact spaces with noncoinciding dimensions*,
       Trans. Moscow Math. Soc. 1991, 199–236.

Chen, Su-shing
[73]   *On subgroups of the noncompact real exceptional Lie group* F_4^*,
       Math. Ann. 204, 271–284.

Christenson, C. O. – Voxman, W. L.
[77]   *Aspects of topology*, New York: Dekker.

Ciftci, S. – Kaya, R. – Ferrar, J. C.
[88]   *On 4-transitivity in the Moufang plane*, J. Geom. 31, 65–68.

Cleary, J. – Morris, S. A.
[88]   *Topologies on locally compact groups*, Bull. Austral. Math. Soc. 38, 105–111.

Coxeter, H. S. M.
[61]   *The real projective plane*, Cambridge Univ. Press (2nd edition).

Curtis, M. L.
[90]   *Abstract linear algebra*, New York etc.: Springer.

Danzer, L. – Laugwitz, D. – Lenz, H.
[57]   *Über das Löwnersche Ellipsoid und sein Analogon unter den einem Eikörper
       einbeschriebenen Ellipsoiden*, Arch. Math. 8, 214–219.

Daverman, R. J.
[80]   *Products of cell-like decompositions*, Topology Appl. 11, 121–139.

Dembowski, P.
[68]   *Finite geometries*, Berlin etc.: Springer.
[71a] *Generalized Hughes planes*, Canad. J. Math. 23, 481–494.
[71b] *Gruppenerhaltende quadratische Erweiterungen endlicher desarguesscher
       projektiver Ebenen*, Arch. Math. 22, 214–220.

Dembowski, P. – Ostrom, T. G.
[68]   *Planes of order n with collineation groups of order* n^2, Math. Z. 103, 239–258.

Deo, S. – Singh, T. B.
[82]   *On the converse of some theorems about orbit spaces*,
       J. London Math. Soc. 25, 162–170.

tom Dieck, T.
[87]   *Transformation groups*, Berlin: de Gruyter.

Dieudonné, J.
[52] *On the structure of unitary groups*, Trans. Amer. Math. Soc. 72, 367–385.
[71] *La géométrie des groupes classiques*, Berlin etc.: Springer (3rd edition).
[73] *Sur les groupes classiques*, Paris: Hermann (3rd edition).
[74] *Sur les automorphismes des corps algébriquement clos*,
     Bol. Soc. Brasileira Mat. 5, 123–126.

Djoković, D.
[76] *A closure theorem for analytic subgroups of real Lie groups*,
     Canad. Math. Bull. 19, 435–439.
[77] *Irreducible connected Lie subgroups of* $GL_n(R)$ *are closed*,
     Israel J. Math. 28, 175–176.

Doignon, J.-P.
[76] *Caractérisations d'espaces de Pasch–Peano*,
     Bull. Acad. Roy. Belg. Cl. Sci. (5) 62, 679–699.

Dold, A.
[63] *Partitions of unity in the theory of fibrations*, Ann. of Math. 78, 223–255.
[72] *Lectures on algebraic topology*, Berlin etc.: Springer.

Dranishnikov, A. N.
[88a] *On the dimension of the product of ANR-compacta*,
      Soviet Math. Dokl. 37, 769–773.
[88b] Mat. Sbornik 135, 551–556, (Russian) translated as:
[89] *On a problem of P. S. Alexandroff*, Math. USSR Sbornik 63, 539–545.

Dress, A.
[69] *Newman's theorems on transformation groups*, Topology 8, 203–207.

Dugundji, J.
[58] *Absolute neighborhood retracts and local connectedness in arbitrary metric spaces*,
     Compositio Math. 13, 229–246.
[66] *Topology*, Boston: Allyn & Bacon.

Dydak, J. – Walsh, J. J.
[87] *Sheaves that are locally constant with applications to homology manifolds*,
     in: S. Mardesic, J. Segal (eds.), Geometric topology and shape theory,
     Proc. Dubrovnik 1986, pp. 65–87, Lecture Notes in Math. 1283,
     Berlin etc.: Springer.
[88] *Sheaves with finitely generated isomorphic stalks and homology manifolds*,
     Proc. Amer. Math. Soc. 103, 655–660.
[93] *Infinite dimensional compacta having cohomological dimension two: an application of the Sullivan conjecture*, Topology 32, 93–104.

Dynkin, E. B.
[52a] Mat. Sbornik 30 (72), 349–462, (Russian) translated as:
[57a] *Semisimple subalgebras of semisimple Lie algebras*,
      Amer. Math. Soc. Transl. (2) 6, 111–244.
[52b] Trudy Moskov. Mat. Obshch. 1, 39–166, (Russian) translated as:

[57b] *The maximal subgroups of the classical groups*,
Amer. Math. Soc. Transl. (2) 6, 245–378.

Ebbinghaus, H.-D. et al.
[90] *Numbers*, New York: Springer.
[92] *Zahlen*, Berlin etc.: Springer (3rd edition).

Eilenberg, S.
[41] *Ordered topological spaces*, Amer. J. Math. 63, 39–45.

Eilenberg, S. – Steenrod, N.
[52] *Foundations of algebraic topology*, Princeton Univ. Press.

Eilenberg, S. – Wilder, R. L.
[42] *Uniform local connectedness and contractibility*, Amer. J. Math. 64, 613–622.

Eisele, E.
[90] *Topological ternary fields not belonging to a topological projective plane*,
Abh. Math. Sem. Univ. Hamburg 60, 257–264.
[91a] *Cartesian groups not belonging to topological projective planes*,
J. Geom. 40, 35–46.
[91b] *Topologische affine Ebenen mit nichtstetigem Parallelismus*,
Geom. Dedicata 40, 297–318.
[92a] *On homeomorphisms and incidence relations of compact projective planes*,
Arch. Math. 58, 615–620.
[92b] *Topologische Divisionsalgebren ohne zugehörige topologische affine Ebene*,
Abh. Math. Sem. Univ. Hamburg 62, 169–177.
[93a] *A class of topological affine translation planes having no topological projective extension*, Results Math. 23, 294–302.
[93b] *Konstruktion topologischer affiner Ebenen mit nichtstetigem Parallelismus durch Knicken von Geraden*, Geom. Dedicata 45, 237–262.

Engelking, R.
[78] *Dimension theory*, Amsterdam: North–Holland.

Fáry, I.
[61] *Dimension of the square of a space*, Bull. Amer. Math. Soc. 67, 135–137.

Faulkner, J. R.
[70] *Octonion planes defined by quadratic Jordan algebras*,
Memoirs Amer. Math. Soc. 104.
[89] *A geometric construction of Moufang planes*, Geom. Dedicata 29, 133–140.

Faulkner, J. R. – Ferrar, J. C.
[85] *Generalizing the Moufang plane*, in: R. Kaya, P. Plaumann, K. Strambach (eds.), Rings and geometry, Proc. Istanbul 1984, pp. 235–288, Dordrecht etc.: Reidel.

Fedorchuk, V. V.
[90] *The fundamentals of dimension theory*,
Part II of A. V. Arkhangel'skiĭ, L. S. Pontryagin (eds.), General topology I,
Encycl. Math. Sci. 17, pp. 91–192, Berlin etc.: Springer.

Fegan, H. D.
[91] *Introduction to compact Lie groups*,
    Singapore etc.: World Scientific Publishing Co.

Franks, J.
[92] *A new proof of the Brouwer plane translation theorem*,
    Ergodic Theory Dynamical Systems 12, 217–226.

Freudenthal, H.
[36] *Einige Sätze über topologische Gruppen*, Ann. of Math. 37, 46–56.
[51] *La structure des groupes à deux bouts et des groupes triplement transitifs*,
    Indag. Math. 13, 288–294.
[55] *Die Bedeutung der topologischen Voraussetzung bei der
    Buckel – van Heemertschen Charakterisierung des Systems der Kreise*,
    J. Reine Angew. Math. 194, 190–192.
[56] *Neuere Fassungen des Riemann–Helmholtz–Lieschen Raumproblems*,
    Math. Z. 63, 374–405.
[57a] *Kompakte projektive Ebenen*, Illinois J. Math. 1, 9–13.
[57b] *Zur Geschichte der Grundlagen der Geometrie. Zugleich eine Besprechung
    der 8. Aufl. von Hilberts "Grundlagen der Geometrie"*,
    Nieuw Arch. Wisk. (3) 5, 105–142.
[64] *Das Helmholtz–Liesche Raumproblem bei indefiniter Metrik*,
    Math. Ann. 156, 263–312.
[65] *Lie groups in the foundations of geometry*, Adv. in Math. 1, 145–190.
[85] *Oktaven, Ausnahmegruppen und Oktavengeometrie*,
    Utrecht 1951; new revised edition 1960; reprinted in: Geom. Dedicata 19, 7–63.

Freudenthal, H. – de Vries, H.
[69] *Linear Lie groups*, New York: Academic Press.

Friberg, B.
[73] *A topological proof of a theorem of Kneser*, Proc. Amer. Math. Soc. 39, 421–426.

Frobenius, F. G.
[1878] *Über lineare Substitutionen und bilineare Formen*,
    J. Reine Angew. Math. 84, 1–63.

Fuchs, L.
[63] *Partially ordered algebraic systems*, Oxford: Pergamon Press.

Gilkey, P. B.
[87] *The eta invariant and non-singular bilinear products on R^n*,
    Canad. Math. Bull. 30, 147–154.

Gleason, A. M.
[56] *Finite Fano planes*, Amer. J. Math. 78, 797–807.

Gleason, A. M. – Palais, R. S.
[57] *On a class of transformation groups*, Amer. J. Math. 79, 631–648.

Gluck, H. – Warner, F. W.
[83] *Great circle fibrations of the three-sphere*, Duke Math. J. 50, 107–132.

Gluck, H. – Warner, F. – Ziller, W.
[86] *The geometry of the Hopf fibrations*, Enseign. Math. 32, 173–198.

Gluškov, V. M.
[57] Uspekhi Mat. Nauk 12, no. 2, 3–41, (Russian) translated as
[60] *The structure of locally compact groups and Hilbert's fifth problem*,
      Amer. Math. Soc. Transl. (2) 15, 55–93.

Godement, R.
[82] *Introduction à la théorie de groupes de Lie*, Université Paris VII.

Goto, M.
[69] *On an arcwise connected subgroup of a Lie group*,
      Proc. Amer. Math. Soc. 20, 157–162.

Gray, B.
[75] *Homotopy theory*, New York: Academic Press.

Greenberg, M. J.
[67] *Lectures on algebraic topology*, New York: Benjamin.

Griffiths, P. – Harris, J.
[78] *Principles of algebraic geometry*, New York: Wiley.

Gröger, D.
[92] *Homomorphe Kopplungen auf dem Körper der reellen Quaternionen*,
      Arch. Math. 58, 354–359.

Groh, H.
[71] *1-dimensional orbits in flat projective planes*, Math. Z. 122, 117–124.
[73] *Moebius planes with locally euclidean circles are flat*, Math. Ann. 201, 149–156.
[76] *Point homogeneous flat affine planes*, J. Geom. 8, 145–162.
[77] *Flat projective planes whose automorphism group contains $\mathbb{R}^2$*,
      in: Arnold–Benz–Wefelscheid (eds.), Beiträge zur Geometrischen Algebra,
      Proc. Duisburg 1976, pp. 129–131, Basel: Birkhäuser.
[79] *$\mathbb{R}^2$-planes with 2-dimensional point transitive automorphism group*,
      Abh. Math. Sem. Univ. Hamburg 48, 171–202.
[81] *Pasting of $\mathbb{R}^2$-planes*, Geom. Dedicata 11, 69–98.
[82a] *$\mathbb{R}^2$-planes with point transitive 3-dimensional collineation group*,
      Indag. Math. 44, 173–182.
[82b] *Isomorphism types of arc planes*, Abh. Math. Sem. Univ. Hamburg 52, 133–149.
[86a] *Geometric lattices with topology*, J. Combin. Theory Ser. A. 42, 111–125.
[86b] *Embedding geometric lattices with topology*,
      J. Combin. Theory Ser. A. 42, 126–136.

Groh, H. – Lippert, M. F. – Pohl, H.-J.
[83] *$\mathbb{R}^2$-planes with 3-dimensional automorphism group fixing precisely a line*,
      J. Geom. 21, 66–96.

Grundhöfer, T.
[81] *Eine Charakterisierung ableitbarer Translationsebenen,*
Geom. Dedicata 11, 177–185.
[83] *Über Projektivitätengruppen affiner und projektiver Ebenen unter besonderer Berücksichtigung von Moufangebenen,* Geom. Dedicata 13, 435–458.
[86] *Automorphism groups of compact projective planes,* Geom. Dedicata 21, 291–298.
[87a] *Ternary fields of compact projective planes,*
Abh. Math. Sem. Univ. Hamburg 57, 87–101.
[87b] *Compact disconnected Moufang planes are desarguesian,*
Arch. Math. 49, 124–126.
[88] *Compact disconnected planes, inverse limits and homomorphisms,*
Monatsh. Math. 105, 261–277.
[90a] *Sharply transitive linear groups and nearfields over p-adic fields,*
Forum Math. 1, 81–101.
[90b] *Non-associative division algebras admitting many derivations,*
Forum Math. 2, 585–601.

Grundhöfer, T. – Hähl, H.
[90] *Fibrations of spheres by great spheres over division algebras and their differentiability,* J. Differential Geom. 31, 357–363.

Grundhöfer, T. – Löwen, R.
[95] *Linear topological geometries,* Chapter 23 of: F. Buekenhout (ed.), Handbook of incidence geometry, Amsterdam: North Holland.

Grundhöfer, T. – Salzmann, H.
[90] *Locally compact double loops and ternary fields,*
Chap. XI of: O. Chein, H. Pflugfelder, J. Smith (eds.), Quasigroups and loops: theory and applications, pp. 313–355, Berlin: Heldermann.

Grundhöfer, T. – Strambach, K.
[86] *Die affinen Projektivitätengruppen der lokalkompakten zusammenhängenden Translationsebenen,* Arch. Math. 47, 274–278.

Grundhöfer, T. – Stroppel, M.
[92] *On restrictions of automorphism groups of compact projective planes to subplanes,* Results Math. 21, 319–327.

Hähl, H.
[75a] *Automorphismengruppen von lokalkompakten zusammenhängenden Quasikörpern und Translationsebenen,* Geom. Dedicata 4, 305–321.
[75b] *Vierdimensionale reelle Divisionsalgebren mit dreidimensionaler Automorphismengruppe,* Geom. Dedicata 4, 323–331.
[75c] *Geometrisch homogene vierdimensionale reelle Divisionsalgebren,*
Geom. Dedicata 4, 333–361.

[76] *Automorphismengruppen achtdimensionaler lokalkompakter Quasikörper*, Math. Z. 149, 203–225.
[77] *Eine Klassifikation achtdimensionaler lokalkompakter Translationsebenen nach ihrer Kollineationsgruppe*, Habilitationsschrift, Tübingen.
[78] *Zur Klassifikation von 8- und 16-dimensionalen lokalkompakten Translationsebenen nach ihren Kollineationsgruppen*, Math. Z. 159, 259–294.
[79] *Lokalkompakte zusammenhängende Translationsebenen mit großen Sphärenbahnen auf der Translationsachse*, Resultate Math. 2, 62–87.
[80a] *Achtdimensionale lokalkompakte Translationsebenen mit großen Streckungsgruppen*, Arch. Math. 34, 231–242.
[80b] *Achtdimensionale lokalkompakte Translationsebenen mit großen kompakten Kollineationsgruppen*, Monatsh. Math. 90, 207–218.
[81] *Homologies and elations in compact, connected projective planes*, Topology Appl. 12, 49–63.
[82a] *Achtdimensionale lokalkompakte Translationsebenen mit zu* $SL_2\mathbb{C}$ *isomorphen Kollineationsgruppen*, J. Reine Angew. Math. 330, 76–92.
[82b] *Kriterien für lokalkompakte topologische Quasikörper*, Arch. Math. 38, 273–279.
[83] *Zur Kollineationsgruppe von achtdimensionalen lokalkompakten Translationsebenen*, Abh. Math. Sem. Univ. Hamburg 53, 84–102.
[84] *Eine Klasse von achtdimensionalen lokalkompakten Translationsebenen mit großen Scherungsgruppen*, Monatsh. Math. 97, 23–45.
[86a] *Achtdimensionale lokalkompakte Translationsebenen mit mindestens 17-dimensionaler Kollineationsgruppe*, Geom. Dedicata 21, 299–340.
[86b] *Charakterisierung der kompakten, zusammenhängenden Moufang-Hughes-Ebenen anhand ihrer Kollineationen*, Math. Z. 191, 117–136.
[87a] *Sechzehndimensionale lokalkompakte Translationsebenen mit* Spin(7) *als Kollineationsgruppe*, Arch. Math. 48, 267–276.
[87b] *Eine Kennzeichnung der Oktavenebene*, Indag. Math. 49, 29–39.
[87c] $SU_4(\mathbb{C})$ *als Kollineationsgruppe in sechzehndimensionalen lokalkompakten Translationsebenen*, Geom. Dedicata 23, 319–345.
[87d] *Differentiable fibrations of the* $(2n-1)$-*sphere by great* $(n-1)$-*spheres and their coordinatization over quasifields*, Results Math. 12, 99–118.
[88] *Die Oktavenebene als Translationsebene mit großer Kollineationsgruppe*, Monatsh. Math. 106, 265–299.
[90] *Sechzehndimensionale lokalkompakte Translationsebenen, deren Kollineationsgruppe* G_2 *enthält*, Geom. Dedicata 36, 181–197.

Hahn, A. J. – O'Meara, O. T.
[89] *The classical groups and K-theory*, Berlin etc.: Springer.

Halder, H.-R.
[71] *Dimension der Bahnen lokal kompakter Gruppen*, Arch. Math. 22, 302–303.
[73] *Über Bahnen lokal kompakter Gruppen auf Flächen*, Geom. Dedicata 2, 101–109.

Hall, M.
[59] *The theory of groups*, New York: Macmillan.

Hartmann, P.
[88] *Die Stellentopologie projektiver Ebenen und Lenz-topologische Ebenen*,
Geom. Dedicata 26, 259–272.
[89] *Zur Stetigkeit in kompakten projektiven Ebenen*,
Abh. Math. Sem. Univ. Hamburg 59, 35–38.

Harvey, F. R.
[90] *Spinors and calibrations*, Boston: Academic Press.

Harvey, R. – Lawson, H. B.
[82] *Calibrated geometries*, Acta Math. 148, 47–157.

Hausdorff, F.
[14] *Grundzüge der Mengenlehre*, Leipzig: Veit & Comp.
[27] *Mengenlehre*, 2. Auflage, Berlin und Leipzig: de Gruyter.

Hausner, M. – Schwartz, J. T.
[68] *Lie groups; Lie algebras*, New York: Gordon and Breach.

Hefendehl-Hebeker, L.
[83] *Isotopieklassen vierdimensionaler quadratischer Divisionsalgebren über Hilbert-Körpern*, Habilitationsschrift, Duisburg.

Hein, W.
[90] *Einführung in die Struktur- und Darstellungstheorie der klassischen Gruppen*,
Berlin etc.: Springer.

Helgason, S.
[62] *Differential geometry and symmetric spaces*, New York: Academic Press.
[78] *Differential geometry, Lie groups, and symmetric spaces*,
New York: Academic Press.

Hering, Ch.
[70] *A new class of quasifields*, Math. Z. 118, 56–57.

Herstein, I. N.
[64] *Topics in algebra*, Waltham (Mass.): Blaisdell.

Herzer, A.
[72] *Dualitäten mit zwei Geraden aus absoluten Punkten in projektiven Ebenen*,
Math. Z. 129, 235–257.

Hewitt, E. – Ross, K. A.
[79] *Abstract Harmonic Analysis I*, Berlin etc.: Springer (2nd edition).

Hilbert, D.
[1899] *Grundlagen der Geometrie*, Leipzig: Teubner.
[30] *Grundlagen der Geometrie*,
Leipzig und Berlin: Teubner (7th edition); Stuttgart: Teubner (13th edition 1987).

Hilgert, J. – Hofmann, K. H.
[85] *Old and new on Sl(2)*, Manuscr. Math. 54, 17–52.

Hilgert, J. – Neeb, K.-H.
[91] *Lie-Gruppen und Lie-Algebren*, Braunschweig: Vieweg.
[93] *Lie Semigroups and their Applications*,
      Lecture Notes in Mathematics 1552, Berlin etc.: Springer.

Hirsch, G.
[49] *La géométrie projective et la topologie des espaces fibrés*,
      in: Topologie Algébrique, Proc. Conf. Paris, CNRS XII, 35–42.
      (German translation in H. Freudenthal (ed.), Raumtheorie, pp. 241–249,
      Darmstadt: Wissenschaftliche Buchgesellschaft 1978).

Hirschfeld, J.
[90] *The nonstandard treatment of Hilbert's fifth problem*,
      Trans. Amer. Math. Soc. 321, 379–400.

Hochschild, G.
[65] *The structure of Lie groups*, San Francisco: Holden–Day.

Hocking, J. G. – Young, G. S.
[61] *Topology*, Reading (Mass.): Addison–Wesley.

Hofmann, K. H.
[58] *Topologische Doppelloops*, Math. Z. 70, 213–230.
[61] *Ein komplexer Neokörper ohne reellen Unterkörper*, Math. Z. 75, 295–298.
[62] *Über die topologische und algebraische Struktur topologischer Doppelloops und einiger topologischer projektiver Ebenen*,
      in: Algebraical and topological foundations of geometry,
      Proc. Colloq. Utrecht 1959, pp. 57–67, Oxford: Pergamon Press.
[65] *Lie algebras with subalgebras of co-dimension one*, Illinois J. Math. 9, 636–643.
[90] *Hyperplane subalgebras of real Lie algebras*, Geom. Dedicata 36, 207–224.
[93] Review of: R. W. Bagley, T. S. Wu, J. S. Yang,
      On connected groups and related topics, Zbl. Math. 765, 22002.

Hofmann, K. H. – Mostert, P. S.
[63] *Splitting in topological groups*, Mem. Amer. Math. Soc. 43, 75 pp.
[68] *One dimensional coset spaces*, Math. Ann. 178, 44–52.

Hofmann, K. H. – Terp, Ch.
[94] *Compact subgroups of Lie groups and locally compact groups*,
      Proc. Amer. Math. Soc. 120, 623–634.

Hohti, A.
[87] *Another alternative proof of Effros' theorem*, Topology Proc. 12, 295–298.

Hopf, H.
[35] *Über die Abbildungen von Sphären auf Sphären niedrigerer Dimension*,
      Fund. Math. 25, 427–440.

Hopf, H. – Samelson, H.
[41] *Ein Satz über die Wirkungsräume geschlossener Liescher Gruppen*,
      Comment. Math. Helv. 13, 240–251.

Hsiung, C.
[86]   *Nonexistence of a complex structure on the six-sphere*,
       Bull. Inst. Math. Acad. Sinica 14, 231–247.

Hu, S.-T.
[59]   *Homotopy theory*, New York: Academic Press.

Hubbuck, J.
[69]   *On homotopy commutative H-spaces*, Topology 8, 119–126.

Hubig, M.
[87]   *Zweidimensionale stabile Ebenen mit auflösbarer, nicht zu $L_2 \times \mathbb{R}$ lokalisomorpher, mindestens dreidimensionaler Kollineationsgruppe*, Diplomarbeit, Tübingen.
[90]   *Sechzehndimensionale kompakte projektive Ebenen mit großer Automorphismengruppe*, Dissertation, Tübingen.

Hughes, D. R.
[60]   *On homomorphisms of projective planes*,
       Proc. Sympos. Appl. Math. 10, Columbia Univ. 1958, 45–52.

Hughes, D. R. – Piper, F. C.
[73]   *Projective planes*, Berlin etc.: Springer.

Humphreys, J. E.
[72]   *Introduction to Lie algebras and representation theory*,
       Berlin etc.: Springer (3rd revised edition 1980).
[90]   *Reflection groups and Coxeter groups*, Cambridge Univ. Press.

Huppert, B.
[67]   *Endliche Gruppen I*, Berlin etc.: Springer.

Hurewicz, W. – Wallman, H.
[48]   *Dimension theory*, Princeton Univ. Press.

Hurwitz, A.
[1898] *Über die Composition der quadratischen Formen von beliebig vielen Variabeln*,
       Nachr. Göttingen 309–316.

Husain, T.
[66]   *Introduction to topological groups*, Philadelphia: Saunders.

Husemoller, D.
[66]   *Fibre bundles*, New York: McGraw–Hill; New York etc.: Springer (3rd ed. 1994).

Itzkowitz, G. – Rothman, S. – Strassberg, H.
[91]   *A note on the real representations of* $SU(2, \mathbb{C})$,
       J. Pure Appl. Algebra 69, 285–294.

Iversen, B.
[92]   *Hyperbolic geometry*, Cambridge Univ. Press.

Iwahori, N.
[59]   *On real irreducible representations of Lie algebras*, Nagoya Math. J. 14, 59–83.

Iwasawa, K.
[49]   *On some types of topological groups*, Ann. of Math. 50, 507–558.
[51a] *Topological groups with invariant compact neighborhoods of the identity*,
       Ann. of Math. 54, 345–348.
[51b] *Über die Einfachheit der speziellen projektiven Gruppen*,
       Proc. Imp. Acad. Tokyo 17, 57–59.

Jacobson, N.
[59]   *Some groups of transformations defined by Jordan algebras. I*,
       J. Reine Angew. Math. 201, 178–195.
[60]   *Some groups of transformations defined by Jordan algebras. II. Groups of type* F_4,
       J. Reine Angew. Math. 204, 74–98.
[61]   *Some groups of transformations defined by Jordan algebras. III.*
       Groups of type E_{6I}, J. Reine Angew. Math. 207, 61–85.
[62]   *Lie algebras*, New York: Interscience.
[80]   *Basic Algebra II*, San Francisco: Freeman.

Jacobson, N. – Taussky, O.
[35]   *Locally compact rings*, Proc. Nat. Acad. Sci. USA 21, 106–108.

Jakobsche, W.
[80]   *The Bing–Borsuk conjecture is stronger than the Poincaré conjecture*,
       Fund. Math. 106, 127–134.

Jha, V. – Johnson, N. L.
[85]   *Translation planes of order n which admit a collineation group of order n*,
       in: C. A. Baker, L. M. Batten (eds.), Finite geometries, Proc. Winnipeg 1984,
       pp. 149–162, New York: Dekker.

Jordan, P.
[49]   *Über eine nicht-desarguessche ebene projektive Geometrie*,
       Abh. Math. Sem. Univ. Hamburg 16, 74–76.

Jungnickel, D.
[87]   *On a theorem of Ganley*, Graphs and Combin. 3, 141–143.

Kalhoff, F.
[86]   *Über Unterebenen affin-archimedischer Ebenen*, Arch. Math. 47, 185–186.
[89]   *Ebenen der Lenz-Barlotti Klasse III.2, die keine Moulton-Ebenen sind*,
       Arch. Math. 53, 99–104.
[90]   *Anordnungsräume unter der Moulton Konstruktion und Ebenen der*
       Lenz Klasse III, J. Geom. 38, 59–77.

Kalhoff, F. – Prieß-Crampe, S.
[90]   *Ordered loops and ordered planar ternary rings*,
       Chap. XIV of: O. Chein, H. Pflugfelder, J. Smith (eds.), Quasigroups and loops:
       theory and applications, pp. 445–465, Berlin: Heldermann.

Kallman, R. R. – Simmons, F. W.
[85] *A theorem on planar continua and an application to automorphisms of the field of complex numbers*, Topology Appl. 20, 251–255.

Kalscheuer, F.
[40] *Die Bestimmung aller stetigen Fastkörper über dem Körper der reellen Zahlen als Grundkörper*, Abh. Math. Sem. Hansische Univ. (Hamburg) 13, 413–435.

Kaplansky, I.
[71] *Lie algebras and locally compact groups*, Univ. of Chicago Press.

Karno, Z. – Krasinkiewicz, J.
[90] *On some famous examples in dimension theory*, Fund. Math. 134, 213–220.

Karzel, H. – Kroll, H.-J.
[88] *Geschichte der Geometrie seit Hilbert*,
      Darmstadt: Wissenschaftliche Buchgesellschaft.

Kawakubo, K.
[91] *The theory of transformation groups*, Oxford Univ. Press.

Kegel, O. H. – Schleiermacher, A.
[73] *Amalgams and embeddings of projective planes*, Geom. Dedicata 2, 379–395.

Keller, H. A.
[86] *On valued, complete fields and their automorphisms*,
      Pacific J. Math. 121, 397–406.

Kerékjártó, B. von
[19] *Über die periodischen Transformationen der Kreisscheibe und der Kugelfläche*,
      Math. Ann. 80, 36–38.

Kervaire, M. A.
[58] *Non-parallelizability of the n-sphere for $n > 7$*,
      Proc. Nat. Acad. Sci. USA 44, 280–283.
[60] *Some nonstable homotopy groups of Lie groups*, Illinois J. Math. 4, 161–169.

Kestelman, H.
[51] *Automorphisms of the field of complex numbers*,
      Proc. London Math. Soc. (2) 53, 1–12.

Kiltinen, J. O.
[73] *On the number of field topologies on an infinite field*,
      Proc. Amer. Math. Soc. 40, 30–36.

Klein, F.
[1871] *Über die sogenannte Nicht-Euklidische Geometrie*, Math. Ann. 4, 573–625.

Kleinfeld, E.
[63] *A characterization of the Cayley numbers*,
      in: A. A. Albert (ed.), Studies in modern algebra, pp. 126–143, Math. Assoc. America, Englewood Cliffs (N.J.): Prentice–Hall.

Knarr, N.
[83] *Topologische Differenzenflächenebenen*, Diplomarbeit, Kiel.
[86] *Topologische Differenzenflächenebenen mit nichtkommutativer Standgruppe*, Dissertation, Kiel.
[87a] *Unstetige Kollineationen 4-dimensionaler Ebenen*, Arch. Math. 48, 548–549.
[87b] *4-dimensionale projektive Ebenen vom Lenz-Barlotti-Typ II.2*,
      Results Math. 12, 134–147.
[88a] *4-dimensionale projektive Ebenen mit großer abelscher Kollineationsgruppe*,
      J. Geom. 31, 114–124.
[88b] *Group preserving extensions of skew parabola planes*,
      in: A. Barlotti et al. (eds), Combinatorics '86, Proc. Passo della Mendola (Trento),
      Ann. Discrete Math. 37, pp. 225–229, Amsterdam: North-Holland.
[90] *The nonexistence of certain topological polygons*, Forum Math. 2, 603–612.
[91] *Konstruktionsverfahren für Translationsebenen unter besonderer Berücksichtigung topologischer Translationsebenen*, Habilitationsschrift, Braunschweig.
[95] *Translation planes: foundations and construction principles*,
      Lecture Notes in Math. 1611, Berlin etc.: Springer.

Knarr, N. – Weigand, C.
[86] *Ein Kriterium für topologische Ternärkörper*, Arch. Math. 46, 368–370.

Kneser, H.
[26] *Die Deformationssätze der einfach zusammenhängenden Flächen*,
      Math. Z. 25, 362–372.

Kolmogoroff, A.
[32] *Zur Begründung der projektiven Geometrie*, Ann. of Math. 33, 175–176.

Komrakov, B. P.
[90] *Maximal subalgebras of real Lie algebras and a problem of Sophus Lie*,
      Soviet Math. Doklady 41, 269–273.

Koornwinder, T. H. (ed.)
[82] *The structure of real semi-simple Lie groups*,
      Amsterdam: Mathematisch Centrum Syllabus 49.

Köthe, G.
[39] *Unendliche abelsche Gruppen und Grundlagen der Geometrie*,
      Jahresber. Deutsch. Math.-Verein. 49, 97–113.

Kramer, L.
[94a] *The topology of smooth projective planes*, Arch. Math. 63, 85–91.
[94b] *Compact polygons*, Dissertation, Tübingen.

Kreh, F.
[92] *Gruppen von Moufang-Ebenen*, Diplomarbeit, Tübingen.

Krupski, P.
[90] *Homogeneity and Cantor manifolds*, Proc. Amer. Math. Soc. 109, 1135–1142.

Kühne, R. – Löwen, R.
[92] *Topological projective spaces*, Abh. Math. Sem. Univ. Hamburg 62, 1–9.

Kuiper, N. H.
[57] *A real analytic non-desarguesian plane*, Nieuw Arch. Wisk. (3) 5, 19–24.

Kuratowski, K.
[68] *Topology, Vol. II*, New York: Academic Press.

Kuzmin, E. N.
[66] *Certain classes of division algebras*, Algebra i Logika Sem. 5, 57–102 (Russian).

Lacher, R. C.
[83] Trudy Mat. Inst. Steklov. 154, 137–150, (Russian) translated as
[84] *Resolutions of generalized manifolds*, Proc. Steklov Inst. Math. 154, 147–159.

Lenz, H.
[65] *Vorlesungen über projektive Geometrie*, Leipzig: Geest & Portig.
[92] *Konvexität in Anordnungsräumen*, Abh. Math. Sem. Univ. Hamburg 62, 255–285.

Lewis, D. W.
[82] *The isometry classification of Hermitian forms over division algebras*,
Lin. Alg. Appl. 43, 245–272.

Lex, W.
[73] *Zur Theorie der Divisionsalgebren*, Mitt. Math. Sem. Gießen 103.

Lie, S.
[1890] *Über die Grundlagen der Geometrie. I. Abhandlung*, Leipziger Berichte, Heft II,
284–321 (also in: Gesammelte Abhandlungen, Zweiter Band, Zweite Abteilung,
I. Teil, pp. 380–413, Leipzig/Oslo: Teubner/Aschehong, 1935).
[1893] *Theorie der Transformationsgruppen, Dritter Abschnitt*, Leipzig: Teubner.

Lippert, M. F.
[86] *Flat projective planes with two-dimensional non commutative automorphism group fixing precisely a semioval*, Dissertation, Darmstadt.

Loos, O.
[69] *Symmetric Spaces I, II*, New York: Benjamin.

Loveland, L. D. – Loveland, S. M.
[95] *Closed generalized Mazurkiewicz sets are curves*, Topology Appl. 61, 151–158.

Löwe, H.
[94] *Zerfallende symmetrische Ebenen mit großem Radikal*,
Dissertation, Braunschweig.
[95] *Symmetric planes with non-classical tangent translation planes*,
Geom. Dedicata, to appear.

Löwen, R.
[76a] *Locally compact connected groups acting on euclidean space with Lie isotropy groups are Lie*, Geom. Dedicata 5, 171–174.
[76b] *Vierdimensionale stabile Ebenen*, Geom. Dedicata 5, 239–294.

[77] *Schleiermachers Starrheitsbedingung für Projektivitäten in der topologischen Geometrie*, Math. Z. 155, 23–28.
[78] *Halbeinfache Automorphismengruppen von vierdimensionalen stabilen Ebenen sind quasi-einfach*, Math. Ann. 236, 15–28.
[79a] *Symmetric planes*, Pacific J. Math. 84, 367–390.
[79b] *Classification of 4-dimensional symmetric planes*, Math. Z. 167, 137–159.
[79c] *Weakly flag homogeneous stable planes of low dimension*, Arch. Math. 33, 485–491.
[81a] *Central collineations and the parallel axiom in stable planes*, Geom. Dedicata 10, 283–315.
[81b] *Characterization of symmetric planes in dimension at most 4*, Indag. Math. 43, 87–103.
[81c] *Equivariant embeddings of low dimensional symmetric planes*, Monatsh. Math. 91, 19–37.
[81d] *Homogeneous compact projective planes*, J. Reine Angew. Math. 321, 217–220.
[81e] *Projectivities and the geometric structure of topological planes*, in: P. Plaumann, K. Strambach (eds.), Geometry — von Staudt's point of view, Proc. Bad Windsheim 1980, pp. 339–372, Dordrecht etc.: Reidel.
[82a] *Stable planes of low dimension admitting reflections at many lines*, Resultate Math. 5, 60–80.
[82b] *A local "Fundamental Theorem" for classical topological projective spaces*, Arch. Math. 38, 286–288.
[83a] *Stable planes with isotropic points*, Math. Z. 182, 49–61.
[83b] *Topology and dimension of stable planes: On a conjecture of H. Freudenthal*, J. Reine Angew. Math. 343, 108–122.
[83c] *Zweidimensionale stabile Ebenen mit nicht-auflösbarer Automorphismengruppe*, Arch. Math. 41, 565–571.
[84a] *Compact projective planes with homogeneous ovals*, Monatsh. Math. 97, 55–61.
[84b] *Ebene stabile Ebenen mit vielen Zentralkollineationen*, Mitt. Math. Sem. Gießen 165, 63–67.
[86a] *Actions of* SO(3) *on 4-dimensional stable planes*, Aequationes Math. 30, 212–222.
[86b] *Stable planes admitting a classical motion group*, Results Math. 9, 119–130.
[86c] *A criterion for stability of planes*, Arch. Math. 46, 275–278.
[86d] *Actions of* $Spin_3$ *on 4-dimensional stable planes*, Geom. Dedicata 21, 1–12.
[89] *Compact spreads and compact translation planes over locally compact fields*, J. Geom. 36, 110–116.
[90] *Four-dimensional compact projective planes with a nonsolvable automorphism group*, Geom. Dedicata 36, 225–234.
[94] *Topological pseudo-ovals, elation Laguerre planes, and elation generalized quadrangles*, Math. Z. 216, 347–369.
[95] *Geometries on surfaces, revisited*, Geom. Dedicata, to appear.

Löwen, R. – Salzmann, H.
[82] *Collineation groups of compact connected projective planes*, Arch. Math. 38, 368–373.

Lüneburg, H.
[66] *Über die Struktursätze der projektiven Geometrie*, Arch. Math. 17, 206–209.
[76] *Characterizations of the generalized Hughes planes*, Canad. J. Math. 28, 376–402.
[80] *Translation planes*, Berlin etc.: Springer.

Lüneburg, M.
[88] *Kompakte, 4-dimensionale projektive Ebenen mit 8-dimensionaler Kollineationsgruppe*, Diplomarbeit, Tübingen.
[92] *Involutionen, auflösbare Gruppen und die Klassifikation topologischer Ebenen*, Mitt. Math. Sem. Gießen 209.

Magyar, Z.
[89] *On the classification of real semi-simple Lie algebras*, Acta Math. Hungar. 54, 99–134.

Maier, P. – Stroppel, M.
[95] *Pseudo-homogeneous coordinates for Hughes planes*, Canad. Math. Bull., to appear.

Mann, L. N.
[67] *Dimensions of compact transformation groups*, Michigan Math. J. 14, 433–444.

Marin, A.
[86] $\mathbb{CP}^2/\sigma$ *ou Kuiper et Massey au pays des coniques*, in: L. Guillou, A. Marin (eds.), A la recherche de la topologie perdue, Progress in Math. Vol. 62, pp. 141–152, Boston: Birkhäuser.

Massey, W. S.
[67] *Algebraic topology: an introduction*, New York: Harcourt, Brace & World, Inc.
[73] *The quotient space of the complex projective plane under conjugation is a 4-sphere*, Geom. Dedicata 2, 371–374.
[91] *A basic course in algebraic topology*, New York etc.: Springer.

Mäurer, H.
[73] *Eine Kennzeichnung der Kugeln*, Math. Z. 133, 41–52.

Mazurkiewicz, S.
[14] C. R. Varsovie 7, 382–384, (Polish) translated as
[69] *Sur un ensemble plan qui a avec chaque droite deux et seulement deux points communs*, in: S. Mazurkiewicz, Travaux de topologie et ses applications, pp. 46–47, Warszawa: PWN-Éditions scientifiques de Pologne.

McCord, M. C.
[66] *Spaces with acyclic point complements*, Proc. Amer. Math. Soc. 17, 886–890.

McKay, W. G. – Patera, J.
[81] *Tables of dimensions, indices, and branching rules for representations of simple Lie algebras*, New York: Dekker.

Milnor, J.
[58] *Some consequences of a theorem of Bott*, Ann. of Math. 68, 444–449.

Mimura, M.
[67] *The homotopy groups of Lie groups of low rank*, J. Math. Kyoto Univ. 6, 131–176.

Mimura, M. – Toda, H.
[64] *Homotopy groups of* SU(3), SU(4) *and* Sp(2), J. Math. Kyoto Univ. 3, 217–250.

Misfeld, J.
[68] *Topologische projektive Räume*, Abh. Math. Sem. Univ. Hamburg 32, 232–263.

Mitchell, W. J. R.
[78] *Absolute suspensions and cones*, Fund. Math. 101, 241–244.

Moise, E. E.
[77] *Geometric topology in dimensions* 2 *and* 3, New York etc.: Springer.

Montgomery, D.
[50] *Simply connected homogeneous spaces*, Proc. Amer. Math. Soc. 1, 467–469.
[57] *Finite dimensionality of certain transformation groups*, Illinois J. Math. 1, 28–35.

Montgomery, D. – Samelson, H.
[43] *Transformation groups of spheres*, Ann. of Math. 44, 454–470.

Montgomery, D. – Zippin, L.
[55] *Topological transformation groups*, New York: Interscience.

Morayne, M.
[87] *Measure preserving analytic diffeomorphisms of countable dense sets in* $\mathbb{C}^n$ *and* $\mathbb{R}^n$, Colloq. Math. 52, 93–98.

Morita, K.
[53] *On the dimension of product spaces*, Amer. J. Math. 75, 205–223.

Mortimer, B.
[75] *A geometric proof of a theorem of Hughes on homomorphisms of projective planes*, Bull. London Math. Soc. 7, 267–268
(cp. Math. Reviews 52 #4120 for a completion).

Mostert, P. S.
[53] *Local cross sections in locally compact groups*,
Proc. Amer. Math. Soc. 4, 645–649.
[56] *Sections in principal fibre spaces*, Duke Math. J. 23, 57–71.
[57] *On a compact Lie group acting on a manifold*,
Ann. of Math. 65, 447–455. Errata ibid. 66 (1957) 589.

Mostow, G. D.
[50] *The extensibility of local Lie groups of transformations and groups on surfaces*, Ann. of Math. 52, 606–636.
[57] *On a conjecture of Montgomery*, Ann. of Math. 65, 513–516.
[61] *On maximal subgroups of real Lie groups*, Ann. of Math. 74, 503–517.

Moulton, F. R.
[02] *A simple non-desarguesian plane geometry*, Trans. Amer. Math. Soc. 3, 192–195.

Myung, H. C. – Sagle, A. A.
[87]   *On Lie-admissible mutations of associative algebras*, Hadronic J. 10, 35–51.

Nagami, K.
[62]   *Dimension-theoretical structure of locally compact groups*,
       J. Math. Soc. Japan 14, 379–396.
[63]   *Cross sections in locally compact groups*, J. Math. Soc. Japan 15, 301–303.
[70]   *Dimension theory*, New York: Academic Press.

Nagano, T.
[59]   *Homogeneous sphere bundles and the isotropic Riemann manifolds*,
       Nagoya Math. J. 15, 29–55.

Nagata, J.
[83]   *Modern dimension theory*, Berlin: Heldermann (revised edition).

Naumann, H.
[54]   *Stufen der Begründung der ebenen affinen Geometrie*, Math. Z. 60, 120–141.

Newman, M. H. A.
[31]   *A theorem on periodic transformations of spaces*, Quart. J. Math. Oxford 2, 1–8.

Niemann, K.
[90]   *Geometrie und Topologie der sechzehndimensionalen Moufang-Hughes-Ebenen*,
       Staatsexamensarbeit, Universität Kiel.

Noll, W.
[65]   *Proof of the maximality of the orthogonal group in the unimodular group*,
       Arch. Rational Mech. Anal. 18, 100–102.

Oniščik, A. L. (= Onishchik, A. L.)
[63]   Math. Sb. 60, 447–485, (Russian) translated as:
[66]   *Transitive compact transformation groups*,
       Amer. Math. Soc. Transl. (2) 55, 153–194.
[94]   *Topology of transitive transformation groups*, Leipzig etc.: J.A. Barth.

Onishchik, A. L. – Vinberg, E. B.
[90]   *Lie groups and algebraic groups*, Berlin etc.: Springer.
[93]   *Foundations of Lie theory*, Part I of A. L. Onishchik (ed.), Lie groups and Lie
       algebras I, Encycl. Math. Sci. 20, pp. 1–94, Berlin etc.: Springer.

Ortega, J. M. – Rheinboldt, W. C.
[70]   *Iterative solution of nonlinear equations in several variables*,
       New York: Academic Press.

Ott, U.
[75]   *Eine neue Klasse endlicher Translationsebenen*, Math. Z. 143, 181–185.

Otte, J.
[93]   *Differenzierbare Ebenen*, Dissertation, Kiel.
[95]   *Smooth projective translation planes*, Geom. Dedicata, to appear.

Palais, R. S.
[60]   *The classification of G-spaces*, Mem. Amer. Math. Soc. 36, 72 pp.
[68]   *The classification of real division algebras*, Amer. Math. Monthly 75, 366–368.

Pasynkov, B. A.
[62]   *On the coincidence of varions definitions of dimensionality for factor spaces of locally bicompact groups*, Uspekhi Mat. Nauk 17, no. 5, 129–135. (Russian)

Pears, A. R.
[75]   *Dimension theory of general spaces*, Cambridge Univ. Press.

Pickert, G.
[59]   *Bemerkungen über die projektive Gruppe einer Moufang-Ebene*, Illinois J. Math. 3, 169–173.
[75]   *Projektive Ebenen*, Berlin etc.: Springer (2nd edition).
[81]   *Projectivities in projective planes*,
       in: P. Plaumann, K. Strambach (eds.), Geometry — von Staudt's point of view, Proc. Bad Windsheim 1980, pp. 1–49, Dordrecht etc.: Reidel.
[94]   *Pasch- oder Veblen-Axiom?*, Geom. Dedicata 50, 81–86.

Plaumann, P. – Strambach, K.
[70]   *Zusammenhängende Quasikörper mit Zentrum*, Arch. Math. 21, 455–465.
[74]   *Hurwitzsche Ternärkörper*, Arch. Math. 25, 129–134.
[81]   (editors) *Geometry — von Staudt's point of view*,
       Proc. Bad Windsheim 1980, Dordrecht etc.: Reidel.
[90]   *Partitionen Liescher und algebraischer Gruppen*, Forum Math. 2, 523–578.

Poguntke, D.
[92]   *The theorem of Lie and hyperplane subalgebras of Lie algebras*,
       Geom. Dedicata 43, 83–91.

Pohl, H.-J.
[90]   *Flat projective planes with 2-dimensional collineation group fixing at least two lines and more than two points*, J. Geom. 38, 107–157.

Polley, C.
[68]   *Lokal desarguessche Salzmann-Ebenen*, Arch. Math. 19, 553–557.
[72a] *Lokal desarguessche Geometrien auf dem Möbiusband*, Arch. Math. 23, 346–347.
[72b] *Zweidimensionale topologische Geometrien, in denen lokal die dreifache Ausartung des desarguesschen Satzes gilt*, Geom. Dedicata 1, 124–140.

Polster, B.
[93]   *Continuous planar functions*, Dissertation, Erlangen.

Polster, B. – Steinke, G. F.
[94]   *Criteria for two-dimensional circle planes*,
       Beiträge Algebra Geom. 35, 181–191.

Poncet, J.
[59]   *Groupes de Lie compacts de transformations de l'espace euclidien et les sphères comme espaces homogènes*, Comment. Math. Helv. 33, 109–120.

Pontrjagin, L. S. (= Pontryagin, L. S.)
[30] *Sur une hypothèse fondamentale de la theorie de la dimension*,
C. R. Acad. Sci. Paris 190, 1105–1107.
[32] *Über stetige algebraische Körper*, Ann. of Math. 33, 163–174.
[34] *The theory of topological commutative groups*, Ann. of Math. 35, 361–388.
[57] *Topologische Gruppen, Teil 1*, Leipzig: Teubner.
[66] *Topological groups*, New York: Gordon and Breach.
[86] *Topological groups*, (3rd edition),
Vol. 2 of L. S. Pontryagin, Selected works, New York etc.: Gordon and Breach.

Porteous, I.
[81] *Topological geometry*, Cambridge Univ. Press (2nd edition).

Prieß-Crampe, S.
[83] *Angeordnete Strukturen: Gruppen, Körper, projektive Ebenen*, Berlin etc.: Springer.

Priwitzer, B.
[94a] *Large automorphism groups of 8-dimensional projective planes are Lie groups*,
Geom. Dedicata 52, 33–40.
[94b] *Kompakte 16-dimensionale Ebenen mit großen halbeinfachen Gruppen*,
Dissertation, Tübingen.

Quinn, F.
[83] *Resolutions of homology manifolds, and the topological characterization of manifolds*, Invent. Math. 72, 267–284, Corrigendum: Invent. Math. 85, 653.
[87] *An obstruction to the resolution of homology manifolds*,
Michigan Math. J. 34, 285–291.

Ranicki, A. – Yamasaki, M.
[95] *Controlled K-Theory*, Topology Appl. 61, 1–59.

Raymond, F.
[60] *Separation and union theorems for generalized manifolds with boundary*,
Michigan Math. J. 7, 7–21.
[65a] *Some remarks on the coefficients used in the theory of homology manifolds*,
Pacific J. Math. 15, 1365–1376.
[65b] *Local triviality for Hurewicz fiberings of manifolds*, Topology 3, 43–57.

Rees, D.
[50] *The nuclei of non-associative division algebras*,
Math. Proc. Cambridge Philos. Soc. 46, 1–18.

Richardson, R. W.
[61] *Groups acting on the 4-sphere*, Illinois J. Math. 5, 474–485.

Riesinger, R.
[91] *Beispiele starrer, topologischer Faserungen des reellen projektiven 3-Raumes*,
Geom. Dedicata 40, 145–163.

Rotman, J. J.
[88] *An introduction to algebraic topology*, New York etc.: Springer.

Rutar, H.-J.
[82]   *Geometrie der Sphären in Cayley-Algebren mit Anwendung auf Kollineationen und Polaritäten von Moufang-Ebenen*, Staatsexamensarbeit, Kiel.

Salzmann, H.
[55]   *Über den Zusammenhang in topologischen projektiven Ebenen*,
       Math. Z. 61, 489–494.
[57]   *Topologische projektive Ebenen*, Math. Z. 67, 436–466.
[58]   *Kompakte zweidimensionale projektive Ebenen*, Arch. Math. 9, 447–454.
[59a] *Homomorphismen topologischer projektiver Ebenen*, Arch. Math. 10, 51–55.
[59b] *Topologische Struktur zweidimensionaler projektiver Ebenen*,
       Math. Z. 71, 408–413.
[59c] *Viereckstransitivität der kleinen projektiven Gruppe einer Moufang-Ebene*,
       Illinois J. Math. 3, 174–181.
[62a] *Kompakte zweidimensionale projektive Ebenen*, Math. Ann. 145, 401–428.
[62b] *Kompakte Ebenen mit einfacher Kollineationsgruppe*, Arch. Math. 13, 98–109.
[63a] *Characterization of the three classical plane geometries*,
       Illinois J. Math. 7, 543–547.
[63b] *Zur Klassifikation topologischer Ebenen*, Math. Ann. 150, 226–241.
[64]   *Zur Klassifikation topologischer Ebenen. II*,
       Abh. Math. Sem. Univ. Hamburg 27, 145–166.
[65]   *Zur Klassifikation topologischer Ebenen. III*,
       Abh. Math. Sem. Univ. Hamburg 28, 250–261.
[66]   *Polaritäten von Moulton-Ebenen*, Abh. Math. Sem. Univ. Hamburg 29, 212–216.
[67a] *Kollineationsgruppen ebener Geometrien*, Math. Z. 99, 1–15.
[67b] *Topological Planes*, Adv. in Math. 2, 1–60.
[69a] *Geometries on surfaces*, Pacific J. Math. 29, 397–402.
[69b] *Kompakte vier-dimensionale Ebenen*, Arch. Math. 20, 551–555.
[69c] *Homomorphismen komplexer Ternärkörper*, Math. Z. 112, 23–25.
[70]   *Kollineationsgruppen kompakter, vier-dimensionaler Ebenen*,
       Math. Z. 117, 112–124.
[71]   *Kollineationsgruppen kompakter 4-dimensionaler Ebenen. II*,
       Math. Z. 121, 104–110.
[72a] *4-dimensional projective planes of Lenz type III*, Geom. Dedicata 1, 18–20.
[72b] *Homogene 4-dimensionale affine Ebenen*, Math. Ann. 196, 320–322.
[72c] *Baer-Unterebenen 4-dimensionaler Ebenen*, Arch. Math. 23, 337–341.
[73a] *Elations in four-dimensional planes*,
       Gen. Top. Appl. (=Topology Appl.) 3, 121–124.
[73b] *Kompakte, vier-dimensionale projektive Ebenen mit 8-dimensionaler Kollineationsgruppe*, Math. Z. 130, 235–247.
[73c] *Reelle Kollineationen der komplexen projektiven Ebene*,
       Geom. Dedicata 1, 344–348.
[74]   *Compact planes of Lenz type III*, Geom. Dedicata 3, 399–403.
[75a] *Homogene kompakte projektive Ebenen*, Pacific J. Math. 60, 217–234.
[75b] *Homogene affine Ebenen*, Abh. Math. Sem. Univ. Hamburg 43, 216–220.

[79a] *Compact 8-dimensional projective planes with large collineation groups,*
Geom. Dedicata 8, 139–161.
[79b] *Automorphismengruppen 8-dimensionaler Ternärkörper,* Math. Z. 166, 265–275.
[81a] *Kompakte, 8-dimensionale projektive Ebenen mit großer Kollineationsgruppe,*
Math. Z. 176, 345–357.
[81b] *Projectivities and the topology of lines,*
in: P. Plaumann, K. Strambach (eds.), Geometry — von Staudt's point of view,
Proc. Bad Windsheim 1980, pp. 313–337, Dordrecht etc.: Reidel.
[82a] *Baer-Kollineationsgruppen der klassischen projektiven Ebenen,*
Arch. Math. 38, 374–377.
[82b] *Compact 16-dimensional projective planes with large collineation groups,*
Math. Ann. 261, 447–454.
[83] *Compact 16-dimensional projective planes with large collineation groups. II,*
Monatsh. Math. 95, 311–319.
[84] *Compact 16-dimensional projective planes with large collineation groups. III,*
Math. Z. 185, 185–190.
[85] *Homogeneous translation groups,* Arch. Math. 44, 95–96.
[87] *Compact 16-dimensional projective planes with large collineation groups. IV,*
Canad. J. Math. 39, 908–919.
[90] *Compact 8-dimensional projective planes,* Forum Math. 2, 15–34.

Samelson, H.
[52] *Topology of Lie groups,* Bull. Amer. Math. Soc. 58, 2–37.
[90] *Notes on Lie algebras,* New York etc.: Springer (2nd edition).

Samuel, P.
[67] *Qu'est-ce qu'une quadrique?,* Enseign. Math. 13, 129–130.
[88] *Projective geometry,* New York etc.: Springer.

Schafer, R. D.
[45] *On a construction for division algebras of order* 16,
Bull. Amer. Math. Soc. 51, 532–534.
[66] *An introduction to nonassociative algebras,* New York: Academic Press.

Schellhammer, I.
[81] *Einige Klassen von ebenen projektiven Ebenen,* Diplomarbeit, Tübingen.

Schnor, B.
[92] *Involutions in the group of automorphisms of an algebraically closed field,*
J. Algebra 152, 520–524.

Schröder, E. M.
[91] *Vorlesungen über Geometrie 1–3,* Mannheim: BI (1991/92).

Schur, F.
[09] *Grundlagen der Geometrie,* Leipzig: Teubner.

Segre, B.
[47] *Gli automorphismi del corpo complesso, ed un problema di Corrado Segre,*
Atti Accad. Naz. Lincei Rend. Cl. Sci. Fis. Mat. Nat. (8) 3, 414–420.

[56]   *Plans graphiques algébriques réels non desarguésiens et correspondences crémoniennes topologiques*, Rev. Math. Pures Appl. 1, no. 3, 35–50.

Seidel, H.-P.
[85]   *Locally homogeneous ANR-spaces*, Arch. Math. 44, 79–81.
[90a] *Generalized symmetric planes*, Geom. Dedicata 33, 337–354.
[90b] *Connected 4-dimensional stable planes with many central collineations*,
       Geom. Dedicata 36, 375–388.
[91]   *Classification of 4-dimensional generalized symmetric planes*,
       Forum Math. 3, 35–59.

Seidenberg, A.
[76]   *Pappus implies Desargues*, Amer. Math. Monthly 83, 190–192.

Seitz, G. M.
[87]   *The maximal subgroups of classical algebraic groups*,
       Mem. Amer. Math. Soc. 365.
[91]   *Maximal subgroups of exceptional algebraic groups*, Mem. Amer. Math. Soc. 441.

Shakhmatov, D. B.
[89]   *A problem of coincidence of dimensions in topological groups*,
       Topology Appl. 33, 105–113.

Siebenmann, L. – Guillou, L. – Hähl, H.
[73]   *Les voisinages ouverts réguliers*, Ann. Sci. École Norm. Sup. 6, 253–293.

Skljarenko, E. G.
[63]   Mat. Sbornik 60, 63–88, (Russian) translated as:
[64]   *On the topological structure of locally bicompact groups and their quotient spaces*,
       Amer. Math. Soc. Transl. (2) 39, 57–82.
[71a] Izv. Akad. Nauk SSSR Ser. Mat. 35, 831–843, (Russian) translated as:
[71b] *On the theory of generalized manifolds*, Math. USSR Izvestija 5, 845–857.

Skornjakov, L. A.
[54]   *Topological projective planes*, Trudy Moskov. Mat. Obshch. 3, 347–373. (Russian)
[57]   *Systems of curves on a plane*, Trudy Moskov. Mat. Obshch. 6, 135–164. (Russian)

Smith, Paul. A.
[41]   *Fixed-point theorems for periodic transformations*, Amer. J. Math. 63, 1–8.
[60]   *New results and old problems in finite transformation groups*,
       Bull. Amer. Math. Soc. 66, 401–415.

Spanier, E. H.
[66]   *Algebraic topology*, New York: McGraw–Hill.

Spencer, J. C. D. (= Yaqub-Spencer, J. C. D.)
[60]   *On the Lenz–Barlotti classification of projective planes*,
       Quart. J. Math. 11, 241–257.

Spera, A. G.
[86]   *Piani di Betten–Walker infiniti ed alcune applicazioni*,
       Boll. Un. Mat. Ital. (6) 5-B, 623–638.

Sperner, E.
[34] *Über die fixpunktfreien Abbildungen der Ebene*,
Abh. Math. Sem. Hamburgische Univ. 10, 1–48.

Sperner, P.
[90] *Vierdimensionale $\mathbb{C}^* \cdot SU_2\mathbb{C}$-Ebenen*, Geom. Dedicata 34, 301–312.

Springer, T. A.
[60] *The projective octave plane. I, II*, Indag. Math. 22, 74–101.
[62] *On the geometric algebra of the octave planes*, Indag. Math. 24, 451–468.

Springer, T. A. – Veldkamp, F. D.
[63] *Elliptic and hyperbolic octave planes. I–III*, Indag. Math. 25, 413–451.

Stauffer, H. B.
[75] *The multiplication problem for spheres*, Math. Stud. 43, 31–38.

Steenrod, N.
[51] *Topology of fibre bundles*, Princeton Univ. Press.

Steinke, G. F.
[85] *Topological affine planes composed of two desarguesian half planes and projective planes with trivial collineation group*, Arch. Math. 44, 472–480.
[89] *Topological properties of locally compact connected Minkowski planes and their derived affine planes*, Geom. Dedicata 32, 341–351.
[95] *Topological circle geometries*, Chapter 24 of: F. Buekenhout (ed.),
Handbook of incidence geometry, Amsterdam: North-Holland.

Stöcker, R. – Zieschang, H.
[88] *Algebraische Topologie*, Stuttgart: Teubner.

Strambach, K.
[67a] *Salzmann-Ebenen mit hinreichend vielen Punkt- oder Geradenspiegelungen*,
Math. Z. 99, 247–269.
[67b] *Eine Charakterisierung der klassischen Geometrien*, Arch. Math. 18, 539–544.
[68] *Zur Klassifikation von Salzmann-Ebenen mit dreidimensionaler Kollineationsgruppe*, Math. Ann. 179, 15–30.
[70a] *Zur Klassifikation von Salzmann-Ebenen mit dreidimensionaler Kollineationsgruppe II*, Abh. Math. Sem. Univ. Hamburg 34, 159–169.
[70b] *Salzmann-Ebenen mit puncttransitiver dreidimensionaler Kollineationsgruppe*,
Indag. Math. 32, 253–267.
[70c] *Zentrale und axiale Kollineationen in Salzmannebenen*, Math. Ann. 185, 173–190.
[71] *Gruppentheoretische Charakterisierungen klassischer desarguesscher und moultonscher Ebenen*, J. Reine Angew. Math. 248, 75–116.
[72] *Vierdimensionale affine Ebenen*, Arch. Math. 23, 342 - 345.
[75] *Algebraische Geometrien*, Rend. Sem. Mat. Univ. Padova 53, 165–210.
[77] *Der von Staudtsche Standpunkt in lokal kompakten Geometrien*,
Math. Z. 155, 11–21.
[86] *Projektivitätengruppen in angeordneten und topologischen Ebenen*,
Arch. Math. 47, 560 - 567.

Stroppel, M.
[90] *A characterization of quaternion planes*, Geom. Dedicata 36, 405–410.
[91] *Achtdimensionale stabile Ebenen mit quasieinfacher Automorphismengruppe*, Dissertation, Tübingen.
[92a] *Planar groups of automorphisms of stable planes*, J. Geom. 44, 184–200.
[92b] *Reconstruction of incidence geometries from groups of automorphisms*, Arch. Math. 58, 621–624.
[92c] *Solvable groups of automorphisms of stable planes*, Seminar Sophus Lie 2, 69–74.
[92d] *Endomorphisms of stable planes*, Seminar Sophus Lie 2, 75–81.
[93a] *Quaternion Hermitian planes*, Results Math. 23, 387–397.
[93b] *Embedding a non-embeddable stable plane*, Geom. Dedicata 45, 93–99.
[93c] *Quasi-perspectivities in stable planes*, Monatsh. Math. 115, 183–189.
[93d] *A note on Hilbert and Beltrami systems*, Results Math. 24, 342–347.
[93e] *Stable planes with large groups of automorphisms: the interplay of incidence, topology, and homogeneity*, Habilitationsschrift, Darmstadt.
[94a] *Compact groups of automorphisms of stable planes*, Forum Math. 6, 339–359.
[94b] *Locally compact Hughes planes*, Canad. Math. Bull. 37, 112–123.
[94c] *Stable planes*, Discrete Math. 129, 181–189.
[94d] *The skew hyperbolic motion group of the quaternion plane*, THD-Preprint No. 1650, Technische Hochschule Darmstadt.
[95a] *Actions of almost simple groups on eight-dimensional stable planes*, Math. Z., to appear.
[95b] *Actions of almost simple groups on compact eightdimensional projective planes are classical, almost*, Geom. Dedicata, to appear.

Suzuki, M.
[82] *Group theory I*, Berlin etc.: Springer.

Switzer, R. M.
[75] *Algebraic topology — homotopy and homology*, Berlin etc.: Springer.

Szambien, H.
[89] *Minimal topological projective planes*, J. Geom. 35, 177–185.

Szenthe, J.
[74] *On the topological characterization of transitive Lie group actions*, Acta Sci. Math. (Szeged) 36, 323–344.

Szőnyi, T.
[90] *Note on planar functions over the reals*, Note di Matematica 10, 59–65.

Taylor, D. E.
[92] *The geometry of the classical groups*, Berlin: Heldermann.

Tecklenburg, H.
[92] *Stufen der Anordnung in Geometrie und Algebra*, Deutscher Universitätsverlag, Wiesbaden.

Tetzlaff, K.
[89] *Einige differentialgeometrische Aspekte der projektiven Ebenen über $\mathbb{R}$, $\mathbb{C}$, $\mathbb{H}$ und $\mathbb{O}$*, Diplomarbeit, Kiel.

Thorbergsson, G.
[91] *Isoparametric foliations and their buildings*, Ann. of Math. 133, 429–446.
[92] *Clifford algebras and polar planes*, Duke Math. J. 67, 627–632.

Timmesfeld, F. G.
[94] *Moufang planes and the groups E_6^K and $SL_2(K)$, K a Cayley division algebra*, Forum Math. 6, 209–231.

Tits, J.
[52] *Sur les groupes doublement transitifs continus*,
Comment. Math. Helv. 26, 203–224 (see also Tits [56]).
[53] *Le plan projectif des octaves et les groupes de Lie exceptionnels*,
Bull. Acad. Roy. Belg. Cl. Sci. 39, 309–329.
[54] *Le plan projectif des octaves et les groupes exceptionnels E_6 et E_7*,
Bull. Acad. Roy. Belg. Cl. Sci. 40, 29–40.
[55] *Sur certaines classes d'espaces homogènes de groupes de Lie*,
Mém. de l'Académie Royale de Belgique, Classe des Sciences XXIX, Fasc. 3.
[56] *Sur les groupes doublement transitifs continus: correction et compléments*,
Comment. Math. Helv. 30, 234–240.
[57] *Transitivité des groupes de mouvements*, J. Naas and K. Schröder (eds.),
Proc. Riemann-Tagung Berlin 1954, pp. 98–111, Berlin: Akademie-Verlag.
[67] *Tabellen zu den einfachen Lie Gruppen und ihren Darstellungen*,
Lecture Notes in Math. 40, Berlin etc.: Springer.
[71] *Représentations linéaires irréductibles d'un groupe réductif sur un corps quelconque*, J. Reine Angew. Math. 247, 196–220.
[77] *Endliche Spiegelungsgruppen, die als Weylgruppen auftreten*,
Invent. Math. 43, 283–295.

Tkachenko, M. G.
[90] *On dimension of locally pseudocompact groups and their quotients*,
Comment. Math. Univ. Carolin. 31, 159–166.

Tondeur, P.
[65] *Introduction to Lie groups and transformation groups*,
Lecture Notes in Math. 7, Berlin etc.: Springer.

Ursul, M. I.
[87] *On the dimension theory of topological fields*,
Izv. Akad. Nauk Moldav. SSR Ser. Fiz.-Tekhn. Mat. Nauk, 47–48, 79. (Russian)

Varadarajan, V. S.
[74] *Lie groups, Lie algebras, and their representations*,
Englewood Cliffs (N.J.): Prentice-Hall; Berlin etc.: Springer 1984.

Veldkamp, F. D.
[68] *Unitary groups in projective octave planes*, Compositio Math. 19, 213–258.

[91]   *Freudenthal and the octonions*, Nieuw Arch. Wisk. (4) 9, 145–162.

Völklein, H.
[81]   *Transitivitätsfragen bei linearen Liegruppen*, Arch. Math. 36, 23–34.

Wähling, H.
[87]   *Theorie der Fastkörper*, Essen: Thales Verlag.
[88]   *Lokalkompakte Fastkörper*, J. Geom. 31, 194–201.

Walker, M.
[76]   *A class of translation planes*, Geom. Dedicata 5, 135–146.

Wan, Z.-X.
[75]   *Lie algebras*, Oxford: Pergamon Press.

Wang, H.-C.
[49]   *A new characterisation of spheres of even dimension*, Indag. Math. 11, 286–295.

Warner, F. W.
[83]   *Foundations of differentiable manifolds and Lie groups*, New York etc.: Springer. (Corrected reprint of the 1971 edition).

Warner, S.
[89]   *Topological fields*, Amsterdam: North-Holland.

Weber, C.
[68]   *Quelques théorèmes bien connus sur les A.N.R. et les C.W. complexes*, Enseign. Math. 13, 211–222.

Weigand, C.
[87]   *Konstruktion topologischer projektiver Ebenen, die keine Translationsebenen sind*, Mitt. Math. Sem. Gießen 177.

Weil, A.
[51]   *L'intégration dans les groupes topologiques et ses applications*, Paris: Hermann (2nd edition).
[67]   *Basic number theory*, Berlin etc.: Springer.

Weiss, E. – Zierler, N.
[58]   *Locally compact division rings*, Pacific J. Math. 8, 369–371.

West, J. E.
[77]   *Mapping Hilbert cube manifolds to ANR's: a solution of a conjecture of Borsuk*, Ann. of Math. 106, 1–18.

Whitehead, J. H. C.
[48]   *On the homotopy type of ANR's*, Bull. Amer. Math. Soc. 54, 1133–1145.

Whyburn, G. T.
[42]   *Analytic topology*,
       New York: Amer. Math. Soc. Colloq. Publ. 28 (3rd edition 1963).

Whyburn, G. – Duda, E.
[79]   *Dynamic topology*, New York etc.: Springer.

Wiener, H.
[1891] *Über Grundlagen und Aufbau der Geometrie*,
Jahresber. Deutsch. Math.-Verein. 1, 45–48.

Wilder, R. L.
[49] *Topology of manifolds*, Providence (RI): Amer. Math. Soc. Colloq. Publ. 32.

Wilker, J. B.
[81] *Inversive geometry*, in: The geometric vein,
the Coxeter Festschrift, pp. 379–442, New York etc.: Springer.

Wolf, J. A.
[84] *Spaces of constant curvature*,
Wilmington: Publish or Perish (first edition New York etc.: McGraw–Hill 1967).

Wonenburger, M. J.
[69] *Automorphisms of Cayley algebras*, J. Algebra 12, 441–452.

Yang, C. T.
[76] *Hilbert's fifth problem and related problems on transformation groups*,
Proc. Sympos. Pure Math. 28, 142–146, Proc. DeKalb (Ill.) 1974.

Zanella, C.
[89] *On topological projective spaces and their Grassmannians*,
Abh. Math. Sem. Univ. Hamburg 59, 125–142.

Zhevlakov, K. A. – Slin'ko, A. M. – Shestakov, I. P. – Shirshov, A. I.
[82] *Rings that are nearly associative*, New York: Academic Press.

Zieschang, H.
[81] *Finite groups of mapping classes of surfaces*,
Lecture notes in Math. 875, Berlin etc.: Springer.

Zippin, M.
[50] *Two-ended topological groups*, Proc. Amer. Math. Soc. 1, 309–315.

Zorn, M.
[31] *Theorie der alternativen Ringe*,
Abh. Math. Sem. Hamburgische Univ. 8, 123–147.
[33] *Alternativkörper und quadratische Systeme*,
Abh. Math. Sem. Hamburgische Univ. 9, 395–402.

Notation

In the sequel, we list some of the symbols that are used within the text, and occasionally give references to places in the text where the definition or further explanation may be found. We have not included symbols that have only 'local' meaning.

General notation.

Aut X	automorphism group of X
$\langle X \rangle$	group generated by X; or point set of the closed subplane generated by X, see the remarks after (41.6), and also (44.1); or closed subternary generated by X, see (83.4)
ker	kernel of a homomorphism
$K^\times$	group of non-zero elements of the field K
End V	ring of endomorphisms of the vector space V
diag $(a_1, \ldots, a_n)$	diagonal matrix with entries $a_1, \ldots, a_n$
$\mathbb{1}$	trivial element (identity) or trivial (sub)group
$\cong$	isomorphic
$\hat{=}$	corresponds to (in some natural identification)
$\stackrel{\circ}{=}$	locally isomorphic (94.1)

As usual, $\mathbb{N}, \mathbb{Z}, \mathbb{Q}, \mathbb{R}, \mathbb{C}$ denote the natural, integer, rational, real, and complex numbers, respectively. We use the convention that $0 \notin \mathbb{N}$ and write $\mathbb{N}_0 = \mathbb{N} \cup \{0\}$. By $\mathbb{H}$ we denote the (skew) field of Hamilton's quaternions, and by $\mathbb{O}$ we denote the division algebra of octonions, as defined in (11.1).

$\mathbb{F}_m$	real Cayley–Dickson algebra of dimension 2^m, see (11.1)
$\|x\|^2$	norm form (11.3)
$\langle x \mid y \rangle$	bilinear form (11.3)
Re x	real part of x
Pu x	pure part of x, see (11.6)
Pu $\mathbb{F}_m$	set of all pure elements of $\mathbb{F}_m$
Ker Q	kernel of the quasifield Q, see (25.3)

Our **geometric notation** is explained mainly in Chapters 1 and 2.

$\mathcal{P} = (P, \mathcal{L}) = (P, \mathcal{L}, \mathbf{F})$	usually a projective plane (21.1)
$\mathbf{F} \subset P \times \mathcal{L}$	set of flags, see (41.5a) and the remarks preceding (21.1)
$p \vee q = pq$	line joining the points p and q
$L \wedge M$	intersection point of the lines L and M
$\mathcal{L}_p$	pencil of lines through p, see (21.1)
$\mathcal{P}^W$	affine part of the projective plane $\mathcal{P}$, see (21.9)
$\overline{\mathcal{A}}$	projective completion of the affine plane $\mathcal{A}$, see (21.9)
$\tau(s, x, t)$	ternary operation (22.2)
$\mathcal{A}_2 \mathbb{F}$	affine plane over $\mathbb{F}$, see (12.1)
$\overline{\mathcal{A}_2 \mathbb{F}}$	projective completion of $\mathcal{A}_2 \mathbb{F}$, see (12.2)
$[s, t]$, $[s]$	lines (12.1), (22.3)
$\mathcal{P}_2 \mathbb{F} = (P_2 \mathbb{F}, \mathcal{L}_2 \mathbb{F})$	projective plane over $\mathbb{F}$, see (13.1)
$\mathcal{P}_2 \mathbb{O} = (P_2 \mathbb{O}, \mathcal{L}_2 \mathbb{O})$	(projective) octonion plane (16.1)
$P_d \mathbb{F}$	point space of the projective d-space over $\mathbb{F}$, see (13.1) and (14.1)
$\mathrm{Ker}(a_1 \ a_2 \ a_3)$	homogeneous coordinates for lines (13.2)
$[A]$	collineation induced by the linear map A, see (13.4), (14.2)
$\pi = \pi^+$	standard elliptic polarity (13.12), (16.7), (18.0)
π^-	standard hyperbolic polarity (13.12), (18.0)
$\mathcal{M}_k$, $\mathcal{M}(s)$	Moulton planes (31.25b), (34.2)
$\mathcal{E}_f$	shift plane (74.1)
$\mathcal{H}_\alpha^\mathbb{F}$	Hughes plane (86.4)
$\Gamma_{[c]}$, $\Gamma_{[A]}$, $\Gamma_{[c,A]} = \Gamma_{[c]} \cap \Gamma_{[A]}$	groups of collineations with center c, axis A, see (23.3)
$\Gamma_{[A,A]}$	group of elations with axis A, see (23.10)
Σ	group of continuous collineations of a compact projective plane (44.2)

Group theoretic notation is standard. For specific groups, like $\mathrm{GL}_n \mathbb{F}$, $\mathrm{SL}_n \mathbb{F}$, $\mathrm{SU}_n(\mathbb{F}, r)$, see (94.32) and the table at the end of the introduction to Chapter 1. We write $\mathfrak{gl}_n \mathbb{F}$, $\mathfrak{sl}_n \mathbb{F}$, $\mathfrak{su}_n(\mathbb{F}, r)$ for the corresponding Lie algebras. Notation for permutation groups is introduced at the beginning of Section 91, and for transformation groups in Section 96.

$\mathbb{Z}_n = \mathbb{Z}/n\mathbb{Z}$	cyclic group
$\mathbb{T} = \mathbb{R}/\mathbb{Z} \cong \mathrm{SO}_2 \mathbb{R}$	torus group
L_2	two-dimensional non-abelian Lie group (94.40)

Notation

$GO_n\mathbb{R}$, $GO_n^+\mathbb{R}$	group of similitudes and of direct similitudes of Euclidean space $\mathbb{R}^n$, see (11.21)
$\gamma^\alpha = \alpha^{-1}\gamma\alpha$	conjugation
$\Delta \trianglelefteq \Gamma$	Δ is normal in Γ
$N \rtimes \Gamma$, $\Gamma \ltimes N$	semidirect products, N is the normal factor
$Cs_\Gamma \Delta$	centralizer of Δ in Γ
$Ns_\Gamma \Delta$	normalizer of Δ in Γ
$C(\Gamma)$	center of the group Γ
$\Gamma' = [\Gamma, \Gamma]$	commutator subgroup
Γ^1	connected component of $\mathbb{1}$ in the topological group Γ, see the beginning of Section 93
$\tilde{\Gamma}$	universal covering group of Γ, see (94.12)
$\mathfrak{l}\Gamma$	Lie algebra of Γ, see (94.5)
$[\,,]$	Lie bracket, see the beginning of Section 94
exp	exponential function (94.7)
$\sqrt{\Gamma}$	(solvable) radical of Γ, see (94.18)
sg $\mathfrak{l}$	signature of the Lie algebra $\mathfrak{l}$, see (94.32c)
rk Γ	torus rank of Γ, see the remark after (94.31)
Γ/Δ	coset space (93.16), (94.3a)
X/Γ	orbit space (96.2b)

We write the action of a group Γ on a set X as $(x, \gamma) \mapsto x^\gamma$, compare the beginning of Section 91. However, matrices (and linear maps) are usually applied to vectors from the left; hence scalars operate from the right.

$x^\Gamma = \{x^\gamma \mid \gamma \in \Gamma\}$	orbit of x under Γ
$\Gamma_x = \{\gamma \in \Gamma \mid x^\gamma = x\}$	stabilizer of x in Γ
$\Gamma_{[X]}$	pointwise stabilizer of X in Γ
$\gamma\|_X$	restriction of γ to X
$\Gamma\|_X = \{\gamma\|_X \mid \gamma \in \Gamma\}$ $\cong \Gamma/\Gamma_{[X]}$	group induced by Γ on X

Topological Notation. Let X be a topological space.

$\mathbb{S}_n$	sphere of dimension n
$G_{n,k}$	Grassmann manifold (64.3)
$X * X$	symmetric square (55.8)
∂X	boundary of X
dim X	covering dimension of X, see (92.1c)
ind X, Ind X	inductive dimensions of X, see (92.3)
$\approx$	homeomorphic
$\simeq$	homotopic (maps), homotopy equivalent (spaces)
$\pi_n X$	homotopy group

$\pi_1 X$	fundamental group
$H_n X$	homology group
$H_n(X; R)$	homology group with coefficients in R
$H_n(X, A)$	relative homology group
$H^n(X)$	cohomology group
χX	Euler characteristic of X

Index

References to definitions are given in **bold** type.

(p, L)-homogeneous (**23.21**)
(p, L)-transitive (**23.21**), (64.1)
2-dimensional compact projective plane (**32.6**)

abelian group (of automorphisms) (55.27), (55.41)f, (61.23), (61.25), (71.11), (72.12)f, (74.26)
absolute neighbourhood retract (ANR) (**51.11**)ff
absolute point, line (**13.16**)
  (of standard hyperbolic polarity) (13.17), (18.21)
  (of standard planar polarity) (18.32)
action of a group on a plane (**71.5**)f
affine part (**21.9**), (41.2)
affine plane (**21.8**)
  (classical) (**12.1**)
almost effective (**96.1**)
almost homogeneous elation group (61.12), (61.14)
almost simple factor (**94.25**)
almost simple group (**94.24**), (94.33)
  (of automorphisms) (84.19), (85.11)
alternative (**11.8**), (11.18)
alternative field (11.13), (24.8)
analytic plane (75.1)f
analytic subgroup (**94.14**), (94.15), (32.21), (95.6)
ANR (absolute neighbourhood retract) (**51.11**)ff
anticompact p. **620**
approximation theorem (93.8)
arc plane (31.25d), (38.5)
associative loop (53.4)

automorphism (**44.2**)
  (of ternary field) (23.5), (44.4), (55.21), (81.5), (83.15), (83.23)
  axial (collineation or group) (**23.7**), (12.4), (23.11), (32.19), (33.4), (44.7)f, Sections 24 and 61
  (of translation plane) (25.4)
axis (of collineation) (**23.7**), (12.4), (44.8)

Baer collineation (**23.14**)
Baer involution (**23.17**), (55.21), (55.29), (55.38)ff, (83.5)
Baer subplane (**21.7**), (41.11), (44.4), (53.10), (55.4)f, (72.3), (74.4b), (86.31)
Baer subternary (**83.4**)
biassociative (**11.9**), (11.18)
Brouwer's theorem (96.30)
bundle involution p. **292**

Cantor manifold (**51.15**)ff, (51.21c)
Cartesian field (**24.4**), (43.3)
Cayley numbers cf. octonions
Cayley triple (**11.16**)
Cayley-Dickson algebra (**11.1**), (11.18), (24.8)
center (of collineation) (**23.7**), (12.4), (44.8)
center map (**61.7**)ff
  (projective) (**61.10**)
central projection (**41.2**)
centralizer (95.4), (95.10)
character group p. **600**
characteristic map (**53.11**)f

classification
  (of almost simple Lie groups) (94.32)
  (of doubly transitive groups) (96.15), (96.17)
  (of 4-dimensional translation planes) (73.22)f
  (of 4-dimensional shift planes) (74.23)
  (of 4-dimensional projective planes) (74.27)
  (of 8-dimensional translation planes) (81.9), (81.15), (82.2), (82.15)ff
  (of 16-dimensional translation planes) (81.10), (81.17), (81.18), (82.5), (82.26)ff
  (of 16-dimensional projective planes) p. 585f
clc (cohomology locally connected) (**54.1**)
collinear p. **131**
collineation (**23.1**), (12.3)
  (of classical planes) (12.10), (13.6), (17.8), (17.10)
  (of octonion plane) (12.14)ff, (17.8), (17.10)
collineation group (23.3), (44.3)
commutative group (94.38), (94.39)
commutative multiplication (52.9)f, (53.9)
compact connected group (93.11)
compact group (of automorphisms) (61.26), (71.9)f, (72.6), (84.9)
compact linear group (95.3)
compact-open topology (**96.3**), (96.6)
completely reducible (**95.1**), (95.2), (95.3)
compression subgroup (**81.8**), (82.7)
conformal transformation (15.2)
conjugation (of Cayley-Dickson algebra) (**11.1**), (11.15)
contractible (**51.3**)f, (51.27)
convex (**31.4**)
convex quadrangle (**31.8**)
convex triangle (**31.6**)
coordinatize (**22.1**), (22.3), (43.2), (43.5)

countable basis (41.8)
covering (**94.2**)
covering dimension Section 92
coverings of Lie groups (94.30)
critical dimension p. **368**, (73.1), p. 449
CW-complex (51.13)
cyclic group (of automorphisms) (55.23)f, (55.27), (55.31)

defect (**92.13**)
deformation retract (**51.25**)f
derivative of a shift plane (**74.14**), (74.31)
derived group p. **607**
Desargues configuration (21.3)
desarguesian (**21.3**)
differentiable plane ($\mathscr{C}$-plane) (**75.2**)ff
differentiable quasifield p. **275**
dimension
  (topological) Section 92
  (of planes) (41.8), (51.9)f, (52.5), (52.10), (54.8), (54.10)f, (55.14), (74.6)
dimension formula (96.10)
discontinuous collineation (55.22)
distributor (**25.4**)
division algebra = semifield
domain invariance (**51.18**)ff
dominant weight (95.7)
dominate (homotopy) (**51.14**)
domination by classical planes p. **283**
doubly transitive group p. **591**, (96.15), (96.17)
dual plane p. **132**, (13.10)
duality (**23.1**), (13.11)
duality principle (13.10)

elation (**23.7**), (12.5), (23.11), (32.14), (44.8), (55.28), Sections 61 and 64
  (of translation plane) (25.4)
equivalent permutation groups p. **591**
equivalent representations p. **616**
Euler characteristic (52.12)ff
exponential (94.7)

exterior point (of standard hyperbolic polarity) (13.17), (18.21)

fiber bundle, fibration (**51.22**)ff
final topology (**31.18**), (31.19)
finite type (**51.14**)
fixed point (55.19), (55.23)f, (55.34), (55.38), (71.4), (73.4)
flag p. **131**
flag space (**41.5**), (32.3), Section 51, (52.14d), (54.10)
flag transitive (63.8)
flexible algebra (**11.10**)
flexible plane p. 213, (**63.11**), (73.3), (73.23), (74.21)
fractional linear transformation (12.12), (15.6)
Frattini argument (91.2)
Fundamental Theorem
  (of affine geometry) (12.10)
  (of projective line) (12.12), (15.6)
  (of projective geometry) (13.6)
fundamental weight (95.7)

generate p. **221**
germ p. 598
Grassmann coordinates (**64.3b**), (64.8)
Grassmann manifold (**64.3**)

H-space p. **261**, (52.8)f
Hadamard criterion (74.19)
half plane (31.2), (31.4)
Hamilton triple (**11.10**), (11.16)
Hausdorff convergence (**31.13**)
Hausdorff product (94.8)
Hilbert's 5th problem p. 597
HLC (homology locally connected) (**54.1**)
homogeneous coordinates (**13.2**), (13.3), (16.8)
homogeneous elation group (61.13)
homology (collineation) (**23.7**), (12.13), (23.11), (32.14), (44.8), Sections 61 and 64

(of translation plane) (25.4), (64.6), (72.11), (81.0b), (81.20)
homology (invariants) (52.13)f
homology manifold (**54.7**)f, (54.10)
homothety (**23.7**), (12.13)
homotopy (**51.3**)
homotopy groups (52.13)f, p. 614
homotopy sequence (96.12)
homotopy sphere (52.5)ff, (52.12)
Hopf fibration (14.11), (16.15), (51.24)
Hughes plane (**86.4**), (86.12), (86.31), (86.34)ff
Hughes situation (72.3), (86.33)f
hyperbolic plane (13.17), (18.23), (31.2)

identification (**92.19**)
incidence structure p. 131
inductive dimension Section 92
integral curve (**34.3**)
integral function (of a spread) (**74.14**)ff
integral plane (of a translation plane) (**74.14**)
interior hyperbolic plane (13.17), (18.23)
interior point (of standard hyperbolic polarity) (13.17), (18.21)
interval (**31.4**)
interval join topology (**31.18**), (31.19)
involution (23.17)ff, (55.29)ff, (83.5), (83.10)ff
irreducible representation (95.4), (95.6)
isomorphic permutation groups p. **591**
isomorphic planes (**23.1**)
isomorphism of topological planes p. 236
isotropic (**62.7**)ff, (63.10)

Kalscheuer nearfield (**64.19**), (64.20), (64.22), (82.24)
kernel (of quasifield) (**25.3**), (11.20), (25.4), (42.6)
Killing form (94.32c)

Lefschetz number (**55.19**)
Lenz-Barlotti type (24.1), (64.17), (74.9)
Lenz type (**24.2**), (24.3), (64.17)
Lenz type II (**24.4**), (43.3)
Lenz type III (**24.5**), (34.8), (63.5), (64.18)
Lenz type IV = translation plane (up to duality)
Lenz type V (**24.7**), (64.15), (81.0e), (82.2)
(examples) (82.2), (82.17)
Levi complement (**94.28**)
Levi subgroup = Levi complement
Lie algebra (94.5)
Lie group (**93.1**), Section 94, (32.21), (44.6), (53.1)ff, (71.2), (87.1)
line at infinity (**21.9**), (12.2)
line homogeneous affine plane (63.1)
linear Lie group (95.9)
linear representation Section 95
linearized version (**95.9**)
linearly transitive (**23.21**), (61.4)
little projective group (**23.7**), (17.1), (17.7)
local homology presheaf (**54.2**)
local homology sheaf (**54.5**)
local homology (**52.4**), (54.6)
local Lie group p. 598
local structure theorem (93.4)
locally contractible p. **595**, (42.8), (51.4), (51.12), (51.27),
locally homeomorphic p. **245**
locally homogeneous (**51.11**), p. **595**
locally isomorphic (**94.1**)
locally trivial fiber bundle (**51.22**)f

Mal'cev-Iwasawa theorem (93.10)
manifold (**51.15**), (52.1)ff, (53.1)ff
maximal compact subgroups (93.10), (94.33)
metrizable (41.8), (41.13)
module function (**81.3**)
mono-associative (11.7)

motion group Section 62
(Euclidean) (**62.1**), (**62.4**)
(elliptic) (13.13), (18.2), (18.5)ff, (18.15)
(hyperbolic) (13.13), (18.2), (18.20)ff, (18.26)
(planar) (18.32)
Moufang identity (12.15)
Moufang plane (**24.8**), (17.3), (42.7), (63.3), (64.24)
Moulton plane (**31.25b**), (34.2), (34.4), (34.8–11), (38.1), (38.3), (64.18)
mutation (**64.16**), (82.21)

natural topology (**31.15**), (31.19), (31.20), (32.2)
nearfield (**25.7**), (24.6), (64.19)ff
(see also: Kalscheuer nearfield)
nearfield plane (**24.6**), (**25.7**), (64.22)
Newman's theorem (96.26), (96.27)
non-semi-simple group (of automorphisms) (84.21), (85.11)
non-solvable group (of automorphisms) (72.1)ff, (73.12)ff, Sections 62 and 63
norm (of Cayley-Dickson algebra) (**11.3**), (11.14)
normalizer (33.12), (37.5)
nucleus (**25.4**), (25.8)

octonion plane (affine) (12.1)
(projective) (16.1), (16.3), (16.11), (16.14)
octonions (**11.1**), (11.13)
one-parameter groups (**93.20**)
open action (96.8)
open join topology (**31.18**), (31.19)
open meet topology (**31.18**), (31.19)
opposite (**31.8**)
orbit (53.1)f
orbit space (**96.2b**)
oval (**55.7**)ff, (38.8)
topological (**55.9**)ff, (74.5), (74.17)

pappian (**21.5**)
parabola model (31.2), (74.2a)

parallel (**21.8**), (12.1)
parallel axiom (12.1)
Pasch axiom (31.7)
Peano continuum (**51.7**)f
pencil p. **132**
pencil intersection topology (**32.1**), (32.2)
perspectivity (**21.6**), (41.2)
pigeon-hole principle (51.20)
planar collineation (**23.14**)
planar function (**74.1**)ff, (74.16)f, (74.19)
point at infinity (**21.9**), (12.2)
point homogeneous projective plane (63.8)
point row p. **132**
point space, topology of (51.29), (52.15), (53.12)ff
polar (**13.11**)
polarity (**23.1**), (13.11), (34.11), (35.5), (36.4), (37.8), (38.8), (74.5)
 (standard elliptic) (13.12), (16.7)
 (standard hyperbolic) (13.12), (18.0)
 (standard planar) (18.28)
 (of classical plane) (13.18), (18.29)
polar triangle (18.20), proof of (18.29)
pole (**13.11**)
polyhedron (51.13)f
presheaf (54.2)
primitive permutation group p. **591**
projective completion (**21.9**), (12.2), (32.3), (43.7), (43.8)
projective group (**23.7**)
projective plane (**21.1**)
 (classical) (12.2), (13.1), (14.4), (14.9), (16.1), (16.3), (16.11), (16.14)
projectivity (**21.6**), (41.2), (66.1)f
pseudo-isotopy (**51.3**)f, (51.27)
pure element (of a Cayley-Dickson algebra) (**11.6**)

quadrangle (non-degenerate) (**21.1**), (13.4), (23.5), (42.3), (44.4)
 (standard) (13.4)

quadrangle, stabilizer of (12.9), (23.5), (44.4c), (44.5), (55.21), (71.7), (81.5), Section 83
quadratic algebra (11.5)
quadric
 (classical projective line as a $\sim$) (15.1), (16.2)
quasi-equivalent (**95.11**)
quasifield (**24.6**), (**25.2**), (42.6), (64.13)
 (examples) (82.1), (82.4), (82.6), (82.16), (82.18), (82.21)
quaternions (**11.1**), (11.9), (11.10), (11.11)
quotient topology (**92.19**)
quotients (93.16)

$\mathbb{R}^2$-plane (**31.1**)
radical (**94.18**)
rank (**94.31**)
reduced stabilizer (73.2d), (81.0c)
reductive (**95.2**)
reflection (**23.20**), (32.12), (32.13), (55.29), (55.34)f
 (of octonion plane) (18.11)
Richardson's theorem (96.34)
rigid (38.7), (64.17)

semi-simple group (**94.20**)
 (of automorphisms) (71.4), (71.8), (84.16), (85.6)
semifield (**24.7**), (**25.8**), (64.14)f
 (examples) (64.16), (82.1), (82.16), (82.21)
sheaf (**54.2**)
 constant (**54.3**)f
 locally constant (**54.3**)
shear (**23.7**), (12.5), (81.0e)
shear point (**81.0e**)
shift (**74.1**)
shift group (**74.1**), (74.7)
shift plane (**74.1**), (31.25c), (74.8)
 proper p. **426**
signature (**94.32c**)
similitude (11.21)

simple group (of automorphisms) (71.8),
    Sections 62, 63, and 86
simply connected (51.28)f
skew hyperbolic plane (**35.1**), (35.2),
    (35.3), (35.4), (35.5), (38.1), (38.3)
skew parabola plane (**36.1**), (36.2),
    (36.3), (36.4), (38.1), (74.2b),
    (74.30)f
slope (12.1), (**22.3**)
solenoid p. **600**
solvable group p. **607**, (94.17)
    (of automorphisms) (73.4)ff, (74.22)
split type (**95.8**)
spread (**25.6**), (64.4)ff, (64.8)ff,
    Section 73, (74.14)ff
square root (55.20)
stable (plane) p. **159**, (31.22), (31.26)
stiffness p. 368, (71.7), (74.13), (81.5),
    (81.8), Section 83
straight (**23.15**)
strain (**23.7**)
strong deformation retract (**51.25**)f
subplane (**21.7**), (41.6), (41.11),
    (41.12), (42.3), (44.1), (44.4), (55.1),
    (83.1)
subternary (**83.4**)
sum theorem (92.9)
symmetric square (**55.8**), (55.12),
    (55.15)
Szenthe's theorem (96.14)

ternary = ternary field
ternary field (**22.2**), (41.3), (43.2),
    (43.4), (43.5), (43.6), (44.4)
Theorem of Lie and Palais (34.3)
topological affine plane (**41.3**), (43.7),
    (43.8)

topological projective plane (**41.1**)
topological ternary field (**41.3**)
topologically generated p. **221**
torus action (55.37), (96.28)
torus (group) p. **600**, (32.17)
torus rank p. **610**, (55.36), (94.32e)
transformation group (**96.1**)
transformation groups on spheres
    (96.22)
transitive transformation group (96.19)
translation (**23.7**), (12.5)
translation line (**24.6**), (25.1), (64.4)
translation plane (**24.6**), (**25.1**),
    (44.6), (52.15), (53.16)ff, (61.22),
    (62.13), (64.4)ff, (64.19)ff, (72.10)ff,
    Sections 73, 81, and 82
    (examples) (64.22), (73.10)ff, (73.13),
    (73.19)ff, (73.23)ff, (82.2), (82.5),
    (82.6), (82.17–25), (82.27–29)
transposition (of spread) (64.10), (82.2)
triality automorphism (**17.12**), (17.16)
triality collineation (**16.6**)
triality principle (12.18)
two-ended group (93.13), (61.2)
type (**94.32b**)

uniform convergence (96.6)
universal covering (**94.2**)

vector field (34.3)
Veronese coordinates (**16.1**), (16.3),
    (16.8)
von Staudt group p. **370**

weight (95.7), (95.10)
Weyl formula p. 619
Weyl group (63.8) (second proof)
Weyl's theorem (94.29)